Sustainable Postharvest Technologies for Fruits and Vegetables

Fruits and vegetables, commonly termed as "fresh produce," are an important component of the human diet, as they provide various beneficial and essential health-related compounds. Nevertheless, fresh produce is susceptible to postharvest deterioration and decay along with loss of certain nutrients due to inappropriate storage conditions and lack of standard postharvest technologies. In addition, a short shelf life is considered another major constraint that must be extended after harvest to ensure a wider availability window of fresh produce for consumers. From this perspective, the use of postharvest approaches is considered imperative to reduce the deterioration of harvested fresh produce in order to extend their storage and shelf life potential on a sustainable basis. *Sustainable Postharvest Technologies for Fruits and Vegetables* covers various aspects of postharvest technologies, with major developments over the recent past, and provides a way forward for the future.

The sustainable use of various technologies and elicitors could be adapted from farm to fork in order to conserve the eating quality of fresh produce. Therefore, this book covers various sustainable postharvest treatments and technologies that could be considered highly effective for the delay of postharvest senescence and deterioration. Among the various technologies, the use of preharvest treatments, controlled atmosphere, dynamic control atmosphere, modified atmosphere and hypobaric conditions has tremendous potential for the fresh fruits and vegetables industry. In the same way, cold plasma, pulsed light, ultraviolet light, ultrasound technology, nanoemulsions, nano-packaging, electrolyzed water, high pressure processing, ozone gas, irradiations, edible coatings, vacuum packaging and active packaging with slow releasing compounds along with nanotechnology are highly practicable and possess tremendous potential to be used in the maintenance of overall eating quality and storage life extension of fresh produce.

Key Features:

- Overviews the major factors affecting postharvest physiology and shelf life potential of fresh produce.
- Focuses on major sustainable technologies having the potential to maintain postharvest quality and extend shelf life of fruits and vegetables.
- Describes practical and recent advances of various approaches indispensable for the maintenance of overall eating quality and food safety attainment for fresh produce on a sustainable basis.
- Covers how quality maintenance and shelf life rely on preharvest practices, nonthermal treatments, storage atmospheres, packaging materials, active packaging, edible packaging, coating application techniques, nanotechnology and ecofriendly plant extracts and natural antagonists.

Sustainable Postharvest Technologies for Fruits and Vegetables

Edited by
Sajid Ali, Shabir Ahmad Mir, B.N. Dar, and Shaghef Ejaz

CRC Press
Taylor & Francis Group
Boca Raton London New York

CRC Press is an imprint of the
Taylor & Francis Group, an **informa** business

First edition published 2025
by CRC Press
2385 NW Executive Center Drive, Suite 320, Boca Raton FL 33431

and by CRC Press
4 Park Square, Milton Park, Abingdon, Oxon, OX14 4RN

CRC Press is an imprint of Taylor & Francis Group, LLC

ISBN: 978-1-032-42601-3 (hbk)
ISBN: 978-1-032-44086-6 (pbk)
ISBN: 978-1-003-37037-6 (ebk)

DOI: 10.1201/9781003370376

Typeset in Times
by Apex CoVantage, LLC

Contents

Preface

Fresh fruits and vegetables are susceptible to postharvest decay and deterioration. The higher decay and deterioration are considered major constraints in the extension of the postharvest shelf and storage life of horticultural produce. Therefore, these factors are considered leading causes of postharvest losses of fresh fruits and vegetables in the world. In this perspective, the use of sustainable postharvest technologies is considered imperative to extend the storage- and shelf-life potential of the fresh produce with maintained quality. The sustainable use of various technologies could be adapted to conserve the eating quality of fresh fruits and vegetables. Therefore, this book covers various sustainable postharvest treatments and technologies which are considered highly effective for the delay of postharvest senescence and deterioration. This book also covers various applications of the potential sustainable technologies in the postharvest physiology and handling of fresh fruits and vegetables. The use of controlled atmosphere, modified atmosphere, hypobaric storage, cold plasma, ultraviolet light, ultrasound technology, pulsed light, vacuum packaging, modified atmosphere, active packaging, nano-packaging, nanoemulsions, electrolyzed water, ozone gas, irradiations, edible coatings, natural plant extracts and antagonistic microorganisms along with nanotechnology are highly practicable and have tremendous potential for the storage-life extension and quality management of fresh fruits and vegetables. In addition, the recent advances in each technology, potential health hazards and major challenges in the application of the said technologies have also been covered. So, this book covers these aspects with recent developments over the recent past and will provide a way forward for future research needs. Overall, *Sustainable Postharvest Technologies for Fresh Fruits and Vegetables* is an informative and imperative reference material for students, professionals, food technologists, the fresh produce industry and exporters of fruits and vegetables.

Dr. Sajid Ali (PhD) is an assistant professor in the Department of Horticulture, Faculty of Agricultural Sciences and Technology, Bahauddin Zakariya University, Multan, Pakistan. Dr. Ali acquired his BSc (Honors) Agriculture with Major Horticulture from Bahauddin Zakariya University, Multan, Pakistan. He received his MSc (Honors) Horticulture and PhD Horticulture (Postharvest Physiology) from University of Agriculture, Faisalabad, Pakistan. He received advanced training in postharvest physiology of horticultural crops from US Horticultural Research Laboratory, USDA-ARS, Fort Pierce, Florida, USA. Dr. Ali has authored and co-authored around 120 international papers and around 25 book chapters. He has edited a book titled *Postharvest Physiological Disorders of Fruits*. He is Editor of *BMC Plant Biology*, Associate Editor of *Frontiers in Sustainable Food Systems* and Academic Editor of *Plos One*. Dr. Ali is an active reviewer for various international scientific journals. He has worked as a research officer and as a project development officer in two research projects at University of Agriculture, Faisalabad, Pakistan. He has won several research projects and awards. He teaches different graduate and postgraduate courses on horticulture.

Shabir Ahmad Mir (PhD) is an assistant professor at the Government College for Women, M.A. Road, Srinagar, Jammu and Kashmir, India. He obtained his MSc (Food Technology) from Islamic University of Science and Technology, Awantipora, India, and PhD (Food Technology) from Pondicherry University, Puducherry, India. He received the best PhD Thesis Award for outstanding research in the field of food technology. He has organized several conferences and workshops in food science and technology. Dr. Mir has published numerous international papers and book chapters and has edited 11 books.

Dr. B.N. Dar (PhD) is a distinguished expert in food technology, recognized for his extensive contributions to the field. As an academic and researcher, he has significantly advanced the understanding of food science, focusing on byproduct utilization and nutritional enhancement. His academic journey is marked by prestigious positions and collaborations with esteemed institutions. Dr. Dar's prolific output includes numerous publications, patents and projects, reflecting his deep commitment to research and innovation. He has been a technical expert member of the Joint FAO/WHO Expert Meeting on Microbiological Risk Assessment (JEMRA) since July 2021. He has presented his research to a wide variety of audiences cutting across government and private organizations to much appreciation. He has published more than 150 research and review papers in some of the top journals in the discipline and has editorship of 10 books. Dr. Dar's work not only contributes to academic knowledge but also has practical implications, bridging the gap between research and industry applications.

Dr. Shaghef Ejaz (PhD) is an associate professor and researcher at the Department of Horticulture, Bahauddin Zakariya University, Multan, Pakistan. Dr. Ejaz acquired his BSc (Honors) Agriculture with Major Horticulture and MSc (Honors) Horticulture from the Department of Horticulture, Bahauddin Zakariya University. He was awarded a PhD scholarship from the Higher Education Commission, Islamabad, Pakistan. He acquired his PhD from the University of Natural Resources and Life Sciences, Vienna, Austria, with a specialization in postharvest quality assessment of fresh horticultural commodities. Dr. Ejaz has research experience of more than 15 years and teaching experience of 10 years. He has published more than 60 research and review articles, with a cumulative Web of Science impact factor (IF) of almost 200. He has also contributed 25 book chapters in books published by various international publishers. Dr. Ejaz has also presented his research work and delivered lectures at more than 15 national and international seminars and conferences. Dr. Ejaz, as Associate Editor and Review Editor, is also on the editorial board of many well-reputed journals. Recently, Dr. Ejaz's research work has been mainly focused on the minimal postharvest processing of fresh fruits and vegetables. Besides being an active researcher, he has been a dynamic teacher and supervisor, which is recognizable from the research publications of his students during his supervision.

Contributors

Khalina Abdan
Department of Biological and Agricultural Engineering
Faculty of Engineering
Universiti Putra Malaysia
Serdang, Selangor, Malaysia

Rehaba Abid
Department of Horticulture
Faculty of Agricultural Sciences and Technology
Bahauddin Zakariya University
Multan, Punjab, Pakistan

Rafiq Ahmad
Department of Microbiology
The University of Haripur
Haripur, Kyber Pakhtunkhwa, Pakistan

Shabbir Ahmad
Department of Food Science and Technology
MNS University of Agriculture
Multan, Punjab, Pakistan

Zienab F.R. Ahmed
Integrative Agriculture Department
College of Agriculture and Veterinary Medicine
United Arab Emirates University
Al Ain, UAE

Shafi Ahmed
Department of Agro Product Processing Technology
Jashore University of Science and Technology
Jashore, Bangladesh

Aqsa Akhtar
School of Food and Agricultural Sciences
University of Management and Technology
Lahore, Punjab, Pakistan

Sharmin Akther
Department of Agro Product Processing Technology
Jashore University of Science and Technology
Jashore, Bangladesh

Maratab Ali
School of Agricultural Engineering and Food Science
Shandong University of Technology
Zibo, Shandong, P.R. China
and
School of Food and Agricultural Sciences
University of Management and Technology
Lahore, Punjab, Pakistan

Muhammad Moaaz Ali
College of Horticulture
Fujian Agriculture and Forestry University
Fuzhou, Fujian, P.R. China

Qasid Ali
Department of Horticulture
Faculty of Agriculture
Akdeniz University
Antalya, Turkey

Sajid Ali
Department of Horticulture
Faculty of Agricultural Sciences and Technology
Bahauddin Zakariya University
Multan, Punjab, Pakistan

Shinawar Waseem Ali
Department of Food Sciences
Faculty of Agricultural Sciences
University of the Punjab
Lahore, Punjab, Pakistan

Laraib Amjad
Key Laboratory of Horticultural Plant Biology of Ministry of Education
Huazhong Agricultural University
Wuhan, Hubei, P.R. China

Muhammad Naveed Arshad
Institute of Biological
Environmental and Rural Sciences
Aberystwyth University
Penglais, Aberystwyth, United Kingdom

Habat Ullah Asad
Centre for Agriculture and Bioscience International
Rawalpindi, Pakistan

Sadia Aslam
School of Food and Agricultural Sciences
University of Management and Technology
Lahore, Punjab, Pakistan

Pedram Assar
Department of Horticultural Science
College of Agriculture
Jahrom University
Jahrom, Fars Province, Iran

Kanza Aziz Awan
Department of Food Science and Technology
University of Central Punjab
Lahore, Punjab, Pakistan

Aliza Batool
Department of Food Science and Technology
MNS University of Agriculture
Multan, Punjab, Pakistan

Meerasabihalli Rangaswami Chandana
Division of Fruits and Horticulture Technology
ICAR-Indian Agricultural Research Institute
New Delhi, Delhi, India

Soheila Aghaei Dargiri
Department of Horticultural Sciences
Faculty of Agriculture and Natural Resources
University of Hormozgan
Bandar Abbas, Hormozgan Province, Iran

Huertas M. Díaz-Mula
Department of Applied Biology
Centro de Investigación e Innovación Agroalimentaria y Agroambiental (CIAGRO-UMH)
University Miguel Hernández
Orihuela, Alicante, Spain

Adem Dogan
Department of Horticulture
Faculty of Agriculture
Akdeniz University
Antalya, Turkey

Arturo Duarte-Sierra
Food Science Department
Center for Research in Plant Innovation (CRIV)
Laval University
Quebec, Canada,
and
Institute on Nutrition and Functional Foods (INAF)
Laval University
Quebec, Canada

Shaghef Ejaz
Department of Horticulture
Faculty of Agricultural Sciences and Technology
Bahauddin Zakariya University
Multan, Punjab, Pakistan

Mustafa Erkan
Department of Horticulture
Faculty of Agriculture
Akdeniz University
Antalya, Turkey

Umar Farooq
Department of Food Science and Technology
MNS University of Agriculture
Multan, Punjab, Pakistan

Nida Firdous
Department of Food Science and Technology
MNS University of Agriculture
Multan, Punjab, Pakistan

María E. García-Pastor
Department of Applied Biology
Centro de Investigación e Innovación Agroalimentaria y Agroambiental (CIAGRO-UMH), University Miguel Hernández
Orihuela, Alicante, Spain

Fernando Garrido-Auñón
Department of Food Technology
Centro de Investigación e Innovación Agroalimentaria y Agroambiental (CIAGRO-UMH), University Miguel Hernández
Orihuela, Alicante, Spain

Mostafa Gouda
College of Biosystems Engineering and Food Science
Zhejiang University
Hangzhou, P.R. China
and
Department of Nutrition & Food Science
National Research Centre
Dokki, Giza, Egypt

Prasoon Gunjan
Division of Food Science and Postharvest Technology
ICAR-Indian Agricultural Research Institute
New Delhi, Delhi, India

Yaodong Guo
College of Health Management
Shangluo University
Shangluo, Shaanxi, P.R. China

Muhammad Wasim Haider
Department of Horticultural Sciences
The Islamia University of Bahawalpur
Bahawalpur, Punjab, Pakistan

Ali Hamza
Department of Food Science & Technology
Faculty of Food and Home Sciences
MNS University of Agriculture
Multan, Punjab, Pakistan

Mahmood Ul Hasan
Horticulture, School of Science
Edith Cowan University
Joondalup, Western Australia, Australia

Norhashila Hashim
Department of Biological and Agricultural Engineering, Faculty of Engineering
Universiti Putra Malaysia
Serdang, Selangor, Malaysia
and
SMART Farming Technology Research Centre (SFTRC), Faculty of Engineering
Universiti Putra Malaysia
Serdang, Selangor, Malaysia

Majid Suhail Hashmi
Department of Food Science and Technology
The University of Agriculture, Peshawar
Peshawar, Khyber Pakhtunkhwa, Pakistan

Muhammad Hassan
Department of Horticulture
Faculty of Agricultural Sciences and Technology
Bahauddin Zakariya University
Multan, Punjab, Pakistan

Martín Ernesto Tiznado Hernández
Coordinación de Tecnología en Alimentos de Origen Vegetal
Centro de Investigación en Alimentación y Desarrollo (CIAD), A.C. Carretera Gustavo Enrique Astiazarán Rosas
Hermosillo, Sonora, Mexico

Md. Sakib Hossain
Bangladesh Food Safety Authority
Ministry of Food
Dhaka, Bangladesh

Bushra Hussain
Department of Horticulture
Faculty of Agriculture
University of Selcuk
Konya, Türkiye

Syed Bilal Hussain
School of Agriculture and Food Science
University College Dublin
Dublin, Ireland

Tanveer Hussain
Department of Horticulture
PMAS Arid Agriculture University
Rawalpindi, Punjab, Pakistan

Ayesha Iftikhar
Department of Agricultural
Environmental and Food Sciences (DiAAA)
University of Molise
Via De Sanctis, Campobasso, Italy

Yoshihiro Imahori
Graduate School of Life and Environmental Sciences
Osaka Prefecture University
Sakai, Osaka, Japan

Aamir Iqbal
Institute of Food Science
Cornell University
Ithaca, New York

Shahid Iqbal
Horticultural Science Department
North Florida Research and Education
Center, University of Florida/IFAS
Quincy, Florida

Fareena Jamil
Department of Food Science and Technology
MNS University of Agriculture
Multan, Punjab, Pakistan

Hafiz Umer Javed
College of Food Engineering
Beibu Gulf University
Qinzhou, Guangxi, P.R. China

Hafiz Muhammad Rashad Javeed
Department of Environmental Sciences
COMSATS University
Islamabad, Vehari, Pakistan

Seema Kanwal
Institute of Plant Protection
MNS University of Agriculture
Multan, Punjab, Pakistan

Nauman Khalid
School of Food and Agricultural Sciences
University of Management and Technology
Lahore, Punjab, Pakistan

Samina Khalid
Department of Biotechnology
COMSATS University
Islamabad, Vehari, Pakistan

Uzman Khalil
Horticultural Sciences Department
IFAS, University of Florida
Gainesville, Florida

Ghulam Khaliq
Department of Horticulture
Lasbela University of Agriculture, Water and Marine Sciences
Uthal, Balochistan, Pakistan

Muhammad Rafiullah Khan
College of Health Management
Shangluo University
Shangluo, Shaanxi, China
and
Department of Food Engineering
Pak-Austria Fachhochschule: Institute of Applied Sciences and Technology
Haripur, Pakistan

Emad Hamdy Khedr
Department of Pomology
Faculty of Agriculture
Cairo University
Giza, Egypt

Neha Khizar
Department of Horticulture
Faculty of Agricultural Sciences and Technology
Bahauddin Zakariya University
Multan, Punjab, Pakistan

Brijesh Kumar
Division of Food Science and Postharvest Technology
ICAR-Indian Agricultural Research Institute
New Delhi, Delhi, India

Ernesto Alonso Lagarda-Clark
Food Science Department
Laval University
Quebec, Canada
and
Center for Research in Plant Innovation (CRIV)
Laval University
Quebec, Canada
and
Institute on Nutrition and Functional Foods (INAF)
Laval University
Quebec, Canada

Xiaoli Li
College of Biosystems Engineering and Food Science
Zhejiang University
Hangzhou, P.R. China

Aman Ullah Malik
The University of Faisalabad
Faisalabad, Punjab, Pakistan

M. Sajid Manzoor
Department of Food Science and Technology
University of Central Punjab
Lahore, Punjab, Pakistan

Romina Alina Marc
Food Engineering Department
Faculty of Food Science and Technology
University of Agricultural Science and Veterinary Medicine of Cluj-Napoca
Cluj-Napoca, Romania

Bernard Maringgal
Faculty of Resource Science and Technology
Universiti Malaysia Sarawak
Samarahan, Sarawak, Malaysia

Nirmal Kumar Meena
Division of Food Science and Postharvest Technology
ICAR-Indian Agricultural Research Institute
New Delhi, Delhi, India
and
Department of Fruit Science
CH&F, Jhalawar, Agriculture University
Kota, India

Munishami Menaka
Division of Food Science and Postharvest Technology
ICAR-Indian Agricultural Research Institute
New Delhi, Delhi, India

Shabir Ahmed Mir
Department of Food Science and Technology
Government College for Women
Srinagar, Jammu and Kashmir, India

Farid Moradinezhad
Department of Horticultural Science
Faculty of Agriculture
University of Birjand
Birjand, South Khorasan, Iran

Ali Muhammad
Department of Food Science and Technology
The University of Agriculture, Peshawar
Peshawar, Khyber Pakhtunkhwa, Pakistan

Ayesha Murtaza
Department of Food Science and Technology
University of Central Punjab
Lahore, Punjab, Pakistan

Muhammad Nafees
Department of Horticultural Sciences
The Islamia University of Bahawalpur
Bahawalpur, Punjab, Pakistan

Ambreen Naz
Department of Home Sciences
Faculty of Food and Home Sciences
MNS University of Agriculture
Multan, Punjab, Pakistan

Aamir Nawaz
Department of Horticulture
Faculty of Agricultural Sciences and Technology
Bahauddin Zakariya University
Multan, Punjab, Pakistan

Safina Naz
Department of Horticulture
Faculty of Agricultural Sciences and Technology
Bahauddin Zakariya University
Multan, Punjab, Pakistan

Israel Ogwuche Ogra
UNESCO International Centre for Biotechnology
Nsukka, Enugu State, Nigeria

Umezuruike Linus Opara
SARChI Postharvest Technology Research Laboratory, Africa Institute for Postharvest Technology
Faculty of AgriSciences
Stellenbosch University
Stellenbosch, South Africa
and
UNESCO International Centre for Biotechnology
Nsukka, Enugu State, Nigeria

Jenifer Puente-Moreno
Department of Food Technology
Centro de Investigación e Innovación Agroalimentaria y Agroambiental (CIAGRO-UMH), University Miguel Hernández
Orihuela, Alicante, Spain

Farzana Rafeeq
Department of Horticulture
MNS University of Agriculture
Multan, Punjab, Pakistan

Asghar Ramezanian
Department of Horticultural Science
School of Agriculture
Shiraz University
Shiraz, Fars, Iran

Somayeh Rastegar
Department of Horticultural Sciences
Faculty of Agriculture and Natural Resources
University of Hormozgan
Bandar Abbas, Hormozgan, Iran

Kashif Razzaq
Department of Horticulture
MNS University of Agriculture
Multan, Punjab, Pakistan

Azri Shahir Rozman
Department of Biological and Agricultural Engineering, Faculty of Engineering
Universiti Putra Malaysia
Serdang, Selangor, Malaysia

Akhmad Sabarudin
Department of Chemistry
Faculty of Science
Brawijaya University
Malang, East Java, Indonesia

Irfan Ali Sabir
College of Horticulture
South China Agricultural University
Guangzhou, Guangdong, P.R. China

Ayesha Sarker
Agricultural and Environmental Research Station
West Virginia State University
Institute, West Virginia

María Serrano
Department of Applied Biology
Centro de Investigación e Innovación Agroalimentaria y Agroambiental (CIAGRO-UMH), University Miguel Hernández
Orihuela, Alicante, Spain

Hossein Shadmanfard
Department of Horticultural Science
School of Agriculture
Shiraz University
Shiraz, Fars, Iran

Waqar Shafqat
Department of Forestry, College of Forest Resources
Mississippi State University
Mississippi State, Mississippi

Hafiz Muhammad Shoaib Shah
Horticulture, School of Science
Edith Cowan University
Joondalup, Western Australia, Australia

Md Rayhan Shaheb
Department of Plant and Agroecosystem Sciences
University of Wisconsin-Madison
Madison, Wisconsin

Misbah Sharif
Department of Home Sciences
Faculty of Food and Home Sciences
MNS University of Agriculture
Multan, Punjab, Pakistan

Muhammad Sibt e Abbas
Department of Food Science and Technology
MNS University of Agriculture
Multan, Punjab, Pakistan

Zora Singh
Horticulture, School of Science
Edith Cowan University
Joondalup, Western Australia, Australia

Leila Taghipour
Department of Horticultural Science, College of Agriculture
Jahrom University
Jahrom, Fars, Iran

Intan Syafinaz Mohamed Amin Tawakkal
Department of Process and Food Engineering
Faculty of Engineering
Universiti Putra Malaysia
Serdang, Selangor, Malaysia

Vijay Yadav Tokala
The Postharvest Education Foundation
La Pine, Oregon

Amber Tufail
Department of Food Science and Technology
University of Central Punjab
Lahore, Punjab, Pakistan

Sami Ullah
Department of Horticulture
MNS University of Agriculture
Multan, Punjab, Pakistan

Muhammad Usman
Department of Food Science and Technology
MNS University of Agriculture
Multan, Punjab, Pakistan

Daniel Valero
Department of Food Technology
Centro de Investigación e Innovación Agroalimentaria y Agroambiental (CIAGRO-UMH), University Miguel Hernández
Orihuela, Alicante, Spain

Mohammad Valipour
Department of Engineering and Engineering Technology
Metropolitan State University of Denver
Denver, Colorado

Bandemuth Renukaradhya Vinod
Division of Food Science and Postharvest Technology
ICAR-Indian Agricultural Research Institute
New Delhi, Delhi, India

Shoaib Younas
School of Food and Biological Engineering
Hefei University of Technology
P.R. China

Xinhua Zhang
School of Agricultural Engineering and Food Science
Shandong University of Technology
Zibo, Shandong, P.R. China

PART I

Postharvest and Sustainable Technologies Applications

An Introduction to Postharvest Handling Technology of Fresh Fruits and Vegetables

1

Umezuruike Linus Opara* and Israel Ogwuche Ogra
*Corresponding Author: opara@sun.ac.za and unesco.icb.nigeria@gmail.com

1.1 INTRODUCTION

Fruits and vegetables (FAVs) constitute a vital component of the human diet and serve as a significant source of biologically active components such as secondary metabolites and vitamins (Poiroux-Gonord et al., 2010). Numerous health benefits can be derived from consumption of FAVs such as extended lifespan (Bellavia et al., 2013), enhanced mental well-being (Conner et al., 2017), improved heart health (Oyebode et al., 2013), lowered chances of cancer (Boffetta et al., 2010), and effective weight control (Rolls et al., 2004), among other positive outcomes (Van Duyn & Pivonka, 2000). Greater overall consumption of FAVs has also been correlated with a decreased likelihood of cognitive decline, thereby establishing its advantageous impact on mental well-being (Payne et al., 2012; Mc Martin et al., 2013). He et al. (2004) reported that healthy middle-aged women who include FAVs in their diet exhibited a decreased likelihood of obesity. Adequate consumption of FAVs has been linked through epidemiological evidence to a decreased likelihood of various non-communicable diseases (Pem & Jeewon, 2015). FAVs have distinct color that is linked with their unique health benefits. The antioxidant attributes of FAVs are indicated by their purple/blue color, which showcases their potential to lower the risks of cancer, stroke, and heart disease. Beetroot and eggplant serve as illustrative examples.

Similarly, the red tint in FAVs has been associated with lower cancer susceptibility and contributes to cardiovascular well-being, as evident in tomatoes, watermelons, and red grapes (Hoejskov, 2014; Amao, 2018; Liu et al., 2000). FAVs with an orange/yellow color contain carotenoids, which play a role in sustaining healthy eyesight, exemplified by carrots, lemons, and pineapples. Brown/white ones like bananas, garlic, onions, and ginger, among others, contain phytochemicals that boast antiviral and antibacterial properties. Additionally, green-colored FAVs, such as green apples, broccoli, spinach, green peppers, lettuce, and cucumbers, encompass phytochemicals renowned for their anticancer attributes.

Furthermore, the fibers present in FAVs have demonstrated the ability to slow down the movement of waste through the intestines by creating a larger mass, resulting in a more gradual absorption of nutrients (Anderson et al., 2010). This process contributes to the prevention of constipation. Additionally, these fibers can undergo fermentation in the colon, leading to an improved presence of short-chains fatty acid contents that possess anti-carcinogenic characteristics (Lattimer & Haub, 2010) and contribute to the well-being of the gastrointestinal tract. Furthermore, FAVs have been proposed as a means to avert osteoporosis in adults, primarily due to their abundant reservoirs of calcium and other essential vitamins that contribute significantly to maintaining bone health (Park et al., 2011). The substantial fiber content found in FAVs could potentially influence the absorption of calcium and help decrease the dietary "acid load" (New, 2001), ultimately enhancing the process of bone formation and curtailing bone resorption. This cascade of effects subsequently contributes to enhanced bone strength (Shen et al., 2012).

Most fruits contain considerable amounts of potassium, which plays a role in mitigating bone loss and the occurrence of kidney stones (Amao, 2018). Fruits contribute to optimal brain function by enhancing memory recall (Mintah et al., 2012) and furnishing the body with the necessary fiber for a well-functioning digestive system (Ridgewell, 1998; Amao, 2018). Moreover, fruits are abundant in vital dietary elements like potassium, antioxidants, and folic acid (Ness & Powles,

DOI: 10.1201/9781003370376-2

1997; Law & Morris, 1998; Tribble, 1999), ensuring not only overall health but also immediate energy supply and the provision of essential vitamins and minerals for bodily functions (Amao, 2018). Likewise, vegetables hold significance by enhancing general well-being, safeguarding crucial bodily organs, aiding in weight management, and fostering healthy skin and hair. Additionally, they supply copious antioxidants that bolster the body's defense against diseases while aiding digestion by preventing issues such as constipation, hemorrhoids, and diarrhea (Amao, 2018). The recommended guideline is to include a diverse range of FAVs in one's diet, as research indicates that their combination offers greater potential advantages compared to the consumption of individual fruits or vegetables alone (Agudo, 2004; Pem & Jeewon, 2015).

1.2 GLOBAL SURGE IN FRUIT AND VEGETABLE PRODUCTION

The appeal of venturing into horticultural crops is gaining traction among impoverished farmers worldwide. The global production of FAVs has shown a more rapid growth rate compared to cereal crops, although originating from a substantially lower starting point (Lumpkin et al., 2005). Between 1960 and 2000, cultivated area for horticultural produce on a global scale has more than doubled. A remarkable illustration of this trend is observed in China, where the horticulture area has increased over fivefold, accounting for about 20% of arable land within the past 25 years (Lumpkin et al., 2005). The surge in worldwide trade of FAVs has been equally impressive (FAO, 2004; Huang, 2004), and the collective value of traded horticultural crops presently surpasses that of cereal crops, constituting almost 21% of the total value exported from developing nations (Weinberger et al., 2005). According to the Food and Agriculture Organization (FAO), the proportionate representation of FAVs (including tree nuts and pulses) in global agricultural exports surged from 11.7% during 1977–1981 to 15.1% in 1987–1991. This share reached an unprecedented peak of 16.5% between 1997 and 2001. In a parallel trend, FAV juices saw their contribution to the total global export value of these products more than double, escalating from 3.6% in the years 1967–1971 to 8.7% during 1997–2001. Similarly, the portion attributed to vegetables and their derivatives grew from 26% to 32.7%, whereas that of fruits and their derivatives (excluding juices) experienced a decline from 48.5% to 39.1% (Huang, 2004).

Several factors underpin the worldwide upsurge in the production and trade of FAVs. Horticultural endeavors prove economically lucrative, with farmers engaged in such cultivation typically enjoying notably higher incomes in contrast to their counterparts involved in cereal production. On a per-capita basis, farm income has been documented to be as much as five times higher (Lumpkin et al., 2005). The cultivation of FAVs provides a platform for productive employment, especially in regions where the labor-to-land ratio is substantial, given that horticultural practices tend to be labor-intensive. Depending on the specific crop, horticultural production necessitates a minimum of twice the labor input and can extend to as much as five times the labor days per hectare when compared to cereal crops (Lumpkin et al., 2005).

China stands as the foremost contributor to global FAV production, with the cultivation of these crops representing substantial sources of revenue for its farmers. The impact of Chinese farmers and exporters on FAV production and trade is evident, anticipated, and positively influential. This influence stems from the surge in consumer demands, rising incomes, and China's accession to the World Trade Organization (WTO). Remarkably, China holds the distinction of being the largest worldwide producer of FAVs, with both markets constituting significant segments of the country's trade in the food sector (Azam & Shafique, 2018). Specifically, China's vegetable production reached approximately 484 million metric tons, out of which 4.4 million metric tons were exported, while fruit production amounted to 220 million metric tons, with exports totaling 3 million metric tons in 2012. Notably, around half of China's commercial gains originate from a mere 30 percent of its total production, enabling it to rank as the primary exporter globally. In the context of fresh vegetables, China holds the fourth position in terms of export value, while for fresh fruits it secures the seventh spot (Azam & Shafique, 2018). The expansion of horticultural production not only contributes to the commercialization of rural economies but also generates numerous employment opportunities beyond the realm of farming activities.

The global expansion of FAV systems presents distinct sustainability complexities compared to other commodities. FAVs encompass an extensive variety of crop types and cultivars cultivated under varying ecological and social circumstances worldwide. Over the past two decades, global FAV production has already surged by 50 percent, particularly concentrated in East and Central Asia, and a majority of this production is attributed to small-scale farmers across different regions (Herrero et al., 2017; FAO, 2020). Anticipated consumption escalation is foreseen in emerging economies where dietary diversification is unfolding, although upholding affordability for consumers may prove challenging (Mason-d'Croz et al., 2019; Stratton et al., 2021). The increased demand could potentially strain FAV producers who are already grappling with resource-intensive practices, pest challenges, changing biophysical conditions, and limited institutional support (Schreinemachers et al., 2018; Myers et al., 2017). Among all food groups, FAVs exhibit the maximum percentages of loss and waste both before and after harvest, with estimates ranging from 35% to 55% across different regions (Gustavsson et al., 2011; FAO, 2019).

1.3 INSUFFICIENT AVAILABILITY OF FRUITS AND VEGETABLES

Throughout history, FAV availability is often insufficient to meet consumption of the recommended levels. By examining the interplay between demand and supply in over 150 countries from 1961 to 2050, Mason-D'Croz et al. (2019) quantified the gap between future FAV supply and the levels of consumption recommended. It was reported that by the year 2015, 81 countries, encompassing 55% of the population globally, managed to attain FAV availability exceeding the minimum target set by WHO. When applying strict age-specific recommendations, only 40 countries, constituting 36% of the global population, achieved satisfactory availability. Despite anticipated economics growth fostering improved FAV availability, predominantly in low-income nations, this alone will prove inadequate. Even when considering highly optimistic socioeconomic forecasts (eliminating food wastage), numerous nations fall short of reaching the minimum recommended availability of FAVs. Notably, sub-Saharan Africa emerges as an area of particular concern, with estimates indicating that by 2050, between 0.8 and 1.9 billion individuals could inhabit states with FAV accessibility below 400 g/person per day.

The challenge of food waste poses a significant impediment that might undermine the anticipated advancements. Assuming a 33% waste rate and socioeconomic trend similar to historical patterns, the global normal availability by 2050 falls short of the age-dependent recommendation, thereby leading to an increase of 1.5 billion people present in countries with insufficient access to FAVs as compared with a scenario of zero waste at the global level. Raising the intake of FAVs forms a crucial element in transitioning toward more healthful and ecologically balanced diets. According to economic projections, it appears that in numerous countries, the forthcoming supply will not suffice to attain the suggested consumption levels, even under optimistic socioeconomic scenarios (Mason-D'Croz et al. 2019). Therefore, comprehensive and strategic public policy efforts aimed at addressing the obstacles related to both producing and consuming FAVs will be imperative. This entails implementing a range of measures and investments, including boosting FAV cultivation, advancing technologies and methods to curtail waste without imposing additional costs on consumers, and enhancing ongoing endeavors to educate the public about healthy dietary choices.

Siegel et al. (2014) utilized worldwide data on agricultural output and size of population to compare the availability of FAVs in the year 2009 with the nutritional requirements of the global population. The analysis indicated that, on average, the global supply of FAV falls short of the population's nutritional needs by approximately 22% based on recommended dietary guidelines (supply-to-need ratio: 0.78 [Range: 0.05–2.01]). This ratio demonstrates considerable variation depending on the economic status of individual countries, with low-income countries exhibiting a median supply-to-need ratio of 0.42, while high-income countries show a ratio of 1.02. Incorporating a sensitivity analysis that factors in food wastage from the perspective of nutritional needs resulted in a similar but slightly more pronounced inadequacy (global supply-to-need ratio: 0.66, varying from 0.37 [low-income countries] to 0.77 [high-income countries]). Furthermore, by employing projections of agricultural production and population growth, estimations for the supply and requirement of FAV for the years 2025 and 2050 were made. Assuming a moderate fertility rate and anticipated increases in agricultural output, the global supply-to-need ratio for FAV experiences a slight rise to 0.81 by 2025 and further to 0.88 by 2050, with similar trends observed across various income levels of countries.

Conversely, in a sensitivity analysis supposing no alteration in current levels of FAV production, the global supply-to-need ratio for FAV declines to 0.66 by 2025 and further to 0.57 by 2050. Furthermore, fresh FAVs generally have high moisture content, which predisposes them to rapid weight loss, mechanical damage, and attack by microbial pathogens, which lead to high incidences of postharvest loss and waste. The primary objectives of postharvest handling practices are to preserve quality attributes and fresh produce safety, reduce incidence of losses and waste, and facilitate trade and market access, which will result in their increasing availability for consumption. This chapter examines the problem of postharvest losses of FAV, discusses the physiological processes which underpin quality evolution and spoilage, and highlights the range of postharvest technologies which can be applied to maintain quality and reduce wastage, thereby increasing the supply of FAVs for consumption.

1.4 FRUIT AND VEGETABLE INTAKE/CONSUMPTION

On a global level, the average consumption of FAVs falls short of the recommended levels, with approximately half of the suggested vegetable intake and only a quarter of the suggested fruit intake being met. This trend is more pronounced in low-income countries and rural populations (Miller et al., 2016). Envisioning further increments in their consumption within future diets, the prominent analysis conducted by the EAT-Lancet Commission underscores that a substantial increase of 50% to 150% in global FAVs production will be necessary to adequately provide a nutritious diet to the anticipated 10 billion individuals by 2050 (Willett et al., 2019). Despite various critiques regarding the EAT-Lancet Commission's report and its growth predictions for various categories of foods (Hirvonen et al., 2019; Mason-d'Croz et al., 2019; Drewnowski, 2020; Hollis et al., 2020; Adesogan et al., 2020; Kim et al., 2020), the consensus to improve the dietary FAVs intake beyond present levels is almost unanimous among both international and local dietary plans and assessments of sustainable diets (Drewnowski et al., 2020).

Previous researchers have analyzed the intake of FAVs across diverse population segments in different regions. The consumption of FAVs by school-age adolescents was investigated by Nago et al. (2012) in Benin (Cotonou). Parental impact emerged as a primary influencer of FAV consumption patterns among adolescent students. Factors at a personal level that shaped their choices included taste preferences, knowledge about the health- and nutrition-related benefits of FAVs, as well as cultural beliefs. The majority the students did not incorporate FAVs into their daily diet except for commonly used ingredients like tomatoes and onions, which found their way into regular meals. Additional determinants encompassed considerations like food safety, the financial aspect of purchasing FAVs, medical recommendations, and the influence of media. It was concluded that to heighten the intake of FAVs among school-age adolescents, the availability of these items should be ensured through nearby food vendors. Furthermore, parental involvement was highlighted as pivotal, with specific attention needed for adolescents from economically disadvantaged backgrounds.

The awareness and intake of FAVs among high school students in Lagos state, Nigeria, was assessed by Silva et al. (2017). The majority of students described a sound understanding of the nutritional and health advantages associated with FAVs. Nevertheless, merely 5.45% of the students adhered to the recommended daily consumption of 400 g. As highlighted by the students, elements fostering sufficient FAV intake encompass parental role modeling, support, and oversight, along with the presence and convenience of these foods at home. Consequently, the research suggests that promoting healthy dietary habits during childhood is crucial for these students as they advance in age.

In South Africa, Peltzer and Phaswana-Mafuya (2012) reported a study on the use of FAVs within the older population, aged 50 years and above, based on a national survey of 3,840 participants. The results showed that the intake of FAVs was inadequate among 68.5% of older adults, evident in their average daily consumption of four servings. Factors which contributed to this insufficient consumption of FAVs included male gender, limited educational attainment, overweight status, minimal religious engagement, poor quality of life, daily tobacco usage, and skin color. As a result, the research recommended the implementation of public awareness campaigns and educational initiatives aimed at augmenting the intake of FAVs among the older adult population in Southern Africa.

Similarly, Mintah et al. (2012) conducted a study investigating the barriers to fruit utilization among students at a public university in Ghana. The investigations revealed that the primary deterrents to consuming FAVs were their high cost and the sense of satiety they induced. The limited intake of fruit was attributed to the fact that eating fruit did not sufficiently alleviate hunger for the students. Approximately 65%of the students failed to fulfill the daily recommended quota of 2–4 servings of FAVs, as stipulated by the USDA 1992 guidelines. The study proposed the implementation of awareness campaigns and educational initiatives regarding fruit consumption to enhance public health, particularly targeting university students and the general population alike.

1.5 POSTHARVEST LOSSES AND WASTES OF FRUITS AND VEGETABLES—AN OVERVIEW

Postharvest losses and waste of FAVs are massive all over the world (Munhuweyi et al., 2016; Santos et al., 2019; Blackenberg et al., 2018, 2021, 2022a, 2022b). Losses and wastage can occur at various points within the entire supply and handling chain, encompassing occurrences during harvesting, transfer to packhouse facilities or markets, grading, sorting, categorization, storage, marketing, processing, as well as at homes prior or after preparations (Figure 1.1). To elaborate further, postharvest losses manifest across the supply chain from the point of harvesting through all subsequent postharvest phases prior to use. These losses emerge inadvertently due to the inherent functioning of production and the systems governing the supply chain, encompassing technical, institutional, and legal aspects (Yahia et al., 2019; Opara et al., 2022). Conversely, waste refers to produce that is suitable for consumption but goes unconsumed and is discarded; this is commonly linked to the actions of consumers or retailers. While losses and waste can be differentiated, primarily due to their distinct origins and resolutions, they are nevertheless interconnected and can sometimes be challenging to distinguish, which leads to their synonymous usage (Yahia et al., 2019).

Postharvest losses are described as a measurable decrease in quality and quantity of a specific product at any point during the postharvest process. This encompasses alterations in the product's availability, edibility, safety, or overall quality, rendering it unsuitable for consumption (De Lucia & Assennato, 1994; FAO & UNEP, 1981; Hailu & Derbew, 2015). The loss can be qualitative or quantitative. Measuring qualitative losses, which encompass reductions in factors like edibility, nutritional value, calorie content, and consumer approval of products, proves to be more challenging than evaluating quantitative losses (Kader, 2005).

Quantitative losses refer to the reduction in mass or volume, diminishing the overall quantity of produce accessible for consumption. These types of losses are often quantified using measurements such as weight units, monetary valuation (currency), and energy content (calories) (Yahia et al., 2019). According to Gustavsson et al. (2011), 50% of cultivated FAVs are wasted. To put it differently, nearly half of all cultivated FAVs are lost even before reaching the end user for consumption. Moreover, this percentage is even more pronounced in developing and underdeveloped nations. A reduction in these postharvest losses plays a significant role in ensuring the sustainable nourishment of the world's population in the coming years (Gustavsson et al., 2011; Opara et al., 2021a, b).

Similarly, Kader et al. (2012) found that roughly a third of produced food is wasted in both developing and developed nations, totaling 1.3 billion tons each year. In higher-income states, a significant amount of food is disposed of before it becomes inedible (over 40% of losses occur during retail and consumer stages), whereas in developing countries, food losses occur early in the postharvest and processing phases of the supply chains. Salami et al. (2010) indicated that around 30–40% of FAVs are lost or discarded once they leave farms. Kader (2002) approximated losses during postharvest for fresh FAVs at 5–35% in developed nations and 20–50% in developing nations. The wastage of fresh FAVs spans the entirety of the food supply chain, from primary agriculture production to eventual household use.

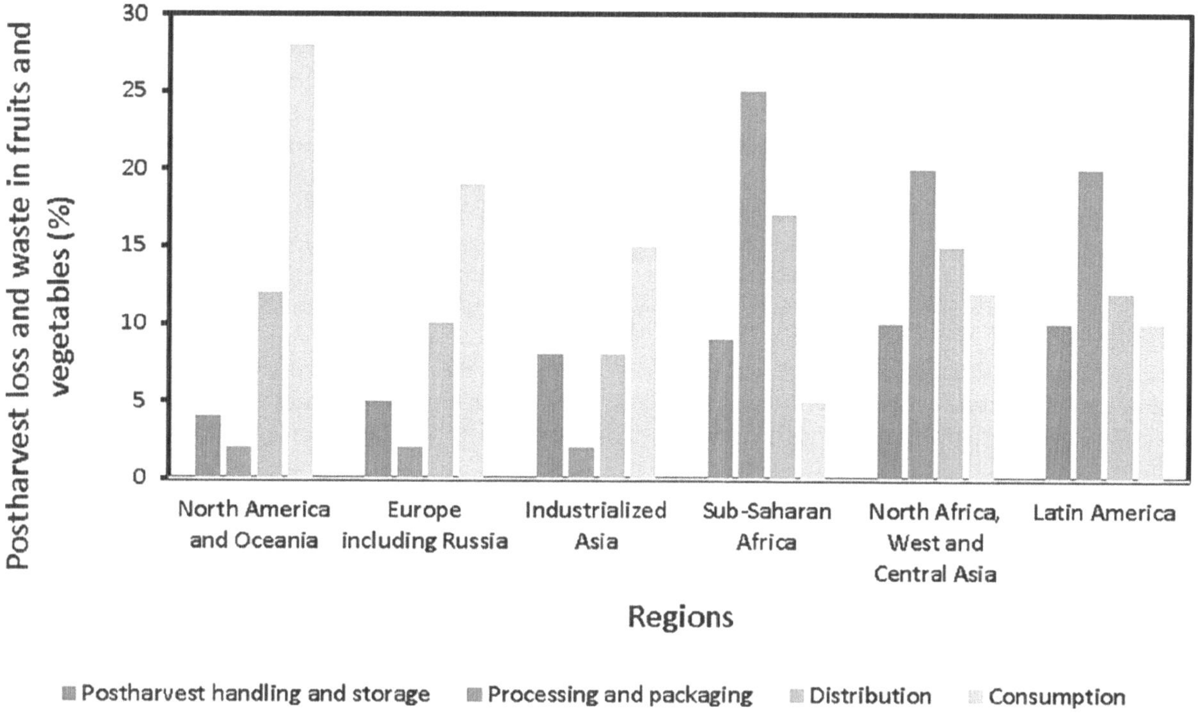

FIGURE 1.1 Percentage of postharvest loss and waste in fruits and vegetables at different stages across different regions (Gustavsson et al., 2011).

The repercussions of food loss and wastage extend to both the economy and the environment. From an economic perspective, they translate into wasted investments that diminish the financial prosperity of participants along the food supply chain. For instance, food wastage during consumption leads to an average annual cost of US$1,600 for a four-person family in the USA and approximately US$1,000 per year for an average household in the UK (Lipinski et al., 2013). In China, the annual value of discarded food reaches about US$32 billion (WRAP, 2011; Zhou, 2013). In sub-Saharan Africa, where numerous farmers earn less than US$2 daily, postharvest losses amass to a value of up to US$4 billion annually (World Bank, NRI & FAO, 2011), exacerbating the economic challenges. Thus, the influences of food loss and wastage extend to areas such as addressing hunger and poverty, enhancing nutrition, generating income, and fostering economic expansion. The occurrence of food loss signifies inadequately operating and ineffective value chains and food systems, thereby constituting losses of economic values for those involved in the said supply chain. The wastage of food resources that are traded globally and discarded in one region can potentially impact the availability and costs of food in other regions.

Food waste and loss also imposes adverse effects on the environment due to greenhouse gases, energy use, biodiversity, water, soil, and others invested in unconsumed food items. The production of food that ultimately goes unused results in redundant CO_2 emission, along with forfeiting the economic values of foods that were produced (Gustavsson et al., 2011). Food waste and loss generally contributes to around 173 billion cubic meters of annual water utilization, accounting for 24% of the total agricultural water use (Kummu et al., 2012). The land used for cultivating the lost and discarded food equates to 198 million hectares annually, an expanse comparable to the size of Mexico (Kummu et al., 2012). Moreover, the production of this lost and wasted food requires the application of 28 million tons of fertilizers each year (Kummu et al., 2012). In addition to these quantifiable repercussions, the natural landscape and vital ecosystem-related services these offer are detrimentally impacted by the resources expended on producing food that ultimately goes to waste (Hailu & Derbew, 2015).

The factors contributing to food loss exhibit variations between developing and developed nations, driven by differences in economic progress. As outlined by Gustavsson et al. (2011), food waste and loss in developed states are primarily attributed to the behaviors of consumers and insufficient coordination among different participants along the supply chain. Sales agreements between farmers and buyers might lead to the wastage of considerable amounts of harvested crops. Stringent quality standards can result in the discarding of food items that do not meet precise appearance or shape criteria. At the consumer level, substantial wastage stems from inadequate purchase planning, food expiration, and the indifferent attitude of those who can afford to dispose of food.

In contrast, the causes of food losses and waste in developing nations are predominantly tied to managerial, financial, and technical restrictions in harvest methods, pre-cooling and cooling facilities, especially in challenging climates, as well as deficiencies in packaging, handling, infrastructure, and marketing systems. Physical and quality losses primarily arise from insufficient temperature controls, use of poor packaging materials, poor handling, and a general lack of awareness about the necessity of preserving the quality and safety of perishable goods at the producer, wholesaler, and retailer levels (Opara & Al-Jufaili, 2006a, 2006b; Kitinoja et al., 2010). Therefore, the ensuing factors stand out as some of the most significant drivers of losses in fresh FAVs, particularly in developing countries and notably in sub-Saharan Africa (Hailu & Derbew, 2015).

Postharvest losses in FAVs are majorly due to four factors, including poor temperature management, microbial action, mechanical injuries, and poor packaging (Figure 1.2). Temperature, whether excessively high or extremely low, stands as the primary factor responsible for impacting the postharvest phase of perishable horticultural goods. Consequently, the temperature levels experienced by horticultural products during harvesting, handling, transportation, and marketing surpass the recommendations for preserving their quality. This is due to the ongoing and rapid processes of biochemical and respiration-related reactions in produce. Elevated temperatures are widely recognized to accelerate respiration rates, spoilage, and moisture loss in fresh produce, resulting in diminished market worth and reduced nutritional content (Hailu & Derbew, 2015). Recorded air and pulp temperatures in regions, for instance sub-Saharan Africa (SSA) and India, significantly exceed the ideal temperatures suggested for optimal postharvest handling, as prescribed for maintaining prime quality (Hardenburg et al., 1986). As a result, the projected shelf life would theoretically be far below the potential duration. Table 1.1 outlines the postharvest percentage losses in FAVs from different countries.

Agrios (2005) noted that postharvest diseases lead to the destruction of 10–30% of the overall crop yield, with some perishable crops, particularly in developing nations, experiencing losses exceeding 30% of their yields. Particularly vulnerable are fresh FAVs, which are highly susceptible to deterioration during postharvest operations. Insufficient treatments, packaging, storage, and transportation can result in spoilage and the proliferation of microorganisms, triggered by shifts in the physiological state of the produce (Wilson et al., 1995). Fungi emerge as the foremost and widespread pathogens, infecting a diverse array of host plants, and producing substantial and economically significant damage to the majority of fresh FAVs throughout storage and transport (Sommer, 1985). Fruits, due to their low pH, elevated moisture contents, and nutritional compositions, exhibit heightened susceptibility to fungal infections. These pathogens, aside from inducing decay, can render FAVs unsuitable for consumption by generating mycotoxins (Moss, 2002).

Mechanical injury is another imperative cause responsible for losses of FAVs. The primary challenge lies in maintaining the freshness of FAVs throughout the distribution process. This is owing to the higher water content, susceptibility to spoilage, and limited shelf life of fresh FAVs, which greatly restrict both transportation and transaction durations (Hailu & Derbew, 2015). Consequently, their transportation demands efficient safety measures. Nevertheless, owing to the inadequate state

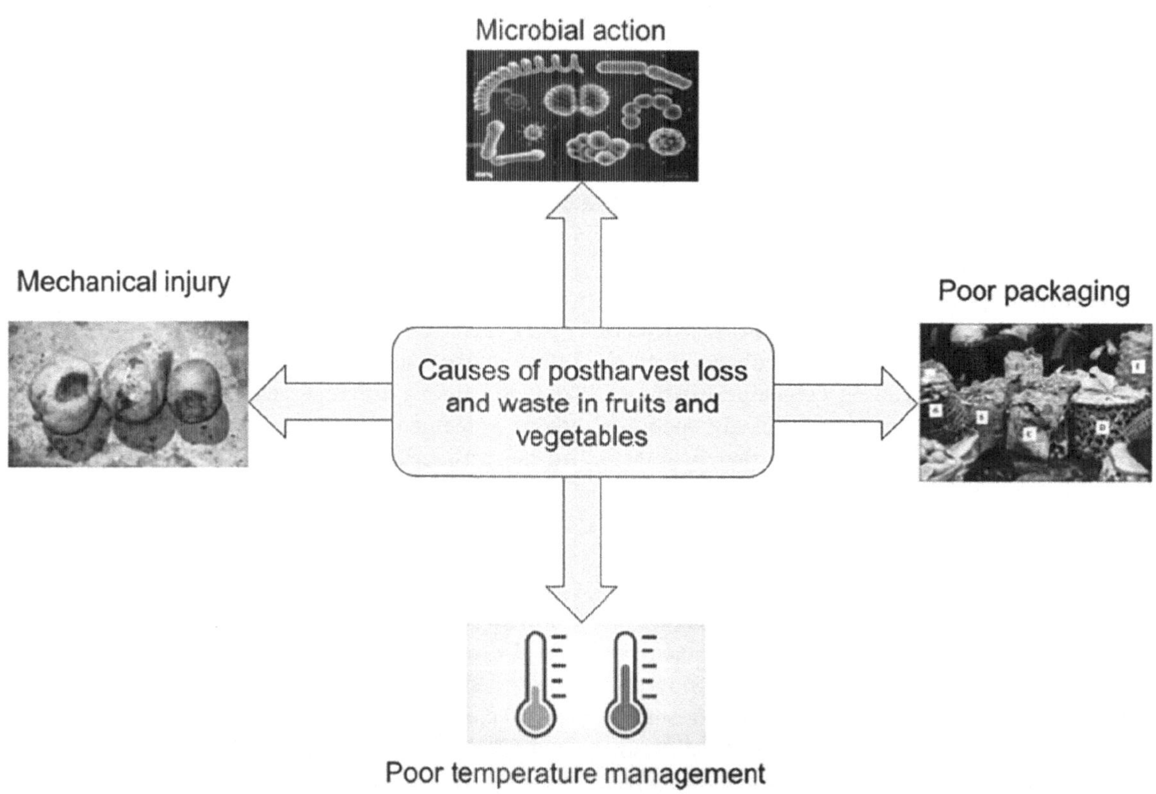

FIGURE 1.2 Major causes of postharvest loss and waste in fruits and vegetables.

TABLE 1.1 Postharvest Losses of Selected Fruits and Vegetables in Different Countries

FRUITS/VEGETABLES	*COUNTRY*	*POSTHARVEST LOSS (%)*	*REFERENCES*
Pomegranate	South Africa	15.3–20.1	Opara et al. (2021a)
Table grape	South Africa	5.51–23.3	Blackenberg et. al. (2021)
Pear	South Africa	18	Blackenberg et. al. (2022a)
Cabbage	South Africa	12 (cold storage), 46 (ambient conditions)	Munhuweyi et al. (2016)
Mango	Benin	17 (early April)	Vayssieres et al. (2008)
		70 (Mid-June, caused by fruit flies)	
	Brazil	28	Choudhury and da Costa (2004)
	Costa Rica	14.1 (dry season)	Arauz et al (1994)
		84.4 (rainy season, caused by anthracnose)	
	Mexico	< 10	Baez-Sanudo et al (1994)
	Pakistan	31–36.1	Mushtaq et al. 2005, Malik and Mazhar (2008)
Banana	Brazil	22.2	Santos et al. (2020)
Papaya		22.2	
Bell pepper		33.3	
Tomato		58.3	
Tomato	Benin	28	IITA (2008)
	Ghana	20–25	Bani et al. (2006); WFLO (2010)
Yams	Ghana	25–63	Bancroft et al. (1998)
Banana	Kenya	18.2–45.8	George and Mwangangi (1994)
Dessert banana		11.2	Save Food (2014)
Plantain		4.6	
Tomato	Saudi Arabia	17	Al-Kahtani and Kaleefah (2011)
Cucumber		21.3	
Fig		19.8	
Grape		15.9–22.8	
Date		15	
Fresh produce	Oman	3–19	Opara (2003)
Tomato	Jordan	18	El-Assi (2002)
Eggplant		19.4	
Pepper		23	
Squash		21.9	
Grape	Iran	13	Jowkar (2005)
Orange	Eqypt	14	El Shazly et al. (2009)
Tomato		15	
Pomegranate		23	Kitinoja and Kader (2015)
Onion		19	
Banana	Sri Lanka	20	Wasala et al. (2014)
Tomato		54	Rupasinge et al. (1991)
Tomato	Pakistan	20–22	Mujib et al. (2007); Zulfiqar et al. (2005)
Potato		12	Zulfiqar et al. (2005)
Onion		9	
Litchi	Bangladesh	4.6–8	Molla et al. (2010)
Fruits & Vegetables		23.6–43.5	Kamrul Hassan et al. (2010)

(Continued)

TABLE 1.1 (Continued)

FRUITS/VEGETABLES	*COUNTRY*	*POSTHARVEST LOSS (%)*	*REFERENCES*
Fruits	Fiji	0.07–2.44	Underhill and Kumar (2014)
Vegetables		4.07–10	
Chili pepper	Laos	10.7	Weinberger et al. (2008)
	Vietnam	16.9	
Yard-long bean	Cambodia	21.8	
	Laos	12.2	
Tomato	Cambodia	24.6	
	Laos	16.9	
	Vietnam	19.1	
Tomato	Rwanda	7.8–14.7	WFLO (2010)
Sweet potato	Tanzania	32.5–35.8	Rees et al. (2001)
Yam	Nigeria	10.5–12.4	Okah (1997)
Tomato		20–28	Olayemi et al. (2010)
Bell pepper		12–15	
Hot pepper		8–10	
Dried onions and tomatoes	Niger	15	Tröger et al. (2007)

of facility distribution and equipment, the quality degradation during distribution and sale is substantial, often resulting from mechanical damage. According to a previous report, the loss rate for FAVs ranges from 25% to 35% in developing countries, a significantly higher figure compared to the loss rates in developed nations, which typically stay below 5%. For instance, in the United States, the loss rate is as low as 1–2% (Fang, 2002). Another study conducted on supermarkets' fresh logistic distributions centers in 2006 highlighted that the majority of FAV losses occur during the transportation, storage, sorting, and shipping stages (Zhang & Deng, 2012).

Poor packaging is also a major contributing factor to postharvest loss of FAVs. According to Kitinoja & AlHassan (2012), the packaging employed for produce in SSA and India displayed notable shortcomings such as being excessively large, inadequately sturdy, and lacking in protective capacity for the produce during both handling and transportation. Moreover, a mere 15% of crops were found to be packed in wooden or plastic crates with poor quality. For instance, the plastic crates utilized in India for mangoes and tomatoes were unclean and coarse on the inner surfaces, leading to abrasions. Similarly, the wooden crates utilized for tomatoes in Ghana lacked ventilation and were designed to hold 60–70 kg of ripened fruit. These oversized wooden crates proved ineffective in safeguarding the produce, resulting in considerable damage to fruits at the bottom of the crate, much of which was discarded prior to being sold.

Minimizing postharvest losses serves as a complementary approach to enhancing the yield of FAVs. This significance is derived not only from the ethical duty to prevent wastage but also due to the fact that the cost of mitigating losses is usually lower compared to expenses involved in producing an equivalent additional quantity of FAVs with the same quality. Consequently, evaluating the postharvest losses incurred during the various stages of handling FAVs can aid in pinpointing the causative factors and their extent (Gajanana et al., 2011). This information, in turn, can contribute to the formulation of appropriate measures at different stages to prevent or diminish such losses. The outcome would be an improved availability of FAVs, both for local consumption and exportation.

Postharvest loss of FAVs is a global challenge that requires the employment of different technological advancements in order to ensure maximum availability and utilization of FAVs. Elik et al. (2019) noted that improvement in postharvest technologies, including effective harvesting methods and advanced packaging systems, plays an essential role in decreasing losses during postharvest, enhancing the quality attributes of fresh produce, and ensuring that a greater quantity of fresh produce is consumed. To reduce postharvest loss in fresh FAVs, Hailu & Derbew, (2015) suggested that improved as well as standard containers and packages should be utilized along with improved and standardized field packaging methods during harvest. They also suggested the utilization of low-energy storage, which does not require electricity but is capable of lowering temperatures by 10–15°C and conserve elevated humidity levels of approximately 95%, thereby extending the shelf life and preserving the quality of fresh produce. Other strategies that could be beneficial include solar drying in tropical and subtropical regions as well as hot water treatment, calcium treatment, and utilization of ethylene action inhibitors.

Gajanana et al. (2011) observed that most studies on the estimation of postharvest loss often tend to consider physical loss alone, which is misleading. It was noted that to accurately assess postharvest losses, it is critical to calculate the economic loss at various stages. This calculated loss should then be employed to estimate the total postharvest loss, instead of relying on the physical loss multiplied by market price, which tends to exaggerate the losses. The estimation process should also incorporate the actual prices received for the produce at different stages, rather than a singular price. Moreover, it is crucial to consider discrepancies in different quality grades when striving for an accurate estimation of postharvest loss.

1.6 MATURITY INDEXING—TECHNIQUES FOR ASSESSING CROP READINESS FOR HARVEST

The principles governing the appropriate timing of FAV harvest play a pivotal role in determining its successive storage capacity, marketing duration, and overall quality. Postharvest physiologists recognize three distinct phases in the lifecycle of FAVs: maturation, ripening, and senescence. The maturity signifies the point at which the fruit or vegetable is prepared for harvesting. During this phase, the edible portion of the produce has attained its full size, while it might not yet be suitable for instant consumption. Ripening, which follows or occurs concurrently with maturation, signifies the stage at which the produce becomes palatable and ready for eating based on its taste. Senescence marks the last phase, characterized by the ultimate deterioration of the fruit or vegetable, leading to changes such as loss of texture, flavor, and other attributes (Barbosa-Cánovas et al., 2003).

Maturity can be classified into two distinct groups: physiological maturity and horticultural maturity. Physiological maturity denotes the point at which a fruit is capable of further ripening and developing if harvested. Horticultural maturity indicates the stage of development at which a plant or its components meet the necessary conditions for consumer-specific purposes, signifying that it is ready for harvesting (Dhatt & Mahajan, 2007; Selvakumar, 2014, Prasad et al., 2018). Maturity indexing techniques help in ensuring the sensory quality and adequate postharvest life of FAVs. They also help to enable the planning of harvesting and packing as well as facilitate internet and phone marketing (Dhatt & Mahajan, 2007). Parameters which are important for maturity index determination are: chronological age, shape, color, surface characteristics, firmness, compositional factors (soluble solids), abscission layer development, tenderness, solidity, sugar content, starch presence, sugar-to-acid ratio, and oil content (Figure 1.3) (Irtwange, 2006).

Diverse maturity indexing techniques are employed in evaluating crop readiness for harvest. These include the optical method, which entails measuring maturity based on light transmission properties. These techniques rely on chlorophyll levels within fruits, which decline as the fruit matures. The fruit is subjected to intense illumination, followed by the cessation of light to envelop the fruit in complete darkness. Consequently, sensors assess the light quantity produced by fruit, which directly correlates to its chlorophyll pigments, thereby indicating its level of maturation (Barbosa-Cánovas et al., 2003; Prasad et al., 2018).

Skin color assessment is another maturity indexing technique which commonly applies to fruit since their skin color changes as they mature or ripen. Though certain fruit do not display noticeable changes in color as these mature with time, a characteristic is influenced by the specific nature of the fruit or vegetable. At the outset, there is a steady fading in the vibrancy of the color, transitioning from a deep green shade to a light green. In numerous cases, this progression culminates in the complete disappearance of the green color, giving way to the emergence of yellow, red, or purple pigments (Dhatt & Mahajan, 2007). This is important to mention that the

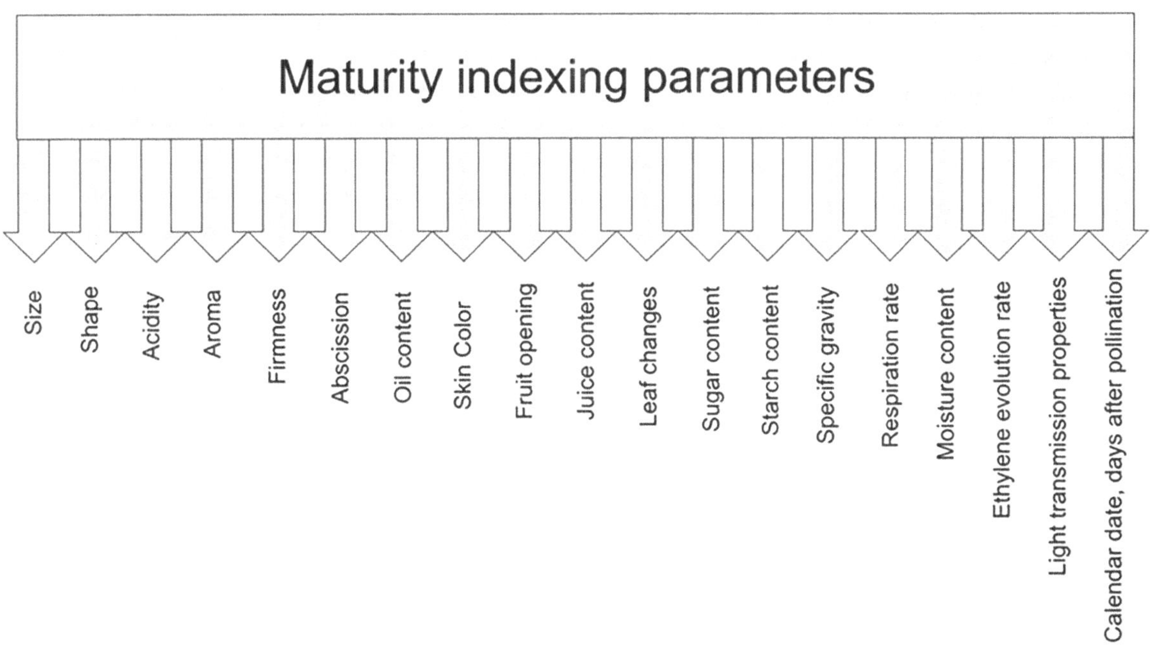

FIGURE 1.3 Parameters for maturity determination in fruits and vegetables.

determination of harvest maturity based on skin color rests on the judgment of the harvesters. Nevertheless, color charts tailored to particular cultivars, such as apples, tomatoes, peaches, chili peppers, and others, are accessible to aid in this process (Barbosa-Cánovas et al., 2003; Prasad et al., 2018).

The shape of a fruit changes throughout its maturation, offering a distinctive feature to ascertain harvest maturity. For example, bananas become progressively rounder in the cross-section and less defined in their angles as they grow. Similarly, mangoes also undergo shape modifications as they mature. During the maturation process of a mango on the tree, alterations in the alignment between the upper part of the fruit (shoulders) and the juncture where the stem is connected can be observed. In less mature mangoes, the shoulders slope away from the stem; however, in more developed mangoes, the shoulders align with the point of attachment, and with further advancement in maturity, the shoulders may even rise above this point (Barbosa-Cánovas et al., 2003; Dhatt & Mahajan, 2007). Changes in crop size during growth are often employed to establish the optimal harvest timing. In bananas, the width of individual segments can serve as an indicator of harvest maturity. Typically, a segment located midway within the cluster is selected, and its greatest width is measured using calipers, a method termed the caliper grade (Barbosa-Cánovas et al., 2003; Dhatt & Mahajan, 2007; Prasad et al., 2018). Gil et al. (2012) reported that a mature lettuce weighs 300–1,500 g and has a maximum core length of 50–80 mm.

Aroma assessment is another maturity indexing technique. Most fruits produce volatile compounds during the ripening process. These compounds contribute to the distinctive aroma of fruits and can serve as an indicator of its ripeness. Nevertheless, these aromas might only become perceptible to humans when the fruits are fully ripe, which restricts its practical applicability at a commercial level (Barbosa-Cánovas et al., 2003; Prasad et al., 2018). Fruit opening can be used to assess maturity in some fruits which have the potential to generate harmful substances during the ripening process, like the ackee tree fruit that accumulates toxic amounts of hypoglycine. The fruit naturally opens up upon full maturation, exposing black seeds encased in yellow arils. It has been demonstrated that at this stage, the presence of hypoglycine is negligible or even absent (Barbosa-Cánovas et al., 2003). The said results align with the conclusions of Brown et al. (1991), who advised against forcibly opening fully developed, bright red ackee fruit for human consumption.

The timing of FAV harvest is frequently influenced by the quality of the leaves. Similarly, in root crops, the state of the leaves can serve as an indicator of the state of the subterranean crop. For instance, when considering the storage of potatoes, the ideal time for harvest is shortly after the leaves and stems have withered. Harvesting them earlier could lead to reduced skin resilience against damage during handling and storage, making them more susceptible to storage-related diseases (Barbosa-Cánovas et al., 2003; Prasad et al., 2018). Similarly, Gil et al. (2012) reported a leaf length of 6–20 cm and a leaf width of 150–250 mm for matured spinach. During the natural growth process of a fruit, an abscission layer forms in the stem. For instance, in the case of cantaloupes, harvest before this abscission layer is entirely developed leads to fruit with less desirable flavor, in contrast to those that remain on the vine until the entire growth period is completed (Barbosa-Cánovas et al., 2003).

The fruit undergoes textural changes progressively as it matures, principally during the phase of ripening when it might become noticeably soft. Additionally, an excessive reduction in moisture content can also impact the crop's texture. The changes in texture can be perceived through touch, and the harvester can often determine the readiness for harvest by gently applying pressure to the fruit. Nowadays, advanced tools like texture analyzers and pressure testers have been designed to measure the texture of FAVs. These devices come in various forms and are currently accessible for assessing the texture of various types of produce (Barbosa-Cánovas et al., 2003; Dhatt & Mahajan, 2007; Prasad et al., 2018).

The juice content in different fruit types tends to increase as fruit undergoes maturation on trees. For juice contents determination, a typical fruit sample is selected, and juice is extracted using a predefined and standardized procedure. The juice volume extracted is then correlated with the mass of initial juice, which is indicative of its level of maturity (Barbosa-Cánovas et al., 2003).

Moisture and oil content can serve as an indicator of fruit maturity, as seen in avocados. According to California's Agricultural Code, avocados must have a minimum oil content of 8% per avocado (excluding seed and skin) at the time of harvest and thereafter, particularly for cultivars of Guatemalan or Mexican races. This oil content is interrelated with the moisture content of the avocado. The determination of oil content involves weighing a small portion (5–10 g) of the avocado's flesh and subsequently extracting the oil using solvents in distillation columns. This technique has proven effective for different cultivars that possess higher oil content (Nagy & Shaw, 1980; Barbosa-Cánovas et al., 2003; Prasad et al., 2018). Moisture content decreases in avocado fruits as it matures, which corresponds to an increase in the oil content (Olaeta-Coscorroza & Undurraga-Martinez, 1995).

Starch and sugar content assessment is another maturation indexing technique. In climacteric fruits, starch accumulates as carbohydrates during maturation. As the fruit undergoes ripening, the starch gets converted into sugar. Conversely, non-climacteric fruits tend to amass sugar throughout the maturation process. A rapid technique to quantify sugar content in fruits involves using a Brix refractometer or a hydrometer. A few drops of sample juice are placed in the sample holder of the refractometer, and a reading is taken to determine the total soluble solids or sugar content. This parameter is commonly used worldwide to indicate maturity (Barbosa-Cánovas et al., 2003; Dhatt & Mahajan, 2007; Prasad et al., 2018).

However, to determine maturity in pear cultivars, one reliable method involves measuring starch contents. This technique entails cutting the fruits in half and dipping the cut portions in solutions having 1% iodine and 4% potassium iodide. Starch presence is shown by blue-black colorations on the cut surface. It has been reported that when the time of harvest approaches, starch changes into sugar. The harvest process begins when about 65–70% of the cut surface exhibits a blue-black color (Barbosa-Cánovas et al., 2003; Prasad et al., 2018) since the energy contents of fruit diminishes as it undergoes the maturation process, leading to the transformation of water-insoluble starch, which holds a calorific value of 4,200 cal/g, into soluble monosaccharide contents with a value of 3,750 cal/g. Bajcar et al. (2016) reported a calorimetric method for evaluating the maturity of harvested fruit, focusing on specific apple, tomato, and strawberry varieties. The calorimetric benchmarks for maturity were established for these fruits as follows: apple—3,930 cal/g dry weight, strawberry—3,880 cal/g dry weight, and tomato—3,910 cal/g dry weight (Bajcar et al., 2016).

In numerous fruits, changes in acidity occur throughout the process of maturation and ripening. For citrus and various other fruits, acidity diminishes steadily as the fruit progresses in maturity on the tree. Collecting samples of these fruits, extracting the juice, and titrating it with standardized alkaline solutions provides a quantity that can be linked to the optimal harvest period. Typically, acidity is not considered a standalone indicator of fruit maturation, but rather in conjunction with soluble solids, resulting in the Brix-to-acid ratio (Barbosa-Cánovas et al., 2003; Prasad et al., 2018). Specific gravity refers to the relative weight of solids or liquids in comparison to distilled water at a temperature of 62 °F (16.7 °C), which serves as the baseline measurement. This method involves associating the weight of equivalent volumes of different constituents with water weight. To determine specific gravity, the produce is first weighed in the air and then in pure water. The ratio of the weight in the air to the weight in water yields the specific gravity values. This process ensures a dependable assessment of fruit maturation. As a fruit undergoes maturation, its specific gravity tends to rise (Barbosa-Cánovas et al., 2003; Dhatt & Mahajan, 2007).

Quamruzzaman et al. (2022) established maturity indices for some FAVs. Tomato having a suitable size (6.2 cm diameter and 6.5 cm length), TSS (4.5%), weight (84 g), pH (4.3), transition to a "red" color, and a "delicious" taste at the fifth week stage is deemed fit for harvest. Similarly, at the fifth week stage, broccoli having appropriate size (12.0 cm length and 13.0 cm diameter), weight (360 g), and "green" coloration is ready for harvest. At the sixth week stage, the netted melon having a suitable size (14.5 cm diameter and 15.2 cm length), weight (800 g), pH (6.3), TSS (10.8%), "fully developed netting" on the skin of the fruit, and a strong flavor is fit for harvest. Concurrently, cucumbers of size (8.8–10.8 cm length and 2.2–2.9 cm diameter), TSS (3.8–4.1%), weight (61–88 g), pH (6.3), and exhibiting "reduced powdery" characteristics are ready for harvest. Other maturity indexing methods include assessment of respiration rate and ethylene evolution rate, flowering and bolting, calendar date, and days after pollination. Harvesting FAVs should be done at the right maturity stage because it not only affects the postharvest quality of product but also enhances its potential shelf life (Ramjan et al., 2017).

1.7 POSTHARVEST PHYSIOLOGY OF FRESH FRUITS AND VEGETABLES

The postharvest physiology of FAVs is affected by factors like ethylene production, respiration rate, and water loss. The effect of the said processes depends on postharvest factors such as humidity, temperature, light, mechanical injury, and postharvest diseases (Kahramanoğlu, 2017; Mishra & Gamage, 2018; Saltveit, 2018). Respiration serves as an indicator for the metabolic functioning of all living produce and holds notable importance in postharvest physiology and degradation of FAV quality. The pace of quality decline typically aligns with the rate of respiration, often serving as an indicative measure of a crop's storage capability. FAVs with elevated respiration rates tend to have shorter shelf lives, while those with lower rates exhibit longer preservation potential. The respiration rate can be employed as a parameter to assess and compare the perishability of different FAVs (Mishra & Gamage, 2018; Saltveit, 2018).

The impact of respiration rates on spoilage can be elucidated by (i) linking the generated reducing power and ATP production from the rate of respiration with deteriorative biochemical processes that contribute to quality decline; (ii) depletion of food reserves; (iii) harmful consequences arising from the increase of CO_2; and (iv) elevation of product temperature caused by the heat produced through respiration. This temperature increase also affects the calculation of cooling requirements during storage (Haard, 1995; 1998). The respiration intensity depends on various factors (such as quantity of sugars, shape, size, cell morphology, and maturation) and external factors (such as light, temperature, water stress, growth regulators, availability of ethylene, oxygen, and CO_2). Among these external factors, temperature, atmospheric compositions, and physical stresses exert the most significant influence on respiration activity. The postharvest regulation of respiration involves managing the said aspects to mitigate the degradation of quality (Mishra & Gamage, 2018).

During the process of ripening, fruits exhibit two distinctive respiration patterns and are categorized as (i) non-climacteric and (ii) climacteric. In climacteric produce, there is a marked surge in respiration rate as ripening progresses due to the production of ethylene, unlike non-climacteric fruits where very little ethylene is produced (Phan et al., 1975; Kader & Barrett, 2003; Saltveit, 2018). The rate of respiration is contingent upon temperature, and the extent of this temperature-related influence varies both among and within different cultivars (Mishra & Gamage, 2018). Physical stresses encountered during growing, harvesting practices, and postharvest handling significantly affect respiration rate patterns. Damage to tissues escalates respiration rates and triggers the production of ethylene, which in turn can amplify respiratory activity, leading to a subsequent decline in quality. When investigating the connection between respiration, vulnerability to bruising, and temperature in sweet cherries, Crisosto et al. (1993) observed that impact-induced bruising had the most pronounced impact on the tested cultivars when the fruit's flesh was below 10 °C. Moreover, a nearly threefold rise in respiration rate resulted from impact-induced bruising when fruits were maintained at 20 °C for a period of two days (MacLeod et al., 1976b).

However, reducing the level of oxygen and elevating the concentration of CO_2 in the ambient atmosphere around fresh FAVs leads to a decrease in respiration rate. The degree of impact varies based on variables like temperature, types of produce, cultivars, age, and maturity level at harvest time (Mishra & Gamage, 2018). This is important that lowering the concentration of oxygen levels below 2–3% results in advantageous decreases in respiration rates and other metabolic activities in the majority of produce. However, it is advised not to completely remove oxygen, as an anaerobic environment can be detrimental to the produce's quality. Such conditions can prompt fermentation, decay, the emergence of undesirable flavors, and alterations in color and texture (Mishra & Gamage, 2018).

Transpiration is a mass transfer phenomenon where water vapor migrates from produce surfaces into the adjacent atmospheres. The rate of transpiration is directly linked to the partial pressures gradient across the transfer surfaces and the surface area, while being inversely associated with the cumulative resistance factors, including surface characteristics and the presence of surface waxes (Sastry et al., 1978; Díaz-Pérez, 2018). The water loss from FAVs leads to a state of water stress within their tissues. This stress could potentially expedite the senescence process, likely due to an elevated rate of breakdown in cellular membranes and the release of solutes (Díaz-Pérez, 2018). If the water loss becomes excessive, it causes the produce to become soft, shriveled, and lose its glossy appearance, leading to calyx browning in eggplants (Risse & Miller, 1983).

Transpiration is majorly responsible for water loss, resulting in a reduction in marketable weight, visual appeal, texture, and nutritional quality (Kader & Barrett, 2003). For the majority of FAVs, moisture loss of 5–10% becomes evident through visible signs of withering and wilting owing to plasmolysis of cells (Salunkhe & Desai, 1984). These signs become concerning when weight loss reaches approximately 5% of the initial harvested weight (Pantastico, 1975; Salunkhe & Desai, 1984). FAVs can potentially be categorized into three general ranges of transpiration rate when stored at low temperatures: high (500–850 mg/kg h mmHg; e.g., carrot and parsnip), intermediate (100–250 mg/kg h mmHg; e.g., cabbage and rutabagas), and low (10–80 mg/kg h mmHg; e.g., potatoes and onions) (Van den Berg & Lenz, 1971). The loss of water through transpiration after harvesting leads to a reduction in cell rigidity, resulting in undesirable changes such as the softening of produce like bell peppers and eggplants; the shriveling of crops like citrus, cucumber, lettuce, spinach, mushrooms, and potatoes; and the diminished shine of eggplants, mangoes, and cucumbers. These effects also extend to alterations in the color of the outer layer of produce, such as the whitening of carrots, among various other factors that negatively impact the overall quality of the harvested goods (Díaz-Pérez, 2018).

Transpiration rate is influenced by a combination of factors related to the product itself and the surrounding environment. These factors include (i) the structure of the skin; (ii) size, shape, and surface area; (iii) differences in water vapor pressure; (iv) air circulation, temperature, relative humidity; (v) respiration heat; (vi) stage of maturity; (vii) endothermic influences of evaporation; and (viii) the concentration of solute found in the produce (Sastry et al., 1978; Sastry & Buffington, 1982; Díaz-Pérez, 2018). Among environmental conditions, relative humidity (RH), temperature, and air circulation have the most notable impact on water loss through transpiration. Generally, elevated surface temperature and reduced RH lead to an increase in transpiration rate (Mishra & Gamage, 2018). Reducing water loss can be achieved by retaining pressure levels higher than atmospheric, employing lower temperatures and high relative humidity levels during storage, selecting appropriate packaging materials, optimizing loading density and thickness (Lenz et al., 1971, Díaz-Pérez, 2018), and implementing techniques such as applying water-resistant coatings like waxes onto surfaces or utilizing suitable plastic film packaging (Pantastico, 1975; Salunkhe & Desai, 1984; Wills et al., 1998; Baldwin, 2003; Ben-Yehoshua & Rodov, 2003, Díaz-Pérez, 2018).

Ethylene gas (C_2H_4) is a natural hormone generated by certain climacteric produce and vegetables during their maturation process, although it lacks color and scent. Its presence can accelerate the ripening of exposed produce, potentially leading to premature maturation within storage facilities. Ethylene plays a pivotal role in stimulating various reactions within plant tissues and plays an important role in the natural ripening of numerous fruits, like apples and bananas. However, its effects can also be detrimental, causing untimely ripening and skin damage in fruits (Kahramanoğlu, 2017). Ethylene serves as a catalyst for the ripening of climacteric and certain non-climacteric fruit, as well as for processes like anthocyanin synthesis, chlorophyll degradation (degreening), seed germination, the development of adventitious roots, senescence, abscission, flower initiations, and both respiratory and phenylpropanoid metabolic activities (Saltveit, 2002). It is worthy of note that ethylene is also capable of accelerating its own production in the ripening of climacteric fruits. Commercially, ethylene is utilized as a "ripening hormone" for climacteric produce such as bananas and mangoes, and as a "degreening hormone" for citrus fruits. The favorable or unfavorable outcomes of ethylene action hinge on various factors, including produce type, specific cultivar, maturity status at harvesting, temperature, and the influence of other hormones (Haard, 1998).

Factors such as the presence of CO_2, insufficient O_2 levels, and low temperatures can hinder the production of ethylene in fruits. Conversely, when produce is injured, ethylene production tends to be higher. Additionally, ethylene can be artificially produced and harnessed as an environmental factor to induce ripening (Kahramanoğlu, 2017). Managing the effects of ethylene is immensely significant from a commercial standpoint for FAVs, as the unfavorable consequences can lead to substantial economic losses. An essential consideration is

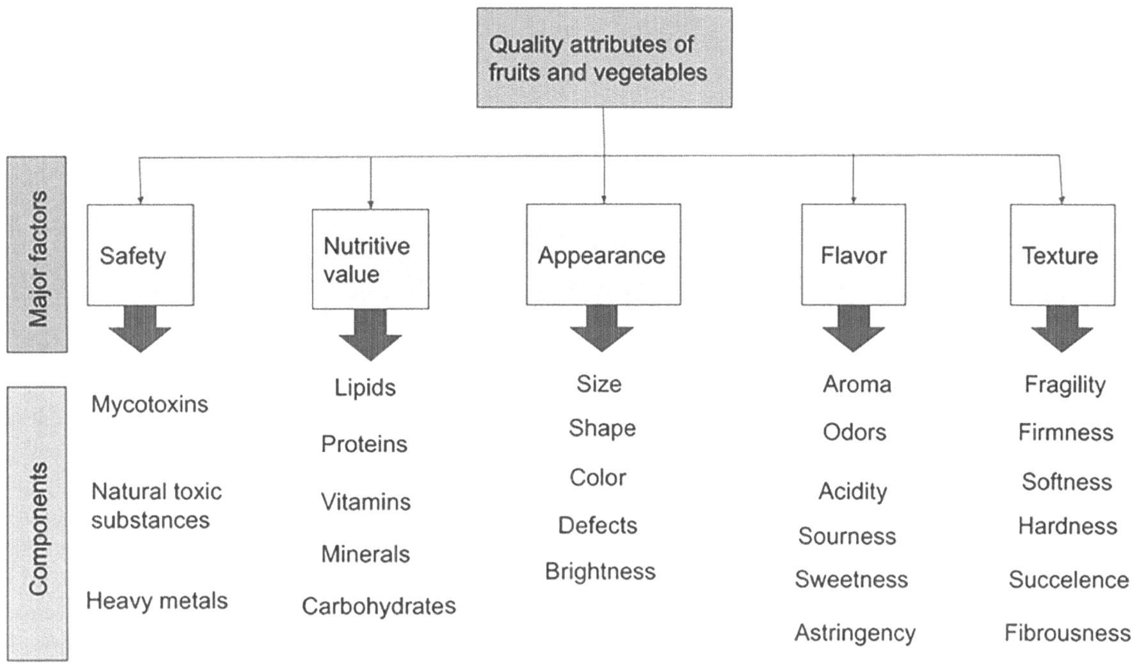

FIGURE 1.4 Quality attributes of fruits and vegetables.

to avoid storing plants that produce ethylene, such as apples, alongside fruits, vegetables, or flowers that are susceptible to its effects, like cabbage. Such co-storage could result in harm to the produce, a decline in quality-related attributes, and a decrease in shelf-life potential (Kahramanoğlu, 2017).

Furthermore, the production rate of ethylene by fresh fruits can be diminished through strategies such as storing at lower temperatures, decreasing oxygen levels (below 8%) while elevating carbon dioxide concentrations above 1%, evading stressors like fruit damage, disease incidence, and water scarcity, and employing cooling to curtail respiration rates (Phan et al., 1975; Salunkhe & Desai, 1984; Kader & Barrett, 2003). Saltveit (2002) outlines three primary approaches to control ethylene's impact, involving the prevention of tissue exposure, its perception, and its responses. To avert tissue exposure, methods like eliminating ethylene from the surroundings through scrubbing, hindering synthesis, and implementing appropriate ventilation are employed. Techniques that function by obstructing ethylene perception include the utilization of silver, CO_2, and 1-methyl cyclopropene (1-MCP), as well as lowering temperatures and adopting ethylene-insensitive cultivars (Beudry & Watkins, 2001; Mishra & Gamage, 2018). Research on 1-MCP has led to its approval for use in certain fruits within the United States (Beudry & Watkins, 2001; Kader & R. Rolle, 2004; Mishra & Gamage, 2018). Reducing plant responsiveness to ethylene can be achieved by lowering temperatures, employing controlled or modified atmospheres, applying genetic engineering, or utilizing chemical-based treatments to impede enzyme activities and alter the synthesis of proteins (Mishra & Gamage, 2018).

Postharvest physiology of FAVs is majorly affected by processes like respiration, transpiration, and ethylene production (Figure 1.4). These three processes are dependent on each other as they not only respond in similar manners to environmental changes like decreasing storage temperatures, but these also affect one another. For instance, low respiration rate reduces ethylene production and transpiration rate. Thus, overlooking the effect of any of these three factors will lead to quality deterioration of FAVs. Similarly, extremes of any of these factors can lead to the development of physiological disorders which will affect the market acceptability of produce. Moreover, some physiological processes resulting in enhanced quality are strongly associated with respiration, transpiration, and ethylene production of FAVs.

1.8 QUALITY INDICATORS AND MEASUREMENT TECHNIQUES

The concept of quality refers to the level of excellence of a product or its suitability for a specific purpose. Quality is a human-created concept that encompasses numerous attributes or features. When applied to produce, quality encompasses sensory characteristics (for instance, texture, appearance, taste, and aroma), nutrition content, chemical compositions, mechanical attributes, functional properties, and certain defects (Abbott, 1999). The ideal quality of a product is associated with a specific stage of development or ripening, at which time the combination of physical features and chemical components achieves the highest level of consumer satisfaction. An essential concept for preserving quality attributes and prolonging shelf life is recognizing that various portions of FAVs remain biologically active, both during their attachment to the parent plant and after being harvested (Chitarra & Alves, 2001). Quality evaluations can take the form of either objectivity or subjectivity. Objective assessments entail quantifiable measurements, whereas subjective evaluations are based on personal judgments. Objective quality attributes are those that can be determined such as size, weight, and color, as well as internal factors associated with quality like sugar and acid concentrations and nutritional values.

Fruit quality assessment encompasses both intrinsic and extrinsic characteristics (Figure 1.4). The intrinsic attributes primarily involve aroma, taste, flavor, texture, nutritional quality (such as levels of soluble sugars, organic acids, starch, total flavonoids, carotenoids content, total phenolics, and antioxidative capacity), mesocarp firmness, disease infestations, and chemical residues. Meanwhile, extrinsic quality pertains to the outward aspects like appearance, size, color, and bruises (Aaby et al., 2002; Awad & De Jager, 2002; Boboeva, 2023). Product quality assessment can be conducted either through a sensory panel or by employing instruments, which may or may not be destructive (Chitarra & Alves, 2001).

Quality criteria fall into three main categories: sensory, nutritional, and safety attributes. These categories should be collectively considered to meet both consumer preferences and public health standards. According to Kader and Rolle (2004), the significance attributed to a specific quality characteristic varies based on the particular product and the perspectives of individuals involved (consumer, producer, and postharvest handler) or the markets concerned with quality evaluation (Brasil & Siddiqui, 2018). For producers, crucial quality attributes encompass high yields, favorable appearance, ease of harvesting, and the ability to withstand extended shipping distances to reach markets. Wholesalers and retailers prioritize attributes like firmness, appearance, and shelf life. In contrast, consumers primarily assesses the quality of FAVs based on their initial visual impression (including perceived "freshness") at purchase time. Subsequent purchases are influenced by the consumer's contentment with the flavor and eating experience of the edible portions of the produce (Brasil & Siddiqui, 2018).

The quality of FAVs can be assessed by evaluating factors such as appearance, texture, flavor, nutritional content, and safety. Appearance is an essential quality factor in the marketing of FAVs. This encompasses characteristics like color, size, shape, glossiness, and absence of defects. Postharvest defects can be categorized as physical, morphological, physiological, or pathological (Kader & Rolle, 2004; Chitarra & Chitarra, 2005). Textural attributes encompass characteristics such as mealiness, firmness, juiciness, crispness, and toughness, which can vary based on the specific product. The texture of FAVs carries significance not only for their edibility and culinary applications but also for their resilience during transportation. Fruits with a delicate texture are prone to significant damage when shipped over extended distances. Often, the transportation of such tender fruits requires harvesting them before they reach their optimal maturity in terms of flavor quality (Kader & Rolle, 2004).

Flavor encompasses characteristics such as sweetness, acidity (sourness), astringency, bitterness, scent, and undesirable tastes. Evaluating flavor quality involves perceiving the tastes and fragrances of various compounds. Combining an objective analytical assessment of crucial components with subjective evaluation from a taste panel is necessary to provide valuable and meaningful insights into the flavor quality of fresh produce. This methodology can establish a baseline level of acceptability. To determine consumer preferences for the flavor of a particular product, it is essential to conduct extensive testing involving a representative cross-section of consumers (Kader & Rolle, 2004). Nutritional quality constitutes a vital aspect within the FAV supply chain (Chitarra & Chitarra, 2005). Fresh produce plays a substantial role in human nutrition, notably as a source of essential minerals, vitamins, and dietary fiber. Additional constituents found in fresh FAVs that potentially contribute to reducing the risk of ailments like cancer include carotenoids, flavonoids, phytosterols, isoflavones, and various other phytochemicals, often referred to as phytonutrients (Kader & Rolle, 2004). The paramount quality concern is safety, and it is imperative that FAVs are devoid of any chemical substance that could potentially harm human well-being (Chitarra & Chitarra, 2005).

Individuals utilize all their senses such as vision, taste, smell, touch, and even hearing to assess quality. The consumer combines all these sensory inputs, including visual appearance, aroma, flavor, tactile sensation, oral sensation, and chewing sounds to form a conclusive assessment of the suitability of a particular fruit or vegetable. While sensory evaluations remain crucial, instrumental measurements are favored in numerous research and commercial contexts. Instruments mitigate disparities among different individuals, offer greater precision, and establish a shared language that connects researchers, industry professionals, and consumers (Shewfelt, 1999). Instrumental detection of appearance involves the measurement of electromagnetic characteristics (typically optical), while texture is assessed through mechanical attributes. Flavor, encompassing both taste and aroma, is evaluated based on chemical properties (Abbott, 1999). Some modern tools used in FAV quality assessment include chromameters, texture analyzers, automated titrators, ethylene gas analyzers, O_2/CO_2 gas analyzers, and refractometers. These quality properties can be assessed by non-destructive and destructive methods.

1.8.1 Destructive Methods

A solvent extraction method using the Soxhlet extractor and refractive index technique is commonly used to measure oil content, which is a quality attribute of fruits (Lee, 1981; Lee et al., 1983; Duduzile Buthelezi et al., 2019). Although Soxhlet extraction provides precise results, it is a laborious process that necessitates extremely high temperatures for operation (Meyer and Terry, 2008). In contrast, the refractive index technique, which is globally acknowledged, raises doubts about its precision. An examination of the statistical relationship between the established Soxhlet and refractive index methods revealed a correlation coefficient of 0.90, highlighting disparities in the accuracies of these approaches (Magwaza & Tesfay, 2015).

Dry matter or moisture content can be evaluated using a domestic food dehydrator, a conventional oven, freeze drying, or a microwave. Typically, the oven method is employed to assess dry matter as described by Lee et al. (1983). In this approach, a fresh sample extracted from fruit tissue is placed in a petri dish, then subjected to drying within an oven at 60 °C for a duration of 2–3 days (or until a consistent mass is achieved), followed by reweighing. Arpaia et al. (2001) described a faster methodology utilizing microwave oven drying as a proficient instrument. This microwave-based approach offers relatively faster drying for avocado samples, spanning from 15 to 40 minutes, contingent upon the specific microwave configurations and model in use (Magwaza & Tesfay, 2015). Additionally, industrially efficient moisture content determination techniques include the microwave method for FAVs, vacuum drying for grapes, and freeze-drying for skim milk (Duduzile Buthelezi et al., 2019).

The textural quality of FAVs can be assessed using a penetrometer (Magness-Taylor), which is the oldest and most widely accepted method. Various versions of the Magness-Taylor (MT) tester exist, varying in terms of shape, instrument size, manual or mechanical operation, and analog or digital display (Abbott, 2004). Similarly, a texture analyzer is also used to evaluate the firmness of FAVs and has a wider application compared with a penetrometer. Similarly, refractometry is a destructive method used to assess the soluble solid content of FAVs. Refractometers are widely used tools found in many laboratories; they come equipped with standardized references and offer high levels of precision and dependability. However, a drawback lies in their lack of specificity (Valero & Ruiz-Altisent, 2000). Similarly, the total acidity of FAVs can be determined via titration using automatic titrators (Valero & Ruiz-Altisent, 2000). This is a destructive method of quality assessment since the produce need to be squeezed to extract juice which will be used for titration.

1.8.2 Non-destructive Methods

Infrared (IR) spectroscopy is a non-destructive method for assessing quality in FAVs. Infrared technology operates on the principle of interactions between substances possessing molecular bonds and electromagnetic radiations within near- and mid-infrared wavelengths. NIR and MIR spectroscopy encompass a range of electromagnetic frequencies spanning from 12,500 to 4,000 cm^{-1} or 800 to 2,500 nm (NIR), and 4,000 to 400 cm^{-1} or 2,500 to 25,000 nm (MIR), respectively (Zerbini, 2006; Nicolai et al., 2007; Arendse et al., 2018a, 2018b). Recently, the utilization of near-infrared spectroscopy (NIRS) has been progressively expanding within the agricultural domain. This expansion involves advancements in instrumentation and spectral analysis techniques aimed at measuring attributes such as fruit firmness, pH, SSC, and TA in fruit (Mouazen et al., 2014; Nicolai et al., 2014; Su & Sun, 2018). Attributes such as TA, TSS, and juice pH have been accurately predicted with NIRS measurements (Arendse et al., 2017, 2018a, 2018b).

Similarly, nuclear magnetic resonance (NMR) and magnetic resonance imaging (MRI) are non-destructive methods in evaluating produce quality (Zerbini, 2006; Okere et al., 2022). The fundamental principle of NMR spectroscopy is based on scientific observation that a majority of elements possess at least one isotope, rendering them magnetic. For instance, isotopes like ^{1}H, ^{13}C, and ^{31}P possess a magnetic moment and can absorb resonance energy when subjected to a powerful magnetic field (Eberle, 2005). Multiple investigations have demonstrated the capacity of NMR and MRI to quantify various physical and chemical attributes of pomegranate fruit (Khoshroo et al., 2009; Zhang & McCarthy, 2012). In a specific study utilizing MRI, Khoshroo et al. (2009) effectively categorized pomegranates into stages of semi-ripeness, ripeness, and over-ripeness while successfully identifying internal defects.

Raman spectroscopy is another non-destructive method for assessing the quality of FAVs (Yang & Ying, 2011; Nikbakht et al., 2011; Zhen et al., 2020). This technique was formulated based on the phenomenon of light's inelastic scattering. It operates on the premise that when samples are exposed to irradiation, inelastic collisions occur between an incident photon and a molecule within the sample (Raman & Krishnan, 1928). In comparison to other methods, Raman spectroscopy offers great advantages in food analysis (Yang & Ying, 2011; Su & Sun, 2018; Xu et al., 2020). More recently, pattern recognitions based on Raman spectroscopy techniques were reportedly used to investigate quality pomegranates (Khodabakhshian & Abbaspour-Fard, 2020; Okere et al., 2022).

Similarly, machine vision is also a non-destructive method which has found extensive utilization in agriculture (Bhargava & Bansa, 2021). One of the driving factors behind this trend is the progress made in digital imaging and data processing techniques, which have fostered the development of intelligent control approaches. In the agricultural sector, evaluations of quality often hinge on visual attributes like color, introducing subjectivity into the process. Machine vision is extensively employed for grading and sorting FAVs (Kumar et al., 2019; Bhargava & Bansa, 2021). For instance, machine vision was utilized by Fashi et al. (2020) to determine the pH of pomegranate fruit.

Moreover, X-ray computed tomography (CT) is another non-destructive method that transforms two-dimensional images into three-dimensional representations to assess and characterize horticultural products. CT presents notable benefits compared to alternative imaging methods, primarily due to its ability to encompass a substantial field of observation (Schoeman et al., 2016; Du Plessis et al., 2017; Arendse et al., 2020). This capability allows for the complete scanning of an entire sample surface without requiring any preparatory steps (Léonard et al., 2008; Arendse et al., 2014).

Electronic nose (e-nose) is a non-destructive method used for the quality assessment of fruits. This technology is engineered to mimic the human sense of smell, enabling the detection and recognition of intricate fruit aromas through the utilization of an array of chemical sensors (Zerbini, 2006; Sanaeifar et al., 2016). A standard e-nose configuration includes components like data capture, an assortment of metal oxide semiconductor (MOS) sensors, gas-based sensors, and a power supply unit (Okere et al., 2022).

Furthermore, a chromameter is utilized to assess the color of FAVs without destroying the food produce. The parameter of color quality of commodities was assessed in redness (a*), lightness (L*), yellowness (b*), and subsequently using equation to calculate for color change (ΔE*) and hue angle (h°) (Sitanggang & Machfoedz, 2023). Sitanggang & Machfoedz, (2023) reported the utilization of a chromameter to determine the color of strawberry products.

Meanwhile, hyperspectral imaging (HSI) and multispectral imaging (MSI) are recent non-destructive methods for obtaining spatial distributions of physical and chemical attributes, facilitating objective evaluations of fruit quality (Elmasry et al., 2012; Kamruzzaman et al., 2012). MSI and HSI techniques share similarities, but their distinction lies in the number and narrowness of bands utilized. Multispectral imaging usually encompasses 3–10 bands (Baiano et al., 2012; Feng & Sun, 2012), whereas hyperspectral imaging can encompass numerous bands, reaching the hundreds or thousands. With its superior spectral resolution, HSI provides enhanced capability to detect imperceptible features. Nonetheless, handling hyperspectral data poses significant challenges in terms of acquisition, processing, and analysis (Feng & Sun, 2012). A typical HSI system typically comprises a detector, source of illumination, a spectrograph, and a power supply unit (Gowen et al., 2015). Gas analyzers are also utilized in the non-destructive measurement of respiration rate and ethylene generation in FAVs.

1.9 MECHANICAL DAMAGE OF FRESH FRUITS AND VEGETABLES

Mechanical damage to FAVs primarily transpires while they are being collected in the field, although it also takes place while being sorted and packed, during transportation, and in the final stage of the supply chain. This could involve instances such as when they are showcased for purchase and chosen by both consumers and retailers (Ruiz-Altisent, 1996). The quality of the said agricultural products might face significant deterioration due to poor handling and management, particularly if they are not promptly consumed. This predicament carries weight as both a food safety and economic concern (Li & Thomas, 2014). Mechanical damage of FAVs arising due to inappropriate postharvest handling is a common defect of fresh horticultural products; it affects appearance attributes and creates sites for pathogen infections and mass loss. Mechanical damage further stimulates the production of ethylene and respiration in plant tissues, leading to an accelerated senescence (Wu, 2010).

Fruits can experience mechanical damage from various sources during transportation, including compression, impact, and vibration. This damage, which can affect both the peel and flesh of fruits like bananas, may stem from forces like impacts or compression. Additionally, friction-related damage can arise when fruits rub against other surfaces. For example, bananas can experience abrasions due to vibration or when they are handled in clusters, such as when they come into contact with conveyor belts on a packing line (Banks et al., 1991; Banks & Joseph, 1991; Li & Thomas, 2014; Opara & Pathare, 2014; Fernado et al., 2018). Numerous studies have highlighted the significant role of vibration during transportation in causing mechanical damage to fruits. Prolonged exposure to vibration can exacerbate the likelihood of such damage. Research on other packaged fruits has demonstrated that loosely packed fruits tend to suffer more severe vibration-related damage compared to tightly packed ones (Dadzie & Orchard, 1997; Wasala et al., 2015; More et al., 2015).

1.9.1 Types of Mechanical Damage

Mechanical damage in FAVs can be classified based on the causative factor into compression damage, impact damage, puncture damage, abrasion damage, and vibration damage.

1.9.1.1 Impact Damage

Impact damage arises when produce strikes a surface with enough force to rupture or potentially separate the cell. This is visually manifested by bruises or cracks. Ordinary instances of impact-induced damages typically occur during the free fall of fruit from trees to the ground during harvesting, as well as in instances of dynamic collisions between individual fruits and between fruit and their packaging or containers. The latter scenario arises due to vibrations, such as those generated by transporting vehicles and the vibration experienced by fruit containers during transport. Among the various mechanisms of mechanical damage in fruit handling, impact damage is the most severe (Van Zeebroeck et al., 2007b; Li & Thomas, 2014). Reducing impact damage can be achieved by harvesting fruits when they are halfway ripe, as their stiffness at this stage surpasses that of fully ripe fruits. This is due to the fact that the level or size of impact-induced damage to fruits is inversely related to their stiffness (Abedi & Ahmadi, 2013; Armstrong, Stone, & Brusewitz, 1997; Zarifneshat et al., 2010).

Models developed to predict bruises resulting from impact test reveal that variables such as impact force, variety of the fruit, point of impact, level of ripeness, temperature of storage, and curvature radius at the impact site considerably influence the vulnerability of tomatoes to bruising (Van Zeebroeck et al., 2007a). Additionally, the duration of the impact emerges as a critical factor. For impacts characterized by low to medium energy, the potential for bruising is predominantly determined by the texture of the fruit, whereas the consequences of impacts with high and very high energy predominantly hinge on the fruit's level of ripeness and the specific location of impact (Linden et al., 2006).

1.9.1.2 Compression Damage

The damage resulting from compression mostly takes place during or after the packaging process when there's an attempt to fit too much product into a container that's too small. Compression leads to the occurrence of bruises and fractures. Furthermore, compression-related harm can arise during mechanical harvestings if the gripping forces surpass a threshold for failure of tissues. Studies on compressions have revealed that the degree of compression, surface curvatures applying pressure, and the internal structural properties all contribute to the mechanical damage sustained by tomato fruits (Li, 2013; Li et al., 2010; Li et al., 2013a; 2013b). The susceptibility of packaged fruits to compression damage rises with advanced fruit ripening or increased vibration levels. Ripeness and vibration are interconnected factors that collectively determine the severity of compression-induced damages in packaged fruit cultivars. This underscores the preference for utilizing semi-ripe tomato fruits for road transportation in packaging containers (Babarinsa & Ige, 2012).

Furthermore, Aliasgarian et al. (2013) reported that activities associated with picking, packaging, and transporting fruits to the market, the arrangement of fruits within containers, and the positioning of containers on trucks all had substantial impacts on the degree of mechanical damage experienced by strawberry fruits. The manner in which fruit bruising occurs due to gradual compression differs significantly from that resulting from impact. In a study by Chen et al. (1987), it was noted that cross-sectional appearance of a bruise resulting from compressive force resembled a parabolic shape. This was consistent with findings for apple and peach fruits. In contrast, the cross-sectional profile of an impact-induced bruise frequently exhibited elongated spikes radiating outward from the point of impact and into the fruit's flesh. The irregular configuration of impact-related bruises renders damage quantification more challenging compared to the scenario of compression-induced damage (Li & Thomas, 2014).

1.9.1.3 Abrasion Damage

Abrasion refers to the erosion of surface layers that occurs when one object moves against another. This process can manifest in the following ways: during the harvesting of roots or tubers, when fruit cultivars are transported at higher speeds, or through the friction-induced rubbing of fruit surfaces against dirt, sand, or containers during the packing phase. Abrasion necessitates a certain amount of energy absorption, enough to remove surface material without inducing tissue failure. Santana et al. (1998) demonstrated that mechanical abrasions, whether occurring at rates of 1 or 4 cm^2 per fruit, had a substantial influence on the physiological development of banana fruits. This was evidenced by changes in respiration and ethylene production, leading to reduced overall quality. The severity of these effects seemed to correlate negatively with air humidity, influencing factors like the fruit's green life and commercial quality (Santana Liado & Marrero Dominguez (1998).

In a study by Macleod et al. (1976a), it was anticipated that the manifestation of symptoms related to tomato abrasion was influenced by the level of ripeness. The occurrence of abrasion hinges on the type of surface coming into contact with the fruit, as well as the configuration of the containers or equipment used for handling. Timm et al. (1996) proposed that the apple fruit industry in the United States could effectively minimize abrasion-related damage by utilizing carefully designed plastic bins and semi-trailers equipped with air-cushioned suspensions for bulk apple transportation. Abrasion-induced damage triggers the release of anthocyanin/phenolic pigments situated within the skin cells. This subsequently initiates various physiological responses that have an impact on the quality of FAVs (Li & Thomas, 2014).

1.9.1.4 Puncture Damage

Puncture damage may arise during the process of harvesting and managing loose fruit, particularly when the stem of harvested fruit penetrates the skin of adjacent fruits. The stems of some FAVs are generally left intact, as these contribute to their distinctive aroma and help preclude postharvest pathogen invasions. The wounds induced by the puncture have negative implications for fruit value, leading to substantial economic losses, a reduction in shelf life, heightened vulnerability to different diseases, and increased water loss (Li & Thomas, 2014). Some studies have indicated that firmness and elasticity of a fruit, along with the toughness of its skin, play a role in determining its vulnerability to puncture injuries (Desmet et al., 2002). This might explain why certain cultivars exhibit reduced susceptibility and why tomatoes, after storage for several days, have been observed to be less prone to puncture damage (Desmet et al., 2003). Furthermore, machine-harvested fruits tend to suffer a considerably increased incidence of puncture-based injuries compared to those harvested by hand. Notably, tomato fruits are highly sensitive to puncture-based damages than to abrasion when subjected to machine-based harvesting (Studer et al., 1981). Puncture, along with abrasion, has been highlighted as a more significant concern than impact injuries for papayas during postharvest handling (Quintana & Paull, 1993). In a mechanical harvesting experiment comprising five different apple cultivars, cuts and punctures emerged as the most prominent issues, as reported by Peterson and Bennedsen (2005).

1.9.1.5 Vibration Damage

Vibration during transportation is the main cause of mechanical damage to FAVs. Li and Thomas (2014) emphasized that the movement and impact experienced by fresh produce on truck beds during transportation can lead to noticeable and direct visible damage. The complexity of the vibration encountered within vehicles during transit significantly impacts the degree of damage that fresh produce sustains along the distribution chain (Ríos-Mesa et al., 2020; Ranathunga et al., 2010). The vibration intensity is contingent upon various factors such as load positioning within the truck, payload weight, truck size, vehicle suspension, road conditions, and speed of the truck (Singh et al., 2006; Soleimani & Ahmadi, 2014, 2015; Zhou et al., 2015). In a recent work conducted by Walkowiak-Tomczak et al. (2021), the alterations in physical and chemical attributes of four apple cultivars subjected to simulated transport vibrations at a frequency of 28 Hz were evaluated. Their findings indicated that the overall color differences of the apple skin were influenced by vibrations, particularly in individual cultivars. Additionally, vibration led to reductions in hardness while increasing antioxidative capacity and polyphenol content across all examined cultivar types.

Similarly, Jung & Park (2012) reported that vibration-induced stress accelerated the deterioration of apple quality characteristics during storage. This resulted in heightened SSC, mass loss, ethylene production, CO_2 concentrations, and diminished firmness and O_2 concentration. Correspondingly, Zhou et al. (2007) demonstrated that mechanical damage from varying vibration levels impacted the integrity of pear cell membranes and polysaccharide components in the cell wall. Therefore, this has contributed to more pronounced color changes and softening of the fruits during subsequent commercial phases of the storage.

1.10 ASSESSMENT OF MECHANICAL DAMAGE

The prevention of mechanical damage will contribute much in averting postharvest loss of FAVs. Fruit susceptibility tests give insights on the best postharvest handling practices that will prevent mechanical damage. Such fruit susceptibility tests include drop tests, spherical impactor tests, pendulum impactor tests, simulation experiments, in-transit experiments, and transit-simulation experiments (Lin et al., 2022; Opara & Pathare, 2014). Mechanical damage can be measured indirectly by evaluating indexes that are linked with it, such as respiration rate, firmness, ethylene production, and microorganism invasion. These indexes have been reported as good parameters for the indirect evaluation of fruit damage (Lin et al., 2022). Bunsiri, et al. (2012) perceived an upsurge in the firmness of mangosteen shortly after impact, but this phenomenon wasn't observed in the undamaged pericarp tissue. Damaged apples exhibited considerably heightened respiration rates and ethylene production in comparison to unaffected apples (Lu et al., 2019). Scalia et al. (2016) investigated the microbial changes in two strawberries subjected to simulated vibration testing. The study demonstrated that volatile organic compounds produced by predominant microorganisms could indirectly indicate the presence of bruise damages (Scalia et al., 2016). The extent of mechanical damage of fresh produce may be assessed based on surface or internal measurement of the affected tissue.

1.10.1 Surface Damage Measurement

Surface damage refers to any defects present on the outer skin of a fruit. The exertion of external forces can result in mechanical damage occurring either on the surface or within the fruit's interior. In certain instances, the internal damage might coincide with the region just below the affected exocarp, while in other cases, the internal damage could be in tissue not directly in contact with the exocarp. Regardless, any observable indications of damage will be limited to the surface level. When it comes to fresh fruits, surface damage, which can encompass browning or ruptures, is easily detectable and hence can be measured (Li & Thomas, 2014). Surface damage can be measured by diameter. Timm et al. (1996) and Vursavus and Ozguven (2004) measured the extent of surface damage on apples by considering the diameters of brown areas that emerged on the fruits' surface. This attribute was employed to assess the impacts of packing techniques and conditions of transportation. It was demonstrated that the degree of bruising was notably influenced by factors such as acceleration, vibration frequencies, and exposure period, along with the method of packaging.

The measurement of injury diameters is a straightforward approach, but it assumes that visible damages take a circular form, essentially constituting a one-dimensional parameter. Conversely, the concept of damage areas is particularly apt for describing abrasions, as it represents two-dimensional parameters that align well with nature of damages. For instance, Puchalski and Brusewitz (1996) categorized the surface abrasions of watermelons by correlating sensory score in the range of 1–20 with the area occupied by wax spots and removed skin. They further explored the influence of the abrasive surfaces and the absorbed energy per unit contact area on the sensory score through a friction test. These innovative methods laid the foundation for quantitatively measuring surface abrasions of fruits. Moreover, damaged areas serve as a rapid means of evaluating bruising.

Pang et al. (1996) and Idah and Yisa (2007) measured the severity of surface damages on apples and tomatoes, respectively, by computing the damage area under the assumption that the damaged tissue's surface shape was either circular or elliptical. Researchers have taken steps to measure the volume of damage that is directly related with surface-related damages. This can be achieved by making assumptions about the internal damage's shape, which isn't visible, or by quantifying the depth of a bruise. In numerous investigations, width, diameter, and sometimes the depth of the bruised tissue were measured using digital calipers. Subsequently, the volume of the damage was computed, with the presumption that the shape of the affected tissue was either spherical, elliptical cone-like, or ellipsoidal (Ahmadi et al., 2010; Maness et al., 1992; Lu et al., 2010: Opara, 2006b; Opara et al., 2007; Opara & Al-Jufaili 2006b).

While manually assessing the damaged areas of fresh fruits is common, it can also be quantified using automated machine vision systems (Blasco et al., 2009; Cubero et al., 2011). Recent advancements in near-infrared hyperspectral and thermal imaging technology have proven to be groundbreaking for the non-destructive detection of subtle or early surface damage in fruits. This method generates minimal waste, requires little to no sample preparation, incurs lower operational costs compared to traditional techniques, and is applicable to a wide array of fruit types (Li & Thomas, 2014). Lu (2003) explored the potential of near-infrared hyperspectral imaging (NIHI) to identify surface damage on apples, focusing on the spectral range between 900 nm and 1,700 nm. The study indicated that the range from 1,000 nm to 1,340 nm was the most suitable for detecting damage. The NIHI system successfully detected both recent and older surface damage, achieving accurate detection rates ranging from 62% to 88% for Red Delicious apples and from 59% to 94% for Golden Delicious apples. Zhao et al. (2008) suggested that subtle damages on apple fruits could be identified through hyperspectral imaging at a wavelength of 547 nm, yielding an approximate success rate of 89%. Baranowski et al. (2009), along with Baranowski et al. (2012), documented that thermal imaging and hyperspectral data could effectively differentiate early damage in apple tissues.

1.10.2 Internal Damage Measurement

Internal damage refers to any defects within the tissue of a fruit, excluding its outer skin. The internal damages that are not directly linked to outer bruising pose a critical challenge to the quality of the fruit, impacting factors like sugar content, firmness, and organic acid content (Alfatni et al., 2013; Montero et al., 2009). Detecting and quantifying this type of damage is exceedingly complex. Moreover, damaged tissue sections typically have irregular shapes and are often confined within the fruit. Damaged fruit tissues tend to exhibit browning during storage, a visual color that can be discerned by cutting the fruit and inspecting it for indentation and discoloration (Pang et al., 1996). Alternatively, permitting the fruit to rot can also reveal the damage on its surface (Van Zeebroeck et al., 2007a, 2007b). However, these methods are inherently destructive and are predominantly employed for quantifying damages in order to comprehend the mechanism causing it.

The visual assessment of damage extent is based on the degree of browning, which is influenced by the presence of phenolics and activities of enzymes like polyphenol oxidases and peroxidases within the tissues (Li & Thomas, 2014). A statistical method can also be used to evaluate the susceptibility of fruit to damage. Using a sensory score that relied on subjective assessment and involved evaluating the damage area of fruits following two-day incubation subsequent to an impact test, Van Linden et al. (2006a, 2006b) documented instances of bruising. They further employed a logistic regression model to effectively forecast the likelihood of different tomatoes sustaining bruises under diverse conditions of impact.

NMR imaging presents a potentially more objective and non-destructive technology for distinguishing between intact and damages on fruit tissues. Because fruits contain a significant amount of water, the way water is distributed between sections of the fruit that are damaged and those that are undamaged varies. By utilizing NMR imaging, it becomes possible to visually represent how water particles are dispersed within the fruit. Subsequently, this technique allows for a non-destructive evaluation of the fruit's quality and any defects it may have. This is achieved by characterizing the behavior of water protons within plant tissue (Li & Thomas, 2014; Hou et al., 2023). Hernandez-Sanchez et al. (2004) and (2007), delved into the variations in MRI between healthy and damaged pear tissues. Their research indicated that disordered tissues exhibited higher transverse relaxation rates, and the proton pools in disordered tissue were clustered into fewer populations compared to healthy tissue. In a similar vein, Milczarek et al. (2009) devised an in-line technique for detecting damaged exocarp tissues in tomatoes using multivariate analysis of NMR images.

Zhou et al. (2012) utilized NMR image processing techniques to detect and classify three distinct defects in pear fruits: these included bruising, damages from falling, and internal discoloration. They developed a relationship between the firmness of pears and texture coefficient derived from magnetic resonance data. Qiao et al. (2019) employed low-field NMR and imaging techniques to examine MRI data concerning the movement of water within blueberries during storage at room temperature. They fed this data into a back propagation neural network (BPNN) model for validation and recognition purposes. The outcomes highlighted that both LF-NMR and MRI were appropriate for analyzing and identifying decay-related diseases in fruits. This could establish a theoretical foundation for the non-destructive assessment of fruit diseases. The utilization of NMR imaging technology for identifying fruit damage primarily centers on the dispersion and mobility of water. The predominant detection methods currently in use encompass signal amplitude, imagery, relaxation time, and assessment of free diffusion coefficients. Among these, relaxation time–based analysis and detection are extensively employed. However, due to the expensive nature of the equipment, widespread practical application is limited.

Hyperspectral imaging detection technology is another method that can be used to assess internal damage in fruits. Liu et al. (2018) utilized a hyperspectral reflectance imaging setup to detect damage in strawberries, employing a fusion of image processing techniques and spectral analysis. They formed a correlation between attributes of quality and distinct spectral characteristics. Furthermore, they devised both linear and nonlinear algorithms for distinguishing various defect categories (Liu et al., 2018). Li et al. (2022) also introduced an approach that merged hyperspectral image attributes with spectral features to forecast the mechanical properties of yellow peaches. Furthermore, visible/near-infrared spectroscopy detection equipment is also utilized in the detection of internal damages in fruit (Arendse et al., 2018b).

Zhao et al. (2016) employed near-infrared spectroscopy as a non-invasive method to detect plum fruit browning. Similarly, in a study by Moscetti et al. (2016), near-infrared spectroscopy was employed to classify olives as damaged or undamaged by hail. Wang et al. (2020) also utilized visible-NIR interaction spectroscopy to evaluate cold damage in kiwifruits without causing harm. Visible/near-infrared spectroscopy can be categorized into quantitative and qualitative measurement. Qualitative detection primarily serves to differentiate the spectral attributes of fruits that are damaged from those that are healthy. On the other hand, quantitative detections establish a correlation between spectral characteristics and features associated with damage. The precision of this method is notably influenced by the algorithms used and the models established, underscoring the need for adept pre-processing techniques and model formulation strategies (Hou et al., 2023).

Acoustic and electrical characteristics detection technology is another method that could be employed to detect internal damages in fruits. Defects within fruit are identified through an analysis of the vibrational response exhibited by fruit samples subjected to specific excitations within applications of the acoustic technique. Zhang and Wu (2020) assessed the susceptibility of distinct parameters to detect internal discoloration in pears. This was achieved by employing a custom-built non-destructive testing apparatus for acoustic vibration. Models were formulated for early and mild internal browning in pears, resulting in accuracies of 91.84% and 81.82%, respectively. These findings offer insights into potential avenues for real-time online detection and automated grading technologies for internal browning in pears through acoustic vibration methods. An alternative approach for non-destructive assessment employs electrical properties to gauge fruit quality, including damage and disease. This technique offers the merits of straightforward operational steps and instruments with heightened sensitivity (Hou et al., 2023). Fan et al. (2021) developed a mathematical model for electrical parameter values that can determine the damaged region of Korla fragrant pears. The study introduced a quantifiable method for evaluating the extent of damage grounded in electrical attributes.

Moreover, X-ray computed tomography (CT) and MRI have been reported to be more suited for evaluating internal damage within fruits (Lin et al., 2022). Diels et al. (2017) devised a technique to automatically recognize and quantify volumes of bruising in the equatorial section of apples using X-ray CT images. Zhou et al. (2015), on the other hand, utilized MRI to non-destructively evaluate changes in the internal structural attributes of Hami melon caused by simulated vibrations. The MRI images acquired clearly distinguished the damaged regions within the affected melons. Optical coherence tomography (OCT) has also been reported as a recent, rapid, and sensitive tool for assessing horticultural produce (Lin et al., 2022). OCT, an optical imaging method devoid of invasiveness and contact, has the capability to generate three-dimensional images of plant tissues with depths reaching 2 mm and resolutions ranging from 5 to 20 m (Li et al. 2019). In the context of identifying mechanical damages in fruits, Zhou et al. (2018) employed OCT to quantify changes in cellular structure in loquats, revealing that the total surface area and quantity of cells were reliable markers for identifying bruises.

1.11 CONTROL STRATEGIES IN REDUCING MECHANICAL DAMAGE

The potential strategies for reducing mechanical damage include decreasing the occurrence of damage through the investigation of the consequences of external forces applied during harvesting, packing, and transportation. Research examining the occurrence and measurement of mechanical injuries to fruit contributes to a better grasp of the mechanisms causing damage and aids in refining packaging solutions. This will also enable the recommendation of enhanced handling techniques to cultivators and other parties involved in the supply chain (Van Linden et al., 2006a; Li & Thomas, 2014; Lin et al., 2022).

Secondly, evaluating both the external and internal damage of harvested fruits in a non-destructive manner, followed by categorizing them into varying levels of damage for immediate treatment or potential handling (i.e., either immediately usable or storable), necessitate an objective and quantifiable assessment of the mechanical damage inflicted on the fruit's surface and interior. While surface damage is readily noticeable and can be objectively determined, internal damage may require some advanced technology for proper assessment (Van Linden et al., 2006a; Li & Thomas, 2014).

Thirdly, packaging methods are the most important control strategy in reducing mechanical damage to FAVs. Various packaging designs and methods are utilized for handling and transporting fresh FAVs after harvesting (Berry et al. 2015). Packaging serves as a buffer against the energy generated by mechanical forces, playing a crucial role in preserving product quality and prolonging the shelf life of the fruit. It stands as a pivotal component in safeguarding fruit from mechanical harm throughout the supply chain (Jarimopas et al., 2007, 2008; Thompson et al., 2008). Fruit packaging typically encompasses both external packaging and interior packaging. Examples of external packaging are cartons and plastic containers, while those of interior packaging are paper-based and foam materials (Lin et al., 2022).

1.12 PACKHOUSE OPERATIONS FOR QUALITY CONTROL AND ASSURANCE

Packhouses are collection points for gathering FAVs before they are sent out for distribution and sale. A packhouse is a structure where harvested FAVs are brought together and readied for transportation and delivery to marketplaces (Opara et al., 2021b). These facilities can range from basic packing sheds with limited equipment and minimal activities to extensive complexes that are fully equipped and capable of handling specialized tasks. The specific tasks conducted within these facilities differ depending on the type of produce and the demands of the market. Items intended for export or supermarket shelves often undergo more intricate processes compared to those meant for local markets (Acedo et al., 2016). A packhouse facilitates quality control measures to guarantee that both the quality and quantity of the products adhere to market standards, while also minimizing losses that might occur during the transportation and delivery to markets (Yaptenco & Esguerra, 2012; Acedo et al., 2016). Packhouses also aid in arranging FAVs in suitable packaging materials for the purpose of marketing. They provide refrigerated storage as a buffer between the production and packaging stages as well as between packaging and marketing (Berk, 2016). Packhouses are usually situated close to farms so they can provide services to the largest number of farmers. They should also be situated such that they facilitate pick-up and drop-off of containers and produce as well as being easily accessible to markets. Packhouse operations vary based on the FAVs. Basic packhouse operations include receiving, sorting and grading, cleaning, treatments, drying, and packaging (Acedo et al., 2016).

1.12.1 Receiving

Receiving entails the arrival of farm produce to the packhouse. The source of the produce and its weight are noted upon arrival. The produce is also inspected for severity of damage, which will facilitate the sorting process. It is necessary to examine the produce for any harm, insect or rodent infestation, decay, foreign substances, and observable traces of chemicals. Additionally, each item should be allocated a code that specifies the supplier, harvest or delivery date, and the location of production (Boonyakiat & Janchamchoi, 2007).

1.12.2 Sorting

In the packing facility, fruits harvested from the orchard undergo sorting and categorization based on commonly employed standards for evaluating fruits. These standards include factors like size, shape, skin color, and the presence of any cuts, marks, insect-related damage, or remnants of sprays (Bill et al., 2014). The sorting process involves removing products that are damaged, diseased, or fail to meet the stipulated quality standards. Typically, this is the first task carried out in the packhouse (Acedo et al., 2016).

1.12.3 Cleaning

Cleaning FAVs can influence market appeal and prices. It also reduces microbial contamination and transport cost. Trimming denotes the elimination of undesirable sections of plants, which could either be rejected by consumers or have the potential to contribute to spoilage. Cleaning involves trimming the stems of tomatoes or eggplants, the roots of leaf mustard, as well as the leaves and lower part of cauliflower, cabbage, or Chinese kale. When dealing with cabbage, it's important to preserve 3–4 outer leaves for protective purposes. Cleaning FAVs also involves wiping tomatoes, eggplants, or cucumbers with a soft, clean cloth or employing clean water to wash away any attached soil and additional debris. Following the washing process, it is necessary to allow the produce to air-dry before the packing stage. Dry brushing is utilized in the cleaning of water-sensitive produce like ginger to remove clods of soil (Yaptenco & Esguerra, 2012; Acedo et al., 2016).

1.12.4 Grading

Grading involves categorizing agricultural products into different clusters based on predetermined quality and size standards that are acknowledged and endorsed by both governmental authorities and the industry. Quality grades, functioning as a globally understood trading language and catalyst for the adoption of technology, play a crucial role. Market orders can be established according to these quality grades, subsequently informing activities both in the packhouse and on the farms. Grading occurs when the sorted produce, devoid of defects, is categorized into distinct grades or classes based on specific weights, sizes (sizing), and the stage of maturity. This grading process can take place either subsequent to sorting or just before the actual packing procedure (Yaptenco & Esguerra, 2012; Acedo et al., 2016). Each group of products is identified by an approved name and size range, such as Extra Class, Class I, or Class II, as outlined in the Codex Alimentarius Commission (CAC) regulations for fresh produce. Despite differences in grading criteria depending on the specific commodity, this practice remains consistent. Some commonly used criteria in grading include the stage of maturity/ripeness, appearance, texture, presence of damage/defects, safety, and wholesomeness (Yaptenco & Esguerra, 2012).

Moreover, the grading of fruit can be carried out manually or with the assistance of electronic sorting machines. At this phase, fruit with higher processing quality might be separated, destined for the factory alongside the fruit discarded during the initial presorting stage. Manual grading requires significant labor and is less precise. To achieve better outcomes, the flow of fruit is divided into multiple conveyors, reducing the number of fruits inspected by all workers. The grades of quality are established based on the prevailing market requirements. Various electronic sorting and grading systems are available. Cameras or electronic "eyes" of computer vision system individually examine the fruits and categorize them based on attributes such as color, defects, and diameter (Berk, 2016). Most computer vision systems assess fruit in the visible light spectrum, mimicking the human eye (Blasco et al., 2007). Systems employing the near-infrared and ultraviolet regions of the electromagnetic spectrum, along with fluorescence, offer more discriminating inspection and can detect defects invisible to the naked eye. According to Blasco et al. (2007), the use of near-infrared technology increased the identification of anthracnose, while fluorescence enhanced the detection of green mold.

1.13 PHYSICAL TREATMENTS

De-sapping or de-latexing is a practice employed to avert the occurrence of staining caused by latex after the removal of a fruit's pedicel. Among the most apparent postharvest quality issues in mangoes, bananas, and papayas are stains resulting from latex and the subsequent burns, often referred to as sapburn. If not effectively addressed, these problems can lead to inferior fruit quality. Upon detachment of a fruit's pedicel, latex is discharged from the point of attachment. If this latex enters the lenticels of the fruit, burn marks are likely to manifest several days postharvest, and these effects are irreversible. Alternatively, latex staining emerges as shiny streaks or patches on the fruit's skin. To prevent latex staining in mangoes, a strategy can be employed during the harvesting process. By using wire-meshed trays positioned alongside the tree rows, latex staining can be averted. After the pedicel is cut, fruits are positioned with their stem ends facing downward within the wire-meshed trays for around 30 minutes, allowing the latex to drip off. Then the fruits are gathered into plastic crates and transported to the packing facility for cleaning (Yaptenco & Esguerra, 2012).

Curing is a physical treatment applied to some FAVs during packhouse operations. Bulb crops like onions necessitate a curing procedure to hinder the intrusion of microorganisms. This step enables the closure of the bulb neck and the drying of the outer layers. Similarly, freshly harvested citrus fruits undergo curing to preclude the incidence of oleocellosis. In the case of Solo papayas, it's beneficial to cure newly harvested fruits for 24 hours within a well-ventilated environment before proceeding with washing and packaging, as this can reduce the occurrence of latex stains on the fruit. For a swifter curing process, the use of heated air can be employed by directing it through the produce placed in curing bins or on racks for a few hours. In instances where forced air movement is absent, the curing process is comparatively slower and might extend over several days (Yaptenco & Esguerra, 2012).

Waxing is another physical treatment applied on some fruits. Waxing serves multiple purposes, including the replacement of natural waxes that might be lost during the harvesting and handling stages, as well as enhancing gloss, diminishing moisture loss, and retarding the ripening process. Moreover, waxes infused with antifungal compounds can be utilized for disease control. The appropriate coating types should confer a glossy appearance; be clear, devoid of scent and flavor; biodegradable; impervious to water; and semi-permeable to both O_2 and CO_2. Commonly treated items comprise citrus fruits like mandarins, oranges, pomelos, apples, and pineapples. In the case of pineapples, waxing proves particularly advantageous in mitigating chilling injuries, especially when the fruits are transported under refrigerated conditions. Various approaches for wax application include immersion, foaming, brushing, and spraying. It's worth noting that certain products might not be well-suited to specific waxing methods due to their structural characteristics (Esguerra & Bautista, 2007; Berk, 2016).

1.14 CHEMICAL TREATMENTS

Acedo et al. (2016) revealed that immersing tomatoes in a chlorine solution of 100–200 ppm for 1–3 minutes can effectively lower microbial presence and decay. Additionally, the utilization of a 0.01% solution of calcinated calcium extracted from scallop powder, applied through a 3–5-minute dipping process, resulted in improved food safety. This alternative sanitizer was developed to address health worries linked to chlorine, which interacts with organic substances in the produce, forming highly carcinogenic compounds called trihalomethanes. It is recommended that the produce undergoes air-drying prior to the packing phase.

In humid tropical regions, bacterial soft rot poses a significant challenge for cabbage cultivation. When 10% alum, lime paste, or an extract derived from guava leaves was administered to the base of cabbages, the loss incurred due to soft rot during trimming was diminished from the range of 20–44% (without treatment) to a much lower 0–20%. This intervention led to a notable improvement in returns, yielding a net profit of approximately US$0.09–0.16 per kilogram of produce, as evidenced by research conducted across Cambodia, Laos, and Vietnam (Acedo et al., 2016). Similarly, chemical treatment like dipping in anionic commercial detergent and 1% alum solution aids in the removal of latex stains on fruit (Yaptenco & Esguerra, 2012). Immersion in a 1% chitosan solution (10 g per liter of water) for a 5-minute duration resulted in a delay of tomato ripening and an extension of shelf life by six days. Moreover, this treatment led to a 50% reduction in weight loss compared to untreated fruits. These effects collectively contributed to a significant net profit of US$0.20 per kilogram of produce (Acedo et al., 2016). Other antimicrobial treatments include treatment with hydrogen peroxide, chlorine, peroxyacetic acid, ozone, and electrolyzed water (Yaptenco & Esguerra, 2012).

Ripening refers to the procedure of artificially hastening the transition of a fruit or fruit-type vegetable, such as tomatoes, from their initial mature-green state to a partially or fully ripe condition. During this phase, the item has developed all the visual and sensory attributes desired by consumers. Often, products are harvested when they are still green because this enhances their resilience during transportation, prolongs their shelf life in the market, permits controlled color development, and offers flexibility in retailing. Ethylene (C_2H_4) is the hormone responsible for triggering the onset of ripening. A frequently used ethylene analogue is acetylene (C_2H_2), which is utilized in artificial ripening processes (Yaptenco & Esguerra, 2012; Berk, 2016). To achieve ripening and degreening, mature and high-quality produce is exposed to ethylene or its chemical counterpart, acetylene, at an optimal ripening temperature with sufficient air circulation. Commonly subjected to artificial ripening are bananas, mangoes, papayas, and tomatoes. Meanwhile, degreening is applied to citrus fruits like pomelos and mandarins (Yaptenco & Esguerra, 2012). Commodities can alternatively undergo ripening using ethephon, which is combined with caustic soda to release ethylene. Approximately 200 mL of ethephon generates 0.028 m^3 of ethylene (Reid, 1992).

1.15 PACKAGING TECHNOLOGIES FOR BULK, RETAIL, AND CONSUMER HANDLING

FAVs have a limited shelf life and are prone to spoilage. The process of handling and marketing these agricultural products often leads to significant postharvest losses. The primary reason for these losses is the inadequate infrastructure for processing and packaging. To prolong the shelf life of fresh produce, it's crucial to minimize physical damage. Effective packaging is key to achieving this goal (Thompson, 1996). Packaging plays vital functions, such as organizing the produce into manageable units for ease of handling; aiding in the containment, movement, and logistical aspects of both fresh and processed goods (Opara, 2009d; Opara et al., 2007; Opara & Mditshwa, 2013); prolonging the postharvest viability of crops; safeguarding the produce throughout distribution, storage, and the marketing process; and carrying information about the products, producer, content, and legal requirements.

The ideal packaging material should possess certain characteristics. These include availability, ease of handling, affordability, proper ventilation for the produce, and environmental friendliness. Additionally, packaging for FAVs should offer features such as protection against physical damage and contaminants, protection from bad smells and external toxicants, non-toxicity, easy opening, resistance to moisture and oxygen, UV light filtering, transparency, low cost, easy disposability, and offering incentives for sales and customer satisfaction (Adhikary & Kumar, 2022).

Due to their perishable nature, the majority of fresh FAVs are still retailed or wholesaled without packaging, often placed in perforated polyethylene (PE) bags or nets. To prevent brushing and subsequent decay of these loosely wrapped items, extended transportation of such goods must be limited. In cases where packaging is necessary at the source or for prolonged storage, packaging materials should possess higher gas permeability and anti-fog characteristics. Printed PE bags find their predominant application in packaging fresh FAVs. The practice of pre-packaging items like carrots, potatoes, onions, beets, parsnips, and radishes for market sales has grown significantly in food distribution. Given the limited water vapor permeability of both PE and polypropylene (PP), these films and bags are sometimes punctured to enable the product to "breathe." Allowing permeation is the fundamental approach to extending the shelf life of fresh produce (Ščetar et al., 2010).

Bags, cartons, crates, baskets, bulk boxes, hampers, and pallet containers are practical receptacles for the management, movement, and sale of fresh produce. Plastics have assumed a crucial function as an effective packaging choice. Beyond being cost-efficient, they safeguard the fruits against microbial contamination, moisture, weight reduction, and even manage the concentration of ethylene within the packaging (Morya & Sharma, 2020). The primary type of packaging commonly employed in this sector involves the utilization of fiberboard cartons. Nevertheless, for the majority of products, additional internal packaging is necessary to minimize harm resulting from friction. This supplementary packaging includes materials like beets, tissue paper wraps, or pads. When dealing with extremely delicate fruits, smaller packages containing fewer layers of fruits are utilized to mitigate damage from compression. In certain cases, molded trays are implemented to isolate each individual piece of produce. Alternatively, individual fruit might be separately wrapped using tissue or waxed paper. This approach enhances physical protection and concurrently diminishes the potential spread of disease organisms within a package (Ščetar et al., 2010). Packaging used for FAVs include modified atmosphere packaging, logistic packaging, and edible coatings and films.

1.15.1 Modified Atmosphere Packaging

Modified atmosphere packaging (MAP) for fresh produce involves sealing the products within a packaging system that modifies the standard air composition, thus creating an ideal gas environment surrounding the product (Opara et al., 2019). The MAP concept involves replacing the air inside a package with a fixed mixture of gases, without further control over their composition once introduced. This method is also referred to as "gas packaging" or "gas exchange packaging." The key benefit of MAP is the extension of the shelf life for products. The primary gases employed in MAP for food preservation include O_2, N_2, and CO_2 (Caleb et al., 2013; Fellows, 2016; Morya & Sharma, 2020). Table 1.2 presents recommended gas percentage ranges for selected FAVs. MAP with elevated CO_2 levels and reduced O_2 content compared to regular air has the potential to effectively slow down physiological and biochemical activities, thereby delaying the aging process in FAVs (Antmann et al., 2008). MAP serves as a valuable alternative to low-temperature storage, especially for conserving fresh tropical FAVs. This is important, as these items have a high respiration rate and are susceptible to chilling injuries (Fellows, 2016; Morya & Sharma, 2020).

Polymeric films are commonly utilized due to their benefits and wide availability. They are especially effective in regulating the movement and levels of gases. By decreasing O_2 levels and increasing CO_2 concentration, these films reduce the respiration rate of products and extend their shelf life. Temperature control is pivotal in MAP, as it directly impacts the respiration rate, thereby influencing the product's shelf life. However, a significant drawback of MAP is the excessive reduction of oxygen levels, which could potentially lead to tissue fermentation and the development of unpleasant off-flavors (Adhikary & Kumar, 2022). The optimal method of

TABLE 1.2 Recommended gas mixtures for MAP of selected fruits and vegetables

COMMODITY		*O_2 (%)*	*CO_2 (%)*	*N_2 (%)*
Fruits	Apple	1–2	1–3	95–98
	Apricot	2–3	2–3	94–96
	Avocado	2–5	3–10	85–95
	Banana	2–5	2–5	90–96
	Grape	2–5	1–3	92–97
	Grapefruit	3–10	5–10	80–92
	Kiwifruit	1–2	3–5	93–96
	Lemon	5–10	0–10	80–95
	Mango	3–7	5–8	85–92
	Orange	5–10	0–5	85–95
	Papaya	2–5	5–8	87–93
	Peach	1–2	3–5	93–96
	Pear	2–3	0–1	96–98
	Pineapple	2–5	5–10	85–93
	Strawberry	5–10	15–20	70–80
	Pomegranate	3–5	0–10	–
Vegetables	Artichoke	2–3	2–3	94–96
	Beans, snap	2–3	5–10	87–93
	Broccoli	1–2	5–10	88–94
	Brussels sprouts	1–2	5–10	91–94
	Cabbage	2–3	3–6	81–95
	Carrot	5	3–4	91–95
	Cauliflower	2–5	2–5	90–96
	Chili peppers	3	5	92
	Corn, sweet	2–4	10–20	76–88
	Cucumber	3–5	0	95–97
	Lettuce (Leaf)	1–3	0	97–99
	Mushrooms	3–21	5–15	65–92
	Spinach	Air	10–20	–
	Tomatoes	3–5	0	95–97
	Onion	1–2	0	98–99

Sources: (Powrie & Skura, 1991; Day, 1993, 1999; Exama et al., 1993; Moleyar & Narasimham, 1994; Smith & Ramaswamy, 1996; Ščetar et al., 2010; Caleb et al., 2012; Adhikary & Kumar, 2022)

packaging for highly respiring products involves a combination of high-oxygen MAP and low-oxygen MAP. This approach proves effective and efficient in preventing the formation of off-flavors and unpleasant odors caused by fermentation, due to the elevated oxygen levels (Day, 1998; Day, 1999).

The term modified atmospheres (MA) refers to altering the composition of the surrounding air around the produce from its normal state by adding or removing gases. This involves manipulating the levels of O_2, CO_2, N_2, and ethylene. In practical terms, MA typically involve lowering O_2 levels to below 5% and/or raising CO_2 levels above 3%. The key distinction between controlled atmosphere (CA) and MA lies in the precision of gas regulation, with CA being more precise than MA (Kader, 2002; Kader & Saltveit, 2003; Wills et al., 2007). The ability of produce to tolerate or be susceptible to the effects of reduced O_2 and increased CO_2 concentrations is a critical factor for the successful advancement of MA technologies (Wu, 2010).

MAP for fresh produce depends on altering the environment within the packaging. This alteration is typically brought about through the natural interplay of two processes: the respiration of the product and the movement of gases through the packaging. It utilizes specialized plastic films that allow selective passage of gases. This interplay results in an elevation of CO_2 levels and a reduction in O_2 concentrations. This adjusted atmosphere holds the potential to effectively decrease the rate of respiration, sensitivity to ethylene, and physiological transformations. Accurate temperature regulation is of paramount significance in MAP due to its impact on both the gas permeability characteristics of the film and the product's respiration rate (Wu, 2010; Scetar et al., 2010; Opara et al., 2019).

The method of MAP involves enclosing perishable goods in polymer films where the gaseous environment is actively or passively adjusted to slow down respiration, minimize moisture loss and spoilage, and/or prolong the shelf life of the items. While many individual films used in MAP lack all the necessary properties for a modified atmosphere package on their own, a variety of these films are combined through techniques such as lamination and co-extrusion to achieve a broad spectrum of physical characteristics in packaging films (Scetar et al., 2010). MAP films can be grouped into various categories. Polyethylene is commonly utilized to create an airtight seal and to manage features like anti-fog capabilities, ease of opening, and the ability to seal effectively even when some level of contamination is present. The extent of atmospheric modification that occurs within packages depends on multiple factors, including the film's permeability to O_2 and CO_2, the respiration rate of the product, and how temperature affects these processes (Beaudry et al., 1992; Beaudry, 1999; Cameron et al., 1994).

Generally, MAP can be categorized as either passive or active (Wu, 2010; Opara et al., 2019). Passive MAP can be established by utilizing the natural air compositions and depending on the respiration of the produce to achieve the targeted gas combination. Successful alteration of the atmosphere is achieved through the respiration rate of the items and the gas permeability of the packaging film. This combination leads to the establishment of a passive and stable gas composition over an extended period, following an initial period of transition (Mahajan et al., 2008). Active MAP entails processes such as gas flushing, gas scavenging, or the utilization of emitters. In active gas flushing, the existing air is replaced with the desired gas mixture. This accelerates the attainment of the optimal gas level and prevents exposure to unfavorable gas environments, all aimed at prolonging shelf life and enhancing safety while preserving the quality of the product (Fonseca et al., 2005).

In the context of active packaging utilizing gas-scavenging or emitting systems, one notable form is the utilization of small sachets containing iron compounds of an oxidizable nature, known as O_2 scavengers (Kartal et al., 2012). These scavengers effectively prevent fruit discoloration and reduce the risk of chilling injuries. O_2 scavengers have demonstrated significant efficacy in decreasing spoilage, prolonging freshness, and minimizing issues like discoloration and mold development in fresh produce (Lee et al., 2014). Likewise, CO_2 absorbers play a role in averting the accumulation of CO_2 to harmful levels (Brody et al., 2001). Similarly, ethylene absorbers contribute to delaying the climacteric increase in respiration and the related ripening process for certain fruits by eliminating ethylene within the packaging. The implementation of scavenger systems has been proven effective for various fruits such as bananas, avocados, kiwifruit, mushrooms, and persimmons (Ozdemir & Floros, 2004).

Opara et al. (2019) further described an intelligent and smart MAP system. This MAP is equipped with the capability to perform advanced functions like detection, sensing, recording, tracking, communication, and the application of scientific reasoning, and these systems are designed to aid decision-making processes. Their aim is to prolong the shelf life, increase safety, elevate quality, deliver valuable information, and alert about potential issues (Yam et al., 2005; Sandhya, 2010). The conceptual framework outlining the information flow within an intelligent and smart system consists of four key elements: intelligent packaging devices, layers of data, data processing, and communication pathways across the food supply chain (Ahvenainen, 2003; Urmila et al., 2015). Smart packaging has the capability to alter its color, signaling to consumers the freshness level of the food and indicating potential spoilage caused by temperature fluctuations during storage or packaging leaks. Time-temperature integrators (TTIs) are instruments designed to exhibit an unalterable alteration in a physical attribute, often manifested through color or shape modification, as a response to the temperature history. These TTIs are anticipated to replicate the change in a specific quality aspect of the food product, mirroring the impact of identical temperature exposure (Scetar et al., 2010).

For instance, Kuswandi et al. (2013) formulated a bromophenol blue–based dye to observe the freshness of guavas. MAP has found broader application in preserving fresh fruits compared to its limited use with freshly cut and minimally processed fruits. This includes fruits like strawberries (Almenar et al., 2007), loquats (Amoros et al., 2008), and mandarins (Del-Valle et al., 2009). Similarly, MAP with reduced O_2 levels of 10% and elevated carbon dioxide (CO_2) levels of 5% reduces deterioration in quality attributes as well as a decline in levels of functional compounds in comparison to storage in normal air conditions over the course of storage lasting 12 days in Parthenon broccoli (Fernández-León et al., 2013). Other applications of MAP include long-term storage of fresh produce such as apples, kiwifruit, pears, oranges, cabbage, and short-term storage of strawberries, bananas, cherries, tomatoes, guavas, etc. (Mangaraj & Goswami, 2009).

1.15.2 Ventilated Packaging for Bulk Handling and Logistics

Shipping or transportation packages are purpose-built for conveying goods over long distances, with varying storage capacities, yet they do not serve a marketing role. These packages are required to withstand the rigors of impact, vibration, and compression during their transit. Different types of packaging materials find application in the transport packaging of horticultural products. The most prevalent form of shipping containers for loads ranging from 5 to 20 kg are corrugated fiberboard cases. However, these are gradually being substituted by corrugated trays that are either stretch-wrapped or shrink-wrapped. Additional types of shipping containers encompass wooden or metal cases, crates, barrels, drums, and sacks. More recently, intermediate bulk containers (IBCs), including combi-bins, which are large boxes constructed from metal, plastic, or corrugated fiberboard, as well as sizable bags crafted from woven plastic fabric, have been introduced to improve handling efficiency and have largely displaced wooden crates. Regarding transport packaging for fresh produce, it can be categorized into two size ranges. The predominant size range, intended for manual carrying, falls between 15 and 25 kg. The other group, which has gained increasing popularity recently and falls within the range of 200–500 kg, is suitable for forklift handling and is referred to as a pallet container (Morya & Sharma, 2020).

1.15.3 Edible Films and Coatings

Another type of packaging for FAVs is edible films and coatings. The utilization of edible coatings seems to be a highly inventive method for prolonging the marketable lifespan of FAVs. This technique operates by serving as a gas transport barrier and inducing effects comparable to those seen in controlled atmosphere storage (Park, 1999). Biopolymer-based materials known as edible films and coatings have been utilized extensively to uphold the quality of fresh produce and extend its shelf life. Many fruits, including apples, pears, plums, citrus, cucumbers, avocados, and tomatoes, are routinely covered prior to being sold. A range of materials that can form films are employed in these edible coatings, and they can be categorized into proteins, lipids, polysaccharides, and their combinations. Examples of such materials encompass corn zein, soy protein, peanut protein, wheat gluten, cottonseed protein, casein, alginates, milk whey proteins, and collagen (Scetar et al., 2010; Morya & Sharma, 2020; Adhikary & Kumar, 2022).

The materials employed for coating must possess specific qualities, including acceptable sensory attributes, suitable barrier characteristics, strong mechanical resilience, reasonable stability against microbes, biochemical changes, and physicochemical shifts. Moreover, they should ensure safety, cost-effectiveness, and uncomplicated production technologies. Edible films and coatings serve several roles for FAVs, including the inhibition or reduction of moisture loss and the regulation of gas exchange. These functions are facilitated by the selective permeability of these edible materials to water vapor and gases, including O_2 and CO_2. Additionally, these coatings can serve as carriers for active substances like antimicrobials, antioxidants, browning agents, colorants, and flavors. This serves to enhance both the quality and safety of the produce (Diab et al., 2001; Chonhenchob et al., 2017; Morya & Sharma, 2020; Adhikary & Kumar, 2022). Diab et al. (2001) reported the utilization of pullulan as a coating hydrocolloid for kiwifruit and strawberries.

Utilizing edible coatings as carriers of antimicrobial agents represents an additional potential strategy for enhancing the safety of pre-cut produce. These antimicrobial coatings have the potential to offer heightened inhibitory effects against both spoilage and harmful bacteria by sustaining effective concentration of active substances on the surfaces of the food. Various classes of antimicrobial agents can potentially be integrated into edible coatings. These encompass organic acids (such as benzoic, lactic, acetic, propionic, and sorbic acids), fatty acid esters (like glyceryl monolaurate), polypeptides (including lysozyme, peroxidase, lactoferrin, and nisin), essential oils derived from plants (such as cinnamon, oregano, and lemongrass), as well as substances like sulphites and nitrites, among others (Franssen & Krochta, 2003; Scetar et al., 2010).

1.16 REQUIREMENTS FOR REFRIGERATED AIR STORAGE AND CONTROLLED ATMOSPHERE STORAGE

A cold storage facility is a sizable, thermally insulated enclosure equipped with appropriate doors and internal cooling systems. Cold storage units designed for FAVs come with specific requirements distinct from regular refrigerated storage spaces. These requirements encompass a substantial cooling capability that demands precise temperature control, and the relative humidity within must exceed 90%. A fundamental criterion for such storage is its capacity to cool a daily amount of produce equivalent to 10% of the total storage capacity, achieving an initial cooling rate of no less than 0.5 °C per hour. For smaller cold storage spaces, about 1 ton of refrigeration capacity is necessary for every 18 tons of produce. Various construction methods can be employed for cold storage facilities, if they adhere to the aforementioned conditions (Shakeel et al., 2022).

Primarily, cold storage is employed to prolong shelf life and preserve the availability of produce. Over time, the freshness, taste, and nutritional content of perishable foods can undergo changes, leading to spoilage. By preserving perishable foods, they can be safeguarded from spoilage and remain accessible year round. It's challenging to enhance the quality of a harvested crop, so proper storage conditions encompassing temperature and humidity are crucial for elongating storage duration and sustaining quality once the crop has been cooled to the optimal storage temperature. Fresh FAVs necessitate lower temperatures (ranging from 0 to 23 °C) and elevated relative humidity (ranging from 80% to 95%) to decrease respiration rates and decelerate metabolic and transpiration processes (Suraj, 2018; Wankhede et al., 2020).

1.16.1 Refrigerated Storage Requirements

FAVs are typically harvested within an ambient temperature ranging from 25 to 35 °C. During these conditions, the respiration rate is elevated, leading to a shorter storage life. The rate of deterioration in stored FAVs is significantly influenced by temperature. Temperature control stands out as the foremost factor in upholding product quality during the span between harvesting and consumption. Respiration and metabolic rates exhibit a direct correlation with ambient room temperatures within a specific range. A higher respiration rate corresponds to hastened deterioration of the produce. One effective approach to reducing this process and extending the storage duration of FAVs is to lower the temperature to a suitable level (Lal Basediya et al., 2013). The act of promptly subjecting harvested produce to lower temperatures for storage purposes reduces the respiration rate. Consequently, this curtails

TABLE 1.3 Refrigerated Storage Requirements of Selected Fruits and Vegetables

COMMODITY		*TEMPERATURE (°C)*	*RELATIVE HUMIDITY (%)*	*APPROXIMATE STORAGE LIFE*
Fruits	Strawberries	0	90–95	3–7 days
	Pomegranates	0–5	90–95	2–3 months
	Apples	0–4	90–95	2–6 months
	Oranges	0–4	85–90	3–4 months
	Pears	0	90–95	2–5 months
Vegetables	Artichokes, Jerusalem	(−1)–0	90–95	4–5 months
	Asparagus	0–3	95–100	2–3 weeks
	Broccoli	0	95–100	10–14 days
	Spinach	0	95–100	10–14 days
	Squashes, summer	9–18	95	1–2 weeks

Sources (Ashrae, 1998; Gast, 2008; Lal Basediya et al., 2013)

respiration-generated heat, thermal degradation, and microbial spoilage and contributes to sustaining the quality and freshness of the stored produce for an extended period (Chopra et al., 2003).

Another crucial consideration in the handling of FAVs is the relative humidity prevalent in the storage environment. The rate of transpiration, denoting the loss of water from produce, is influenced by the moisture content of the air, often denoted as relative humidity. At elevated relative humidity levels, the produce retains its marketable weight, visual appeal, nutritional value, and flavor, while attributes like wilting, softening, and excessive juiciness are minimized (Lal Basediya et al., 2013). When the relative humidity is low, the air holds only a small portion of its potential water vapor capacity, and it's also capable of taking on additional moisture. This condition results in increased transpiration rates. Conversely, in situations of high relative humidity, the rate of water transpiration is low, consequently leading to reduced cooling efficiency (Odesola & Onyebuchi, 2009). Keeping elevated humidity levels around harvested produce serves to curtail water loss, which otherwise could lead to diminished quality and a loss in marketable weight, thus affecting returns. Nonetheless, it's important to combine high humidity with low-temperature storage to also avoid the proliferation of fungi and bacteria. Table 1.3 outlines the recommended refrigerated storage conditions for selected FAVs.

1.16.2 Controlled Atmosphere Storage Requirements

Controlled atmosphere (CA) technology stands out as one of the most successful strategies introduced by the postharvest industry during the 20th century (Falagán & Terry, 2018). The conventional employment of CA involves elevating carbon dioxide levels while simultaneously reducing the concentration of oxygen. It has been demonstrated that CA alters the immediate environment surrounding the produce, thus influencing the internal gas composition. This adjustment curtails the metabolic activity of fruits or vegetables and retards the onset of senescence. The practical outcome is an elongated availability of produce throughout the year, the preservation of physicochemical and functional quality, and a potential decrease in consumer expenses. In addition to these benefits, CA's ability to mitigate storage-related issues, such as chilling injuries, contributes to a reduction in food wastage, which in turn has positive economic, social, and environmental effects (Rama & Narasimham, 2003; Falagán & Terry, 2018).

CA storage offers advantageous outcomes, such as extending the shelf life of perishable commodities by decreasing respiration and senescence, reducing ethylene production and the sensitivity of produce to it, diminishing the occurrence and severity of decay, and managing fungi, bacteria, and pests in FAVs. Nevertheless, if employed improperly, CA technology can lead to adverse effects like exacerbating physiological issues, uneven fruit ripening, the development of undesirable flavors, and heightened susceptibility to decay (Kader, 2002; Kader & Saltveit, 2003; Wills et al., 2007). The boundary between beneficial and detrimental outcomes from CA combinations is often quite narrow. It's important to note that CA is intended as a supplement to refrigerated storage and shouldn't be used as a replacement for appropriate temperature- and humidity-control management (Kader, 2002; Kader & Saltveit, 2003; Wu, 2010).

The effectiveness of CA hinges on various factors: the specific crop variety, whether it is climacteric or non-climacteric, the storage temperature, the carefully chosen gas concentrations, and the maturity stage of the commodity, its quality at harvest, and any pre-storage treatments applied. Under optimal conditions for a given crop, CA accomplishes the delay of senescence through mechanisms like diminishing respiration rates and substrates oxidation, postponing the ripening

TABLE 1.4 Controlled Storage Requirements of Selected Fruits and Vegetables

COMMODITY		O_2 *(%)*	CO_2 *(%)*	*TEMPERATURE (°C)*
Climateric fruits	Apple	2–3	1–5	0–2
	Mango	5	5	10–15
	Papaya	2–5	5–8	10–15
Non-climateric fruits	Pomegranate	3–5	0–10	1–8
	Blackberry	5–10	10–19	0–5
	Orange	5–10	0–5	5–10
Vegetables	Beetroot	10	3	0
	Broccoli	1–2	5–10	0–5
	Lettuce	2–5	0	0–5
	Okra	3–5	0	8–10
	Sweet potato	2–4	5–10	0–5

Sources (Kader, 1992; Bishop, 1996; Rama & Narasimham, 2003; Caleb et al., 2012)

of climacteric fruits, and decreasing the pace of ethylene production. Additionally, CA brings down the respiration rate of pathogens and can sustain the natural diseases resistance of the produce (Rama & Narasimham, 2003; Thompson, 2010; Falagán & Terry, 2018). Table 1.4 outlines the recommended controlled storage conditions for selected FAVs.

1.17 FUTURE PERSPECTIVES

Raising the intake of FAVs forms a crucial element in transitioning toward more healthful and ecologically balanced diets. According to future economic projections, availability is insufficient in several countries to enable the populations to attain the recommended consumption levels. Therefore, comprehensive and strategic public policy efforts aimed at addressing the obstacles related to producing, handling, and consuming FAVs will be imperative. This entails implementing a range of measures and investments, including boosting FAV cultivation, advancing technologies and methods to curtail waste without imposing additional costs on consumers, and enhancing ongoing endeavors to educate the public about healthy dietary choices. Furthermore, the international nutrition and agricultural sectors must collaboratively devise inventive approaches to enhance the production and consumption of FAVs, addressing the imperative of meeting the health requirements of the population, particularly in less developed parts of the world.

Postharvest waste and loss of FAVs are high, especially in developing countries. Further research is needed to come up with innovative ways to curb postharvest loss along the FAV supply chain (Opara, 2006a, 2009a, 2012, 2015; Fawole & Opara, 2018). Similarly, further investigations are necessary to gain a comprehensive knowledge of the dynamic physiological reactions of FAVs to CA and MAP technologies. This understanding is vital for pinpointing the optimal conditions suited for each cultivar and scenario. A particularly essential aspect worthy of further research is the effect of postharvest handling and processing on produce flavor, which is a major quality concern among consumers.

Furthermore, integrating biosensors into MAP presents fresh opportunities for swiftly identifying and diagnosing biological risks, thereby enabling timely management of food safety concerns, including instances of contamination. Likewise, there is a need for enhancements in the structures and functionality of film polymers, alongside the development and application of novel sustainable materials for MAP. Cost reduction is pivotal to expanding the accessibility of these technologies to a broader range of companies in the sector. This cost reduction would foster the wider adoption of MAP, subsequently curbing FAV loss and wastage.

The need to assure food and nutrition security for a growing global population is recognized, and with high incidence of food losses and waste along the value chain, saving the harvest through postharvest technologies has become paramount (Opara, 2006a, 2006b, 2010a, 2010b, 2015). Postharvest technologies also link production to markets through quality control and assurance and other value addition techniques and procedures (Opara, 2006b, 2009a). To succeed in harnessing the prospects of postharvest technology in food systems transformation and modernization, more investments in research are needed (Opara, 2009b, 2012), and especially to build capacity (human, infrastructure, and local institutions) in places where these are deficient (Opara, 2009c, 2012).

1.18 CONCLUSIONS

FAVs constitute a vital component of human diets and serve as a significant source of biologically active constituents, like secondary metabolites and vitamins. Despite this huge importance, the average consumption of FAVs falls short of the recommended levels on a global scale. Parental impact emerged as a primary influencer of the FAV consumption patterns among adolescents in developing nations. Promoting and encouraging FAV consumption by parents can change low consumption among children and adolescents. Public awareness campaigns and educational initiatives aimed at augmenting the intake of FAVs among older people is also recommended. FAV cultivation and trade have witnessed an upward surge due to the increasing awareness of their health and economic benefits. Postharvest loss and waste of FAVs are massive across the globe and pose a serious challenge to FAV availability for maximum consumption. A reduction in these postharvest losses plays a significant role in ensuring the sustainable nourishment of the world's population in the coming years. Maturity indexing techniques helps in ensuring sensory quality and adequate postharvest life of FAVs. Some identified maturity indices for FAVs includes skin color, shape and size of fruit and leaf, aroma, texture, acidity, moisture and oil content, starch, and sugar content. Harvesting FAVs should be done at the right stage of maturity because maturity status influences the postharvest quality of product and shelf life.

Factors affecting postharvest physiology of FAVs include respiration rate, ethylene production, and transpiration/water loss. These three processes are dependent on each other, as they not only respond in a similar manner to environmental changes like reduced storage temperature, but they also influence one another. For instance, a low respiration rate reduces ethylene production and transpiration rates. Thus, overlooking the effect of any of these three factors will lead to quality deterioration of FAVs. Quality indicators of FAVs include intrinsic aroma, taste, flavor, texture, nutritional quality, flesh firmness, diseases, appearance, size, color, and bruises. FAV quality can be assessed using destructive and non-destructive methods. Non-destructive methods like infrared spectroscopy, electronic nose, nuclear magnetic resonance, machine vision, X-ray computed tomography, and magnetic resonance imaging are finding wider application in FAV quality assessment. Mechanical damage is a major factor influencing the quality of FAVs. FAVs experience mechanical damage from various sources during handling and transportation, including compression, impact, abrasion, puncture, and vibration. Packaging methods are the most important control strategy in reducing mechanical damage to FAVs. Diverse packaging methods and designs are utilized for handling and transporting fresh FAVs.

A packhouse facilitates quality control measures to guarantee that both the quality and quantity of the FAVs adhere to market standards while also minimizing losses that might occur during transportation and delivery to markets. Packaging is an important packhouse operation. Packaging plays vital functions, such as organizing the produce into manageable units for ease of handling; aiding in the containment, movement, and logistical aspects of FAVs; prolonging the postharvest viability of FAVs; and safeguarding them throughout distribution, storage, and the marketing process. The common packaging for FAVs is logistic/bulk packaging, MAP, and edible coatings and films. Cold storage is critical in extending the shelf life of FAVs. Storage at low temperatures and very high relative humidity alongside controlled atmosphere technology is an effective approach in preserving the quality of FAVs and averting postharvest loss and waste.

ACKNOWLEDGMENT

This work is based on the research supported in part by the National Research Foundation of South Africa (Grant Numbers: 64813). The opinions, findings, and conclusions or recommendations expressed are those of the author(s) alone, and the NRF accepts no liability whatsoever in this regard. Professor UL Opara acknowledges the Department of Science and Innovation of South Africa for this grant.

REFERENCES

Aaby, K., Haffner, K., Skrede, G. (2002). Aroma quality of Gravenstein apples influenced by regular and controlled atmosphere storage. *LWT-Food Science and Technology*, *35*, 254–259.

Abbott, J. A. (1999). Quality measurement of fruits and vegetables. *Postharvest Biology and Technology*, *15*(3), 207–225. https://doi.org/10.1016/S0925-5214(98)00086-6

Abbott, J. A. (2004). Textural quality assessment for fresh fruits and vegetables. *Advances in Experimental Medicine and Biology*, *542*, 265–279. https://doi.org/10.1007/978-1-4419-9090-7_19

Abedi, G., & Ahmadi, E. (2013). Design and evaluation of a pendulum device to study postharvest mechanical damage in fruits: bruise modeling of Red Delicious apple. *Australian Journal of Crop Science*, *7*, 962e968.

Acedo, A. L., Rahman, M. A., Buntong, B., Gautam, D. M. (2016). *Establishing and managing smallholder vegetable packhouses to link farms and markets* (Publication No. 16–801). AVRDC—The World Vegetable Center, p. 46.

Adesogan, A. T., Havelaar, A. H., McKune, S. L., Eilittä, M., Dahl, G. E. (2020). Animal source foods: sustainability problem or malnutrition and sustainability solution? Perspective matters. *Global Food Security*, *25*, 100325. https://doi.org/10.1016/j.gfs.2019.100325

Adhikary, T., Kumar, D. (2022). Advances in postharvest packaging systems of fruits and vegetable. *Postharvest Technology—Recent Advances, New Perspectives and Applications*, 1–15. https://doi.org/10.5772/intechopen.101124

Agrios, G. N. (2005). *Plant pathology* (5th ed.). Elsevier Inc and Academic Press, p. 553.

Agudo, A. (2004). *Measuring intake of fruits and vegetables* (Background paper for the joint FAO/WHO workshop on fruit and vegetables for health). Kobe, pp. 1–3

Ahmadi, E., Ghassemzadeh, H. R., Sadeghi, M., Moghaddam, M., Neshat, S. Z. (2010). The effect of impact and fruit properties on the bruising of peach. *Journal of Food Engineering*, *97*, 110–117.

Ahvenainen, R. (2003). Active and intelligent packaging: an introduction. In: Ahvenainen, R. (Ed.), *Novel food packaging techniques*. Woodhead, pp. 5–21.

AL-Kahtani, S. H., Kaleefah, A. M. (2011). Postharvest technology impact on marketing loss and economic resources losses for important vegetables and fruit crops in Saudi Arabia. A technical project summary funded by King Abdul-Aziz city for science and technology, titled: Agricultural marketing in Kingdom of Saudi Arabia: Current situation, challenges, and solutions. Report number: White Paper No. 15-02. https://doi.org/10.13140/RG.2.1.3921.6402

Alfatni, M. S. M., Shariff, A. R. M., Abdullah, M. Z., Marhaban, M. H. B., Saaed, O. M. B. (2013). The application of internal grading system technologies for agricultural products e review. *Journal of Food Engineering, 116*, 703–725.

Aliasgarian, S., Ghassemzadeh, H. R., Moghaddam, M., Ghaffari, H. (2013). Mechanical damage of strawberry during harvest and postharvest operations. *World Applied Sciences Journal, 22*, 969–974.

Almenar, E., Del-Valle, V., Hernández-Muñoz, P., Lagarón, J. M., Catalá, R., Gavara, R., (2007). Equilibrium modified atmosphere packaging of wild strawberries. *Journal of the Science of Food and Agriculture, 87*(10), 1931–1939.

Amao, I. (2018). *Health benefits of fruits and vegetables: review from Sub-Saharan Africa*. InTech.

Amoros, A., Pretel, M. T., Zapata, P. J., Botella, M. A., Romojaro, F., Serrano, M. (2008). Use of modified atmosphere packaging with microperforated polypropylene films to maintain postharvest loquat fruit quality. *Food Science and Technology International, 14*(1), 95–103.

Anderson, J., Perryman, S., Young, L., Prior, S. (2010). *Dietary fibre fact (Sheet No. 9.333)*. Colorado State University. www.ext.colostate.edu/pubs/foodnut/ 09333.

Antmann, G., Lema, P., Lareo, C. (2008). Influence of modified atmosphere packaging on sen- sory quality of shiitake mushrooms. *Postharvest Biology and Technology, 49*, 164–170.

ARAUZ, L. F., Wang, A., Duran, J. A., Monterrey, M. (1994). Causas de pérdidas poscosecha de mango a nivel mayorista en Costa Rica (Factors causing postharvest losses of mango fruit at central distribution market in Costa Rica). *Agronomia Costarricense, 18*, 47–51.

Arendse, E., Amos, O., Samukelo, L., Linus, U. (2018a). Non-destructive prediction of internal and external quality attributes of fruit with thick rind: a review. *Journal of Food Engineering, 217*, 11–23.

Arendse, E., Fawole, O. A., Magwaza, L. S., Nieuwoudt, H. H., Opara, U. L. (2017). Development of calibration models for the evaluation of pomegranate aril quality by Fourier-transform near infrared spectroscopy combined with chemometrics. *Biosystem Engineering, 159*, 22–32.

Arendse, E., Fawole, O. A., Magwaza, L. S., Nieuwoudt, H. H., Opara, U. L. (2018b). Fourier transform near infrared diffuse reflectance spectroscopy and two spectral acquisition modes for evaluation of external and internal quality of intact pomegranate fruit. *Postharvest Biology and Technology, 138*, 91–98

Arendse, E., Fawole, O. A., Opara, U. L. (2014). Discrimination of pomegranate fruit quality by instrumental and sensory measurements during storage at three temperature regimes. *Journal of Food Processing and Preservervation, 39*, 1745–4549.

Arendse, E., Nieuwoudt, H., Magwaza, L. S., Fredric, J., Nturambirwe, I., Fawole, O. A., Opara, U. L. (2020). Recent advancements on vibrational spectroscopic techniques for the detection of authenticity and adulteration in horticultural products with a specific focus on oils, juices and powders. *Food and Bioprocess Technology, 14*, 1–22.

Armstrong, P. R., Stone, M. L., Brusewitz, G. H. (1997). Nondestructive acoustic and compression measurements of watermelon for internal damage detection. *Transactions of ASAE, 13*, 641–645.

Arpaia, M. L., Boreham, D., Hofshi, R. (2001). Development of a new method for measuring minimum maturity of avocados. *California Avocado Society Yearbook, 85*, 153–178.

Ashrae (1998). *ASHRAE handbook, refrigeration*. American Society of Heating, Refrigeration and Air-Conditioning Engineers, Inc. SI edn

Awad, M. A., De Jager, A. (2002). Influences of air and controlled atmosphere storage on the concentration of potentially healthful phenolics in apples and other fruits *Postharvest Biology and Technology, 27*, 53–58.

Azam, A., Shafique, M. (2018). An overview of fruits and vegetables trade of China. *International Journal of U- and e-Service, Science and Technology, 11*(1), 33–44. https://doi.org/10.14257/ijunesst.2018.11.1.03

Babarinsa, F., Ige, M. (2012). Strength parameters of packaged Roma tomatoes at break point under compressive loading. *International Journal of Scientific and Engineering Research, 3*, 1–8.

Baez-Sanudo, R., Rodriguez-Felix, A., Siller, J. H., Bringas-Taddei, E., Baez, M. A., Camarena-Gomez, G., Martinez-Antunez, R. (1994). Habitos de compra, consume y perdidas de mango a nivel consumidor (Purchasing habits, consumption and loss of mango at the consumer level). *Proceedings of the Interamerican Society for Tropical Horticulture, 38*, 28–31.

Baiano, A., Terracone, C., Peri, G., Romaniello, R. (2012). Application of hyperspectral imaging for prediction of physico-chemical and sensory characteristics of table grapes. *Computers and Electronics in Agriculture, 87*, 142–151.

Bajcar, M., Saletnik, B., Zardzewiały, M., Drygaś, B., Czernicka, M., Puchalski, C., Zaguła, G. (2016). Method for determining fruit harvesting maturity. *Journal of Microbiology, Biotechnology and Food Sciences, 6*(2), 773–776. https://doi.org/10.15414/jmbfs.2016.6.2.773-776

Baldwin, E. A. (2003). Coatings and other supplemental treatments to maintain vegetable quality. In: J. A. Bartz and J. K. Brecht (Eds.) *Postharvest physiology and pathology of vegetables*. Marcel Dekker, p. 413

Bancroft, R. et al. (1998). The marketing system for fresh yams in Techiman, Ghana and associated postharvest losses. *Tropical Agriculture 75*(1/2), 115–119.

Bani, R. J., Josiah, M. N., Kra, E. Y. (2006). Postharvest losses of tomatoes in transit. *Agricultural Mechanization in Asia, Africa and Latin America, 37*(2), 84–86.

Banks, N. H., Borton, C. A., Joseph, M. (1991). Compression bruising test for bananas. *Journal of the Science of Food and Agriculture, 56*, 223–226.

Banks, N. H., Joseph, M. (1991). Factors affecting resistance of banana fruit to compression and impact bruising. *Journal of the Science of Food and Agriculture, 56*, 315–323.

Baranowski, P., Mazurek, W., Witkowska-Walczak, B., Slawin´ski, C. (2009). Detection of early apple bruises using pulsed-phase thermography. *Postharvest Biology and Technology, 53*, 91–100.

Baranowski, P., Mazurek, W., Wozniak, J., Majewska, U. (2012). Detection of early bruises in apples using hyperspectral data and thermal imaging. *Journal of Food Engineering, 110*, 345–355.

Barbosa-Cánovas, G. V., Fernández-Molina, J. J., Alzamora, S. M., Tapia, M. S., López-Malo, A., Chanes, J. W. (2003). *Handling and preservation of fruits and vegetables by combined methods for rural areas* (Technical Manual: FAO Agricultural Services Bulletin 149). Food and Agriculture Organization of the United Nations

Beaudry, R. M. (1999) Effect of O_2 and CO_2 partial pressure on selected phenomena affecting fruit and vegetable quality. *Postharvest Biology and Technology, 15*, 293–303.

Beaudry R. M., Cameron A. C., Shirazi A., Dostal Lange, D. L. (1992) Modified atmosphere packaging of blueberry fruit: effect of temperature on package O_2 and CO_2. *Journal of the American Society for Horticultural Science, 117*(3), 436–441.

Bellavia, A., Larsson, S. C., Bottai, M., Wolk, A., Orsini, N. (2013). Fruit and vegetable consumption and all-cause mortality: a dose-response analysis. *The American Journal of Clinical Nutrition, 98*(2), 454–459.

Ben-Yehoshua, S., Rodov, V. (2003). Transpiration and water stress. In: J. A. Bartz and J. K. Brecht (Eds.) *Postharvest physiology and pathology of vegetables*. Marcel Dekker, p. 111.

Berk, Z. (2016). Packing house operations. In *Citrus fruit processing*. Elsevier, pp. 107–125. https://doi.org/10.1016/B978-0-12-803133-9.00007-2

Berry, T. M., Delele, M. A. Griessel, H., Opara, U. L. (2015). Geometric design characterisation of ventilated multi-scale packaging used in the South African pome fruit industry. *Agricultural Mechanization in Asia, Africa, and Latin America*, *46*, 34–42.

Beudry, R., Watkins, C. (2001). Use of 1-MCP on apples. *Perishable Handling Q*, *108*, 12.

Bhargava, A., Bansal, A. (2021). Fruits and vegetables quality evaluation using computer vision: a review. *Journal of King Saud University—Computer and Information Sciences*, *33*, 243–257

Bill, M., Sivakumar, D., Thompson, A. K., Korsten, L. (2014). Avocado fruit quality management during the postharvest supply chain. *Food Reviews International*, *30*(3), 169–202. https://doi.org/10.1080/87559129.2014.907304

Bishop, D. (1996). Controlled atmosphere storage. In: C. V. Dellino (Ed.) *Cold and chilled storage technology*. Blackie.

Blackenberg, A., Fawole, O.A., Opara, U.L. (2018). Quantifying postharvest losses of 'Crimson Seedless' table grapes along the supply chain. *Acta Horticulturae*, *1201*, 29–34. https://doi.org/10.17660/ActaHortic.2018.1201.5

Blackenberg, A., Opara, U.L., Fawole, A.O. (2021). Postharvest losses in quantity and quality of Table Grape (cv. Crimson Seedless) along the supply chain and associated economic, environmental and resource impacts. *Sustainability, 13*, 4450. https://doi.org/10.3390/su13084450

Blackenberg, A., Opara, U.L., Fawole, O.A. (2022b). Quantifying the magnitude of postharvest fruit losses: a case study of industry packhouse data in the Western Cape, South Africa. *Acta Horticulturae*, *1349*, 243–250. https://doi.org/10.17660/ActaHortic.2022.1349.34

Blackenburg, A., Fawole, O.A., Opara, U.L. (2022a). Postharvest losses in quantity and quality of pear (cv. Packham's Triumph) along the supply chain and associated economic, environmental and resource impacts. *Sustainability, 14*(2), 603. https://doi.org/10.3390/su14020603

Blasco, J., Aleixos, N., Cubero, S., Go´mez-Sanchis, J., Molto, E. (2009). Automatic sorting of satsuma segments using computer vision and morphological features. *Computers and Electronics in Agriculture*, *66*, 1–8.

Blasco, J., Aleixos, N., Gómez, J.R., & Moltó, E. (2007). Citrus sorting by identification of the most common defects using multispectral computer vision. *Journal of Food Engineering*, *83*, 384–393.

Boboeva, K. (2023). Study on the quality indicators of apple varieties grown at low level under-grafted M-9. *IOP Conference Series: Earth and Environmental Science*, *1142*(1). https://doi.org/10.1088/1755-1315/1142/1/012059

Boffetta, P., Couto, E., Wichmann, J., Ferrari, P., Trichopoulos, D., Bueno-de-Mesquita, H. B. (2010). Fruit and vegetable intake and overall cancer risk in the European Prospective Investigation into Cancer and Nutrition (EPIC). *Journal of the National Cancer Institute*, *102*(8), 529–537. doi: 10.1093/jnci/djq072. PMID: 20371762

Boonyakiat, D., Janchamchai, K. (2007). The Royal Project packinghouse quality assurance. [In: A. N. Fardous, W. Schnitzler, M. Qaryouti (Eds.) Proceedings of the First International Symposium on Fresh Food Quality Standards: Better Food by Quality and Assurance]. *Acta Horticulturae*, *741*, 41–47.

Brasil, I. M., Siddiqui, M. W. (2018). Postharvest quality of fruits and vegetables: an overview. In *Preharvest modulation of postharvest fruit and vegetable quality*. Elsevier Inc. https://doi.org/10.1016/B978-0-12-809807-3.00001-9

Brody, A. L., Strupinsky, E. P., Kline, L. R. (2001). *Active packaging for food applications*. CRC press.

Brown, M., Bates, R. P., Mcgowan, C., Cornell, J. A. (1991). Influence of fruit maturity on the hypoglycin a level in Ackee (*Blighia Sapida*). *Journal of Food Safety*, *12*(2), 167–177. https://doi.org/10.1111/j.1745-4565.1991.tb00075.x

Bunsiri, A., Paull, R. E., Ketsa, S. (2012). Increased activities of phenyalanine ammonia lyase, peroxidase, and cinnamyl alcohol dehydrogenase in relation to pericarp hardening after physical impact in mangosteen (*Garcinia mangostana* L.). *The Journal of Horticultural Science and Biotechnology 87*(3), 231–236. https://doi.org/10.1080/14620316.2012.11512857.

Caleb, O.J., Mahajan, P.V., Al-Said, F. A., Opara, U. L. (2013). Modified atmosphere packaging technology of fresh and fresh-cut produce and the microbial consequences—a review. *Food and Bioprocess Technology*, *6*, 303–329. https://doi.org/10.1007/s11947-012-0932-4

Caleb, O.J., Opara, U.L., Witthuhn, C.R. (2012). Modified atmosphere packaging of pomegranate fruit and arils: a review. *Food and Bioprocess Technology*, *5*, 15–30. https://doi.org/10.1007/s11947-011-0525-7

Cameron, A. C., Beaudry, R. M., Banks, N. H., Yelanich, M. V. (1994) Modified atmosphere packaging of blueberry fruit: modelling respiration and package oxygen partial pressures as a function of temperature. *Journal of the American Society for Horticultural Science*, *119*, 534–539.

Chen, P., Ruiz, M., Lu, F., Kader, A. A. (1987). Study of impact and compression damage on Asian pears. *Transactions of ASABE*, *30*, 1193–1197.

Chitarra, A. B., Alves, R. E. (2001). *Tecnologia de pós-colheita para frutas tropicais*. Instituto Frutal/Sindifruta, pp. 1, 314.

Chitarra, M. I. F., Chitarra, A. B. (2005). *Pós-colheita de frutas e hortaliças: fisiologia e manuseio* (Vol. 1). Universidade Federal de Lavras.

Chonhenchob, V., Singh, P., Singh, J. (2017). *Packaging & distribution of fresh fruits & vegetables*. DEStech Publication, pp. 4, 82.

Chopra, S., Baboo, B., Alesksha, Kudo, S. K., Oberoi, H. S. (2003). An effective on farm storage structure for tomatoes. Proceedings of the International Seminar on Downsizing Technology for Rural Development held at RRL, Bhubaneswar, Orissa, India, October 7–9, pp. 591–598.

Choudhury, M. M., da Costa, T. S. (2004). *Perdas na cadeia de comercializacao da manga (Losses in the commercialization network of mango)*. Documentos-da-Embrapa-Semi-Arido 186, p. 41.

Conner, T. S., Brookie, K. L., Carr, A. C., Mainvil, L. A., Vissers, M. C. M. (2017). Let them eat fruit! The effect of fruit and vegetable consumption on psychological well-being in young adults: a randomized controlled trial. *PLoS One*, *12*(2), e0171206. https://doi.org/10.1371/journal.pone.0171206

Crisosto, C. H., Garner, D., Doyle, J., Day, K. R. (1993). Relationship between fruit respiration, bruising susceptibility, and temperature in sweet cherries. *HortScience, 28*, 132.

Cubero, S., Aleixos, N., Molto´, E., Go´mez-Sanchis, J., Blasco, J. (2011). Advances in machine vision applications for automatic inspection and quality evaluation of fruits and vegetables. *Food and Bioprocess Technology*, *4*, 487–504.

Dadzie, B. K., Orchard, J. E. (1997). *Routine postharvest screening of banana/plantain hybrids: criteria and methods*. International Plant Genetic Resources Institute (IPGRI).

Day, B. P. F. (1993), Fruit and vegetables. In: R.T. Parry (Eds.) *Principles and applications of MAP of foods*. Blackie Academic and Professional, pp. 114–133.

Day, B. P. F. (1998). Novel MAP. A brand-new approach. *Food Manufacture*, *73*, 22–24.

Day, B. P. F. (1999). Recent developments in active packaging. *South African, Food & Beverage Manufacturing Review*, *26*(8), 21–27.

De Lucia, M., Assennato, D. (1994). *Agricultural engineering in development: Postharvest operations and management of food grains* (FAO Agricultural Services Bulletin No. 93). FAO.

Del-Valle, V., Hernández-Muñoz, P., Catalá, R., Gavara, R. (2009). Optimization of an equi- librium modified atmosphere packaging (EMAP) for minimally processed mandarin segments. *Journal of Food Engineering, 91*(3), 474–481.

Desmet, M., Lammertyn, J., Scheerlinck, N., Verlinden, B. E., Nicolaï, B. M. (2003). Determination of puncture injury susceptibility of tomatoes. *Postharvest Biology and Technology*, *27*(3), 293–303. doi: 10.1016/s0925-5214(02)00115-1

Desmet, M., Lammertyn, J., Verlinden, B. E., Nicolaï, B. M. (2002). Mechanical properties of tomatoes as related to puncture injury susceptibility. *Journal of Texture Studies*, *33*(5), 415–429. doi: 10.1111/j.1745-4603.2002.tb01357.x

Dhatt, A. S., Mahajan, B. V. C. (2007). Harvesting, handling and storage of horticultural crops. *Journal of Experimental Psychology: General*, *136*(1), 23–42.

Diab, T., Biliaderis, C., Gerasopoulos, D., Sfakiotakis, E. (2001). Physicochemical properties and application of pullulan edible films and coatings in fruit preservation. *Journal of the Science of Food and Agriculture*, *81*, 988–1000

Díaz-Pérez, J. C. (2018). Transpiration. In *Postharvest physiology and biochemistry of fruits and vegetables*. Elsevier Inc. https://doi.org/10.1016/B978-0-12-813278-4.00008-7

Diels, E., van Dael, M., Keresztes, J., Vanmaercke, S., Verboven, P., Nicolai, B., Saeys, W., Ramon, H., Smeets, B. (2017). Assessment of bruise volumes in apples using X-ray computed tomography. *Postharvest Biology and Technology*, *128*, 24–32. doi: 10.1016/j.postharvbio.2017.01.013.

Drewnowski, A. (2020). Analysing the affordability of the EAT–Lancet diet. *Lancet*, *8*, 6–7.

Drewnowski, A., Finley, J., Hess, J. M., Ingram, J., Miller, G., Peters, C. (2020). Toward healthy diets from sustainable food systems. *Current Developments in Nutrition*, *4*, 1–12

Du Plessis, A., Broeckhoven, C., Guelpa, A., le Roux, S. G. (2017). Laboratory x-ray micro-computed tomography: A user guideline for biological samples. *Gigascience*, *6*, 1–11.

Duduzile Buthelezi, N. M., Tesfay, S. Z., Ncama, K., Magwaza, L. S. (2019). Destructive and non-destructive techniques used for quality evaluation of nuts: A review. *Scientia Horticulturae*, *247*, 138–146. https://doi.org/10.1016/j.scienta.2018.12.008

Eberle, K. (2005). Evaluation of near infrared and nuclear magnetic resonance spectroscopy for rapid quality control of south african extra virgin olive oils (Master's Thesis, Stellenbosch University, Stellenbosch).

EL-ASSI, N. (2002). Postharvest losses of peppers and squashes produced for local markets in Jordan. *Mutah Lil-Buhuth Wad-Dirasat*, *17*, 35–45 (in Arabic).

Elik, A., Yanik, D. K., Istanbullu, Y., Guzelsoy, N. A., Yavuz, A., Gogus, F. (2019). Strategies to reduce post-harvest losses for fruits and vegetables. *International Journal of Scientific and Technological Research*, *5*(3), 29–39. https://doi.org/10.7176/JSTR/5-3-04

Elmasry, G., Kamruzzaman, M., Sun, D. W., Allen, P. (2012). Principles and applications of hyperspectral imaging in quality evaluation of agro-food products: A review. *Critical Reviews in Food Science and Nutrition*, *52*, 8398

Elshazly, F. A. et al. (2009). *A study of losses in some important agricultural crops. Unpublished report of the agricultural economics research institute, ministry of agriculture and land reclamation*. Giza, p. 177 (in Arabic).

Esguerra, E. B., Bautista, O. K. (2007). Waxing. In: O. K. Bautista, E. B. Esguerra (Eds.) *Postharvest technology for Southeast Asian perishable crops*. University of the Philippines at Los Baños—Department of Agriculture, Bureau of Agricultural Research, pp. 127–133.

Exama, A., Arul, J., Lencki, R. W., Lee, L. Z., Toupin C. (1993). Suitability of plastic fi lms for modifi ed atmosphere packaging of fruits and vegetables. *Journal of Food Science*, *58*, 1365–1370.

Falagán, N., Terry, L. A. (2018). Recent advances in controlled and modified atmosphere of fresh produce. *Johnson Matthey Technology Review*, *62*(1), 107–117. https://doi.org/10.1595/205651318X696684

Fan, X., Yu, S., Lan, H., Zhang, H., Zhang, Y., Liu, Y. (2021). Based on the electrical properties of korla fragrant pear static damage degree of quantitative research. *Journal of Agricultural Mechanization Research*, *43*(9), 194–198. http://dx.doi.org/10.13427/j.cnki.njyi.2021.09.035.

Fang, X. (2002). *Why traders laugh in middle and cry at both ends in Agricultural Logistics*. International Trade News, p. 001.

FAO (2019). *The state of food and agriculture: moving forward on food loss and waste reduction*. FAO.

FAO (2020). *Fruit and vegetables—your dietary essentials*. The International Year of Fruits and Vegetables. https://doi.org/10.4060/cb2395en

FAO (Food and Agriculture Organization of the United Nations) (2004). *FAOSTAT*. FAO.

FAO & UNEP (1981). *Food loss prevention in perishable crops* (FAO Agricultural Services Bulletin No. 43). FAO.

Fashi, M., Naderloo, L., Javadikia, H. (2020). Pomegranate grading based on pH using image processing and artificial intelligence. *Journal of Food Measurement and Characterization*, *14*, 3112–3121.

Fawole, O.A., Opara, U.L. (2018). Value-addition of sunburned pomegranate fruit to reduce postharvest losses: A cosmeceutical perspective. *Acta Horticulturae*, *1225*, 221–226. https://doi.org/10.17660/ActaHortic.2018.1225.30

Fellows P. J. (2016). *Food processing technology: principles and practice* (4th ed.). FAO, pp. 462–508.

Feng, Y. Z., Sun, D. W. (2012). Application of hyperspectral imaging in food safety inspection and control: A review. *Critical Reviews in Food Science and Nutrition*, *52*, 1039–1058.

Fernández-León, M. F., Fernández-León, A. M., Lozano, M., Ayuso, M. C., Amodio, M. L., Colelli, G., (2013). Retention of quality and functional values of broccoli 'Parthenon' stored in modified atmosphere packaging. *Food Control*, *31*(2), 302–313.

Fernando, I., Fei, J., Stanley, R., Enshaei, H. (2018). Measurement and evaluation of the effect of vibration on fruits in transit—Review. *Packaging Technology and Science*, *31*, 1–16.

Fonseca, S. C., Oliveira, F. A., Brecht, J. K., Chau, K. V., (2005). Influence of low oxygen and high carbon dioxide on shredded Galega kale quality for development of modified atmosphere packages. *Postharvest Biology and Technology*, *35*(3), 279–292.

Franssen L. R., Krochta J. M. (2003) Edible coatings containing natural antimicrobials for processed foods. In: S. Roller (Ed.) *Natural antimicrobials for the minimal processing of foods*. CRC Press.

Gajanana, T. M., Sreenivasa Murthy, D., Sudha, M. (2011). Postharvest losses in fruits and vegetables in South India—A review of concepts and quantification of losses. *Indian Food Packer*, *65*(6), 178–187.

Gast, K. L. B. (2008). Storage conditions fruits & vegetables. *Postharvest Management of Commercial Horticultural Crops, Bulletin*, *4*, 1–8.

George, J. B., Mwangangi, B. M. (1994). Some factors affecting banana storage and ripening: a case study of banana handling and ripening in Kenya. *Acta Horticulturae*, *368*, 628–633.

Gil, M. I., Tudela, J. A., Martínez-Sánchez, A., Luna, M. C. (2012). Harvest maturity indicators of leafy vegetables. *Stewart Postharvest Review*, *8*(1), 1–9. https://doi.org/10.2212/spr.2012.1.2

Gowen, A. A., Feng, Y., Gaston, E., Valdramidis, V. (2015). Recent applications of hyperspectral imaging in microbiology. *Talanta*, *137*, 43–54.

Gustavsson, J., Cederberg, C., Sonesson, U., van Otterdijk, R., Meybeck, A. (2011), *Global food losses and food waste: extent causes and prevention*. Food and Agriculture Organization (FAO) of the United Nations.

Haard, N. E. (1995). Foods as cellular systems: impact on quality and preservation. A review. *Journal of Food Biochemistry*, *19*, 191–238.

Haard, N. E. (1998). Foods as cellular systems: impact on quality and preservation. In: I. A. Taub, R. P. Singh (Ed.) *Food storage and stability*. CRC Press, p. 39.

Hailu, G., Derbew, B. (2015). Extent, causes and reduction strategies of postharvest losses of fresh fruits and vegetables –a review. *Journal of Biology, Agriculture and Healthcare*, *5*(5), 49–64.

Hardenburg, R. E., Watada, A. T., Wang, C. Y. (1986). *The commercial storage of fruits, vegetables and nursery stocks* (Agriculture Handbook No 66). US Dept. of Agriculture.

He, K., Hu, F. B., Colditz, G. A., Manson, J. E., Willett, W. C., Liu, S. (2004). Changes in intake of fruits and vegetables in relation to risk of obesity and weight gain among middle-aged women. *International Journal of Obesity*, *28*, 1569–1574. doi: 10.1038/sj.ijo.0802795

Hernandez-Sanchez, N., Hills, B. P., Barreiro, P., Marigheto, N. (2007). An NMR study on internal browning in pears. *Postharvest Biology and Technology*, *44*, 260–270.

Hernandez-Sanchez, N., Barreiro, P., Ruiz-Altisent, M., Ruiz-Cabello, J., Fernandez-Valle, M. E. (2004). Detection of freeze injury in oranges by magnetic resonance imaging of moving samples. *Applied Magnetic Resonance*, *26*, 431–445.

Herrero, M. et al (2017). Farming and the geography of nutrient production for human use: a transdisciplinary analysis. *The Lancet Planetary Health*, *1*, 33–42.

Hirvonen, K., Bai, Y., Headey, D., Masters, W. A. (2019). Affordability of the EAT-Lancet reference diet: a global analysis. *The Lancet Global Health*, *19*, 1–8.

Hoejskov, P. S. (2014). Importance of fruit and vegetables for public health and food safety. Presentation at the Pacific Regional Workshop on Fruit and Vegetables for Health; PROFAV, Naji, Fiji, pp. 20–23.

Hollis, J. L., Collins, C. E., deClerck, F., Chai, L., McColl, K., Demaio, A. R. (2020). Defining healthy and sustainable diets for infants, children and adolescents. *Global Food Security, 27*, 100401. https://doi.org/10.1016/j.gfs.2020.100401

Hou, J., He, Z., Liu, D., Zhu, Z., Long, Z., Yue, X., Wang, W. (2023). Mechanical damage characteristics and nondestructive testing techniques of fruits: a review. *Food Science and Technology*, (Brazil), *43*, https://doi.org/10.1590/fst.001823

Huang, S. W. (2004). *An overview of global trade patterns in fruits and vegetables*. Global Trade Patterns in Fruits and Vegetables, pp. 3–15.

Idah, P., Yisa, M. (2007). An assessment of impact damage to fresh tomato fruits. *AU Journal of Technology*, *10*, 271–275.

IITA. (2008). *Pertes Post-récoltes de Légumes Frais dans le Sud du Bénin (Piments, Laitues et Tomates)*. IITA, p. 74 (in French)

Irtwange, S. (2006). Maturity, quality and marketing of fruits and vegetables. *Agricultural Engineering International, 8*.

Jarimopas, B., Rachanukroa, D., Singh, S. P., Sothornvit, R. (2008). Post-harvest damage and performance comparison of sweet tamarind packaging. *Journal of Food Engineering*, *88*(2), 193–201. https://doi.org/10.1016/j.jfoodeng.2008.02.015

Jarimopas, B., Singh, S. P., Sayasoonthorn, S., Singh. J. (2007). Comparison of package cushioning materials to protect post-harvest impact damage to apples. *Packaging Technology and Science*, *20*(5), 315–24. doi: 10.1002/pts.760.

Jowkar, M. M., Mohammadpour, H., Farshadfar, Z., Jowkar, A. (2005). A look at postharvest in Iran. *Acta Horticulturae*, *682*, 2177–2182.

Jung, H., Park, J. G. (2012). Effects of vibration stress on the quality of packaged apples during simulated transport. *Journal of Biosystems Engineering*, *37(1)*, 44–50.

Kader, A. A. (1992). *Post harvest technology of horticultural crops* (2nd ed.). Division of Agriculture and Natural Resources

Kader, A. A. (2002). Postharvest biology and technology: An overview. In: A. A. Kader (Ed.) *Postharvest technology of horticultural crops* (3rd ed.). University of California, Agriculture & Natural Resources, Publication #3311, pp. 39–47

Kader, A. A. (2005). Increasing food availability by reducing postharvest losses of fresh produce. *Proc. 5th Int. Postharvest Symp* (Eds. F. Mencarelli and P. Tonutti Acta Hort), *682*.

Kader, A. A. Barrett, D. M. (2003). Classification, composition of fruits, and postharvest maintenance of quality. In: L. P. Somogyi, H. S. Ramaswamy, Y. H. Hui (Eds.) *Processing fruits: science and technology, Vol. 1, biology, principles, and applications*. Technomic Publishing Co., p. 1.

Kader, A. A., Kitinoja, L., Hussein, A. M., Abdin, O., Jabarin, A., Sidahmed, A. E. (2012), Role of agroindustry in reducing food losses in the Middle East and North Africa Region. UNFAO Report NE2012234004. https://ucanr.edu/datastoreFiles/234-2297.pdf

Kader, A. A. Rolle, R. S. (2004). *The role of post-harvest management in assuring the quality and safety of horticultural produce* (FAO Agricultural Services Bulletin 152), FAO of the UN.

Kader, A. A., Saltveit, M. E. (2003). Atmosphere modification. In: J. A. Bartz, J. K. Brecht (Eds.) *Postharvest physiology and pathology of vegetables* (2nd ed.). Marcel Dekker, pp. 229–246.

Kahramanoğlu, İ. (2017). Introductory chapter: postharvest physiology and technology of horticultural crops. *Postharvest Handling*, 1–6. https://doi.org/10.5772/intechopen.69466

Kamrul Hassan, M. et al (2010). *Postharvest loss assessment: a study to formulate policy for loss reduction of fruits and vegetables and socioeconomic uplift of the stakeholders USAID Report*. USAID (Bangladesh).

Kamruzzaman, M., Barbin, D., Elmasry, G., Sun, D. W., Allen, P. (2012). Potential of hyperspectral imaging and pattern recognition for categorization and authentication of red meat. *Innovative Food Science and Emerging Technologies*, *16*, 316–325.

Kartal, S., Aday, M. S., Caner, C. (2012). Postharvest biology and technology use of microper-forated films and oxygen scavengers to maintain storage stability of fresh strawberries. *Postharvest Biology and Technology*, *71*, 32–40.

Khodabakhshian, R., Abbaspour-Fard, M. H. (2020). Pattern recognition-based Raman spectroscopy for non-destructive detection of pomegranates during maturity. Spectrochim. *Acta-Part A Mol. Biomol. Spectrosc*, *231*, 118127.

Khoshroo, A., Keyhani, A., Zoroofi, R. A., Rafiee, S., Zamani, Z., Alsharif, M. R. (2009) Classification of pomegranate fruit using texture analysis of MR images. *Agricultural Engineering International: The CIGR Ejournal*, *XI*.

Kim, B. F. et al (2020). Country-specific dietary shifts to mitigate climate and water crises Glob. *Environment Change*, *62*, 101926

Kitinoja, L, AlHassan, H. A., Saran, S., Roy, S. K. (2010). Identification of appropriate postharvest technologies for improving market access and incomes for small horticultural farmers in sub-Saharan Africa and South Asia. Invited paper in three parts for the IHC Postharvest Symposium Lisbon, August 23. Acta Hortic (in press), p. 597.

Kitinoja, L., AlHassan, H. Y. (2012), Identification of appropriate postharvest technologies for small scale horticultural farmers and marketers in Sub-Saharan Africa and South Asia—Part 1. In: M.I. Cantwell, D.P.F. Almeida (Eds.). *Postharvest losses and quality assessments* (Proc. XXVIIIth IHC—IS on Postharvest Technology in the Global Market Acta Hort. 934). ISHS, pp. 31–40

Kitinoja, L., Kader, A. A. (2015). *Measuring postharvest losses of fresh fruits and vegetables in developing countries.* The Postharvest Education Foundation.

Kumar, A., Rajpurohit, V. S., Bidari, K. Y. (2019). Multi class grading and quality assessment of pomegranate fruits based on physical and visual parameters. *International Journal of Fruit Science, 19*, 372–396.

Kummu, M., de Moel, H., Porkka, M., Siebert, S., Varis, O., Ward, P.J. (2012). Lost food, wasted resources: global food supply chain losses and their impacts on freshwater, cropland, and fertilizer use. *Science of the Total Environment, 438*, 477–489.

Kuswandi, B., Maryska, C., Abdullah, A., Heng, L.Y. (2013). Real time on-package freshness indicator for guavas packaging. *Journal of Food Measurement and Characterization, 7*(1), 29–39.

Lal Basediya, A., Samuel, D. V. K., Beera, V. (2013). Evaporative cooling system for storage of fruits and vegetables—A review. *Journal of Food Science and Technology, 50*(3), 429–442. https://doi.org/10.1007/s13197-011-0311-6

Lattimer, J. M., Haub, M. D. (2010). Effects of dietary fibre and its components on metabolic health. *Nutrients, 2*, 1266–1289.

Law, M. R., Morris, J. K. (1998). By how much does fruit and vegetable consumption reduce the risk of ischemic heart disease? *European Journal of Clinical Nutrition, 52*, 549–553.

Lee, J., Rudell, D. R., Watkins, C. B. (2014). Metabolic changes in 1-methylcyclopropene (1-MCP)-treated 'Empire' apple at different storage temperatures. *Acta Horticulturae, 1048*, 113–120.

Lee, S., Young, R. E., Schiffman, P. M., Coggins Jr., C. W. (1983). Maturity studies of avocado fruit based on picking dates and dry weight. *Journal of the American Society for Horticultural Science, 108*, 390–394.

Lee, S. K. (1981). Methods for percent oil analysis of avocado fruit. *California Avocado Society Yearbook, 65*, 133–141.

Lenz, C. P., Van den Berg, L., McCullough, R. S. (1971). Study of factors affecting temperature relative humidity and moisture loss in fresh fruit and vegetable storage. *Canadian Institute of Food Technology, 4*, 146.

Léonard, A., Blacher, S., Nimmol, C., Devahastin, S. (2008). Effect of far-infrared radiation assisted drying on microstructure of banana slices: an illustrative use of X-ray microtomography in microstructural evaluation of a food product. *Journal of Food Engineering, 85*, 154–162

Li, M., Landahl, S. East, A. R., Verboven, P., Terry, L. A. (2019). Optical coherence tomography—a review of the opportunities and challenges for postharvest quality evaluation. *Postharvest Biology and Technology, 150*, 9–18. https://doi.org/10.1016/j.postharvbio.2018.12.005.

Li, B., Zhang, F., Liu, Y., Yin, H., Zou, J., Ou-yang, A. (2022). Quantitative study on impact damage of yellow peach based on hyperspectral image information combined with spectral information. *Journal of Molecular Structure, 1272*, 134176. https://doi.org/10.1016/j.molstruc.2022.134176

Li, Z. (2013). The effect of compressibility, loading position and probe shape on the rupture probability of tomato fruits. *Journal of Food Engineering, 119(3)*, 471–476.

Li, Z. G., Li, P. P., Liu, J. Z. (2010). Effect of tomato internal structure on its mechanical properties and degree of mechanical damage. *African Journal of Biotechnology, 9*, 1816–1826.

Li, Z., Li, P., Yang, H. (2013a). Stability tests of grasping tomato with two fingers for harvesting robot. *Biosystems Engineering, 116*, 163–170.

Li, Z., Li, P., Yang, H., Liu, J. (2013b). Internal mechanical damage prediction in tomato compression using multiscale finite element models. *Journal of Food Engineering, 116*, 639–647.

Li, Z., Thomas, C. (2014). Quantitative evaluation of mechanical damage to fresh fruits. *Trends in Food Science and Technology, 35(2)*, 138–150. https://doi.org/10.1016/j.tifs.2013.12.001

Lin, M., Fawole, O. A., Saeys, W., Wu, D., Wang, J., Opara, U. L., Nicolai, B., Chen, K. (2022). Mechanical damages and packaging methods along the fresh fruit supply chain: A review. *Critical Reviews in Food Science and Nutrition.* https://doi.org/10.1080/10408398.2022.2078783

Linden, V. V., Scheerlinck, N., Desmet, M. (2006). Factors that affect tomato bruise development as a result of mechanical impact. *Postharvest Biology and Technology, 42*, 260–270.

Lipinski, B., Hanson, C., Lomax, J., Kitinoja, L., Waite, R., Searchinger, T. (2013). Reducing food loss and waste (Working Paper, Installment 2 of Creating a Sustainable Food Future). World Resources Institute, pp. 8–9.

Liu, S., Manson, J. E., Lee, I. M., Cole, S. R., Hennekens, C. H., Willett, W. C. (2000). Fruit and vegetable intake and risk of cardiovascular disease: The Women's health study. *American Journal of Clinical Nutrition, 72*(4), 922–928.

Liu, Q., Sun, K., Peng, J., Xing, M., Pan, L., Tu, K. (2018). Identification of bruise and fungi contamination in strawberries using hyperspectral imaging technology and multivariate analysis. *Food Analytical Methods, 11*(5), 1518–1527. http://dx.doi.org/10.1007/s12161-017-1136-3

Lu, F., Ishikawa, Y., Kitazawa, H., Satake, T. (2010). Measurement of impact pressure and bruising of apple fruit using pressure-sensitive film technique. *Journal of Food Engineering, 96*(4), 614–620. https://doi.org/10.1016/j.jfoodeng.2009.09.009

Lu, F., Xu, F., Li, Z., Liu, Y., Wang, J., Zhang, L. (2019). Effect of vibration on storage quality and ethylene biosynthesis-related enzyme genes expression in harvested apple fruit. *Scientia Horticulturae, 249*, 1–6. https://doi.org/10.1016/j.scienta.2019.01.031

Lu, R. (2003). Detection of bruises on apples using near infrared hyperspectral imaging. *Transactions of the ASAE, 46*, 1–8.

Lumpkin, T. A., Weinberger, K., Moore, S. (2005). Increasing income through fruit and vegetable production opportunities and challenges. *Annual General CGIAR Meeting*, 1–10.

MacLeod, R. F., Kader, A. A., Morris, L. L. (1976a). Stimulation of ethylene and CO_2 production of mature green tomatoes by impact bruising. *HortScience, 11*, 604.

Macleod, R. F., Kader, A. A., Morris, L. L. (1976b). Damage to fresh tomatoes can be reduced. *California Agriculture, 11*, 10–17.

Magwaza, L. S., Tesfay, S. Z. (2015). A review of destructive and non-destructive methods for determining avocado fruit maturity. *Food and Bioprocess Technology, 8(10)*, 1995–2011. https://doi.org/10.1007/s11947-015-1568-y

Mahajan, P. V., Oliveira, F. A. R., Macedo, I. (2008). Effect of temperature and humidity on the transpiration rate of the whole mushrooms. *Journal of Food Engineering, 84(2)*, 281–288.

Malik, A.U., Mazhar, M. S. (2008). *Evaluation of postharvest losses in mango.* Report to the Australian Center for International Agricultural Research, p. 8.

Maness, N. O., Brusewitz, G. H., McCullum, T. G. (1992). Impact bruise resistance comparison among peach cultivars. *HortScience*, *27*, 1008–1011.

Mangaraj, S., Goswami, T. K. (2009). Modified atmosphere packaging of fruits and vegetables for extending shell-life: Review. *Fresh produce*, *3(1)*, 1–31.

Mason-D'Croz, D., Bogard, J. R., Sulser, T. B., Cenacchi, N., Dunston, S., Herrero, M., Wiebe, K. (2019). Gaps between fruit and vegetable production, demand, and recommended consumption at global and national levels: an integrated modelling study. *The Lancet Planetary Health*, *3(7)*, 318–329. https://doi.org/10.1016/S2542-5196(19)30095-6

Mc Martin, S. E., Jacka, F. N., Colman, I. (2013). The association between fruit and vegetable consumption and mental health disorders: Evidence from five waves of a national survey of Canadians. *Preventive Medicine*, *56*, 225–230.

Meyer, M. D., Terry, L. A. (2008). Development of a rapid method for the sequential extraction and subsequent quantification of fatty acids and sugars from avocado mesocarp tissue. *Journal of Agricultural and Food Chemistry*, *56*(16), 7439–7445. https://doi.org/10.1021/jf8011322

Milczarek, R. R., Saltveit, M. E., Garvey, T. C., McCarthy, M. J. (2009). Assessment of tomato pericarp mechanical damage using multivariate analysis of magnetic resonance images. *Postharvest Biology and Technology*, *52*, 189–195.

Miller, V. et al (2016). Availability, affordability, and consumption of fruits and vegetables in 18 countries across income levels: Findings from the prospective urban rural epidemiology (PURE) study. *The Lancet Global Health*, *4*, 695–703.

Mintah, B. K., Eliason, A. E., Nsiah, M., Baah, E. M., Hagan, E., Ofosu, D. B. (2012). Consumption of fruits among students: A case of Public University in Ghana. *African Journal of Food, Agriculture, Nutrition and Development*, *12*(2), 5979–5993.

Mishra, V. K., Gamage, T. V. (2018). Postharvest physiology of fruit and vegetables. Handbook of food preservation [Rahman, M. S. Saltveit, M. E. (2018). Respiratory metabolism]. *Postharvest Physiology and Biochemistry of Fruits and Vegetables, 1*, 73–91. https://doi.org/10.1016/B978-0-12-813278-4.00004-X

Moleyar, V., Narasimham, P. (1994). Modified atmosphere packaging of vegetables: An appraisal. *Journal of Food Science*, *31*, 267–278.

Molla, M. M. et al. (2010). Survey on postharvest practices and losses of litchi in selected areas of Bangladesh. *Bangladesh Journal of Agricultural Research*, *35*(3), 439–451.

Montero, C. R. S., Schwarz, L. L., Santos, L. C. D., Andreazza, C. S., Kechinski, C. P., Bender, R. J. (2009). Postharvest mechanical damage affects fruit quality of Montenegrina and Rainha tangerines. *Pesquisa Agropecuaria Brasileira*, *44*, 1630–1640.

More, P., Behere, D., Housalmal, S., Khodke, S., Jadhav, S., Kakade, A. (2015). Assessment of vibration damage to banana bunches simulated by vibration tester. *International Journal of Tropical Agriculture*, *33*, 1419–1422.

Morya, S., Sharma, A. (2020). *Advances in horticultural crop management and value addition editors*. Professional University of Punjab, p. 10.

Moscetti, R., Haff, R. P., Monarca, D., Cecchini, M., Massantini, R. (2016). Near-infrared spectroscopy for detection of hailstorm damage on olive fruit. *Postharvest Biology and Technology*, *120*, 204–212. http://dx.doi.org/10.1016/j.postharvbio.2016.06.011

Moss, M. O. (2002), Mycotoxin review. 1. Aspergillus and Penicillium. *Mycologist*, *16*, 116–119.

Mouazen, A. M., Nicolaï, B., Terry, L. A. (2014). The use of Vis/NIRS and chemometric analysis to predict fruit defects and postharvest behaviour of 'Nules Clementine' mandarin fruit. *Food Chemistry*, *163*, 267–274.

Mujib U. R., Naushad, K., Inayatullah, J. (2007). Postharvest losses in tomato crop (a case of Peshawar valley). *Sarhad Journal Agriculture*, *23*(4), 1279–1284.

Munhuweyi, K., Sigge, G.O., Opara, U.L. (2016). Postharvest losses of cabbages from retail to consumer and the socio-economic and environmental impacts. *British Food Journal*, *118(2)*, 286–300. http://dx.doi.org/10.1108/BFJ-08-2014-0280

Mushtaq, K., Javed, M. S., Bari, A. (2005). Postharvest losses in mango: the case of Pakistani Punjab. Proceedings-of-the-International-Conference-on-Postharvest-Technology-and-Quality-Management-in-Arid-Tropics,-Sultanate-of-Oman,-31-January–2-February, pp. 89–94.

Myers, S. S., Smith, M. R., Guth, S., Golden, C. D., Vaitla, B., Mueller, N. D., Dangour, A. D., Huybers, P. (2017). Climate change and global food systems: potential impacts on food security and undernutrition. *The Annual Review* of *Public Health*, *38*, 259–77.

Nago, E. S., Verstraeten, R., Lachat, C. K., Dossa, R. A., Kolsteren, P. W. (2012). Food safety is a key determinant of fruit and vegetable consumption in urban Beninese adolescents. *Journal of Nutrition Education and Behaviour*, *44*(6), 548–555. doi: 10.1016.j.jneb.2011.06.006

Nagy, S., Shaw, P. E. (1980). *Tropical and subtropical fruits—composition, properties and uses*. AVI Publishing, Inc., pp. 15–22. https://books.google.com.ng/books?id=ue4fAQAAIAAJ

Ness, A. R., Powles, J. W. (1997). Fruit and vegetable and cardiovascular disease: a review. *International Journal of Epidemiology*, *26*, 1–12.

New, S. (2001). Fruit and vegetable consumption and skeletal health: is there a positive link? Nutrition Foundation. *Nutrition Bulletin*, *26*, 121–125.

Nicolaï, B. M., Bulens, I., De Baerdemaker, J., De Ketelaere, B., Hertog, M.L.A.T.M., Verboven, P., Lammertyn, J. (2014). Non-destructive evaluation: detection of external and internal attributes frequently associated with quality and damage. In: *Postharvest handling* (3rd ed.) *A systems approach*. Academic Press, pp. 363–385.

Nicolaï, B. M., Beullens, K., Bobelyn, E., Peirs, A., Saeys, W., Theron, K. I., Lammertyn, J. (2007). Nondestructive measurement of fruit and vegetable quality by means of NIR spectroscopy: a review. *Postharvest Biology Technology*, *46*, 99–118.

Nikbakht, A. M., Hashjin, T. T., Malekfar, R., Gobadian, B. (2011). Nondestructive determination of tomato fruit quality parameters using raman spectroscopy. *Journal of Agricultural Science and Technology*, *13*, 517–526.

Odesola, I. F., Onyebuchi, O. (2009). A review of porous evaporative cooling for the preservation of fruits and vegetables. *The Pacific Journal of Science and Technology, 10*(2), *935–941.*

Okah, R. N. (1997). Economic evaluation of losses in yams stored in traditional barns. *Tropical Science*, *38*, 125–127.

Okere, E. E., Arendse, E., Ambaw Tsige, A., Perold, W. J., Opara, U. L. (2022). Pomegranate quality evaluation using non-destructive approaches: a review. *Agriculture (Switzerland)*, *12*(12), 1–25. https://doi.org/10.3390/agriculture12122034

Olaeta-Coscorroza, J. A., Undurraga-Martinez, P. (1995). Estimacion del Indice de Madurez en Paltas. In: *Harvest and post-harvest technologies for fresh fruits and vegetables. proceedings of the international conference*. Guanajuato, pp. 521–525.

Olayemi, F.F. et al. (2010). Assessment of post-harvest challenges of small-scale farm holders of tomatoes, bell pepper and hot pepper in some local government areas of Kano State, Nigeria. *Bayero Journal of Pure and Applied Sciences*, *3*(2), 39–42.

Opara, I.K., Fawole, O.A., Kelly, C., Opara, U.L. (2021a). Quantification of on-farm pomegranate fruit postharvest losses and waste, and implications on sustainability indicators:

South African case study. *Sustainability, 13*, 5168. https://doi.org/10.3390/su13095168

Opara, I. K., Fawole, O. A., Opara, U. L. (2021b). Postharvest losses of pomegranate fruit at the packhouse and implications for sustainability indicators. *Sustainability, 13*, 5187. https://doi.org/10.3390/su13095187

Opara, I. K., Fawole, O. A., Opara, U. L. (2022). Pomegranate production and export during the past decade in South Africa and incidence of postharvest losses—a review. *Acta Horticulturae, 1349*, 325–332. https://doi.org/10.17660/ActaHortic.2022.1349.45

Opara, L. U. (2003). Postharvest losses at the fresh produce retail chain in the Sultanate of Oman. Australian-postharvest-horticulture-conference, -Brisbane,-Australia,-1–3-October,-2003, pp 248–249 (abstract).

Opara, L. U. (2006a). Editorial: a new era in postharvest technology. *International Journal of Postharvest Technology & Innovation, 1*(1), 1–3.

Opara, L. U. (2006b). Editorial: Postharvest technology for linking production to markets. *International Journal of Postharvest Technology & Innovation, 1*(2), 139–141.

Opara, L. U. (2009a). Editorial: Harnessing postharvest technology for development. *International Journal of Postharvest Technology & Innovation, 1*(4), 357–359.

Opara, L. U., Al-Jufaili, S. M. (2006a). Status of fisheries postharvest industry in the Sultanate of Oman: Part 2 quantification of fresh fish losses. *Journal of Fisheries International, 1*, 150–156.

Opara, L. U., Al-Jufaili, S. M. (2006b). Status of fisheries postharvest industry in the Sultanate of Oman: Part 3. Regression models of quality loss in fresh tuna fish. *Journal of Fisheries International, 1*, 141–143.

Opara, L.U., Al-Jufaili, S.M., Rahman, M.S. (2007). Postharvest handling and preservation of fresh fish and seafoods. Chapter 6 In: M. S. Rahman (Ed.), *Handbook of food preservation* (2nd ed.). CRC Press, pp. 151–172.

Opara, U. L. (2009b). Challenging our postharvest technology research and innovation system. *South African Fruit Journal, 8*(5), 9.

Opara, U. L. (2009c). Building human capacity for our future. *South African Fruit Journal, 8*(4), 25–26.

Opara, U.L., (2009d). Quality management: An industrial approach to produce handling. Chapter 8 In: W. J. Florkowski, R. L. Shewfelt, B. Brueckner, S. E. Prussia (Eds.), *Postharvest handling: a systems approach* (2nd ed.). Elsevier, pp. 154–204.

Opara, U. L. (2010a). Editorial: High incidence of postharvest food losses is worsening global food and nutrition security. *International Journal of Postharvest Technology & Innovation, 2*(1), 1–3.

Opara, U.L. (2010b). Making every harvest count! The role of postharvest research in development. *South African Fruit Journal, 8*(6), 33–35.

Opara, U.L. (2012). More investment and participation needed in postharvest research and innovation. *South African Fruit Journal, 10*(1), 25–26.

Opara, U.L. (2015). *To feed the world, we must save the harvest.* Resource, p. 7.

Opara, U. L., Caleb, O. J., Belay, Z. A. (2019). Modified atmosphere packaging for food preservation. In: *Food quality and shelf life*. Elsevier Inc. https://doi.org/10.1016/B978-0-12-817190-5.00007-0

Opara, U. L., Mditshwa, A. (2013). The role of packaging in securing the food system: adding value to food products and reducing losses and waste: a review. *African Journal of Agricultural Research, 8*(22), 2621–2630. doi:10.5897/AJAR2013.6931.

Opara U. L., Pathare, P. B. (2014). Bruise damage measurement and analysis of fresh horticultural produce—A review. *Postharvest Biology and Technology, 91*, 9–24.

Oyebode, O., Gordon-Dseagu, V., Walker, A., Mindell, J. S. (2013). Fruit and vegetable consumption and all-cause, cancer and CVD mortality: Analysis of Health Survey for England data. *Journal of Epidemiology and Community Health, 68*(9), 856–62.

Ozdemir, M., Floros, J. D. (2004). Active food packaging technologies. *Critical Reviews in Food Science and Nutrition, 44*(3), 185–193.

Pantastico, E.B. (1975). General introduction: structure of fruits and vegetables. In Postharvest Physiology, Handling, and Utilization of Tropical and Subtropical Fruits and Vegetables (E. B. Pantastico, Ed.), AVI Publishing, Westport, CT, p. 1.

Pang, D. W., Studman, C. J., Banks, N. H., Baas, P. H. (1996). Rapid assessment of the susceptibility of apples to bruising. *Journal of Agricultural Engineering Research, 64*, 37–48.

Park, H. J. (1999) Development of advanced edible coatings for fruits. *Trends in Food Science & Technology, 10*, 254–260.

Park, H. M., Heo, J., Park, Y. (2011). Calcium from plant sources is beneficial to lowering the risk of osteoporosis in postmenopausal Korean women. *Nutrition Research, 31*, 27–32.

Payne, M. E., Steck, S. E., George, R. R., Steffens, D. C. (2012). Fruit, Vegetable, and Antioxidant Intakes Are Lower in Older Adults with Depression. *The Journal of the* Academy *of* Nutrition *and Dietetics, 112*, 2022–2027.

Peltzer, K., Phaswana-Mafuya, N. (2012). Fruit and vegetable intake and associated factors in older adults in South Africa. *Global Health Action, 5*, 18668. doi: 10.3402/gha.v5i0.18668

Pem, D., Jeewon, R. (2015). Fruit and Vegetable Intake: Benefits and Progress of Nutrition Education Interventions- Narrative Review Article. *Iran Journal of Public Health, 44*(10), 1309–1321.

Peterson, D. L., Bennedsen, B. S. (2005). Isolating damage from mechanical harvesting of apple. *Transactions of the ASAE, 21*, 31–34.

Phan, C. T., Pantastico, E. B., Ogata, K., Chachin, K. (1975). Respiration and respiratory climacteric. In Postharvest Physiology, Handling, and Utilization of Tropical and Subtropical Fruits and Vegetables (E. B. Pantastico, Ed.), AVI Publishing, Westport, CT, 1975, p. 86.

Poiroux-Gonord, F., Bidel, L. P. R., Fanciullino, A. L., Gautier, H., Lauri-Lopez, F., Urban, L. (2010). Health benefits of vitamins and secondary metabolites of fruits and vegetables and prospects to increase their concentrations by agronomic approaches. *Journal of Agricultural and Food Chemistry, 58*(23), 12065–12082. https://doi.org/10.1021/jf1037745

Powrie, W. D., Skura, B. J. (1991). Modified atmosphere packaging of fruits and vegetables. In: Ooraikul B, Stiles ME, Horwood E, editors. Modified Atmosphere Packaging of Food. New York, USA; pp. 169–245

Prasad, K., Jacob, S., Siddiqui, M. W. (2018). Fruit Maturity, Harvesting, and Quality Standards. Preharvest Modulation of Postharvest Fruit and Vegetable Quality, February 2020, 41–69. https://doi.org/10.1016/B978-0-12-809807-3.00002-0

Puchalski, C., Brusewitz, G. (1996). Watermelon surface abrasion e a sensory method. *International Agrophysics, 10*, 117–122.

Qiao, S., Tian, Y., Song, P., He, K., Song, S. (2019). Analysis and detection of decayed blueberry by low field nuclear magnetic resonance and imaging. *Postharvest Biology and Technology, 156*, 110951. http:// dx.doi.org/10.1016/j.postharvbio.2019.110951.

Quamruzzaman, A. K. M., Islam, F., Akter, L., Mallick, S. R. (2022). Effect of Maturity Indices on Growth and Quality of High Value Vegetables. *American Journal of Plant Sciences, 13*, 1042–1062. https://doi.org/10.4236/ajps.2022.137069 R

Quintana, M. E. G., Paull, R. E. (1993). Mechanical injury during postharvest handling of Solo papaya fruit. *Journal of the American Society for Horticultural Science, 118*, 618–622.

Rama, M. V., Narasimham, P. (2003). Controlled-atmosphere storage. Effects on fruit and vegetables. *Encyclopedia of Food Sciences and Nutrition, 1978*, 1607–1615. https://doi.org/10.1016/b0-12-227055-x/00292-3

Raman, C. V., Krishnan, K.S. (1928). A new type of secondary radiation. *Nature, 121*, 501–502.

Ramjan, M., Pandey, A. K., Angami, T. (2017). Assessment of maturity indices in vegetables. Biomolecules Reports, December, 1–5.

Ranathunga, C., et al. (2010). Vibration effects in vehicular road transportation. In IPSL Annual Technical Sessions, I.o. Physics. Sri Lanka: Colombo.

Rees, D. et al. (2001). Effect of damage on market value and shelf life of sweetpotato in urban markets of Tanzania. *Tropical Science, 41*(3), 1–9.

Reid, M.S. (1992). Ethylene in postharvest technology. In A.A. Kader, ed. Postharvest technology of horticultural crops, pp. 97–98. Davis, California, Univ. of California.

Ridgewell, J. (1998). Examining Food and Nutrition. London: Oxford University Press; p. 58

Ríos-Mesa, A. F., et al. (2020). Effect of vehicle vibration on the mechanical and sensory properties of avocado (Persea Americana Mill. Cv. Hass) during road transportation. *International Journal of Fruit Science, 20*, S1904–S1919.

Risse, L. A., Miller, W. R. (1983). Film wrapping and decay of eggplant. *Proceedings of the Florida State Horticultural Society, 96*, 350–352.

Rolls, B. J., Ello-Martin, J. A., Tohill, B. C. (2004). What can intervention studies tell us about the relationship between fruit and vegetable consumption and weight management? *Nutrition Review, 62*(1),1–17.

Ruiz-Altisent, M. (1996). Engineering research to improve fruit quality. *Land Technology*, 8–9.

Rupasinghe, H.P.V. et al (1991). A case study on identification and assessment of postharvest losses of tomato. University College, NW Province (Sri Lanka).

Salami, P., Ahmadi, H., Keyhani, A., Sarsaifee, M. (2010). Strawberry post-harvest energy losses in Iran. *Researcher, 2*(4), 67–73.

Saltveit, M. E. (2002). Ethylene effects. In The Commercial Storage of Fruits, Vegetables, and Florist and Nursery Stocks—A Draft Version of the Revision to USDA Agricultural Handbook Number 66 (K. C. Gross, C. Y. Wang, and M. Saltveit, Eds), (revised in 2004) (*www.ba.ars.usda.gov/hb66/index.html*)

Saltveit, M. E. (2018). Respiratory metabolism. Postharvest Physiology and Biochemistry of Fruits and Vegetables, 1, 73–91. https://doi.org/10.1016/B978-0-12-813278-4.00004-X

Salunkhe, D. K., Desai, B. B. (1984). Postharvest Biotechnology of Vegetables, CRC Press, Boca Raton, FL.

Sanaeifar, A., Mohtasebi, S. S., Ghasemi-Varnamkhasti, M., Shafie, M. M. (2016). Evaluation of an electronic nose system for characterization of pomegranate varieties. *Agricultural Engineering International, 18*, 317–323.

Sandhya, (2010). Modified atmosphere packaging of fresh produce: current status and future needs. *Food Science Technology, 43*, 381–392.

Santana Liado, J., Marrero Dominguez, A. (1998). The effects of peel abrasion on the postharvest physiology and commercial life of banana fruits. *Acta Horticulturae, 490*, 547e554.

Santos, S. F. dos, Cardoso, R. de C. V., Borges, Í. M. P., Almeida, A. C. e., Andrade, E. S., Ferreira, I. O., & Ramos, L. C. (2020). Post-harvest losses of fruits and vegetables in supply centers in Salvador, Brazil: Analysis of determinants, volumes and reduction strategies. *Waste Management, 101*, 161–170. https://doi.org/10.1016/j.wasman.2019.10.007

Sastry, S. K., Baird, C. & Buffington, D. E. (1978). Transpiration rates of certain fruits and vegetables. ASHRAE Trans., 84(1):237.

Sastry, S.K., Buffington, D. K. (1982). Transpiration rates of stored perishable commodities: mathematical model and experiments on tomatoes. *ASHRAE Transactions, 88*(1), 159.

Save Food, (2014). Food Loss Assessments: causes and solutions. Kenya case studies (Bananas).

Scalia, G. L., Aiello, G., Miceli, A., Nasca, A., Alfonzo, A., Settanni, L. (2016). Effect of vibration on the quality of strawberry fruits caused by simulated transport. *Journal of Food Process Engineering, 39*(2), 140–56. doi: 10.1111/jfpe.12207.

Ščetar, M., Kurek, M., Galić, K. (2010). Trends in fruit and vegetable packaging – a review. *Croatian Journal of Food Technology, Biotechnology and Nutrition, 5*, 69–86.

Schoeman, L., Williams, P., du Plessis, A., Manley, M. (2016). X-ray micro-computed tomography (μCT) for non-destructive characterisation of food microstructure. *Trends Food Science and Technology, 47*, 10–24.

Schreinemachers, P., Simmons, E. B., Wopereis, M. C. S. (2018). Tapping the economic and nutritional power of vegetables. *Global Food Security, 16*, 36–45.

Selvakumar, R., (2014). A text book of Glaustus Olericulture. New Vishal Publications. New Delhi.

Shakeel, Q., Shaheen, M. R., Ali, S., Ahmad, A., Raheel, M., Bajwa, R. T. (2022). Postharvest management of fruits and vegetables. Applications of Biosurfactant in Agriculture, September, 1–16. https://doi.org/10.1016/B978-0-12-822921-7.00001-5

Shen, C. L., Bergen, V. V., Chyu, M. C. et al. (2012). Fruits and dietary phytochemicals in bone protection. Nutr Res, 32: 897–910.

Shewfelt, R. L., (1999). 'What is quality?' Postharvest Biol. Technol. (in press).

Siegel, K. R., Ali, M. K., Srinivasiah, A., Nugent, R. A., Narayan, K. M. V. (2014). Do we produce enough fruits and vegetables to meet global health need? *PLoS ONE, 9*(8), https://doi.org/10.1371/journal.pone.0104059

Silva, O. O., Ayankogbe, O. O., Odugbemi, T. O. (2017). Knowledge and consumption of fruits and vegetables among secondary school students of Obele Community Junior High School, Surulere, Lagos State, Nigeria. *Journal of Clinical Sciences, 14*, 68–73.

Singh, J., Singh, S. P., Joneson, E. (2006). Measurement and analysis of US truck vibration for leaf spring and air ride suspensions, and development of tests to simulate these conditions. *Packaging Technology and Science: International Journal, 19*, 309–323.

Sitanggang, F. A., Machfoedz, M. M. (2023). Proceedings of the 2nd International Conference for Smart Agriculture, Food, and Environment (ICSAFE 2021). In N. Huda, I. Jaswir, Y. Romdhonah, A. Alimuddin, T. Ahamed, & N. Nasser (Eds.), Proceedings of the 2nd International Conference for Smart Agriculture, Food, and Environment (ICSAFE 2021). Atlantis Press International BV. https://doi.org/10.2991/978-94-6463-090-9

Smith J. P., Ramaswamy H. S. (1996) Packaging of fruits and vegetables. In L. Somogyi, H. S. Ramaswamy, & Y. H. Hui (Eds.), Processing fruits: Science and Technology. Lancaster, PA: Technomic Publishing Co., pp. 379–427

Soleimani, B., Ahmadi, E. (2014). Measurement and analysis of truck vibration levels as a function of packages locations in truck bed and suspension. *Computers and Electronics in Agriculture, 109*, 141–147.

Soleimani, B., Ahmadi, E. (2015). Evaluation and analysis of vibration during fruit transport as a function of road conditions, suspension system and travel speeds. *Engineering in Agriculture, Environment and Food, 8*(1), 26–32.

Sommer, N. F. (1985), Strategies for control of post-harvest disease of selected commodities. In: Post-harvest Technology of Horticultural Crops. University of California Press. 83–98p.

Stratton, A. E., Finley, J. W., Gustafson, D. I., Mitcham, E. J., Myers, S. S., Naylor, R. L., Otten, J. J., Palm, C. A. (2021). Mitigating sustainability tradeoffs as global fruit and vegetable systems expand to meet dietary recommendations. *Environmental Research Letters*, *16*(5), https://doi.org/10.1088/1748-9326/abe25a

Studer, H., Chen, P., Kader, A. (1981). Damage evaluation of machine harvested fresh market tomatoes. *Transactions of the ASAE*, *24*, 284e287.

Su, W. H., Sun, D. W. (2018). Fourier transform infrared and Raman and hyperspectral imaging techniques for quality determinations of powdery foods: A review. *Comprehensive Reviews in Food Science and Food Safety*, *17*, 104–122

Suraj, M. S. (2018). Design of Cold Storage. *International Research Journal of Engineering and Technology*, *5*, 2395–0056.

Thompson, A. K. (1996). Postharvest Technology of Fruit and Vegetables. Oxford: Blackwell.

Thompson, A. K. (2010). "Controlled Atmosphere Storage of Fruits and Vegetables", CABI International, London, UK, 288 pp

Thompson, J. F., Slaughter, D. C., Arpaia, M. L. (2008). Suspended tray package for protecting soft fruit from mechanical damage. *Applied Engineering in Agriculture 24*, 71–5. doi: 10.13031/2013. 24149.

Timm, E. J., Brown, G. K., Armstrong, P. R. (1996). Apple damage in bulk bins during semi-trailer transport. *Transactions of the ASABE*, *12*, 369–377.

Tribble, D. L. (1999). AHA Science Advisory. Antioxidant consumption and risk of coronary heart disease, emphasis on vitamin C, vitamin E and beta-carotene; a statement for health care professionals from the American Heart Association. *Circulation*, *99*(4), 591–595. https://doi.org/10.1161/01.cir.99.4.591

Troger, K., Henselb, O., Bürkert, A. (2007). Conservation of Onion and Tomato in Niger—Assessment of Post-Harvest Losses and Drying Methods. Tropentag 2007 University of KasselWitzenhausen and University of Göttingen, Conference on International Agricultural Research for Development

Underhill, S. J. R., Kumar, S. (2014). Quantifying horticulture postharvest wastage in three municipal fruit and vegetable markets in Fiji. Int'l J. Postharvest Technology and Innovation, Vol. 4, Nos. 2/3/4.

Urmila, K., Li, H., Chen, Q., Hui, Z., Zhao, J. (2015). Quantifying of total volatile basic nitrogen (TVB-N) content in chicken using a colorimetric sensor array and nonlinear regression tool. *Analytical Methods, 7*(13), 5682–5688.

Valero, C., Ruiz-Altisent, M. (2000). Design guidelines for a quality assessment system of fresh fruits in fruit centers and hypermarkets. *Agricultural Engineering International: The CIGR Journal of Scientific Research and Development, 2*, 1–20.

Van den Berg, L., Lenz, C. P. (1971). Moisture loss of vegetables under refrigerated storage conditions. *Canadian Institute of Food Science and Technology*, *4*, 143.

Van Duyn, M. A., Pivonka, E. (2000). Overview of the health benefits of fruit and vegetable consumption for the dietetics professional: selected literature. *Journal* of the *American Dietetic Association*, 100(12), 1511–1521.

Van Linden, V., Scheerlinck, N., Desmet, M., De Baerdemaeker, J. (2006b). Factors that affect tomato bruise development as a result of mechanical impact. *Postharvest Biology and Technology*, *42*, 260–270.

Van Linden, V., De Ketelaere, B., Desmet, M., Baerdemaeker, D. J. (2006a). Determination of bruise susceptibility of tomato fruit by means of an instrumented pendulum. *Postharvest Biology and Technology*, *40*, 7–14.

Van Zeebroeck, M. Van Iinden, V. van, Darius, P., Ketelaere, B. de, Ramon, H., Tijskens, E. (2007a). The effect of fruit properties on the bruise susceptibility of tomatoes. *Postharvest Biology and Technology*, *45*, 168–175. doi: 10.1016/j.postharvbio.2006.12.022

Van Zeebroeck, M., Van Linden, V., Ramon, H., Baerdemaeker, J. D., Nicolai, B.M., Tijskens, E. (2007b). Impact damage to apples during transport and handling. *Postharvest Biology and Technology*, *45*, 157–167.

Vayssieres, J., Korie, S., Coulibaly, O., Temple, L., Boueyi, S. P. (2008). The mango tree in central and northern Benin: Cultivar inventory, yield assessment, infested stages and loss due to fruit flies (diptera tephritidae). *Fruits*, *63*, 335–348.

Vursavus, K., Ozguven, F. (2004). Determining the effects of vibration and packaging method on mechanical damage in golden delicious apples. *Turkish Journal of Agriculture and Forestry*, *28*, 311–320.

Walkowiak-Tomczak, D., et al. (2021). The effect of mechanical vibration during transport under model conditions on the shelf-life, quality and physico-chemical parameters of four apple cultivars. *Agronomy*, *11*(1), 81.

Wang, Z., Künnemeyer, R., McGlone, A., Burdon, J. (2020). Potential of Vis-NIR spectroscopy for detection of chilling injury in kiwifruit. *Postharvest Biology and Technology*, *164*, 111160. http://dx.doi.org/10.1016/j.postharvbio.2020.111160

Wankhede, S. A., Kale, V. S., Shaligram, A. D. (2020). Review on cold storage system for vegetables. *International Multidisciplinary E-Research Journal*, *230*(special issue). ISSN-2348-7143.

Wasala, W., Dharmasena, D., Dissanayake, T., Thilakarathne, B. (2015). Vibration simulation testing of banana bulk transport packaging systems. Tropical Agricultural Research, vol. 26.

Wasala, W.M.C.B. et al. (2014). Postharvest Losses, Current Issues and Demand for Postharvest Technologies for Loss Management in the Main Banana Supply Chains in Sri Lanka. *Journal of Postharvest Technology*, *2*(1), 080–087, January' 2014.

Waste and Resource Action Programme (WRAP), (2011), New estimates for household food and drink waste in the UK. [Online] Accessible: www.wrap.org.uk/content/new-estimateshousehold-food-and-drink-waste-uk.

Weinberger, K., Genova II, C., Acedo, A. (2008). Quantifying postharvest loss in vegetables along the supply chain in Vietnam, Cambodia and Laos. *International Journal of Postharvest Technology and Innovation*, *1*, 288–297.

Weinberger, K., Lumpkin, T. A. (2005). Horticulture for Poverty Alleviation: The Unfunded Revolution. Working Paper No. 15. Shanhua: AVRDC—The World Vegetable Center.

Wflo. (2010). Identification of Appropriate Postharvest Technologies for Improving Market Access and Incomes for Small Horticultural Farmers in Sub-Saharan Africa and South Asia. WFLO Grant Final Report to the Bill & Melinda Gates Foundation, March 2010. 318 pp.

Willett, W. et al (2019). Food in the Anthropocene: the EAT–lancet commission on healthy diets from sustainable food systems. *Lancet*, *393*, 1–49.

Wills, R. B. H., McGlasson, W. B., Graham, D., Joyce, D. C. (2007). Postharvest—An Introduction to the Physiology and Handling of Fruits, Vegetables and Ornamentals (5th ed.). CAB International, Oxfordshire, UK. 227 pp.

Wills, R. H. H., McGlasson, W. B., Graham, D., Joyce, D. (1998). Postharvest: An Introduction to the Physiology and Handling of Fruit Vegetables and Ornamentals, New South Wales University Press, Sydney

Wilson, L. G., Boyette, M. D., Estes, G. A. (1995). Postharvest handling and cooling of fresh fruits, vegetables and flowers for small farms leaflets 800–804. North Carolina cooperative extension service. 17p.

World Bank, Natural Resources Institute, and FAO (2011), Missing Food: The Case of Postharvest Grain Losses in Sub-Saharan Africa. Washington, DC: The World Bank.

Wu, C. (2010). An Overview of Postharvest Biology and Technology of Fruits and Vegetables. Technology, 2–11.

Xu, Y., Zhong, P., Jiang, A., Shen, X., Li, X., Xu, Z., Shen, Y., Sun, Y., Lei, H. (2020). Raman spectroscopy coupled with chemometrics for food authentication: A review. *Trends in Analytical Chemistry*, *131*, 116017.

Yahia, E. M., Fonseca, J. M., Kitinoja, L. (2019). Postharvest losses and waste. In Postharvest Technology of Perishable Horticultural Commodities. Elsevier Inc. https://doi.org/10.1016/B978-0-12-813276-0.00002-X

Yam, K. L., Takhistov, P. T., Miltz, J., (2005). Intelligent packaging: concepts and applications. *Journal of Food Science, 70*(1), R1–R10.

Yang, D. & Ying, Y. (2011). Applications of Raman spectroscopy in agricultural products and food analysis: A review. *Applied Spectroscopy Reviews*, *46*, 539–560.

Yaptenco, K., Esguerra, E. (2012). Good practice in the design, management and operation of a fresh produce packing-house (Issue January).

Zarifneshat, S., Ghassemzadeh, H. R., Sadeghi, M., Abbaspour-Fard, M. H., Ahmadi, E., Javadi, A., Shervani-Tabar, M. T. (2010). Effect of impact level and fruit properties on golden delicious apple bruising. *American Journal of Agricultural and Biological Sciences*, *5*, 114–121.

Zerbini, P. E. (2006). Emerging Technologies for Non- Destructive Quality Evaluation of Fruit. *Journal of Fruit and Ornamental Plant Research*, *14*, 13–22.

Zhang, L., McCarthy, M. J. (2012). Black heart characterization and detection in pomegranate using NMR relaxometry and MR imaging. *Postharvest Biology and Technology*, *67*, 96–101.

Zhang, H., Wu, J. (2020). Detection of early browning in pears using vibro-acoustic signals. *Transactions of the Chinese Society of Agricultural Engineering*, *36*(17), 264–271. http://dx.doi.org/10.11975/j. issn.1002-6819.2020.17.031

Zhang, S., Deng, X. (2012), Fresh fruits and vegetables distribution system in China—Analysis on the feasibility of Agricultural super-docking. 39–41p.

Zhao, J., Liu, J., Chen, Q., Saritporn, V. (2008). Detecting subtle bruises on fruits with hyperspectral imaging. *Transactions of the CSAM*, *39*, 106–109.

Zhao, Z., Wang, Y., Guo, D., Niu, X., Cheng, W., Gu, Y. (2016). Identification of plum fruit Browning by near-infrared spectroscopy. *Spectroscopy and Spectral Analysis*, *36*(7), 2089–2093. http://dx.doi. org/10.3964/j.issn.1000-0593(2016)07-2089-05.

Zhen, O. P., Hashim, N., Maringgal, B. (2020). Quality evaluation of mango using non-destructive approaches: A review. *Journal of Agricultural and Food Engineering*, *1*, 0003

Zhou, R., et al. (2007). Effect of transport vibration levels on mechanical damage and physiological responses of Huanghua pears (Pyrus pyrifolia Nakai, cv. Huanghua). *Postharvest Biology and Technology*, *46*(1), 20–28.

Zhou, R., Wang, X. C., Hu, Y. S., Zhang, G. X., Yang, P. Q., Huang, B. L. (2015). Reduction in Hami melon (Cucumis melo var. saccharinus) softening caused by transport vibration by using hot water and shellac coating. *Postharvest Biology and Technology*, *110*, 214–23. doi: 10.1016/j.postharvbio.2015.08.022

Zhou, S., Ying, Y., Shang, D. (2012). Morphology based noninvasive detection for fragrant pears browning with magnetic resonance imaging. *Journal of Zhejiang University (Engineering Science)*, *46*(12), 2141–2145. http://dx.doi.org/10.3785/j. issn.1008-973X.2012.12.002

Zhou, W. (2013), Food Waste and Recycling in China: A Growing Trend? [Online] Accessible: www.worldwatch.org/food-waste-and-recyclingchina-growing-trend-1

Zhou, Y., Wu, D., Hui, G., Mao, J., Liu, T., Zhou, W., Zhao, Y., Chen, Z., Chen, F. (2018). Loquat bruise detection using optical coherence tomography based on microstructural parameters. *Food Analytical Methods, 11*(10), 2692–2698. doi: 10.1007/s12161-018-1246-6

Zulfiqar, M., Khan, D., Bashir, M. (2005). An assessment of marketing margins and physical losses at different stages of marketing channels for selected vegetable crops of Peshawar Valley. *Journal of Applied Sciences*, *5*(9), 1528–1532.

Sustainable Technologies Prospects and Potential Applications in Postharvest Storage of Fruits and Vegetables

2

Nirmal Kumar Meena*, Bandemuth Renukaradhya Vinod*, Munishami Menaka, Prasoon Gunjan, Brijesh Kumar, and Meerasabihalli Rangaswami Chandana

**Corresponding Authors:* nirmalchf@gmail.com; vinodbr0026@gmail.com

2.1 INTRODUCTION

Fruits and vegetables (FVs) are widely recognized as crucial components of a healthy diet. They serve as vital sources of essential minerals, vitamins, antioxidants, phenols, and dietary fibers that contribute to overall human well-being. The consumption of FVs has been consistently linked to a multitude of positive effects on human health, as they can cure several cardiovascular diseases. The field of FV production is gaining momentum within the horticulture sector. Despite huge production, a significant portion, about one-third, of the produced FVs is lost prior to reaching market. This represents a substantial challenge for the horticulture sector, as the aim is to provide high-quality fresh produce to increasingly competitive markets.

Efforts are required to address this issue and improve the post-harvest handling, storage, and distribution processes to minimize losses and ensure the delivery of superior FVs to consumers. Such advancements are crucial to meet the rising demand for nutritious and high-quality produce in a sustainable manner. The FVs are prone to higher perishability owing to their higher moisture contents and accelerated metabolic activities, such as ethylene production and respiration rates. Various factors, for instance improper handling practices, certain physiological disorders, microbial spoilage, and selected storage issues, significantly contribute to losses throughout the supply chain (Meena & Choudhary, 2020; Asrey et al., 2023; Meena & Asrey, 2018). Following harvest, the FVs continue to undergo respiration, leading to the consumption of O_2 and the release of CO_2 and water. As a consequence, lipids, organic acids, proteins, and carbohydrates undergo stimulated metabolism, resulting in compromised energy replacement as the FVs are detached from their parent plants. Chilling-related stresses further exacerbate the problem by disrupting metabolic processes and inducing detrimental alterations in membrane fluidity. The causes of substantially higher losses in FVs can be ascribed into various pathological, physiological, and physical factors, including combinations of these (Holt et al., 1983). In addition, selected physiological factors encompassing the inherent characteristics and responses of the produce also contribute to the higher losses. The pathological factors involve the detrimental impact of microorganisms and pathogens, thus leading to the deterioration of FVs in general. The physical factors include mechanical damages, temperature fluctuations, and inappropriate packaging materials. Thus, it is critical to manage these factors appropriately to collectively impede the extent of postharvest losses.

Proper postharvest management practices and operation as well as implementation of some novel and green technologies play a crucial role in mitigating postharvest losses and reducing global food wastes. There is an increasing consideration regarding the application of cutting-edge technologies

DOI: 10.1201/9781003370376-3

and in-depth understandings of postharvest physiology to curb the said losses (Al-Dairi et al., 2023; Asrey et al., 2023). Application of selected chemical- and non-chemical-oriented technologies can notably decrease the losses of FVs after harvest. Use of precooling methods, irradiation, ozonation, edible coatings, plant extracts, light-emitting diodes (LEDs), hot water treatments (HWT), brassinosteroids, melatonin, salicylic acid (SA), jasmonates, and advanced storage technologies could help in minimizing storage losses by modifying the FVs' physiology (Zhou et al., 2007; Baghel et al., 2018; Zhu et al., 2019; Ghasil et al., 2022; Ali et al., 2020, 2022; Vinod et al., 2023). Recently substantial advancements have been made to minimize the ripening and metabolic aspects through the use of innovative hormonal approaches for postharvest FVs (Baghel et al., 2018; Meena et al., 2023). A study by Ghasil et al. (2022) revealed that the use of novel molecules such as nitric oxide and sodium nitroprusside could considerably preserve the food value of sugar apples and enhance their storage life. Another work by Ali et al. (2023) demonstrated that the application of SA enhances the antioxidant enzymes and disease tolerance of stored fruits. Controlled atmospheric storage helps increase the storage life of various climacteric fruits by lowering O_2 levels and elevating CO_2 levels. Despite that, "green technology" implementation and safer methodologies are in the infant stage and need to be further investigated (Bhan et al., 2022; Meena et al., 2023).

To effectively address these challenges, a postharvest physiology and technology perspective is essential, as it offers the necessary knowledge to develop innovations and technologies that preserve the quality of FV products while minimizing losses and waste. The present chapter describes the role of advanced postharvest interventions in the storage physiology and shelf life of FVs.

2.2 SUSTAINABLE TECHNOLOGIES FOR POSTHARVEST MANAGEMENT

2.2.1 Precooling

The FVs, even after being removed from the parent plant, endure respiration and transpiration as living organisms. When freshly harvested, they undergo a significant loss of water owing to these processes, possibly due to the stresses of harvesting. Field heat specifies the difference in temperature between harvested produce and ideal storage temperatures. It is crucial to promptly and effectively remove field heat after harvesting to prevent water loss, wilting, and shriveling, which can cause severe damage to the appearance of the produce. Additionally, field heat promotes the growth of decay-causing microorganisms and increases ethylene production. A general rule is that an hour's delay at a field temperature of 35 °C can decrease the shelf life to approximately a day, even under optimal storage conditions (Makule et al., 2022).

Research has demonstrated that without precooling, commercial FVs experience postharvest losses of about 25–30%, whereas with precooling, these losses are reduced to 5–10% (Zhou et al., 2007). Temperature plays a critical role in determining the postharvest life and quality of FVs. Among various available postharvest technologies, precooling is considered indispensable. Elansari et al. (2019) define precooling as "the rapid removal of field heat from freshly harvested perishable produce to slow down metabolism and minimize quality degradation before transportation or storage". Precooling was first developed by Powell and his colleagues in 1904 at the U.S. Department of Agriculture (Ryall & Pentzer, 1982). The major objective of precooling is to reduce the biochemical and physiological activity of perishables by eliminating field heat, preparing them for long-term storage (Duan et al., 2020). Furthermore, precooling can prevent the onset of certain disorders of a physiological nature in FVs, delay ripening or aging, decrease postharvest decays, and preserve overall quality (Duan et al., 2020), as mentioned in Table 2.1. It also helps in reducing the essential refrigeration capacity of cold-storage rooms.

2.2.1.1 Factors Affecting Precooling Techniques

Different factors affect the precooling process of FVs after harvest. Among these factors, the characteristics of the produce, including their shape, surface-to-volume ratio, size, porosity, internal structure, density, thermal property, and respiration heat production rates, play a significant role (Pathare et al., 2012). Additionally, velocity and temperature of the cooling mediums, along with the relative humidity percentage of surrounding environments, are crucial variables that can impact the effectiveness of precooling in general (Pathare et al., 2012).

2.2.1.2 Types of Precooling Systems

Pre-cooling techniques commonly include forced air cooling, room cooling, vacuum cooling, hydrocooling, package icing, and cryogenic cooling. While these techniques technically differ, they all aim to quickly remove heat from the produce by using cooling mediums like air and water (Wang & Sun, 2003; Duan et al., 2020).

2.2.1.2.1 Room Cooling

Room cooling refers to the practice of placing containerized goods or bulk in a low-temperature room for a period of time to cool them down (Prusky, 2011). This method has been applied for a long time and involves exposing the goods to cold air. Typically, the chilled air is provided by coolers equipped with evaporator coils and fans that circulate the air over the coils. Nevertheless, the efficacy of room cooling is influenced by several factors, such as air circulation, packaging of the

TABLE 2.1 Effect of Precooling Methods on FVs

CROP	*PRECOOLING METHODS*	*KEY FINDINGS*	*REFERENCE*
Baby cos lettuce	Vacuum cooling (0.6 kPa; 25 min)	Extended shelf life by 7 d, high cells intact and better chloroplast integrity, high phenolics and flavonoids content	Kongwong et al. (2019)
Barhee dates (Khalal stage)	Hydrocooling with ice-water mix	Extended shelf life, maintained	Elansari et al. 2008
Button mushrooms	FAC with refrigerated storage (2 °C) plus modified atmosphere packaging	Reduced water loss. Color change and browning index was minimal	Salamat et al. (2020)
Cashew apple	Hydrocooled by dipping in chlorinated water at 1, 3, 5, and 7 °C	Retained higher total phenolic contents, vitamin C, and firmness. Activity of polyphenol oxidase and peroxidase are lower than control	Sena et al. (2019)
Litchi	Hydrocooling (30 min) followed by drying for 1 h	Reduced weight loss and maintained fruit freshness	Liang et al. (2013)
Mango	Aluminum-cladded burnt-clay-brick evaporative cooling	Tissue breakdown, color changes, pH and titratable acidity were lower	Balogun et al. (2020)
	Chilled water dipping (5° C; 26 min)	Highest shelf life of 35 d, minimum PLW, and less spoilage	Khanbarad et al. (2013)
Pomegranate	FAC system in conjunction with cold storage (7 ± 1.2 °C; 91.4 ± 6.3%) + Plastic wrapping of pomegranates in 10 μm thick HDPE plastic film	Plastic lining delayed cooling rates	Ambaw et al. (2017)
Okra, tomato, green bell pepper	FAC (air speed: 1200 rev/min) with ice	Good sensory quality, good microstructure quality of fruits	Alabi et al. (2023)

FAC: Forced air cooling system; **HDPE**: High density polyethylene

goods, venting, initial temperature, and wrapping materials. To achieve proper cooling, it is important to ensure homogeneous air distribution with air velocities of at least 60–120 m/min around the packages (Elansari et al., 2019). Stacking the packages in a way that allows airflow between them and ensuring good ventilation of the packages are also essential. These measures create the necessary turbulence to remove heat and achieve adequate cooling.

One advantage of room cooling is that it reduces the need for additional handling since the goods can be cooled and stored in the same space. It is a comparatively simple precooling method. Room cooling is suitable for produce that is sensitive to surface moisture or moisture and for perishable goods that do not deteriorate quickly after harvesting, such as dry beans, sprouts, beets, cucumbers, peppers, dry onions, tomatoes, garlic, parsnips, potatoes, pumpkins, radishes, rhubarb, herbs, and turnips (Brosnan & Sun, 2001). However, one drawback of room cooling is that it is not space-efficient. Another challenge is the defrosting process, which can be problematic. Since room cooling relies on conduction, it is a slower cooling process compared to other methods (Ambaw et al., 2017).

2.2.1.2.2 Forced Air Cooling

Forced air cooling is a method used to precool packaged commodities by directing cold air to flow rapidly through the containers, promoting efficient convective heat transfer. This technique, also known as pressure cooling, relies on a small difference of pressure between the two sides of the containers to generate effective air movement and heat exchange. The temperature and velocity of the air play crucial roles in forced air cooling (Alabi et al., 2023). To prevent excessive water loss from commodities and minimize the possibility of chilling injury, it is important to avoid extremely low air temperatures and excessive airflow (Delele et al., 2013). Forced air cooling is commercially used for precooling bulk and palletized produce due to its affordability, versatility, and effectiveness (Dehghannya et al., 2010; Ambaw et al., 2017). In contrast with the traditional room cooling, forced air cooling offers faster cooling times because the cooling air directly contacts the product. In fact, this method can cool faster by tenfold or more. However, it should be noted that forced air cooling is generally slower than hydrocooling, which uses water for cooling purposes (Kitinoja & Thompson, 2010; Salamat et al., 2020).

According to Rao (2015a), forced air cooling can be accomplished through three main arrangements: circulating higher-velocity air in a refrigerated room, passing cold air through the commodity on a conveyor in a cooling tunnel, or utilizing pressure differentials to force air through packed containers. The third arrangement, which involves forced air serpentine cooling, tunnel-type cooling, or cold wall forced air cooling, is the most commonly used method. Forced air cooling is particularly suitable for strawberries and tomatoes, as they cannot be submerged in water. However, one limitation of forced air cooling is that it requires specific stacking patterns, and therefore skilled operators are necessary to achieve the desired loading patterns for a satisfactory cooling rate.

Alabi et al. (2023) demonstrated that forced air cooling combined with ice effectively removed field heat from okra, tomatoes, and green bell peppers, preserving their sensory quality and microstructures better than room cooling. Salamat et al. (2020) observed that combining forced air cooling, MAP, and refrigerated storage at 2°C for button mushrooms resulted in minimal water loss (less than 5%), maintained firmness, and reduced color changes and browning incidence in contrast with other storage methods. In the study by Ambaw et al. (2017), using a 10 μm thick HDPE plastic liner in CFB boxes during forced air cooling resulted in delayed cooling rates for pomegranates, extending the cooling time from 2.5 h to 8 h.

2.2.1.2.3 Hydrocooling

Hydrocooling is one of the most well-established and efficient cooling techniques, utilized for its exceptional heat-transfer capabilities through water. The said cooling method entails the application of cold water to FVs using cool shower systems or immersing selected produce in a vigorously agitated mixture of cold water or ice water (Liang et al., 2013). It is imperative to highlight that the hydrocooling technique may not be universally suitable for all cultivars of produce, mainly those which are susceptible to water-induced damages, such as strawberries. The commodities suitable for hydrocooling should be water-tolerant and must not adversely respond to sanitizing agents present in cold water. Some of the major FV commodities suitable for hydrocooling include litchis, apples, and cherries (Prusky, 2011). Nevertheless, hydrocooling applications are limited owing to its physiological and pathological effects on selected produce, such as chilling-associated injury, prompt softening, and increased growth of bacteria and fungi-based infestations. Therefore, vigilant consideration should be given to specific characteristics of FV-based produce before using hydrocooling after harvest.

One benefit of hydrocooling is its rapid cooling ability in contrast to other precooling methods. According to Rao (2015b), hydrocooling can remove field heat from produce in just 20–30 min, whereas forced air cooling typically takes several hours. Sena et al. (2019) noted that hydrocooling cashew apple fruits in chlorinated water at 5 °C delayed rapid softening and inhibited weight loss along with reduced total phenols and vitamin C losses. On the other hand, cold-stored cashew apples generally exhibited higher activity of peroxidase and polyphenol oxidase than hydrocooled ones. Litchi fruits which were subjected to hydrocooling followed by suitable drying (to rule out excessive water) had a higher weight than both the initial weight and the control groups (Liang et al., 2013). Another related study conducted by Elansari et al. (2008) observed that hydrocooling with an ice-water mixture successfully prolonged shelf life and maintained Barhee date quality throughout the distribution supply chain.

2.2.1.2.4 Vacuum Cooling

The main principle behind vacuum cooling (VC) depends on the concept of latent heat of vaporization. This means that cooling of produce is achieved by allowing moisture to evaporate from the surface and inside the commodity itself (Zhu et al., 2019). To further enhance the evaporation process, atmosphere pressure is reduced from 760 mm Hg to 4.6 mm Hg, causing water to boil at a lower temperature of 0 °C (32 °F). This rapid evaporation normally leads to desiccation and the cooling of produce, resulting in weight reduction of about 2–4%. To efficiently minimize moisture loss and inhibit wilting and shrinkage of FVs, water could be sprayed on commodities before it is kept in VC chambers (Zhu et al., 2019). Because water is the principal cooling agent, it can be assumed that the quantity of heat excluded from FVs is directly proportional to the quantity of evaporated water from their surface.

The VC is predominantly well-suited for certain leafy produce such as lettuce, as its numerous leaves provide a large surface area, which enables a rapid cooling process. This aforementioned method is commonly used in commercial applications for mushrooms and other selected leafy produce like spinach, broccoli, cabbage, celery, and cauliflower (Zhu et al., 2019; Garrido et al., 2015; Kongwong et al., 2019). Balogun et al. (2020) worked on mango storage in an aluminum-cladded burnt-clay-brick evaporative cooler system, which resulted in decreased metabolic rates, reduced tissue breakdown, and suppressed color changes, pH, and acidity as compared to ambient storage conditions. In another study by Kongwong et al. (2019), the VC prolonged shelf life of baby lettuce for about 16 days and improved antioxidative potential and phenolics concentration under storage.

2.2.1.2.5 Contact or Package Icing

Package icing stands as a cooling technique used during transit by enveloping the produce with crushed or flaked ice (Elansari et al., 2019). Diverging from alternate cooling procedures, ice not only swiftly eliminates heat upon application but also persistently absorbs heat during its melting process. Multiple approaches exist to administer ice for the sought-after cooling outcome. The approach of package icing entails the direct application of slush, flakes, or crushed ice directly onto the produce within shipping containers. Nonetheless, this approach might yield irregular cooling effects due to ice retention until complete liquefaction. In contrast, liquid icing offers an alternative avenue, utilizing ice slurry for a more sustained cooling performance. Additionally, there is the technique of top icing, where ice is positioned atop packaged containers, often serving as a supplementary measure to complement other cooling methods.

The main benefit of icing is that it prevents the commodity from drying out during cooling. However, there are drawbacks. Icing requires packaging that can handle constant water contact, adds significant weight to the package, and can create an environment favorable for postharvest disease and soft rot incidence due to the meltwater (Wills et al., 1998; Brosnan & Sun, 2001). Moreover, icing involves high capital and operating costs. Contact icing is limited to cooling specific commodities like kale, Brussels sprouts, collards, broccoli, carrots, radishes, and onions (Hardenburg et al., 1986; Elansari et al., 2019).

2.2.1.2.6 Cryogenic Cooling

Cryogenic precooling is based on utilizing the latent heat of evaporation from liquid nitrogen (N_2) or solid CO_2 (dry ice) to achieve extremely low "boiling" temperatures of −196 ℃ and −78 ℃, respectively (Brosnan & Sun, 2001). In this method, the commodity is cooled as it passes through a tunnel where the liquid N_2 or solid CO_2 evaporates. Nevertheless, the said temperature can cause the commodity to freeze, rendering it unsuitable for the fresh market. This issue is addressed by carefully controlling the conveyor speed and evaporation rate.

The use of liquid N_2 as a refrigerant for precooling fruits is particularly interesting due to the inert nature of N_2 gas. Cryogenic cooling systems are comparatively affordable to install but generally expensive to operate since these rely on costly cryogens and non-toxic refrigerants. The said technique is mainly employed for the cooling of selected soft fruits that have specific production seasons (Brosnan & Sun, 2001). The higher costs of liquid N_2, dry ice, and certain other non-toxic refrigerants make this process additionally appropriate for comparatively expensive products.

2.3 OZONE IN POSTHARVEST HANDLING OF FRUITS AND VEGETABLES

The demand for selected non-thermal, unconventional, and eco-friendly technologies to effectively manage FVs is on the rise. Ozone treatments have emerged as progressive methods for preserving quality attributes of fresh produce due to its potent disinfectant and antimicrobial potential. Ozone, a distinct form of O_2, facilitates the oxidation of various organic compounds (Pandiselvam et al., 2019). Notably, the utilization of ozone to impede microorganism growth is considered safer for food storage and processing, encompassing FVs, as recognized by the U.S. FDA (Food and Drug Administration). Moreover, the U.S. EPA (Environmental Protection Agency) acknowledges ozone's application in wastewater treatments as a decontamination process. Additionally, applying aqueous ozone to cleanse food-contact surfaces or items is deemed a sanitization measure, yielding no harmful by-products. Since ozone naturally decomposes into O_2 post-application, it leaves no residues and poses no risk to consumers, thus presenting an environmentally friendly and safe option for curbing postharvest fruit diseases (Vinod et al., 2023).

Ozone is an extremely reactive, naturally occurring gas that has strong oxidizing qualities. According to reports, the reducing potential of ozone is 1.5 times greater than that of chlorine and 3000 times greater than that of hypochlorous acid. Ozone is a potent oxidizer that has been recommended as an environmentally responsible and secure replacement for conventional chemical treatments. Depending on the commodity, it can be applied in either a gaseous or aqueous form (Crowe et al., 2007). Ozone can reduce pesticides and chemical residues as well as change non-biodegradable organic materials into forms which can be broken down by bacteria and other microorganisms (Rodgers et al., 2004). The bactericidal impacts of ozone extend to both gram-positive bacteria (including *B. cereus*, *S. aureus*, *L. monocytogenes*, and *E. faecalis*) and gram-negative bacteria (such as *P. aeruginosa* and *Y. enterocolitica*). This strong ozone effect applies to both bacterial spores and certain other actively growing cells. Ozone is effective in killing yeast cells, including *C. albicans* and *Z. Bacilli*. Furthermore, it has the ability to chemically suppress the effects of ethylene gas, which thereby reduces the ripening process and eradicates undesirable flavors created by bacteria (Rice et al., 1982).

2.3.1 Applications of Ozone

2.3.1.1 Regulate Ripening of Fruits and Vegetables

By preventing the biosynthesis of ethylene and immediately oxidizing it while in storage, ozone treatment controls the ripening of fruits. To preserve fresh produce quality, it is crucial to reduce ethylene biosynthesis by using ozone to block the actions of enzymes that help the process of ethylene generation. This is due to the fact that an increase in ethylene production promotes several biochemical processes, including respiration, which causes softening, nutrient loss, and a general decline in quality. It was known as a rate-determining phase by Wang et al. (2002). As outlined in Minas et al.'s study (2014), ozone treatment exhibited inhibitory effects on ACC synthase, an enzyme involved in ethylene biosynthesis, across the storage duration. Consequently, ozone-treated kiwifruit cv. "Hayward" displayed reduced levels of ACC compared to untreated fruit, leading to a postponement in the initiation of ripening. Ozone also had the effect of diminishing the activity of ACC oxidase, which is responsible for converting ACC to ethylene, CO_2, and cyanide. Notably, during storage, ozone quickly reacts with ethylene gas, forming CO_2 and water, as indicated by Yin et al. (2013). Research by Skog and Chu (2001) supports the idea that gaseous ozone facilitates ethylene oxidation, leading to the creation of CO_2 and water. Ozone's substantial oxidizing capacity is evident in aqueous environments, where a lone oxygen molecule released by ozone can attach to various compounds, triggering oxidation and the formation of new compounds. In this process, both organic and inorganic substances react and oxidize, leading to the precipitation of oxidized inorganic molecules out of the water.

2.3.1.2 Inhibition of Cell Wall–Degrading Enzymes

The activities of enzymes that break down cell walls have been found to be inhibited by ozone, which prevents the prompt softening of fruits. By blocking the action of cell wall–degrading enzymes, mainly pectin methylesterase (PME) and polygalacturonase (PG) activity, ozone treatments have been found to considerably slow down fruit softening and aid in maintaining fresh FV firmness. Toti et al. (2018) explained that melons exposed to ozone exhibited a decline in the activity of enzymes responsible for the degradation of cell walls, including PG, α-arabinopyranosidase, and β-galactopyranosidase. Furthermore, Minas et al. (2014) found a substantial role of ozone in modulating the softening of kiwifruit, reporting that this was governed by PG activity. Through the suppressions of the translation cascades of genes like AdACS1 and AdACO1, which code for enzyme activity essential for ethylene biosynthesis, ozone indirectly constrains the activities of cell wall–degrading enzymes. Reduced ethylene synthesis follows an inhibition of ACS and ACO activity, which then postpones the rapid ripening and softening. By slowing down the solubilization mechanism of pectins, ozone directly impedes the activities of the said enzymes. Galactosidase (β-Gal) and polygalacturonase (PG) are then suppressed to slow down prompt softening. This thereafter decreases PME activity and the rate of depolymerization, which results in a slower rate of cell wall disintegration in fruits that have been exposed to ozone.

2.3.1.3 Antimicrobial Action of Ozone on Postharvest Pathogens

Ozone is potentially reactive in nature and has the strong oxidizing property of free radicals, which gives rise to a diverse range of antimicrobial action. The influence of ozone-based treatments on numerous microorganisms is intricate due to its interaction with various components of cells (including respiratory enzymes, proteins, and unsaturated fatty acid contents), such as cell membranes, cell envelopes (peptidoglycans), cytoplasm (enzymes and nucleic acids), spores coats of pathogens, and virus capsids (comprising proteins and peptidoglycans). These complex interplays lead to bacterial inactivation in response to the exposure of treatments of ozone (Greene et al., 2012). Numerous researchers have suggested two main principal mechanisms by which ozone exerts its microbial inactivation influences. The first mechanism involves sulfhydryl groups' oxidation and the degradation of amino acids found in peptides, enzymes, and proteins. The said oxidation process results in the formation of small peptides as a result of exposure to ozone gas. In contrast, the second mechanism centers on the oxidation base reactions of polyunsaturated fatty acids, resulting in the creation of acid peroxides. Both these processes contribute to the multifaceted cascades through which it inactivates selected microorganisms on fresh FVs. It is believed that harm to cell envelopes or their disintegration causes cellular contents to leak out, and cell lysis occurs as a result, eventually causing microorganism inactivation (Greene et al., 2012). Luo et al. (2019) carried out research to investigate the efficiency of ozone treatments in preventing postharvest kiwifruit pathogens. According to the findings, kiwifruit's shelf life was enhanced for about 8 days and postharvest diseases were prominently reduced in response to ozone treatment. Similar to these results, Liang et al. (2019) studied the impact of ozone treatment on cherry tomato postharvest diseases. Their findings demonstrated that cherry tomato postharvest pathogens were inhibited, which ultimately showed to suppressed disease incidence after ozone treatment, resulting in cherry tomato shelf life increase of around 6 days. The main objective of using ozone treatments was to eliminate or reduce gray molds caused by *Botrytis cinerea*, which is a very prevalent postharvest disease in FVs such as susceptible strawberries, grapes, blackberries, and peaches, as well as plums, carrots, and tomatoes, among others.

2.4 EDIBLE COATINGS

Edible coating is defined as a uniform and thin layer which is directly applied on the surface of selected commodities either by spraying or dipping as per good manufacturing practices (de Vasconcellos Santos Batista et al., 2020; Gupta et al., 2023). It is odorless, colorless, and tasteless, comprising hydrocolloids, proteins, polysaccharides, lipids, and their composites (Sanchis et al., 2016). These compounds, derived from natural origins like carnauba, beeswax, shellac, chitosan, paraffin, gums, gelatin, and so on, are the most common edible coating agents which do not impose any adverse influence on human health (Vargas et al., 2008). Edible coatings not only maintain the postharvest quality and extend the shelf life of fresh FVs but also find broader applications as antimicrobial agents, modifiers of texture, dyes, antioxidants, flavor enhancers, sources of nutrients, surfactants, emulsifiers, and even plasticizers. Furthermore, these substances demonstrate excellent adhesion to the surface of products, especially when utilized in combination with other materials.

Commonly utilized polysaccharides in the formulation of edible films and coatings encompass starch, dextrin, pectin, tapioca, cellulose, corn, and its derivatives like hydroxypropyl methyl cellulose, carboxymethylcellulose and methylcellulose along with chitosan, pullulan, alginate, gellan, carrageenan, quince seed mucilage, and policaju gum (Hashemi et al., 2017). Among these, polysaccharide-based coating materials such as gum acacia, guar gum and carboxymethyl cellulose have shown the potential to extend the shelf life of numerous fruits, including Kinnow oranges (Ali et al., 2021; Bhan et al., 2022; Gupta et al., 2023).

These are known to control various pre- and postharvest diseases of fresh fruit. The formation of an extra layer around the fruit covers the stomata, leading to a decrease in the respiration rate by creating a modified atmosphere around the fruits that acts as a semi-permeable barrier to gases (O_2, CO_2), moisture (Alali et al., 2018), and volatile compound movement (Kumari et al., 2017), thus retarding ripening and senescence and prolonging shelf life (Parven et al., 2020). It also minimizes chemical-based adverse reactions by decreasing microbial contamination and deterioration (Thakur et al., 2018). Polysaccharides-based coatings are characterized by reduced water loss, whereas their gas barrier property induces required atmospheric modifications and enhanced shelf life without severe anaerobic conditions development.

2.4.1 Chitosan

Chitosan, a biopolymer of β-(1–4)-2-acetamido-D-glucose and β-(1–4)-2-amino-D-glucose units is derived from the deacetylation of crustaceans' chitin exoskeleton in an alkaline medium (Reddy et al., 2021). It is a biocompatible, biodegradable, and nontoxic compound exhibiting antimicrobial activity (Vilaplana et al., 2020) and good film-forming properties. It also possesses inherent anti-fungal properties and makes it efficient against the growth of several pathogens (Bautista-Banos et al., 2006).

2.4.2 Carrageenan

Carrageenan, a naturally occurring hydrophilic polymer, comprises a linear arrangement of partly sulfated galactan units. This biodegradable film serves as an antioxidant and contributes to prolonging the shelf life of FVs by conserving their sensorial and organoleptic attributes. Notably, these films are non-toxic and economically viable for consumption.

2.4.3 Essential Oils

Essential oils are aromatic plant extracts and composites of natural and volatile compounds that evaporate at room temperature. The major plant sources of EOs are roots, stems, leaves, flowers, fruits, and seeds (Rodrigues et al., 2021).

2.4.4 Plant Extracts

Plant extracts are concentrated derivatives obtained from dried raw materials through solvent extraction. These extracts often possess potent antifungal properties, capable of inhibiting the proliferation of pathogenic microorganisms responsible for causing diseases (Matrose et al., 2021).

2.4.5 Aloe Vera

Aloe mucilage exhibits dual properties of antifungal action against postharvest fungal diseases and antibacterial effects against both gram-positive and gram-negative bacteria (Mendy et al., 2019). Application of an aloe gels–based coating plays a substantial role in preserving textural properties, retarding softening and changes in peel color, regulating respiration rate, preventing moisture loss and maturity progression, delaying enzymatic browning, and suppressing the growth of certain microbial pathogens in FVs. Parven et al. (2020) further elucidated the multifaceted mechanisms of aloe coatings, including a protective barrier formation against O_2 and moisture, as well as inhibitory effects against microbes.

2.4.6 Edible Gums

Gums, a collective term designated for naturally occurring polysaccharides, are generally valued for their potential to create gels, produce thick solutions, and stabilize emulsion-based formulations (Salehi, 2020). Water-soluble gums, also known as "hydrocolloids", have a wide range of applications, for instance coating agents, textural enhancers, packaging films, thickeners, emulsifiers, stabilizers, gelling agents, and sources of dietary fiber (Dick et al., 2015).

2.4.7 Agar

Agar, a universally used food additive, is designated Generally Recognized as Safe (GRAS) by the FDA. Agar-derived polysaccharides have the potential to function as functional foods, offering protection against inflammatory diseases. It is notable that enzymatic hydrolysis is of minimal concern due to the scarcity of agarases (enzymes that degrade agaroses). These enzymes are found primarily in marine bacteria and a few bacilli not commonly present in food products (Ziedan et al., 2018).

2.5 IRRADIATION-BASED TREATMENTS

Ionizing radiations, such as gamma rays from cobalt-60 or cesium-137 sources, electrons produced by machine sources (e-beam), or X-rays, are used to irradiate food, including fresh FVs. This radiation damages insects, plant cells, or microorganisms or in the food by producing ions or free radicals that react with cellular components (Yahia, 2016). Irradiation may thereby alter the metabolism, ripening, and senescence of products while also controlling insects and microbes. Due to the expense, size, and need for irradiation plants, as well as the lack of certainty regarding whether or not consumers will accept irradiated food, commercial applications of irradiation are still not very widespread (Yahia, 2016).

Radiation doses are quantified in Grays (Gy), with 1 Gy being equivalent to 100 rads. A dose of 1 Gy signifies the absorption of 1 joule of energy per kilogram of food material. In various regions, irradiation has found practical applications. For instance, in Australia, it is employed to generate sterile male Queensland fruit flies, while in Hawaii irradiation is utilized in papaya to address issues related to the papaya fruit fly. The sprouting in potatoes, onions, garlic, and yams can be controlled by a dose of 0.05–0.3 kGy. Most insects are sterilized at a dose of 0.1–1.0 kGy. An irradiation dose of 1.0–1.5 kGy can control fungal infection.

Excessive doses of irradiation can lead to undesired outcomes, such as discoloration, softening, loss of phytochemicals, and alterations in taste. Research has shown that e-beam irradiation had minimal effects on the phytochemical constituents of pecan kernels (Villarreal-Lozoya et al., 2009). Pecan kernels from Kanza and Desirable cultivars, subjected to e-beam irradiation doses of 1.5 and 3.0 kGy and stored in accelerated conditions (40 °C and 55–60% RH for 134 days), exhibited reduced levels of phenolics and condensed tannins. However, this reduction did not significantly impact antioxidant capacity. While progress has been made, comprehensive studies are required, particularly regarding the molecular mechanisms that influence the phytochemical content of irradiated FVs (Alothman et al., 2009).

Irradiation effectively reduces the risk of foodborne ailments, prevents spoilage, inhibits maturation or sprouting, and eliminates pests. During irradiation, the radiations penetrate the fruits, leading to the ionization of atoms in its path, causing changes in essential macromolecules of harmful microorganisms and ultimately resulting in their elimination. The extent of ionization is determined by the quantity of energy transferred per unit mass or radiation dosage administered. Microbial cells are impacted by irradiation through mechanisms such as DNA disruption and heightened generation of free radicals, leading to disruptions in membranes, impaired protein function, growth inhibition, and genetic mutations (Reddy et al., 2018). Moreover, irradiation has the prospective to slow down the natural ripening process of fruits, thereby prolonging their shelf life. This delay in ripening is attributed to down-regulated ACC oxidase enzyme activity, which is pivotal in the production of ethylene. In irradiated fruits, enzyme activities of PG, PME, and β-GAL are notably reduced compared to control fruits (D'innocenzo & Lajolo, 2001).

For climacteric fruits like papaya, applying gamma radiation doses ranging from 0.25 to 1.00 kGy can effectively extend their shelf life by slowing down the ripening process. In the context of controlling fungal infections, irradiation has been found effective at levels between 1.0 and 1.5 kGy. Irradiation serves as a prevalent method for eliminating insect infestation, with a dose of 0.15 Gy being authorized for papaya imports into the USA to eliminate Tephritid fruit fly incidence. However, in practical applications, a dose of around 0.40 kGy is commonly employed to meet the strict requirements of zero tolerance for surface pests (Rashid et al., 2015).

2.5.1 Application of Irradiation

2.5.1.1 Sprouting Inhibition

Rotting and sprouting in potatoes, garlic, onions, ginger, and yams can be prevented by irradiation (Acheson & Steele, 2001). This method offers a benefit over refrigeration, which is costly, particularly in developing nations with limited access to energy. Also, it is superior to potentially inexpensive non-chemical-based therapy (Kalyani & Manjula, 2014).

2.5.1.2 Disinfestations

Food irradiation is employed as a method to counter insect and pest infestations in grains, pulses, and dried fruits. This technique holds an advantage over the use of chemicals that could potentially pose risks to human health and the environment, as it does not leave any chemical residues in the food (Hallman & Blackburn, 2016). Furthermore, it proves to be more efficacious compared to conventional practices such as airtight storage containers, which might not completely eliminate all insects.

2.5.1.3 Food-Borne Pathogens

Food pathogens are eliminated using the food irradiation process, which lowers the prevalence of food-borne illnesses (Bintsis, 2017). While overcooking vegetables can damage certain essential vitamins, preparing and drying spices exposes them to microorganisms. Irradiation will eliminate the bacteria in spices and prevent the vitamin destruction in overcooked vegetables.

2.5.1.4 Shelf-Life Extension

Using irradiation, the shelf life of many food products may be significantly increased. This holds true for fruits, meat, poultry, and fish. According to the latest reports, irradiation to fruits before these mature can considerably enhance their potential shelf life period (Mostafavi et al., 2010).

2.6 LIGHT-EMITTING DIODES

The utilization of light-emitting diodes (LEDs) in the context of fruit physiology has recently exhibited a noteworthy influence within various storages and growing settings (Gong et al., 2015; Perera et al., 2022). These LEDs, renowned for their ability to manipulate the illumination conditions, have been employed as an eco-conscious approach to enhance plant resilience against a wide range of stresses. Furthermore, they have demonstrated efficacy in extending the storage duration and preserving the postharvest characteristics of horticultural crop yields throughout the past few decades (Poonia et al., 2022; Zhang et al., 2022).

In the context of microbial adhesion to surfaces of animal-derived food, both LED irradiations within the range of 400–460 nm, functioning on the principle of photodynamics inactivations, and UV-LED application, operating on an ultraviolet sterilizations principles basis, exhibit substantial bactericidal effectiveness. However, a standardized framework to ascertain the optimal LED irradiation dosage alongside specific wavelengths remains absent. Yet it is plausible that an enhanced LED irradiations system will find extensive utility in the food industry in the foreseeable future. Noteworthy is that the pre-treatment of mature green tomatoes with blue light encompassing the 440–450 nm range has exhibited the potential to impede softening and maintain firmness throughout storage (Dhakal & Baek, 2014). Similarly, the application of blue light has demonstrated the ability to enhance PpLOXs expression and ethylene biosynthesis–related genes like PpACS3 and PpACO1 in peaches (Gong et al., 2015). It's worth noting, however, that there is a limited body of research addressing LED applications in postharvest treatments of FVs.

2.7 HYDROTHERMAL TREATMENTS

The increasing awareness of sustainable resource utilization and the protection of human health and the ecosystem have led to the growing popularity of heat treatments in postharvest practices. These treatments effectively manage insect pests; prevent fungus-based rots, and influence ripening and temperature responses of harvested commodities. The intake of FVs reduces disease occurrence related to oxidative stresses (Petruzzi et al., 2017). Heat treatments are given for a short duration of up to 1 h and for a long-term duration of up to 4 days. In order to preclude overripe flavor development in cantaloupes and melon fruits, as well as to prolong the shelf life of bean sprouts, plums, grapes, and peaches, heat treatments have been applied to a variety of FVs, including firm potatoes, tomatoes, carrots, and strawberries, among others. These treatments also reduce fruit decay and preserve color in broccoli, green beans, asparagus, kiwifruit, lettuce, and celery (Schirra et al., 2000; Lurie, 1998; Fallik, 2004). Hot water treatment, vapor heat treatments, and hot air treatment are main heat treatments commercially available for the preservation of FVs by altering their physiology and metabolism.

2.7.1 Hot Water Treatment

Hot water treatment involves the transfer of heat from higher temperatures to a low-temperature medium of FV tissues. It is one of the cheapest and most cost-effective methods used in horticultural crops. The potential of HWT to control the enzyme activities which influence the qualitative traits of fresh FVs is directly connected to its efficacy. Although many FVs can withstand 50–60°C water exposure for up to 10 min, postharvest plant pathogens can be controlled by shorter exposures at these temperatures (Barkai-Golan & Phillips, 1991). Contrarily, fruit dips in hot water necessitate 60 min of exposure to 46 °C. HWT is frequently used to manage diseases like anthracnose and to get rid of fruit fly larvae on fruits like mangoes, citrus fruits, and avocados.

Luengwilai et al. (2012) reported that hot water treatment in green tomatoes from 30 to 50 °C for 3–9 min to green tomato at 40 °C for 7 min was the most suitable combination, showing good visual quality and progressed ripening after storage. Mango cv. Tainong subjected to hot water treatment at 55 °C for 10 min exhibited suppressed respiration (Zhang et al., 2012). Mango cv. Keitt at 46 °C for 90 min showed inhibition of the ACC oxidase enzyme (Bender et al., 2003). According to Prakash and Pandey (2000), controlling mango postharvest anthracnose with hot water dips at 52°C for 5, 15, and 30 min is successful. According to the research, HWT at 50°C for 5 min effectively reduced the occurrence of anthracnose in mango by 96%. Pomegranate in hot water dipped at 45 °C for 4 min showed the results of reduced occurrence of chilling injury (Mirdehghan et al., 2007). Hot water is known to stimulate enzyme β-amylase genes involved in hydrolyses of starch and increased TSS with increasing fruit ripening in mango (Bambalele et al., 2021). Setagane et al. (2021) observed that hot water application for Hass cultivar of avocado at 42 °C for 24 min helps in mitigating chilling injury and better development of purple color. The combined 1-MCP application with hot water showed long distant transportation of FVs (Ngamchuachit et al., 2014). Recently the use of hot water forced convection (HWFC) application in zucchini showed effective results in terms of fruit quality, reducing CI and membrane injury, increasing treatment of HWFC at a temperature of 40 °C and water flow rate of 2 m/s for 20 min (Zhang et al., 2019). Recently HWT in mango cv. Chenab Gold at 48°C for 60 min after harvesting showed reduced mass loss and improved surface color at ambient conditions (Javed et al., 2022). Hot water dips can not only prevent pests and diseases but can also enhance the grade of horticultural perishables.

2.7.2 Vapor Heat Treatment

Vapor heat treatment (VHT) is a technique involving the use of moist air at temperatures ranging from 40 to 50°C to eliminate insect eggs and larvae, making it a valuable quarantine measure for ensuring the safety of fresh market shipments (Vinod et al., 2023). The transmission of heat occurs through the condensation of water vapor on the cooler fruit surface. VHT has been applied successfully in addressing Mediterranean and Mexican fruit fly infestations, without requiring forced air (Le et al., 2010). In the case of guava, VHT maintained at a core temperature of 47.5 °C for 25 min considerably decreased fruit fly infestation, resulting in enhanced marketability index, preserved weight, increased color, and reduced disease occurrence under regular storage conditions (Malik et al., 2021). The VHT has also been studied for its impact on aroma profiles in mango fruits. Treatment for 20 min at 48 °C led to accelerated fruit ripening, showing an edible-soft stage within 4 days post-treatment due to reversible inhibition of aroma volatile emission at different ripening stages (Singh & Saini, 2014). Ethylene treatment along with VHT for papayas at 20 °C for 24 h revealed that ethylene did not interfere with the ripening process of the fruits (Hasan et al., 2018). For mangoes infested with *B. dorsalis*, VHT at 47 °C for 40 min effectively reduced the quantity of eggs and larvae without compromising fruit quality (Lestari et al., 2017). The versatility of VHT extends to fulfilling phytosanitary prerequisites for the export of horticultural produce. In line with export mandates for pest-free products, VHT serves as an effective method to meet these requirements. For example, VHT disinfestation as an approved protocol for Carabao mango from the Philippines and Irwin and Haden mangoes from Taiwan was effective fruit temperature treatment at 46 °C for 10 min and 46.5 °C for 30 min, respectively. This highlights the role of VHT in enabling access to a high-value market, such as Japan.

2.7.3 Hot Air Treatments

The hot air treatment is a progressive heating method compared with other methods such as hot water dipping or vapor heat treatment. It involves subjecting FVs to elevated temperatures for longer durations, typically ranging from 12 to 96 h at temperatures between 38 °C and 46 °C. In case of hot air treatments, FVs are placed within a heating chamber where hot air is introduced at a temperature of 40 °C for 2 days, followed by a 40-day storage period at 10 °C. This approach was found to maintain higher sugar concentrations and lower citric acid content (Chen et al., 2012). The treatment based on hot air at 38 °C for 12 h exhibited significant results in enhancing the quality of cherry tomato fruits. The said treatment was associated with the stimulation of the phenylpropanoid pathway, thus leading to increased levels of certain secondary metabolites. At the same time, genes associated with metabolite production were substantially up-regulated, which thereby contributed to substantial reduction in disease incidence such as black and gray mold (*Botrytis cineraria*) (Wei et al., 2017).

2.8 USE OF PHYTOHORMONES

Application of novel plant-based biomolecules has been found effective in modifying the physiology and delaying the senescence of FVs. Plant growth regulators or phytohormones are signaling molecules that significantly affect plants' regulatory mechanisms at lower concentrations (Wani et al., 2016). Phytohormones play important roles in postharvest storage and the physiology of FVs by activating the expression of various defense genes, the inhibition of senescence-related enzymes, regulating the integrity of plasma membranes, and stimulating the activity of various antioxidative enzymes (Xiang et al., 2021). Use of brassinosteroids (BRs), jasmonates, salicylic acid (SA), polyamines, nitric oxide, and melatonins have exerted positive effects in preserving bioactive compounds by delaying ripening, softening enzymatic activities, and elevating defense-related enzymes (Baghel et al., 2018, Meena et al., 2023, Ghasil et al., 2022). The various hormone-associated impacts have been mentioned in Table 2.2.

TABLE 2.2 Effects of Various Plant Hormones on Postharvest Parameters of FVs

CROP	*PHYTOHORMONE*	*RECOMMENDED CONCENTRATION*	*INFERENCE*	*REFERENCE*
Avocado	Salicylic acid	0.1 mM MeSA	Alleviates chilling injury by maintaining cell integrity achieved by alteration of fatty acid composition and content.	Glowacz et al. (2017)
Apricot	Polyamines	1 mM putrescine or spermidine	Delays ripening and chilling injury by enhancing the antioxidant system.	Saba et al. (2012)
Broccoli	Cytokinins	0.89 mM 6-BA	Reduces POD activity, MDA content and enhanced SOD, APX, and CAT activities, therefore enhancing the antioxidant system.	Xu et al. (2012)
Mangosteen	Jasmonates	0.20 mM MeJA	Enhances antioxidant system and delays pericarp hardening and weight loss	Mustafa et al. (2018)
Jujube	Brassinosteroids	5 μM brassinolide	Inhibits development of blue mold rot caused by *P. expansum* and enhances the activity of defense-related enzymes, including PPO, PAL, SOD, and CAT.	Zhu et al. (2010)

(Continued)

TABLE 2.2 (Continued)

CROP	PHYTOHORMONE	RECOMMENDED CONCENTRATION	INFERENCE	REFERENCE
Tomato	Brassinosteroids	3μmol L^{-1} brassinolide	Increases ascorbic acid, lycopene content, soluble sugars, respiration rate, and ethylene production but leads to decrease in chlorophyll content in fruits.	Zhu et al. (2015b)
Satsuma mandarin (*Citrus unshiu*)	Brassinosteroids	5 mg L^{-1} 24-epibrassinolide	Significantly enhances the postharvest biotic stress tolerance. Increase in H_2O_2 content and proline, GABA, ornithine, D-xylose, and D-galactose.	Zhu et al. (2015a)
Pineapple	Salicylic acid	2.0 mM	Reduces the incidence of internal browning or blackheart by inhibiting the activity of PAL and PPO enzymes.	Lu et al. (2010)
Pineapple	Abscisic acid	380 μM ABA	Reduces incidence of internal browning by 23.4–86.3% by enhancing the antioxidative system. Decreased the activity of phenolic compounds, PPO, and PAL.	Zhang et al. (2022)
Strawberry	Salicylic acid	1–2 mmol L^{-1}	Controls fungal decay, reduces ethylene production, and maintains overall quality of fruits.	Babalar et al. (2007)
Grapes	Salicylic acid	1–4 mmol L^{-1}	Reduces water loss and controls fungal decay caused by *Botrytis cinerea*. Increases berry firmness and reduces rachis browning rate (2 mmol L^{-1}). No effect on TSS, titratable acidity, or pH	Ranjbaran et al. (2011)
Cassava	Abscisic acid	–	Alleviates postharvest physical deterioration (PPD), increases endogenous levels, and reduces endogenous H_2O_2 levels.	Yan et al. (2022)

2.8.1 Brassinosteroids

Brassinosteroids are natural steroid phytohormones comprising nearly 70 polyhydroxylated sterol derivatives and were first isolated from *Brassica napus* pollen, hence named "brassins" (Rao et al., 2002). The brassinolide (BL), 24-epibrassinolide (24-EBR), and 28-homobrassinolide (28-HBL) are commonly used BRs at varying levels to modulate pre- and postharvest aspects (Nawaz et al., 2017). Brassinolide is biosynthesized from campesterol through either the early or later C-6 oxidations pathway involving an early C-22 oxidation branch in it (Fujioka & Yokota, 2003). Exogenous BR application impedes cell wall degradations and suppresses softening catalyzed by several enzymes such as PME and PG (He et al., 2018), regulates sugar and energy metabolism (Lu et al., 2019), inhibits lipid peroxidation, maintains higher membrane integrity (Aghdam & Mohammadkhani, 2014), regulates ripening and color metabolism (Tavallali, 2018), inhibits POD and PPO enzymes, delays enzymatic browning (Gao et al., 2015), and inhibits production of ROS (O-, H_2O_2) species. Gao et al. (2016) influences a variety of physiological responses during postharvest life and maintaining the quality of different FVs.

2.8.2 Salicylic Acid

Salicylic acid (SA) or *ortho*-hydroxybenzoic acid, a phenylpropanoid compound, is an endogenous signaling molecule responsible for higher physiological and metabolic activities regulating the growth and development of plants. Salicin is the best-known SA derivative occurring naturally in *Salix alba* (white willow). In plants, SA is synthesized through distinct enzymatic routes, i.e., either from chorismate via isochorismate (chloroplast-localized) (Wildermuth et al., 2001) or through phenylalanine (cytoplasm-localized) (Leon et al., 1993). Recently, SA has been utilized in postharvest technologies for various horticultural crops. The SA and acetyl salicylic acid (ASA) have been found to suppress the biosynthesis of ethylene in pears, apple and pear discs, and a variety of horticultural crops, implying a role for SA as an ethylene antagonist; decrease lipid oxidation and membrane senescence (Kazemi et al., 2011); reduce chilling injuries and browning reactions (Sayyari et al., 2009); enhance firmness, anthocyanin synthesis, and antioxidant capacity (Wei et al., 2011); controls postharvest decay; and reduces respiratory rate (Babalar et al., 2007) in plants.

2.8.3 Jasmonates

Methyl jasmonate (MeJA) and its free acid jasmonic acid (JA) are together known as jasmonic acid. Jasmonic acid is biosynthesized in plant tissues via an octadecanoid's pathway through activation of membrane lipids by phospholipase to release α-linolenic acid (Seo et al., 2001). Methyl jasmonate (MeJA) was one of the plant volatiles first identified from the flowers of *Jasminum grandiflorum* (Cheong & Choi, 2003). Postharvest application of MeJA alters volatile patterns, improves innate resistance to pathogenic fungi in postharvest crops, and enhances heat shock proteins (HSPs) and C-repeat binding factors (CBFs) (Aghdam & Bodbodak, 2013). Dhami et al. (2023) demonstrated that both pre- and postharvest application of 0.5 mM MeJA was found effective in preserving bioactive compounds and food value in Kinnow mandarins. Cai et al. (2020) found that MeJA enhanced the release of peach aroma–associated lactone and carotenoid contents, thereby improving the sensorial attributes, such as taste and odor, of postharvest fruit. Similarly, it prompted aromatic volatiles biosynthesis in ripe pear fruits (Qin et al., 2017). Jasmonates play an important role in alleviating chilling injuries during postharvest storage of FVs through various mechanisms, including enhancement of membrane integrity and antioxidant system, and alterations in PAL and PPO activity (Dhami et al., 2023).

2.8.4 Abscisic Acid

ABA affects the shelf life of various horticultural crops after harvest, but little is known about the signaling mechanism involved in postharvest physiological processes modulation by the ABA. It enhances ripening-related activities like sugar accumulation, softening, and degradation of chlorophyll and enhanced color, especially anthocyanin in colored grapes (Fuentes et al., 2019). However, ABA applications are reported to reduce cell wall rigidity loosening, resulting in the softening and cracking of berries (Gambetta et al., 2010). The application of ABA to table grapes resulted in cell wall–degrading enzymes genes induction, such as PG, that stimulate pectin depolymerization and solubilization, leading to reduced firmness (Koyama et al., 2010). Internal browning or blackheart causes huge economic losses in pineapples, and 380 µM ABA application decreased internal browning incidence by 23.4–86.3% (Zhang et al., 2015). In cassava, exogenous ABA application reduced endogenous H_2O_2 content (Yan et al., 2022).

2.9 STORAGE TECHNOLOGIES

Storage of fresh FVs is an important operation after harvesting the produce. Owing to high moisture content and metabolic activities, they are more liable to spoilage. As fresh commodities are live entities, respiration and ripening continue even after harvest. In middle-income countries, more so over in tropical countries like India, storage is a major problem where sufficient store houses are not available (Meena & Choudhary, 2022). There are some traditional ways for storing FVs. In recent years, commercial use of advanced storage techniques, for example controlled atmosphere (CA) storage and modified atmospheric (MAP) storage, have attained great attention from stakeholders (Meena & Choudhary, 2022). Temperature, relative humidity, gaseous combinations, and pressure modifications are major interconnected things for storage, which need to be modified as per the nature of commodities and structure.

2.9.1 Traditional Methods of Storage

These methods are also known as low-cost storage methods which have been traditionally used for fresh FV preservation. This method does not require refrigeration and includes in situ, sand, clamps, cellars, barns, coir, pits, windbreaks, evaporative cooling, zero energy cool chambers, and night ventilation systems. Moreover, several structures like cellars, clamps, and ventilated structures have been used for many years (Kale et al., 2016).

2.9.2 Zero Energy Cool Chambers (ZECCs)

ZECCs work on the basis of an evaporation cooling system which helps in the short-term farm storage of FVs and has proven a milestone in hot and dry regions of India (Jha & Chopra, 2006). In 1982, Roy and Khurdiya developed ZECCs at the Indian Agricultural Research Institute in New Delhi, India, intended for on-farm storage using locally sourced materials like bricks, bamboo, and jute cloth. It is suited for short-term storage of hot-season crops like mangoes, tomatoes, and peppers. Effective operation requires outdoor temperatures exceeding 32 °C and low wet bulb temperatures below 2 °C. Chilling effectiveness relies on ambient air's initial humidity and evaporative surface efficiency. This system can lower crop temperature by 10–15°C while maintaining about 90% humidity, thus enhancing produce quality and shelf life (Kaur et al., 2021; Mishra et al., 2020).

2.9.3 Advanced Methods of Storage

2.9.3.1 Low-Temperature Storage (Refrigerated or Cold Storage)

For fresh FVs, the lower the storage temperature, the longer the storage life will be. The metabolic functions of crops are responsible for the accelerated pace of degradation. The respiration rate of crops undergoes an exponential increase as the temperature rises within the crop's physiological temperature range. This phenomenon, known as the Q10 effect, signifies that for every 10 °C alteration in temperature, the metabolic rate escalates by a factor of two to three. The Van't Hoff rule aids in predicting this change. It is important to define the applicable temperature range for this prediction, as many crops are susceptible to chilling injuries at temperatures higher than their freezing point. This temperature sensitivity differs not only among different crop species but also among various cultivars of the same species. When compared to storage at 0 °C, numerous crops exhibit significantly prolonged marketable lifespans when stored at or slightly above their freezing point. Notably, this phenomenon has been observed in crops like pears and broccoli. Originating in Japan, the low-temperature technique was devised for increasing the preservation of FVs throughout the entire distribution chain, including the end consumer's refrigerator.

2.9.3.2 Modified Atmosphere Storage

Modified atmospheres (MA) pertain to the composition of the surrounding air differing from that of typical atmospheric conditions, characterized by approximately 21% O_2, approximately 0.03% CO_2, and approximately 79% N_2. The MAP is a method employed to modify the atmosphere in the vicinity of a product, accomplished through the utilization of a polymeric film possessing selective penetrability to O_2 and CO_2, or by introducing a precisely controlled gas mixture into the packaging system. The initial implementation of MAP can be traced back to 1927 when it was employed to prolong the shelf life of apples by subjecting them to reduced O_2 levels and increased CO_2 concentrations (Davies, 1995). Subsequent research has validated the effectiveness of MAP for both climacteric (Sellamuthu et al., 2013; Villalobos et al., 2014 and non-climacteric (Finnegan et al., 2013; Barrios et al., 2014) fruits. The MAP has also been found to contribute to the accumulation of phenolic compounds in numerous fruits, such as pomegranates, plums, sweet cherries, and nectarines. This effect may be attributed to the suppression of major enzymes involved in phenol biosynthesis, including chalcone synthase, PAL, and anthocyanidins synthase (Desjardins, 2008), or the reduction in activities of phenols-degrading enzymes such as PPO and POD (Pourcel et al., 2007).

Selcuk and Erkan (2014) revealed that MAP enabled storage of pomegranates for up to 100 days without detrimentally decreasing fruit quality. Furthermore, MAP significantly impeded weight loss and decay development and preserved the visual appearance of pomegranate fruits. Similarly, Rao and Shivashankara (2018) noted that MAP consisting of 3–8% O_2 and 7–11% CO_2 successfully maintained the peel color, moisture content, and extended the shelf life of Bhagwa pomegranates for 90 days at a temperature of 8°C. In the context of avocado fruits, the utilization of lower O_2 levels (2–5%) and elevated CO_2 concentrations (3–10%) through MAP was found to suppress the rate of respiration and ethylene synthesis, thereby delaying undesirable changes in texture characteristics, color, and susceptibility to chilling-related injuries. Similar effects have also been observed in bell pepper and certain other tropical fruit types (Singh et al., 2014).

2.9.3.3 Controlled Atmosphere Storage

Controlled atmosphere (CA) also refers to a different atmosphere than air which is continuously controlled and adjusted over time. In the horticultural industry, CA storage is the most successful method to enhance the shelf life of fresh commodities (Bodbodak & Moshfeghifar, 2016). The controlled admission of some air into the storage room prevented the O_2 level from becoming too low during storage. Current technologies used to generate CA include hollow fiber systems and pressure swing adsorption systems that provide a consistent supply of low O_2 for continuous ventilation of produce (Hoehn et al., 2009). An advantage of these techniques is that the starting raw material, regular air, is freely available and no hazardous gaseous by-products are generated. However, to avoid unfavorable low and high levels of O_2 and CO_2, respectively, the atmosphere must be continuously monitored and corrective action must be taken. These advancements enable the quick establishment and long-term maintenance of an efficient atmosphere, particularly for many apple cultivars, containing 2–5% CO_2 and 2–3% O_2.

Weight loss in CA-stored FVs can be primarily ascribed to processes of respiration and transpiration, which result in the consumption of organic matter (Li et al., 2014; Xin et al., 2017). In case of fruits stored in CA conditions, the suppressed ripening and respiration rates collectively contribute to lower losses in fresh weight (Sivakumar & Korsten, 2010). The appropriate level of O_2 and CO_2 within a CA storage system plays an essential role in maintaining internal fruit atmosphere, thereby promoting a favorable storage environment for lemon fruit and considerably extending their storage duration (Ma et al., 2019). In addition, protocols such as initial low O_2 stress followed by ordinary CA storage has been shown to impede superficial scald—O_2 is set at 0.5% for 14–21 days at the start of storage and then raised to 1.5%, with the temperature raised to 2 °C for 3–4 days. The O_2 content is then increased to 1.5% and the temperature is reduced to 1 °C for the remainder of the 8-month storage.

Awad and Jager (2003) studied the effect of air and CA storage on concentrations of phenolic contents in apple fruit and noted that long-term storage can be retained with no loss of flavonoid concentration during storage. Sánchez-Mata et al. (2003) stored green beans at 8 °C under different atmospheric conditions, i.e., normal atmosphere air, 3% CO_2 + 5% O_2, 3% CO_2 + 3% O_2, and 3% CO_2 + 1% O_2 and found that 3% CO_2 + 3% O_2 at 8 °C was best to retain the nutritive constituents of the beans with reduced anaerobic process.

2.9.3.4 Hypobaric (Sub-Atmosphere) Storage

Hypobaric storage is a controlled atmosphere (CA) storage method that involves placing produce in refrigerated containers under a partial vacuum. The container is sealed and pressure is reduced to the desired level using a vacuum pump, typically ranging from 1.3 to 13 kPa. To avoid water loss and continue appropriate O_2 levels, humid air is continuously removed from the container. This technique effectively retards the ripening process of FVs by reducing the partial pressure of oxygen, and in some cases, the concentrations of ethylene. This slower ripening is a result of decreased ethylene biosynthesis from the produce and reduced respiratory rates. The application of hypobaric storage has historically been limited due to the high costs associated with creating low internal pressures. However, advancements have led to the availability of commercial hypobaric shipping containers, and the engineering and construction expenses for hypobaric chambers have decreased over time. Research by Gao et al. (2006) demonstrated successful storage of loquat fruit for 48 days under controlled air or 40–50 kPa pressure at 2–4 °C, maintaining fruit quality and preventing decay. Similarly, Li et al. (2006) observed that hypobaric storage of green asparagus preserved different biochemical attributes along with reduced malondialdehyde accumulation. Maryam et al. (2022) explored the effects of pre-storage hypobaric treatment (20 kPa for 2 h) on strawberry fruit postharvest storage and found that it maintained better fruit quality compared to control conditions.

2.10 USE OF MINERALS AND ANTIOXIDANTS

Calcium is an important nutrient mineral affecting the quality of horticultural crops, and its deficiency has been associated with the development of a range of postharvest physiological disorders. These disorders are associated with either the presence of excess of calcium or its inadequate concentration. Golden specks in *Solanum lycopersicum* are developed due to excess calcium in the fruits, resulting in reduced shelf life and quality (Mestre et al., 2012). Blossom end rot in tomatoes (Ho & White, 2005) and bitter pit and superficial scald in apples (Saure, 1996) are related with calcium deficiency (Table 2.3). Calcium-related physiological disorders are usually associated with loss of membrane integrity and ultimately lead to cell death. Also, resistance to softening of tissues due to application of calcium is due to the formation of calcium pectate (increase the rigidity of the middle portion of the cell wall) and stabilization of membrane systems.

Therefore, Ca^{2+} are potential regulators of postharvest physiology of FVs at the cellular level and an imperative constituent of cell walls binding together the strands of pectin, therefore maintaining fruit firmness. In pears, postharvest application of 4% calcium chloride maintains fruit firmness and reduces core browning up to 75 days of storage at 0–1 °C and 90–95% RH (Mahajan & Dutt, 2004). Fungal decay during postharvest storage greatly limits the economic value of FVs such as gray mold and soft rot of strawberries, blue mold of citrus fruits, anthracnose in mangoes, bacterial soft rot of tomatoes, dry rot of potatoes, and black rot of carrots. Dipping of strawberry fruits in 1% calcium gluconate prevents both loss of firmness and fungal decay (Hernández-Muñoz et al., 2006). In tomatoes, application of 2% $CaCl_2$ + 3 along with 4.5 kJ m^{-2} UV-C results in reduced ethylene production and has lesser fungal decay than the control (Mansourbahmani et al., 2017).

Among all the plant mineral nutrients, K^+ has the strongest influence on fruit quality and marketability. In addition to the regulation of stomatal closure and its potential role in stress alleviation in plants, it is also involved in the transportation of

TABLE 2.3 Effects of Ca on Postharvest Quality of FVs

CROP	*RECOMMENDED CONCENTRATION*	*INFERENCE*	*REFERENCE*
Peach	62.5 mM calcium chloride, calcium lactate and calcium propionate	Maintains tissue firmness and less susceptible to physiological disorders after extended cold storage.	Manganaris et al. (2007)
Strawberry	1% calcium chloride solution	Increases Ca content of fruits, maintains fruit firmness and soluble solids, and does not affect the sensorial quality.	García et al. (1996)
Apple	2–4% calcium solution	Increases fruit firmness. Reduces peroxidase activity, pH, and TSS/TA ratio.	Rabiei et al. (2011)
Broccoli	2% $CaCl_2$ + 2mM SA	Reduces loss of chlorophyll, phenolic compounds, carotenoids, flavonoids, and vitamin C.	El-Beltagi et al. (2022)
Sweet pepper	0.5–1.5 g L^{-1}	Reduces weight loss and exhibits higher TSS, chlorophyll, TA, and carotenoid content. Also exhibit higher antioxidant antioxidative properties.	Barzegar et al. (2018)
Tomato	2% $CaCl_2$ + 3 and 4.5 kJ m^{-2} UV-C	Maintains firmness and higher phenolic content. Reduced ethylene production and decay of fruits.	Mansourbahmani et al. (2017)

photoassimilates, enzyme activation, and turgor maintenance. In horticultural crops, adequate K^+ concentration is associated with improved fruit color, increased TSS, ascorbic acid, and also shelf life. In bananas, potassium silicate application during postharvest improved the keeping quality (Nikagolla et al., 2019). Also, preharvest application of K_2O (4–20g/plant) reduced the incidence of postharvest incidence of internal browning in pineapple fruits stored at 7°C and 95% RH for 15 days (Soares et al., 2005). Biofortification of potassium (15 mM) in KCl form improved the quality of cherry tomatoes during postharvest storage at 4°C (Constán-Aguilar et al., 2014). In long sweet peppers (*Capsicum annum* L.), preharvest foliar application of KCl and K_2SO_4 reduced decay and weight loss of fruits along with increased vitamin C and total sugars (Shehata et al., 2019).

2.10.1 Antioxidants

Ascorbic acid, also known as vitamin C is the major natural antioxidant and is produced via the Smirnoff-Wheeler pathway or the L-galactose pathway (Wheeler et al., 1998). Ascorbic and its derivatives are widely used as antibrowning compounds in various edible coating types to retain the quality of FVs because of their high antioxidative properties. In strawberries, ascorbic acid incorporation with chitosan coatings improved quality by delaying senescence and softening of fruits, thereby improving their storability for 15 days as the coated fruits showed higher antioxidant activity (POD, SOD, APX, CAT) than non-coated ones (Saleem et al., 2021). In combination with organic acids and calcium salts, application of ascorbic acid prevents discoloration and membrane disintegration by reducing PPO activity. Application of ascorbic acid at 200 ppm concentration during postharvest results in reduced physical weight loss, fruit decay, along with higher concentration of POD and CAT activities (Azam et al., 2020). Besides its antibrowning properties, ascorbic acid also has antibacterial properties in different fruit crops, such as in fresh-cut bananas, apples, and papayas. It is also an important ROS scavenger, and endogenous ascorbic acid content can improve potato resistance against *Phytophthora infestans* (Chung et al., 2019).

2.11 CONCLUSIONS

FVs are an integral part of the human diet due to their richness in bioactive compounds. Continuous respiration, ethylene evolution, enzymatic changes, etc. are common physiological activities which lead to certain postharvest changes and quality deterioration. Improper postharvest handling triggers postharvest losses. Use of certain novel sustainable postharvest technology in an integrated manner could minimize this loss. Use of advanced storage techniques not only delay the ripening process but also reduce the incidence of fungal infection. Certain smart packaging technologies could enhance the shelf life inside the packaging itself. However, there is need to understand proper mechanisms, compatibility, and suitability of upcoming technologies such as nano-bio composite coatings, use of LEDs in storage, etc.

REFERENCES

Acheson, D., Steele, J.H. (2001). Food irradiation: a public health challenge for the 21st century. *Clinical Infectious Diseases, 33*(3), 376–377.

Aghdam, M.S., Bodbodak, S. (2013). Physiological and biochemical mechanisms regulating chilling tolerance in fruits and vegetables under postharvest salicylates and jasmonates treatments. *Scientia Horticulturae, 156*, 73–85.

Aghdam, M.S., Mohammadkhani, N. (2014). Enhancement of chilling stress tolerance of tomato fruit by postharvest brassinolide treatment. *Food and Bioprocess Technology, 7*, 909–914.

Alabi, K.P., Fadeyibi, A., Musbaudeen, U. (2023). The forced-air cooling system: effect of fan speeds on the cooling characteristics of some horticultural produce. *Agricultural Engineering International: CIGR Journal, 25*(2).

Alali, A.A., Awad, M.A., Al-Qurashi, A.D., Mohamed, S.A. (2018). Postharvest gum Arabic and salicylic acid dipping affect quality and biochemical changes of 'Grand Nain' bananas during shelf life. *Scientia Horticulturae, 237*, 51–58.

Al-Dairi, M., Pathare, P.B., Al-Yahyai, R., Jayasuriya, H., Al-Attabi, Z. (2023). Postharvest quality, technologies, and strategies to reduce losses along the supply chain of banana: a review. *Trends in Food Science & Technology, 134*, 177–191.

Ali, S., Anjum, M.A., Ejaz, S., Hussain, S., Ercisli, S., Saleem, M.S., Sardar, H. (2021). Carboxymethyl cellulose coating delays chilling injury development and maintains eating quality of 'Kinnow' mandarin fruits during low temperature storage. *International Journal of Biological Macromolecules, 168*, 77–85.

Ali, S., Anjum, M.A., Nawaz, A., Ejaz, S., Anwar, R., Khaliq, G., Hasan, M.U. (2022). Postharvest γ-aminobutyric acid application mitigates chilling injury of aonla (*Emblica officinalis* Gaertn.) fruit during low temperature storage. *Postharvest Biology and Technology, 185*, 111803.

Ali, S., Anjum, M.A., Ullah, S., Nawaz, A., Ejaz, S., Khaliq, G. (2023). Salicylic acid mediated postharvest chilling and disease stress tolerance in horticultural crops. In Ghorbanpour, M., Shahid, M.A. (Eds.), *Plant Stress Mitigators* (pp. 69–85). Academic Press.

Ali, S., Ejaz, S., Anjum, M.A., Nawaz, A., Ahmad, S. (2020). Impact of climate change on postharvest physiology of edible plant products. In Hasanuzzaman, M. (ed.), *Plant Ecophysiology and Adaptation Under Climate Change: Mechanisms and Perspectives I: General Consequences and Plant Responses* (pp. 87–115).

Alothman, M., Bhat, R., Karim, A.A. (2009). Effects of radiation processing on phytochemicals and antioxidants in plant produce. *Trends in Food Science & Technology, 20*(5), 201–212.

Ambaw, A., Mukama, M., Opara, U.L. (2017). Analysis of the effects of package design on the rate and uniformity of cooling of stacked pomegranates: numerical and experimental studies. *Computers and Electronics in Agriculture, 136*, 13–24.

Asrey, R., Sharma, S., Barman, K., Prajapati, U., Negi, N., Meena, N.K. (2023). Biological and postharvest interventions to manage the ethylene in fruit: a review. *Sustainable Food Technology*, *1*, 803–826.

Awad, M.A., de Jager, A. (2003). Influences of air and controlled atmosphere storage on the concentration of potentially healthful phenolics in apples and other fruits. *Postharvest Biology and Technology*, *27*(1), 53–58.

Azam, M., Hameed, L., Qadri, R., Ejaz, S., Aslam, A., Khan, M.I., . . . Anjum, N. (2020). Postharvest ascorbic acid application maintained physiological and antioxidant responses of Guava (*Psidium guajava* L.) at ambient storage. *Food Science and Technology*, *41*, 748–754.

Babalar, M., Asghari, M., Talaei, A., Khosroshahi, A. (2007). Effect of pre-and postharvest salicylic acid treatment on ethylene production, fungal decay and overall quality of Selva strawberry fruit. *Food Chemistry*, *105*(2), 449–453.

Baghel, M., Nagaraja, A., Srivastav, M., Meena, N.K., Senthil Kumar, M., Kumar, A., Sharma, R.R. (2018). Pleiotropic influences of brassinosteroids on fruit crops: a review. *Plant Growth Regulation*, *87*, 375–388.

Balogun, A.A., Ariahu, C.C., Alakali, J.S. (2020). Quality evaluation of mango stored in evaporative coolers. *Current Journal of Applied Science and Technology*, *39*(6), 1–10.

Bambalele, N.L., Mditshwa, A., Magwaza, L.S., Tesfay, S.Z. (2021). Recent advances on postharvest technologies of mango fruit: a review. *International Journal of Fruit Science*, *21*(1), 565–586.

Barkai-Golan, R., Phillips, D.J., 1991. Postharvest heat treatment of fresh fruits and vegetables for decay control. *Plant Disease*, *75*, 1085–1089.

Barrios, S., De Aceredo, A., Chao, G., De Armas, V., Ares, G., Martín, A., . . . Lema, P. (2014). Passive modified atmosphere packaging extends shelf life of enzymatically and vacuum-peeled ready-to-eat valencia orange segments. *Journal of Food Quality*, *37*(2), 135–147.

Barzegar, T., Fateh, M., Razavi, F. (2018). Enhancement of postharvest sensory quality and antioxidant capacity of sweet pepper fruits by foliar applying calcium lactate and ascorbic acid. *Scientia Horticulturae*, *241*, 293–303.

Bautista-Baños, S., Hernandez-Lauzardo, A.N., Velazquez-Del Valle, M.G., Hernández-López, M., Barka, E.A., Bosquez-Molina, E., Wilson, C.L. (2006). Chitosan as a potential natural compound to control pre and postharvest diseases of horticultural commodities. *Crop Protection*, *25*(2), 108–118.

Bender, R.J., Seibert, E., Brecht, J.K. (2003). Heat treatment effects on ACC oxidase activity of 'Keitt' mangoes. *Brazilian Journal of Plant Physiology*, *15*, 145–148.

Bhan, C., Asrey, R., Meena, N.K., Rudra, S.G., Chawla, G., Kumar, R., Kumar, R. (2022). Guar gum and chitosan-based composite edible coating extends the shelf life and preserves the bioactive compounds in stored Kinnow fruits. *International Journal of Biological Macromolecules*, *222*, 2922–2935.

Bintsis, T. (2017). Foodborne pathogens. *AIMS Microbiology*, *3*(3), 529.

Bodbodak, S., Moshfeghifar, M. (2016). Advances in controlled atmosphere storage of fruits and vegetables. In Siddiqui, M.W. (Ed.), *Eco-Friendly Technology for Postharvest Produce Quality* (pp. 39–76). Academic Press.

Brosnan, T., Sun, D.W. (2001). Precooling techniques and applications for horticultural products—a review. *International Journal of Refrigeration*, *24*(2), 154–170.

Cai, H., Han, S., Yu, M., Ma, R., Yu, Z. (2020). The alleviation of methyl jasmonate on loss of aroma lactones correlated with ethylene biosynthesis in peaches. *Journal of Food Science*, *85*(8), 2389–2397.

Chen, M., Jiang, Q., Yin, X.R., Lin, Q., Chen, J.Y., Allan, A.C., Xu, C.J., Chen, K.S. (2012). Effect of hot air treatment on organic acid-and sugar-metabolism in Ponkan (*Citrus reticulata*) fruit. *Scientia Horticulturae*, *147*, 118–125.

Cheong, J.J., Do Choi, Y. (2003). Methyl jasmonate as a vital substance in plants. *Trends in Genetics*, *19*(7), 409–413.

Chung, I.M., Venkidasamy, B., Upadhyaya, C.P., Packiaraj, G., Rajakumar, G., Thiruvengadam, M. (2019). Alleviation of *Phytophthora infestans* mediated necrotic stress in the transgenic potato (*Solanum tuberosum* L.) with enhanced ascorbic acid accumulation. *Plants*, *8*(10), 365.

Constán-Aguilar, C., Leyva, R., Romero, L., Soriano, T., Ruiz, J.M. (2014). Implication of potassium on the quality of cherry tomato fruits after postharvest during cold storage. *International Journal of Food Sciences and Nutrition*, *65*(2), 203–211.

Crowe, K.M., Bushway, A.A., Bushway, R.J., Davis-Dentici, K., Hazen, R.A. (2007). A comparison of single oxidants versus advanced oxidation processes as chlorine-alternatives for wild blueberry processing (*Vaccinium angustifolium*). *International Journal of Food Microbiology*, *116*, 25–31.

Davies, A.R. (1995). Advances in modified-atmosphere packaging. In *New Methods of Food Preservation* (pp. 304–320). Boston, MA: Springer US.

Dehghannya, J., Ngadi, M., Vigneault, C. (2010). Mathematical modeling procedures for airflow, heat and mass transfer during forced convection cooling of produce: a review. *Food Engineering Reviews*, *2*, 227–243.

Delele, M.A., Ngcobo, M.E.K., Getahun, S.T., Chen, L., Mellmann, J., Opara, U.L. (2013). Studying airflow and heat transfer characteristics of a horticultural produce packaging system using a 3-D CFD model. Part I: model development and validation. *Postharvest Biology and Technology*, *86*, 536–545.

Desjardins, Y. (2008). Physiological and ecological functions and biosynthesis of health-promoting compounds in fruit and vegetables. *Tomás BFA y Gil MI, Improving the Health-Promoting Properties of Fruit and Vegetable Products, Edit. Elsevier, USA*, *23*, 201–247.

de Vasconcellos Santos Batista, D., Reis, R.C., Almeida, J.M., Rezende, B., Bragança, C.A.D., da Silva, F. (2020). Edible coatings in post-harvest papaya: impact on physical–chemical and sensory characteristics. *Journal of Food Science and Technology*, *57*(1), 274–281.

Dhakal, R., Baek, K.H. (2014). Short period irradiation of single blue wavelength light extends the storage period of mature green tomatoes. *Postharvest Biology and Technology*, *90*, 73–77.

Dhami, K.S., Asrey, R., Vinod, B.R., Meena, N.K. (2023). Postharvest methyl jasmonate alleviates chilling injury and maintains quality of 'kinnow'(*citrus nobilis* lour *x C. deliciosa* tenora) fruits under differential storage temperature. *Erwerbs-Obstbau*, 1–8.

Dick, M., Costa, T.M.H., Gomaa, A., Subirade, M., de Oliveira Rios, A., Flôres, S.H. (2015). Edible film production from chia seed mucilage: effect of glycerol concentration on its physicochemical and mechanical properties. *Carbohydrate Polymers*, *130*, 198–205.

D'innocenzo, M., Lajolo, F.M. (2001). Effect of gamma irradiation on softening changes and enzyme activities during ripening of papaya fruit. *Journal of Food Biochemistry*, *25*(5), 425–438.

Duan, Y., Wang, G.B., Fawole, O.A., Verboven, P., Zhang, X.R., Wu, D., Opara, U.L., Nicolai, B., Chen, K. (2020). Postharvest precooling of fruit and vegetables: a review. *Trends in Food Science & Technology*, *100*, 278–291.

Elansari, A.M. (2008). Hydrocooling rates of Barhee dates at the Khalal stage. *Postharvest Biology and Technology*, *48*(3), 402–407.

Elansari, A.M., Fenton, D.L., Callahan, C.W. (2019). Precooling. In E.M. Yahia (Ed.), *Postharvest Technology of Perishable*

Horticultural Commodities (pp. 161–207). Woodhead Publishing.

El-Beltagi, H.S., Ali, M.R., Ramadan, K.M., Anwar, R., Shalaby, T.A., Rezk, A.A., . . . El-Mogy, M.M. (2022). Exogenous postharvest application of calcium chloride and salicylic acid to maintain the quality of broccoli florets. *Plants*, *11*(11), 1513.

Fallik, E. (2004). Prestorage hot water treatments (immersion, rinsing and brushing). *Postharvest Biology and Technology*, *32*(2), 125–134.

Finnegan, E., Mahajan, P.V., O'Connell, M., Francis, G.A., O'Beirne, D. (2013). Modelling respiration in fresh-cut pineapple and prediction of gas permeability needs for optimal modified atmosphere packaging. *Postharvest Biology and Technology*, *79*, 47–53.

Fuentes, L., Figueroa, C.R., Valdenegro, M. (2019). Recent advances in hormonal regulation and cross-talk during non-climacteric fruit development and ripening. *Horticulturae*, *5*(2), 45.

Fujioka, S., Yokota, T. (2003). Biosynthesis and metabolism of brassinosteroids. *Annual Review of Plant Biology*, *54*(1), 137–164.

Gambetta, G.A., Matthews, M.A., Shaghasi, T.H., McElrone, A.J., Castellarin, S.D. 2010. Sugar and abscisic acid signaling orthologs are activated at the onset of ripening in grape. *Planta*, *232*, 219–234.

Gao, H., Kang, L., Liu, Q., Cheng, N., Wang, B., Cao, W. (2015). Effect of 24-epibrassinolide treatment on the metabolism of eggplant fruits in relation to development of pulp browning under chilling stress. *Journal of Food Science and Technology*, *52*, 3394–3401.

Gao, H., Zhang, Z., Lv, X., Cheng, N., Peng, B., Cao, W. (2016). Effect of 24-epibrassinolide on chilling injury of peach fruit in relation to phenolic and proline metabolisms. *Postharvest Biology and Technology*, *111*, 390–397.

Gao, H.Y., Chen, H.J., Chen, W.X., Yang, Y.T., Song, L.L., Jiang, Y.M., Zheng, Y.H. (2006). Effect of hypobaric storage on physiological and quality attributes of loquat fruit at low temperature. *Acta Horticulturae*, *712*, 269–274.

García, J.M., Herrera, S., Morilla, A. (1996). Effects of postharvest dips in calcium chloride on strawberry. *Journal of Agricultural and Food Chemistry*, *44*(1), 30–33.

Garrido, Y., Tudela, J.A., Gil, M.I. (2015). Comparison of industrial precooling systems for minimally processed baby spinach. *Postharvest Biology and Technology*, *102*, 1–8.

Ghasil, I., Meena, N.K., Bhatnagar, P., Singh, H., Kumar, A., Jain, S.K., Bhateshwar, M. (2022). Effect of postharvest treatments on biochemical and bioactive compounds of custard apple (*Annona Squamosa* L.) Cv. Balanagar. *International Journal of Fruit Science*, *22*(1), 826–836.

Glowacz, M., Roets, N., Sivakumar, D. (2017). Control of anthracnose disease via increased activity of defence related enzymes in 'Hass' avocado fruit treated with methyl jasmonate and methyl salicylate. *Food Chemistry*, *234*, 163–167.

Gong, D., Cao, S., Sheng, T., Shao, J., Song, C., Wo, F., Chen, W., Yang, Z. (2015). Effect of blue light on ethylene biosynthesis, signalling and fruit ripening in postharvest peaches. *Scientia Horticulturae*, *197*, 657–664.

Greene, A.K., Güzel-Seydim, Z.B., Seydim, A.C. (2012). Chemical and physical properties of ozone. In O'Donnell, C., Tiwari, B.K., Cullen, P.J., Rice, R.G. (Eds.), *Ozone in Food Processing* (1st ed., pp. 19–32). Blackwell Publishing Ltd.

Gupta, V., Thakur, R., Barik, M., Das, A.B. (2023). Effect of high amylose starch-natural deep eutectic solvent based edible coating on quality parameters of strawberry during storage. *Journal of Agriculture and Food Research*, *11*, 100487.

Hallman, G.J., Blackburn, C.M. (2016). Phytosanitary irradiation. *Foods*, *5*(1), 8.

Hardenburg, R.E., Watada, A.E., Wang, C.Y. (1986). The commercial storage of fruits, vegetables, and florist and nursery stocks. *Agriculture Handbook*, *66*.

Hasan, F.H., Tajidin, N.E., Ahmad, S.H., Mohamed, M.T.M., Shukor, N.I.A. (2018). Ripening characteristics of vapour heat treated 'Frangi' papaya (*Carica papaya* L. cv. Frangi) as affected by maturity stages and ethylene treatment. *Bragantia*, *77*, 372–384.

Hashemi, S.M.B., Mousavi Khaneghah, A., Ghaderi Ghahfarrokhi, M., Eş, I. (2017). Basil-seed gum containing *Origanum vulgare* subsp. viride essential oil as edible coating for fresh cut apricots. *Postharvest Biology and Technology*, *125*, 26–34.

He, Y., Li, J., Ban, Q., Han, S., Rao, J. (2018). Role of brassinosteroids in persimmon (*Diospyros kaki* L.) fruit ripening. *Journal of Agricultural and Food Chemistry*, *66*(11), 2637–2644.

Hernández-Muñoz, P., Almenar, E., Ocio, M.J., Gavara, R. (2006). Effect of calcium dips and chitosan coatings on postharvest life of strawberries (*Fragaria* x *ananassa*). *Postharvest Biology and Technology*, *39*(3), 247–253.

Ho, L.C., White, P.J. (2005). A cellular hypothesis for the induction of blossom-end rot in tomato fruit. *Annals of Botany*, *95*(4), 571–581.

Hoehn, E., Prange, R.K., Vigneault, C. (2009). Storage technology and applications. In Yahia, E.M. (Ed.), *Modified and Controlled Atmospheres for the Storage, Transportation, and Packaging of Horticultural Commodities* (pp. 35–68). CRC Press.

Holt, J.E., Schoorl, D., Muirhead, I.F. (1983). Post-harvest quality control strategies for fruit and vegetables. *Agricultural Systems*, *10*(1), 21–37.

Javed, S., Fu, H., Ali, A., Nadeem, A., Amin, M., Razzaq, K., Hussain, S.B. (2022). Comparative response of mango fruit towards pre- and post-storage quarantine heat treatments. *Agronomy*, *12*(6), 1476.

Jha, S.N., Chopra, S. (2006). Selection of bricks and cooling pad for construction of evaporatively cooled storage structure. *Inst Engineers (I)(AG)*, *87*, 25–28.

Kale, S.J., Nath, P., Jalgaonkar, K.R., Mahawar, M.K. (2016). Low cost storage structures for fruits and vegetables handling in Indian conditions. *Indian Horticulture Journal*, *6*(3), 376–379.

Kalyani, B., Manjula, K. (2014). Food irradiation–technology and application. *International Journal of Current Microbiology and Applied Sciences*, *3*(4), 549–555.

Kaur, J., Aslam, R., Saeed, P.A. (2021). Storage structures for horticultural crops: a review. *Environment Conservation Journal, 22(SE)*, 95–105.

Kazemi, M., Hadavi, E., Hekmati, J. (2011). Role of salicylic acid in decreases of membrane senescence in cut. *American Journal of Plant Physiology*, *6*, 106–112.

Khanbarad, S.C., Patil, N., Sutar, R.F., Joshi, D.C. (2013). Studies on pre-cooling of mango for extension of shelf-life. *Journal of Agricultural Engineering*, *50*(4), 23–31.

Kitinoja, L., Thompson, J.F. (2010). Pre-cooling systems for small-scale producers. *Stewart Postharvest Review*, *6*(2), 1–14.

Kongwong, P., Boonyakiat, D., Poonlarp, P. (2019). Extending the shelf life and qualities of baby cos lettuce using commercial precooling systems. *Postharvest Biology and Technology*, *150*, 60–70.

Koushesh saba, M., Arzani, K., Barzegar, M. (2012). Postharvest polyamine application alleviates chilling injury and affects apricot storage ability. *Journal of Agricultural and Food Chemistry*, *60*(36), 8947–8953.

Koyama, K., Sadamatsu, K., Goto-Yamamoto, N. (2010). Abscisic acid stimulated ripening and gene expression in berry skins of the Cabernet Sauvignon grape. *Functional and Integrative Genomics*, *10*, 367–381.

Kumari, M., Mahajan, H., Joshi, R., Gupta, M. (2017). Development and structural characterization of edible films for improving fruit quality. *Food Packaging and Shelf Life*, *12*, 42–50. https://doi.org/10.1016/j.fpsl.2017.02.003

Le, T., Shiesh, C., Lin, H., Lee, E. (2010). Vapour heat quarantine treatment for Taiwan native mango variety fruits infested with fruit fly. *Journal of Applied Horticulture*, *12*(2), 107–112.

Leon, J., Yalpani, N., Raskin, I., Lawton, M.A. (1993). Induction of benzoic acid 2-hydroxylase in virus-inoculated tobacco. *Plant Physiology*, *103*(2), 323–328.

Lestari, T.W.W., Wijonarko, A., Murdita, W., Suputa, S. (2017). Effect of vapor heat treatment on the mortality of *Bactrocera dorsalis* (Diptera: Tephritidae) and the quality of mango cv. Arumanis. *Jurnal Perlindungan Tanaman Indonesia*, *21*(1), 38–46.

Li, L., Guo, Y., Nian, B. (2014). Responses of postharvest broccoli (*Brassica oleracea* L. var. italica) florets to controlled atmospheres with varying CO_2/O_2 levels at different temperature. *Journal of Applied Botany and Food Quality*, *87*.

Li, W., Zhang, M., Yu, H.Q. (2006). Study on hypobaric storage of green asparagus. *Journal of Food Engineering*, *73*(3), 225–230.

Liang, Y., Ji, L., Chen, C., Dong, C., Wang, C. (2018). Effects of ozone treatment on the storage quality of post-harvest tomato. *International Journal of Food Engineering*, *14*(7–8), 20180012.

Liang, Y.S., Wongmetha, O., Wu, P.S., Ke, L.S. (2013). Influence of hydrocooling on browning and quality of litchi cultivar Feizixiao during storage. *International Journal of Refrigeration*, *36*(3), 1173–1179.

Lu, X.H., Sun, D.Q., Mo, Y.W., Xi, J.G., Sun, G.M. (2010). Effects of post-harvest salicylic acid treatment on fruit quality and anti-oxidant metabolism in pineapple during cold storage. *The Journal of Horticultural Science and Biotechnology*, *85*(5), 454–458.

Lu, Z., Wang, X., Cao, M., Li, Y., Su, J., Gao, H. (2019). Effect of 24-epibrassinolide on sugar metabolism and delaying postharvest senescence of kiwifruit during ambient storage. *Scientia Horticulturae*, *253*, 1–7.

Luengwilai, K., Beckles, D.M., Saltveit, M.E. (2012). Chilling-injury of harvested tomato (*Solanum lycopersicum* L.) cv. Micro-Tom fruit is reduced by temperature pre-treatments. *Postharvest Biology and Technology*, *63*, 123–128.

Luo, A., Bai, J., Li, R., Fang, Y., Li, L., Wang, D., . . . Kou, L. (2019). Effects of ozone treatment on the quality of kiwifruit during postharvest storage affected by *Botrytis cinerea* and *Penicillium expansum*. *Journal of Phytopathology*, *167*(7–8), 470–478.

Lurie, S. (1998). Postharvest heat treatments. *Postharvest Biology and Technology*, *14*, 257–269.

Ma, Y., Li, S., Yin, X., Xing, Y., Lin, H., Xu, Q., . . . Chen, C. (2019). Effects of controlled atmosphere on the storage quality and aroma compounds of lemon fruits using the designed automatic control apparatus. *BioMed Research International*, *2019*.

Mahajan, B.V.C., Dhatt, A.S. (2004). Studies on postharvest calcium chloride application on storage behaviour and quality of Asian pear during cold storage. *Journal of Food, Agriculture and Environment*, *2*(3–4), 157–159.

Makule, E., Dimoso, N., Tassou, S.A. (2022). Precooling and cold storage methods for fruits and vegetables in Sub-Saharan Africa-A review. *Horticulturae*, *8*(9), 776.

Malik, A.U., Hasan, M.U., Hassan, W.U., Khan, A.S., Shah, M.S., Rajwana, I.A., Anwar, R. (2021). Postharvest quarantine vapour heat treatment attenuates disease incidence, maintains eating quality and improves bioactive compounds of 'Gola' and 'Surahi' guava fruits. *Journal of Food Measurement and Characterization*, *15*, 1666–1679.

Manganaris, G.A., Vasilakakis, M., Diamantidis, G., Mignani, I.J.F.C. (2007). The effect of postharvest calcium application on tissue calcium concentration, quality attributes, incidence of flesh browning and cell wall physicochemical aspects of peach fruits. *Food Chemistry*, *100*(4), 1385–1392.

Mansourbahmani, S., Ghareyazie, B., Kalatejari, S., Mohammadi, R.S., Zarinnia, V. (2017). Effect of post-harvest UV-C irradiation and calcium chloride on enzymatic activity and decay of tomato (*Lycopersicon esculentum* L.) fruit during storage. *Journal of Integrative Agriculture*, *16*(9), 2093–2100.

Maryam, A., Anwar, R., Malik, A.U., Khan, A.S., Ali, S., Waris, F., Hasan, M.U., El-Mogy, M.M. (2022). Pre-storage hypobaric treatment reduces microbial spoilage and maintains eating quality of strawberry fruits during low temperature conditions. *Journal of Food Processing and Preservation*, *46*(11), e17121.

Matrose, N.A., Obikeze, K., Belay, Z.A., Caleb, O.J. (2021). Plant extracts and other natural compounds as alternatives for postharvest management of fruit fungal pathogens: a review. *Food Bioscience*, *41*, 100840.

Meena, N.K., Asrey, R. (2018). Tree age affects physicochemical, functional quality and storability of Amrapali mango (*Mangifera indica* L.) fruits. *Journal of the Science of Food and Agriculture*, *98*(9), 3255–3262.

Meena, N.K., Asrey, R., Singh, S., Singh, D., Paul, V. (2023). Exogenous application of brassinolide affects ripening, phenotypic traits, jelly seed and sensory properties of 'Dashehari' mango at ambient storage. *Annals of Plant and Soil Research*, *25*(2), 205–210.

Meena, N.K., Choudhary, K. (2019). Harnessing the recent approaches in postharvest quality retention of fruits. In *Modern Fruit Industry*. IntechOpen.

Meena, N.K., Choudhary, K. (2022). Storage methods of horticultural crops. In *Postharvest Management of Horticultural Crops*. Jaya Publishing House.

Mendy, T.K., Misran, A., Mahmud, T.M.M., Ismail, S.I. (2019). Antifungal properties of Aloe vera through in vitro and in vivo screening against postharvest pathogens of papaya fruit. *Scientia Horticulturae*, *257*, 108767.

Mestre, T.C., Garcia-Sanchez, F., Rubio, F., Martinez, V., Rivero, R.M. (2012). Glutathione homeostasis as an important and novel factor controlling blossom-end rot development in calcium-deficient tomato fruits. *Journal of Plant Physiology*, *169*(17), 1719–1727.

Minas, I.S., Vicente, A.R., Dhanapal, A.P., Manganaris, G.A., Goulas, V., Vasilakakis, M., . . . Molassiotis, A. (2014). Ozone-induced kiwifruit ripening delay is mediated by ethylene biosynthesis inhibition and cell wall dismantling regulation. *Plant Science*, *229*, 76–85.

Mirdehghan, S.H., Rahemi, M., Martínez-Romero, D., Guillén, F., Valverde, J.M., Zapata, P.J., . . . Valero, D. (2007). Reduction of pomegranate chilling injury during storage after heat treatment: role of polyamines. *Postharvest Biology and Technology*, *44*(1), 19–25.

Mishra, A., Jha, S.K., Ojha, P. (2020). Study on Zero energy cool chamber (ZECC) for storage of vegetables. *International Journal of Scientific and Research Publications*, *10*(1), 427–433.

Mostafavi, H.A., Fathollahi, H., Motamedi, F. (2010). Food irradiation: applications, public acceptance and global trade. *African Journal of Biotechnology*, *9*(20).

Mustafa, M.A., Ali, A., Seymour, G., Tucker, G. (2018). Delayed pericarp hardening of cold stored mangosteen (*Garcinia mangostana* L.) upon pre-treatment with the stress hormones methyl jasmonate and salicylic acid. *Scientia Horticulturae*, *230*, 107–116.

Nawaz, F., Naeem, M., Zulfiqar, B., Akram, A., Ashraf, M.Y., Raheel, M., Aurangzaib, M. (2017). Understanding brassinosteroid-regulated mechanisms to improve stress tolerance in plants: a critical review. *Environmental Science and Pollution Research*, *24*(19), 15959–15975.

Ngamchuachit, P., Barrett, D.M., Mitcham E.J. (2014). Effects of 1-methylcyclopropene and hot water quarantine treatment on quality of "Keitt" mangos. *Journal of Food Science*, *79*, 505–509.

Nikagolla, N.G.D.N., Udugala-Ganehenege, M.Y., Daundasekera, W.A.M. (2019). Postharvest application of potassium silicate improves keeping quality of banana. *The Journal of Horticultural Science and Biotechnology*, *94*(6), 735–743.

Pandiselvam, R., Subhashini, S., Banuu Priya, E.P., Kothakota, A., Ramesh, S.V., Shahir, S. (2019). Ozone based food preservation: a promising green technology for enhanced food safety. *Ozone: Science & Engineering*, *41*(1), 17–34.

Parven, A., Sarker, Md. R., Megharaj, M., Md. Meftaul, I. (2020). Prolonging the shelf life of Papaya (*Carica papaya* L.) using Aloe vera gel at ambient temperature. *Scientia Horticulturae*, *265*, 109228. https://doi.org/10.1016/j.scienta.2020.109228

Pathare, P.B., Opara, U.L., Vigneault, C., Delele, M.A., Al-Said, F.A.J. (2012). Design of packaging vents for cooling fresh horticultural produce. *Food and Bioprocess Technology*, *5*, 2031–2045.

Perera, W.P.T.D., Navaratne, S., Wickramasinghe, I. (2022). Impact of spectral composition of light from light-emitting diodes (LEDs) on postharvest quality of vegetables: a review. *Postharvest Biology and Technology*, *191*, 111955.

Petruzzi, L., Campaniello, D., Speranza, B., Corbo, M.R., Sinigaglia, M., Bevilacqua, A. (2017). Thermal treatments for fruit and vegetable juices and beverages: a literature overview. *Comprehensive Reviews in Food Science and Food Safety*, *16*(4), 668–691.

Poonia, A., Pandey, S., Vasundhara. (2022). Application of light emitting diodes (LEDs) for food preservation, post-harvest losses and production of bioactive compounds: a review. *Food Production, Processing and Nutrition*, *4*(1), 8.

Pourcel, L., Routaboul, J.M., Cheynier, V., Lepiniec, L., Debeaujon, I. (2007). Flavonoid oxidation in plants: from biochemical properties to physiological functions. *Trends in Plant Science*, *12*(1), 29–36.

Prakash, O.M., Pandey, B.K. (2000). Control of mango anthracnose by hot water and fungicides treatment. *Indian Phytopathology*, *53*(1), 92–94.

Prusky, D. (2011). Reduction of the incidence of postharvest quality losses, and future prospects. *Food Security*, *3*, 463–474.

Qin, G.H., Wei, S.W., Tao, S.T., Zhang, H.P., Huang, W.J., Yao, G.F., Zhang, S.L. (2017). Effects of postharvest methyl jasmonate treatment on aromatic volatile biosynthesis by 'Nanguoli'fruit at different harvest maturity stages. *New Zealand Journal of Crop and Horticultural Science*, *45*(3), 191–201.

Rabiei, V., Shirzadeh, E., Sharafi, Y., Mortazavi, N. (2011). Effects of postharvest applications of calcium nitrate and acetate on quality and shelf-life improvement of "Jonagold" apple fruit. *Journal of Medicinal Plants Research*, *5*(19), 4912–4917.

Ranjbaran, E., Sarikhani, H., Wakana, A., Bakhshi, D. (2011). Effect of salicylic acid on storage life and postharvest quality of grape (Vitis vinifera L. cv. Bidaneh Sefid). *Journal of the Faculty of Agriculture*, Kyushu University, *56*(2), 263–269. https://doi.org/10.5109/20319

Rao, C.G. (2015a). *Engineering for Storage of Fruits and Vegetables: Forced-Air Cooling of Fruits and Vegetables* (pp. 71–83). Academic Press.

Rao, C.G. (2015b). *Engineering for Storage of Fruits and Vegetables: Hydrocooling of Fruits, Vegetables, and Cut Flowers* (pp. 85–110). Academic Press.

Rao, S.S.R., Vardhini, B.V., Sujatha, E., Anuradha, S. (2002). Brassinosteroids–a new class of phytohormones. *Current Science*, 1239–1245.

Rashid, M.H.A., Grout, B.W.W., Continella, A., Mahmud, T.M.M. (2015). Low-dose gamma irradiation following hot water immersion of papaya (*Carica papaya* Linn.) fruits provides additional control of postharvest fungal infection to extend shelf life. *Radiation Physics and Chemistry*, *110*, 77–81.

Reddy, P.H., Yin, X., Manczak, M., Kumar, S., Pradeepkiran, J.A., Vijayan, M., Reddy, A.P. (2018). Mutant APP and amyloid beta-induced defective autophagy, mitophagy, mitochondrial structural and functional changes and synaptic damage in hippocampal neurons from Alzheimer's disease. *Human Molecular Genetics*, *27*(14), 2502–2516.

Reddy, V.R.S., Sharma, R.R., Raghavendra, H.R., Prajapati, U. (2021). Microbiological efficacy of decontamination methodologies for fresh produce. In *Food Security and Plant Disease Management* (pp. 323–355). Elsevier.

Rice, R.G., Farquhar, J.W., Bollyky, L.J. (1982). Reviewof the applications of ozone for increasing storage times of perishables foods. *Ozone-Science and Engineering*, *4*, 147–163.

Rodgers, S.L., Cash, J.N., Siddiq, M., Ryser, E.T. (2004). A comparison of different chemical sanitizers for inactivating *Escherichia coli* O157:H7 and *Listeria monocytogenes* in solution and on apples, lettuce, strawberries, and cantaloupe. *Journal of Food Protection*, *67*(4), 721–731.

Rodrigues, J.P., de Souza Coelho, C.C., Soares, A.G., Freitas-Silva, O. (2021). Current technologies to control fungal diseases in postharvest papaya (*Carica papaya* L.). *Biocatalysis and Agricultural Biotechnology*, *36*, 102128.

Ranjbaran, E., Sarikhani, H., Wakana, A., Bakhshi, D. (2011). Effect of salicylic acid on storage life and postharvest quality of grape (Vitis vinifera L. cv. Bidaneh Sefid). Journal of the Faculty of Agriculture, Kyushu University, 56(2), 263-269. https://doi.org/10.5109/20319

Ryall, A.L., Pentzer, W.T. (1982). *Handling, Transportation and Storage of Fruits and Vegetables. Volume 2. Fruits and Tree Nuts*. AVI Publishing Co., Inc.

Salamat, R., Ghassemzadeh, H.R., Ranjbar, F., Jalali, A., Mahajan, P., Herppich, W.B., Mellmann, J. (2020). The effect of additional packaging barrier, air moment and cooling rate on quality parameters of button mushroom (*Agaricus bisporus*). *Food Packaging and Shelf Life*, *23*, 100448.

Saleem, M.S., Anjum, M.A., Naz, S., Ali, S., Hussain, S., Azam, M., Ejaz, S. (2021). Incorporation of ascorbic acid in chitosan-based edible coating improves postharvest quality and storability of strawberry fruits. *International Journal of Biological Macromolecules*, *189*, 160–169.

Salehi, F. (2020). Edible coating of fruits and vegetables using natural gums: a review. *International Journal of Fruit Science*, *20*(Suppl. 2), S570–S589. https://doi.org/10.1080/15538362.2020.1746730

Sánchez-Mata, M.C., Camara, M., Dıez-Marques, C. (2003). Extending shelf-life and nutritive value of green beans (*Phaseolus vulgaris* L.), by controlled atmosphere storage: macronutrients. *Food Chemistry*, *80*(3), 309–315.

Sanchís, E., González, S., Ghidelli, C., Sheth, C.C., Mateos, M., Palou, L., Pérez-Gago, M.B. (2016). Browning inhibition and microbial control in fresh-cut persimmon (*Diospyros kaki* Thunb. cv. Rojo Brillante) by apple pectin-based edible coatings. *Postharvest Biology and Technology*, *112*, 186–193.

Saure, M.C. (1996). Reassessment of the role of calcium in development of bitter pit in apple. *Functional Plant Biology*, *23*(3), 237–243.

Sayyari, M., Babalar, M., Kalantari, S., Serrano, M., Valero, D. (2009). Effect of salicylic acid treatment on reducing chilling injury in stored pomegranates. *Postharvest Biology and Technology*, *53*(3), 152–154.

Schirra, M., D'Hallewin, G., Ben-Yehoshua, S., Fallik, E. (2000). Host–pathogen interactions modulated by heat treatment. *Postharvest Biology and Technology*, *21*, 71–85.

Selcuk, N., Erkan, M. (2014). Changes in antioxidant activity and postharvest quality of sweet pomegranates cv. Hicrannar under modified atmosphere packaging. *Postharvest Biology and Technology*, *92*, 29–36.

Sellamuthu, P.S., Mafune, M., Sivakumar, D., Soundy, P. (2013). Thyme oil vapour and modified atmosphere packaging reduce anthracnose incidence and maintain fruit quality in avocado. *Journal of the Science of Food and Agriculture*, *93*(12), 3024–3031.

Sena, E.D.O.A., da Silva, P.S.O., de Araujo, H.G.S., de Aragão Batista, M.C., Matos, P.N., Sargent, S A., Carnelossi, M.A.G. (2019). Postharvest quality of cashew apple after hydrocooling and coold room. *Postharvest Biology and Technology*, *155*, 65–71.

Seo, H.S., Song, J.T., Cheong, J.J., Lee, Y.H., Lee, Y.W., Hwang, I., . . . Choi, Y.D. (2001). Jasmonic acid carboxyl methyltransferase: a key enzyme for jasmonate-regulated plant responses. *Proceedings of the National Academy of Sciences*, *98*(8), 4788–4793.

Setagane, L., Mafeo, T.P., Mathaba, N., Shikwambana, K. (2021). Mitigation of chilling injury with hot water treatment to improve early-season 'HASS'avocado (*Persea americana*) fruit peel colour. *Research on Crops*, *22*(1), 60–67.

Shehata, S.A., El-Mogy, M.M., Mohamed, H.F. (2019). Postharvest quality and nutrient contents of long sweet pepper enhanced by supplementary potassium foliar application. *International Journal of Vegetable Science*, *25*(2), 196–209.

Singh, R., Giri, S.K., Kotwaliwale, N. (2014). Shelf-life enhancement of green bell pepper (*Capsicum annuum* L.) under active modified atmosphere storage. *Food Packaging and Shelf Life*, *1*(2), 101–112.

Singh, S.P., Saini, M.K. (2014). Postharvest vapour heat treatment as a phytosanitary measure influences the aroma volatiles profile of mango fruit. *Food Chemistry*, *164*, 387–395.

Sivakumar, D., Korsten, L. (2010). Fruit quality and physiological responses of litchi cultivar McLean's Red to 1-methylcyclopropene pre-treatment and controlled atmosphere storage conditions. *LWT-Food Science and Technology*, *43*(6), 942–948.

Skog, C.L., Chu, L.J. (2001). Effect of ozone on qualities of fruits and vegetables in cold storage. *Canadian Journal of Plant Science*, *81*(4), 773–778.

Soares, A.G., Trugo, L.C., Botrel, N., da Silva Souza, L.F. (2005). Reduction of internal browning of pineapple fruit (*Ananas comusus* L.) by preharvest soil application of potassium. *Postharvest Biology and Technology*, *35*(2), 201–207.

Sudhakar Rao, D.V., Shivashankara, K.S. (2018). Effect of modified atmosphere packaging on the extension of storage life and quality maintenance of pomegranate (cv.'Bhagwa') at ambient and low temperatures. *Journal of Food Science and Technology*, *55*, 2103–2113.

Tavallali, V. (2018). Vacuum infiltration of 24-epibrassinolide delays chlorophyll degradation and maintains quality of lime during cold storage. *Acta Scientiarum Polonorum Hortorum Cultus*, *17*(1), 35–48.

Thakur, R., Pristijono, P., Golding, J.B., Stathopoulos, C.E., Scarlett, C.J., Bowyer, M., . . . Vuong, Q.V. (2018). Development and application of rice starch based edible coating to improve the postharvest storage potential and quality of plum fruit (*Prunus salicina*). *Scientia Horticulturae*, *237*, 59–66.

Toti, M., Carboni, C., Botondi, R. (2018). Postharvest gaseous ozone treatment enhances quality parameters and delays softening in cantaloupe melon during storage at 6 °C. *Journal of the Science of Food and Agriculture*, *98*(2), 487–494.

Vargas, M., Pastor, C., Chiralt, A., McClements, D.J., Gonzalez-Martinez, C. (2008). Recent advances in edible coatings for fresh and minimally processed fruits. *Critical Reviews in Food Science and Nutrition*, *48*(6), 496–511.

Vilaplana, R., Chicaiza, G., Vaca, C., Valencia-Chamorro, S. (2020). Combination of hot water treatment and chitosan coating to control anthracnose in papaya (*Carica papaya* L.) during the postharvest period. *Crop Protection*, *128*, 105007. https://doi.org/10.1016/j.cropro.2019.105007

Villalobos, M.D.C., Serradilla, M.J., Martín, A., Ruiz-Moyano, S., Pereira, C., Córdoba, M.D.G. (2014). Use of equilibrium modified atmosphere packaging for preservation of 'San Antonio' and 'Banane' breba crops (*Ficus carica* L.) *Postharvest Biology and Technology*, *98*, 14–22.

Villarreal-Lozoya, J.E., Lombardini, L., Cisneros-Zevallos, L. (2009). Electron-beam irradiation effects on phytochemical constituents and antioxidant capacity of pecan kernels [*Carya illinoinensis* (Wangenh.) K. Koch] during storage. *Journal of Agricultural and Food Chemistry*, *57*(22), 10732–10739.

Vinod, B.R., Asrey, R., Sethi, S., Prakash, J., Meena, N.K., Menaka, M., Shivaswamy, G. (2023). Recent advances in physical treatments of papaya fruit for postharvest quality retention: a review. *eFood*, *4*(2), e79.

Wang, K.L.C., Li, H., Ecker, J.R., 2002. Ethylene biosynthesis and signaling networks. *Plant Cell*, *14*, S131–S151.

Wang, L., Sun, D.W. (2003). Recent developments in numerical modelling of heating and cooling processes in the food industry-a review. *Trends in Food Science & Technology*, *14*(10), 408–423.

Wani, S.H., Kumar, V., Shriram, V., Sah, S.K. (2016). Phytohormones and their metabolic engineering for abiotic stress tolerance in crop plants. *The Crop Journal*, *4*(3), 162–176.

Wei, Y., Liu, Z., Su, Y., Liu, D., Ye, X. (2011). Effect of salicylic acid treatment on postharvest quality, antioxidant activities, and free polyamines of asparagus. *Journal of Food Science*, *76*(2), S126–S132.

Wei, Y., Zhou, D., Peng, J., Pan, L., Tu, K. (2017). Hot air treatment induces disease resistance through activating the phenylpropanoid metabolism in cherry tomato fruit. *Journal of Agricultural and Food Chemistry*, *65*(36), 8003–8010.

Wheeler, G.L., Jones, M.A., Smirnoff, N. (1998). The biosynthetic pathway of vitamin C in higher plants. *Nature*, *393*(6683), 365–369.

Wildermuth, M.C., Dewdney, J., Wu, G., Ausubel, F.M. (2001). Isochorismate synthase is required to synthesize salicylic acid for plant defence. *Nature*, *414*(6863), 562–565.

Wills, R., McGlasson, B., Graham, D., Joyce, D. (1998). *Postharvest: An Introduction to the Physiology and Handling of Fruit, Vegetables, and Ornamentals*. UNSW Press.

Xiang, W., Wang, H.W., Sun, D.W. (2021). Phytohormones in postharvest storage of fruit and vegetables: mechanisms and applications. *Critical Reviews in Food Science and Nutrition*, *61*(18), 2969–2983.

Xin, Y., Chen, F., Lai, S., Yang, H. (2017). Influence of chitosan-based coatings on the physicochemical properties and pectin nanostructure of Chinese cherry. *Postharvest Biology and Technology*, *133*, 64–71.

Xu, F., Yang, Z., Chen, X., Jin, P., Wang, X., Zheng, Y. (2012). 6-Benzylaminopurine delays senescence and enhances health-promoting compounds of harvested broccoli. *Journal of Agricultural and Food Chemistry*, *60*(1), 234–240.

Yahia, E.M. (2016). *Irradiation of Produce: What for, How and Is It Safe? Scientists Speak, Report* (pp. 57–63). World Food Logistics Organization.

Yan, Y., Zhao, S., Ye, X., Tian, L., Shang, S., Tie, W., . . . Hu, W. (2022). Abscisic acid signaling in the regulation of postharvest physiological deterioration of sliced cassava tuberous roots. *Journal of Agricultural and Food Chemistry*, *70*(40), 12830–12840.

Yin, X.R., Zhang, Y., Zhang, B., Yang, S.L., Shi, Y.N., Ferguson, I.B., Chen, K.S. (2013). Effects of acetylsalicylic acid on kiwifruit ethylene biosynthesis and signaling components. *Postharvest Biology and Technology*, *83*, 27–33.

Zhang, M., Liu, W., Li, C., Shao, T., Jiang, X., Zhao, H., Ai, W. (2019). Postharvest hot water dipping and hot water forced convection treatments alleviate chilling injury for zucchini fruit during cold storage. *Scientia Horticulturae*, *249*, 219–227.

Zhang, Q., Liu, Y., He, C., Zhu, S. (2015). Postharvest exogenous application of abscisic acid reduces internal browning in pineapple. *Journal of Agricultural and Food Chemistry*, *63*(22), 5313–5320.

Zhang, X., Zhang, M., Xu, B., Mujumdar, A.S., Guo, Z. (2022). Light-emitting diodes (below 700 nm): improving the preservation of fresh foods during postharvest handling, storage, and transportation. *Comprehensive Reviews in Food Science and Food Safety*, *21*(1), 106–126.

Zhang, Z., Gao, Z., Li, M., Hu, M., Gao, H., Yang, D., Yang, B. (2012). Hot water treatment maintains normal ripening and cell wall metabolism in mango (*Mangifera indica* L.) fruit. *HortScience*, *47*(10), 1466–1471.

Zhou, R., Su, S., Yan, L, Li, Y. (2007). Effect of transport vibration levels on mechanical damage and physiological responses of Huanghua pears (*Pyrus pyrifolia* Nakai, cv. Huanghua). *Postharvest Biology and Technology*, *46*(1), 20–28.

Zhu, F., Yun, Z., Ma, Q., Gong, Q., Zeng, Y., Xu, J., . . . Deng, X. (2015a). Effects of exogenous 24-epibrassinolide treatment on postharvest quality and resistance of Satsuma mandarin (*Citrus unshiu*). *Postharvest Biology and Technology*, *100*, 8–15.

Zhu, T., Tan, W.R., Deng, X.G., Zheng, T., Zhang, D.W., Lin, H.H. (2015b). Effects of brassinosteroids on quality attributes and ethylene synthesis in postharvest tomato fruit. *Postharvest Biology and Technology*, *100*, 196–204.

Zhu, Z., Geng, Y., Sun, D.W. (2019). Effects of operation processes and conditions on enhancing performances of vacuum cooling of foods: a review. *Trends in Food Science & Technology*, *85*, 67–77.

Zhu, Z., Zhang, Z., Qin, G., Tian, S. (2010). Effects of brassinosteroids on postharvest disease and senescence of jujube fruit in storage. *Postharvest Biology and Technology*, *56*(1), 50–55.

Ziedan, E.S.H., El Zahaby, H.M., Maswada, H.F., Zoeir, H.A. (2018). Agar-agar a promising edible coating agent for management of postharvest diseases and improving banana fruit quality. *Journal of Plant Protection Research*, *58*(3), 234–240.

PART II

Gaseous Atmosphere Applications

Controlled Atmosphere Storage for Fruits and Vegetables

3

Yoshihiro Imahori*

**Corresponding Author:* imahori@omu.ac.jp

3.1 INTRODUCTION

Horticultural commodities contain various important nutrients such as carbohydrates, proteins, minerals, vitamins, and dietary fiber for the human diet that benefit human health. As fruits and vegetables after harvesting are also living organisms, physiological and biochemical changes continually happen. Some are desirable alterations, but most are not. Also, horticultural commodities include much water content and are highly susceptible to dehydration and to mechanical damage. Thus, as fruits and vegetables are fresh products, they are perishable and subject to widespread quality losses through microbial effects, physical damage, and senescence after harvesting (Imahori, 2012, 2014; Madani et al., 2019). The losses in quality and quantity after harvesting significantly affect the relationship between harvest and consumption of horticultural commodities. Their irreversible alterations cannot be stopped, but they are able to be slowed down within certain thresholds (Imahori, 2014; Madani et al., 2019). Therefore, the preservation of quality and the improvement of shelf life are important in horticultural commodities. Improving and prolonging the shelf life of commodities is crucial for reducing the loss of economical values (Imahori, 2014; Madani et al., 2019).

Up to the present, postharvest technologies have widely developed to sustain the postharvest quality of horticultural commodities (Imahori et al., 2021; Endo et al., 2019). It became possible to construct the worldwide transport and supply chain of horticultural commodities and contribute to the year-round supply system with these technologies. Together with low-temperature storage, controlled atmosphere (CA) storage, that is, gaseous storage, is an important postharvest technology.

CA storage is postharvest technology that utilizes decreased O_2 and/or increased CO_2 atmospheres and is known for gaseous storage that maintains the quality and extends the shelf life of commodities. To expose horticultural commodities to reduced O_2 and/or enhanced CO_2 gases can either bring benefits or harmful effects dependent on the concentration of these gases, storage temperature, exposing duration, and kind of horticultural commodity. The advantage of CA storage is the delay of fruit ripening, reduction of physiological and pathological disorders, and decrease of the need for exterminating horticultural commodities (Imahori, 2012).

CA technology is a relatively old preservation technique. The advantage of gaseous storage on harvested crops was known for centuries from ancient practices in ancient China, ancient Greece, and ancient Egypt (Yahia et al., 2019). In 200 BC Egyptians had preserved crops in sealed limestone crypts. Romans also had utilized sealed underground pits for storage of grain in 100 BC (Yahia et al., 2019). In 1821, Jacques Etienne Berard performed the first scientific studies about CA storage and found that harvested apples stored in a low-oxygen atmosphere (0.5–1.0 kPa) did not ripen for more than 6 months (Beaudry, 1999; Dilley, 2006; Thompson, 2010). In 1927, Franklin Kidd and Cyril West indicated that a suppression of respiratory action was related with the storage duration of apples and demonstrated that decreased O_2 and elevated CO_2 in gaseous conditions of storage extended the shelf life of apples. They called this innovative technology "gas storage", and this observation became the first study of CA storage in the United Kingdom (Beaudry, 1999; Thompson, 2010; Falagán & Terry, 2018). After that, studies in CA storage progressively proceeded to obtain the influence of CA storage and the optimal gaseous conditions of horticultural commodities (Falagán & Terry, 2018). Kidd and West's study led to the development of the CA industry in the early 1930s. Robert Smock in the United States coined the term "controlled atmosphere storage" (Thompson, 2010). Later CA research have expanded in various regions in the world and have also extended to various commodities

DOI: 10.1201/9781003370376-5

beyond apples (Yahia, 2006). The successful practices of CA storage also inspired the development of modified atmosphere (MA) packaging using plastic film packages (Beaudry, 2010).

In 1950–2000 many CA technologies have been developed and refined, and consequently technical innovation realized and expanded the construction of CA facilities and its utilization worldwide (Prange et al., 2005; Bodbodak & Moshfeghifar, 2016).

This chapter reviews some scientific and technological information as well as recent approaches on the research and application of CA storage as a sustainable postharvest technology for horticultural commodities.

3.2 TYPES OF CONTROLLED ATMOSPHERES

The technology of CA has developed steadily since its beginning. In its early stage, research for methods to constant storage temperature and storage atmosphere, for development of mechanical refrigerant system and gastight rooms, and for utilization of respiratory action by horticultural commodities progressed. As a result, advancements in the construction of CA storage rooms progressed. The better utilization of efficient equipment led to more precise regulation of O_2 and CO_2 concentrations in the CA of storage rooms. Exact regulation of low O_2 levels became easy with the introduction of automatic measurement and control systems (Hoehn et al., 2009). In the next stage, more precise gas measurements were developed and became possible with the control of the gas level at a constant concentration with the percent level. This led to more rapid establishment of CA conditions.

3.2.1 Controlled Ventilation

This type is the least basial CA storage by buildup of CO_2 produced by the respiratory action of horticultural commodities stored in a gastight storage room. The level of CO_2 is kept in 5–10 kPa (approximately 5–10% of concentration in the total gas) or higher and is regulated by ventilating with outside air only. The total CO_2 and O_2 concentrations are equal to 20–21%. The range of O_2 concentrations is approximately 10–16 kPa, but it is not enough to act as a sufficient suppressing effect against respiratory metabolism (Hoehn et al., 2009, Bodbodak & Moshfeghifar, 2016).

As many apple and pear cultivars are more delicate to increased CO_2 concentration and cause the physiological disorders related to elevated CO_2, it is necessary that CO_2 levels are kept under 3 kPa and O_2 levels at 18 kPa in controlled ventilation storage. As a result, there are uninfluenced or moderate effects on shelf life compared with refrigeration (Hoehn et al., 2009; Bodbodak & Moshfeghifar, 2016). In contrast, as cherries and berries have tolerance to higher CO_2, controlled ventilation storage enhances the storability and maintains the fruit quality. Developing controlled ventilation storage requires a gastight room or a device monitoring CO_2, pallet packaging systems, and an installation that enables ventilation (Hoehn et al., 2009; Bodbodak & Moshfeghifar, 2016).

3.2.2 Conventional Controlled Atmosphere

This CA is the traditional system and is commercially used in the preservation of apples and pears. This system commonly maintains low O_2 concentrations of 2–3 kPa and high CO_2 concentrations of 2–3 kPa in an airtight room, efficiently decreasing the respiration of apples and pears and keeping their qualities (Hoehn et al., 2009; Bodbodak & Moshfeghifar, 2016, Yahia et al., 2019). Reduced O_2 and elevated CO_2 concentrations in conventional CA are generated by CO_2 production by the respiration of fruits and vegetables until the purpose atmosphere is established. The O_2 level is kept by ventilation with outside air, and the CO_2 level is kept by removal (scrubbing) to maintain constant levels through the storage duration, because of CO_2 excess by respiration (Figure 3.1).

To remove excessive CO_2 in a CA storage room, a scrubber is used and chemically or physically removes CO_2 in the storage atmosphere (Hoehn et al., 2009; Bodbodak & Moshfeghifar, 2016; Yahia et al., 2019). An appropriate measuring system is necessary for monitoring concentrations of O_2 and CO_2 in the storage atmosphere. The storage room is necessary to be relatively airtight and equipped with dependable remote temperature control within ± 0.5°C. The O_2 and CO_2 concentrations in the storage atmosphere are also controlled within ± 0.5 kPa with gas measurement and CO_2 adsorbers (Markarian et al., 2003). A CA storage room and apple fruits in storage are shown in Figure 3.2A and B.

3.2.3 Modified Atmosphere

Gaseous conditions in this system are generated by the respiration of fruits and vegetables within gastight environments, or by flushing single or mixed gases into the container; thus, this system is also named "biological CA". Atmospheric balance in MA establishes the positive effects in points of reducing physiological metabolism and delaying the senescence of fruits and vegetables.

As this method passively decreases O_2 levels and increases CO_2 levels and the atmosphere composition is determined by the balance between the respiratory action of commodities and the statical gas diffusion, it needs several weeks to achieve the desired atmosphere and cannot be further adjusted once the storage atmosphere is established (Hoehn et al., 2009; Bodbodak & Moshfeghifar, 2016; Yahia et al., 2019).

MA packaging applies MA technology and uses the changes of packaging atmosphere composition surrounding the horticultural commodities by the interaction between the respiration of commodities and gas diffusion through plastic films with specific gas permeability (Figure 3.2 C and D).

FIGURE 3.1 Corridor with controlled atmosphere room.

FIGURE 3.2 Controlled atmosphere storage room (A) and apples stored in storage room (B), modified atmosphere packaging (MAP) on broccoli under refrigerated storage (C) and MAP and refrigerated storage (D).

As respiration enhances rapidly with temperature rather than the gas permeability of films, O_2 levels in packaging decrease and CO_2 levels increase; consequently, anaerobic metabolism under higher temperatures than designed MA conditions is induced. Thus, the regulation of temperature is significant together with the choice of films for appropriate atmosphere composition. As MA condition is usually difficult to keep for long durations, short-term storage and transport is utilized (Hoehn et al., 2009).

3.3 BENEFICIAL EFFECTS OF CONTROLLED ATMOSPHERE STORAGE

CA storage is widely utilized around the world to store horticultural commodities and is one of the most beneficial postharvest technologies that has been introduced to the horticultural commodity industry. There are the synergistic effects of two gases with low O_2 and high CO_2 that are contained in storage atmosphere. As this CA storage also combines with low-temperature storage, CA technology can bring some benefits during postharvest handling, transport, and storage (Thompson, 2010).

CA reduces the rates in respiration and ethylene production in horticultural commodities and results in the retardation of ripening and senescence with its physiological and biochemical alterations (Kader, 1992; Yahia et al., 2019). Generally, low O_2 concentrations (less than 2 kPa) and/or high CO_2 concentrations (over 10 kPa) cause the decrease of respiratory metabolism in plants, but the adaptability to these gases is dependent on the species and cultivar of horticultural commodities (Table 3.1) (Yahia et al., 2019; Imahori, 2021).

Ethylene is related to senescence and plant ripening, and its biosynthesis and ethylene action need oxygen. Thus, the lower O_2 levels enable reduced ethylene production. High CO_2 atmospheres also work as a competitive inhibitor by competing with ethylene to bind with ethylene receptors (Yahia et al., 2019). Generally, low O_2 concentrations (less than 8 kPa) and/or high CO_2 concentrations (over 1 kPa) cause a decrease in ethylene sensitivity in fruits (Kader, 1992).

CA, especially high CO_2 levels, alleviates physiological disorders of fruits and vegetables, such as chilling injuries, russet spotting that occurs in lettuce, and several storage disorders found in apples (Kader, 1992; Yahia et al., 2019).

CA can control postharvest pathogens and, consequently, the incidence and severity of decay that occurs during storage. Elevated levels of CO_2 (10–15 kPa) inhibit Botrytis rot development on strawberries and cherries (Kader, 1992). CO_2 permeates the bio membrane of microorganisms and changes the pH within the cell by producing carbonate, resulting in the mortality of the microorganisms (Yahia et al., 2019).

CA is used for insect control. Insect mortality depends on a low O_2 level (below 5 kPa) and a high CO_2 level (above 50 kPa), relative humidity level, and temperature. The exposure period is usually short, from a few hours to a few days (Yahia et al., 2019).

TABLE 3.1 Threshold Level of Oxygen and Carbon Dioxide Required to Cause Injury to Fruits and Vegetables and Injury Symptoms

	LOW O_2 INJURY		*HIGH CO_2 INJURY*	
PRODUCE	*O_2 INJURY LEVEL (KPA)*	*O2 INJURY SYMPTOMS*	*CO2 INJURY LEVEL (KPA)*	*CO2 INJURY SYMPTOMS*
Strawberry	<2	Off-flavor, discoloration	>25	Off-flavor
Orange	<5	Off-flavor	>5	Off-flavor
Persimmon	<3	Off-flavor	>10	Off-flavor
Kiwifruit	<1	Off-flavor	>7	Softening of flesh
Blueberry	<2	Off-flavor	>25	Off-flavor, skin browning
Japanese pear	<1	Off-flavor	>8	Off-flavor
Banana	<1	Off-flavor, discoloration	>7	Off-flavor, softening of flesh
Apple	<2	Off-flavor	>1	Off-flavor, softening of flesh
Cauliflower	<2	Off-flavor, discoloration, water-soaked tissue	>5	Off-odor
Cucumber	<1	Off-odor, discoloration, surface pitting	>10	Discoloration, pitting
Onion	<1	Off-odor	>1	Softening
Tomato	<3	Off-flavor, softening	>2	Softening, uneven ripening
Chinese chive	<1	Off-odor, discoloration	>30	Off-odor
Sweet pepper	<1	Off-odor discoloration	>5	Off-odor, discoloration

3.4 PHYSIOLOGICAL AND BIOCHEMICAL EFFECTS OF CONTROLLED ATMOSPHERES

Reduced O_2 levels and/or elevated CO_2 levels that are constituted in CA atmospheres can bring either beneficial or harmful effects to fruits and vegetables dependent upon gas composition, gas concentrations, exposure temperature, and exposure period. Up to now, the evaluation of CA storage on the storability and quality keeping for many commodities has been conducted. But exposure of decreased O_2 and/or increased CO_2 for sustaining the quality of horticultural commodities is very risky, as extremely low O_2 and/or extremely high CO_2 may have harmful effects on them. Thus, to understand the physiological and biochemical influences of decreased O_2 and increased CO_2 atmospheres to fruits and vegetables is incredibly significant (Imahori, 2021).

3.4.1 Physiological and Biochemical Responses to a Low-Oxygen Atmosphere

Exposure of a horticultural commodity to a lower O_2 atmosphere can have beneficial or harmful effects dependent upon O_2 concentrations, exposure temperature, and exposure duration. Horticultural commodities exposed to stressful levels of O_2 concentrations for prolonged periods may accumulate ethanol and acetaldehyde, leading to browning of the plant tissues and abnormal ripening (Imahori et al., 2005; Imahori, 2012; Madani et al., 2019). Oxygen levels below 0.2 kPa result in anaerobic metabolism in plant cells (Kader, 1986). As O_2 levels decrease in plant cells, mitochondria respiration becomes lower and fermentative metabolism associated with low-oxygen consequences is activated, resulting in the increase of fermentation products and of overall CO_2 levels, and occurring in the activation of the glycolysis pathway, known as the Pasteur effect (Madani et al., 2019; Imahori, 2021). When oxygen levels decrease within the plant cell, fermentation metabolism accumulates fermentation products such as acetaldehyde, ethanol, and lactate, as well as lower intracellular pH and ATP contents in the plant cells. In the fermentation pathway, pyruvate changes to acetaldehyde by the decarboxylation of pyruvate with the action of pyruvate decarboxylase (PDC); furthermore, acetaldehyde changes to ethanol by the catalysis of alcohol dehydrogenase (ADH) utilizing NADH. Also, lactate is directly produced from pyruvate in a single step by lactate dehydrogenase (LDH) utilizing NADH (Figure 3.3).

Fermentative end products, ethanol and lactate, are synthesized with a varying level within plant tissues, if oxidative phosphorylation and electron transport are limited and glycolysis is enhanced under low O_2 (Imahori, 2012; Imahori, 2021). O_2 levels in CA are usually recommended to be over the

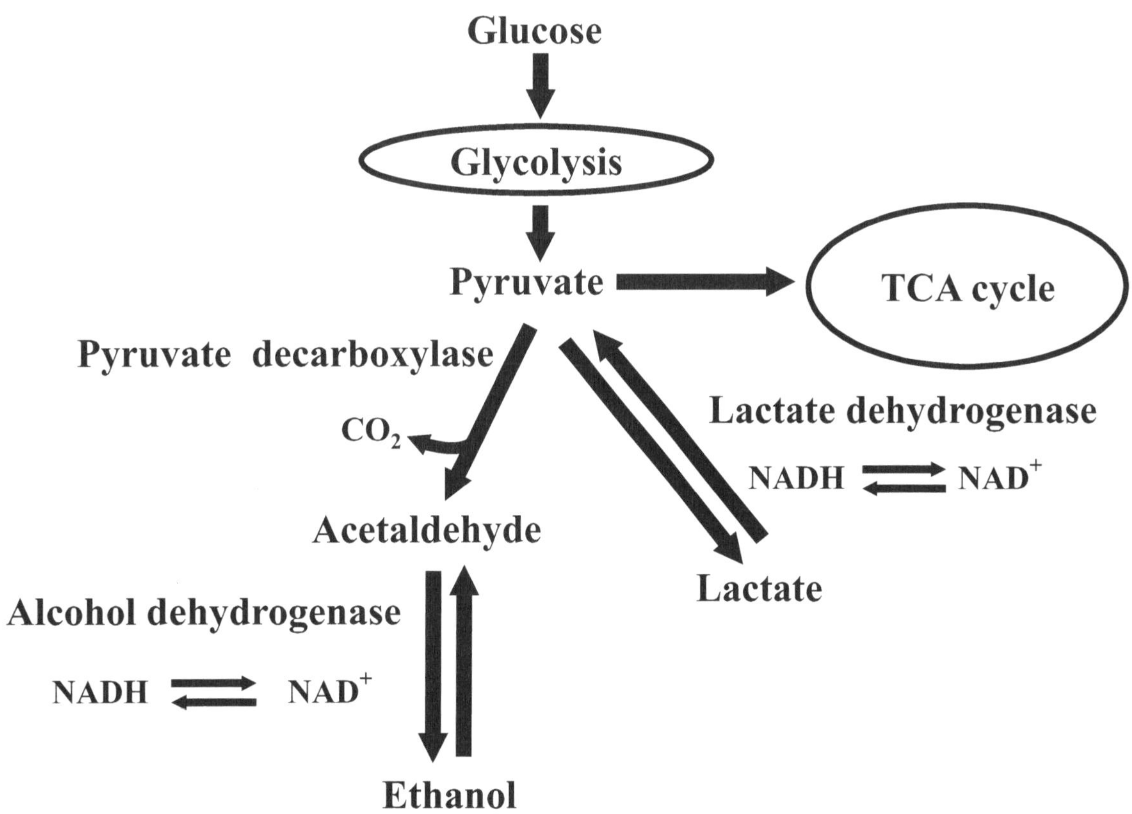

FIGURE 3.3 Schematic pathway of fermentation metabolism.

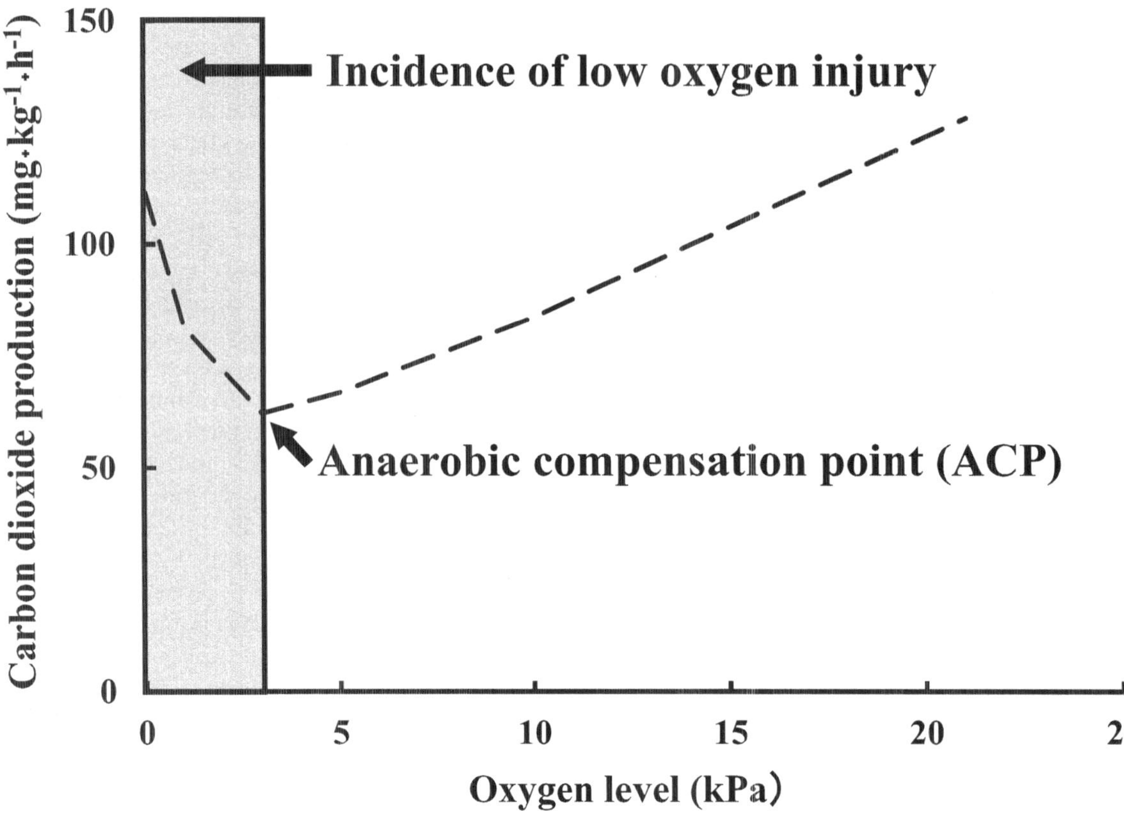

FIGURE 3.4 Effect of oxygen level on carbon dioxide production in banana fruit.

anaerobic compensation point (ACP) to stay safe during CA storage (Figure 3.4). ACP indicates a close relationship with the lowest oxygen limit (LOL) (Thewes et al., 2015).

The O_2 levels that CO_2 produce in fruits and vegetables is minimal. In O_2 levels below ACP, the fermentation metabolism activates and storage disorders occur (Thewes et al., 2015; Bessemans et al., 2016). Generally, at least 1 kPa O_2 in CA storage is needed to avoid anaerobic metabolism (Madani et al., 2019; Imahori, 2021).

Fermentative metabolism is regulated by two mechanisms in plants; one is the control of the molecular level of fermentative enzymes such as LDH, PDC, and ADH in a low O_2 atmosphere, and the other is the control of the metabolic level of these enzymes. Usually, activation of these enzymes via low O_2 is thought to be largely caused by the activation of transcription and the translation levels of related genes, consequently processing new mRNA synthesis of their related genes and *de novo* synthesis of these enzyme proteins (Imahori, 2012). However, molecular regulation by the gene expression of fermentative enzymes do not become the major mechanism, although the regulation of fermentation enzymes through the induction of the molecular level (transcription and/or translation) is essential for the accumulation of fermentative products (Imahori, 2012). It is reported that in bell pepper fruits stored in 0 kPa O_2, the accumulation of transcription and the increase of activity in ADH is not correlated with the changes in the production of acetaldehyde and ethanol in fermentative products (Imahori et al., 2000).

The cytoplasmic pH changes in plants are also thought to be the regulator that controls fermentative metabolism. A self-controlling system called the pH-stat hypothesis has been proposed regarding lactate and ethanol production in fermentative metabolism. At the beginning of anaerobic metabolism, LDH is activated by cytoplasm pH that indicates alkaline and catalyzes the reaction of lactate production from pyruvate; consequently, lactate accumulation reduces cytoplasmic pH and inhibits LDH activity and promotes PDC activity, leading to ethanol production (Imahori et al., 2003; Imahori, 2012). However, the reduction of cytoplasmic pH, besides lactate accumulation, is caused by the potential passive H^+ and K^+ leakage from the vacuole and protoplasm for the limitation of available ATP levels and the suppression of vacuolar H^+-ATPase activity (Benkeblia, 2021).

Levels of substrates and cofactors relate to the metabolic regulation for fermentative metabolism. The difference in *Km*s for pyruvate between pyruvate dehydrogenase (PDH) and PDC becomes the regulation factor that controls pyruvate flux into the TCA cycle or the ethanolic fermentation pathway, as the *Km* of PDH is in the μM range, while that of PDC is in the mM range (Tadege et al., 1999). Indeed, as the *Km* of PDC is too low, pyruvate levels become the limiting factor. However, when respiration alters to anaerobic respiration due to a decrease in O_2 levels, pyruvate levels considerably increase. As a result, pyruvate becomes utilized in the reaction of PDC due to the change in conformation of the allosteric enzyme together with binding with the substrate. Thus, the lag phase in the production of ethanol at the beginning of anaerobic respiration is not due to the requirement for a lower intrasellar pH; rather, the lag phase needs to accumulate pyruvate (Tadege et al., 1999). PDC activity, therefore, becomes a key controller of ethanolic fermentation under O_2 limiting conditions, and ethanolic fermentation may actually be controlled based on the PDH/PDC stat hypothesis under a low O_2 atmosphere (Tadege et al., 1999; Imahori et al., 2003; Imahori, 2012).

3.4.2 Physiological and Biochemical Responses to a High Carbon Dioxide Atmosphere

The reactions of horticultural commodities to high CO_2 atmospheres differ significantly among or within organ types, developmental stages, species, and cultivars, and such atmospheres can have both beneficial and harmful physiological and biochemical effects (Beaudry, 1999). The influence of CO_2 is dependent on its concentration and other environmental situations, such as temperature. CO_2 functions as either an inducer or a suppressor of respiratory metabolism dependent on its exposure concentration in situ, the exposure period, the kind of horticultural commodity, and the exposure temperature (Imahori et al., 2007). The influences of a high CO_2 atmosphere on horticultural commodities are activation of the glycolytic pathway, development of fermentative metabolism, buildup of succinate and/or alanine, and reduction of pH and ATP concentration (Mathooko, 1996). A CO_2 atmosphere above 20 kPa activates the ethanol fermentative pathway and accumulates ethanol in plant tissues, depending on the kind of commodity and the concentrations of O_2 in the CA atmosphere (Kader, 1986; Imahori, 2012). In a 20 kPa or higher CO_2 atmosphere containing atmospheric oxygen, ethanol accumulation was found in lettuce, strawberries, and figs (Mathooko, 1996). These phenomena indicate that as a fermentative product, several substrates in energy-related metabolism are flowing through the fermentative pathway (Imahori et al., 2007; Imahori, 2012).

A high CO_2 atmosphere inhibits the activity of succinate dehydrogenase (SDH) that converts succinate into fumarate in the tricarboxylic acid (TCA) cycle and accumulates succinate in the plant (Imahori et al., 2007). Relating to the inhibition of SDH activity by elevated CO_2, the reduction of succinate oxidation to fumarate leads to the accumulation of toxic compounds and the depletion of malate in plant tissues (Mathooko, 1996). The primary effect of high CO_2 on fruits and vegetables seems to be on the kinetic reversible reaction catalyzed by SDH in the TCA cycle (Mathooko, 1996). High CO_2 leads to the reduction of enzyme synthesis and the alteration of the enzymatic structure or conformation, resulting in the diminution of the SDH protein caused by the increase of degradation or the inactivation of the SDH enzyme. While regulating the TCA cycle by reducing SDH activity due to high CO_2, fermentation metabolism is influenced by the activation of ADH, accumulating ethanol in plant tissues. Therefore, the reaction of the high CO_2 atmosphere to horticultural commodities indicates the molecular primary action, which is dependent on the kinetics of a reversible reaction within the TCA cycle acted on by SDH (Mathooko, 1996).

3.5 STORAGE FACILITIES FOR CONTROLLED ATMOSPHERE STORAGE

Storage facilities for CA in horticultural commodities basically consist of refrigerated and controlled gas storage.

These facilities have refrigeration equipment, gastight and thermally insulated rooms or enclosures, fabricate and maintenance systems of gas in the predetermined concentrations under a specific environment, and measurement and controlling systems (Hoehn et al., 2009; Bodbodak & Moshfeghifar, 2016).

Essentially, a CA storage room is a refrigerated storage room with an effective sealing system that minimizes gas exchange. Each room generally has a bladder called a breather bag, subject to keep equal pressure inside and outside the room to prevent incidence of structural damage in the CA room. Also, each room has an automatic monitoring system to check the atmosphere composition and correct the gas exchange. It is necessary for the walls, ceiling, floor, and doors in the storage room to be airtight (Yahia et al., 2019). The storage room is made from metal insulated panels and expanded polyurethane foam for thermal and gaseous insulation. Panels are fitted together with gastight locking devices (Thompson, 2010). Elastomer resins are applied on all walls and the ceiling to maintain the gas barrier properties and gas tightness in the CA room (Yahia et al. 2019). Connections between the storage room and machinery equipment external to the room are connected via hermetically sealed passages (Yahia et al., 2019).

3.6 GENERATION OF CONTROLLED ATMOSPHERE

Several capabilities are needed for establishing and maintaining CA (Figure 3.5). They consist of O_2 removal, CO_2 removal, and ethylene removal, as well as the supplementation of air to supply O_2 used by respiratory action, and in some cases CO_2 supplement (Hoehn et al., 2009; Bodbodak & Moshfeghifar, 2016). Choice of the suitable functions and devices depends on the kinds of horticultural commodities and required storage conditions (Bodbodak & Moshfeghifar, 2016).

3.6.1 Oxygen Removal

It is common to reduce O_2 concentrations by the respiration of fruits and vegetables at the beginning of CA storage. The rapid decrease of O_2 concentrations in the storage room is

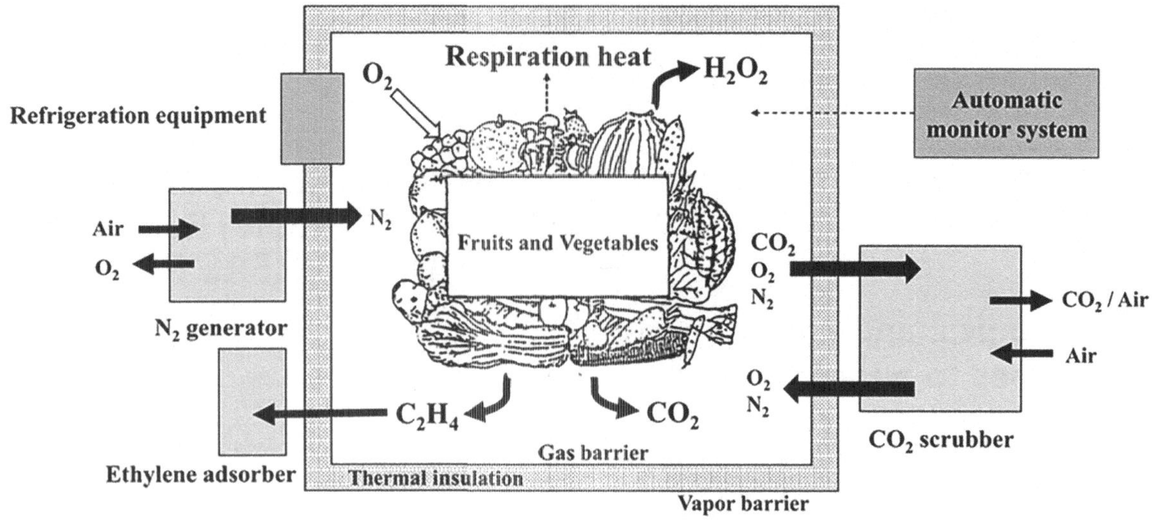

FIGURE 3.5 Scheme of a controlled atmosphere (CA) system.

advantageous. The concentration of O_2 rapidly reduces by flash with N_2. N_2 is commonly obtained as a liquid or gas, or by production via gas separator systems (Hoehn et al., 2009; Bodbodak & Moshfeghifar, 2016). Gas separator systems include a hollow fiber membrane based on membrane technology and a pressure swing adsorber based on adsorption technology (Malcom, 2005; Hoehn et al., 2009; Bodbodak & Moshfeghifar, 2016; Yahia et al., 2019).

A propane burner was utilized to burn propane due to converting O_2 to CO_2 until around 1990. Operation of gas generators is conducted by open flame or catalytic burner. The use of a catalytic burner is desirable, as it can be utilized on a recycling mode, reducing to 3–5 kPa O_2 (Hoehn et al., 2009; Bodbodak & Moshfeghifar, 2016; Yahia et al., 2019). However, incomplete burning of propane raised a safety concern. The combustion products of this system include CO, which is toxic to humans; therefore, ventilation is necessary before entering the storage room (Yahia et al., 2019). A CO_2 scrubber and an explosive gas detector are also necessary for removing much of the CO_2 from the CA room. The burning of propane raises ethylene. As the ripening and senescence of horticultural commodities is promoted by the accumulation of ethylene gas, ethylene gas removal is necessary (Hoehn et al., 2009; Bodbodak & Moshfeghifar, 2016; Yahia et al., 2019).

In alternative O_2 removal, ammonia acts with the O_2 in the CA room and produces H_2 and N_2 by cracking ammonia at a high temperature via ammonia burners (Hoehn et al., 2009; Bodbodak & Moshfeghifar, 2016; Yahia et al., 2019). This system is advantageous for not producing CO or ethylene, but the leakage of ammonia is a disadvantage because it can cause damage to plants (Yahia et al., 2019).

Since 1986 a new method of rapid removal of O_2 has been used by flushing N_2 into the CA storage room (Malcolm, 2005; Dilley, 2006). This equipment separates air into O_2 and N_2 and is generally called a N_2 generator. There are two types that are commercially available (Hoehn et al., 2009). One type is the hollow fiber membrane (HFM) type based on membrane technology, and the other is the pressure swing adsorption (PSA) type based on adsorption technology. In both, compressed air of 800–1,300 kPa is necessary for the feed stream of the equipment (Hoehn et al., 2009). The PSA type separates air into N_2 and O_2, utilizing a carbon molecular sieve that selectively adsorbs O_2 and CO_2 in the adsorber bed and passes N_2 through it. As pressure decreases, these gases adsorbed in the adsorber bed are released and the adsorber bed is regenerated (Dilley, 2006; Hoehn et al., 2009). There is also a vacuum pressure swing adsorber that uses a vacuum during the regeneration (Malcolm, 2005; Hoehn et al., 2009; Bodbodak & Moshfeghifar, 2016). The HFM type uses the difference in pressure between the inside and outside of the membrane fiber. That pressure difference becomes the driving force that gases permeate across the wall of the membrane fiber. The rate of gas permeation through the membrane fiber is different for each type of gas. As O_2, CO_2, and H_2O are all called fast gases because they diffuse at fast speeds and N_2 is called a slow gas because it diffuses at a low speed, N_2 can be continuously separated from other gases in the air and regeneration of membrane fibers is not necessary (Hoehn et al., 2009; Bodbodak & Moshfeghifar, 2016).

3.6.2 Carbon Dioxide Removal

It is necessary to remove excess CO_2 from CA rooms. The control and removal of CO_2 in the CA room is commonly accomplished by scrubber systems.

The caustic soda (NaOH) solution that is dissolved by water is the chemical reagent used in CO_2 scrubbers in commercial CA storage. That solution cycles within open tubes and absorbs CO_2. Caustic soda reacts with CO_2 and produces $NaCO_3$. However, as utilization of NAOH was discontinued due to its corrosiveness and potential for danger for its handling and disposal, now lime is used instead (Hoehn et al., 2009; Bodbodak & Moshfeghifar, 2016).

Lime is most commonly used in CO_2 scrubber systems and the simplest method uses hydrated lime (Ca $(OH)_2$), which changes to limestone ($CaCO_3$) during CO_2 absorption (Hoehn et al., 2009). Lime bags are located directly inside the storage room or in a layer attached to it through pipes (Yahia et al., 2019).

Activated charcoal (activated carbon) has much more space per volume and higher porosity than normal carbon and has many microscopic pores that can absorb CO_2 (Yahia et al., 2019). Storage room air is cycled through a scrubber that comprises chambers or two cylindrical beds packed with activated charcoal. The CO_2-free air is again circulated to the storage room (Hoehn et al., 2009; Bodbodak & Moshfeghifar, 2016). CO_2-saturated charcoal is regenerated by flowing low CO_2 air (freshly outside air) into packed activated charcoal (Hoehn et al., 2009; Bodbodak & Moshfeghifar, 2016; Yahia et al., 2019).

A molecular sieve scrubber is another CO_2 removal method. A molecular sieve is made of crystalline metal, sodium, or aluminum silicate zeolites. It is regenerated by heating to produce uniform tiny pores that trap smaller molecules while permitting larger molecules to pass (Yahia et al., 2019).

Excess CO_2 in a storage room can also be removed by flushing the room with N_2. N_2 generators are utilized for independently controlling CO_2 and O_2 concentrations in CA storage rooms (Hoehn et al., 2009).

3.6.3 Ethylene Removal

Ethylene gas is a plant hormone that reduces the shelf life of horticultural products by promoting ripening and senescence and contributes to physiological disorders (Hoehn et al., 2009; Bodbodak & Moshfeghifar, 2016; Yahia et al., 2019). There are two ethylene scrubber systems that are used commercially. One is the catalytic oxidation of ethylene to water and CO_2 by ozone gas or ultraviolet radiation. The other involves the use of ethylene-absorbing beads. Some CO_2 scrubbers also enable the absorption of ethylene. Flushing with N_2 to remove CO_2 also reduces ethylene concentrations in storage rooms (Hoehn et al., 2009; Bodbodak & Moshfeghifar, 2016).

Ethylene-absorbing beads are made from smaller globular particles infiltrated by potassium permanganate. Several porous and inactive matrix materials like alumina, silica gel, and zeolite are used for globular particles. Air in the storage room is cycled through a sealed cartridge filled by beads and ethylene is absorbed (Hoehn et al., 2009; Bodbodak & Moshfeghifar, 2016).

3.7 CONTROL OF STORAGE ATMOSPHERES

Successful CA storage depends on achieving the exact levels of atmospheric gas and retaining a proper and stable composition of atmospheric gas within the CA storage room. Control of the atmosphere in storage room is influenced by the physiological characteristics of horticultural commodities stored in the CA room, the O_2 and CO_2 controlling equipment, the refrigeration system, and environmental conditions. Thus, the influences on CA storage are not simple and rely on other environmental factors (Hoehn et al., 2009; Thompson, 2010; Bodbodak & Moshfeghifar, 2016).

3.7.1 Temperature Control

Use of refrigeration is commonly the most effectual technique for keeping the quality of products and elongating their shelf life after harvest. It is also one of the greatest extensively utilized postharvest technologies for storage and transport (Endo et al., 2019; Imahori et al., 2021; Kantakhoo et al., 2022). CA storage is the effective postharvest technique, together with refrigeration, to enhance the retention of quality and prolongation of shelf life of horticultural commodities due to its synergistic effects (Brizzolara et al., 2017). Therefore, a refrigeration system is commonly included in the integral components of CA storage (Thompson, 2010; Bodbodak & Moshfeghifar, 2016). The refrigeration system is mainly constituted with an evaporator, compressor, condenser, expansion valve, and refrigerant that cycles into each part (Bodbodak & Moshfeghifar, 2016).

The control of temperature regulates with pipes that refrigerant flows inside the storage room (Thompson, 2010). The refrigerants used are generally ammonia, dichlorodifluoromethane (R-12), or chlorofluorocarbon (R-22). But with the destruction of the ozone layer, new and less detrimental refrigerants, such as hydrofluorocarbon (R-410A) that is mixed 50:50 with difluoromethane (R-32) and pentafluoroethane (R-125), have been introduced. The liquid refrigerant that flows into the pipes is cooled and passed through the storage room to decrease the atmospheric temperature (Thompson, 2010).

The evaporator is arranged inside the storage room as part of the system. The liquid refrigerant that became at low pressure and low temperature is cycled, and then is vaporized in the pipes of the evaporator due to absorbing heat. The resulting vapor is pumped to the compressor through the pipe and is compressed into a hot and high-pressure vapor. This compressed gas is pumped continuously to the condenser and is cooled by passing through a radiator and changed to a high-pressure liquid. Generally, a radiator composed of a network of pipes is opened to the atmosphere. Then the high-pressure liquid passes through an expansion valve to control the flow of the refrigerant and reduce the refrigerant pressure. Consequently, the refrigerant vaporizes and the refrigerant temperature decreases. That chilled mix of vaporized and liquid refrigerant passes through the evaporator and then a series of refrigeration cycles completes (Thompson, 2010; Bodbodak & Moshfeghifar, 2016).

3.7.2 Humidity Control

Water loss causes unfavorable quality loss such as shriveling, wilting, softening, and peel damage and impairs the quality and shelf life of horticultural commodities during transport and storage (Madani et al., 2019). Therefore, it is significant to maintain high humidity within the storage room during the storage period. Humidity within the storage room usually must maintain closer to water saturation, as long as moisture within the room does not condense on the surface of the horticultural commodities. As the heat that is absorbed with cooling coils in the refrigeration system is related with refrigerant temperature, if the temperature of the refrigerant is below the temperature of the storage room, water within the room will condense on the surface of the evaporator. Also, if the temperature of the refrigerant is lower than water's freezing point, the water will freeze, and the cooling effect will reduce (Thompson, 2010; Bodbodak & Moshfeghifar, 2016). As the moisture is removed from the air of the storage room, the water loss of the commodities is caused by vapor evapotranspiration. Therefore, it is necessary to keep the temperature of the refrigerant near to the room temperature. As the temperature of horticultural commodities should be maintained, the temperature of the refrigerant should be considered with the removal of respiration heat caused by commodities, heat caused by fans, and leakage of chilled air through the insulation of the room and doors (Thompson, 2010; Bodbodak & Moshfeghifar, 2016).

As another system maintaining high humidity within the storage room, there is "jacketed storage". The storage room consists of a metal inner wall and an outer insulation wall, and the refrigeration pipes can chill the air space between the inner and outer walls. Therefore, as cooling coils indirectly contact the room air, cold temperatures can be maintained by the cooling pipes without water loss occurring in the horticultural commodities, and the whole face of the wall in the room functions as the cooling surface (Thompson, 2010; Bodbodak & Moshfeghifar, 2016).

Ice-bank cooling is a cooling system in which refrigerant pipes are soaked in a water tank and then the water is frozen, and the produced ice is used to chill the water. Cooled water converts to a fine mist and then is utilized to chill and moisten the air within the storage room (Thompson, 2010; Bodbodak & Moshfeghifar, 2016).

A hollow-fiber membrane contactor system can exactly retain relative humidity at 90.5 ± 0.1% in ventilated 500 L containers. The membrane in the system is made by polyetherimide and coated inside the membrane with a thin, non-porous silicone layer (Dijkink et al., 2004; Thompson, 2010; Bodbodak & Moshfeghifar, 2016). The liquid desiccant that is made by the dilute aqueous glycerol solution is pumped at a low speed through the hollow fibers. By this method, water vapor can sufficiently transfer between the atmosphere in the room and the liquid desiccant (Dijkink et al., 2004; Dijkink et al., 2004; Thompson, 2010; Bodbodak & Moshfeghifar, 2016).

3.7.3 Oxygen Control

In a CA storage system, control and maintenance of O_2 levels during storage is the most important procedure. The determination of O_2 levels is measured by chemical analyzers. The reactions with a pyrogallol solution of O_2 and with alkaline solutions of CO_2 are used in the measurement of the chemical analyzer. These concentrations are measured by changes in the gas volume of a sample collected from the CA storage room. There are two kinds of chemical analyzer systems. One is the Fyrite analyzer, and the other is the Orsat analyzer. The Fyrite analyzer system is not expensive and is handy, but its accuracy is inadequate for O_2 levels lower level than 2 kPa. The Orsat analyzer system is generally placed in the middle part of a CA storage room, and samples are collected by directly aspirating air from the CA storage room with a pump via a permanent tube. The cost of an Orsat analyzer is cheap and its accuracy is permitted at higher O_2 concentrations. But this analyzer is not suitable as an automated system, as the sampling of the gas is time consuming and requires a skilled operator (Hoehn et al., 2009; Bodbodak & Moshfeghifar, 2016).

An O_2 analyzer with a paramagnetic sensor is dependable, accurate, and longevity. O_2 has highly specific paramagnetic property and its property is utilized for analysis (Hoehn et al., 2009; Bodbodak & Moshfeghifar, 2016).

Recently electrochemical cells have improved and used for O_2 analysis. Its analyzer has similar accuracy and resolution of parametric sensor (Hoehn et al., 2009; Bodbodak & Moshfeghifar, 2016).

In traditional CA system, as O_2 reached at required level for supply, fresh air had frequently been introduced from outside of storage room (Thompson, 2010). Recently, with continuous equipment development, the accuracy that O_2 and CO_2 levels can keep is enhancing. Control of O_2 levels near the theoretical minimum is possible, because recent systems can accomplish much lower variations in gas levels, and ultra-low oxygen storage is recently common (Thompson, 2010).

3.7.4 Carbon Dioxide Control

The levels of CO_2 can be analyzed chemically by Fyrite or Orsat analysis similar to O_2 measurement. But their methods are not suitable for automatic operation due to their time-consuming nature and need for precision. An infrared CO_2 sensor is suitable and commonly utilized. Infrared absorbance wavelength at 4,260 ± 20 nm, and 15,000 nm is specific against CO_2, and absorbance strength is proportionate for CO_2 levels. For warranting good storage practices, regularly measuring O_2 and CO_2 concentrations utilizing a constant analyzer system is important (Hoehn et al., 2009; Bodbodak & Moshfeghifar, 2016).

A scrubber accomplishes the removal of excess CO_2 from a CA storage room. Many types of scrubbers have been developed, but they can generally be categorized into two types. One is type that chemicals react with O_2 and O_2 is removed is called passive scrubbing. This method places bags or pallets of CO_2-absorbing chemicals inside the storage room and maintains a lower CO_2 level and is referred to as 'product generated' scrubbing, as CO_2 is produced by the respiration of the commodities. The other type is called active scrubbing, which comprises two containers that the absorbing material of CO_2 enters. When reducing CO_2 from the air, the air passes through one of these containers and CO_2 is removed from the other one. When one container is full of CO_2, the containers reverse and CO_2 is removed from one while the other absorbs CO_2 from the air (Thompson, 2010).

3.7.5 Ethylene Control

The measurement of ethylene is difficult for low concentrations of CO_2 and is not common in CA storage rooms. But as ethylene concentrations that are reactive to fruits and vegetables is extremely low, ethylene control is necessary for effective operation of CA storage. There is a practiced method that measures ethylene levels via disposable tubes and color-changing chemicals. Recently, ethylene analyzers or detectors have been developed. They equip electrochemical sensors measuring 0–100 ppm for ethylene concentration within a CA room. Gas chromatography is the best measurement method for ethylene analysis (Hoehn et al., 2009; Bodbodak & Moshfeghifar, 2016).

3.7.6 Gas Control Equipment

The atmosphere within a CA storage room must be maintained at appropriate and stable conditions for the achievement of successful CA storage. However, its conditions always change due to several factors, and it is difficult to achieve an appropriate CA composition. The control of the atmosphere is always influenced by the physiological response of horticultural commodities stored in the storage room, the regulation equipment of CO_2 and O_2 in the atmosphere, the operation cycle of the refrigeration system and weather alterations (Hoehn et al., 2009; Bodbodak & Moshfeghifar, 2016). Measurement of O_2 and CO_2 concentrations at intervals of time is necessary for confirming and correcting whether these data shift from the set points. Control in conventional CA can be performed by manual measurements and controls once per day, whereas control in modern and large CA are conducted by automated control systems. At present, these systems have been important for the regulation of storage conditions and have ensured the optimal quality and efficiency of CA storage (Raghavan et al., 2005; Hoehn et al., 2009).

Automated control systems decrease the variations between set points and measuring values and determine whether the particular control equipment is operational. Devices for controlling equipment such as scrubbers consist of proportional, proportional-integrated, or proportional–integral–derivative (PID) controllers; on/off switches; and personal computer (PC)–based systems (Bodbodak & Moshfeghifar, 2016). PID controllers are more developed systems and more intelligent and are frequently installed in CA operations as a function of self-tuning or self-learning systems. Additionally, PC-based systems directly regulate online controlling of O_2 and CO_2 concentrations within a CA storage room. They are configured with a computer, a data acquisition system, communication ports, control software, and switching devices (Hoehn et al., 2009; Bodbodak & Moshfeghifar, 2016).

3.8 ADVANCED TECHNIQUES IN CONTROLLED ATMOSPHERE STORAGE

With the utilization of CA technologies, the storability of fruits and vegetables extended by leaps and bounds. But CA conditions of fruits and vegetables are usually determined by experiential observation and recommended conditions of static and particular O_2 and CO_2 concentrations for horticultural commodities are commonly used (Delele et al., 2019). These CA conditions do not correctly consider seasonal variation, differences in origin between commodities, or dynamic alterations during the storage period (Saltveit, 2003; Delele et al., 2019).

In many cases, the recommended concentration of O_2 is set up at concentrations above the anaerobic compensation point and decided minimum possible O_2 levels so that physiological disorders do not occur during the storage period (Bessemans et al., 2016).

Even using CA technology, as low O_2 levels cause the acceleration of quality loss, and extremely low O_2 levels and/or CO_2 accumulation lead to physiological disorders, sub-optimal CA conditions may cause quality loss, such as physiological disorders and fruit softening during long periods of storage (Delele et al., 2019; Thewes et al., 2022).

Chemical treatment can efficiently control the progress of physiological disorders. For instance, the chemical treatment of apples with anti-scald agents such as diphenylamine and 1-methylcyclopropene is utilized for suppressing physiological disorders and prolonging the storability of fruits by decreasing the production of ethylene and harmful volatile substances (Mditshwa et al., 2018). However, as these treatments use synthetic and environmentally unfriendly chemicals, there are negative influences. Chemical treatments are limited in their application in many countries. Therefore, non-chemical treatments have been sought after for preventing physiological disorders that occur during storage (Prange et al., 2011; Mditshwa et al., 2018).

Much evidence has been indicated that a strictly controlled low O_2 atmosphere reduces the occurrence of physiological disorders and is safe. Several advanced CA technologies that can precisely control oxygen levels associated with a low oxygen atmosphere have been recently developed to prevent physiological disorders (Mditshwa et al., 2018).

3.8.1 Rapid Controlled Atmosphere

Rapid CA rapidly establishes CA conditions after sealing the CA room. The O_2 level in the CA room is lowered below 5 kPa by purging the CA room with N_2 as quickly as possible after harvest and then by achieving further decrease by respiration (Wright et al., 2015; Bodbodak & Moshfeghifar, 2016). This technology is mostly used for many horticultural crops such as pears, apples, and cabbages and improves firmness and keeping of titratable acidity (Wright et al., 2015; Yahia et al., 2019).

3.8.2 Delayed Controlled Atmosphere

Delayed CA is a CA method in which the O_2 level in a CA room is reduced in a delayed or gradual fashion, as opposed to rapid CA (Wright et al., 2015). Rapid CA causes higher rates of internal browning in some apple cultivars, such as Fuji, Elstar, Pink Lady, and Braeburn, as well as in Conference pears (Wright et al., 2015). But delaying the establishment of CA conditions and the accumulation of CO_2 by several weeks after harvest indicated prevention of the incidence of their physiological disorders due to higher concentrations of adenosine triphosphate in fruit tissues during the entirety of CA storage (Saquet et al., 2003; Bodbodak & Moshfeghifar, 2016).

3.8.3 Ultra-Low Oxygen

The lowest recommendation of O_2 levels for traditional CA storage is almost 2 kPa. In contrast, the O_2 levels in ultra-low oxygen are set below conventional CA guidelines, generally below 1 kPa (Bodbodak & Moshfeghifar, 2016; Yahia et al., 2019). The purpose of ultra-low oxygen is to maximize the benefits of CA (Mditshwa et al., 2018). This CA technology is based on rapid CA. The O_2 levels are rapidly decreased to approximately 1–1.3 kPa after harvest, while the CO_2 level is required to regulate below 2–2.5 kPa (Yahia et al., 2019). Therefore, more airtight room is necessary than in traditional CA to keep a lower level of O_2, and the concentration of the gas must be adjusted frequently to keep it within the acceptable range (Bodbodak & Moshfeghifar, 2016). Excellent quality of fruit is found in Granny Smith and Golden Reinders apples stored at ultra-low oxygen compared with traditional CA (Mditshwa et al., 2018). But the maturity of the fruit influences the effectiveness of this technology (Mditshwa et al., 2018). This technology is utilized in pears, kiwifruit, nectarines, grapefruits, lettuce, onions, and other commodities besides apples (Ekman et al., 2005; Bodbodak & Moshfeghifar, 2016).

3.8.4 Initial Low-Oxygen Stress

This CA technology is similar to low-oxygen stress (Yahia et al., 2019) and has an exposure process at low O_2 levels below 0.5 kPa for a limited period up to 2 weeks (Bodbodak & Moshfeghifar, 2016; Mditshwa et al., 2018). This method induces a transient induction of low-oxygen stress and fermentative metabolism at the beginning of storage and is based on the concept of utilizing biological feedback with the production of a certain ethanol level during long periods of storage (Wright et al., 2015; Mditshwa et al., 2018). It is necessary to store commodities under ultra-low oxygen conditions after initial low-oxygen stress treatment (Yahia et al., 2019).

3.8.5 Repeated Low-Oxygen Stress

This method simply repeats low-oxygen stress 2–3 times based on the evaluation of ethanol during storage (Mditshwa et al., 2018). Until a particular level of ethanol that does not cause physiological disorders, low-oxygen stress treatment takes place and then the O_2 level is increased repeatedly throughout storage (Wright et al., 2015).

3.8.6 Dynamic Controlled Atmosphere

Dynamic controlled atmosphere (DCA) is a CA technology that realizes and maintains a low oxygen atmosphere variably with a dynamic form as a method that can minimally customize oxygen levels at the start and during the storage period without inducing unfavorable fermentative metabolism (Beaudry, 2010; Yahia et al., 2019; Thewes et al., 2021). A DCA system is constructed with the monitoring of gas levels in accordance with the physiology of the horticultural commodities, the adjustment of gases in storage, and the processing of data in the control system (Bodbodak & Moshfeghifar, 2016). In this technique there have been outstanding characters, including real-time monitoring by bio-sensers, non-chemical and non-destructive analysis, and rapid and frequent measurement (Tran et al., 2015). The application and development of this technology strongly binds the development of machinery technology and the information collection of plant physiological phenomenon of each horticultural commodity under excessive conditions (Yahia et al., 2019).

The changes in gas concentrations during CA storage are regularly monitored by sensors. There are several kinds of sensors used in this technology (Bodbodak & Moshfeghifar, 2016). These monitor the low oxygen limit by quantifying different metabolic processes, chlorophyll fluorescence, ethanol, carbon dioxide, and respiration quotient during CA storage (Mditshwa et al., 2018).

Various stresses, including extremely low O_2 in storage, high temperature, extreme radiation, and drought occur in chlorophyll fluorescence. These stresses decrease the transmission of excitation energy to the reaction center in photosystems (Thewes et al., 2021). Chlorophyll fluorescence can predict the level of stress response and development of physiological disorder during storage of horticultural commodities (Tran et al., 2015). O_2 concentrations change according to physiological responses in plants. Low O_2 stress extremely affects the minimum fluorescence (F_0) of photosynthesis (Wright et al., 2015; Mditshwa et al., 2018). When O_2 concentration lowers to critical levels, F_0 greatly increases as electron flux through photosystem II is suppressed (Beaudry, 2010; Yahia et al., 2019). HarvestWatch™ and FruitObserver®, known as commercially available systems, are globally utilized for measuring relative levels of low O_2 stress (Beaudry, 2010; Mditshwa et al., 2018; Thewes et al., 2021). By the monitoring system of chlorophyll fluorescence, it is possible to store fruits and vegetables at a level slightly above LOL at which commodities can be stored without incurring undesirable changes, decreasing metabolism of horticultural commodities, and maintaining quality compared with static CA (Wright et al., 2015; Mditshwa et al., 2018; Thewes et al., 2021). The effect of chlorophyll fluorescence in DCA storage has been evaluated. Using chlorophyll fluorescence to monitor the response against extremely low oxygen as a non-destructive indicator, storage atmosphere could be optimized and maintain quality in Greenstar apples (Tran et al., 2015). DCA using chlorophyll fluorescence effectively maintained the quality after storage of Royal Gala and Galaxy apples during long periods of storage (Thewes et al., 2015).

Ethanol is the product of ethanolic fermentation metabolism. LOL is determined by measuring ethanol concentrations (Wright et al., 2015; Mditshwa et al., 2018). Measuring ethanol levels was first adopted in the DCA system to monitor fruits' response to variations of O_2 levels in the storage room (Thewes et al., 2021). O_2 levels in the storage room are gradually reduced until the quality of horticultural commodities are considered to be unaffected by ethanol concentration. As ethanol in the storage room is detected, O_2 levels are increased. DCS™ (Dynamic controlled system) is known as a commercially available system (Thewes et al., 2021). This system is placed inside the CA room and the ethanol concentration of headspace gas in the CA room is determined (Wright et al., 2015; Mditshwa et al., 2018). DCA using ethanol sensors decreased the incidence of skin spots and maintained higher firmness in Elstar apples under the lowest possible O_2 concentrations (Veltman et al., 2003).

The production of carbon dioxide is an important indicator in metabolism during storage. When O_2 partial pressure (pO_2) reduces until ACP at aerobic condition, the CO_2 production rate also decreases. But when pO_2 is decreased below ACP, the CO_2 production rate sharply increases. Estimating LOL during storage in real time is enabled by measuring CO_2 using a gas analyzer (Thewes et al., 2020). DCA based on CO_2 production could correctly estimate LOL, have lower incidence in flesh breakdown, maintain higher flesh firmness and decrease the incidence of decay in Imperial Gala and Golden Delicious apples (Thewes et al., 2020). Similarly, this DCA safely induced anaerobic metabolism, consequently increasing the availability of substrate to synthesis aroma compounds (Bűchele et al., 2023).

Respiratory quotient (RQ) in fruits and vegetables during storage can estimate LOL. RQ is the ratio of CO_2 production to O_2 consumption of horticultural commodities (Bessemans et al., 2016; Wright et al., 2015; Mditshwa et al., 2018). In DCA based on RQ, O_2 and CO_2 levels during storage are dynamically measured and the RQ value is calculated in real time (Delele et al., 2019). The RQ value is approximately 1.0 during aerobic conditions, but it increases when O_2 levels during storage fall below acceptable concentrations (Mditshwa et al., 2018; Delele et al., 2019; Thewes et al., 2020). The optimal value in RQ for stored apples is 1.3–1.5. Fruit with an RQ value of 1.3 generally maintains better firmness in flesh and decreases decay and incidences of physiological disorders (Thewes et al., 2021). The increase of the RQ value detects the switching point between aerobic and anaerobic respiration and the critical point of ethanol production (Weber et al., 2023). Advanced Control Respiration (ACR), SafePod®, and RQ-SToreFresh are utilized commercially to monitor RQ during storage (Mditshwa et al., 2018). DCA based on RQ for Maxi Gala apples maintained better quality at a constant 2 °C for 9 months of storage plus 7 and 14 days of shelf life (Wendt et al., 2022). When comparing responses of low O_2 by monitors in apples, the response of RQ measured by the SafePod® coincided with the increase in yield of chlorophyll fluorescence yield measured by HarvestWatch™ (Rees et al., 2021). RQ monitoring has an advantage with the direct measurement of LOL due to the ratio of CO_2 production to O_2 consumption, whereas ethanol and chlorophyll monitors are systems to determine LOL due to secondary phenomenon and cannot directly monitor respiration (Thewes et al., 2021).

3.8.7 Insecticidal Controlled Atmosphere

Controlled atmospheres that combine extremely low O_2 levels and remarkably high CO_2 levels indicate insecticidal effects. The insecticidal atmospheric composition should be below 1 kPa O_2 and/or over 50 kPa CO_2. Exposure with CA is short, from a few hours to a few days, depending on the temperature (Yahia, 2019). These insecticidal CA atmospheres are the optimum concentrations used for transport or storage of almost all horticultural commodities, and thus usually evoke stress in these commodities. The potential problem for development and application of this system is the possible fermentation in crop tissue. Conquering this problem usually becomes the restrictive factor in the development of an effective insecticidal CA technology (Prange et al., 2005; Yahia, 2006).

Insect mortality by insecticidal CA depends on the species and the development stage of the insects. For mechanisms of insecticidal CA, it is thought that extreme high CO_2 causes the spiracles of the pest to remain open, and consequently they become dehydrated (Yahia, 2019).

3.9 QUALITY OF FRUIT AND VEGETABLES DURING AND AFTER CONTROLLED ATMOSPHERE STORAGE

Technology in CA storage that utilizes reduced O_2 and/or elevated CO_2 atmospheres maintains quality and prolongs storability of horticultural products. The potential advantages of CA storage for horticultural products are to delay ripening and senescence, to hold attributes of quality and appearance, to maintain nutritional compounds, and to control physiological disorders and postharvest diseases (Imahori, 2012; Madani et al., 2019). Whereas, if excessive modification of the atmosphere occurs beyond the appropriate range of gas concentrations, extremely low O_2 and/or very high CO_2 induce detrimental effects in horticultural commodities (Imahori, 2012; Plotto et al., 2020).

3.9.1 Effect of Controlled Atmosphere on Quality Attributes

The quality in fruits and vegetables consists of numerous attributes that complexly interact with each other. Generally, these attributes are decided by either the subjective or objective decisions of the consumer. Horticultural products are utilized not only for their nutritional value but also for their appearance, such as color, flavor, and aroma (Yahia, 2019). CA storage can maintain quality attributes in fruits and vegetables. However, it is shown that CA storage can have harmful effects on the quality attributes of horticultural products (Madani et al., 2019).

Organic acids and sugars and are related to sourness and sweetness in fresh horticultural products. Acidity is determined by pH, individual acid levels, or titratable acidity (TA), whereas sugars are determined by the quantification of individual sugars or soluble solids content (SSC). SSC, TA, and SSC/TA are measured as a quality index of fresh horticultural products (Plotto et al., 2020).

Extending cold storage of Rich Lady peaches under air for 2 weeks lowered sensory acceptance scores, but CA storage improved their acceptability due to the enhancement of sweetness and flavor by high SSC, low TA, and the release of specific volatiles (Ortiz et al., 2009).

Apples stored in CA storage maintained high sugar contents containing sucrose, fructose, and glucose by increasing sucrose synthesis and delaying sucrose hydrolysis, by enhancing the gene expression and activity of sucrose synthesis–related enzymes (Zhu et al., 2013).

CA prevented the loss of acidity in 1-methylcyclopropene pretreated Abbé Fétel pears after long-term storage (Rizzolo et al., 2014). Similarly, CA maintained the acidity in kiwifruits during cold storage (Latocha et al. 2014). Apples stored in DCA reduced the respiration and ethylene synthesis, and consequently maintained TA compared with those in regular atmosphere and static CA (Bűchele et al., 2023).

The most obvious effect of CA has been keeping the firmness of horticultural products. Fruit softening related with ripening and senescence causes the destruction of cohesiveness and structure in cell wall tissue. Cell walls consist of cellulose and hemicellulose microfibrils imbedded in a matrix of pectin polysaccharides and proteins. Fruit softening is caused by the structural loss of tissues due to the decomposition and solubilization of hemicellulose and pectin in the middle lamella (Plotto et al., 2020). DCA indicated better quality than static CA by reducing the fruit firmness loss in apples (Weber et al., 2023). High CO_2 treatment improved the fruit firmness in strawberries, but its effect was dependent on cultivar, CO_2 concentration, and exposure period (Plotto et al., 2020), whereas CO_2 concentrations higher than 6–12 kPa in MA packaging caused fruit softening in blueberries (Rodriguez & Zoffoli, 2016).

Color changes in fruits and vegetables are strongly influenced by atmospheric compositions during storage, especially changes from green to yellow during fruit ripening and senescence. Usually, CA storage decreases the loss of chlorophyll contents and similarly reduces the production of other pigments such as carotenoids, lycopene, anthocyanin, and xanthophylls (Mattheis & Fellman, 2000). Snow pea pods stored at 5°C by MA packaging equilibrated with CA atmosphere concentrations of 5 kPa O_2 and 5 kPa CO_2 maintained chlorophyll contents better than those stored at air (Pariasca et al., 2001).

3.9.2 Effect of Controlled Atmosphere on Phytochemical Compounds

Horticultural commodities have an incredibly significant role in a healthy human diet and are sources of minerals, vitamins, dietary fiber, and phytochemicals. Phytochemicals contain carotenoids and flavonoids such as anthocyanins, phenolic acids, and polyphenols and can reduce the risk of cancer, heart disease, and other diseases. The level of antioxidant capacity associates with ascorbic acid, phenolic compounds, anthocyanins, carotenoids, and tocopherols (Kader, 2009).

Snow pea pods stored at 5°C by MA packaging equilibrated with CA atmosphere concentrations of 5 kPa O_2 and 5 kPa CO_2 maintained more ascorbate contents than those stored at air (Pariasca et al., 2001). Similarly, broccoli heads by MA packaging prevented the loss of ascorbate and the reduction of total antioxidant capacity better than those stored at 1°C for 28 days (Serrano et al., 2006). In Bohnapfel apples stored for 7 months by CA storage of 1 kPa O_2 and 3 kPa CO_2 atmosphere conditions, ascorbate content was decreased but total antioxidant capacity was maintained (Trierweiler et al., 2004).

External (skin) and internal (flesh) color in strawberries changed to a darker red color and anthocyanin contents increased after storage at 5°C in 2 kPa O_2 for 10 days (Holcroft & Kader, 1999). In table grapes, anthocyanin contents in a CA atmosphere of 20 kPa O_2 and 20 kPa CO_2 are higher than those in non-treatment, although low temperatures (0°C) induced accumulation of anthocyanin in both CA and non-treatment (Romero et al. 2008). Blueberries stored at 5°C in a CA atmosphere of elevated O_2 (60–100 kPa) for up to 35 days encouraged the increase of total phenolics and total anthocyanins (Zheng et al., 2003).

Carotenoid synthesis such as lycopene of tomatoes or β-carotene of mangoes is suppressed by the delay of fruit ripening due to CA storage (Bodbodak & Moshfeghifar, 2016).

3.9.3 Effect of Controlled Atmosphere on Flavor and Aroma

Flavor and aroma in horticultural commodities are significant quality attributes that determine their acceptability by consumers (Sing et al., 2013). The flavor is a combination of taste and aroma and depends on their degrees. Taste depends on a balance of sweetness, acidity, or sourness, and low or no astringency. Aroma depends on odor active concentrations of volatile compounds (Kader, 2008; Sing et al., 2013). Volatile compounds consist of esters, alcohols, aldehydes, and ketones (Kader, 2008). Specific flavors in fruits and vegetables are produced from various primary and secondary metabolites, while aroma volatiles are synthesized from various metabolic pathways in amino acids, fatty acids, terpenoids, glucosinolates, and pigmented compounds (Sing et al., 2013). The overall aroma of horticultural commodities is dictated by a balance between all compounds and by the interaction between volatile compounds of a recognized aroma. The influence of individual volatiles is dependent on its odor threshold and the odor active concentration of compounds (Cukrov et al., 2019).

In general, CA delays ripening and senescence of horticultural commodities, consequently suppressing aroma production and reducing aroma development and accumulation. However, CA influences the flavor of fruits and vegetables both positively and negatively (Plotto et al., 2020; Bodbodak & Moshfeghifar, 2016). Apples stored in CA experienced reduced development of volatile compounds such as alcohols, esters, and 1-methoxy-4(2-propenyl) benzene, which was recognized by a sensory test (Bai et al., 2005). Royal Gala apples stored under regular CA for 9 months at 1 °C plus 7 days at 20 °C contained higher amounts of key compounds such as butyl acetate, 2-methylbutyl acetate, and hexyl acetate than fruits stored under DCA storage based on chlorophyll fluorescence. This DCA storage reduced ester production, especially 2-methylbutyl acetate. But DCA storage based on respiratory quotient contained higher levels of key volatiles with increased ethanol and ethyl acetate, although further below the threshold due to the development of anaerobic metabolism (Both et al., 2017). In peaches stored at three different CA conditions for 40 days at −0.5 °C, atmosphere composition affected the flavor perception of fruits, and lower O_2 CA condition in 2 kPa O_2 and 5 kPa CO_2 reduced flavor perception related to acids and several volatiles such as lactones, esters, and aldehyde (Cano-Salazar et al., 2013). Suppression of aroma production is particularly caused by ultra-low oxygen storage or long-term CA. Usually, low O_2 levels, higher CO_2 levels, and longer storage period result in decreased production of aroma volatile (Plotto et al., 2020).

The development of off-flavor is a major issue in the storage and handling of horticultural commodities after harvest. The appearance and perception of off-flavor relates to the development of fermentation metabolism and the accumulation of its products, acetaldehyde and ethanol (Porat & Fallik, 2008). Optimal gas conditions for CA are altered according to cultivar, maturity of ripening stage, storage temperature, and exposure duration. These sudden changes enhance fermentative metabolism and result in fermentation damage (Porat & Fallik, 2008). The off-flavor is the most general and significant harmful symptom that occurs beyond the tolerance of fruits and vegetables to low O_2 (Imahori et al., 2005).

3.9.4 Effect of Controlled Atmosphere on Physiological Disorders

The commercial utilization of CA in horticultural commodities involves delaying the deterioration of horticultural commodities, inhibiting the damage of pests, and killing the pest entirely. The beneficial effect of CA is the regulation of various metabolic processes that influence the appearance of physiological disorders.

Physiological disorder or physiological injury is a visible event of disturbance in the normal metabolism of horticultural commodities and is not concerned with infection of microorganisms (Prange & DeLong, 2006). Physiological disorders are associated with nutrition, temperature, respiration, senescence, and miscellaneous other factors. Regarding physiological disorders of fruits and vegetables, CA indicates an alleviative effect in some cases and an exacerbated effect in others, as well as occasional induction (Schotsmans et al., 2009). Abnormal fruit ripening, browning, and destruction of tissues and accumulation of toxic fermentation products occurs in horticultural commodities during or after long-term CA or unsuitable CA storage conditions (Cukrov et al., 2019). The appearance and extent of physiological disorders are affected by CA atmosphere composition and its concentration. Additionally, cultivar, growing environment, harvest maturity of commodities, CA storage temperature, and CA storage humidity affect physiological disorders. As cultivars and individual products have differences in intercellular space size and gas diffusion rate passing between the surface and cells individually, the susceptibility to physiological disorders differs among each horticultural commodity (Bodbodak & Moshfeghifar, 2016).

When horticultural commodities are transported to or stored at a temperature below a certain threshold, they cannot conduct their normal metabolic activities, and consequently physiological and biochemical alterations occur. If these crops are exposed to the damaging low temperature for too long, changes to several metabolic processes occur, followed by changes in appearance and cell death. This physiological disorder is called chilling injury (CI) and takes place in several horticultural commodities (Endo et al., 2019; Imahori et al., 2021; Kantakhoo et al., 2022). The advantageous efficacies of CA storage and MA treatment include delay of ripening, reduction of physiological and pathological disorders such as CI, and the possibility for disinfesting (Imahori, 2012). The CA condition with elevated CO_2 (5 kPa) decreases CI levels and increases shelf life via maintenance of the membrane functionality of guava fruit (Alba-Jiménez et al., 2018). The MA condition with reduced O_2 (14.9–16.7 kPa) and elevated CO_2 (4.2–7.3 kPa) for tomato fruit prevents the CI symptoms and maintains quality at 4°C for 14 days (Park et al., 2018). The hypoxia pretreatment with 3 kPa O_2 at room temperature for 24 h decreased CI symptoms of avocado fruit stored at 2°C for 3 weeks. The pretreated fruit indicated lower respiration and ethylene levels, and fewer symptoms of electrolyte leakage than the control. The pretreatment with 40 kPa CO_2 for 3, 6, or 9 days decreased CI symptoms of lemon fruit during storage at 0°C after 30, 60, or 90 days (Sevillano et al., 2009). In contrast, green Chondrolia table olives stored at 5°C in high CO_2 (2 or 5 kPa) or CA (2 kPa O_2 and 2 or 5 kPa CO_2) increased CI symptoms, such as the occurrence of brown-black areas on the fruit's surface (Nanos et al., 2002).

Russet spotting is an ethylene injury that is a senescence-based physiological disorder caused by ethylene concentrations as low as 0.1 ppm and occurs in crisphead lettuce, appearing as small brown spots on both sides of the midrib (Prange & DeLong, 2006, Schotsmans et al., 2009). Occurrence of this disorder was reduced by O_2 levels below 8 kPa or CO_2 levels of 5 kPa or more (Prange & DeLong, 2006).

Superficial scald or storage scald is a lightly unequal browning on the surface, or more exactly the hypodermis, of apples and pears after harvest and results in a loss of marketability as fresh fruit (Prange & DeLong, 2006; Schotsmans et al., 2009). The occurrence of storage scald can suppress by CA methods of initial low oxygen stress or DCA due to reducing respiration, the production of ethylene, and metabolic activity (Schotsmans et al., 2009). In Gem pears stored in CA of 1 kPa O_2 and 0.05 kPa CO_2 at –1.1°C for 8–10 months, the incidence of superficial scald and internal breakdown was reduced due to the delay of ethylene production's peak (Dong et al., 2023).

In apples and pears, internal browning, also called core browning, brown core, or brown heat of core flesh, is generally used to refer to several disorders characterized by the browning of the cortex and/or core and subsequent or concurrent tissue breakdown (Prange & DeLong, 2006; Schotsmans et al., 2009). Core browning and soft scald occur in apples due to extreme cold storage temperatures (Prange & DeLong, 2006). Apples stored in DCA reduced internal browning incidences compared with those in regular atmosphere and static CA (Bűchele et al., 2023).

Bitter pit, Jonathan spot, cork spot, or lenticel blotch pit that occur in apples, or cork spot in pears, are called pitting disorders. These disorders occur due to mineral imbalance caused by calcium deficiency against relatively high levels of magnesium and potassium (Prange & DeLong, 2006; Schotsmans et al., 2009). Their symptoms similarly appear on the surface of fruit and in the outer cortex under the epidermis of the fruit (Schotsmans et al., 2009). A symptom of bitter pit is a tiny, brownish, dry, and slightly bitter tasting lesion (3–5 mm in cross-section) in the flesh tissues (Prange & DeLong, 2006; Schotsmans et al., 2009). CA storage and MA storage can prevent the incidence and development of physiological disorders such as Jonathan spot and bitter pit due to a delay of senescence in apples (Prange & DeLong, 2006, Schotsmans et al., 2009).

3.9.5 Effect of Controlled Atmosphere on Pathological Diseases

Chemical treatment can efficiently control pathological diseases and pest infestation in postharvest storage and is widely utilized. However, as these chemicals are environmentally unfriendly, there are negative effects. Non-chemical treatment has been sought due to reducing the applied amounts and residues of chemicals in horticultural commodities (Mditshwa et al., 2018). Atmosphere environment that changed O_2 and CO_2 levels can influence the metabolism of microorganisms and regulate them during the storage period of horticultural commodities. Reduced O_2 and especially elevated CO_2 in CA treatment indicate negative influences on the growth and development of microorganisms. Additionally, the infestation of diseases is prevented in fruits ejected from high CO_2 CA storage than those previously stored under air (Thompson, 2010). However, high CO_2 levels effective to disease are rather injurious to the quality of horticultural commodities, and CA treatment with ultra-low oxygen is necessary. Physiological conditions such as respiration also influence the tolerance of horticultural commodities to CA treatment (Thompson, 2010; Liu, 2020).

On *in vitro* experiment at 1.5 °C, Chinese cabbage infected with fungi (*Phytophthora brassicae*) showed lower disease development in CA storage of 1.5 kPa O_2 and 0.5 kPa CO_2, or 3 kPa O_2 and 3 kPa CO_2 than control (Hermansen & Hoftun, 2005). Higher CO_2 (15–20 kPa) delayed incidences of decay in cherries, blackberries, blueberries, raspberries, strawberries, figs, and grapes (Thompson, 2010). In control of *Botrylis cinerea* rot in Hayward kiwifruits inoculated with *Botrylis cinerea* and stored in air for 12 weeks at 0 °C after higher CO_2 CA treatment, due to increases in antifungal compound concentrations by CA treatment, fruits that were exposed to 60 kPa CO_2 at 40 °C completely inhibited spore germination and growth and decreased the incidence of disease, while fruits that were exposed to 60 kPa CO_2 at 30 °C were partly reduced (Thompson, 2010).

3.9.6 Effect of Controlled Atmosphere on Pest Infestation

Atmospheres in extremely decreasing O_2 and highly increasing CO_2 act on the metabolism in insect pests and show insecticidal effects for the control of insects (Yahia, 2006). Insecticidal CA treatment is a safe and environmentally friendly physical treatment compared to chemical fumigants. It is possible that not all pests will be controlled with CA treatment under a tolerance condition that can be accepted by horticultural commodities. Susceptibility to CA in pests is different across pest groups and species, as well as variations in the life stages of the pest (Liu, 2020). Also, the efficacy in CA storage depends on O_2 and CO_2 concentrations, treatment temperature, exposure period, relative humidity, the species of the insect, and the development stage of the pest. The extremely reduced O_2 concentration, higher CO_2 concentration, high temperature, low relative humidity, and short time are significant for the extermination of pests. Usually, low O_2 concentrations under 1 kPa and high CO_2 concentrations over 20 kPa are necessary for insecticidal CA atmospheres. Pest control can be completed within 2–4 days at room temperature or within 2–4 hours at a high temperature (Yahia, 2006).

Tolerance to CA treatment varies with the difference of horticultural commodities. Extreme O_2 and CO_2 concentrations cannot be tolerated by many fruits and vegetables due to the development of anaerobiosis. Green onions and broccoli can tolerate high CO_2, whereas lettuce is sensitive to CO_2 (Liu, 2020). Thus, the development of anaerobic metabolism due to the mixture of very low O_2 and very high CO_2 results in the accumulation of fermentation products, and it limits the utilization of insecticidal CA for horticultural commodities (Yahia, 2006).

Ultra-low O_2 treatments for 5 days with 0.003 kPa O_2, and for 10 days with 0.03 kPa O_2, perfectly controlled western flower thrips on broccoli at 1°C (Liu, 2007), while higher CO_2 (60 kPa) treatment effectively controlled coding moth larvae better than 0.5 kPa O_2 treatment (Soderstrom et al., 1990). The combination of insecticidal CA and high temperature can eliminate many pests within 2–4 hours instead of days. Higher temperatures above 40°C accomplished the faster mortality of pests under insecticidal CA, although horticultural commodities are susceptible to fermentation and heat injury (Yahia, 2006; Mitcham, 2007). Cherries treated for 41 min at 45°C or for 27 min at 47°C in 1 kPa O_2, 15 kPa CO_2 could be protected from western cherry fruit flies and codling moths (Retamales et al. 2003). Similarly, treatments for 45 min at 45°C or for 25 min 47°C by an atmosphere of 1 kPa O_2 + 15 kPa CO_2 in sweet cherries could control all development stages of codling moths and maintain market quality (Neven, 2005).

3.10 POSTHARVEST TREATMENT OF FRUITS AND VEGETABLES FOR CONTROLLED ATMOSPHERE STORAGE

In many cases, practical application of CA technology needs the combination of other postharvest technologies to store and transport horticultural commodities without problems in the logistics stage. Thus, the combined treatment with other postharvest treatments can preserve the quality after harvest and prolonging the storage life of commodities.

3.10.1 High-Temperature Treatment

Root vegetables like sweet potatoes, potatoes, and yams have a cork layer on their surface. The cork layer acts as protection from microorganism infection and excess water loss. But it subjects root vegetables to damage in harvest and handling. Exposing root vegetables to high temperatures and humidity for a few days can repair the damaged parts and decrease postharvest loss. This treatment is known as curing and is also utilized with citrus fruits (Thompson, 2010; Bodbodak & Moshfeghifar, 2016).

Drying is also a high-temperature treatment and helps the storage of bulb onions and garlic. It is a simple treatment and dries the outer layer to reduce microorganism infection and excess water loss (Thompson, 2010; Bodbodak & Moshfeghifar, 2016).

Immersion in hot water, or rinsing or brushing with hot water, is a pretreatment for reducing microorganism infection (Fallik, 2004). Visible symptoms by anthracnose in ripening was effectively prevented by hot water immersion treatment for 75 min at 46 °C and storage for 2 weeks at 10 °C in CA condition of 3 kPa O_2 or 3 kPa O_2 and 10 kPa CO_2 on mature green mangoes (Kim et al., 2007). Shortly exposing citrus fruit to a higher CO_2 or O_2 atmosphere at curing temperatures of 37 °C controls the infection of blue and green molds (Montesions-Herro et al., 2012).

Heat treatment can effectively alleviate ripening heterogeneity of horticultural commodities. Hass avocado fruits experienced significantly reduced ripening heterogeneity for early and middle season fruits by hot water immersing for 1 h at 38 °C followed by storage for 30 days at 5 °C in CA of 4 kPa O_2 and 6 kPa CO_2 (Hernández et al., 2017).

3.10.2 1-Methylcyclopropene Treatment

1-Methylcyclopropene (1-MCP) is a potent inhibitor that prevents the ethylene action at extremely low concentrations by inhibiting ethylene receptor binding in horticultural crops. The 1-MCP treatment efficiently controls ripening and senescence of horticultural products and maintains the quality of horticultural commodities and decreases the appearance of physiological disorders (Steffens et al., 2021; Weber et al., 2023). The 1-MCP delays the peak appearance of 1-aminocyclopropane-1-carboxylic acid levels and the corresponding expressions of transcript in ethylene receptors (Gang et al., 2009). 1-MCP is used as a gaseous fumigant in extremely low concentrations (ppb to ppm) for a short time directly after harvest in various horticultural crops, but mainly in apples. The utilization of 1-MCP became possible via aqueous dipping or being sealed in plastic bags. With the development of commercial chemical agents, 1-MCP was available to use in storage rooms and gastight tents or in containers before loading into storage rooms or gastight rooms (Thompson, 2010). Gastight curtains were temporarily used to partition part of the storage room by a zippered access door to independently fumigate a relatively small volume of commodities. Gastight pallet covers were also used for 1-MCP fumigation (Thompson, 2010; Bodbodak & Moshfeghifar, 2016).

CA storage inhibits the biosynthesis in ethylene and the ethylene action and reduces the metabolic activity in horticultural commodities. Although meaningful extensions in shelf life are also accomplished by 1-MCP treatment only, the greatest beneficial effects by 1-MCP treatment were derived when combined with CA storage of apples, (Johnson, 2008). After 1-MCP treatment, long-term CA storage for apples indicated the most effective quality retention (Prange et al., 2005). Extremely low O_2 and/or very high CO_2 concentrations in CA storage often results in a high occurrence of physiological disorders. In the CA storage of pears, CA extended or strengthened the influences of 1-MCP treatment, while 1-MCP treatment could not replace CA storage but could enhance the effects of CA (Rizzolo et al., 2005).

DCA storage with 1-MCP treatment, and subsequently ultra-low O_2 storage, is indicated as an effective replacement method for DPA treatment on apple cultivars that are susceptible to symptoms of scald that occur during long periods of storage (Raffo et al., 2009).

3.11 CONCLUSIONS

Controlled atmosphere (CA) storage is a system of postharvest technology that has widely developed to maintain the postharvest quality of horticultural commodities. It became possible to build the worldwide transport and supply chain of commodities and to provide the year-round supply system with this technology. CA is one of most successful postharvest technologies that has been introduced to the horticultural commodity industry. CA storage utilizes decreased O_2 and/or increased CO_2 atmospheres and is a form of gaseous storage, keeping the postharvest quality and extending the storage life of horticultural commodities. Exposing commodities to decreased O_2 and/or increased CO_2 gases brings either benefits or detrimental effects. It depends on the levels of these gases, treatment temperature, exposure period, and the species of horticultural commodity. Exposure of horticultural commodities to reduced O_2 and/or increased CO_2 is very risky, as it may contribute to harmful effects in fruits and vegetables. Therefore, understanding the physiological and biochemical effects of reduced O_2 and elevated CO_2 atmospheres to horticultural commodities is incredibly significant for research and development of CA technology.

Several functions are necessary for establishing and maintaining CA storage. Their functions are composed of O_2 removal, CO_2 removal, and ethylene removal, as well as the supplementation of air for the replacement of O_2 used by respiration and the addition of CO_2. Selection of the suitable functions and devices depends on the kinds of horticultural commodities and storage conditions, and the success of CA storage depends on achieving accurate levels of atmosphere gas and keeping a proper and steady composition of atmosphere gas within the CA storage room. Control of the atmosphere in the storage room is influenced by physiological conditions of the commodities that are stored in the CA room, the O_2 and CO_2 controlling equipment, the refrigeration system, and weather conditions. Thus, it needs to be considered that the effects of CA are not simple and rely on other environmental factors.

Although the storability of fruits and vegetables is extended by leaps and bounds with the utilization of CA technologies, CA conditions of fruits and vegetables are usually established on experiential observation and recommended conditions of static and crop-specific O_2 and CO_2 levels do not consider seasonal variation and differences in origin between batches of crops and the dynamic alterations occurring during the storage period. Consequently, low O_2 levels can cause the acceleration of quality loss, and extremely low O_2 levels and/or CO_2 accumulation cause physiological disorders, such as physiological disorders and fruit softening. Chemical treatment can efficiently control the development of physiological disorders. However, as these treatments use synthetic and environmentally unfriendly chemicals, there are negative effects. Although much evidence has been indicated that a strictly controlled low O_2 atmosphere can decrease the occurrence of physiological disorders and is safe, several advanced CA technologies that can precisely control oxygen levels associated with a low-oxygen atmosphere have been recently developed to prevent physiological disorders and have been applied to CA storage of apples and pears.

Recently many progressive research and development studies have been advanced and permitted for the diversity of utilization for CA technology. However, research and development are still necessary to raise the utilization of the technology and applications of storage, transport, and packaging for more horticultural commodities.

REFERENCES

Alba-Jiménez, J.E., Benito-Bautista, P., Nava, G.M., Rivera-Pastrana, D.M., Vázquez-Barrios, M.E., Mercado-Silva, E.M. (2018). Chilling injury is associated with changes in microsomal membrane lipids in guava fruit (*Psidium guajava* L.) and the use of controlled atmospheres reduce these effects. *Scientia Horticulturae*, 240, 94–101.

Bai, J., Baldwin, E.A., Goodner, K. L., Mattheis, J.P., Brecht, J.K. (2005). Response of four apple cultivars to 1-methylcyclopropene treatment and controlled atmosphere storage. *HortScience*, 40, 1534–1538.

Beaudry, R.M. (1999). Effect of O_2 and CO_2 partial pressure on selected phenomena affecting fruit and vegetable quality. *Postharvest Biology and Technology*, 15, 293–303.

Beaudry, R.M. (2010). Future trends and innovation in controlled atmosphere storage and modified atmosphere packaging technologies. *Acta Horticulturae*, 876, 21–28.

Benkeblia, N. (2021). Physiological and biochemical response of tropical fruits to hypoxia/anoxia. *Frontiers in Plant Sciences*, 12, Article 670803.

Bessemans, N., Verboven, P., Verlinden, B.E., Nicolai, B.M. (2016). A novel type of dynamic controlled atmosphere storage based on the respiratory quotient (RQ-DCA). *Postharvest Biology and Technology*, 115, 91–102.

Bodbodak, S., Moshfeghifar, M. (2016). Advances in controlled atmosphere storage of fruits and vegetables. In *Eco-friendly technology for postharvest produce quality*, ed. M.W. Siddiqui, 39–76. Amsterdam: Elsevier.

Both, V., Thewes, F.R., Brackmann, A., de Oliveira Anese, R., de Freitas Ferreira, D., Wagner, R. (2017). Effects of dynamic controlled atmosphere by respiratory quotient on some quality parameters and volatile profile of 'Royal Gala' apple after long-term storage. *Food Chemistry*, 215, 483–492.

Brizzolara, S., Santucci, C., Tenori, L., Hertog, M., Nicolai, B., Stürz, S., Zanella, A., Tonutti, P. (2017). A metabolomics approach to elucidate apple fruit responses to static and dynamic controlled atmosphere storage. *Postharvest Biology and Technology*, 127, 76–87.

Bűchele, F., Khera, K., Thewes, F.R., Kittemann, D., Neuwald, D.A. (2023). Dynamic control of atmosphere and temperature based on fruit CO_2 production: practical application in apple storage and effects on metabolism, quality, and volatile profiles. *Food and Bioprocess Technology*, https://doi.org/10.1007/s11947-023-03079-0

Cano-Salazar, J., López, M.L., Echeverría, G. (2013). Relationships between the instrumental and sensory characteristics of four peach and nectarine cultures stored under air and CA atmosphere. *Postharvest Biology and Technology*, 75, 58–67.

Cukrov, D., Brizzolara, S., Tonutti, P. (2019). Physiological and biochemical effects of controlled and modified atmosphere. In *Postharvest physiology and biochemistry of fruits and vegetables*, ed. E. M. Yahia, 425–4414. Duxford: Elsevier.

Delele, M.A., Bessemans, N., Gruyters, W., Rogge, S., Janssen, S., Verlinden, B.E., Smeets, B., Ramon, H., Verboven, P., Nicolai, B.M. (2019). Spatial distribution of gas concentrations and RQ in a controlled atmosphere storage container with pear fruit in very low oxygen conditions. *Postharvest Biology and Technology*, 156, 110903.

Dijkink, B.H., Tomassen, M.M., Willemsen, J.H.A., van Doon, W.G. (2004). Humidity control during bell pepper storage, using a hollow fiber membrane contactor system. *Postharvest Biology and Technology*, 32, 311–320.

Dilley, D.R. (2006). Development of controlled atmosphere storage technologies. *Stewart Postharvest Review*, 6, 5.

Dong, Y., Zhi, H., Leisso, R. (2023). Physiological disorders, textural property, and cell wall metabolism of 'Gem' pears (*Pyrus communis* L.) affected by oxygen regimes and harvest maturity under controlled atmosphere storage. *Scientia Horticulturae*, 317, 112067.

Ekman, J.H., Golding, J.B., McGlasson, W.B. (2005). Innovation in cold storage technologies. *Stewart Postharvest Review*, 3, 1–14.

Endo, H., Miyazaki, K., Ose, K., Imahori, Y., (2019). Hot water treatment to alleviate chilling injury and enhance ascorbate-glutathione cycle in sweet pepper fruit during postharvest cold storage. *Scientia Horticulturae*, 257, 108715.

Falagán, N., Terry, L.A. (2018). Recent advances in controlled and modified atmosphere of fresh produce. *Johnson Matthey Technology Review*, 62, 107–117.

Fallik, E. (2004). Prestorage hot water treatments immersion, rinsing and brushing. *Postharvest Biology and Technology*, 32, 125–134.

Gang, M., Wang, R., Cheng-Rong, W., Kato, M., Yamawaki, K., Fei-Fei, Q., Hui-Lian, X. (2009). Effect of 1-methylcyclopropene on expression of genes for ethylene biosynthesis enzymes and ethylene receptors in post-harvest broccoli. *Plant Growth Regulation*, 57, 223–232.

Hermansen, A., Hoftun, H. (2005). Effect of storage in controlled atmosphere on post-harvest infections of Phytophthora brassicae and chilling injury in Chinese cabbage (*Brassica rapa* L pekinensis (Lour) Hanelt). *Journal of the Science of Food and Agriculture*, 85(8), 1365–1370.

Hernández, I., Fuentealba, C., Olaeta, J.A., Poblete-Echeverría, C., Defilippi, B.G., González-Agüero, M., Campos-Vargas, R., Lurie, S., Pedreschi, R. (2017). Effects of heat shock and nitrogen shock pre-treatments on ripening heterogeneity of Hass avocados stored in controlled atmosphere. *Scientia Horticulturae*, 225, 408–415.

Hoehn, E., Prange, R.K., Vigneault, C. (2009). Storage technology and application. In *Modified and controlled atmospheres for the storage, transportation, and packaging of horticultural commodities*, ed. E. M.Yahia, 17–50. Boca Raton: CRC Press.

Holcroft, D., Kader, A.A. (1999). Controlled atmosphere-indued changes in pH and organic acid metabolism may affect color of stored strawberry fruit. *Postharvest Biology and Technology*, 17, 19–32.

Imahori, Y. (2012). Postharvest stress treatments in fruits and vegetables. In *Abiotic stress responses in plants, metabolism, productivity and sustainability*, ed. P. Ahmad and M.N.V. Prasad, 347–358. New York: Springer Science1business media.

Imahori, Y. (2014). Role of ascorbate peroxidase in postharvest treatment of horticultural crops. In *Oxidative damage to plants: Antioxidant networks and signaling*, ed. A. Ahmad, 425–451. San Diego: Elsevier.

Imahori, Y. (2021). Gas damage. In *Engeiriyougaku*, ed. N. Yamauch, Y. Imahori, 223–228. Tokyo: Buneido.

Imahori, Y., Bai, J., Ford, B.L., Baldwin, E.A. (2021). Effect of storage temperature on chilling injury and activity of antioxidant enzymes in carambola "Arkin" fruit. *Journal of Food Processing and Preservation*, 45, e15178.

Imahori, Y., Kota, M., Ueda, Y., Ishimaru, M., Chachin, K. (2000). Ethanolic fermentaion enzymes, their products and transcription of alcohol dehydrogenase from bell pepper fruit held under various low oxygen atmospheres. *Journal of the Japanese Society for Horticultural Science*, 69, 266–272.

Imahori, Y., Matushita, K., Kota, M., Ueda, Y., Ishimaru, M., Chachin, K. (2003). Regulation of fermantive metabolism in tomato fruit under low oxygen stress, *Journal of Horticultural Science and Biotechnology*, 78, 386–393.

Imahori, Y., Suzuki, Y., Kawagishi, M., Ishimaru, M., Chachin, K. (2007). Physiological responses and quality attributes of Chinese chive leaves exposed to CO_2-enriched atmospheres. *Postharvest Biology and Technology*, 46, 160–166.

Imahori, Y., Uemura, K., Kishioka, I., Fujiwara, H., Tulio, A. Z., Ueda, Y., Chachin, K. (2005). Relationship between low-oxygen injury and ethanol metabolism in various fruits and vegetables. *Acta Horticulturae*, 682, 1103–1108.

Johnson, D.S. (2008). Factors affecting the efficacy of 1-MCP applied to retard apple ripening. *Acta Horticulturae*, 796, 59–67.

Kader, A.A. (1986). Biochemical and physiological basis for effects of controlled and modified atmospheres on fruit and vegetables. *Food Technology*, 40, 54–60.

Kader, A.A. (1992). Modified atmospheres during transport and storage. In *Postharvest technology of horticultural crops*. ed. A. A. Kader, 85–95. Oakland: University of California.

Kader, A.A. (2008). Flavor quality of fruits and vegetables. *Journal of the Food and Agriculture*, 88, 1863–1868.

Kader, A.A. (2009). Effect on nutritional quality. In *Modified and controlled atmospheres for the storage, transportation, and packaging of horticultural commodities*, ed. E. M. Yahia, 111–118. Boca Raton: CRC Press.

Kantakhoo, J., Ose, K., Imahori, Y. (2022). Effects of hot water treatment to alleviate chilling injury and enhance phenolic metabolism in eggplant fruit during low temperature storage. *Scientia Horticulturae*, 304, 111325.

Kim Y., Brecht, J. Talcott, S.T. (2007). Antioxidant phytochemical and fruit quality changes in mango (*Mangifera indica* L.) following hot water immersion and controlled atmosphere storage. *Food Chemistry*, 105, 1327–1334.

Latocha, P., Krupa, T., Jankowski, P., Radzanowska, J. (2014). Changes in postharvest physicochemical and sensory characteristics of hardy kiwifruit (*Actinidia arguta* and its hybrid) after cold storage under normal versus controlled atmosphere. *Postharvest Biology and Technology*, 88, 21–33.

Liu, Y.B. (2007). Ultralow oxygen treatment for postharvest control of western flower thrips, *Frankliniella occidentalis*, on broccoli. *Journal of Economic Entomology*, 100, 717–722.

Liu, Y.B. (2020). CA requirements for postharvest pest control. In *Controlled and modified atmospheres for fresh and fresh-cut produce*, ed. M. I. Gil and R. Beaudry, 65–74. London: Academic Press.

Madani, B., Mirshekari, A., Imahori, Y. (2019). Physiological responses to stress. In *Postharvest physiology and biochemistry of fruits and vegetables*, ed. E. M. Yahia, 405–423. Duxford: Elsevier.

Malcolm, G.I. (2005). Advancements in the implementation of CA technology for storage of perishable commodities. *Acta Horticulturae*, 682, 1593–1597.

Markarian, N.R., Vigneault, C., Gariepy, Y., Rennie, T.J. (2003). Computerized monitoring and control for a research controlled-atmosphere storage facility. *Computers and Electronic in Agriculture*, 39, 23–37.

Mathooko, F.M. (1996). Regulation of respiratory metabolism in fruits and vegetables by carbon dioxide. *Postharvest Biology and Technology*, 9, 247–264.

Mattheis, J., Fellman, J.K. (2000). Impacts of modified atmosphere packaging and controlled atmospheres on aroma, flavor, and quality of horticultural commodities. *HortTechnology*, 10, 507–510.

Mditshwa, A, Fawole, O.A., Opara, U.L. (2018). Recent developments on dynamic controlled atmosphere storage of apples-A review. *Food Packaging and Shelf Life*, 16, 59–68.

Mitcham, E.J. (2007). Heat with controlled atmospheres. In *Heat treatments for postharvest pest control: theory and practice*, ed. J. Tang, E. Mitcham, S. Wang, S. Lurie, 251–268. Oxon: CABI.

Montesinos-Herrero, C., Angel del Rio, M., Rojas-Argudo, C., Palou, L. (2012). Short exposure to high CO_2 and O_2 at curing temperature to control postharvest diseases of citrus fruit. *Plant Disease*, 96, 423–430.

Nanos, G.D., Kirisakis, A.K., Sfakiotakis, E.M. (2002). Preprocessing storage conditions for green 'Conservolea' and 'Chondrolia' table olive. *Postharvest Biology and Technology*, 25, 109–115.

Neven, L.G. (2005). Combined heat and controlled atmosphere quarantine treatments for control of codling moth in sweet cherries. *Journal of Economic Entomology*, 98, 709–715.

Ortiz, A., Echeverría, G., Graell, J., Lara, I. (2009). Overall quality of 'Rich Lady' peach fruit after air- or CA storage. The importance of volatile emission. *LWT—Food Science and Technology*, 42, 1520–1529.

Pariasca, J.A.T., Miyazaki, T., Hisaka, H., Nakagawa, H., Sato, T. (2001). Effects of modified atmosphere packaging (MAP) and controlled atmosphere (CA) storage on the quality of snow pea pods (*Pisumsativum* L. var. *saccharatum*). *Postharvest Biology and Technology*, 21, 213–223.

Park, M.H., Sangwanangkul, P., Choi, J.W. (2018). Reduced chilling injury and delayed fruit ripening in tomatoes with modified atmosphere and humidity packaging. *Scientia Horticulturae*, 231, 66–72.

Plotto, A., Bai, J., Baldwin, E. (2020). Effect of CA/MA on sensory quality. In *Controlled and modified atmospheres for fresh and fresh-cut produce*, ed. M. I. Gil, R. Beaudry, 109–130. London: Academic Press.

Porat, R., Fallik, E. (2008). Production of off-flavours in fruit and vegetables under fermentative conditions. In *Fruits and vegetables flavour*, ed. B. Brückner, S. G. Wyllie, 150–164. Boca Raton, FL: CRC Press.

Prange, R., DeLong, J.M. (2006). Controlled-atmosphere related disorders of fruits and vegetables. *Stewart Postharvest Review*, 5, 7.

Prange, R.K., DeLong, J.M., Daniels-Lake, B.J., Harrison, P.A. (2005). Innovation in controlled atmosphere technology. *Stewart Postharvest Review*, 1, 6.1–6.11.

Prange, R.K., DeLong, J.M., Wright, A.H. (2011). Storage of pears using dynamic controlled-atmosphere (DCA), a non-chemical method. *Acta Horticulturae*, 909, 707–717.

Raffo, A., Kelderer, M., Paoletti, F., Zanella, A. (2009). Impact of innovative controlled atmosphere storage technologies and postharvest treatments on volatile compound production in cv. Pinova Apples. *Journal of Agricultural and Food Chemistry*, 57, 915–923.

Rees, D., Bishop, D., Schaefer, J., Colgan, R., Thurston, K., Fisher, R., Duff, A. (2021). SafePod: A respiration chamber to characterise apple fruit response to storage atmospheres. *Postharvest Biology and Technology*, 181, 111674.

Reghavan, G.S.V., Vigneault, C., Gariépy, Y., Markarian, N.R., Alvo, P. (2005). Refrigerated and controlled/modified atmosphere. In *Processing fruits*. ed. D. M. Barett, L. Somogy, H. Ramaswamy, 23–52. Boca Raton: CRC Press.

Retamales, J., Manríquez, D., Castillo, P., Defilippi, B. (2003). Controlled atmosphere in Bing cherries from Chile and problems caused by quarantine treatments for export to Japan. *Acta Horticulturae*, 600, 149–153.

Rizzolo, A., Cambiaghi, P., Grassi, M., Zerbini, P. E. (2005). Influence of 1-methylcyclopropene and storage atmosphere on changes in volatile compounds and fruit quality of conference pears. *Journal of Agricultural and Food Chemistry*, 53, 9781 – 9789.

Rizzolo, A., Grassi, M., Vanoli, M. (2014). 1-Methylcyclopropene application, storage temperature and atmosphere modulate sensory quality changes in shelf-life of 'Abbé Fétel' pears. *Postharvest Biology and Technology*, 92, 87–97.

Rodriguez, J., Zoffoli, J.P. (2016). Effect of sulfur dioxide and modified atmosphere packaging on blueberry postharvest quality. *Postharvest Biology and Technology*, 117, 230–238.

Romero, I., Sanchez-Ballesta, M.T., Maldonado, R., Escribano, M.I., Mereodio, C. (2008). Anthocyanin, antioxidant activity and stress-induced gene expression in high CO_2-treated table grapes stored at low temperature. *Journal of Plant Physiology*, 165, 522–530.

Saltiveit, M.E. (2003). Is it possible to find an optimal controlled atmosphere? *Postharvest Biology and Technology*, 27, 3–13.

Saquet, A.A., Streif, J., Bangerth, F. (2003). Reducing internal browning disorders in "Braeburn" apples by delayed controlled atmosphere storage and some related physiological and biochemical changes. *Acta Horticulturae*, 682, 453–458.

Schotsmans, W.C., DeLong, J.M., Larrigaudière, C., Prange, R.K. (2009). Effects on physiological disorders. In *Modified and controlled atmospheres for the storage, transportation, and packaging of horticultural commodities*, ed. E. M.Yahia, 159–192. Boca Raton: CRC Press.

Serrano, M., Martínez-Romero, D., Guillén, F., Castillo, S., Valero, D. (2006). Maintenance of broccoli quality and functional properties during cold storage as affected by modified atmosphere packaging. *Postharvest Biology and Technology*, 39, 61–68.

Sevillano, L., Sanchez-Ballesta, M.T., Romojaro, F., Flores, F.B. (2009). Physiological, hormonal and molecular mechanisms regulating chilling injury in horticultural species. Postharvest technologies applied to reduce its impact. *Journal of the Science of Food and Agriculture*, 89, 555–573.

Singh, R.J., Srivastava, S., Sane, V.A. (2013). Biology and biotechnology of fruit flavor and aroma volatiles. *Stewart Postharvest Review*, 4, 2.

Soderstrom, E.L., Brandl, D.G., Mackey, B.E. (1990). Responses of codling moth (*Lepidoptera*: *Tortricidae*) life storge to high carbon dioxide or low oxygen atmospheres. *Journal of Economic Entomology*, 83, 472–475.

Steffens, C.A., Soardi, K., Heinzen, A.S., Amaral Vignali Alves, J., da Silva, J.C., Talamini do Amarante, C.V., Brackmann, A. (2021). Quality of "Cripps Pink" apples following the application of 1-MCP, ethanol vapor and nitric oxide as pretreatments for controlled atmosphere storage. *Journal of Food Processing and Preservation*, 46(1), e16121.

Tadege, M., Dupuis, I., Kuhlemeier, C. (1999). Ethanolic fermentation: New functions for an old pathway. *Plant Science*, 4, 320–325.

Thewes, F.R., Both, V., Brackmann, A., Weber, A., Anese, R.Q. (2015). Dynamic controlled atmosphere and ultralow oxygen storage on 'Gala' mutants quality maintenance. *Food Chemistry*, 188, 62–70.

Thewes, F.R., Brackmann, A., Both, V., de Oliveira Anese, R., Ludwig, V., Wendt, L. M., Berghetti, M.R.P., Thewes, F.R., Rossato, F.P. (2022). Dynamic controlled atmosphere: Does the frequency of respiratory quotient determination during storage affect apple fruit metabolism and quality?. *Postharvest Biology and Technology*, 194, 112097.

Thewes, F.R., Brackmann, A., Both, V., de Oliveira Anese, R., Schultz, E.E., Ludwig, V., Wendt, L.M., Berghetti, M.R.P., Thewes, F. R. (2020). Dynamic controlled atmosphere based on carbon dioxide production (DCA–CD): Lower oxygen limit establishment, metabolism and overall quality of apples after long-term storage. *Postharvest Biology and Technology*, 168, 111285.

Thewes, F.R., Wood, R.M., Both, V., Keshri, N., Geyer, M., Pansera-Espíndola, B., Neuwald, D.A. (2021). Dynamic controlled atmosphere: A review of methods for monitoring fruit responses to low oxygen. *Comunicata Scientiae*, 12, e3782.

Thompson, A.K. (2010). *Controlled atmosphere storge of fruits & vegetables*. Oxfordshire: CABI.

Tran, D.T., Verlinden, B.E., Hertog, M., Nicolai, B.M. (2015). Monitoring of extremely low oxygen control atmosphere storage of 'Greenstar' apples using chlorophyll fluorescence. *Scientia Horticulturae*, 184, 18–22.

Trierweiler, B., Krieg, M., Tauscher, B. (2004). Antioxidative capacity of different apple cultivars after long-time storage. *Journal of applied Botany and Food Quality*, 78, 117–119.

Veltman, R.H., Verschoor, J.A., Ruijsch, J.H., van Dugteren, R. (2003). Dynamic control system (DCS) for apples (*Malus domestica* Borkh. cv 'Elstar'): optimal quality through storage based on product response. *Postharvest Biology and Technology*, 27, 79–86.

Weber, A. Soldateli, F.J., Thewes, F.R., Both, V., Brackmann, A. (2023). Dynamic controlled atmosphere isolated or associated with 1-MCP and elevated storage temperature affect the overall quality and anaerobic metabolism of 'Royal Gala' apples. *Erwerbs-Obstbau*, 65.

Wendt, L.M., Ludwig, V., Rossato, F.P. (2022). Combined effects of storage temperature variation and dynamic controlled atmosphere after long-term storage of 'Maxi Gala' apples. *Food Packaging and Shelf Life*, 31, 100770.

Wright, A.H., DeLong, J.M., Arul, J., Prange, R.K. (2015). The trend toward lower oxygen levels during apple (*Malus* × *domestica* Borkh) storage – A review. *Journal of Horticultural Science & Biotechnology*, 90, 1–13.

Yahia, E.M. (2006). Modified and controlled atmospheres for tropical fruits. *Stewart Postharvest Review*, 2, 123–183.

Yahia, E.M. (2019). Introduction, In *Postharvest physiology and biochemistry of fruits and vegetables*, ed. E. M. Yahia, 1–17. Duxford: Elsevier.

Yahia, E.M., Fadanelli, L., Mattè, P., Brecht, J.K. (2019). Controlled atmosphere storage. In *Postharvest technology of perishable horticultural commodities*, ed. E. M. Yahia, 439–479. Duxford: Elsevier.

Zheng, Y., Wang, C.Y., Wang, S., Zheng, W. (2003). Effect of high-oxygen atmospheres on blueberry phenolics, anthocyanins, and antioxidant capacity. *Journal of Agricultural and Food Chemistry*, 51, 7162–7169.

Zhu, Z., Liu, R., Li, B., Tian, S. (2013). Characterization of genes encoding key enzymes involved in sugar metabolism of apple fruit in controlled atmosphere storage. *Food Chemistry*, 141, 3323–3328.

Dynamic Controlled Atmosphere Technology for Fruits and Vegetables

4

Muhammad Rafiullah Khan*, Ali Muhammad, Yaodong Guo, Majid Suhail Hashmi, and Rafiq Ahmad
*Corresponding Author: rafiullah192@yahoo.com

4.1 INTRODUCTION

Consumer preferences for fresh fruits and vegetables are increasing day by day, as fresh produce are the constituents of health-essential compounds. Therefore, the preservation of fresh-like characteristics of these produce is very important. As fresh produce are living organisms and respire after harvesting or detaching from parent plants, alteration of the gas's environment surrounding this produce is an effective preservative technology. Some types of fruit show a limited harvest window which lasts only a few months during the year. To ensure fruit availability throughout the growing season, a part of the harvest should be stored. To minimize fruit metabolism and extend postharvest life, storage temperatures are normally kept as low as feasible (Brackmann et al., 2008; Steffens et al., 2007). A variety of methods can be employed to prolong the shelf life of horticultural commodities such as fruits and vegetables during storage and market distribution following harvest. Other services such as coating, modified atmosphere packing and controlled environment storage are available. Scientists, particularly postharvest technologists, have developed new and safe preservative techniques. Among these technologies, controlled atmosphere (CA) is an advanced technology in which fruits are kept in static CA environments where oxygen, temperature and CO_2 concentrations are sustained at a particular condition during storage seasons.

For the storage of fruit, controlled atmosphere systems (CA) are combined with low temperatures that alter pCO_2, which is partial pressure of CO_2, and pO_2 partial pressure of oxygen (Brackmann et al., 2008; Both et al., 2014). In fruit metabolism these partial pressures vary and depend upon various factors such as growing season, fruit storage, maturity and temperature, which can affect optimal pO_2 values during controlled environment storage (Weber et al., 2011, Kitteman et al., 2015; Bekele et al., 2016; Thewes et al., 2020).

However, some limitations were observed in static CA storage. Among these limitations, the first is the maintenance of the lowest possible oxygen (O_2) levels to minimize the losses of the stored products, which is difficult due to the great variability of the tolerance levels of the cultivars to these gases. The second important limitation of static CA is that it is challenging to access fruits and all the dynamic information on particular product-based quality aspects, as sometimes for inspection purposes or quality checks, the stored fruits are taken out of storage rooms (Schouten et al., 1997).

Another technology known as ultra-low oxygen (ULO) emerged in response to the limitation of CA in the complete control of certain physiological disorders in fruits during storage to maximize the beneficial effects of CA. For example, in comparison to the traditional CA system, superior fruit quality of apples such as Golden Reinders and Granny Smith (Zanella, 2003; Altisent et al., 2009; Weber et al., 2011) were obtained when stored in ULO (0.7–1 kPa). However, for long-term storage fruit maturity is the major influencing factor in the ULO system (Zanella, 2003). De Long et al. (2007) and Zanella (2003) reported that above the anaerobic threshold level the concentration of low oxygen cannot be detected by static ULO and similarly increased off flavor (alcoholic) due to the slow synthesis of aromatic volatile compounds.

Another approach is initial low oxygen stress (ILOS), which induces low O_2 stress at the initial stage of storage, for example, 0.5% O_2 for 10 days (Wang & Dilley, 2000; Van der Merwe et al, 2001). Another approach, RLOS (repeated low oxygen stress), was established in which low O_2 stress was

DOI: 10.1201/9781003370376-6

used repeatedly depending on ethanol assessment under CA storage (Fadanelli et al., 2013; Prange et al., 2013).

Another approach was designed as dynamic control atmosphere (DCA), in which the controlled atmosphere condition is modified dynamically and the metabolic responses of the stored product are optimized in the respective storage environment (Tran et al., 2015). This is an emerging technology in the postharvest treatment and preservation of fresh produce quality; therefore, this chapter will cover different aspects and recent advances in DCA for fresh fruits.

4.2 DYNAMIC CONTROLLED ATMOSPHERE (DCA)

Several technologies have been developed over the last 25 years to establish the optimum pO_2 to monitor the metabolism of fruit during storage at controlled atmosphere (CA). For decades, researchers have been intrigued by the promise that during long-term storage DCA is used as a non-chemical postharvest therapy for preserving apple fruits. DCA storage can preserve fruit quality without the use of extra pesticides and has the potential to save significant energy through the utilization of increased storage temperatures. Thus, the DCA system is an adaption of the static controlled atmosphere system in which apple fruits are stored under the lowest oxygen limits (LOL) permitted by fruits prior to the beginning of anaerobic metabolism (Prange, 2018; Wright et al., 2015). Zanella (2003) stated that DCA storage exploring reduced oxygen limits by changing oxygen levels dynamically as a storage time function, which relies on a measurement of metabolic responses of fruits in response to low oxygen concentrations. According to Zanella (2003), DCA is a cutting-edge storage method that allows fruits to be preserved at low O_2 concentrations without developing off-flavor or storage problems. Mditshwa et al. (2018) characterized DCA as an imperative system which permits the standardization of O_2 levels at the start and throughout storage, and due to its non-chemical nature this technique has sparked widespread attention in farmers and also researchers worldwide in the field of postharvest technology.

Under DCA, the pO_2 (oxygen partial pressure) continuously changes during storage based on the requirements of fruit physiology (Prange, 2018; Wright et al., 2015). The OOPP (optimum oxygen partial pressure) varies due to some factors such as growing season, cultivar, temperature, species storage period and harvest maturity. Storage techniques for detecting low oxygen limit (LOL) and fruits that are storing under LOL were developed based on these variables. A proper technology/method is required to continuously monitor the LOL in an extremely low partial pressure during storage. To monitor such a low partial pressure of gas, four methods are generally used, which are also known as DCA sensor technologies.

4.3 TYPES OF DCA SENSOR TECHNOLOGIES

Sensors can be used to detect gases. There are four types of sensors/systems that are commonly utilized in DCA; each is briefly detailed in the following sections.

4.3.1 Respiration Quotient Sensor

Respiration quotient is the measurement ratio of CO_2 production over the utilization of O_2. The DCA based on the respiration quotient is a recently designed method in low oxygen storage, mostly used for apples. This method monitors the critical point at which fermentative metabolism is about to start. Temperature is always a key parameter to be considered in postharvest storage. The DCA depends on the respiration quotient effectively maintaining the postharvest quality of apples during storage. Different cultivars of apple have been tested by different researchers. Weber et al. (2015) and Weber et al. (2017) reported that DCA based on the respiration quotient conserved the quality attributes of Royal Gala and Fuji Suprema apple cultivars. Overall quality retention of the Idared cultivar was reported by Gasser et al. (2005), while results in Galaxy and Granny Smith cultivars were reported by Brackmann et al. (2014) and Bessemans et al. (2016), respectively.

4.3.2 Chlorophyll Fluorescence Sensor

Another effective and widely used technology for preserving apple fruit quality is the DCA based on chlorophyll fluorescence. The concentration of gases varies due to the physiological behaviors of the fruits, and the sensitivity to O_2 can be monitored using the chlorophyll fluorescence sensor. Harvest Watch™ (Satlantic L.P., Halifax, Nova Scotia, Canada) is a widely used sensor for measuring the chlorophyll fluorescence of apples during storage (Mditshwa et al., 2018). Firmness in various cultivars was preserved by DCA-CF, and this method has been found effective in optimal and pre-optimal harvested fruit types (Zanella et al., 2005; Mditshwa et al., 2017b) during storage in the chlorophyll fluorescence–based DCA. Prange et al. (2003) retained the quality of Summer 1 and McIntosh cultivars, while DeLong et al. (2007) preserved the overall quality of Cortland and Delicious cultivars. This DCA technology has been widely used for other cultivars and commendably preserved the quality of various cultivars of apple fruits (Thewes et al., 2015; Weber et al., 2015; Tran et al., 2015). The affecting characteristics for the chlorophyll fluorescence–based dynamic regulated environments are cultivar and fruit maturity status (Mditshwa et al., 2018).

4.3.3 Ethanol Sensor

Fermentative metabolism is a general concept in the CA storage of fresh fruits. Therefore, monitoring this threshold or status is important to avoid the adverse effects of ethanol and prevent economic loss. DCA-ET (ethanol-based DCA) is another advancement in the low oxygen storage environment for various fruits and vegetables in storage rooms.

According to Wright et al. (2015), fruit quality in the storage chamber is based on ethanol concentrations, when the ethanol reaches approximately 1 ppm due to reduction in oxygen levels, this concentration in the chamber is considered to be undesirable due to the application of the dynamic control environment, which is more effective in the long-term storage of apple fruits of various cultivars. Most ethanol levels can be monitored by a DCS™ sensor in store rooms Prange et al. (2013).

4.3.4 Dynamic Controlled Atmosphere System Based on CO_2 Production (DCA-CD)

Fresh produce are living organisms and produce CO_2 during respiration. This CO_2 reduces the O_2 level in the atmosphere surrounding the fresh produce. Therefore, measuring CO_2 production is another method used for determining LOL. Commercially, the DCA-CD system is available with a brand name of FruitAtmo® by the company Frigotec in Germany. Basically, CO_2 production rate decreases as O_2 pressure in storage rooms decreases until a certain oxygen partial pressure is reached, at which point CO_2 production begins to increase due to anaerobic metabolism (Thewes et al., 2021). Because of the reaction of CO_2 production to oxygen fluctuation, it is possible to monitor the LOL of fruit only based on CO_2 measurements (Thewes, 2019). DCA based on CO_2 has precisely measured the LOL for certain cultivars of apple fruits and maintained quality better than DCA based on the respiration quotient and the CA-ULO (Thewes et al., 2020). Thewes et al. (2020) further mentioned that one advantage of DCAs over other systems is that the LOL estimation in the DCA-CD system does not require the intake of oxygen measurement, which is a major problem due to leakage in commercial storage rooms (Bessemans et al., 2016; 2018).

In commercial storage rooms the DCA-CD system is easily applicable for the estimation of LOL via the production of CO_2. The DCA-CD system is also helpful to maintain fruit quality through the initiation of anaerobic metabolism at a safe level (Thewes et al., 2020). The quality of apple cultivars was maintained by the application of DCA-CD, CA+1MCP and DCA-CF storage systems, which showed that higher concentrations of volatile compounds such as ester were higher than the other two systems (Thewes, 2019). Similarly, in the storage area, the amount of CO_2 is very little affected compared to oxygen uptake by the external atmosphere (Bessemans et al., 2018).

4.4 DCA INFLUENCES ON PHYSICOCHEMICAL PROPERTIES OF FRESH PRODUCE

After harvesting, fresh food is vulnerable to a range of physiological disorders, although these can be reduced to reduce losses. The list that follows includes typical physiological problems that affect apples as well as how a changing, controlled environment affects them.

4.4.1 Firmness

When evaluating apple fruit quality, firmness is an important quality attribute to consider. During storage firmness is directly related to the ripening process, water content, fruit maturity and storage condition of fruit and vegetable crops (Bessemans et al., 2016; Mditshwa et al., 2017b).

Ethylene is one of the major contributors to decreases in firmness, as is the associated process like internal cell space, increase in soluble pectin, volume, loss of cellulosic sugar, arabinose and galactose (De Ell et al., 2001; Wei et al., 2010). The correlation of fruit softening and firmness is with a poly-galacturonase gene which depends upon ethylene production; similarly firmness is also associated with cell wall–degrading enzymes such as β-xylosidase and polygalacturose (Costa et al., 2010).

Postharvest treatments like heat treatment and 1-MCP can help to maintain the firmness of apples to some extent (Moran & McManus, 2005; Rupasinghe et al., 2000). However, non-chemical methods such as DCA have been developed to maintain the firmness of apple fruits. DeLong et al. (2007) found higher firmness values of Delicious apples stored in DCA than the apples stored in CA. Studies showed higher firmness of Golden Delicious and Granny Smith apple cultivars, respectively, in DCA than in normal air or traditional CA conditions (Zanella et al., 2005; Bessemans et al., 2016). Mattheis et al. (1998) reported high firmness values of Gala apple fruits kept under DCA than static CA. Therefore, these studies suggest that DCA has a positive effect on controlling the fruits' metabolism, delaying the ripening process, slowing down ethylene biosynthesis, controlling the activity of associated enzymes and consequently maintaining the firmness of the stored products.

According to De Belie et al. (2000), fruit and vegetable commodities under postharvest storage firmness losses can be monitored by automated monitoring systems. One such approach to predict hardness in apple fruits during storage is the multispectral imaging approach (Pend & Lu, 2006). Heat treatment and 1-MCP are the postharvest therapy of fruits and vegetables for slowing firmness loss during storage, especially in apple fruits (Rupasinghe et al., 2000; Moran & McManus, 2005). For example, DeLong et al. (2007) found that the hardness of Delicious apples was substantially higher when stored in DCA than regulated atmosphere. Higher firmness values were also found in DCA than the normal air or CA in apples. Higher firmness in stored fruits was observed in DCA systems (0.7% oxygen level) as compared to static CA, which contains 2% oxygen (Tran et al., 2015; Bessemans et al., 2016).

According to Costa et al. (2010), the ethylene-dependent polygalacturonase gene is linked to fruit firmness and softening. Ortiz et al. (2011) and Wei et al. (2010) reported a relationship of cell wall disintegration of preserved fruit with polygalacturose and β-Xylosidase enzymes for significant effect on fruit firmness during long-term storage. Mattheis et al. (1998) found that DCA-stored apples exhibited increased firmness and lower ethylene generation compared with static CA during 120-day storage in cold storage and 7-day shelf life storage. It has been demonstrated that apples preserved with low oxygen technology are firmer than fruit stored in a RA system (Rebeaud, et al., 2015; Zanella & Rossi, 2015). Throughout the ripening process, tissue disintegration, respiratory metabolism and ethylene generation are restricted by LOT, resulting in less firmness loss (Mditshwa et al., 2018; Graell et al., 1997). Therefore, these studies suggest that DCA had a positive effect on the activities of these enzymes.

4.4.2 Biochemical Quality–Based Attributes

Total soluble solids (TSS) and total acidity (TA) and its ratio are essential quality parameters of apples at both temperatures. Therefore, these quality parameters must be maintained while extending the postharvest shelf life of fruits. Fruits can be conserved for longer periods of time (7 days shelf life and 8 months cold storage) with higher levels of TA and TSS by the application of DCA technologies. Similarly this technology is very effective to maintain the quality (TSS, TA and its ratio) of apple fruits for 6 months and a shipment period of 6 weeks (De long et al., 2007; Mditshwa et al., 2017a). This could be accredited to the decrease of respiration and thereby delaying the degradation of organic acid in DCA, which consequently maintain the TSS and TA content and their ratio in the stored products (Zanella, 2003; Mditshwa et al., 2017a). Loss of TA occurred due to the increase in cellular activity when fruit and vegetable commodities are stored in open air (Mditshwa et al., 2017a).

DCA-CF is more effective than RA-stored apple fruits in maintaining higher TSS content and TA levels in long-term cold storage for 36 and 16 weeks (Rebeaud & Gasser, 2015; Mditshwa et al., 2015). The DCA-CF system also maintained the quality of the Uta cultivar of pears in terms of TA and firmness compared to the CA system (Lafer, 2011). Similarly Magazin et al. (2013) observed that quality parameters such as TA decreased as the storage period increased in apple fruits. Mostly decreases in TA can be observed during extended storage periods due to the use of organic acids in respiration (Melgarejo et al., 2000), but this decline can be controlled by the application of DCA-CF and also to some extent by CA and ULO or RLOS (Both et al., 2016).

4.4.3 Antioxidant Activities

Fruits are a rich source of antioxidant compounds. The decrease in total antioxidant capacity in fruits is one of the effectiveness of LO technologies during storage (Mditshwa et al. 2015) in which DCA-CF is more effective than CA in store rooms at 0 °C. According to Shaham et al. (2003), the behavior of the antioxidant capacity of 1-MCP-treated apple fruits are similar in retaining antioxidant capacity by the application of reduced O_2 technology such as ULO and CA, while these technologies also have a positive effect on fruit during regular storage conditions.

Water-soluble antioxidants decreased under long-term storage in Delicious and Cortland apple cultivars at 0 °C (Barden & Bramlage, 1994). A decrease in antioxidant activity was also observed in the Granny Smith variety of apple fruit (Mditshwa et al., 2015). According to Leja et al. (2003), radical scavenging activities can be increased by changing the storage condition to 1 °C for 120 days of storage with CA, with 2% O_2 and CO_2. DCA-CF and RLOS are the most effective methods for retaining maximum antioxidant capacity in fruits in storage rooms and in open environment and can prolong the shelf life of fruit for 7 days.

4.4.4 Sensory Characteristics

Postharvest treatments should attempt to maintain or improve customer acceptability throughout long-term cold storage by sensory attributes. Consumer satisfaction is a final goal, and therefore sensory analysis must be conducted to evaluate the sensory attributes of the stored produce in a DCA. In accordance with sensory tests and consumer expectations surveys, fruits like apples kept at CA or other LOS systems should have high consumer acceptability in terms of firmness, crispness, and juiciness with exceptional flavor and high sugar–acid balance compared to in open air or on the shelf (Lopez et al., 2007; Streif, 2008).

According to their findings, reduced flesh firmness was highly associated with lower customer approval of air-stored fruit. On the other hand, higher soluble solid concentration was highly connected to higher consumer approval of CA-kept fruit. According to Kittemann et al. (2015), consumer acceptance was higher in DCA-treated fruits (Green Star and Jonagold apples) than ULO alone or in combination with MCP (Tran et al., 2015), while low mealiness and higher crispness were the characteristics of DCA-treated fruit compared to CA storage after 6 months (Aubert et al., 2015). Environmental conditions of storage rooms have great effect on the sensory attributes of fruits. One of the characteristics of DCA technology is that it maintains the quality of fruit for a long time compared to CA. DCA technology also retained lower 2-methylbutyl acetate levels at 1 °C for 9 months than CA (Both et al., 2017). LO technologies are very effective in maintaining the sensory characteristics for a long time, but sometimes ethanol and acetaldehydes accumulate in certain apple cultivars, resulting in off-flavor. Pesis (2005) reported that off-flavor is due to the accumulation of acetaldehydes and ethanol and a technology which attracts the consumers, on ever-changing O_2 levels in fruits during storage is the DCA technology, which has a great advantage for ensuring high-quality fruits and vegetables with acceptable overall sensory attributes during and after storage.

4.5 POSITIVE EFFECTS OF DYNAMIC CONTROLLED ATMOSPHERE

The loss of aroma during postharvest storage of fruits is mainly concerned with consumer dissatisfaction of fruit quality (Crisosto & Crisosto, 2005). Under control atmosphere (CA) treatment some physiological disorders like internal browning of fruit flesh, ethylene emission and defects in softening capacity were observed during storage. During CA treatments increased aroma volatile content accumulation derived from fatty acids increased consumer acceptability (Liu et al., 2022).

DCA is a non-chemical therapy during long-term storage of fruit- and vegetable-based produce. Physiological disorders such as superficial scalds are mainly suppressed by the DCA system, while retention of color and firmness in fruits were high during storage due to decreased ethylene biosynthesis and respiration rate. Fruit contains a high concentration of bioactive substances such as antioxidants, phenolics, organic acids and vitamins (Zanella, 2003; Jung & Watkins, 2008). Due to their high perishability fruits and vegetables are extremely exposed to environmental conditions. To decrease the postharvest loss, cold storage is one of the options for short-term storage; however, long-term storage can lead to physiological disorders such as internal browning, superficial scald and water core. To maximize marketability chemicals such as 1-MCP and diphenylamine are used to handle this problem in fruits, especially in apples (Jung & Watkins, 2008; Mditshwa et al., 2018). These chemicals are used to prolong the postharvest life through reduction in the formation of ethylene by producing abnormalities within the fruits and producinf some toxic volatile compounds (Methyl Hepten One, conjugated trienols and farnesene) which affect the quality of stored fruits. Jung and Watkin (2008) concluded that these chemical compounds are excellent at preserving the quality of fruits in terms of titratable acidity, color and fruit firmness. Furthermore, there is growing concern that DPA (diphenylamine) may cause cancer in humans. Due to investigations on its broad spectrum toxic effects and potential risk to human beings, aquatic habitats and some animals, its usage is limited in some countries (Drzyzga, 2003; Zanella, 2003).

As an outcome, recent postharvest research has concentrated on developing non-chemical apple preservation methods. The pome industry commonly employs non-chemical postharvest treatments such as CA and ILOS (Sabban-Amin et al., 2011; Zanella, 2003). These technologies have a reputation for lowering respiration rate and ethylene accumulation, two crucial biochemical activities during fruit storage (Wright et al., 2015). Fruit maintained in low-oxygen environments has a lower incidence of greasiness and surface scald (DeLong et al., 2004). Although low-oxygen storage (LOS) provides several advantages, there are certain disadvantages of this technology. This includes minimal O_2 damages and higher prevalence of off-flavor in apple fruits (Wright et al., 2015). DCA technology has been established to improve the efficacy of LOS by managing physiological problems and lowering the prevalence of off-flavors (Wright et al., 2015; Weber et al., 2015).

4.6 NEGATIVE EFFECTS OF THE DCA SYSTEM

Despite the fact that CA technology is rapidly being employed and its performance has substantially increased over the years, the technology still has several shortcomings. Such issues must be addressed as soon as possible by focused and dedicated studies. While DCA is quite appropriate for controlling physiological diseases, studies have shown that high CO_2 and low O_2 damage is common in DCA-preserved apple fruits. For example, a study by DeLong et al. (2007) showed that Delicious apple fruits stored in DCA had a higher rate of low O_2 injury incidence than fruit kept in normal CA for 8 months. This injury is characterized by epidermal purpling. Surprisingly, the Cortland apples kept in the same settings in the same investigation showed no low O_2 damage. The disparate results could be attributable to the variations in the apple cultivars' biochemical, genetic, physical and physiological characteristics. Low O_2 damage has also been demonstrated to be influenced by the interaction between storage temperature and DCA type. Weber et al. (2015) found 0% low O_2 damage in DCA-CF-kept Royal Gala apple against 49% impact in DCA-RQ-preserved apples. They suggested that an increase from 0.5 °C to 1.0 °C in storage temperature be applied for DCA-RQ-stored Royal Gala apples to lessen the occurrence of low O_2 damage. Overall, it might be argued that DCA-CF sensors outperform DCA-RQ technology in detecting crucial O_2 levels during long-term periods of storage. The recent invention of a flexible sensor–based respirometer capable of monitoring the real-time respiration quotient and LOL for fresh commodities, on the other hand, may be advantageous to the apple business (Mahajan et al., 2016). Its performance at variable conditions of storage and O_2 level may overcome some of the problems that presently employed, technology-based approaches present at diverse storage circumstances. Another common issue in DCA-kept apples is CO_2 injury, which shows as internal discoloration combined with the occurrence of cavity development (Lafer, 2007; Zanella et al., 2005). In extreme situations, the condition results in skin damage that appears as partially sunken, rough, brown lesions on the affected fruits. In their investigation into DCA potential as a non-chemical-based strategy during postharvest. Moreover, Zanella et al. (2005) found that Granny Smith apples stored in DCA had a condition approaching CO_2 injury. In a similar study, Lafer (2007) found that Braeburn apples that were kept in DCA for an 8-month period as opposed to CA experienced an external CO_2 harm of more than 3.6%. The primary cause of this disease has been discovered as higher partial pressures of CO_2 along with very low concentrations of O_2 in DCA storage (Lafer, 2007; Zanella et al., 2005). Cultivars, harvest times, storage periods and range of the storage temperatures are also highly associated with CO_2-induced injuries (Streif et al., 2001). Before the fruits are stored in the DCA system, prior and effective risk assessment for CO_2-induced injury occurrence should be considered an imperative approach to reduce the quantitative and qualitative losses of the commodity.

4.7 SYSTEM OF DCA AND RELATED TECHNOLOGIES

Apple storage technology has evolved over time. In recent decades, there has been a determined attempt to adopt non-chemical treatments rather than synthetic and environmentally unfriendly chemicals. The recent technologies which have evolved include static CA [usually > 1% O_2 (v/v)] and ultra-low O_2 (ULO) [typically 1% O_2 (v/v)]. However, the inability of CA, ULO and ILOS to regulate some physiological problems prompted additional investigation. To preserve fruit quality, these types of technologies are often supplemented with certain treatments based on chemical elicitors such as 1-MCP (Chervin et al., 2001; Watkins et al., 2000). DCA is the most recent technology which has achieved worldwide acceptance among the different researchers and apple fruit growers and stakeholders.

DCA is a technology which permits the customization of O_2 concentrations at the start of and throughout storage. Recent research has emphasized the ability of the DCA system to reduce certain disorders of a physiological nature with the objective to preserve better postharvest quality of apple fruits (Weber et al., 2015; Wright et al., 2015; Mditshwa et al., 2017a, 2017b). DCA can be utilized as a pure non-chemical postharvest approach, unlike other LOS methods that require chemical treatments. The CA-based technique has been used to store and preserve better quality of fruit and vegetable produce globally for around two centuries (Thompson, 2010).

The two strategies for establishing CA conditions are delayed CA and quick CA (Wright et al., 2015). Delayed CA is extremely beneficial, as noted in apple fruits, for the preservation of general quality under postharvest storage. This is owing to its capacity to inhibit internal discoloration and CO_2 damage, as opposed to quick CA (Argenta et al., 2000; Saquet et al., 2002). Due to CA's inability to totally regulate certain physiological problems, ultra-low oxygen (ULO) storage was developed. Particularly, O_2 concentrations in ULO technology are lower than those recommended by conventional CA recommendations (Zanella, 2003; Watkins, 2008). The objective of ULO is to maximize the benefits of CA. Higher maintenance of fruit quality, for example, was observed under 0.7 kPa^{-1} kPa based ULO for Granny Smith (Zanella, 2003; Weber et al., 2011) and Golden Reinders apples (Altisent et al., 2009) in contrast to fruit groups kept under normal CA conditions. However, the maturity status of the fruits is the primary element regulating ULO efficacy throughout prolonged storage periods. Zanella (2003) described that ULO is more efficient for inhibiting core flush disorder in properly matured Granny Smith apples. An additional drawback of ULO is the significant likelihood of alcoholic off-flavors due to decreased synthesis of aroma volatiles (Watkins, 2008). Static ULO is incapable of detecting low O_2 concentration slightly above the anaerobic threshold level (Zanella, 2003; DeLong et al., 2007). Despite the fact that the ULO environment is efficient in preserving quality attributes of apples, the drawbacks of this technique demand additional investigation. Another new LOS technology is initial low oxygen stress (ILOS). The scientific foundation for ILOS's ability to preserve the quality of apples dates back to the early 1980s. At the start of storage, ILOS generates a temporary stress of low oxygen concentration (about 0.5% O_2 for a 10-day period) (Wang & Dilley, 2000). ILOS is thus founded on the idea that a particular quantity of ethanol is favorable to the quality of apples during long-term preservation (Zanella & Stürz, 2012).

Another new and innovative postharvest technology which has been recently used by some producers of fruit in Italy is repeated low oxygen stress (RLOS) (Fadanelli et al., 2013). RLOS basically uses repetitive low oxygen concentration during CA storage based on ethanol assessment (Prange et al., 2013). According to Fadanelli et al. (2013), RLOS should be performed 2–3 times during postharvest cold storage. RLOS has the capacity to sustain quality for 6 months of storage at 1.3 °C, according to research on Red Delicious apples (Zanella & Stürz, 2012).

4.8 CONCLUSIONS

This chapter stated the potential of dynamic controlled atmosphere storage technology, which is one of the safest recently developednon-chemical-treatment technologies used for fresh produce, particularly apple fruits. Different types or sensor technologies of DCA systems have been detailed. The effect of DCA on various quality attributes like firmness, total soluble solid, color, antioxidant properties and sensory characteristics have been explained. DCA is a recommended technology for the preservation of fresh produce; however, care should be taken while selecting the oxygen and carbon dioxide compositions to avoid possible injury to the fruits. Therefore, it is recommended to study the tolerance levels of each commodity to oxygen and carbon dioxide before designing a DCA system.

REFERENCES

Altisent, R., Echeverría, G., Lara, I., López, M., Graell, J. (2009). Shelf-life of 'Golden Reinders' apples after ultra low oxygen storage: Effect on aroma volatile compounds, standard quality parameters, sensory attributes and acceptability. *Food Science and Technology International, 15 (5)*, 481–493.

Argenta, L., Fan, X., Mattheis, J. (2000). Delaying establishment of controlled atmosphere or CO_2 exposure reduces 'Fuji' apple CO_2 injury without excessive fruit quality loss. *Postharvest Biology and Technology, 20 (3)*, 221–229.

Aubert, C., Mathieu-Hurtiger, V., Vaysse, P. (2015). Effects of dynamic atmosphere on volatile compounds, polyphenolic content, overall fruit quality, and sensory evaluation of pink lady apples. *Acta Horticulturae, 1071*, 275–280.

Barden, C.L., Bramlage, W.J. (1994). Relationships of antioxidants in apple peel to changes in α-farnesene and conjugated trienes during storage, and to superficial scald development after storage. *Postharvest Biology and Technology, 4*, 23–33.

Bekele, E.A., Ampofo-Asiama, J., Alis, R.R., Hertog, M.L.A.T.M., Nicolai, B.M., Geeraerd, A.H. (2016). Dynamics of metabolic adaptation during initiation of controlled atmosphere storage of 'Jonagold' apple: Effects of storage gas concentrations and conditioning. *Postharvest Biology and Technology, 117*, 9–20. https://doi.org/10.1016/j.postharvbio.2016.02.003.

Bessemans, N., Verboven, P., Verlinden, B.E., Nicolaï, B.M., (2016). A novel type of dynamic controlled atmosphere storage based on the respiratory quotient (RQ-DCA). *Postharvest Biology and Technology, 115*, 91–102. https://doi.org/10.1016/j.postharvbio.2015.12.019.

Bessemans, N., Verboven, P., Verlinden, B.E., Nicolaï, B.M. (2018). Model based leak correction of real-time RQ measurement for dynamic controlled atmosphere storage. *Postharvest Biology and Technology, 136*, 31–41. https://doi.org/10.1016/j.postharvbio.2017.09.011.

Both, V., Brackmann, A., Thewes, F.R., de Freitas Ferreira, D., Wagner, R. (2014). Effect of storage under extremely low oxygen on the volatile composition of 'Royal Gala' apples. *Food Chemistry, 156*, 50–57.

Both, V., Thewes, F.R., Brackmann, A., de Freitas Ferreira, D., Pavanello, E.P., Wagner, R. (2016). Effect of low oxygen conditioning and ultralow oxygen storage on the volatile profile, ethylene production and respiration rate of 'Royal Gala' apples. *Scientia Horticulturae, 209*, 156–164.

Both V., Thewes F.R., Brackmann A., Anese R.O., Ferreira D.F., Wagner R. (2017). Effects of dynamic controlled atmosphere by respiratory quotient on some quality parameters and volatile profile of 'Royal Gala' apple after long-term storage. *Food Chemistry, 215*, 483–492. doi.org/10.1016/j.foodchem.2016.08.009. https://www.sciencedirect.com/science/article/pii/S0308814616312390.

Brackmann, A., Weber, A., Both, V. (2014). CO_2 partial pressure for respiratory quotient and Harvest Watch™ dynamic controlled atmosphere for 'Galaxy' apples storage. *Acta Horticulturae* 1079(1079), 435–440. https://doi.org/ 10.17660/ActaHortic.2015.1079.56.

Brackmann, A., Weber, A., Pinto, J.A.V., Neuwald, D.A., Steffens, C.A. (2008). Manutenção da qualidadepós-colheita de maçãs "Royal Gala" e "Galaxy" sobarmazenamentoematmosferacontrolada. *Ciência Rural, 38*, 2478–2484. English abstract. https://doi.org/10.1590/S0103-84782008000900010. https://www.scielo.br/j/cr/a/MrV75bH7hVcrpwSRGTtjm8F/?format=pdf&lang=pt (accessed 5April 2024).

Chervin, C., Raynal, J., Andre, N., Bonneau, A., Westercamp, P. (2001). Combining controlled atmosphere storage and ethanol vapors to control superficial scald of apple. *Hortscience, 36 (5)*, 951–952.

Costa, F., Peace, C.P., Stella, S., Serra, S., Musacchi, S., Bazzani, M., Van de Weg, W. (2010). QTL dynamics for fruit firmness and softening around an ethylene-dependent polygalacturonase gene in apple (Malus × domesticaborkh.). *Journal of Experimental Botany, 61 (11)*, 3029–3039. https://doi.org/10.1093/jxb/erq130.

Crisosto, C.H., Crisosto, G.M. (2005). Relationship between ripe soluble solids concentration (RSSC) and consumer acceptance of high and low acid melting flesh peach and nectarine (*Prunus persica* (L.)Batsch) cultivars. *Postharvest Biology and Technology, 38*, 239–246.

De Belie, N., Schotte, S., Coucke, P., De Baerdemaeker, J. (2000). Development of an automated monitoring device to quantify changes in firmness of apples during storage. *Postharvest Biology and Technology, 18 (1)*, 1–8.

De Ell, J.R., Khanizadeh, S., Saad, F., Ferree, D.C. (2001). Factors affecting apple fruit firmness—A review. *Journal of American Pomological Society, 55*, 8–27.

DeLong, J.M., Prange, R., Harrison, P. (2007). Chlorophyll fluorescence-based low-O_2 CA storage of organic 'Cortland' and 'Delicious' apples. *Acta Horticulturae, 737*, 31–37.

DeLong, J.M., Prange, R.K., Harrison, P.A. (2004). The influence of 1-methylcyclopropene on 'Cortland' and 'McIntosh' apple quality following long-term storage. *Hortscience, 39 (5)*, 1062–1065.

Drzyzga, O. (2003). Diphenylamine and derivatives in the environment: A review. *Chemosphere, 53 (8)*, 809–818.

Fadanelli, L., Turrini, L., Zeni, F., Buglia, L. (2013). Apples: DCA storage with repeated gas stress-the experience in the management of commercial cells. XI international controlled and modified atmosphere research conference, 1071.

Gasser, F., Dätwyler, D., Schneider, K., Naunheim, W., Hoehn, E. (2005). Effects of decreasing oxygen levels in the storage atmosphere on the respiration and production of volatiles of 'Idared' apples. *Acta Horticulturae, 682*, 1585–1592.

Graell, J., Larrigaudiere, C., Vendrell, M. (1997). Effect of low-oxygen atmospheres on quality and superficial scald of 'Top red' apples. *Food Science and Technology International, 3*, 203–211.

Jung, S., Watkins, C.B. (2008). Superficial scald control after delayed treatment of apple fruit with diphenylamine (DPA) and 1-methylcyclopropene (1-MCP). *Postharvest Biology and Technology, 50 (1)*, 45–52.

Kittemann, D., McCormick, R., Neuwald, D.A. (2015). Effect of high temperature and 1-MCP application or dynamic controlled atmosphere on energy savings during apple storage. *European Journal of Horticultural Science, 80 (1)*, 33–38.

Lafer, G. (2007). Storability and fruit quality of 'Braeburn' apples as affected by harvest date, 1-MCP treatment and different storage conditions. International conference on ripening regulation and postharvest fruit quality, 796, 179–184.

Lafer, G. (2011). Effect of different CA storage conditions on storability and fruit quality of organically grown 'Uta' pears. *Acta Horticulturae, 909*, 757–760.

Leja, M., Mareczek, A., Ben, J. (2003). Antioxidant properties of two apple cultivars during long-term storage. *Food Chemistry, 80*, 303–307.

Liu, H., He, H., Liu, C., Wang, C., Qiao, Y., Zhang, B. (2022). Changes of sensory quality, flavor-related metabolites and gene expression in peach fruit treated by controlled atmosphere (CA) under cold storage. *International Journal of Molecular Sciences, 23*, 7141. https://doi.org/10.3390/ijms23137141.

López, M., Villatoro, C., Fuentes, T., Graell, J., Lara, I., Echeverría, G. (2007). Volatile compounds, quality parameters and consumer acceptance of 'Pink Lady' apples stored in different conditions. *Postharvest Biology and Technology, 43 (1)*, 55–66.

Magazin, N., Keserović, Z., Milić, B., Dorić, M. (2013). Fruits quality of granny smith apples treated with 1-methylcyclopropene or diphenylamine and stored under ULO conditions. *Acta Horticulturae, 981*, 619–624.

Mahajan, P.V., Luca, L., Edelenbos, M. (2016). Development of a small and flexible sensor-based respirometer for real-time determination of respiration rate, respiratory quotient and low O_2 limit of fresh produce. *Computers and Electronics in Agriculture, 121*, 347–353.

Mattheis, J., Buchanan, D., Fellman, J. (1998). Volatile compounds emitted by 'Gala' apples following dynamic atmosphere storage. *Journal of the American Society for Horticultural Science, 123*, 426–432.

Mditshwa, A., Fawole, O.A., Opara, U.L. (2018). Recent developments on dynamic controlled atmosphere storage of apples—A review. *Food Packaging and Shelf Life, 16*, 59–68.

Mditshwa, A., Fawole, O.A., Vries, F., van der Merwe, K., Crouch, E., Opara, U.L. (2017a). Minimum exposure period for dynamic controlled atmospheres to control superficial scald in Granny Smith apples for long distance supply chains. *Postharvest Biology and Technology, 127*, 27–34.

Mditshwa, A., Fawole, O.A., Vries, F., van der Merwe, K., Crouch, E., Opara, U.L. (2017b). Repeated application of dynamic controlled atmospheres reduced superficial scald incidence in 'Granny Smith' apples. *Scientia Horticulturae, 220*, 168–175.

Mditshwa, A., Vries, F., van der Merwe, K., Crouch, E., Opara, U.L. (2015). Antioxidant content and phytochemical properties of apple 'Granny Smith' at different harvest times. *South African Journal of Plant and Soil, 32*, 221–226.

Melgarejo, P., Salazar, D.M., Artes, F. (2000). Organic acids and sugars composition of harvested pomegranate fruits. *European Food Research and Technology, 211*, 185–190.

Moran, R.E., McManus, P. (2005). Firmness retention, and prevention of coreline browning and senescence in 'Macoun' apples with 1-methylcyclopropene. *Hortscience, 40 (1)*, 161–163.

Ortiz, A., Graell, J., Lara, I. (2011). Cell wall-modifying enzymes and firmness loss in ripening Golden Reinders apples: A comparison between calcium dips and ULO storage. *Food Chemistry, 128 (4)*, 1072–1079.

Peng, Y., Lu, R. (2006). Improving apple fruit firmness predictions by effective correction of multispectral scattering images. *Postharvest Biology and Technology, 41 (3)*, 266–274.

Pesis, E. (2005). The role of the anaerobic metabolites, acetaldehyde and ethanol, in fruit ripening, enhancement of fruit quality and fruit deterioration. *Postharvest Biology and Technology, 37 (1)*, 1–19.

Prange, R.K. (2018). Dynamic Controlled Atmosphere (DCA) Storage of fruits and vegetables. In: *Reference Module in Food Science*, Elsevier, p. B9780081005965214000. https://doi.org/10.1016/B978-0-08-100596-5.21349-8.

Prange, R.K., DeLong, J.M., Harrison, P.A., Leyte, J.C., McLean, S.D. (2003). Oxygen concentration affects chlorophyll fluorescence in chlorophyll-containing fruit and vegetables. *Journal of the American Society for Horticultural Science, 128 (4)*, 603–607.

Prange, R., Wright, A., DeLong, J., Zanella, A. (2013). A review on the successful adoption of dynamic controlled-atmosphere (DCA) storage as a replacement for diphenylamine (DPA), the chemical used for control of superficial scald in apples and pears. *XI International Controlled and Modified Atmosphere Research Conference, 1071*, 389–396.

Rebeaud, S.G., Gasser, F. (2015). Fruit quality as affected by 1-MCP treatment and DCA storage—A comparison of the two methods. *European Journal of Horticultural Science, 80*, 18–24.

Rupasinghe, H., Murr, D., Paliyath, G., Skog, L. (2000). Inhibitory effect of 1-MCP on ripening and superficial scald development in 'McIntosh' and 'Delicious' apples. *Journal of Horticultural Science and Biotechnology, 75 (3)*, 271–276.

Sabban-Amin, R., Feygenberg, O., Belausov, E., Pesis, E. (2011). Low oxygen and 1- MCP pretreatments delay superficial scald development by reducing reactive oxygen species (ROS) accumulation in stored 'Granny Smith' apples. *Postharvest Biology and Technology, 62 (3)*, 295–304.

Saquet, A., Streif, J., Bangerth, F. (2002). Reducing internal browning disorders in 'Braeburn' apples by delayed controlled atmosphere storage and some related physiological and biochemical changes. *XXVI International Horticultural Congress: Issues and Advances in Postharvest Horticulture, 628*, 453–458.

Schouten, S.P., Prange, R.K., Verschoor, J., Lammers, T.R., Oosterhaven, J., (1997). Improvement of quality of Elstar apples by dynamic control of ULO conditions. In: Mitcham, E.J. (Ed.), *Seventh International Controlled Atmosphere Research Conference. Proceedings Volume 2: Apples and Pears*. University of California, pp. 71–78.

Shaham, Z., Lers, A., Lurie, S. (2003). Effect of heat or 1-methylcyclopropene on antioxidative enzyme activities and antioxidants in apples in relation to superficial scald development. *Journal of the American Society of Horticultural Sciences, 128*, 761–766.

Steffens, C.A., Brackmann, A., Pinto, J.A. V., Eisermann, A.C. (2007). Taxa respiratória de frutas de climatemperado. *Pesquisa Agropecuária Brasileira, 42*, 313–321. Abstract in English. https://doi.org/ 10.1590/S0100-204X2007000300003. https://www.scielo.br/j/pab/a/bfPxVP4SxDHxFyCpjzsVcbc/?lang=pt (accessed 5 April 2024).

Streif, J. (2008). Ripening management and postharvest fruit quality. *III International Conference Postharvest Unlimited, 858*, 121–129.

Streif, J., Saquet, A., Xuan, H. (2001). CA-related disorders of apples and pears. *VIII International Controlled Atmosphere Research Conference, 600*. www.actahort.org/books/600/.

Thewes, F.R. (2019). Dynamic controlled atmosphere methods × 1-MCP: Metabolism and quality of stored apples. Ph.D Thesis. Federal University of Santa Maria, Santa Maria, p. 239.

Thewes, F.R., Both, V., Brackmann, A., Weber, A., de Oliveira Anese, R. (2015). Dynamic controlled atmosphere and ultralow oxygen storage on 'Gala' mutants quality maintenance. *Food Chemistry, 188*, 62–70.

Thewes, F.R., Brackmann, A., Both, V., Anese, R. de O., Schultz, E.E., Ludwig, V., Wendt, L.M., Berghetti, M.R.P., Thewes, F.R. (2020). Dynamic controlled atmosphere based on carbon dioxide production (DCA-CD): Lower oxygen limit establishment, metabolism and overall quality of apples after long-term storage. *Postharvest Biology and Technology, 168*, 111285. https://doi.org/10.1016/j.postharvbio.2020.111285.

Thewes, F.R., Rachael. M.W., Vanderlei. B., Nandita. K., Martin. G., Bruno, P.E., Michael, H.H., Auri, B., Jens, N.W., Daniel, A.N. (2021). Dynamic controlled atmosphere: A review of methods for monitoring fruit responses to low oxygen Comunicata Scientiae. *Horticultural Journal*, 1–13. https://doi.org/10.14295/CS.v12.3782.

Thompson, A.K. (2010). *Controlled Atmosphere Storage of Fruits and Vegetables*. CABI. www.cabidigitallibrary.org/doi/10.1079/9781845936464.0000.

Tran, D.T., Verlinden, B.E., Hertog, M., Nicolaï, B.M. (2015). Monitoring of extremely low oxygen control atmosphere storage of 'Greenstar' apples using chlorophyll fluorescence. *Scientia Horticulturae*, *184*, 18–22.

Van der Merwe, J., Combrink, J., Calitz, F. (2001). Effect of controlled atmosphere storage after initial low oxygen stress treatment on superficial scald development on South African-grown Granny Smith and Top red apples. *VIII International Controlled Atmosphere Research Conference*, *600*. https://doi.org/10.17660/ACTAHORTIC.2003.600.34.

Wang, Z., Dilley, D.R. (2000). Initial low oxygen stress controls superficial scald of apples. *Postharvest Biology and Technology*, *18 (3)*, 201–213.

Watkins, C.B. (2008). Dynamic controlled atmosphere storage: A new technology for the New York storage industry. *New York Fruit Quarterly*, *16 (1)*, 23–26.

Watkins, C.B., Nock, J.F., Whitaker, B.D. (2000). Responses of early, mid and late season apple cultivars to postharvest application of 1-methylcyclopropene (1-MCP) under air and controlled atmosphere storage conditions. *Postharvest Biology and Technology*, *19 (1)*, 17–32.

Weber, A., Brackmann, A., Both, V., Pavanello, E.P., Anese, R.D.O., Thewes, F.R. (2015). Respiratory quotient: Innovative method for monitoring Royal Gala apple storage in a dynamic controlled atmosphere. *Scientia Agricola*, *72 (1)*, 28–33.

Weber, A., Brackmann, A., Anese, R.D.O., Both, V., Pavanello, E.P. (2011). 'Royal Gala'apple quality stored under ultralow oxygen concentration and low temperature conditions. *Pesquisa Agropecuária Brasileira*, *46 (12)*, 1597–1602.

Weber, A., Thewes, F.R., de Oliveira Anese, R., Both, V., Pavanello, E.P., Brackmann, A. (2017). Dynamic controlled atmosphere (DCA): Interaction between DCA methods and 1-methylcyclopropene on 'Fuji Suprema' apple quality. *Food Chemistry*, *235*, 136–144.

Wei, J., Ma, F., Shi, S., Qi, X., Zhu, X., Yuan, J. (2010). Changes and postharvest regulation of activity and gene expression of enzymes related to cell wall degradation in ripening apple fruit. *Postharvest Biology and Technology*, *56 (2)*, 147–154.

Wright, A., Delong, J., Arul, J., Prange, R. (2015). The trend toward lower oxygen levels during apple (*Malus × domestica* Borkh) storage. *The Journal of Horticultural Science and Biotechnology*, *90 (1)*, 1–13.

Zanella, A. (2003). Control of apple superficial scald and ripening—a comparison between 1-methylcyclopropene and diphenylamine postharvest treatments, initial low oxygen stress and ultralow oxygen storage. *Postharvest Biology and Technology*, *27*, 69–78. https://doi.org/10.1016/s0925-5214(02) 00187–4.

Zanella, A., Cazzanelli, P., Panarese, A., Coser, M., Chistè, C., Zeni, F. (2005). Fruit fluorescence response to low oxygen stress: Modern storage technologies compared to 1-MCP treatment of apple. *Acta Horticulturae*, *682*, 1535–1542.

Zanella, A., Rossi, O. (2015). Post-harvest retention of apple fruit firmness by 1-methylcyclopropene (1-MCP) treatment or dynamic CA storage with chlorophyll fluorescence (DCA-CF). *European Journal of Horticultural Sciences*, *80*, 11–17.

Zanella, A., Stürz, S. (2012). Replacing DPA postharvest treatment by strategical application of novel storage technologies controls scald in 1/10th of EU's apples producing area. VII international postharvest symposium, 1012.

Ozone Gas Technology for Fresh Fruits and Vegetables

5

Vijay Yadav Tokala*
*Corresponding Author: vijayyadav.t@postharvest.org

5.1 INTRODUCTION

Demand for fresh fruits and vegetables has increased globally over the past few decades, which can be attributed to dietary patterns changing, a rise in taste preferences, health awareness, and consumers' present way of life (Pollock, 2001). But due to physiological, pathological, and physical factors, fresh fruits and vegetables are highly perishable compared to grain crops and therefore are vulnerable to significant postharvest losses. About 13.2% of foods, with a net worth of US$400 billion, are lost annually between harvest and the retail market, while the losses in fresh fruit and vegetables as high as 40% of total production (FAO, 2019). Fresh horticulture produce has a limited shelf life, contributing to the losses due to its high perishability (Kaczmarek et al., 2019). The three primary reasons for fresh product losses are physiological, pathological, and physical factors. Consumer acceptability decreases due to physiological changes brought on by ripening, senescence, and the development of postharvest physiological disorders leading to losses. Postharvest deterioration brought on by microbiological contamination (mostly by fungus and bacteria) may account for a sizable portion of losses. Severe losses also occur due to mechanical damage caused by improper harvesting, handling, and shipping practices. Improper handling during production and postharvest leads to contamination of fresh produce with pathogens that also increase the chances of foodborne diseases in humans. The evidence of increased outbreaks caused by foodborne infections grew along with the rise in the consumption of fresh fruits and vegetables (Sivapalasingam et al., 2004). Food safety has equal attention as the reduction in food losses, especially by the elimination of pathogens from fresh produce (Xu, 1999). To minimize the microbial burden from the produce, traditionally they are dipped in water treated with chlorine or fungicides (Garcia et al., 2003). Any residues of the fungicides on the produce could pose serious health issues in the long run. Chlorine is good at cleaning surface microorganisms; however, it can remove the original aroma and taste (Hassenberg et al., 2008). At high pH levels, chlorine is ineffective against spore-forming microorganisms and can create possibly dangerous chlorinated by-products such as trihalomethanes or halo acetic acids (Akbas & Olmez, 2007). Ozone has proved to be one of the most effective substitutes in the food industry to reduce the microbial load from the surface of fresh produce with no detectable residues in/on treated produce (Parish et al., 2003). Ozone-treated produce has no residues, as ozone easily converts to molecular oxygen while converting non-biodegradable pesticides and chemical constituents to safe types (Selma et al., 2008a). In addition to receiving full US FDA (Food and Drug Administration) clearance for usage in the food industry, ozone has been certified as Generally Recognized as Safe (GRAS) since 1997 (Tzortzakis et al., 2007a). Applications of ozone in the food processing industry have also been approved by organic certification due to ozone's quick breakdown into respirable oxygen and lack of chemical residues on the processed produce (Selma et al., 2008b). By employing ozone in the right quantities in the storage environment, fruit and vegetable commodities can be kept away from diseases, producing the least physiological stress (Tokala et al., 2018; 2020). According to Nichols and Varas (1992), ozone may oxidize a variety of organic compounds, allowing it to reduce pesticide residues in treated water. The fruit washed in ozonated water had low levels of harmful chemicals such as mycotoxins and pesticide residues (McKenzie et al., 1997). When exposed to high concentrations of ozone the fresh horticulture produce exhibited physiological stress symptoms (Tokala et al., 2020). When compared to fresh fruit, vegetables are typically more vulnerable to ozone-induced injuries, and gaseous ozone has been considered not suitable for commercial application to leafy vegetables (Sarron et al., 2021). Several researchers have been examining the application of ozone to extend the shelf life and/or enhance the quality of fresh horticulture produce in recent years. Furthermore, postharvest applications of ozone to fresh food in closed storage rooms were found to be efficient even at low doses. These elements, along with ozone's rapid decomposition, make it impossible for ozone's postharvest application to have any negative

DOI: 10.1201/9781003370376-7

effects on the environment. This chapter discusses the different physiological effects of the applied treatments of gaseous ozone on fresh horticultural produce.

5.2 PROPERTIES OF OZONE

Ozone has three oxygen molecules arranged at a bond angle of 116° 49' (Beltran, 2003). It naturally occurs in the atmosphere and is produced by lightning and ultraviolet light. Ozone has a vapor density of 24, making it 1.5 times heavier than normal air. In smaller proportions, pure ozone gas has a sky-blue color with a pleasing smell. But when inhaled in larger amounts it has a pungent, irritating, acrid smell and causes headaches and nausea (Bocci, 2013). Ozone is a strong, highly reactive oxidant with a larger oxidation potential than commonly used sanitizers. It may be used as a gas or in the form of a solution by being dissolved in water (Smilanick, 2003). Ozone gas rapidly decomposes into oxygen, and due to its instability, it is hard to store it. Ozone gas has a half-life as short as 25 min at 30 °C. The standard oxidizing power of ozone is much higher than chlorine or hypochlorous acid (Bocci, 2011).

Ozone has the potential to oxidize a wider range of compound concentrations such as alcohols, amines, heterocycles, arenes, alkenes, alkanes, and sulfides (Manousaridis et al., 2005). According to several authors, ozone treatment significantly lowers microbial growth and prevents pathogenic infections on fruits (Horvitz & Cantalejo, 2014). According to Skog and Chu (2001), ozone-based treatments can also decelerate the ripening process by oxidizing the ethylene released by the fruits and vegetables in storage rooms. The best concentrations for various horticulture produce to prolong storage without compromising marketable quality is yet to be standardized.

Ozone can be applied in gaseous form in a closed storage room or in aqueous form by bubbling gas into water. When compared to a solution of an aqueous nature, ozone in its gaseous form has a longer half-life, lasting 12 hours compared to 20 min in water at 20 °C.

5.3 OZONE GENERATION AND ITS APPLICATION

Due to its residual-free nature and possibility for treatment in aqueous as well as gaseous forms, the ozone applications have gained significant commercial attention in postharvest horticulture. The precursors for ozone synthesis are also widely available and economically viable, and the effectiveness and durability of ozone depend on the quality of the input gas (Crowe et al., 2007; Guzel-Seydim et al., 2004). As it is unstable, ozone cannot be kept in a cylinder similar to other gases and is typically produced for a single application. Schonbein reported three different processes for producing ozone: electrolyzation of aqueous acids solution, arcing oxygen, and exposing phosphorus to wet air. Compared to all other approaches, the electrolysis procedure produced greater amounts of ozone from oxygen (Rubin, 2001). Siemens (1857) described a thorough procedure for producing ozone from air or oxygen using a silent discharge device. This work paved the way for the creation of commercial ozone-producing machinery. The corona discharge process involves the introduction of significant concentrations of dry oxygen through low-power electrical discharge, which cause high-energy electrons to break apart oxygen molecules and results in the production of ozone. Numerous ozone-generating devices have been created utilizing the corona discharge theory, but establishing the factors to define the exact concentration range of ozone produced by this method is yet to be researched. In order to avoid the production of hazardous nitrogen oxides, commercial ozonizers utilizing this concept must utilize air that has had the nitrogen removed or has had pure oxygen added (Chang et al., 1991).

The corona discharge technique for producing ozone requires gases containing significant amounts of dry oxygen, which starts the dissociation of oxygen molecules by very energetic electrons. Ozone is created by a three-body collision process with additional oxygen molecules after the dissociation event. High gas temperatures in the corona further encourage breakdown events, and ozone combines with atomic oxygen to produce oxygen (Carlins & Clark, 1982). The corona discharge concept was used to create several ozone-generating devices, but due to the existence of complicated chemical interactions that are still poorly understood, no attempt has been made to predict the ozone concentrations caused by corona discharges in the air. To prevent the production of harmful nitrogen oxides, commercial ozonizers utilizing this concept must utilize pure oxygen or air that has had the nitrogen removed (Chang et al., 1991).

According to Horvitz and Cantalejo (2014), gaseous ozone may be effectively utilized to sanitize storage areas; break down mycotoxins; prevent the formation of bacteria, mold, and yeasts on edible foods; and control the infestation of insect pests. The gaseous ozone fumigation method is relatively less expensive, simple to use, and can also oxidize ethylene gas in storage rooms to delay ripening as well as remove off-odors created by bacterial activity (Rice et al., 1982; Glowacz & Rees, 2016). As the ozone gas is highly oxidizing, safety considerations are essential while it is applied. It is important to take precautions to accurately measure ozone levels often and maintain safe exposure intervals. Ozone gas is generally used in an enclosed room with the commodity inside (Figure 5.1). A range of sensor-based ozone gas detectors are available for real-time estimation of ozone gas concentrations (Figure 5.2). The ideal time for ozone treatment relies on several variables, including physiological and physical characteristics of treated

FIGURE 5.1 Ozone generator installed inside cold storage with apple fruit. (Source: Vijay Yadav Tokala)

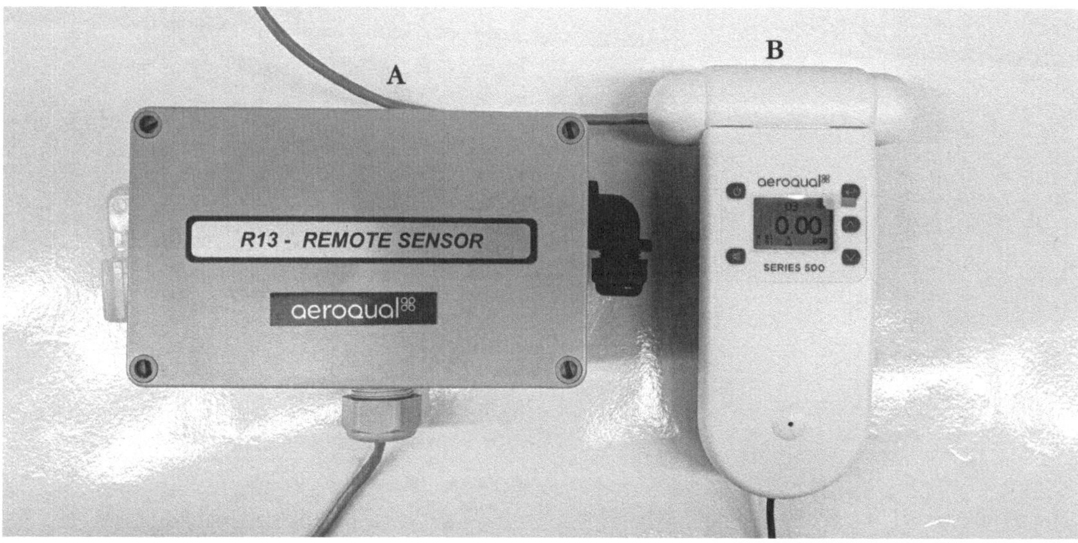

FIGURE 5.2 Gas detector for real-time estimation of ozone gas. (Source: Vijay Yadav Tokala)

commodities, the treatment equipment, the air quality, and temperature (Kim et al., 1999).

Higher ozone concentrations must be applied to successfully lower viable microbe counts on fruit surfaces, but doing so can also put the produce under oxidative stress and cause irreversible damage to fruit quality (Wani et al., 2015; Tokala et al., 2021). According to Forney (2003), fruits are often more tolerant of greater concentrations of ozone than vegetables. Several researchers have examined the effects of varying ozone concentrations on the nutritional value, microbiological development, and freshness of fresh fruits and vegetables (Table 5.1).

5.4 EFFECTS OF OZONE APPLICATION

5.4.1 Ethylene Production and Rates of Respiration

Upon extended exposure to the storage environment, ozone oxidizes ethylene into carbon dioxide, carbon monoxide, and water (Bailey, 1978). By using a proteomic technique, Minas et al.

TABLE 5.1 Effects of Different Concentrations of Gaseous Ozone Application on the Quality of Horticulture Produce

	CROP	*OZONE CONCENTRATION*	*INFERENCES*	*REFERENCES*
1	Apple	1 μL.L^{-1}	Reduced fungicidal residues; increased decay rate	Antos et al. (2018)
		0.1 μL.L^{-1}	Increased levels of sugars, ethylene, and respiration	Tokala et al. (2021)
		0.4 μL.L^{-1}	No effect on internal ethylene production; oxidized ethylene present in storage environment; fruit firmness, total soluble solids (TSS), and total acidity were not affected; no phytotoxicity symptoms	Skog and Chu (2001)
		0.5 μL.L^{-1}	Suppressed the growth of *Penicillium expansum*	Yaseen et al. (2015)
		2–10 mg.m^{-3}	Higher fruit texture maintained	Li et al. (1989)
		0.05–0.087 μL.L^{-1}	Reduced *Listeria innocua* count	Sheng et al. (2018)
2	Apricot	0.5–1.5 μL.L^{-1}	Rise in weight loss, SSC, and firmness	Panou et al. (2018)
3	Banana	8 μL.L^{-1}	Increased phenol and flavonoid levels; reduced ascorbic acid	Alothman et al. (2010)
4	Bell pepper	0.1–0.3 μmol.mo L^{-1}	Higher concentrations increased weight loss; peel color and fruit texture had no significant effect	Glowacz et al. (2015)
5	Bell pepper (fresh-cut)	1–9 μL.L^{-1}	Reduced *Salmonella typhimurium* and *L. monocytogenes*	Alwi and Ali (2014)
6	Blueberry	15 μL.L^{-1}	Increased antioxidant activity	Piechowiak, et al. (2020a)
		216 μL.L^{-1}; 433 μL.L^{-1}	Reduction of *E. coli* O157:H7, *Salmonella*, and *L. monocytogenes*	Bridges et al. (2018)
7	Broccoli	200 nL. L^{-1}	Rates of ethylene production and respiration were not affected; weight loss not affected	Forney et al. (2003)
		0.4 μL.L^{-1}	Delayed changes in color to yellow from green; improved base browning	Skog and Chu (2001)
8	Cantaloupe	0.15–0.3 μL.L^{-1}	Improved firmness; reduced fruit softening enzymes; lowered ethylene	Toti et al. (2018)
		10000 μL.L^{-1}	Reduced *Salmonella*	Selma et al. (2008b)
9	Cantaloupe (fresh-cut)	5000 μL.L^{-1}; 20000 μL.L^{-1}	Reduced total coliforms, yeast, and bacteria	Selma et al. (2008b)
		5000; 20000 μL.L^{-1}	Reduced yeast and bacteria count; maintained firmness, aroma, and visual quality	Selma et al. (2008b)
10	Carrot	15 μL. L^{-1}	At high concentrations microbial load decreased but respiration rates increased	Liew and Prange (1994)
		1000 nL L^{-1}	Respiration rates increased; firmness reduced; stress volatile production increased	Forney et al. (2007)
11	Clementine	0.1μmol.mol^{-1}	*Botrytis cineraria* spores reduced	Tzortzakis et al. (2007a)
12	Cucumber	0.5 μmol.mol^{-1}	Stomatal conductance and polyamine concentration increased	Agrawal et al. (1993)
		0.1–0.3 μL.L^{-1}	Reduced weight loss; maintained texture	Glowacz et al. (2015)
13	Fig	1.0–9.0 μL. L^{-1}	Sweetness not affected; reduced *Escherichia coli* and *Bacillus cereus*	Akbas & Ozdemir (2008)
		5 mg. L^{-1}	*Escherichia coli* count reduced	Oztekin et al. (2006)

14	Grape	0.3 μL.L^{-1}	Maintained levels of cyprodinil, fenhexamid, and pyrimethanil	Karaca et al. (2012)
		0.3 μL.L^{-1}	No effect on SSC or pH; rise in weight loss; reduced pesticide residue	Karaca (2019)
		60 μL.L^{-1}	Trans-resveratrol and trans-piceatannol levels increased	Segade et al. (2019)
		200 μL.L^{-1}	Decreased load of *Botrytis cinerea*	Ozkan et al. (2011)
		10000 μL.L^{-1}	Reduced pesticide residues	Gabler et al. (2010)
		2500–5000 μL.L^{-1}	Reduced gray mold (*B. cinerea*), phytotoxicity not observed	Gabler et al. (2010)
		0.1 μL.L^{-1}	Maintained high stilbenoid, total flavonols, and anthocyanin levels	Artes-Hernandez et al. (2003)
		0.1 μL.L^{-1}	Enhanced total phenolic and flavonoid concentrations	Artes-Hernandez et al. (2007)
		2 μL.L^{-1}	Increased resveratrol levels and weight loss; reduced sensory values	Cayuela et al. (2009)
		0.1 μL.L^{-1}	Titratable acidity (TA), SSC, aroma, and texture were not affected	Artes-Hernandez et al. (2004)
		200–350 μL.L^{-1}	More effective in controlling fungal growth at a higher relative humidity	Ozkan et al. (2011)
15	Guava	8 μL.L^{-1}	Reduced levels of total phenols, flavonoids, and ascorbic acid	Alothman et al. (2010)
16	Kiwifruit	4 mg.hr^{-1}	Reduced incidence of *Botrytis cineraria*; maintained high levels of reducing sugars	Barboni et al. (2010)
		0.3 μL.L^{-1}	Higher SSC; reduced microbial growth; induced ripening-associated changes	Goffi et al. (2019)
		0.3 μL.L^{-1}	Reduced malondialdehyde accumulation; inhibited lipoxygenase activity and enhanced catalase activity	Goffi et al. (2020)
		0.3 μL.L^{-1}	Decreased disease incidence; maintained phenol levels	Minas et al. (2010)
		0.3 μL.L^{-1}	Delayed ripening; reduced ethylene production; enhanced antioxidant activity	Minas et al. (2012)
		0.3 μL.L^{-1}	Reduced ethylene production and fruit-softening enzyme activity	Minas et al. (2014)
		0.3 μL.L^{-1}	Reduced ripening and ethylene production	Minas et al. (2018)
17	Lettuce	36000 μL.L^{-1}	Reduced surface contamination by norovirus	Predmore et al. (2015)
18	Mandarin	2–10 mg.m^{-3}	Weight loss and rotting reduced; retarded activity of fruit softening enzymes and senescence	Li et al. (1989)
19	Mango	2–10 μL.L^{-1}	Reduced ethylene production, respiration rate, weight loss; delayed color changes	Tran et al. (2013)

(Continued)

TABLE 5.1 (Continued)

	CROP	*OZONE CONCENTRATION*	*INFERENCES*	*REFERENCES*
20	Mulberry	0.3, 2.42 $\mu L.L^{-1}$	Reduced aerobic mesophilic bacteria; increased weight loss	Tabakoglu and Karaca (2018)
		2 $\mu L.L^{-1}$	Maintained SSC, firmness, color, and titratable acidity; reduced respiration rate and the decay rate	Han et al. (2017)
21	Onion	50 $nL.L^{-1}$ to 200 $nL.L^{-1}$	Reduced physiological weight loss; fungal infections decreased; no significant effect on internal decay and sprouting	Fan et al. (2001)
		50 $nL.L^{-1}$ to 250 $nL.L^{-1}$	No substantial impact on sprouting and firmness; mass loss and fungal infection reduced	Song et al. (2000)
22	Papaya	1.5–5 $\mu L.L^{-1}$	> 3.5 $\mu L.L^{-1}$ caused the injury; increased PAL, POD, and PPO enzymes	Ong et al. (2014)
		1.5–5 $\mu L.L^{-1}$	Reduced weight loss; increased SSC, lycopene, and antioxidant activity	Ali et al. (2014)
23	Papaya (fresh-cut)	9.2 $\mu L.L^{-1}$	Increased phenolic content; reduced microbial load	Yeoh et al. (2014)
24	Peach	0.3 $\mu L.L^{-1}$	Rates of respiration and ethylene production are not affected; higher weight loss	Palou et al. (2002)
25	Pear	100 $\mu L.L^{-1}$	Reduced weight loss; maintained high SSC and titratable acidity	Alencar et al. (2014)
26	Pepper	0.1–0.3 $\mu L.L^{-1}$	No substantial influence on quality and weight loss	Glowacz et al. (2015)
		1019 $\mu L.L^{-1}$	Increased activity of PPO; activity of PAL and tyrosine ammonia-lyase reduced	Sachadyn-Krol et al. (2016)
		1–9 $\mu L.L^{-1}$	Increased weight loss, respiration rates, and membrane permeability; reduced firmness and acidity	Alwi and Ali (2014)
27	Persimmon	0.1 $\mu L.L^{-1}$	Stimulated mass loss and electrolytes leakage	Salvador et al. (2006)
28	Pineapple	8 $\mu L.L^{-1}$	Increased phenol and flavonoid levels; reduced ascorbic acid	Alothman et al. (2010)
29	Pomegranate	5.2 and 10.4 $mg.h^{-1}.m^{-3}$	Developed off-odor and lower content of acetaldehyde	Ozon and Kafkas (2015)
30	Plum	0.1 $\mu mol.mol^{-1}$	Suppressed gray mold incidence–related spores	Tzortzakis et al. (2007a)
31	Raspberry	4–5 $\mu L.L^{-1}$	Reduced enzymes involved in phenolic degradation	Piechowiak, et al. (2020b)
		8–10 $\mu L.L^{-1}$	Higher activity of SOD and PAL	Piechowiak & Balawejder (2019)
		30000 $\mu L.L^{-1}$	*Salmonella* and *E. coli* reduced	Bialka & Demirci (2007)
32	Red pitaya (fresh-cut)	10 $\mu L.L^{-1}$	Conserved high fruit firmness, SSC, and titratable acidity	Li et al. (2023)
33	Spinach	1–10 $\mu L.L^{-1}$	Induced ozone injury (discoloration and dark spots on leaves) at higher concentrations	Calatayud et al. (2004)

34	Strawberry	1–5 μL.L^{-1}	Enhanced anthocyanins, phenols, and flavonoids	Chen et al. (2019)
		36000 μL.L^{-1}	Decreased surface contamination by virus	Predmore et al. (2015)
		30000 μL.L^{-1}	Reduced *Salmonella* and *E. coli*	Bialka and Demirci (2007)
		4 mg.L^{-1}	High ascorbic acid levels maintained; reduced respiration rates, senescence, and weight loss	Zhang et al. (2011)
		36000 μL.L^{-1}	Reduced *Salmonella* count	Zhou et al. (2018)
		40 mg.L^{-1}	High biothiol levels maintained	Demirkol et al. (2008)
35	Tangerine	200mg.L^{-1}	*Penicillium digitatum* spores reduced; no significant effect on weight loss, peel color, or acidity	Whangchai et al (2010)
		300 μL.L^{-1}	Increased SOD and catalase activities; delayed disease incidence	Boonkorn et al. (2012)
36	Tomato	0.05–1.0 μL.L^{-1}	Higher firmness was maintained; no effect on ethylene production	Tzortzakis (2007)
		0.05 μL.L^{-1}	Inhibited expression of signal transduction and defense-related genes	Tzortzakis et al. (2011)
		10 μL.L^{-1}	Delayed softening and PME activity; induced phenolics	Rodoni et al. (2010)
		800; 1600; 3200 μL.L^{-1}	Reduced firmness and lycopene; negatively impacted the appearance	Wang et al. (2019)
		12.82; 23.08 μL.L^{-1}	Enhanced shelf life	Venta et al. (2010)
		216; 433 μL.L^{-1}	Reduction of *E. coli*, *Salmonella*, and *L. monocytogenes*; surface bleached	Bridges et al. (2018)
		2500; 15000 μL.L^{-1}	Reduced *Salmonella* count; discoloration; no softening	Das et al. (2006)
37	Zucchini	0.1–0.3 μL.L^{-1}	Maintained firmness	Glowacz et al. (2015)

(2012) have demonstrated that ozone can prevent the synthesis of ethylene and postpone the ripening of kiwifruits. In cold storage containers, gaseous ozone treatment has lowered ethylene levels in apple and pear fruits (Skog & Chu, 2001). The ethylene exposure responses, such as the color change from green to yellow and the rate of floret opening, were significantly reduced in ozone-treated broccoli florets. According to Palou et al. (2001), the use of ozone gas efficiently reduced the levels of ethylene in sweet orange fruit loaded in temperature-controlled containers for export purposes. It was also proposed that ozone generators might be useful in storage rooms where mixed produce containing both ethylene-producing (apples) and ethylene-sensitive (broccoli) produce is kept. Cold-stored strawberries treated with ozone showed noticeably lowered respiration rates and delayed senescence (Zhang et al., 2011). In fresh-cut carrots, the simultaneous application of ozone and controlled atmosphere (CA) lowered respiration rate and ethylene biosynthesis (Chauhan et al., 2011).

Some researchers opined that high ozone levels in storage environments may accelerate rates of C_2H_4 breakdown and hence slow down the ripening process (Gane, 1937; Kuprianoff, 1953; Dickson et al., 1992). In contrast, the treatment of ozone had no discernible impact on the rate of respiration or the synthesis of ethylene in fresh commodities (Palou et al., 2002; Tzortzakis et al., 2007a). On the other hand, Liew and Prange (1994) noticed a rise in respiratory intensity when carrots were exposed to increased ozone concentrations. Even in highbush blueberries, ozone application resulted in methanol and ethanol synthesis, altered metabolism, and increased respiration (Song et al., 2003). Similarly, Tokala et al. (2021) found increased levels of respiration and ethylene production in apple cultivars stored in cold storage with ozone. Variations in ozone concentrations may account for the different ways that various fruits and vegetables react in terms of ethylene production and respiration rates. The rate of respiration generally serves as a scale for physiological activity, and increased respiration rate indicates that increased ozone concentrations have caused damage at the cellular level (Forney et al., 2003).

5.4.2 Biochemical Quality Parameters

Several studies have shown that gaseous ozone application can boost the antioxidants in fruits and vegetables. Following ozone treatment papaya fruit exhibited a rise in beta-carotene and lycopene content. It was demonstrated that ozone application caused higher detoxification of the reactive oxidative species (ROS) in raspberry fruit (Piechowiak & Balawejder, 2019). In table grapes, ozone application significantly increased the amounts of bioactive substances such as stilbenoid, viniferins, and phytoalexins (resveratrol and pterostilbene) (Gonzalez-Barrio et al., 2006). The treatment with ozone markedly increased levels of total flavonoid and total phenolics in fresh-cut banana and pineapple fruits, improving their nutritional value. However, when the same treatments were applied to guava fruit, ascorbic acid levels and related attributes of the eating quality were decreased (Alothman et al., 2010).

After treatment with ozone, high soluble sugar concentrations of tomato fruits were maintained (Tzortzakis et al., 2007b), while Shalluf (2012) found that the levels of glucose and fructose significantly increased in tomatoes after ozone treatment. In another work, it was found that higher amounts of glucose and fructose sugars in red bell pepper fruits were noted after their continuous ozone exposure (Glowacz et al., 2015). In the same way, Tokala et al. (2021) also reported increased levels of sugars in apple cultivars following ozone treatment in cold storage.

Fruit softening significantly affects postharvest life, and it is the result of several cell wall degradation processes such as solubilization of middle lamella, pectin, and hemicellulose depolymerization (Wang et al., 2018). Numerous genes and enzymes that break down cell walls have been functionally characterized. Although there was no substantial difference in the activity of polygalacturonase (PG), and β-Galactosidase (β-Gal), short-term ozone application suppressed fruit softening in tomatoes at ambient temperature by reducing pectin methyl esterase (PME) activity (Rodoni et al. 2010). While Toti et al. (2018) observed no effect on PME, they found reduced activity of PG, α-arabinosidase, and β-Gal when cantaloupe was treated with ozone.

According to Palou et al. (2001; 2002), ozone treatment and low temperatures had a synergistic impact on delaying ethylene-associated changes and sustaining fruit quality with the absence of any phytotoxic effects. In another work, it was also observed that the fruit quality characteristics of fresh blueberries were preserved throughout storage at appropriate refrigeration temperatures when combined with ozone gas (Concha-Meyer et al., 2015). According to Perez et al. (1999), applying gaseous ozone to strawberries during cold storage decreased their aroma of volatile ester compounds. On the other side, Nadas et al. (2003) also found that the use of gaseous ozone during cold storage was successful in preserving fruit quality; it also reversibly decreased the volatile compound production essential for the distinctive scent of strawberry fruit.

According to Cayuela et al. (2009), gaseous ozone caused increased weight loss and reduced organoleptic score in table grapes. Table grapes exposed to higher concentrations of ozone shocks under low temperature conditions retained high phenolics, flavonoids, and anthocyanin content throughout the length of time after cold storage (Artes-Hernandez et al., 2003, 2007). On the other side, it was observed that long-term preserved arils of pomegranates when subjected to ozone treatment exhibited reduced the levels of acetaldehyde and off-odor volatile content (Ozon & Kafkas, 2015). Strawberry fruit exposed to very high concentrations of ozone in cold storage had decreased phenolic levels, mostly because of a considerable decrease in the quantities of procyanidin contents (Allende et al., 2007).

Total anthocyanins and total flavonoids of table grapes packed in polypropylene packaging were relatively higher when treated with the use of ozone gas (Artes-Hernandez et al., 2003). The carrots treated with ethylene antagonist 1-methylcyclopropene (1-MCP) and then kept in cold ozonated storage exhibited a transient escalation of respiration rate and stress-related volatile compounds (such as hexanal and ethanol) levels (Forney et al., 2007). Similarly, Tokala et al. (2021) also reported a rise in the rates of ethylene production and respiration in apple cultivars treated with ethylene antagonists 1*H*-cyclopropabenzene (BC), 1*H*-cyclopropa[*b*]naphthalene (NC), and 1-MCP and cold stored with gaseous ozone application.

5.4.3 Microbial Population

Ozone exerts its anti-microbial effect through two mechanisms. Microbes such as bacteria and viruses are mostly destroyed by ozone by direct oxidation (Fan, 2021). The strong oxidant properties of ozone damage and destroy bacterial cell structures, mainly the membranes of microorganisms, which results in cell death (Alwi & Ali, 2014). Ozone has also been shown to affect the structure of viral particles and destroy viral proteins on virus surfaces while leaving RNA intact (Predmore et al., 2015). The second pathway includes the generation of ROS from the breakdown of ozone, including OH, OOH, and H_2O_2 (Forney, 2003). High levels of ROS produced by ozone gradually destroy cellular constituents such as proteins, lipids, and DNA by upsetting the equilibrium between oxidative stress and defense enzymes (Alwi & Ali, 2014).

Ozone therapy significantly slows down microbial development and prevents diseased fruit regions from spreading. When compared to treatments for isolated fungi, ozone treatments given to fruits that had been infected with a pathogen had a more inhibitory effect on the growth of mold (Tzortzakis et al., 2007a, 2008). As a result of the major changes that ozone treatment has brought about in fruit-pathogen interactions, pathogen growth has been effectively suppressed without any unfavorable chemicals (Tzortzakis et al., 2008; Smilanick et al., 1999). Additionally, they mentioned that ozone had a major impact even at low concentrations, but the larger concentrations would have a bigger impact. Ozone's significant biocidal properties result from a synergistic interaction between its increased oxidizing efficacy and its ability to cross biological membranes.

At the right concentrations, ozone can significantly lower the ethylene levels and decrease the microbial load on the surface, which can delay senescence and significantly prolong the shelf life of fresh produce. In combination with cold, controlled atmosphere conditions, modified atmosphere packing could significantly enhance the storage life of fresh horticulture commodities while preserving quality attributes (Skog & Chu, 2001; Zhang et al., 2011).

5.5 LIMITATIONS

Despite being a powerful surface sanitizer, gaseous ozone was not suitable in decreasing decay-causing microorganisms present in wounds of the fruit (Smilanick et al., 1999). Due to ozone's limited ability to penetrate wound entries and possible reactions with reactive compounds viz., dirt, fruit residues, or organic materials other than pathogen inoculum in wounds, its anti-microbial capacity is reduced (Karaca, 2010). Even though ozone has been linked to several instances of microbial decrease, spore development returns when ozone-treated fruits are exposed to normal air (Smilanick, 2003). Ozone spray is not a substitute for the use of fungicides since it is ineffective at stopping microbial development in wounds and less effective at stopping latent infections that are already present on produce (Glowacz & Rees, 2016).

Since not all fruit and vegetable commodities react the same way, even at the same dose, phytotoxicity in ozone-sensitive crops is a major problem (Skog & Chu, 2001). Prior to its commercial deployment, it is crucial to understand the acceptable level of ozone exposure for certain fruits and vegetables.

Ozone gas is not hazardous at low doses, but at high concentrations it can be lethal to people, particularly harming the respiratory system. To protect employees from the harmful effects of ozone, a workplace managing ozone must have an efficient ozone monitoring and alarm system as well as appropriate personal protection gear. To prevent any threats to one's health when utilized in enclosed buildings like cold storage, safety measures must be strictly considered. Ozone levels must be closely monitored, and the treated rooms must be opened at proper safety intervals. (Tokala et al., 2018)

5.6 CONCLUSIONS

Ozone has great oxidation capability and a broad range of antimicrobial properties. It breaks down quickly into oxygen and leaves no residues. Although gaseous ozone has a longer half-life than an aqueous solution, it must be applied at larger concentrations or for longer periods to be beneficial. The internal ethylene level in vegetables and fruits was not significantly reduced with gaseous ozone treatment in closed storage containers. In the majority of cases, the ozone treatment raised concentrations of sugars, antioxidants, phenolic contents, and anthocyanin contents but decreased levels of organic acids (citric acid, ascorbic acid, etc.). However, it is necessary to standardize the safe ozone concentration for various cultivars of several fruits and vegetables as they had exhibited oxidative stress symptoms at high concentrations of ozone. Future research should focus on standardizing safe doses of ozone for various fruits and vegetables while examining their impact on endogenous enzymes and phytohormones involved in produce quality and storage life. The effectiveness of applying ozone in combination with different storage conditions and postharvest treatments in prolonging storage life and preserving the quality of fresh produce warrants further research in a variety of fruits and vegetables.

REFERENCES

Agrawal, M., Donald T. Krizek, Shashi B. Agrawal, George F. Kramer, Edward H. Lee, Roman M. Mirecki, Randy A. Rowland. (1993). Influence of inverse day/night temperature on ozone sensitivity and selected morphological and physiological responses of cucumber. *Journal of the American Society for Horticultural Science* 118, 5, 649–654.

Akbas, M. Y., Ö. Hülya. (2007). Effectiveness of organic acid, ozonated water and chlorine dippings on microbial reduction and storage quality of fresh-cut iceberg lettuce. *Journal of the Science of Food and Agriculture* 87, 14, 2609–2616.

Akbas, M. Y., M. Ozdemir. (2008). Application of gaseous ozone to control populations of *Escherichia coli*, *Bacillus cereus* and *Bacillus cereus* spores in dried figs. *Food Microbiology* 25(2), 386–391.

Alencar, E. R., L. R. Faroni, M. S. Pinto, A. R. da Costa, A. F. Carvalho. (2014). Effectiveness of ozone on postharvest conservation of pear (*Pyrus communis* L). *Journal of Food Processing and Technology* 5, 317, 3–5. doi: 10.4172/2157-7110.1000317

Ali, A., K. O. Mei, Charles F. Forney. (2014). Effect of ozone preconditioning on quality and antioxidant capacity of papaya fruit during ambient storage. *Food Chemistry* 142, 19–26.

Allende, Ana, Alicia Marín, Begoña Buendía, Francisco Tomás-Barberán, Maria I. Gil. (2007). Impact of combined postharvest treatments (UV-C Light, Gaseous O_3, Superatmospheric O_2 and High CO_2) on health promoting compounds and shelf-life of strawberries. *Postharvest Biology and Technology* 46, 3, 201–211.

Alothman, Mohammad, Bhupinder Kaur, Ariffin Fazilah, Rajeev Bhat, Alias A. Karim. (2010). Ozone-induced changes of antioxidant capacity of fresh-cut tropical fruits. *Innovative Food Science & Emerging Technologies* 11, 4, 666–671.

Alwi, Nurul A., Asgar Ali. (2014). Reduction of *Escherichia coli* O157, *Listeria monocytogenes* and *Salmonella enterica* sv. *Typhimurium* populations on fresh-cut bell pepper using gaseous ozone. *Food Control* 46, 304–311.

Antos, P., B. Piechowicz, J. Gorzelany, N. Matłok, D. Migut, R. Józefczyk, M. Balawejder. (2018). Effect of ozone on fruit quality and fungicide residue degradation in apples during cold storage. *Ozone: Science & Engineering* 40, 6, 482–486

Artés-Hernández, Francisco, Encarna Aguayo, Francisco Artés, Francisco A. Tomás-Barberán. (2007). Enriched ozone atmosphere enhances bioactive phenolics in seedless table grapes after prolonged shelf life. *Journal of the Science of Food and Agriculture* 87, 5, 824–831.

Artés-Hernández, F., E. Aguayo, F. Artés. (2004). Alternative atmosphere treatments for keeping quality of 'autumn seedless' table grapes during long-term cold storage. *Postharvest Biology and Technology* 31, 59–67.

Artés-Hernández, Francisco, Francisco Artés, Francisco A. Tomás-Barberán. (2003). Quality and enhancement of bioactive phenolics in cv. Napoleon table grapes exposed to different postharvest gaseous treatments. *Journal of Agricultural and Food Chemistry* 51, 18, 5290–5295.

Bailey, P. S. (1978). *Organic Chemistry, vol. 39, pt. 1: Ozonation in Organic Chemistry, vol. 1: Olefinic Compounds*. New York: Academic Press, 272.

Barboni, Toussaint, Magali Cannac, Nathalie Chiaramonti. (2010). Effect of cold storage and ozone treatment on physicochemical parameters, soluble sugars and organic acids in *Actinidia deliciosa*. *Food Chemistry* 121, 946–951.

Beltran, Fernando J. (2003). *Ozone Reaction Kinetics for Water and Wastewater Systems*. New York: CRC Press.

Bialka, Katherine L., Ali Demirci. (2007). Utilization of gaseous ozone for the decontamination of *Escherichia coli* O157: H7 and *Salmonella* on raspberries and strawberries. *Journal of Food Protection* 70, 5, 1093–1098.

Bocci, Velio. (2011). *Ozone: A New Medical Drug*. Dordrecht: Springer.

Bocci, Velio. (2013). *Oxygen-ozone Therapy. A Critical Evaluation*. New York: Springer Science & Business Media, 1–37.

Boonkorn, P., H. Gemma, S. Sugaya, S. Setha, J. Uthaibutra, K. Whangchai. (2012). Impact of high-dose, short periods of ozone exposure on green mold and antioxidant enzyme activity of tangerine fruit. *Postharvest Biology and Technology* 67, 25–28.

Bridges, D. F., B. Rane, V. C. Wu. (2018). The effectiveness of closed-circulation gaseous chlorine dioxide or ozone treatment against bacterial pathogens on produce. *Food Control* 91, 261–267.

Calatayud, A., D. J. Iglesias, M. Talon, E. Barreno. (2004). Response of spinach leaves (*Spinacia oleracea* L.) to ozone measured by gas exchange, chlorophyll a fluorescence, antioxidant systems, and lipid peroxidation. *Photosynthetica* 42, 23–29.

Carlins, James J., Richard G. Clark. (1982). Ozone generation by corona discharge. In *Handbook of Ozone Technology and Applications*. Ann Arbor: Ann Arbor Science, 41–75.

Cayuela, José Antonio, A. Vazquez, A. G. Perez, J. M. Garcia. (2009). Control of table grapes postharvest decay by ozone treatment and resveratrol induction. *Revista de Agaroquimica y Tecnologia de Alimentos* 15, 5, 495–502.

Chang, J-S., Phil A. Lawless, Toshiaki Yamamoto. (1991). Corona discharge processes. *IEEE Transactions on Plasma Science* 19, 1152–1166.

Chauhan, O. P., P. S. Raju, N. Ravi, Asha Singh, A. S. Bawa. (2011). Effectiveness of ozone in combination with controlled atmosphere on quality characteristics including lignification of carrot sticks. *Journal of Food Engineering* 102, 1, 43–48.

Chen, C., H. Zhang, C. Dong, H. Ji, X. Zhang, L. Li, Z. Ban, N. Zhang, W. Xue. (2019). Effect of ozone treatment on the phenylpropanoid biosynthesis of postharvest strawberries. *RSC Advances* 9, 44, 25429–25438.

Concha-Meyer, Anibal, Joseph D. Eifert, Robert C. Williams, Joseph E. Marcy, Gregory E. Welbaum. (2015). Shelf life determination of fresh blueberries (*Vaccinium corymbosum*) stored under controlled atmosphere and ozone. *International Journal of Food Science*, http://dx.doi.org/10.1155/2015/164143

Crowe, Kristi M., Alfred A. Bushway, Rodney J. Bushway, Katherine Davis-Dentici, Russell A. Hazen. (2007). A comparison of single oxidants versus advanced oxidation processes as chlorine-alternatives for wild blueberry processing (*Vaccinium angustifolium*). *International Journal of Food Microbiology* 116, 1, 25–31.

Daş, E., G. C. Gürakan, A. Bayındırlı. (2006). Effect of controlled atmosphere storage, modified atmosphere packaging and gaseous ozone treatment on the survival of Salmonella Enteritidis on cherry tomatoes. *Food Microbiology* 23, 5, 430–438. https://doi.org/10.1016/j.fm.2005.08.002

Demirkol, Omca, Arzu Cagri-Mehmetoglu, Zhimin Qiang, Nuran Ercal, Craig Adams. (2008). Impact of food disinfection on beneficial biothiol contents in strawberry. *Journal of Agricultural and Food Chemistry* 56, 21, 10414–10421.

Dickson, Russell Garland, S. J. Kays, S. E. Law, M. A. Eiteman. (1992). Abatement of ethylene by ozone treatments in controlled atmosphere storage of fruits and vegetables. In *American Society of Agricultural Engineers. Meeting (USA)*. New York: ASAE, pp. 1–9.

Fan, L. H., J. Song, P. D. Hildebrand, C. F. Forney. (2001). Corona discharge reduces mold on commercially stored onions. *Acta Horticulturae* 553, 427–428.

Fan, X. (2021). Gaseous ozone to preserve quality and enhance microbial safety of fresh produce: Recent developments and research needs. *Comprehensive Reviews in Food Science and Food Safety*, 1–22. https://doi.org/10.1111/1541-4337.12796

Food and Agriculture Organization (FAO). (2019). *The State of Food and Agriculture 2019. Moving Forward on Food Loss and Waste Reduction*. London: Author

Forney, C. F. (2003). Postharvest Response of Horticultural Products to Ozone. In: *Postharvest Oxidative Stress in Horticultural Crops*, edited by D. M. Hodges, 13–54. New York: CRC Press.

Forney, Charles F., Jun Song, Lihua Fan, Paul D. Hildebrand, Michael A. Jordan. (2003). Ozone and 1-methylcyclopropene alter the postharvest quality of broccoli. *Journal of the American Society for Horticultural Science* 128, 3, 403–408.

Forney, Charles F., Jun Song, Paul D. Hildebrand, Lihua Fan, Kenneth B. McRae. (2007). Interactive effects of ozone and 1-methylcyclopropene on decay resistance and quality of stored carrots. *Postharvest Biology and Technology* 45, 3, 341–348.

Forney, Charles F., Jun Song, Paul D. Hildebrand, Lihua Fan, Kenneth B. McRae. (2007). Interactive effects of ozone and 1-methylcyclopropene on decay resistance and quality of stored carrots. *Postharvest Biology and Technology* 45, 3, 341–348.

Gabler, Franka Mlikota, Joseph L. Smilanick, Monir F. Mansour, Hakan Karaca. (2010). Influence of fumigation with high concentrations of ozone gas on postharvest gray mold and fungicide residues on table grapes. *Postharvest Biology and Technology* 55, 2, 85–90.

Gane, R. (1937). The respiration of bananas in presence of ethylene. *New Phytologist* 36, 2, 170–178.

Garcia, A., J. R. Mount, P. M. Davidson. (2003). Ozone and chlorine treatment of minimally processed lettuce. *Journal of Food Science* 68, 9, 2747–2751.

Glowacz, Marcin, Richard Colgan, Deborah Rees. (2015). Influence of continuous exposure to gaseous ozone on the quality of red bell peppers, cucumbers and zucchini. *Postharvest Biology and Technology* 99, 1–8.

Glowacz, Marcin, Deborah Rees. (2016). The practicality of using ozone with fruit and vegetables. *Journal of the Science of Food and Agriculture* 96, 14, 4637–4643.

Goffi, V., A. Magri, R. Botondi, M. Petriccione. (2020). Response of antioxidant system to postharvest ozone treatment in 'Soreli' kiwifruit. *Journal of the Science of Food and Agriculture* 100, 3, 961–968.

Goffi, V., L. Zampella, R. Forniti, M. Petriccione, R. Botondi. (2019). Effects of ozone postharvest treatment on physicochemical and qualitative traits of *Actinidia chinensis* 'Soreli' during cold storage. *Journal of the Science of Food and Agriculture* 99, 13, 5654–5661

González-Barrio, Rocío, David Beltrán, Emma Cantos, María I. Gil, Juan Carlos Espín, Francisco A. Tomás-Barberán. (2006). Comparison of ozone and UV-C treatments on the postharvest stilbenoid monomer, dimer, and trimer induction in Var. 'superior' white table grapes. *Journal of Agricultural and Food Chemistry* 54, 12, 4222–4228.

Guzel-Seydim, Zeynep B., Annel K. Greene, A. C. Seydim. (2004). Use of ozone in the food industry. *LWT-Food Science and Technology* 37, 4, 453–460.

Han, Q., H. Gao, H. Chen, X. Fang, W. Wu. (2017). Precooling and ozone treatments affects postharvest quality of black mulberry (*Morus nigra*) fruits. *Food Chemistry* 221, 1947–1953.

Han, Y., J. D. Floros, R. H. Linton, S. S. Nielsen, P. E. Nelson. (2002). Response surface modeling for the inactivation of *Escherichia coli* O157: H7 on Green Peppers (*Capsicum annuum*) by ozone gas treatment. *Journal of Food Science* 67, 3, 1188–1193.

Hassenberg, K., A. Fröhling, M. Geyer, O. Schlüter, W. B. Herppich. (2008). Ozonated wash water for Inhibition of *Pectobacterium carotovorum* on carrots and the effect on the physiological behaviour of produce. *European Journal of Horticultural Science* 73, S37–S42.

Horvitz, S., M. J. Cantalejo. (2014). Application of ozone for the postharvest treatment of fruits and vegetables. *Critical Reviews in Food Science and Nutrition* 54, 3, 312–339.

Kaczmarek, M., Avery, S. V., Singleton, I. (2019). Microbes associated with fresh produce: Sources, types and methods to reduce spoilage and contamination. *Advances in Applied Microbiology* 107, 29–82.

Karaca, H. (2019). The effects of ozone-enriched storage atmosphere on pesticide residues and physicochemical properties of table grapes. *Ozone: Science & Engineering* 41, 5, 404–414.

Karaca, H., Walse, S. S., Smilanick, J. L. (2012). Effect of continuous 0.3 μL/L gaseous ozone exposure on fungicide residues on table grape berries. *Postharvest Biology and Technology* 64, 1, 154–159.

Karaca, Hakan. (2010). Use of ozone in the citrus industry. *Ozone: Science & Engineering* 32, 2, 122–129.

Kim, Jin-Gab, Ahmed E. Yousef, Sandhya Dave. (1999). Application of ozone for enhancing the microbiological safety and quality of foods: A review. *Journal of Food Protection* 62, 9, 1071–1087.

Kuprianoff, J. (1953). The use of ozone for the cold storage of fruit. *Z. Kaltentech* 10, 1–4.

Li, Jin, Xiaoyu Wang, Honglin Yao, Zonggan Yao, Jiaxun Wang, Yaguang Luo. (1989). Influence of discharge products on postharvest physiology of fruit. In *Sixth International Symposium on High Voltage Engineering*, New Orleans, LA, August 28–September 1, p. 4.

Li, Chen, Shan Wang, Jiaqi Tao, Jiayi Wang, Zhaoxia Wu. (2023). Monitoring quality parameters and antioxidant potential of fresh-cut red pitaya fruit treated with gaseous ozone using kinetic models. *Journal of Food Measurement and Characterization* 1–17.

Liew, Chiam L., Robert K. Prange. (1994). Effect of ozone and storage temperature on postharvest diseases and physiology of carrots (*Daucus carota* L.). *Journal of the American Society for Horticultural Science* 119, 3, 563–567.

lowacz, Marcin, Richard Colgan, Deborah Rees. (2015a). The use of ozone to extend the shelf-life and maintain quality of fresh produce. *Journal of the Science of Food and Agriculture* 95, 4, 662–671.

Manousaridis, G., A. Nerantzaki, E. K. Paleologos, A. Tsiotsias, I. N. Savvaidis, M. G. Kontominas. (2005). Effect of ozone on microbial, chemical and sensory attributes of shucked mussels. *Food Microbiology* 22, 1, 1–9.

McKenzie, K. S., A. B. Sarr, K. Mayura, R. H. Bailey, D. R. Miller, T. D. Rogers, W. P. Norred et al. (1997). Oxidative degradation and detoxification of mycotoxins using a novel source of ozone. *Food and Chemical Toxicology* 35, 8, 807–820.

Minas, I. S., Tanou, G., Krokida, A., Karagiannis, E., Belghazi, M., Vasilakakis, M., Papadopoulou, K. K., Molassiotis, A. (2018). Ozone-induced inhibition of kiwifruit ripening is amplified by 1- methylcyclopropene and reversed by exogenous ethylene. *BMC Plant Biology* 18, 1, 1–19.

Minas, I. S., Vicente, A. R., Dhanapal, A. P., Manganaris, G. A., Goulas, V., Vasilakakis, M., Crisosto, C. H., Molassiotis, A. (2014). Ozone-induced kiwifruit ripening delay is mediated by ethylene biosynthesis inhibition and cell wall dismantling regulation. *Plant Science* 229, 76–85.

Minas, Ioannis S., Georgia Tanou, Maya Belghazi, Dominique Job, George A. Manganaris, Athanassios Molassiotis, Miltiadis Vasilakakis. (2012). Physiological and proteomic approaches to address the active role of ozone in kiwifruit post-harvest ripening. *Journal of Experimental Botany* 63, 7, 2449–2464.

Minas, Ioannis S., George S. Karaoglanidis, George A. Manganaris, Miltiadis Vasilakakis. (2010). Effect of ozone application during cold storage of kiwifruit on the development of stem-end rot caused by *Botrytis cinerea*. *Postharvest Biology and Technology* 58, 3, 203–210.

Nadas, A., M. Olmo, J. M. Garcia. (2003). Growth of *Botrytis cinerea* and strawberry quality in ozone-enriched atmospheres. *Journal of Food Science* 68, 5, 1798–1802.

Nickols, D., A. J. Varas. (1992). Ozonation. *Disinfection Alternatives for Safe Drinking Water*. New York: Bryant, EA, Fulton, GP, and Budd, GC Van Nostrand Reinhold, pp. 197–258.

Ong, K. C., J. N. Cash, M. J. Zabik, M. Siddiq, A. L. Jones. (1996). Chlorine and ozone washes for pesticide removal from apples and processed apple sauce. *Food Chemistry* 55, 2, 153–160.

Ong, M. K., A. Ali, P. G. Alderson, C. F. Forney. (2014). Effect of different concentrations of ozone on physiological changes associated to gas exchange, fruit ripening, fruit surface quality and defence-related enzymes levels in papaya fruit during ambient storage. *Scientia Horticulturae* 179, 163–169.

Ozkan, Ragip, Joseph L. Smilanick, Ozgur Akgun Karabulut. (2011). Toxicity of ozone gas to conidia of *Penicillium digitatum, Penicillium italicum*, and *Botrytis cinerea* and control of gray mold on table grapes. *Postharvest Biology and Technology* 60, 1, 47–51.

Ozon, A. T., Kafkas, E. (2015). Determination of aroma volatile composition of ozone treated arils from longterm-stored whole pomegranate fruits using HS-SPME GC/MS technique. *Hacettepe Journal of Biology and Chemistry* 43, 3, 153–157.

Öztekin, Serdar, Bülent Zorlugenç, Feyza Kırogˇlu Zorlugenç. (2006). Effects of ozone treatment on microflora of dried figs. *Journal of Food Engineering* 75, 3, 396–399.

Palou, Lluıs, Carlos H. Crisosto, Joseph L. Smilanick, James E. Adaskaveg, Juan P. Zoffoli. (2002). Effects of continuous 0.3 ppm ozone exposure on decay development and physiological responses of peaches and table grapes in cold storage. *Postharvest Biology and Technology* 24, 1, 39–48.

Palou, Lluís, Joseph L. Smilanick, Carlos H. Crisosto, Monir Mansour. (2001). Effect of gaseous ozone exposure on the development of green and blue molds on cold stored citrus fruit. *Plant Disease* 85, 6, 632–638.

Panou, A. A., Karabagias, I. K., Riganakos, K. A. (2018). The effect of different gaseous ozone treatments on physicochemical characteristics and shelf life of apricots stored under refrigeration. *Journal of Food Processing and Preservation* 42, 5, e13614.

Parish, M. E., L. R. Beuchat, T. V. Suslow, L. J. Harris, E. H. Garrett, J. N. Farber, F. F. Busta. (2003). Methods to reduce/eliminate pathogens from fresh and fresh-cut produce. *Comprehensive Reviews in Food Science and Food Safety* 2, 161–173.

Pérez, Ana G., Carlos Sanz, José J. Ríos, Raquel Olias, José M. Olías. (1999). Effects of ozone treatment on postharvest strawberry quality. *Journal of Agricultural and Food Chemistry* 47, 4, 1652–1656.

Piechowiak, T., M. Balawejder. (2019). Impact of ozonation process on the level of selected oxidative stress markers in raspberries stored at room temperature. *Food Chemistry* 298, 125093.

Piechowiak, T., K. Grzelak-Błaszczyk, M. Sójka, M. Balawejder. (2020b). Changes in phenolic compounds profile and glutathione status in raspberry fruit during storage in ozone-enriched atmosphere. *Postharvest Biology and Technology* 168, 111277.

Piechowiak, T., B. Skóra, M. Balawejder. (2020a). Ozone treatment induces changes in antioxidative defense system in blueberry fruit during storage. *Food and Bioprocess Technology* 13, 1240–1245.

Pollack, Susan L. (2001). Consumer demand for fruit and vegetables: the US example. *Changing Structure of Global Food Consumption and Trade* 6, 49–54.

Predmore, A., G. Sanglay, J. Li, K. Lee. (2015). Control of human norovirus surrogates in fresh foods by gaseous ozone and a proposed mechanism of inactivation. *Food Microbiology* 50, 118–125.

Rice, Rip G., John W. Farquhar, L. Joseph Bollyky. (1982). Review of the applications of ozone for increasing storage times of perishable foods. *Ozone-Science & Engineering* 4, 3, 147–163.

Rodoni, Luis, Natalia Casadei, Analía Concellón, Alicia R. Chaves Alicia, Ariel R. Vicente. (2010). Effect of short-term ozone treatments on tomato (*Solanum lycopersicum* L.) fruit quality and cell wall degradation. *Journal of Agricultural and Food Chemistry* 58, 1, 594–599.

Rubin, Mordecai B. (2001). The history of ozone. The Schönbein Period, 1839–1868. *Bulletin for the History of Chemistry* 26, 1, 40–56.

Sachadyn-Król, M., M. Materska, B. Chilczuk, M. Karaś, A. Jakubczyk, I. Perucka, I. Jackowska. (2016). Ozoneinduced changes in the content of bioactive compounds and enzyme activity during storage of pepper fruits. *Food Chemistry* 211, 59–67.

Salvador, Alejandra, Isabel Abad, Lucía Arnal, J. M. Martínez-Jávega. (2006). Effect of ozone on postharvest quality of persimmon. *Journal of Food Science* 71, 6, S443–S446.

Sarron, E., P. Gadonna-Widehem, T. Aussenac. (2021). Ozone treatments for preserving fresh vegetables quality: A critical review. *Foods* 10, 3, 605.

Segade, S. R., S. Vincenzi, S. Giacosa, L. Rolle. (2019). Changes in stilbene composition during postharvest ozone treatment of 'Moscato bianco' winegrapes. *Food Research International* 123, 251–257.

Selma, María V., Ana M. Ibáñez, Ana Allende, Marita Cantwell, Trevor Suslow. (2008a). Effect of gaseous ozone and hot water on microbial and sensory quality of cantaloupe and potential transference of *Escherichia coli* O157: H7 during cutting. *Food Microbiology* 25, 1, 162–168.

Selma, María Victoria, Ana María Ibáñez, Marita Cantwell, Trevor Suslow. (2008b). Reduction by gaseous ozone of salmonella and microbial flora associated with fresh-cut cantaloupe. *Food Microbiology* 25, 4, 558–565.

Shalluf, Milad A. (2012). Study of effect different of ozone doses on sugars content in tomatoes at different stages of ripening. *World Academy of Science, Engineering and Technology, International Journal of Biological, Biomolecular, Agricultural, Food and Biotechnological Engineering* 6, 4, 182–184.

Sheng, L., I. Hanrahan, X. Sun, M. H. Taylor, M. Mendoza, M.-J. Zhu. (2018). Survival of *Listeria innocua* on Fuji apples under commercial cold storage with or without low dose continuous ozone gaseous. *Food Microbiology* 76, 21–28.

Siemens, Werner. (1857). Ueber die elektrostatische Induction und die Verzögerung des Stroms in Flaschendrähten. *Annalen der Physik* 178, 9, 66–122.

Sivapalasingam, Sumathi, Cindy R. Friedman, Linda Cohen, Robert V. Tauxe. (2004). Fresh produce: a growing cause of outbreaks of foodborne illness in the United States, 1973 through 1997. *Journal of Food Protection* 67, 10, 2342–2353.

Skog, L. J., C. L. Chu. (2001). Effect of ozone on qualities of fruits and vegetables in cold storage. *Canadian Journal of Plant Science* 81, 4, 773–778.

Smilanick, J. L. (2003). Postharvest use of ozone on citrus fruit. *Packinghouse Newsletter* 199, 1–6.

Smilanick, Joseph L., Carlos Crisosto, Franka Mlikota. (1999). Postharvest use of ozone on fresh fruit. *Perishables Handling Q* 99, 10, 10–14.

Song, Jun, Lihua Fan, Paul D. Hildebrand, Charles F. Forney. (2000). Biological effects of corona discharge on onions in a commercial storage facility. *Hort Technology* 10, 3, 608–612

Song, J., Fan, L., C. F. Forney, M. A. Jordan, P. D. Hildebrand, W. Kalt and D. A. J. Ryan. (2003). Effect of ozone treatment and controlled atmosphere storage on quality and phytochemicals in highbush blueberries. *Acta Horticulturae* 600, 417–423.

Tabakoglu, N., H. Karaca. (2018). Effects of ozone-enriched storage atmosphere on postharvest quality of black mulberry fruits (*Morus nigra* L.). *LWT-Food Science and Technology* 92, 276–281.

Tokala, V. Y., Z. Singh, P. N. Kyaw. (2021). 1*H*-cyclopropabenzene and 1*H*-cyclopropa [*b*] naphthalene fumigation downregulates ethylene production and maintains fruit quality of controlled atmosphere stored 'Granny Smith' apple. *Postharvest Biology and Technology* 176, 111499.

Tokala, V. Y., Z. Singh, A. D. Payne. (2018). Postharvest Uses of Ozone Application in Fresh Horticultural Produce. In: *Postharvest Biology and Nanotechnology,* edited by G. Paliyath, J. Subramanian, L.-T. Lim, K. Subramanian, A. K. Handa, & A. K. Mattoo, 129–170. Hoboken, NJ: John Wiley & Sons.

Tokala, V. Y., Z. Singh, P. N. Kyaw. (2020). Fumigation and dip treatments with 1*H*-cyclopropabenzene and 1*H*-cyclopropa[*b*]naphthalene suppress ethylene production and maintain fruit quality of cold-stored 'Cripps Pink' apple. *Scientia Horticulturae* 272, 109597.

Toti, M., C. Carboni, R. Botondi. (2018). Postharvest gaseous ozone treatment enhances quality parameters and delays softening in cantaloupe melon during storage at 6°C. *Journal of the Science of Food and Agriculture* 98, 2, 487–494.

Tran, T. T. L., S. Aimla-Or, V. Srilaong, P. Jitareerat, C. Wongs-Aree, A. Uthairatanakij. (2013). Fumigation with ozone to extend the storage life of mango fruit cv. Nam Dok Mai No. *4. Agricultural Science Journal* 44, 4, 663–72.

Tzortzakis, Nikos, Anne Borland, Ian Singleton, Jeremy Barnes. (2007b). Impact of atmospheric ozone-enrichment on quality-related attributes of tomato fruit. *Postharvest Biology and Technology* 45, 3, 317–325.

Tzortzakis, N., T. Taybi, R. Roberts, I. Singleton, A. Borland, J. Barnes. (2011). Low-level atmospheric ozone exposure induces protection against *Botrytis cinerea* with down-regulation of ethylene-, jasmonate-and pathogenesis-related genes in tomato fruit. *Postharvest Biology and Technology* 61, 2–3, 152–159.

Tzortzakis, Nikos, Ian Singleton, Jeremy Barnes. (2007a). Deployment of low-level ozone-enrichment for the preservation of chilled fresh produce. *Postharvest Biology and Technology* 43, 2, 261–270.

Tzortzakis, Nikos, Ian Singleton, Jeremy Barnes. (2008). Impact of low-level atmospheric ozone-enrichment on black spot and anthracnose rot of tomato fruit. *Postharvest Biology and Technology* 47, 1, 1–9.

Tzortzakis, N. G. (2007). Maintaining postharvest quality of fresh produce with volatile compounds. *Innovative Food Science & Emerging Technologies* 8(1), 111–116.

Venta, M. B., S. S. C. Broche, I. F. Torres, M. G. Pérez, E. V. Lorenzo, Y. R. Rodriguez, S. M. Cepero. (2010). Ozone application for postharvest disinfection of tomatoes. *Ozone: Science & Engineering* 32, 5, 361–371

Wang, L., X. Fan, K. Sokorai, J. Sites. (2019). Quality deterioration of grape tomato fruit during storage after treatments with gaseous ozone at conditions that significantly reduced populations of Salmonella on stem scar and smooth surface. *Food Control* 103, 9–20.

Wang, D., T. H. Yeats, S. Uluisik, J. K. Rose, G. B. Seymour. (2018). Fruit softening: Revisiting the role of pectin. *Trends in Plant Science* 23, 4, 302–310.

Wani, S., J. K. Maker, J. R. Thompson, J. Barnes, I. Singleton. (2015). Effect of ozone treatment on inactivation of *Escherichia coli* and *Listeria* sp. on spinach. *Agriculture* 5(2), 155–169.

Whangchai, K., K. Saengnil, C. Singkamanee, J. Uthaibutra. (2010). Effect of electrolyzed oxidizing water and continuous ozone exposure on the control of *Penicillium digitatum* on tangerine cv.'Sai Nam Pung' during storage. *Crop Protection* 29(4), 386–389.

Xu, Liang ji. (1999). Use of ozone to improve the safety of fresh fruits and vegetables. *Food Technology* 53, 58–61.

Yaseen, T., A. Ricelli, B. Turan, P. Albanese, A. M. D'onghia. (2015). Ozone for post-harvest treatment of apple fruits. *Phytopathologia Mediterranea* 54, 1, 94–103.

Yeoh, W. K., A. Ali, C. F. Forney. (2014). Effects of ozone on major antioxidants and microbial populations of fresh-cut papaya. *Postharvest Biology and Technology* 89, 56–58.

Zhang, Xiaona, Zide Zhang, Lei Wang, Zhenliang Zhang, Jing Li, Congzhi Zhao. (2011). Impact of ozone on quality of strawberry during cold storage. *Frontiers of Agriculture in China* 5, 3, 356–360.

Zhou, Z., S. Zuber, F. Cantergiani, I. Sampers, F. Devlieghere, M. Uyttendaele. (2018). Inactivation of foodborne pathogens and their surrogates on fresh and frozen strawberries using gaseous ozone. *Frontiers in Sustainable Food Systems* 2, 51.

PART III

Light, Gas and Ultrasonic Applications

Pulsed Light for Fresh Quality Preservation of Fruits and Vegetables

6

Maratab Ali*, Nauman Khalid, Aqsa Akhtar, Sadia Aslam, Zienab F.R. Ahmed, and Xinhua Zhang*

*Corresponding Authors: Maratab Ali (maratab_ali@sdut.edu.cn); Xinhua Zhang (zxh@sdut.edu.cn)

6.1 INTRODUCTION

Fresh fruits and vegetables (F&V) contain health-promoting nutrients, including carbohydrates, vitamins, antioxidants, dietary fibers, and minerals, making them vital to human diets (Irfan et al., 2023; Tang et al., 2022). However, as perishable commodities, both fresh and fresh-cut F&V are potentially prone to quality depletion unless novel methods of preservation are being employed. It has been reported that fresh produce quality losses during the postharvest cycle are also frequently caused by the ineffective application of preservation technologies such as thermal treatments (Liu et al., 2022; Obileke et al., 2022). In this perspective, non-thermal postharvest procedures have gained substantial interest in previous years as an alternative to the conventional thermal approaches owing to their capability to retain essential nutritional and sensory characteristics of fresh produce. In the last decade, emerging postharvest physical technologies, such as PL, have become potential treatments for maintaining the nutritive value and microbial safety of F&V during postharvest storage (Palumbo et al., 2022; Macedo et al., 2023; Salazar-Zúñiga et al., 2022; Salehi, 2022). PL technology is one of the most recent approaches investigated for processing and preserving food commodities. This method of preserving food was revealed in the 1930s and granted its initial patent in 1984 (Patent No. US4464336 A, 1984) (Gomez-Lopez et al., 2007). Since 1996, the Food and Drug Administration (FDA) has permitted the application of pulsed light treatments in food, with exposures limited to 12 J cm^{-2} maximum UV dosage, spectral lines ranges between 200 and 1100 nm, and pulse length less than 2 ms (Salazar-Zúñiga et al., 2022). In addition, the scientific literature has used many nomenclatures to represent this technology, including higher-intensity pulse UV light, wide-spectrum pulsed light, intense light pulse, pulse whitish light, and pulsed UV light (Vargas-Ramella et al., 2021).

The benefiting effects of PL treatment include quality maintenance by delaying ripening and microbial inactivation in different F&V-based commodities such as acerola (Macedo et al., 2023), bayberries (Wang et al., 2023), apricots (Hua et al., 2022), mushrooms (Zhang et al., 2021), tomatoes (Pirozzi et al., 2021a), pennywort (Zaharah et al., 2020), persimmons (Denoya et al., 2020), melons (Filho et al., 2020), strawberries (Cao et al., 2019), fresh-cut lettuce (Tao et al., 2019), fresh-cut tomatoes (Valdivia-Nájar et al., 2018a, b), fresh-cut mangoes (de Almeida Lopes et al., 2017), etc. In particular, the PL technique has been examined as a trustworthy alternative for microbiological inactivation of the pathogens without impairing the nutritional, physicochemical, or sensorial properties of fresh food at certain levels. The effectiveness of PL, however, is solely reliant on the equipment factors that can be optimized, involving the quantity of flashes, employed voltage, distance between the specimen and lamps, the spectrum of light flashes, treatment duration, and sample characteristics (Mahendran et al., 2019). To explore an in-depth PL effect, researchers are contemplating developments in its application forms and obtaining its intended effects through different approaches, from individual PL treatment to its combinational application with other technologies such as cold plasma, edible coatings, etc. (Hua et al., 2022; Pirozzi et al., 2021a; Taştan et al., 2017).

Thus, for a comprehensive understanding, this chapter discusses the recent literature on PL technology, expanding on its possible mechanism of action, optimizing equipment parameters, postharvest effects, and limitations observed to date in order to ensure reliability among the horticultural produce industry, scientists, and consumers in PL technology as a sustainable postharvest treatment that can preserve nutrient contents, sensorial quality, and reduce the incidence of

DOI: 10.1201/9781003370376-9

microbiological deteriorations in F&V produce under storage during postharvest.

6.2 MECHANISM OF ACTION

PL treatment has a greater impact on both physiological mechanisms and microbial decontamination in fresh and fresh-cut F&V produce. Regarding the physiological mechanisms, several studies have reported the PL action on the gas respiratory pathway, antioxidant system, and color mechanism (Macedo et al., 2023; Hua et al., 2022; Lopes et al., 2016). Meanwhile, photochemical, photothermal, and photophysical phenomena are involved in the PL technology process for inactivating microorganisms (Salazar-Zúñiga et al., 2022). These mechanisms are able to operate independently or in combination with one another. It has been established that the primary factor in PL treatment's potential to inactivate microorganisms is the photochemical action. In terms of photochemical influence, the key factor in PL germicidal activity is UV penetration on microbial DNA. The antibacterial effect is a result of pyrimidine dimer formation, which is triggered by UV light as reported by Pollock et al. (2017). However, the photochemical transformation of target molecules like thymine prevents the synthesis of new DNA, which inactivate microflora via clonogenic cell death (Salazar-Zúñiga et al., 2022). Moreover, the production of 5-thymine and 5,6-dihydro-thymine, as well as the breakdown of single and double pyrimidine dimers, can both promote and impede the growth of yeast cells and spores (Bhavya & Umesh Hebbar, 2017; Gómez-López et al., 2012). As for the photothermal influence, once the final temperature reaches the pasteurization temperature, its impacts can result in the inactivation of microorganisms (Obileke et al., 2022). In order to destroy superficial pathogenic organisms like bacteria, mold, and yeasts, PL can cause elevated temperature and matrix overheating that can reach pasteurization conditions (Mandal et al., 2020). In agreement, Nicorescu et al. (2013) demonstrated that photothermal stress was involved in the disruption of the *B. subtilis* cell wall during the PL treatment. It was also found that *E. coli* underwent structural changes after being subjected to PL, likely as a result of the overheating, interstitial water evaporation, and membrane disintegration, as reported by Xu and Wu (2016). However, it has also been shown that the production of heat had a relatively small impact on microbial inactivation compared to the photochemical effect (Mandal et al., 2020). As the third pathway for PL-triggered inactivation of microflora, however, less research has been conducted. In this regard, physical changes in bacteria, molds, and yeast cells have all been associated with cell wall distortion, structural alterations, and cell form modifications (Ramos-Villarroel et al., 2012; Obileke et al., 2022; Salazar-Zúñiga et al., 2022).

6.3 EQUIPMENT DESIGN AND WORKING PARAMETERS

The PL system's core components include a high-voltage power source, storing capacitor, pulse-generating circuit used for investigating the pulse shape and spectral parameters, gas-discharging flashed lamp, and an inductor for commencing the electrical energy discharge to the flash lamps (Mahendran et al., 2019). The inert gas inside the lamp is subjected to an electrical pulse with high voltage and current, and as a result of the extreme collisions among electrons and gas molecules, the gas molecules are excited and release an intense light pulse over a very short period of time (Kautkar & Pandey, 2018). The PL equipment is compatible with batch and continuous modes. The most common systems are batch systems (Figure 6.1a), which are employed for initial laboratory-scale studies to know the consequence of the primary PL impact on the disinfection of both solid and liquid products. The device

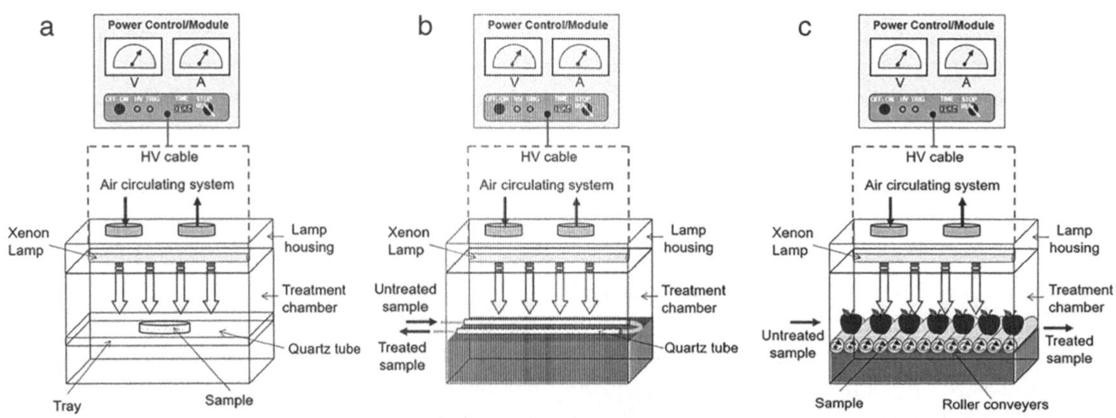

FIGURE 6.1 Pulsed light equipment design (a) batch system equipment, (b) continuous treatment system for liquid foods, and (c) continuous system for solid foods (Reprinted with permission from Pirozzi et al., 2021b).

comprises a container in which a liquid or solid sample is deposited on a tray that can be rotated to alter the sample's distance to the light source. Typically, the upper portion of the treatment chamber is equipped with a covering lamp that contains a xenon lamp. To guarantee more consistent irradiation of the surface of the product, multiple lamps can be employed. A high-voltage connection attached a power unit to the lamp, allowing the treatment time or amount of flash provided to the sample to be regulated. A cooling system that uses water or purified airflow reduces the heat that the bulb inevitably generates (Pirozzi et al., 2021b). Continuous flow systems can be considered to treat both solid and liquid materials (Figure 6.1b and c). To process liquid substances, liquids are pumped into the processed zone via a quartz tube that allows light flashes to pass throughout. The flow velocity and the volume of the quartz tubes both influence how long the exposure lasts and, as a result, how many flash or fluence are ultimately delivered to the sample. To guarantee optimal disinfection effectiveness with the least amount of energy consumption, optimization of geometry, the sum of quartz hoses, and their placement in relation to the light is necessary. Instead, solid foodstuffs can be delivered into the treatment compartment using roller systems that move the sample through the irradiation zone at a rate that ensures the required exposure time (Pirozzi et al., 2021b). The efficacy of PL application is highly dependent on the optimization of equipment factors such as flash numbers, employed voltage, distance between the sample and lamps, luminous intensity of light flashes, treatment duration, sample characteristics, and the microflora contamination threshold (Vargas-Ramella et al., 2021; Obileke et al., 2022). Additionally, PL efficiency is subjected to the sample composition and structure, as well as to the food's microbial diversity (Valdivia-Nájar et al., 2017).

6.4 EFFECTS OF PL ON THE QUALITY REGULATION OF FRESH FRUITS AND VEGETABLES

PL has the potential to improve fresh produce storage quality particularly by delaying ripening and inhibiting microbial growth. In accordance, Macedo et al. (2023) found that PL deferred ripening and enhanced Okinawa acerola fruit quality by decreasing the rate of respiration, ethylene synthesis, weight loss, and ascorbic acid degradation; increasing firmness and total antioxidant activity; and decreasing H_2O_2, membrane lipid peroxidation, and cell wall hydrolysis. Wang et al. (2023) investigated the effects of PL treatment on Chinese bayberries and found that it decreased the number of pathogenic fungal microbiotas, increased PAL activity, and had no impact on TA, TSS, MDA content, and SOD activity. Hua et al. (2022) summed up research showing that treatments with intense pulse light (IPL) prevented deterioration in apricots that had been exposed to vibration and mechanical damage during transport. This was accomplished by decreasing microbial growth, enhancing non-enzymatic antioxidant attributes like flavonoid content, ABTS, total phenols, and DPPH scavenging activity under upregulated enzyme activity (POD, PPO, SOD, and CAT), and decreasing cellular structure damage and water migration. It has also been described that IPL can significantly reduce PPO activity to lower the browning index; this effect is dose-dependent, with higher PPO concentrations requiring more intense and fluent treatments (Zhang et al., 2021). Further, Pirozzi et al. (2021a) noticed that optimum PL treatment conditions (4 J/cm^2) in combination with edible coatings enriched with oregano essential oil nano-emulsion enhanced the quality of tomato fruit by inhibiting the endogenous flora growth and regulating quality characteristics such as pH, TSS, and color index during a 15-day storage period at ambient conditions. Moreover, it has been speculated that applying PL could reduce the astringency of persimmons (Denoya et al., 2020). Furthermore, Lopes et al. (2016) reported that PL treatment enhanced the postharvest quality of Tommy Atkins mangoes, particularly by regulating the fruit color. This was accomplished by simultaneously increasing total carotenoid and phenolic contents and total antioxidant and PAL activity. Exposure to PL, as shown by Rock et al. (2015), not only increased the antioxidant activity and phenolic contents of fresh blueberries in comparison with the control, but also drastically enhanced the total anthocyanin content under increased PAL activity. Additionally, it has been noted that PL treatment may promote the storage of tomatoes' lycopene, carotenoid, and phenolic compounds, all of which are considered health-promoting compounds in humans, as reported by Pataro et al. (2015).

In recent years, it has been proven that PL can inactivate microorganisms present naturally or inoculated on the surfaces of fresh produce. According to Filho et al. (2020), PL technology protects melons from postharvest disease by directly restraining *Fusarium pallidoroseum* growth via upregulating particular fruit defense system–related markers like saponarin (7), pipecolic acid (11), and orientin (3) against pathogens. According to Zaharah et al. (2020), pennywort leaves treated with PL at 6.9 J/cm^2 retained their physical characteristics while inactivating the *E. coli* population, extending shelf life by around 4 days. As per Cao et al. (2019), PL treatments have the ability to prolong strawberries' shelf life by hindering the growth of *Salmonella* and visible mold while having little to no effect on quality parameters. Additionally, Duarte-Molina et al. (2016) noted that the strawberry postharvest storage life under PL application was prolonged due to a reduction in fungal population and a softening depletion. Moreover, Huang and Chen (2019) found inhibited *Salmonella* growth on blueberries, tomatoes, and iceberg lettuce under PL and UV treatment aligned with washing solution (WPL and WUV). Aguilo-Aguayo et al. (2014) found that fresh tomatoes treated with PL had lower levels of microbiological contaminants like *Saccharomyces cerevisiae* and were of higher quality due to the retention of higher total lycopene and ß-carotene contents during storage. In addition, Luksiene et al. (2013) observed that higher-power pulsed light (HPPL) treatments inactivated yeasts/microfungi distributed on the strawberries' surface by 1.0 log, while inoculated *B. cereus* and *L. monocytogenes* were deactivated by 1.5 and 1.1 log. *E. coli* growth on green onion stems and leaves were reduced by 1.4 and 3.1 log, respectively, under the combined treatment of PL and 1000 ppm of the surfactant SDS, as reported by Xu et al. (2013). By preventing the growth of naturally occurring microbial populations, including coliforms, listeria spp., mesophilic aerobic bacteria, psychrotrophic bacteria, yeasts, and molds in spinach, PL treatment was found to be efficient in enhancing storage quality (Agüero et al., 2016). In another work, PL application induced the rapid inactivation of murine norovirus and uulane virus, *Salmonella* and *E. coli*, on strawberries and blueberries, as reported by Huang et al. (2017). Similar observations were noted in blueberries associated with the decontamination of *Salmonella* under water-assisted PL treatment (Cao et al., 2017). Overall effects of PL treatment on F&V have been listed in Table 6.1.

TABLE 6.1 Effects of PL Treatment on Fresh Fruits and Vegetables

FRESH PRODUCE	*DOSE CONDITIONS*	*INTERFERENCE*	*REFERENCE*
Okinawa acerola	0.6 J/cm^2	↓Respiration rate, ethylene production, weight loss, AsA degradation, H_2O_2 levels, membrane lipid peroxidation, and cell wall hydrolysis; ↑firmness and total antioxidant activity	Macedo et al. (2023)
Chinese bayberry	9 kJ/m^2	↓Decay incidence, pathogenic fungal microbiotas, PPO, and PAL activity; ↑firmness	Wang et al. (2023)
Apricot	200 J/min	↓Microbial growth, cell structural damage, and water migration; ↑total phenols, flavonoids, ABTS, DPPH scavenging activity, SOD, PPO, POD, and CAT activities	Hua et al., (2022)
Tomato	4 J/cm^2 + edible coating enriched with oregano essential oil nano-emulsion	↓Microbial growth	Pirozzi et al., (2021a)
Melon	9 kJ m^{-2}	↓Growth of F. *pallidoroseum*	Filho et al., (2020)
Pennywort	6.9 J/cm^2	↓Total aerobic mesophilic count and inactivation of *E. coli* population	Zaharah et al., (2020)
Strawberry	3–5 J/cm^2	↓*Salmonella* and visible mold growth	Cao et al., (2019)
Blueberry, tomato, and iceberg lettuce	WUV (13–28 mW/cm^2), and WPL (0.15–0.30 J/cm^2)	↓*Salmonella* growth	Huang and Chen, (2019)
Strawberry and blueberry	22.5 J/cm^2	↓*Murine norovirus* (MNV-1), *E. coli*, and *Salmonella* growth	Huang et al., (2017)
Blueberry	0.066 J/cm^2	↓*Salmonella* growth	Cao et al., (2017)
Tommy Atkins mango	0.6 J/cm^2	↑Color intensity, total carotenoids content, total antioxidant activity, PAL activity, and phenolic contents	Lopes et al., (2016)
Strawberry	2.4–47.8 J/cm^2	↓Fungal incidence and depletion of softening	Duarte-Molina et al., (2016)
Spinach	20, and 40 kJ m^{-2}	↓Mesophilic aerobic bacteria, psychrotrophic bacteria, coliforms, listeria, spp. Yeasts, and mold populations	Agüero et al., (2016)
Blueberry	–	↑Antioxidant activity (ORAC), total phenolics contents, total anthocyanin content, and PAL activity	Rock et al., (2015)
Tomato	1–8 J/cm^2	↑Lycopene contents, total carotenoids, phenolic compounds, and antioxidative activity	Pataro et al., (2015)
Tomato	1–30 J/cm^2	↓*Saccharomyces cerevisiae* growth; ↑total lycopene and α, and ß-carotene contents	Aguilo-Aguayo et al., (2014)
Strawberry	0–1.95 J/cm^2	↓*Bacillus cereus*, *Listeria monocytogenes*, and yeasts/microfungi growth	Luksiene et al., (2013)
Green onion leaves and stem	1.27 J/cm^2 + 1000 ppm (SDS)	↓*E. coli* population	Xu et al., (2013)

6.5 EFFECTS OF PL ON THE QUALITY REGULATION OF FRESH-CUT FRUITS AND VEGETABLES

Recent work has also linked the use of PL to control the postharvest quality of fresh-cut F&V produce. For example, Tao et al. (2019) exhibited that PL treatment (8.2–12.5 J/cm²) preserved the postharvest quality of fresh-cut lettuce by delaying weight loss and total bacteria, mold, and yeast count; conserving color; maintaining significantly increased soluble solid contents, chlorophyll, and AsA; and inactivating foodborne pathogens like *E. coli*, *S. enteritidis*, *L. monocytogenes* and *S. aureus* on the surface of the lettuce. Furthermore, Valdivia-Nájar et al. (2018a) and Valdivia-Nájar et al. (2018b) found that PL treatment of 8 J/cm² reduced microbial growth by 2 log CFU/g, increased lycopene and total phenolic content, and reduced antioxidative losses in fresh-cut tomatoes. It has been indicated that fresh-cut mango slices maintained their firmness, color index, TSS and carotenoids levels, and antioxidant activity when treated with 4 pulses of PL (0.7 J/cm²), provided in short progression, as reported by de Almeida Lopes et al. (2017). Furthermore, Taştan et al. (2017) used a combination of high-intensity PL treatments (12 J/cm²) and antimicrobial attributed coatings, comprising chitosan-incorporating carvacrol emulsions, to preserve the qualities of fresh-cut cucumber pieces by reducing *E. coli* growth >5 log cycles than the control. Moreover, Moreira et al. (2017) found that PL application and dipping fresh-cut apples in an AsA/$CaCl_2$ solution before applying a pectin coating was helpful in reducing browning index and firmness loss. Adding fiber to the pectin coating had no effect on the microbiological loads or the sensory quality of apple cubes. However, pectin-coated, PL-treated apple pieces, in contrast to untreated and PL control samples, showed considerably increased levels of antioxidant activity. At the same time, the application of both treatments together led to a nearly 2 log CFU/g drop in microflora counts at the end of storage. In another study, Moreira et al. (2015) exhibited that combined treatment of both coating and PL simultaneously reduced the microbial spoilage and retained sensorial characteristic scores above the refusal thresholds even after extended storage in fresh-cut apples. In addition, the findings indicated that a gellan-gum coating comprising apple fiber as well, after a PL treatment effectively decreased the firmness loss and browning index of apple pieces during storage. Salinas-Roca et al. (2016) showed that combined treatment of malic acid and PL, as well as PL, alginates coatings, and malic acid application, synergistically inhibited *L. innocua* levels by 4.5 and 3.9 logs in fresh-cut mangoes, respectively. Meanwhile, all treated mango slices experienced a drop in color characteristics and total soluble solids content during the storage period.

According to Charles et al. (2013), 8 J/cm² PL treatment boosted PPO and PAL activity while maintaining the softening, color intensity, and carotenoids content in fresh-cut mangoes. Koh et al. (2016) reported that PL treatment (7.8 J/cm²) is a promising method for prolonging the shelf life of fresh-cut cantaloupe by inactivating microorganism-based infestation without reducing nutritive values. Moreover, Aguilo-Aguayo et al. (2014) found that exposing fresh-cut avocado to the highest PL dose (3.6, 6.0, and 14 J/ cm²) reduces total aerobic mesophilic bacteria (1.20 log CFU/g) growth and inhibits yeast and mold proliferation for 3 days, prolonging the fruit's shelf life potential to 15 days. PL treatments also improved fresh-cut avocado color values, possibly because of enhanced chlorophyll retention. For 15 days, the lipidic component of fresh-cut avocados treated with PL showed negligible peroxide accumulation and constant distinct extinctions coefficient at 272 and 232 nm, indicating that the treatment may considerably preserve avocado quality throughout storage. In another study, combined application of PL and anti-browning solutions associated with AsA and $CaCl_2$ were effective in preventing browning on apple slices subjected to a PL dose of 71.6 J/cm², along with reduced microflora count (Gómez et al. 2012), Moreover, Oms-Oliu et al. (2010) also showed that PL treatment (4.8 J/cm²) allowed fresh-cut mushrooms to have a longer shelf life by lowering microbiological growth and browning index in comparison to untreated samples. Overall effects of PL treatments on F&V have been summarized in Table 6.2.

TABLE 6.2 Effects of PL Treatment on Fresh-Cut Fruits and Vegetables

FRESH-CUT PRODUCE	*DOSE CONDITIONS*	*INTERFERENCE*	*REFERENCE*
Lettuce	8.2–12.5 J/cm²	↓Weight loss and total bacteria, mold, and yeast count; ↑color intensity, chlorophyll, and AsA; ↓activation of *S. enteritidis*, *E. coli*, *S. aureus*, and *L. monocytogenes* foodborne pathogenic strains	Tao et al. (2019)
Tomato	8 J/cm²	↓Microbial growth (*E. coli* and *L. innocua* growth), AsA, and firmness; ↑lycopene and total phenolic contents	Valdivia-Nájar et al. (2017); Valdivia-Nájar et al. (2018a, b)

(Continued)

TABLE 6.2 (Continued)

FRESH-CUT PRODUCE	DOSE CONDITIONS	INTERFERENCE	REFERENCE
Mango	0.7 J/cm^2 (4 pulses)	↓Mass loss, respiratory rate, and decline in yellow color; ↑Firmness, soluble solids, total carotenoids, and antioxidant activity	de Almeida Lopes et al. (2017)
Cucumber	12 J/cm^2 + antimicrobial coating (chitosan suspension containing 0.08% carvacrol)	↓*E. coli* growth >5 log cycles	Taştan et al. (2017)
Apple	12 J/cm^2 + pectin-based edible coatings enriched with dietary fiber	↓Browning, softening and microbial count (2 log CFU g^{-1}); ↑antioxidant activity	Moreira et al. (2017)
Mango	12 J/cm^2 + malic acid + alginate coating, or 8 J/cm^2	↓*L. innocua* counts, mold, yeast, psychrophilic bacteria, color, and TSS; ↑PPO and PAL activity	Charles et al. (2013); Salinas-Roca et al. (2016)
Cantaloupe	7.8 J/cm^2, or 0.9 J/cm^2 repetitive pulsed light (RPL)	↓Total plate count, aerobic mesophilic microflora, yeast and mould populations, and CO_2; ↑AsA and firmness	Koh et al. (2016a); Koh et al., (2016b)
Apple	12 J/cm^2 + gellan-gum (0.5% w/v) enriched with apple fiber	↓Microbiological deterioration, softening and browning; ↑firmness	Moreira et al. (2015)
Avocado	3.6, 6.0 and 14 J/cm^2	↓Aerobic mesophilic microflora, yeast and mold; ↑hue value, chlorophyll, peroxide formation	Aguilo-Aguayo et al. (2014)
Apple	71.6 J/cm^2 + anti-browning agent (ascorbic acid + $CaCl_2$ solution)	↓Browning, and native microflora count	Gómez et al. (2012)

6.6 LIMITATIONS

The PL approach is not without significant limitations. The efficiency of the PL technology is largely impacted by the opacity and composition of the sample matrix (Ren et al., 2021; Mandal et al., 2020). Additional concerns include sample overheating, pH and color changes at higher fluences, and the risk of ozone formation (Mahendran et al., 2019). It is critical to note that, in such cases, PL treatments have detrimental effects on specific components, which can result in changes in sensory aspects of fresh produce, such as accelerating the breakdown of some natural coloring pigments and the development of an unpleasant flavor and smell. In accordance, Ignat et al. (2014) found that PL-treated apple slices had an unacceptable flavor and a darker brown color than the untreated samples. To overcome these types of drawbacks, combined application of PL with other technological interventions, or modification of the PL system must be explored to conserve the quality characteristics of F&V during postharvest while reducing the negative impact on quality characteristics. Nevertheless, the PL research field is primarily concerned with foodborne pathogen inactivation, and comprehensive investigation and analyses of PL have not been conducted on the other quality traits of F&V.

6.7 CONCLUSIONS

PL technology appeared as a promising non-thermal approach to conserve the postharvest quality of fresh and fresh-cut fruit and vegetable commodities, particularly by inactivating natural or inoculated microflora and delaying ripening. However, PL effectiveness is significantly dependent on optimized system factors, including the number of flashes, applied voltage, spectral range of light flashes, distance between the sample and lamps, treatment duration, sample matrix, and microbial strains and their contamination level. Meanwhile, it is imperative to note that, in some situations, PL treatments have negative effects on specific fresh produce components, including coloring pigments and flavor compounds, which can result in changes to the sensory qualities of fresh produce. Therefore, more research is recommended on PL technology via its combined application with other strategies or its modification in equipment design for the better conservation of sensory and nutritional attributes along with the microbial safety of F&V.

REFERENCES

Agüero, M. V., Jagus, R. J., Martín-Belloso, O., Soliva-Fortuny, R. (2016). Surface decontamination of spinach by intense pulsed light treatments: Impact on quality attributes. *Postharvest Biology and Technology*, *121*, 118–125.

Aguilo-Aguayo, I., Oms-Oliu, G., Martin-Belloso, O., Soliva-Fortuny, R. (2014). Impact of pulsed light treatments on quality characteristics and oxidative stability of fresh-cut avocado. *LWT-Food Science and Technology*, *59(1)*, 320–326.

Bhavya, M., Umesh Hebbar, H. (2017). Pulsed light processing of foods for microbial safety. *Food Quality and Safety*, *1(3)*, 187–202.

Cao, X., Huang, R., Chen, H. (2017). Evaluation of pulsed light treatments on inactivation of *Salmonella* on blueberries and its impact on shelf-life and quality attributes. *International Journal of Food Microbiology*, *260*, 17–26.

Cao, X., Huang, R., Chen, H. (2019). Evaluation of food safety and quality parameters for shelf life extension of pulsed light treated strawberries. *Journal of Food Science*, *84(6)*, 1494–1500.

Charles, F., Vidal, V., Olive, F., Filgueiras, H., Sallanon, H. (2013). Pulsed light treatment as new method to maintain physical and nutritional quality of fresh-cut mangoes. *Innovative Food Science & Emerging Technologies*, *18*, 190–195.

de Almeida Lopes, M. M., Silva, E. O., Laurent, S., Charles, F., Urban, L., de Miranda, M. R. A. (2017). The influence of pulsed light exposure mode on quality and bioactive compounds of fresh-cut mangoes. *Journal of Food Science and Technology*, *54*, 2332–2340.

Denoya, G. I., Pataro, G., Ferrari, G. (2020). Effects of postharvest pulsed light treatments on the quality and antioxidant properties of persimmons during storage. *Postharvest Biology and Technology*, *160*, 111055.

Duarte-Molina, F., Gómez, P. L., Castro, M. A., Alzamora, S. M. (2016). Storage quality of strawberry fruit treated by pulsed light: Fungal decay, water loss and mechanical properties. *Innovative Food Science & Emerging Technologies*, *34*, 267–274.

Filho, F. O., Silva, E. d. O., Lopes, M. M. d. A., Ribeiro, P. R. V., Oster, A. H., Guedes, J. A. C., Zampieri, D. d. S., Bordallo, P. d. N., Zocolo, G. J. (2020). Effect of pulsed light on postharvest disease control-related metabolomic variation in melon (*Cucumis melo*) artificially inoculated with *Fusarium pallidoroseum*. *PloS One*, *15(4)*, e0220097.

Gómez, P. L., García-Loredo, A., Nieto, A., Salvatori, D. M., Guerrero, S., Alzamora, S. M. (2012). Effect of pulsed light combined with an antibrowning pretreatment on quality of fresh cut apple. *Innovative Food Science & Emerging Technologies*, *16*, 102–112.

Gómez-López, V. M., Koutchma, T., Linden, K. (2012). Ultraviolet and pulsed light processing of fluid foods. In *Novel thermal and non-thermal technologies for fluid foods* (pp. 185–223). Elsevier.

Gomez-Lopez, V. M., Ragaert, P., Debevere, J., Devlieghere, F. (2007). Pulsed light for food decontamination: A review. *Trends in Food Science & Technology*, *18(9)*, 464–473.

Hua, X., Li, T., Wu, C., Zhou, D., Fan, G., Li, X., Cong, K., Yan, Z., Wu, Z. (2022). Novel physical treatments (Pulsed light and cold plasma) improve the quality of postharvest apricots after long-distance simulated transportation. *Postharvest Biology and Technology*, *194*, 112098.

Huang, R., Chen, H. (2019). Comparison of water-assisted decontamination systems of pulsed light and ultraviolet for salmonella inactivation on blueberry, tomato, and lettuce. *Journal of Food Science*, *84(5)*, 1145–1150.

Huang, Y., Ye, M., Cao, X., Chen, H. (2017). Pulsed light inactivation of murine norovirus, *Tulane virus*, *Escherichia coli O157: H7* and *Salmonella* in suspension and on berry surfaces. *Food Microbiology*, *61*, 1–4.

Ignat, A., Manzocco, L., Maifreni, M., Bartolomeoli, I., Nicoli, M. C. (2014). Surface decontamination of fresh-cut apple by pulsed light: Effects on structure, colour and sensory properties. *Postharvest Biology and Technology*, *91*, 122–127.

Irfan, M., Kumar, P., Ahmad, M. F., Siddiqui, M. W. (2023). Biotechnological interventions in reducing losses of tropical fruits and vegetables. *Current Opinion in Biotechnology*, *79*, 102850.

Kautkar, S., Pandey, J. P. (2018). An elementary review on principles and applications of modern non-conventional food processing technologies. *Journal homepage*, *7*(5). www. ijcmas. com

Koh, P. C., Noranizan, M. A., Karim, R., Nur Hanani, Z. A. (2016a). Microbiological stability and quality of pulsed light treated cantaloupe (*Cucumis melo* L. reticulatus cv. Glamour) based on cut type and light fluence. *Journal of Food Science and Technology*, *53*, 1798–1810.

Koh, P. C., Noranizan, M. A., Karim, R., Nur Hanani, Z. A. (2016b). Repetitive pulsed light treatment at certain interval on fresh-cut cantaloupe (*Cucumis melo* L. reticulatus cv. Glamour). *Innovative Food Science & Emerging Technologies, 36*, 92–103.

Liu, X., Le Bourvellec, C., Yu, J., Zhao, L., Wang, K., Tao, Y., Renard, C. M., Hu, Z. (2022). Trends and challenges on fruit and vegetable processing: Insights into sustainable, traceable, precise, healthy, intelligent, personalized and local innovative food products. *Trends in Food Science & Technology*, *125*, 12–25.

Lopes, M. M., Silva, E. O., Canuto, K. M., Silva, L. M., Gallão, M. I., Urban, L., Ayala-Zavala, J. F., Miranda, M. R. A. (2016). Low fluence pulsed light enhanced phytochemical content and antioxidant potential of 'Tommy Atkins' mango peel and pulp. *Innovative Food Science & Emerging Technologies*, *33*, 216–224.

Luksiene, Z., Buchovec, I., Viskelis, P. (2013). Impact of high-power pulsed light on microbial contamination, health promoting components and shelf life of strawberries. *Food Technology and Biotechnology*, *51(2)*, 284.

Macedo, J. J., Sanches, A. G., Rabelo, M. C., Lopes, M. M., Freitas, V. S., Silveira, A. G., Moura, C. F., Silva, E. O., Gallão, M. I., Gomes-Filho, E. (2023). Pulsed light influences several metabolic routes, delaying ripening and improving the postharvest quality of acerola. *Scientia Horticulturae*, *307*, 111505.

Mahendran, R., Ramanan, K. R., Barba, F. J., Lorenzo, J. M., López-Fernández, O., Munekata, P. E., Roohinejad, S., Sant'Ana, A. S., Tiwari, B. K. (2019). Recent advances in the application of pulsed light processing for improving food safety and increasing shelf life. *Trends in Food Science & Technology*, *88*, 67–79.

Mandal, R., Mohammadi, X., Wiktor, A., Singh, A., Pratap Singh, A. (2020). Applications of pulsed light decontamination technology in food processing: An overview. *Applied Sciences*, *10(10)*, 3606.

Moreira, M. R., Álvarez, M. V., Martín-Belloso, O., Soliva-Fortuny, R. (2017). Effects of pulsed light treatments and pectin edible coatings on the quality of fresh-cut apples: a hurdle technology approach. *Journal of the Science of Food and Agriculture, 97(1)*, 261–268.

Moreira, M. R., Tomadoni, B., Martín-Belloso, O., Soliva-Fortuny, R. (2015). Preservation of fresh-cut apple quality attributes by pulsed light in combination with gellan gum-based prebiotic edible coatings. *LWT-Food Science and Technology, 64(2)*, 1130–1137.

Nicorescu, I., Nguyen, B., Moreau-Ferret, M., Agoulon, A., Chevalier, S., Orange, N. (2013). Pulsed light inactivation of *Bacillus subtilis* vegetative cells in suspensions and spices. *Food Control, 31(1)*, 151–157.

Obileke, K., Onyeaka, H., Miri, T., Nwabor, O. F., Hart, A., Al-Sharify, Z. T., Al-Najjar, S., Anumudu, C. (2022). Recent advances in radio frequency, pulsed light, and cold plasma technologies for food safety. *Journal of Food Process Engineering, 45(10)*, e14138.

Oms-Oliu, G., Aguiló-Aguayo, I., Martín-Belloso, O., Soliva-Fortuny, R. (2010). Effects of pulsed light treatments on quality and antioxidant properties of fresh-cut mushrooms (*Agaricus bisporus*). *Postharvest Biology and Technology, 56(3)*, 216–222.

Palumbo, M., Attolico, G., Capozzi, V., Cozzolino, R., Corvino, A., de Chiara, M. L. V., Pace, B., Pelosi, S., Ricci, I., Romaniello, R. (2022). Emerging postharvest technologies to enhance the shelf-life of fruit and vegetables: An overview. *Foods, 11(23)*, 3925.

Pataro, G., Sinik, M., Capitoli, M. M., Donsì, G., Ferrari, G. (2015). The influence of post-harvest UV-C and pulsed light treatments on quality and antioxidant properties of tomato fruits during storage. *Innovative Food Science & Emerging Technologies, 30*, 103–111.

Pirozzi, A., Del Grosso, V., Ferrari, G., Pataro, G., Donsì, F. (2021a). Combination of edible coatings containing oregano essential oil nanoemulsion and pulsed light treatments for improving the shelf life of tomatoes. *Chemical Engineering Transactions, 87*, 61–66.

Pirozzi, A., Pataro, G., Donsì, F., Ferrari, G. (2021b). Edible coating and pulsed light to increase the shelf life of food products. *Food Engineering Reviews, 13*, 544–569.

Pollock, A. M., Singh, A. P., Ramaswamy, H. S., Ngadi, M. O. (2017). Pulsed light destruction kinetics of *L. monocytogenes*. *LWT, 84*, 114–121.

Ramos-Villarroel, A. Y., Aron-Maftei, N., Martín-Belloso, O., Soliva-Fortuny, R. (2012). Influence of spectral distribution on bacterial inactivation and quality changes of fresh-cut watermelon treated with intense light pulses. *Postharvest Biology and Technology, 69*, 32–39.

Ren, M., Yu, X., Mujumdar, A. S., Yagoub, A. E.-G. A., Chen, L., Zhou, C. (2021). Visualizing the knowledge domain of pulsed light technology in the food field: A scientometrics review. *Innovative Food Science & Emerging Technologies, 74*, 102823.

Rock, C., Guner, S., Yang, W., Gu, L., Percival, S., Salcido, E. (2015). Enhanced antioxidant capacity of fresh blueberries by pulsed light treatment. *Journal of Food Research, 4(5)*, 89.

Salazar-Zúñiga, M., Lugo-Cervantes, E., Rodríguez-Campos, J., Sanchez-Vega, R., Rodríguez-Roque, M., Valdivia-Nájar, C. (2022). Pulsed light processing in the preservation of juices and fresh-cut fruits: A review. *Food and Bioprocess Technology*, 1–16.

Salehi, F. (2022). Application of pulsed light technology for fruits and vegetables disinfection: A review. *Journal of Applied Microbiology, 132(4)*, 2521–2530.

Salinas-Roca, B., Soliva-Fortuny, R., Welti-Chanes, J., Martín-Belloso, O. (2016). Combined effect of pulsed light, edible coating and malic acid dipping to improve fresh-cut mango safety and quality. *Food Control, 66*, 190–197.

Tang, T., Zhang, M., Mujumdar, A. S. (2022). Intelligent detection for fresh-cut fruit and vegetable processing: Imaging technology. *Comprehensive Reviews in Food Science and Food Safety, 21(6)*, 5171–5198.

Tao, T., Ding, C., Han, N., Cui, Y., Liu, X., Zhang, C. (2019). Evaluation of pulsed light for inactivation of foodborne pathogens on fresh-cut lettuce: Effects on quality attributes during storage. *Food Packaging and Shelf Life, 21*, 100358.

Taştan, Ö., Pataro, G., Donsì, F., Ferrari, G., Baysal, T. (2017). Decontamination of fresh-cut cucumber slices by a combination of a modified chitosan coating containing carvacrol nanoemulsions and pulsed light. *International Journal of Food Microbiology, 260*, 75–80.

Valdivia-Nájar, C. G., Martín-Belloso, O., Giner-Seguí, J., Soliva-Fortuny, R. (2017). Modeling the inactivation of *Listeria innocua* and *Escherichia coli* in fresh-cut tomato treated with pulsed light. *Food and Bioprocess Technology, 10*, 266–274.

Valdivia-Nájar, C. G., Martín-Belloso, O., Soliva-Fortuny, R. (2018a). Impact of pulsed light treatments and storage time on the texture quality of fresh-cut tomatoes. *Innovative Food Science & Emerging Technologies, 45*, 29–35.

Valdivia-Nájar, C. G., Martín-Belloso, O., Soliva-Fortuny, R. (2018b). Kinetics of the changes in the antioxidant potential of fresh-cut tomatoes as affected by pulsed light treatments and storage time. *Journal of Food Engineering, 237*, 146–153.

Vargas-Ramella, M., Pateiro, M., Gavahian, M., Franco, D., Zhang, W., Khaneghah, A. M., Guerrero-Sánchez, Y., Lorenzo, J. M. (2021). Impact of pulsed light processing technology on phenolic compounds of fruits and vegetables. *Trends in Food Science & Technology, 115*, 1–11.

Wang, Y., Zheng, Y., Shang, J., Wu, D., Zhou, A., Cai, M., Gao, H., Yang, K. (2023). Pulsed light reduces postharvest losses of Chinese bayberries by affecting fungal microbiota during cold storage. *Food Control, 146*, 109524.

Xu, W., Chen, H., Huang, Y., Wu, C. (2013). Decontamination of Escherichia coli O157: H7 on green onions using pulsed light (PL) and PL–surfactant–sanitizer combinations. *International Journal of Food Microbiology, 166(1)*, 102–108.

Xu, W., Wu, C. (2016). The impact of pulsed light on decontamination, quality, and bacterial attachment of fresh raspberries. *Food Microbiology, 57*, 135–143.

Zaharah, R. S., Noranizan, M., Son, R., Roselina, K., Yusof, N., Koh, P., Hasni, H. N. (2020). Microbiological and physical properties of pennywort (*Centella asiatica*) leaves using pulsed light technology. *International Food Research Journal, 27(1)*, 16–27.

Zhang, J., Yu, X., Xu, B., Yagoub, A. E. A., Mustapha, A. T., Zhou, C. (2021). Effect of intensive pulsed light on the activity, structure, physico-chemical properties and surface topography of polyphenol oxidase from mushroom. *Innovative Food Science & Emerging Technologies, 72*, 102741.

Cold Plasma Technology for Fruit and Vegetable Quality Preservation during Postharvest Storage

7

Ernesto Alonso Lagarda-Clark,
Martín Ernesto Tiznado Hernández, and Arturo Duarte-Sierra*
*Corresponding Author: arturo.duarte-sierra@fsaa.ulaval.ca

Abbreviations

APCP: Atmospheric pressure cold plasma
APX: Ascorbate peroxidase
CAP: Cold atmospheric plasma
CAT: Catalase
CP: Cold plasma
DBD: Dielectric barrier discharge
DCSBD: Diffuse coplanar surface barrier discharge
LPCP: Low-pressure cold plasma
LTP: Low-temperature plasma
MHD: Magnetohydrodynamics
OAUGDP: One atmosphere uniform glow discharge plasma
POD: Peroxidase
PPO: Polyphenol oxidase
RF: Radiofrequency
ROS: Reactive oxygen species
ROSN: Reactive oxygen-nitrogen species
SOD: Superoxide dismutase

7.1 INTRODUCTION

Plasma is a state of matter in which a gas is ionized, giving rise to a collection of free electrons and positively charged ions. It is a distinctive state of matter with unique properties and is commonly found in natural and man-made phenomena. CP (i.e., a type of plasma technology that operates at or near room temperature, unlike other forms of plasma that require high temperatures) consists of excited and ground-state molecules, electrons, ions, ultraviolet photons and reactive neutrals, such as radicals, and is a partially ionized gas (Simmons, 2012). The medical field of low-temperature plasma (LTP) began in the mid-1990s, with a proof-of-principle experiment demonstrating its effective bactericidal properties. Such experiments highlighted that reactive oxygen and nitrogen species (ROSN) played a vital role in the biological outcomes of LTP and found that it could disinfect both abiotic surfaces and biological tissues, with potential applications for wound healing. As more experimental data emerged, the LTP research community recognized the potential of these applications, and 2005 saw a surge of research interest in the field (Laroussi, 2020).

CP applications in agriculture and food science have shown great promise in various areas, including growth improvement, seed sterilization, soil remediation, food preservation, sterilization and decontamination (Ansari et al., 2022; Domonkos et al., 2021; Hashizume et al., 2021; Niemira, 2012; Puač et al., 2018). Indeed, surface decontamination was one of the first applications of plasma, and there is a record of a patent from 1968 created by Wilson P. Menashi. Although the first work with oxygen plasma was carried out in 1989, ever since there has been a great deal of research on the mechanism of microbial inactivation by plasma-based agents (Misra et al., 2011). Aside from food decontamination, other plasma technology applications in food processing have emerged. Such applications include the modification of starch to enhance its functional properties. A further area of application is the utilization of plasma-activated water as a nitrite source in the curing of meat. Moreover, cold atmospheric plasma (CAP) has

DOI: 10.1201/9781003370376-10

revealed the potential for accelerating oxidation in the analysis of oils and food products containing them (Puač et al., 2018).

In contrast, prolonging the shelf life of fresh produce is critical for the financial success and competitiveness of the industry. Unfortunately, many factors can cause produce to spoil prematurely, including mishandling, insufficient humidity, exposure to incorrect temperatures, and contact with ethylene gas (Pan et al., 2019). Therefore, the use of plasma technology has also awakened great interest in postharvest biology and technology as a tool for extending the shelf life of fresh produce, minimizing spoilage and preserving quality (Jin et al., 2022; Palumbo et al., 2022). Furthermore, plasma technology has been reported to be suitable in decreasing pesticide residues in fresh produce, which is a significant concern in the postharvest industry (Misra et al., 2014). The use of plasmas for the decontamination of produce surfaces started around the end of the 2010s. As such, Critzer et al. (2007) explored the impact of one atmosphere constant glow discharge plasma (OAUGDP) on the elimination of *Escherichia coli* O157:H7, *Salmonella* and *Listeria monocytogenes* from apples, melons and lettuce, respectively. Niemira & Sites (2008) also pioneered the use of CP produced in a sliding arc to inactivate strains of *Escherichia coli* O157:H7 and *Salmonella Stanley* on the surface of Golden Delicious apple fruits. That same year in the United Kingdom, CP was used to decrease microbial growth on the surface of melons and mangoes (Perni et al., 2008).

Although the main postharvest application of plasma is still the elimination of fungi and bacteria, other trends stand out, such as the increase of phytochemical composition in fruits and vegetables, and the treatment of material to enhance their barricade property and increase the shelf life of fresh produce. Therefore, plasma technology has the ability to be an effective tool for extending the shelf life of fresh produce, reducing spoilage, and maintaining quality during storage. Further research is needed to optimize plasma technology for use in the postharvest industry, but the results so far suggest that it is a promising area of research for the future in this field.

7.2 HISTORY OF PLASMA

In the mid-19th century, the Czech physiologist Jan Evangelista Purkinje first introduced the use of the Greek word plasma (i.e., plasma, meaning "formed or molded") as a designation for the pale fluid that is left after the elimination of all corpuscular material from the blood (Bellan, 2008; Smirnov, 2011). However, the study of plasma as a distinct state of matter can be traced to the late 19th and early 20th centuries. Significant discoveries in the field of plasma physics were made at that time, which helped to lay the foundation for its study. In 1879, Sir William Crookes performed experiments with cathode rays in vacuum tubes, which were later recognized as plasma (Gates, 2018). The electron, later recognized as a fundamental particle of plasma, was discovered by J.J. Thomson in 1897 (Downard, 2009). Irving Langmuir, in the 1920s, coined the concept of plasma and conducted studies on ionized gases (Gates, 2018), for which he received the Nobel Prize in Chemistry in 1932. Tonks (1967), who was working with Langmuir at the time the term plasma was first used described how it happened:

> Dr. Langmuir approached me for a term to describe the main part of the gas discharges, as opposed to the region near the wall or electrode, which is commonly referred to as the "sheath". Since the high conductivity and complete neutralization of the space charge, no potential difference can be applied to this main part since it is in the sheath. Dr. Langmuir was not looking to create a new term, but rather a term that would accurately describe this specific region. I was asked to suggest a suitable word, and I replied that I would think about it. The following day, Langmuir came and declared that he had a name for the main part of the gas discharge—"plasma". The word evoked the mental image of blood plasma, and Langmuir even referenced it. Given what was known at the time, the selection of the term "plasma" appeared highly suitable.

Then, a rapid expansion of plasma physics research occurred in the 1930s and 1940s due to advances in vacuum technology and high-frequency radio waves, and researchers such as Hannes Alfvén made important contributions to the field, including the discovery of magnetohydrodynamics (MHD) and the development of MHD waves (Fälthammar, 2007). Research in plasma physics expanded further in the 1950s and 1960s due to the development of nuclear fusion technology (Meade, 2010). Ultimately, the development of higher-power lasers in the 1960s opened up the field of laser plasma physics (Fitzpatrick, 2022). Ever since, plasma physics has become a multidisciplinary field encompassing a wide range of research areas, for example fusion energy, space physics, astrophysics and plasma processing. Nowadays, plasma physicists continue to make important contributions to a variety of fields, including food science, and its study remains a thriving and active research area.

7.2.1 What Is Plasma?

Plasma is a state of matter in which some or all of the particles in a gas are ionized. In other words, plasma is a gas that has been partially or completely ionized, creating a mixture of positively charged ions and negatively charged electrons that behave collectively in response to external electric and magnetic fields (Laroque et al., 2022; Smirnov, 2011). A more technical definition of plasma comes from the field of plasma physics, which defines plasma as a quasineutral (i.e., the number of negative and positively charged particles are equal), conductive fluid that comprises charged particles (i.e., ions and electrons) that interact with each other through electromagnetic forces (Bittencourt, 2004; Fitzpatrick, 2022). This definition emphasizes the collective behavior of charged particles in plasma and the importance of electromagnetic fields in governing their behavior. Furthermore, some of the most important plasma parameters include electron density, temperature, Debye length, plasma collision as well as collision frequency (Bellan, 2008; Bittencourt, 2004; Chen, 2015; Fitzpatrick,

2022; Gibbon, 2020), and three additional criteria must be satisfied for a plasma to be defined (Chen, 2015):

$\lambda_D << L$ An ionized gas can be regarded as plasma if the Debye length, λ_D, of the Debye sphere formed when an external potential is introduced is significantly lower than the dimension of the plasma system.

$N_D >> 1$ The numbers of charged particle, N_D, in the Debye sphere, which is given by $N_D = \frac{4}{3}\pi\lambda_D^3 \times n$, is significantly greater than 1, where n is the number density of charge in the Debye sphere.

$\omega\tau > 1$ The product of plasma frequency ω and the mean time of collision with the neutral particles τ is greater than 1.

Electron density: The electron density of a plasma is a measure of the number of free electrons in the plasma per unit volume. The range of electron densities in plasma can also vary widely, from 10^6 electrons per cubic centimeter in some low-density plasmas to 10^{20} electrons per cubic centimeter in high-density plasmas, such as those found in fusion experiments (Gibbon, 2020; Wiesemann, 2014). The electron density of a plasma can affect its electrical conductivity, the Debye length and other properties.

Temperature: When the temperature of plasma increases, the charged particles in the plasma gain more kinetic energy, which causes a shift in the distribution of energy among the particles. This shift in the distribution affects the interactions between the particles and can cause changes in various plasma properties, such as its density and electrical conductivity (Chen, 2015). However, a higher temperature does not essentially mean a large amount of heat. For instance, the temperature of electrons inside fluorescent light bulbs can be around 20,000°K, but the density of electrons is much lower than that of gas at atmospheric pressure, so the total heat transmitted to the bulb's wall is not significant. Similarly, although laboratory plasmas can have temperatures of around 1,000,000°K, their low densities mean that the heating of the wall does not require serious attention (Chen, 2015).

Debye length (λ_D): Within a plasma, which is a gas of charged particles (electrons and ions), the charges in the plasma tend to shield each other's electric fields. Consequently, the electric field around a charged particle in the plasma is somewhat shielded by the other charged particles in its proximity. The extent of this shielding will depend on temperature and the density of the plasma (Chen, 2015). The Debye length is an imperative parameter in plasma physics, as it decides the behavior of the charged particles in the plasma. For example, the Debye length affects how plasma waves propagate and how particles interact with each other. In general, the Debye length is a measure of how strongly the plasma is coupled, or how easily charges can move through the plasma (Chen, 2015; Gibbon, 2020; Wiesemann, 2014).

7.2.2 Types of Plasma

Plasma can be classified into two types, lower-temperature plasma ($T_e = 10^4$ to 10^{5}°K) and higher-temperature plasma ($T_e = 10^6$ to 10^{8}°K), depending on the electron temperature (T_e) (Figure 7.1) (Pan et al., 2019). Lower-temperature plasma can, in turn, be divided into thermal plasma and non-thermal (cold)

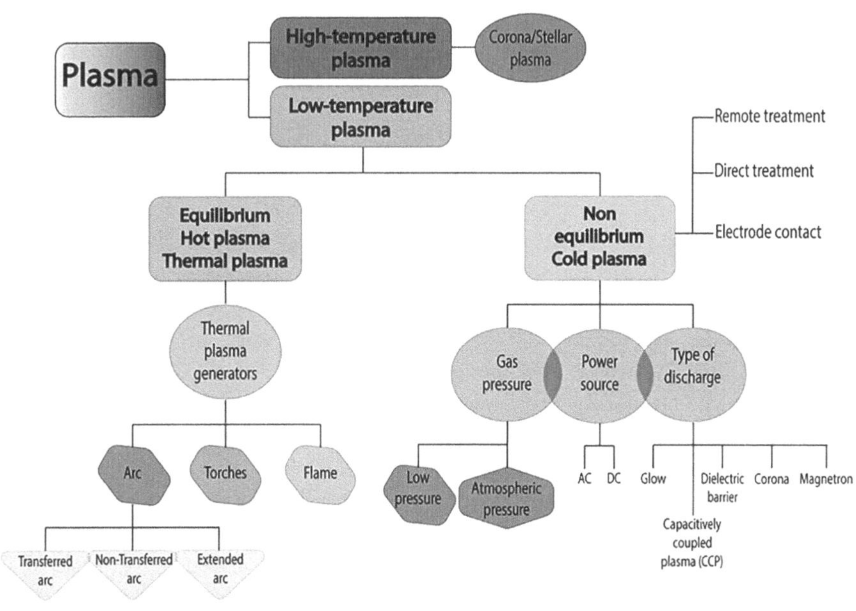

FIGURE 7.1 Types of plasma. The two main categories of plasma are those created at high temperatures and those created at relatively low temperatures. Within the latter class are those that are thermodynamically equilibrated and those that are not. Essentially, plasmas in the former category are used in scientific and commercial applications related to the agri-food sector. The figure here has been adapted from Samal (2017).

plasma depending on the thermodynamic equilibrium. The CP, by contrast, is in a state of thermodynamic disequilibrium (i.e., the temperature of ions, electrons and neutrals within the plasma is not equal), with ions and neutrals remaining at low temperatures (i.e., near room temperature), with electrons reaching a temperature of around 10^{4}°K. Therefore, the CP system is used in food-associated industries, as it maintains at relatively low temperatures (Hashizume et al., 2021; Samal, 2017).

Furthermore, various types of electrical discharges, such as corona discharge, radiofrequency (RF) discharge, glow discharge, dielectric barrier discharge (DBD), pulsed corona discharge, microwave discharge and plasma jet, are used for generating nonthermal or CP (Pan et al., 2019; Samal, 2017). Among these, DBD and plasma jet are the most popular types of CP generation methods due to their straightforward design and flexibility to adjust to different treatment requirements and targets. All the mechanisms are described in Figure 7.1.

7.3 COLD PLASMA APPLICATIONS IN POSTHARVEST MANAGEMENT

7.3.1 Advantages of Cold Plasma in Postharvest Management

Fresh products are handled, and postharvest shelf life is one of the main variables to guarantee commercial success. Using CP treatments within the industry offers benefits in the decontamination process, as traditional sanitization may lose efficacy against microorganisms caused by biofilm build-up as well as by the chemical residues on raw products (Ziuzina et al., 2014; Bourke et al., 2018). A further positive aspect of CP treatments is that they prolong the shelf life of fresh produce. Fruit and vegetable commodities are highly perishable due to their susceptibility to various pathogenic organisms, enzymatic activity and ripening, which reduces their shelf life and quality (Wu et al., 2023). The primary use of CP treatments in the postharvest field is the inactivation of microorganisms. That is why several authors have mentioned the significance of CP in enhancing food safety. Won et al. (2017) have reported using cold microwave plasma to improve the storability of mandarins through the inhibition of *Penicillium italicum*. Siddique et al. (2018) describe fungal species inhibition in fresh products by CP treatments.

Further, the direct elimination of mycotoxin was also reported (Misra et al., 2019). Bacterial inhibition by CP dielectric barrier treatment and cold microwave plasma has been successfully utilized to control pathogens such as *Escherichia coli* O157:H7, *Salmonella*, and *Listeria monocytogenes* (Min et al., 2017; Kim et al., 2017). We define the mechanism of the antimicrobial properties of CP in Section 7.3.6.

Consequently, several authors have described the improvement of the shelf life of fresh produce by means of CP treatments. Abbaszadeh et al. (2018) reported CP treatments in unpackaged and packaged fig fruit, improving the quality through dielectric barrier discharge (DBD) plasma, with direct application for 90 s (unpackaged) and 30 s (in packaged), reaching enhancement in sensorial characteristics such as color, texture, odor and acceptability. These results were shown to be statistically different compared with the controls.

Fresh-cut mango was treated with 75 kV for 3 min to observe the effects on shelf life for 4 days at 15°C. It was found that a reduction in the generation of ROS and malondialdehyde (MDA), browning-related phenolic oxidation inhibition, led to a higher content of phenolics and flavonoids and lower enzymatic activity (Yi et al., 2022).

7.3.2 Effects of Cold Plasma Generators and Their Effect on Fresh Product Quality

Cold plasma treatments induce various chemical and physical processes in fresh produce, and these reactions are associated with changes in the quality parameters of such produce. Despite the advantages of using CP, such as decontamination and bioactive induction, the use of CP may involve brief disadvantages, such as lipid oxidation (Gavahian et al., 2018). Due to the degradation of lipids, the alteration of volatile organic compounds has been described, for instance the conversion of limonene and p-cymene into y-terpene, a-terpineol, terpinene-4-ol and linalool after treatment with atmospheric CP treatment processing for 15, 30, 45 and 60 s with a discharge at 70 Kv. This conversion may change the flavor of citric fruits and juices (Alves Filho et al., 2020). Likewise, the color of fresh products is altered after the application of CP derived from enzymatic changes. For example, the interaction of the endogenous enzymatic activity of fruits and vegetables, such as polyphenol oxidase (PPO), directly influences enzymatic browning. In potato slices, microwave CP treatment of 400 W after 40 min reached a 72.4% PPO activity decrement (Han et al., 2019; Kang et al., 2019; Wang et al., 2022). Pipliya et al. (2022) described the use of dielectric barrier discharge (DBD) to decrease enzymatic browning through the reduction of PPO enzymatic activity and amount of peroxide after 10 min of treatment at 25 kV.

Texture/firmness is another critical parameter in the handling of fresh produce during postharvest. This is a consequence of physical damage and the involvement of enzymes in the degradation process of pectin and other cell wall polysaccharides as well as gene expression (Bangar et al., 2022). The use of CP has proved to be a viable solution to this postharvest challenge. The improvement of the firmness and color of Namwa bananas, for example, resulted in the extension of their shelf life by decreasing ethylene production after treatment with a DBD system, which involved the plate electrodes connected to 220 V/50 Hz power after 10 min (Udtachee et al., 2023).

Likewise, promising results in Cavendish banana fruit have been reported by Sandanuwan et al. (2020) after 30 s of treatment through a DBD system at 15 kV. The treatment revealed that it decreased the weight loss compared to the control by 2.8%; however, the result didn't show a statistical difference. Further CP treatments are described in the following sections.

7.3.3 Low-Pressure Cold Plasma (LPCP)

LPCP is carried out through three reactors: inductively coupled, capacitively coupled and microwave reactions; these reactors are sealed from room air and include a vacuum pump to reduce the pressure in the process (Pedrow et al., 2020). During the procedure, gas is decomposed into radicals and atoms, which could be partially ionized. Typically, the radiofrequency range is 40 kHz or 13.36 MHz, with pressure around 10^{-4} to 10^{-2} kPa (Tendero et al., 2006).

The use of LPCP has been effectively tested in fresh produce decontamination without changing its organoleptic properties. Basaran et al. (2008) described the use of LPCP treatment for surface decontamination of the nut surface using sulfur hexafluoride and air; the treatment was carried out for 20 min. The results revealed the efficacy of sulfur hexafluoride and LPCP against *Aspergillus parasiticus* with a reduction in the initial population by 5 logs. Another treatment mixing citric acid followed by CP at low pressure has been described on sliced apples by Adam et al. (2022). The experiment reached a significant reduction in the *Salmonella spp* population (5.68 log CFU/g) and a reduction in polyphenol oxidase activity (78%), after dipping cut apples in citric acid (5%), followed by 3 min of LPCP.

Kashfi et al. (2020) studied that the use of LPCP in peppermint reduced microbial loads, improved antioxidative potential and total phenolics concentration and maintained color. The experiment involved operating the reactor at a radiofrequency of 13.56 MHz in a parallel plate capacitive coupled configuration. The chambers were then vacuumed with rotary and root pumps and maintained at 40 mTorr. The results demonstrated substantial differences in antioxidant activity between the controls and the plasma treatment when 50 W were used for 20 min. Specifically, the DPPH scavenging percentage increased by 9% due to an enhancement in phenolic concentration (29.72 mg GAE/g). In addition, the treatment inhibited *E. coli* O157:H7 growth. However, the peppermint treated with CP resulted in a dark color that may lead to customer rejection.

7.3.4 Atmospheric Pressure Cold Plasma (APCP)

APCP has higher reactivity due to the formation of excited ionized species. The characteristics of APCP depend on the charged particle densities, which in turn involve many factors, such as the power of the power generator, the conditions of the working gas (molecular, atomic, chemical composition) and the discharge parameters (Bárdos & Bránková, 2010). The generation of APCP does not need vacuum facilities compared to an LPCP system, which must be developed under specific pressure conditions (10^{-4} to 10^{-2} kPa) (Keidar et al., 2021). Using APCP treatments in fresh produce has many advantages, such as plant growth promotion through ROS and RONS regulation in the intracellular redox state. Furthermore, its great potential from the economic and eco-friendly view has been defined (Cui et al., 2022). The generation of APCP can be carried out with different plasma sources. Atmospheric pressure plasma jets (Figure 7.2a) comprise a plasma needle, radiofrequency, electrode, quartz capillary and gas feeding source. In combination with argon, this kind of plasma source has been reported as a successful treatment to inhibit mycotoxin release by *Aspergillus niger* in palm fruit (Ouf et al., 2015).

Moreover, the design of APCP using a jet as a source has shown a reduction of 6 logs of *E. coli* in cherry juice after a voltage of 19.6 kV, an oxygen level of 1% and a gas flow rate of 4.5 L/min (Hosseini et al., 2021). Atmospheric pressure plasma jets have also been tested *in vitro* to eradicate pathogens. The inactivation of *E. coli* has been described in quartz tubes by Fan et al. (2021). During the study, a frequency of 9–17 kHz and higher-purity argon achieved the elimination of *E. coli* after 13 min. Complementary treatments are presented in Table 7.1.

A second source of APCP is the dielectric barrier discharge (DBD) (Figure 7.2b). The definition of DBD is described at length in Chapter 2 of *Applications of cold plasma in food safety* by Ding et al. (2022). This discharge employs a layer of dielectric materials (quartz, polymer, glass, ceramics) to cover one or both electrodes, which discharge alternative current voltages over a wider frequency range (50 Hz-1 MHz) along with pulsed voltage. DBD has already proven to be a promising treatment strategy for fresh produce against pathogenic bacteria, fungi and spoilage yeasts; these treatments are described in Table 7.1.

Several authors have assessed the impact of corona discharge plasma jet (CDPJ) treatments in improving the microbial quality and shelf life of treated produce (Figure 7.2c). Most of the literature characterizes the origin of plasma atmospheric pressure as a discharge that involves a gap measuring between 10 and 30 mm and a voltage of 60–100 kV. In this process, the voltage is applied using one or more electrode heads, which direct gas onto a plate, leading to the generation of plasma that subsequently comes into contact with the sample (Laroque et al., 2022; Warne et al., 2021) it is described in Figure 7.2.

These treatments show the result of CP decontamination. Principally, researchers have described the inactivation or reduction of bacterial and fungal species. However, the present authors suggest that obtaining ideal conditions must be proved with several treatments.

7.3.5 Improving Shelf Life

Fresh fruit and vegetable shelf life refers to the time after harvest when produce is of acceptable quality and remains safe for

TABLE 7.1 Cold Plasma Decontamination Treatments

FOOD MATRIX	PLASMA TREATMENT	RESULTS	REFERENCE
Fresh lettuce	Exposition to N_2-CP 400 W and 900 W, 10 min	Bacteriostatic effect against the growth of *E. coli* O157:H7.	Song et al. (2015)
Cherry tomatoes	Exposition to 30, 60 and 180 kV for 300 s	Inhibition of *Botrytis cinerea* growth.	Misra et al. (2014)
Lettuce, tomatoes, and carrots	Atmospheric pressure cold plasma 12.83 KV (60 Hz) in argon, 10 min	Inactivation 1.6 log against *E. coli* O157:H7	Bermúdez-Aguirre et al. (2013)
Strawberries	DBD-60 kV pulsed at 50 Hz, 5 min	Mesophiles load reduction of 3.3 Log cycles.	Misra et al. (2014)
Bulk grape tomatoes	DBD-35 kV, 3 min	*Salmonella spp* reduction of 0.9–3.3 log CFU	Min et al. (2018)
Almonds	Jet plasma, distance for emitter 6 cm, 20 s	*E. coli* O157:H7 reduction of 1.37 CFU/g	Niemira (2012)
Blueberries	Jet plasma, distance for emitter 7.5 cm, 60 s	Reduction of mold and yeast 1.5 log CFU/g	Lacombe et al. (2015)

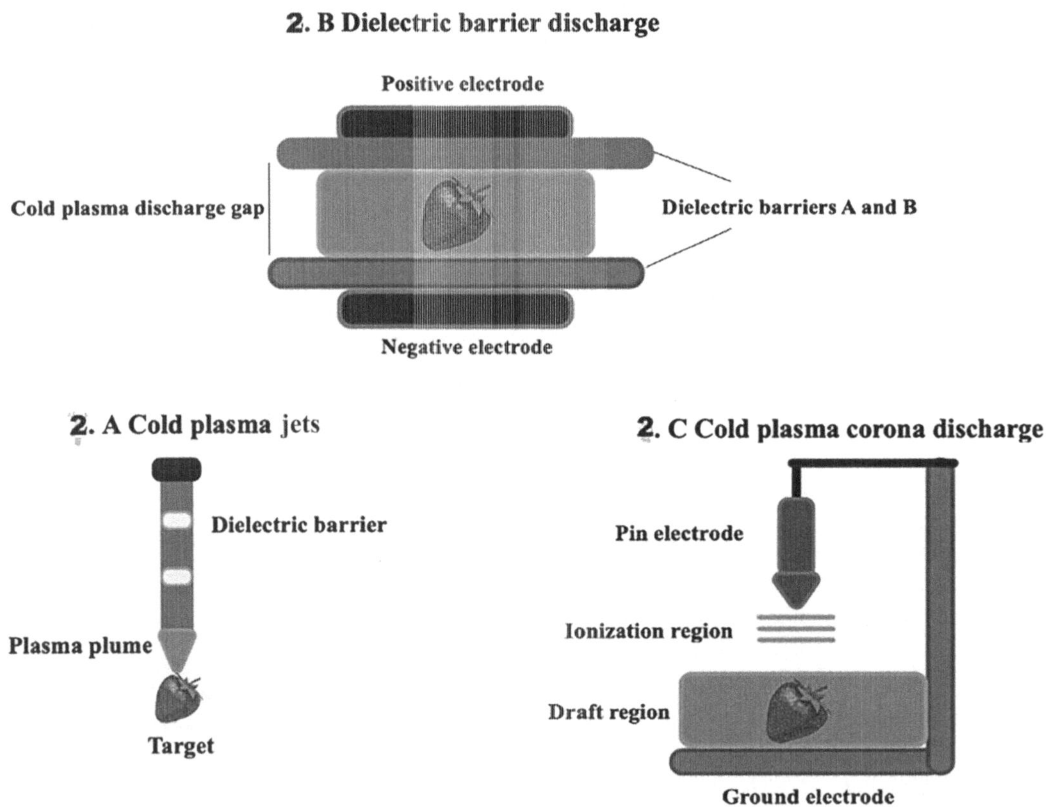

FIGURE 7.2 Cold plasma treatment sources. Adapted from Shen et al. (2022), and Laroque et al. (2022). Systems for the generation of cold plasma are described. These treatments are designed through adaptation, sample stability and effect. (a) Cold plasma jet increases the possibility of modifying the distance between the discharge and the target, offering more control during the discharge. (b) Dielectric barrier discharge allows the possibility to generate more uniformity of discharge in the target. (c) Cold plasma corona discharge operates under discharge gaps up to 7 cm driven by low-frequency AC. During the treatment in this source it is hardly possible to get a uniform discharge; however, it allows the discharge in a specific zone.

consumption. Fresh produce shelf life can vary depending on several factors such as relative humidity, temperature, genotype and the maturity of the product at the time of harvest. Fresh produce is susceptible to the proliferation of pathogens due to their composition. Cold plasma treatments are a tool that has proven to be efficient in improving the shelf life of produce by inactivating phytopathogenic organisms, as well as altering biochemical and physiological processes (Pan et al., 2019; Asl et al., 2022).

Atmospheric air pressure cold plasma treatments in postharvest have been shown to enhance the quality of fresh produce by delaying ripening, enhancing texture and reducing the loss of color and weight (Dong & Yang, 2019; Chen et al., 2020). As an example, Jia et al. (2022) reported an improvement in tomato quality when atmospheric cold plasma using air as a working gas, a voltage treatment was performed of 60 kV, induced an inhibition of chlorophyll degradation. The use of CP treatments to improve the shelf life involves the capacity of this kind of matter state to increase the induction of bioactive components in fruit and microorganism inactivation that could reduce the shelf life of fresh products. These points are described further in the following section.

7.3.6 Pathogen and Microorganism Inactivation

Cold plasma treatments are effective in reducing microorganism populations on fruit and vegetable surfaces. In this way, the shelf life of fresh produce is prolonged and the risk of foodborne diseases is reduced. Numerous other authors have developed CP decontamination treatments, which are included in Table 7.1.

Ongoing research has demonstrated the possibility for the development of antimicrobial treatments with CP (Niveditha et al., 2021; Pant et al., 2022; Hernández-Torres et al., 2022).

Specific parameter modifications and applications have been developed to generate these treatments, creating atmospheric conditions with CP, dielectric barrier charges or using CP jets (Cheng et al., 2006; Di et al., 2018; Feizollahi et al., 2021). CP treatment used as a disinfectant against microorganisms affecting the quality of fresh produce exhibit three distinct modes of action. These include inducing cell wall leakage, causing damage to intracellular proteins and harming nucleic acids (Hernández-Torres et al., 2022). Cell leakage occurs due to damage to membrane integrity resulting from various interactions between ROS, UV radiation and energetically charged particles with the cell walls of microorganisms. These processes are strongly associated with membrane lipid oxidation, reducing surface microbial load by cell wall damage (Mahnot et al., 2019; Han et al., 2016).

Lu et al. (2014) described cell leakage in *E. coli* ATCC25922, *E. coli* NCTC 12900 and *L. monocytogenes*. Due to reactions between microorganisms and RONS, the cell leakage mechanism is described in Figure 7.3.

7.3.7 Induction of Secondary Metabolites

CP treatments can help preserve the nutritive value of fresh produce by slowing the ripening process and by reducing ethylene production. For example, it was found that CP treatment of 2 min at 8 kV direct power supply and at 2.0–4.0 A by intermittent corona discharge plasma jet effectively slowed the ripening of tomatoes and preserved their vitamin C content (Lee et al., 2018). Foods of plant origin are the main sources of bioactive compounds. However, one of the most pursued objectives in postharvest research is the conservation of these compounds throughout storage, since bioactive compounds are directly correlated with organoleptic

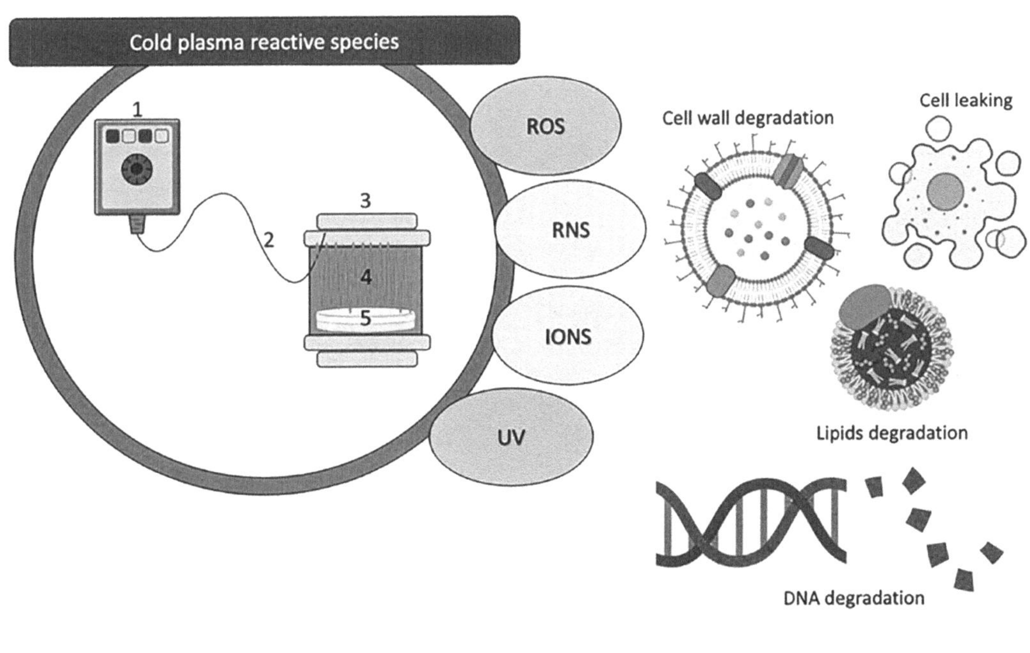

FIGURE 7.3 Cold plasma decontamination mechanism. 1: power source, 2: electrode, 3: dielectric barriers; 4: plasma; 5: sample. Adapted from Liao et al. (2018) and Hernández-Torres et al. (2022). Interactions between DBD, RONS, UV and ions with cell walls result in cell leakage by lipid oxidation and DNA damage resulting from nucleic acid decomposition. Plasma species cause damage to the DNA molecules present in the microbial cells. RONS are formed by CP interaction and cause damage in the cell nucleus.

TABLE 7.2 Cold Plasma Treatments as a Bioactive Inductor

FOOD MATRIX	TREATMENT CONDITIONS	BIOACTIVE ENHANCEMENT	REFERENCE
Banana slices	Atmospheric plasma—6.9 kV for 46 s using air as working gas	Increment of total phenolic compounds of 19.86 mg/ 100 g	Pour et al. (2022)
Strawberries	Atmospheric plasma—60 kV for 15 min using air as working gas	Increment of total phenolic compounds of 12.6 mg/ g	Rana et al. (2020)
Rocket-salad leaves	Cold plasma activated water	Increment of vitamin B2 of 0.21 mg/ 100g and vitamin B3 of 0.65 mg/ 100 g	Abouelenein et al. 2023
Shiitake mushrooms	Cold plasma activated water	Increment of total phenolic compounds of 303.2 mg/100 g	Shishir et al. (2020)
Jujubes	Dielectric barrier discharge—5 kV, 40 kHz, using air as working gas.	Increment of total phenolic, procyanidins, and flavonoids content in 13.85 %, 53.81 %, 33.89 %	Bao et al. (2021)
Blueberries	Dielectric barrier discharge—45 kV for 50 s	Increment of total anthocyanins of 0.26 mg/100 g	Zhou et al. (2020)

Source: All the increments are compared with the negative control.

quality while having potential positive impacts on human health (Saremnezhad et al., 2021). The enhancement of bioactive compounds by CP treatments under different conditions has been defined by several authors (Table 7.2). Nevertheless, the impact and mechanism of CP treatments to improve bioactive compound contents in fruit and vegetable produce are not well understood (López-Gámez et al., 2021; Dantas et al., 2021).

Recently, CP treatments have been investigated for the induction of bioactive constituents in fruits, vegetables and cereals (Sirgedaitė-Šėžienė et al., 2021). Yodpitak et al. (2019) reported the influence of CP to increase vitamin E (14%) content after one day of the treatment in brown rice and a constant distance between two electrodes was kept constant at 0.5 cm. The power supply was radiofrequency and the samples trayed between the two electrodes at the potency of 100 W, 135 W, 170 W and 200 W, using argon as the working gas.

Li et al. (2019) reported an increase of phenolics in cut pitahayas, such as protocatechuic acid (132.3%), gallic acid (106%) and *p*-coumaric acid (108.8%) after a CP treatment with a power of 60 kV for 5 min. In contrast, p-hydroxybenzoic acid and caffeic acid were reduced by 6.8% and 16.7%, respectively. Yet in blueberries, phenols, flavonoids and anthocyanins increased after DBD at 45 kV for 50 s (Zhou et al., 2023). Likewise, Ji et al. (2020) reported increased content of phenols and anthocyanins in blueberries with atmospheric cold plasma treatment for 60 s at 12 kV and 5 kHz. Moreover, it was found that the enhancement was statistically significant after 20 days of storage. Additionally, Li et al. (2019) reported total phenolic enhancement in strawberries after 5 days of DBD at 45 kV for 1 min. An increase in total phenolic content has also been demonstrated in mango pulp after 4 days of treatment; however, ascorbic acid concentration diminished during this period. These findings confirmed that CP treatments alter the bioactive profile of plant products. Treatments are described in Table 7.2.

7.4 COLD PLASMA TREATMENT OF SPECIFIC FRUITS AND VEGETABLES

7.4.1 Fresh Apples

Apples are one of the most consumed fresh products worldwide, with an annual production of almost 86 million tons (FAO, 2018). Further, the apple is recognized as a high nutritional value fruit and used as an ingredient in many gastronomic dishes. One of the most important aspects of CP treatments is the alteration of the sensory quality of fresh produce. In contrast, glow discharge plasma improves the sweetness of apple products. It has also been stated that DBD reduces the total soluble solids content of apple products, which could also lead to the acidification of apple juices (Wang et al., 2023; Farias et al., 2021).

Disinfection of apple surfaces using CP treatments reduces pathogenic microorganisms such as *E. coli* O157:H7 and *Salmonella Stanley* (Niemira & Sites et al., 2008). CP treatments are largely used in apple by-products. In apple juice, the use of these treatments has been reported. *Citrobacter freundii* was reduced by 5 logs after 480 s using argon and 0.1% oxygen (Surowsky et al., 2014). Also, a decrease of 3.98–4.34 logs CFU/mL in apple juice after 40 s by dielectric barrier discharge-atmosphere CP was reported (Liao et al., 2018).

A variety of effects on the nutritional and nutraceutical content have been reported in CP-treated apples. After the use of a DBD system with a oscillation frequency of 12 kHz, the supply power was in the range of 150 W for 10 minutes; after the treatment, a 20% increase in polyphenols was observed compared to untreated fruits (Tappi et al., 2018). Also, the increment of polyphenols in apple cubes has been reported during DBD CP treatment (600 Hz for 15 min), which may be due to lower polyphenol oxidase and peroxidase activity (Farias et al., 2022). Nevertheless, contrasting effects were also observed in apple cubes, in which polyphenols were not enhanced, and the antioxidant activity was decreased by 10% after dielectric barrier discharge at 12.7 kHz of frequency oscillation for 120 min (Ramazzina et al., 2016).

7.4.2 Berries

Berries are becoming increasingly popular in new diet trends due to their high levels of nutrients and phytochemicals. These fruits are primarily composed of phenolics, flavonoids, anthocyanins, tannins and phenolic acids. Despite their many health benefits, berries are vulnerable to contaminations in response to different biological hazards (Golovinskaia & Wang, 2021; Lacombe et al., 2017). One of the major limitations of consuming berries is their short postharvest shelf life, which can pose a challenge for growers and suppliers. However, various treatments can be applied to prolong their shelf life. These treatments include the use of edible oils, coatings and controlled atmospheres, among others (Abugoch et al., 2016; Huynh et al., 2019).

Emergent technologies such as CP treatments are described as novel techniques involving enhancing quality and decontamination in berries. Recent research has shown promising results in using DCSBD to reduce *Salmonella Enteritidis* contamination in berries. In particular, after 300 s of exposure, there was a decrease of 6.64 logs in their population. Also, *Bacillus subtilis* endospores were reduced by 2.85 logs CFU/g (Medvecká et al., 2020), and *A. niger* was partially inactivated after 15 s of plasma irradiation (Wiktor et al., 2020). A different CP treatment showed total mesophilic aerobes of 1.25 logs CFU/g in blueberries after 10 days of storage (Pathak et al., 2020). In red currants, the reduction of 1.1 logs CFU/g mesophilic aerobic bacteria was similar after CP treatment (Limnaios et al., 2021). CP application on strawberries has been the subject of numerous reports. Ahmadnia et al. (2021) described the dielectric barrier discharge of CP to enhance the shelf life of strawberries; the results showed shelf-life prolongation with a treatment of 20 min and 14% duty cycle. During the treatment, total mesophilic aerobic bacterial count and yeasts/molds were decreased by 1.46 and 2.75 logs CFU/g, respectively. Further treatments revealed that a CP atmosphere was capable of reducing the pathogenic load of *Listeria innocua* by 3.8 logs CFU/g, as reported by Ziuzina et al. (2020); moreover, the treatment did not show a substantial difference in color, pH, firmness or soluble solids content.

Alternatively, other authors describe CP as a tool to increase the bioactive content of fresh fruits and vegetables. In blueberries, the increment of ascorbic acid has been reported after 80 kV for 1 min with 14.01 mg/100 g. This is a higher range than the control (8.91 mg/100 g). Likewise, the flavonoid content increased after CP treatment at 60 kV for 5 min. Yet anthocyanin content significantly decreased when the voltage was increased. In addition, the use of DBD CP use at a discharge voltage of 12 kV and frequency of 5 kHz for 60 s, after 10 days of storage, induced an increment of anthocyanins, and phenols showed an increment in blueberries from 0.66 to 0.94 OD_{280}/ g, and 0.62 to 0.80 OD_{530}-OD_{600}/g, respectively. The increment of DPPH antioxidant activity did not increase until after 40 days of storage (2.87 to 3.55 mmol TE/g) (Ji et al., 2020). Moreover, the use of atmospheric CP at 60 kV for 1 min revealed an increment of polyphenolic content from 210.12 to 250.12 mg GAE/100 g, and the increment of total flavonoid content showed an increase from 89.22 to 98.22 mg CEQ/100 g (Sarangapani et al., 2017).

7.4.3 Leafy Greens

Postharvest handling of leafy greens represents a significant challenge because of their susceptibility to microbial loads through environmental contamination, which involves wild animal contamination or contaminated agricultural water (Ceuppens et al., 2015). Leafy green vegetables are sources of foodborne disease outbreaks caused by associated microorganisms such as *E. coli* O157:H7, *L. monocytogenes*, *Salmonella spp* and viruses such as norovirus (Liu et al., 2013; Shah et al., 2019; Kyere et al., 2020; Herman et al., 2015). Leafy green vegetables are susceptible to microbial colonization; therefore, the use of CP treatments is an alternative to reduce these infections. Romaine lettuce was subjected to DBD atmospheric CP for 5 min at 34.8 kV, resulting in a significant decrease in bacterial count. Specifically, there was a decrease of 4.7 logs CFU/g of *E. coli* O157:H7 and a decrease of 6 logs CFU/g of *Salmonella* in lettuce (Min et al., 2017). Several studies have reported on the efficacy of CP treatments in reducing bacterial contamination in various types of leafy green vegetables. In a study conducted by Klockow & Keener (2008) that focused on spinach, they utilized DBD to reduce the microbial load. This study found that the treatment generated ozone and exhibited 3–5 log cycle reduction of *E. coli* O157:H7. Reduction of *E. coli* O157:H7 has also been reported in other commodities. Min et al. (2017) defined the effect of DBD during storage at 4°C for 7 days; in the study, CP treatment led to a bactericidal effect in Romaine lettuce.

The use of CP techniques in leafy greens has also been described by Perinban et al. (2022). The research involved a treatment method that utilized a DBD nonthermal plasma system and water to create plasma-activated water conditions. The water was subjected to plasma at a voltage of 10 kV and a frequency of 20 kHz for varying durations of 10, 20, 30, 45 and 60 min, after which kale and spinach leaves were immersed for 10 min. The results indicated that the antioxidative activity in spinach significantly improved after 20 min of treatment, as demonstrated by a reduction in DPPH radical (17.65%) than the controls. Nevertheless, the treatments resulted in a reduction of FRAP radical (1.63 mM TE/g) after 20 min.

7.4.4 Tomatoes

Tomatoes are consumed worldwide as the second most important commercialized vegetable crop, with annual production of 163 million tons (Ali et al., 2020). The high antioxidant content, minerals, vitamins, essential amino acids, and phytosterols are beneficial compounds against cancer, cardiovascular diseases and cognitive functions (Dominguez et al., 2020). Despite the high production of tomatoes, postharvest losses can reach 25–42% worldwide (Arah et al., 2015). Due to this, the use of novel treatments such as CP has been implemented (Ali et al., 2021). Decontamination in tomatoes by CP treatments has shown promising results. A reduction of 4.08 logs CFU/g after 20 min and a power of 51.7 W was reported by Han et al. (2020). In addition, the use of an atmospheric-pressure plasma jet (APPJ) was able to inactivate the tomato pathogen *Cladosporium fulvum* during the assay, plasma discharge was developed at 7 kV and 9 kHz for 10, 20, 30, 40, 50, 60, 120, 180, 240 and 300 s; therefore, the bactericidal effect was reached in all treatments after 60 s. This was achieved by disrupting the cellular membrane of the pathogen (Lu et al., 2014).

E. coli load has been reduced by CP treatments in tomatoes by using a CAP at 60 kV and 5-, 10-, 15- and 30-min assays. The best results reached 6 logs CFU/mL of the pathogen after 15 min of treatment (Prasad et al., 2017).

The use of CP treatments in tomatoes has not only been defined for disinfection. For instance, in tomato seeds, CP improves germination and enhances seed growth after 10 min of treatment, and the reactive oxygen species showed an increase (Adihkari et al., 2020). Previous reports have explored the application of CP to enhance nutrient absorption in tomato seeds. As an example, three variations were tested by exposing the seeds to inductive helium coupled with cold plasma. The plasma frequency used was 13.56 MHz for 15 s, while the radiofrequency discharge was set at 60 W, 80 W and 100 W. Results indicated that the germination potential significantly improved after exposure to an 80 W discharge (from 67.7% to 75.5%). Moreover, this discharge led to an increase in the phosphorus and nitrogen content by 19.1% and 23.7%, respectively (Jiang et al., 2018).

Studies have shown a notable increase in tomato shelf life. During storage, the corona discharge plasma jet at 3.0 A and 4.0 A induced a significant extension of 10–15 days compared with a 6-day period for controls (Lee et al., 2018). Similarly, the utilization of ACP, including as a working gas air, and three different voltages (40 kV, 60 kV and 80 kV) were tested for a duration of 5 min in tomatoes. Thereafter, tomatoes were monitored for 32 days. After the evaluation, 60 kV treatment was most effective in improving quality parameters such as color and firmness. Moreover, a reduction in weight loss rate was observed and the use of 60 kV treatment reached the lowest reduction (Jia et al., 2022).

7.5 COLD PLASMA MECHANISM OF ACTION ON FRUITS AND VEGETABLES DURING POSTHARVEST

There are several reviews available about the effects of CP application on fruits and vegetables focused on nutritional quality (Chen et al., 2020), shelf life (Sousa-Gallagher et al., 2016), enzyme activity (Mayookha et al., 2022), microbiological quality (Mao et al., 2021; Wang et al., 2012), bioactive compounds (Boateng, 2022; Yu & Fan, 2021) and elimination of hazardous pesticides (Gracy et al., 2022), among others. In this section, we will focus on the mode of action underlying the technology. The effects of CP during postharvest of fruits had been observed on shelf life, microbial safety, nutritional quality and organoleptic quality. It is important to state that, to our knowledge, no efforts to elucidate the CP mode of action had been carried out using next-generation technology.

The effect of CP on fruits and vegetables can be explained by the inhibition of enzyme activity and the generation of ROS, which can damage DNA, lipid membranes, carbohydrates and proteins, causing cell death (Karuppanapandian et al., 2011).

7.6 ORGANOLEPTIC QUALITY

The effect of the DBD system with the level of 45 kV on fresh-cut strawberries with 90% of fruit color development was tested. The fruit was stored at 4°C for 1 week afterward. An increase in flavonoids and anthocyanins biosynthesis was found, which correlates with the increase in gene expression of the genes encoding for phenylalanine ammonium lyase, cinnamate-4-hydroxylase and 4-coumarate coenzyme A ligase. Also, an increase in enzyme activity was found. This data explains the increase in the metabolites mentioned, because these enzymes catalyze the first three steps in their biosynthesis.

Furthermore, an increase in the radical scavenging capacity was recorded, which correlates with the observed improvement in the total phenolic content. An increase in ROS, O_2 and H_2O_2 was also recorded, which explains the observed increased activity of the enzyme catalase (CAT), ascorbate peroxidase (APX) and copper/zinc-dependent and manganese-dependent superoxide dismutase (SOD). Furthermore, an increase in gene activity during storage after 12 h of SOD and after 3 days of both CAT and APX was found. Higher firmness in plasma-treated strawberries under storage compared with the controls was observed; however, no analysis of plant cell wall enzymes was carried out. A decrease in total aerobic bacteria in plasma-treated strawberries was also found, which is most likely due to the impact of the ROS on the cell structures, although no analyses were carried out to support this statement (Li et al., 2019).

Corn salad was subjected to a combination of an atmospheric pressure plasma jet for 60 s at 8 W of power and an atmosphere of argon mixed with 0.1% of oxygen. A 4.1 log units reduction of artificially inoculated *Escherichia coli* strain DSM 1116 was found. Also, a reduction in the leaves' photosynthetic performance by 20% was observed and was shown to have a large effect on organoleptic quality and, consequently, on shelf life (Baier et al., 2014).

With the goal of reducing the number of *Escherichi coli* inoculated by hand, the air was ionized with a 2.45 GHz microwave-driven air plasma torch with a power of 1.2 kW and a 20 $Lmin^{-1}$ of gas flow to treat apples, cucumbers, tomatoes and carrots. Photosynthetic efficiency was measured in apples and cucumbers. It recorded different levels of effect on the number of bacteria. Further, it was found to have no negative effect on apple quality, whereas the cucumbers showed irreversible damage after 5 min of treatment, suggesting a negative effect on fruit physiological function. Also, in cucumbers, a reduction in the elastic modulus was recorded, which was explained by an increase in softening due to injuries in the exocarp. Tomatoes only showed changes to a darker red color, whereas carrots developed a brown color along with small exudation of an orange-colored liquid. It was concluded that there is a need to develop specific treatments for every commodity to reduce the number of bacteria without inducing damage to the commodity's organoleptic quality (Baier et al., 2014).

Highbush blueberries (*Vaccinium* O'Neal) were treated with atmospheric CP with an energy of 12 kV for 30, 60 and 90 s. A more significant reduction in total bacteria as well in decay rate with a larger duration of the treatment was found. Higher fruit firmness during the storage time with 60 s of treatment was recorded, although no efforts to measure the plant cell wall enzymes were carried out. Further, an increase in the enzyme peroxidase, catalase and SOD was found, which explains the reduction in ROS, O_2 and H_2O_2. The results of this experiment clearly demonstrate that it is possible to increase the shelf life of fruits. In this article, a model to explain the mode of action of atmospheric CP is provided (Ji et al., 2020).

Shiitake mushrooms were treated with CP set at 650 W of power while immersed in CP-activated water. It was found that this treatment reduced the time to dry the product at all temperatures tested. Further, analysis of the microstructural changes by scanning electron microscopy exhibited that CP application increased both the size of the intracellular spaces and the number of cavities, which explains why the drying of the product was faster. Also, mushroom powder showed better quality in terms of swelling index and water retention percentage (Karim et al., 2021). In agreement with this data, the treatment of button mushrooms pre-soaked in plasma-activated water, with a CP set at 50 kV, showed a reduction in the browning index and color change. It was found that this visual effect was due to polyphenol oxidase enzyme activity inactivation (Zheng et al., 2022).

An experiment to analyze the effect of CP on browning-inducing enzymes was tested on an *in vitro* system with the enzymes polyphenol oxidase and peroxidase, dissolved on pH 6.5 phosphate-buffered saline buffer. It was found that enzyme activity was reduced by 90% after 180 s and 85% after 240 s on PPO and POD, respectively. It was shown that the reduction was due to a reduction in the alpha-helix and an increase in the beta-sheet structural domain content (Surowsky et al., 2013). This experiment provides evidence of the plasma mode of action to maintain the organoleptic quality of fresh produce under postharvest by inactivating enzymes.

Lysozyme of hen egg white was treated with a low-frequency plasma jet with a maximum power level of 5.0 kV to know the mechanism of enzyme inactivation. It was shown that a 30 min treatment induced more than 90% of enzyme activity. This effect was due to a change in the chemical structures of the enzyme tryptophan residues as well as other unknown amino acid residues. It was also shown to have a negative effect on the secondary structure. Further, it was suggested that the influences observed were due to ROS effect rather than UV radiation (Takai et al., 2012).

Apples and potatoes cut into cubes were subjected to an air treatment plasma process generated by microwave at 1.2 kW power and a flow rate of 20 L/min. After 10 min, polyphenol oxidase enzyme activity was declined to 62% and 77% in apple and potato, respectively. On another side, peroxidase enzymatic activity was suppressed by 65% and 89% in apple and potato, respectively. The effect on enzymatic activity was able to inhibit the browning development in potatoes completely, although in apple tissues a browning development was observed, although this was different compared with the usual apple browning color. After the treatment, the tissue was freeze-dried as well as dried by using warm air. After these treatments, apple tissues were stored for 19 days, and polyphenol oxidase enzymatic activity of about 10% in both treatments was recorded, whereas the enzymatic activity of peroxidase was close to zero in both treatments. In the case of potatoes, after 19 days of storage, peroxidase enzymatic activity close to zero for freeze-dried and warm air treatments was recorded, whereas, in the case of polyphenol oxidase, the freeze-dried tissues showed activity close to zero and tissues warm air–dried showed about 10%. These data clearly indicate that the reduction in enzymatic activity remained during several days of storage, maintaining the organoleptic quality of the tissues (Bußler et al., 2017).

7.7 BACTERIA

A mixture of *Listeria monocytogenes*, *Salmonella enterica* subsp. Enterica, Shiga toxin-producing *Escherichia coli* collocated in a coverslip was treated with CP designed to be generated by a surface dielectric barrier discharge set with a 13.5 V input power. A reduction in CFU/mL between 2.34 and 3.99, depending upon the equipment arrangement, was observed. Observation with transmission electron microscopy showed physical damage in membranes, leading to cell lysis as the cause of bacteria death (Timmons et al., 2018).

The bacteria *E. coli* strain K12 was treated with a device generating nonthermal atmospheric pressure plasma, which included argon and a power level of 125 W. The effects on gene expression were studied with a microarray containing a large density of oligonucleotides. It was found that the activation of genes plays a role in the SOS DNA response. Also, an increase in genes that remove superoxide radicals was recorded. The data mentioned clearly suggest a harmful effect of either UV or ROS on DNA and, consequently, in bacterial death (Sharma et al., 2009).

An experiment to test the mode of action of high-voltage atmospheric plasma in Gram-positive and Gram-negative bacteria was carried out by treating *Escherichia coli* and *Staphylococcus aureus* with atmospheric CP with a power level of 120 kV. The results showed that in the case of Gram-negative bacteria, the cell walls and plasma membranes are the main target of the generated ROS. This causes a disruption and cell leakage. On the other side, in the case of Gram-positive bacteria, the reactive oxygen species react with intracellular components such as DNA with no induction of cell leakage (Han et al., 2016).

7.8 FOOD SAFETY

An imperative application of non-thermal plasma is the elimination of toxic compounds present in fruits. In this regard, the pesticides chlorpyrifos and cypermethrin were collocated in the mango skin by using a solution of acetone nitrile with Tween 20. Mango fruits were treated with different flow rates of argon gas treated with a CP device developed with a discharge system driven by a neutron generator and DC transformer at 8 kV and 0.6 A to generate plasma at 600 W. It recorded a reduction of 74% and 62.6% for chlorpyrifos and cypermethrin, respectively. It was suggested that the generation of ROS induced the degradation of pesticides. Furthermore, no adverse effects on color, texture profile, total soluble solids, total carotenoids and total phenolics were recorded. However, a slight reduction in titratable acidity percentage was observed with the highest flow rate of argon gas (Phan et al., 2018). Other authors have reported enhancing bioactive properties through the induction of biosynthesis of the polyphenols and retarding microorganisms and enzymatic degradation (Munekata et al., 2020). Moreover, the use of CP in degrading residual pesticides on lettuce has been described; malathion and chlorpyrifos have been reduced by 79.6% and 74.1% after 80 kV for 180 s (Cong et al., 2021). Likewise, the use of CP to reduce pesticides in fresh products has been observed by Mahbubeh et al. (2017). During the experiment the device was constant at 13 kHz and 10 kV for 10 min, and specific quantities of diazinon (1,000 ppm, 500 ppm). The results showed the highest reduction of diazinon in apple, decreasing 87.38% for a 500 ppm test; likewise, the reduction of diazinon (82.24%) in cucumbers was reported.

7.9 FUTURE DIRECTIONS AND ONGOING RESEARCH

Many authors have defined future directions for researching postharvest handling enhancement through CP treatments. Generally, these treatments allow the research community to seek new environmentally friendly strategies to decontaminate fresh products or increase their quality through secondary metabolite induction; however, to scale up fresh products, companies may need to revise the set up process because all the parameters must be checked before the transition from laboratory to industry. CP treatments during postharvest may decrease even millionaires' losses. Effects such as reducing microbial load, degradation of pesticides and allergens and increment of bioactive compounds are straightly related to prolonging shelf life and preserving quality over an extended time (Gavahian & Khaneghah, 2020; Zhang et al., 2019). Although increasing the shelf life of fresh products has been a big challenge for food scientists, CP technology has brought the possibility of obtaining better treatments without altered organoleptic characteristics. Future directions aim to develop more profound research on CP to achieve better results through parameter modifications. Due to promising results, CP is currently hugely investigated (Xiang et al., 2021). An important area of future research for plasma treatments is the better generation of scientific information to elucidate this technology's mechanism of action. This data will help to design and optimize plasma treatment by using a step-wise approach and not strial and error.

7.10 INDUSTRIAL APPLICATION

Industrial applications of CP aim to eliminate bacterial loads because efficiency as a decontaminant has been hugely reported. Nevertheless, the food research community constantly seeks standard protocols to reach repeatable results (Keener & Misra, 2016). CP has been reported on extensively at the laboratory level, and the standardization of protocols on a large scale is only partially understood. However, many authors describe the possibility of using CP in the food industry (Umair et al., 2022). This emerging nonthermal technology may minimize the undesirable effects of conventional thermal technologies. Due to these results, using CP in the fresh products industry is becoming more important for reaching high-quality and safe values in fruits and vegetables.

7.11 CONCLUSIONS

Cold plasma treatments provide new strategies for the food industry and related research to guarantee enhanced quality and safety of fruit and vegetable products. This technology has the possibility to increase the shelf life of fresh products by overcoming challenges that seemed impossible in the past. Looking forward, CP treatments will allow the possibility of benefiting the producers and customers economically by reducing postharvest losses and foodborne diseases. However, the use of CP and scaling it up to industrial applications need to be more deeply studied to understand how all the mechanisms are carried out. The standardization of CP treatments in fresh products is still a challenge for the postharvest research community; nevertheless, laboratory scale results show promising results to utilize the technology at the industrial scale for fresh products and subproducts.

REFERENCES

Abbaszadeh, R., Alimohammad, K., Zarrabi Ekbatani, R. (2018). Application of cold plasma technology in quality preservation of fresh fig fruit (Siyah): a feasibility study. *International Journal of Horticultural Science and Technology*, *5*(2), 165–173. https://doi.org/10.22059/ijhst.2018.258024.240

Abouelenein, D., Mustafa, A. M., Nzekoue, F. K., Caprioli, G., Angeloni, S., Tappi, S., . . . Vittori, S. (2023). The impact of plasma activated water treatment on the phenolic profile, vitamins content, antioxidant and enzymatic activities of rocket-salad leaves. *Antioxidants*, *12*(1). https://doi.org/10.3390/antiox12010028

Abugoch, L., Tapia, C., Plasencia, D., Pastor, A., Castro-Mandujano, O., Lopez, L., Escalona, V. H. (2016). Shelf-life of fresh blueberries coated with quinoa protein/chitosan/sunflower oil edible film. *Journal of the Science of Food and Agriculture*, *96*(2), 619–626. https://doi.org/10.1002/jsfa.7132

Adam, A. M., Jeganathan, B., Vasanthan, T., Roopesh, M. S. (2022). Dipping fresh-cut apples in citric acid before plasma-integrated low-pressure cooling improves Salmonella and polyphenol oxidase inactivation. *Journal of the Science of Food and Agriculture*, *102*(8), 3425–3434. https://doi.org/10.1002/jsfa.11690

Adhikari, B., Adhikari, M., Ghimire, B., Adhikari, B. C., Park, G., Choi, E. H. (2020). Cold plasma seed priming modulates growth, redox homeostasis and stress response by inducing reactive species in tomato (*Solanum lycopersicum*). *Free Radical Biology and Medicine*, *156*, 57–69. https://doi.org/10.1016/j.freeradbiomed.2020.06.003

Ahmadnia, M., Sadeghi, M., Abbaszadeh, R., Ghomi Marzdashti, H. R. (2021). Decontamination of whole strawberry via dielectric barrier discharge cold plasma and effects on quality attributes. *Journal of Food Processing and Preservation*, *45*(1), e15019.

Ali, M., Cheng, J. H., Sun, D. W. (2021). Effect of plasma activated water and buffer solution on fungicide degradation from tomato (*Solanum lycopersicum*) fruit. *Food Chemistry*, *350*, 129195. https://doi.org/10.1016/j.foodchem.2021.129195

Ali, M. Y., Sina, A. A. I., Khandker, S. S., Neesa, L., Tanvir, E. M., Kabir, A., Gan, S. H. (2020). Nutritional composition and bioactive compounds in tomatoes and their impact on human health and disease: a review. *Foods*, *10*(1), 45. https://doi.org/10.3390/foods10010045

Alves Filho, E. G., de Brito, E. S., Rodrigues, S. (2020). Effects of cold plasma processing in food components. In *Advances in cold plasma applications for food safety and preservation* (pp. 253–268). Academic Press. https://doi.org/10.1016/B978-0-12-814921-8.00008-6

Ansari, A., Parmar, K., Shah, M. (2022). A comprehensive study on decontamination of food-borne microorganisms by cold plasma. *Food Chemistry: Molecular Sciences*, *4*, 100098. https://doi.org/10.1016/j.fochms.2022.100098

Arah, I. K., Amaglo, H., Kumah, E. K., Ofori, H. (2015). Preharvest and postharvest factors affecting the quality and shelf life of harvested tomatoes: a mini review. *International Journal of Agronomy*, *2015*. https://doi.org/10.1155/2015/478041

Asl, P. J., Rajulapati, V., Gavahian, M., Kapusta, I., Putnik, P., Khaneghah, A. M., Marszałek, K. (2022). Non-thermal plasma technique for preservation of fresh foods: a review. *Food Control*, *134*, 108560. https://doi.org/10.1016/j.foodcont.2021.108560

Baier, M., Görgen, M., Ehlbeck, J., Knorr, D., Herppich, W. B., Schlüter, O. (2014). Non-thermal atmospheric pressure plasma: screening for gentle process conditions and antibacterial efficiency on perishable fresh produce. *Innovative Food Science & Emerging Technologies*, *22*, 147–157. https://doi.org/10.1016/j.ifset.2014.01.011

Bangar, S. P., Trif, M., Ozogul, F., Kumar, M., Chaudhary, V., Vukic, M., Changan, S. (2022). Recent developments in cold plasma-based enzyme activity (browning, cell wall degradation, and antioxidant) in fruits and vegetables. *Comprehensive Reviews in Food Science and Food Safety*, *21*(2), 1958–1978. https://doi.org/10.1111/1541-4337.12895

Bao, T., Hao, X., Shishir, M. R. I., Karim, N., Chen, W. (2021). Cold plasma: an emerging pretreatment technology for the drying of jujube slices. *Food Chemistry*, *337*. https://doi.org/10.1016/j.foodchem.2020.127783

Bárdos, L., Baránková, H. (2010). Cold atmospheric plasma: sources, processes, and applications. *Thin Solid Films*, *518*(23), 6705–6713. https://doi.org/10.1016/j.tsf.2010.07.044

Basaran, P., Basaran-Akgul, N., Oksuz, L. (2008). Elimination of Aspergillus parasiticus from nut surface with low pressure cold plasma (LPCP) treatment. *Food Microbiology*, *25*(4), 626–632. https://doi.org/10.1016/j.fm.2007.12.005

Bellan, P. M. (2008). *Fundamentals of plasma physics* (1st pbk. ed.). Cambridge University Press.

Bermúdez-Aguirre, D., Wemlinger, E., Pedrow, P., Barbosa-Cánovas, G., Garcia-Perez, M. (2013). Effect of atmospheric pressure cold plasma (APCP) on the inactivation of Escherichiacoli in fresh produce. *Food Control*, *34*(1), 149–157. https://doi.org/10.1016/j.foodcont.2013.04.022

Bittencourt, J. A. (2004). Introduction. In J. A. Bittencourt (Ed.), *Fundamentals of plasma physics* (pp. 1–32). Springer. https://doi.org/10.1007/978-1-4757-4030-1_1

Boateng, I. D. (2022). Recent processing of fruits and vegetables using emerging thermal and non-thermal technologies. A critical review of their potentialities and limitations on bioactives, structure, and drying performance. *Critical Reviews in Food Science and Nutrition*. https://doi.org/10.1080/10408398.2022.2140121

Bourke, P., Ziuzina, D., Boehm, D., Cullen, P. J., Keener, K. (2018). The potential of cold plasma for safe and sustainable food production. *Trends in Biotechnology*, *36*(6), 615–626. https://doi.org/10.1016/j.tibtech.2017.11.001

Bußler, S., Ehlbeck, J., Schlüter, O. K. (2017). Pre-drying treatment of plant related tissues using plasma processed air: impact on enzyme activity and quality attributes of cut apple and potato. *Innovative Food Science & Emerging Technologies*, *40*, 78–86.

Ceuppens, S., Johannessen, G. S., Allende, A., Tondo, E. C., El-Tahan, F., Sampers, I., Uyttendaele, M. (2015). Risk factors for Salmonella, shiga toxin-producing Escherichia coli and Campylobacter occurrence in primary production of leafy greens and strawberries. *International Journal of Environmental Research and Public Health, 12*(8), 9809–9831. https://doi.org/10.3390/ijerph120809809

Chen, F. (2015). *Introduction to plasma physics and controlled fusion*. Springer International Publishing. https://books.google.ca/books?id=mFg-CwAAQBAJ

Chen, Y. Q., Cheng, J. H., Sun, D. W. (2020). Chemical, physical and physiological quality attributes of fruit and vegetables induced by cold plasma treatment: mechanisms and application advances. *Critical Reviews in Food Science and Nutrition, 60*(16), 2676–2690. https://doi.org/10.1080/10408398.2019.1654429

Cheng, C., Liu, P., Xu, L., Zhang, L. Y., Zhan, R. J., Zhang, W. R. (2006). Development of a new atmospheric pressure cold plasma jet generator and application in sterilization. *Chinese Physics, 15*(7), 1544–1548. https://doi.org/10.1088/1009–1963/15/7/028

Cong, L., Huang, M., Zhang, J., Yan, W. (2021). Effect of dielectric barrier discharge plasma on the degradation of malathion and chlorpyrifos on lettuce. *Journal of the Science of Food and Agriculture, 101*(2), 424–432. https://doi.org/10.1002/jsfa.10651

Critzer, F. J., Kelly-Wintenberg, K., South, S. L., Golden, D. A. (2007). Atmospheric plasma inactivation of foodborne pathogens on fresh produce surfaces. *Journal of Food Protection, 70*(10), 2290–2296. https://doi.org/10.4315/0362-028x-70.10.2290

Cui, D. J., Yin, Y., Sun, H., Wang, X. J., Zhuang, J., Wang, L., Jiao, Z. (2022). Regulation of cellular redox homeostasis in *Arabidopsis thaliana* seedling by atmospheric pressure cold plasma-generated reactive oxygen/nitrogen species. *Ecotoxicology and Environmental Safety, 240* https://doi.org/10.1016/j.ecoenv.2022.113703

Dantas, A. M., Batista, J. D. F., dos Santos Lima, M., Fernandes, F. A., Rodrigues, S., Magnani, M., Borges, G. D. S. C. (2021). Effect of cold plasma on açai pulp: enzymatic activity, color and bioaccessibility of phenolic compounds. *LWT, 149*, 111883. https://doi.org/10.1016/j.lwt.2021.111883

Di, L., Zhang, J., Zhang, X. (2018). A review on the recent progress, challenges, and perspectives of atmospheric-pressure cold plasma for preparation of supported metal catalysts. *Plasma Processes and Polymers, 15*(5), 1700234. https://doi.org/10.1002/ppap.201700234

Ding, T., Cullen, P. J., Yan, W. (2022). *Applications of cold plasma in food safety*. Springer.

Dominguez, R., Gullon, P., Pateiro, M., Munekata, P. E. S., Zhang, W. G., Lorenzo, J. M. (2020). Tomato as potential source of natural additives for meat industry. A review. *Antioxidants, 9*(1). https://doi.org/10.3390/antiox9010073

Domonkos, M., Tichá, P., Trejbal, J., Demo, P. (2021). Applications of cold atmospheric pressure plasma technology in medicine, agriculture and food industry. *Applied Sciences, 11*(11).

Dong, X. Y., Yang, Y. L. (2019). A novel approach to enhance blueberry quality during storage using cold plasma at atmospheric air pressure. *Food and Bioprocess Technology, 12*(8), 1409–1421. https://doi.org/10.1007/s11947-019-02305-y

Downard, K. M. (2009). J. J. Thomson goes to America. *Journal of the American Society for Mass Spectrometry, 20*(11), 1964–1973. https://doi.org/10.1016/j.jasms.2009.07.008

Fälthammar, C.-G. (2007). The discovery of magnetohydrodynamic waves. *Journal of Atmospheric and Solar-Terrestrial Physics, 69*(14), 1604–1608. https://doi.org/10.1016/j.jastp.2006.08.021

Fan, Z. Q., Zhong, J. Y., Li, Z. W., Zheng, Y. C., Wang, Z. Z., Bai, S. P. (2021). Inactivation of *Escherichia coli* using atmospheric pressure cold plasma jet with thin quartz tubes. *Journal of Physics D-Applied Physics, 54*(45). doi:ARTN 455204 10.1088/1361–6463/ac1d6f

FAO. (2018). *Guide: preventing post-harvest losses in the apple supply chain in Lebanon* (Vol. 1). FAO.

Farias, T. R. B., Filho, E. G. A., Silva, L. M. A., De Brito, E. S., Rodrigues, S., Fernandes, F. A. N. (2021). NMR evaluation of apple cubes and apple juice composition subjected to two cold plasma technologies. *LWT-Food Science and Technology, 150*. doi: https://doi.org/10.1016/j.lwt.2021.112062

Farias, T. R., Rodrigues, S., Fernandes, F. A. (2022). Comparative study of two cold plasma technologies on apple juice antioxidant capacity, phenolic contents, and enzymatic activity. *Journal of Food Processing and Preservation, 46*(10), e16871 (Surowsky, B., Fischer, A., Schlueter, O., & Knorr, D. (2013). Cold plasma effects on enzyme activity in a model food system. *Innovative Food Science & Emerging Technologies, 19*, 146–152. https://doi.org/10.1016/j.ifset.2013.04.002)

Feizollahi, E., Misra, N. N., Roopesh, M. S. (2021). Factors influencing the antimicrobial efficacy of dielectric barrier discharge (DBD) atmospheric cold plasma (ACP) in food processing applications. *Critical Reviews in Food Science and Nutrition, 61*(4), 666–689. https://doi.org/10.1080/10408398.2020.1743967

Fitzpatrick, R. (2022). *Plasma physics: an introduction*. CRC Press. https://books.google.ca/books?id=8RmbEAAAQBAJ

Gates, D. A. (2018). Plasma: an international open access journal for all of plasma science. *Plasma, 1*(1), 45–46.

Gavahian, M., Chu, Y. H., Khaneghah, A. M., Barba, F. J., Misra, N. N. (2018). A critical analysis of the cold plasma induced lipid oxidation in foods. *Trends in Food Science & Technology, 77*, 32–41. https://doi.org/10.1016/j.tifs.2018.04.009

Gavahian, M., Khaneghah, A. M. (2020). Cold plasma as a tool for the elimination of food contaminants: recent advances and future trends. *Critical Reviews in Food Science and Nutrition, 60*(9), 1581–1592. https://doi.org/10.1080/10408398.2019.1584600

Gibbon, P. (2020). Introduction to plasma physics. *arXiv preprint arXiv:2007.04783*.

Golovinskaia, O., Wang, C. K. (2021). Review of functional and pharmacological activities of berries. *Molecules, 26*(13). https://doi.org/10.3390/molecules26133904

Gracy, T. K. R., Sharanyakanth, P. S., Radhakrishnan, M. (2022). Non-thermal technologies: solution for hazardous pesticides reduction in fruits and vegetables. *Critical Reviews in Food Science and Nutrition, 62*(7), 1782–1799. https://doi.org/10.1080/10408398.2020.1847029

Han, L., Patil, S., Boehm, D., Milosavljević, V., Cullen, P. J., Bourke, P. (2016). Mechanisms of inactivation by high-voltage atmospheric cold plasma differ for escherichia coli and staphylococcus aureus. *Applied and Environmental Microbiology, 82*(2), 450–458. https://doi.org/doi:10.1128/AEM.02660-15

Han, Y. X., Cheng, J. H., Sun, D. W. (2019). Activities and conformation changes of food enzymes induced by cold plasma: a review. *Critical Reviews in Food Science and Nutrition, 59*(5), 794–811. https://doi.org/10.1080/10408398.2018.1555131

Hashizume, H., Kitano, H., Mizuno, H., Abe, A., Yuasa, G., Tohno, S., Tanaka, H., Ishikawa, K., Matsumoto, S., Sakakibara, H., Nikawa, S., Maeshima, M., Mizuno, M., Hori, M. (2021). Improvement of yield and grain quality by periodic cold plasma treatment with rice plants in a paddy field *Plasma Processes and Polymers, 18*(1), 2000181. https://doi.org/10.1002/ppap.202000181

Herman, K. M., Hall, A. J., Gould, L. H. (2015). Outbreaks attributed to fresh leafy vegetables, United States, 1973–2012. *Epidemiology & Infection, 143*(14), 3011–3021. https://doi.org/10.1017/S0950268815000047

Hernández-Torres, C. J., Reyes-Acosta, Y. K., Chávez-González, M. L., Dávila-Medina, M. D., Verma, D. K., Martínez-Hernández, J. L., Aguilar, C. N. (2022). Recent trends and technological development in plasma as an emerging and promising technology for food biosystems. *Saudi Journal of Biological Sciences*, *29*(4), 1957–1980. https://doi.org/10.1016/j.sjbs.2021.12.023

Hosseini, S. M., Samani, B. H., Rostami, S., Lorigooini, Z., Gavahian, M., Barba, F. J. (2021). Design and characterisation of jet cold atmospheric pressure plasma and its effect on Escherichia coli, colour, pH, and bioactive compounds of sour cherry juice. *International Journal of Food Science and Technology*, *56*(10), 4883–4892. https://doi.org/10.1111/ijfs.15220

Huynh, N. K., Wilson, M. D., Eyles, A., Stanley, R. A. (2019). Recent advances in postharvest technologies to extend the shelf life of blueberries (Vaccinium sp.), raspberries (*Rubus idaeus* L.) and blackberries (Rubus sp.). *Journal of Berry Research*, *9*(4), 687–707. https://doi.org/10.3233/JBR-190421

Ji, Y., Hu, W., Liao, J., Jiang, A., Xiu, Z., Gaowa, S., Liu, C. (2020). Effect of atmospheric cold plasma treatment on antioxidant activities and reactive oxygen species production in postharvest blueberries during storage. *Journal of the Science of Food and Agriculture*, *100*(15), 5586–5595. https://doi.org/10.1002/jsfa.10611

Jia, S. T., Zhang, N., Ji, H. P., Zhang, X. J., Dong, C. H., Yu, J. Z., Liang, L. Y. (2022). Effects of atmospheric cold plasma treatment on the storage quality and chlorophyll metabolism of postharvest tomato. *Foods*, *11*(24). https://doi.org/10.3390/foods11244088

Jiang, J. F., Li, J. G., Dong, Y. H. (2018). Effect of cold plasma treatment on seedling growth and nutrient absorption of tomato. *Plasma Science & Technology*, *20*(4). ARTN 044007. https://doi.org/10.1088/2058–6272/aaa0bf

Jin, T., Dai, C., Xu, Y., Chen, Y., Xu, Q., Wu, Z. (2022). Applying cold atmospheric plasma to preserve the postharvest qualities of winter jujube (*Ziziphus jujuba* Mill. cv. Dongzao) during cold storage [Original Research]. *Frontiers in Nutrition*, *9*. https://doi.org/10.3389/fnut.2022.934841

Kang, J. H., Roh, S. H., Min, S. C. (2019). Inactivation of potato polyphenol oxidase using microwave cold plasma treatment. *Journal of Food Science*, *84*(5), 1122–1128. https://doi.org/10.1111/1750-3841.14601

Karim, N., Shishir, M. R. I., Bao, T., Chen, W. (2021). Effect of cold plasma pretreated hot-air drying on the physicochemical characteristics, nutritional values and antioxidant activity of shiitake mushroom. *Journal of the Science of Food and Agriculture*, *101*(15), 6271–6280. https://doi.org/10.1002/jsfa.11296

Karuppanapandian, T., Moon, J.C., Kim, C., Manoharan, K., Kim, W. (2011). Reactive oxygen species in plants: their generation, signal transduction, and scavenging mechanisms [other journal article]. *Australian Journal of Crop Science*, *5*(6), 709–725. https://search.informit.org/doi/10.3316/informit.282079847301776

Kashfi, A. S., Ramezan, Y., Khani, M. R. (2020). Simultaneous study of the antioxidant activity, microbial decontamination and color of dried peppermint (Mentha piperita L.) using low pressure cold plasma. *LWT-Food Science and Technology*, *123*. https://doi.org/10.1016/j.lwt.2020.109121

Keidar, M., Weltmann, K. D., Macheret, S. (2021). Fundamentals and applications of atmospheric pressure plasmas. *Journal of Applied Physics*, *130(8)*. https://doi.org/10.1063/5.0065750

Kim, J. H., Min, S. C. (2017). Microwave-powered cold plasma treatment for improving microbiological safety of cherry tomato against Salmonella. *Postharvest Biology and Technology*, *127*, 21–26. https://doi.org/10.1016/j.postharvbio.2017.01.001

Keener, K. M., Misra, N. N. (2016). Chapter 14–future of cold plasma in food processing. In N. N. Misra, O. Schlüter, & P. J. Cullen (Eds.), *Cold plasma in food and agriculture* (pp. 343–360). Academic Press. https://doi.org/10.1016/B978-0-12-801365-6.00014-7

Klockow, P. A., Keener, K. M. (2008). Quality and safety assessment of packaged spinach treated with a novel atmospheric, non-equilibrium plasma system. In *2008 Providence, Rhode Island, June 29–July 2, 2008* (p. 1). American Society of Agricultural and Biological Engineers. https://doi.org/10.13031/2013.25061

Kyere, E. O., Foong, G., Palmer, J., Wargent, J. J., Fletcher, G. C., Flint, S. (2020). Biofilm formation of Listeria monocytogenes in hydroponic and soil grown lettuce leaf extracts on stainless steel coupons. *LWT*, *126*, 109114. https://doi.org/10.1016/j.lwt.2020.109114

Lacombe, A., Niemira, B. A., Gurtler, J. B., Fan, X. T., Sites, J., Boyd, G., Chen, H. Q. (2015). Atmospheric cold plasma inactivation of aerobic microorganisms on blueberries and effects on quality attributes. *Food Microbiology*, *46*, 479–484. https://doi.org/10.1016/j.fm.2014.09.010

Lacombe, A., Niemira, B. A., Gurtler, J. B., Sites, J., Boyd, G., Kingsley, D. H., Chen, H. Q. (2017). Nonthermal inactivation of norovirus surrogates on blueberries using atmospheric cold plasma. *Food Microbiology*, *63*, 1–5. https://doi.org/10.1016/j.fm.2016.10.030

Laroque, D. A., Seó, S. T., Valencia, G. A., Laurindo, J. B., Carciofi, B. A. M. (2022). Cold plasma in food processing: design, mechanisms, and application. *Journal of Food Engineering*, *312*, 110748. https://doi.org/10.1016/j.jfoodeng.2021.110748

Laroussi, M. (2020). Cold plasma in medicine and healthcare: the new frontier in low temperature plasma applications [mini review]. *Frontiers in Physics*, *8*. https://doi.org/10.3389/fphy.2020.00074

Lee, T., Puligundla, P., Mok, C. (2018). Intermittent corona discharge plasma jet for improving tomato quality. *Journal of Food Engineering*, *223*, 168–174. https://doi.org/10.1016/j.jfoodeng.2017.11.004

Li, M. L., Li, X. A., Han, C., Ji, N. N., Jin, P., Zheng, Y. H. (2019). Physiological and metabolomic analysis of cold plasma treated fresh-cut strawberries. *Journal of Agricultural and Food Chemistry*, *67*(14), 4043–4053. https://doi.org/10.1021/acs.jafc.9b00656

Li, X., Li, M., Ji, N., Jin, P., Zhang, J., Zheng, Y., Li, F. (2019). Cold plasma treatment induces phenolic accumulation and enhances antioxidant activity in fresh-cut pitaya (*Hylocereus undatus*) fruit. *LWT*, *115*, 108447. https://doi.org/10.1016/j.lwt.2019.108447

Liao, X., Li, J., Muhammad, A. I., Suo, Y., Chen, S., Ye, X., Ding, T. (2018). Application of a dielectric barrier discharge atmospheric cold plasma (Dbd-Acp) for Eshcerichia coli inactivation in apple juice. *Journal of Food Science*, *83*(2), 401–408. https://doi.org/10.1111/1750-3841.14045

Limnaios, A., Pathak, N., Bovi, G. G., Frohling, A., Valdramidis, V. P., Taoukis, P. S., Schluter, O. (2021). Effect of cold atmospheric pressure plasma processing on quality and shelf life of red currants. *LWT-Food Science and Technology*, *151*, ARTN 112213. https://doi.org/10.1016/j.lwt.2021.112213

Liu, C., Hofstra, N., Franz, E. (2013). Impacts of climate change on the microbial safety of pre-harvest leafy green vegetables as indicated by Escherichia coli O157 and Salmonella spp. *International Journal of Food Microbiology*, *163*(2–3), 119–128. https://doi.org/10.1016/j.ijfoodmicro.2013.02.026

López-Gámez, G., Elez-Martínez, P., Martin-Belloso, O., Soliva-Fortuny, R. (2021). Enhancing carotenoid and phenolic contents in plant food matrices by applying non-thermal technologies: bioproduction vs improved extractability. *Trends in Food*

Science & Technology, *112*, 622–630. https://doi.org/10.1016/j.tifs.2021.04.022

Lu, H., Patil, S., Keener, K. M., Cullen, P. J., Bourke, P. (2014). Bacterial inactivation by high-voltage atmospheric cold plasma: influence of process parameters and effects on cell leakage and DNA. *Journal of Applied Microbiology*, *116*(4), 784–794. https://doi.org/10.1111/jam.12426

Mahbubeh Mousavi, S., Imani, S., Dorranian, D., Larijani, K., Shojaee, M. (2017). Effect of cold plasma on degradation of organophosphorus pesticides used on some agricultural products. *Journal of Plant Protection Research*, *57*(1). https://doi.org/10.1515/jppr-2017-0004

Mahnot, N. K., Mahanta, C. L., Farkas, B. E., Keener, K. M., Misra, N. N. (2019). Atmospheric cold plasma inactivation of Escherichia coli and Listeria monocytogenes in tender coconut water: inoculation and accelerated shelf-life studies. *Food Control*, *106*. https://doi.org/10.1016/j.foodcont.2019.06.004

Mao, L. L., Mhaske, P., Zing, X., Kasapis, S., Majzoobi, M., Farahnaky, A. (2021). Cold plasma: microbial inactivation and effects on quality attributes of fresh and minimally processed fruits and ready-to-eat vegetables. *Trends in Food Science & Technology*, *116*, 146–175. https://doi.org/10.1016/j.tifs.2021.07.002

Mayookha, V. P., Pandiselvam, R., Kothakota, A., Ishwarya, S. P., Khanashyam, A. C., Kutlu, N., Abd El-Maksoud, A. A. (2022). Ozone and cold plasma: emerging oxidation technologies for inactivation of enzymes in fruits, vegetables, and fruit juices. *Food Control*,109399. https://doi.org/10.1016/j.foodcont.2022.109399

Meade, D. (2010). 50 years of fusion research. *Nuclear Fusion*, *50*(1), 014004. https://doi.org/10.1088/0029-5515/50/1/014004

Medvecká, V., Mošovská, S., Mikulajová, A., Valík, Ľ., Zahoranová, A. (2020). Cold atmospheric pressure plasma decontamination of allspice berries and effect on qualitative characteristics. *European Food Research and Technology*, *246*, 2215–2223. https://doi.org/10.1007/s00217-020-03566-0

Min, S. C., Roh, S. H., Niemira, B. A., Boyd, G., Sites, J. E., Fan, X. T., . . . Jin, T. Z. (2018). In-package atmospheric cold plasma treatment of bulk grape tomatoes for microbiological safety and preservation. *Food Research International*, *108*, 378–386. https://doi.org/10.1016/j.foodres.2018.03.033

Min, S. C., Roh, S. H., Niemira, B. A., Boyd, G., Sites, J. E., Uknalis, J., Fan, X. T. (2017). In-package inhibition of E. coli O157:H7 on bulk Romaine lettuce using cold plasma. *Food Microbiology*, *65*, 1–6. https://doi.org/10.1016/j.fm.2017.01.010

Misra, N. N., Brijesh kumar, T., Raghavarao, K., Cullen, P. J. (2011). Nonthermal plasma inactivation of food-borne pathogens. *Food Engineering Reviews*,*3*, 159–170. https://doi.org/10.1007/s12393-011-9041-9

Misra, N. N., Keener, K. M., Bourke, P., Mosnier, J. P., Cullen, P. J. (2014). In-package atmospheric pressure cold plasma treatment of cherry tomatoes. *Journal of Bioscience and Bioengineering*, *118*(2), 177–182. https://doi.org/10.1016/j.jbiosc.2014.02.005

Misra, N. N., Yadav, B., Roopesh, M. S., Jo, C. (2019). Cold plasma for effective fungal and mycotoxin control in foods: mechanisms, inactivation effects, and applications. *Comprehensive Reviews in Food Science and Food Safety*, *18*(1), 106–120. https://doi.org/10.1111/1541-4337.12398

Munekata, P. E. S., Dominguez, R., Pateiro, M., Lorenzo, J. M. (2020). Influence of plasma treatment on the polyphenols of food products-a review. *Foods*, *9*(7). https://doi.org/10.3390/foods9070929

Niemira, B. A. (2012). Cold plasma reduction of Salmonella and Escherichia coli O157:H7 on almonds using ambient pressure gases. *Journal of Food Science*, *77*(3), M171–M175. https://doi.org/10.1111/j.1750-3841.2011.02594.x

Niemira, B. A., Sites, J. (2008). Cold plasma inactivates Salmonella stanley and Escherichia coli O157: H7 inoculated on Golden Delicious apples. *Journal of Food Protection*, *71*(7), 1357–1365. doi: https://doi.org/10.4315/0362-028X-71.7.1357

Niveditha, A., Pandiselvam, R., Prasath, V. A., Singh, S. K., Gul, K., Kothakota, A. (2021). Application of cold plasma and ozone technology for decontamination of Escherichia coli in foods—A review. *Food Control*, *130*. https://doi.org/10.1016/j.foodhyd.2014.11.010

Ouf, S. A., Basher, A. H., Mohamed, A. A. H. (2015). Inhibitory effect of double atmospheric pressure argon cold plasma on spores and mycotoxin production of *Aspergillus niger* contaminating date palm fruits. *Journal of the Science of Food and Agriculture*, *95*(15), 3204–3210. https://doi.org/10.1002/jsfa.7060

Palumbo, M., Attolico, G., Capozzi, V., Cozzolino, R., Corvino, A., de Chiara, M. L. V., Pace, B., Pelosi, S., Ricci, I., Romaniello, R., Cefola, M. (2022). Emerging postharvest technologies to enhance the shelf-life of fruit and vegetables: an overview. *Foods*, *11*(23). https://doi.org/10.3390/foods11233925

Pan, Y., Cheng, J.H., Sun, D.W. (2019). Cold plasma-mediated treatments for shelf life extension of fresh produce: a review of recent research developments. *Comprehensive Reviews in Food Science and Food Safety*, *18*(5), 1312–1326. https://doi.org/10.1111/1541-4337.12474

Pant, K., Thakur, M., Nanda, V. (2022). Application of cold plasma techniques for the fruit and vegetable processing industry. In *Non-thermal processing technologies for the fruit and vegetable industry* (pp. 33–56). CRC Press.

Pathak, N., Bovi, G. G., Limnaios, A., Frohling, A., Brincat, J. P., Taoukis, P., Schluter, O. (2020). Impact of cold atmospheric pressure plasma processing on storage of blueberries. *Journal of Food Processing and Preservation*, *44*(8). https://doi.org/10.1111/jfpp.14581

Pedrow, P., Hua, Z., Xie, S., Zhu, M. J. (2020). Engineering principles of cold plasma. In *Advances in cold plasma applications for food safety and preservation* (pp. 3–48). Academic Press. https://doi.org/10.1016/B978-0-12-814921-8.00001-3

Perinban, S., Orsat, V., Lyew, D., Raghavan, V. (2022). Effect of plasma activated water on disinfection and quality of kale and spinach. *Food Chemistry, 397*. https://doi.org/10.1016/j.foodchem.2022.133793

Perni, S., Liu, D. W., Shama, G., Kong, M. G. (2008). Cold atmospheric plasma decontamination of the pericarps of fruit. *Journal of Food Protection, 71*(2), 302–308. https://doi.org/10.4315/0362-028x-71.2.302

Phan, K. T. K., Phan, H. T., Boonyawan, D., Intipunya, P., Brennan, C. S., Regenstein, J. M., Phimolsiripol, Y. (2018). Nonthermal plasma for elimination of pesticide residues in mango. *Innovative Food Science & Emerging Technologies*, *48*, 164–171.

Pipliya, S., Kumar, S., Srivastav, P. P. (2022). Inactivation kinetics of polyphenol oxidase and peroxidase in pineapple juice by dielectric barrier discharge plasma technology. *Innovative Food Science & Emerging Technologies*, *80*. https://doi.org/10.1016/j.ifset.2022.103081

Pour, A. K., Khorram, S., Ehsani, A., Ostadrahimi, A., Ghasempour, Z. (2022). Atmospheric cold plasma effect on quality attributes of banana slices: its potential use in blanching process. *Innovative Food Science & Emerging Technologies*, *76*, ARTN 102945. https://doi.org/10.1016/j.ifset.2022.102945

Prasad, P., Mehta, D., Bansal, V., Sangwan, R. S. (2017). Effect of atmospheric cold plasma (ACP) with its extended storage on the inactivation of Escherichia coli inoculated on tomato. *Food Research International*, *102*, 402–408. https://doi.org/10.1016/j.foodres.2017.09.030

Puač, N., Gherardi, M., Shiratani, M. (2018). Plasma agriculture: a rapidly emerging field. *Plasma Processes and Polymers*, *15*(2), 1700174. https://doi.org/10.1002/ppap.201700174

Ramazzina, I., Tappi, S., Rocculi, P., Sacchetti, G., Berardinelli, A., Marseglia, A., Rizzi, F. (2016). Effect of cold plasma treatment on the functional properties of fresh-cut apples. *Journal of Agricultural and Food Chemistry*, *64*(42), 8010–8018. https://doi.org/10.1021/acs.jafc.6b02730

Rana, S., Mehta, D., Bansal, V., Shivhare, U. S., Yadav, S. K. (2020). Atmospheric cold plasma (ACP) treatment improved in-package shelf-life of strawberry fruit. *Journal of Food Science and Technology-Mysore*, *57*(1), 102–112. https://doi.org/10.1007/s13197-019-04035-7

Samal, S. (2017). Thermal plasma technology: the prospective future in material processing. *Journal of Cleaner Production*, *142*, 3131–3150. https://doi.org/10.1016/j.jclepro.2016.10.154

Sandanuwan, T., Attygalle, D., Amarasinghe, S., Weragoda, S. C., Ranaweera, B., Rathnayake, K., Alankara, W. (2020). Shelf life extension of cavendish banana fruit using cold plasma treatment. *Mercon 2020: 6th International Multidisciplinary Moratuwa Engineering Research Conference (Mercon)*, 182–186. Retrieved from <Go to ISI>://WOS:000788397000031 https://doi.org/10.1080/07373937.2019.1683860

Sarangapani, C., O'Toole, G., Cullen, P.J., Bourke, P. (2017). Atmospheric cold plasma dissipation efficiency of agrochemicals on blueberries. *Innovative Food Science & Emerging Technologies*, *44*, 235–241. https://doi.org/10.1016/j.ifset.2017.02.012

Saremnezhad, S., Soltani, M., Faraji, A., Hayaloglu, A. A. (2021). Chemical changes of food constituents during cold plasma processing: a review. *Food Research International*, *147*. https://doi.org/10.1016/j.foodres.2021.110552

Shah, U., Ranieri, P., Zhou, Y., Schauer, C. L., Miller, V., Fridman, G., Sekhon, J. K. (2019). Effects of cold plasma treatments on spot-inoculated Escherichia coli O157: H7 and quality of baby kale (Brassica oleracea) leaves. *Innovative Food Science & Emerging Technologies*, *57*, 102104. https://doi.org/10.1016/j.ifset.2018.12.010

Sharma, A., Collins, G., Pruden, A. (2009). Differential gene expression in Escherichia coli following exposure to nonthermal atmospheric pressure plasma. *Journal of Applied Microbiology*, *107*(5), 1440–1449.

Shen, J., Cheng, C., Xu, Z., Lan, Y., Ni, G., Sui, S. (2022). Principles and characteristics of cold plasma at gas phase and gas-liquid phase. *Applications of Cold Plasma in Food Safety*, 1–36.

Shishir, M. R. I., Karim, N., Bao, T., Gowd, V., Ding, T., Sun, C. D., Chen, W. (2020). Cold plasma pretreatment—a novel approach to improve the hot air drying characteristics, kinetic parameters, and nutritional attributes of shiitake mushroom. *Drying Technology*, *38*(16), 2134–2150. https://doi.org/10.1080/07373937.2019.1683860

Siddique, S. S., Hardy, G. E. S., Bayliss, K. L. (2018). Cold plasma: a potential new method to manage postharvest diseases caused by fungal plant pathogens. *Plant Pathology*, *67*(5), 1011–1021. https://doi.org/10.1111/ppa.12825

Simmons, A. (2012). 11—Future trends for the sterilisation of biomaterials and medical devices. In S. Lerouge & A. Simmons (Eds.), *Sterilisation of biomaterials and medical devices* (pp. 310–320). Woodhead Publishing. https://doi.org/10.1533/9780857096265.310

Sirgedaitė-Šėžienė, V., Mildažienė, V., Žemaitis, P., Ivankov, A., Koga, K., Shiratani, M., Baliuckas, V. (2021). Long-term response of Norway spruce to seed treatment with cold plasma: dependence of the effects on the genotype. *Plasma Processes and Polymers*, *18*(1), 2000159. https://doi.org/10.1002/ppap.202000159

Smirnov, B. M. (2011). General concepts in physics of excited and ionized gases. In *Fundamentals of ionized gases* (pp. 1–52). https://doi.org/10.1002/9783527637102.ch1

Song, A. Y., Oh, Y. J., Kim, J. E., Bin Song, K., Oh, D. H., Min, S. C. (2015). Cold plasma treatment for microbial safety and preservation of fresh lettuce. *Food Science and Biotechnology*, *24*(5), 1717–1724. https://doi.org/10.1007/s10068-015-0223-8

Sousa-Gallagher, M. J., Tank, A., Sousa, R. (2016). Emerging technologies to extend the shelf life and stability of fruits and vegetables. In P. Subramaniam (Ed.), *Stability and shelf life of food* (2nd ed., Vol. 297, pp. 399–430). https://doi.org/10.1016/b978-0-08-100435-7.00014-9

Surowsky, B., Fischer, A., Schlueter, O., Knorr, D. (2013). Cold plasma effects on enzyme activity in a model food system. *Innovative Food Science & Emerging Technologies*, *19*, 146–152. https://doi.org/10.1016/j.ifset.2013.04.002

Surowsky, B., Fröhling, A., Gottschalk, N., Schlüter, O., Knorr, D. (2014). Impact of cold plasma on *Citrobacter freundii* in apple juice: inactivation kinetics and mechanisms. *International Journal of Food Microbiology*, *174*, 63–71. https://doi.org/10.1016/j.ijfoodmicro.2013.12.031

Takai, E., Kitano, K., Kuwabara, J., Shiraki, K. (2012). Protein inactivation by low-temperature atmospheric pressure plasma in aqueous solution. *Plasma Processes and Polymers*, *9*(1), 77–82.

Tappi, S., Ramazzina, I., Rizzi, F., Sacchetti, G., Ragni, L., Rocculi, P. (2018). Effect of plasma exposure time on the polyphenolic profile and antioxidant activity of fresh-cut apples. *Applied Sciences-Basel*, *8*(10). https://doi.org/10.3390/app8101939

Tendero, C., Tixier, C., Tristant, P., Desmaison, J., Leprince, P. (2006). Atmospheric pressure plasmas: a review. *Spectrochimica Acta Part B-Atomic Spectroscopy*, *61*(1), 2–30. https://doi.org/10.1016/j.sab.2005.10.003

Timmons, C., Pai, K., Jacob, J., Zhang, G., Ma, L. M. (2018). Inactivation of Salmonella enterica, Shiga toxin-producing Escherichia coli, and Listeria monocytogenes by a novel surface discharge cold plasma design. *Food Control*, *84*, 455–462. https://doi.org/10.1016/j.foodcont.2017.09.007

Tonks, L. (1967). The birth of "Plasma". *American Journal of Physics*, *35*(9), 857–858. https://doi.org/10.1119/1.1974266

Udtachee, K., Yatongchai, C., Sarapirom, S. (2023). Utilization of DBD plasma in shelf-life extension for Namwa banana. *Journal of Physics: Conference Series*, *2431*(1), 012033. https://doi.org/10.1088/1742-6596/2431/1/012033

Umair, M., Jabbar, S., Ayub, Z., Aadil, R. M., Abid, M., Zhang, J. H., Zhao, L. Q. (2022). Recent advances in plasma technology: influence of atmospheric cold plasma on spore inactivation. *Food Reviews International*, *38*, 789–811. doi:10.1080/87559129.2021.1888972

Wang, J., Yu, Y. D., Zhang, Z. G., Wu, W. C., Sun, P. L., Cai, M., Yang, K. (2022). Formation of sweet potato starch nanoparticles by ultrasonic-assisted nanoprecipitation: effect of cold plasma treatment. *Frontiers in Bioengineering and Biotechnology*, *10*, 986033. https://doi.org/10.3389/fbioe.2022.986033

Wang, R. X., Nian, W. F., Wu, H. Y., Feng, H. Q., Zhang, K., Zhang, J., Zhu, W. D., Becker, K. H., Fang, J. (2012). Atmospheric-pressure cold plasma treatment of contaminated fresh fruit and vegetable slices: inactivation and physiochemical properties evaluation. *European Physical Journal D*, *66*(10), Article 276. https://doi.org/10.1140/epjd/e2012-30053-1

Wang, Z. W., Jia, H., Yang, J. Y., Hu, Z. Q., Wang, Z. L., Yue, T. L., Yuan, Y. H. (2023). Inactivation of *Alicyclobacillus contaminans* in apple juice by dielectric barrier discharge plasma. *Food Control*, *146*. https://doi.org/10.1016/j.foodcont.2022.109475

Warne, G. R., Williams, P. M., Pho, H. Q., Tran, N. N., Hessel, V., Fisk, I. D. (2021). Impact of cold plasma on the biomolecules and organoleptic properties of foods: a review. *Journal of Food Science*, *86*(9), 3762–3777. https://doi.org/10.1111/1750-3841.15856

Wiesemann, K. (2014). A short introduction to plasma physics. *arXiv preprint arXiv:1404.0509*.

Wiktor, A., Hrycak, B., Jasiński, M., Rybak, K., Kieliszek, M., Kraśniewska, K., Witrowa-Rajchert, D. (2020). Impact of atmospheric pressure microwave plasma treatment on quality of selected spices. *Applied Sciences*, *10*(19), 6815. https://doi.org/10.3390/app10196815

Won, M. Y., Lee, S. J., Min, S. C. (2017). Mandarin preservation by microwave-powered cold plasma treatment. *Innovative Food Science & Emerging Technologies*, *39*, 25–32. https://doi.org/10.1016/j.ifset.2016.10.021

Wu, Y., Cheng, J. H., Keener, K. M., Sun, D. W. (2023). Inhibitory effects of dielectric barrier discharge cold plasma on pathogenic enzymes and anthracnose for mango postharvest preservation. *Postharvest Biology and Technology*, *196*, 112181. https://doi.org/10.1016/j.postharvbio.2022.112181

Xiang, Q. S., Huangfu, L. L., Dong, S. S., Ma, Y. F., Li, K., Niu, L. Y., Bai, Y. H. (2021). Feasibility of atmospheric cold plasma for the elimination of food hazards: recent advances and future trends. *Critical Reviews in Food Science and Nutrition*. https://doi.org/10.1080/10408398.2021.2002257

Yi, F., Wang, J., Xiang, Y., Yun, Z., Pan, Y., Jiang, Y., Zhang, Z. (2022). Physiological and quality changes in fresh-cut mango fruit as influenced by cold plasma. *Postharvest Biology and Technology*, *194*, 112105. https://doi.org/10.1016/j.postharvbio.2022.112105

Yodpitak, S., Mahatheeranont, S., Boonyawan, D., Sookwong, P., Roytrakul, S., Norkaew, O. (2019). Cold plasma treatment to improve germination and enhance the bioactive phytochemical content of germinated brown rice. *Food Chemistry*, *289*, 328–339. https://doi.org/10.1016/j.foodchem.2019.03.061

Yu, Q., Fan, L. P. (2021). Improving the bioactive ingredients and functions of asparagus from efficient to emerging processing technologies: a review. *Food Chemistry*, *358*, Article 129903. https://doi.org/10.1016/j.foodchem.2021.129903

Zhang, Z. H., Wang, L. H., Zeng, X. A., Han, Z., Brennan, C. S. (2019). Non-thermal technologies and its current and future application in the food industry: a review. *International Journal of Food Science and Technology*, *54*(1), 1–13. https://doi.org/10.1111/ijfs.13903

Zheng, Y., Zhu, Y. F., Zheng, Y. H., Hu, J. J., Chen, J., Deng, S. G. (2022). The effect of dielectric barrier discharge plasma gas and plasma-activated water on the physicochemical changes in button mushrooms (*Agaricus bisporus*). *Foods*, *11*(21), Article 3504. https://doi.org/10.3390/foods11213504

Zhou, D. D., Sun, R., Zhu, W. X., Shi, Y. L., Ni, S. J., Wu, C. E., Li, T. T. (2023). Impact of dielectric barrier discharge cold plasma on the quality and phenolic metabolism in blueberries based on metabonomic analysis. *Postharvest Biology and Technology*, *197*. https://doi.org/10.1016/j.postharvbio.2022.112208

Zhou, Y. H., Vidyarthi, S. K., Zhong, C. S., Zheng, Z. A., An, Y., Wang, J., Xiao, H. W. (2020). Cold plasma enhances drying and color, rehydration ratio and polyphenols of wolfberry via microstructure and ultrastructure alteration. *LWT*, *134*, 110173. https://doi.org/10.1016/j.postharvbio.2022.112208

Ziuzina, D., Misra, N. N., Han, L., Cullen, P. J., Moiseev, T., Mosnier, J. P., Bourke, P. (2020). Investigation of a large gap cold plasma reactor for continuous in-package decontamination of fresh strawberries and spinach. *Innovative Food Science & Emerging Technologies*, *59*, 102229.

Ziuzina, D., Patil, S., Cullen, P. J., Keener, K. M., Bourke, P. (2014). Atmospheric cold plasma inactivation of Escherichia coli, Salmonella enterica serovar Typhimurium and Listeria monocytogenes inoculated on fresh produce. *Food Microbiology*, *42*, 109–116. https://doi.org/10.1016/j.fm.2014.02.007

Ultraviolet (UV) Light Technology for Postharvest Fruits and Vegetables

8

Muhammad Wasim Haider*, Muhammad Nafees, Mohammad Valipour, Habat Ullah Asad, and Romina Alina Marc

*Corresponding Author: wasim.haider@iub.edu.pk

8.1 INTRODUCTION

Food waste is one of the major postharvest problems impacting modern food systems (Santeramo & Lamonaca, 2021). A lot of food products, both edible and non-edible, are unnecessarily lost or wasted every year through inappropriate transportation and storage facilities, unexpected weather changes, incorrect handling, consumer behavior, or rejection in relation to market standards (Areche et al., 2022). Less food loss or waste means more food for everyone, less pollution, decreased greenhouse gas emissions, improved production levels, economic growth, and more sustainable communities (Ali et al., 2023; Elik et al., 2019). This can be attained by adopting cutting-edge, eco-friendly techniques to achieve the goal of preventing, reusing, and recycling food waste (De Simone et al., 2020).

New techniques are continuously emerging to minimize food waste and prolong the shelf life of perishable produce (Spada et al., 2018). Ultraviolet (UV) light, a non-ionizing radiation, has currently emerged as a non-chemical and non-thermal technique with promising characteristics to preserve the nutritional quality of freshly harvested commodities, including fruits and vegetables (Darré et al., 2022; Zhang et al., 2017; Usall et al., 2016). The use of UV treatment has been very common for water purification, surface decontamination, and air disinfection; however, its use is still limited in the field of postharvest food treatment (Prasad et al., 2020; Ribeiro & Alvarenga, 2012).

UV light possesses wavelengths within the spectrum of electromagnetic radiation, which is shorter than that of visible light (Singh et al., 2021). This method involves exposing the commodity to specific wavelengths within the range of 200–280 nm in the UV-C spectrum (Kaczmarek et al., 2019). UV light offers numerous benefits in the postharvest management of fresh produce in comparison to conventional methods. One of the notable aspects of this novel treatment is its complete avoidance of chemicals and reduction of any potential risk they may pose to humans and the environment (Kahramanoğlu et al., 2020). This makes UV light a desirable option for organic produce and for customers looking for non-chemical alternatives. Secondly, the use of UV light is an effective means to eliminate various microorganisms, even those that have become resistant to traditional antimicrobial treatments (Zhang et al., 2017; Palou et al., 2008).

The broad-spectrum effectiveness of UV light is crucial to preserve the quality of treated horticultural commodities throughout the supply chain (D'Souza et al., 2015). Furthermore, the UV light treatment procedure is handy and swift, and it can be seamlessly integrated into present postharvest practices without any modifications or disruptions (Ribeiro & Alvarenga, 2012).

UV light application has shown a remarkable reduction in microbial load and preserving quality and freshness in horticultural commodities (Ali et al., 2018). UV, in particular UV-C, has germicidal properties that can deactivate or kill different kinds of microbes, such as viruses, bacteria, fungi, and molds (Shahi et al., 2021; Deepshikha et al., 2017). Previous reports have indicated that this technique is efficient in minimizing spoilage microorganisms present on the surface of fruits like apples, grapes, strawberries, and citrus, including *Penicillium expansum*, *Botrytis cinerea*, and different species of *Alternaria* (Leng et al., 2022; Dwiastuti et al., 2021; Temur & Tiryaki, 2013). UV treatment in fresh vegetables like lettuce, spinach, and tomatoes has been successfully used to kill bacteria like *Salmonella enterica*, *Listeria monocytogenes*, and *Escherichia coli* (Prado-Silva et al., 2015; Mahmoud, 2010).

In addition to having antimicrobial effects, UV light has proven to encourage a variety of physiological and biochemical changes in the treated produce (Charles & Arul, 2007).

DOI: 10.1201/9781003370376-11

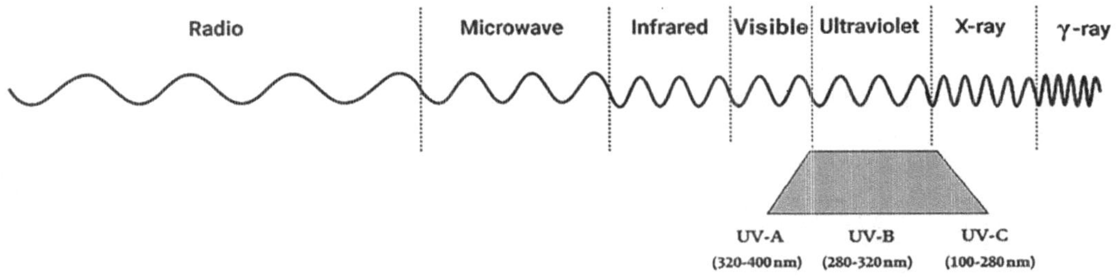

FIGURE 8.1 The wavelength range of ultraviolet radiations in the electromagnetic spectrum.

These responses can enhance postharvest quality attributes such as antioxidant capacity, color development, vitamin content, and flavor. Furthermore, UV light can modulate ripening, ethylene metabolism, and senescence in certain commodities (Martínez-Romero et al., 2007). This ability to control fruit ripening can be particularly advantageous for lengthening the shelf life of highly perishable commodities.

Even though UV treatment has shown significant potential in postharvest management, numerous factors must be considered before its successful implementation. These include optimizing UV wavelength, dose, and exposure times for various commodities and assuring uniformity and consistency of treatment across large-scale operations as well as addressing potential side effects, including skin browning or damage. Additionally, the viability of UV light systems for large-scale use in the industry must also be assessed.

8.2 DEFINITION, TYPES, AND PRINCIPLES OF UV LIGHT

UV light is a particular form of electromagnetic radiation that falls within a wavelength range that is smaller than the visible spectrum but greater than both γ- and X-rays (Figure 8.1), and its application to harvested produce results in the synthesis of bioactive compounds, the prevention of postharvest diseases, and the modification of ripening and senescence.

The efficacy of UV radiation is dependent on its wavelength, intensity, and exposure duration (Darré et al., 2022). A few UV sources available in the market include pulsed light, mercury lamps, and light-emitting diodes (LEDs).

The UV light sources are grouped into three main categories: UV-A, B, and C (Rocha et al., 2015). UV-A light possesses the most extended wavelength among the ultraviolet spectrum (320–400 nm) and is the least harmful type of UV radiation. It is commonly used in insect traps. UV-B light is characterized by a relatively lower wavelength range (280–320 nm) in comparison to UV-A light and can cause sunburn, skin damage, and skin cancer. UV-B light is especially effective in inducing changes in plant gene expression, promoting photomorphogenesis and increasing plant resilience to environmental stressors. UV-C radiation has the smallest wavelength (100–280 nm) and is commonly found to produce resistance to postharvest storage rots in fruits and vegetables (Figure 8.1).

The following sections explore the four primary principles of UV light technology for postharvest fruits and vegetables.

8.2.1 Preservation of Nutritional Quality

UV light treatment has been found to preserve the nutritional content of horticultural commodities (Cano-Lamadrid et al., 2022). Some food components like polyphenols (e.g., phenolics, flavonoids, anthocyanins) and vitamins (e.g., vitamin A, vitamin C, and folate) are sensitive to thermal processing methods, which can result in their degradation (Barrett & Lloyd, 2012). UV light treatment, being a non-thermal method, helps retain the nutritional value by minimizing the loss or degradation of these compounds (Chandra et al., 2021).

Heat- or cold-sensitive nutrients, such as vitamins, can be preserved more effectively with UV light (Zhang et al., 2021). This boosts the produce's dietary benefits and nutritional quality. UV light also has little impact on the physical characteristics of fresh commodities, such as texture, color, and taste. Studies have shown that UV-treated fruits and vegetables can retain higher levels of vitamin A, vitamin C, folate, and various other phytochemicals compared to those subjected to conventional cold or thermal treatments (Bisht et al., 2021; Ma et al., 2017; Barrett et al., 2010).

8.2.2 Inhibition of Enzyme Activity

UV radiation treatment can delay the period of ripening and senescence in freshly harvested produce by reducing the enzyme activities that break down cell walls, pigments, and other components (Zhang et al., 2022). This delay in ripening helps to maintain color, firmness, and nutritional quality, thus extending the storage life of horticultural commodities (Mahajan et al., 2014).

The role of enzymes is substantial in the biological functions of fruits and vegetables (Ali et al., 2023; Haider et al., 2023). During ripening and senescence, enzymes like proteases, pectinases, and cellulases become activated, leading to the disintegration of proteins, pigments, and cell wall structures (Temiz & Ayhan, 2017). This enzymatic breakdown contributes to nutrient losses, color changes, and the softening of tissues, resulting in a shorter shelf life. UV radiation treatment effectively inhibits the activity of these enzymes, thereby slowing down the degradation process and preserving the qualitative traits of the treated produce (Abdipour et al., 2019).

8.2.3 Germicidal Effect

UV light in an appropriate spectrum (UV-C range) has germicidal effects, effectively killing or inactivating pathogenic microorganisms such as fungi, molds, bacteria, and viruses (Shahi et al., 2021). It alters the DNA structure of these pathogens, inhibiting their reproduction and spread (Gmurek et al., 2022).

Viruses like tobacco mosaic virus and tomato-spotted wilt virus can infect a wide range of fruits and vegetables (Ong et al., 2020). UV light treatment damages the viral nucleic acids, rendering them non-infectious, thereby lowering the incidence of viral diseases. The treatment also stimulates the production of antimicrobial compounds, such as phytoalexins, which further inhibit pathogen proliferation (Gómez-López et al., 2021).

Bacterial pathogens, including species of Erwinia, Xanthomonas, and Pseudomonas can cause postharvest bacterial soft rot and other diseases (Zhao et al., 2013). UV light technology effectively reduces the population of these bacteria, reducing the chances of contamination and decay (Fan et al., 2017). Fungal pathogens, such as *Penicillium* spp. (blue mold) and *Botrytis cinerea* (gray mold), are major concerns in the postharvest storage of fresh produce (Joel et al., 2019). UV light treatment disrupts the cell walls and membranes of these fungi, inhibiting their growth and preventing the development of visible decay symptoms (Papoutsis et al., 2019).

8.2.4 Activation of Defense Mechanisms

UV light triggers various defense systems in the exposed produce, including the synthesis of secondary metabolites (Santin et al., 2019; Turtoi, 2013). These substances have antioxidative and antibacterial properties that help keep food safe from oxidative stress and hinder microbial development (Bensid et al., 2022). The treatment of fresh produce with UV radiations also causes several genes involved in defense mechanisms to come into expression (Pombo et al., 2011).

The synthesis of antibacterial substances, proteins associated with pathogenesis (PR), and other defensive compounds is controlled by these genes (Zhou et al., 2015). The activation of these genes improves the ability of produce to combat pathogens and other stresses during postharvest storage. Moreover, the application of postharvest UV light treatment enhances the functionality of enzymatic antioxidants in harvested commodities. These enzymes, including POD, SOD, and CAT, assist in the stabilization of reactive oxygen species and protect against oxidization during after-harvest storage. By improving antioxidant capacity, the produce's treatment with UV light improves its quality and prolongs its shelf life.

Postharvest UV light treatment also promotes the production of flavonoids and phenolics in fruits and vegetables. These compounds possess antioxidant, antimicrobial, and anti-inflammatory properties. Their increased production contributes to the defense mechanisms of the produce, hindering the proliferation of pathogens and reducing postharvest decay. Blueberries, strawberries, and raspberries, for instance, have been subjected to UV-C light to prevent postharvest deterioration. As a result, the berries' shelf life was increased while fungal pathogen growth was inhibited (Brazaitytė et al., 2019; Huynh et al., 2019).

8.3 IMPACT OF POSTHARVEST UV LIGHT APPLICATION ON FRUITS AND VEGETABLES

The impact of UV radiation on fresh produce has been extensively studied due to its potential in improving postharvest quality, shelf life, and safety (De Corato, 2020). The results of multiple studies related to the influence of UV light on postharvest fruits and vegetables are summarized in Table 8.1. UV light has been observed to have plenty of positive effects, including increased antioxidant activity (Murray et al. 2017), reduced microbial load (Mditshwa et al., 2017), delayed ripening (Ali et al., 2018), and increased defense mechanisms against pathogens (Bensid et al., 2022). The following are the most obvious UV light effects on fruits and vegetables.

8.3.1 Improved Defense Mechanisms

UV light exposure stimulates defense systems in fruits and vegetables, enhancing their resilience against diseases and external stresses. It increases the synthesis of phytoalexins and other defense-related compounds, including antioxidants and pathogenesis-related proteins (Urban et al. 2016). These defense mechanisms help protect the produce from microbial attacks and improve their ability to withstand postharvest stresses (Huang et al., 2021).

8.3.2 Stimulation of Antioxidant Activities

Antioxidants play a crucial role in lowering oxidative stress and safeguarding against cellular damage caused by free radicals (Cömert et al., 2020). UV light can enhance the synthesis and accumulation of antioxidants, thereby improving the overall nutritional value and health benefits of the irradiated horticultural commodities (Gonzalez-Aguilar et al., 2010). UV-B and UV-C can activate specific molecular pathways associated with the production of phenolics and flavonoids in the produce (Figure 8.2). These compounds act as antioxidants, preserving the produce from oxidative damage and maintaining its nutritious quality (Avena-Bustillos et al., 2012).

Recent studies have revealed greater antioxidant activity in UV-treated horticultural produce, leading to extended storage time and improved health benefits (Sonntag et al., 2023; Zhang et al., 2022; Sethi et al., 2018). In the past, reports by Charles et al. (2008) on tomatoes and Hagen et al. (2007) on apples have also shown substantial improvement in flavonoid and phenolic content after exposure to UV-B and UV-C radiation, respectively. Similarly, studies on grapes by Maurer et al. (2017) have demonstrated that UV-C exposure increased the accumulation of flavonoids. UV light treatment has also been found to enhance the flavonoid content in many other fruits and vegetables (Table 8.1).

TABLE 8.1 Impact of Postharvest UV Radiation on the Physicochemical Attributes of Various Fruits and Vegetables

NO.	FRUIT/VEGETABLE	UV LIGHT EXPOSURE		EFFECTS ON POSTHARVEST QUALITY	REFERENCE
		WAVELENGTH	DURATION		
1	Strawberries	Pulsed light (47.8 J cm^{-2})	10 sec	Postponed decay, extended shelf life	Duarte-Molina et al. (2016)
		UV-C (1.0 kJ m^{-2})	5 sec		Idzwana et al. (2020)
2	Apples	UV-C (65.5 J cm^{-2})	5 min	Enhanced color development, increased phenolic content	Chen et al. (2016)
		UV-C (0.5 kJ m^{-2})	7 hr		Ribeiro & Alvarenga (2012)
		UV-B (0.17 W m^{-2})	12 hr		Hagen et al. (2007)
3	Mangoes and pineapples	UV-C (2.06 kJ m^{-2})	10 s	Microbial count lowered; shelf life enhanced	Romero et al. (2017)
		UV-C (75 J cm^{-2})	5 min		George et al. (2015)
		UV-C (4.93 kJ m^{-2})	10 s		González-Aguilar et al. (2007)
4	Peaches	UV-B (219 kJ m^{-2})	12 hr	Increased anthocyanin content, improved color	Scattino et al. (2014)
5	Grapes	UV-A (80 μmol m^{-2} s^{-1})	10 min		Azuma et al. (2019)
		UV-C (2.0 kJ m^{-2})			Maurer et al. (2017)
6	Sweet cherries	UV-C (4 kJ m^{-2})	45 s	Increased phenolic content, flavonoids, and anthocyanins; prolonged shelf life	Martínez-Hernández et al. (2020)
7	Blueberries	UV-C (4.0 kJ m^{-2})	10 s	Elevated anthocyanins and total phenolics levels, reduced decay	Xu et al. (2016)
8	Lemons	UV-B (22 kJ m^{-2})	3 min	Elevated content of flavonols, anthocyanins, and flavanones-dihydro flavonols; enhanced antifungal efficacy for *Penicillium digitatum*	Ruiz et al. (2017)
9	Bananas	UV-B (5 kJ m^{-2})	4 hr	Chilling injury decreased and reactive oxygen species (ROS) increased	Ruan et al. (2015)
		UV-C (0.03 kJ m^{-2})	10 hr		Pongprasert et al. (2011)
10	Sweet oranges	UV-C 100 μmol m^{-2} s^{-1}	6 days	Higher carotenoid content	Hu et al. (2019)
11	Pomegranate arils	UV-C (4.54 kJ m^{-2})	3.5 min	Preserved enzymatic antioxidants	Maghoumi et al. (2013)
12	Pears	UV-B (6.4 W·m^{-2})	15 mins	Highest phenolic content	Ortega-Hernández et al. (2019)

NO.	FRUIT/VEGETABLE	UV LIGHT EXPOSURE		EFFECTS ON POSTHARVEST QUALITY	REFERENCE
		WAVELENGTH	DURATION		
13	Watermelons	UV-C (2.8 kJ m^{-2})	10 min	The total antioxidant activity enhanced	Artés-Hernández et al. (2010)
14	Lettuce	Pulsed light (10.5 J cm^{-2})	10 s	Reduced microbial load, maintained quality	Mukhopadhyay et al. (2021)
		UV-C 7.5 W cm^{-2}	2 hr		Lee et al. (2021)
15	Tomatoes	2.16 J m^{-2}	60 min	Increased lycopene content	Bhat (2016)
		UV-C (4.83 kJ m^{-2})	15 min		Pinheiro et al. (2015)
		UV-C (12.2 kJ m^{-2})	3 hr		Bravo et al. (2012)
		UV-C (3.7 kJ m^{-2})	3 min	Increased accumulation of lignin and phenolic compounds	Charles et al. (2008)
16	Spinach and Chinese kale	UV-C (1.82 kJ m^{-2})	60 min	Delayed senescence, extended shelf life	Tavanikone and Pranamornkith (2019)
		UV-C (24 kJ m^{-2})	3 min		Escalona et al. (2010)
17	Carrots	UV-C (23 μW cm^{-2})	3 min	Reduced microbial load, extended shelf life	Lee et al. (2021)
		UV-C (2 kJ m^{-2})	5 min		Li et al. (2021)
18	Bell peppers	UV-C (9.0 kJ m^{-2})	14 hr	Delayed ripening, reduced microbial growth	Castillejo et al. (2022)
		UV-C (20 kJ m^{-2})	5 min		Rodoni et al. (2012)
19	Broccoli	White light (19.0 W m^{-2})	3 hr	Enhanced antioxidant activity, extended shelf life	Pintos et al. (2020)
		Fluorescent light (24 μmol m^{-2} s^{-1})	5 min		Zhan et al. (2012)
20	Cauliflower and cabbage	UV-C (23 μW cm^{-2})	3 min	Delayed browning, improved microbial safety	Lee et al. (2021)
21	Brussels sprouts	UV-A and UV-B	16 hr	Reduction in infestation of aphids	Acharya et al. (2016)
22	Garlic	UV-C (2.0 kJ m^{-2})	3 min	Increased total phenolics, reduced microbial loads	Park and Kim (2015)

8.3.3 Lowered Microbial Count

Microbial contamination poses a significant risk to the quality and safety of perishable produce (Balali et al., 2020). Traditional antimicrobial treatments may raise concerns about residual chemical residues and their potential health and environmental impacts (Chen et al., 2019). UV radiation, UV-C in particular, offers a non-chemical alternative that addresses these concerns and provides an effective means of microbial control. These disrupt the DNA structure of viruses, bacteria, fungi, and molds, preventing their proliferation and lowering postharvest disease development (Shin et al., 2016; Temur & Tiryaki, 2013; Ribeiro & Alvarenga, 2012).

8.3.4 Delayed Ripening

UV light treatment can slow the ripening process in fruits and vegetables. It inhibits the enzymes' activities responsible for the breakdown of pigments, flavor compounds, and cell wall components, hence retaining firmness and extending shelf life (De Corato, 2020).

Numerous studies have indicated the efficacy of UV light treatment in delaying ripening and enhancing the postharvest life span of various fruits and vegetables. For example, UV light treatment of strawberries has been shown to delay color changes and reduce the softening rate during storage (Miao et al., 2016). UV light treatment has also been shown to slow down the ripening process in bananas by suppressing the activity of enzymes like polygalacturonase and amylase, which are involved in the breakdown of cell wall components and starch (Ruan et al., 2015). Continuous research and optimization of UV light treatment for different fruits and vegetables will further contribute to its effective implementation in the horticulture industry.

8.3.5 Regulation of Phytochemicals

UV radiation can alter fruit and vegetable phytochemical biosynthesis. It may elevate the levels of phytochemicals like carotenoids, polyphenols, and ascorbic acid, which are critically important for human health (Al-Juhaimi et al., 2018). UV light exposure has been shown to boost the organoleptic and nutritional attributes of freshly harvested fruits and vegetables, making them more appealing to consumers (Vargas-Ramella et al., 2021; Ribeiro & Alvarenga, 2012).

8.4 COMPARISON OF UV LIGHT TECHNOLOGY WITH OTHER POSTHARVEST TREATMENTS

The previous research proved that UV light technology, UV-C in particular, has evolved as a potential treatment after harvesting fruits and vegetables, with significant advantages for shelf life, nutritional quality, and food safety (Esua et al., 2020; Mditshwa et al., 2017; Pataro et al., 2015; Pinheiro et al., 2015). It is mainly effective against surface pathogens and has proved to have antimicrobial effects by destroying the DNA or RNA of microorganisms (Anderson et al., 2000). It eliminates a broad spectrum of microbes, like viruses, bacteria, and fungi, without leaving any chemical residues on the produce. This makes it a good alternative to hot water and chemical treatments, which can be harmful for the environment and mankind (Ribeiro & Alvarenga, 2012).

UV-C radiation also improves the nutritive value of fresh produce. For instance, the treatment of fresh produce with UV-C radiation can elevate the levels of antioxidants like phenolics and ascorbic acid (Bisht et al., 2021; Xu et al., 2016; Zhan et al., 2012). Also, UV treatment, when properly applied, has minimal influence on the nutritional and organoleptic attributes of the produce (Ma et al., 2017). However, the effects of UV light on fruits and vegetables can be complex and multifaceted, based on the type of produce, the intensity and duration of the UV treatment, and the stage of ripeness or maturity of the produce (Meneses-Espinosa et al., 2023; Darré et al., 2022). For example, some studies have reported that UV-C radiation can modify the color and texture of produce, which may affect consumer acceptance (Pandiselvam et al., 2022; Esua et al., 2020; Fan et al., 2017). Another potential limitation of UV light technology is its penetration depth, which may limit its effectiveness in treating produce with thick or dense surfaces (Pinela & Ferreira, 2017). However, this can be overcome by using multiple UV lamps or by combining UV treatment with other postharvest treatments such as cold storage, controlled atmosphere (CA) storage, or modified atmosphere packaging (MAP). Despite these challenges, UV light technology has shown great potential for improving the safety, quality, and shelf life of harvested produce. Additional research should be carried out to discover the most effective doses for the postharvest treatment of different fruits and vegetables in the food industry (Figure 8.2).

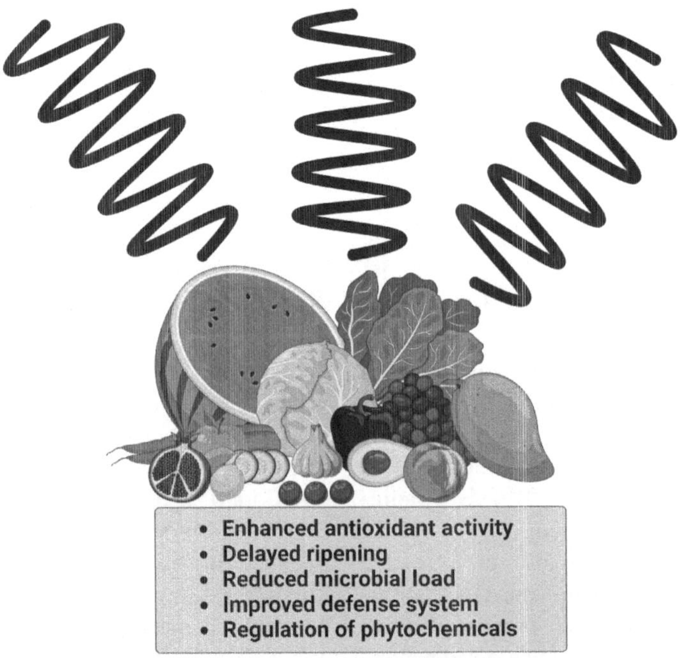

FIGURE 8.2 Postharvest changes in fruits and vegetables exposed to ultraviolet radiation.

8.5 POTENTIAL AREAS FOR FUTURE RESEARCH

The implementation of UV light as a postharvest treatment has shown considerable expansion in the horticulture industry, particularly in fruits and vegetables. This technology involves exposing produce to UV-C light, which can decrease microbial development and lengthen its life. Nevertheless, further investigation is required to comprehensively understand the potential of UV light technology in postharvest horticulture.

One potential area for future research is to study the consequences of UV radiation on the nutritional profile of produce. While UV light can reduce the growth of harmful microbes, it can also have an adverse effect on the nutritional quality of produce. For instance, a study on strawberries found that UV light treatment lowered the content of ascorbic acid and total phenolics in the fruits (Idzwana et al., 2020). However, the effects of UV light can vary depending on various factors, including variation in produce type, treatment intensity and duration, and maturation stage. Hence, additional investigation is required to comprehensively understand the impact of UV radiation on the nutrient values of different fruits and vegetables.

Another area for future research is the potential of combining UV light treatment with other after-harvest applications, such as edible coatings, plant extracts, cold storage, CA storage, and MAP. These treatments are often employed to extend the storage life of fresh commodities, and combining them with UV light treatment could further enhance their effectiveness. For example, a study conducted on blueberries found that combining UV light treatment with MAP led to a substantial decline in mold growth and an extension in the fruit's shelf life (Perkins-Veazie et al., 2008).

Furthermore, the potential limitations of UV light technology for postharvest horticulture should also be explored. One limitation is that UV-C light can only penetrate the surface of produce, so it may not be effective in treating internal microorganisms. Additionally, UV light treatment can cause damage to the surface of produce, which can lead to a higher deterioration rate and reduced shelf life. Therefore, more investigation is required to develop methods for effective UV light treatment that minimizes damage to the surface of produce and maximizes the penetration of UV-C radiation.

8.6 MERITS AND DEMERITS OF USING UV RADIATION IN FOOD

The following are the major merits and demerits of UV radiation in foods.

MERITS	*DEMERITS*
• This is a simple treatment. • Non-ionizing in nature. • Approved by food control agencies. • UV-C is a strong germicide that controls a wide range of microbes. • Little changes in physicochemical attributes. • Capable of eliciting hermetic reactions, resulting in the accumulation of phytochemicals in metabolically active foods. • Fewer restrictions than other irradiation methods. • There are no byproducts or waste produced. • Energy-efficient and cost-effective. • Could be combined with other preservation techniques. • No water requirement.	• Insufficient surface penetration efficacy in solid foods. • Direct exposure is necessary to achieve maximal efficacy and germicidal action. • Challenging to implement commercially. • Dangerous for workers if not appropriately safeguarded. • Consumers are concerned even though the UV light is non-ionizing.

8.7 CONCLUSIONS

In conclusion, the postharvest exposure of fresh produce to UV radiation is a promising method due to its broad-spectrum antimicrobial efficacy, minimal impact on quality attributes, and non-chemical nature. It offers an alternative to chemical treatments and provides an effective means to control surface pathogens and extend shelf life. However, UV light treatment has some limitations, such as its surface treatment–only capability, dependence on treatment time and intensity, and the associated costs of equipment and implementation. To further enhance the application of UV light technology in postharvest treatments, future research can focus on the following areas: (a) Research should be focused on optimization of wavelength, intensity, and exposure time of UV rays for different kinds of produce. This will increase the deactivation of pathogens and the preservation of the nutritional quality of the produce. (b) Further studies are needed to evaluate the long-term effects of UV light treatment on the nutritional composition, sensory attributes, and overall quality of treated fruits and vegetables. This will help to ensure that UV light treatment does not compromise the nutritional value or consumer acceptance of the product. (c) Investigating the synergistic effects of combining UV light treatment with other postharvest treatments, such as MAP or edible coatings, could lead to enhanced microbial control and shelf-life extension. (d) Efforts should be made to explore techniques that allow UV light to penetrate deeper into the produce to target internal pathogens. This could involve innovative approaches such as surface modifications or the development of UV-activated compounds. (e) Studies can be conducted to find the efficacy of UV light against pathogens in a variety of fruits and vegetables. Understanding the variability in UV sensitivity of various fruits and vegetables will help in tailoring treatment protocols for maximum effectiveness. (f) Researchers should concentrate on discovering cost-effective UV light treatment systems that can be easily integrated into existing postharvest operations. This will make the technology more accessible to a wider range of producers.

REFERENCES

Abdipour, M., M. Hosseinifarahi, N. Naseri. (2019). Combination method of UV-B and UV-C prevents post-harvest decay and improves organoleptic quality of peach fruit. *Scientia Horticulturae*, 256,108564. https://doi.org/10.1016/j.scienta.2019.108564.

Acharya, J., O. Rechner, S. Neugart, M. Schreiner, H. M. Poehling. (2016). Effects of light-emitting diode treatments on *Brevicoryne brassicae* performance mediated by secondary metabolites in Brussels sprouts. *Journal of Plant Diseases and Protection*, 123, 321–330. https://doi.org/10.1007/s41348-016-0029-9.

Ali, A., W. K. Yeoh, C. Forney, M. W. Siddiqui. (2018). Advances in postharvest technologies to extend the storage life of minimally processed fruits and vegetables. *Critical Reviews in Food Science and Nutrition*, 58, 2632–2649. https://doi.org/10.1080/10408398.2017.1339180.

Ali, S., A. S. Khan, A. Nawaz, S. Naz, S. Ejaz, A. A Shah, M. W. Haider. (2023). The combined application of Arabic gum coating and γ-aminobutyric acid mitigates chilling injury and maintains eating quality of 'Kinnow' mandarin fruits. *International Journal of Biological Macromolecules*, 236, 123966. https://doi.org/10.1016/j.ijbiomac.2023.123966.

Al-Juhaimi, F., K. Ghafoor, M. M. Özcan, M. H. A. Jahurul, E. E. Babiker, S. Jinap, F. Sahena, M. S. Sharifudin, I. S. M. Zaidul. (2018). Effect of various food processing and handling methods on preservation of natural antioxidants in fruits and vegetables. *Journal of Food Science and Technology*, 55, 3872–3880. https://doi.org/10.1007/s13197-018-3370-0.

Anderson, J. G., N. J. Rowan, S. J. MacGregor, R. A. Fouracre, O. Farish. (2000). Inactivation of food-borne enteropathogenic bacteria and spoilage fungi using pulsed-light. *IEEE Transactions on Plasma Science*, 28, 83–88. https://doi.org/10.1109/27.842870.

Areche, F. O., A. H. Gondal, O. T. Landeo, D. D. C. Flores, A. R. Rodríguez, P. L. Pérez, J. T. Huaman, A. V. Roman, R. J. M. Y. Correo. (2022). Innovative trends in reducing food waste and ensuring a more sustainable food system and environment. *CABI Reviews*, 2022, 202217053. https://doi.org/10.1079/cabireviews202217053

Artés-Hernández, F., P. A. Robles, P. A. Gómez, A. Tomás-Callejas, F. Artés. (2010). Low UV-C illumination for keeping overall quality of fresh-cut watermelon. *Postharvest Biology and Technology*, 55, 114–120. https://doi.org/10.1016/j.postharvbio.2009.09.002.

Avena-Bustillos, R. J., W. X. Du, R. Woods, D. Olson, A. P. Breksa III, T. H. McHugh. (2012). Ultraviolet-B light treatment increases antioxidant capacity of carrot products. *Journal of the Science of Food and Agriculture*, 92, 2341–2348. https://doi.org/10.1002/jsfa.5635.

Azuma, A., H. Yakushiji, A. Sato. (2019). Postharvest light irradiation and appropriate temperature treatment increase anthocyanin accumulation in grape berry skin. *Postharvest Biology and Technology*, 147, 89–99. https://doi.org/10.1016/j.postharvbio.2018.09.008.

Balali, G. I., D. D. Yar, V. G. Afua Dela, P. Adjei-Kusi. (2020). Microbial contamination, an increasing threat to the consumption of fresh fruits and vegetables in today's world. *International Journal of Microbiology*, 2020, 1–10. https://doi.org/10.1155/2020/3029295.

Barrett, D. M., J. C. Beaulieu, R. Shewfelt. (2010). Color, flavor, texture, and nutritional quality of fresh-cut fruits and vegetables: desirable levels, instrumental and sensory measurement, and the effects of processing. *Critical Reviews in Food Science and Nutrition*, 50, 369–389. https://doi.org/10.1080/10408391003626322.

Barrett, D. M., B. Lloyd. (2012). Advanced preservation methods and nutrient retention in fruits and vegetables. *Journal of the Science of Food and Agriculture*, 92, 7–22. https://doi.org/10.1002/jsfa.4718.

Bensid, A., N. El Abed, A. Houicher, J. M. Regenstein, F. Özogul. (2022). Antioxidant and antimicrobial preservatives: Properties, mechanism of action and applications in food–a review. *Critical Reviews in Food Science and Nutrition*, 62, 2985–3001. https://doi.org/10.1080/10408398.2020.1862046.

Bhat, R. (2016). Impact of ultraviolet radiation treatments on the quality of freshly prepared tomato (*Solanum lycopersicum*) juice. *Food Chemistry*, 213, 635–640. https://doi.org/10.1016/j.foodchem.2016.06.096.

Bisht, B., P. Bhatnagar, P. Gururani, V. Kumar, M. S. Tomar, R. Sinhmar, N. Rathi, S. Kumar. (2021). Food irradiation: Effect of ionizing and non-ionizing radiations on preservation of fruits and vegetables–a review. *Trends in Food Science and Technology*, 114, 372–385. https://doi.org/10.1016/j.tifs.2021.06.002.

Bravo, S., J. García-Alonso, G. Martín-Pozuelo, V. Gómez, M. Santaella, I. Navarro-González, M. J. Periago. (2012). The influence of post-harvest UV-C hormesis on lycopene, β-carotene, and phenolic content and antioxidant activity of breaker tomatoes. *Food Research International*, 49, 296–302. https://doi.org/10.1016/j.foodres.2012.07.018.

Brazaitytė, A., V. Vaštakaitė-Kairienė, N. Rasiukevičiūtė, A. Valiuškaitė. (2019). UV LEDs in postharvest preservation and produce storage. In *Ultraviolet LED technology for food applications*. Cambridge, MA: Academic Press, pp. 67–90.

Cano-Lamadrid, M., F. Artés-Hernández. (2022). Thermal and nonthermal treatments to preserve and encourage bioactive compounds in fruit-and vegetable-based products. *Foods*, 11, 3400. https://doi.org/10.3390/foods11213400.

Castillejo, N., L. Martínez-Zamora, F. Artés–Hernández. (2022). Postharvest UV radiation enhanced biosynthesis of flavonoids and carotenes in bell peppers. *Postharvest Biology and Technology*, 184, 111774. https://doi.org/10.1016/j.postharvbio.2021.111774.

Chandra, R. D., M. N. U. Prihastyanti, D. M. Lukitasari. (2021). Effects of pH, high pressure processing, and ultraviolet light on carotenoids, chlorophylls, and anthocyanins of fresh fruit and vegetable juices. *eFood*, 2, 113–124. https://doi.org/10.2991/efood.k.210630.001.

Charles, M. T. J. Arul. (2007). UV treatment of fresh fruits and vegetables for improved quality: a status report. *Stewart Postharvest Review*, 3, 1–8. https://doi.org/10.2212/spr.2007.3.6.

Charles, M.T., J. Mercier, J. Makhlouf, J. Arul. (2008). Physiological basis of UV-C induced resistance to *Botrytis cinerea* in tomato fruit. I. Role of pre- and post-challenge accumulation of the phytoalexin-rishitin. *Postharvest Biology and Technology*, 47, 10–20. https://doi.org/j.postharvbio.2007.05.013.

Chen, C., W. Hu, Y. He, A. Jiang, R. Zhang. (2016). Effect of citric acid combined with UV-C on the quality of fresh-cut apples. *Postharvest Biology and Technology*, 111,126–131. https://doi.org/10.1016/j.postharvbio.2015.08.005.

Chen, J., G. G. Ying, W. J. Deng. (2019). Antibiotic residues in food: extraction, analysis, and human health concerns. *Journal of Agricultural and Food Chemistry*, 67, 7569–7586. https://doi.org/10.1021/acs.jafc.9b01334.

Cömert, E. D., B. A. Mogol, V. Gökmen. (2020). Relationship between color and antioxidant capacity of fruits and vegetables. *Current Research in Food Science*, 2,1–10. https://doi.org/10.1016/j.crfs.2019.11.001.

Darré, M., A. R. Vicente, L. Cisneros-Zevallos, F. Artés-Hernández. (2022). Postharvest ultraviolet radiation in fruit and vegetables: Applications and factors modulating its efficacy on bioactive compounds and microbial growth. *Foods*, 11, 653. https://doi.org/10.3390/foods11050653.

De Corato, U. (2020). Improving the shelf-life and quality of fresh and minimally-processed fruits and vegetables for a modern food industry: A comprehensive critical review from the traditional technologies into the most promising advancements. *Critical Reviews in Food Science and Nutrition*, 60, 940–975. https://doi.org/10.1080/10408398.2018.1553025.

De Simone, N., B. Pace, F. Grieco, M. Chimienti, V. Tyibilika, V. Santoro, V. Capozzi, G. Colelli, G. Spano, P. Russo. (2020). *Botrytis cinerea* and table grapes: A review of the main physical, chemical, and bio-based control treatments in post-harvest. *Foods*, 9, 1138. https://doi.org/10.3390/foods9091138.

Deepshikha, B. Kumari, E. P. Devi, G. Sharma, S. Rawat, J. P. Jaiswal. (2017). Irradiation as an alternative method for post-harvest disease management: An overview. *International Journal of Agriculture Environment and Biotechnology*, 10, 625–633. http://dx.doi.org/10.5958/2230-732X.2017.00077.8.

D'Souza, C., H.G. Yuk, G.H. Khoo, W. Zhou. (2015). Application of light-emitting diodes in food production, postharvest preservation, and microbiological food safety. *Comprehensive Reviews in Food Science and Food Safety*, 14, 719–740. https://doi.org/10.1111/1541-4337.12155.

Duarte-Molina, F., P. L. Gómez, M. A. Castro, S. M. Alzamora. (2016). Storage quality of strawberry fruit treated by pulsed light: Fungal decay, water loss and mechanical properties. *Innovative Food Science and Emerging Technologies*, 34, 267–274. https://doi.org/10.1016/j.ifset.2016.01.019.

Dwiastuti, M. E., L. Soesanto, T. G. Aji, N. F. Devy. (2021). Biological control strategy for postharvest diseases of citrus, apples, grapes and strawberries fruits and application in Indonesia. *Egyptian Journal of Biological Pest Control*, 31, 1–12. https://doi.org/10.1186/s41938-021-00488-1.

Elik, A., D. K. Yanik, Y. Istanbullu, N. A. Guzelsoy, A. Yavuz, F. Gogus. (2019). Strategies to reduce post-harvest losses for fruits and vegetables. *Strategies*, 5, 29–39. https://doi.org/10.7176/JSTR/5-3-04.

Escalona, V. H., E. Aguayo, G. B. Martínez-Hernández, F. Artés. (2010). UV-C doses to reduce pathogen and spoilage bacterial growth in vitro and in baby spinach. *Postharvest Biology and Technology*, 56, 223–231. https://doi.org/10.1016/j.postharvbio.2010.01.008.

Esua, O. J., N. L. Chin, Y. A. Yusof, R. Sukor. (2020). A review on individual and combination technologies of UV-C radiation and ultrasound in postharvest handling of fruits and vegetables. *Processes*, 8, 1433. https://doi.org/10.3390/pr8111433.

Fan, X., R. Huang, H. Chen. (2017). Application of ultraviolet C technology for surface decontamination of fresh produce. *Trends in Food Science and Technology*, 70, 9–19. https://doi.org/10.1016/j.tifs.2017.10.004.

George, D. S., Z. Razali, V. Santhirasegaram, C. Somasundram. (2015). Effects of ultraviolet light (UV-C) and heat treatment on the quality of fresh-cut Chokanan mango and Josephine pineapple. *Journal of Food Science*, 80, S426–S434. https://doi.org/10.1111/1750-3841.12762.

Gmurek, M., E. Borowska, T. Schwartz, H. Horn. (2022). Does light-based tertiary treatment prevent the spread of antibiotic resistance genes? Performance, regrowth and future direction. *Science of The Total Environment*, 153001. https://doi.org/10.1016/j.scitotenv.2022.153001.

Gómez-López, V. M., E. Jubinville, M. I. Rodríguez-López, M. Trudel-Ferland, S. Bouchard, J. Jean. (2021). Inactivation of foodborne viruses by UV light: A review. *Foods*, 10, 3141. https://doi.org/10.3390/foods10123141.

Gonzalez-Aguilar, G. A., J. A. Villa-Rodriguez, J. F. Ayala-Zavala, E. M. Yahia. (2010). Improvement of the antioxidant status of tropical fruits as a secondary response to some postharvest treatments. *Trends in Food Science and Technology*, 21, 475–482. https://doi.org/10.1016/j.tifs.2010.07.004.

González-Aguilar, G. A., R. Zavaleta-Gatica, M. E. Tiznado-Hernández. (2007). Improving postharvest quality of mango 'Haden'by UV-C treatment. *Postharvest Biology and Technology*, 45, 108–116. https://doi.org/10.1016/j.postharvbio.2007.01.012.

Hagen, S. F., G. I. A. Borge, G. B. Bengtsson, W. Bilger, A. Berge, K. Haffner, K. A. Solhaug. (2007). Phenolic contents and other health and sensory related properties of apple fruit (*Malus domestica* Borkh., cv. Aroma): Effect of postharvest UV-B irradiation. *Postharvest Biology and Technology*, 45, 1–10. https://doi.org/10.1016/j.postharvbio.2007.02.002.

Haider, M. W., M. Nafees, R. Iqbal, S. Ali, H. U. Asad, F. Azeem, M. Arslan, M. H. U. Rahman, A. R. Z. Gaafar, M. S. Elshikh. (2023). Combined application of hot water treatment and eucalyptus leaf extract postpones senescence in harvested green chilies by conserving their antioxidants: A sustainable approach. *BMC Plant Biology*, 23, 576. https://doi.org/10.1186/s12870-023-04588-y.

Hu, L., C. Yang, L. Zhang, J. Feng, W. Xi. (2019). Effect of light-emitting diodes and ultraviolet irradiation on the soluble sugar, organic acid, and carotenoid content of postharvest sweet oranges (*Citrus sinensis* (L.) Osbeck). *Molecules*, 24, 3440. https://doi.org/10.3390/molecules24193440.

Huang, X., J. Ren, P. Li, S. Feng, P. Dong, M. Ren. (2021). Potential of microbial endophytes to enhance the resistance to postharvest diseases of fruit and vegetables. *Journal of the Science of Food and Agriculture*, 101, 1744–1757. https://doi.org/10.1002/jsfa.10829.

Huynh, N.K., M.D. Wilson, A. Eyles, R.A. Stanley. (2019). Recent advances in postharvest technologies to extend the shelf life of blueberries (Vaccinium spp.), raspberries (Rubusidaeus L.) and blackberries (Rubus sp.). *Journal of Berry Research*, 9, 687–707. https://doi.org/10.3233/JBR-190421.

Idzwana, M. I. N., K. S. Chou, R. M. Shah, N. C. Soh. (2020). The Effect of ultraviolet light treatment in extend shelf life and preserve the quality of strawberry (*Fragaria × ananassa*) cv. Festival. *International Journal on Food Agriculture and Natural Resources*, 1, 15–18. https://doi.org/10.46676/ij-fanres.v1i1.4.

Joel, A.A., D. Percival, L. Abbey, S.K. Asiedu, B. Prithiviraj, A. Schilder. (2019). Biofungicides as alternative to synthetic fungicide control of grey mould (*Botrytis cinerea*)—prospects and challenges. *Biocontrol Science and Technology*, 29, 207–228. https://doi.org/10.1080/09583157.2018.1548574.

Kaczmarek, M., S. V. Avery, I. Singleton. (2019). Microbes associated with fresh produce: Sources, types and methods to reduce spoilage and contamination. In *Advances in Applied Microbiology*, ed. P. Shukla. Cambridge, MA: Academic Press, pp. 29–82.

Kahramanoğlu, İ., M. F. Nisar, C. Chen, S. Usanmaz, J. Chen, C. Wan. (2020). Light: An alternative method for physical control of postharvest rotting caused by fungi of citrus fruit. *Journal of Food Quality*, 2020, 1–12. https://doi.org/10.1155/2020/8821346.

Lee, J. Y., S. Y. Yang, K. S. Yoon. (2021). Control measures of pathogenic microorganisms and shelf-life extension of fresh-cut vegetables. *Foods*, 10, 655. https://doi.org/10.3390/foods10030655.

Leng, J., L. Yu, Y. Dai, Y. Leng, C. Wang, Z. Chen, M. Wisniewski, X. Wu, J. Liu, Y. Sui. (2022). Recent advances in research on biocontrol of postharvest fungal decay in apples. *Critical Reviews in Food Science and Nutrition*, 1–14. https://doi.org/10.1080/10408398.2022.2080638.

Li, L., C. Li, J. Sun, M. Xin, P. Yi, X. He, J. Sheng, Z. Zhou, D. Ling, F. Zheng, J. Li, G. Liu, Z. Li, Y. Tang, Y. Yang, J. Tang. (2021). Synergistic effects of ultraviolet light irradiation and high-oxygen modified atmosphere packaging on physiological quality, microbial growth and lignification metabolism of fresh-cut carrots. *Postharvest Biology and Technology*, 173, 111365. https://doi.org/10.1016/j.postharvbio.2020.111365.

Ma, L., M. Zhang, B. Bhandari, Z. Gao. (2017). Recent developments in novel shelf-life extension technologies of fresh-cut fruits and vegetables. *Trends in Food Science and Technology*, 64, 23–38. https://doi.org/10.1016/j.tifs.2017.03.005.

Maghoumi, M., P. A. Gomez, F. Artés-Hernández, Y. Mostofi, Z. Zamani, F. Artés. (2013). Hot water, UV-C and superatmospheric oxygen packaging as hurdle techniques for maintaining overall quality of fresh-cut pomegranate arils. *Journal of the Science of Food and Agriculture*, 93, 1162–1168. https://doi.org/10.1002/jsfa.5868.

Mahajan, P.V., O.J. Caleb, Z. Singh, C.B. Watkins, M. Geyer. (2014). Postharvest treatments of fresh produce. *Philosophical Transactions of the Royal Society A: Mathematical, Physical and Engineering Sciences*, 372, 1–19. https://doi.org/10.1098/rsta.2013.0309.

Mahmoud, B. S. (2010). The effects of X-ray radiation on *Escherichia coli* O157: H7, *Listeria monocytogenes*, *Salmonella enterica* and *Shigella flexneri* inoculated on whole Roma tomatoes. *Food Microbiology*, 27, 1057–1063. https://doi.org/10.1016/j.fm.2010.07.009.

Martínez-Hernández G.B., V. Blanco, P.J. Blaya-Ros, R. Torres, R. Domingo, F. Artés-Hernández. (2020). Effects of UV-C on bioactive compounds and quality changes during shelf life of sweet cherry grown under conventional or regulated deficit irrigation. *Scientia Horticulturae*, 269, 109398. https://doi.org/10.1016/j.scienta.2020.109398.

Martínez-Romero, D., G. Bailén, M. Serrano, F. Guillén, J. M. Valverde, P. Zapata, S. Castillo, D. Valero. (2007). Tools to maintain postharvest fruit and vegetable quality through the inhibition of ethylene action: a review. *Critical Reviews in Food Science and Nutrition*, 47, 543–560. https://doi.org/10.1080/10408390600846390.

Maurer, L. H., A.M. Bersch, R. O. Santos, S. C. Trindade, E. L. Costa, M.M. Peres, C. A. Malmann, M. Schneider, V. C. Bochi, C. K. Sautter et al. (2017). Postharvest UV-C irradiation stimulates the non-enzymatic and enzymatic antioxidant system of 'Isabel' hybrid grapes (Vitis labruscax Vitis vinifera L.). *Food Research International*, 102, 738–747. https://doi.org/10.1016/j.foodres.2017.09.053.

Mditshwa, A., L. S. Magwaza, S. Z. Tesfay, N. C. Mbili. (2017). Effect of ultraviolet irradiation on postharvest quality and composition of tomatoes: A review. *Journal of Food Science and Technology*, 54, 3025–3035. http://dx.doi.org/10.1007/s13197-017-2802-6.

Meneses-Espinosa, E., D. Gálvez-López, R. Rosas-Quijano, L. Adriano-Anaya, A. Vázquez-Ovando. (2023). Advantages and disadvantages of using emerging technologies to increase postharvest life of fruits and vegetables. *Food Reviews International*, 1–26. https://doi.org/10.1080/87559129.2023.2212061.

Miao, L., Y. Zhang, X. Yang, J. Xiao, H. Zhang, Z. Zhang, Y. Wang, G. Jiang. (2016). Colored light-quality selective plastic films affect anthocyanin content, enzyme activities, and the expression of flavonoid genes in strawberry (*Fragaria× ananassa*) fruit. *Food Chemistry*, 207, 93–100. https://doi.org/10.1016/j.foodchem.2016.02.077.

Mukhopadhyay, S., K. Sokorai, D. O. Ukuku, T. Jin, X. Fan, O. M. Olanya, J. Leng, V. Juneja. (2021). Effects of direct and in-package pulsed light treatment on inactivation of *E. coli* O157: H7 and reduction of microbial loads in Romaine lettuce. *LWT-Food Science and Technology*, 139, 110710. https://doi.org/10.1016/j.lwt.2020.110710.

Murray, K., F. Wu, J. Shi, S. Jun Xue, K. Warriner. (2017). Challenges in the microbiological food safety of fresh produce: Limitations of post-harvest washing and the need for alternative interventions. *Food Quality and Safety*, 1, 289–301. https://doi.org/10.1093/fqsafe/fyx027.

Ong, S.N., S. Taheri, R.Y. Othman, C.H. Teo. (2020). Viral disease of tomato crops (*Solanum lycopesicum* L.): An overview. *Journal of Plant Diseases and Protection*, 127, 725–739. https://doi.org/10.1007/s41348-020-00330-0.

Ortega-Hernández, E., V. Nair, J. Welti-Chanes, L. Cisneros-Zevallos, D.A. Jacobo-Velázquez. (2019). Wounding and UVB light synergistically induce the biosynthesis of phenolic compounds and ascorbic acid in red prickly pears (*Opuntia ficus-indica* cv. Rojo Vigor). *International Journal of Molecular Sciences*, 20, 5327. https://doi.org/10.3390/ijms20215327.

Palou, L., J. L. Smilanick, S. Droby. (2008). Alternatives to conventional fungicides for the control of citrus postharvest green and blue moulds. *Stewart Postharvest Review*, 2, 1–16. http://dx.doi.org/10.2212/spr.2008.2.2.

Pandiselvam, R., S. Barut Gök, A.N. Yüksel, Y. Tekgül, G. Çalişk an Koç, A. Kothakota. (2022). Evaluation of the impact of UV radiation on rheological and textural properties of food. *Journal of Texture Studies*, 53, 800–808. https://doi.org/10.1111/jtxs.12670.

Papoutsis, K., M.M. Mathioudakis, J.H. Hasperué, V. Ziogas. (2019). Non-chemical treatments for preventing the postharvest fungal rotting of citrus caused by *Penicillium digitatum* (green mold) and *Penicillium italicum* (blue mold). *Trends Food Science and Technology*, 86, 479–491. https://doi.org/10.1016/j.tifs.2019.02.053.

Park, M.H., J.G. Kim. (2015). Low-dose UV-C irradiation reduces the microbial population and preserves antioxidant levels in peeled garlic (*Allium sativum* L.) during storage. *Postharvest Biology and Technology*, 100, 109–112. https://doi.org/10.1016/j.postharvbio.2014.09.013.

Pataro, G., M. Sinik, M. M. Capitoli, G. Donsì, G. Ferrari. (2015). The influence of post-harvest UV-C and pulsed light treatments on quality and antioxidant properties of tomato fruits during storage. *Innovative Food Science and Emerging Technologies*, 30, 103–111. https://doi.org/10.1016/j.ifset.2015.06.003.

Perkins-Veazie, P., J. K. Collins, L. Howard. (2008). Blueberry fruit response to postharvest application of ultraviolet radiation. *Postharvest Biology and Technology*, 47, 280–285.

Pinela, J., and I. C. Ferreira. (2017). Nonthermal physical technologies to decontaminate and extend the shelf-life of fruits and vegetables: Trends aiming at quality and safety. *Critical Reviews in Food Science and Nutrition*, 57, 2095–2111. https://doi.org/10.1080/10408398.2015.1046547.

Pinheiro, J., C. Alegria, M. Abreu, E. M. Gonçalves, C. L. Silva. (2015). Use of UV-C postharvest treatment for extending fresh whole tomato (*Solanum lycopersicum*, cv. Zinac) shelf-life. *Journal of Food Science and Technology*, 52, 5066–5074. https://doi.org/10.1007/s13197-014-1550-0.

Pintos, F. M., J. H. Hasperué, A. R. Vicente, L. M. Rodoni. (2020). Role of white light intensity and photoperiod during retail in broccoli shelf-life. *Postharvest Biology and Technology*, 163, 111121. https://doi.org/10.1016/j.postharvbio.2020.111121.

Pombo, M.A., H.G. Rosli, G.A. Martínez, P.M. Civello. (2011). UV-C treatment affects the expression and activity of defense genes in strawberry fruit (Fragaria × ananassa, Duch.). *Postharvest Biology and Technology*, 59, 94–102. https://doi.org/10.1016/j.postharvbio.2010.08.003.

Pongprasert, N., Y. Sekozawa, S. Sugaya, H. Gemma. (2011). A novel postharvest UV-C treatment to reduce chilling injury (membrane damage, browning and chlorophyll degradation) in banana peel. *Scientia Horticulturae*, 130, 73–77. https://doi.org/10.1016/j.scienta.2011.06.006.

Prado-Silva, L., V. Cadavez, U. Gonzales-Barron, A. C. B. Rezende, A. S. Sant'Ana. (2015). Meta-analysis of the effects of sanitizing treatments on *Salmonella*, *Escherichia coli* O157: H_7, and *Listeria monocytogenes* inactivation in fresh produce. *Applied and Environmental Microbiology*, 81, 8008–8021. https://doi.org/10.1128/AEM.02216-15.

Prasad, A., L. Du, M. Zubair, S. Subedi, A. Ullah, M. S. Roopesh. (2020). Applications of light-emitting diodes (LEDs) in food processing and water treatment. *Food Engineering Reviews*, 12, 268–289. https://doi.org/10.1007/s12393-020-09221-4.

Ribeiro, C., B. Alvarenga. (2012). Prospects of UV radiation for application in postharvest technology. *Emirates Journal of Food and Agriculture*, 24, 586–597. https://doi.org/10.9755/ejfa.v24i6.14677.

Rocha, A. B., S. L. Honório, C. L. Messias, M. Otón, P. A. Gómez. (2015). Effect of UV-C radiation and fluorescent light to control postharvest soft rot in potato seed tubers. *Scientia Horticulturae*, 181, 174–181. http://dx.doi.org/10.1016/j.scienta.2014.10.045.

Rodoni, L. M., A. Concellón, A. R. Chaves, A. R. Vicente. (2012). Use of UV-C treatments to maintain quality and extend the shelf life of green fresh-cut bell pepper (*Capsicum annuum* L.). *Journal of Food Science*, 77, C632–C639. https://doi.org/10.1111/j.1750-3841.2012.02746.x.

Romero, L., J. Colivet, N. M. Aron, A. Ramosvillarroel. (2017). Impact of ultraviolet light on quality attributes of stored fresh-cut mango. The Annals of the University of Dunarea de Jos of Galati. Fascicle VI. *Food Technology*, 41, 62–80.

Ruan, J., M. Li, H. Jin, L. Sun, Y. Zhu, M. Xu, J. Dong. (2015). UV-B irradiation alleviates the deterioration of cold-stored mangoes by enhancing endogenous nitric oxide levels. *Food Chemistry*, 169, 417–423. https://doi.org/10.1016/j.foodchem.2014.08.014.

Ruiz, V.E., L. Cerioni, I.C. Zampini, S. Cuello, M.I. Isla, M. Hilal, V.A. Rapisarda. (2017). UV-B radiation on lemons enhances antifungal activity of flavedo extracts against *Penicillium digitatum*. *LWT—Food Science and Technology*, 85, 96–103. https://doi.org/10.1016/j.lwt.2017.07.002.

Santeramo, F.G., E. Lamonaca. (2021). Food loss–food waste–food security: A new research agenda. *Sustainability*, 13, 4642. https://doi.org/10.3390/su13094642.

Santin, M., L. Lucini, A. Castagna, G. Rocchetti, M. T. Hauser, A. Ranieri. (2019). Comparative "phenol-omics" and gene expression analyses in peach (*Prunus persica*) skin in response to different postharvest UV-B treatments. *Plant Physiology and Biochemistry*, 135, 511–519. https://doi.org/10.1016/j.plaphy.2018.11.009.

Scattino C., A. Castagna, S. Neugart, H. M. Chan, M. Schreiner, C. H. Crisosto, A. Ranieri. (2014). Post-harvest UV-B irradiation induces changes of phenol contents and corresponding biosynthetic gene expression in peaches and nectarines. *Food Chemistry*, 163, 51–60. https://doi.org/10.1016/j.foodchem.2014.04.077.

Sethi, S., A. Joshi, B. Arora. (2018). UV treatment of fresh fruits and vegetables. In *Postharvest Disinfection of Fruits and Vegetables*, ed. M. W. Siddiqui. Cambridge: Academic Press, pp. 137–157.

Shahi, S., R. Khorvash, M. Goli, S. M. Ranjbaran, A. Najarian, A. Mohammadi Nafchi. (2021). Review of proposed different irradiation methods to inactivate food-processing viruses and microorganisms. *Food Science and Nutrition*, 9, 5883–5896. https://doi.org/10.1002/fsn3.2539.

Shin, J. Y., S. J. Kim, D. K. Kim, D. H. Kang. (2016). Fundamental characteristics of deep-UV light-emitting diodes and their application to control foodborne pathogens. *Applied and Environmental Microbiology*, 82, 2–10. https://doi.org/10.1128/AEM.01186-15.

Singh, H., S.K. Bhardwaj, M. Khatri, K.H. Kim, N. Bhardwaj. (2021). UVC radiation for food safety: An emerging technology for the microbial disinfection of food products. *Chemical Engineering Journal*, 417, 128084. https://doi.org/10.1016/j.cej.2020.128084.

Sonntag, F., H. Liu, S. Neugart. (2023). Nutritional and physiological effects of postharvest UV radiation on vegetables: A review. *Journal of Agricultural and Food Chemistry*, 71, 9951 – 9972. https://doi.org/10.1021/acs.jafc.3c00481.

Spada, A., A. Conte, M. A. Del Nobile. (2018). The influence of shelf life on food waste: A model-based approach by empirical market evidence. *Journal of Cleaner Production*, 172, 3410–3414. https://doi.org/10.1016/j.jclepro.2017.11.071.

Tavanikone, C., T. Pranamornkith. (2019). Effects of hot water combined with UV-C treatment on Chinese kale (*Brassica oleracea* var. Alboglabra) during storage. *Journal of Food Science and Agricultural Technology*, 5, 172–176. http://rs.mfu.ac.th/ojs/index.php/jfat.

Temiz, A., D.K. Ayhan. (2017). Enzymes in Minimally Processed Fruits and Vegetables. In *Minimally Processed Refrigerated Fruits and Vegetables*, eds. Yildiz, F., and R.C. Wiley. New York, NY: Springer, pp. 93–153.

Temur, C., O. Tiryaki. (2013). Irradiation alone or combined with other alternative treatments to control postharvest diseases. *African Journal of Agricultural Research*, 8, 421–434. https://doi.org/10.5897/AJAR11x.004.

Turtoi, M. (2013). Ultraviolet light treatment of fresh fruits and vegetables surface: A review. *Journal of Agroalimentary Processes and Technologies*, 19, 325–337.

Urban, L., F. Charles, M. R. A. de Miranda, J. Aarrouf. (2016). Understanding the physiological effects of UV-C light and exploiting its agronomic potential before and after harvest. *Plant Physiology and Biochemistry*, 105, 1–11. https://doi.org/10.1016/j.plaphy.2016.04.004.

Usall, J., A., Ippolito, M. Sisquella, F. Neri. (2016). Physical treatments to control postharvest diseases of fresh fruits and vegetables. *Postharvest Biology and Technology*, 122, 30–40. https://doi.org/10.1016/j.postharvbio.2016.05.002.

Vargas-Ramella, M., M. Pateiro, M. Gavahian, D. Franco, W. Zhang, A. M. Khaneghah, Y. Guerrero-Sánchez, J. M. Lorenzo. (2021). Impact of pulsed light processing technology on phenolic compounds of fruits and vegetables. *Trends in Food Science and Technology*, 115, 1–11. https://doi.org/10.1016/j.tifs.2021.06.037.

Xu, F., S. Wang, J. Xu, S. Liu, G. Li. (2016). Effects of combined aqueous chlorine dioxide and UV-C on shelf-life quality of blueberries. *Postharvest Biology and Technology*, 117, 125–131. https://doi.org/10.1016/j.postharvbio.2016.01.012.

Zhan, L., J. Hu, Y. Li, L. Pang. (2012). Combination of light exposure and low temperature in preserving quality and extending shelf-life of fresh-cut broccoli (*Brassica oleracea* L.). *Postharvest Biology and Technology*, 72, 76–81. https://doi.org/10.1016/j.postharvbio.2012.05.001.

Zhang, W., H. Jiang, J. Cao, W. Jiang. (2021). Advances in biochemical mechanisms and control technologies to treat chilling injury in postharvest fruits and vegetables. *Trends in Food Science and Technology*, 113, 355–365. https://doi.org/10.1016/j.tifs.2021.05.009.

Zhang, H., G. K. Mahunu, R. Castoria, M. T. Apaliya, Q. Yang. (2017). Augmentation of biocontrol agents with physical methods against postharvest diseases of fruits and vegetables. *Trends in Food Science and Technology*, 69, 36–45. https://doi.org/10.1016/j.tifs.2017.08.020.

Zhang, X., M. Zhang, B. Xu, A. S. Mujumdar, Z. Guo. (2022). Light-emitting diodes (below 700 nm): Improving the preservation of fresh foods during postharvest handling, storage, and transportation. *Comprehensive Reviews in Food Science and Food Safety*, 21, 106–126. https://doi.org/10.1111/1541-4337.12887.

Zhao, Y., P. Li, K. Huang, Y. Wang, H. Hu, Y. Sun. (2013). Control of postharvest soft rot caused by *Erwinia carotovora* of vegetables by a strain of *Bacillus amyloliquefaciens* and its potential modes of action. *World Journal of Microbiology and Biotechnology*, 29, 411–420. https://doi.org/10.1007/s11274-012-1193-0.

Zhou, G., Q.S. Shi, X.M. Huang, X.B. Xie. (2015). The three bacterial lines of defense against antimicrobial agents. *International Journal of Molecular Sciences*, *16*, 21711–21733. https://doi.org/10.3390/ijms160921711.

9 Ultrasound Technology for Fresh Fruits and Vegetables

Mostafa Gouda* and Xiaoli Li*
*Corresponding Authors: mostafa-gouda@zju.edu.cn; goudarowing@yahoo.com (M. Gouda); xiaolili@zju.edu.cn (X. Li)

9.1 INTRODUCTION

The integration of physical and chemical methodologies for the evaluation and alteration of organic flora, such as vegetables and fruits, has been employed for numerous years with the aim of enhancing their efficacy and excellence (Gouda et al., 2021b; Li et al., 2019). The medicinal applications of vegetables and fruits have been extensively recorded through their traditional usage and contemporary scientific research, particularly in the context of China's rich history. There has been a notable focus among scientists on investigating the underlying scientific principles and mode of operation of vegetables and fruits, with a specific emphasis on their chemical components (Gouda et al., 2016; Zhao et al., 2022). Particular consideration has been in investigating various methods for processing and preparation of fruit commodities to capitalize on the potential of their potent bioactivity and to apprehend the changes in their chemical constituents (Dang et al., 2019).

Ultrasound (US) treatment is an acoustic technology that possesses the capability to be employed for the finding and/or alteration of bioactive compounds without the need for an invasive approach. Functioning as a physical treatment, it has the potential to manipulate the physical and chemical characteristics of biological systems to different degrees contingent upon the specific conditions of the process (such as frequency, intensity, and duration), as well as the composition and structure of the plant tissues (Hsieh et al., 2019). The aforementioned technology is environmentally friendly and provides prospects for the development of novel functional products or the extraction of more potent functional bioactive compounds from vegetables and fruits. Additionally, it is a technology characterized by low energy consumption and maintenance costs, thereby yielding multiple economic advantages (Ojha et al., 2020).

However, US does not possess a comprehensive standardized technology that can be readily adjusted on a commercial scale to modify vegetables, fruits, and other food products. Enhancing our comprehension of the intricate relationships between conditions of processing (including power, frequency, and duration) and the structural and biochemical constituents of vegetables and fruits can provide valuable support for future industrial applications (Zhang et al., 2019). Conversely, the generation of free radicals through cavitation induced by the shock wave energy of the US system can lead to phytochemical modifications. These alterations in phytochemicals pose a significant challenge to the efficacy of US technology. Modern research has indicated that US may be proven as a worthy source to protect the food quality and safety (Chandrapala et al., 2012), so it may be added on to current complex and lengthy preservation methods (Gouda et al., 2021b).

9.2 VEGETABLES AND FRUITS DEFINITION

Horticultural produce usually contains such beneficial composites that act as a source of improvement in human health (Ahmed et al., 2017; Fouad et al., 2015; Nguyen et al., 2019). Many untransferable diseases like neurological, cancerous, and heart diseases can be controlled by the simple or complex compounds extracted from fruits (Ghazzawy et al., 2022; Lv et al., 2022; Venkatakrishnan et al., 2019; Zong et al., 2021). There is a bulk of beneficial constituents (carotenoids, phenolics, flavonoids, etc.) that act as a source of human growth and development (Gouda et al., 2016; Giacometti et al., 2018). Prevailing usage of fruits and vegetables minimizes the concern about quality parameters, so when techniques are applied to produce it will be a key for more safe, nutritionally active, and ecofriendly fruits and vegetables (Gouda et al., 2023a, 2023b; Tadda et al., 2023).

DOI: 10.1201/9781003370376-12

9.3 ULTRASOUND TECHNOLOGY USES AS EXTRACTION AND MEASUREMENT OF IMPORTANT COMPOUNDS IN FRUITS AND VEGETABLES

Ultrasounds are characterized as mechanically induced sound waves having a frequency of more than 20 kHz, inaudible to humans. Ultrasound usage on substrate produce minute holes after the overlapping of two different sound waves (Bu & Alheshibri, 2021). Cavities produced in plant material act as a source of losing heat waves and pressure, preventing microbial manifestation (Yusoff et al., 2022; Gouda et al., 2021b). Free radicals (OH^-, O^+, H_2O_2) are produced by the breakdown of water molecules due to exposure to sound waves. Efficiency of sound technology depends on sound rarefication, velocity, and penetration power of the sound waves. Ultrasonics can be differentiated based on frequency (kHz) in addition to the potency (W) of the produced power. Additionally, it may be classified as corresponding to sound concentration W/m^2 and density Ws/m^3 (Knorr et al., 2004). US may be distinguished as low or high frequency. High frequency is stated as 100 kHz, while high potency varies from 10 to 1000 W/cm^2 (Sukor et al., 2020).

The current instruments that use the piezoceramic-based transducers are highly regarded due to their exceptional acoustic characters, including the generation of higher sound pressure–based amplitudes while utilizing minimal power inputs (Kim et al., 2020). The applications of US technology in postharvest preservation of vegetables and fruits are significantly progressing with time. US proved to be an imperative measuring tool for the quantification of compositions of food owing to its non-destructive nature, rapid operational capacity, and adaptableness to the optical opaque system. For example, lower-intensity ultrasound (< 1 W/cm^2) or higher-frequency ultrasound (> 100 kHz) has been employed to acquire comprehensive information regarding the dimensions, structure, and compositions of plant-based products during storage (Arvanitoyannis et al., 2017). The chemical composition of these products results in distinct responses to ultrasound properties such as velocity, attenuation, frequency, and power. When it comes to determining particle size, US utilizes the same principle as light scattering in suspensions or emulsions, whereby the attenuation and velocity coefficients of US depend on concentrations of the particles (Figure 9.1).

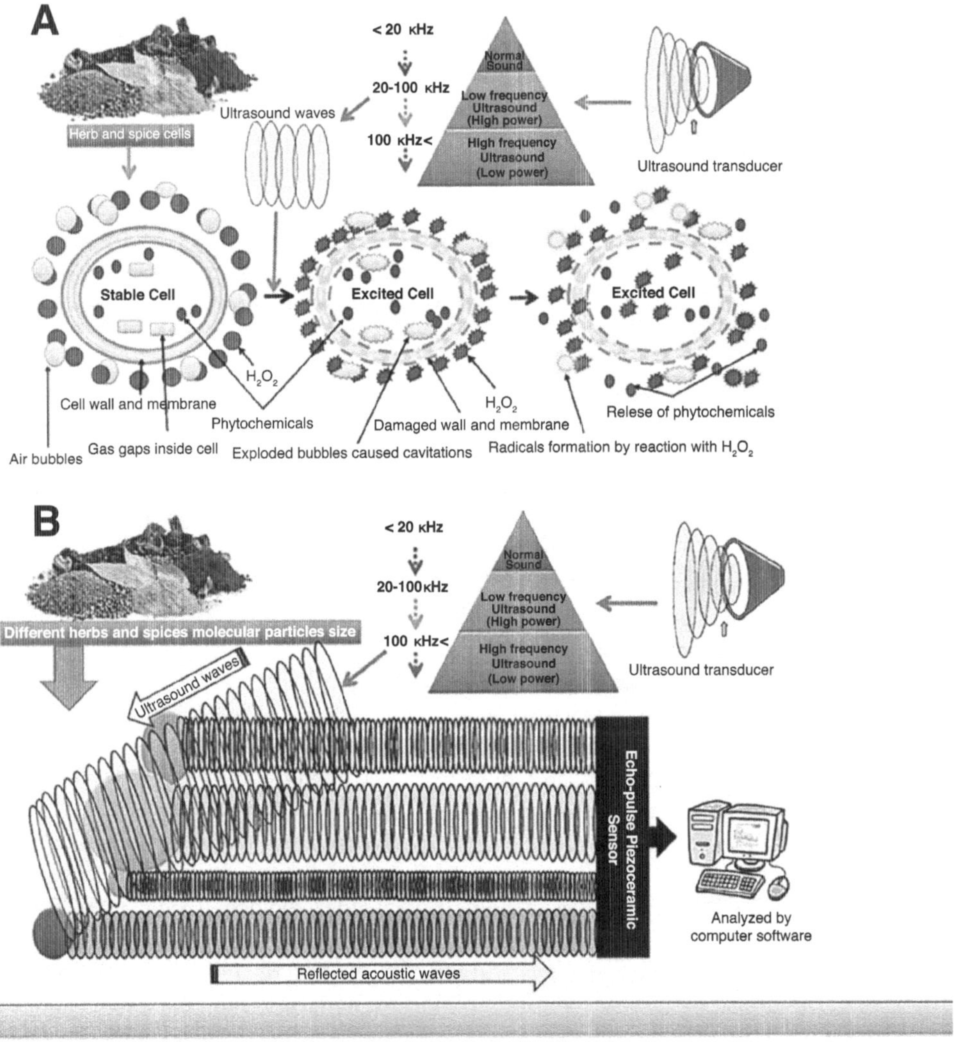

FIGURE 9.1 The major effect of lower-frequency and higher-intensity US vs higher-frequency and lower-intensity US as assistant (A) and analytical (B) technology. Adapted from (Gouda et al., 2021a).

Despite the advantageous influences of US, it has been witnessed that it can induce various oxidation reactions and inhibit the activity of numerous enzymes in food, including glucosidases and peroxidases (Nguyen et al., 2021). This phenomenon is likely attributed to intensive pressure, temperature, and shear force produced by US waves, leading to the denaturation of proteins (Chandrapala et al., 2012). Additionally, the vigorous agitation caused by microstreaming has the potential to alter hydrogen bonds and Van der Waals interactions in polypeptides, leading to protein denaturation, as noted in *Dolichos lablab* (Zhao et al., 2021).

The effects of US are reliant on changes in chemicals structures of materials enduring (US-based) treatments. The mechanical effects of US have been shown to prompt heterogeneous responses involving unfragmentable substrate types within US bubbles, thereby improving chemical reactivity (Duan et al., 2021). These relationships, known as mechanochemistry, may be utilized to control transformations at the molecular stages by readjusting or bonds cleaving at specific breaking sites (Huo et al., 2021). In an investigation, the influence of US on camptothecin content, a plant alkaloid monoterpene generated by *C. acuminata*, was explored, and it was observed that US transformed the disulfide bonds of the molecules into thiol bonds. Consequently, US modified the molecules by converting covalently attached chains of linear polymers in the β-position to a disulfide moiety (Table 9.1) (Huo et al., 2021).

TABLE 9.1 Examples of Certain Vegetable Plants Treated with Ultrasound (US) for Extraction of Important Compounds

SAMPLES NAME	*SCIENTIFIC NAME*	*PROCESS*	*US METHOD*	*COMPOUNDS OF INTEREST*	*FREQ. (KHZ)*	*INT. (W)*	*TIME (MIN)*	*T. (°C)*	*US INFLUENCES*	*REFERENCES*
Arrowhead	*Sagittaria sagittifolia*	Sonication chamber with ultrasonic generator	Thermosonication treatment	Bioactive proteins	28–40	60 W/L	10–60	30–50	The findings indicated that the utilization of ultrasound therapy exerted a significant influence on the configuration of proteins, thereby enhancing their vulnerability to the activity of digestive enzymes such as pepsin and trypsin. In this context, ultrasound treatment emerges as an influential factor for promoting protein breakdown.	Wen et al. (2018)
Lettuce	*Lactuca sativa*	Ultrasonic probe sonotrode system	Sonication extraction for antimicrobial evaluation	Phytochemical mixtures	26	200, 400	5,10	40,45	Ultrasound examination revealed the utmost degree of covetable phenolic compounds, which demonstrate the most elevated degree of antimicrobial and antiviral capabilities, specifically targeting murine and hepatitis A viruses. Additionally, it enhanced the purification of lettuce herb.	Abdelkebir et al. (2019) Millan-Sango et al. (2016)
Garlic	*Allium sativum*	Ultrasonic disperser	Thermosonication treatment	Polysaccharide	40	500	0–30 min	65	Ultrasound enhanced the C=C bond and c/o ratio, which is advantageous for enhancing the materials' electrochemicals performance.	Teng et al. (2020)
Parsley	*Petroselinum crispum*	Ultrasound bath	Ultrasonic drying preprocess	Phytochemical mixture	21	100–300	20	30	The US pretreatment cut the drying time and energy consumption by up to 29.8% and 33.6%, respectively. Additionally, it made parsley more resistant to drying in terms of chlorophylls	(Sledz, Wiktor, Rybak, Nowacka, & Witrowa-Rajchert, 2016)

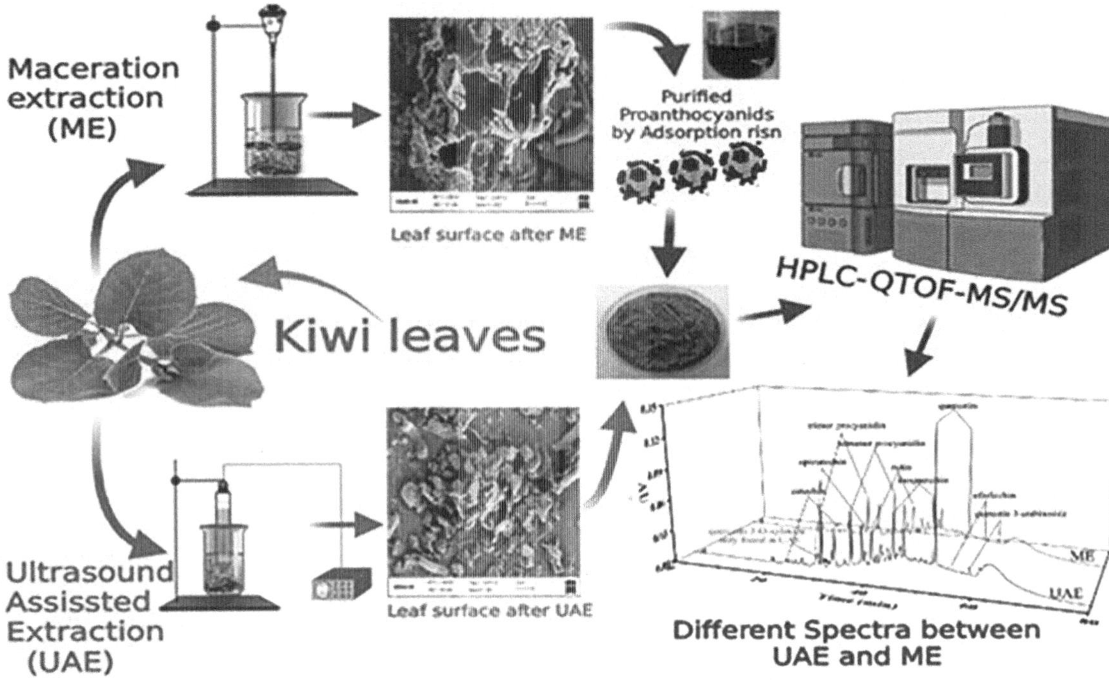

FIGURE 9.2 Flowchart of proanthocyanidin (PA) extraction by US-assisted extraction (UAE) and maceration extraction (ME) techniques (Lv et al., 2021).

According to Giacometti et al. (2018), US significantly affects plant phytochemicals, particularly polyphenols. To extract proanthocyanidin (PA) from kiwifruit (*Actinidia chinensis*) leaves, for instance, we optimized the US method in a recent study (Lv et al., 2021). The results included a detailed chemical composition and bioactivity relationship of purified kiwifruit leaf samples PAs (PKLPs) (Figure 9.2). The best conditions for extracting PAs were found to be 40% US-amplitude, 30 mL/g dry weight (DW) solvent to solids ratio (S/S), and 70 °C for 15 min, according to modeling of the PA-affected parameters identified in the data. Moreover, Rashed et al. (2016) said that an increased extraction of total phenolic content was made.

Additionally, ultrasonic therapy modifies certain secondary bonds of β-turns and β-sheets, affecting the hydrophilic groups and hydrophobic cores of the protein. This has an impact on the protein structure in fruits and vegetables (Tang et al., 2020). Research on the protein of the arrowhead (*Sagittaria sagittifolia*) revealed that treatment with US (28–40 kHz, 30–50 °C) resulted in protein unfolding, decreased contents of α-helix and β-turn, and increased contents of β-sheet and random coil, whereas, in comparison to the control samples, the concentration of free sulfhydryl (SH) rose by 38.87% at 40 kHz and 40 °C (Gouda et al., 2022b). Furthermore, it has the ability to promote the synthesis of peptide subunits that arise from the enzymatic breakdown of plant proteins (Wen et al., 2018). The non-covalent connections that give the folded macromolecule Boulatov (2021) its three-dimensional structure are disrupted when it is overstretched, which also impacts the macromolecule's capacity to participate in interactions (such as changes in enzyme active sites) (Figure 9.2).

The majority of research generally said that many different types of enzymes can be chemically and physically inactivated by high-intensity US (Wen et al., 2018). The molecular weight of the enzyme has a significant impact on how sensitive the enzymes are to US. Polymeric enzymes, for instance, break down into monomeric subunits. These monomeric enzymes may then break down even more or clump together after receiving prolonged US treatment. Free radicals and shear stresses brought on by cavitation encourage the denaturation of enzyme proteins, which inactivates the enzymes (Liu et al., 2020). As per Liu et al. (2020), US-inactivated pectin methylesterase (PME) hydrolyzes pectins and yields a compound with limited stability. In contrast with thermosonication or sonication alone at equal intensity level, a combination of US, pressures, and heat-based treatments (manothermosonication) demonstrated maximum inactivation of this enzyme. According to Tiwari et al. (2009), the high rates of inactivation caused by temperature and pressure may be attributed to their effect on pectin and other molecules that interact with PME. Conversely, their absence results in lower enzyme resistance to the effects of temperature and pressure. Nakayama and Imai (2013) discovered that the enzyme-based hydrolysis of the kenaf plant (*Hibiscus cannabinus*) celluloses to create glucose sugar can be accelerated by applying a pretreatment of US (200 W, 20 kHz, 10 min). The authors claimed that by eliminating the covering components on the cellulose from kenaf, US created a stronger connection between cellulase and cellulose.

9.3.1 Ultrasound Application as an Effective Measurement Technology

A crucial scientific and technological concern is the advancement of US acoustic sensors (Gouda et al., 2022a). Acoustic sensors are extensively utilized in a broad range of technological systems, including signal processing devices, biological and chemical safety systems, environmental monitoring systems, and many more. According to Gouda et al. (2021) and Lakshmanan et al. (2020), US is the foundation of this technology and has grown to be one of the most significant new technologies (Figure 9.3). Furthermore, for several decays, acoustic wave devices have been employed commercially in a variety of sectors. This is because of the inherent dependability and sensitivity of these sensors. Piezoelectric materials are used in almost all acoustic wave devices and sensors to produce the acoustic wave (Figure 9.3). The quartz materials used to create this piezoelectricity may be able to resonantly stabilize electronic oscillators (de Jong et al., 2015).

Taha et al. (2020) synthesized zinoxide nano-coating by usage of Roselle plant extract under US conditions to ossify the material. In the study both the structural and photocatalysis changes were obsereved. It was observed under X-ray diffraction analysis that ZnO paritical size was reduced (40 to 31 nm) due to the presence of biochemcials in Roselle. Moreover, it was exposed that the Roselle plant extract enhanced photocatalysis degradation capacity while resulting in enlargement of the area and a decrease in the energy gap. Tostado-Plascencia et al. (2018) performed an experiment to see the amalgamation of layer by layer carbon nanomaterials by use of US and Roselle plant content. The interaction potential and the creation of covalent bonds were shown to rise with increased US time, according to the authors, and this resulted in a 160% increase in the thickness of the nanotube. Therefore, one of the potential methods for creating nanoparticles is using hibiscus phytochemicals (Tostado-Plascencia et al., 2018).

9.3.2 Electrophysical Characterization of Ultrasound-Based Acoustic Sensors

Acoustic wave devices are characterized by the way the piezoelectric substrate propagates waves. Because of the wavelengths' infiltration depth, most of the energy density is concentrated in areas close to the surface (Mujahid & Dickert, 2017). As a result, any changes in the physical or chemical composition of the surface or its surroundings will affect the waves and, in turn, the devices that depend on them (Casalinuovo et al., 2006). For example, Kiontke et al. (2021) significantly

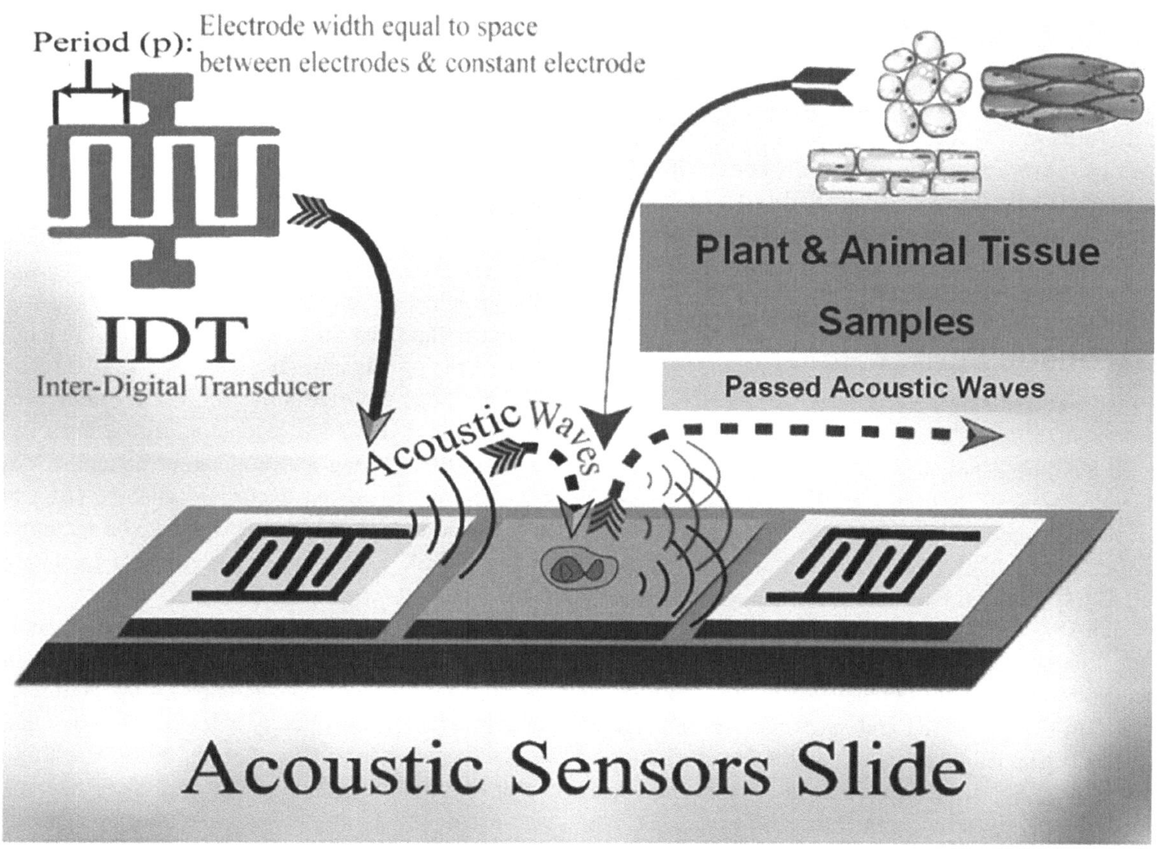

FIGURE 9.3 Schematic flow chart of SAW-based chemical sensors: a two-port delay line and a resonator with sensing over layers of target analytes (Gouda et al., 2023a).

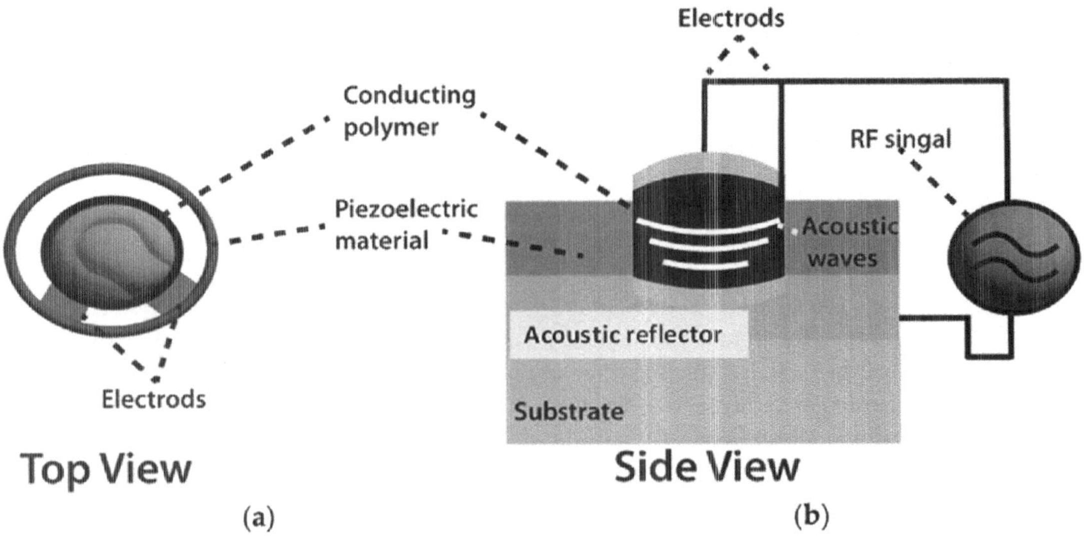

FIGURE 9.4 Schematics of bulk acoustic wave (BAW) chemical sensors. (a) Vertical top view. (b) Side view (Khodagholy et al., 2012; Löbl et al., 2001).

enhanced the signal using SAW nebulization help. According to that study, because SAW enhanced the accessible droplet surface area, it was able to boost the sensitivity response of phenylenediamines and aminophenols up to eightfold without warming the material under examination. Additionally, SAW devices provide a wide range of adjustable operating frequencies (from MHz to GHz) that allow for wireless operation and sensitivity tuning. Furthermore, a wide range of velocity and displacement direction may be employed to differentiate between acoustic waves, based on materials and boundary circumstances. Bulk waves, which travel through the substrate, are another type of characteristic wave (Löbl et al., 2001). The two major popular bulk acoustic wave (BAW) devices are the shear-horizontal acoustic plates mode (SH-APM) sensors and the thickness-shear modes (TSM) resonators (Figure 9.4).

9.3.3 Acoustic Sensor Rate for Analytes

Temperature has been shown to affect the reaction time of a SAW chemical sensor (Balashov et al., 2021). As the temperature rises, the analytes may disperse or dissociate more quickly, which would shorten the reaction time. For example, Chen et al. (2014) reported that 4.44 × 106 cm^2/g of IgE antibody sensitivity was measured using a mounted SAW sensor with an isolated chamber. Based on the frequency of the Sezawa acoustic waves, this should have specific requirements for the substrate being utilized and situations for enhancing the rate of allergen detection in tissues and fluids of a biological nature (Figure 9.3). Furthermore, according to Roche-Gaso et al. (2009), all sonic wave sensors are sensitive to changes in a broad range of physical parameters. The rates of layer-analytes interactions and, in turn, the SAW sensor's reaction time are influenced by several circumstances (Nicolae et al., 2022). The response and recovery periods of SAW sensors in the case of a mass-based sensing layer is significantly influenced by the diffusion rate of adsorbed masses into films, to piezoelectric substrates, and back to the films' surfaces (Lange & Rapp, 2008).

According to Lange et al. (2013), chemical analysis technology should work in conjunction with other associated technological studies to accomplish the addition of necessary, advantageous components to increase accuracy and stability of acoustic-based sensors in order to increase its viability as a globally distributed functional chemical analysis technology. For instance, acoustic sensors enhanced with cellulose from powdered bacteria have shown advantages in several areas, including high sensitivity (Wang et al., 2020). The mass loads enable these types of sensors to be used in particle and film thickness sensors, among other uses. If the sensor is coated with an adhesive material, it becomes a particulate sensor; any particulate that falls on the surface will remain there and obstruct the propagation of waves. A 200 MHz ST-cut quartz SAW has been reported to be 1000 times more sensitive than the tested 10 MHz TSM resonator, with a mass resolution of 3 pg. In 2020, Wang and colleagues created SAW, a sensor that uses quartz-coated bacterial cellulose (BC) to measure medium humidity. They said that for the medium humidity feeling, the BC-SAW sensor demonstrated good long-term stability and short-term repeatability.

The purpose of acoustic-based biosensors is to link the analyte adsorption–measuring process to a change in the physical properties of the acoustic wave (Jiang et al., 2020b), such as its frequency and velocity, which may be connected to the analyte concentration (Fogel et al., 2016). Existing molecular biosensors based on fluorescent emissions are restricted in their use due to light scattering and interference with the fluorescents of their compounds. By using US, deep tissue may be scanned with ease and at great spatiotemporal resolution. In general, acoustic sensors have the potential to develop into strong instruments for even more varied cellular investigation. This has also been utilized in assessment of the cytocompatibility of artificial surfaces. Additionally, a number of studies have been conducted on the tracking of single-cell microalgae using sound sensor–based investigations (Gouda et al., 2022b; Tekaya et al., 2013) and have revealed a strong correlation between the kinds of acoustic sensors and their effects on microalgae chemical composition investigations.

9.4 POTENTIAL OF ULTRASONIC TECHNOLOGY FOR HARVESTED FRUIT AND VEGETABLE PRESERVATION

9.4.1 Pesticide Residual Removal and Cleaning

External pollutants, including synthetic, natural, and physical ones, can impact the intake of fruits and vegetables. Pesticides are one such pollutant that have gained popularity in recent times (Azam et al., 2020; Jankowska et al., 2019; Elibariki & Maguta, 2017). Frequent use of horticultural produce containing pesticides deposits lead to health complications and compromise human life (Valcke et al., 2017; Chiu et al., 2019). Different deposits of field dirt, muck, microbes, and other pollutants on the surface of fruit and vegetable produce results in reduced shelf life of the produce. Scientists have been trying to introduce a way to minimize the traces of pesticides and sterilize produce to minimize the laborious procedures of washing and stripping to get desired outcomes (Pandiselvam et al., 2020). To overcome these issues ultrasonic technology is being practiced. Zhu et al. (2019) employed an ultrasonic technique to eliminate pesticide remnants (such as carbendazim, chlorothalonil, and pyrazophos) from pak choi leaves. The residuals of all the said herbicides were dramatically reduced by applying ultrasonic treatment. Pesticide residues can be removed more effectively with ultrasonic cleaning than with conventional water. The outcomes further showed that the elimination impact of chemicals improved with an increased ultrasonic frequency. In order to get rid of 16 pesticide contaminants on the skin of strawberry fruits, Lozowicka et al. (2016) investigated the effects of many techniques, including boiling, ultrasonic cleansing, and washing with normal and ozonized water. Out of the several approaches tested, ultrasonic-based cleaning proved to be the most successful in eliminating all traces of the substances under investigation. Zhou et al. (2019) found similar outcomes when they investigated low ultrasonic technology–based treatment and eliminated 5 pesticides from grapes and strawberry, as it was observed that ultrasonic treatment removed the pesticides more effectively than standard water cleaning.

Ultrasonic cleaning is based on the concept of cavitation produced by ultrasonic-based sound waves that remove the particulate matter (Zhou et al., 2019). The cavitation mechanism comprised the breaking of particles of contamination along with the breakdown of water molecules. There will be production by hydrolysis of reactive oxygen species such as hydroxyl radicle during the oxidation of organic contaminants (Cengiz et al., 2018). Water treated by sonication is being employed for the eradication of contaminants like 2,4-dihydroxybenzoic acid (Jiang et al., 2020b), p-ni-trophenol (PNP) (de Lima Leite et al., 2002), substituted phenols (Xie et al., 2016), diuron (N-(3,4-di-chlorophenyl)-N (Kim et al., 2015), and N-dimethyl-urea) (Bringas et al., 2011). The ultrasonic technique, when employed alone on different fruits and vegetables, resulted in better results in reducing pesticide and field residues. Cengiz et al. (2018) conducted an experiment of both US and electric current alone and in combination on tomato crops to remove captan, thiamethoxam, and metalaxyl deposits. The combination of both techniques showed improved outcomes in reducing all three pesticidal residues. Investigations similarly revealed that the breakdown rate of methomyl was accelerated by the use of US in conjunction with the Fenton (H_2O_2/Fe^{2+}), H_2O_2, and UV-Fenton ($H_2O_2/Fe^{2+}/UV$) processes. The method that used hydrogen peroxide and ultrasonic bubbles proved to be the most successful (Raut-Jadhav et al., 2016). In comparison to standard procedures, ultrasonic sanitation and pesticide residual removal is a highly effective, environmentally friendly, and energy-efficient approach. In particular, using ultrasonic techniques in combination with other techniques yields more effective results. Future studies in this field can be strengthened by scientists (Jiang et al., 2020a).

9.4.2 Sterilization of Fruits and Vegetables

Harvested fruits and vegetables are mainly damaged by microbial activity. Food-based diseases from microbes like *salmonella* and *E. coli* (Neto et al., 2019; Olaimat et al., 2012) are common. So there is a need to develop some effective techniques to prevent these microbial activities (Xin et al., 2009). Control of microbe activity and epidemics of foodborne illnesses is very critical. Ultrasonic technology is a successful supplemental sterilization method employed in cleaning wastewater (Xin et al., 2009), water for consumption and sanitation (Zou & Wang, 2017), and other commercial uses. Ultrasonic disinfection is used in the food sector to sterilize liquid and packaged foodstuffs (Samani et al., 2016; Wang et al., 2020; Khandpur & Gogate, 2016; Chen et al., 2020). It is widely assumed that microbial killing by ultrasound occurs mostly through cavitation, which may induce the development and breakdown of bubbles as well as greater pressure and heat in liquids (Chen et al., 2020; Wordon et al., 2012). The resulting enormous hydraulic wave force may cause damage to cell exteriors and cytoplasmic walls (Brilhante São José & Dantas Vanetti, 2012; Sivagamasundari et al., 2020). Aside from direct injury to cells, an additional inactive process is that reactive oxygen species created by bubble explosions infiltrate the inside of cells and interact with their inner parts to kill the cells (Liao et al., 2018, Stanley et al., 2004). Researchers are paying more attention to ultrasonic sterilization as a non-thermal sterilization method (Lin et al., 2019), particularly in fruit and vegetable storage after harvesting. Experiments have been done on the ultrasonic therapy of newly gathered strawberries to lower the number of microbes. Tomatoes were treated via US to investigate its influence on the postharvest microbiological loads and storage quality. It was found that US significantly lowered microbiological populations (Table 9.2). The molds, yeasts, and mesophilic populations in the untreated control were noticeability higher compared to US-treated tomatoes (Pinheiro et al., 2015).

The combination of ultrasound with gibberellic acid (GA_3) and acetic acid (AA) showed a significantly lower population of bacterial pathogens in green asparagus compared with the untreated groups (Wang & Fan, 2019). The effect of US in combination with modified atmospheric packaging (MAP) was explored on microbial counts and fresh-cut quality maintenance of cucumbers (Fan et al., 2019). This was found that the combination of MAP and US efficiently acted to inhibit the growth of yeasts and molds and overall plate count during storage. All the US-based treatments have not been found effective in maintaining the quality of the treated produce. Some researchers have discovered negative impacts when US was used in combination with other treatments (Kilicli et al., 2019).

Immersion, spraying, and drenching are some ineffective fungal treatment methods currently in use. These techniques generally result in increased residual concentrations in fruits and wastewater that are higher than the statutory restrictions. The application of the ultrasonic technique can enhance the efficiency of conventional fungicides and lower pesticide residual levels. It was discovered that intense ultrasonography increased the impact of imazalil on the preservation of postharvest citrus. Citrus and mandarins absorb more antifungal agents when immersed in an imazalil combination treated with US fruits that are treated with ultrasonic technology more rapidly and profoundly, therefore re-lowering the amount of time and dose used. Furthermore, neither the fruit surface nor its overall quality score will be impacted by the US technique (Dore et al., 2013). Further research has demonstrated that US may accelerate the rate of liquid–solid mass transfer and encourage the diffusion of liquids (Knorr et al., 2004), hence the use of the ultrasonic technique in disinfection has created new opportunities.

9.4.3 Inactivation of Enzyme Activities

Enzymes are proteins in nature that very specifically catalyze biological processes in living organisms (Palmer & Bonner, 2007). Vegetable and fruit commodities in storage respond differently depending on the actions of the enzymes. In nearly all living things, enzymes catalyze the completion of metabolic processes. Normal enzyme action may swiftly lower the overall quality after harvesting of fruits and vegetables and lead to a variety of metabolic changes. Several methods, both heating and non-heating, are being employed in the food industry to regulate enzymatic action (Jiménez-Sánchez et al., 2017, Tao et al., 2007; Xu et al., 2020). The non-thermal technique is more appealing than standard heating since it reduces the duration of the processing, saves energy, protects the ecosystem, and helps fruits and vegetables conserve their nutrition (Ali et al., 2023). The use of ultrasonics over non-heating methods for food preservation has recently attracted a lot of interest (Guerrero et al., 2017). The cavitation generated by ultrasonic waves is likely to modify various aspects of the enzymes related to their surroundings, including pH, temperature, stress, and pressure. The enzyme ultimately gets denatured and inactivated as a result of these modifications (Sánchez-Rubio et al., 2016). However, residual amino acids are impacted by the interaction between free radicals generated by the hydrolysis by US ($H_2O \rightarrow$ ROS species) and enzyme action. Ultrasonic rate, ultrasonic intensity, denseness, volume, amount of gas dissolved, and kind of medium through which the ultrasonic waves are delivered are some of the variables that affect how ultrasonic waves affect enzyme activity. Furthermore, studies have demonstrated that moderate US improves enzyme activity (Gouda et al., 2021a). In practical situations, ultrasonic waves may include both pH and temperature conditions to inhibit enzyme action; this would have noticeably higher efficacy than simply changing the temperature or pH alone, which can also induce the degradation of proteins (Jiang et al., 2020b).

The primary enzymes involved in the vegetable and fruit browning process are polyphenol-oxidase (PPO) and peroxidase (POD), which negatively affect the cosmetic quality (Oliveira & Sulaiman, 2019; Chemat & Khan, 2011; Ali et al., 2021). When cell integrity is disrupted, the enzymes get in touch with substrates, and in the existence of oxygen, browning occurs (Ali et al., 2021). PPO can catalyze the hydroxylation procedure of monophenols and oxidation-based conversion of o-diphenol into o-quinones, as well as the ultimate polymer synthesis to brownish pigments (Jang & Moon, 2011). In the existence of H_2O_2, POD may oxidize a variety of substrates. Furthermore, it can generate ROS, which are abundant in most of the newly harvested produce and are linked to the creation of brown pigmentation and undesirable tastes (Jiménez-Sánchez et al., 2017; Tao et al., 2007; Zhang et al., 2018). Bananas are particularly vulnerable to chilling injury during storage. Alteration in cell membrane structure produces initial damage, while PPO oxidation of phenolic substances causes subsequent reaction, promoting browning and blackening of banana peels (Pongprasert et al., 2011). Khademi et al. (2019) studied the impact of salicylic acid and ultrasonic waves on chilling-induced damages in bananas. Studies have demonstrated that, in contrast to the control, US at an intensity of 40 kHz, either alone or in combination with a salicylic acid treatment, resulted in higher PPO inactivation in banana fruits throughout storage. Zhang et al. (2019) studied the effects of ultrasonic waves (30 kHz) and MAP on postharvest quality control of pak choi. Both PPO and POD activities of pak choi were successfully inactivated throughout a 30-day duration of storage at 4 °C due to the combined effect of US for 10 min (US-10) and MAP. At 15 days of preservation, the activity level of PPO in the samples subjected to UT-10 + MAP was the lowest. Chen et al. (2012) demonstrated that ultrasonics suppressed the activity of PPO and POD, thereby considerably reducing epicarp browning in litchi fruit.

Fruit and vegetable cell walls are complex structures that have a role in cell development, shape, and relations with the atmosphere. They are mostly made up of cellulose, hemicellulose, and pectin, along with proteinous units (Micheli, 2001). Certain vegetables and fruits lose their firmness because of alterations to the cell wall, including polysaccharide breakdown and a reduction in cell wall constituents. Turgidity reduction is connected with middle-lamella breakdown and a drop in water-soluble pectin in deformation, which not only lowers the fruit's value but also impacts shelf life, transportation, and resistance to infection, rendering the fruit susceptible (Chen & Zhu, 2011; Mohebbi et al., 2020). The breakdown of carbohydrates by enzymes that degrade cell walls, such as poly-galacturonase (PG), pectin-methyleesterase (PME), cellulase, mannosidase, and galactosidase (Gamboa-Santos et al., 2012), is one of the primary causes of these changes. According to the literature, the US technique, as well as combining US and thermal treatments (thermosensation), might be beneficial in inactivating such enzymes in food products (Gouda et al., 2022b; Paniwnyk, 2014; Terefe et al., 2009). Gamboa-Santos et al. (2012) observed that the US technique might be a great substitute for limited PME inhibition for improved carrot structural retention. Zhi et al. (2017) discovered that combining ultrasonic and calcium chloride treatments may better limit the production of PG and pectic-lyase in jujube fruit while maintaining decreased activities of the cell wall responsible for enzymatic alterations. Vercet et al. (1999) investigated the inactivating impact of manothermosonication (MTS) on temperature-resistant PME from *Citrus sinensis* and discovered that ultrasonic applications and temperature simultaneously increased the deactivation ratio of these enzymes.

The MTS is a blend of ultrasonic and a mild temperature with slight pressure. Though heat has a powerful detrimental impact on enzymes, whenever employed solely it frequently necessitates heat at higher levels for an extended period, leading to a considerable loss in the nutrition of food and extreme sensory alterations. The MTS considerably increased the impact of temperature on enzymes, thereby resulting in their inactivation (Sala et al., 1995). Vercet et al. (2002) investigated the impacts of MTS on tomatoes' pectin enzyme and discovered that MTS treatment deactivated 62% of the PG activities, while heat treatments had no impact. Lopez et al. (1998) reported the same outcome, that MTS deactivated tomato pectic enzymes significantly and effectively (Lopez et al., 1998). Terefe et al. (2009) used heat and ultrasonic waves on tomatoes to differentiate the impact of both heat and ultrasonic waves on PG and PME enzymes in the juice. The results demonstrated that ultrasonography increased the efficacy of PG and PME activity inhibition compared with the untreated control.

9.4.4 Impact of Ultrasound Application on Physicochemical Attributes

Slow variations in biochemical attributes is key to longer shelf life of fresh produce during cold storage conditions. Biochemical properties observed are soluble solids, vitamin C, pH, titrable acid, phenolics, fruit turgidity, and a few others. There is a long list of US procedures as prospective treatments to maintain the biochemical properties of fresh produce. MAP with the addition of US was observed by Zhang et al. (2019) and revealed that such a treatment suppressed the decline in TSS, chlorophyll, and ascorbic acid content in bok choy leaves. The same results were observed when bok choy leaves were treated with a chlorine dioxide solution in addition to US (180 W) (Wu et al., 2019). Ling et al. (2018) observed that soluble solids and ascorbic acid were much higher when treated with peracetic acid (PA) and US compared to untreated; moreover, in combined form (US+PA), the phenolic and flavonoid content were improved. There was significant difference between treated and Non-treated mushrooms both phenotypically and chemically during storage at 4 °C (Lagnika et al., 2013). US-treated mushrooms had better phenolic content compared to non-treated mushrooms. Esua et al. (2019) examined the effect of sonication on tomatoes and found that it enhanced the ratio of red carotene content (90%), phenolics (30%), and vitamin C (60%) during a 4-week storage period. After 8 days of cold storage, Cao et al. (2010) found that the TSS, TA, and VC levels in strawberries treated with ultrasonication were considerably higher than those of untreated ones. It was discovered that the US method was more efficient in maintaining the nutritional value of strawberries. According to Muzaffar et al. (2016), US treatment for cherries proved successful in preserving and enhancing their juice pH, TSS, and color retention during their 14-day storage. According to the results of Xu et al.'s (2019) work, ultrasonic application or 1-MCP-based treatment efficiently inhibited the increase in TSS of apple fruits, and combination treatment of 1-MCP and US was the most effective in maintaining the overall quality compared with their individual treatments. Wang and Fan (2019) reported that exposure of *Asparagus altilis* mushrooms to US treatment in combination with AA and GA_3 exhibited lower weight loss; increased TSS; and higher ascorbic acid, chlorophyll pigments, and TPC, along with better sensory quality. In another work, Gani et al. (2016) reported that strawberries treated with ultrasonography for 30–40 min led to higher preservation of most of the biochemical attributes in contrast with the untreated group. In the same way, Pinheiro et al. (2015) reported that US application efficiently enhanced TPC in tomatoes, reduced the deterioration of cosmetic appearance, and conserved the higher sensory quality of the US-treated fruits compared with the untreated group.

The cell walls of fruits change during the maturation process and after harvest. The concentration of pectins degrade after harvest, which leads to enhanced textural changes of the fruits, thereby leading to shorter shelf life for fruits of deteriorated quality (Ali et al., 2023). Hence, it is a worldwide truth that pectin substances are a key to conserve the turgidity of fruits (Naveed et al., 2024). In several studies sonication has been proved as a key source to reduce the softness of fresh produce, including tomatoes (Pinheiro et al., 2015), strawberries (Cao et al., 2010), plums (Chen & Zhu, 2011), jujubes (Zhi et al., 2017), and others. Sonication usage may enhance the storage life, maintaining turgidity and general acceptability of vegetables and fruits. Likewise, an investigation has disclosed that sonication has an optimistic impact on the polycarbohydrate content of cell walls. Day et al. (2012) studied the influence of sonication on the physiology of *docus carota*. He proposed that ultrasonics (60c, 10 min) improved cell wall turgidity better than boiling for the same amount of time. Moreover, it was discovered that a CaCl2 (0.5%) solution as a washing dip proved more helpful. Calcium ion plays an important role in the conservation of pectic ingredients by developing a bond among Ca and pectin. Soluble pectin, pectic lyase, and others are major contributors that play an important role in the firmness of fruit. Mohebbi et al. (2020) revealed that calcium ion may enhance the shelf life of fruit by overcoming the firmness issue. It has been proved that US enhances calcium ion absorption and establishes strong cell structure. Zhi et al. (2017) discovered that, together, calcium chloride and US (10 g-l, 350 W, 40 kHz) could be a helpful tool in disseminating calcium ion at the cellular level, delaying sodium carbonate–based pectin, and avoiding the development of hydro-soluble pectates. According to the findings of Zhang et al. (2018), calcium pretreatment and ultrasonic waves may efficiently maintain higher chelate-soluble pectins from degrading and thereby keep strawberries from becoming excessively soft in storage, preserving the quality of the treated fruits.

It has been proposed that in the initial stages of heat, the material of the cell passes through the cell walls, explaining the process of the action of ultrasonic waves on cell walls. Cell turgidity gradually declines, thereby affecting the rapid reduction in pressure on cell wall composition (Day et al., 2012). As demonstrated by the impact of ultrasonic waves in combined treatment with Ca application, some have proposed that ultrasonic waves can improve mass flow where dispersion occurs (Yildirim et al., 2011). Fruit and vegetable texture has been found to be modified by US-based treatments, thereby producing several small pores that increase the tissue's mass movement (Mothibe et al., 2011). It's imperative to describe that ultrasonic-based treatments are effective and can alter cell wall composition if not used properly. According to the findings of Zhang et al. (2020), different US-associated attributes, such as ultrasonics level and temperature duration, have a major influence on pectin concentrations. According to one study, the volume of waves produced by ultrasound cavities is negatively related to the rate of US. Low-powered ultrasonics generate enormous cavity bubbles, and the levels of heat and pressure in the site rise, leading to cell wall destruction (Cao et al., 2010). As a result, appropriate US settings must be chosen to increase the beneficial impact of cell wall constituents while maintaining fruit quality. Although there has been little research in this sector, the impact of ultrasonography on cell wall chemistry may be investigated further in the years to come. This could be observed that the US technique may preserve the physiochemical indices of produce after harvesting, lengthen shelf life, and maintain the nutritional value of fruits and vegetables. The collective use of ultrasonic and other preservation technologies, in particular, has been documented in recent years and has shown a highly beneficial

TABLE 9.2 The Impact of Ultrasound (US) in Combination with Other Treatments on Postharvest Quality Attributes of Different Fruits and Vegetables

CROPS	*TREATMENTS*	*FREQUENCY*	*TIME*	*INFLUENCES*	*REFERENCES*
Plums	US + ClO_2	40 kHz at 100 W + 40 mg L^{-1}	10 min	The primary microorganisms reduced effectively using the single-step approach of liquid ClO_2 coupled with US sterilization.	Chen and Zhu (2011)
Cherry tomatoes	US + PA	45 kHz + 40 mg L^{-1}	10 min	Cherry tomatoes could be effectively sanitized by using the ultrasonic method in conjunction with an industrial sanitizer.	Brilhante São José and Dantas Vanetti (2012)
Lettuce and Carrots	US + Tween-20	40 kHz at 30 W + 0.1%	5 min	The most efficient technique to prevent *Bacillus cereus* spores is to combine US with Tween 20. This prevented the deterioration of lettuce and carrots quality.	Sagong et al. (2013)

CROPS	TREATMENTS	FREQUENCY	TIME	INFLUENCES	REFERENCES
Watercress, Parsley, and Strawberry	US + PA	45 kHz + 40 mg L^{-1}	10 min	US combined with PA were a viable treatment for these vegetable disinfection.	de São José et al. (2015)
Head lettuce	US + NaOCl	37 kHz at 380 W + 200 mg L^{-1}	100 min	The combined treatment was most effective to preserve head lettuce quality.	Park et al. (2016)
Strawberries	US + PA	40 kHz at 500 W + 40 mg L^{-1}	5 min	Out of all the treatments, the combination of US and PA acid had the greatest detoxification result.	do Rosário et al. (2017)
Purple cabbage	US + *sodium Dichloroisocyanurate*	40 kHz at 500 W + 100 mg L^{-1}	5 min	Combining sodium dichloroisocyanurate with US proved efficient to reduce microorganisms with better quality.	Duarte et al. (2018)
Green asparagus	US + AA and GA_3	40 kHz at 360 W + 50 mg kg^{-1} + 2%	10 min	Freshness of asparagus was preserved the asparagus's physicochemical characteristics.	Wang and Fan (2019)
Tomatoes	US 50% power		1.5 min	After 150 s of US treatment at 50% intensity and 90 s at 75% strength, 100% peroxidase inhibition was noted.	Ercan and Soysal (2011)
Litchi	US 40 kHz at 120 W		10 min	After harvest litchi fruit, US postponed the browning of epicarp and reduced the activity of POD and PPO during the early stages of storage.	Chen et al. (2012)
Mushrooms	US 20 kHz at 400 W		10 min	The use of US prevented PPO from reacting and improved the preservation of the white coloration in the mushrooms.	Lagnika et al. (2013)
Tomatoes	US 45 kHz		19 min	With the use of US, the overall tomato was preserved.	Pinheiro et al. (2015)
Cherries	US + PA 40 kHz at 400 W and 0.4%		20–40 min	A 30- to 40-min US exposure improved color stability with better quality.	Muzaffar et al. (2016)
Jujubes	US + $CaCl_2$ 40 kHz at 350 W and 10 g L^{-1}		10 min	Polygalacturonase and pectate lyase activities were inhibited by the combination of US and $CaCl_2$.	Zhi et al. (2017)
Loquats	US + PA 40 kHz at 400 W + 0.4%		6 min	Total phenolics and flavonoids were found to be higher in the US-PA combination along with preserved TSS and ascorbic acid concentration. Compared to the control, loquat fruits underwent the combination treatment retained superior quality.	Ling et al. (2018)

(Continued)

TABLE 9.2 (Continued)

CROPS	TREATMENTS	FREQUENCY	TIME	INFLUENCES	REFERENCES
Bok choy	US + MAP with 30 kHz at 2.4 W g^{-1}		10 min	The combined effects of packing in a modified environment and US effectively lowered activity of POD and PPO.	Zhang et al. (2019)
Pak choy	US + aqueous ClO_2 80 kHz at 180 W + 50 ppm		10 min	The effects of US in conjunction with ClO_2 treatment on ascorbic acid, total phenolic levels, membrane leakage, and the breakdown of chlorophyll were reduced.	Wu et al. (2019)
Apples	US + 1-MCP 0.9 μL/L 33 kHz at 60 W		20 hr	1-MCP + US treatment increased firmness, inhibited the increase of SSC, and raised MdETR1 and MdERS1 gene expression levels in comparison to untreated group.	Xu et al. (2019)
Strawberries	US 40 kHz at 350 W		10 min	Biochemical attributes were maintained, and the firmness was effectively maintained by US treatment. The shelf-life of strawberries was increased by using the combinational treatment.	Cao et al. (2010)

Source: PA = Peracetic acid, AA = Acetic acid, GA3 = Gibberellic acid, NaOCl = Sodium hypochlorite.

impact, which may be utilized to further improve the possibility of application.

9.5 POSSIBLE NEGATIVE INFLUENCES AND LIMITATIONS OF ULTRASONIC APPLICATION ON FRESH FRUIT AND VEGETABLE COMMODITIES

Because fresh produce possesses a significant amount of water and a crispy texture, the power of US frequently damages their cells and tissues. When US is applied, it is occasionally discovered that the intended preservation impact does not occur and that it can actually have an adverse impact on the treated produce. Though treatment can eliminate traces of pesticides on pak choi leaves, physical damage and a drop in vitamin C levels were noted in the US-exposed leaves (Zhu et al., 2019). Furthermore, research revealed that altering the intensity of the ultrasonic waves lowered the proportion of injured leaves, and no visible harm to the leaves was observed while the ultrasound's wavelength was set at 135 kHz. Nonetheless, vitamin C reduction was seen in such circumstances, demonstrating that pak choi leaf injury wasn't the sole cause of vitamin depletion. Neto et al. (2019) discovered that ultrasonic treatment reduces the physiological texture of lettuce. Muzaffar et al. (2016) demonstrated that the hardness of cherries was considerably reduced after 20 min of US treatment. According to Vivek et al. (2017), the ultrasonic treatment lowered the firmness and AA level of *Docynia indica* by 3% and 5%, respectively, compared to the untreated sample. Pieczywek et al. (2017) discovered that US waves on apples improved total pectin dissolution in fruit while creating extracellular gaps in fruit cells, thereby resulting in textural changes.

Occasionally, US treatment can result in a substantial temperature rise (Leong et al., 2017), which isn't ideal for the preservation of fresh produce since low-frequency ultrasonic application may produce cavity bubbles. Neto et al. (2019)'s research demonstrated that the antibacterial impact of sonic treatment was reduced and the number of bacteria in the specimens grew quickly at higher temperatures. It was demonstrated by Zhang et al. (2020) that as the ultrasonic heat increased, the cherry tomato's cell wall integrity became lower. Regarding the decrease of enzymatic functions by ultrasonics, experiments revealed that because the majority of enzymes are highly resistant to US, strong US treatment takes an extended period and negatively affects treated produce, reducing its taste and texture. Even shorter wave ultrasonography is available that facilitates enzymatic interactions with reagents by dissolving bigger particles, potentially boosting the performance of the enzymes (Wang et al., 2018). Furthermore, several ultrasonic procedures call for the use of extremely high sound frequency ranges, which, if safety measures are not followed, may cause the operator to lose their auditory sense. In factories, it can occasionally be challenging to produce sonic fields with enough intensity (Mothibe et al., 2011). There are a few restrictions on the use of ultrasonography that must be taken into account while using this type of equipment in fruits and vegetables at higher doses.

9.6 CONCLUSIONS AND FUTURE PERSPECTIVES

The effectiveness of US for the substitution, augmentation, and advancement of many traditional processing techniques in the field has been supported by numerous studies. According to the vast majority of publications, US (25–50 kHz) enhances the yields of polyphenols, carotenoids, flavonoids, and essential oils depending on the temperature, pressure, and time parameters that are applied. The effects of cavitation (20–40 kHz, 300 W) on various granules enhance the effectiveness of physical, chemical, and enzyme-based reactions for bioactive polysaccharide and protein extraction. Several phytochemicals, such as carotenoids, were severely oxidized and destroyed by higher power intensity (400–600 W). Thus, further work should be performed to learn more about a product's chemical composition and use of high-frequency US (>100 kHz). Fruits and vegetables can be preserved using non-heating methods such as US application. In addition, fresh fruits and vegetables may be made safer and greener by using ultrasonic disinfection to remove traces of pesticides, cleaning fruit and vegetable surfaces more effectively, and using fewer chemical cleansers. Additionally, PPO, POD, and cell wall breakdown of enzyme function may be inhibited; sterilization can be achieved; and better physical and chemical attributes of fruits and vegetables can be maintained with the use of ultrasonics. Moreover, the use of ultrasonic processing actively contributes to the stabilization of sugars, thereby in the cell wall, enhancement of the physical characteristics of cell integrity and preservation of fruit firmness can be ensured. On the other side, highly intense ultrasonics, for instance, deteriorate the tissue of fresh produce, reducing its firmness and increasing nutritional losses. Additionally, more costly apparatus and increased power consumption are restricting the adoption of ultrasonography at a commercial scale. Due to these challenges, further studies are required in the coming years to explore the potential applications of ultrasonics in the postharvest of fresh produce such as fruits and vegetables. These studies should focus on integrating US as an additional conservation technique with conventional methods of preservation or some other suitable alternative and effective treatments.

ACKNOWLEDGMENTS

The authors would like to thank the "Belt and Road" joint project fund between Zhejiang University, China, and the National Research Centre, Egypt (SQ2023YFE0103360) for funding the current chapter. Additionally, this research work was supported by the National Natural Science Foundation of China (32171889) and the Key R&D Projects in Zhejiang Province (2022C02044, 2023C02009).

REFERENCES

Abdelkebir, R., Alcántara, C., Falcó, I., Sánchez, G., Garcia-Perez, J. V., Neffati, M., Lorenzo, J. M., Barba, F. J., Collado, M. C. (2019). Effect of ultrasound technology combined with binary mixtures of ethanol and water on antibacterial and antiviral activities of *Erodium glaucophyllum* extracts. *Innovative Food Science & Emerging Technologies*, *52*, 189–196.

Ahmed, F. E., Gouda, M. M., Hussein, L. A., Ahmed, N. C., Vos, P. W., Mohammad, M. A. (2017). Role of melt curve analysis in interpretation of nutrigenomics' MicroRNA expression data. *Cancer Genomics & Proteomics*, *14*(6), 469–481.

Ali, S., Khan, A. S., Malik, A. U., Anwar, R., Anjum, M. A., Nawaz, A., Naz, S. (2021). Combined application of ascorbic and oxalic acids delays postharvest browning of litchi fruits under controlled atmosphere conditions. *Food Chemistry*, *350*, 129277. https://doi.org/10.1016/j.foodchem.2021.129277

Ali, S., Khan, A. S., Nawaz, A., Naz, S., Ejaz, S., Shah, A. A., Haider, M. W. (2023). The combined application of Arabic gum coating and γ-aminobutyric acid mitigates chilling injury and maintains eating quality of 'Kinnow' mandarin fruits. *International Journal of Biological Macromolecules*, *236*, 123966. https://doi.org/10.1016/j.ijbiomac.2023.123966

Arvanitoyannis, I. S., Kotsanopoulos, K. V., Savva, A. G. (2017). Use of ultrasounds in the food industry-Methods and effects on quality, safety, and organoleptic characteristics of foods: A review. *Critical Reviews in Food Science and Nutrition*, *57*(1), 109–128.

Azam, S. R., Ma, H., Xu, B., Devi, S., Siddique, M. A. B., Stanley, S. L., Zhu, J. (2020). Efficacy of ultrasound treatment in the removal of pesticide residues from fresh vegetables: A review. *Trends in Food Science & Technology*, *97*, 417–432.

Balashov, S. M., Rocha, J. M., Hurtado, M. R. F., Prestes, J. A. L., de Campos, A. F. M., Moshkalev, S. A. (2021). Improved stability and performance of surface acoustic wave nanosensors using a digital temperature compensation. *Frontiers in Sensors*, *2*.

Boulatov, R. (2021). The liberating force of ultrasound. *Nature Chemistry*, *13*(2), 112–114.

Brilhante São José, J. F., Dantas Vanetti, M. C. (2012). Effect of ultrasound and commercial sanitizers in removing natural contaminants and *Salmonella enterica* Typhimurium on cherry tomatoes. *Food Control*, *24*(1).

Bringas, E., Saiz, J., Ortiz, I. (2011). Kinetics of ultrasound-enhanced electrochemical oxidation of diuron on boron-doped diamond electrodes. *Chemical Engineering Journal*, *172*(2–3), 1016–1022.

Bu, X., Alheshibri, M. (2021). The effect of ultrasound on bulk and surface nanobubbles: A review of the current status. *Ultrasonics Sonochemistry, 76*, 105629.

Cao, S., Hu, Z., Pang, B. (2010). Optimization of postharvest ultrasonic treatment of strawberry fruit. *Postharvest Biology and Technology*, *55*(3), 150–153.

Cao, S., Hu, Z., Pang, B., Wang, H., Xie, H., Wu, F. (2010). Effect of ultrasound treatment on fruit decay and quality maintenance in strawberry after harvest. *Food Control*, *21*(4), 529–532.

Casalinuovo, I. A., Pierro, D., Bruno, E., Francesco, P., Coletta, M. (2006). Experimental use of a new surface acoustic wave sensor for the rapid identification of bacteria and yeasts. *Letters of Applied Microbiology*, *42*(1), 24–29.

Cengiz, M. F., Başlar, M., Basançelebi, O., Kılıçlı, M. (2018). Reduction of pesticide residues from tomatoes by low intensity electrical current and ultrasound applications. *Food Chemistry*, *267*, 60–66.

Chandrapala, J., Oliver, C., Kentish, S., Ashokkumar, M. (2012). Ultrasonics in food processing—Food quality assurance and food safety. *Trends in Food Science & Technology*, *26*(2), 88–98.

Chemat, F., Khan, M. K. (2011). Applications of ultrasound in food technology: Processing, preservation and extraction. *Ultrasonics Sonochemistry*, *18*(4), 813–835.

Chen, Y.C., Chang, W.T., Cheng, C.C., Shen, J.Y., Kao, K.S. (2014). Development of human IgE biosensor using Sezawa-mode SAW devices. *Current Applied Physics*, *14*(4), 608–613.

Chen, F., Zhang, M., Yang, C. H. (2020). Application of ultrasound technology in processing of ready-to-eat fresh food: A review. *Ultrasonics Sonochemistry*, *63*, 104953.

Chen, Y., Jiang, Y., Yang, S., Yang, E. N., Yang, B., Prasad, K. N. (2012). Effects of ultrasonic treatment on pericarp browning of postharvest litchi fruit. Journal of Food Biochemistry, *36*(5), 613–620.

Chen, Z., Zhu, C. (2011). Combined effects of aqueous chlorine dioxide and ultrasonic treatments on postharvest storage quality of plum fruit (Prunus salicina L.). *Postharvest Biology and Technology, 61*(2–3), 117–123.

Chiu, Y. H., Sandoval-Insausti, H., Ley, S. H., Bhupathiraju, S. N., Hauser, R., Rimm, E. B., Chavarro, J. E. (2019). Association between intake of fruits and vegetables by pesticide residue status and coronary heart disease risk. *Environment International*, *132*, 105113.

Dang, Z., Liu, X., Wang, X., Li, M., Jiang, Y., Wang, X., Yang, Z. (2019). Comparative effectiveness and safety of traditional Chinese medicine supporting Qi and enriching blood for cancer related anemia in patients not receiving chemoradiotherapy: A meta-analysis and systematic review. *Drug Des Devel Ther*, *13*, 221–230.

Day, L., Xu, M., Øiseth, S. K., Mawson, R. (2012). Improved mechanical properties of retorted carrots by ultrasonic pre-treatments. *Ultrasonics Sonochemistry*, *19*(3), 427–434.

de Jong, M., Chen, W., Geerlings, H., Asta, M., Persson, K. A. (2015). A database to enable discovery and design of piezoelectric materials. *Scientific Data*, *2*, 150053.

de Lima Leite, R. H., Cognet, P., Wilhelm, A. M., Delmas, H. (2002). Anodic oxidation of 2, 4-dihydroxybenzoic acid for wastewater treatment: Study of ultrasound activation. *Chemical Engineering Science*, *57*(5), 767–778.

de São José, J. F. B., Vanetti, M. C. D. (2015). Application of ultrasound and chemical sanitizers to watercress, parsley and strawberry: Microbiological and physicochemical quality. *LWT-Food Science and Technology*, *63*(2), 946–952.

do Rosário, D. K. A., da Silva Mutz, Y., Peixoto, J. M. C., Oliveira, S. B. S., de Carvalho, R. V., Carneiro, J. C. S., Bernardes, P. C. (2017). Ultrasound improves chemical reduction of natural contaminant microbiota and *Salmonella enterica subsp. enterica* on strawberries. *International Journal of Food Microbiology*, *241*, 23–29.

Dore, A., Molinu, M. G., Venditti, T., D'hallewin, G. (2013). Use of high-intensity ultrasound to increase the efficiency of imazalil in postharvest storage of citrus fruits. *Food and Bioprocess Technology*, *6*, 3029–3037.

Duan, B., Shao, X., Han, Y., Li, Y., Zhao, Y. (2021). Mechanism and application of ultrasound-enhanced bacteriostasis. *Journal of Cleaner Production*, *290*, 125750.

Duarte, A. L. A., do Rosário, D. K. A., Oliveira, S. B. S., de Souza, H. L. S., de Carvalho, R. V., Carneiro, J. C. S., Bernardes, P. C. (2018). Ultrasound improves antimicrobial effect of sodium dichloroisocyanurate to reduce *Salmonella Typhimurium* on purple cabbage. *International Journal of Food Microbiology*, *269*, 12–18.

Elibariki, R., Maguta, M. M. (2017). Status of pesticides pollution in Tanzania–A review. *Chemosphere*, *178*, 154–164.

Ercan, S. Ş., Soysal, Ç. (2011). Effect of ultrasound and temperature on tomato peroxidase. *Ultrasonics Sonochemistry*, *18*(2), 689–695.

Esua, O. J., Chin, N. L., Yusof, Y. A., Sukor, R. (2019). Combination of ultrasound and ultraviolet-C irradiation on kinetics of color, firmness, weight loss, and total phenolic content changes in tomatoes during storage. *Journal of Food Processing and Preservation, 43*(10), e14161.

Fan, K., Zhang, M., Jiang, F. (2019). Ultrasound treatment to modified atmospheric packaged fresh-cut cucumber: Influence on microbial inhibition and storage quality. *Ultrasonics Sonochemistry*, *54*, 162–170.

Fogel, R., Limson, J., Seshia, A. A. (2016). Acoustic biosensors. *Essays in Biochemistry*, *60*(1), 101–110.

Fouad, M. T., Moustafa, A., Hussein, L., Romeilah, R., Gouda, M. (2015). In-vitro antioxidant and antimicrobial activities of selected fruit and vegetable juices and fermented dairy products commonly consumed in Egypt. *Research Journal of Pharmaceutical Biological and Chemical Sciences*, *6*(2), 541–550.

Gamboa-Santos, J., Montilla, A., Soria, A. C., Villamiel, M. (2012). Effects of conventional and ultrasound blanching on enzyme inactivation and carbohydrate content of carrots. *European Food Research and Technology*, *234*, 1071–1079.

Gani, A., Baba, W. N., Ahmad, M., Shah, U., Khan, A. A., Wani, I. A., Gani, A. (2016). Effect of ultrasound treatment on physico-chemical, nutraceutical and microbial quality of strawberry. *LWT-Food Science and Technology*, *66*, 496–502.

Ghazzawy, H. S., Gouda, M. M., Awad, N. S., Al-Harbi, N. A., Alqahtani, M. M., Abdel-Salam, M. M., Abdein, M. A., Al-Sobeai, S. M., Hamad, A. A., Alsberi, H. M., Gabr, G. A., Hikal, D. M. (2022). Potential bioactivity of Phoenix dactylifera fruits, leaves, and seeds against prostate and pancreatic cancer cells. *Frontiers in Nutrition*, *9*, 998929.

Giacometti, J., Bursac Kovacevic, D., Putnik, P., Gabric, D., Bilusic, T., Kresic, G., Stulic, V., Barba, F. J., Chemat, F., Barbosa-Canovas, G., Rezek Jambrak, A. (2018). Extraction of bioactive compounds and essential oils from Mediterranean herbs by conventional and green innovative techniques: A review. *Food Research International*, *113*, 245–262.

Gouda, M., El-Din Bekhit, A., Tang, Y., Huang, Y., Huang, L., Li, X., He, Y. (2021a). Recent innovations of ultrasound green technology in herbal phytochemistry: A review. *Ultrasonics Sonochemistry*, 105538.

Gouda, M., Ghazzawy, H. S., Alqahtani, N., Li, X. (2023a). The recent development of acoustic sensors as effective chemical detecting tools for biological cells and their bioactivities. *Molecules*, *28*(12).

Gouda, M., He, Y., Bekhit, A. E., Li, X. (2022a). Emerging technologies for detecting the chemical composition of plant and animal tissues and their bioactivities: An editorial. *Molecules*, *27*(9).

Gouda, M., Moustafa, A., Hussein, L., Hamza, M. (2016). Three week dietary intervention using apricots, pomegranate juice or/and fermented sour sobya and impact on biomarkers of antioxidative activity, oxidative stress and erythrocytic glutathione transferase activity among adults. *Nutrition Journal*, *15*(1), 52.

Gouda, M., Nassarawa, S. S., Gupta, S. D., Sanusi, N. I., Nasiru, M. M. (2023b). Evaluation of carbon dioxide elevation on phenolic compounds and antioxidant activity of red onion (*Allium cepa* L.) during postharvest storage. *Plant Physiology and Biochemistry, 200*.

Gouda, M., Tadda, M. A., Zhao, Y., Farmanullah, F., Chu, B., Li, X., He, Y. (2022b). Microalgae bioactive carbohydrates as a novel sustainable and eco-friendly source of prebiotics: Emerging health functionality and recent technologies for extraction and detection. *Frontiers in Nutrition*, *9*, 806692.

Gouda, M., Chen, K., Li, X., Liu, Y., He, Y. (2021b). Detection of microalgae single-cell antioxidant and electrochemical potentials by gold microelectrode and Raman micro-spectroscopy combined with chemometrics. *Sensors and Actuators B: Chemical*, *329*, 129229.

Guerrero, S. N., Ferrario, M., Schenk, M., Carrillo, M. G. (2017). Hurdle technology using ultrasound for food preservation. In *Ultrasound: Advances for food processing and preservation* (pp. 39–99). Academic Press.

Hsieh, Y. H., Li, Y., Pan, Z., Chen, Z., Lu, J., Yuan, J., Zhu, Z., Zhang, J. (2019). Ultrasonication-assisted synthesis of alcohol-based deep eutectic solvents for extraction of active compounds from ginger. *Ultrasonics Sonochemistry*, *63*, 104915.

Huo, S., Zhao, P., Shi, Z., Zou, M., Yang, X., Warszawik, E., Loznik, M., Gostl, R., Herrmann, A. (2021). Mechanochemical bond scission for the activation of drugs. *Nature Chemistry*, *13*(2), 131–139.

Jang, J. H., Moon, K. D. (2011). Inhibition of polyphenol oxidase and peroxidase activities on fresh-cut apple by simultaneous treatment of ultrasound and ascorbic acid. *Food Chemistry*, *124*(2), 444–449.

Jankowska, M., Łozowicka, B., Kaczyński, P. (2019). Comprehensive toxicological study over 160 processing factors of pesticides in selected fruit and vegetables after water, mechanical and thermal processing treatments and their application to human health risk assessment. *Science of the total Environment*, *652*, 1156–1167.

Jiang, X., Jin, H., Gui, R. (2020a). Visual bio-detection and versatile bio-imaging of zinc-ion-coordinated black phosphorus quantum dots with improved stability and bright fluorescence. *Biosensors and Bioelectronics*, *165*, 112390.

Jiang, Q., Zhang, M., Xu, B. (2020b). Application of ultrasonic technology in postharvested fruits and vegetables storage: A review. *Ultrasonics Sonochemistry*, *69*, 105261. https://doi.org/10.1016/j.ultsonch.2020.105261

Jiménez-Sánchez, C., Lozano-Sánchez, J., Segura-Carretero, A., Fernandez-Gutierrez, A. (2017). Alternatives to conventional thermal treatments in fruit-juice processing. Part 1: Techniques and applications. *Critical Reviews in Food Science and Nutrition*, *57*(3), 501–523.

Khademi, O., Ashtari, M., Razavi, F. (2019). Effects of salicylic acid and ultrasound treatments on chilling injury control and quality preservation in banana fruit during cold storage. *Scientia Horticulturae*, *249*, 334–339.

Khandpur, P., Gogate, P. R. (2016). Evaluation of ultrasound based sterilization approaches in terms of shelf life and quality parameters of fruit and vegetable juices. *Ultrasonics Sonochemistry*, *29*, 337–353.

Khodagholy, D., Malliaras, G. G., Owens, R. M. (2012). 8.05-Polymer-Based Sensors. In *Polymer science: A comprehensive reference*, vol. 8 (pp. 101–128). Elsevier.

Kilicli, M., Baslar, M., Durak, M. Z., Sagdic, O. (2019). Effect of ultrasound and low-intensity electrical current for microbial safety of lettuce. *LWT—Food Science and Technology*, *116*, 108509.

Kim, H.-P., Kang, W.-S., Hong, C.-H., Lee, G.-J., Choi, G., Ryu, J., Jo, W. (2020). Advanced Ceramics for Energy Conversion and Storage. In Olivier Guillon (ed.) *Piezoelectrics*. 157–206.

Kim, K., Cho, E., Thokchom, B., Cui, M., Jang, M., Khim, J. (2015). Synergistic sonoelectrochemical removal of substituted phenols: Implications of ultrasonic parameters and physicochemical properties. *Ultrasonics Sonochemistry*, *24*, 172–177.

Kiontke, A., Roudini, M., Billig, S., Fakhfouri, A., Winkler, A., Birkemeyer, C. (2021). Author correction: Surface acoustic wave nebulization improves compound selectivity of low-temperature plasma ionization for mass spectrometry. *Scientific Reports*, *11*(1), 11620.

Knorr, D., Zenker, M., Heinz, V., Lee, D. U. (2004). Applications and potential of ultrasonics in food processing. *Trends in Food Science & Technology*, *15*(5), 261–266.

Lagnika, C., Zhang, M., Mothibe, K. J. (2013). Effects of ultrasound and high pressure argon on physico-chemical properties of white mushrooms (*Agaricus bisporus*) during postharvest storage. *Postharvest Biology and Technology*, *82*, 87–94.

Lagnika, C., Zhang, M., Mothibe, K. J. (2013). Effects of ultrasound and high pressure argon on physico-chemical properties of white mushrooms (*Agaricus bisporus*) during postharvest storage. *Postharvest Biology and Technology*, *82*, 87–94.

Lakshmanan, A., Jin, Z., Nety, S. P., Sawyer, D. P., Lee-Gosselin, A., Malounda, D., Swift, M. B., Maresca, D., Shapiro, M. G. (2020). Publisher correction: Acoustic biosensors for ultrasound imaging of enzyme activity. *Nat Chem Biol*, *16*(9), 1035.

Lange, K., Gruhl, F. J., Rapp, M. (2013). Surface Acoustic Wave (SAW) biosensors: Coupling of sensing layers and measurement. *Methods Mol Biol*, *949*, 491–505.

Lange, K., Rapp, M. (2008). Influence of intermediate aminodextran layers on the signal response of surface acoustic wave biosensors. *Analytical Biochemistry*, *377*(2), 170–175.

Leong, T., Juliano, P., Knoerzer, K. (2017). Advances in ultrasonic and megasonic processing of foods. *Food Engineering Reviews*, *9*(3), 237–256.

Li, X. L., Sha, J. J., Chu, B. Q., Wei, Y. Z., Huang, W. H., Zhou, H., Xu, N., He, Y. (2019). Quantitative visualization of intracellular lipids concentration in a microalgae cell based on Raman micro-spectroscopy coupled with chemometrics. *Sensors and Actuators B-Chemical*, *292*, 7–15.

Liao, X., Li, J., Suo, Y., Chen, S., Ye, X., Liu, D., Ding, T. (2018). Multiple action sites of ultrasound on *Escherichia coli* and Staphylococcus aureus. *Food Science and Human Wellness*, *7*(1), 102–109.

Lin, L., Wang, X., Li, C., Cui, H. (2019). Inactivation mechanism of E. coli O157: H7 under ultrasonic sterilization. *Ultrasonics Sonochemistry*, *59*, 104751.

Ling, C., Xu, J., Shao, S., Wang, L., Jin, P., Zheng, Y. (2018). Effect of ultrasonic treatment combined with peracetic acid treatment reduces decay and maintains quality in loquat fruit. *Journal of Food Quality*, *2018*, 1–8.

Liu, J., Bi, J., McClements, D. J., Liu, X., Yi, J., Lyu, J., Zhou, M., Verkerk, R., Dekker, M., Wu, X., Liu, D. (2020). Impacts of thermal and non-thermal processing on structure and functionality of pectin in fruit- and vegetable- based products: A review. *Carbohydrate Polymers*, *250*, 116890.

Löbl, H. P., Klee, M., Milsom, R., Dekker, R., Metzmacher, C., Brand, W., Lok, P. (2001). Materials for bulk acoustic wave (BAW) resonators and filters. *Journal of the European Ceramic Society*, *21*(15), 2633–2640.

Lopez, P., Vercet, A., Sanchez, A. C., Burgos, J. (1998). Inactivation of tomato pectic enzymes by manothermosonication. *Zeitschrift für Lebensmitteluntersuchung und-Forschung A*, *207*, 249–252.

Lozowicka, B., Jankowska, M., Hrynko, I., Kaczynski, P. (2016). Removal of 16 pesticide residues from strawberries by washing with tap and ozone water, ultrasonic cleaning and boiling. *Environmental Monitoring and Assessment*, *188*, 1–19.

Lv, J.M., Gouda, M., El-Din Bekhit, A., He, Y.K., Ye, X.Q., Chen, J.C. (2022). Identification of novel bioactive proanthocyanidins with potent antioxidant and anti-proliferative activities from kiwifruit leaves. *Food Bioscience*, *46*.

Lv, J. M., Gouda, M., Zhu, Y. Y., Ye, X. Q., Chen, J. C. (2021). Ultrasound-assisted extraction optimization of proanthocyanidins from kiwi (*Actinidia chinensis*) leaves and evaluation of its antioxidant activity. *Antioxidants (Basel)*, *10*(8).

Micheli, F. (2001). Pectin methylesterases: Cell wall enzymes with important roles in plant physiology. *Trends in Plant Science*, *6*(9), 414–419.

Millan-Sango, D., Garroni, E., Farrugia, C., Van Impe, J. F. M., Valdramidis, V. P. (2016). Determination of the efficacy of ultrasound combined with essential oils on the decontamination of Salmonella inoculated lettuce leaves. *LWT—Food Science and Technology*, *73*, 80–87.

Mohebbi, S., Babalar, M., Zamani, Z., Askari, M. A. (2020). Influence of early season boron spraying and postharvest calcium dip treatment on cell-wall degrading enzymes and fruit firmness in 'Starking Delicious' apple during storage. *Scientia Horticulturae*, *259*, 108822.

Mothibe, K. J., Zhang, M., Nsor-atindana, J., Wang, Y.C. (2011). Use of ultrasound pretreatment in drying of fruits: Drying rates, quality attributes, and shelf life extension. *Drying Technology*, *29*(14), 1611–1621.

Mujahid, A., Dickert, F. (2017). Surface Acoustic Wave (SAW) for chemical sensing applications of recognition layers. *Sensors*, *17*(12).

Muzaffar, S., Ahmad, M., Wani, S. M., Gani, A., Baba, W. N., Shah, U., Wani, T. A. (2016). Ultrasound treatment: Effect on physicochemical, microbial and antioxidant properties of cherry (*Prunus avium*). *Journal of Food Science and Technology*, *53*, 2752–2759.

Nakayama, R.I., Imai, M. (2013). Promising ultrasonic irradiation pretreatment for enzymatic hydrolysis of Kenaf. *Journal of Environmental Chemical Engineering*, *1*(4), 1131–1136.

Naveed, F., Nawaz, A., Ali, S., Ejaz, S. (2024). Xanthan gum coating delays ripening and softening of jujube fruit by reducing oxidative stress and suppressing cell wall polysaccharides disassembly. *Postharvest Biology and Technology*, *209*, 112689. https://doi.org/10.1016/j.postharvbio.2023.112689

Neto, L., Millan-Sango, D., Brincat, J. P., Cunha, L. M., Valdramidis, V. P. (2019). Impact of ultrasound decontamination on the microbial and sensory quality of fresh produce. *Food Control*, *104*, 262–268.

Nguyen, L., Duong, L. T., Mentreddy, R. S. (2019). The U.S. import demand for spices and herbs by differentiated sources. *Journal of Applied Research on Medicinal and Aromatic Plants*, *12*, 13–20.

Nguyen, T. M. C., Gavahian, M., Tsai, P.J. (2021). Effects of ultrasound-assisted extraction (UAE), high voltage electric field (HVEF), high pressure processing (HPP), and combined methods (HVEF+UAE and HPP+UAE) on Gac leaves extraction. *LWT—Food Science and Technology*, *143*, 111131.

Nicolae, I., Viespe, C., Miu, D., Marcu, A. (2022). Analyte discrimination by SAW sensor variable loop amplification probing. *Sensors and Actuators B: Chemical*, *358*.

Ojha, K. S., Aznar, R., O'Donnell, C., Tiwari, B. K. (2020). Ultrasound technology for the extraction of biologically active molecules from plant, animal and marine sources. *TrAC Trends in Analytical Chemistry*, *122*, 115663.

Olaimat, A. N., Holley, R. A. (2012). Factors influencing the microbial safety of fresh produce: A review. *Food Microbiology*, *32*(1), 1–19.

Oliveira, M., Sulaiman, A. (2019). Polyphenoloxidase in fruit and vegetables: Inactivation by thermal and non-thermal processes. *Encyclopedia of Food Chemistry*.

Palmer, T., Bonner, P. L. (2007). *Enzymes: Biochemistry, biotechnology, clinical chemistry*. Elsevier.

Pandiselvam, R., Kaavya, R., Jayanath, Y., Veenuttranon, K., Lueprasitsakul, P., Divya, V., Ramesh, S.V. (2020). Ozone as a novel emerging technology for the dissipation of pesticide residues in foods–a review. *Trends in Food Science & Technology*, *97*, 38–54.

Paniwnyk, L. (2014). Application of ultrasound. In *Emerging technologies for food processing* (pp. 271–291). Academic Press.

Park, S. Y., Mizan, M. F. R., Ha, S. D. (2016). Inactivation of Cronobacter sakazakii in head lettuce by using a combination of ultrasound and sodium hypochlorite. *Food Control*, *60*, 582–587.

Pieczywek, P. M., Kozioł, A., Konopacka, D., Cybulska, J., Zdunek, A. (2017). Changes in cell wall stiffness and microstructure in ultrasonically treated apple. *Journal of Food Engineering*, *197*, 1–8.

Pinheiro, J., Alegria, C., Abreu, M., Gonçalves, E. M., Silva, C. L. M. (2015). Influence of postharvest ultrasounds treatments on tomato (*Solanum lycopersicum*, cv. Zinac) quality and microbial load during storage, *Ultrasonics Sonochemistry*, *27*, 552–559.

Pongprasert, N., Sekozawa, Y., Sugaya, S., Gemma, H. (2011). A novel postharvest UV-C treatment to reduce chilling injury (membrane damage, browning and chlorophyll degradation) in banana peel. *Scientia Horticulturae*, *130*(1), 73–77.

Rashed, M. M., Tong, Q., Abdelhai, M. H., Gasmalla, M. A., Ndayishimiye, J. B., Chen, L., Ren, F. (2016). Effect of ultrasonic treatment on total phenolic extraction from Lavandula pubescens and its application in palm olein oil industry. *Ultrasonics Sonochemistry*, *29*, 39–47.

Raut-Jadhav, S., Pinjari, D. V., Saini, D. R., Sonawane, S. H., Pandit, A. B. (2016). Intensification of degradation of methomyl (carbamate group pesticide) by using the combination of ultrasonic cavitation and process intensifying additives. *Ultrasonics Sonochemistry*, *31*, 135–142.

Rocha-Gaso, M. I., March-Iborra, C., Montoya-Baides, A., Arnau-Vives, A. (2009). Surface generated acoustic wave biosensors for the detection of pathogens: A review. *Sensors (Basel)*, *9*(7), 5740–5769.

Sagong, H. G., Cheon, H. L., Kim, S. O., Lee, S. Y., Park, K. H., Chung, M. S., Kang, D. H. (2013). Combined effects of ultrasound and surfactants to reduce Bacillus cereus spores on lettuce and carrots. *International Journal of Food Microbiology*, *160*(3), 367–372.

Sala, F. J., Burgos, J., Condon, S., Lopez, P., Raso, J. (1995). Effect of heat and ultrasound on microorganisms and enzymes. In *New methods of food preservation* (pp. 176–204). Springer US.

Samani, B. H., Khoshtaghaza, M. H., Minaei, S., Zareifourosh, H., Eshtiaghi, M. N., Rostami, S. (2016). Design, development and evaluation of an automatic fruit-juice pasteurization system using microwave–ultrasonic waves. *Journal of Food Science and Technology*, *53*(1), 88.

Sánchez-Rubio, M., Taboada-Rodríguez, A., Cava-Roda, R., López-Gómez, A., Marín-Iniesta, F. (2016). Combined use of thermo-ultrasound and cinnamon leaf essential oil to inactivate Saccharomyces cerevisiae in natural orange and pomegranate juices. *LWT—Food Science and Technology*, *73*, 140–146.

Sivagamasundari, V., Sannasiraj, S. A. (2020). Experimental study of vertical wave-in-deck force and pressure on a thin plate due to regular and focused waves. *Journal of Ocean Engineering and Marine Energy*, *6*(2), 199–210.

Sledz, M., Wiktor, A., Rybak, K., Nowacka, M., Witrowa-Rajchert, D. (2016). The impact of ultrasound and steam blanching pre-treatments on the drying kinetics, energy consumption and selected properties of parsley leaves. *Applied Acoustics*, *103*, 148–156.

Stanley, K. D., Golden, D. A., Williams, R. C., Weiss, J. (2004). Inactivation of Escherichia coli O157: H7 by high-intensity ultrasonication in the presence of salts. *Foodbourne Pathogens & Disease*, *1*(4), 267–280.

Sukor, N. F., Jusoh, R., Kamarudin, N. S., Abdul Halim, N. A., Sulaiman, A. Z., Abdullah, S. B. (2020). Synergistic effect of probe sonication and ionic liquid for extraction of phenolic acids from oak galls. *Ultrasonics Sonochemistry*, *62*, 104876.

Tadda, M. A., Gouda, M., Shitu, A., Yu, Q., Zhao, X., Ying, L., Zhu, S., Liu, D. (2023). Baobab fruit powder promotes denitrifiers' abundance and improves poly(butylene succinate) biodegradation for a greener environment. *Journal of Environmental Chemical Engineering*, *11*(3).

Taha, K. K., Modwi, A., Elamin, M. R., Arasheed, R., Al-Fahad, A. J., Albutairi, I., Arasheed, H.A., Alfaify, M., Anojaidi, K., Algethami, F. K., Bagabas, A. (2020). Impact of Hibiscus extract on the structural and activity of sonochemically fabricated ZnO nanoparticles. *Journal of Photochemistry and Photobiology A: Chemistry*, *390*, 112263.

Tang, S. Q., Du, Q. H., Fu, Z. (2020). Ultrasonic treatment on physicochemical properties of water-soluble protein from Moringa oleifera seed. *Ultrasonics Sonochemistry*, *71*, 105357.

Tao, F., Zhang, M., Yu, H. Q. (2007). Effect of vacuum cooling on physiological changes in the antioxidant system of mushroom under different storage conditions. *Journal of Food Engineering*, *79*(4), 1302–1309.

Tekaya, N., Gammoudi, I., Braiek, M., Tarbague, H., Morotė, F., Raimbault, V., Sakly, N., Rebière, D., Ben Ouada, H., Lagarde, F., Ben Ouada, H., Cohen-Bouhacina, T., Dejous, C., Jaffrezic Renault, N. (2013). Acoustic, electrochemical and microscopic characterization of interaction of Arthrospira platensis biofilm and heavy metal ions. *Journal of Environmental Chemical Engineering*, *1*(3), 609–619.

Teng, Z., Han, K., Li, J., Gao, Y., Li, M., Ji, T. (2020). Ultrasonic-assisted preparation and characterization of hierarchical porous carbon derived from garlic peel for high-performance supercapacitors. *Ultrasonics Sonochemistry*, *60*, 104756.

Terefe, N. S., Gamage, M., Vilkhu, K., Simons, L., Mawson, R., Versteeg, C. (2009). The kinetics of inactivation of pectin methylesterase and polygalacturonase in tomato juice by thermosonication. *Food Chemistry*, *117*(1), 20–27.

Tiwari, B. K., Muthukumarappan, K., O'Donnell, C. P., Cullen, P. J. (2009). Inactivation kinetics of pectin methylesterase and cloud retention in sonicated orange juice. *Innovative Food Science & Emerging Technologies*, *10*(2), 166–171.

Tostado-Plascencia, M. M., Sanchez-Tizapa, M., Zamudio-Ojeda, A. (2018). Synthesis and characterization of multiwalled carbon nanotubes functionalized with chlorophyll-derivatives compounds extracted from Hibiscus tiliaceus. *Diamond and Related Materials*, *89*, 151–162.

Valcke, M., Bourgault, M. H., Rochette, L., Normandin, L., Samuel, O., Belleville, D., Phaneuf, D. (2017). Human health risk assessment on the consumption of fruits and vegetables containing residual pesticides: A cancer and non-cancer risk/benefit perspective. *Environment International*, *108*, 63–74.

Venkatakrishnan, K., Chiu, H.F., Wang, C.K. (2019). Popular functional foods and herbs for the management of type-2-diabetes mellitus: A comprehensive review with special reference to clinical trials and its proposed mechanism. *Journal of Functional Foods*, *57*, 425–438.

Vercet, A., Lopez, P., Burgos, J. (1999). Inactivation of heat-resistant pectinmethylesterase from orange by manothermosonication. *Journal of Agricultural and Food Chemistry*, *47*(2), 432–437.

Vercet, A., Sánchez, C., Burgos, J., Montañés, L., Buesa, P. L. (2002). The effects of manothermosonication on tomato pectic enzymes and tomato paste rheological properties. *Journal of Food Engineering*, *53*(3), 273–278.

Vivek, K., Mishra, S., Sasikumar, R. (2017). Effect of ultra-sonication on postharvest quality parameters and microbial load on. *Scientia Horticulturae*, *225*, 163–170.

Wang, D., Yan, L., Ma, X., Wang, W., Zou, M., Zhong, J., Liu, D. (2018). Ultrasound promotes enzymatic reactions by acting on different targets: Enzymes, substrates and enzymatic reaction systems. *International Journal of Biological Macromolecules*, *119*, 453–461.

Wang, J., Fan, L. (2019). Effect of ultrasound treatment on microbial inhibition and quality maintenance of green asparagus during cold storage. *Ultrasonics Sonochemistry*, *58*, 104631.

Wang, J. L., Guo, Y. J., Li, D. J., Long, G. D., Tang, Q. B., Zu, X. T., Ma, J. Y., Du, B., Tang, Y. L., Torun, H., Fu, Y. Q. (2020). Bacterial cellulose coated ST-cut quartz surface acoustic wave humidity sensor with high sensitivity, fast response and recovery. *Smart Materials and Structures*, *29*(4).

Wen, C., Zhang, J., Zhou, J., Duan, Y., Zhang, H., Ma, H. (2018). Effects of slit divergent ultrasound and enzymatic treatment on the structure and antioxidant activity of arrowhead protein. *Ultrasonics Sonochemistry*, *49*, 294–302.

Wordon, B. A., Mortimer, B., McMaster, L. D. (2012). Comparative real-time analysis of Saccharomyces cerevisiae cell viability, injury and death induced by ultrasound (20 kHz) and heat for the application of hurdle technology. *Food Research International*, *47*(2), 134–139.

Wu, W., Gao, H., Chen, H., Fang, X., Han, Q., Zhong, Q. (2019). Combined effects of aqueous chlorine dioxide and ultrasonic treatments on shelf-life and nutritional quality of bok choy (*Brassica chinensis*). *LWT-Food Science and Technology*, *101*, 757–763.

Xie, F., Xu, Y., Xia, K., Jia, C., Zhang, P. (2016). Alternate pulses of ultrasound and electricity enhanced electrochemical process for p-nitrophenol degradation. *Ultrasonics Sonochemistry*, *28*, 199–206.

Xin, Q., Zhang, X., Li, Z., Lei, L. (2009). Sterilization of oil-field re-injection water using combination treatment of pulsed electric field and ultrasound. *Ultrasonics Sonochemistry*, *16*(1), 1–3.

Xu, B., Yuan, J., Wang, L., Lu, F., Wei, B., Azam, R. S., Bhandari, B. (2020). Effect of multi-frequency power ultrasound (MFPU) treatment on enzyme hydrolysis of casein. *Ultrasonics Sonochemistry*, *63*, 104930.

Xu, F., Liu, S., Xiao, Z., Fu, L. (2019). Effect of ultrasonic treatment combined with 1-methylcyclopropene (1-MCP) on storage quality and ethylene receptors gene expression in harvested apple fruit. *Journal of Food Biochemistry*, *43*(8), e12967.

Yildirim, A., Öner, M. D., Bayram, M. (2011). Fitting Fick's model to analyze water diffusion into chickpeas during soaking with ultrasound treatment. *Journal of Food Engineering*, *104*(1), 134–142.

Yusoff, I. M., Mat Taher, Z., Rahmat, Z., Chua, L. S. (2022). A review of ultrasound-assisted extraction for plant bioactive compounds: Phenolics, flavonoids, thymols, saponins and proteins. *Food Research International, 157*, 111268.

Zhang, L., Wang, P., Sun, X., Chen, F., Lai, S., Yang, H. (2020). Calcium permeation property and firmness change of cherry tomatoes under ultrasound combined with calcium lactate treatment. *Ultrasonics Sonochemistry*, *60*, 104784.

Zhang, M., De Baerdemaeker, J., Schrevens, E. (2003). Effects of different varieties and shelf storage conditions of chicory on deteriorative color changes using digital image processing and analysis. *Food Research International*, *36*(7), 669–676.

Zhang, X. T., Zhang, M., Devahastin, S., Guo, Z. (2019). Effect of combined ultrasonication and modified atmosphere packaging on storage quality of pakchoi (*Brassica chinensis* L.). *Food and Bioprocess Technology*, *12*, 1573–1583.

Zhang, L., Zhao, S., Lai, S., Chen, F., Yang, H. (2018). Combined effects of ultrasound and calcium on the chelate-soluble pectin and quality of strawberries during storage. *Carbohydrate Polymers*, *200*, 427–435.

Zhao, Y., Wen, C., Feng, Y., Zhang, J., He, Y., Duan, Y., Zhang, H., Ma, H. (2021). Effects of ultrasound-assisted extraction on the structural, functional and antioxidant properties of Dolichos lablab L. Protein. *Process Biochemistry*, *101*, 274–284.

Zhao, Y., Zhang, J., Gouda, M., Zhang, C., Lin, L., Nie, P., Ye, H., Huang, W., Ye, Y., Zhou, C., He, Y. (2022). Structure analysis and non-invasive detection of cadmium-phytochelatin2 complexes in plant by deep learning Raman spectrum. *Journal of Hazardous Materials*, *427*, 128152.

Zhi, H., Liu, Q., Xu, J., Dong, Y., Liu, M., Zong, W. (2017). Ultrasound enhances calcium absorption of jujube fruit by regulating the cellular calcium distribution and metabolism of cell wall polysaccharides. *Journal of the Science of Food and Agriculture*, *97*(15), 5202–5210.

Zhou, Q., Bian, Y., Peng, Q., Liu, F., Wang, W., Chen, F. (2019). The effects and mechanism of using ultrasonic dishwasher to remove five pesticides from rape and grape. *Food Chemistry*, *298*, 125007.

Zhu, Y., Zhang, T., Xu, D., Wang, S., Yuan, Y., He, S., Cao, Y. (2019). The removal of pesticide residues from pakchoi (*Brassica rape* L. ssp. chinensis) by ultrasonic treatment. *Food Control*, *95*, 176–180.

Zong, W., Gouda, M., Cai, E., Wang, R., Xu, W., Wu, Y., Munekata, P. E. S., Lorenzo, J. M. (2021). The Antioxidant Phytochemical Schisandrin A Promotes Neural Cell Proliferation and Differentiation after Ischemic Brain Injury. *Molecules*, *26*(24).

Zou, H., Wang, L. (2017). The disinfection effect of a novel continuous-flow water sterilizing system coupling dual-frequency ultrasound with sodium hypochlorite in pilot scale. *Ultrasonics Sonochemistry*, *36*, 246–252.

PART IV

Pressure Manipulations-Based Applications

Advances in Hypobaric Storage of Fresh Horticultural Produce

10

Samina Khalid*, Hafiz Muhammad Rashad Javeed, and Hafiz Umer Javed
*Corresponding Author: saminakhalid@cuiatd.edu.pk

10.1 INTRODUCTION

The atmosphere is made up of many gases in varying concentrations, with nitrogen accounting for 78% of the total and being mixed with 21% O_2, 0.9% argon, 0.04% CO_2, water vapor (variable), and trace gases. The composition of these gases has a profound impact on living beings, including plants and animals. Horticultural product is still alive and respiring after harvest, and the gaseous composition of the storage atmosphere might alter their postharvest life. Changing the gaseous composition around fresh fruit may reduce its storage life.

Hypobaric storage, often known as sub-atmospheric pressure or low-pressure storage (pressure less than 101 kPa), is a relatively new postharvest method for keeping food fresh. The principle of Dalton's law asserts that the overall gas mixture pressure is equal to the sum of the partial pressures of each of the constituent gases (Thompson, 2015). When overall pressure is reduced, the partial pressure of all gases and water vapor is reduced as well. A low-oxygen atmosphere is achieved in hypobaric storage by lowering the total pressure of air, which lowers the partial pressure of O_2. Burg and Burg in 1966 established the notion of hypobaric storage for horticultural goods under low-pressure settings, typically at 50 kPa. The partial pressure of each gas component in air is decreased by drawing air and creating a vacuum in direct proportion to the overall pressure. The O_2 partial pressure at 10.1 kPa total pressure, for instance, is comparable to 2.1 kPa at atmospheric pressure. Lower oxygen levels slow down the rate of aerobic respiration, reducing the breakdown of respiratory substrate and production of energy prerequisite for biochemical events related with the ripening process of fruits.

Hypobaric storage has been acknowledged for a long time and is a re-emerging technology with consistent usage (Vigneault et al., 2012). It can accurately and regularly manage the inner temperature and air configuration (Wenxiang et al., 2006). This storage strategy not only swiftly removes heat in the vacuum cooling method (Vigneault et al., 2008) and harmful gases that may be produced by the respiration of fresh product in storage (Wang et al., 2001), but it also reduces the oxygen level. Because oxygen concentrations can fall to anaerobic levels, the containers are opened, fresh air circulates on a regular basis, and hypobaric conditions are re-established. Hypobaric storage of fresh produce improves the self-defense system against microbial attack and external pressure, reduces metabolic reactions, and extends shelf life.

Excessive moisture loss in hypobaric storage has frequently been reported incorrectly as limiting its utility for fresh produce. Many modern hypobaric treatment systems can now maintain high humidity levels within the treatment compartment, reducing moisture loss and shriveling in fresh produce as well as respiration and endogenous ethylene production (Burg, 2004). The transformation of 1-aminocyclopropane carboxylic acid (ACC) to ethylene requires oxygen during the ethylene synthesis process. Low oxygen levels in hypobaric conditions may reduce ethylene generation, thereby extending fresh produce ripening (Beaudry, 1999).

10.2 HYPOBARIC STORAGE MECHANISM

Air is made up of various gases as well as moisture in the form of water vapors. When the partial pressure of air is decreased in hypobaric storage, the partial pressure of all of its constituents is also reduced. Burg and Burg studied gas exchange in fruits in 1966 and discovered that the internal concentration of CO_2

DOI: 10.1201/9781003370376-14

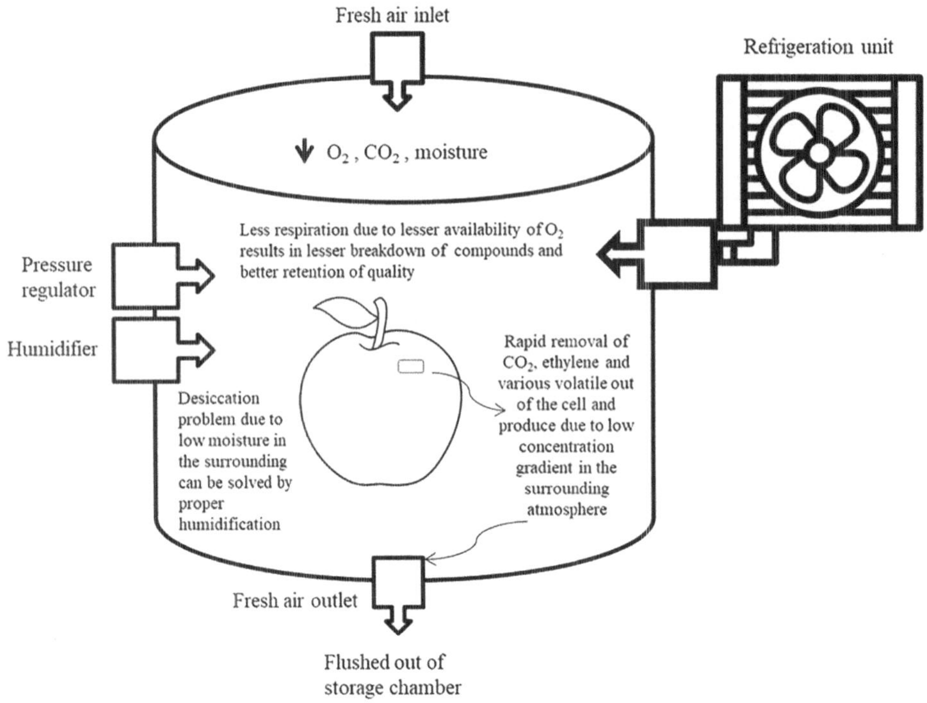

FIGURE 10.1 Schematic diagram of hypobaric storage chamber.

and ethylene in fruits varies directly as a function of atmospheric pressure. Commodities are stored in hypobaric storage systems under low atmospheric pressure with continuous fresh air ventilation, high relative humidity, and low temperature to prevent desiccation (Figure 10.1). Lower oxygen levels in hypobaric storage slow respiration and metabolic degradation of various compounds, thereby resulting in better produce quality retention. Because of their lower availability in the surrounding environment, ethylene, CO_2, and volatile compounds produced inside fresh produce cells rapidly move outside (Figure 10.1). When the air is flushed out, these compounds are removed from the storage atmosphere, reducing the harmful effects on fresh produce. Lower moisture content in the surrounding atmosphere causes moisture to move outward from fresh produce, resulting in drying and desiccation. This issue can be solved by keeping the storage environment at a higher relative humidity (Figure 10.1).

10.3 EFFECTS OF HYPOBARIC STORAGE ON DIFFERENT ATTRIBUTES OF FRUITS AND VEGETABLES

10.3.1 Effect of Hypobaric Storage on Respiration

At atmospheric pressure (101 pKa), air is composed of 78.9% nitrogen, 20.9% oxygen, and 0.03% CO_2, with partial pressures of 78.9, 20.9, and 0.03 kPa, respectively. Dalton's law states that "The partial pressure of each gas component in air is altered in direct proportion to the total pressure as a result of properties of gas". As a result, lowering of total absolute pressure from 101 kPa to 10.13494 kPa reduces the O_2 partial pressure from 21% to around 2.1%. This slows aerobic respiration, reducing carbohydrate catabolism (respiration-related attributes) and production of energy prerequisite for biochemical processes related with fruit ripening. In comparison to normal atmospheric conditions, hypobaric storage of Chinese bayberry reduced the fruit's respiration rate (Chen et al., 2013b). Loquat respiration rate was also reduced by hypobaric storage (Gao et al., 2006). In blueberry cv. Bluecrop, hypobaric storage had little effect on respiratory rate (Li et al., 2019). Hypobaric storage of green asparagus considerably suppressed respiration; prohibited the breakdown of chlorophyll, ascorbic acid, and acidity; enhanced sensory traits; and postponed senescence during postharvest (Wenxiang et al., 2006). In contrast to these findings, Tovar et al. (2011) discovered an increased mango respiration under hypobaric conditions of 34 kPa for 20 minutes.

10.3.2 Ethylene Synthesis in Hypobaric Storage

The transformation of ACC to ethylene requires oxygen in the ethylene synthesis. Low oxygen levels in hypobaric conditions may decrease ethylene synthesis, postponing fresh produce ripening (Beaudry, 1999) (Figure 10.2). The use of $KMnO_4$ in conjunction with low pressure storage (50 kPa) reduced ethylene production in tomato fruit (Muhammad et al., 2022).

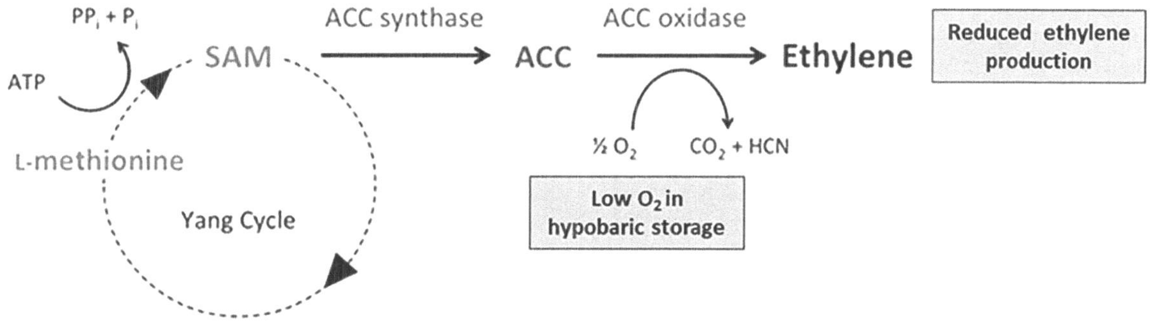

FIGURE 10.2 Ethylene production in hypobaric storage adopted from Chang (2016) with some modifications.

10.3.3 Gas Diffusion (O_2, CO_2, and Ethylene Volatiles)

As previously stated, one of the most significant effects of low-pressure conditions is a decrease in O_2 partial pressure. The effects of storing fruits and vegetables in low-oxygen environments have been shown to improve their postharvest life. Burg (2010) stated that "the higher rate of gas diffusion at a lower pressure excludes the commodities surface to central O_2 gradient produced in response to consumption of O_2 related respiration, resulting different commodity categories to produce closely identical lower O_2 tolerance, near 0.1%". Burg (2004) described that fermentation does not occur under hypobaric conditions at O_2 concentration as low as 0.06–0.15%, whereas fermentation can happen at atmospheric pressures.

During respiration, CO_2 is produced and diffuses outward in intracellular spaces, eventually moving out through the cuticle or other openings, while O_2 moves inward through an inverse process (Burton, 1982). The partial pressure of gases (CO_2 and O_2) in sub-atmospheric storage is low. Fick's law states that

$$J = A.\frac{\Delta C}{R}$$

where J is the gas flux (m^2s^{-1}), A is the surface area (m^2), C is the concentration gradient of gas across the barrier, and R is the resistance to gas diffusion (sm^{-1}).

This indicates that in a plant system gas movement is proportional to the surface area and concentration difference of the gas across the barrier and inversely proportional to barrier resistance (Ben-Yehoshua & Rodov, 2003). In storage, low-pressure conditions have been found to speed up the outward movement of gas from horticultural crop inner tissues (Goszczynska & Ryszard, 1988). CO_2 formation increases in plant tissue during the respiration process, resulting in CO_2 release into the intercellular spaces. Finally, CO_2 moves from the intercellular spaces to the storage chamber (where the concentration of CO_2 is low) to sustain the gaseous equilibrium between plant tissue and the adjacent environment (Burton, 1982). Burg (2004) discovered that low-pressure conditions can reduce ambient and within cell CO_2, which is a significant benefit that cannot be reproduced by accumulating CO_2 to concentrations in some CA storage recommendations. The higher CO_2 was required for effective storage of Waldin avocados, even in low-pressure storage (Spalding & Reeder, 1976). In avocados the CO_2 was assumed to be vital in managing rotting and alleviating chilling injury (Yahia, 2011), and CO_2 may not be included in a hypobaric system. According to Burg (2004), the influences of hypobaric conditions on CO_2 removal from intercellular and cellular spaces may lead to reduced fungal and bacterial growth, improved fruit ascorbic acid preservation, inactivation of the ethylene-forming enzyme, and inhibition of succinate formation. Wells (1974) discovered that under hypobaric conditions, postharvest pathogens such as *Erwinia atroseptica*, *E. carotovora*, and *P. fluorescens* were incapable of proliferating in the very low CO_2 concentrations as well as lower O_2 concentrations found in the cells of vegetables, validating the influence of CO_2 on bacterial proliferation. Nevertheless, Enfors and Molin (1980) discovered that when *P. fragi* was allowed to grow under low O_2 (0.0025 atmosphere O_2) and uncovered to 0.99 atmosphere CO_2, the CO_2 impeding effects were added to the O_2 limitation. Cucumbers when kept under low-pressure conditions of 532 mm Hg for 6 hours may display an indirect stress response of prolonged open stomata that happens only when the fruit is moved to ambient air conditions may be due to CO_2 deficiency in hypobaric storage, according to Laurin et al. (2006).

Little ethylene is synthesized inside the cell in a low O_2 storage environment. Because this ethylene concentration is greater than that of the surrounding atmosphere, ethylene is radially removed from the cells and into the surrounding environment. Because the hypobaric chamber is persistently ventilated and depleted of air, C_2H_4 is synthesized by fresh produce and will be expelled out. This C_2H_4 removal may decrease its impact on fruit ripening as well as the progression of some physiological disorders related with ethylene buildup. According to Burg and Kosson (1983), low-pressure storage reduces inner concentrations of volatiles such as C_2H_4. Chen et al. (2013a) also found that fresh bamboo stems kept at 2 °C under 50 kPa for 35 days produced less C_2H_4 than those kept at atmospheric condition, delaying textural changes. Chen et al. (2013b) found no substantial differences in ethylene synthesis between hypobaric and normal atmospheric storage in Chinese bayberries.

The CO_2, ethylene, and numerous volatile by-products generated during various metabolic activities instantly move out of produce and are produced out of storage chambers in hypobaric storage at constantly ventilated partial pressure (Burg, 2004). Ripening and senescence of fresh produce are postponed and storage life is extended as a result of the hypobaric conditions and low levels of ethylene in the storage atmosphere. Other volatiles synthesized by fresh produce throughout storage may be removed by hypobaric storage. Bangerth (1984) discovered that apples kept below 50–75 mm Hg for extended periods did not produce characteristic flavor and aroma upon the exposure to ambient air condition and allowed to ripen. This could be owing to reduced partial O_2 pressure, as the similar effects have been described for apple fruits kept under CA conditions (Burg, 2004).

10.3.4 Hypobaric Storage on Moisture Loss

The rate of moisture loss from fresh fruits and vegetables is higher at lower air pressure, given that the speed of evaporation is inversely proportional to pressure. The reduced store pressure reduces water's boiling point from 100 °C at normal pressure conditions to 0 °C at 4.6 mm Hg, resulting in rapid water loss from fresh produce. Furthermore, partial vacuum decreases the amount of water molecules in the surrounding air, which raises the vapor pressure deficit and thus the driving force of transpiration (Burg & Kosson, 1983).

The problem of inadequate relative humidity in hypobaric storage of fresh produce could be solved by spraying water during storage (Li et al., 2006). Mass loss can also be affected by specific hypobaric pressures used for the storage of fresh produce. Apelbaum et al. (1977) discovered that a pressure less than 50 mm Hg in hypobaric storage increases mango fruit desiccation. According to Cicale and Jamieson (1978), as cited by Burg (2004), lower storage pressures below 61 mm Hg result in higher moisture loss in avocados. According to Patterson and Melsted (1977), cherries exhibited more shriveling when stored in hypobaric conditions. Capsicum kept in hypobaric conditions lost more moisture than capsicum stored at atmospheric conditions (Hughes et al., 1981). An et al. (2009) described that curled lettuce kept in hypobaric conditions had a higher desiccation rate than those kept at normal atmospheric conditions. Burg (2004) discovered that spring onions at 0–3 °C, dissipate less mass during low-pressure storage than under atmospheric conditions. In general, radishes lost less weight when stored under low pressure (56 mm Hg) at 1 °C and high relative humidity than when kept at ambient air conditions (McKeown & Lougheed, 1980). According to Burg's (2004) study, Cicale and Jamieson (1978) have shown that for 35-day storage at 6 °C avocados dissipated 1.2% of their mass at pressure from 61 to 203 mm Hg than with 5.7% in ambient air conditions. Spalding and Reeder (1976) concluded that humidity conditions did not seem to be a reason in avocado postharvest life as the acceptability of avocados kept in low-pressure storage at 80–85% RH and 98–100% RH did not differ considerably.

The humidification of air as it comes into the chambers is part of any successful hypobaric system, though Burg (2004) noted that this was not always sufficient in a number of cases. McKeown and Lougheed (1980) reported in the literature that the maximum mass loss in asparagus happened when the pressure was decreased and the humidity was not near saturation. Laurin et al. (2006) described that spraying with water can solve the desiccation problem in cucumbers in low-pressure storage. In his hypobaric patent application Burg stated that, "Even if relative humidity of 80% is usefully compatible for storing fruits and vegetables, a preferable relative humidity to around 90% shall be maintained in storage chamber" (Burg, 1976). According to Burg and Kosson (1983), lower air pressure near the plant tissues reduces the cells' hydrostatic pressure. This decrease in pressure may result in a decrease in cellular water potential, but it will only have a minor effect on cellular activity, and changes of this magnitude within 20 atmospheres are likely to produce no more than 3% or 4% reductions. Burton (1989) noted the vapor pressures gradient that persists between the surface of fresh produce and the surrounding air. A considerable portion of fresh produce is composed of water, which is transported from cell to cell and then to the commodity's surface, where it evaporates into the atmosphere by the latent heat of the commodity. The vacuum cooling of fresh produce is based on this principle. Some fresh produce have a waxy cuticle, while certain root crops have a periderm that prevents moisture loss. The permeability of fresh produce tissues to gaseous exchange is influenced by intracellular space, cuticular composition, and stomata structure, lenticel, and hydrothode. According to Ben-Yehoshua and Rodov (2003) and Laurin et al. (2006), the dehydration of fresh produce under hypobaric storage is anticipated due to a rise in transpiration rate assisted by the properties and functions of stomata, lenticels, cuticle, and epidermal cells. Because grapes' tissues do not have a higher number of useful stomatas, moisture loss happens mainly via the cuticles, limiting water loss. Stomata seem to be affected by storage atmosphere. After 96 hours, cucumbers kept under hypobaric storage showed considerably higher opened stomata compared to those deposited at normal atmospheric conditions (Laurin et al., 2006). Dehydration under hypobaric conditions, they added, most likely occurs at stimulated rates of transpiration, which is assisted by stomatal action and properties, in addition to moisture content being relatively more volatile in nature at lower atmospheric pressures.

10.3.5 Disease Development

Because of the negative impact on consumers, synthetic methods for controlling postharvest diseases are discouraged. To control disease development in fruits and vegetables after harvest, many alternative techniques, including low-pressure storage, are being used. Hypobaric storage (0.1–0.25 0.008% partial pressure of O_2) inhibits spore germination and mycelial growth while also acting as a bactericidal agent (Burg, 2004). In hypobaric storage, however, Bangerth (1974) discovered a higher incidence of decay in tomatoes, peppers, and cucumbers. *P. digitatum* spore germination was inhibited at 100 mm Hg, *Alternaria alternata* and *B. cinerea* spore germination happened only at 50 mm Hg, but *Geotrichum candidum* spore germination could not be stopped even at 25 mm Hg, according to Barkai-Golan (1990). Romanazzi et al. (2001) described that hypobaric storage decreased fungal infestations in sweet cherries, table grapes, and strawberries. Adams et al. (1976) found that pressure reduced mycelial growth and patulin production when he explored the effects of diverse pressure (760–122 mm Hg) on the development of *P. patulum* and *P. expansum*, as well as production of toxic patulins. In oranges Archer et al. (2021) found that exposure to both low oxygen and low pressure (0.9 kPa O_2 and 6.6 kPa, respectively) treatments reduced in vivo growth of *P. digitatum* and *P. italicum*. Archer et al. (2019) also reported that oranges kept at 4 kPa pressure for 3 and 6 days at 10 °C reduced *P. digitatum* in vitro growth. Pristijono et al. (2017) investigated whether low pressure of 4 kPa decreased calyx rots in tomato fruits when compared to fruits stored at ambient atmospheres of 101 kPa. Strawberries kept under low pressure conditions of 50.7 kPa had reduced bacterial development (An et al. 2009). The use of $KMnO_4$ in conjunction with low-pressure storage (50 kPa) prevents decay in tomato fruit (Muhammad et al., 2022). Loquat fruit kept in hypobaric storage of 0.04–0.05 MPa had 87% reduced decay when compared to a control (Gao et al., 2006).

The hypobaric pressure in the range of 190, 380, and 570 mm Hg was applied for 4 hours at 20 °C and strawberries were kept at 20 or 5 °C; the lower pressure of 380 mm Hg was proved effective as it steadily postponed rot growth at both 20 and 5 °C, but it had no effect on mass loss, firmness, or an initial rise in respiration rate (Hashmi et al., 2013b). In another study, it was found that keeping strawberries at 380 mm Hg for 4 hours decreased rot development. The same treatment was effective on fruit inoculated with spores of *R. stolonifer* or *B. cinerea*. They discovered increased PAL, chitinase, and POD activity, but not PPO activity (Hashmi et al., 2013c).

10.3.6 Membrane Lipid Peroxidation

Malondialdehyde (MDA) is produced as a result of membrane lipid peroxidation. MDA levels and electrolyte leakages are indicators of cell membrane damage, which worsens with low-temperature storage (Zhao et al., 2006), resulting in chilling injuries to chilling-sensitive crops. Oxidative stress causes lipid peroxidation and sub-atmospheric pressure, which reduces oxidative stress and results in lower MDA content production. Li et al. (2008) found that hypobaric storage of asparagus could inhibit membrane lipid peroxidation. It also suppressed MDA accumulation in Chinese bayberries cv. Dongkui (Chen et al., 2013b). Li et al. (2006) reported similar findings when they investigated whether hypobaric storage could decrease MDA production and slow aging (Figure 10.3).

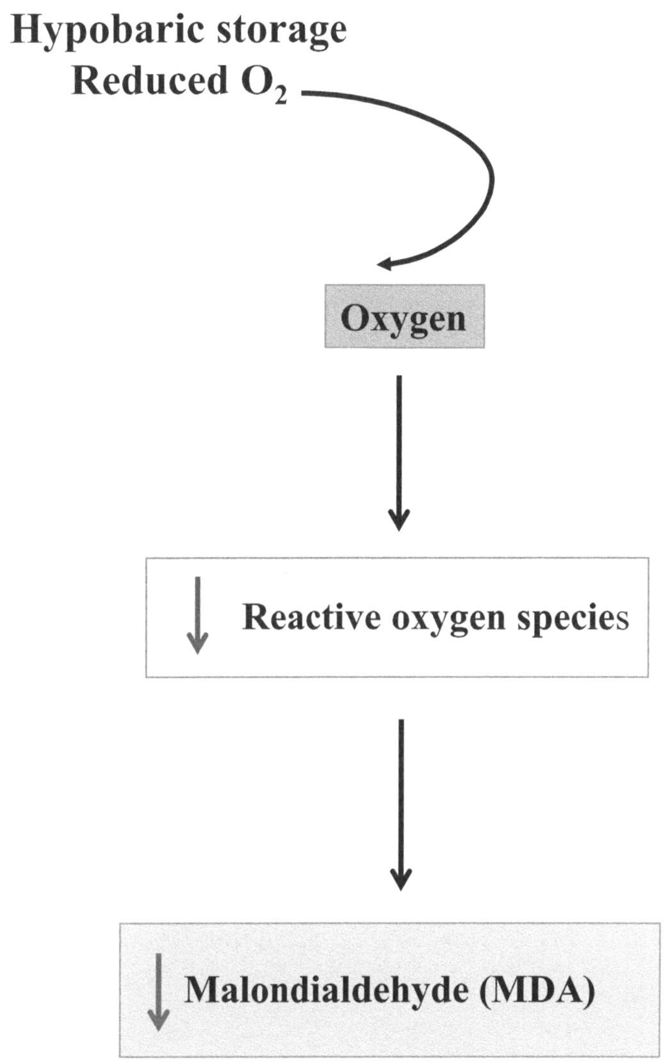

FIGURE 10.3 Effect of hypobaric storage on malondialdehyde (MDA).

10.3.7 Plant Defense Mechanisms and Bioactive Compounds

When respiration occurs reactive oxygen species (ROS) such as hydrogen peroxide, superoxide, and the hydroxyl radical are produced, which causes tissue damage and impairs the storage quality and marketability of fresh produce (Hodges et al., 2004). Because of the low partial pressure of oxygen in hypobaric storage, respiration slows down, resulting in less ROS accumulation, which delays the commencement of senescence in fresh produce (Hashmi et al., 2013b; Wang et al., 2015). Higher antioxidative enzyme activity such as catalase (CAT) and peroxidase (POD), which scavenge ROS, indicates that fresh produce has a greater storage potential. In peaches, low-pressure storage sustained high CAT activity (Chen et al. 2010). According to Chen et al. (2013b), low-pressure storage improves CAT and POD activity in Chinese bayberries. Hypobaric storage stimulates the activity of oxidative enzymes such as superoxide dismutase (SOD) and CAT, as well as controlling the rise

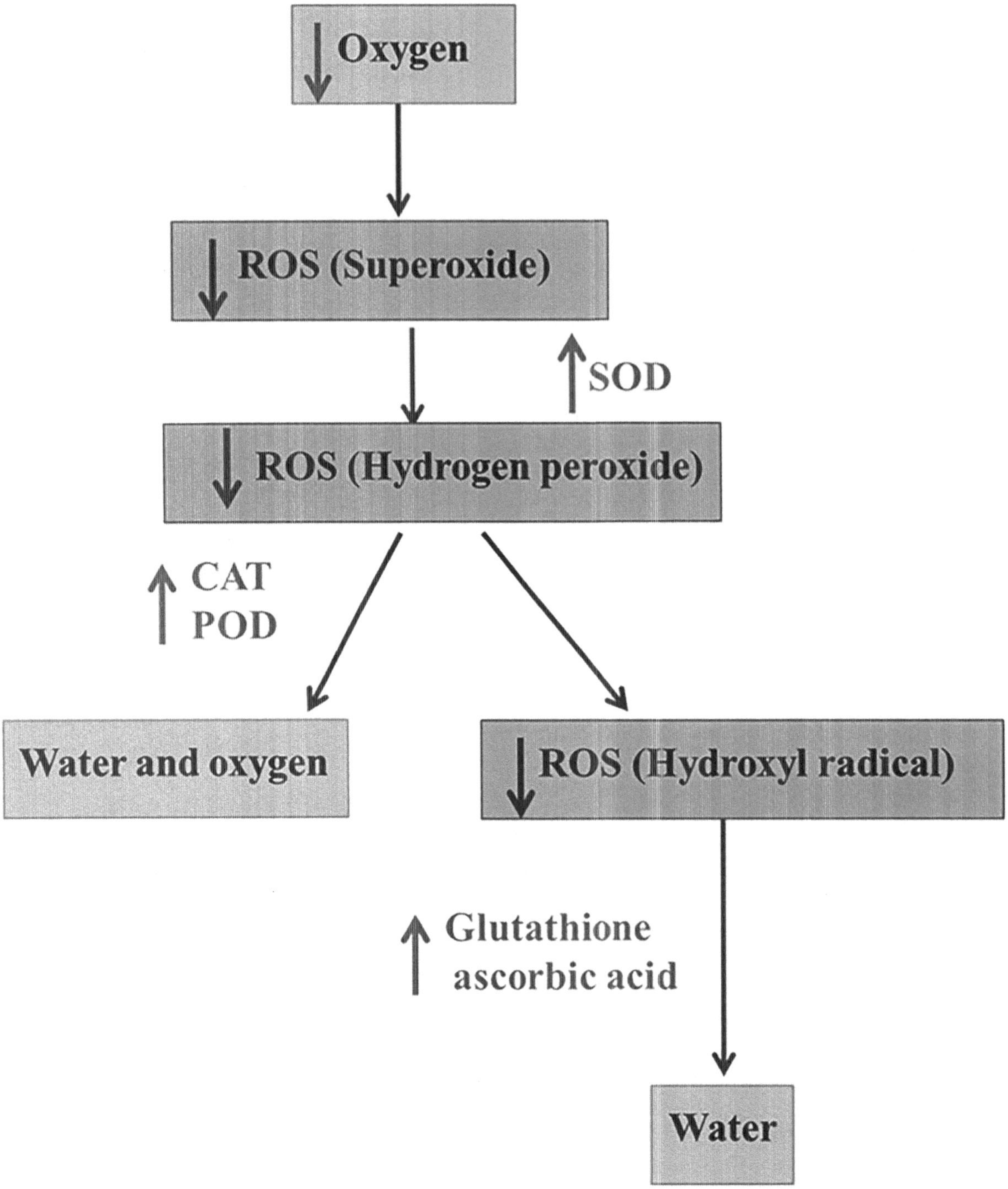

FIGURE 10.4 Effect of hypobaric storage on bioactive compounds and defense enzymes activity (Merksamer et al., 2013).

of POD activity (Li et al., 2008). Antioxidant capacity in fruits and vegetables are thought to be attributable to bioactive compounds such as anthocyanins, total flavonoids, and total phenols. According to Li et al. (2019), the effects of low-pressure storage on blueberry bioactive compounds was achieved by (i) promoting ROS breakdown via CAT, (ii) decreasing ROS and bioactive compound breakdown via PPO and POD, and (iii) increasing blueberry ROS scavenging and antioxidant capacity. Low-pressure storage decreased POD and PPO activities in loquat (Gao et al., 2006) and stimulated ascorbate peroxidase (APX) and glutathione reductase (GR) activity in peaches (Figure 10.4) (Song et al., 2016).

10.3.8 Color Characteristics

Peel color is a significant indicator of fresh produce quality. One of the major quality issues that occurs during transportation and storage is the degradation of green color due to chlorophyll (Chl) breakdown in green horticultural produce.

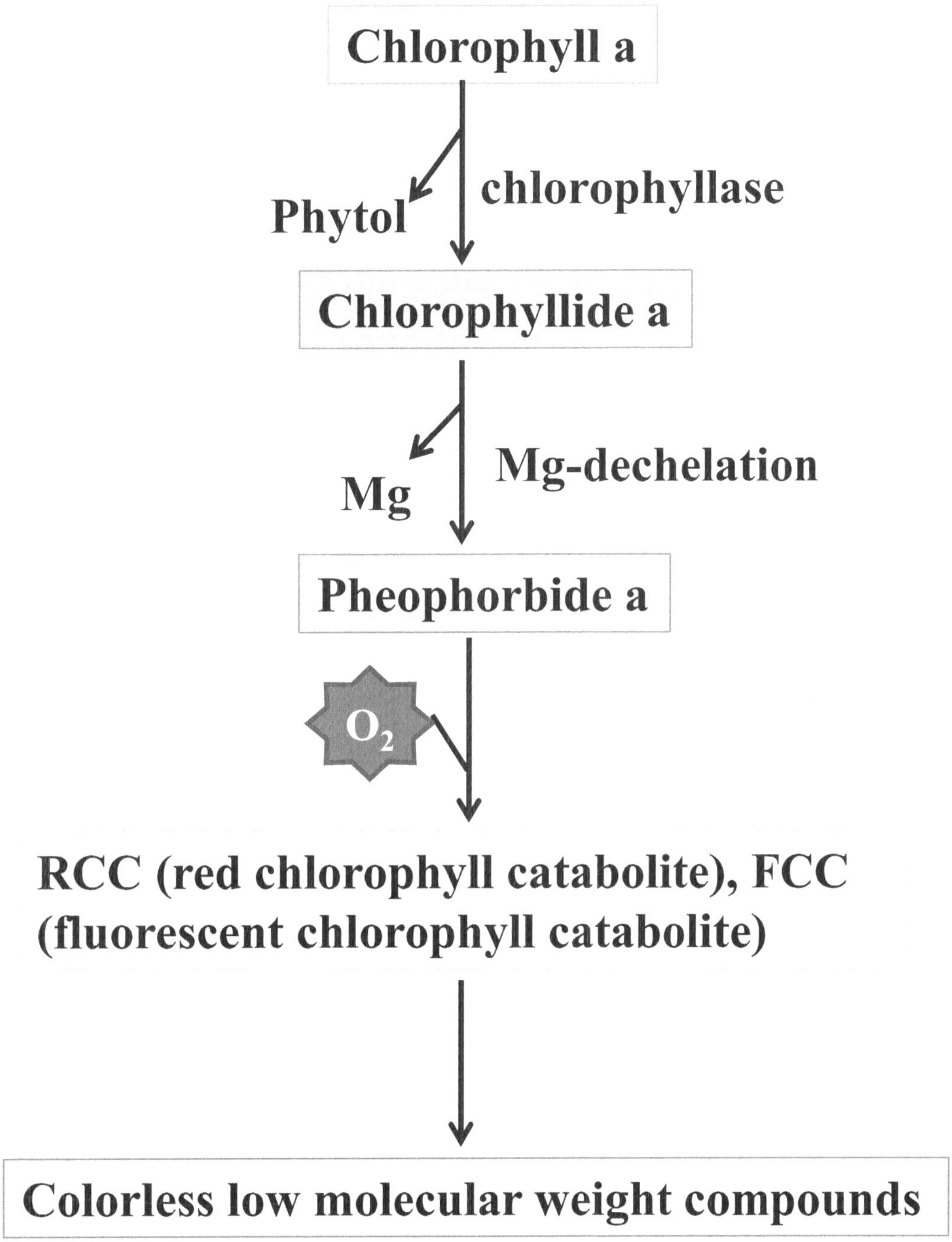

FIGURE 10.5 Chlorophyll degradation pathway in horticultural crops (Yamauchi, 2015).

After harvest, ROS production causes oxidative stress, which leads to senescence and chlorophyll loss. Tsuchiya et al. (1997) found that the chlorophyllase (Chlase) enzyme removes phytol from Chl a to produce chlorophyllide (Chlide) in the chlorophyll degradation pathway (Figure 10.5). Mg-dechelating substances (MDS) remove the Mg atom from Chlide a, forming pheophorbide (Pheide) a (Suzuki et al. 2005). In the presence of oxygen, Pheide a is transformed to red chlorophyll catabolite (RCC) and fluorescent chlorophyll catabolite (FCC). Lower oxygen levels in the sub-atmospheric environment reduce chlorophyll compound degradation (Hörtensteiner & Lee, 2007) (Figure 10.5).

Asparagus surface color degeneration was delayed by hypobaric storage (Li et al., 2008). Burg (2004) demonstrated that the skin of Acorn squash stayed green after 11 days of storage at 7 °C under hypobaric conditions of 7.33–8 kPa. There was no substantial change in skin color between fruit stored under hypobaric conditions of 4 kPa at 10 °C and fruit kept in normal atmospheric pressure of 101 kPa at either 10 °C or 20 °C. Under hypobaric conditions, slow chlorophyll degradation was observed in Marmande tomatoes (*Solanum lycopersicum*) (del Carmen Salas-Sanjuán et al., 2023). Cucumbers underwent the smallest changes in total color difference under hypobaric storage (10 kPa) (Zapotoczny &

Markowski, 2014). Color values were kept in apple cv. Royal Gala (Rahman et al. 2022) in a hypobaric chamber. Under low-pressure storage conditions, no chlorophyll degradation was observed in *Hibiscus-rosa-sinensis* (Skytt Andersen & Kirk, 1986). Green asparagus postharvest senescence was significantly delayed after hypobaric storage (Wenxiang et al., 2006). Zhang et al., (2021) investigated whether short-term hypobaric storage preserved the peel color of 'Taishanhong' pomegranate.

10.3.9 Insect Attack

According to Back and Cotton (1925), the effect of low hypobaric storage on insect mortality has been studied since the 1880s. Many researchers reported that exposing fruits and vegetables to low-pressure conditions disinfested them during export (Burg, 2010; Davenport et al., 2006; Johnson & Zettler, 2009; Mbata & Philips, 2001). Burg (2004) found that low-pressure conditions of 15–20 mm Hg effectively control the majority of green peach aphids on lettuce heads. Low O_2 contents (Navarro & Calderon, 1979) and low relative humidity (Jiao et al. 2012, 2013, Navarro, 1978) are the primary causes of insect mortality in hypobaric conditions. In hypobaric environments adenosine triphosphate synthesis was decreased, resulting in increased membrane phospholipid hydrolysis, instigating cell damage or mortality (Herreid, 1980; Mitcham et al., 2006). Davenport et al. (2006) stated that in mango fruits kept at 13 °C at 15–20 mm Hg with ≥98% relative humidity all the eggs and larvae of Caribbean fruit flies were killed. The percent mortality of eggs of *Rhagoletis pomonella* (Walsh) was higher when infested apples were kept under low-pressure conditions (Hulasare et al., 2013).

10.4 ADVANTAGES OF HYPOBARIC STORAGE ON FRESH HORTICULTURAL PRODUCE

Fresh produce are living entities that respond to relatively low pressure changes, i.e., absolute pressures ranging from 0 to 0.5 MPa. Hypobaric (low-pressure) storage can extend the storage and shelf life of perishable horticultural commodities (Figure 10.6). Reduced O_2 partial pressure is known to be advantageous to the storage and/or shelf life of various perishables.

10.4.1 Fruits

Romanazzi et al. (2003) applied short-term hypobaric storage treatments of 25 and 50 kPa for 4 h to sweet cherries and reported that 50 kPa treatments for 4 h reduced fungal decay compared to produce kept at normal air pressure. Romanazzi et al. (2001) reported that a 4-hour application of 50 kPa hypobaric pressure was the most useful treatment against mold and rot on sweet cherries and strawberries and a 24-hour application of 25 kPa reduced the occurrence of mold on table grapes. The investigators also discovered that after hypobaric treatment, wounded and inoculated table grapes had significantly lower infection and lesion diameter than controls. When compared to the control, the LP had no influence on fruit TSS, TA, or fruit respiration rates, ethylene synthesis, or fruit firmness while significantly increasing ethanol concentration in fruits and decreasing weight loss (Archer et al., 2021). Jinhua Wang et al. (2015) observed that the extended

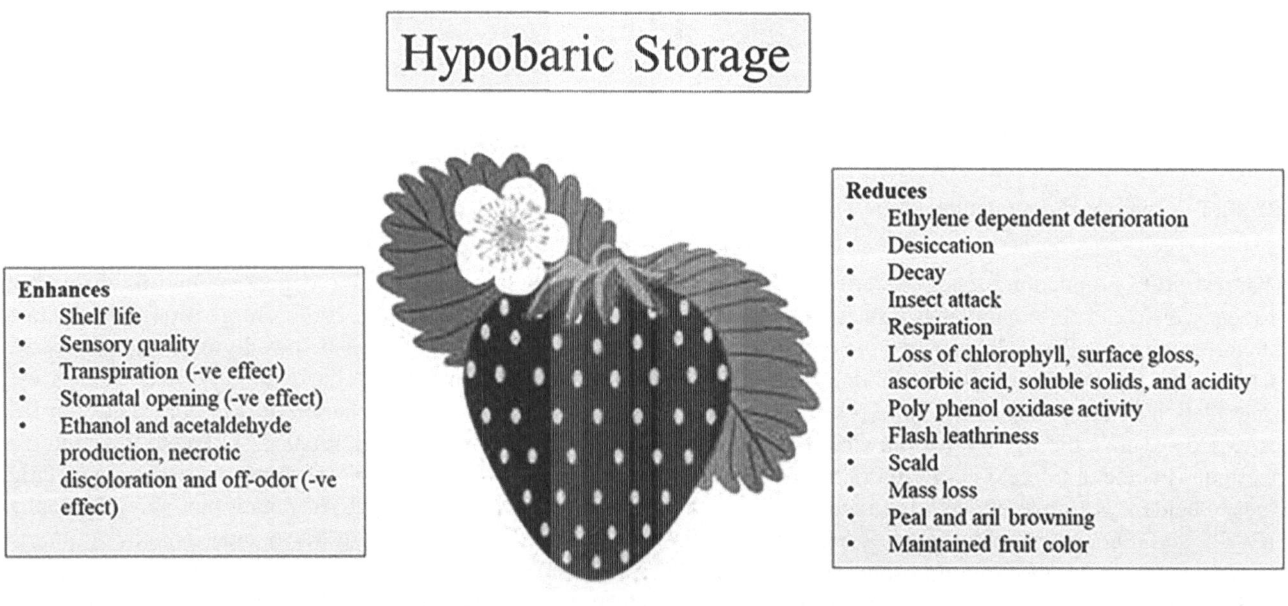

FIGURE 10.6 Effect of hypobaric storage on fresh produce quality.

shelf life of honey peaches in low-pressure storage might be due to improved energy status, superior antioxidative capacity, and reduced membrane leakage. In the highly vulnerable apple cultivars Law Rome and Granny Smith, hypobaric storage (at 5 kPa) reduces the development of superficial scald within one month of harvest, while it happens when kept in ambient air or in controlled atmosphere (Wang & Dilley, 2000). This treatment also decreases the formation of oxidative substances in fresh apples after 0.5 to 6 months of storage (Wang & Dilley, 2000). Wall et al. (2017) investigated the postharvest behavior of tropical fruits cherimoya, papaya, and mango stored under hypobaric conditions at 2.7 kPa pressure, 100% RH, and 13 °C; 2.7 kPa pressure, 100% RH, and 10 °C; and 2.7 kPa, 13 °C. They discovered that hypobaric storage increases the storage life of cherimoya by 10–12 days, papaya by 3–4 days, and mango by 7–10 days when compared to ambient storage. They also reported that hypobaric conditions delayed the ripening of these tropical fruits, but that hypobaric treatments did not reduce disease incidence after ripening. Huan et al. (2021) reported that hypobaric storage of kiwifruit at room temperature successfully reduced fruit rot, mass loss, alcoholic flavor development, acetaldehyde, ethanol, and enzymatic activities of pyruvate decarboxylase (PDC) and alcohol dehydrogenase (ADH), sustained greater levels of total phenolic, quinic acid, and citric acid. Hypobaric pressure (50 kPa) with 1-MCP treatment on Royal Gala apple fruit for 4 hours at low temperature preserved fruit quality parameters such as mass loss, pH, TSS, spoilage percentage, firmness, and juiciness for 120 days (Rehman et al., 2022). According to Huan et al. (2021), low-pressure treatment reduced ethylene generation and hence enhanced apple shelf life. Zhou et al. (2010) found that in Datuanmilu peach fruit, low temperature (2–3 °C) and hypobaric conditions (1.450–1 550 kPa) could significantly reduce respiration rate, decrease respiration peak, delay the onset of peak (P0.05), retain color, and decrease MDA content, as well as reduce free radical production rates and membrane leakage (Table 10.1).

Poor postharvest storage conditions cause physiological disorders in horticultural crops (Kader & Ben-Yehoshua, 2000). Hypobaric storage (40–50 kPa) was found to cause a prominent reduction in enzyme activities of PPO and POD, respiration and ethylene production, reduction in TA content and ascorbic acid, fruit browning, flesh leatheriness, and rottening of loquat fruits by 87% when compared to a control (Gao et al., 2006).

According to Chen et al. (2013b), Chinese bayberry fruit kept in low-pressure atmospheres (85, 55, and 155) had less decay, lower malonaldehyde accumulation and respiratory rate, higher catalase, peroxidase activities, titratable acidity, total phenolics, and ferric reducing antioxidant power than those stored under normal atmospheric conditions (101 kPa). Short-term hypobaric treatment reduced fruit decay rate, peel browning, maintained peel color, and increased titratable acidity of Taishanhong pomegranate, according to Zhang et al. (2021).

Zhang et al. (2021) reported that short-term low-pressure storage was beneficial in lowering the CI index and preserving fruit quality in peaches. The fruit's respiration rate was suppressed, and the climacteric peak was postponed. Peachess also had greater disaccharide contents and lower monosaccharide contents, as well as higher antioxidant enzyme activity, effectively suppressed membrane degradation by preventing phospholipase D (PLD) and lipoxygenase (LOX) activity, and inhibited MDA accumulation.

Some climacteric fruits, such as apples, asparagus, avocados, bananas, mangoes, peaches, sweet cherries, and tomatoes, ripening was delayed by hypobaric treatment (Li et al., 2006; Chen et al., 2010). Hashmi et al. (2016) described that 6 hours of low-pressure treatment was more beneficial against fungal decay in strawberry fruits than 4 hours. Reactive oxygen species are produced during fruit ripening. Hypobaric storage influenced the antioxidant system, which protects fruits from reactive oxygen species; as a result, reactive oxygen species accumulation decreased and senescence was slowed (Hashmi et al., 2013b; Wang et al., 2015). According to Chen et al. (2008), programmed hypobaric storage of pear cv. Cuiguan considerably decreased the respiration rate and ethylene synthesis, inhibited softening, and preserved ascorbic acid and soluble solid contents. Northland blueberry hypobaric storage (25 kPa) extends fruit storage life by lowering oxygen levels and lowering carbon dioxide production, which slows respiration (James et al., 2022). Furthermore, in tomatoes, hypobaric storage promotes the preservation of bioactive compounds like anthocyanins, flavonols, and phenols (Kou et al., 2016). Hypobaric storage (25 kPa) at 0 °C significantly reduced shrinkage, mass loss, and improved persimmon fruit firmness (Bibi et al., 2021). Hashmi et al. (2013a) demonstrated that pre-storage hypobaric treatment of blueberries reduced the occurrence of postharvest rot. However, hypobaric storage had no significant influence on the fruit's firmness or respiration rate. Wang et al. (2015) discovered that Honey peaches had a lower decay index, total soluble solids (TSS), ascorbic acid, MDA, and membrane permeability in hypobaric storage. Hypobaric treatment effectively delayed the production of O_2 and H_2O_2, increased CAT and SOD activity, and increased ATP and ADP content while decreasing LOX activity and AMP content.

Low pressure storage technique has been used in asparagus (Li et al., 2008), lettuce and strawberries (An et al., 2009), pears (Li et al., 2017), tomatoes (Sohn et al., 1988; Melgar et al., 2015), other horticultural products (Burg, 2004; Vithu & Moses, 2017), and some tropical fruits such as mangoes (Spalding & Reeder, 1977). The use of this technique has been shown to delay senescence in a variety of fresh produce. When mango fruit was kept at 13 kPa, ripening was delayed, extending its shelf life (Apelbaum et al., 1977). Pressure was found to be inversely proportional to shelf-life extension, though pressures below 7 kPa instigated dehydration, resulting in fruit Wilting (Romanazzi et al., 2001). Cui (2008) demonstrated that hypobaric storage (20.3, 50.7, or 101.3 kPa) reduces respiration, retains ascorbic acid content, reduces superoxide anion synthesis rate, and increases the postharvest life of the jujube. Jin et al. (2006) discovered that a hypobaric pressure of 55.7 kPa led to preservation of the best jujube quality.

When strawberries and cut lettuce leaves were placed in sub-atmospheric conditions, the rate of respiration significantly decreased (An et al., 2009). Tovar et al. (2011) testified that hypobaric treatment improves the uniform ripening of mango fruit by applying a vacuum of 34 kPa for 20 minutes and ethylene exposure for 30 minutes to mango fruits. Short-term hypobaric storage of pomegranate reduced the degree of peel and aril browning, slowed fruit decay, and preserved the peel color index (Zhang et al., 2021). When compared to CK, HT sustained higher firmness and lesser MDA contents, and markedly higher rate of respiration, but delayed respiration peaks. Enzymes like PPO and SOD were considerably higher at the enzyme and gene levels.

HT clearly repressed CAT activity but had no effect on PAL or APX activity (p 0.05). To further investigate the prompted resistance mechanism of low-pressure treatment, studies of defense-related genes are mandatory. Wu et al. (2022) found that peach fruit stored under hypobaric conditions (55 ± 5 kPa and 4 ± 0.5 °C) retained greater fruit firmness and lesser MDA contents, as well as increased respiration rate but delayed peak respiration. It also increased SOD and PPO enzymes while inhibiting CAT activity and had no effect on Pal or APX activity.

10.4.2 Vegetables

In fresh-cut broccoli, hypobaric storage inhibited the buildup of malondialdehyde, reduced the degradation of ascorbic acid, and preserved the firmness and chewiness at comparatively higher levels, but had no effect on chlorophyll content or membrane permeability when compared to controlled atmospheric conditions (Xin-guang et al., 2014).

Asparagus kept at low pressures ranging from 0.035 to 0.045 MPa had a 50-day shelf life, compared to 25 days in low-temperature storage and 6 days at ambient temperature (Wenxiang et al., 2006). Low-pressure storage slows the rate of respiration; prevents chlorophyll, ascorbic acid, and acidity degradation; improves organoleptic quality; and delays senescence in green asparagus (Wenxiang et al., 2006). To prevent desiccation, the asparagus was sprinkled with water during storage (Table 10.1).

Hypobaric storage enhanced the quality of Acorn squash and cucumbers (McKeown & Lougheed, 1980; Burg, 2004). Burg (2004), on the other hand, discovered no quality enhancement for Yellow crookneck squash stored under hypobaric conditions. Sun and Li (2017) studied the postharvest behavior of broccoli in micro-vaccum storage and found that it halted chlorophyll degradation; improved the activity of CAT, SOD, and POD;

TABLE 10.1 Major Effects of Hypobaric Conditions on Various Fruits and Vegetables

S. NO.	*PARAMETER*	*RESULTS*	*REFERENCE*
1	**Physiological disorder**	Reduced core browning in pear cv Yali	Li et al. (2017)
		Decreased fruit browning and flesh leatheriness in loquat	Gao et al. 2006
		Reduced superficial scald in apple cv Law Rome and Granny Smith	Wang and Dilley (2000)
		Reduced internal browning in peach	Song et al. (2016)
		Reduced in stem-end browning; lower incidence of postharvest rot in zucchini cv Cylindrica by 50%	Pristijono et al. (2018a)
2	**Respiration rate & ethylene production**	Decreased respiration rate in loquat	Gao et al. (2006)
		Repressed respiration rate of peaches, and delayed climacteric peak	Zhang et al. (2021)
		Little influence on respiratory rate in blueberry cv. Bluecrop	Li et al. (2019)
		Hypobaric storage did not affect fruit respiration rate of blue berries	Hashmi et al. (2013a)
		Hypobaric conditions (55 ± 5 kPa and 4 ± 0.5 °C) increased respiration rate but delayed respiration peak in peaches.	Wu et al. (2022)
		Low-pressure storage of 20.3, 50.7, or 101.3 kPa restrain rate of respiration in jujube	Cui (2008)
		Reduced ethylene production in Marmande tomatoes (*Solanum lycopersicum*) cv. Rojito	del Carmen Salas-Sanjuán et al. (2023)
		Decreased respiration in apples	Huan et al. (2021)
		Reduced respiratory rate in Chinese bayberries	Chen et al. (2013b)
		Reduced respiratory intensity and rate of ethylene production in Dongzao jujube fruit.	Min 2013
		No influence on fruit ethylene synthesis and rate of respiration in oranges	Archer et al. (2021)
3	**Disease/decay**	Reduced pathogen growth in oranges	Archer et al. (2021)
		Lower decay, no flesh rot in green capsicum	Pristijono et al. (2018b)
		Lower decay in Chinese bayberries (Myrica rubra Sieb. & Zucc.)	Chen et al. (2013)

S. NO.	PARAMETER	RESULTS	REFERENCE
		Lower decay in apples	Rehman et al. (2022)
		Reduced fungal rot in blueberries	Hashmi et al. (2013a)
		Decreased fungal decay in strawberries, sweet cherries, and table grapes	Romanazzi et al. (2001)
		Reduced bacterial and fungal growth	Burg (2004)
		Little influence on decay rate in blueberry cv. Bluecrop	Li et al. (2019)
		Decreased fruit decay in loquat	Gao et al. (2006)
4	**Enzyme activity**	Decreased POD and PPO activities in loquat	Gao et al. (2006)
		Deactivation of the ethylene generating enzymes	Burg (2004)
		Fruits showed higher antioxidant enzyme activity, suppressed membrane degradation efficiently by preventing phospholipase D (PLD) and lipoxygenase (LOX) activity and inhibit MDA buildup.	Zhang et al. (2021)
		Increased catalase and peroxidase activities	Chen et al. (2013b)
		High mitochondrial cytochrome C oxidase and succinic dehydrogenase activity. Increased APX, GR, and MDHAR activity in peach	Song et al. (2016)
		Improve the activities of superoxide dismutase (SOD) and catalase (CAT) and suppress the increase of peroxidase activity in asparagus	Li et al. (2008)
		Repressed ascorbic acid oxidase and alcohol dehydrogenase activities, in dongzao jujube fruit	Min (2013)
		Hypobaric conditions (55 ± 5 kPa and 4 ± 0.5 °C) lower MDA contents and increased SOD and PPO enzymes suppressed CAT activity and had no prominent influence on Pal and APX activites.	Wu et al. (2022)
		Short-term hypobaric storage improved the activity of fatty acid synthetase (FAS), and withdrawn the activity of lipoxygenase (LOX), reduced the production of MDA and electrolyte leakage.	Zhang et al. (2022)
		Retained CAT and SOD activity, repressed ROS, MDA and membrane leakage in persimmons	Zhou et al. (2008)
		Reduced the rate of superoxide anion production,	Cui (2008)
5	**Fruit quality**	LP storage had significant influence on juice ethanol contents and fruit mass loss whereas no effect on TSS, TA and fruit firmness	Archer et al. 2021
		Hypobaric conditions (55 ± 5 kPa and 4 ± 0.5 °C) retain higher fruit firmness of peach.	Wu et al. (2022)
		Fruits also maintained higher disaccharide and lower monosaccharide concentrations.	Zhang et al. (2021)
		Hypobaric storage (20.3, 50.7 or 101.3 kPa), retain ascorbic acid contents, and extended the postharvest life of the jujube cultivar	Cui (2008)
		Hypobaric condition of 55.7 kPa maintained jujube quality	Jin et al. (2006)
		Peaches (*Prunus persica* L. Batsch cv. Yingshuanghong) showed higher fruit quality and lower chilling injury in short-term hypobaric storage.	Zhang et al. (2021)
		LP storage conditions of 50.7 kPa reserved higher levels of ascorbic acid in strawberries	An et al. (2009)
		Low pressure of 4 kPa had no prominent effect on firmness and SSC/TA ratio of tomato fruit	Pristijono et al. (2017)
		Hypobaric storage considerably kept hardiness and ascorbic acid content, decreased acetaldehyde and ethanol contents in flesh and in dongzao jujube fruit	Min (2013)
		Delayed the degeneration of surface color, reduced the buildup of MDA contents, superoxide anion (O^{2-}) and hydrogen peroxide (H_2O_2), repressed the surge of relative leakage rate in green asparagus	Li et al. (2008)
		Before storage treatment of 20 kPa for 2 h without postharvest delay efficiently reserved fruit quality and prolonged storage life of strawberries	Maryam et al. (2022)

(Continued)

S. NO.	PARAMETER	RESULTS	REFERENCE
		Higher levels of acceptability in zucchinis	Pristijono et al. (2018a)
		Smallest changes in total color difference and highest force at bio yield and energy to bio yield were observed in 10 kPa. Reduced changes in fiber, pectins, and sugar contents in cucumber	Zapotoczny and Markowski (2014)
		Higher overall acceptability in green capsicum	Pristijono et al. (2018b)
		Repressed increase in membrane fluidity, while it reserved mitochondrial permeability transition pore (MPTP) concentration high. Reduced the degradation of vitamin C and glutathione in peaches	Song et al. (2016)
		Extend shelf life in apples	Huan et al. (2021)
		Reduced MDA accumulation, greater TA and total phenolics compared with those kept under normal air pressure. High ferric reducing antioxidant power in Chinese bayberries cv. Dongkui	Chen et al. (2013b)
		Retained the firmness and maintained the color values in apple cv. Royal Gala	Rehman et al. (2022)
		α-farnesene and MHO production was low in apple cv. Law Rome and Granny Smith	Wang and Dilley (2000)
		Double the storage life compared to normal atmosphere in rose	Bredmose (1981)
		No chlorophyll degradation and proline accumulation but more open stomata at normal pressure after storage, unsatisfactory rooting in *Hibiscus-rosa-sinensis* L cv. Moesiana	Skytt Andersen and Kirk (1986)
		Hindering the buildup of malondialdehyde, decreasing the degradation of ascorbic acid, and preserving the hardness and chewiness in fresh-cut broccoli	Xin-guang et al. (2014)
		Weight loss, firmness in blueberries	Hashmi et al. (2013a)
		Increased weight loss due to prolonged stomatal openion upto 96 h at normal pressure in cucumbers	Laurin et al. (2006)
		Better ascorbic acid retention, prevention of succinate formation	Burg (2004)
		101 kPa chlorophyll degradation and fast carotinoid production, negative sensory analysis fruit softening 75, 50 kPa slow degardation and slow carotinoid production, fruit softening, positive sensory analysis in Marmande tomatoes (*Solanum lycopersicum*,)	del Carmen Salas-Sanjuán et al. (2023)
		Inhibit loss of titratable acidity and ascorbic acid in loquat	Gao et al. (2006)
		pH total soluble solid content, bioactive components such as anthocyanins, total phenol, and total flavonoid improved in blueberry cv. Bluecrop	Li et al. (2019)

and reduced free radicals such as super oxide and hydrogen peroxide generation, MDA content, and membrane permeability.

Li and Zhang (2006) and Li et al. (2006) observed positive effects of low-pressure storage on the shelf life and quality of asparagus (Table 10.1).

10.5 DISADVANTAGES OF HYPOBARIC STORAGE ON FRESH PRODUCE

Under low pressure, stomata opened and rooting ability was reduced in rose cultivars compared to those under normal atmospheric pressure. During storage, no ethylene generation, chlorophyll loss, proline buildup, or respiratory rise was observed (Skytt Andersen & Kirk, 1986).

The chlorophyll content and membrane permeability of fresh-cut broccoli were not considerably different between the two packagings (Xin-guang et al., 2014). Cucumber fruits subjected to 70 kPa for 6 hours in darkness to simulate air flight conditions and later kept in a cold storage facility at normal pressure for 96 hours had considerably higher open stomatas than controls (Laurin et al., 2006). Desiccation of fresh produce exposed to low-pressure conditions may be triggered by better transpiration due to the actions and properties of the stomata (Laurin et al., 2006), lenticel, cuticle, and cells of the epidermis (Ben-Yehoshua & Rodov, 2003).

Pressures less than 100 kPa tend to extend shelf life; nevertheless, dehydration frequently appears to be caused by low-pressure storage due to a decrease in water vapor partial pressure. Mango dehydrated at pressures lower than 7 kPa did not produce the appropriate orange and red colors upon ripening (Apelbaum et al., 1977).

10.6 LIMITATIONS OF HYPOBARIC TECHNOLOGY

Storage under hypobaric conditions is rarely used. Despite the advantages of hypobaric storage as well as some advantages over control atmosphere storage, it has not been accepted in commercial fruit and vegetable storage. Fresh horticultural produce exposed to hypobaric conditions are likely to desiccate due to enhanced transpiration, which is caused by characteristics and opening of stomatas (Laurin et al., 2006), cuticle, lenticels, and epidermal cells (Ben-Yehoshua & Rodov, 2003).

Insufficient control of temperature, cold areas on the surface of the vacuum chamber, moistening at atmosphere pressures compared with lower pressures, insufficient air variations, leaked vacuum chamber, and a failure to recognize that great turgor pressures of plant cells inhibits this type of storage from producing volatiles to boil and outgas (Burg, 2014). Excessive low oxygen levels cause toxic ethanol and acetaldehyde concentrations to accumulate in plant tissues (Imahori et al., 2005). The main limitation of using this technique with low pressure treatment is generating low pressure in a vessel safely and repeatedly. Furthermore, the initial investment in equipment is relatively high. Nonetheless, the use of pressure treatments and the range of products to which they are applied expands year after year.

Solution: Water can be sprinkled to solve the problem of low RH, which causes dehydration during low-pressure storage (Laurin et al., 2006).

Caution: It is important to remember that playing with pressure always contains a safety aspect as well as worker risk.

10.7 CONCLUSIONS

In conclusion, it is evident that hypobaric storage is useful technology which could be used and is being used for various fresh fruits and vegetables. It affects various attributes of fresh produce, such as suppression of diseases, color changes, desiccation inhibition, delay of membrane disintegration, and maintenance of characteristic qualities of selected fruits and vegetables. At the same time, it also shows some negative effects. For example, fresh horticultural produce exposed to hypobaric conditions are likely to desiccate due to enhanced transpiration, which is caused by characteristics and opening of stomata. In addition, insufficient control of temperature, cold vacuum chamber surface, moistening at atmosphere pressures compared with lower pressures, insufficient air control, leaked vacuum chambers, and a failure to recognize that higher turgor pressures of plant cells further limits this type of storage utilization at the commercial level. Similarly, excessive low oxygen levels cause toxic ethanol and acetaldehyde concentrations to accumulate in tissues of fresh produce, which should be carefully managed.

REFERENCES

Adams, K.B., Wu, M.T., Salunkhe, D.K. (1976). Effects of subatmospheric pressure on the growth and patulin production of *P. expansum* and *P. patulum*. *LWT-Food Science and Technology*, 9, 153–155.

An, D.S., Park, E., Lee, D.S. (2009). Effect of hypobaric packaging on respiration and quality of strawberry and curled lettuce. *Postharvest Biology and Technology*, 52, 78–83.

Apelbaum, A., Zauberman, G., Fuchs, Y. (1977). Subatmospheric pressure storage of mango fruits. *Scientia Horticulturae*, 7, 153–160.

Archer, J., Pristijono, P., Gallien, Q., Houizot, L., Bullot, M., Palou, L., Golding, J.B. (2021). Preliminary investigations on the effect of low-pressure treatment on in vitro and in vivo growth of *Penicillium* sp. in oranges. *Acta Horticulturae*, 1325, 55–58. https://doi.org/10.17660/ActaHortic.2021.1325.9

Archer, J., Pristijono, P., Vuong, Q. V., Palou, L., Golding, J.B. (2021). Effect of low pressure and low oxygen treatments on fruit quality and the in vivo growth of *Penicillium digitatum* and *Penicillium italicum* in oranges. *Horticulturae*, 7(12), 582.

Back, E.A., Cotton, R.T. (1925). The use of vacuum for insect control. *Journal of Agriculture Research*, 31, 1035–1041.

Bangerth, F. (1974). Hypobaric storage of vegetables. *Acta Horticulturae*, 38, 23–32.

Bangerth, F. (1984). Changes in sensitivity for ethylene during storage of apple and banana fruits under hypobaric conditions. *Scientia Horticulturae*, 24, 151–163.

Barkai-Golan, R. (1990). Postharvest disease suppression by atmospheric modifications. In *Food Preservation by Modified Atmospheres* (pp. 237–264), ed. M. Calderon and R. Barkai-Golan. CRC Press.

Beaudry, R.M. (1999). Effect of O_2 and CO_2 partial pressure on selected phenomena affecting fruit and vegetable quality. *Postharvest Biology and Technology*, 15(3), 293–303.

Ben-Yehoshua, S., Rodov, V. (2003). Transpiration and water stress. In *Postharvest Physiology and Pathology of Vegetables* (pp. 111–161), ed. J. A. Bartz and J. K. Brecht. Marcel Dekker.

Bibi, F., Muhammad, A., Khalid, M. U., Shaukat, M., Ahmad, U., Batool, Z., Nishat, Z. (2021). Effect of hypobaric treatment and storage temperature on shelf life of persimmon fruit. *Life Science Journal*, 18(2), 55–60.

Bredmose, N.B. (1981). Effects of low pressure on storage life and subsequent keeping quality of cut roses. In *Acta Horticulturae* (Vol. 113, pp. 73–79). http://www.pubhort.org/actahort/books/113/113_11.htm

Burg, S.P. (1976). Low temperature hypobaric storage of metabolically active matter. US patents 3,958,028 and, 061,483.

Burg, S.P. (2004). *Postharvest Physiology and Hypobaric Storage of Fresh Produce*. CABI Publisher.

Burg, S.P. (2010). Experimental errors in hypobaric laboratory research. *Proceeding of 9th International Controlled Atmosphere Research Conference*, 1, 45–61.

Burg, S.P. (2014). *Hypobaric Storage in Food Industry*. Academic Press.

Burg, S.P., Burg, E.A. (1965). Gas exchange in fruits. *Plant Physiology*, 18, 870–884.

Burg, S.P., Burg, E.A. (1966). Fruit storage at subatmospheric pressures. *Science*, 153(3733), 314–315.

Burg, S.P., Kosson, R. (1983). Metabolism, heat transfer and water loss under hypobaric conditions. In *Postharvest Physiology and Crop Preservation*, ed. M. Lieberman (pp. 399–424). Plenum Corp.

Burton, W. (1982). *Post-Harvest Physiology of Food Crops*. Longman.

Burton, W.G. (1989). *The Potato*, 3rd ed. Longmans.

Chang, C. (2016). Q & A. How do plants respond to ethylene and what is its importance? *BMC Biology*, 14, 1–7.

Chen, H., Ling, J., Wu, F., Zhang, L., Sun, Z., Yang, H. (2013a). Effect of hypobaric storage on flesh lignification, active oxygen metabolism and related enzyme activities in bamboo shoots. *LWT-Food Science and Technology*, 51, 190–195.

Chen, H., Yang, H., Gao, H., Long, J., Tao, F., Fang, X., Jiang, Y. (2013b). Effect of hypobaric storage on quality, antioxidant enzyme and antioxidant capability of the Chinese bayberry fruits. *Chemistry Central Journal*, 7(4), 1–7.

Chen, W., Gao, H., Chen, H., Mao, J., Song, L., Ge, L. (2010). Effects of hypobaric storage on postharvest physiology and quality of flesh-melting textured juicy peach. *Transactions of the Chinese Society of Agricultural Machinery*, 41(9), 108–112. (in Chinese)

Chen, W.X., Gao, H.Y., Zhou, Y.J., Chen, H.J., Mao, J.L., Song, L.L., Tao, F., Ge, L.M., Zheng, Y.H. (2008). Study of programmed hypobaric storage of 'Cuiguan' pears. *Acta Horticulturae*, 804, 645–650.

Cicale, D., Jamieson, W. (1978). *Preliminary Investigation of the Storage of Haas Avocados under Hypobaric Conditions*. Report for Grumman Allied Industries, Inc., Garden City, New York.

Cui, Y. (2008). Effects of hypobaric conditions on physiological and biochemical changes of Lizao jujube. *Journal Anhui Agricultural Science*, 36, 12900–12901.

Davenport, T.L., Burg, S.P., White, T.L. (2006). Optimal lowpressure conditions for long-term storage of fresh commodities kill Caribbean fruit fly eggs and larvae. *HortTechnology*, 16(1), 98–104.

del Carmen Salas-Sanjuán, M., del Mar Rebolloso, M., del Moral, F., Valenzuela, J.L. (2023). Use of sub-atmospheric pressure storage to improve the quality and shelf-life of marmande tomatoes cv. Rojito. *Foods*, 12(6).

Enfors, S.O., Molin, G. (1980). Effect of high concentrations of carbon dioxide on growth rate of *Pseudomonas fragi*, *Bacillus cereus* and *Streptococcus cremoris*. *Journal of Applied Bacteriology*, 48, 409–416.

Gao, H.Y., Chen, H.J., Chen, W.X., Yang, Y.T., Song, L.L., Jiang, Y.M., Zheng, Y.H. (2006). Effect of hypobaric storage on physiological and quality attributes of loquat fruit at low temperature. *IV International Conference on Managing Quality in Chains–The Integrated View on Fruits and Vegetables Quality*, 712, 269–274.

Goszczynska, D.M., Ryszard, M.R. (1988). Storage of cut flowers. *Horticultural Reviews*, 10, 35–62.

Hashmi, M.S., East, A.R., Palmer, J.S., Heyes, J.A. (2013a). Hypobaric treatment reduces fungal rots in blueberries. *VII International Postharvest Symposium*, 1012, 609–614.

Hashmi, M.S., East, A.R., Palmer, J.S., Heyes, J.A. (2013b). Pre-storage hypobaric treatments delay fungal decay of strawberries. *Postharvest Biology and Technology*, 77, 75–79.

Hashmi, M.S., East, A. R., Palmer, J. S., Heyes, J.A. (2013c). Hypobaric treatment stimulates defence-related enzymes in strawberry. *Postharvest Biology and Technology*, 85, 77–82.

Hashmi, M.S., East, A.R., Palmer, J.S., Heyes, J.A. (2016). Hypobaric treatments of strawberries: A step towards commercial application. *Science Horticulturae*, 198, 407–413.

Herreid, C.F. (1980). Hypoxia in invertebrates. *Comparative Biochemistry and Physiology*, 67, 311–320.

Hodges, D.M., Lester, G.E., Munro, K.D., Toivonen, P.T.A. (2004). Oxidative stress: importance for postharvest quality. *HortScience*, 39(5), 924.

Hörtensteiner, S., Lee, D.W. (2007) Chlorophyll catabolism and leaf coloration. In *Senescence Processes in Plants. Annual Plant Reviews* (vol 26, pp. 12–38), ed. S. Gan. Blackwell.

Huan, C., Li, H., Jiang, Z., Shen, S., Zheng, X. (2021). Effect of hypobaric treatment on off-flavour development and energy metabolism in 'Bruno' kiwifruit. *LWT—Food Science and Technology*, 136, 110349.

Hughes, P.A., Thompson, A.K., Plumbley, R.A., Seymour, G.B. (1981) Storage of capsicums (*Capsicum annum* cultivar Bellboy) under controlled atmosphere, modified atmosphere and hypobaric conditions. *Journal of Horticultural Sciences*, 56, 261–266.

Hulasare, R., Payton, M.E., Hallman, G.J., Phillips, T.W. (2013). Potential for hypobaric storage as a phytosanitary treatment: mortality of *Rhagoletis pomonella* (Diptera: Tephritidae) in apples and effects on fruit quality. *Journal of Economic Entomology*, 106(3), 1173–1178.

Imahori, Y., Uemura, K., Kishioka, I., Fujiwara, H., Tilio, A.Z., Ueda, Y., et al. (2005). Relationship between low-oxygen injury and ethanol metabolism in various fruits and vegetables. In *Acta Horticulturae* (Vol. 682). *Proceedings of the Fifth International Postharvest Symposium*, Verona, Italy, June 6–11, 2004.

James, A., Yao, T., Ma, G., Gu, Z., Cai, Q., Wang, Y. (2022). Effect of hypobaric storage on Northland blueberry bioactive compounds and antioxidant capacity. *Scientia Horticulturae*, 291, 110609.

Jiao, S., Johnson, J.A., Fellman, J.K., Mattinson, D.S., Tang, J., Davenport, T.L., Wang, S. (2012). Evaluating the storage environment in hypobaric chambers used for disinfesting fresh fruits. *Biosystem Engineering*, 111, 271–279.

Jiao, S., Johnson, J.A., Tang, J., Mattinson, D.S., Fellman, J.K., Davenport, T.L., Wang, S. (2013). Tolerance of codling moth, and apple quality associated with low pressure/low temperature treatments. *Postharvest Biology and Technology*, 85, 136–140.

Jin, A.X., Wang, Y.P., Liang, L.S. (2006). Effects of atmospheric pressure on the respiration and softening of DongZao jujube fruit during hypobaric storage. *Journal of the Northwest Forestry University*, 21, 143–146.

Johnson, J.A., Zettler, J.L. (2009). Response of postharvest tree nut lepidopoteran pests to vacuum treatments. *Journal of Economic Entomology*, 102(5), 2003–2010.

Kader A.A. Ben-Yehoshua S. (2000). Effects of superatmospheric oxygen levels on postharvest physiology and quality of fresh fruits and vegetables *Postharvest Biology and Technology*, 20, 1–13.

Kou, X., Wu, J.Y., Wang, Y., Chen, Q., Xue, Z., Bai, Y., Zhou, F. (2016). Effects of hypobaric treatments on the quality, bioactive compounds, and antioxidant activity of tomato. *Journal of Food Science*, 81, H1816–H1824.

Laurin, É., Nunes, M.C.N., Émond, J.P., Brecht, J.K. (2006). Residual effect of low-pressure stress during simulated air transport on Beit Alpha-type cucumbers: Stomata behavior. *Postharvest Biology and Technology*, 41, 121–127.

Li, H., James, A., He, X., Zhang, M., Cai, Q., Wang, Y. (2019). Effect of hypobaric treatment on the quality and reactive oxygen species metabolism of blueberry fruit at storage. *CyTA-Journal of Food*, 17(1), 937–948.

Li, J., Bao, X., Xu, Y., Zhang, M., Cai, Q., Li, L., Wang, Y. (2017). Hypobaric storage reduced core browning of Yali pear fruits. *Science Horticulturae*, 225, 547–552.

Li, W., Zhang, M. (2006). Effect of three-stage hypobaric storage on cell wall components, texture and cell structure of green asparagus. *Journal of Food Engineering*, 77(1), 112–118.

Li, W., Zhang, M., Yu, H. (2006). Study on hypobaric storage of green asparagus. *Journal of Food Engineering*, 73(3), 225–230.

Li, W.X., Zhang, M., Wang, S.J. (2008) Effect of three-stage hypobaric storage on membrane lipid peroxidation and activities of defense enzyme in green asparagus. *LWT—Food Science and Technology*, 41, 2175–2181.

Maryam, A., Anwar, R., Malik, A. U., Khan, A. S., Ali, S., Waris, F., . . . & El-Mogy, M. M. (2022). Pre-storage hypobaric treatment reduces microbial spoilage and maintains eating quality of strawberry fruits during low temperature conditions. *Journal of Food Processing and Preservation*, 46(11), e17121.

Mbata, G.N., Philips, T.W. (2001). Effects of temperature and exposure time on mortality of stored-product insects exposed to low pressure. *Journal of Economic Entomology*, 94(5), 1302–1307.

McKeown, A., Lougheed, E. (1980). Low pressure storage of some vegetables. *Acta Horticulturae*, 116, 83–100.

Melgar, D., Conde, E., Megías, Z., Rebolloso, M.M., Valenzuela, J.L., Jamilena, M. (2015). Efecto de la conservación hipobárica sobre la calidad de fruto en tomate Raf. *Acta Horticulturae*, 71, 352–355.

Merksamer, P.I., Liu, Y., He, W., Hirschey, M.D., Chen, D., Verdin, E. (2013). The sirtuins, oxidative stress and aging: an emerging link. *Aging (Albany NY),* 5(3), 144–150.

Min, C. Z. (2013). The study on hypobaric storage mechanism and technology in dongzao jujube fruit (Doctoral dissertation, Master's thesis, Tianjin University of Science and Technology. http:// www. dissertationtopic. net/doc/1089396. Accessed October).

Mitcham, E.J., Martin, T., Zhou, S. (2006). The mode of action of insecticidal controlled atmospheres. *Bulletin of Entomological Research*, 96, 213–222.

Muhammad, A., Dayisoylu, K. S., Khan, H., Khan, M.R., Khan, I., Hussain, F., Idrees, M. (2022). An integrated approach of hypobaric pressures and potassium permanganate to maintain quality and biochemical changes in tomato fruits. *Horticulturae*, 9(1), 9.

Navarro, S. (1978). The effects of low oxygen tensions on three stored-product insect pests. *Phytoparasitica*, 6(2), 51–58

Navarro, S., Calderon, M. (1979). Mode of action of low atmospheric pressures on Ephestia cautella (Wlk.) pupae. *Experientia*, 35, 620–621.

Patterson, M.E., Melsted, J.L. (1977) Sweet cherry handling and storage alternatives. In *Controlled Atmospheres for the Storage and Transport of Perishable Agricultural Commodities. Proceedings of the 2nd National Controlled Atmosphere Research Conference* (p. 29), ed. D.H. Dewey. Michigan State University.

Pristijono, P., Scarlett, C.J., Bowyer, M.C., Vuong, Q.V., Stathopoulos, C.E., Jessup, A.J., Golding, J.B. (2017). Use of low-pressure storage to improve the quality of tomatoes. *The Journal of Horticultural Science and Biotechnology*, 92(6), 583–590.

Pristijono, P., Bowyer, M.C., Scarlett, C.J., Vuong, Q.V., Stathopoulos, C.E., Golding, J.B. (2018a). The application of low pressure storage to maintain the quality of zucchinis. *New Zealand Journal of Crop and Horticultural Science*, 46(3), 254–263.

Pristijono, P., Bowyer, M.C., Scarlett, C.J., Vuong, Q.V., Stathopoulos, C.E., Golding, J.B. (2018b). Effect of low-pressure storage on the quality of green capsicums (*Capsicum annum* L.). *The Journal of Horticultural Science and Biotechnology*, 93(5), 529–536.

Rehman, W.U., Majid, S.H., Yasser, D., Sadiq, S., Ayaz, A., Sahib, A., Waqar, A. (2022). Hypobaric treatment augments the efficacy of 1-MCP in apple fruit. *Journal of Food Science and Technology*, 59, 4221–4229.

Romanazzi G., Nigro F., Ippolito A. (2003). Short hypobaric treatments potentiate the effect of chitosan in reducing storage decay of sweet cherries. *Postharvest Biology and Technology*, 29, 73–80.

Romanazzi, G., Nigro, F., Ippolito, A., Salerno, M. (2001). Effect of short hypobaric treatments on postharvest rots of sweet cherries, strawberries and table grapes. *Postharvest Biology and Technology*, 22, 1–6.

Skytt Andersen, A., Kirk, H. (1986). Low pressure storage of herbaceous cuttings. *III International Symposium on Postharvest Physiology of Ornamentals,* 181, 305–312.

Sohn, T.H., Cheon, S.H., Choi, S.W., Moon, K.D. (1988) Changes of flavor components in tomato fruits during subatmospheric pressure storage. *Applied and Biological Chemistry*, 31, 298–307

Song, L., Wang, J., Shafi, M., Liu, Y., Wang, J., Wu, J., Wu, A. (2016). Hypobaric treatment effects on chilling injury, mitochondrial dysfunction, and the ascorbate–glutathione (AsA-GSH) cycle in postharvest peach fruit. *Journal of Agricultural and Food Chemistry*, 64(22), 4665–4674.

Spalding, D.H., Reeder, W.F. (1976). Low pressure (hypobaric) storage of avocados. *HortScience*, 11(5), 491–492.

Spalding, D.H., Reeder, W.F. (1977). Low pressure (hypobaric) storage of mangos. *Journal of American Society of Horticultural Sc*iences, 102, 367–369.

Sun, Y., Li, W. (2017). Effects the mechanism of micro-vacuum storage on broccoli chlorophyll degradation and builds prediction model of chlorophyll content based on the color parameter changes. *Scientia Horticulturae*, 224, 206–214.

Suzuki, T., Kunieda, T., Murai, F., Morioka, S., Shioi, Y. (2005). Mg-dechelation activity in radish cotyledons with artifi cial and native substrates, Mg-chlorophyllin a and chlorophyllide A. *Plant Physiology and Biochem*istry, 43(5), 459–464.

Thompson, A.K. (2010). *Controlled Atmosphere Storage of Fruits and Vegetables*, 2nd ed. CAB International.

Thompson, A.K. (2015). *Fruit and Vegetable Strage: Hypobaric, Hyperbaric and Controlled Atmosphere*. Springer.

Tovar, B., Montalvo, E., Damián, B.M., García, H.S., Mata, M. (2011). Application of vacuum and exogenous ethylene on Ataulfo mango ripening. *LWT-Food Science and Technology*, 44(10), 2040–2046.

Tsuchiya, T., Ohta, H., Masuda, T., Mikami, B., Kita, N., Shioi, Y., Takamiya, K. (1997). Purifi cation and characterization of two isozymes of chlorophyllase from mature leaves of Chenopodium album. *Plant and Cell Physiology*, 38(9), 1026–1031.

Vigneault, C., Leblanc, D.I., Goyette, B., Jenni, S. (2012). Invited review: Engineering aspects of physical treatments to increase fruit and vegetable phytochemical content. *Canadian Journal of Plant Science*, 92(3), 373–397.

Vigneault, C., Rennie, T.J., Toussaint, V. (2008). Cooling of freshly cut and freshly harvested fruits and vegetables. *Stewart Postharvest Reviewes*, 4(3), 4.1–4.10

Vithu, P., Moses, J.A. (2017). Hypobaric storage of horticultural products: A review. In *Engineering Practices for Agricultural Production and Water Conservation* (pp. 155–170). CRC Press.

Wall, M.M., Silva, S.T., Sanxter, S.S., Follett, P.A. (2017). Hypobaric storage of fresh tropical fruit. In *HortScience. Proceedings 2017 ASHS Annual Conference*, Waikoloa, HI.

Wang, J.H., You, Y.L., Chen, W.X., Xu, Q.Q., Wang, J., Liu, Y.K. Wu, J.S. (2015). Optimal hypobaric treatment delays ripening of honey peach fruit via increasing endogenous energy status and enhancing antioxidant defence systems during storage. *Postharvest Biology and Technology*, 101, 1–9. doi:10.1016/j.postharvbio.2014.11.004.

Wang, L.P., Zhang, P., Wang, S.J. (2001). Advances in research on theory and technology for hypobaric storage of fruit and vegetable. *Storage and Process*, 5, 3–6.

Wang, Z., Dilley D.R. (2000). Hypobaric storage removes scald-related volatiles during the low temperature induction of superficial scald of apples. *Postharvest Biology and Technol*ogy, 18, 191–199.

Wells, J.M. (1974). Growth of *Erwinia carotovora, E. atroseptica,* and *Pseudomonas fluorescence* in low-oxygen and high-carbon dioxide atmospheres. *Phytopathology*, 64, 1012–1015.

Wenxiang, L., Zhang, M., Han-qing, Y. (2006). Study on hypobaric storage of green asparagus. *Journal of Food Engineering*, 73, 225–230.

Wu, X., Wu, H., Yu, M., Ma, R., Yu, Z. (2022). Effect of combined hypobaric and cold storage on defense-related enzymes in postharvest peach fruit during ripening. *Acta Physiologiae Plantarum*, 44(9), 93.

Xin-guang, F., Lu, X. Zhen-fu, Z., Feng-jun, G., Mei-lan, W., Chang-feng, Z., Xian-zhang, Z. (2014). Comparative analysis between hypobaric storage and controlled atmosphere storage in the preservation of fresh-cut broccoli. *Food Science*, 35(2), 277–281.

Yahia, E.M. (2011). *Post-Harvest Biology and Technology of Tropical and Subtropical Fruits. Volume 3: Cocona to Mango*. Woodhead Publishing Ltd.

Yamauchi, N. (2015). Postharvest chlorophyll degradation and oxidative stress. *Abiotic stress Biology in Horticultural Plants*, 101–113.

Zapotoczny, P., Markowski, M. (2014). Influence of hypobaric storage on the quality of greenhouse cucumbers. *Bulgarian Journal of Agricultural Science*, 20(6), 1406–1412.

Zhang, X., Liu, T., Zhu, S., Wang, D., Sun, S., Xin, L. (2022). Short-term hypobaric treatment alleviates chilling injury by regulating membrane fatty acids metabolism in peach fruit. *Journal of Food Biochemistry*, 46(7), e14113.

Zhang, X., Sun, S., Yang, J., Zou, M., Xin, L. (2021). Improving pomegranate fruit quality by short-term hypobaric treatment combined with modified atmosphere packaging storage. *Acta Horticulturae*, 1319, 217–222.

Zhao, Z., Jiang, W., Cao, J., Zhao, Y., Gu, Y. (2006). Effect of cold-shock treatment on chilling injury in mango (*Mangifera indica* L. cv. Wacheng) fruit. *Journal of the Science of Food and Agriculture*, 86, 2458–2462.

Zhou, Y., Gao, H., Chen, H., Mao, J., Chen, W., Song, L. (2008). Effects of hypobaric storage on active oxygen metabolism of persimmon (*Diospyros Kaki* L.) fruit. *Acta Horticulturae*, 804, 523–526.

Zhou, H., Qiao, Y., Zhang, S., Wang, H., Chen, Z., Zheng, X. (2010). Effects of low temperature and hypobaric on softening and membrane injury physiology of Datuanmilu peach. *Jiangsu Journal of Agricultural Sciences*, 26(4), 802–807.

Impact of High-Pressure Carbon Dioxide on the Processing of Fruits and Vegetables

11

Ayesha Murtaza*, Aamir Iqbal, Kanza Aziz Awan, Amber Tufail, Shinawar Waseem Ali, Shoaib Younas, and M. Sajid Manzoor
*Corresponding Author: ayesha.murtaza@ucp.edu.pk

11.1 INTRODUCTION

Fruits and vegetables are considered a great source of minerals, fiber, vitamins, and bioactive compounds, and these are utilized in fresh and processed forms. According to FAO, 40–50% of food wastage is due to fruits and vegetables (Zhu et al., 2023). Discoloration is mainly due to enzymatic browning in fresh fruit and vegetable products caused by different enzymatic activities, such as phenylalanine ammonia-lyase (PAL), polyphenol oxidase (PPO), and peroxidase (POD), which affects the appearance and nutritional value of fruits and vegetables and reduces their economical value (Wang et al., 2022). For prolonged storage of fruits and vegetables and their products, browning is a restrictive factor which shortens its shelf life. Different browning enzymes are found to be a major threat for food industries, and thus a major cause of quality degradation. This is one of the major issues in the food industry, as it depreciates the overall acceptability and quality of produce in general.

Due to the demand of high living standards, consumers are mainly focused on improving their lives, and for this purpose, fruits and vegetables are essential due to their high nutritional status and essential dietary fiber, vitamins, and minerals. The quality and shelf life of fruits and vegetables are mainly compromised due to processing and storage conditions which result in discoloration, spoilage, lignification, nutritional losses, and chilling injuries (De Jaegere et al., 2022; Wang et al., 2022). Different processing conditions and discoloration phenomena can cause a reduction in the sensory attributes of fruits and vegetables, eventually reducing their economic value. Improved quality, safety of food, and nutritional content are increasingly in demand from consumers. Traditional thermal pasteurization techniques, however, have the potential to contaminate food. Additionally, traditional thermal pasteurization techniques always change the color of food, which can lower the value of food products (Du et al., 2023). As a result, researchers have been studying a growing number of non-thermal techniques lately (Benito-Román et al., 2022). These techniques include usage of HHP, ultrasound, ultraviolet irradiation (UV), pulse electric field, high-pressure homogenization, HPCD, and microfiltration technologies (Figure 11.1A). Among all techniques, HPCD is a non-thermal food processing technique that uses high pressure and CO_2 to preserve the food. This technique has been used in fruit and vegetable preservation for several reasons, including improving their shelf life, reducing spoilage, and preserving their nutritional value. A brand-new non-thermal HPCD that implements pressurized CO_2 at 0.1 MPa (1 bar) pressure and temperature that's not too high (lower than thermal pasteurization) (Briongos et al., 2016). At various pressures and temperatures, carbon dioxide act in different phases. Above critical conditions, CO_2 is always a supercritical fluid with the characteristics of both a gas or a liquid (7.38 MPa, 31.1 °C or 73.8 bar, 304.25 K). Up to this point, studies on HPCD inactivation have typically been carried out on temperatures between room temperature (25 °C) and pressure between 0.1 and 50 MPa. Consequently, CO_2 usually form as a gas or supercritical fluid for HPCD inactivation technology.

In the food industry, HPCD is employed in bacterial inactivation, spore forming, and viruses, among other technologies. Insufficient explanation and specific processes of microorganism inactivation are a significant barrier for the commercial application of HPCD (Liu et al., 2013). Additionally, the

DOI: 10.1201/9781003370376-15

impacts of HPCD on the nutrients and food quality have been introduced. This chapter helps to demonstrate that by inactivating some enzymes and microbes, additionally providing theoretical support for subsequent work on the influence of HPCD and could be a viable alternative pasteurization method that increases nutritional facts and aids in improving the quality of fruits and vegetables and their discloration mechanism and control measures, which helps researchers gain a better understanding of this novel technology and its influence on the mechanism of fruits and vegetables.

11.2 HIGH-PRESSURE CARBON DIOXIDE (HPCD)

HPCD is a type of high-pressure processing (HPP) that uses carbon dioxide gas as the pressure-transmitting medium (Zhao et al., 2019). During HPCD processing, the selected foods are placed in sealed containers, and then high pressure–based treatment is applied by using a high-pressure vessel or pumps. The CO_2 gas is thereafter introduced into the containers, which then further increases the applied pressure and helps to conserve the foods. The CO_2 gas used in HPCD can exhibit several effects on the foods being treated. Firstly, it can efficiently inactivate or eliminate microorganisms of a harmful nature, such as bacteria, viruses, and yeasts that can cause deterioration and/or foodborne infections. Secondly, it can denature certain enzymes that can cause foods degradation, thus help in extending the shelf life of the foods. Thirdly, it can lessen the O_2 concentration (owing to increased CO_2 levels) in the containers, which can further slow down oxidative reactions that can cause food spoilage (Krupa & Tomala, 2021). It is a non-thermal conservation method, thus it does not rely on a heat source to preserve the targeted food products (Marszałek et al., 2015). This is in comparison to traditional preservation approaches such as canning and pasteurization, in which heat is used to inactivate microorganisms and enzyme-based reactions. Therefore, HPCD can better preserve the nutritional qualities, along with flavor and texture properties, of the foods compared to conventional methods of a thermal nature, making HPCD a promising alternative for fresh and minimally processed foods preservation.

HPCD has also been found to be an appropriate preservation approach for freshly harvested fruit and vegetable commodities, as it offers a wider range of potential benefits (Figure 11.1B). It can preserve the natural nutritive contents of fruits and vegetables, increase their probable shelf life, improve their desired safety status, reduce the requirements for artificial preservatives, and retain their characteristic color and texture-based properties. These benefits can improve access to highly nutritious foods and reduce food waste production, as well as protect public health with no major side effects on the environment and the treated commodities in general. HPCD is one of the innovative approaches to preserve fruit and vegetable products that has the prospective to positively transform the foods-related industry and support a healthier and sustainable food system. The HPCD could be used to preserve a wide variety of fruit and vegetable commodities. The sources of HPCD utilized in this method are typically high-pressure-based equipment, such as a high-pressure vessel or pump, and CO_2 gas. Here are some more specific sources of HPCD in fruits and vegetables.

The HPCD processing technique explains the mechanism in which it contains one pressure vessel to have high pressure (>50 MPa) and one temperature vessel containing two thermocouples to observe the temperature of the carbon dioxide vessel as well as the temperature of the targeted commodity. The vessel is pressurized with the plunger pump and commercially available 99.5% of carbon dioxide used in the HPCD system (Hu et al., 2013). The pressure vessel is sanitized and heated to the essential temperature and then the sample is placed inside the vessel where anticipated pressures for the particular period are generally given to the food product. After given a particular duration of treatments, by opening the pressure relieving valve, the vessel is depressurized and the product is refrigerated until final usage.

The numerous factors, for instance, food ingredients, pH, temperature, time, and pressure, impart an important role in the inactivation of enzyme effectiveness by the HPCD technique for different food products. Several studies revealed the effect of PPO inactivation in different food commodities. In carrot juice, PPO inactivation improved with enhancing pressure (Park et al., 2002), whereas in apples juice activity of PPO declined by 38.5% upon increased pressure intensity

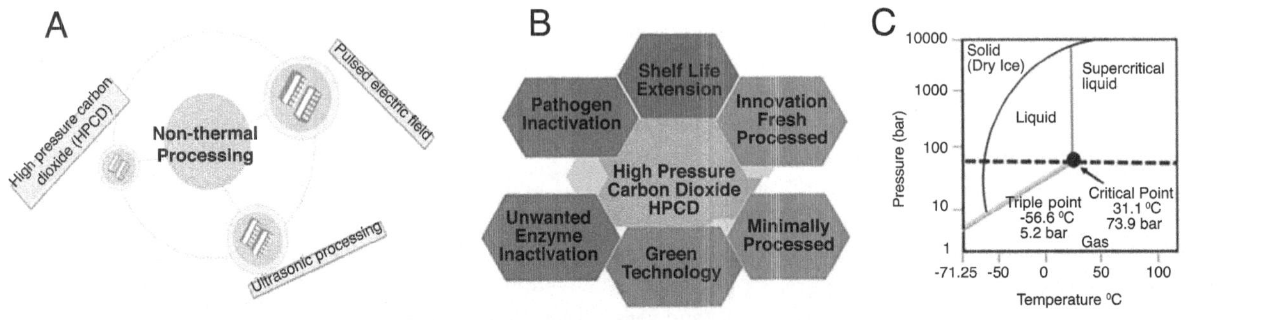

FIGURE 11.1 (A) Non-thermal technologies; (B) benefits of HPCD technology; (C) phase diagram of carbon dioxide demonstrated schematically.

to 30 MPa at persistent temperature and time duration (Gui et al., 2007). Different conditions and elevated temperature stimulated PPO inactivation, preferably due to the enhanced diffusibility of CO_2 at a higher temperature that instigated the collision of CO_2 with enzyme molecules, leading to enzyme inactivation (Yu et al., 2020).

In contrast to the HHP technique, HPCD abolishes the spoilage enzymes and microorganisms at a relatively lower pressure and key advantages of the methods over the traditional technologies of processing. It also has many advantages in contrast to other methods, such as CO_2 gas is easily available, less costly, nontoxic, and non-flammable; additionally, after processing conditions, this can be eliminated with release of pressure (Briongos et al., 2016). However, several reports suggested the HPCD effectiveness to deactivate different enzymes such as PME (Hu et al., 2013), PAL (Ma et al., 2022), and PPO (Kong et al., 2021; Zhang et al., 2022). Based on the reported literature, HPCD considered an effective method for fruit and vegetable processing techniques. However, the exact mechanism of suppression of enzyme activity under HPCD is still not clear. The leading theory illustrated that enzyme inactivation is due to the conformational and structural changes of enzymes. Many scientists have summed up the influence of supercritical carbon dioxide ($SSCO_2$) on the inactivation of different enzymes (Hu et al., 2013), but this chapter summarizes the potential of HPCD in the inactivation of enzymes in association with enzymes' structural modifications and microbial inactivation in different fruits and vegetables.

11.2.1 Carbon Dioxide (CO_2) Gas

CO_2 is extensively utilized in the food and medical sectors due to its inertness, nonflammability, non-corrosiveness, affordability, lack of odor and taste, and its Generally Recognized as Safe (GRAS) status. It plays a cardinal role in HPCD because it possesses a critical point at 31.1 °C (304.2 K) and 7.3 MPa (72.8 bar), enabling operations at ambient temperatures and mild pressures, which is particularly beneficial for preserving the quality of thermo-labile and oxidizable natural food products (Reverchon & de Marco, 2006). In the subcritical or supercritical fluid phase, CO_2 exhibits properties that lie between liquids and gases, with density similar to that of a liquid. The manipulation of temperature and pressure can modify its diffusivity, density, dielectric constant, and viscosity. Changes in pressure and temperature affect the phase equilibrium between the liquid and gas states, with increased pressure favoring the liquid phase and increased temperature favoring the gas phase. Understanding of CO_2 solubility and phase behavior in different solution types under variable pressures and temperature conditions is critical for applications in different food processes and designs of process factors. The solutbility of CO_2 in water, which decreases with elevated pressure regimes and temperatures, serves as a vital guideline for change predictions in solubility and phase behavior when compared with some highly complex liquids systems. The presence of other materials in the solutions can either improve or diminish CO_2 solubility potential, with higher solids content thus reducing solubility with the addition of some polar co-solvent agents increasing the overall solubility (Spilimbergo et al., 2013).

11.2.2 Working Principle of HPCD

The HPCD working principle involves exposing a substance or material, such as foods or beverages, to elevated pressure levels using CO_2 as the pressurizing medium. HPCD generally operates on the principle of increasing the pressure of CO_2 and can lead to numerous effects, such as microbial pathogen inactivation, enzyme inactivation, preservation of food quality, and modifications of certain properties of the produce being treated. This is important such that when HPCD is applied to a substance, for instance either fresh fruits and vegetables or their juices (Plazzotta & Manzocco, 2019), it involves placing the said objects in a pressure vessel or chamber and introducing CO_2 at high pressures (Hu et al., 2013). The pressure is typically raised gradually to several hundred megapascals (MPa), which is considerably higher than atmospheric pressure. The elevated pressure therefore alters the behavior and characteristics of the CO_2, leading to certain interactions with the materials being treated with HPCD. The high pressure in HPCD can disrupt the cellular structures and functions of infestation-causing microorganisms, including bacteria, yeast, and molds, thus effectively reducing their populations to ensure food safety. This can contribute to extending the shelf life and improving the safety of the treated product. HPCD can also induce enzyme inactivation. The elevated pressure can cause structural changes in enzymes, leading to their denaturation or loss of activity. This is particularly useful in maintaining the quality of food products, as certain types of the enzymes play an imperative role in various biochemical reactions that can cause spoilage or degradation as well as discoloration of the fresh produce over time. Furthermore, HPCD can impact other properties of the substance, such as the preservation of nutritional content, textural nature, and color. The higher pressure can help to retain vitamins, antioxidative compounds, and other beneficial nutrients in the treated produce such as fruit and vegetable commodities. It can also affect sensory attributes of the treated produce by modifying the structure of proteins, carbohydrates, and certain other components (Hu et al., 2013).

11.3 HPCD MECHANISM

11.3.1 Microbial Inactivation by HPCD

11.3.1.1 Vegetative Cells' Inactivation by HPCD

The higher pressure of CO_2 causes the hydrophobicity of bacterial cells to rise further, which can make the treated cells more likely to clump together. The higher cell concentration causes more cells to clump together, which thereby lowers the HPCD inactivation rates (Ding et al., 2022). Cells tend to clump under the pressure applied during HPCD to separate the cells, and this could produce a more severe shear force. Therefore, a higher inactivation rate is caused by higher pressure. Pressured CO_2 has the potential to harm cell surfaces and disturb intracellular structure. Using scanning electron microscopes and transmission electron microscopes (SEM and TEM), it is possible to see how HPCD treatment changes the morphology of microbial cells. Longer treatment times may intensify these changes. The extracellular surface of HPCD-treated cells had a significant number of bulges, which suggests cytoplasmic leakage. However, because inactive cells can appear to be integral under TEM, morphological alteration is not always the cause of cell death.

High pressure results in increased CO_2 diffusivity and decreased extracellular pH. HCO_{3-} and CO_3^{2-} can be produced from dissolved CO_2 (Wang et al., 2019). H^+ can be liberated from H_2O in the interim. Due to H+ release in the aqueous extracellular environment, which can increase the permeabilization of the membrane and permit more CO_2 to enter the cytoplasm, the pH will decrease. This explains why dry strains' HPCD inactivation efficiency was significantly reduced. The pH that is caused by pressured CO_2 will not, however, continuously reduce with the pressure and temperature regardless of the liquid matrix (Yu et al., 2020).

The permeability and fluidity of the membrane are improved. Approximately 81% of the cells were permeable after 15 minutes of CO_2 application under pressure (12 MPa and 35 °C), while only 18% of the cells were only partially permeable. Furthermore, when exposed to pressured CO_2 at a pressure of about 12 MPa, the ratio of phosphoglyceride to phosphatidylethanolamine decreased, indicating a decrease in the stability of the *E. coli* membrane as well (Illera et al., 2018). The potential for microbial growth-related increases in cell membrane permeability and intracellular cytoplasmic solute leakage, were both confirmed by this study. Another study found that HPCD can result in a reduced quantity of saturated fatty acid contents and an increased amount of unsaturated fatty acids (Yu et al., 2020). Cell membrane permeabilization and stability are subsequently changed. Membrane damage was also seen in the majority of cells. Some membrane proteins may become less active and lose the ability to perform their biological functions as a result of the lower pH that pressured CO_2 produces. Pressured CO_2 can alter the electrostatic interactions between proteins and lipids by producing HCO_{3-} ions (Hama & Fitzsimmons-Thoss, 2022). For instance, it is thought that negatively charged phospholipids, which may be altered by HCO_{3-} ions, have an impact on the topology of the inner membrane protein. The HCO_{3-} ions may also have an impact on where the charged amino acids are located within the membrane. It should be noted that the makeup and distribution of membrane lipids affect the conformation and function of membrane proteins.

11.3.1.2 Spore Inactivation by HPCD

Due to their distinctive structure, spores typically have a high ability to resist physical and chemical treatment. The exosporium, germ cell wall, outer membrane, coat layer, cortex, inner membranes, and core are all components of spore structure (Garcia-Gonzalez et al., 2009). The exosporium, coat layer, and cortex that make up the permeability barrier help to protect the spores from pressure or attack by chemicals and lytic enzymes. Spore deactivation using HPCD typically goes like this. The enzymes within the spores may activate by heat that results in modifications. That gives CO_2 a chance to enter the cell and cause harm to its metabolic and structural components. The HPCD then causes the inner membrane's fluidity and permeability to increase. The HPCD treatment decreases the spore's ability to withstand heat because that improves the permeability of the inner membrane (Xu et al., 2013). It should be noted that because the damage to the germinant receptors in HPCD-treated spores prevented them from undergoing typical germination, germination is not the cause of the reduced heat resistance of spores. The absence of germination then prevents spore outgrowth, which results in spore death (Figure 11.2).

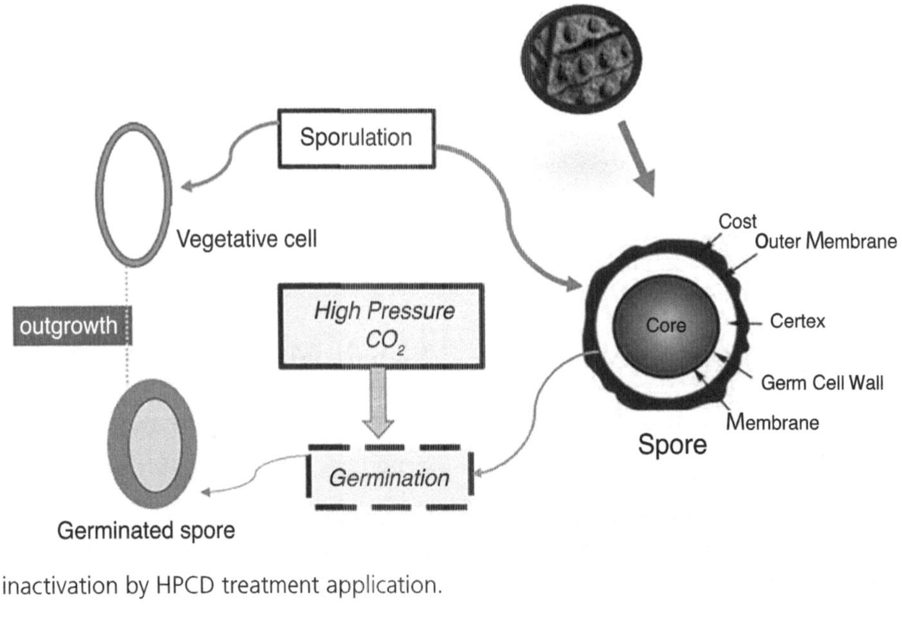

FIGURE 11.2 Spore inactivation by HPCD treatment application.

11.3.2 Enzyme Inactivation by HPCD

HPCD may help enzymes become inactive; POD and PPO are the most significant enzymes to adversely affect food quality by the process of browning, off-flavor formation, and vitamin and pigments loss, particularly in some fruits and vegetables and other products (Ma et al., 2022). Temperature, pressure, and time all affect how well HPCD inhibits the activity of peroxidase and polyphenol oxidase. After 20 minutes of CO_2 treatment at 45 °C and 20 MPa, 61.39% of the original activity of peroxidase was still present.

However, as soon as the temperature reached 65 °C, only 29.32% of the original peroxidase activity was still present. POD activity remained constant at 95.57%, even when the temperature was raised to 66 °C without HPCD (Xu et al., 2011). For 30 minutes, a pressure of CO_2 increase to 31 MPa effectively rendered 84% of peroxidase inactive. The polyphenol oxidase is significantly more susceptible to HPCD when compared to peroxidase. Polyphenol oxidase was inactivated under 31 MPa and 46 °C HPCD conditions for 31 minutes; therefore, this enzyme will be completely inactivated. For inactivating enzymes, the level of CO_2 is more significant than pressure. Without applying pressure, 70% CO_2 inactivated 51% of the protease. Without CO_2, the inactivation was only 20% when 150 MPa was applied. The collagenase activity yielded the same results. Collagenase inactivation did not increase at 150 MPa pressure compared to 0 pressure and CO_2-treated collagenase 100%. In addition to these instances, HPCD has successfully inactivated numerous other enzymes.

11.4 FACTORS INFLUENCING THE MICROBIAL AND ENZYMATIC INACTIVATION BY HPCD

A century ago, compressed CO_2 was found to have bactericidal properties. It has been demonstrated that a variety of factors affect how effective HPCD is at inactivating microorganisms (Illera et al., 2018). This discussion of HPCD inactivation factors is necessary because many of these factors, to varying degrees, affect pasteurization effectiveness. We anticipate this will also offer practical advice for business operations.

11.4.1 Species of Microorganisms

Scientists have thoroughly investigated how HPCD kills a variety of bacteria. Different bacteria exhibit various reactions to HPCD therapy. Aerobic psychrophilic microorganisms have been found to be significantly sensitive to HPCD compared to aerobic mesophilic microorganisms. Additionally, *L. monocytogenes* is highly sensitive to HPCD treatments as compared to *E. coli*.

According to the most recent research, no differences in bactericidal effects between gram-positive and gram-negative bacteria or fungi were discovered. This implies that HPCD bactericidal effects have little to do with the cell wall (Casas et al., 2012). However, more research must be done to fully understand the bactericidal effects of HPCD on gram-positive and gram-negative bacteria, as well as on fungi. This research must vary and control the treatment conditions (pressure, temperature, time, and medium).

While HPCD can easily inactivate bacterial vegetative cells under the same pressure, time, and temperature conditions, it is more challenging to do so with their spore forms. For instance, to effectively inactivate *Bacillus subtilis* spores, temperature that is generally high (85 °C) for 60 minutes, 20 MPa HPCD is needed. The same species' vegetative cells, however, can experience a more than 7 log reduction after only 2.5 minutes at 39 °C and 7.5 MPa (Yu et al., 2020). The following is a ranking of spore resistance to HPCD: *B. licheniformis*, *G. stearothermophilus*, *B. cereus*, and *B. coagulans* are all better than *B. subtilis*. Additionally, it was found that HPCD can render bacteriophage T4 inactive. Twenty-five minutes at 0.7 MPa and 22 °C, it was possible to achieve log reductions of approximately 4.0 for bacteriophage Q, just under 3.0 bacteriophage X174, > 3.1 bacteriophage T4, > 3.4 bacteriophage MS2, > 3.0 for bacteriophage T (Table 11.1 and Table 11.2).

11.4.2 Concentration of Cells

Compressed carbon dioxide is used to inactivate bacteria; samples with a higher cell concentration perform less effectively than samples with a lower cell concentration. More cells clumped together when pressurized CO_2 was applied at higher concentrations, making it more difficult for the gas to enter the cells and have the desired bactericidal effect. The initial cell concentration should be considered when applying HPCD to real-world industrial applications.

11.4.3 Water Content and pH

Some researchers discovered that lowering the initial environmental pH could improve the efficacy of microbe inactivation by HPCD. However, the pH of the endpoint after the treatment of carbon dioxide did not differ significantly depending on the treatment of pressure, microorganism strain, and exposure time. While treating the bacteriophage virus MS2 with HCl (hydrochloric acid) at the same pressure as for HPCD, CO_2 treatment showed a greater sterilization effect than HCl treatment. Therefore, microbial inactivation is not caused by the decrease in pH caused by HPCD; however, a reduced initial environment pH can still increase inactivation effectiveness.

The effectiveness of HPCD inactivation is significantly influenced by water. Increased water content might improve bactericidal effectiveness. According to some research, wet cells are more susceptible to HPCD than dry cells. For instance, cell colonies decreased by more than 3 log (CFU/mL)

TABLE 11.1 HPCD Inhibition on Microbial Cells (Garcia-Gonzalez et al., 2007)

MICROORGANISMS	FOOD MEDIA	PRESSURE (MPA)	TIME (MIN)	TEMP. (°C)	REDUCTION/ INACTIVATION RATE	REFERENCE
Listeria monocytogenes	Carrot juice	6	720	45	6.4 decimal reduction	Erkmen (2000)
Brochotrix thermosphacta	Apple juice	6.05	120	45	4.7 decimal reduction	Erkmen (2000)
Natural aerobic microflora	Peach juice	6.05	180	45	4.6 decimal reduction	Erkmen (2000)
Enterococcus faecalis	Orange juice	15	45	55	5.9 decimal reduction	Erkmen (2000)
Yersinia enterocolitica	Peach juice	6	240	45	4.7 decimal reduction	Erkmen (2001)
E. coli	Nutrient broth	10	50	30	7.5 decimal reduction	Özcan and Erkmen (2001)
Bacteria	Onions	5.9	120	23	90% inactivation	Haas et al. (1989)
Bacteria	Strawberries	6.2	120	−22	99% inactivation	Haas et al. (1989)

TABLE 11.2 HPCD Impact on Fruits and Vegetables (Ferrentino & Spilimbergo, 2011)

FOOD SUBSTRATES	TARGETED MICROORGANISMS	PRESSURE (MPA)	TIME (MIN)	TEMPERATURE (°C)	EFFECT
Fresh-cut pears	*Saccharomyces cerevisiae*	10	10	55	Total inactivation
Celery and leafstalks	TPC	62.8	30	40	Reduction in TPC by 140 cfu/g
Kimchi	Lactic acid bacteria	7.0	1440	–	Inadequate reduction of LAB
Spinach	*E. coli Ke12*	7.5–10	40	40	No detectable level of reduction

when moist *E. coli* in lysogeny broth was subjected to a 15-minute HPCD treatment at 35 °C and 10 MPa. However, when the same conditions were applied to dry *E. coli*, the number of cell colonies decreased by about 0.5 log (CFU/mL). Additional researchers have demonstrated that the water content is crucial to HPCD's ability to kill bacteria. In general, liquid materials are more effectively inactivated by HPCD technology than solids.

11.4.4 Conditions

The longer the exposure time, the more effectively CO_2 treatment destroyed cells. For instance, a 15-minute treatment can reduce *E. coli* CFU/g by 1.4 logs, whereas a 45-minute treatment can result in a log reduction. The longer treatment times may not always be worth the effort to boost inactivation, though. The efficiency rose over time up until a certain point in the treatment period; even when the treatment was lasting, there was progress at 51 °C.

Using high pressure would result in a higher CO_2 inactivating effect. The higher inactivation efficiency may be explained by the fact that higher pressure can improve the solubilization of CO_2, facilitating its contact with cell penetration (Yu et al., 2020). At lower temperatures, high pressure can produce the same inactivation efficacy as low pressure can at higher temperatures. Although the pressure has a favorable impact on HPCD inactivation effectiveness, temperature is more crucial. Temperature is thought to be the most crucial factor for inactivating bacteria among all other considerations. Temperature, however, should not be taken into account separately for the inactivation of bacteria. Both 38 °C and 50 °C produced comparable images of cell cytoplasmic damage; however, only the high temperature may result in complete inactivation.

11.5 IMPACT OF HPCD ON FRUITS AND VEGETABLES

11.5.1 HPCD Impact on Nutritional Factors

The goal of HPCD, a type of cutting-edge nonthermal pasteurization technology, is to enhance nutrition preservation following pasteurization processing. However, it is necessary to talk about how HPCD affects food nutrition (Cappelletti et al., 2015). It was demonstrated that the remaining Ca^{2+} in orange juice could remain greater than 90% after HPCD treatment.

Longer treatment times and greater pressure can produce more Ca^{2+} loss rather than higher temperatures.

HPCD can increase the content of sugar, according to Marszałek et al. (2020). Though, when HPCD treatment was used on apple juice that had been supplemented with inulin and subjected to lower pressure, lower temperature, and shorter treatment times, the sugar content was not significantly affected. Pressure of 60 MPa can result in loss of 33%, and 30 MPa can result in loss of 30%, the preserved content of vitamin C is not significantly affected by pressure. However, the temperature can be lowered by using pressure. Pressure has less of an impact on vitamin deterioration than temperature; 71–90% of vitamin C was protected before the temperature dropped from 45 to 38 °C. Additionally, ultrasound can be used to preserve more vitamin C at lower HPCD temperatures.

Organic acids, a class of nutritional components with antioxidant activity in several fruits, showed no significant effects from HPCD treatment at pressures below 20 MPa. The antioxidant activity–rich phenolic compounds are not significantly impacted by HPCD treatment either. Higher temperatures, however, can result in an even greater loss of phenolic content; according to Murtaza et al. (2022), about 30% of the polyphenols were degraded at higher temperatures and pressures (60 MPa, 55 °C).

Since there is no reduction after a 30-minute HPCD treatment at 30 MPa and 45 °C, anthocyanins are protected during HPCD. The anthocyanin content only decreased by 3% when the pressure was raised to 60 MPa (Garcia-Gonzalez et al., 2007). Betaxanthins, which are responsible for the color of plants and have antioxidant properties, were also exposed to HPCD treatment. It was discovered that betaxanthin degradation can be accelerated by both higher pressure and temperature. Particularly when temperature and pressure were raised to 55 °C for 30 minutes, 32.1% betaxanthin degradation was found. Betacyanins appear highly sensitive to HPCD when compared to betaxanthins. However, these analyses suggest that reduced pressure and temperature are preferable in industrial applications because high pressure and temperature can cause nutrition degradation.

11.5.2 HPCD Impact on Organoleptic Qualities of Food

Another crucial task is to increase consumer acceptance of the products. It was discovered that the HPCD technique can lower food's volatile compound concentrations and may have an impact on their aroma. Food titratable acidity, total soluble solids, and sweetness are all influenced by pH (fructose, glucose, and sucrose). HPCD has little impact on food pH because of the intricate food matrix. However, the pH of various foods varies greatly (Roobab et al., 2021). Apple juice and strawberry juice always have pH values of 3 and 4, no matter the temperature, pressure, or length of treatment. HPCD can nonetheless cause a very slight drop in apple juice's pH. Lower pH values are caused by higher pressure, which suggests that high pressure and temperature may cause higher CO_2 to dissolve into water (Yu et al., 2020).

Investigations were done into how HPCD affected food color. Other sources of food and HPCD treatment circumstances caused the yellowness to vary. Apple, carrot, and Hami melon juice all exhibit less browning. Total color differences have been used in recent studies to examine changes in food color. Food color changed more when the temperature was higher, and HPCD can intensify the color change in apple juice at the same temperature. However, after HPCD treatment, strawberry juice color did not significantly change. Orange juice was exposed to 30 MPa, 40 °C for 40 minutes and experienced a clear color change. This implies that HPCD has different effects on food color (Table 11.3).

Another study found that pressure had a greater impact on food color than CO_2 did. When pressure is applied without CO_2 (Yu et al., 2020), Perez-Won et al. (2020) demonstrated that significant color changes could be seen. With CO_2 alone, without pressure treatment, the color change was less pronounced (Perez-Won et al., 2020).

11.5.3 HPCD Impact on Viscosity

Juices treated with HPCD have greater viscosity, average particle size, and cloudiness, which can offer customers high-quality juice. For instance, orange juice that had undergone HPCD treatment contained a greater percentage of smaller particles, which enhanced the juice's quality by adding to its cloudiness and stability. Furthermore, it was found that HPCD at low pressure and temperature had no impact on the viscosity of inulin-enriched apple juice, whereas increased viscosity was seen in peach juice that had undergone the treatment. Bovine milk's turbidity and average particle size can increase with high temperature and long treatment time, but HPCD can reduce the milk's viscosity and total solids content, which is not the case with fruit juices (Table 11.4).

TABLE 11.3 HPCD Inactivation on the Quality of Food

ATTRIBUTES	*FOOD TYPES*	*CONDITIONS OF HPCD*	*RESULTS*
pH	Apple juice	20 min, 20 MPa, (65, 25 °C)	2.5%, 11.4% decreased (3.58, 3.25)
Color	Orange juice	40 °C, 30 MPa, and 40 min	ΔE = 5.1

TABLE 11.4 Impact of HPCD on Flavor Profile

FRUITS	PRESSURE (MPA)	TIME (MIN)	TEMP. °C	CO_2 / SAMPLE (W/W)	CHANGES OCCURRING IN FLAVOR	REFERENCE
Apple juice	10	10	36	36	Partially significant	Gasperi et al. (2009)
Orange juice	38,72,107	10	Continuous	0.04–1.18	Insignificant within 2 weeks, but significant from the 3rd week	Kincal et al. (2006)
Beer	27.6, 20.7	5	21	10%	No changes	Dagan and Balaban (2006)
Grape juice	48.3	35	Continuous	35	Significant	Gunes et al. (2005)

11.5.4 Effects of HPCD on Flavor of Fruits and Vegetables

The sensory impression of flavor in food is primarily influenced by the chemical senses of taste and smell. A food's flavor can change if the substances that give it its sourness, sweetness, bitterness, or odor change. There has been some progress in our understanding of HPCD-induced flavor changes so far.

11.5.5 HPCD Impact on Taste of Fruits and Vegetables

Food sourness was found to have a close relationship with pH and titratable acidity. At 4.90 MPa, HPCD reduced the initial carrot juice pH from 6.5–4.4 to 6.5–4.4. After HPCD, the apple juice pH dropped from 3.56 to 3.52. According to Gunes et al. (2006), treated juices had a high acidic or carbonated taste after HPCD, which was undesirable. HPCD's pH-lowering effect was caused by the dissolution of CO_2 into liquid foods, which dissociated into carbonate, bicarbonate, and hydrogen ions (Table 11.5).

Polyphenols found in plant foods and beverages were thought to be responsible for the astringent mouthfeel. Because polyphenolic compounds and proteins can interact to form soluble complexes, complexes that are soluble in water can precipitate HPCD-denatured proteins can bind polyphenolic compounds, and HPCD reduced the total polyphenols in peach juice significantly.

11.5.6 Aroma of Fruits and Vegetables

Food aromas were typically produced by volatile flavor compounds. Normally, HPCD reduced volatile compound concentrations in foods. Following HPCD, there was a 35% decrease in volatile components in fresh apple juice, with high variation

TABLE 11.5 HPCD Effect on the Taste of Fruits and Vegetables Juices

SPECIES OF FOOD	PARAMETERS OF TASTE	PRESSURE (MPA)	TIME (MIN)	TEMPERATURE (• C)	CHANGE	REFERENCE
Carrot juice	pH	4.9	5	5	Reduction	Park et al. (2002)
Orange juice	TA, pH, and phenols	40	10–60	55	There is no change	Niu et al. (2010)
Apple juice	pH	16.0	150	60	Reduction	Ferrentino et al. (2009a)
Peach juice	pH	30	60	55	There is no change	Zhou et al. (2010)
Apple slices	pH	20	20	45, 55, 65	There is no change	Niu et al. (2010)
Nutrient broth	pH	10	50, 80, 90	30	Reduction	Erkmen (2001)
Apple cider	pH	6.9–48.3	–	25–45	There is no change	Gunes et al. (2006)
Red grapefruit juice	pH, TA, and total phenolics	34.5	7	40	There is no change	Ferrentino et al. (2009b)
Coconut water	pH, TA, and soluble phenolics	34.5	6	25	Increment	Damar et al. (2009)

TABLE 11.6 HPCD Effect on Flavor of Juices

SPECIES OF FOOD	*PRESSURE (MPA)*	*TIME (MIN)*	*TEMPERATURE (°C)*	*SAMPLE OF CO_2*	*CHANGE*	*REFERENCE*
Apple juice	10	5, 10, 20	36	–	Overall depletions of 35%	Gasperi et al. (2009)
Orange juice	600	2.17	25	–	Particularly relevant for low molecular weight	Boff et al. (2003)
Hami melon juice	35	60	55	–	Insignificant	Chen et al. (2010)
Beer	27.6	5	21%	10%	Depletions for ethyl hexanoate of 49%	Dagan and Balaban (2006)

TABLE 11.7 HPCD Effect on Flavor by Sensory Analysis

SPECIES OF FOOD	*PRESSURE (MPA)*	*TIME (MIN)*	*SAMPLE OF CO_2*	*TEMP. (°C)*	*RESULTS*	*REFERENCE*
Apple juice	10	10	–	36	Partially important	Gasperi et al. (2009)
Coconut water beverage	34.7	7	14%	41	Minimal changes	Damar et al. (2009)
Beer	20.8	6	11%	22	Insignificant	Dagan and Balaban (2006)
Orange juice	71, 108	10	0.41–1.19	–	Minimal changes	Kincal et al. (2006)
Muscadine grape juice	35	6.2	8 or 16%	29	Minimal changes	Pozo-Insfran et al. (2007)
Grape juice	48.3	–	0.17	35	Minimal changes	Gunes et al. (2005)

being seen for the esters (particularly isopentyl acetate and hexyl acetate), which are characterized by lower threshold values and are in charge. The concentrations of phellandrene, trans-2-hexenol, ethyl butyrate, and limonene in orange juice were significantly reduced by HPCD (between 50% and 83%). Even though concentrations in orange juice of ethyl butyrate and trans-2-hexenol were decreased after HPCD treatment, they remained higher than in juice that had undergone thermal processing (Table 11.6).

Furthermore, HPCD-induced homogenization reduced the size of larger particles in foods, promoted the release of volatile components in foods, and altered mouth feel. According to Niu et al., the volatile components in HPCD-treated orange juice was improved. The chemical reaction in food is highly dependent on temperature. Lower temperatures during HPCD could help to prevent degradation of heat-sensitive components caused by thermal processing. The biochemical reactions in foods were caused by native enzymes. Because of their residual activity, foods containing indigenous enzymes catalyzed the substrates, and HPCD-treated foods produced off-flavors during storage.

Lipoxygenase and hydroperoxide lyase catalyzed the oxidation of polyunsaturated fatty acids, and the oxidation metabolisms pathway from linolenic acids to volatile aldehydes was partially responsible for the rancid taste. HPCD-treated red cabbage's residual myrosinase catalyzed the conversion of glucosinolates into pungent byproducts like isothiocyantes. The activity of lipoxygenase affected the volatile flavor of Hami melon juice, particularly the 4 weeks of posttreatment storage (Table 11.7).

11.5.7 Effects of HPCD on Food Texture

Food textural parameters are perceived through touch and when it is chewed after being placed in the mouth. These parameters are related to the processes of clarification, cloud formation, turbidity, and the rheological process for liquid foods like purees and juices (Hu et al., 2013). These terms were frequently used interchangeably, regardless of the fact that they referred to slightly different characteristics.

11.6 CONCLUSIONS

High-pressure carbon dioxide (HPCD) processing is quickly expanding in the field of food processing, and the prospects for technology appear excellent. High-pressure processing (HPP) and HPCD stand for high pressure and carbon dioxide. HPCD offers a method for preserving food that does not involve the use of heat or chemical preservatives, both of which have the potential to lower the food's overall quality as well as its nutritional value. This is one of the most significant advantages offered by these two processes. It is expected that there will be greater demand for items that have been treated with HPP and HPCD as consumers develop a greater interest in natural and minimally processed meals. HPCD can help meet consumers' rising expectations for safety and quality. It has significant advantages over HHP technology, such as lower equipment costs and its large scale due to its lower pressure requirements (less than 50 MPa). Furthermore, because CO_2 is non-toxic, it is known as a "green technology". HPCD process parameters include treatment time, pressure, and temperature for various food products. Research is required to produce data that is credible for upcoming commercial-scale applications.

The current situation of HPCD pasteurization technology knowledge has been explained. HPCD's bactericidal effect on various microorganisms has been widely tested, and this process has been validated for use in the food industry as an innovative, non-thermal technique. HPCD's impact on food quality was also taken into account. Although HPCD has been extensively studied for microorganism inactivation, the detailed mechanisms remain unknown. Until now, research has suggested that the primary cause of bactericidal activity is a change in the cell membrane. As a result, in future studies, scientists should pay closer attention to membrane alteration caused by HPCD treatment. Research on HPCD's inactivation of spores and viruses has not been sufficient, particularly for spore-forming bacteria like *Clostridium*, which results in spoilage and foodborne diseases.

11.7 LIMITATIONS AND FUTURE PROSPECTS

HPCD technology is effective in inactivating enzymes and may be an efficacious technique but still possesses some limitations. The equipment and maintenance required to carry out HPCD may be costly, and highly trained personnel are required to operate it. Sensory properties, such as aroma, taste, and color, of liquid foodstuffs, i.e., juices and fresh produce, are affected negatively. Sensory quality changes such as pH, particle size, and corresponding properties are changed. The process must be critically monitored, as vitamin and antioxidant quantities can be affected by the use of HPCD. The process may not be suitable for all kinds of juices and enzymes, so the working parameters need to be scrutinizingly tailored to the specific substrate being tested (Benito-Roman et al., 2019).

More research into mathematical analysis for the deactivation of different microorganisms is needed to give information for commercial processes. Quite systematic and numeric studies of the effect of HPCD on nutrient content are required because this is important for nutrient content and efficiency, features which affect a customer's willingness to purchase a product. Despite the reality that no poisonous chemicals were discovered in HPCD-treated bread, further investigations into the safe operation of HPCD-treated food are required because HPCD is a new pasteurization technology.

Technology is continuously advancing, which results in the development of new pieces of processing equipment and methods that are intended to increase the process's overall efficacy as well as its efficiency. For instance, there is a continuing study into improving the processing parameters, such as pressure and temperature, to achieve greater preservation while limiting the influence on the sensory and nutritional quality of the food that is altered as a result of the procedure (Yang et al., 2021).

REFERENCES

Benito-Román, Ó., Alonso-Riaño, P., De Cerio, E. D., Sanz, M. T., & Beltrán, S. (2022). Semi-continuous hydrolysis of onion skin wastes with subcritical water: Pectin recovery and oligomers identification. *Journal of Environmental Chemical Engineering*, *10*(3), 107439.

Boff, J. M., Truong, T. T., Min, D. B., & Shellhammer, T. H. (2003). Effect of thermal processing and carbon dioxide-assisted high-pressure processing on pectinmethylesterase and chemical changes in orange juice. *Journal of Food Science*, *68*(4), 1179–1184.

Briongos, H., Illera, A. E., Sanz, M. T., Melgosa, R., Beltrán, S., & Solaesa, A. G. (2016). Effect of high pressure carbon dioxide processing on pectin methylesterase activity and other orange juice properties. *LWT—Food Science and Technology*, *74*, 411–419.

Casas, J., Valverde, M. T., Marín-Iniesta, F., & Calvo, L. (2012). Inactivation of *Alicyclobacillus acidoterrestris* spores by high pressure CO_2 in apple cream. *International Journal of Food Microbiology*, *156*(1), 18–24. https://doi.org/10.1016/j.ijfoodmicro.2012.02.015

Chen, J. L., Zhang, J., Song, L., Jiang, Y., Wu, J., & Hu, X. S. (2010). Changes in microorganism, enzyme, aroma of hami melon (*Cucumis melo* L.) juice treated with dense phase carbon dioxide and stored at 4 °C. *Innovative Food Science and Emerging Technologies*, *11*(4), 623–629. https://doi.org/10.1016/j.ifset.2010.05.008

Dagan, G. F., & Balaban, M. O. (2006). Pasteurization of beer by a continuous dense-phase CO_2 system. *Journal of Food Science*, *71*(3), E164–E169.

Damar, S., Balaban, M. O., & Sims, C. A. (2009). Continuous dense-phase CO_2 processing of a coconut water beverage. *International Journal of Food Science & Technology*, *44*(4), 666–673.

De Jaegere, I., Cornelis, Y., De Clercq, T., Goossens, A., & Van de Poel, B. (2022). Overview of Witloof Chicory (Cichorium intybus L.) discolorations and their underlying physiological and biochemical causes. *Frontiers in Plant Science*, *13*. https://doi.org/10.3389/fpls.2022.843004

Ding, T., Liao, X., & Feng, J. (2022). *Stress Responses of Foodborne Pathogens*. Springer Nature.

Du, L., Sun, Y., Han, L., & Su, S. (2023). Inactivation of *Saccharomyces cerevisiae* by combined high pressure carbon dioxide and high pressure homogenization. *The Journal of Supercritical Fluids*, *193*, 105816. https://doi.org/10.1016/j.supflu.2022.105816

Erkmen, O. (2000). Effect of carbon dioxide pressure on *Listeria monocytogenes* in physiological saline and foods. *Food Microbiology*, *17*(6), 589–596.

Erkmen, O. (2001). Effects of high-pressure carbon dioxide on *Escherichia coli* in nutrient broth and milk. *International Journal of Food Microbiology*, *65*(1–2), 131–135.

Ferrentino, G., Bruno, M., Ferrari, G., Poletto, M., & Balaban, M. O. (2009a). Microbial inactivation and shelf life of apple juice treated with high pressure carbon dioxide. *Journal of Biological Engineering*, *3*, 1–9.

Ferrentino, G., Plaza, M. L., Ramirez-Rodrigues, M., Ferrari, G., & Balaban, M. O. (2009b). Effects of dense phase carbon dioxide pasteurization on the physical and quality attributes of a red grapefruit juice. *Journal of Food Science*, *74*(6), E333–E341.

Ferrentino, G., & Spilimbergo, S. (2011). High pressure carbon dioxide pasteurization of solid foods: Current knowledge and future outlooks. *Trends in Food Science & Technology*, *22*(8), 427–441.

Garcia-Gonzalez, L., Geeraerd, A. H., Spilimbergo, S., Elst, K., Van Ginneken, L., Debevere, J., & Devlieghere, F. (2007). High pressure carbon dioxide inactivation of microorganisms in foods: The past, the present and the future. *International Journal of Food Microbiology*, *117*(1), 1–28. https://doi.org/10.1016/j.ijfoodmicro.2007.0

Gasperi, F., Aprea, E., Biasioli, F., Carlin, S., Endrizzi, I., Pirretti, G., & Spilimbergo, S. (2009). Effects of supercritical CO_2 and N_2O pasteurisation on the quality of fresh apple juice. *Food Chemistry*, *115*(1), 129–136. https://doi.org/10.1016/j.foodchem.2008.11.078

Gui, F., Wu, J., Chen, F., Liao, X., Hu, X., Zhang, Z., & Wang, Z. (2007). Inactivation of polyphenol oxidases in cloudy apple juice exposed to supercritical carbon dioxide. *Food Chemistry*, *100*(4), 1678–1685. https://doi.org/10.1016/j.foodchem.2005.12.048Gunes, G., Blum, L. K., & Hotchkiss, J. H. (2005). Inactivation of yeasts in grape juice using a continuous dense phase carbon dioxide processing system. *Journal of the Science of Food and Agriculture*, *85*(14), 2362–2368.

Gunes, G., Blum, L. K., & Hotchkiss, J. H. (2005). Inactivation of yeasts in grape juice using a continuous dense phase carbon dioxide processing system. *Journal of the Science of Food and Agriculture*, *85*(14), 2362–2368.

Gunes, G., Blum, L. K., & Hotchkiss, J. H. (2006). Inactivation of Escherichia coli (ATCC 4157) in diluted apple cider by dense-phase carbon dioxide. *Journal of Food Protection*, *69*(1), 12–16.

Haas, G. J., Prescott, H. E., Jr., Dudley, E., Dik, R., Hintlian, C., & Keane, L. (1989). Inactivation of microorganisms by carbon dioxide under pressure. *Journal of Food Safety*, *9*(4), 253–265.

Hama, J. R., & Fitzsimmons-Thoss, V. (2022). Determination of unsaturated fatty acids composition in Walnut (Juglans regia L.) oil using NMR spectroscopy. *Food Analytical Methods*, *15*(5), 1226–1236. https://doi.org/10.1007/s12161-021-02203-0

Hu, W., Zhou, L., Xu, Z., Zhang, Y., & Liao, X. (2013). Enzyme inactivation in food processing using high pressure carbon dioxide technology. *Critical Reviews in Food Science and Nutrition*, *53*(2), 145–161.

Illera, A. E., Sanz, M. T., Beltrán, S., Melgosa, R., Solaesa, A. G., & Ruiz, M. O. (2018). Evaluation of HPCD batch treatments on enzyme inactivation kinetics and selected quality characteristics of cloudy juice from Golden delicious apples. *Journal of Food Engineering*, *221*, 141–150. https://doi.org/10.1016/j.jfoodeng.2017.10.017

Kincal, D., Hill, W. S., Balaban, M., Portier, K. A., Sims, C., Wei, C. I., & Marshall, M. (2006). A continuous high-pressure carbon dioxide system for cloud and quality retention in orange juice. *Journal of Food Science*, *71*. https://doi.org/10.1111/j.1750-3841.2006.00087.x

Krupa, T., & Tomala, K. (2021). Effect of oxygen and carbon dioxide concentration on the quality of minikiwi fruits after storage. *Agronomy*, *11*(11), 2251.

Liu, Y., Hu, X. S., Zhao, X. Y., & Zhang, C. (2013). Inactivation of polyphenol oxidase from watermelon juice by high pressure carbon dioxide treatment. *Journal of Food Science and Technology*, *50*, 317–324.

Ma, W., Li, J., Murtaza, A., Iqbal, A., Zhang, J., Zhu, L., & Hu, W. (2022). High-pressure carbon dioxide treatment alleviates browning development by regulating membrane lipid metabolism in fresh-cut lettuce. *Food Control*, *134*, 108749.

Marszałek, K., Skąpska, S., Woźniak, L., & Sokołowska, B. (2015). Application of supercritical carbon dioxide for the preservation of strawberry juice: Microbial and physicochemical quality, enzymatic activity and the degradation kinetics of anthocyanins during storage. *Innovative Food Science & Emerging Technologies*, *32*.

Murtaza, A., Xu, X., Pan, S., & Hu, W. (2022). A review of discoloration in fruits and vegetables: Formation mechanisms and inhibition. *Food Reviews International*, 1–22. https://doi.org/10.1080/87559129.2022.2119997

Niu, S., Xu, Z., Fang, Y., Zhang, L., Yang, Y., Liao, X., & Hu, X. (2010). Comparative study on cloudy apple juice qualities from apple slices treated by high pressure carbon dioxide and mild heat. *Innovative Food Science & Emerging Technologies*, *11*(1), 91–97.

Özcan, M., & Erkmen, O. (2001). Antimicrobial activity of the essential oils of Turkish plant spices. *European Food Research and Technology*, *212*, 658–660.

Park, S. J., Lee, J. I., & Park, J. (2002). Effects of a combined process of high-pressure carbon dioxide and high hydrostatic pressure on the quality of carrot juice. *Journal of Food Science*, *67*(5), 1827–1834.

Perez-Won, M., Lemus-Mondaca, R., Herrera-Lavados, C., Reyes, J. E., Roco, T., Palma-Acevedo, A., . . . & Aubourg, S. P. (2020). Combined treatments of high hydrostatic pressure and CO2 in coho salmon (Oncorhynchus kisutch): Effects on enzyme inactivation, physicochemical properties, and microbial shelf life. *Foods*, *9*(3), 273.

Plazzotta, S. & Manzocco, L. (2019). High-Pressure Carbon Dioxide Treatment of Fresh Fruit Juices. *Value-Added Ingredients and Enrichments of Beverages*, *431*. https://doi.org/10.1016/B978-0-12-816687-1.00013-8

Pozo-Insfran, D., Balaban, M. O., & Talcott, S. T. (2007). Inactivation of polyphenol oxidase in muscadine grape juice by dense phase-CO2 processing. *Food Research International*, *40*, 894–899. https://doi.org/10.1016/j.foodres.2007.03.002

Roobab, U., Shabbir, M. A., Khan, A. W., Arshad, R. N., Bekhit, A. E. D., Zeng, X. A., & Aadil, R. M. (2021). High-pressure treatments for better quality clean-label juices and beverages: Overview and advances. *LWT-Food Science and Technology*, *149*, 111828. https://doi.org/10.1016/j.lwt.2021.111828

Spilimbergo, S., Komes, D., Vojvodic, A., Levaj, B., & Ferrentino, G. (2013). High pressure carbon dioxide pasteurization of fresh-cut carrot. *The Journal of Supercritical Fluids*, *79*, 92–100.

Wang, H., Iqbal, A., Murtaza, A., Xu, X., Pan, S., & Hu, W. (2022). A review of discoloration in fruits and vegetables: Formation mechanisms and inhibition. *Food Reviews International*, 1–22. https://doi.org/10.1080/87559129.2022.2119997

Wang, S., & Springerlink (Online Service). (2019). *Chemical Hazards in Thermally-Processed Foods*. Springer Singapore.

Yang, G., Li, C., Zhu, X., Yan, J., & Liu, J. (2021). Prevalence of and risk factors associated with sleep disturbances among HPCD exposed to COVID-19 in China. *Sleep Medicine*. https://doi.org/10.1016/j.sleep.2020.12.034

Yu, T., Niu, L., & Iwahashi, H. (2020). High-pressure carbon dioxide used for pasteurization in food industry. *Food Engineering Reviews*, *12*(3), 364–380. https://doi.org/10.1007/s12393-020-09240-1

Zhao, W., Sun, Y., Cheng, Y., Ma, Y., & Zhao, X. (2019). Effect of high-pressure carbon dioxide on the quality of cold-and hot-break tomato pulps. *Journal of Food Processing and Preservation*, *43*(7), e13959.

Zhou, L., Zhang, Y., Leng, X., Liao, X., & Hu, X. (2010). Acceleration of precipitation formation in peach juice induced by high-pressure carbon dioxide. *Journal of Agricultural and Food Chemistry*, *58*(17), 9605–9610.

Zhu, Y., Luan, Y., Zhao, Y., Liu, J., Duan, Z., & Ruan, R. (2023). Current technologies and uses for fruit and vegetable wastes in a sustainable system: A review. *Foods*, *12*(10), 1949.

High-Pressure Processing of Fresh Fruits and Vegetables

12

Ayesha Sarker* and Md Rayhan Shaheb
*Corresponding Author: ayesha.sarker@wvstateu.edu

12.1 INTRODUCTION

High-pressure processing (HPP) is a science-based, innovative food safety and conservation method. It is a commercialized, nonthermal processing–oriented technology that allows food processors to pasteurize and sterilize foods at ambient temperature. It is a viable substitute to thermal processing for making shelf-stable foods, including low-acid shelf-stable foods (Patterson, 2005; Trejo Araya et al., 2009; Parker, 2020). HPP is considered the best commercial processing–based methods compared to other nonthermal technologies, for example, pulsed electrics fields and light, cold plasma, electron beams, and modified atmosphere packaging (Farkas, 2016). Conventional thermal pasteurization technology may affect organoleptic properties, flavors, and nutritional value and consume higher energy related to heating and successive cooling. The benefit of HPP over traditional thermal pasteurization is it eliminates food pathogens and prolongs the shelf life of the treated food commodities and food products circulated through the entire cold chain while maintaining the foods' sensory attributes and nutritive values (Huang et al., 2017). By eliminating foodborne pathogens, bacteria, and molds; prolonging shelf life; and retaining the organoleptic and nutritional potentials of foods, HPP exhibits an environmentally friendly food processing technology for the modern food industry. The commercial usage of HPP technology has been widely seen in several food and beverage categories globally. Most examples are juices (e.g., premium), smoothies and beverages, avocado products (e.g., guacamole), ready-to-cook and ready meal (e.g., burger patties) products, seafood products, dips, spreads, sandwich fillings, baby foods, dairy products, and even some pet foods (Huang et al., 2017). Nowadays, many manufacturers, including Avure (USA), Stansted Fluid Power (UK), Hiperbaric (Spain), Kobelco (Japan), Multivac (Germany), Baotou Kefa (China), and Toyo Koatsu (Japan) manufacture HPP equipment with various specifications, capabilities, and abilities to satisfy experimental needs. For example, leading manufacturers Avure, based in OH, USA, and Hiperbaric, based in Burgos, Spain, developed HPP equipment with a capacity of 525 L and annual production of almost 60 million tons (Balasubramaniam et al., 2008). HPP jam in Japan is considered the earliest commercial product, while various HPP-based food products have also been introduced in many parts of Europe, North America, and, in recent years, Asia. Considering the high investment costs involved, an industrial alliance was established in 2015, where companies such as American Pasteurization and Universal Pure in the USA, in cooperation with Avure and Hiperbaric and several food beverage manufacturers, provide outsourcing services for food manufacturers to access the advantages of this technology (Huang et al., 2017).

The widespread application of HPP technology to diverse food applications, including different brands, has boosted significant development in HPP equipment and its adoption in food markets over the past decades. However, challenges still exist, including improvement of processing equipment, production conditions, pressure-resistant properties, transmitting fluids, dynamic parameters, and quantifications of indicator organisms; additionally, hygienic safety assessments for HPP products should be thoroughly implemented and strengthened. Emerging technologies, such as intelligent production lines, next-generation cloud-based sensors, automation, robotics, and artificial intelligence, have created new opportunities for the food processing industry like any other industry in today's world. Applying these smart technologies for evaluating, analyzing, and monitoring the systems and products and data-driven decisions can enormously impact the food processing industry. This chapter aims to highlight the principles and practices of HPP; their effect on the quality, safety, and shelf life of fruit and vegetable products; possible challenges and strategies; and future research needs, including food process models for the evaluation of microbial inactivation. Further, it outlines the necessity of university–industry collaboration for advancing the sector and effective extension programming, which are crucial for strengthening the sustainability of food and agriculture.

DOI: 10.1201/9781003370376-16

12.2 A BRIEF HISTORY OF HPP TECHNOLOGY ADVANCEMENT

HPP has evolved for many decades with various levels of effort from scientists worldwide, especially in Japan, Europe, the United States, and more recently, Asia. The rapid progress and success of HPP and pasteurization have played a significant role in advancing food safety technology. A brief history of the HPP development has been outlined under four different periods and described here:

I. Before 1900:

1894: The benefits of HPP were revealed in early 1894 at West Virginia University (Hite et al., 1914). Pressures ranging from 200 to 680 MPa exhibited inactivated yeasts, molds (fungus), and bacterial spoilage in foods.

1899: The first report in food application indicated that HPP (680 MPa) could reduce bacterial levels by 5 to 6 logs in milk (Hite, 1899).

II. Between 1900 and 1950:

1900–1923: Key contributions are research on engineering aspects of HPP, such as compressibility under pressure, polymorphic transformations, change of phase, and thermal conductivity of liquids (Bridgman, 1909, 1912, 1914, 1923). The water phase diagram under pressure was developed.

1920: Phenol manufactured from chlorobenzene under pressure (Brown, 1920).

1924: McGraw-Hill, New York, published the book *Commercial Fruit and Vegetable Products* (Cruess, 1924). Hite et al.'s (1914) findings were cited and reported that HPP is effective for fruit juice preservation. HPP studies on microorganisms, starches, different proteins, and enzyme activity during these periods were observed (Johnson et al., 1954). Despite these efforts, HPP for commercial food processing was very slow. The main barriers were the lack of pressure vessels, necessary pumps, low gas transmission, instrumentation, and food-grade and flexible packaging unavailability. Frozen foods, with their preservation and packaging technologies, became popular.

Early 1950s: Ionizing radiation became available as a commercial "cold sterilization" technology for food pasteurization and sterilization. Several decades of research have contributed to moving HPP to the next stage of advancement.

III. HPP technology advancements between 1950 and 2000:

1960: The HACCP concept was first developed by NASA and Pillsbury.

1970: *Clostridium botulinum* spores were first reported as resistant to inactivation at 680 MPa (Sale et al., 1970). pH < 4.5 found potential for commercial sterilization in HPP foods. Supercritical pressures of 30–50 MPa were seen for coffee decaffeination (King, 2014).

1970–1977: High pressure of 140 MPa applied on red meats pre- and post-rigor mortis helped tenderize cuts of beef (Macfarlane, 1973; Bouton et al., 1977).

1974: In the 34th Annual Meeting, The Institute of Food Technologists (IFT) recognized HPP as a tool to preserve fruits. Packaging of apricots in hermetically sealed flexible pouches was an example of using the HPP.

1980: In-vivo proteolytic enzyme inactivation was reported in the beef muscle (Kennick et al., 1980).

The 1980s: Several technical problems, however, exhibited as barriers to the commercial application of HPP. Modified atmosphere–packaged salads are introduced to extend shelf life.

The late 1980s: The commercial applications of higher hydrostatic pressures were introduced in Japan. It helped preserve acid food products, such as yogurt and strawberry jam. A consortium of 25 companies was established to exploit the potential application of HPP nonthermal technology to pasteurize and sterilize foods with minimal heat damage.

1984: Commercialization efforts of HPP research and development led by Dan Farkas, Dallas Hoover, and Dietrich Knorr were started at the University of Delaware, USA. Results highlighted that HPP (340–680 MPa) preserves covalent bonds in biological materials and maintains product quality but would disrupt hydrogen bonds. This caused disruption of enzymes and membranes activities of protein contents and subsequent death of microbes, such as yeast, molds, bacteria, and parasites.

1990: Research efforts expanded, and the Industry-University High-Pressure Processing Consortium was launched among the University of Delaware, Oregon State University, and Combat Feeding Directorate of the US Army at the Natick Soldier Research, Development and Engineering Command, MA. Marcia Walker and D. Farkas played a significant role.

The 1990s: Japan commercially introduced HPP-treated (400–900 MPa) fruit jams and jellies in the mid-1990s.

1992: A new book, *High-pressure Science and Biotechnology* (Hayashi & Balny, 1992) was published. It covered the topics of the joint conference held in Germany in 1989 and highlighted Japanese and European research works on HPP.

1994: The University of Reading hosted a symposium on March 28–29, 1994. A new book, *High-Pressure Processing of Foods* (Ledward et al., 1995) was published. This served as a milestone and led the early history of R&D to the widespread commercial application of HPP.

1996 and late 1990s: Outbreak of *E. coli* O157:H7 food poisoning in apple juice reported in 1996. The Food and Drug Administration (FDA) proposed pasteurizing all apple juice products with heat. Studies in the late 1990s confirmed that HPP is an effective process in preserving apple and orange juices that ensure safety without damaging nutrients and flavor caused by heat. The United States fully embraced HACCP Final Rule in 1996.

1997: Pressure-treated guacamole products from avocado were commercialized in the United States.

IV. HPP technology after 2000:

Several years into the development process, HPP technology has now entered the most advanced stages of research and development and commercial applications.

2003: The USDA Food Safety and Inspection Service (FSIS) accepted HPP as a post-lethality treatment approved for controlling *L. monocytogenes* in processed RTE meat (Simonin et al., 2012). Regulatory agencies such as Health Canada, the European Commission, and others recognized HPP as a pasteurization process to warrant food safety.

2009: The FDA approved HPP's commercial application for low-acid foods sterilization (Stewart et al., 2015). The FDA further approved pressure-assisted thermal processings (PATP).

2011: The first pressure-treated food product, 'Spanish sliced cooked ham', was introduced in Europe (Tonello, 2011).

2013–2014: Synchronous applications of pressures, heats, and electric fields were adopted for the elimination of microbes such as bacterial spores (Park et al., 2013, 2014).

2015: An industrial alliance was established among companies American Pasteurization and Universal Pasteurization, Avure and Hiperbaric, and numerous foods and beverages manufacturing companies with an aim to encourage use of HPP nonthermal-based technology for future foundations.

2015 to date: Extensive research and development works on food processing, preservation, food quality and safety, processing equipment, production conditions, etc., using state-of-the-art tools and models, are being investigated. Efforts of academia–industry partnerships, equipment manufacturers, food processors, and the government's initiative toward the HPP-based food processing industry are noticeable.

12.3 BASIC PRINCIPLES OF HPP

Food products in the high-pressure system experience a reduction in volume as a function of applied 0 pressures. In this system, food products' behavior follows the three basic principles: i) Le Chatelier's principle, ii) microscopic ordering, and iii) the isostatic principle (Balasubramaniam et al., 2008). In Le Chatelier's principle, phenomena like changes in a phase transition, molecular configuration, and chemical reaction are led by a decrease in volume escalated by pressure. The microscopic ordering principle explains that an increase in pressure at a certain temperature enhances the degree of ordering of molecules of a given substance. The key working principle in the HPP system is the isostatic principle, which describes that pressure is uniformly distributed throughout the entire sample irrespective of the shape and size. Isostatic pressure does not impact molecular bonding and configuration; thus, food products get their original shape once de-pressurized (Howson, 2009). Therefore, HPP technology enables the pasteurization of foods and inactivates pathogenic and spoilage organisms with no or minimal effects on texture, flavor, taste, shape, and appearance; extends shelf life; and maintains food safety and quality.

A standard HPP system includes both batch (mainly for solid and liquid foods) and semi-continuous equipment (used for pumpable foods) lines. Key components of the HPP batch design systems are a pressure vessel (thick wall cylinder), closure(s) for sealing the vessel, a yoke or device for holding the closure, HP pump and intensifier(s), a pressure controlling and monitoring system, and a temperature and product-handling system for loading and unloading the products (Ting, 2011). A simple flowchart of the high-pressure processing food pasteurization method is shown in Figure 12.1.

12.4 BENEFITS OR ADVANTAGES AND DISADVANTAGES OF HPP

HPP technology has multiple food safety benefits, including reducing foodborne pathogen issues, bacteria, and molds, extending shelf life, reconditioning contaminated foods, and

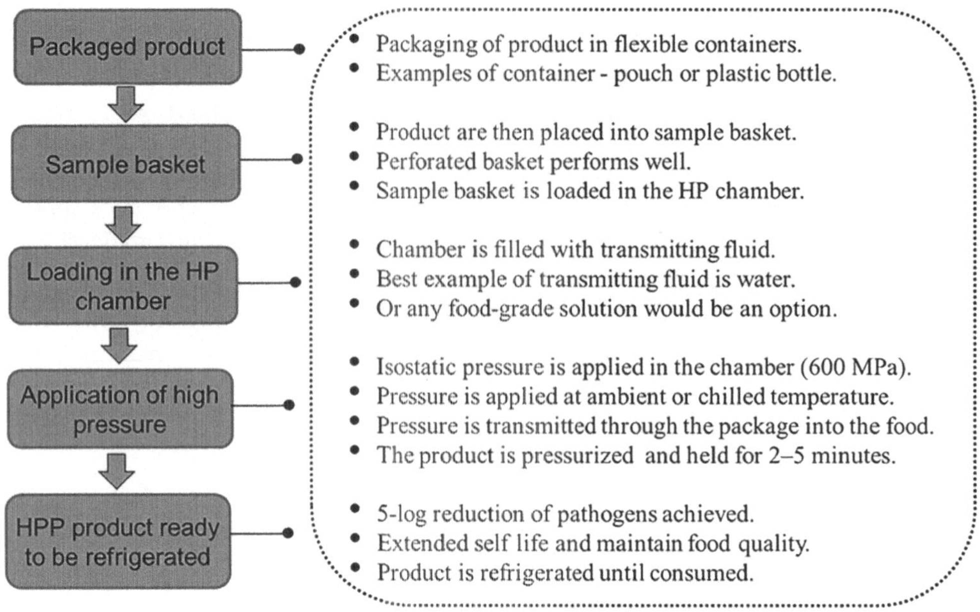

FIGURE 12.1 A simple flowchart of the high-pressure processing food pasteurization method.

maintaining the nutritional qualities of the products. As a minimal processing technology for foods, HPP technology ensures the quality, safety, and flavor of the food products (Daryaei & Balasubramaniam, 2012). This technology's key benefits and/or advantages are shown in Figure 12.2.

Based on the available literature, the key advantages/benefits and limitations of HPP technology are described in the following sections.

12.4.1 Key Benefits and/or Advantages of HPP

1 HPP technology enables food processing at room or chilled temperatures.
2 Regardless of geometry and size, HPP technology allows prompt pressure transmittance through the systems, thereby making size reduction optional, which can be a great advantage.
3 It eliminates microbes while virtually eliminating heat damage, thus resulting in overall food quality improvements.
4 HPP incorporation into the production process offers high-quality and healthy products.
5 It can further eliminate the necessity for chemical-based additives/preservatives and provide an opportunity to develop constituents with innovative functional characteristics.
6 The technology also prevents secondary contamination after the pasteurization of foods.
7 HPP maintains the quality and microbiological safety of foods without added preservative contents. It further allows food processors to maintain sensory attributes such as natural color, flavors, taste, acceptability, and nutritive values of the processed food products.
8 It extends the food products' shelf life. As a rule of thumb of HPP, an improved shelf life of foods means lesser waste and greater sustainability.
9 It further allows beverages manufacturers and brands to have a greater geographic and distribution reach.
10 Low energy consumption and low contamination risk are two other significant advantages of HPP. Thus, it can be recognized as an environmentally friendly food processing technology.

12.4.2 Limitations or Disadvantages of HPP

HPP technology has some limitations and/or disadvantages, which are as follows:

1 Most HPP products are required to be stored and transported via a refrigeration-based system. Pressure treatments alone are insufficient to eliminate harmful pathogens. For example, lower-acid HPP food products may exhibit probable microbiological threats due to the existence of harmful *Clostridium botulinum* bacterial spores.
2 HPP requires water to act as a pressure-transmitting medium and air bubbles of the products to be deformed under HPP treatment. Various types of food, such as flour, powder, paste, flour with the least amount of water, and products containing many air bubbles, are generally unsuitable for HPP treatment.

FIGURE 12.2 Key interactive benefits of the high-pressure food processing approach.

3 The compressibility of the packaging material must be at least 15%; thus, only plastic-based packaging material is generally appropriate for HPP.
4 Technical barriers and effectiveness in HPP equipment production are considerable.
5 Large capital investment is another barrier for many food manufacturers and companies that causes a conservative approach toward HPP.
6 HPP is still performed batch-wise, making it difficult to use HPP on higher-speed production lines. So, shortening the decontamination time is an option to enhance the production capacity; however, it can result in insufficient pasteurization of the products.

12.5 CURRENT KNOWLEDGE OF HPP IN FRUITS AND VEGETABLES IN BRIEF

HPP is an antimicrobial pasteurization process for foods, and pasteurization with dry ingredients is currently a key focus of manufacturers. HPP offers an opportunity to develop food products of high-quality attributes, greater safety, and extended shelf life of fruit and vegetable products (Gould, 1996; Fernandez Garcia et al., 2001). HPP processing is sometimes recognized as 'high hydrostatic pressure' (HHP) and 'ultra-high pressure' (UHP) processing in the foods processing industry (Balasubramaniam et al., 2008). UHP technology-based juice products from apples to sweet oranges to peaches, and including carrots, mixed citrus juices, and frozen raspberries, are well-adopted in the food industry. When treated with UHP, products from these fruits can retain sucrose, vitamin C, carotenoid contents, and anti-mutagenic and antioxidative attributes (Butz et al., 2003). Besides reducing pathogens, it extends the shelf life of juices and fresh-cut fruits and products by more than threefold (Huang et al., 2017). Consumers widely accept HPP fruit beverages, such as fresh and safe juices, indicating prompt growths in the output values in the HPP-based food markets. Besides the high-quality fruit juices, HPP-packaged coconut water and HPP avocado products have similar nutrients and flavors to the original (Huang et al., 2017; Podolak et al., 2020), indicating greater demands for HPP technology and developing various HPP foods.

Several Centers of Excellence have been established through the partnership between the university and industry in the USA to promote advanced research on HPP, packaging, and food safety (Balasubramaniam & Farkas, 2008). The Center for Advanced Processing and Packaging Studies (Columbus, OH), the Center for Nonthermal Processing of Foods (Pullman, WA), the National Center for Food Safety and Technology (Chicago, IL), and the HPP Lab (Blacksburg, VA) are some examples where multi-university, industry, and US Army Natick Soldier RDEC, MA, and the food industry have collaborated. This kind of initiative and involvement in strengthening the HPP technology-based food sector requires more from manufacturers and food processing industries, outsourcing industries, universities, and private and public partnerships.

12.6 APPLICATION OR EFFECT OF HPP ON THE QUALITY AND SAFETY OF FRUIT- AND VEGETABLE-BASED FOOD PRODUCTS

The demand for higher-quality, nutritious, fresh-like, and healthy additive-free and value-added food products is high among consumers in modern society (Sarker, 2022; Sarker et al., 2023). Thermally treated foods generally tend to lose quality features. Pressure treatment inactivates microorganisms without changing nutritional and organoleptic attributes. Therefore, retaining colors and texture in HPP fruits and vegetables in their original forms (Rastogi et al., 2007; Norton & Sun, 2008), eliminating pathogens, and extending shelf life are the most beneficial outputs for the food processing industry and consumers (Huang et al., 2017). HPP technology is widely accepted in commercial applications for processing several food and beverage categories. HPP-based food products are, for instance, fruit juices, smoothies, and beverages, avocado products (e.g., guacamole), ready-to-cook and -eat food products (e.g., burger patties), seafood products, sandwich fillings, and dairy-based food products. Minimally processed vegetables and fruits are also popular in the food industry (Podolak et al., 2020; Sarker et al., 2021). Consumers' preference is being increased to minimally or nonthermally processed foods with greater safety and quality, an increased shelf life, and higher nutritive values (Mohideen et al., 2015; Sarker & Siddiqui, 2023). The application and/or effect of HPP on food product quality attributes have been described in the following sections.

12.6.1 Effect of HPP on Organoleptic Attributes, Functionality, and Nutritional Quality

Research on HPP demonstrated that HPP treatment has no or minimal impact on food chemistry and physical-chemical changes in fruits and vegetables. For maximum reduction in microflora and retention of greater quality, higher pressure and shorter time (600 MPa, 5 min) were recommended for mango pulp (Kaushik et al., 2014). Despite the HPP effect on pigments, color changes are negligible in HPP-processed fruits (Patras et al., 2009a, 2009b; Perera et al., 2010). Cultivar is an important consideration in selecting raw materials for HPP. Cultivars with greater retention ability of vitamin C indicates better suitability for HPP (Wolbang et al., 2008). Further, HPP plus refrigeration was found to be promising and efficient for retaining high-quality cucumber juice with a 20-day shelf life. The findings could help processors commercialize HPP nonthermal technology to replace traditional thermal processing in developing clear vegetable juices (Liu et al., 2016). Based on the commodity, an optimum pressure was recommended for microbial and bioactive compounds' stability during storage (Błaszczak et al., 2017). However, the stability of HP-processed mango pulp was not confirmed because of the challenge associated with the complete inactivation of enzymes using HPP technology (Kaushik et al., 2014). Optimal thermal processing is effective for the complete inactivation of tissue enzymes, such as from the oxidoreductase group; in contrast, HPP is ineffective for such a purpose (Szczepańska et al., 2020). Therefore, there is a faster dilapidation of bioactive compound contents in HPP-treated fruit and vegetable products due to higher residual activity of tissue enzymes, which eventually leads to the shortening of the product's shelf life (Marszałek et al., 2018). Besides, total polyphenol content was enhanced in carrot juice with static high pressure of 450 MPa and 600 MPa over lower pressure with pulses (Szczepańska et al., 2020). In another study, ultrafiltration (UF) treatment followed by HPP was recommended as an efficient approach for the overall quality retention of stored pear juice (Zhao et al., 2016). However, texture alteration of fruits and vegetables is an issue for industrial-scale processing using HPP. HPP-induced modification of textural properties is simultaneously influenced by altering cell morphology, membrane integrity, and depolymerization and solubility of water-insoluble pectin (Sun et al., 2019). In fresh horticultural produce, the structure and functionality of biomembranes and proteins are affected by HPP; the applied pressure initially affects cell permeability and turgor, which, above a threshold, leads to tissue damage (Rux et al., 2020). This effect was demonstrated as a function of pressure intensity, and if a low pressure (100 MPa) is induced, the changes in fresh tissue properties can be reversible (Rux et al., 2020). Research revealed that HPP treatment (160 MPa) coupled with enzyme immobilization at a temperature of 37 °C for 20 min decreases bitterness by decreasing the naringin contents of citrus fruit juices (Luís Ferreira et al., 2008).

12.6.2 Effects of HPP on the Safety of Fruit and Vegetable Products

Consumers prefer processed fruits and fruit products that retain high quality, fresh flavor, and are microbiologically safe with zero additives (Patterson, 2005; Perera et al., 2010). Essentially, the HPP effectiveness on microbial growth in food products depend on the different magnitudes of thermal pressure combined practice. HPP with 400–600 MPa pressure alone inactivates spoilage and microorganisms (e.g., yeasts, molds, and bacterial cells) and, combined with heat (>70 °C), was more effective in eliminating pathogenic spores (Khaliq et al., 2021). Application of pressure-assisted thermal processing (PATP) or ultrahigh-pressure thermal processing sterilization (500–600 MPa and 90–120 °C for 3–10 min) to preheated food materials (70–90 °C) inactivates strains of *C. botulinum* (Juliano et al., 2010; Stewart et al., 2015). HPP has also been seen to inactivate vegetative microorganisms. The HPP and heat synergies with less thermal exposure eliminate bacterial spores, for example, *C. botulinum* and varieties of *Clostridia* and *Bacillus* (Margosch et al., 2006; Stewart et al., 2015). Studies show that gram-negative bacteria exhibit lower pressure resistance than gram-positive bacteria, while vegetative microbe cells demonstrate reduced pressure resistance in exponential phases in contrast with the stationary phase (Khaliq et al., 2021). Pressure with ambient temperature triggers nutrient germinant receptors and starts the microbial spore germination process (Farkas & Hoover, 2000). Mild temperature (20 °C and 50 °C) and pressure (100–400 MPa) can also start the germination of spores (Black et al., 2007). The effects of mild pressure and heat treatments in combination inactivates the germinated spores. The usage of pressure and exposure time in the HPP system affects food-borne pathogens and spoilage. For instance, the *E. coli* bacteria (strain ATCC 25922) in HPP-treated apple juice decreased by 0.12, 0.18, and 1.31 log CFU/mL, respectively, from the initial cell concentration (approximately 10^6 CFU/mL) with the increase in pressures of 250 MPa, 300 MPa, and 400 MPa after 1.5 min (Lavinas et al., 2008). Peach and orange juices inoculated by *L. monocytogenes* (10^7 CFU/mL) were subjected to HP of 400 MPa and 600 MPa and showed that the reduction of microbial populations was 2.76 log CFU/mL under 400 MPa pressure than 6.47 log CFU/mL at 600 MPa (Dogan & Erkmen, 2004). The effect of HPP on the quality attributes of some fruit and vegetable products is described in Table 12.1.

TABLE 12.1 Effect of High-Pressure Processing on the Quality Attributes of Some Fruit and Vegetable Products

FOOD GROUP	*PROCESSING CONDITIONS MPA/MIN/°C*	*IMPACT OF HPP ON FOODS AND FOOD PRODUCTS OVER CONTROLLED TRIAL*	*REFERENCES*
HPP fruit products			
Orange milk juice beverage	400 MPa, 9 min at 26.6 °C	• Retained the product's ascorbic acid contents • No changes in color	Khaliq et al. (2021)
Orange juice	350 MPa, 1 min at 30 °C	• Better quality • Shelf life > two months under refrigeration	Donsì et al. (1996)
Orange juice	600 MPa, 5 min at 25 or 80 °C	• Greater folates retention at 25 °C • No significant losses in folate at 80 °C	Butz et al. (2004)
Orange cubes	100–400 MPa, 5–60 min at 20 °C	• Firmness decreased	Basak and Ramaswamy (1998)
Apple puree	400 MPa, 5 min at 20 °C	• No effect on phenolic content • Retained greater vitamin C	Khaliq et al. (2021)
Kiwifruit	400 MPa, 30 min at 5 20 °C	• Fruit became yellow under HPP • Texture was acceptable • Greater activities of PPO and POD enzymes at 20 °C	Prestamo and Arroyo (2000)
Lemon juice	450 MPa, 2, 5, or 10 min	• No fungus growth was detected • Yeast caused spoilage of control sample • Minimal effects on constituents and physicochemical properties	Donsi et al. (1998)
Peaches	400 MPa, 30 min at 5 or 20 °C	• Peaches turned brown but were prevented with the addition of ascorbic acid • Texture was acceptable • Greater activities of PPO and POD enzymes at 20 °C	Prestamo and Arroyo (2000)
Raspberry puree	200–800 MPa, 15 min at 18–22 °C	• Anthocyanin stability was higher at 4 °C	Suthanthangjai et al. (2005)
Apple cubes	100–400 MPa, 5–60 min at 20 °C	• The quick loss of firmness during compression	Basak and Ramaswamy (1998)
Pears	400 MPa, 30 min at 5 or 20 °C	• Browning of pears for HPP • Ascorbic acid addition prevented such a change • Texture was acceptable • Activities of PPO and POD enzymes lower at 5 °C	Prestamo and Arroyo (2000)

(Continued)

TABLE 12.1 (Continued)

FOOD GROUP	*PROCESSING CONDITIONS MPA/MIN/°C*	*IMPACT OF HPP ON FOODS AND FOOD PRODUCTS OVER CONTROLLED TRIAL*	*REFERENCES*
Strawberry puree	800 MPa, 20 min at 20 °C	• Greater reduction of concentrations of volatile compounds (e.g., funeral, linalool, nerolidol, and some esters) than unprocessed food	Khaliq et al. (2021)
Strawberries	200–700 MPa, 0–30 min at 80–130 °C and 36–93 °C (initial).	• Degradation of anthocyanins • Pressure reduced the degradation rate	Verbeyst et al. (2010)
Pear cubes	100–400 MPa, 5–60 min at 20 °C	• Quick decrease of firmness during compression	Basak and Ramaswamy (1998)
Grapefruit juice	160 MPa, 20 min at 37 °C	• Reduced bitterness (narington content) 75%	Luís Ferreira et al. (2008)
Melon	400 MPa, 30 min at 5 °C or 20 °C	• Melon was suitable • Texture was acceptable • Decreased microorganisms	Prestamo and Arroyo (2000)
Pineapple cubes, Aronia juice, Korla pear juice	100–400 MPa, 5–60 min at 20 °C 200–600 MPa,15 min at ambient temperature Ultrafiltration (UF) followed by HPP. 500 MPa, 10 min at room temperature)	• Firmness reduced substantially • Reduced microorganism growth • Improved antioxidant capacity • Reduction in total phenolic concentration • Enhanced juice clarity • Greater retention of vitamin C, total phenols, and aroma compounds • Increased antioxidant capacity	Basak and Ramaswamy (1998) Błaszczak et al. (2017) Zhao et al. (2016)
HPP vegetable products			
Broccoli juice	850 MPa at 30 °C –90 °C	• Color changed, e.g., greenness	Weemaes et al. (1999)
Tomato puree	50–400 MPa, 15 min at 25 °C	• Inactivation of PPO, POD, and pectin methylesterase was increased	Plaza et al. (2003)
Tomato puree	100–600 MPa, 12 min at 20 °C	• Greater losses of total lycopene and 13-cis isomer (%). Product stability was higher at 500 MPa and storage at 4 ± 1 °C	Qiu et al. (2006)
Tomato	300–400 MPa	• Decreased microorganism and organoleptic properties were affected • Skin became loosened and peeled away • Retained firm flesh. No changes occurred in color and flavor	Arroyo et al. (1997; 1999)
Carrot	700 MPa, 3 min at 124 °C and 90 °C (initial).	• Formation of furfural and retention of nutrient, texture, and color observed • Enhanced the overall quality	Vervoort et al. (2012)
Asparagus, spinach, cauliflower	300–400 MPa	• The population of aerobic fungi, yeasts, and mesophiles was reduced • Sensorial attributes were affected • Slight browning in case of cauliflower • Low temperature and longer treatment time increased microorganisms' inactivation • Retain better sensorial attributes	Arroyo et al. (1997; 1999)

FOOD GROUP	PROCESSING CONDITIONS MPA/MIN/°C	IMPACT OF HPP ON FOODS AND FOOD PRODUCTS OVER CONTROLLED TRIAL	REFERENCES
Green pepper, red pepper, and celery pieces	100–400 MPa, 5–60 min at 20 °C	• Firmness was decreased or regained during compression	Basak and Ramaswamy (1998)
Sweet potato	600 MPa, 5–30 min at 25 °C	• Firmness and maximum cutting force reduced • Significant changes in texture were observed at 600 MPa/5 min	Oliveira et al. (2015)
Lettuce	300–400 MPa	• Aerobic mesophiles, yeasts, and fungi reduced • Retained firmness but underwent browning • Flavor was unaffected • Low temperature and longer treatment time eliminated microorganisms	Arroyo et al. (1997, 1999)
Carrot slices, cucumber juice, carrot juice, carrot, red cabbage leaves	100–400 MPa, 5–60 min at 20 °C 500 MPa, 5 min 300 MPa×3 pulses and 600 MPa in static mode 400–1200 MPa, 2 min 150–200 MPa, 5–20 min at 35–55 °C	• Firmness was reduced or regained • Retention of color and aroma compounds • Enhanced clarity and sensory acceptability • Enhancement of total phenolic content • Color darkening due to partial inactivation of PPO and POD enzymes • Texture softening • Cell turgor and tissue integrity at < 150 MPa and < 45 °C	Basak and Ramaswamy (1998) Liu et al. (2016) Szczepańska et al. (2020) Sun et al. (2019) Rux et al. (2020)

Note: HPP—High pressure processing, PPO—Polyphenol oxidase, POD—Peroxidase, and MPa—Mega pascal.

12.7 MODELING HP PROCESSES IN FOOD PRODUCTS

Researchers attempt to develop mathematical models using thermodynamic and empirical principles to predict the processing impact on food ingredients. Heat and mass transfer models are desirable. However, the lack of proper knowledge of critical thermo-physical properties of materials and/or food constituents (e.g., water) in the high-pressure domains is one of the key obstacles when modeling heat transfer in HP processes (Hogan et al., 2005). Critical variables such as temperatures in the vessel chamber (before processing) and product, uniform distribution of temperature through the products, pressurizing fluids to product ratio, and pressurization time substantially influence the heat transfer in the HPP vessel (Hogan et al., 2005). Thus, a simple model should consider these variables and allow for predicting temperature changes in a given sample under various pressures to control the process and its associated effects. HPP models facilitate the prediction of both temperature and pressure conditions essential for achieving microbial population inactivation in specific HPP-treated food products (Polydera et al., 2004).

Microbial spore inactivation curve changes with processing techniques. Thermal pressure–combined treatment shows a unique tailing for microbial inactivation curves, while a regular linear relationship was revealed for thermal processing (Rajan et al., 2006). Researchers sometimes use irregular kinetic models to explain the interactive effect of thermal pressure microbial inactivation. According to the integrated process lethality model, pressure and temperature exhibited lethal impacts that caused the killing of microbial spores (Nguyen et al., 2014). Several other studies reported that the elimination of bacterial vegetative cell populations (e.g., *E. coli* O157:H7 and *L. monocytogenes*) in HPP-treated juices followed a first-order kinetics model (Ramaswamy et al., 2003; Dogan & Erkmen, 2004; Lavinas et al., 2008; Katsaros et al., 2010; Guerrero-Beltrán et al., 2011) and inactivation of microorganisms is assumed to be log-linear with time (Podolak et al., 2020).

Microbial inactivation curves tend to be exponential, with a dramatic initial reduction in bacterial cells, and follow a first-order kinetics model (Ponce et al., 1998). Tailing is often observed in microbial survival curves, resulting in substantial deviancies from the linear associations between treatment times or pressure levels and the log reduction (Chen & Hoover, 2003). Thus, either the Weibull model or other non-linear models, for instance, the biphasic model (Yoo et al., 2015) for the inactivation kinetics of *E. coli* bacteria (O157:H7) on the surface of oranges treated with 400 MPa HPP (for 1 min) performs well to adjust microbial inactivation. Several non-linear models, namely, the Weibull, biphasic, modified Gompertz, Baranyi, and log-logistic models, were observed to apply for microbiota inactivation in cucumber juice (Zhao et al., 2014). However, researchers' efforts to develop mathematical models for process validation and optimization are ubiquitous, considering all critical variables and conditions.

12.8 CHALLENGES AND FUTURE RESEARCH NEEDS

HPP technology has entered a new era in the global food processing industry, and its advancement to multiple food application in the food processing industries is noticeable. However, the HPP faces several challenges, including processing equipment, production conditions, heat transfer problems, non-uniformity in processing, production capacity, high pressure, pasteurization, microbial safety and product quality, hygiene safety, HPP food storage and preservation, lack of reliable and reproducible data for process validation (Rastogi et al., 2007; Khaliq et al., 2021). Since HPP technology has emerged as a new pasteurization nonthermal technology for the food processing industry, its health and safety aspects, sensory and nutritional attributes, economic value, product outcome, and products form should be thoroughly evaluated. Academia, the industry, and the government's concerned agencies collaboration are crucial to improve, access, and ensure HPP food production conditions, including pressure-resistant characteristics, transmitting fluids (e.g., water), indicator microorganisms, and hygienic safety assessment for HPP products. Further, the establishment of plants, operations, technical guidance, low inputs, and minimal energy investments, shortcomings associated with traditional techniques, microorganisms, limited shelf life, product sensitivity to high temperatures, pressures, and difficulties connected with the storage and transportation of HPP food products needs to be assessed and strengthened.

Many forward-looking concepts or emerging technologies have evolved toward automation, data-driven decision-making, and precise management. These concepts, including intelligent production lines, next-generation cloud-based sensors, automation, robotics, and artificial intelligence (e.g., machine learning), can be effectively applied to monitor, evaluate, and manage the quality and safety of food products in the food processing factory and industry. These emerging tools, along with data-driven management decisions and the power of 'big data' and analytical techniques (Duong et al., 2020; Wauters et al., 2012), can exhibit an enormous positive impact on modern agriculture, food science, and industry. The Internet of Things (IoT) is a powerful tool that has benefited the food industry by monitoring product quality in real time by recognizing influential conditions and process parameters (Panda et al., 2019). The machine vision approach can be applied to real-time monitoring of quality, spoilage, and microbial progression of processed fruit and vegetable produce. This approach has been utilized in cucumber quality monitoring in a recent investigation, and this was found to be a promising tool (Sarker & Grift, 2023).

Furthermore, the development of new machine and equipment designs combined with handling technology through the university–industry partnership is critically important to the advancement of the HPP-based food industry. There is a huge scope to combine HPP technology with other nonthermal technologies and existing trends in the food sector. The improvement in reliability, modeling of processes, product throughput with larger vessel size and modern pumping systems, the line speed of the processing equipment, and cost reduction are instrumental in expanding the industry adoption of the HPP. More food manufacturers' access without higher capital investment to the benefits of HPP to outsourcing companies like Universal Pure (https://universalpure.com/) can boost the food sector substantially. Relevant laws, regulations, or specifications must ensure microbial safety and product quality for HPP food products. Further, translating HPP research into knowledge and disseminating this knowledge through effective extension programming should be strengthened. The key examples of technology disseminating tools for stakeholders include publications of articles and factsheets, webinars, podcasts, YouTube videos, seminars, scientific meetings, and advisory services to stakeholders, including utilizing university expertise for training to the food industry and relevant government and non-government personnel to strengthen HPP-based food industry.

12.9 CONCLUSIONS

Advancements in HPP technology to satisfy consumers' demand for quality food products and their commercial application are notable. The global food industry still considers thermal pasteurization as the key pasteurization technology despite its possible impact on sensory attributes such as taste, texture, flavor, appearance, and nutritional value of food products. HPP technology is an alternative or novel processing technology of food products to heat treatment; it can overcome such limitations without compromising food product safety, shelf life, taste, flavor, and nutritional value by containing fewer additives. By eliminating pathogenic and spoilage organisms while maintaining food nutritional attributes almost intact and ensuring extended shelf life and higher market value in a sustainable manner, HPP technology pasteurizes foods with minimal effects on organoleptic properties and nutritional ingredients and plays a pivotal role in food and nutritional security globally. With the gradual increase in consumer demand for natural, fresh, and aesthetically appealing but affordable and diversified HPP food products, food producers adopting HPP technology are expected to fulfill and satisfy those demands on the local, regional, national, and global levels. However, the advancement of the HPP-based food industry will be accelerated through continuous research and development for addressing newer challenges and issues, including HPP equipment, production conditions, capital investment, food safety and quality, regulations, and monitoring, and thus, collaboration among academia, the industry, and concerned departments of the government is warranted.

CONFLICT OF INTEREST

The authors declare that there are no conflicts of interest.

AUTHOR'S CONTRIBUTION

Conceptualization—Ayesha Sarker (AS) and Md Rayhan Shaheb (MRS); Review of literature collections, data analysis, and interpretation—AS and MRS; Writing—original manuscript, AS; Review and editing—AS and MRS. Both authors have read and agreed to the published version of the chapter.

REFERENCES

Arroyo, G., Sanz, P. D., Préstamo, G. (1997). Effect of high pressure on the reduction of microbial populations in vegetables. *Journal of Applied Microbiology*, *82*(6). https://doi.org/10.1046/j.1365-2672.1997.00149.x

Arroyo, G., Sanz, P. D., Préstamo, G. (1999). Response to high-pressure, low-temperature treatment in vegetables: Determination of survival rates of microbial populations using flow cytometry and detection of peroxidase activity using confocal microscopy. *Journal of Applied Microbiology*, *86*(3). https://doi.org/10.1046/j.1365-2672.1999.00701.x

Balasubramaniam, V. M., Farkas, D. (2008). High-pressure food processing. *Food Science and Technology International*, *14*(5), 413–418. https://doi.org/10.1177/1082013208098812

Balasubramaniam, V. M., Farkas, D., Turek, E. (2008). Preserving foods through high-pressure processing. *Food Technology*, *62*(11), 32–38.

Basak, S., Ramaswamy, H. S. (1998). Effect of high pressure processing on the texture of selected fruits and vegetables. *Journal of Texture Studies*, *29*(5). https://doi.org/10.1111/j.1745-4603.1998.tb00185.x

Black, E. P., Setlow, P., Hocking, A. D., Stewart, C. M., Kelly, A. L., Hoover, D. G. (2007). Comprehensive reviews in food science and food safety response of spores to processing. *Comprehensive Reviews in Food Science and Food Safety*, *6*, 103–119.

Błaszczak, W., Amarowicz, R., Górecki, A. R. (2017). Antioxidant capacity, phenolic composition and microbial stability of aronia juice subjected to high hydrostatic pressure processing. *Innovative Food Science and Emerging Technologies*, *39*, 141–147. https://doi.org/10.1016/j.ifset.2016.12.005

Bouton, P. E., Ford, A. L., Harris, P. V., Macfarlane, J. J., O'Shea, J. M. (1977). Pressure-heat treatment of postrigor muscle: Effects on tenderness. *Journal of Food Science*, *42*, 132–135.

Bridgman, P. W. (1909). An experimental determination of certain compressibilities. *Proceedings of the American Academy of Arts and Sciences*, *44*(10). https://doi.org/10.2307/20022428

Bridgman, P. W. (1912). Water, in the liquid and five solid forms, under pressure. *Proceedings of the American Academy of Arts and Sciences*, *47*(13). https://doi.org/10.2307/20022754

Bridgman, P. W. (1914). Change of phase under pressure I. The phase diagram ofeleven substances with especial reference to melting curve. *Physical Review*, *3*, 153–203.

Bridgman, P. W. (1923). The thermal conductivity of liquids under pressure. *Proceedings of the American Academy of Arts and Sciences*, *59*(7). https://doi.org/10.2307/20026073

Brown, K. (1920). The manufacture of phenol in a continuous high pressure autoclave. *Industrial and Engineering Chemistry*, *12*(3). https://doi.org/10.1021/ie50123a027

Butz, P., Serfert, Y., Fernandez Garcia, A., Dieterich, S., Lindauer, R., Bognar, A., Tauscher, B. (2004). Influence of high-pressure treatment at 25 °C and 80 °C on folates in orange juice and model media. *Journal of Food Science*, *69*(3). https://doi.org/10.1111/j.1365-2621.2004.tb13380.x

Butz, P., Fernández García, A., Lindauer, R., Dieterich, S., Bognár, A., Tauscher, B. (2003). Influence of ultra high pressure processing on fruit and vegetable products. *Journal of Food Engineering*, *56*(2–3), 233–236. https://doi.org/10.1016/S0260-8774(02)00258-3

Chen, H., Hoover, D. G. (2003). Pressure inactivation kinetics of Yersinia enterocolitica ATCC 35669. *International Journal of Food Microbiology*, *87*(1–2). https://doi.org/10.1016/S0168-1605(03)00064-3

Cruess, W. V. (1924). *Commercial fruit and vegetable products*. New York: McGraw-Hill Inc.

Daryaei, H., Balasubramaniam, V. M. (2012). Microbial decontamination of food by high pressure processing. In A. D. and M. O. Ngadi (Ed.), *Novel Methods and Applications. Woodhead Publishing Series in Food Science, Technology and Nutrition* (pp. 370–406). https://doi.org/10.1533/9780857095756.2.370

Dogan, C., Erkmen, O. (2004). High pressure inactivation kinetics of Listeria monocytogenes inactivation in broth, milk, and peach and orange juices. *Journal of Food Engineering*, *62*(1). https://doi.org/10.1016/S0260-8774(03)00170-5

Donsì, G., Ferrari, G., Di Matteo, M. (1996). High pressure stabilization of orange juice: Evaluation of the effects of process conditions. *Italian Journal of Food Science*, *8*(2).

Donsi, G., Ferrari, G., Matteo, M. di., Bruno, M. C. (1998). High-pressure stabilization of lemon juice. *Italian Food & Beverage Technology*, 14–16.

Duong, L. N. K., Al-Fadhli, M., Jagtap, S., Bader, F., Martindale, W., Swainson, M., Paoli, A. (2020). A review of robotics and autonomous systems in the food industry: From the supply chains perspective. *Trends in Food Science and Technology*, *106*(May), 355–364. https://doi.org/10.1016/j.tifs.2020.10.028

Farkas, D. F. (2016). A short history of research and development efforts leading to the commercialization of high-pressure processing of food. *Food Engineering Series*, 19–36. https://doi.org/10.1007/978-1-4939-3234-4_2

Farkas, D., Hoover, D. (2000). High pressure processing: Kinetics of microbial inactivation for alternative food processing technologies. *Journal of Food Science*, 47–64.

Fernandez Garcia, A., Butz, P., Tauscher, B. (2001). Effects of high-pressure processing on carotenoid extractability, antioxidant activity, glucose diffusion, and water binding of tomato puree (*Lycopersicon esculentum* mill.). *Journal of Food Science*, *66*(7), 1033–1038. https://doi.org/10.1111/j.1365-2621.2001.tb08231.x

Gould, G. W. (1996). Methods for preservation and extension of shelf life. *International Journal of Food Microbiology*, *33*, 51–64.

Guerrero-Beltrán, J., Barbosa-Cánovas, G. V., Welti-Chanes, J. (2011). High hydrostatic pressure effect on Saccharomyces cerevisiae, Escherichia coli and Listeria innocua in pear nectar. *Journal of Food Quality*, *34*(6). https://doi.org/10.1111/j.1745-4557.2011.00413.x

Hayashi, R., Balny, C. (1992). *High pressure science and biotechnology*. Amsterdam: Elsevier Science B.V.

Hite, B. H. (1899). The effect of pressure in the preservation of milk: A preliminary report. *West Virginia Agricultural and Forestry Experiment Station Bulletins*, *58*.

Hite, B. H., Giddings, N. J., Weakley, C. (1914). The effect of pressure on certain microorganisms encountered in the preservation of fruits and vegetables. W V Agric Exp Station Bull, *146*, 1–67.

Hogan, E., Kelly, A. L., Sun, D. W. (2005). *High pressure processing of foods. An Overview. Emerging Technologies for Food Processing*. Elsevier Ltd. https://doi.org/10.1016/B978-012676757-5/50003-7

Howson, G. (2009). *High pressure processing for food safety, extended shelf life and all-natural*. New York: Avure Technologies.

Huang, H. W., Wu, S. J., Lu, J. K., Shyu, Y. T., Wang, C. Y. (2017). Current status and future trends of high-pressure processing in food industry. *Food Control*, *72*(12), 1–8. https://doi.org/10.1016/j.foodcont.2016.07.019

Johnson, F. H., Eyring, F. H., Polissar, M. J. (1954). *The kinetic basis of molecular biology*. New York: John Wiley.

Juliano, P., Koutchma, T., Sui, Q., Barbosa-Cánovas, G. V., Sadler, G. (2010). Polymeric-Based Food Packaging for High-Pressure Processing. *Food Engineering Reviews*. https://doi.org/10.1007/s12393-010-9026-0

Katsaros, G. I., Tsevdou, M., Panagiotou, T., Taoukis, P. S. (2010). Kinetic study of high pressure microbial and enzyme inactivation and selection of pasteurization conditions for Valencia orange juice. *International Journal of Food Science and Technology*, *45*(6), 1119–1129. https://doi.org/10.1111/j.1365-2621.2010.02238.x

Kaushik, N., Kaur, B. P., Rao, P. S., Mishra, H. N. (2014). Effect of high pressure processing on color, biochemical and microbiological characteristics of mango pulp (Mangifera indica cv. Amrapali). *Innovative Food Science and Emerging Technologies*, *22*, 40–50. https://doi.org/10.1016/j.ifset.2013.12.011

Kennick, W. H., Elgasim, E. A., Holmes, Z. A., Meyer, P. F. (1980). The effect of pressurisation of pre-rigor muscle on post-rigor meat characteristics. *Meat Science*, *4*(1). https://doi.org/10.1016/0309-1740(80)90021-2

Khaliq, A., Farhan, M., Chughtai, J., Mehmood, T. (2021). High-Pressure Processing; Principle, applications, impact, and future prospective. In *Sustainable Food Processing and Engineering Challenges* (pp. 75–108). Elsevier Inc. https://doi.org/10.1016/B978-0-12-822714-5/00003-6

King, J. W. (2014). Modern supercritical fluid technology for food applications. *Annual Review of Food Science and Technology*, *5*(1). https://doi.org/10.1146/annurev-food-030713-092447

Lavinas, F. C., Miguel, M. A. L., Lopes, M. L. M., Valente Mesquita, V. L. (2008). Effect of high hydrostatic pressure on cashew apple (Anacardium occidentale L.) juice preservation. *Journal of Food Science*, *73*(6). https://doi.org/10.1111/j.1750-3841.2008.00791.x

Ledward, D. A., Johnston, D. E., Earnshaw, R. G., Hasting, A. P. M. (1995). *High pressure processing of foods*. Nottingham: Leicestershire Press and Nottingham University Press.

Liu, F., Zhang, X., Zhao, L., Wang, Y., Liao, X. (2016). Potential of high-pressure processing and high-temperature/short-time thermal processing on microbial, physicochemical and sensory assurance of clear cucumber juice. *Innovative Food Science and Emerging Technologies*, *34*, 51–58. https://doi.org/10.1016/j.ifset.2015.12.030

Luís Ferreira, Afonso, C., Vila-Real, H., Alfaia, A., Ribeiro, M. H. L. (2008). Evaluation of the effect of high pressure on naringin hydrolysis in grapefruit juice with narimgimase immobilised in calcium alginate beads. *Food Technology and Biotechnology*, *46*(2), 146–150.

Macfarlane, J. J. (1973). Pre-rigor pressurization of muscle: Effects on pH, shear value and taste panel assessment. *Journal of Food Science*, *38*(2), 294–229.

Margosch, D., Ehrmann, M. A., Buckow, R., Heinz, V., Vogel, R. F., Gänzle, M. G. (2006). High-pressure-mediated survival of Clostridium botulinum and Bacillus amyloliquefaciens endospores at high temperature. *Applied and Environmental Microbiology*, *72*(5). https://doi.org/10.1128/AEM.72.5.3476-3481.2006

Marszałek, K., Woźniak, Ł., Barba, F. J., Skąpska, S., Lorenzo, J. M., Zambon, A., Spilimbergo, S. (2018). Enzymatic, physicochemical, nutritional and phytochemical profile changes of apple (Golden Delicious L.) juice under supercritical carbon dioxide and long-term cold storage. *Food Chemistry*, *268*, 279–286. https://doi.org/10.1016/j.foodchem.2018.06.109

Mohideen, F. W., Solval, K. M., Li, J., Zhang, J., Chouljenko, A., Chotiko, A., . . . Sathivel, S. (2015). Effect of continuous ultrasonication on microbial counts and physico-chemical properties of blueberry (Vaccinium corymbosum) juice. *LWT*, *60*(1). https://doi.org/10.1016/j.lwt.2014.07.047

Nguyen, L. T., Balasubramaniam, V. M., Ratphitagsanti, W. (2014). Estimation of Accumulated Lethality Under Pressure-Assisted Thermal Processing. *Food and Bioprocess Technology*, *7*(3). https://doi.org/10.1007/s11947-013-1140-6

Norton, T., Sun, D. W. (2008). Recent advances in the use of high pressure as an effective processing technique in the food industry. *Food and Bioprocess Technology*, *1*(1). https://doi.org/10.1007/s11947-007-0007-0

Oliveira, M. M. De, Tribst, A. A. L., Leite Júnior, B. R. D. C., Oliveira, R. A. De, Cristianini, M. (2015). Effects of high pressure processing on cocoyam, Peruvian carrot, and sweet potato: Changes in microstructure, physical characteristics, starch, and drying rate. *Innovative Food Science and Emerging Technologies*, *31*. https://doi.org/10.1016/j.ifset.2015.07.004

Panda, S. K., Blome, A., Wisniewski, L., Meyer, A. (2019). IoT Retrofitting Approach for the Food Industry. *IEEE International Conference on Emerging Technologies and Factory Automation, ETFA, 2019-Septe*, 1639–1642. https://doi.org/10.1109/ETFA.2019.8869093

Park, S. H., Balasubramaniam, V. M., Sastry, S. K. (2014). Quality of shelf-stable low-acid vegetables processed using pressure-ohmic-thermal sterilization. *LWT*, *57*(1). https://doi.org/10.1016/j.lwt.2013.12.036

Park, S. H., Balasubramaniam, V. M., Sastry, S. K., Lee, J. (2013). Pressure-ohmic-thermal sterilization: A feasible approach for the inactivation of Bacillus amyloliquefaciens and Geobacillus stearothermophilus spores. *Innovative Food Science and Emerging Technologies*, *19*. https://doi.org/10.1016/j.ifset.2013.03.005

Parker. (2020). Advances in food safety technology, including high pressure processing (HPP) and pasteurization. Retrieved from www.food-safety.com/articles/1583-advances-in-food-safety-technology-including-high-pressure-processing-hpp-and-pasteurization

Patras, A., Brunton, N., Da Pieve, S., Butler, F., Downey, G. (2009a). Effect of thermal and high pressure processing on antioxidant activity and instrumental colour of tomato and carrot purées. *Innovative Food Science and Emerging Technologies*, *10*(1). https://doi.org/10.1016/j.ifset.2008.09.008

Patras, A., Brunton, N. P., Da Pieve, S., Butler, F. (2009b). Impact of high pressure processing on total antioxidant activity, phenolic, ascorbic acid, anthocyanin content and colour of strawberry and blackberry purées. *Innovative Food Science and Emerging Technologies*, *10*(3). https://doi.org/10.1016/j.ifset.2008.12.004

Patterson, M. F. (2005). Microbiology of pressure-treated foods. In *Journal of Applied Microbiology* (Vol. 98). https://doi.org/10.1111/j.1365-2672.2005.02564.x

Perera, N., Gamage, T. V., Wakeling, L., Gamlath, G. G. S., Versteeg, C. (2010). Colour and texture of apples high pressure processed in pineapple juice. *Innovative Food Science and Emerging Technologies*, *11*(1), 39–46. https://doi.org/10.1016/j.ifset.2009.08.003

Plaza, L., Muñoz, M., De Ancos, B., Cano, M. P. (2003). Effect of combined treatments of high-pressure, citric acid and sodium chloride on quality parameters of tomato puree. *European Food Research and Technology*, *216*(6), 514–519. https://doi.org/10.1007/s00217-003-0689-0

Podolak, R., Whitman, D., Black, D. G. (2020). Factors affecting microbial inactivation during high pressure processing in juices and beverages: A review. *Journal of Food Protection*, *83*(9), 1561–1575. https://doi.org/10.4315/JFP-20-096

Polydera, A. C., Galanou, E., Stoforos, N. G., Taoukis, P. S. (2004). Inactivation kinetics of pectin methylesterase of greek Navel orange juice as a function of high hydrostatic pressure and temperature process conditions. *Journal of Food Engineering*, *62*(3). https://doi.org/10.1016/S0260-8774(03)00242-5

Ponce, E., Pla, R., Capellas, M., Guamis, B., Mor-Mur, M. (1998). Inactivation of Escherichia coli inoculated in liquid whole egg by high hydrostatic pressure. *Food Microbiology*, *15*(3). https://doi.org/10.1006/fmic.1997.0164

Prestamo, G., Arroyo, G. (2000). Preparation of preserves with fruits treatedby high pressure. *Alimentari*, *318*, 25–30.

Qiu, W., Jiang, H., Wang, H., Gao, Y. (2006). Effect of high hydrostatic pressure on lycopene stability. *Food Chemistry*, *97*(3). https://doi.org/10.1016/j.foodchem.2005.05.032

Rajan, S., Ahn, J., Balasubramaniam, V. M., Yousef, A. E. (2006). Combined pressure-thermal inactivation kinetics of Bacillus amyloliquefaciens spores in egg patty mince. *Journal of Food Protection*, *69*(4), 853–860. https://doi.org/10.4315/0362-028X-69.4.853

Ramaswamy, H. S., Riahi, E., Idziak, E. (2003). High-pressure destruction kinetics of E. coli (29055) in apple juice. *Journal of Food Science*, *68*(5). https://doi.org/10.1111/j.1365-2621.2003.tb12323.x

Rastogi, N. K., Raghavarao, K. S. M. S., Balasubramaniam, V. M., Niranjan, K., Knorr, D. (2007). Opportunities and challenges in high pressure processing of foods. *Critical Reviews in Food Science and Nutrition*, *47*(1). https://doi.org/10.1080/10408390600626420

Rux, G., Gelewsky, R., Schlüter, O., Herppich, W. B. (2020). High hydrostatic pressure treatment effects on selected tissue properties of fresh horticultural products. *Innovative Food Science and Emerging Technologies*, *61*, 102326. https://doi.org/10.1016/j.ifset.2020.102326

Sale, A. J., Gould, G. W., Hamilton, W. A. (1970). Inactivation of bacterial spores by hydrostatic pressure. *Journal of General Microbiology*, *60*(3), 323–334. https://doi.org/10.1099/00221287-60-3-323

Sarker, A. (2022). A review on the application of bioactive peptides as preservatives and functional ingredients in food model systems. *Journal of Food Processing and Preservation*, *46*(8), 1–18. https://doi.org/10.1111/jfpp.16800

Sarker, A., Deltsidis, A., Shaheb, M. R., Grift, T. E. (2021). Effect of aloe vera gel-glycerol edible coating on the shelf-life and the kinetics of colour change of minimally processed cucumber during storage. *International Journal of Postharvest Technology and Innovation*, *8*(1), 38–60. https://doi.org/10.1504/IJPTI.2021.116079

Sarker, A., Grift, T. E. (2023). Monitoring postharvest color changes and damage progression of cucumbers using machine vision. *Journal of Food Research*, *12*(2), 37–50. https://doi.org/10.5539/jfr.v12n2p37

Sarker, A., Jung, Y., Siddiqui, R. (2023). Yoghurt fortification with green papaya powder and banana resistant starch: effects on the physicochemical and bioactive properties. *International Journal of Food Science and Technology*, 1–12. https://doi.org/10.1111/ijfs.16672

Sarker, A., Siddiqui, R. A. (2023). Effects of ultrasonic processing on the quality properties of fortified yogurt. *Ultrasonics Sonochemistry*, *98*, 106533. https://doi.org/10.1016/j.ultsonch.2023.106533

Simonin, H., Duranton, F., de Lamballerie, M. (2012). New insights into the high-pressure processing of meat and meat products. *Comprehensive Reviews in Food Science and Food Safety*. https://doi.org/10.1111/j.1541-4337.2012.00184.x

Stewart, C., Dunne, P. C., Keener, L. (2015). Pressure-assisted thermal sterilization validation. In V. Balasubramaniam, G. Barbosa-Canovas, & H. Lelieveld (Eds.), *High pressure processing principles, technology and applications*. New York: Springer.

Sun, Y., Kang, X., Chen, F., Liao, X., Hu, X. (2019). Mechanisms of carrot texture alteration induced by pure effect of high pressure processing. *Innovative Food Science and Emerging Technologies*, *54*(17), 260–269. https://doi.org/10.1016/j.ifset.2018.08.012

Suthanthangjai, W., Kajda, P., Zabetakis, I. (2005). The effect of high hydrostatic pressure on the anthocyanins of raspberry (Rubus idaeus). *Food Chemistry*, *90*(1–2), 193–197. https://doi.org/10.1016/j.foodchem.2004.03.050

Szczepańska, J., Barba, F. J., Skąpska, S., Marszałek, K. (2020). High pressure processing of carrot juice: Effect of static and multi-pulsed pressure on the polyphenolic profile, oxidoreductases activity and colour. *Food Chemistry*, *307*(September 2019), 125549. https://doi.org/10.1016/j.foodchem.2019.125549

Ting, E. (2011). High-pressure processing equipment fundamentals. In *Nonthermal Processing Technologies for Food*. https://doi.org/10.1002/9780470958360.ch2

Tonello, C. (2011). Case studies on high-pressure processing of foods. In *Nonthermal Processing Technologies for Food*. https://doi.org/10.1002/9780470958360.ch4

Trejo Araya, X. I., Smale, N., Zabaras, D., Winley, E., Forde, C., Stewart, C. M., Mawson, A. J. (2009). Sensory perception and quality attributes of high pressure processed carrots in comparison to raw, sous-vide and cooked carrots. *Innovative Food Science and Emerging Technologies*, *10*(4), 420–433. https://doi.org/10.1016/j.ifset.2009.04.002

Verbeyst, L., Oey, I., Van der Plancken, I., Hendrickx, M., Van Loey, A. (2010). Kinetic study on the thermal and pressure degradation of anthocyanins in strawberries. *Food Chemistry*, *123*(2), 269–274. https://doi.org/10.1016/j.foodchem.2010.04.027

Vervoort, L., Van Der Plancken, I., Grauwet, T., Verlinde, P., Matser, A., Hendrickx, M., Van Loey, A. (2012). Thermal versus high pressure processing of carrots: A comparative pilot-scale study on equivalent basis. *Innovative Food Science and Emerging Technologies, 15*. https://doi.org/10.1016/j.ifset.2012.02.009

Wauters, T., Verbeeck, K., Verstraete, P., Vanden Berghe, G., De Causmaecker, P. (2012). Real-world production scheduling for the food industry: An integrated approach. *Engineering Applications of Artificial Intelligence*, *25*(2). https://doi.org/10.1016/j.engappai.2011.05.002

Weemaes, C., Ooms, V., Indrawati, Ludikhuyze, L., Broeck, I. Van den, Loey, A. Van, Hendrickx, M. (1999). Pressure temperature degradation of green color in broccoli juice. *Journal of Food Sci- Ence*, *64*, 504–508.

Wolbang, C. M., Fitos, J. L., Treeby, M. T. (2008). The effect of high pressure processing on nutritional value and quality attributes of Cucumis melo L. *Innovative Food Science and Emerging Technologies*, *9*(2), 196–200. https://doi.org/10.1016/j.ifset.2007.08.001

Yoo, S., Ghafoor, K., Kim, J. U., Kim, S., Jung, B., Lee, D. U., Park, J. (2015). Inactivation of Escherichia coli O157:H7 on orange fruit surfaces and in juice using photocatalysis and high hydrostatic pressure. *Journal of Food Protection*, *78*(6). https://doi.org/10.4315/0362-028X.JFP-14-522

Zhao, L., Wang, Y., Wang, S., Li, H., Huang, W., Liao, X. (2014). Inactivation of naturally occurring microbiota in cucumber juice by pressure treatment. *International Journal of Food Microbiology*, *174*. https://doi.org/10.1016/j.ijfoodmicro.2013.12.023

Zhao, L., Wang, Y., Hu, X., Sun, Z., Liao, X. (2016). Korla pear juice treated by ultrafiltration followed by high pressure processing or high temperature short time. *LWT*, *65*, 283–289. https://doi.org/10.1016/j.lwt.2015.08.011

PART V

Packaging Materials and Emulsions Applications

Nanoemulsions-Based Edible Coatings for Postharvest Quality Preservation of Fruits and Vegetables

13

Sajid Ali*, **Ghulam Khaliq, Ayesha Iftikhar, Shabir Ahmad Mir and Shahid Iqbal**

**Corresponding Author:* sajidali@bzu.edu.pk

13.1 INTRODUCTION

Fresh vegetables and fruits are the leading source of different health-associated phytonutrients which are considered beneficial and essential for humans' normal and healthy life style. These products contain plentiful quantities of vitamins, phenols, organic acids, carotenoids, anthocyanins, minerals and certain other bioactive compounds which have generally been found essential for normal and healthy lives (Veiga et al., 2020). The short shelf life potential of fruits and vegetables is considered the leading hindrance in the widespread consumption of such commodities. The short shelf life has been found to be related to significant losses of fresh fruits and vegetables around the globe, thereby resulting in global food insecurity. Therefore, some appropriate and effective technologies which can maintain the general quality and reduce the postharvest losses with extended storage/shelf life potential are needed.

The fruits and vegetables are also susceptible to postharvest diseases, which generally lead to about 10–30% of the losses during harvesting, storage, transportation and subsequent marketing (Mossa et al., 2021). The packaging of fresh fruits and vegetables is important in order to maintain their quality and to reduce postharvest losses. Nevertheless, the major and substantial challenges of the global food packaging industry are maintenance of the quality and prolongation of shelf life under widely variable environmental conditions. There are various types of packaging materials. The widespread increase and accumulation of plastic-based films has raised many environmental-related challenges. In this context, the technology based on edible coatings has gained much attention because these are generally considered easy to apply, widely available and generally do not create any potential environmental-related issues (Maringgal et al., 2020).

There are many ways to apply edible coatings on the fresh fruits and vegetables in general. In recent times, nanotechnology has been introduced as a new tool for the preparation of various emulsion-based edible coating treatments with enhanced functionalities and properties (Triphati et al., 2021; de Oliveira Filho et al., 2021). It has been reported that the edible coatings can be made from conventional (microemulsion) and nanoemulsion methods (de Oliveira Filho et al., 2021). Nanoemulsions-based edible coatings have gained much attention and are continuously applied to various fruits and vegetables with varying efficacy and outcomes. The droplets of nanoemulsions are on a nanoscale, with a particle size of less than 100 nm in a dispersed aqueous solution (McClements, 2012). These nanoemulsions have certain advantages with practical applications in the food industry. The nanoemulsions can carry certain active compounds, such as antimicrobial agents, oil-soluble vitamins, and nutraceutical and flavoring agents that thereby have the capacity to preserve and/or further enhance the quality attributes of food commodities (Salvia-Trujillo et al., 2017; Hasan et al., 2018; Mushtaq et al., 2023). The application of nanoemulsions-based edible coatings has significantly increased in recent times, and the related literature is still increasing day by day (Pandey et al., 2022; Mushtaq et al., 2023). Although much research has been reported regarding the potential application of nanoemulsions on fresh fruits and vegetables, a lot of work is still needed to explore the potential of these coating types on a crop-by-crop basis. This chapter reviews the potential use of

DOI: 10.1201/9781003370376-18

nanoemulsions-based edible coatings of different fruits and vegetables during postharvest storage.

13.2 DEFINITION AND CHARACTERISTICS OF NANOEMULSIONS

The emulsions are normally made of two immiscible liquids. These frequently contain water and oil, thereby forming comparatively a stable mixture. In general, the emulsion system is that which contains a continuous and dispersed phase and could be differentiated into three-dimensional organization comprising water and oil phases. An oil-based droplet dispersed in water phase is oil-in-water (O/W) emulsion, and the other phase is water-in-oil (W/O) emulsion. The W/O-based coatings are the most common emulsion so far reported (Jin et al., 2016; de Oliveira Filho et al., 2021). The emulsions are categorized into three major classes based on their thermodynamic stabilities, physical properties and stable mechanisms. These include (a) conventional or microemulsion, (b) nanoemulsions and (c) microemulsion. It is important to mention that nanoemulsions and conventional emulsions are thermodynamically unstable (McClements, 2012; Salvia-Trujillo et al., 2015; McClements & Rao, 2011; Zhang et al., 2020). The size of the droplet is the leading factor in nanoemulsions, which affects the capacity of such coatings to enhance the bioavailability and efficacy of the incorporated hydrophobic substances such as carotenoids (Zhang et al., 2020) and improve the antimicrobial potential of the added essential oils (Salvia-Trujillo et al., 2015) and other oil-based compounds (Prakash et al., 2020).

13.3 MATERIALS FOR NANOEMULSIONS-BASED COATINGS FORMULATIONS

Different types of materials have been reported for the preparation of nanoemulsions for fresh fruits and vegetables. Nevertheless, it is important that each type of material possesses its own properties and efficacy as well as safety status to be used as nanoemulsions because fresh fruits and vegetables are edible produce (Table 13.1).

Paraffins, rice bran, waxes, organic acids, resins, neutral lipids, organic fatty acids and acetoglycerides could be used for the preparation of edible films and coatings because these are considered hydrophobic in nature and exhibit appropriate moisture barrier properties, thereby suppressing the physiological and biochemical degradation of the coated edible produce (Umaraw & Verma, 2017; Mushtaq et al., 2023). Lipids-oriented nanoemulsions can also improve the solubilization owing to increased adsorption. On the other side, several lipids have the ability to improve the glossiness of the coated produce. In contrast, resins or lipids used for the preparation of edible coatings and films generally exhibit decreased mechanical strength due to their non-polymer nature (Singh et al., 2017; Mushtaq et al., 2023).

Proteins-based materials can also be used for the preparation of nanoemulsions. Proteins exhibit amphiphilic properties owing to their hydrophobic and hydrophilic nature. Proteins with low water solubility are often less permeable to gases and water vapors. Owing to their superior mechanical, physical and film-forming properties, proteins have been found to be well suited to the requirements of certain food products.

TABLE 13.1 Major Advantages and Disadvantages of Different Materials Used for the Preparation of Potential Edible Coatings. This Information Is Based on the Summary after Modification of the Information Reported by de Oliveira Filho et al. (2021)

MATERIALS	*MAIN MATRICES*	*MAJOR ADVANTAGES*	*MAJOR DISADVANTAGES*	*REFERENCES*
Proteins	Casein, gelatin, zein, whey protein, soy protein, quinoa and myofibrillar's proteins	Suitable barrier property against gases without imparting anaerobic environment	Protein-based coatings are considered brittle and susceptible to cracking	Arnon-Rips and Poverenov (2018)
Lipids	Animals and plant-based resins and waxes. Vegetable-based oils and certain (FA) fatty acids	Suitable barrier property against the moisture along with provision of glossy appearance on the treated produce	Ineffective gas and mechanical barricade properties	Miranda et al. (2020); Dehghani et al. (2018); Eddin et al. (2019)
Polysaccharides	Chitosan, starch, alginates, cellulose, pectins and derivatives	Appropriate gas and mechanical barrier property	Hydrophilic nature of the said polysaccharides may result in a poor moisture barricade	Thakur et al. (2019); Mohamed et al. (2020)
Composites	Combinations of different polysaccharides along with proteins and certain lipid-based compounds	Effective gas and moisture barrier property	Development of potential non-homogeneous emulsions	Galus and Kadzińska (2015); Dhumal and Sarkar (2018); Miranda et al. (2019); De Oliveira Filho et al. (2020)

However, due to their disproportionately higher propensity to trigger allergic reactions, proteins have a finite range of applications in the preparation of nanoemulsions (Mushtaq et al., 2023).

Fruits and vegetables are typically covered with edible coatings made of polysaccharides because of their natural flexibility and capacity to stick to the surface of the food. According to Versino et al. (2016), amphiphilic polysaccharides such as modified alginate, modified celluloses, certain galactomannan compounds, modified starches, pectin, gum arabic or cellulose-based derivative gums, carrageenan and alginates are likely employed as emulsifiers. They are widely accessible and have a reasonably lower price. When polysaccharides are dissolved in water, the majority of them create powerful hydrogen bonds, which have a positive mechanical impact on the films and edible coatings. Because of their hydrophilic nature, polysaccharides are permeable to moisture, which limits their ability to prevent water loss from meals (Kong et al., 2022; Mushtaq et al., 2023). In general, polysaccharides have little to no impact on the flavor or appearance of the treated commodities. In addition, unlike proteins, polysaccharides-based materials do not trigger allergic reactions. Polysaccharide nanoemulsions exhibit a higher water permeability and water solubility. The higher water vapor permeability and water solubility are considered major drawbacks. Due to their poor moisture barrier ability as well, these are unfavorable for liquid-based products (Kong et al., 2022; Mushtaq et al., 2023). In addition to the previously listed materials, the use of surfactants, stabilizers, ionic surfactants, non-ionic surfactants, zwitterionic surfactants, emulsifiers, weighting agents, stabilizers, texture modifiers, ripening inhibitors and plasticizers has also been reported in minor concentrations (Mushtaq et al., 2023).

13.4 PREPARATION TECHNIQUES OF NANOEMULSIONS

It has been reported that for the development of stable nanoemulsions in O/W for edible coatings three major components are need. These include (a) the continuous phase, (b) the disperse phase and (c) emulsifiers for the stabilization of the aforementioned phases (Figures 13.1 and 13.2). The process of dispersion is based on fractions of lipids of the emulsions in which active ingredients of a non-polar nature can be added directly (García-Márquez et al., 2017; Rohasmizah & Azizah, 2022). High and low energy approaches have been used for nanoemulsions preparation (Figure 13.3). On the other hand, lipophilic compounds may also be added by dissolving in the carrier oil prior to emulsification (Luo et al., 2017). Polysaccharides such as carrageenan, alginates, starches, pectins, gums and protein, including soy and whey protein, are the major and most commonly used materials which can be utilized as continuous phases. It has been reported that these polysaccharides and proteins are the most critical, as they provide the structural supporting matrix for the development of nanoemulsions-based edible coating materials (Luo et al., 2017). It is important that humans eat fruits and vegetables, so it is important that the developed nanoemulsions should be based on food-grade materials, as these are directly applied to the produce (fruits and vegetables). For this purpose, spans, tweens and lecithin-based materials have been used in nanoemulsions-based coatings materials (Nash & Erk, 2017; Rohasmizah & Azizah, 2022). In addition, surface active biopolymers, including gums and proteins,

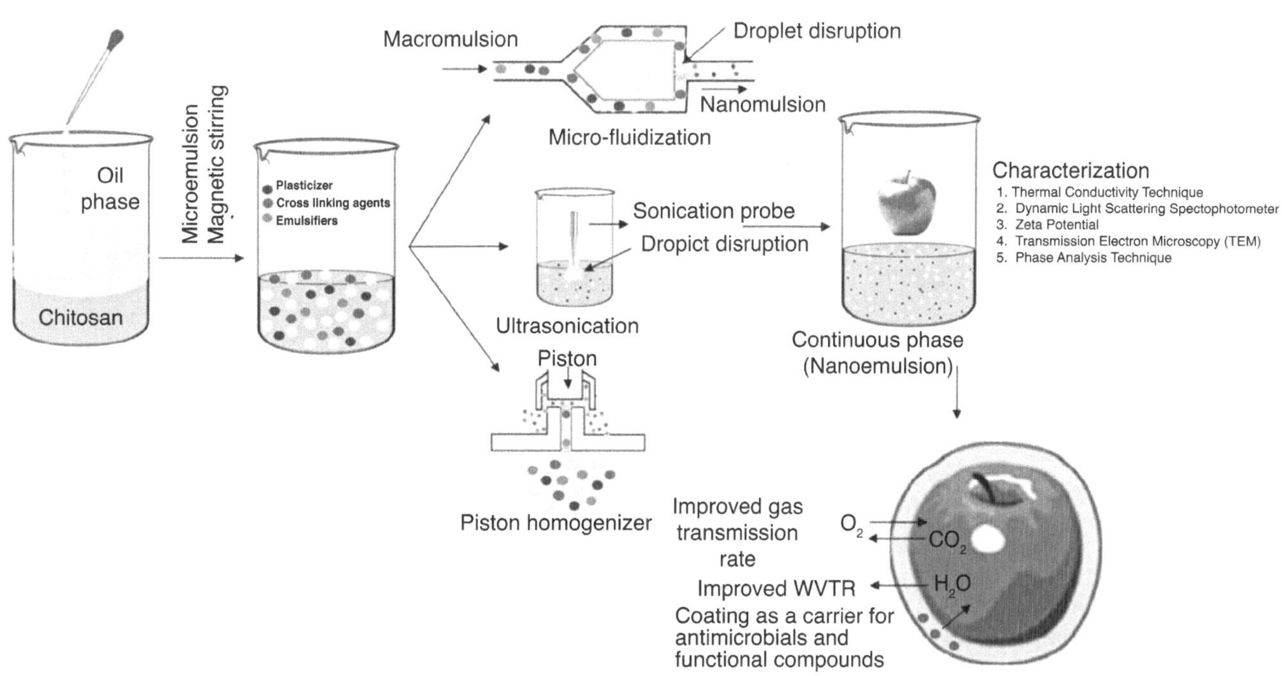

FIGURE 13.1 Schematic presentations of the fabrication, application and impact of nanoemulsions on fresh food. This figure is adapted (with permission) from Chaudhary et al. (2020).

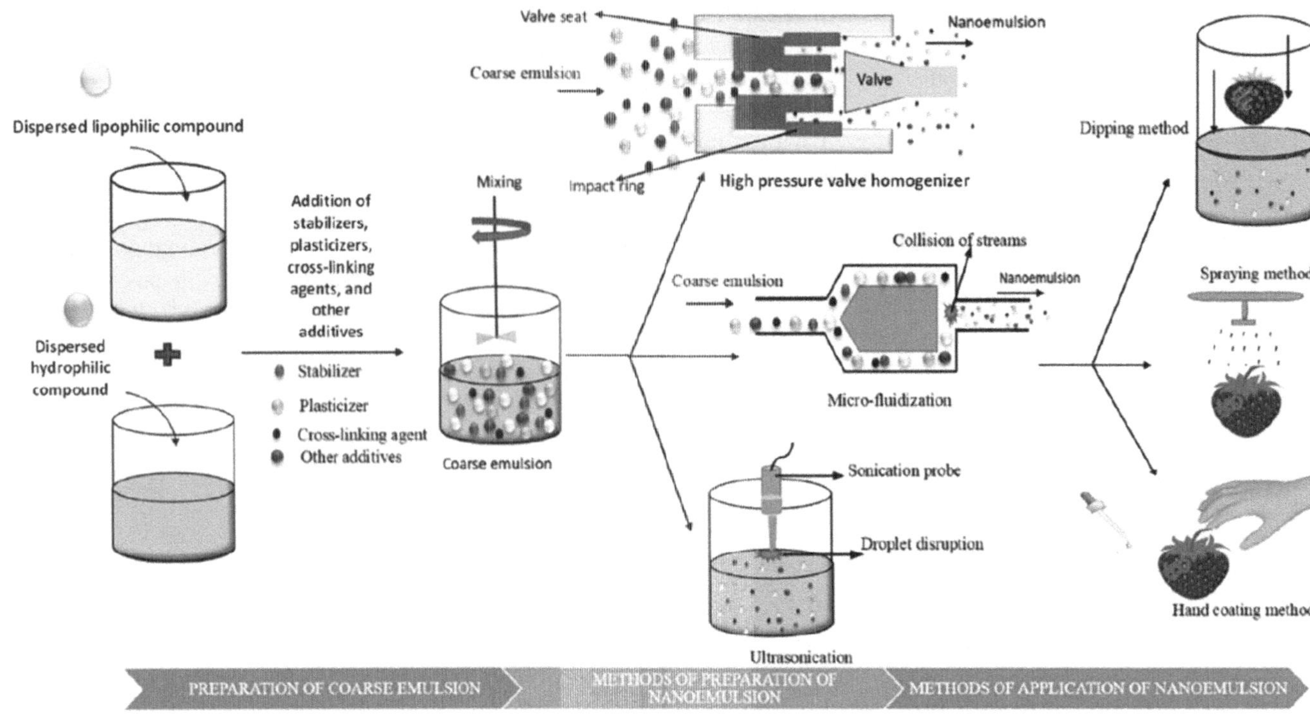

FIGURE 13.2 The nanoemulsion preparation illustration and its methods of application on fresh produce. This figure is adapted (with permission) from Panwar et al. (2024).

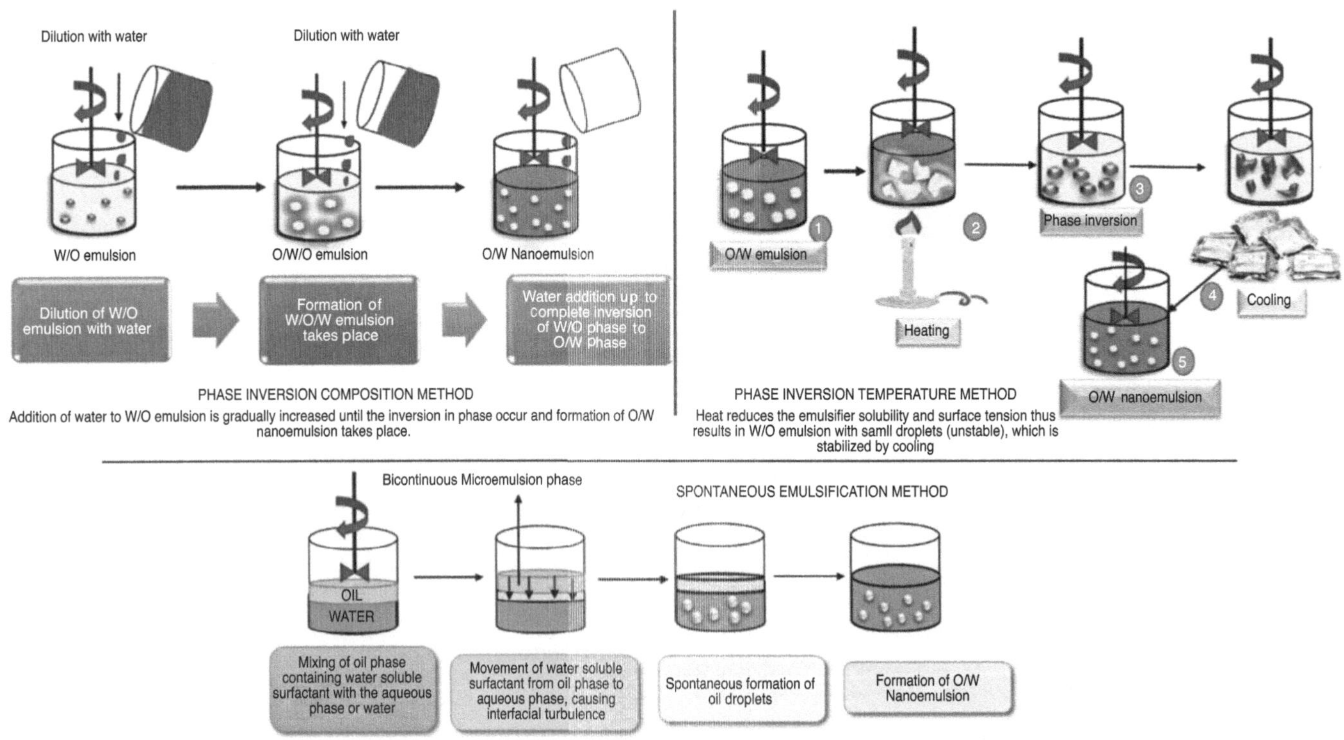

FIGURE 13.3 The nanoemulsion preparation illustration based on low energy approaches (including phase inversion and spontaneous methods of emulsification). This figure is adapted (with permission) from Panwar et al. (2024).

may also be considered for the formation of nanoemulsions (Bai et al., 2016; Rohasmizah & Azizah, 2022). It has been described that nanoemulsions-based coatings could be developed by using two major approaches. One technique generally involves the utilization of an emulsifier, continuous phase and polymers. Thereafter, the aforementioned components form a coarse emulsion and the mixed components are homogenized in order to reduce the size of the droplets into nanoemulsions (Artiga-Artigas et al., 2017; Rohasmizah & Azizah, 2022). The second approach consists of two major steps in which nanoemulsions are prepared directly without incorporating biopolymers in the aqueous phase at the first step. This will then be added to a previously prepared nanoemulsions solution at a later stage (Severino et al., 2015; Rohasmizah & Azizah, 2022). It is important to mention that the encapsulations of the non-polar substance in the edible coatings are essential under both the approaches and situations (Rohasmizah & Azizah, 2022).

13.5 STABILITY CONSIDERATIONS OF NANOEMULSIONS

The formation of nanoemulsions requires energy. This energy is generally provided by physical and chemical systems as well as by mechanical equipment. The methods that need mechanical energy are known as high-energy methods and utilize high-pressure homogenizers, micro-fluidizers and ultrasonic homogenization agents. The energy-dependent methods use the chemical and physical properties of the system and are grouped as lower-energy (phase inversion, spontaneous emulsification and emulsion inversion) methods (McClements, 2021; Zhang et al., 2020). The amphiphilic materials, i.e., phospholipids, surfactants, polysaccharides and proteins, efficiently decrease the interfacial-based tension, thereby significantly maintaining the stability of the prepared droplet. Nanoemulsions such as W/O and O/W are considered the major stable systems. Nevertheless, under certain conditions, emulsions, for instance O/W/O and W/O/W, may develop and are highly unstable for coalescence (Dhandevi & Jeewon, 2015; McClements & Xiao, 2012; McClements, 2021). In this case, the stability of the developed nanoemulsions may be reduced and compromised. It is important that the vegetables and fruits are highly perishable and contain significantly higher water percentages. Therefore, O/W-based emulsion has been reported as the most researched and investigated type of emulsion in the fresh produce industry owing to the possibility of oil droplets loading with lipophilic components generally surrounded by water molecules (Jin et al., 2016; McClements, 2021). The stability may also depend upon the type of polysaccharide used in the preparation of the nanoemulsions. For instance, emulsions are not produced from the chitosan alone, and the emulsions produced at low concentrations are generally considered unstable in contrast to its higher concentrations. The higher concentrations-based emulsion exhibits improved capacity for emulsification along with high stability in nature under certain temperature and aging conditions (Chaudhary et al., 2020).

13.6 TECHNIQUES FOR THE APPLICATION OF NANOEMULSIONS COATINGS

There are several methods or approaches which have been reported for the application of edible coatings. It is important to mention that in the case of edible coatings, two main components are imperative which affect the coating's quality and efficacy. These include (a) the methods of applications and (b) the clinginess potential of the coating type on a commodity (Pandey et al., 2022). The following approaches and methods have been reported for the application of microemulsions and nanoemulsions on different fruits and vegetables.

Dipping produce in an edible coating solution is the most common method of edible coating application. The dipping forms a semipermeable covering on the surface of fruits and vegetables for a specific time period. This type of coating application develops a layer on the produce which subsequently resists the transfer of gases and moisture loss with extended shelf life potential (de Oliveira Filho et al., 2021; Ali et al., 2022a; Ali et al., 2022b; Pandey et al., 2022). The major advantage of this type of application is that the subject (produce) is generally enclosed from all sides even in the case of commodities with a rough surface texture. In some cases, the approach based on multilayer, particularly layer-by-layer, treatment is used in case of fresh-cut products owing to the sticky nature of the commodity (Poverenov et al., 2014; Pandey et al., 2022; Ali et al., 2024).

The spray technique is another approach to edible coating application. This method is generally used in packing houses and provides an attractive and homogenous coating on the produce. Additionally, spraying results in the suppression of possible contamination of the prepared edible coating solution (Dhanapal et al., 2012; de Oliveira Filho et al., 2021). In this application method, the vegetables or fruits are kept on plates or rollers at a specific speed and the coating material is applied under dispersion nozzles whose activation depends on an automatic or manual system. The process is generally repeated until the desired thickness of the coating has been attained. One drawback of this technique is that viscous solutions cannot be applied in the form of spray because it will clog the equipment (Atieno et al., 2019).

Manual edible coating application in the form of filmogenic solutions is another method of edible coating treatment.

The fruits and vegetables can be uniformly coated with the edible coating solution by spreading the prepared coating solution with hands while wearing latex gloves (de Oliveira Filho et al., 2021). This technique has been found to be suitable to rule out the probable contamination of the solution and to minimize waste of the coating solution. Nevertheless, a major negative aspect of this technique is the possibility of non-homogenous thick films forming because the whole surface of the produce is not covered appropriately (Sun et al., 2015; Miranda et al., 2021; de Oliveira Filho et al., 2021).

Vacuum impregnation is another application method for certain treatments, and the most crucial element is pressure. This technique is typically used to improve the vitamin and mineral characteristics of fruits and vegetables. The only variation from the standard dipping-draining process is the pressure. The samples are combined in partial vacuum pumps that are connected to the treatment solutions. The vacuum impregnation procedure includes three key factors: (a) atmospheric restoration time, (b) vacuum pressure and (c) vacuum pressure time period (Senturk et al., 2018). However, it is important that this technique has mostly been reported for the application of some elicitors. The literature about the application of nanoemulsions and/or microemulsions-based edible coatings on fruits and vegetables is very negligible.

13.7 ACTION MECHANISMS OF NANOEMULSIONS COATINGS

In general, edible coatings lead to the development of a modified atmosphere inside the coated produce (fruits and vegetables), normally leading to increased levels of carbon dioxide and lower concentrations of ethylene (Ali et al., 2022a; Ali et al., 2024). This type of modified atmosphere results in the suppressed activities of metabolism, thereby considerably reducing the quantitative and qualitative losses of fresh produce such as fruits and vegetables (Ali et al., 2022b). The reduced metabolic activities thereby also help in the maintenance of the eating quality with minimal losses of produce. In addition, sometimes some types of additives such as antimicrobial and/or antibrowning agents can be incorporated in the nanoemulsions, which slowly release to suppress the browning and conserve the visual quality of the coated commodities (Manzoor et al., 2021; Javanmardi et al., 2023). It is important to mention that the addition of antimicrobial agents increases the efficacy of the nanoemulsions in general. On the other side, the addition of some antioxidant compounds is also possible and the added antioxidative compounds may serve to further increase the efficacy of the nanoemulsions along with improvement of the structural integrity of the developed nanoemulsions (Manzoor et al., 2021). It is also important to report that the addition of some antifungal chemicals or natural agents is also possible in the nanoemulsions (Yang et al., 2021). Similarly, the added antifungal compounds/essential oils and bioactive compounds have been found to be effective in increasing the performance of the developed nanoemulsions (Mossa et al., 2021; Abdalla et al., 2023). Therefore, the action mechanism of the nanoemulsions generally depends upon the type of nanoemulsions, the addition of certain antibrowning agents, antioxidant compounds and antifungal agents on the whole.

13.8 MAJOR ADVANTAGES OF NANOEMULSIONS COATINGS FOR FRUITS AND VEGETABLES

13.8.1 Incorporation of Antimicrobial/Antibrowning Compounds

Fresh fruits and vegetables are highly perishable commodities due to their higher moisture content, and due to this they are susceptible to postharvest disease infestations during storage. Therefore, it is important to protect these commodities from decay and diseases in order to extend their shelf and/or storage life for longer periods of time (Al-Tayyar et al., 2020). For this purpose, different strategies and approaches have been reported. Among these, one is to encapsulate certain bioactive and/or antimicrobial compounds which thereafter are released at a slower rate during storage, and these therefore protect the stored fruits and vegetables from spoilage, ultimately prolonging their storage potential and impeding the disease- and decay-based losses (Kong et al., 2022; Oyom et al., 2022). Therefore, the encapsulation or incorporation of the bioactive and slow-releasing antimicrobial compounds is imperative. For this purpose, the encapsulation of the bioactive compounds and antimicrobial agents can be done with nanoemulsions-based edible coatings (Mahfoudhi et al., 2016; Lohith Kumar & Sarkar, 2018; Oyom et al., 2022). Different types of bioactive compounds and antimicrobial agents' incorporation and encapsulation have been reported for the preservation of fresh fruits and vegetables during postharvest storage. Due to consumer awareness and potential health hazards, the use of natural agents has been much researched and increased in recent years. Essential oils are potentially considered organic and safe, and their utilization as a potential additive or antimicrobial agent has been much increased. They are also considered to assist in decreasing water vapor permeability in certain types of hydrophilic coatings and films (Atarés & Chiralt, 2016; Al-Tayyar et al., 2020). Thyme, lemongrass and sage essential oils in nanoemulsions have been reported as potential antimicrobial agents (Acevedo-Fani et al., 2015). The addition of lemongrass nanoemulsions in a sodium alginate-based coating was found effective in suppressing the spoilage of apple fruit during storage (Salvia-Trujillo et al., 2015). The edible coating based on *Aloe vera* gel incorporated with nanoemulsions of calcium chloride and catechin maintained the eating quality of strawberry fruits along with significantly reduced yeast and mold counts during storage (Amiri et al., 2022). Sodium alginate coating which contained cinnamaldehyde essential oil nanoemulsions resulted in reduced browning, better quality and reduced pseudomona counts on mushrooms (Louis et al., 2021). The incorporation of quercetin in O/W-based nanoemulsions significantly reduced the onset of browning in apple fruits (Mendoza-Wilson et al., 2022).

13.8.2 Texture Preservation Ability of Nanoemulsions

The storage of climacteric and certain non-climacteric fruits generally leads to a decline in firmness owing to the degradation of cell wall integrity in response to certain hydrolase enzyme activity (Hasan et al., 2018). Edible coatings have been found to improve and maintain the postharvest textural characteristics of fresh and fresh-cut fruits and vegetables under storage. Likewise, nanoemulsions-based coatings showed enormous potential to act as texture enhancers when applied on fresh-cut horticultural produce such as fruits and vegetables. Zambrano-Zaragoza et al. (2014) found that nanoemulsions-based coatings on xanthan gum wax was effective in suppressing the loss of firmness in fresh-cut apple slices. Similarly, xanthan gum and lipids-based coating of carnauba wax was also effective in reducing the loss of firmness of guava for a 30-day period at storage conditions of 10°C (Zambrano-Zaragoza et al., 2013). It is important to report that the activity of PPO and PME was decreased in apple and guava when coated with nanoemulsions comprising nopal mucilage, alpha-tocopherol and xanthan gum (Zambrano-Zaragoza et al., 2013; Garcıa-Betanzos et al., 2017). The higher textural preservation was also noted in green beans (Severino et al., 2014), strawberries (Eshghi et al., 2013) and zucchini (Donsi et al., 2014) treated with chitosan-based nanoemulsions.

13.9 APPLICATIONS OF NANOEMULSIONS IN POSTHARVEST STORAGE OF FRESH FRUITS AND VEGETABLES

13.9.1 Preservation of Fruits and Vegetables with Polysaccharides-Based Nanoemulsions

Fresh fruits and vegetables are perishable commodities and should be handled with care and proper attention to minimize the desiccation-based quantitative and qualitative losses in order to ensure global food security. Different types of approaches and treatments have been used and are being used for the postharvest quality preservation of fresh fruits and vegetables during storage. Among the different treatments, the application of polysaccharides-based nanoemulsions coatings has been found effective in reducing the losses of fresh fruits and vegetables after harvest. These include cellulose, pectin, chitosan, pullulan, alginates, gums and other related substances/derivatives (Al-Tayyar et al., 2020). These have been reported as the main and basic coating types that are considered generally effective in the modification of the internal atmosphere of the coated horticultural produce owing to their ability to block the uptake of oxygen and the accumulation of carbon dioxide (Ali et al., 2024).

13.9.2 Preservation of Fruits and Vegetables with Proteins-Based Nanoemulsions

Proteins naturally occur in fibrous or globular forms. The fibrous proteins are not soluble (insoluble) in water and normally play the function of structural elements of the animal's tissues, whereas proteins which are soluble in water are generally present in globular forms and play certain critical roles in living systems (Hasan et al., 2018; Al-Tayyar et al., 2020). Different types of globular proteins, including whey, corn zein, wheat gluten and soy proteins, have been exploited for the formation of edible coatings and edible films. In water, protein interactions create hydrophobic reactions, ionic bonds and hydrogen bonds. The water solubility of the protein depends upon the constituents of the functional group. It has been reported that edible films and coatings normally exhibit smaller barriers against moisture but enormous permeability to different gaseous molecules and mechanical strength when the selected protein tends to develop ionic or hydrogen bonds (Umaraw & Verma, 2017). The lower water-soluble protein is considered normally less permeable to gases and vapor. Due to higher mechanical and physical properties as well as edible film-forming ability, proteins fit in the specific needs for various food-based products. Nevertheless, proteins show reduced applications owing to their comparatively higher chances of leading to certain allergic reactions (Mushtaq et al., 2023). Therefore, the utilization of proteins-based nanoemulsions is negligible so far and needs further comprehensive research work.

A nanoemulsion based on β-carotene which was stabilized with hydrolyzed whey proteins containing pectin exhibited good stability for a period of 30 days (López-Monterrubio et al., 2021). In another work, Vildan et al. (2020) reported that whey protein–based isolate resulted in higher hardness, thereby leading to the suppression of softening under postharvest storage conditions in fruit bars. An antimicrobial edible coating based on nanoemulsion of thymol/quinoa protein/chitosan was developed for the safety attainment, sensory quality preservation and eating quality enhancement of refrigerated strawberry fruits during conditions of commercial storage. It was observed that the nanoemulsion was effective in reducing postharvest decay and overall quality preservation of strawberry fruits (Robledo et al., 2018).

13.9.3 Preservation of Fruits and Vegetables with Lipids-Based Nanoemulsions

Application of lipids-based edible coatings and nanoemulsions have been found effective in reducing the postharvest losses of treated vegetables and fruits after harvest (Miranda et al., 2022a; Duan et al., 2023). The treatments on wax-based coatings have been found effective in increasing the shelf life of coated fruits such as citrus by delaying color development due to retarded biosynthesis of the carotenoids pathway during storage (Duan et al., 2023). The combined treatment based on *Cymbopogon martinii* essential oil and carnauba wax nanoemulsion significantly suppressed the ripening and maintained the general eating quality of the coated papaya fruits (Table 13.2). This treatment also reduced the incidence of rot and did not show any negative impact on the quality of fruits for 12 days (de Oliveira Filho et al., 2022). The application of 18% carnauba wax along with ginger essential oil and hydroxypropyl methylcellulose (HPMC) significantly suppressed the natural fruit decay in treated papaya fruit. In addition, it also conserved the biochemical quality attributes along with better color and aroma volatile compound content (Miranda et al., 2022b). In another work, the influence of carnauba wax nanoemulsions was studied on fresh tomatoes during storage at 23°C for 15 days. The coated tomatoes exhibited better biochemical quality with reduced metabolic activities, having higher sensory evaluation-based scores in contrast with the controls (Miranda et al., 2022a). The application of carnauba wax nanoemulsions along with *Metha spicata* and *Syzigium aromaticum* was tested in improving the postharvest storage life of papaya fruits along with the maintenance of eating quality (de Oliveira Filho et al., 2023). It was observed that the combination of carnauba wax emulsion with *Metha spicata* significantly suppressed postharvest decay severity and showed maintained biochemical attributes throughout storage. The combination probably acted as an antimicrobial approach, which ultimately reduced the disease incidence and extended the storage life by showing the potential of quality preservation in comparison with untreated controls (de Oliveira Filho et al., 2023).

13.9.4 Preservation of Fruits and Vegetables with Essential Oils–Based Nanoemulsions

The essential oils (EOS) obtained from certain plants and spices/herbs show antimicrobial and antioxidative properties (Viuda-Martos et al., 2010). So, these are considered effective and peculiar organic derivatives to be applied in food-related industries with the lowest potential for toxicity and other health-related issues if consumed. This is due to their categorization as Generally Recognized as Safe (GRAS). Nevertheless, the utilization of EOS in the form of preservatives for foods such as fruits and vegetables is often restricted due to the strong flavor of such products. Applying EOS could alter the taste and flavor of the produce. Therefore, to preclude this problem, EOS are generally incorporated into edible coatings and/or films (Donsì et al., 2015). It is also imperative that by keeping in mind the reduction of chemical-based additives researches have paid significant and considerable attention to apply natural alternatives which have certain antimicrobial and antioxidative properties that do not impart any negative or deleterious impact on the health of consumers (Alves-Silva et al., 2013).

The potential application of EOS-based nanoemulsions has been reported on different fruits and vegetables, including fresh-cut or minimally processed products. For instance, an edible film was developed from sodium alginate that contained nanoemulsions of different EOS such as thyme, lemongrass and sage. It was observed that thyme-based nanoemulsions incorporated into sodium alginate films showed significantly higher flexibility, transparency and water vapor resistance. In addition, it also showed markedly higher antimicrobial efficacy by suppressing the growth of *E. coli* (Acevedo-Fani et al., 2015). In another work, nanoemulsions were developed from lemongrass added in sodium alginate for increasing the shelf life of fresh-cut apple slices under refrigeration storage. The treatment showed reduced *E. coli* proliferation for 14 days along with maintenance of general eating quality (Salvia-Trujilo et al., 2015). In another research, the impact of lemongrass nanoemulsion was investigated on changes in the texture and appearance of table grapes. It was observed that the changes in quality were delayed in lemongrass EOS-based nanoemulsions in contrast with traditional emulsions (Oh et al., 2017).

The application of 18% carnauba wax along with ginger essential oil and HPMC significantly suppressed the natural fruit decay in treated papaya fruit. In addition, it also conserved the biochemical quality attributes along with better color and aroma volatile compounds (Miranda et al., 2019). The postharvest gray mold caused by *Botrytis cinerea* significantly increased decay-based losses in tomatoes. Nanothymol-based emulsion markedly suppressed the growth of *Botrytis cinerea* on tomato fruits (Zhang et al., 2022). In a recent study, the impact of citral nanoemulsion coating was investigated on the population of *L. monocytogenes* and the sensorial quality of fresh-cut papaya and melons. It was noted that citral nanoemulsion coating inactivated *L. monocytogenes* and preserved juiciness, firmness, brightness and related aspects compared with controls (Luciano et al., 2023). This is imperative, as it indicates that fresh-cut produce is considered generally vulnerable to microbial infestation after cutting and processing. The EOS-based emulsions have been found effective in reducing microbial infestation and contributing to food safety goals. In a recent study by Guo et al. (2023), a carvacrol-based emulsion was applied on fresh-cut apple slices. It was found that emulsions-based treatment markedly suppressed yeast and mold as well as viable bacterial count (Table 13.2).

TABLE 13.2 Summary of Different Nanoemulsions Used in Postharvest Preservation of Fruits and Vegetables

PRODUCE	NANOEMULSIONS	FUNCTIONS	INFERENCES	REFERENCES
Plums	Carnauba wax and lemongrass essential oil	Antimicrobial	Suppressed respiration and ethylene generation with preserved quality and improved shelf life	Kim et al. (2013)
Apples	Carnauba-shellac wax + lemongrass essential oil	Antimicrobial	Reduced microbial infestation, suppressed loss of quality and improved sensory attributes	Jo et al. (2014)
Grapes	Carnauba wax and lemongrass oil	Antioxidant and antimicrobial	Improved glossiness, inhibited microbe growth and prolonged shelf life	Kim et al. (2014)
Apples	Nopal mucilage	Antibrowning and antioxidant	Reduced browning and preserved texture of fruits	Zambrano-Zaragoza et al. (2014)
Papayas	Carnauba wax	Antimicrobial	Reduced ripening due to decreased mass loss and suppressed decay	Ohashi et al. (2015)
Apples	Sodium alginate and lemongrass essential oil	Antimicrobial	Nanoemulsion-based edible coating with 0.1% (v/v) LEO Decreased microbial growths, diminished enzymatic discoloration and preserved general quality	Salvia-Trujillo et al. (2015)
Okra pods	Sapindus extract and basil oil	Antimicrobial	Reduced microbe growth, extended storage life with preserved quality	Gundewadi et al. (2018)
Mangoes	Thai essential oil and hydroxypropyl methylcellulose	Antimicrobial	Preserved quality and extended storage life	Klangmuang and Sothornvit (2018)
Fresh-cut oranges	Essential oil of orange peel	Antioxidant and antimicrobial	Antibacterial and antifungal activities improved shelf life with better eating properties	Radi et al. (2018)
Tomatoes	*Flourensia cernua* extracts, chitosan and sodium alginate	Antioxidant and antimicrobial	Enhanced shelf life, reduced quality loss and suppressed decay	de Jesús Salas-Méndez et al. (2019)
Strawberries	Nutmeg seed oil and chitosan	Antimicrobial	Microbial growth was suppressed along with preserved quality	Horison et al. (2019)
Melons	Cinnamon essential oil	Antimicrobial	Significantly suppressed the growth of microbes	Paudel et al. (2019)
Guavas	Alginate, ZnO and chitosan	Antioxidant and antimicrobial	Extended storage life with better eating quality	Arroyo et al. (2020)
Apples	Flax seed and fenugreek	Antioxidant and antimicrobial	Suppressed respiration and ethylene with preserved quality	Rashid et al. (2020)
Melons	Citral oil	Antimicrobial	Inactivated *L. monocytogenes* and preserved sensory quality	Luciano et al. (2023)
Papayas	Citral oil	Antimicrobial	Suppressed *L. monocytogenes* growth and preserved sensory attributes	Luciano et al. (2023)
Apples	Carvacrol	Antimicrobial	Improved physicochemical attributes and suppressed microbiological growth	Guo et al. (2023)

13.10 SAFETY AND REGULATORY CONSIDERATIONS OF NANOEMULSIONS COATINGS

Due to their nanometric size, nanoemulsions may partially withstand digestion and provide possible safety issues for the substances used in their formation. Due to their quicker absorption in contrast with traditional emulsion systems, their capacity to raise the bioavailability of bioactive compounds to toxic levels and their potential for enhanced absorption by the epithelial cells, which may result in alterations of the gastrointestinal tract's functions, these can be of serious concern in hormonal and metabolic dysfunctions. Nanoemulsions can also be swiftly digested by gastrointestinal tract enzymes due to their higher surface area, which lessens the potentially harmful effects that could arise in the organ cells from their accumulation (McClements & Xiao, 2012). It is imperative to test the possible toxicity of the nanoemulsions because they can pose some serious health-related concerns if they penetrate human cells. Therefore, to test the possible toxicity of the nanoemulsions, *in vitro* research was carried out by using cell cultures, often representations of normal cells like fibroblasts. According to the findings of Kaur et al. (2017), Hep G2 cells did not reveal any potential toxicity upon exposure to certain nanoemulsions. Bergamot's essential oil–based nanoemulsion revealed a cytotoxic effect against $CaCo_2$ cells at higher doses (Marchese et al., 2020; Wani et al., 2018). Although some research regarding the possible toxicity of the nanoemulsions has been carried out under *in vitro* conditions, there is still a serious need for more comprehensive research on the possible toxicity of nanoemulsions under *in vivo* conditions (Wani et al., 2018). Moreover, there is incomplete information in the recent literature about the probable impact of nanoemulsions-based edible coatings on antimicrobial agents, such as essential oils, on the digestive system. Antimicrobial agents can affect the intestinal microbiota, so this effect needs to be carefully and comprehensively investigated in the near future (de Oliveira Filho et al., 2021).

The nanoemulsions made with GRAS agents do not appear to have substantial cytotoxic effects. The droplet's nanoscale size proposes that these are quickly converted into monoglycerides as they pass through the small intestine and free fatty acids, which are normal byproducts of the digestive system and shouldn't have any harmful effects (McClements & Rao, 2011). Some fruits and vegetables, such as fresh cut, are eaten with edible coatings. So, it is imperative that the addition of antioxidants and antimicrobial and texture enhancers should not influence the nanoemulsions-treated fresh-cut fruits and vegetables on the basis of consumer acceptance (Hasan et al., 2018). The addition of some essential oils into the nanoemulsions-based coatings may sometimes affect the taste and quality of the treated fresh-cut fruits and vegetables at higher concentrations (Sánchez-González et al., 2011). Therefore, the concentrations of the used essential oil nanoemulsions must be selected with great care, and the impact of the nanoemulsions may vary based on the commodity (fruits and vegetables). It is also important that the potential cytotoxicity of the food-grade-based nanoemulsions do not have any defined and standard limits (Hasan et al., 2018). Nevertheless, certain factors such as emulsifier, droplet size and concentration of the incorporated agents and emulsifier may affect the toxicity of the nanoemulsions (Hasan et al., 2018). It has been reported that the essential oils–based nanoemulsions compatible with antimicrobial efficiency were biocompatible with humans' epithelial cells, i.e., WKD, McCoy, HEp-2 and Caco-2 (Bonferoni et al., 2017), or had no impact of potential toxicity on the peripheral mononuclear blood cells (De Godoi et al., 2017). It is important that these are merely preliminary reports and potentially lack the in-depth insights on the toxicity aspects of nanoemulsions-based coatings. Therefore, some more and comprehensive research is needed on these aspects to get more reliable information about the possible toxicity and safety of nanoemulsions coatings (Hasan et al., 2018).

13.11 CONCLUSIONS

Fresh fruits and vegetables could be treated with different types of postharvest treatments. The treatments should be practically viable and sustainable to ensure the effective preservation and maintenance of the characteristic quality of fruits and vegetables. For this purpose, different types of nanoemulsions which could be efficiently applied on fruits and vegetables have been proven effective and ecofriendly. Nanoemulsions could be used for the enhancement of texture and as antimicrobial/antibrowning agents. Nanoemulsions contain suitable barrier, optical, microstructural and mechanical properties which are considered indispensable for the improvement of the visual quality and effective postharvest storage of fresh fruits and vegetables. Future work should focus on the structural stability and further safety regulations of nanoemulsions. The incorporation of some effective and ecofriendly antimicrobials should be further explored. In addition, some health-associated compounds could also be incorporated into nanoemulsions, and this aspect must be further investigated.

REFERENCES

Abdalla, G., Mussagy, C. U., Brasil, G. S. A. P., Scontri, M., da Silva Sasaki, J. C., Su, Y., et al. (2023). Eco-sustainable coatings based on chitosan, pectin, and lemon essential oil nanoemulsion and their effect on strawberry preservation. *International Journal of Biological Macromolecules*, *249*, 126016. https://doi.org/10.1016/j.ijbiomac.2023.126016

Acevedo-Fani, A., Salvia-Trujillo, L., Rojas-Graü, M. A., Martín-Belloso, O. (2015). Edible films from essential-oil-loaded nanoemulsions: Physicochemical characterization and antimicrobial properties. *Food Hydrocolloids*, *47*, 168–177. https://doi.org/10.1016/j.foodhyd.2015.01.032

Ali, S., Ullah, M. A., Nawaz, A., Naz, S., Shah, A. A., Gohari, G., Khaliq, G., Razzaq, K. (2022b). Carboxymethyl cellulose coating regulates cell wall polysaccharides disassembly and delays ripening of harvested banana fruit. *Postharvest Biology and Technology*, *191*, 111978. https://doi.org/10.1016/j.postharvbio.2022.111978

Ali, S., Ishtiaq, S., Nawaz, A., Naz, S., Ejaz, S., Haider, M. W., Shah, A.A., Ali, M.M., Javad, S. (2024). Layer by layer application of chitosan and carboxymethyl cellulose coatings delays ripening of mango fruit by suppressing cell wall polysaccharides disassembly. *International Journal of Biological Macromolecules*, 128429. https://doi.org/10.1016/j.ijbiomac.2023.128429

Ali, S., Zahid, N., Nawaz, A., Naz, S., Ejaz, S., Ullah, S., Siddiq, B. (2022a). Tragacanth gum coating suppresses the disassembly of cell wall polysaccharides and delays softening of harvested mango (*Mangifera indica* L.) fruit. *International Journal of Biological Macromolecules*, *222*, 521–532. https://doi.org/10.1016/j.ijbiomac.2022.09.159

Al-Tayyar, N. A., Youssef, A. M., Al-Hindi, R. R. (2020). Edible coatings and antimicrobial nanoemulsions for enhancing shelf life and reducing foodborne pathogens of fruits and vegetables: A review. *Sustainable Materials and Technologies*, *26*, e00215. https://doi.org/10.1016/j.susmat.2020.e00215

Alves-Silva, J. M., dos Santos, S. M. D., Pintado, M. E., Pérez-Álvarez, J. A., Fernández-López, J., Viuda-Martos, M. (2013). Chemical composition and in vitro antimicrobial, antifungal and antioxidant properties of essential oils obtained from some herbs widely used in Portugal. *Food Control*, *32*(2), 371–378.

Amiri, S., Rezazad Bari, L., Malekzadeh, S., Amiri, S., Mostashari, P., Ahmadi Gheshlagh, P. (2022). Effect of *Aloe vera* gel-based active coating incorporated with catechin nanoemulsion and calcium chloride on postharvest quality of fresh strawberry fruit. *Journal of Food Processing and Preservation*, *46*(10), e15960. https://doi.org/10.1111/jfpp.15960

Arnon-Rips, H., Poverenov, E. (2018). Improving food products' quality and storability by using Layer by Layer edible coatings. *Trends in Food Science & Technology*, *75*, 81–92.

Arroyo, B. J., Bezerra, A. C., Oliveira, L. L., Arroyo, S. J., de Melo, E. A., Santos, A. M. P. (2020). Antimicrobial active edible coating of alginate and chitosan add ZnO nanoparticles applied in guavas (*Psidium guajava* L.). *Food Chemistry*, *309*, 125566.

Artiga-Artigas, M., Acevedo-Fani, A., Martín-Belloso, O. (2017). Improving the shelf life of low-fat cut cheese using nanoemulsion-based edible coatings containing oregano essential oil and mandarin fiber. *Food Control*, *76*, 1–12.

Atarés, L., Chiralt, A. (2016). Essential oils as additives in biodegradable films and coatings for active food packaging. *Trends in Food Science & Technology*, *48*, 51–62. https://doi.org/10.1016/j.tifs.2015.12.001

Atieno, L., Owino, W., Ateka, E. M., Ambuko, J. (2019). Influence of coating application methods on the postharvest quality of cassava. International Journal of Food Science, 2148914, https://doi.org/10.1155/2019/2148914

Bai, L., Huan, S., Gu, J., McClements, D. J. (2016). Fabrication of oil-in-water nanoemulsions by dual-channel microfluidization using natural emulsifiers: Saponins, phospholipids, proteins, and polysaccharides. *Food Hydrocolloids*, *61*, 703–711.

Bonferoni, M. C., Sandri, G., Rossi, S., Usai, D., Liakos, I., Garzoni, A., Fiamma, M., Zanetti, S., Athaassiou, A., Caramella, F., Ferrari, F. (2017). A novel ionic amphiphilic chitosan derivative as a stabilizer of nanoemulsions: Improvement of antimicrobial activity of *Cymbopogon citratus* essential oil. *Colloids and Surfaces B: Biointerfaces*, *152*, 385–392.

Chaudhary, S., Kumar, S., Kumar, V., Sharma, R. (2020). Chitosan nanoemulsions as advanced edible coatings for fruits and vegetables: Composition, fabrication and developments in last decade. *International Journal of Biological Macromolecules*, *152*, 154–170. https://doi.org/10.1016/j.ijbiomac.2020.02.276

de Jesús Salas-Méndez, E., Vicente, A., Pinheiro, A. C., Ballesteros, L. F., Silva, P., Rodríguez-García, R., et al. (2019). Application of edible nanolaminate coatings with antimicrobial extract of *Flourensia cernua* to extend the shelf-life of tomato (*Solanum lycopersicum* L.) fruit. *Postharvest Biology and Technology*, *150*, 19–27.

de Oliveira Filho, J. G., Bezerra, C.C.D.O.N., Albiero, B. R., Oldoni, F. C. A., Miranda, M., Egea, M. B., de Azeredo, H.M.C., Ferreira, M. D. (2020). New approach in the development of edible films: The use of carnauba wax micro-or nanoemulsions in arrowroot starch-based films. *Food Packaging and Shelf Life*, *26*, 100589.

de Oliveira Filho, J. G. D., Duarte, L. G., Silva, Y. B., Milan, E. P., Santos, H. V., Moura, T. C., . . . Ferreira, M. D. (2023). Novel Approach for Improving Papaya Fruit Storage with Carnauba Wax Nanoemulsion in Combination with *Syzigium aromaticum* and *Mentha spicata* Essential Oils. *Coatings*, *13*(5), 847. https://doi.org/10.3390/coatings13050847

de Oliveira Filho, J. G. D., Silva, G. D. C., Oldoni, F. C. A., Miranda, M., Florencio, C., Oliveira, R. M. D. D., Gomes, M.D.P., Ferreira, M. D. (2022). Edible coating based on carnauba wax nanoemulsion and *Cymbopogon martinii* essential oil on papaya postharvest preservation. *Coatings*, *12*(11), 1700. https://doi.org/10.3390/coatings12111700

de Oliveira Filho, J. G., Miranda, M., Ferreira, M. D., Plotto, A. (2021). Nanoemulsions as edible coatings: A potential strategy for fresh fruits and vegetables preservation. *Foods*, *10*(10), 2438.

de Godoi, S. N., Quatrin, P. M., Sagrillo, M. R., Nascimento, K., Wagner, R., Klein, B., Santos, R.C.V., Ourique, A.F. (2017). Evaluation of stability and in vitro security of nanoemulsions containing *Eucalyptus globulus* oil. *BioMed Research International*, *2017*. 2723418. doi: 10.1155/2017/2723418

Dehghani, S., Hosseini, S. V., Regenstein, J. M. (2018). Edible films and coatings in seafood preservation: A review. *Food Chemistry*, *240*, 505–513.

Dhanapal, A., Sasikala, P., Rajamani, L., Kavitha, V., Yazhini, G., Banu, M. S. (2012). Edible films from polysaccharides. *Food Science and Quality Management*, *3*(0), 9.

Dhandevi, P. E. M., Jeewon, R. (2015). Fruit and vegetable intake: Benefits and progress of nutrition education interventions-narrative review article. *Iranian Journal of Public Health*, *44*(10), 1309.

Dhumal, C.V., Sarkar, P. (2018). Composite edible films and coatings from food-grade biopolymers. *Journal of Food Science and Technology*, *55*, 4369–4383.

Donsı, F., Cuomo, A., Marchese, E. Ferrari, G. (2014). Infusion of essential oils for food stabilization: Unraveling the role of nanoemulsion-based delivery systems on mass transfer and antimicrobial activity. *Innovative Food Science & Emerging Technologies*, *22*, 212–220.

Donsì, F., Marchese, E., Maresca, P., Pataro, G., Vu, K. D., Salmieri, S., Lacroix, M., Ferrari, G. (2015). Green beans preservation by combination of a modified chitosan based-coating containing nanoemulsion of mandarin essential oil with high pressure or pulsed light processing. *Postharvest Biology and Technology*, *106*, 21–32.

Duan, B., Wang, Z., Reymick, O. O., Zhang, M., Chen, Y., Meng, K., Tao, N. (2023). Wax treatment delays the coloration of postharvest citrus fruit by retarding the carotenoid biosynthesis pathway. *Scientia Horticulturae*, *321*, 112379. https://doi.org/10.1016/j.scienta.2023.112379

Eddin, A. S., Ibrahim, S. A., Tahergorabi, R. (2019). Egg quality and safety with an overview of edible coating application for egg preservation. *Food Chemistry*, *296*, 29–39.

Eshghi, S., Hashemi, M., Mohammadi, A., Badie, F., Mohammad hosseini, Z. Ahmadi, K. (2013). Effect of nano-emulsion coating containing chitosan on storability and qualitative characteristics of strawberries after picking. *Iranian Journal of Nutrition Sciences & Food Technology*, *8*, 9–19.

Galus, S., Kadzińska, J. (2015). Food applications of emulsion-based edible films and coatings. *Trends in Food Science & Technology*, *45*(2), 273–283.

Garcıa-Betanzos, C.I., Hernandez-S anchez, H., Bernal-Couoh, T.F., Quintanar-Guerrero, D. de la Luz Zambrano-Zaragoza, M. (2017). Physicochemical, total phenols and pectin methylesterase changes on quality maintenance on guava fruit (*Psidium guajava* L.) coated with candeuba wax solid lipid nanoparticles-xanthan gum. *Food Research International*, *101*, 218–227.

García-Márquez, E., Higuera-Ciapara, I., Espinosa-Andrews, H. (2017). Design of fish oil-in-water nanoemulsion by microfluidization. *Innovative Food Science & Emerging Technologies*, *40*, 87–91.

Gundewadi, G., Rudra, S. G., Sarkar, D. J., Singh, D. (2018). Nanoemulsion based alginate organic coating for shelf life extension of okra. *Food Packaging and Shelf Life*, *18*, 1–12.

Guo, Q., Qing, Y., Qiang, L., Du, G., Shi, K., Tang, J., Yan, X., Chen, H., Yue, Y., He, Y., Yuan, Y., Yue, T. (2023). Improving microbiological and physicochemical properties of fresh-cut apples using carvacrol emulsions. *Food Bioscience*, *52*, 102450. https://doi.org/10.1016/j.fbio.2023.102450

Hassan, B., Chatha, S. A. S., Hussain, A. I., Zia, K. M., Akhtar, N. (2018). Recent advances on polysaccharides, lipids and protein based edible films and coatings: A review. *International Journal of Biological Macromolecules*, *109*, 1095–1107.

Horison, R., Sulaiman, F. O., Alfredo, D., Wardana, A. A. (2019). Physical characteristics of nanoemulsion from chitosan/nutmeg seed oil and evaluation of its coating against microbial growth on strawberry. *Food Research*, *3*(6), 821–827.

Javanmardi, Z., Saba, M. K., Nourbakhsh, H., Amini, J. (2023). Efficiency of nanoemulsion of essential oils to control Botrytis cinerea on strawberry surface and prolong fruit shelf life. *International Journal of Food Microbiology*, *384*, 109979. https://doi.org/10.1016/j.ijfoodmicro.2022.109979

Jin, W., Xu, W., Liang, H., Li, Y., Liu, S., Li, B. (2016). Nanoemulsions for food: Properties, production, characterization, and applications. In *Emulsions* (pp. 1–36). Academic Press.

Jo, W. S., Song, H. Y., Song, N. B., Lee, J. H., Min, S. C., Song, K. B. (2014). Quality and microbial safety of 'Fuji'apples coated with carnauba-shellac wax containing lemongrass oil. *LWT-Food Science and Technology*, *55*(2), 490–497.

Kaur, K., Kumar, R., Goel, S., Uppal, S., Bhatia, A., Mehta, S. K. (2017). Physiochemical and cytotoxicity study of TPGS stabilized nanoemulsion designed by ultrasonication method. *Ultrasonics Sonochemistry*, *34*, 173–182.

Kim, I. H., Lee, H., Kim, J. E., Song, K. B., Lee, Y. S., Chung, D. S., Min, S. C. (2013). Plum coatings of lemongrass oil-incorporating carnauba wax-based nanoemulsion. *Journal of Food Science*, *78*(10), E1551–E1559.

Kim, I. H., Oh, Y. A., Lee, H., Song, K. B., Min, S. C. (2014). Grape berry coatings of lemongrass oil-incorporating nanoemulsion. *LWT-Food Science and Technology*, *58*(1), 1–10.

Klangmuang, P., Sothornvit, R. (2018). Active coating from hydroxypropyl methylcellulose-based nanocomposite incorporated with Thai essential oils on mango (cv. Namdokmai Sithong). *Food Bioscience*, *23*, 9–15.

Kong, I., Degraeve, P., Pui, L. P. (2022). Polysaccharide-based edible films incorporated with essential oil nanoemulsions: Physico-chemical, mechanical properties and its application in food preservation-A review. *Foods*, *11*(4), 555. https://doi.org/10.3390/foods11040555

Lohith Kumar, D.H., Sarkar, P. (2018). Encapsulation of bioactive compounds using nanoemulsions. *Environmental Chemistry Letters*, *16*, 59–70. https://doi.org/10.1007/s10311-017-0663-x

López-Monterrubio, D. I., Lobato-Calleros, C., Vernon-Carter, E. J., Alvarez-Ramirez, J. (2021). Influence of 1*β*-carotene concentration on the physico-chemical properties, degradation and antioxidant activity of nanoemulsions stabilized by whey protein hydrolyzate-pectin soluble complexes. *LWT-Food Science and Technology*, *143*, 111148. https://doi.org/10.1016/j.lwt.2021.111148

Louis, E., Villalobos-Carvajal, R., Reyes-Parra, J., Jara-Quijada, E., Ruiz, C., Andrades, P., Gacitua, J., Beldarraín-Iznaga, T. (2021). Preservation of mushrooms (*Agaricus bisporus*) by an alginate-based-coating containing a cinnamaldehyde essential oil nanoemulsion. *Food Packaging and Shelf Life*, *28*, 100662. https://doi.org/10.1016/j.fpsl.2021.100662

Luciano, W. A., Pimentel, T. C., Bezerril, F. F., Barão, C. E., Marcolino, V. A., Carvalho, R. D. S. F., et al. (2023). Effect of citral nanoemulsion on the inactivation of Listeria monocytogenes and sensory properties of fresh-cut melon and papaya during storage. *International Journal of Food Microbiology*, *384*, 109959. https://doi.org/10.1016/j.ijfoodmicro.2022.109959

Luo, X., Zhou, Y., Bai, L., Liu, F., Deng, Y., McClements, D. J. (2017). Fabrication of β-carotene nanoemulsion-based delivery systems using dual-channel microfluidization: Physical and chemical stability. *Journal of Colloid and Interface Science*, *490*, 328–335.

Mahfoudhi, N., Ksouri, R., Hamdi, S. (2016). Nanoemulsions as potential delivery systems for bioactive compounds in food systems: Preparation, characterization, and applications in food industry. In *Emulsions* (pp. 365–403). Academic Press.

Manzoor, S., Gull, A., Wani, S. M., Ganaie, T. A., Masoodi, F. A., Bashir, K., Malik, A.R., Dar, B. N. (2021). Improving the shelf life of fresh cut kiwi using nanoemulsion coatings with antioxidant and antimicrobial agents. *Food Bioscience*, *41*, 101015. https://doi.org/10.1016/j.fbio.2021.101015

Marchese, E., D'onofrio, N., Balestrieri, M. L., Castaldo, D., Ferrari, G., Donsì, F. (2020). Bergamot essential oil nanoemulsions: Antimicrobial and cytotoxic activity. *Zeitschrift für Naturforschung C*, *75*(7–8), 279–290.

Maringgal, B., Hashim, N., Tawakkal, I. S. M. A., Mohamed, M. T. M. (2020). Recent advance in edible coating and its effect on fresh/fresh-cut fruits quality. *Trends in Food Science & Technology*, *96*, 253–267.

McClements, D. J. (2012). Nanoemulsions versus microemulsions: terminology, differences, and similarities. *Soft Matter*, *8*(6), 1719–1729.

McClements, D. J. (2021). Advances in edible nanoemulsions: Digestion, bioavailability, and potential toxicity. *Progress in Lipid Research*, *81*, 101081.

McClements, D. J., Rao, J. (2011). Food-grade nanoemulsions: formulation, fabrication, properties, performance, biological fate, and potential toxicity. *Critical Reviews in Food Science and Nutrition*, *51*(4), 285–330. https://doi.org/10.1080/10408398.2011.559558

McClements, D. J., Xiao, H. (2012). Potential biological fate of ingested nanoemulsions: influence of particle characteristics. *Food & Function*, *3*(3), 202–220.

Mendoza-Wilson, A. M., Balandrán-Quintana, R. R., Valdés-Covarrubias, M. Á., Cabellos, J. L. (2022). Potential of quercetin in combination with antioxidants of different polarity incorporated in oil-in-water nanoemulsions to control enzymatic browning of apples. *Journal of Molecular Structure*, *1254*, 132372. https://doi.org/10.1016/j.molstruc.2022.132372

Miranda, M., Marilene De Mori, M. R., Spricigo, P. C., Pilon, L., Mitsuyuki, M. C., Correa, D. S., Ferreira, M. D. (2022a). Carnauba wax nanoemulsion applied as an edible coating on fresh tomato for postharvest quality evaluation. *Heliyon*, *8*(7). https://doi.org/10.1016/j.heliyon.2022.e09803

Miranda, M., Gozalbo, A. M., Sun, X., Plotto, A., Bai, J., de Assis, O., Baldwin, E. (2019). Effect of mono and bilayer of carnauba wax based nano-emulsion and HPMC coatings on post-harvest quality of 'redtainung' papaya. In *Proceedings of the Embrapa Instrumentação-Artigo em anais de congresso (ALICE)* (pp. 3–5). São Carlos.

Miranda, M., Sun, X., Ference, C., Plotto, A., Bai, J., Wood, D., Assis, O.B.G., Ferreira, M.D., Baldwin, E. (2021). Nano-and micro-carnauba wax emulsions versus shellac protective coatings on postharvest citrus quality. *Journal of the American Society for Horticultural Science*, *146*(1), 40–49.

Miranda, M., Sun, X., Marín, A., Dos Santos, L. C., Plotto, A., Bai, J., Assis, O.B.G., Ferreira, M.D., Baldwin, E. (2022b). Nano-and micro-sized carnauba wax emulsions-based coatings incorporated with ginger essential oil and hydroxypropyl methylcellulose on papaya: Preservation of quality and delay of post-harvest fruit decay. *Food Chemistry: X*, *13*, 100249. https://doi.org/10.1016/j.fochx.2022.100249

Mohamed, S. A., El-Sakhawy, M., El-Sakhawy, M. A. M. (2020). Polysaccharides, protein and lipid-based natural edible films in food packaging: A review. *Carbohydrate Polymers*, *238*, 116178.

Mossa, A. T. H., Mohafrash, S. M., Ziedan, E. S. H., Abdelsalam, I. S., Sahab, A. F. (2021). Development of eco-friendly nanoemulsions of some natural oils and evaluating of its efficiency against postharvest fruit rot fungi of cucumber. *Industrial Crops and Products*, *159*, 113049. https://doi.org/10.1016/j.indcrop.2020.113049

Mushtaq, A., Wani, S. M., Malik, A. R., Gull, A., Ramniwas, S., Nayik, G. A., Ercisli, S., Marc, R.A., Ullah, R., Bari, A. (2023). Recent insights into Nanoemulsions: Their preparation, properties and applications. *Food Chemistry: X*, 100684. https://doi.org/10.1016/j.fochx.2023.100684

Nash, J. J., Erk, K. A. (2017). Stability and interfacial viscoelasticity of oil-water nanoemulsions stabilized by soy lecithin and Tween 20 for the encapsulation of bioactive carvacrol. *Colloids and Surfaces A: Physicochemical and Engineering Aspects*, *517*, 1–11.

Oh, Y. A., Oh, Y. J., Song, A. Y., Won, J. S., Song, K. B., Min, S. C. (2017). Comparison of effectiveness of edible coatings using emulsions containing lemongrass oil of different size droplets on grape berry safety and preservation. *LWT*, *75*, 742–750.

Ohashi, T. L., Pilon, L., Spricigo, P. C., Miranda, M., Corrêa, D. S., Ferreira, M. D. (2015). Postharvest quality of 'golden' Papayas (*Carica papaya*.) coated with carnauba wax nanoemulsions. *Revista Iberoamericana de Tecnología Postcosecha*, *16*(2), 199–209.

Oyom, W., Zhang, Z., Bi, Y., Tahergorabi, R. (2022). Application of starch-based coatings incorporated with antimicrobial agents for preservation of fruits and vegetables: A review. *Progress in Organic Coatings*, *166*, 106800. https://doi.org/10.1016/j.porgcoat.2022.106800

Pandey, V. K., Islam, R. U., Shams, R., Dar, A. H. (2022). A comprehensive review on the application of essential oils as bioactive compounds in Nano-emulsion based edible coatings of fruits and vegetables. *Applied Food Research*, *2*(1), 100042.

Panwar, A., Kumar, V., Dhiman, A., Thakur, P., Sharma, V., Sharma, A., Kumar, S. (2024). Nanoemulsion based edible coatings for quality retention of fruits and vegetables-decoding the basics and advancements in last decade. *Environmental Research*, *240*, 117450.

Paudel, S. K., Bhargava, K., Kotturi, H. (2019). Antimicrobial activity of cinnamon oil nanoemulsion against Listeria monocytogenes and Salmonella spp. on melons. *LWT*, *111*, 682–687.

Poverenov, E., Danino, S., Horev, B., Granit, R., Vinokur, Y., Rodov, V. (2014). Layer-by-layer electrostatic deposition of edible coating on fresh cut melon model: Anticipated and unexpected effects of alginate–chitosan combination. *Food and Bioprocess Technology*, *7*(5), 1424–1432.

Prakash, A., Baskaran, R., Vadivel, V. (2020). Citral nanoemulsion incorporated edible coating to extend the shelf life of fresh cut pineapples. *LWT*, *118*, 108851.

Radi, M., Akhavan-Darabi, S., Akhavan, H. R., Amiri, S. (2018). The use of orange peel essential oil microemulsion and nanoemulsion in pectin-based coating to extend the shelf life of fresh-cut orange. *Journal of Food Processing and Preservation*, *42*(2), e13441.

Rashid, F., Ahmed, Z., Ameer, K., Amir, R. M., Khattak, M. (2020). Optimization of polysaccharides-based nanoemulsion using response surface methodology and application to improve postharvest storage of apple (*Malus domestica*). *Journal of Food Measurement and Characterization*, *14*, 2676–2688.

Robledo, N., Lopez, L., Bunger, A., Tapia, C. Abugoch, L. (2018). Effects of antimicrobial edible coating of thymol nanoemulsion/quinoa protein/chitosan on the safety, sensorial properties, and quality of refrigerated strawberries (*Fragaria × ananassa*) under commercial storage environment. *Food and Bioprocess Technology*, *11*, 1566–1574.

Rohasmizah, H., Azizah, M. (2022). Pectin-based edible coatings and nanoemulsion for the preservation of fruits and vegetables: A review. *Applied Food Research*, 100221. https://doi.org/10.1016/j.afres.2022.100221

Salvia-Trujillo, L., Rojas-Graü, M. A., Soliva-Fortuny, R., Martín-Belloso, O. (2015). Use of antimicrobial nanoemulsions as edible coatings: Impact on safety and quality attributes of fresh-cut Fuji apples. *Postharvest Biology and Technology*, *105*, 8–16.

Salvia-Trujillo, L., Soliva-Fortuny, R., Rojas-Graü, M. A., McClements, D. J., Martín-Belloso, O. (2017). Edible nanoemulsions as carriers of active ingredients: A review. *Annual Review of Food Science and Technology*, *8*, 439–466.

Sánchez-González, L., Vargas, M., González-Martínez, C., Chiralt, A., Chafer, M. (2011). Use of essential oils in bioactive edible coatings: a review. *Food Engineering Reviews*, *3*(1), 1–16.

Senturk Parreidt, T., Schmid, M., Müller, K. (2018). Effect of dipping and vacuum impregnation coating techniques with alginate-based coating on physical quality parameters of cantaloupe melon. *Journal of Food Science*, *83*(4), 929–936.

Severino, R., Ferrari, G., Vu, K. D., Donsì, F., Salmieri, S., Lacroix, M. (2015). Antimicrobial effects of modified chitosan based coating containing nanoemulsion of essential oils, modified atmosphere packaging and gamma irradiation against *Escherichia coli* O157:H7 and *Salmonella Typhimurium* on green beans. *Food Control*, *50*, 215–222. 10.1016/j.foodcont.2014.08.029.

Severino, R., Vu, K.D., Donsı, F., Salmieri, S., Ferrari, G. Lacroix, M. (2014). Antibacterial and physical effects of modified chitosan based-coating containing nanoemulsion of mandarin essential oil and three non-thermal treatments against *Listeria innocua* in green beans. *International Journal of Food Microbiology*, *191*, 82–88.

Singh, Y., Meher, J. G., Raval, K., Khan, F. A., Chaurasia, M., Jain, N. K., Chourasia, M. K. (2017). Nanoemulsion: Concepts, development and applications in drug delivery. *Journal of Controlled Release*, *252*, 28–49. https://doi.org/10.1016/j.jconrel.2017.03.008

Sun, X., Baldwin, E., Ritenour, M., Plotto, A., Bai, J. (2015). Evaluation of natural colorants and their application on citrus fruit as alternatives to Citrus Red No. 2. *HortScience*, *50*(9), 1353–1357.

Thakur, R., Pristijono, P., Scarlett, C. J., Bowyer, M., Singh, S. P., Vuong, Q. V. (2019). Starch-based films: Major factors affecting their properties. *International Journal of Biological Macromolecules*, *132*, 1079–1089.

Tripathi, A. D., Sharma, R., Agarwal, A., Haleem, D. R. (2021). Nanoemulsions based edible coatings with potential food applications. *International Journal of Biobased Plastics*, *3*(1), 112–125. https://doi.org/10.1080/24759651.2021.1875615

Umaraw, P., Verma, A. K. (2017). Comprehensive review on application of edible film on meat and meat products: an eco-friendly approach. *Critical Reviews in Food Science and Nutrition*, *57*(6), 1270–1279. https://doi.org/10.1080/10408398.2014.986563

Veiga, M., Costa, E. M., Silva, S., Pintado, M. (2020). Impact of plant extracts upon human health: A review. *Critical Reviews in Food Science and Nutrition*, *60*(5), 873–886. https://doi.org/10.1080/10408398.2018.1540969

Versino, F., Lopez, O. V., Garcia, M. A., Zaritzky, N. E. (2016). Starch-based films and food coatings: An overview. *Starch-Stärke*, *68*(11–12), 1026–1037. https://doi.org/10.1002/star.201600095

Vildan, E., İsmail, T., Selman, T. (2020). The effect of edible coatings on physical and chemical characteristics of fruit bars. *Journal of Food Measurement & Characterization*, *14*(3), 1775–1783.

Viuda-Martos, M., El Gendy, A. E. N. G., Sendra, E., Fernandez-Lopez, J., Abd El Razik, K. A., Omer, E. A., Perez-Alvarez, J. A. (2010). Chemical composition and antioxidant and anti-Listeria activities of essential oils obtained from some Egyptian plants. *Journal of Agricultural and Food Chemistry*, *58*(16), 9063–9070.

Wani, T. A., Masoodi, F. A., Jafari, S. M., McClements, D. J. (2018). Safety of nanoemulsions and their regulatory status. In *Nanoemulsions* (pp. 613–628). Academic Press.

Yang, R., Miao, J., Shen, Y., Cai, N., Wan, C., Zou, L., Chen, C., Chen, J. (2021). Antifungal effect of cinnamaldehyde, eugenol and carvacrol nanoemulsion against *Penicillium digitatum* and application in postharvest preservation of citrus fruit. *LWT*, *141*, 110924. https://doi.org/10.1016/j.lwt.2021.110924

Zambrano-Zaragoza, M.L., Mercado-Silva, E., Ramirez-Zamorano, P., Cornejo-Villegas, M.A., Gutierrez-Cortez, E. Quintanar-Guerrero, D. (2013). Use of solid lipid nanoparticles (SLNs) in edible coatings to increase guava (*Psidium guajava* L.) shelf-life. *Food Research International*, *51*, 946–953.

Zambrano-Zaragoza, M.L., Mercado-Silva, E., Del Real, L.A., Gutierrez-Cortez, E., Cornejo-Villegas, M.A. Quintanar-Guerrero, D. (2014). The effect of nano-coatings with α-tocopherol and xanthan gum on shelf-life and browning index of fresh-cut "Red Delicious" apples. *Innovative Food Science & Emerging Technologies*, *22*, 188–196.

Zhang, J., Hao, Y., Lu, H., Li, P., Chen, J., Shi, Z., Xie, Y., Mo, H., Hu, L. (2022). Nano-Thymol emulsion inhibits *Botrytis cinerea* to control postharvest gray mold on tomato fruit. *Agronomy*, *12*(12), 2973. https://doi.org/10.3390/agronomy12122973

Zhang, R., Zhang, Z., McClements, D. J. (2020). Nanoemulsions: An emerging platform for increasing the efficacy of nutraceuticals in foods. *Colloids and Surfaces B: Biointerfaces*, *194*, 111202.

14 Nano-Packaging Technology for Conserving the Quality of Fruits and Vegetables

Leila Taghipour* and Pedram Assar
*Corresponding Author: L_taghipoor@yahoo.com; Leilataghipour@jahromu.ac.ir

14.1 INTRODUCTION

The problematic encounter with fruit and vegetable waste and loss, as a part of the food waste dilemma, is increasingly gaining technical momentum within the global scope of food security (Mokrane et al., 2023). The length of time that harvested fruits and vegetables remain fresh is mainly determined by factors such as the type of fruit or vegetable, the maturity of the produce, the storage temperature, the amount of water lost, and the firmness of its texture. These factors work together in correlation with ethylene production, physical damage during transportation, and unforeseen incidents during storage. More importantly, microorganisms and pests can exacerbate postharvest losses, primarily due to the fragile nature of horticultural commodities and their soft, fleshy texture. This results in shorter shelf lives during the expected storage duration (Hosseini et al., 2023; Sau et al., 2021; Videira-Quintela et al., 2022).

As scientific literature has grown, so too has awareness of the attention required by horticultural commodities during pre- and postharvest stages. Sometimes, experts and people involved in postharvest activities can feel inundated by the vast array of postharvest treatments and strategies that claim to optimize the shelf life and quality maintenance of horticultural produce. This has led to various efforts aimed at meeting these challenges through optimal trial-and-error initiatives (Eshghi et al., 2022). It can be challenging to determine where to start when it comes to the practical implementation of postharvest management. The tools available can either make postharvest conduct simpler or more complicated, depending on the range of tools used (Schudel et al., 2023).

Besides different physical and chemical postharvest treatments on fruits and vegetables, food packaging (FP) is essential to preserve the quality and safety of horticultural commodities during transportation and storage until they reach the consumer. Currently, non-degradable materials used in food packaging pose a significant global environmental problem. To address this, biomaterials have successfully been utilized to create edible and eco-friendly films, aiming to increase food quality, prolong potential shelf life, and decrease waste from packaging materials (Tharanathan, 2003). Nevertheless, the limited use of these edible and biodegradable polymeric materials is hindered by performance issues (e.g., brittleness, poor barrier properties against gases and moisture), challenges of processing (e.g., lower heat distortion temperatures), and concerns of costs.

The packaging industry has undergone significant evolution, with advancements in material science and technology driving innovation. Alongside this, consumer demand for convenience, sustainability, and shelf-life extension has played a crucial role in shaping food packaging practices. Notably, nanotechnology has evolved as a major technological breakthrough in twenty-first-century FP development. Recent progress in nanotechnology over the past two decades has enabled researchers to effectively handle postharvest issues, particularly in relation to fruits and vegetables. These advancements provide not only advantageous options but are also increasingly recognized as critical accomplishments for preventing postharvest losses (Arabpoor et al., 2021; Dubey et al., 2023).

Nanotechnology involves manipulating and creating different nanoparticles at the nanometer scale (< 100 nm), which holds immense scientific and commercial importance. This advancement has paved the way for the development of nanomaterials, which are used in sustainable food, fruit, and vegetable packaging. These nanomaterials are highly valued by manufacturers owing to their eco-friendly nature and enhanced characteristics, such as improved barrier properties, mechanical strengths, and antimicrobial activity. By providing an initial line of defense against spoilage caused by environmental factors and storage conditions, these features contribute to the preservation of food quality (Videira-Quintela et al., 2022).

DOI: 10.1201/9781003370376-19

It is imperative to state that although nanotechnology and many of the approaches related to using nanomaterials are still in the experimental stage, they have several potential uses in fruit and vegetable packaging. These uses include improving the potential shelf life of produce, enhancing the safety of produce, and reducing waste. Some specific methods being explored for each of these areas include:

14.1.1 Improving Shelf Life

- Use of nano-coatings to reduce respiratory rate and desiccation in fruits and vegetables (Ashfaq et al., 2022).
- Incorporation of selected nanoparticle contents into packaging material to develop a better barrier nanocomposite against different gases, moisture, and certain other external factors which can cause deterioration of fresh produce (Sharma et al., 2017a).
- Intelligent packaging development which can detect and efficiently respond to certain changes in temperature fluctuations, humidity, and several other associated environmental conditions (Pereda et al., 2019).

14.1.2 Enhancing Food Safety

- Uses of nanoparticles to form antimicrobial coatings/films that can kill or inhibit the growth of bacteria, viruses, and some detrimental pathogens on the surface of produce (Ashfaq et al., 2022).
- Nano-based intelligent packaging development encompassing appropriate nanosensors that can detect contaminations and certain toxin-based compounds in food packages (Sharma et al., 2017a).
- Producing nano-based intelligent packing systems with tags and/or labels consisting of nanoscale features that can identify and track food commodities throughout supply chain operations (Ashfaq et al., 2022).

14.1.3 Reducing Waste

- Nanotechnology applications can help to produce biodegradable packaging materials that can degrade more quickly, thus reducing potential waste occurrence (Pereda et al., 2019).
- Incorporation of nanoparticles into packaging may help to increase their mechanical strength and durability, thus reducing the probability of damage or leakage during transport and storage after harvest (Sharma et al., 2017a).
- Creation of nano-based active packaging systems that may also release selected preservatives or antioxidative constituents to help prolong fresh produce shelf life and limit spoilage under storage (Ashfaq et al., 2022).

This chapter provides comprehensive insights into the growing trends of nano-packaging technology for preserving horticultural produce. It also covers recent advancements in the application of nanomaterials and discusses various approaches for the conservation of quality and monitoring of horticultural produce, categorized as enhanced active packaging, nano reinforced packaging, and certain intelligent packaging types.

14.2 NANO-SCALED ORGANIC, INORGANIC, AND COMBINED MATERIALS IN THE PACKAGING OF HORTICULTURAL COMMODITIES

Based on the aforementioned information, fruit and vegetable nano-packaging has been increasingly explored owing to its potential to improve shelf life, ensure safety, and preserve the quality of fresh produce. Numerous reviews on aspects of nano-packaging material have recently been reported (Sharma et al., 2017a; Ashfaq et al., 2022).

Some of the most commonly used classes of nanomaterials and nanostructure-based components in the field include:

14.2.1 Bio-nanocomposites

As revealed in the review article by Vasile (2018), several polymer types are universally used in food packaging (FP) as substrates or matrices for the development of bioactive coating. These polymers include non-biodegradable synthetic polymer types, such as high-density polyethylenes, low-density polyethylenes, polypropylene, ethylene vinyl alcohol copolymer, polyethylene terephthalate, polyamides, polystyrenes, and degradable polymers such as polyhydroxyalkanoates and poly(hydroxybutyrate-co-hydroxyvalerate), poly(lactic acid), polycaprolactone, and polyvinyl alcohol, as well as biopolymers, such as proteins and polysaccharides. However, biopolymers face challenges related to water migration, resistance, and penetrability. Nonetheless, the said characteristics can be enhanced by incorporating reinforcing nanofillers, also known as nanoparticles. This makes it possible to make bio-nanocomposites that are engineered to have certain barrier properties, like better gas permeability, which increases the storage life of fresh produce.

14.2.2 Nanocellulose

This is a type of nanomaterial that is derived from plant-based cellulose fibers. Nanocellulose has unique mechanical properties and can be used to reinforce packaging materials, making them more resistant to physical damage and reducing the amount of oxygen that can permeate through the package (Ashfaq et al., 2022).

14.2.3 Nanoclay

It encompasses nanoparticles of clay, with montmorillonite (MMT) being the most commonly utilized type. MMT is a naturally occurring clay that is abundantly available and affordable. By incorporating nanoclay into polymer matrices, nanocomposites are developed, possessing improved mechanical, thermal, and barrier characteristics in contrast with conventional biocomposites. The achievement of uniform dispersions of most clay types within organic polymer types presents some challenges owing to their hydrophilic nature on the surface when applied (De Paiva et al., 2008). To address these challenges, organoclays have now emerged as a valuable and sustainable solution in the development of nanocomposite polymers, thus ensuring an essential application in terms of their cost-effectiveness. These are derived from easily accessible sources of natural origins and can also be manufactured in large-scale production facilities. For instance, organo MMT is created by replacing inorganic cations found in MMT with ammonium ions of certain organic types. The process therefore enhances compatibility among the MMT and organic polymers, thereby resulting in much more organized and stable arrangements of layers within their structures (Picard et al., 2007).

14.2.4 Chitosan Nanoparticles

Chitosan is considered an environmentally friendly, biodegradable, and nontoxic polysaccharide with suitable and efficient coating properties. It possesses exceptional characteristics, having an antimicrobial nature, which thus makes it an innovative and promising nanoparticle preference for the development of various types of edible packaging (Divya & Jisha, 2018). Incorporation of chitosan into some other biodegradable polymers, for instance, polylactic acid, further increases its gas and moisture barricade characteristics (Yanat & Schroën, 2021). The nanoparticles made from chitosan thus possess the potential to substantially increase the mechanical, physical, antimicrobial, and barrier attributes of the formed biocomposite films. These exert antimicrobial effects through chelation and ionic interactions with bacterial cell surfaces, impeding nutrient transport and causing cell death. Moreover, the hydrophobic nature of chitosan nanoparticles and the development of covalent and hydrogen bonds with biopolymers reduce moisture diffusion, enhancing moisture impermeability (Garavand et al., 2022).

14.2.5 Metal and Metal Oxide Nanoparticles

Different metal [gold (Au), silver (Ag), zinc (Zn)] and metal oxide [titanium dioxide (TiO_2), silicon oxide (SiO_2), zinc oxide (ZnO), and magnesium oxide (MgO)] nanoparticles have been studied for their use in nanoreinforced and active packages. Their

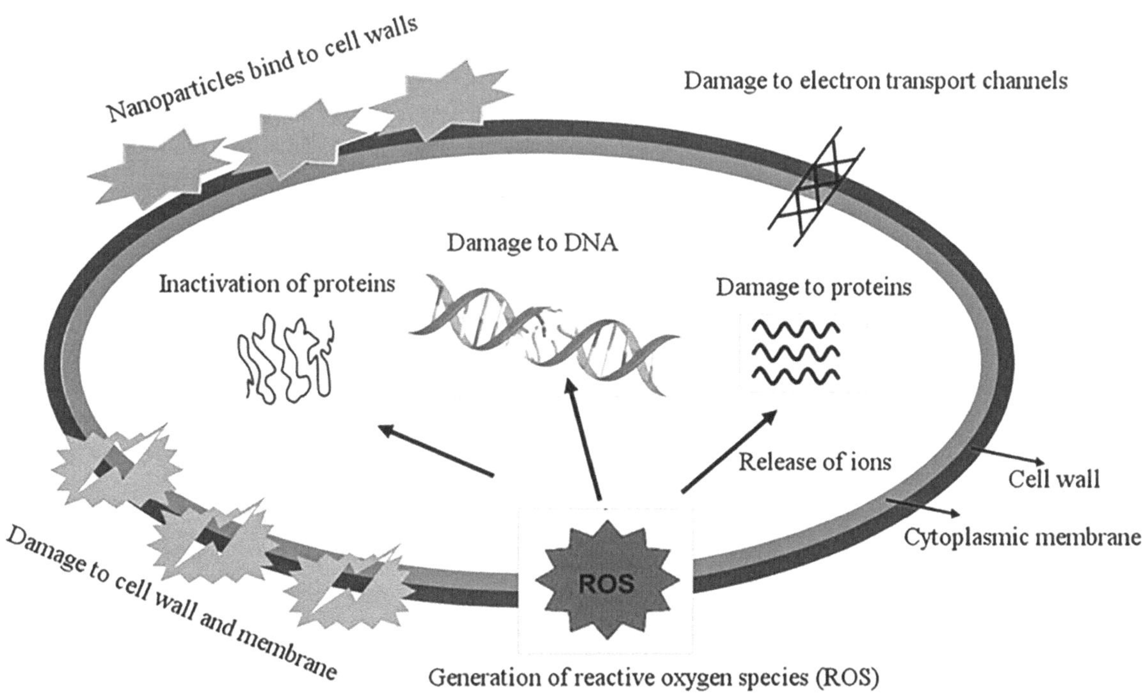

FIGURE 14.1 Mechanisms of action of nanoparticles on microbial cells (adapted from Ashfaq et al. 2022).

excellent antimicrobial properties can be attributed to diverse mechanisms, such as direct association with certain types of microbes (disrupting electron transfer or cell covering), cell component oxidation, and secondary metabolite generation that results in cell damage (Figure 14.1) (Sharma et al., 2017a).

14.2.6 Nanotubes of Carbon

Carbon nanotubes, as carbon allotropes, exist in a cylindrical shape with nanometer-scale diameters. They can be categorized into two types: multi-walled nanotubes and single-walled nanotubes consisting of concentric cylinders (Huang et al., 2015). These nanotubes are employed to enhance the antimicrobial and mechanical characteristics of packaging polymers (Rezić et al., 2017). Furthermore, they find utility in the development of smart sensors for monitoring atmospheric modifications and detecting food spoilage within intelligent packaging systems (Chaudhary et al., 2020; Zhu et al., 2017).

14.2.7 Nanoemulsions

Nanoemulsions are water-in-oil- or oil-in-water-based emulsions in which the droplet size is typically less than 100 nm. It is well reported that nanoemulsions can encapsulate (and later slowly release) several types of bioactive compounds, for instance vitamins, antioxidant constituents, and certain antimicrobials, which may also be combined with packaging materials to safeguard fresh produce such as fruit and vegetable commodities from spoilage and contamination (Aswathanarayan & Vittal, 2019).

14.2.8 Nanofabrication Methods

Nanofabrication is the process of constructing structures or devices with nanometer dimensions. The production of these nanomaterial types involves various intricate techniques and/or methods that require scientific insight and technological proficiency in general (Sardar et al., 2022).

The nanofabrication methods include bottom-up and top-down approaches (Figure 14.2). The top-down methods predominantly serve as the prevailing source of large-scale production of the said nanomaterials. Nanofabrication works by diminishing the size of bulk materials/constituents through various processes, such as nanolithography, milling, and precision engineering. By reducing the dimensions of target materials, their surface areas are substantially increased, resulting in numerous advantageous characteristics and enhanced overall functionalities. On the contrary, the bottom-up approach represents a relatively recent innovation that facilitates the creation of target nanostructures by assembling individual atoms or molecules. This process, commonly known as self-assembly, occurs when distinct molecular entities assemble and arrange themselves into larger functional structures. The self-assembly normally depends on intricately balanced forces of either repulsion or attraction among these constitutive building blocks (Pereda et al., 2019). A review on advances in techniques of nanofabrication has been published by Ashfaq et al. (2022), in which various aspects of nanofabrication were reviewed and discussed.

14.3 NANO PACKAGING APPROACHES FOR FRUITS AND VEGETABLES

The nanotechnology of polymers in food packaging is an emerging aspect that has the capacity to revolutionize the global food packaging industry. In this context, nanotechnology applications involve the development, manufacturing, processing, and modification of polymer nanocomposites that encompass certain types of nanoparticles or devices of nano-range (Sharma et al., 2017a). Polymer nanocomposites have modified, altered, and improved

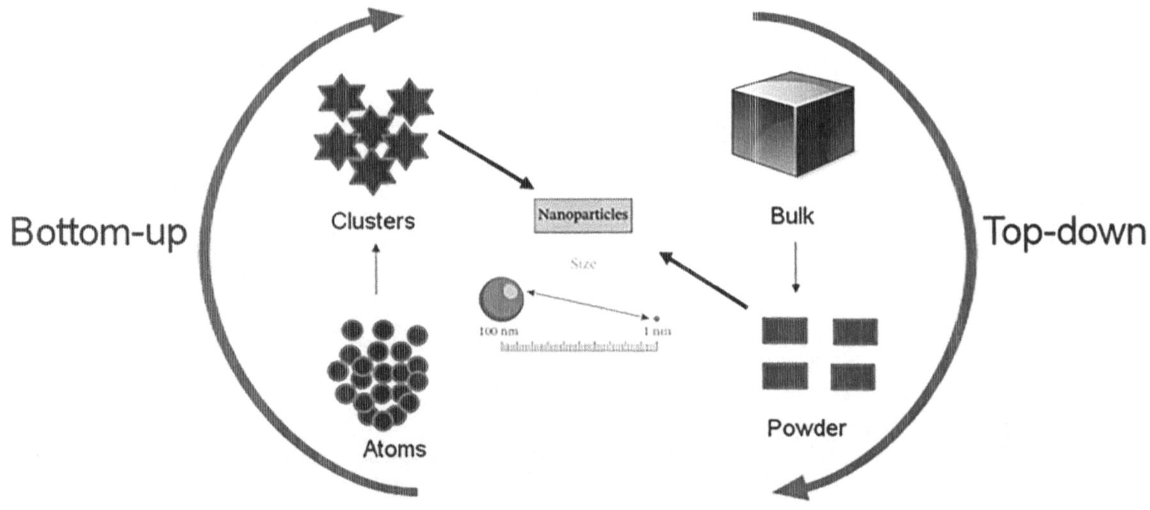

FIGURE 14.2 Top-down and bottom-up approaches of nanofabrication (adapted from Ashfaq et al. 2022).

barrier properties (against moisture, O_2, CO_2, and emission of flavor and ethanol contents), and considerably better physical and chemical quality when compared with polymers used for conventional treatments for preservation of horticultural commodities with films, coatings, and other packaging materials. Furthermore, polymer nanocomposites could be used in producing active packaging (for inhibiting or impeding microbe growth and associated food decay) and intelligent packaging (for preserving and observing food conditions) approaches (Silvestre et al., 2011).

14.3.1 Nanoreinforced Improved Packaging

Nanocomposites are polymer matrices reinforced with selected nanoparticles, specifically engineered to possess characteristics that enhance overall quality maintenance and extend the shelf life of fresh produce. This can include applications of nanoparticles and nanostructures during the formation of polymer films, coatings, and other types of packaging materials to enhance their mechanical strength and barrier properties. For example, the use of nanoscale clay particles has been found to increase the mechanical properties of selected polymers, making them substantially resistant against tearing or punctures (Ghanbarzadeh et al., 2015; Gutiérrez et al., 2017).

14.4 NANO-COATING: AN IMPORTANT HORTICULTURAL NANOREINFORCED PACKAGING TECHNIQUE FOR FRUITS AND VEGETABLES

14.4.1 Principles, Materials, and Approaches to Postharvest Nano-Coating of Horticultural Produce

The application of fruit and vegetable coating involves the deposition of a thin layer or film, which is made from edible or non-edible conventional- or nano-materials designed to impede desiccation under storage. Edible coatings, as thin layers made of biopolymers such as polysaccharides, lipids, and proteins, as well as some other edible ingredients, can be deposited on the surfaces of fruit and vegetable produce through various methods, such as brushing or spraying, immersing, or dipping into nanoemulsions to provide protection against desiccation and general quality deterioration (Eshghi et al., 2022). The edible films are also becoming highly prevalent among the scientific community and in the industrial sector. These differ from edible coatings, as they undergo a pre-forming procedure, requiring particular and specific technical steps in their structural manufacturing processes before use on fresh produce (Eshghi et al., 2022).

The use of conventional coatings and films was very common for preserving coated objects and reducing biochemical degradation before the development of innovative nanoscale coatings. Before this innovation, it was common to observe that the interaction of the stored commodity with the surrounding environment and the impact of gaseous modification were less effective with traditional coatings compared to the influences of nanoscale coatings. This involves the mitigation of stressful conditions of unwanted exchanges between coated fruits and vegetables and their prevailing surroundings to suppress quality deterioration (Gemail et al., 2023). As an example of conventional coating types, Eshghi et al. (2022) studied the influence of chitosan and gum ghatti coatings, both individually and in combination, on preserving bioactive constituents, including phenolic content, flavonoids, and antioxidant activity of Rishbaba grapes for 60 days under cold storage conditions. The results revealed that dipping grapes in chitosan alone or in combined forms with gum ghatti appropriately helped to preserve phenolic acids in grapes and led to higher concentrations of selected beneficial compounds in contrast with uncoated samples. In addition, chitosan and gum ghatti treatments as a combination proved highly effective in preventing fungal decay and stimulating the accumulation of soluble sugars and polyamines, indicating its potential as a postharvest strategy for enhancing fruit quality and preservation.

Although conventional coatings and films show promise in extending the shelf life of horticultural produce, to fully preserve food quality and provide multiple properties simultaneously, composites and nanocomposites are being explored (Corbo et al., 2015). Conventional coatings typically contain hydrocolloids, such as polysaccharides, proteins, lipids, waxes, and fatty acids, and may form composite structures such as protein-polysaccharides or protein-protein blends. Traditional coatings have not been fully acknowledged for their structural properties, antimicrobial activity, and barrier role. Nano-scale materials offer a way to compensate for the limitations of conventional coatings (Oliani et al., 2022). In other words, nano-coatings have more complex structures and formulations compared to conventional coatings, often requiring the latest technologies for their production (Singh & Packirisamy, 2022).

Nano-coatings are now used in different types of FP to enhance their designs. Incorporating nanofillers improves the mechanical properties, barrier properties, and color of edible coatings and films in contrast with simple edible coatings. Nevertheless, additional investigation is needed to make them more competitive with synthetic films, which can further improve their efficacy to prolong the shelf life of fresh horticultural produce (Jeevahan & Chandrasekaran, 2019).

Nanomaterials offer high value due to their flexibility in operating within different micromechanical settings, associated with a wide range of sizes. This feature allows for various implications and the possibility of establishing different dosages depending on the application's case on fruits and vegetables (Garg et al., 2021). More effectively than conventional coating, nano-coating upholds the normal regulation of oxygen levels and ethylene biosynthesis, thereby decelerating metabolic activities, respiration and oxidation rates, and chlorophyll loss. Also, they can efficiently control moisture loss, exhibit antimicrobial effects, protect the produce from postharvest diseases, and help maintain tissue firmness (Anean et al., 2023).

New biopolymers, along with the wide range of nanomaterials used in horticultural and food science research and industry, have emerged as significant contributors to the development of innovative edible coatings. Biodegradable polymer matrices are a useful platform for making antimicrobial nanocomposites while maintaining the nutritional profile and visual appeal of produce (Arabpoor et al., 2021; Eshghi et al., 2022; Gardesh et al., 2016; Nasr et al., 2021; Sreelatha et al., 2022). In horticultural postharvest applications, the solidity of nanocomposites and their compatibility with conversion into liquid, colloidal formulations have expanded future experimentation possibilities for material biosynthesis and conserving horticultural commodities from postharvest damage (Ye et al., 2023). Depending on their specific elements, biopolymers can modify the thermal requirements of nanocomposite applications and regulate moisture retention and gaseous exchange to extend shelf life (Chavan et al., 2023; Salem et al., 2019; Zare et al., 2019).

Nanocomposites are inherently antimicrobial due to their ability to disrupt cellular activity in the cytoplasm and organelles, offering ample time for thorough contact between microbial surface membranes and respective nanoparticles (Ventura-Aguilar et al., 2023). This triggers a chain of reactions that interferes with regular electron transport patterns in microbial cells, ultimately leading to cell death (Mohammadi et al., 2023).

The current trends in postharvest research are giving rise to demands for coatings and films that operate at the nanoscale, with an emphasis on edible formulations. One primary reason for this is that the full potential of nano-coatings has yet to be explored in scientific experiments related to postharvest management. Although nano-coatings are commonly applied to fruit and vegetable produce to maintain their quality, optimizing shelf-life improvements necessitates a case-by-case assessment of each fruit and vegetable species. Only then can researchers generate reliable results and unfailing recommendations for each species. Moreover, storage location variables add complexity to the matter because relevant variables may respond more intricately in nano applications than conventional ones when aiming to counter postharvest damage cost-efficiently (Ruffo Roberto et al., 2019). Meanwhile, it is necessary to pay attention to the fact that nano-coatings are sensitive to even minute changes in the volume of ingredients, which means that production protocols must be continuously optimized and reassessed for each particular situation and application (Singh & Packirisamy, 2022).

14.4.2 Nano-Coating for Preserving Horticultural Produce Quality and Microbial Control

The emergence of nano-enabled crop protection agents offers an eco-friendly, cost-effective, and sustainable alternative to conventional chemical treatments. Significant prevention of fungal and bacterial diseases was observed during the assessment of nanoparticles' potential, both pre- and postharvest, as well as in *in vitro*, *in vivo*, and *ex vivo* studies. Some examples extracted from scientific literature are explained next.

Copper nanoparticles, like other nanoparticles, are a valuable asset for targeting pathogenic specificity. They reportedly eradicated infestations of *Penicillium digitatum* and *Fusarium solani* (Khamis et al., 2017). Noman et al. (2023) investigated the effectiveness of biogenic cu-nanoparticles (bio-CuNPs) in combating bacterial fruit blotch, a disease that affects watermelon. The research examined the antibacterial properties and disease-suppressive efficacy of the bio-CuNPs. These nanoparticles were extracellularly synthesized with *Bacillus altitudinis* strain WM-2/2 and revealed a spherical shape. The foliar application of bio-CuNPs significantly inhibited bacterial blotch in watermelons by prompting biofilm inhibition, increasing oxidative stress, and stimulating cell integrity disruptions. The study exhibited the potential of bio-CuNPs as a tool for increasing overall production and reducing food insecurity. Moreover, CuNPs are considered a viable substitute for chemical-based fungicides against *Fusarium* species in dragon fruits and tomatoes (Van Viet et al., 2016). The effectiveness of CuNPs can be enhanced further by using them in combination with other protective agents. Bouqellah (2023) demonstrated a novel technique for biosynthesizing trimetallic magnetic zinc and copper nanoparticles (CMC-Cu-Zn-FeMNPs) of carboxymethyl cellulose (CMC). The combination showed the highest efficiency in downregulating occurrence and severity of *Fusarium* wilt in tomatoes.

Citrus black rot disease, caused by *Alternaria citri*, poses a significant problem for citrus during both pre- and postharvest periods, resulting in substantial economic losses of up to 30–35% on an annual basis. Despite major efforts to use fungicides, their effectiveness in controlling disease has not substantially improved over the last few decades (Sardar et al., 2022). Sardar et al. (2022) investigated the antifungal efficacy of zinc oxide (ZnO) and copper oxide (CuO) biosynthesized nanoparticles using lemon peel extracts against *Alternaria citri*. It was noted that both types, i.e., CuO and ZnO nanoparticles, markedly inhibited fungal growth, with mixed metal oxide NPs showing the highest antifungal activity. The NPs were effective at concentrations as low as 80 mg mL^{-1} and completely inhibited fungal growth and further spread at higher concentrations. This suggests that green-synthesized NPs could be used as an alternative to chemical fungicides for controlling citrus black rot disease. Moreover, an *in vitro* experiment as reported by Sreelatha et al. (2022) revealed that thymol-based chitosan nanoparticles can successfully suppress *Xanthomonas campestris* pv. Campestris, which is considered another leading cause of black rot in certain other cruciferous crops. Antifungal efficiency has reportedly emanated from silica and silver sulfide nanocomposites against *Aspergillus niger* (Fateixa et al., 2009), as well as nanoferrites against *Fusarium oxysporum*, *Dematophora necatrix*, and *Colletotrichum gloeosporioides* (Sharma et al., 2017b).

Chitosan nanoparticles are famous for their antibacterial properties. Nano-coating materials made from this substance were found effective against a wide range of bacteria, including *Escherichia coli* and *Salmonella typhimurium* (Paomephan et al., 2018), *Listeria monocytogenes*, and *Staphylococcus aureus* (Feyzioglu & Tornuk, 2016), and *Pseudomonas aeruginosa* (Bernal-Mercado et al., 2022). According to the results of Gardesh et al. (2016), an edible coating with 100 nm chitosan nanoparticles at a concentration of 0.5% provided significant improvements in the postharvest attributes of apples under storage. This led to better overall conservation of eating quality than with conventional chitosan. The treatment markedly extended shelf life, impeded weight loss, and retained the qualitative attributes of Golab Kohanz apple fruits. Additionally, nano-coated apples exhibited better resistance against *Penicillium axpansum*. Melo et al. (2018) explored the effects of edible coatings made from chitosan nanoparticles on the postharvest quality of grapes under storage. It was noted that the nanoparticles effectively slowed down the ripening of harvested grapes. The treatments also led to reduced soluble solids concentration, weight loss, and reduced sugar levels compared to controls. Additionally, it also helped to maintain general freshness and preserve titratable acid contents, along with the preserved sensorial attributes of fresh grapes.

In another study, Salem et al. (2019) used AgNPs to observe potency against *Botrytis cinerea* in tomatoes. The AgNPs substantially prolonged the shelf life of tomatoes, and the severity of fungal decay was markedly reduced during 40 days of storage. Recently, Ouzakar et al. (2023) made zinc oxide nanoparticles (ZnONPs) from the culture supernatants of *Phaeodactylum tricornutum* and sprayed them on sweet cherries to see how well they kept fruit quality over a 10-day period of storage. The nanoparticles successfully inhibited *E.coli*, *C. albicans*, and *S. aureus* growth on sweet cherries during storage. Additionally, the use of ZnONPs significantly delayed changes in both the cosmetic and biochemical quality of treated fruits, which thereby resulted in reduced decay occurrence. Cherries coated with 500 mg L^{-1} of ZnONPs showed the lowest changes in moisture and sugar content compared to untreated cherries. Moreover, the ZnONPs coating showed a beneficial impact in preserving ascorbic acid, polyphenols, and flavonoids in contrast with controls. The data indicated an increased concentration of total flavonoids and antioxidative capacity in fruits treated with 500 mg L^{-1} ZnONPs in contrast with the controls. Generally, the study suggests that utilizing a green ZnONPs-based coating could be considered a safe and effective technique for extending the preservation of commercial fruits. *Botrytis cinerea* is the primary cause of gray mold infestation in grapes. Hashim et al. (2019) observed how chitosan and silica nanoparticles sufficiently prevented gray mold development under storage. The fungal DNA was substantially deconstructed in response to the penetrative effects of the applied nanoparticles.

As mentioned earlier, chitosan has wide applicability as a conventional coating. This has resulted in the development of numerous nanocomposites for edible nano-coatings for perishable horticultural produce (Kumar et al., 2020; Nasr et al., 2021). Using AgNPs, chitosan was reportedly used as a component of nanocomposite materials to extend the shelf life of apples and tomatoes (Nisha et al., 2016). Three types of nano-coatings were utilized, namely, glucose-chitosan with chitinase, glucose-chitosan, and chitosan-silver nanocomposite. Moisture loss was minimal, and contamination was highly controlled in response to the silver nanoparticles present in the nanocomposites. The chitosan coating had antimicrobial effects that prevented fruits and vegetables from deteriorating due to microbial activity for a certain period of time during storage. Additionally, the nano-coating minimized the rate of phenolic loss.

The loquat fruit originated in China and is abundant in essential vitamins and minerals. It is necessary to store loquats in cold storage, along with some exogenous treatments. However, the primary obstacle to prolonging their cold storage time is chilling injury (CI), which can be managed with eco-friendly postharvest treatments. Without these treatments, CI ultimately reduces the time period during which fruits can be marketed (Shah et al., 2023). As reported by Song et al. (2016), treatment of loquat fruits with chitosan/nano-silica coatings resulted in a reduction of symptoms associated with CI. The coating helped to maintain the structural integrity of the cell wall, which thus prevented weight loss and internal browning disorder under cold storage. Finally, fruits showed higher concentrations of TSS, TA, and sugar contents. The treatment proved effective in preserving antioxidants and the postharvest quality of loquat fruits.

Similar results were observed in response to the nano-coating of citrus fruits. It was observed that chitosan-clay nanocomposites (CCNC) were applied as coatings to retain the qualitative attributes of lemon fruits (Taghinezhad & Ebadollahi, 2017). Different chemical attributes, such as acidity, firmness, soluble solids content, and peel shear value, were monitored for 63 days during storage. The CCNC effectively conserved fruit firmness and suppressed weight loss, along with higher acidity and soluble solids concentration during the storage period. In the case of citrus fruits, coating Thomson sweet oranges with CCNC helped to maintain fruit quality during cold storage for 90 days at 6 °C. The treated fruits exhibited better eating characteristics compared to the control group. This included a higher juice pH, peel moisture content, surface chroma values, and firmness. In addition, the CCNC coating was also found to enhance the orange fruits' resistance to fungal infection, preserve their color, and maintain a higher texture under cold storage (Ruffo Roberto et al., 2019).

Colletotrichum musae is the leading pathogen responsible for anthracnose incidence in bananas and results in substantial losses of harvested fruit in various developing countries. Treatment of ripe bananas during postharvest can be an appropriate and effective strategy to manage anthracnose disease. It was noted that incorporating chitosan-montmorillonite nanocomposite into free-standing film using a casting technique improved the postharvest longevity of banana fruits by suppressing ethylene production. This, in turn, could indirectly affect the dynamics of pathogenic infestations of bananas (Wantat et al., 2022). Green beans were found to be compatible with modified chitosan layers comprising nanoemulsions of mandarin essential oil, which countered pathogenic infections during storage. Suspension of n-palmitoyl chitosan made in acetic acid allowed the effective incorporation of mandarin essential oil into nanoemulsions. This chitosan-based treatment thus demonstrated strong efficacy against *Listeria innocua* (Donsì et al., 2015).

Tomatoes are one of the most commonly cultivated crops in the world and require suitable storage conditions after harvest to preserve their vitamin C and lycopene contents. The turgor pressure of tomato cells is critical for salad preparation, and the application of nano-coatings can help to maintain this pressure for an extended storage period (Kumar et al., 2020). Treatment of tomato fruits with hydroxypropyl methylcellulose enriched with *Piper betle* leaf essential oil nanoemulsion coating led to a substantial improvement in the shelf life of the treated group compared to those stored without nanoemulsion application (Poovai et al., 2023).

Prunus avium, commonly known as sweet cherry, is rich in various antioxidant compounds that can remain preserved in fruit for longer periods of time during storage if suitable conditions are provided. The ideal postharvest management of sweet cherry involves modified-atmosphere packaging and does not require chemical fungicides for decay control (Cabañas et al., 2023). Nabifarkhani et al. (2015) carried out a study to examine the impact of postharvest edible coating made from chitosan-based nanocomposites (comprising 1% chitosan as the matrix material along with 0.1% nano-cellulose) and thyme essential oil (at a concentration of 1%; dissolved in 25% ethanol) on the fruit characteristics of sweet cherry that were kept at 1 °C and 90% relative humidity for a duration of 6 weeks. The implementation of treatments resulted in the preservation of total soluble solids, anthocyanins, and total sugar content. In comparison to control samples, these treatments have also led to a reduction in fruit weight loss ($P < 0.05$).

Cellulose nanofibrils (CNFs) and nanocrystals (CNCs) play an important role in synthesizing nanocomposites when biopolymers are used for nano-coatings (Jeet et al., 2022, Wang et al., 2023). It was found that the application of cellulose nanofibrils on fresh-cut apple fruits maintained antioxidant properties and aroma volatiles with reduced reactive oxygen species and fruit decay (Wang et al., 2023). Laureth et al. (2018) conducted a study in which it was found that nanocellulose (nanocrystals [NC])–based nanocomposites consisting of polymer filmmaking solutions of citrus pectin (PEC), carboxymethyl cellulose (CMC), and cassava starch were effective in improving the storage life and nutrient contents of Tahiti acid lime fruits. The fruits were treated with different nanocomposite concentrations and pure polymer solutions for 1 min, air-dried, and kept under ambient conditions for 9 days. The NC-PEC nanocomposite was found to be most effective in impeding CO_2 diffusion, retaining chlorophyll pigments, and reducing fresh mass loss. The study concluded that the incorporation of nanocellulose in pectin biopolymers forms a stable coating with a suitable barrier property, which can preserve the postharvest quality of fresh produce and is advantageous for potential commercialization.

Piña-Barrera et al. (2019) investigated how the coating made from polymeric and pullulan nanocapsules containing thyme essential oil influences the quality and shelf life of grapes. It was noted that a multisystem coating led to improvement in the antioxidative properties of the essential oil. The treated grapes retained their characteristic color, flesh firmness, and taste scores for longer than the controls. The treatment also decreased the produce's metabolism, oxidative stress indicators, and microbial decay under postharvest storage. Furthermore, the volatile compounds of the essential oil were found to be higher in treated fruits than in the controls. Finally, it was concluded that NC-EOt coatings can be a viable solution to prolong the storage period of horticultural products.

14.5 MULTIFACETED APPROACHES TO NANO-PACKAGING OF FRUITS AND VEGETABLES

To extend the shelf life of fruits and vegetables during transportation and to prevent contamination from foodborne pathogens, packaging must address several challenges. Protection, convenience, communication, and inhibition are the four primary functions of packaging, which safeguards the product from external environmental factors, communicates marketing messages to the consumer, provides ease of use and time-saving convenience, and accommodates products of varying sizes and shapes. However, traditional packaging no longer satisfactorily meets the increasingly complex needs associated with regulatory requirements, market globalization, and consumer requirements for innovative packaging, minimally processed produce, and enhanced food safety measures. In other words, there is a need for smarter packaging approaches, for instance, active packaging and intelligent packaging, as described next (Figures 14.3, 14.4, and 14.5).

14.5.1 Nano-Based Active Packaging

Active packaging is a form of packaging that purposefully includes bioactive materials in either the packaging materials or headspace. These materials serve as releasers of bioactive compounds or scavengers of undesirable molecules, aiming to improve the overall performance of the packaging system and counteract any detrimental effects (Basavegowda et al. 2020). These systems often involve the absorption or scavenging of undesirable compounds such as O_2, CO_2, ethylene, and excessive water. Other active packaging releases or adds certain compounds, for instance CO_2, as well as antioxidative and antimicrobial compounds, and other additives into headspaces through sachets, labels, or films (Bodbodak & Rafiee, 2016; Yildirim et al., 2018). Hence, this deliberate approach aims to enhance produce, packaging, and environmental interactions, which may result in prolonged shelf life, increased food safety,

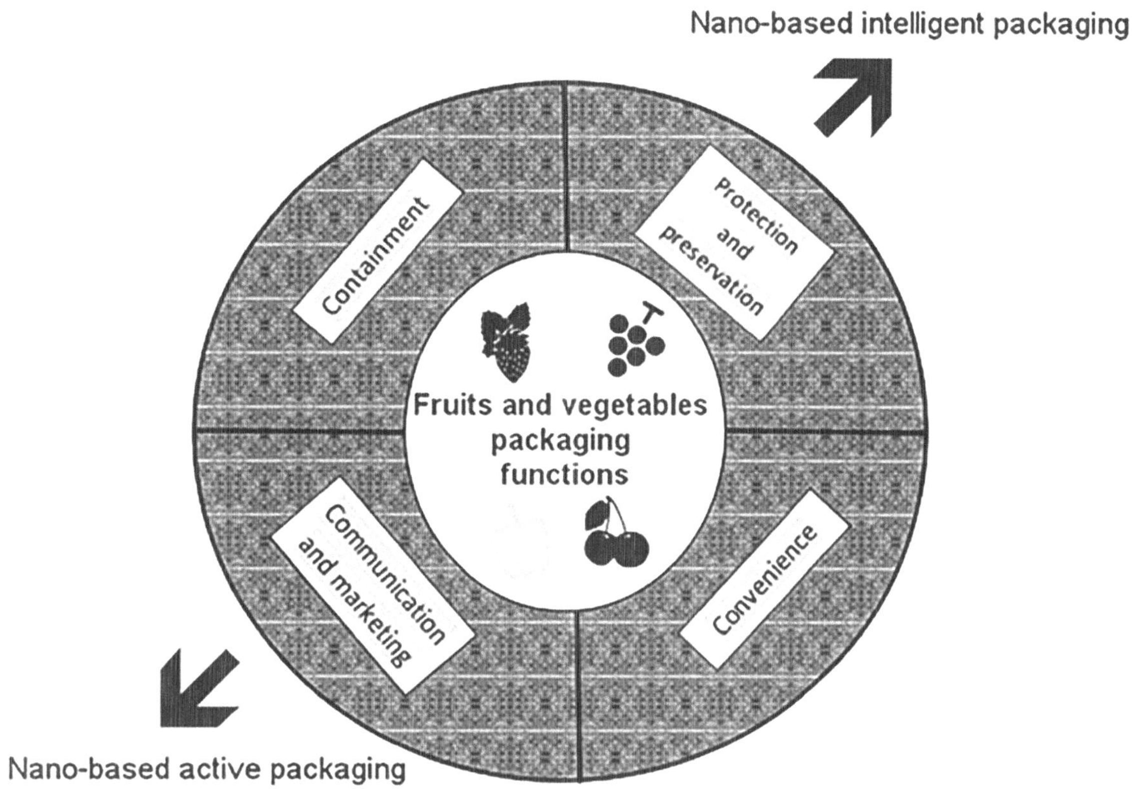

FIGURE 14.3 Fruit and vegetable packaging functions (adapted after modifications from Salgado et al., 2021).

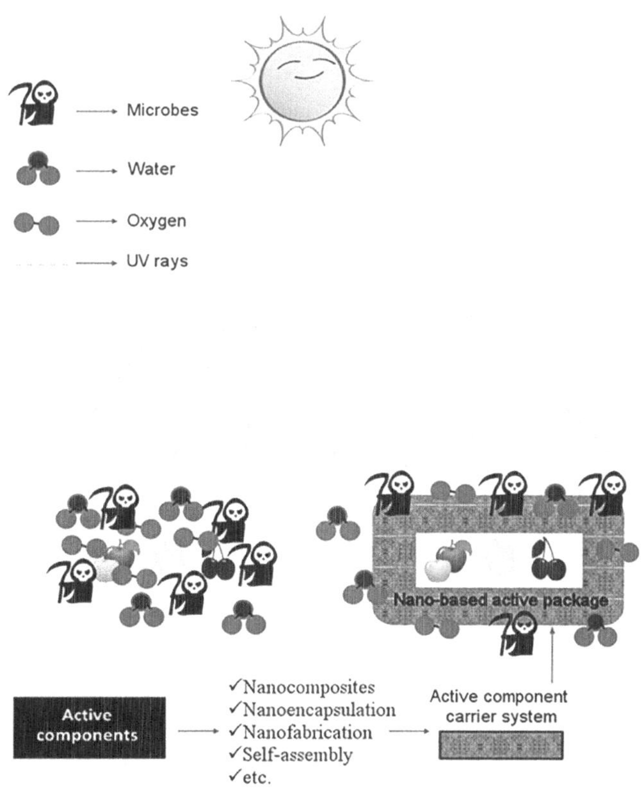

FIGURE 14.4 Nano-based active packaging of fruits and vegetables (adapted from Ashfaq et al., 2022).

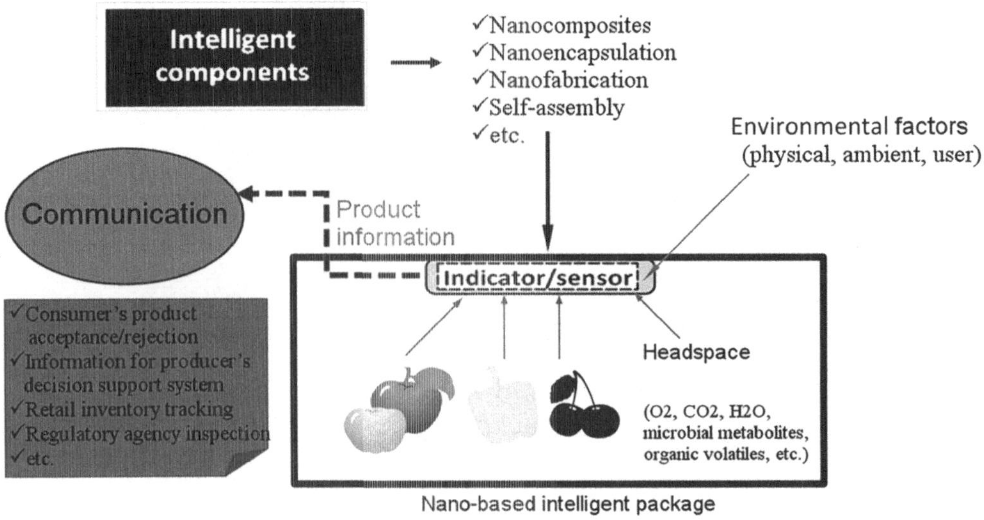

FIGURE 14.5 Nano-based intelligent packaging of fruits and vegetables (adapted after some modifications from Mihindukulasuriya and Lim, 2014; Sharma et al., 2017a).

enhanced sensory characteristics, and maintained product quality, which may not be achieved otherwise.

Anthocyanin pigments have great potential to be used in producing active packaging films for fresh horticultural produce storage purposes. Cheng et al. (2022) reported on an active packaging film made via the addition of red cabbage anthocyanin extracts (RCAE) to acetylated di-starch phosphate (ADSP) as film substrates. The RCAE inclusion eventually helped to enhance the barrier property against light, O_2, and water in the films while also improving their thermal stability and antioxidative potential. The benefits were attributed to the development of hydrogen bonding through electrostatic interactions between RCAE and ADSP constituents. It was thus demonstrated that RCAE incorporation was substantially suitable to strengthen the physical properties of ADSP-based composites, thereby making them highly applicable as packaging films for fruit and vegetable produce.

In the case of nano-packaging, active packaging utilizes selected compounds that actively participate through nanoscale media to protect perishable fruits and vegetables from various damage-associated factors, thereby conserving the characteristic quality of horticultural produce under storage. Antimicrobial nanosystems can be categorized into two distinct groups based on their mode of operation. The first group involves one that releases antimicrobial compounds from nanocapsules into package headspace, thus allowing for interaction with the product surface. In contrast, the other group immobilizes the antimicrobial compound within the packaging material, utilizing materials of a nanocomposite nature (Ayala-Zavala et al., 2014).

Nano-based active packaging systems that release antimicrobial compounds, such as sachets or plastic films containing nanocapsules, are employed to inhibit microbial activity. These systems can be categorized into two groups: indirect and direct antimicrobial activity. Nanosystems with indirect antimicrobial activity scavenge oxygen and moisture, remove ethylene, and absorb/emit CO_2. While their major function is to reduce deterioration owing to enzymatic reactions and altered internal atmospheres, these also have an inhibitory effect on aerobic bacteria growth (Pereda et al., 2019). TiO_2 nanoparticles have been utilized in producing oxygen scavenger films (Xiao-e et al., 2004), while some silver-based nanoparticle treatments exhibit antimicrobial activity and can absorb and decompose ethylene (Li et al., 2009), thereby extending the shelf life of perishable products like fruits and vegetables.

The presence of certain compounds, like metals and essential oils, in the headspace can directly act as antimicrobial agents (Espitia et al., 2012). Ayala-Zavala and González-Aguilar (2010) developed a cyclodextrin-essential oils nanocapsule to extend the freshness duration of fresh-cut tomatoes. They postulated that the release of antimicrobial compounds from the molecular complexes formed between volatile essential oil components and cyclodextrin could be facilitated by internal moisture. Another direct system is using antimicrobial agents incorporated within the bulk polymer of active packaging, residing in nanocavities. These agents move toward the surface to engage with the microorganisms, while synthetic and natural polymers are used as carriers for this functionality (Silvestre et al., 2011). Various compounds, both organic and inorganic, such as silver zeolites, organic acids, peptides, enzymes, and volatile compounds, have been employed as antimicrobial agents (Duncan, 2011; Seil & Webster, 2012). However, a disadvantage of this type of package is that heat-sensitive substances cannot be utilized due to their inactivation during the package's processing. An alternative solution involves the nanoencapsulation of such additives before their incorporation into the polymer extrusion process (Pereda et al., 2019). The controlled release of active compounds into the package's headspace is regulated by different factors, including physical modifications, heat, enzymatic activity, pH, and relative humidity (Ho et al., 2011). Of these factors, relative humidity appears to have the most significant impact on the release of antimicrobial constituents (Mastromatteo et al., 2010; Podsiadlo et al., 2005).

Antimicrobial active packages that incorporate nanoassemblies into the polymer matrix are rare. Metal nanoparticles (such as silver, gold, and zinc), metal oxide nanomaterials (including titanium dioxide, zinc oxide, silicon oxide, and magnesium oxide), and carbon nanotubes can be used and function through direct contact, but they can also slowly migrate and interact with the food matrix (Pereda et al., 2019). Silver has higher temperature stability and low volatility at the nanoscale, making it an effective antifungal and antimicrobial agent toward many microorganisms (Liau et al., 1997). The antimicrobial capacity of silver nanocomposites is attributed to their ability to rupture and adhere to cell surfaces, increase permeability, degrade lipopolysaccharides, and bind with electron donor groups in biological molecules containing oxygen, sulfur, or nitrogen (Damm et al., 2007; 2008). Metal oxide nanoparticles have the ability to act as UV blockers and photo-catalytic disinfectants (Jiang et al., 2015). Carbon nanotubes exhibit antibacterial properties and have been observed to impact the survival of *Escherichia coli* upon direct contact. This effect may be attributed to the piercing action of the nanotubes, which leads to irreversible damage in microbial cells (Pereda et al., 2019).

As mentioned earlier, to efficiently inhibit the growth of microorganisms by immobilizing the antimicrobial compound within the packaging material, direct contact between the polymer and fresh produce is necessary. However, it is possible for nanomaterials to migrate from the packaging into the food, which can lead to inadequate packaging performance and potential health risks for consumers. For instance, there are significant concerns regarding the cytotoxic effects of nanoscale chitosan (de Azeredo, 2009) and carbon nanotubes (Monteiro-Riviere et al., 2005) on human cells if they eventually migrate into produce that has been packed with nanomaterials. As a result, the incorporation of these materials into food packaging is discouraged.

Safakas et al. (2023) used essential oils, thymol, carvacrol, and olive leaf extract in inorganic bentonite and organically modified montmorillonite to render the nanocarriers bioactive and employable in nanofilms. The nanocomposites had substantial antimicrobial properties and acted as barriers against factors that undermined the mechanical integrity of the biofilms in nano-packaging. As mentioned earlier, montmorillonite effectively binds with chitosan and serves as a nanocarrier in biofilms for nano-packaging after preparation through the casting technique (Wantat et al., 2022).

López-Gómez et al. (2023) investigated the efficacy of plant essential oils (EOs) to maintain lemon fruit quality under cold and commercial storage conditions. The antimicrobial and antioxidant properties of EOs have been extensively studied, and this research focused on the effects of EOs released from active packages on the antioxidant activities and quality characteristics of lemon fruits. The results showed that essential oils released from inclusion complexes enclosed in active packaging can control microbial growth. The active packaging induced stress that activated both enzymatic and non-enzymatic antioxidant defense systems in the flavedo tissue, including ascorbic acid and phenolic contents. In the meantime, the active packaging did not affect physicochemical quality attributes such as firmness, color, and weigh loss of samples. In research on the postharvest storage of lettuce, packaging films were impregnated with nano-bentonite particles in MAP conditions for 12 days. The results indicated that nano-packaging was greatly effective in preserving the physicochemical attributes and sensory qualities of lettuce when the developed biofilms contained 1% and 3% nano-bentonite particles. This thus helped to suppress yeast growth and reduce the severity of mold infestations (Farahanian et al., 2023).

Poly(3-hydroxybutyrate) is a distinctive polymer known for its biodegradation and biocompatible nature, and it falls under polyhydroxyalkanoates. Cherpinski et al. (2018) utilized the biodegradable nature and scavenging capacity of poly(3-hydroxybutyrate) in combination with palladium nanoparticles to develop active packaging as biofilms through an electrospinning procedure. The resulting biofilms contained hexadecyltrimethylammonium bromide and revealed promising mechanical integrity and thermal stability, as well as notable scavenging activity, thus showing its significant potential in nano-packaging. In another study, Rusková (2023) utilized poly(3-hydroxybutyrate) as a matrix to encapsulate EOs for the development of nano-packaging films. The antioxidative and antibacterial properties of the films were explored on strawberries, ultimately revealing their strong potential to extend the shelf life of treated fruit. Interestingly, lemongrass-based EO was found to be more effective in contrast with oregano EO when incorporated into the poly(3-hydroxybutyrate) matrix. Oliani et al. (2022) made polyethylene nano-packaging films with the addition of AgNPs while incorporating lower density-based titanium dioxide. The nano-packaging films were less permeable to O_2 and exhibited improved antibacterial activity.

14.5.2 Nano-Based Intelligent Packaging

Intelligent packaging performs various functions, including communicating and detecting information to aid in decision-making, in prolonging shelf life, improving safety, enhancing quality, and providing warnings about potential issues. This system employs a variety of inventive communication techniques, such as sensors to detect time/temperature conditions and history, oxygen and CO_2 levels, spoilage or package leakage, ripeness of commodities and freshness, microbiological growth, and particular foodborne pathogens. Indicator systems can provide qualitative and quantitative information through color changes when incorporated into packaging materials or placed inside packages (Ashfaq et al., 2022; Sharma et al., 2017a).

Fruits and vegetables are among the most essential foods for maintaining a healthy and balanced diet. Nano-based intelligent packaging entails the use of highly sensitive and low-cost nanosensors made of nanomaterials placed in food packages or packaging films to monitor parameters such as temperature, humidity, gas composition, and pathogenic contamination levels. These nanosensors play a vital role in maintaining food quality and preventing the consumption of harmful foods. With their ability to detect and report issues with the product in real time, nanosensors have the potential to revolutionize the way we package and monitor food products, leading to improved efficiency in the agricultural supply chain and a reduction in food waste, benefiting both producers and consumers (Chelliah et al., 2021).

According to recent advancements, moisture absorbers, oxygen scavengers, and barrier packaging items constitute approximately 80% of the total market share in multifaceted approaches to nano-packaging (Chellaram et al., 2014). The primary emphasis on nanosensors lies in scenarios where temperature changes occur or where packaging systems possess micropores or sealing deficiencies that could potentially expose food products to higher-than-desired levels of oxygen. Such exposure can lead to unfavorable alterations in food products. Timestrip® has devised an innovative solution, known as iStrip, specifically designed for refrigerated food products. This system employs gold nanoparticles that exhibit a vivid red coloration when exposed to temperatures surpassing the freezing point. However, if unintended freezing occurs, it triggers an irreversible process of agglomeration among the gold nanoparticles, consequently causing the loss of the distinctive red hue (Pereda et al. 2019). There is growing interest in developing nontoxic and irreversible oxygen nanosensors for O_2-free food packaging systems. Lee et al. (2002) developed a UV-activated colorimetric O_2 indicator using TiO_2 nanoparticles. Upon their exposure to UV light, these remain colorless, and the sensor bleaches until it encounters O_2, reinstating its original blue color. Another sensor, described in a separate study (Mills & Hazafy, 2009), utilizes nanocrystalline SnO_2 and exhibits a gradual color change in response to oxygen.

In the recent time of technological advancements, nanosensors have emerged as an innovative means that utilize polymer conductance to detect and characterize microorganisms through analysis of gas emissions (Ahuja et al., 2007). The nanosensors consist of insulation-based polymeric matrices that integrate conducting particles such as gold, platinum and palladium nanoparticles (Mannino & Scampicchio, 2007). Their functionality lies in their ability to reveal selected changes in resistance upon exposure to gases such as H_2S, SO_2, NH_3, H_2, and NO_2 emitted from certain microorganisms in response to food degradation or spoilage (Arshak et al., 2007; Kang et al., 2007). Furthermore, there have been reports indicating that nanocomposites comprising tungsten trioxide and tin dioxide possesses the ability to detect C_2H_4 gas, which has been reported as an imperative indicator of fruit ripening (Pimtong-Ngam et al., 2007).

Color-based pH indicators have proven highly effective in monitoring pH-related alterations caused by microbial and physiological activity during storage. This is especially relevant in the context of fruit and vegetable produce ripening and spoilage, as indicators serve as valuable tools in detecting changes in pH levels. Spoilage normally occurs when bacteria and fungi ferment sugar, thus leading to a drop in pH and the development of off flavors. On the other hand, normal fruit ripening involves starch conversion to simple sugars, formation of desirable flavors, and an increase in pH (Maftoonazad & Ramaswamy, 2019). A pH biosensor consists of two parts: a dye that is considered sensitive to pH changes and a solid matrix that supports the dye (Mishra et al., 2011). Natural pigments extracted from different fruits and vegetables, such as anthocyanin from RCE, are commonly used for colorimetric pH determination due to their broad range of color changes with varying pH levels (Pourjavaher et al., 2017). The solid matrix used in biosensors can affect sensitivity, response time, reproducibility, and reversibility. Numerous types of polymeric materials have been described in the literature as solid bases (Pourjavaher et al., 2017). Maftoonazad & Ramaswamy (2019) developed a real-time electrospun pH-sensitive biosensor nanofiber mat consisting of RCE incorporated into a polyvinyl alcohol matrix, and it was able to rapidly and sensitively detect pH changes, for instance, spoilage, in date fruit during storage. These types of biosensors could be utilized inside produce packages or in packaging films as bio-indicators to monitor the progression of unfavorable changes in the characteristic quality of produce. In a research report, nanosensors fabricated with engineered platinum nanoparticles were found to exhibit potential detection of pH fluctuations (Martins et al., 2012).

In general, the concept of intelligent packaging in the food industry is a novel idea that is still under development for further advancement and optimization. The field offers promising prospects for the enhancement of food safety and harbors a hopeful outlook for the future owing to the emergence of advanced technologies, i.e., nanotechnology.

14.6 SAFETY CONCERNS AND FUTURE PROSPECTS OF NANO-PACKAGING

Nano–food packaging signifies cutting-edge advancements in the field of nanotechnology and has developed as a potential and revolutionary alternative to traditional food packaging. The incorporation of nanomaterials has emerged as a new generation of packaging technology that has gained substantial momentum and is presently one of the most broadly developed areas within nanotechnology fields, food sciences, and postharvest physiology management of horticultural produce. Nanocomposites offer enormous potential to improve the performance of eco-friendly edible films and coatings by incorporating nanoreinforcements. Even at lower concentrations, the addition of nanoparticles to biopolymers was found to increase their mechanical, thermal, and barricade properties. Therefore, nanoparticle-based technology plays a critical role in improving biopolymer utilization feasibility in numerous applications, including food packaging systems. Furthermore, nanoparticles can confer smart properties to food packaging. For instance, these can exhibit antimicrobial characteristics, enable O_2 scavenging, facilitate enzyme immobilization, and/or serve as critical indicators of exposure regarding degradation-related factors. So, nanocomposite-based packaging not only passively safeguards food items from environment-related factors but also improves their stability or provides reliable indications of their suitability for consumption. However, the safety concerns surrounding nanomaterials cannot be overlooked. Their minute size raises apprehensions about their potential ability to penetrate cells and remain within the body. The characterization of nanomaterials as new or unnatural substances lacks consensus. While the properties and safety profiles of bulk materials are generally well-understood, their nano-sized counterparts often exhibit divergent characteristics.

Currently, scientific data on the migration of different types of NPs from packaging materials into food, as well as their potential toxicological effects, remain limited. Nanoparticles may also accumulate in the environment and enter the food chain, leading to respiratory or dermatological disorders. These knowledge gaps are due to the absence of appropriate analytical methods for tracking nanoparticles with lower contents and smaller sizes. Furthermore, more time is required to arrive at a comprehensive decision on whether specific nanomaterials are safe for large-scale human consumption. The assessment of long-term health impacts resulting from extended exposure to nanoparticles is of paramount significance, and case studies may lead to varying outcomes, particularly in industrial applications. It is therefore essential for regulatory authorities and government organizations around the world, for instance, the FDA in the United States and the European Union, to research the potential risks of NPs and communicate their findings to ensure their safe use. By gathering extensive scientific data and undertaking rigorous toxicity investigations, researchers can elucidate the potential risks associated with nanomaterials and develop suitable safety guidelines and strict regulations.

14.7 CONCLUSIONS

In conclusion, this chapter highlights the importance of nano-packaging technology as a solution to the preservation challenges faced by the fresh produce industry. Incorporating nanomaterials and nanostructures into packaging materials presents promising methods for prolonging shelf life, preserving freshness, and maintaining the general quality and safety of perishable produce. The chapter precisely focuses on three pivotal aspects of nano-packaging: nanoreinforcement, nano-based active packaging, and nano-based intelligent packaging. Nanoreinforcement involves incorporating nanomaterials and nanostructures to improve the strength and barricade properties of packaging matrices, protecting against O_2, moisture, and microbial infestations. Nano-based active packaging integrates active NPs or nanostructures into packaging, enabling interactions with packaged fruit and vegetable commodities to extend shelf life, inhibit spoilage, and maintain overall quality. Nano-based intelligent packaging consists of sensors, indicators, and monitors, and all of these provide real-time information regarding quality and safety aspects, allowing for timely detection of deterioration and decay. As the field of nano-packaging continues to advance, it is imperative to identify the limitations and challenges related to the experimental phase of nanotechnology in packaging. Nanomaterials, due to their exceptionally small size, raise apprehensions about their potential for cellular penetration and bioaccumulation. As a result, there is a need for comprehensive assessments and rigorous toxicity studies addressing the long-term effects of nanoparticle exposure to form an informed consensus on their safety for human consumption and the environment. It means that in the years ahead, it will be crucial to undertake further research and foster collaboration among the scientific community, regulatory agencies, for instance, the FDA and European Union; and the food industry.

Keywords:

Active packaging, Horticultural crops, Food safety, Intelligent packaging, Nanomaterials, Nanoreinforcement, Shelf life

REFERENCES

Ahuja, T., Mir, I. A., & Kumar, D. (2007). Biomolecular immobilization on conducting polymers for biosensing applications. *Biomaterials, 28(5)*, 791–805.

Anean, H. A., Mallasiy, L. O., Bader, D. M., & Shaat, H. A. (2023). Nano edible coatings and films combined with zinc oxide and pomegranate peel active phenol compounds has been to extend the shelf life of minimally processed pomegranates. *Materials, 16(4)*, 1569.

Arabpoor, B., Yousefi, S., Weisany, W., & Ghasemlou, M. (2021). Multifunctional coating composed of *Eryngium campestre* L. essential oil encapsulated in nano-chitosan to prolong the shelf-life of fresh cherry fruits. *Food Hydrocolloids, 111*, 106394.

Arshak, K., Adley, C., Moore, E., Cunniffe, C., Campion, M., & Harris, J. (2007). Characterisation of polymer nanocomposite sensors for quantification of bacterial cultures. *Sensors and Actuators B: Chemical, 126(1)*, 226–231.

Ashfaq, A., Khursheed, N., Fatima, S., Anjum, Z., & Younis, K. (2022). Application of nanotechnology in food packaging: Pros and Cons. *Journal of Agriculture and Food Research, 7*, 100270.

Aswathanarayan, J. B., & Vittal, R. R. (2019). Nanoemulsions and their potential applications in food industry. *Frontiers in Sustainable Food Systems, 3*, 95.

Ayala-Zavala, J. F., & González-Aguilar, G. A. (2010). Optimizing the use of garlic oil as antimicrobial agent on fresh-cut tomato through a controlled release system. *Journal of food science, 75(7)*, M398-M405.

Ayala-Zavala, J. F., González-Aguilar, G. A., Ansorena, M. R., Alvarez-Párrilla, E., & de la Rosa, L. (2014). Nanotechnology tools to achieve food safety. In Bhat, R., & Gomez-Lopez, V.M. (Eds.), *Practical Food Safety: Contemporary Issues and Future Directions* (pp. 341–353). Wiley.

Basavegowda, N., Mandal, T. K., & Baek, K. H. (2020). Bimetallic and trimetallic nanoparticles for active food packaging applications: A review. *Food and Bioprocess Technology*, *13*, 30–44.

Bernal-Mercado, A. T., Juarez, J., Valdez, M. A., Ayala-Zavala, J. F., Del-Toro-Sánchez, C. L., & Encinas-Basurto, D. (2022). Hydrophobic chitosan nanoparticles loaded with carvacrol against *Pseudomonas aeruginosa* biofilms. *Molecules*, *27(3)*, 699.

Bodbodak, S., & Rafiee, Z. (2016). Recent trends in active packaging in fruits and vegetables. In Siddiqui, M.W. (Ed.), *Eco-friendly Technology for Postharvest Produce Quality* (pp. 77–125). Academic Press.

Bouqellah, N. A. (2023). In silico and in vitro investigation of the antifungal activity of trimetallic Cu–Zn-magnetic nanoparticles against *Fusarium oxysporum* with stimulation of the tomato plant's drought stress tolerance response. *Microbial Pathogenesis*, *178*, 106060.

Cabañas, C. M., Hernández, A., Serradilla, M. J., Moraga, C., Martín, A., Córdoba, M. D. G., & Ruiz-Moyano, S. (2023). Improvement of shelf-life of cherry (*Prunus avium* L.) by combined application of modified-atmosphere packaging and antagonistic yeast for long-distance export. *Journal of the Science of Food and Agriculture*, *103(9)*, 4592–4602.

Chaudhary, P., Fatima, F., & Kumar, A. (2020). Relevance of nanomaterials in food packaging and its advanced future prospects. *Journal of Inorganic and Organometallic Polymers and Materials*, *30*, 5180–5192.

Chavan, P., Lata, K., Kaur, T., Jambrak, A. R., Sharma, S., Roy, S., . . . & Rout, A. (2023). Recent advances in the preservation of postharvest fruits using edible films and coatings: A comprehensive review. *Food Chemistry*, 135916.

Chellaram, C., Murugaboopathi, G., John, A. A., Sivakumar, R., Ganesan, S., Krithika, S., & Priya, G. (2014). Significance of nanotechnology in food industry. *APCBEE procedia*, *8*, 109–113.

Chelliah, R., Wei, S., Daliri, E. B. M., Rubab, M., Elahi, F., Yeon, S. J., . . . & Oh, D. H. (2021). Development of nanosensors based intelligent packaging systems: Food quality and medicine. *Nanomaterials*, *11(6)*, 1515.

Cheng, M., Yan, X., Cui, Y., Han, M., Wang, Y., Wang, J., . . . & Wang, X. (2022). Characterization and release kinetics study of active packaging films based on modified starch and red cabbage anthocyanin extract. *Polymers*, *14(6)*, 1214.

Cherpinski, A., Gozutok, M., Sasmazel, H. T., Torres-Giner, S., & Lagaron, J. M. (2018). Electrospun oxygen scavenging films of poly (3-hydroxybutyrate) containing palladium nanoparticles for active packaging applications. *Nanomaterials*, *8(7)*, 469.

Corbo, M. R., Campaniello, D., Speranza, B., Bevilacqua, A., & Sinigaglia, M. (2015). Non-conventional tools to preserve and prolong the quality of minimally-processed fruits and vegetables. *Coatings*, *5(4)*, 931–961.

Damm, C., Münstedt, H., & Rösch, A. (2007). Long-term antimicrobial polyamide 6/silver-nanocomposites. *Journal of Materials Science*, *42*, 6067–6073.

Damm, C., Münstedt, H., & Rösch, A. (2008). The antimicrobial efficacy of polyamide 6/silver-nano-and microcomposites. *Materials Chemistry and Physics*, *108(1)*, 61–66.

De Azeredo, H. M. (2009). Nanocomposites for food packaging applications. *Food Research International*, *42(9)*, 1240–1253.

De Paiva, L. B., Morales, A. R., & Díaz, F. R. V. (2008). Organoclays: properties, preparation and applications. *Applied Clay Science*, *42(1–2)*, 8–24.

Divya, K., & Jisha, M. S. (2018). Chitosan nanoparticles preparation and applications. *Environmental Chemistry Letters*, *16*, 101–112.

Donsì, F., Marchese, E., Maresca, P., Pataro, G., Vu, K. D., Salmieri, S., . . . & Ferrari, G. (2015). Green beans preservation by combination of a modified chitosan based-coating containing nanoemulsion of mandarin essential oil with high pressure or pulsed light processing. *Postharvest Biology and Technology*, *106*, 21–32.

Dubey, N., Chitranshi, S., Dwivedi, S. K., & Sharma, A. (2023). Postharvest physiology, value chain advancement, and nanotechnology. In Goyal, M.R., Mishra, S.K., & Kumar, S. (Eds.), *Fresh-Cut Fruits And Vegetables. Nanotechnology Horizons in Food Process Engineering: Volume 3: Trends, Nanomaterials, and Food Delivery* (p. 99). Apple Academic Press.

Duncan, T. V. (2011). Applications of nanotechnology in food packaging and food safety: barrier materials, antimicrobials and sensors. *Journal of Colloid and Interface Science*, *363(1)*, 1–24.

Eshghi, S., Karimi, R., Shiri, A., Karami, M., & Moradi, M. (2022). Effects of polysaccharide-based coatings on postharvest storage life of grape: Measuring the changes in nutritional, antioxidant and phenolic compounds. *Journal of Food Measurement and Characterization*, *16(2)*, 1159–1170.

Espitia, P. J. P., Soares, N. D. F. F., Coimbra, J. S. D. R., de Andrade, N. J., Cruz, R. S., & Medeiros, E. A. A. (2012). Zinc oxide nanoparticles: Synthesis, antimicrobial activity and food packaging applications. *Food and Bioprocess Technology*, *5*, 1447–1464.

Farahanian, Z., Zamindar, N., Goksen, G., Tucker, N., Paidari, S., & Khosravi, E. (2023). Effects of nano-bentonite polypropylene nanocomposite films and modified atmosphere packaging on the shelf life of fresh-cut iceberg lettuce. *Coatings*, *13(2)*, 349.

Fateixa, S., Neves, M. C., Almeida, A., Oliveira, J., & Trindade, T. (2009). Anti-fungal activity of SiO_2/Ag_2S nanocomposites against *Aspergillus niger*. *Colloids and Surfaces B: Biointerfaces*, *74(1)*, 304–308.

Feyzioglu, G. C., & Tornuk, F. (2016). Development of chitosan nanoparticles loaded with summer savory (*Satureja hortensis* L.) essential oil for antimicrobial and antioxidant delivery applications. *LWT*, 70, 104–110.

Garavand, F., Cacciotti, I., Vahedikia, N., Rehman, A., Tarhan, Ö., Akbari-Alavijeh, S., . . . & Jafari, S. M. (2022). A comprehensive review on the nanocomposites loaded with chitosan nanoparticles for food packaging. *Critical Reviews in Food Science and Nutrition*, *62(5)*, 1383–1416.

Gardesh, A. S. K., Badii, F., Hashemi, M., Ardakani, A. Y., Maftoonazad, N., & Gorji, A. M. (2016). Effect of nanochitosan based coating on climacteric behavior and postharvest shelf-life extension of apple cv. Golab Kohanz. *LWT*, *70*, 33–40.

Garg, R., Rani, P., Garg, R., & Eddy, N. O. (2021). Study on potential applications and toxicity analysis of green synthesized nanoparticles. *Turkish Journal of Chemistry*, *45(6)*, 1690–1706.

Gemail, M. M., Elesawi, I. E., Jghef, M. M., Alharthi, B., Alsanei, W. A., Chen, C., . . . & Gad, M. M. (2023). Influence of wax and silver nanoparticles on preservation quality of murcott mandarin fruit during cold storage and after shelf-life. *Coatings*, *13(1)*, 90.

Ghanbarzadeh, B., Oleyaei, S. A., & Almasi, H. (2015). Nanostructured materials utilized in biopolymer-based plastics for food packaging applications. *Critical Reviews in Food Science and Nutrition*, *55(12)*, 1699–1723.

Gutiérrez, T. J., Ponce, A. G., & Alvarez, V. A. (2017). Nano-clays from natural and modified montmorillonite with and without added blueberry extract for active and intelligent food nanopackaging materials. *Materials Chemistry and Physics, 194*, 283–292.

Hashim, A.F., Youssef, K., & Abd-Elsalam, K.A. (2019). Ecofriendly nanomaterials for controlling gray mold of table grapes and maintaining postharvest quality. *European Journal of Plant Pathology*, *154*, 377–388.

Ho, B. T., Joyce, D. C., & Bhandari, B. R. (2011). Release kinetics of ethylene gas from ethylene–α-cyclodextrin inclusion complexes. *Food Chemistry*, *129(2)*, 259–266.

Hosseini, A., Koushesh Saba, M., & Watkins, C. B. (2023). Microbial antagonists to biologically control postharvest decay and preserve fruit quality. *Critical Reviews in Food Science and Nutrition*, 1–13.

Huang, J. Y., Li, X., & Zhou, W. (2015). Safety assessment of nanocomposite for food packaging application. *Trends in Food Science & Technology*, *45(2)*, 187–199.

Jeet, K., Kaur, R., & Kaur, M. (2022). Nanocellulose for food packaging applications. In Dar, A. H., & Nayik, G. A. (Eds.), *Nanotechnology Interventions in Food Packaging and Shelf Life* (pp. 39–60). CRC Press.

Jeevahan, J., & Chandrasekaran, M. (2019). Nanoedible films for food packaging: A review. *Journal of Materials Science*, *54(19)*, 12290–12318.

Jiang, Y., O'Neill, A. J., & Ding, Y. (2015). Zinc oxide nanoparticle-coated films: fabrication, characterization, and antibacterial properties. *Journal of Nanoparticle Research*, *17*, 1–9.

Kang, S., Pinault, M., Pfefferle, L. D., & Elimelech, M. (2007). Single-walled carbon nanotubes exhibit strong antimicrobial activity. *Langmuir*, *23(17)*, 8670–8673.

Khamis, Y., Hashim, A. F., Margarita, R., Alghuthaymi, M. A., & Abd-Elsalam, K. A. (2017). Fungicidal efficacy of chemically-produced copper nanoparticles against *Penincillium digitatum* and *Fusarium solani* on citrus fruit. *Philippine Agricultural Scientist (Philippines)*, *100(1)*.

Kumar, S., Mukherjee, A., & Dutta, J. (2020). Chitosan based nanocomposite films and coatings: Emerging antimicrobial food packaging alternatives. *Trends in Food Science & Technology*, *97*, 196–209.

Laureth, J. C. U., Moraes, A. J. D., Franca, D. L. B. D., Flauzino Neto, W. P., & Braga, G. C. (2018). Physiology and quality of 'Tahiti' acid lime coated with nanocellulose-based nanocomposites. *Food Science and Technology*, *38*, 327–332.

Lee, S. W., Mao, C., Flynn, C. E., & Belcher, A. M. (2002). Ordering of quantum dots using genetically engineered viruses. *Science*, *296(5569)*, 892–895.

Li, X., Xing, Y., Jiang, Y., Ding, Y., & Li, W. (2009). Antimicrobial activities of ZnO powder-coated PVC film to inactivate food pathogens. *International Journal of Food Science & Technology*, *44(11)*, 2161–2168.

Liau, S. Y., Read, D. C., Pugh, W. J., Furr, J. R., & Russell, A. D. (1997). Interaction of silver nitrate with readily identifiable groups: relationship to the antibacterialaction of silver ions. *Letters in Applied Microbiology*, *25(4)*, 279–283.

López-Gómez, A., Navarro-Martínez, A., & Martínez-Hernández, G. B. (2023). Effects of essential oils released from active packaging on the antioxidant system and quality of lemons during cold storage and commercialization. *Scientia Horticulturae*, *312*, 111855.

Maftoonazad, N., & Ramaswamy, H. (2019). Design and testing of an electrospun nanofiber mat as a pH biosensor and monitor the pH associated quality in fresh date fruit (Rutab). *Polymer Testing*, *75*, 76–84.

Mannino, S., & Scampicchio, M. (2007). Nanotechnology and food quality control. *Veterinary Research Communications*, *31*, 149–151.

Martins, A. J., Benelmekki, M., Teixeira, V., & Coutinho, P. J. G. (2012). Platinum nanoparticles as pH sensor for intelligent packaging. *Journal of Nano Research*, *18*, 97–104.

Mastromatteo, M., Mastromatteo, M., Conte, A., & Del Nobile, M. A. (2010). Advances in controlled release devices for food packaging applications. *Trends in Food Science & Technology*, *21(12)*, 591–598.

Melo, N. F. C. B., de MendonçaSoares, B. L., Diniz, K. M., Leal, C. F., Canto, D., Flores, M. A., . . . & Stamford, T. C. M. (2018). Effects of fungal chitosan nanoparticles as eco-friendly edible coatings on the quality of postharvest table grapes. *Postharvest Biology and Technology*, *139*, 56–66.

Mihindukulasuriya, S. D. F., & Lim, L. T. (2014). Nanotechnology development in food packaging: A review. *Trends in Food Science & Technology*, *40*(2), 149–167.

Mills, A., & Hazafy, D. (2009). Nanocrystalline SnO_2-based, UVB-activated, colourimetric oxygen indicator. *Sensors and Actuators B: Chemical*, *136(2)*, 344–349.

Mishra, A. K., Mishra, S. B., & Tiwari, A. (2011). Nanocomposites and their biosensor applications. *Biosensor Nanomaterials*, 255–268.

Mohammadi, P., Ali, A. A., & Ghadam, P. (2023). Mycogenic nanoparticles and their applications as antimicrobial and antibiofilm agents in postharvest stage. In Abd-Elsalam, K. A. (Ed.), *Fungal Cell Factories for Sustainable Nanomaterials Productions and Agricultural Applications* (pp. 635–655). Elsevier.

Mokrane, S., Buonocore, E., Capone, R., & Franzese, P. P. (2023). Exploring the global scientific literature on food waste and loss. *Sustainability*, *15(6)*, 4757.

Monteiro-Riviere, N. A., Nemanich, R. J., Inman, A. O., Wang, Y. Y., & Riviere, J. E. (2005). Multi-walled carbon nanotube interactions with human epidermal keratinocytes. *Toxicology Letters*, *155(3)*, 377–384.

Nabifarkhani, N., Sharifani, M., Daraei Garmakhany, A., Ganji Moghadam, E., & Shakeri, A. (2015). Effect of nano-composite and Thyme oil (*Tymus Vulgaris* L) coating on fruit quality of sweet cherry (Takdaneh Cv) during storage period. *Food Science & Nutrition*, *3(4)*, 349–354.

Nasr, F., Pateiro, M., Rabiei, V., Razavi, F., Formaneck, S., Gohari, G., & Lorenzo, J. M. (2021). Chitosan-phenylalanine nanoparticles (Cs-Phe Nps) extend the postharvest life of persimmon (*Diospyros kaki*) fruits under chilling stress. *Coatings*, *11(7)*, 819.

Nisha, V., Monisha, C., Ragunathan, R., & Johney, J. (2016). Use of chitosan as edible coating on fruits and in micro biological activity- an ecofriendly approach. *International Journal of Pharmaceutical Science Invention*, *5(8)*, 7–14.

Noman, M., Ahmed, T., White, J. C., Nazir, M. M., Azizullah, Li, D., & Song, F. (2023). *Bacillus altitudinis*-stabilized multifarious copper nanoparticles prevent bacterial fruit blotch in watermelon (*citrullus lanatus* l.): direct pathogen inhibition, in planta particles accumulation, and host stomatal immunity modulation. *Small*, *19(15)*, 2207136.

Oliani, W. L., Pusceddu, F. H., & Parra, D. F. (2022). Silver-titanium polymeric nanocomposite non ecotoxic with bactericide activity. *Polymer Bulletin*, *79(12)*, 10949–10968.

Ouzakar, S., Senhaji, N. S., Saidi, M. Z., El Hadri, M., El Baaboua, A., El Harsal, A., & Abrini, J. (2023). Antibacterial and antifungal activity of zinc oxide nanoparticles produced by *Phaeodactylum tricornutum* culture supernatants and their potential application to extend the shelf life of sweet cherry (*Prunus avium* L.). *Biocatalysis and Agricultural Biotechnology*, *49*, 102666.

Paomephan, P., Assavanig, A., Chaturongakul, S., Cady, N. C., Bergkvist, M., & Niamsiri, N. (2018). Insight into the antibacterial property of chitosan nanoparticles against *Escherichia coli* and *Salmonella* Typhimurium and their application as vegetable wash disinfectant. *Food Control*, *86*, 294–301.

Pereda, M., Marcovich, N. E., & Ansorena, M. R. (2019). Nanotechnology in food packaging applications: Barrier materials, antimicrobial agents, sensors, and safety assessment. In Martínez, L. M. T., Kharissova, O. V., & Kharisov, B. I. (Eds.), *Handbook of Ecomaterials* (pp. 2035–2056). Springer, Cham International Publishing.

Picard, E., Gauthier, H., Gérard, J. F., & Espuche, E. (2007). Influence of the intercalated cations on the surface energy of montmorillonites: Consequences for the morphology and gas barrier properties of polyethylene/montmorillonites nanocomposites. *Journal of Colloid and Interface Science, 307(2)*, 364–376.

Pimtong-Ngam, Y., Jiemsirilers, S., & Supothina, S. (2007). Preparation of tungsten oxide–tin oxide nanocomposites and their ethylene sensing characteristics. *Sensors and Actuators A: Physical, 139(1–2)*, 7–11.

Piña-Barrera, A. M., Álvarez-Román, R., Báez-González, J. G., Amaya-Guerra, C. A., Rivas-Morales, C., Gallardo-Rivera, C. T., & Galindo-Rodríguez, S. A. (2019). Application of a multisystem coating based on polymeric nanocapsules containing essential oil of *Thymus vulgaris* L. to increase the shelf life of table grapes (*Vitis vinifera* L.). *IEEE Transactions on NanoBioscience, 18(4)*, 549–557.

Podsiadlo, P., Choi, S. Y., Shim, B., Lee, J., Cuddihy, M., & Kotov, N. A. (2005). Molecularly engineered nanocomposites: layer-by-layer assembly of cellulose nanocrystals. *Biomacromolecules, 6(6)*, 2914–2918.

Poovai, P. D., Kumaran, N., Iyengar, A., Kalpana, P., & Ramasubramaniyan, M. R. (2023). A study on coating of Hydroxypropyl methylcellulose incorporated with a nano-emulsion of Piper betel leaf essential oil to enhance shelf-life and improve postharvest quality of Tomato (*Solanum lycopersicum* L.). *Journal of Applied and Natural Science, 15(1)*, 252–261.

Pourjavaher, S., Almasi, H., Meshkini, S., Pirsa, S., & Parandi, E. (2017). Development of a colorimetric pH indicator based on bacterial cellulose nanofibers and red cabbage (*Brassica oleraceae*) extract. *Carbohydrate Polymers, 156*, 193–201.

Rezić, I., Haramina, T., & Rezić, T. (2017). Metal nanoparticles and carbon nanotubes—perfect antimicrobial nano-fillers in polymer-based food packaging materials. In Grumezescu, A. M. (Ed.), *Food Packaging* (pp. 497–532). Academic Press.

Ruffo Roberto, S., Youssef, K., Hashim, A. F., & Ippolito, A. (2019). Nanomaterials as alternative control means against postharvest diseases in fruit crops. *Nanomaterials, 9(12)*, 1752.

Rusková, M., Opálková Šišková, A., Mosnáčková, K., Gago, C., Guerreiro, A., Bučková, M., . . . & Antunes, M. D. (2023). Biodegradable active packaging enriched with essential oils for enhancing the shelf life of strawberries. *Antioxidants, 12(3)*, 755.

Safakas, K., Giotopoulou, I., Giannakopoulou, A., Katerinopoulou, K., Lainioti, G. C., Stamatis, H., . . . & Ladavos, A. (2023). Designing antioxidant and antimicrobial polyethylene films with bioactive compounds/clay nanohybrids for potential packaging applications. *Molecules, 28(7)*, 2945.

Salem, E. A., Nawito, M. A. S., & Abd El-Raouf, A. E. -R. (2019). Effect of silver nano-particles on gray mold of tomato fruits. *Journal of Nanotechnology Research, 1(4)*, 108–118.

Salgado, P. R., Di Giorgio, L., Musso, Y. S., & Mauri, A. N. (2021). Recent developments in smart food packaging focused on biobased and biodegradable polymers. *Frontiers in Sustainable Food Systems, 5*, 630393.

Sardar, M., Ahmed, W., Al Ayoubi, S., Nisa, S., Bibi, Y., Sabir, M., . . . & Qayyum, A. (2022). Fungicidal synergistic effect of biogenically synthesized zinc oxide and copper oxide nanoparticles against *Alternaria citri* causing citrus black rot disease. *Saudi Journal of Biological Sciences, 29(1)*, 88–95.

Sau, S., Sarkar, S., Mitra, M., & Gantait, S. (2021). Recent trends in agro-technology, post-harvest management and molecular characterisation of pomegranate. *The Journal of Horticultural Science and Biotechnology, 96(4)*, 409–427.

Schudel, S., Shoji, K., Shrivastava, C., Onwude, D., & Defraeye, T. (2023). Solution roadmap to reduce food loss along your postharvest supply chain from farm to retail. *Food Packaging and Shelf Life, 36*, 101057.

Seil, J. T., & Webster, T. J. (2012). Antimicrobial applications of nanotechnology: methods and literature. *International Journal of Nanomedicine*, 2767–2781.

Shah, H. M. S., Khan, A. S., Singh, Z., & Ayyub, S. (2023). Postharvest biology and technology of loquat (*Eriobotrya japonica* Lindl.). *Foods, 12(6)*, 1329.

Sharma, C., Dhiman, R., Rokana, N., & Panwar, H. (2017a). Nanotechnology: an untapped resource for food packaging. *Frontiers in Microbiology, 8*, 1735.

Sharma, P., Sharma, A., Sharma, M., Bhalla, N., Estrela, P., Jain, A., . . . & Thakur, A. (2017b). Nanomaterial fungicides: in vitro and in vivo antimycotic activity of cobalt and nickel nanoferrites on phytopathogenic fungi. *Global Challenges, 1(9)*, 1700041.

Silvestre, C., Duraccio, D., & Cimmino, S. (2011). Food packaging based on polymer nanomaterials. *Progress in Polymer Science, 36(12)*, 1766–1782.

Singh, D. P., & Packirisamy, G. (2022). Biopolymer based edible coating for enhancing the shelf life of horticulture products. *Food Chemistry: Molecular Sciences, 4*, 100085.

Song, H., Yuan, W., Jin, P., Wang, W., Wang, X., Yang, L., & Zhang, Y. (2016). Effects of chitosan/nano-silica on postharvest quality and antioxidant capacity of loquat fruit during cold storage. *Postharvest Biology and Technology, 119*, 41–48.

Sreelatha, S., Kumar, N., Yin, T. S., & Rajani, S. (2022). Evaluating the antibacterial activity and mode of action of thymol-loaded chitosan nanoparticles against plant bacterial pathogen *Xanthomonas campestris* pv. *campestris*. *Frontiers in Microbiology, 12*, 792737.

Taghinezhad, E., & Ebadollahi, A. (2017). Potential application of chitosan-clay coating on some quality properties of agricultural product during storage. *Agricultural Engineering International: CIGR Journal, 19(3)*, 189–194.

Tharanathan, R. N. (2003). Biodegradable films and composite coatings: Past, present and future. *Trends in Food Science & Technology, 14(3)*, 71–78.

Vasile, C. (2018). Polymeric nanocomposites and nanocoatings for food packaging: A review. *Materials, 11(10)*, 1834.

Ventura-Aguilar, R. I., Bautista-Baños, S., Mendoza-Acevedo, S., & Bosquez-Molina, E. (2023). Nanomaterials for designing biosensors to detect fungi and bacteria related to food safety of agricultural products. *Postharvest Biology and Technology, 195*, 112116.

Videira-Quintela, D., Guillén, F., Martin, O., & Montalvo, G. (2022). Antibacterial LDPE films for food packaging application filled with metal-fumed silica dual-side fillers. *Food Packaging and Shelf Life, 31*, 100772.

Viet, P. V., Nguyen, H. T., Cao, T. M., & Hieu, L. V. (2016). *Fusarium* antifungal activities of copper nanoparticles synthesized by a chemical reduction method. *Journal of Nanomaterials, 2016*.

Wang, Y., Zhang, J., Wang, D., Wang, X., Zhang, F., Chang, D., . . . & Wang, X. (2023). Effects of cellulose nanofibrils treatment on antioxidant properties and aroma of fresh-cut apples. *Food Chemistry, 415*, 135797.

Wantat, A., Seraypheap, K., & Rojsitthisak, P. (2022). Effect of chitosan coatings supplemented with chitosan-montmorillonite nanocomposites on postharvest quality of 'Hom Thong' banana fruit. *Food Chemistry, 374*, 131731.

Xiao-e, L., Green, A. N., Haque, S. A., Mills, A., & Durrant, J. R. (2004). Light-driven oxygen scavenging by titania/polymer nanocomposite films. *Journal of Photochemistry and Photobiology A: Chemistry, 162(2–3)*, 253–259.

Yanat, M., & Schroën, K. (2021). Preparation methods and applications of chitosan nanoparticles; with an outlook toward reinforcement of biodegradable packaging. *Reactive and Functional Polymers, 161*, 104849.

Ye, S., Chen, M., Liu, Y., Gao, H., Yin, C., Liu, J., . . . & Zhang, Y. (2023). Effects of nanocomposite packaging on postharvest quality of mushrooms (*Stropharia rugosoannulata*) from the perspective of water migration and microstructure changes. *Journal of Food Safety*, e13050.

Yildirim, S., Röcker, B., Pettersen, M. K., Nilsen-Nygaard, J., Ayhan, Z., Rutkaite, R., . . . & Coma, V. (2018). Active packaging applications for food. *Comprehensive Reviews in Food Science and Food Safety, 17(1)*, 165–199.

Zare, M., Namratha, K., Ilyas, S., Hezam, A., Mathur, S., & Byrappa, K. (2019). Smart fortified PHBV-CS biopolymer with ZnO–Ag nanocomposites for enhanced shelf life of food packaging. *ACS Applied Materials & Interfaces, 11(51)*, 48309–48320.

Zhu, R., Desroches, M., Yoon, B., & Swager, T. M. (2017). Wireless oxygen sensors enabled by Fe (II)-polymer wrapped carbon nanotubes. *ACS Sensors, 2(7)*, 1044–1050.

Modified Atmosphere Packaging and Individual Shrink Wrapping for Fresh Fruits and Vegetables

15

Mahmood Ul Hasan, Zora Singh*,
Hafiz Muhammad Shoaib Shah, and Aman Ullah Malik
*Corresponding Author: z.singh@ecu.edu.au

15.1 INTRODUCTION

Fruits and vegetables are an integral part of the human diet and are extensively growing worldwide to feed the continuously increasing population. Fruits and vegetables contribute significantly to global food and nutritional security (Singh, 2022). Even with the rising global demand, about one billion people are still undernourished and facing several nutritional shortcomings (Bond et al., 2013). Nevertheless, the wastage in fruit and vegetable supply chains is the main scourge of recent times. Overall, food loss and wastage in global fresh produce supply chains are approaching 60% depending on the perishability of the crop (Porat et al., 2018). These losses occur at different stages along the supply chain, commencing from crop production through harvesting, storage, processing, transportation, and retail marketing until it reaches the consumer. Major postharvest losses have been reported, accordingly, several technologies are being adopted to minimize losses in fresh horticultural produce and enhance the availability of food for the growing population to combat global hunger.

In various non-chemical approaches, packaging has been a commonly used technique for decades to protect commodities from the external environment, providing a less expensive way to preserve and overcome the concerns of food safety (Mangaraj et al., 2009). Among all such approaches, modified atmosphere packaging (MAP) has been adopted rapidly and widely used at a commercial scale because it offers a great deal of benefits for protecting fresh minimally processed (fresh-cut) products that are highly susceptible to surface browning, water loss, microbial spoilage, higher respiration rates, and ethylene production (Kader & Watkins, 2000; Singh et al., 2009a). The application of MAP was first reported in 1927 to extend the storage life of apples in modified atmosphere with manipulated gas composition (Phillips, 1996). MAP technology has received much attention from consumers because of its nature of preserving fresh fruits and vegetables without any synthetic chemical application or chemical residues on fruit and vegetable surfaces (Belay et al., 2019).

The MAP technique involves the use of semipermeable polymeric films which help in altering the internal atmosphere by modifying the gas composition; elevated CO_2 and decreased O_2 in the package around the fresh commodity significantly reduce the respiration and permeation of gases, delay ripening and senescence, and extend the storage life of the product by restricting oxidative stress and reducing microbial activity and shrivelling during the postharvest period (Qu et al., 2022). MAP technology functions in two ways: active or passive. In passive packaging, gaseous exchange and permeability depends on the natural process of ripening and respiration for achieving the desired gas composition inside the package to delay the postharvest senescence of certain packed commodities (Caleb et al., 2013; Costa et al., 2011). Active packaging is a technique in which the internal modified atmosphere is established by scavenging inappropriate gases produced inside the package via absorbents, gas scavengers, or flushing (Charles et al., 2005). Additionally, individual shrink wrapping (ISW) is a type of packaging which works for individual units (fruit or vegetable) instead of fruit lot, followed by sealing the commodity in a flexible film by removing extra air between the fruit and film, altering the micro atmosphere, a considerably reducing water loss, and preserving quality in the postharvest period (Rao, 2017) (Figure 15.1). This chapter

DOI: 10.1201/9781003370376-20

covers different aspects of MAP technology with its principal functions; packaging materials used for MAP, including biodegradable packaging; and the effects of MAP on nutritional quality; physiological processes such as ethylene biosynthesis, respiration, water loss, and firmness; chilling injury, enzymatic browning, flesh softening, colour degradation, physiological disorders, microbial spoilage, and MAP crosstalk with other postharvest technologies for postharvest preservation.

15.2 MAP TECHNOLOGY—RECENT ADVANCEMENTS

In the last two decade, scientists in collaboration with the food packaging industry have put significant efforts into developing crop-specific MAP bags for their application at a commercial scale. Over time, a rich source of literature has been compiled regarding MAP advancements, fundamental principles, and their effects on variable aspects of fresh whole or fresh-cut fruits and vegetables (Belay et al., 2016; Brecht, 2006; Caleb et al., 2013; Kader & Watkins, 2000; Mahajan & Lee, 2023; Mangaraj et al., 2009; Mattheis & Fellman, 2000; Moradinezhad et al., 2018; Qu et al., 2022; Rennie & Sunjka, 2017; Singh et al., 2009a; Werner & Hotchkiss, 2005; Wilson et al., 2019; Zhuang et al., 2014), covering research efforts approaching its commercial success, and considering a few reports on ISW technology, Ben-Yehoshua (1985) and Rao (2017) have summarized the research findings depicting ISW-relevant information. MAP is a dynamic technology where the interplay between the product respiration and gas permeation via films used for packaging works simultaneously to attain equilibrium for permeable gaseous exchange (Mahajan & Lee, 2023). Accordingly, the equilibrium requirement can only be achieved primarily through identification of appropriate plastic films and their permeability of gas matches to the respiration of the desired crop to be packed for prolonging storage life. The successful interaction between MAP and fresh produce relies on three basic components; the physiology of the crop, polymer engineering, and converting technology comprising raw polymer fabrication and modification in their structure for its desired characteristics (Brandenburg & Zagory, 2009; Caleb et al., 2013). Regarding the physiology of the crop, among extrinsic factors, temperature and relative humidity under storage, levels of CO_2 and O_2, weight loss, and duration of storage influence the rate of respiration, which directly affects the storage quality of fresh produce. Similarly, several intrinsic factors, including genotype, type of tissue, cultural practices, season of production, growing location, harvest maturity of the crop, harvest handling, diseases, and pre-storage treatments, determine the fate of MAP technology for storage life extension and quality maintenance of fresh fruits and vegetables during the postharvest period (Caleb et al., 2013; Mangaraj et al., 2009). Various attempts have been made to develop a protocol for crop-specific models, MAP bags impregnated with certain anti-senescence or antimicrobial compounds, and testing the different type of biodegradable plastic films for ensuring quality maintenance of stored fruits and vegetables and reducing any environmental hazards, such as plastic waste.

15.3 PACKAGING MATERIALS AND TYPES

Several flexible plastic films have been used for manufacturing MAP bags, liners, punnets, and lids for preserving fresh fruits and vegetables. Different types of polymers, including low-density polyethylene (LDPE), linear LDPE, linear medium LDPE, high-density polyethylene (HDPE), ultralow-density polyethylene (ULDPE), ethylene vinyl acetate (EVA), polyvinyl chloride (PVC), polypropylene (PP), polyethyleneterapthalate (PET), and polystyrene (PP) are commonly used in developing MAP with distinct physical, chemical, and gas transmission characteristics. These films can be individual, blended monolayer films, or a combination of different polymers/layers (Brandenburg & Zagory, 2009). Additionally, the micro-perforation in the structure of MAP films is useful in maintaining equilibrium and increasing gas permeability, showing good results for crops with high rates of respiration. The storage temperature has a significant influence on respiration modulation in fresh horticultural produce to be packed in MAP bags. The failure to maintain storage temperature escalates respiration rate, and commodities packed in MAP bags consumed more oxygen following lower levels inside the package, which may result in increased microbial spoilage during the postharvest period. Accordingly, breathable MAPs such as BreatheWay™ were introduced to compensate for the temperature fluctuations (within moderate limits) along with appropriate gas permeability during storage (Poças et al., 2008). Recently, multiple concerns have been raised regarding plastic packaging usage in fresh produce industries, prompting researchers and other industry stakeholders to search for MAP technology specifically made with biodegradable or biobased plastic films, and these films have less gaseous exchange permeability than other polyolefin films (Mahajan & Lee, 2023). For sustainable packaging, new materials with more biodegradability or developed through bio-based compositions such as gelatin, starch, cellulose, or polylactic acid (PLA) offers an alternative to traditional MAPs, but its efficacy may be variable (Abdul-Khalil et al., 2018; Mistriotis et al., 2016). Qin et al. (2016) have reported that PLA-based active packaging of cold stored hot peppers showed significant reduction in weight loss and microbial growth and exhibited higher marketability with retained firmness and preserved ascorbic acid in comparison with PE packaging during 28 days of postharvest storage. Furthermore, the emerging trend of using nanotechnology in packaging materials could enhance the performance of MAPs for preserving fruits and vegetables (Wilson et al., 2019). For

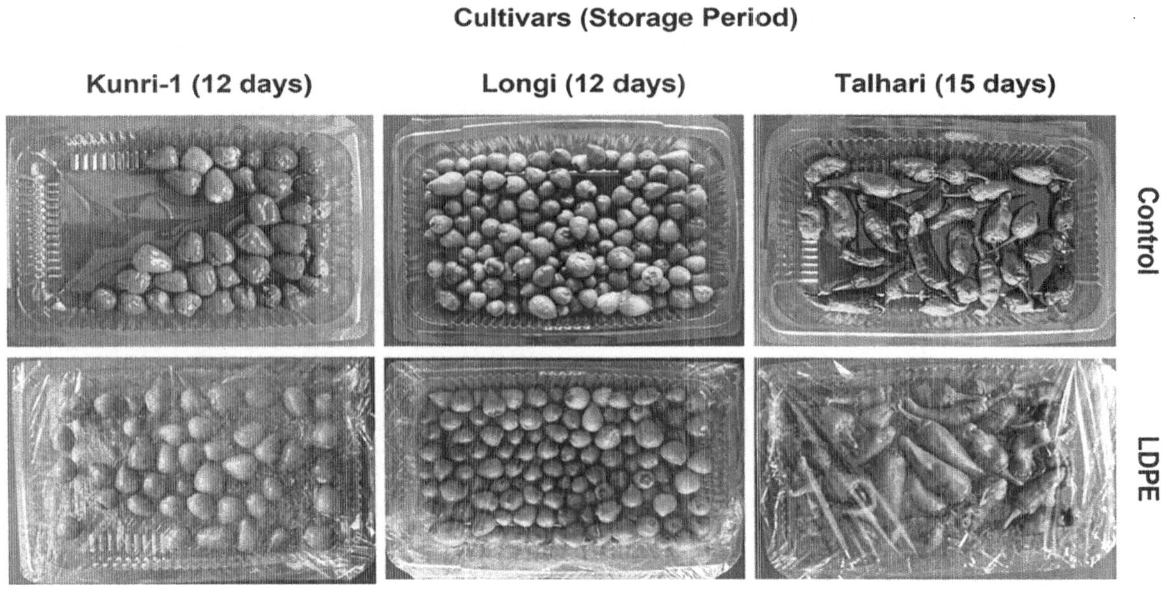

FIGURE 15.1 Effect of low-density polyethylene (LDPE- 12 μ) packaging on the postharvest quality of three chilli cultivars of Pakistan. Source: Mahmood Ul Hasan and Aman Ullah Malik, ACIAR project—strengthening vegetable value chains in Pakistan for greater community livelihood benefits.

instance, PLA-based films along with the release of clove oil loaded on mesoporous silica nanoparticles exhibited lower water loss, restricted the activities of peroxidase (POD) and polyphenol oxidase (PPO), and conserved higher phenolics as well as ascorbic acid in cold stored white button mushrooms (Zhu et al., 2022). The LDPE has been found effective in preserving the green color of chillies (Figure 15.1).

The concept of smart or intelligent packaging had emerged to replace the traditional packaging system which has greatly contributed to providing a solution to the food industry as an early strategy in the distribution chain. Considering the consumers' need, the innovative packaging has been introduced with improved functions of providing information regarding data accumulation of quality detection, traceability, application of scientific technology, help in making decisions, extending storability, and possible issues (Poças et al., 2008). Yam et al. (2005) have stated that advancements in smart/intelligent packaging include the use of barcode labels, TTIs (time temperature indicators), RFID (radio frequency identification)–oriented tags, gas indicators, and biosensors, which helps in monitoring the quality of packed commodities. The physical shock indicators are also aligned with smart packaging to avoid transportation losses in fruits and vegetables (Poças et al., 2008). Additionally, the composition/concentration of CO_2 and O_2 in headspace changes during storage while the suitable devices/sensor technologies can identify and quantify the changes in the gases may help in predicting the shelf life and quality of packed fresh produce (Zhuang et al., 2014). For example, the development of novel O_2 sensor technology possibly could be used to determine the variation in O_2 in packaged produce, as earlier tested by Seefeldt et al. (2012) in broccoli and rocket leaves, which showed the continuous monitoring of O_2 during the entire period of storage. Poças et al. (2008) have revealed that usage of smart packaging technology is more progressive in Japanese and US markets compared to Europe, and their application is more prevalent in other industries, i.e., electronic devices and medical or medicinal products where the products are more sensitive and highly prone to damage along the distribution chain. The potential reasons behind less application at the commercial scale may be due to cost implications, as fruit and vegetable industries have smaller margins compared to other products, a longer period of return, and are slower to make decisions on technical solutions in their supply chain operations (Poças et al., 2008).

Furthermore, the ethylene biosynthesis in fresh fruits and vegetables packed in MAP bags may substantially affect their quality and lower their marketable life. Several techniques have been tested to degrade/downregulate ethylene inside packages. Among all, the use of photocatalytic technology or photooxidative degradation, use of oxidants such as the application of ozone, potassium permanganate ($KMnO_4$), and use of adsorbents (activated carbon, zeolite, cyclodextrin, etc.) and ethylene antagonists, are vital to remove extra ethylene, reduce fruit and vegetable sensitivity to ethylene, and help in downregulating their production and ethylene action during the postharvest period (Zhangh et al., 2022). The application of titanium oxide (TiO_2) has been commonly used for photocatalytic oxidation of ethylene due to its low cost, low toxicity, and good thermal stability, as well as its impact on the suppression of microorganisms during storage (Tian et al., 2019). In addition, the implication of $KMnO_4$ inside the package has great potential of ethylene degradation into CO_2 and H_2O, while its colour turns from purple to brown and prevents fruit quality from damages associated with excessive ethylene production during storage (Zhangh et al., 2022). Similarly, among various ethylene antagonists, the application of 1-MCP and nitric oxide

(NO) outclassed and had a synergistic impact while combining with MAP on the postharvest quality conservation of fruits and vegetables (Vázquez-Celestino et al., 2016).

The condensation risk in MAPs has been regarded as a limitation in perishable fruits and vegetables when subjected to temperature fluctuation at higher RH, which leads to poor cosmetic quality and increased microbial spoilage (Bovi et al., 2016). The water vapour permeability of the packaging is also imperative when selecting the type of packaging for avoiding condensation along with maintaining desired equilibrium of CO_2 and O_2 concentrations. To overcome the condensation risk, macro- and micro-perforated films have been widely used for packaging fresh fruits and vegetables. Nevertheless, the micro- and macro-perforation in packaging films also had application constraints, such as higher rates of transpiration than of vapour transmission in fruits and vegetables, and a higher number of perforations may not adequately develop the internal modified atmosphere (Mahajan & Lee, 2023). The application of moisture absorbers such as potassium chloride, calcium chloride, calcium oxide, and sorbitol in different proportions may help in capturing moisture in packaging, as tested in mushrooms during the postharvest period (Azevedo et al., 2011; Mahajan et al., 2008). Besides its efficiency, consumer acceptability is less for fresh produce containing moisture absorbing pads or sachets (Mahajan & Lee, 2023). Accordingly, multiple polymers (co-extruded and biodegradable) have been introduced with enhanced water vapour permeability as compared to traditional polyethylene or polypropylene films. The blending of co-extruded film materials (e.g. hydrophilic polyamides) with other polymers, or non-polymers, substantially helped in achieving the desired water vapour permeability (Bovi et al., 2016; Rodov et al., 2010). Aharoni et al. (2007) have jointly developed a series of packaging films (Xtend®) with higher water vapour permeability potential. Tested on several fruits and vegetables, the films significantly restricted condensation, maintained in-package RH, inhibited various storage issues (chilling injury, decay incidence, sprouting, discolouration, microbial growth, etc.), delayed postharvest senescence, and expanded the marketable window with extended shelf life.

15.4 EFFECTS OF MAP AND ISW TECHNOLOGIES ON THE POSTHARVEST QUALITY OF FRUITS AND VEGETABLES

15.4.1 Nutritional Quality

Fresh fruits and vegetables are rich source of a variety of phytonutrients with plenty of potential antioxidants playing a vital role in human health by strengthening the immune system and subsequently preventing various chronic diseases. The antioxidant capacity of a fruit or vegetable depends on the availability of different bioactive compounds such as flavonoids, carotenoids, anthocyanins, phenolic acids, polyphenols, ascorbic acid, and other vitamins, and their concentrations vary among commodities (Kader, 2009). MAP technology has been reported to preserve antioxidant compounds in a wide range of fresh horticultural produce, thereby helping extend the storage period. MAP technology has proved to be a saviour of ascorbic acid levels from postharvest degradation in stored fruits and vegetables. For instance, Islam et al. (2022) have reported that jujube fruit packed in MAP bags exhibited significantly higher ascorbic acid level during 35 days of cold storage following 4 days at ambient conditions. The strawberries packed in MAP bags with prior application with *Aloe vera* significantly maintained higher ascorbic acid during storage (Esmaeili et al., 2021). Similar results for the retention of higher levels of ascorbic acid have been widely reported in apricot (Dorostkar et al., 2022), litchi (Ali et al., 2019a), pakchoi (Zhang et al., 2022), cucumber (Wang et al., 2022), and chillies (Sarfraz et al., 2020) packed in different types of MAP either alone or in combined form with other treatments. However, it has also been reported that Sindhri and Sufaid Chaunsa mangoes packed in MAP bags showed non-significant changes to levels of ascorbic acid during cold storage (Hafeez et al., 2016). The phenolic compounds in fresh fruits and vegetables have been preserved in MAP bags during the postharvest period. Öztürk et al. (2019a) have evaluated the effects of preharvest application of gibberellic acid and $CaCl_2$ on the phenolics profile of sweet cherries at harvest and during storage. Preharvest-treated cherries packed in LDPE (22 µm, Xtend®) showed reduced losses of phenolic acids (4-hydroxybenzoic acid, epicatechin, caffeic acid, *p*-coumaric acid, and protocatechuic acid) during 21 days of cold storage. Additionally, the application of MAP technology helps in maintaining colour pigments with high antioxidant properties such as anthocyanins, flavonoids, and carotenoids in stored fruits and vegetables. Gimeno et al. (2021) have revealed that red raspberries packed in MAPs showed higher flavonoids and anthocyanins, in addition to increased phenolics with lower rot incidence during 12 days of cold storage. Similarly, the blueberries pre-treated with gaseous ozone followed by MAP storage significantly preserved individual and total anthocyanins during a postharvest period of 15 days (Pinto et al., 2020). However, the sweet cherries 0900 Ziraat stored in MAP bags showed lower monomeric anthocyanins as compared to control fruit during 21 days of cold storage followed by three days at shelf conditions, which might be due to lowering the senescence by retaining the colour and quality (Aglar et al., 2017). In addition, enzymatic activity has also been reported to vary in different crops under different storage conditions. Ali et al. (2019a) have reported that Gola litchi fruit packed in MAP bags showed significant results for maintaining higher activities of ascorbate peroxidase (APX), superoxide dismutase (SOD), and catalase (CAT) and lower peroxidase (POD) and polyphenol oxidase (PPO) enzymes during 28 days of cold storage, thereby helping reduce pericarp browning. Moreover, the total soluble solids and acid content in fresh horticultural produce are the key biochemical quality markers for determining consumer acceptability in the market (Hasan et al., 2021). The mango, medlar, and jujube fruit packed in MAP bags displayed lower TSS, which may be due to delayed fruit ripening during cold storage (Ozturk et al., 2019; Perumal et al., 2021; Reche et al., 2019). On the other hand, the change in titratable acidity (TA) of fruit may be variable due to the physiology of the crop or cultivar, for instance fruit packed in MAP bags exhibited higher retention of TA in Banganapalli and Totapuri mangoes, while MAP-packed Sindhri mango fruit revealed non-significant results for TA among different treatments (Hafeez et al., 2016; Perumal et al., 2021). However, TA content decreases along the advancement of the storage period of fruit irrespective of treatments. The medlar cv. İstanbul packed in MAP bags showed lower degradation of individual organic acids during the entire period of cold storage. The levels of oxalic acid, citric acid, malic acid, and fumaric acid were noted as decreasing but higher in MAP-packed medlar fruit as compared to control fruit during 60 days of cold storage (Ozturk et al., 2019). Accordingly, the MAP technology preserved the nutritional quality of packed fruit and vegetables, while the successes for quality preservation are variable among different crops.

15.4.2 Respiration Rate and Ethylene Production

In MAP technology, the physiological process of respiration rate is the pivotal parameter to modulate during storage for managing quality and extending the storage life of fruits or vegetables to be packed in MAPs. This is a simultaneous process of gas exchange permeation from the bag and respiration of the packed product. Additionally, MAP manufacturing significantly relies on the respiration rate of the produce, which helps in modifying internal gas atmosphere, delaying senescence, and extending storability (Mangaraj et al., 2009). Nevertheless, the sustainable equilibrium with optimum conditions of storage may guarantee the impressive results in maintaining postharvest quality (Mahajan & Lee, 2023). Furthermore, ethylene production is a key process of ripening in fresh horticultural produce which may be variable according to their physiological classification as climacteric or non-climacteric, which greatly influences the storability of the produce (Hasan et al., 2021). However, the mitigation of ethylene production or perception around the stored fruits and vegetables is the most critical strategy for conserving the end product's quality. Accordingly, the lowering of oxygen levels may assist in reducing ethylene production, though it is imperative to develop MAPs (alleviation of ethylene through diffusion), or inclusion of any technology such as ethylene scrubbers, absorbents, or antagonists to avoid their devastating effects on quality (Qu et al., 2022). Illeperuma and Jayasuriya (2002) have reported that Pollack avocadoes packed in LDPE (0.05 mm) showed suppressed ethylene production and respiration during 32 days of cold storage at 12 °C. In another study, jujube fruit packed in Xtend® MAP bags showed lower respiration along with delayed TSS during the entire period of cold storage (Islam et al., 2022) (Table 15.1). Similarly, Mauritius litchi fruit packed in 30 µm MAP bags showed lower ethylene production, respiration besides its significant effects on peri-carp dry matter, and lower TSS during 9 days of cold storage (Somboonkaew & Terry, 2010). Reduced respiration has also been reported in MA-packed mangoes; however, it seems that response to MAP bags is cultivar dependent. Hafeez et al. (2016) reported that mango cv. Sindhri packed in Xtend® MAP bags showed lower respiration during cold storage at 11 °C, while Sufaid Chaunsa mangoes packed in similar bags depicted non-significant effects for respiration while stored at similar temperature conditions. Moreover, melons and tomatoes packed in MAP bags showed declined ethylene production and rate of respiration, with maintained fruit quality under storage (Bailén et al., 2006; Flores et al., 2004) (Table 15.1).

15.4.3 Water Loss

Water loss is the prime quality indicator for the conservation of fresh fruits and vegetables. It is also known as weight loss, physiological loss of weight, or moisture loss during the postharvest period (Lufu et al., 2020). MAP technology has the potential to lower respiration and ethylene production, which significantly delays the senescence of fresh fruits and vegetables, thereby reducing water loss during storage. Table 15.1 summarizes the percentage of water loss after the application of MAP technology in a wide range of fruits and vegetables. Mpho et al. (2013) reported that avocados cvs. Fuerte, Hass, and Ryan showed higher water loss in the untreated control (between 9% and 10%) compared to fruit packed in MAP bags (0.4–0.6%) after 18 days of cold storage following 5 days of ripening. Nevertheless, the water loss is slightly different among the cultivars even under similar packaging materials and storage conditions, while the variation in water loss may be due to the genotypic change among fruit and the stage of maturity at harvest. Similar findings for change in water loss have been reported in sweet cherry cultivars, including Siah-e Mashhad, Érdi Botermo, and Albaloo Mohallai. Packed in three-layer PE packaging, these cherries exhibited 0.2–0.7% loss of water during 60 days of cold storage (Khorshidi et al., 2011). In addition, a fruit packed in different packaging materials also showed different results. Ding et al. (2002) reported that loquat cv. Mogi packed in MAP bags exhibited lower water loss of about 1.5% compared to fruit packed in perforated PE, which showed approximately 8.9% water loss after 60 days of cold storage at 5 °C (Table 15.1). Hayat et al. (2017) have reported that Kaghzi limes packed in MAP bags experienced substantially reduced water loss by 1.04% compared to unpacked control fruit (33.46%) after 45 days of cold storage at 10 °C. In leafy vegetables, water loss is the major constraint for maintaining quality during the postharvest and marketing supply chains. Siddique et al. (2021) have stated that Desi spinach packed in Xtend® MAP bags showed 1.3% water loss and those packed in Biofresh® MAP bags showed 4.2%, which was a substantial reduction compared to the open control spinach leaves (39.7%) after 20 days of cold storage (Table 15.1).

TABLE 15.1 Effects of MAP on Postharvest Water Loss and Physiochemical Quality of Fresh Fruits and Vegetables during Storage

CROP	*CULTIVAR*	*MAP (TYPE/ SPECIFICATIONS)*	*STORAGE CONDITIONS AND DURATION*	*WATER LOSS (%)*	*PHYSIOCHEMICAL QUALITY*	*REFERENCE*
Avocado	Fuerte	MAP (0.00313%)	10 °C, 85% RH, 14d + 5d ripening at 25 °C	0.6	↓ disease incidence and severity, ↑ firmness and sensory attributes	Mpho et al. (2013)
	Hass			0.4	–	
	Ryan			0.5	–	
	Fuerte	MAP (BOP- 0.00313%)	10 °C, 85% RH, 18d + 5–10d ripening at 25 °C	0.3	↓ disease incidence and severity, grey pulp, vascular browning, firmness loss, maintained ↑ overall acceptability, sensory attributes	Sellamuthu et al. (2013)
	Hass			0.3	–	
	Pollock	LDPE (0.05 mm)	12 °C, 94% RH, 32d	1.1	↓ ethylene, respiration	Illeperuma and Nikapitiya (2002)
Apricot	Roxana	MAP	0 ± 0.5 °C, 45d	8.8	↑ firmness, TSS, TA, ↓ internal browning	Varlı-Yunusoğlu and Ekinci (2023)
Broccoli	Lucky	MAP bags (Xtend®)	0 °C, 95% RH, 28d	10.8	↓ colour degradation, ↑ visual quality	Sabir (2012)
	Parthenon	Polypropylene bags (30 µm)	5 °C, 12d	0.75	↑ overall appearance and colour	Fernández-León et al. (2013)
Blueberry	Brigitta	MAP (60 µm)	0 °C, 45d	0.1	↑ firmness, TSS, ↓ dehydration percentage	Rodriguez and Zoffoli (2016)
Cherries	Siah-e Mashhad	3-layer PE (70 µm)	0 ± 1 °C, 90%, 10% O_2 + 15% CO_2 + 75% N_2, 60d	0.24	↑ firmness	Khorshidi et al. (2011)
	Érdi Botermo			0.70	–	
	Albaloo Mohallai			0.52	--	
	0900 Ziraat	MAP (20 µm LDPE), Xtend®	0 °C with 90–95% RH, 0d, 2d at 20 ± 2 °C	4.50	↓ TSS, pitting percentage, ↑ TA, stem colour score, fruit elasticity, colour values	Özkaya et al. (2015)
Grapes	Flame Seedless	nonperforated polypropylene (N-PP) (20 µm)	1 °C with 90% RH, 53d	2.16	↓ TSS, TA, ↑ berry firmness, crunchiness, sweetness, overall quality	Martínez-Romero et al. (2003)
	Crimson Seedless	eugenol + (N-PP)	1 °C with 90% RH, 35d	0.2	↔ firmness, ↓ ethylene, TSS:TA ratio, fruit decay, ↑ colour values, rachis browning (score)	Valverde et al. (2005)
	Red Globe	ZOEpac-210 bags (25 µm LDPE)	0 °C with 90–95% RH, 120d	0.38	↓ weight loss, rachis browning, ○ fungal decay	Candir et al. (2012)
Grapefruit	Star Ruby	MAP (20 µm) Xtend®	10 °C, 112d + 7d at 20 °C	3.29	↓ TA, ↑ TSS, TSS:TA ratio, ○ taste	Chaudhary et al. (2015)

(Continued)

TABLE 15.1 (Continued)

CROP	*CULTIVAR*	*MAP (TYPE/ SPECIFICATIONS)*	*STORAGE CONDITIONS AND DURATION*	*WATER LOSS (%)*	*PHYSIOCHEMICAL QUALITY*	*REFERENCE*
Jujube	Li	MAP bags (Xtend®)	4 ± 0.5 °C, 90 ± 5 % RH	1.4	↓ respiration, TSS, ↑ firmness, colour values (L*, hue angle), TA	Islam et al. (2022)
Loquat	Mogi	MAP	5 °C, 60d	1.5	↔ organic acids, ○ total sugars, ↑ core browning	Ding et al. (2002)
		Perforated PE		8.9	↑ visual quality (score), chemical compounds, ↓ disorders	
Litchi	Mauritius	PropaFresh™ PFAM (30 μm)	13 °C, 9d	0.37	↓ pericarp dry matter, TSS, ethylene production, respiration, ↔ sugars, organic acids	Somboonkaew and Terry (2010)
	Gola	MAP bags Xtend®	5 ± 1 °C, 28d	3.1	↓ fruit decay, skin browning, relative electrolyte leakage, ↑ TSS, TA, taste, flavour, aroma, overall acceptability	Ali et al. (2019a)
Mango	Karuthacolomban	LDPE (0.05.mm) + scavenger	13 °C with 94% RH, 24d	2.0	↓ respiration, ethylene, ↑ firmness, TA	Illeperuma and Jayasuriya (2002)
	Sindhri	Map bags (Xtend®)	11 °C, 28d	5.79	↓ respiration, TSS, softness, colour score ↑ marketable fruit, reducing sugars, flavour	Hafeez et al. (2016)
	Sufaid Chaunsa	Map bags (Xtend®)	11 °C, 35d	4.78	○ respiration, TSS, softness, disease severity, ↓TA ↑ marketable fruit, reducing sugars, flavour	
	Samar Bahisht Chaunsa	Map bags (Xtend®)	13 ± 1 °C with 85–90% RH, 35d	3.2[a]	↓ peel colour (score), TSS, TA, shrivelling, ethylene production	Ali et al. (2019b)
		MAP bags (Biofresh®)	13 ± 1 °C with 85–90% RH, 35d	4.2[a]	↑ peel colour (score), taste, marketability index, ↓ TSS, TA, shrivelling, ethylene production	
Melon	Antisense	MAP (15 μm)	2 °C with 90–95% RH, 21d + 4d (22 °C)	5.9	↓ respiration, ethylene production, ion leakage	Flores et al. (2004)
Mushroom	Shiitake	Biorientated polypropylene (BOPP) MAP (35 μm)	4 ± 1 °C with 90% RH, 20d	4.4	↓ textural quality	Jiang et al. (2010)

CROP	CULTIVAR	MAP (TYPE/ SPECIFICATIONS)	STORAGE CONDITIONS AND DURATION	WATER LOSS (%)	PHYSIOCHEMICAL QUALITY	REFERENCE
	Button	BOPP MAP (35 μm)	4±1 °C with 90% RH, 16d	2.5	↓ cap opening	Jiang et al. (2011)
Persimmon	Hachiya	PE (35 μm) wrapping	0±1 °C with 90±5% RH, 90d + 7d at 20±1 °C with 60±5% RH	0.76	↑ firmness, acidity, overall acceptability	Ozer et al. (2013)
	Fuyu	LDPE (50 μm)	1±1 °C with 90±5% RH, 84d + 5d at 25 °C with 70% RH	0.9	↔ firmness, colour values, ↓ TSS	Cia et al. (2006)
Pomegranate	Hicrannar	MAP	6±0.5 °C with 90±5% RH, 120d + 3d at 20 °C	1.63	↓ visual quality, decay incidence, ↑ TSS	Selcuk and Erkan (2014)
Plum	Tegan Blue	MAP bags	0±1 °C with 90±5% RH, 60d	0.9	↓ ethylene production, TA ↑ firmness, TSS:TA ratio	Khan and Singh (2008)
	Autumn Giant	PE (30μm)	0 °C with 90–92% RH, 60d	1.58	↓ flesh browning, titratable acidity, ↑ overall quality	Erkan and Eski (2012)
	Black Beauty	PE (30μm)		1.78	-	
	Friar	FF-602 box liner	0 °C with 85%RH, 40d	0.1	↓ TSS:TA ratio, dark skin fruit, decay, red flesh	Cantín et al. (2008)
Strawberry	Benihppe	MAP (H350)	4 ±5 °C with 90–95% RH, 18d	5.3	↑ firmness, TSS, ↓ decay index, respiration	Lei et al. (2022)
Spinach	Desi	MAP bags (Biofresh®)	4±1 °C with 90±5% RH, 20d	4.2	↑ visual quality, marketability index, ↓ decay incidence, ion leakage	Siddique et al. (2021)
		MAP bags (Xtend®)		1.3	–	
Tomato	Liberto	PE (50μ)	13 °C, 60d	2.3	↔ colour values, ↑ firmness, TA	Batu and Thompson (1998)
	Beef	(N-PP) (20 μm) + granular-activated with palladium carbon	8 °C with 90% RH, 28d	0.49	↓ ethylene production, respiration, internal package odour, decay ↔ skin and mesocarp colour, ↑ firmness	Bailén et al. (2006)
	Perla	MAP bags (Xtend®)	12 °C with 90% RH, 21d	1.7	↑ skin colour (hue value), TA, TSS:TA ratio, ↔ elasticity,	Sabir and Agar (2011)

Source: TSS = total soluble solids, TA = titratable acidity, TSS:TA ratio = sugar acid ratio, ↑ = increased, ↓ = decreased, ↔ = maintained, ○ = non-significant, [a] = at removal from storage

However, the spinach leaves packed in perforated PE bags exhibited higher water loss of 26.5% and those subjected to a short hot-water blanching treatment displayed 33.4% water loss after the entire period of cold storage. Furthermore, the effects of ISW on water loss and the postharvest quality of fresh fruits and vegetables are presented in Table 15.2. For instance, the Royal Delicious apples subjected to ISW (Cryovac 9 μ) reported water loss of 10.2%, half the loss noted in the unwrapped control fruit (20.4%) during 60 days of storage under zero-energy cool-room conditions (Table 15.2). ISW reduces colour change and conserves freshness, as noted in cucumbers (Figure 15.2)

TABLE 15.2 Effects of Individual Shrink Wrapping (ISW) on Postharvest Water Loss and Physiochemical Quality of Fresh Fruits and Vegetables during Storage

CROP	*CULTIVAR*	*ISW (TYPE/ SPECIFICATION)*	*STORAGE CONDITIONS AND DURATION*	*WATER LOSS (%)*	*PHYSIOCHEMICAL QUALITY*	*REFERENCE*
Apple	Royal Delicious	Cryovac (9 µ)	18–22 °C, 82–88% RH, 60d	10.2	↓ decay, ↑ firmness, juice recovery, TSS, TA	Sharma et al. (2010)
Banana	Dwarf Cavendish	Cryovac (9 µ)	33 °C, 72% RH, 21d	7.2	↑ firmness, ↓ colour index, TA, ascorbic acid	Surekha et al. (2017)
Bell pepper	Selika	Cryovac D-940	2 °C, 95% RH, 21d	0.5	↓ decay, ion leakage, deformation	Ilić et al. (2012)
Cucumber	Padmini	Cryovac D955 (60 gauge)	12 ± 1 °C, 90–95% RH, 15d	0.6	↑ firmness, sensory quality	Dhall et al. (2012)
Capsicum	Bachata	Cryovac BDF-2001	8 °C, 80 ± 5% RH, 35d	4.3	↑ firmness, ascorbic acid, phenolics	Sudhakar et al. (2021)
Honeydew melon	Inodorus Naud	Cryovac XDR 17 pm	4 °C, 42d	4.6	↓ *Alternaria* rot, *Fusarium* rot, ↔ firmness, ↑ TSS	Edwards and Blennerhassett (1990)
Guava	Hisar Safeda	Polybags	7 ± 3 °C, 21d	3.5	○ firmness, ↓ respiration, ↑ TA	Rana et al. (2015)
	Allahabad Safeda	BOPP (23 µ)	8–12 °C and 88–90% RH, 28d	7.0	↑ ascorbic acid	Sahoo et al. (2015)
Kinnow mandarin	–	LDPE (19µ)	25–40 °C and 31–70% RH, 21d	10.3	↑ firmness, TSS, eating quality, ↓ TA	Mandal (2015)
Lemon	Assam	Shrink film (15 µ)	29–32 °C, 55–78% RH, 32d	3.92	↓ TSS, ↑ TA, marketability	Piloo et al. (2020)
Muskmelon	TAM Uvalde	HDPE (12.7 µm, Clysar 50 EHC-F film)	4 ± 2 °C, 90 ± 5% RH, 21d	–	↑ visual quality, ↓ browning, surface mould	Collins et al. (1990)
Mango	Alphonso	D-955 (14 µm polylefin film, Cryovac)	8 °C, 35d	1.41	↑ firmness, ↓ ethylene, respiration	Rao and Shivashankara (2015)
	Banganaoalli			0.53	–	
Papaya	Taiwan Red Lady	Polyolefin shrink film (15 µ)	32–39 °C, 72–83% RH, 12d	9.9	↓ TSS, moisture content, ↑ ascorbic acid, vitamin A, peel colour (score), taste (score)	Perli et al. (2019)
Peach	Shan-i-Punjab	HeatshrinkableRD-106 (15µ, Cryovac)	18–20 °C, 90–95% RH, 9d	1.10	↑ firmness, sensory quality, TSS, TA, total sugars, ↓ decay	Mahajan et al. (2015)
Pomegranate	Wonderful	Shrink film (0.025 µm)	7 ± 0.5 °C, 92 ± 2% RH	< 2	↑ decay, phenolics; catechin and rutin, total anthocyanins, ↓ visual quality	Mphahlele et al. (2016)
	Mridula	Cryovac® (BDF-2001)	8 °C, 75–80% RH, 90d	0.43	↑ colour values (chroma, L*), TA, peel moisture, sensory attributes, ↓ TA	Sudhakar (2018)
	Bhagwa			0.68	–	
Zucchini	Natura	Shrink film (18 µm)	4 °C, 14d	0.6	↑ firmness, ↓ respiration, ethylene production	Megías et al. (2015)
	Sinatra			0.8	–	

Source: TSS = total soluble solids, TA = titratable acidity, ↑ = increased, ↓ = decreased, ↔ = maintained, ○ = non-significant

FIGURE 15.2 Individual shrink wrapping of cucumbers for extending shelf life and maintaining quality.

15.4.4 Colour Preservation

MAP technology has also been reported for delaying colour degradation, thereby extending the storage life of fresh horticultural produce. The preservation of colour during ripening as well as the postharvest period is highly dependent on the modified atmosphere inside the package and the storage conditions. The modulation of high CO_2 and lower O_2 inside the package significantly helps in delaying chlorophyll degradation in a wide range of fruits and vegetables where green to yellow colours are indicators of quality. MAPs also help in inhibiting the degradation anthocyanins, xanthophylls, carotenoids, and lycopene, resulting in the delay of colour development during storage (Mattheis & Fellman, 2000). Sarfraz et al. (2020) stated that eight green chilli cultivars packed in open-top trays wrapped with Glad® Wrap film packaging experienced significantly delayed postharvest senescence; reduced red chilli percentage, water loss, and wrinkling; and extended shelf life up to 6 days depending on cultivars. Figure 15.1 depicts the retention of green colour and maintained overall quality of three Pakistani green chilli cultivars, including Talhari, Longi, and Kunri-1 packed in LDPE (12 μ) after 15 and 12 days of storage at 18 ± 2 °C. Similarly, Parthenon broccoli florets packed in MAP bags showed significant results for retaining green colour (L*, a*, and b* values) and decreasing the degradation of chlorophyll a and b besides impressive preservation of ascorbic acid and total phenolic content during 12 days of cold storage (Fernández-León et al. 2013). In jujubes, MAP application alone or in combination with an *Aloe vera* gel coating showed similar results for preserving colour quality in terms of L* and hue angle during cold storage and following 4 days at ambient conditions (Islam et al., 2022). Likewise, tomatoes packed in micro-perforated LDPE bags showed delayed colour development with reduced biosynthesis of lycopene during 21 days of cold storage (Olveira-Bouzas et al., 2021). A study on spinach leaves packed in passive MAP bags showed a lower value for lightness (L*), a higher value for hue angle with delayed degradation of chlorophyll content, yellowing, and reduced water loss during the entire period of cold storage (Batziakas et al., 2020). On the other hand, the MAP bags also preserved anthocyanins biosynthesis or delayed its degradation in several red-pigmented fruit such as litchi (Ali et al., 2019a), sweet cherry (Aglar et al., 2017), pomegranate (Banda et al., 2015), and raspberry (Briano et al., 2015) during storage. Similar to MA packaging in fresh produce, ISW also preserved the colour of stored fruit. For instance, pomegranate cvs. Mridula and Bhagwa subjected to different types of wrappings exhibited higher chromaticity values of lightness (L*) and chroma during 90 days of cold storage (Sudhakar, 2018) (Table 15.2). MAP technology could be the suitable strategy for delaying ripening and colour development during storage as exhibited in mango fruit (Figure 15.3).

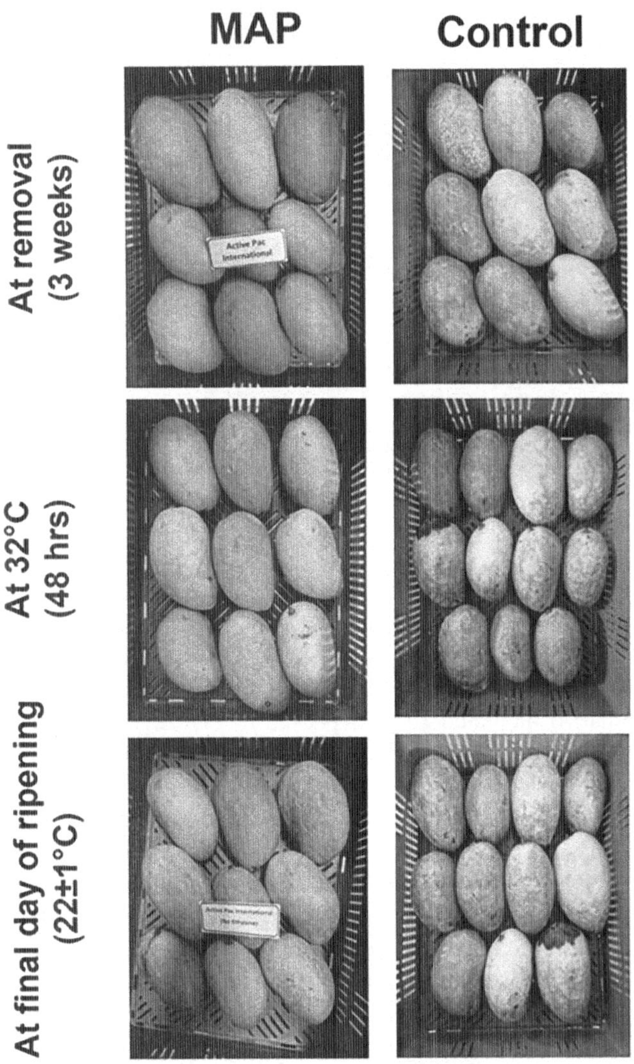

FIGURE 15.3 Effects of MAP bagging on the postharvest quality of mid-season harvest of Samar Bahisht Chaunsa mangoes during storage and shelf periods. Source: Aman Ullah Malik, PARB project—Exploiting controlled atmosphere technology potential for extended storage and shipping of fresh produce to international markets.

15.4.5 Chilling Injury

Chilling injury (CI) is the most prevailing storage disorder in fresh fruits and vegetables, particularly from tropical and subtropical regions when subjected to cold temperature conditions, which leads to less acceptance in the market due to either external or internal undesirable changes (Shah et al., 2023). The use of MAP technology helps to reduce this disorder, as reported in some studies. Roxana apricots packed in MAP bags showed no CI either alone or in combination with other postharvest dip application during 45 days of cold storage (Varlı-Yunusoğlu & Ekinci, 2023). Nguyen et al. (2004) have reported that Kluai Khai bananas packed in MAP bags following 30 days of storage at 10 °C substantially reduced CI symptoms (greyish peel browning) by 1.8-fold compared to unpacked control fruit (Table 15.3). In addition, the MAP bagging of bananas significantly retained better eating quality attributes, with higher phenolic content during the entire period of storage. The chill-injured peels of control bananas showed higher activities of PPO and phenylalanine ammonia lyase (PAL) enzymes held responsible for browning compared to those subjected to MAP bagging. In another study, Porat et al. (2004) revealed that Shamouti oranges and Star Ruby grapefruit packed in Xtend® XF10-μ liners and stored for 5 weeks at 2 °C following 5 days at ambient conditions showed markedly reduced CI percentages compared to the unwrapped control and those packed in macro-perforated liners, respectively (Table 15.3). Furthermore, persimmons are highly sensitive to CI stress. Zhao et al. (2020) have reported that Youhou persimmons subjected to pre-storage 1-methylcyclopropene (1-MCP) treatment following bagging in MAP (PE 0.05 mm) packages showed substantial reduction of CI during the entire cold storage period for 70 days at 1 °C following 5 days at ambient conditions. On the other hand, ISW also showed its potential in alleviating CI in fruits and vegetables. Megías et al. (2015) stated that zucchini cvs. Sinatra and Natura subjected to ISW (LDPE- 18 μm) exhibited substantial reduction of CI and lower ethylene production with their respective genes and respiration, and mitigated the oxidative stress by lowering MDA content and H_2O_2 during the entire period of cold storage. Accordingly, the application of MAP and ISW technologies could be a suitable alternative for attenuating CI and mitigating oxidative stress alone or in combination with other postharvest treatment depending on the crop and genotype.

15.4.6 Browning Incidence

Browning incidence is another crucial problem occurring in whole and fresh-cut fruit and vegetables due to the oxidation of antioxidants during storage. MAP technology has been reported to delay enzymatic browning by preserving the antioxidant system. Litchi is very susceptible to pericarp browning, which limits its acceptability in the market. Ali et al. (2019a) have described that litchi cv. Gola packed in MAP bags (20 μm, Xtend®) significantly delayed pericarp browning by

TABLE 15.3 Effects of MAP on the Reduction of Chilling Injury (CI), Browning Incidence (BI), and Postharvest Quality in Cold-Stored Fruits and Vegetables

CROP	*CULTIVAR*	*STORAGE CONDITIONS AND DURATION*	*CI REDUCTION (FOLD)*	*INFERENCE*	*REFERENCE*
Apricot	Roxana	0 ± 0.5 °C, 45d	-[a]	↓ internal browning	Varlı-Yunusoğlu and Ekinci (2023)
Banana	Kluai Khai	10 °C, 90% RH, 30d	1.85	↑ eating quality (score), TPC, ↓ PPO, PAL activities	Nguyen et al. (2004)
Citrus	Shamouti	2 °C, 35d + 5d at 20 °C	3.84	↓ weight loss, rind breakdown, ageing	Porat et al. (2004)
	Star Ruby		4.33	–	
Nectarine	Maria Aurelia	0 °C, 40d +2d at 20 °C	3.1	↓ ethylene, respiration, PPO, PME, PG activities, ↑ firmness, colour value (hue)	Özkaya et al. (2016)
Pomegranate	Shishe-kab	5 ± 0.5 °C, 85 ± 5% RH, 84d	1.6	↑ anthocyanins, TSS, TA, TSS:TA ratio	Moradinezhad et al. (2018)
Papaya	Solo	13 °C and 85–90% RH, 30d +7d at 20 °C	3.0	↓ electrical conductivity, ↑ colour, ascorbic acid	Singh and Rao (2005)
Plum	BlackAmber	0–1 °C, 56d + 6d at 21 ± 1 °C	1.12	↑ firmness, SOD, POD, CAT activities, ↓ LOX activity, electrical conductivity	Singh and Singh (2013)
Persimmon [b]	Youhou	1 ± 0.5 °C, 90–95% RH, 70d	13.7	↑ firmness, colour value, juice yield, PG activity, ↓ EL, MDA content, LOX, POD, PPO activities, WSP	Zhao et al. (2020)
BI reduction (fold)					
Litchi	Gola	5 ± 1 °C, 28d	2.0	↓ relative electrolyte leakage, MDA content, H_2O_2, superoxide anion	Ali et al. (2019a)
Longan	Daw	5 ± 2 °C and 80 ± 5% RH, 21d	1.27	↑ PPO, POD activities	Khan et al. (2020)
Pear	Yali	0 °C, 100d	2.36	↑ TPC, ↓ *PbPAL1*, *PbPAL2*, and *PbPPO1* gene expression	Cheng et al. (2015)
Cherries [c]	Lapins	0 °C, 42d	1.42	↓ pitting, splitting, decay, anthocyanins, bitter taste, ↑ skin lightness, firmness	Wang et al. (2015)
	Skeena	–	2.52	–	
Mushroom	Shiitake	4 °C, 17d	1.25	↓ respiration, PPO activity, reducing sugar content, ↑ firmness, TPC, SSC, ascorbic acid	Ye et al. (2012)

Source: CAT = catalase, EL = electrolyte leakage, H_2O_2 = hydrogen peroxide, LOX = lipoxygenase, MDA = malondialdehyde, POD = peroxidase, PPO = polyphenol oxidase, PAL = phenylalanine, PME = pectin methylesterase, PG = polygalacturonase, SOD = superoxide dismutase, TPC = total phenolics content, TSS = total soluble solids, TA = titratable acidity, TSS:TA ratio = sugar acid ratio, ↑ = increased, ↓ = decreased, ↔ = maintained, WSP = water soluble pectin, ○ = non-significant, [a] = no chilling injury observed in MAP stored fruit, [b] = MAP + 1-MCP, [c] = pedicel browning

two-fold by mitigating oxidative stress and suppressing the accumulation of malondialdehyde (MDA), hydrogen peroxide (H_2O_2), and superoxide anions during 28 days of cold storage at 5 ± 1 °C (Table 15.3). In addition, MAP technology also reduced relative electrolyte leakage; maintained higher activities of APX, SOD, and CAT enzymes; and lowered PPO and POD activity during the entire period of storage. Similarly, MAPs have been reported to reduce the browning incidence in longans (Khan et al., 2020), pears (Cheng et al., 2015), and mushrooms (Ye et al., 2012). In another study, the MA packaging significantly delayed pedicle browning, pitting, and splitting; reduced decay and bitter taste; increased skin lightness; and maintained the firmness of sweet cherry cvs. Lapins and Skeena fruit during 42 days of cold storage at 0 °C (Wang et al., 2015) (Table 15.3). However, in apricots, MAP bagging displayed pronounced internal browning, which may be due to the crop's physiology and its response to developed internal modified atmosphere during cold storage, which could be alleviated by applying a pre-storage antiripening treatment (Varlı-Yunusoğlu & Ekinci, 2023).

15.4.7 Microbial Growth and Decay

Spoilage due to microbial growth is more prevalent in minimally processed fruits and vegetables, which may be due to the processes involved, such as postharvest washing, peeling, cutting, chopping, trimming, and shredding, prior to packaging and storage, while whole fresh produce is also susceptible during the postharvest period depending on field conditions, harvest handling, the type of crop, its genotype, and storage conditions (Werner & Hotchkiss, 2005; Hasan et al., 2021). MAP technology alone or in combination with other technologies largely employed in the minimally processed fruit and vegetable industry prevents fresh-cut and whole fruits and vegetables from microbial spoilage and subsequent decay, but their success rate depends on several factors. Martínez-Ferrer et al. (2002) have reported that fresh-cut mango and pineapple fruit packed in MAP bags with gas composition of 10% CO_2 and 4% O_2 with balanced N_2 showed substantial reduction in total yeast populations compared to storage under vacuum or air conditions during 30 days of cold storage. In another study, Waghmare and Annapure (2013) revealed that fresh-cut papaya fruit packed in MAP bags with atmospheric composition of 10% CO_2 and 5% O_2 following cold storage at 5 °C showed lower growth of mesophilic aerobics, yeasts, and moulds as compared to controls, and combination with other dip treatments (calcium chloride, citric acid) significantly reduced microbial growth and associated decay and maintained overall quality during storage. Similarly, the Hicrannar pomegranates packed in Xtend® MAP bags showed lower decay incidence of 13.3% as compared to unpacked control fruit (40%) after 120 days of cold storage (Selcuk & Erkan, 2014). Additionally, bell peppers packed in MAP bags (LDPE pouches, 60 µm) showed lower incidence of internal rotting caused by *Fusarium lactis* species in bell pepper fruit during the shelf period at ambient conditions. Furthermore, ISW application in combination with a short hot-water dip treatment significantly inhibited decay incidence due to surface organisms and maintained quality of Topmark netted muskmelons during cold storage (Mayberry & Hartz, 1992). Accordingly, MAP technology could be helpful in reducing microbial spoilage in stored fruits and vegetables while their effectiveness for lowering postharvest decay may be enhanced with any of the pre-storage treatments.

15.5 CROSSTALK OF MAP TECHNOLOGY WITH OTHER POSTHARVEST TREATMENTS

15.5.1 Antiripening Chemicals

In packaging technologies, it has been reported that pre-storage application of antiripening chemicals in combination with MAP showed pronounced effects in maintaining the quality of fruits and vegetables. However, it appears that MAP bagging/wrapping does not always work individually for alleviating storage problems such as CI or browning, premature fruit softening, and non-uniform ripening during the post-storage shelf period. Accordingly, several chemicals have been tested along with MAP bagging and have shown promising results for storage life extension of fresh horticultural produce. For instance, the pre-packaging application of 1-MCP treatment (300 nL L^{-1}) combined with MAP bagging significantly prevented the green ripe disorder in Baxi bananas during high-temperature storage by delaying ethylene production, respiration rate, and suppressing ROS-mediated oxidative stress by lowering the accumulation of MDA, H_2O_2, and superoxide anion content during 17 days of cold storage following 7 days of ripening period (Li et al., 2023). Similarly, the combined application of MAP and 1-MCP treatment has shown significantly mitigated CI in Youhou persimmons during 70 days of cold storage and a 5-day shelf period (Zhao et al., 2020). Khan and Singh (2008) reported that MAP application alone was less effective than the combined application of MAP and 1-MCP in delaying the biosynthesis of ethylene (ethylene production, lowering ACC [1-aminocyclopropane-1-carboxylic acid] content, activities of ACS [1-aminocyclopropane-1-carboxylic acid synthase], and ACO [1-aminocyclopropane-1-carboxylic acid oxidase] enzymes), fruit softening, and extending storage life for 7 weeks in Tegan Blue Japanese plums during cold storage following 8 days of ripening at shelf conditions without compromising its fruit quality. Similar results of combined application of MAP and 1-MCP treatments for delaying postharvest senescence have been reported for mulberries (Kızıldeniz et al., 2023), nectarines (Özkaya et al., 2016), broccoli (Sabir, 2012), and tomatoes (Sabir & Agar, 2011) during storage. Nitric oxide is another promising antiripening chemical which down-regulates the physiological processes in fresh fruits and vegetables (Singh et al., 2009b). Thus, the combined application of nitric oxide and biaxially oriented polypropylene (BOPP) bagging showed positive results in inhibiting browning incidence, maintaining higher firmness, and increasing TPC and ascorbic acid in button mushrooms during cold storage. Similarly, there are multiple reports for other chemicals such as sulphur dioxide (Candir et al., 2012; Rodriguez & Zoffoli, 2016), calcium chloride (Waghmare & Annapure, 2013), and salicylic acid (Bal, 2018) treatments combined with MAP application for preserving quality during the postharvest period.

15.5.2 Edible Coating

The application of pre-storage edible coatings has emerged as a novel approach due to their organically safe nature containing antioxidative compounds and its antimicrobial properties showing potential for storage life extension of fruits and vegetables. In addition, the edible coatings also functioned on the principle of developing modified atmosphere inside the coated product helps in developing a barrier against postharvest senescence (Hasan et al., 2021). The combined application of different coatings followed by packaging in MAPs showed positive results on the storage quality of fruits and vegetables. Islam et al. (2022) have reported that an *Aloe vera* (ALV) gel coating plus MA packaging significantly helps in preserving the colour (L* and hue angle values), phenolics content, ascorbic acid, antioxidant activities (FRAP and DPPH), and total flavonoids in jujube fruit during cold storage and shelf period at ambient conditions. Similarly, the ALV gel coating enriched with other food-safe compounds, such as ascorbic acid and calcium lactate and the addition of essential oils followed by MAP bagging markedly lowered microbial load, retained higher anthocyanins and ascorbic acid in cold-stored strawberries (Esmaeili et al., 2021). Furthermore, the application of chitosan combined with MAP exhibited impressive results in maintaining the organoleptic quality and lowered rachis browning in ready-to-eat Italia grapes during cold storage and shelf period (Liguori et al., 2021). Similar results for the combined application of chitosan coating and MA packaging have been reported in button mushrooms during cold and ambient storage (Gholami et al., 2020). Accordingly, the combined application of edible coating and MA packaging could be a suitable alternative to synthetic chemical application for preserving the quality of stored fruits and vegetables. The coatings for commercial-scale application are quite challenging, but we urge researchers to study the sustainability of combined technology.

15.5.3 Essential Oils

The application of essential oils (EOs) is gaining interest as a substitute for synthetic chemical treatments for attenuating the incidence of postharvest disease and inhibiting rots and spoilage of fruits and vegetables during storage. Several EOs have already been tested for their application in agro-food industries due to their role as antimicrobials, antioxidants, insecticides, and herbicides (El Khetabi et al., 2022). Owing to the role of EOs in preventing decay index in stored fruits and vegetables, it has also been reported that pre-storage fumigation of few EOs have the potential to significantly suppress ethylene production and the respiration rate, thereby mitigating oxidative stress; alleviating cold storage constraints, such as superficial scald in apple fruit; and maintaining postharvest quality during long-term storage (Malekipoor et al., 2022) Accordingly, the application of EOs combined with MAP technology presented synergistic effects on the quality of stored fruits and vegetables. Perumal et al. (2021) have reported that the combined application of thyme oil vapour (TOV) and MAP significantly reduced the postharvest spread of anthracnose and retained higher firmness in mango cvs. Banganapalli and Totapuri. In another study, TOV and MA packaging considerably reduced disease incidence and browning, and expressed increased antioxidants, PAL enzyme activity, TPC, and flavonoids during storage. They suggested that TOV could be a suitable alternative to commercially used prochloraz fungicide application in the postharvest supply chain of avocados (Sellamuthu et al., 2013). The combined application of lemon grass oil (LGO) followed by MA packaging preserved the quality of strawberry cultivars by lowering water loss, loss of firmness, spoilage, and the development of off-odour during storage (Kahramanoğlu, 2019). Likewise, the LGO and MAP bagging showed a declined trend in anthracnose incidence, vascular browning, and grey pulp in addition to lower water loss and loss of firmness in avocados during cold storage and thenripening period (Mpho et al., 2013). Hence, the combined application of EOs and MAP could have a positive impact on the shelf-life extension of fresh fruits and vegetables.

15.6 CONCLUSIONS AND FUTURE RESEARCH

This chapter covers the effects and limitations of MAP and ISW technologies for storage-life extension and quality maintenance of fruits and vegetables. The growing interest of industry stakeholders associated with the marketing and trade of horticultural produce are more inclined towards expanding the marketable window by preserving postharvest quality using MAP technology in their supply chain operations. Accordingly, the innovation in MAP technology along with pre-storage or complementary postharvest technologies are still in the infancy stage to cope with the challenges of cool chain systems, such as chilling injury, browning, and post-storage softening. Manufacturing MAP bags is also a challenge for crops with high respiration rates, as they sometimes fail to develop the desirable atmosphere inside the bag, leading to low-grade quality. Increasing global concerns of plastic waste have also directed researchers to develop a holistic, environmentally safe, and biodegradable packaging for preserving fruits and vegetables.

REFERENCES

Abdul-Khalil, H., Banerjee, A., Saurabh, C. K., Tye, Y., Suriani, A., Mohamed, A., Karim, A., Rizal, S., & Paridah, M. (2018). Biodegradable films for fruits and vegetables packaging application: preparation and properties. *Food Engineering Reviews*, *10*, 139–153. https://doi.org/10.1007/s12393-018-9180-3

Aglar, E., Ozturk, B., Guler, S. K., Karakaya, O., Uzun, S., & Saracoglu, O. (2017). Effect of modified atmosphere packaging and 'Parka' treatments on fruit quality characteristics of sweet cherry fruits (*Prunus avium* L.'0900 Ziraat') during cold storage and shelf life. *Scientia Horticulturae*, *222*, 162–168. https://doi.org/10.1016/j.scienta.2017.05.024

Aharoni, N., Rodov, V., Fallik, E., Porat, R., Pesis, E., & Lurie, S. (2007). Controlling humidity improves efficacy of modified atmosphere packaging of fruits and vegetables. *Acta Horticulturae*, 121–128. https://doi.org/10.17660/ActaHortic.2008.804.14

Ali, S., Khan, A. S., Malik, A. U., Anjum, M. A., Nawaz, A., & Shah, H. M. S. (2019a). Modified atmosphere packaging delays enzymatic browning and maintains quality of harvested litchi fruit during low temperature storage. *Scientia Horticulturae*, *254*, 14–20. https://doi.org/10.1016/j.scienta.2019.04.065

Ali, O., Malik, A. U., Khan, A. S., Amin, M., & Rehman, A. (2019b). Effect of modified atmosphere packaging on the postharvest life and quality of mango cv. Samar Bahisht Chaunsa stored at chilling temperature. *Pakistan Journal of Agricultural Sciences*, *56*(4). https://doi.org/10.21162/PAKJAS/19.7036

Azevedo, S., Cunha, L. M., Mahajan, P. V., & Fonseca, S. C. (2011). Application of simplex lattice design for development of moisture absorber for oyster mushrooms. *Procedia Food Science*, *1*, 184–189. https://doi.org/10.1016/j.profoo.2011.09.029

Bailén, G., Guillén, F., Castillo, S., Serrano, M., Valero, D., & Martínez-Romero, D. (2006). Use of activated carbon inside modified atmosphere packages to maintain tomato fruit quality during cold storage. *Journal of Agricultural and Food Chemistry*, *54*(6), 2229–2235. https://doi.org/10.1021/jf0528761

Bal, E. (2018). Combined treatment of modified atmosphere packaging and salicylic acid improves postharvest quality of nectarine (*Prunus persica* L.) fruit. *Journal of Agricultural Science and Technology*, *18*, 1345–1354. jast.modares.ac.ir/article-23–5197-en.html

Banda, K., Caleb, O. J., Jacobs, K., & Opara, U. L. (2015). Effect of active-modified atmosphere packaging on the respiration rate and quality of pomegranate arils (cv. Wonderful). *Postharvest Biology and Technology*, *109*, 97–105.

Batziakas, K. G., Singh, S., Ayub, K., Kang, Q., Brecht, J. K., Rivard, C. L., & Pliakoni, E. D. (2020). Reducing postharvest losses of spinach stored at nonoptimum temperatures with the implementation of passive modified atmosphere packaging. *HortScience*, *55*(3), 326–335.

Batu, A., & Thompson, A. K. (1998). Effects of modified atmosphere packaging on post harvest qualities of pink tomatoes. *Turkish Journal of Agriculture and Forestry*, *22*(4), 365–372.

Belay, Z. A., Caleb, O. J., & Opara, U. L. (2016). Modelling approaches for designing and evaluating the performance of modified atmosphere packaging (MAP) systems for fresh produce: A review. *Food Packaging and Shelf Life*, *10*, 1–15. https://doi.org/10.1016/j.fpsl.2016.08.001

Belay, Z. A., Caleb, O. J., & Opara, U. L. (2019). Influence of initial gas modification on physicochemical quality attributes and molecular changes in fresh and fresh-cut fruit during modified atmosphere packaging. *Food Packaging and Shelf Life*, *21*, 100359. https://doi.org/10.1016/j.fpsl.2019.100359

Ben-Yehoshua, S. (1985). Individual seal-packaging of fruit and vegetables in plastic film—a new postharvest technique. *HortScience*, *20*(1), 32–37. https://doi.org/10.21273/HORTSCI.20.1.32

Bond, M., Meacham, T., Bhunnoo, R., & Benton, T. G. (2013). *Food waste within global food systems. A Global Food Security Report*. www.foodsecurity.ac.uk

Bovi, G. G., Caleb, O. J., Linke, M., Rauh, C., & Mahajan, P. V. (2016). Transpiration and moisture evolution in packaged fresh horticultural produce and the role of integrated mathematical models: A review. *Biosystems Engineering*, *150*, 24–39. https://doi.org/10.1016/j.biosystemseng.2016.07.013

Brandenburg, J. S., & Zagory, D. (2009). Modified and controlled atmosphere packaging technology and applications. In Y. Elhadi M (Ed.), *Modified and Controlled Atmospheres for the Storage, Transportation, and Packaging of Horticultural Commodities* (pp. 91–110). CRC Press. https://doi.org/10.1201/9781420069587

Brecht, J. K. (2006). Controlled atmosphere, modified atmosphere and modified atmosphere packaging for vegetables. *Stewart Postharvest Review*, *5*(2), 1–6.

Briano, R., Giuggioli, N. R., Girgenti, V., & Peano, C. (2015). Biodegradable and compostable film and modified atmosphere packaging in postharvest supply chain of raspberry fruits (cv. G randeur). *Journal of Food Processing and Preservation*, *39*(6), 2061–2073.

Caleb, O. J., Mahajan, P. V., Al-Said, F. A.-J., & Opara, U. L. (2013). Modified atmosphere packaging technology of fresh and fresh-cut produce and the microbial consequences-a review. *Food and Bioprocess Technology*, *6*, 303–329. https://doi.org/10.1007/s11947-012-0932-4

Candir, E., Ozdemir, A. E., Kamiloglu, O., Soylu, E. M., Dilbaz, R., & Ustun, D. (2012). Modified atmosphere packaging and ethanol vapor to control decay of 'Red Globe' table grapes during storage. *Postharvest Biology and Technology*, *63*(1), 98–106. https://doi.org/10.1016/j.postharvbio.2011.09.008

Cantín, C. M., Crisosto, C. H., & Day, K. R. (2008). Evaluation of the effect of different modified atmosphere packaging box liners on the quality and shelf life of 'Friar' plums. *HortTechnology*, *18*(2), 261–265. https://doi.org/10.21273/HORTTECH.18.2.261

Charles, F., Anchez, J. S., & Gontard, N. (2005). Modeling of active modified atmosphere packaging of endives exposed to several postharvest temperatures. *Journal of Food Science*, *70*(8), e443-e449. https://doi.org/10.1111/j.1365-2621.2005.tb11512.x

Chaudhary, P. R., Jayaprakasha, G., Porat, R., & Patil, B. S. (2015). Influence of modified atmosphere packaging on 'star ruby' grapefruit phytochemicals. *Journal of Agricultural and Food Chemistry*, *63*(3), 1020–1028. https://doi.org/10.1021/jf505278x

Cheng, Y., Liu, L., Zhao, G., Shen, C., Yan, H., Guan, J., & Yang, K. (2015). The effects of modified atmosphere packaging on core browning and the expression patterns of PPO and PAL genes in 'Yali' pears during cold storage. *LWT-Food Science and Technology*, *60*(2), 1243–1248. https://doi.org/10.1016/j.lwt.2014.09.005

Cia, P., Benato, E. A., Sigrist, J. M., Sarantopóulos, C., Oliveira, L. M., & Padula, M. (2006). Modified atmosphere packaging for extending the storage life of 'Fuyu' persimmon. *Postharvest Biology and Technology*, *42*(3), 228–234. https://doi.org/10.1016/j.postharvbio.2006.06.016

Collins, J., Bruton, B., & Perkins-Veazie, P. (1990). Organoleptic evaluation of shrink-wrapped muskmelon. *HortScience*, *25*(11), 1409–1412. https://doi.org/10.21273/HORTSCI.25.11.1409

Costa, C., Lucera, A., Conte, A., Mastromatteo, M., Speranza, B., Antonacci, A., & Del Nobile, M. A. (2011). Effects of passive and active modified atmosphere packaging conditions on ready-to-eat table grape. *Journal of Food Engineering*, *102*(2), 115–121. https://doi.org/10.1016/j.jfoodeng.2010.08.001

Dhall, R. K., Sharma, S. R., & Mahajan, B. (2012). Effect of shrink wrap packaging for maintaining quality of cucumber during storage. *Journal of Food Science and Technology*, *49*, 495–499. https://doi.org/10.1007/s13197-011-0284-5

Ding, C.-K., Chachin, K., Ueda, Y., Imahori, Y., & Wang, C. Y. (2002). Modified atmosphere packaging maintains postharvest quality of loquat fruit. *Postharvest Biology and Technology*, *24*(3), 341–348. https://doi.org/10.1016/S0925-5214(01)00148-X

Dorostkar, M., Moradinezhad, F., & Ansarifar, E. (2022). Influence of active modified atmosphere packaging pre-treatment on shelf life and quality attributes of cold stored apricot fruit. *International Journal of Fruit Science*, *22*(1), 402–413. https://doi.org/10.1080/15538362.2022.2047137

Edwards, M., & Blennerhassett, R. (1990). The use of postharvest treatments to extend storage life and to control postharvest wastage of honeydew melons (*Cucumis melo* L. var. Inodorus Naud.) in cool storage. *Australian Journal of Experimental Agriculture, 30*(5), 693–697. https://doi.org/10.1071/EA9900693

El Khetabi, A., Lahlali, R., Ezrari, S., Radouane, N., Lyousfi, N., Banani, H., . . . & Barka, E. A. (2022). Role of plant extracts and essential oils in fighting against postharvest fruit pathogens and extending fruit shelf life: A review. *Trends in Food Science & Technology, 120*, 402–417.

Erkan, M., & Eski, H. (2012). Combined treatment of modified atmosphere packaging and 1-methylcyclopropene improves postharvest quality of Japanese plums. *Turkish Journal of Agriculture and Forestry, 36*(5), 563–575. https://doi.org/10.3906/tar-1111-34

Esmaeili, Y., Zamindar, N., Paidari, S., Ibrahim, S. A., & Mohammadi Nafchi, A. (2021). The synergistic effects of aloe vera gel and modified atmosphere packaging on the quality of strawberry fruit. *Journal of Food Processing and Preservation, 45*(12), e16003. https://doi.org/10.1111/jfpp.16003

Fernández-León, M., Fernández-León, A., Lozano, M., Ayuso, M., Amodio, M. L., Colelli, G., & González-Gómez, D. (2013). Retention of quality and functional values of broccoli 'Parthenon' stored in modified atmosphere packaging. *Food Control, 31*(2), 302–313. https://doi.org/10.1016/j.foodcont.2012.10.012

Flores, F. B., Martínez-Madrid, M. C., Ben Amor, M., Pech, J. C., Latché, A., & Romojaro, F. (2004). Modified atmosphere packaging confers additional chilling tolerance on ethylene-inhibited cantaloupe Charentais melon fruit. *European Food Research and Technology, 219*, 614–619. https://doi.org/10.1007/s00217-004-0952-z

Gimeno, D., Gonzalez-Buesa, J., Oria, R., Venturini, M. E., & Arias, E. (2021). Effect of modified atmosphere packaging (MAP) and UV-C irradiation on postharvest quality of red raspberries. *Agriculture, 12*(1), 29. https://doi.org/10.3390/agriculture12010029

Gholami, R., Ahmadi, E., & Ahmadi, S. (2020). Investigating the effect of chitosan, nanopackaging, and modified atmosphere packaging on physical, chemical, and mechanical properties of button mushroom during storage. *Food Science & Nutrition, 8*(1), 224–236.

Hafeez, O., Malik, A. U., Khalid, M. S., Amin, M., Khalid, S., & Umar, M. (2016). Effect of modified atmosphere packaging on postharvest quality of mango cvs. Sindhri and Sufaid Chaunsa during storage. *Turkish Journal of Agriculture-Food Science and Technology, 4*(12), 1104–1111. https://doi.org/10.24925/turjaf.v4i12.1104-1111.818

Hasan, M. U., Riaz, R., Malik, A. U., Khan, A. S., Anwar, R., Rehman, R. N. U., & Ali, S. (2021). Potential of *Aloe vera* gel coating for storage life extension and quality conservation of fruits and vegetables: An overview. *Journal of Food Biochemistry, 45*(4), e13640. https://doi.org/10.1111/jfbc.13640

Hayat, F., Nawaz Khan, M., Zafar, S., Balal, R., Azher Nawaz, M., Malik, A., & Saleem, B. (2017). Surface coating and modified atmosphere packaging enhances storage life and quality of 'Kaghzi lime'. *Journal of Agricultural Science and Technology, 19*(5), 1151–1160. jast.modares.ac.ir/article-23-863-en.html

Ilić, Z. S., Trajković, R., Pavlović, R., Alkalai-Tuvia, S., Perzelan, Y., & Fallik, E. (2012). Effect of heat treatment and individual shrink packaging on quality and nutritional value of bell pepper stored at suboptimal temperature. *International Journal of Food Science & Technology, 47*(1), 83–90. https://doi.org/10.1111/j.1365-2621.2011.02810.x

Illeperuma, C. K., & Nikapitiya, C. (2002). Extension of the postharvest life of Pollock' avocado using modified atmosphere packaging. *Fruits, 57*(5–6), 287–295. https://doi.org/10.1051/fruits:2002025

Illeperuma, C. K., & Jayasuriya, P. (2002). Prolonged storage of 'Karuthacolomban' mango by modified atmosphere packaging at low temperature. *The Journal of Horticultural Science and Biotechnology, 77*(2), 153–157. https://doi.org/10.1080/14620316.2002.11511472

Islam, A., Acıkalın, R., Ozturk, B., Aglar, E., & Kaiser, C. (2022). Combined effects of *Aloe vera* gel and modified atmosphere packaging treatments on fruit quality traits and bioactive compounds of jujube (*Ziziphus jujuba* Mill.) fruit during cold storage and shelf life. *Postharvest Biology and Technology, 187*, 111855. https://doi.org/10.1016/j.postharvbio.2022.111855

Jiang, T., Zheng, X., Li, J., Jing, G., Cai, L., & Ying, T. (2011). Integrated application of nitric oxide and modified atmosphere packaging to improve quality retention of button mushroom (*Agaricus bisporus*). *Food Chemistry, 126*(4), 1693–1699. https://doi.org/10.1016/j.foodchem.2010.12.060

Jiang, T., Luo, S., Chen, Q., Shen, L., & Ying, T. (2010). Effect of integrated application of gamma irradiation and modified atmosphere packaging on physicochemical and microbiological properties of shiitake mushroom (*Lentinus edodes*). *Food Chemistry, 122*(3), 761–767. https://doi.org/10.1016/j.foodchem.2010.03.050

Kader, A. A. (2009). Effects on nutritional quality. In E. Yahia (Ed.), *Modified and Controlled Atmospheres for the Storage, Transportation, and Packaging of Horticultural Commodities* (pp. 129–136). CRC Press. https://doi.org/10.1201/9781420069587

Kader, A. A., & Watkins, C. B. (2000). Modified atmosphere packaging—toward 2000 and beyond. *HortTechnology, 10*(3), 483–486. https://doi.org/10.21273/HORTTECH.10.3.483

Kahramanoğlu, İ. (2019). Effects of lemongrass oil application and modified atmosphere packaging on the postharvest life and quality of strawberry fruits. *Scientia Horticulturae, 256*, 108527.

Khan, A. S., & Singh, Z. (2008). 1-Methylcyclopropene application and modified atmosphere packaging affect ethylene biosynthesis, fruit softening, and quality of 'Tegan Blue' Japanese plum during cold storage. *Journal of the American Society for Horticultural Science, 133*(2), 290–299. https://doi.org/10.21273/JASHS.133.2.290

Khan, M. R., Chinsirikul, W., Sane, A., & Chonhenchob, V. (2020). Combined effects of natural substances and modified atmosphere packaging on reducing enzymatic browning and postharvest decay of longan fruit. *International Journal of Food Science & Technology, 55*(2), 500–508. https://doi.org/10.1111/ijfs.14293

Khorshidi, S., Davarynejad, G., Tehranifar, A., & Fallahi, E. (2011). Effect of modified atmosphere packaging on chemical composition, antioxidant activity, anthocyanin, and total phenolic content of cherry fruits. *Horticulture, Environment, and Biotechnology, 52*, 471–481. https://doi.org/10.1007/s13580-011-0027-6

Kızıldeniz, T., Hepsağ, F., & Hayoğlu, İ. (2023). Improving mulberry shelf-life with 1-methylcyclopropene and modified atmosphere packaging. *Biochemical Systematics and Ecology, 106*, 104578. https://doi.org/10.1016/j.bse.2022.104578

Lei, T. t., Qian, J., & Yin, C. (2022). Equilibrium modified atmosphere packaging on postharvest quality and antioxidant activity of strawberry. *International Journal of Food Science & Technology, 57*(11), 7125–7134. https://doi.org/10.1111/ijfs.16052

Li, X., Xiong, T., Zhu, Q., Zhou, Y., Lei, Q., Lu, H., Chen, W., Li, X., & Zhu, X. (2023). Combination of 1-MCP and modified atmosphere packaging (MAP) maintains banana fruit quality under high temperature storage by improving antioxidant system and cell wall structure. *Postharvest Biology and Technology*, *198*, 112265. https://doi.org/10.1016/j.postharvbio.2023.112265

Liguori, G., Sortino, G., Gullo, G., & Inglese, P. (2021). Effects of modified atmosphere packaging and chitosan treatment on quality and sensorial parameters of minimally processed cv.'Italia' table grapes. *Agronomy*, *11*(2), 328.

Lufu, R., Ambaw, A., & Opara, U. L. (2020). Water loss of fresh fruit: Influencing pre-harvest, harvest and postharvest factors. *Scientia Horticulturae*, *272*, 109519. https://doi.org/10.1016/j.scienta.2020.109519

Mahajan, B., Dhillon, W., Kumar, M., & Singh, B. (2015). Effect of different packaging films on shelf life and quality of peach under super and ordinary market conditions. *Journal of Food Science and Technology*, *52*, 3756–3762. https://doi.org/10.1007/s13197-014-1382-y

Mahajan, P. V., & Lee, D. S. (2023). Modified atmosphere and moisture condensation in packaged fresh produce: Scientific efforts and commercial success. *Postharvest Biology and Technology*, *198*, 112235. https://doi.org/10.1016/j.postharvbio.2022.112235

Mahajan, P. V., Rodrigues, F. A., Motel, A., & Leonhard, A. (2008). Development of a moisture absorber for packaging of fresh mushrooms (*Agaricus bisporous*). *Postharvest Biology and Technology*, *48*(3), 408–414. https://doi.org/10.1016/j.postharvbio.2007.11.007

Malekipoor, R., Singh, Z., & Payne, A. (2022). Fumigation with lemon and cinnamon oils suppresses ethylene production and maintains the fruit quality of controlled atmosphere-stored organic apples. *Food Packaging and Shelf Life*, *34*, 100958.

Mandal, G. (2015). Effect of lac-wax, citrashine and individual shrink wrapping of fruits on storage life of late harvested kinnow under ambient conditions. *International Journal of Bio-Resource Environment And Agricultural Sciences*, *1*(3), 84–89.

Mangaraj, S, Goswami, TK, Mahajan, & PV. (2009). Applications of plastic films for modified atmosphere packaging of fruits and vegetables: a review. *Food Engineering Reviews*, *1*, 133–158. https://doi.org/10.1007/s12393-009-9007-3

Martínez-Ferrer, M., Harper, C., Pérez-Muntoz, F., & Chaparro, M. (2002). Modified atmosphere packaging of minimally processed mango and pineapple fruits. *Journal of Food Science*, *67*(9), 3365–3371.

Martínez-Romero, D., Guillén, F., Castillo, S., Valero, D., & Serrano, M. (2003). Modified atmosphere packaging maintains quality of table grapes. *Journal of Food Science*, *68*(5), 1838–1843.

Mattheis, J., & Fellman, J. K. (2000). Impacts of modified atmosphere packaging and controlled atmospheres on aroma, flavor, and quality of horticultural commodities. *HortTechnology*, *10*(3), 507–510. https://doi.org/10.21273/HORTTECH.10.3.507

Mayberry, K., & Hartz, T. (1992). Extension of muskmelon storage through the use of hot water life treatment and polyethylene wraps. *HortScience*, *27*(4), 324–326. https://doi.org/10.21273/HORTSCI.27.4.324

Megías, Z., Martínez, C., Manzano, S., García, A., Rebolloso-Fuentes, M. d. M., Garrido, D., Valenzuela, J. L., & Jamilena, M. (2015). Individual shrink wrapping of zucchini fruit improves postharvest chilling tolerance associated with a reduction in ethylene production and oxidative stress metabolites. *PLoS One*, *10*(7), e0133058. https://doi.org/10.1371/journal.pone.0133058

Mistriotis, A., Briassoulis, D., Giannoulis, A., & D'Aquino, S. (2016). Design of biodegradable bio-based equilibrium modified atmosphere packaging (EMAP) for fresh fruits and vegetables by using micro-perforated poly-lactic acid (PLA) films. *Postharvest Biology and Technology*, *111*, 380–389. https://doi.org/10.1016/j.postharvbio.2015.09.022

Moradinezhad, F., Khayyat, M., Ranjbari, F., & Maraki, Z. (2018). Physiological and quality responses of Shishe-Kab pomegranates to short-term high CO_2 treatment and modified atmosphere packaging. *International Journal of Fruit Science*, *18*(3), 287–299. https://doi.org/10.1080/15538362.2017.1419399

Mphahlele, R. R., Fawole, O. A., & Opara, U. L. (2016). Influence of packaging system and long term storage on physiological attributes, biochemical quality, volatile composition and antioxidant properties of pomegranate fruit. *Scientia Horticulturae*, *211*, 140–151. https://doi.org/10.1016/j.scienta.2016.08.018

Mpho, M., Sivakumar, D., Sellamuthu, P. S., & Bautista-Baños, S. (2013). Use of lemongrass oil and modified atmosphere packaging on control of anthracnose and quality maintenance in avocado cultivars. *Journal of Food Quality*, *36*(3), 198–208.

Nguyen, T. B. T., Ketsa, S., & van Doorn, W. G. (2004). Effect of modified atmosphere packaging on chilling-induced peel browning in banana. *Postharvest Biology and Technology*, *31*(3), 313–317. https://doi.org/10.1016/j.postharvbio.2003.09.006

Olveira-Bouzas, V., Pita-Calvo, C., Vázquez-Odériz, M. L., & Romero-Rodríguez, M. Á. (2021). Evaluation of a modified atmosphere packaging system in pallets to extend the shelf-life of the stored tomato at cooling temperature. *Food Chemistry*, *364*, 130309.

Ozer, M., Akbudak, B., & Altioglu, I. (2013). Postharvest quality of'Hachiya'astringent persimmons (*Diospyros kaki* L.) as affected by hot water treatment and modified atmosphere packaging. *Italian Journal of Food Science*, *25*(1).

Özkaya, O., Yildirim, D., Dündar, Ö., & Tükel, S. S. (2016). Effects of 1-methylcyclopropene (1-MCP) and modified atmosphere packaging on postharvest storage quality of nectarine fruit. *Scientia Horticulturae*, *198*, 454–461. https://doi.org/10.1016/j.scienta.2015.12.016

Özkaya, O., Şener, A., Saridaş, M. A., Ünal, Ü., Valizadeh, A., & Dündar, Ö. (2015). Influence of fast cold chain and modified atmosphere packaging storage on postharvest quality of early season-harvested sweet cherries. *Journal of Food Processing and Preservation*, *39*(6), 2119–2128. https://doi.org/10.1111/jfpp.12455

Ozturk, A., Yildiz, K., Ozturk, B., Karakaya, O., Gun, S., Uzun, S., & Gundogdu, M. (2019). Maintaining postharvest quality of medlar (*Mespilus germanica*) fruit using modified atmosphere packaging and methyl jasmonate. *LWT-Food Science and Technology*, *111*, 117–124. https://doi.org/10.1016/j.lwt.2019.05.033

Öztürk, B., Ağlar, E., Karakaya, O., Saracoğlu, O., & Sefa, G. (2019a). Effects of preharvest GA3, CaCl2 and modified atmosphere packaging treatments on specific phenolic compounds of sweet cherry. *Turkish Journal of Food and Agriculture Sciences*, *1*(2), 44–56. https://doi.org/10.14744/turkjfas.2019.009

Perli, Monisha, Nama, S., Chandana Yasoda, B., Singh, H., & Yarramsetty, S. M. (2019). Shelf life extension of carica papaya by shrink wrapping. *International Journal of Chemical Studies*, *7*(5), 817–825.

Perumal, A. B., Nambiar, R. B., Sellamuthu, P. S., & Emmanuel, R. S. (2021). Use of modified atmosphere packaging combined with essential oils for prolonging post-harvest shelf life of mango (cv. Banganapalli and cv. Totapuri). *LWT-Food Science and Technology*, *148*, 111662. https://doi.org/10.1016/j.lwt.2021.111662

Phillips, C. A. (1996). Modified atmosphere packaging and its effects on the microbiological quality and safety of produce. *International Journal of Food Science & Technology*, *31*(6), 463–479. https://doi.org/10.1046/j.1365-2621.1996.00369.x

Piloo, N., Singh, S., Messar, O., & Pandey, A. (2020). Effect of heat shrinkable film on storability of Citrus cv. Assam Lemon [*Citrus limon* Burm. f.] under ambient conditions. *Journal of Pharmacognosy and Phytochemistry*, *9*(3), 797–800.

Pinto, L., Palma, A., Cefola, M., Pace, B., D'Aquino, S., Carboni, C., & Baruzzi, F. (2020). Effect of modified atmosphere packaging (MAP) and gaseous ozone pre-packaging treatment on the physico-chemical, microbiological and sensory quality of small berry fruit. *Food Packaging and Shelf Life*, *26*, 100573. https://doi.org/10.1016/j.fpsl.2020.100573

Poças, M. F., Delgado, T., & Oliveira, F. A. (2008). Smart packaging technologies for fruits and vegetables. In *Smart packaging technologies for fast moving consumer goods* (pp. 151–166). https://doi.org/10.1002/9780470753699.ch9

Porat, R., Weiss, B., Cohen, L., Daus, A., & Aharoni, N. (2004). Reduction of postharvest rind disorders in citrus fruit by modified atmosphere packaging. *Postharvest Biology and Technology*, *33*(1), 35–43. https://doi.org/10.1016/j.postharvbio.2004.01.010

Porat, R., Lichter, A., Terry, L. A., Harker, R., & Buzby, J. (2018). Postharvest losses of fruit and vegetables during retail and in consumers' homes: Quantifications, causes, and means of prevention. *Postharvest Biology and Technology*, *139*, 135–149. https://doi.org/10.1016/j.postharvbio.2017.11.019

Qin, Y., Zhuang, Y., Wu, Y., & Li, L. (2016). Quality evaluation of hot peppers stored in biodegradable poly (lactic acid)-based active packaging. *Scientia Horticulturae*, *202*, 1–8. https://doi.org/10.1016/j.scienta.2016.02.003

Qu, P., Zhang, M., Fan, K., & Guo, Z. (2022). Microporous modified atmosphere packaging to extend shelf life of fresh foods: A review. *Critical Reviews in Food Science and Nutrition*, *62*(1), 51–65. https://doi.org/10.1080/10408398.2020.1811635

Rana, S., Siddiqui, S., & Goyal, A. (2015). Extension of the shelf life of guava by individual packaging with cling and shrink films. *Journal of Food Science and Technology*, *52*, 8148–8155. https://doi.org/10.1007/s13197-015-1881-5

Rao, D. S. (2017). Individual shrink wrapping technology. In P. Sunil (Ed.), *Novel postharvest treatments of fresh produce* (pp. 557–578). CRC Press. https://doi.org/10.1201/9781315370149

Rao, D. S., & Shivashankara, K. (2015). Individual shrink wrapping extends the storage life and maintains the antioxidants of mango (cvs.'Alphonso'and 'Banganapalli') stored at 8 C. *Journal of Food Science and Technology*, *52*, 4351–4359. https://doi.org/10.1007/s13197-014-1468-6

Reche, J., García-Pastor, M., Valero, D., Hernández, F., Almansa, M., Legua, P., & Amorós, A. (2019). Effect of modified atmosphere packaging on the physiological and functional characteristics of Spanish jujube (*Ziziphus jujuba* Mill.) cv'Phoenix'during cold storage. *Scientia Horticulturae*, *258*, 108743. https://doi.org/10.1016/j.scienta.2019.108743

Rennie, T. J., & Sunjka, P. S. (2017). Modified atmosphere for storage, transportation, and packaging. In P. Sunil (Ed.), *Novel Postharvest Treatments of Fresh Produce* (pp. 433–480). CRC Press. https://doi.org/10.1201/9781315370149

Rodov, V., Ben-Yehoshua, S., Aharoni, N., & Cohen, S. (2010). Modified humidity packaging of fresh produce. In *Horticultural Reviews* (Vol. 37, pp. 281–329). https://doi.org/10.1002/9780470543672.ch5

Rodriguez, J., & Zoffoli, J. P. (2016). Effect of sulfur dioxide and modified atmosphere packaging on blueberry postharvest quality. *Postharvest Biology and Technology*, *117*, 230–238. https://doi.org/10.1016/j.postharvbio.2016.03.008

Sabir, F. K. (2012). Postharvest quality response of broccoli florets to combined application of 1-methylcyclopropene and modified atmosphere packaging. *Agricultural and Food Science*, *21*(4), 421–429.

Sabir, F. K., & Agar, I. T. (2011). Effects of 1-methylcyclopropene and modified atmosphere packing on postharvest life and quality in tomatoes. *Journal of Food Quality*, *34*(2), 111–118.

Sahoo, N. R., Panda, M. K., Bal, L. M., Pal, U. S., & Sahoo, D. (2015). Comparative study of MAP and shrink wrap packaging techniques for shelf life extension of fresh guava. *Scientia Horticulturae*, *182*, 1–7. https://doi.org/10.1016/j.scienta.2014.10.029

Sarfraz, S., Hasan, M. U., Malik, A. U., Anwar, R., & Riaz, R. (2020). Low cost glad wrap film packaging delays postharvest senescence and maintains fruit quality of green chilies. *Multidisciplinary Digital Publishing Institute Proceedings*, *36*(1), 43. https://doi.org/10.3390/proceedings2019036043

Seefeldt, H. F., Løkke, M. M., & Edelenbos, M. (2012). Effect of variety and harvest time on respiration rate of broccoli florets and wild rocket salad using a novel O_2 sensor. *Postharvest Biology and Technology*, *69*, 7–14. https://doi.org/10.1016/j.postharvbio.2012.01.010

Selcuk, N., & Erkan, M. (2014). Changes in antioxidant activity and postharvest quality of sweet pomegranates cv. Hicrannar under modified atmosphere packaging. *Postharvest Biology and Technology*, *92*, 29–36. https://doi.org/10.1016/j.postharvbio.2015.05.018

Sellamuthu, P. S., Mafune, M., Sivakumar, D., & Soundy, P. (2013). Thyme oil vapour and modified atmosphere packaging reduce anthracnose incidence and maintain fruit quality in avocado. *Journal of the Science of Food and Agriculture*, *93*(12), 3024–3031.

Shah, H. M. S., Singh, Z., Afrifa-Yamoah, E., Hasan, M. U., Kaur, J., & Woodward, A. (2023). Insight into the role of melatonin in mitigating chilling injury and maintaining the quality of cold-stored fruits and vegetables. *Food Reviews International*, 1–27. https://doi.org/10.1080/87559129.2023.2212042

Sharma, R., Pal, R., Singh, D., Samuel, D., Kar, A., & Asrey, R. (2010). Storage life and fruit quality of individually shrink-wrapped apples (*Malus domestica*) in zero energy cool chamber. *Indian Journal of Agricultural Sciences*, *80*(4), 338–341.

Siddique, W., Hasan, M., Shah, M., Ali, M., Hayat, F., & Mehmood, A. (2021). Impact of blanching and packaging materials on postharvest quality and storability of fresh spinach. *Journal of Horticultural Science & Technology*, *4*, 7–12. https://doi.org/10.46653/jhst2141007

Singh, S., & Rao, D. S. (2005). Effect of modified atmosphere packaging (MAP) on the alleviation of chilling injury and dietary antioxidants levels in 'Solo'papaya during low temperature storage. *European Journal of Horticultural Science*, *70*(5), 246–252.

Singh, S. P., & Singh, Z. (2013). Controlled and modified atmospheres influence chilling injury, fruit quality and antioxidative system of J apanese plums (*Prunus salicina* L indell). *International Journal of Food Science & Technology*, *48*(2), 363–374. https://doi.org/10.1111/j.1365-2621.2012.03196.x

Singh, S., Singh, Z., & Swinny, E. (2009b). Postharvest nitric oxide fumigation delays fruit ripening and alleviates chilling injury during cold storage of Japanese plums (*Prunus salicina* Lindell). *Postharvest Biology and Technology*, *53*(3), 101–108. https://doi.org/10.1016/j.postharvbio.2009.04.007

Singh, Z. (2022). Interventions to minimise postharvest losses, ensure the quality and safety of tropical and subtropical fruits: an overview. *Acta Horticulturae*, *1340*, 1–12. https://doi.org/10.17660/ActaHortic.2022.1340.1

Singh, Z., Singh, S., & Yahia, E. M. (2009a). Subtropical fruits. In E. M. Yahia (Ed.), *Modified and Controlled Atmospheres for the Storage, Transportation, and Packaging of Horticultural Commodities* (pp. 317–361). https://doi.org/10.1201/9781420069587

Somboonkaew, N., & Terry, L. A. (2010). Physiological and biochemical profiles of imported litchi fruit under modified atmosphere packaging. *Postharvest Biology and Technology*, *56*(3), 246–253. https://doi.org/10.1016/j.postharvbio.2010.01.009

Sudhakar, R. (2018). Individual shrink wrapping extends the storage life and maintains the quality of pomegranates (cvs.'Mridula' and 'Bhagwa') at ambient and low temperature. *Journal of Food Science and Technology*, *55*(1), 351–365. https://doi.org/10.1007/s13197-017-2945-5

Sudhakar, Rao, D., Hebbar, S., & Narayana, C. (2021). CFB box wrapping: A new shrink wrapping technology for extension of storage life of colour capsicum (cv. Bachata). *Journal of Food Science and Technology*, *58*, 3039–3048. https://doi.org/10.1007/s13197-020-04807-6

Surekha, D., Edukondalu, L., Smith, D., & Kumar, K. R. (2017). Quality evaluation of shrink wrapped bananas. *International Journal of Current Microbiology and Applied Sciences*, *6*(10), 2076–2084.

Tian, F., Chen, W., Cai'E, W., Kou, X., Fan, G., Li, T., & Wu, Z. (2019). Preservation of Ginkgo biloba seeds by coating with chitosan/nano-TiO2 and chitosan/nano-SiO2 films. *International Journal of Biological Macromolecules*, *126*, 917–925. https://doi.org/10.1016/j.ijbiomac.2018.12.177

Valverde, J. M., Guillén, F., Martínez-Romero, D., Castillo, S., Serrano, M., & Valero, D. (2005). Improvement of table grapes quality and safety by the combination of modified atmosphere packaging (MAP) and eugenol, menthol, or thymol. *Journal of Agricultural and Food Chemistry*, *53*(19), 7458–7464. https://doi.org/10.1021/jf050913i

Varlı-Yunusoğlu, S., & Ekinci, N. (2023). The Effect of post-harvest salicylic acid and modified atmosphere packaging treatments on the storage of 'Roxana' apricots. *Erwerbs-Obstbau*, 1–9. https://doi.org/10.1007/s10341-023-00829-4

Vázquez-Celestino, D., Ramos-Sotelo, H., Rivera-Pastrana, D. M., Vázquez-Barrios, M. E., & Mercado-Silva, E. M. (2016). Effects of waxing, microperforated polyethylene bag, 1-methylcyclopropene and nitric oxide on firmness and shrivel and weight loss of 'Manila'mango fruit during ripening. *Postharvest Biology and Technology*, *111*, 398–405. https://doi.org/10.1016/j.postharvbio.2015.09.030

Waghmare, R., & Annapure, U. (2013). Combined effect of chemical treatment and/or modified atmosphere packaging (MAP) on quality of fresh-cut papaya. *Postharvest Biology and Technology*, *85*, 147–153. https://doi.org/10.1016/j.postharvbio.2013.05.010

Wang, F., Mi, S., Chitrakar, B., Li, J., & Wang, X. (2022). Effect of cold shock pretreatment combined with perforation-mediated passive modified atmosphere packaging on storage quality of cucumbers. *Foods*, *11*(9), 1267. https://doi.org/10.3390/foods11091267

Wang, Y., Bai, J., & Long, L. E. (2015). Quality and physiological responses of two late-season sweet cherry cultivars 'Lapins' and 'Skeena'to modified atmosphere packaging (MAP) during simulated long distance ocean shipping. *Postharvest Biology and Technology*, *110*, 1–8. https://doi.org/10.1016/j.postharvbio.2015.07.009

Werner, B., & Hotchkiss, J. (2005). Modified atmosphere packaging. In *Microbiology of fruits and vegetables* (pp. 453–476). CRC Press. https://doi.org/10.1201/9781420038934

Wilson, M. D., Stanley, R. A., Eyles, A., & Ross, T. (2019). Innovative processes and technologies for modified atmosphere packaging of fresh and fresh-cut fruits and vegetables. *Critical Reviews in Food Science and Nutrition*, *59*(3), 411–422. https://doi.org/10.1080/10408398.2017.1375892

Yam, K. L., Takhistov, P. T., & Miltz, J. (2005). Intelligent packaging: concepts and applications. *Journal of Food Science*, *70*(1), 1–10. https://doi.org/10.1111/j.1365-2621.2005.tb09052.x

Ye, J.-j., LI, J.-r., HAN, X.-x., Zhang, L., JIANG, T.-j., & Miao, X. (2012). Effects of active modified atmosphere packaging on postharvest quality of shiitake mushrooms (Lentinula edodes) stored at cold storage. *Journal of Integrative Agriculture*, *11*(3), 474–482. https://doi.org/10.1016/S2095-3119(12)60033-1

Zhang, X.-j., Zhang, M., Chitrakar, B., Devahastin, S., & Guo, Z. (2022). Novel combined use of red-white LED illumination and modified atmosphere packaging for maintaining storage quality of postharvest pakchoi. *Food and Bioprocess Technology*, *15*(3), 590–605. https://doi.org/10.1007/s11947-022-02771-x

Zhangh, A., Han, M., Xie, Y., Wang, M., & Cao, C. (2022). Application of ethylene-regulating packaging in post-harvest fruits and vegetables storage: A review. *Packaging Technology and Science*, *35*(6), 461–471. https://doi.org/10.1002/pts.2644

Zhao, Q., Jin, M., Guo, L., Pei, H., Nan, Y., & Rao, J. (2020). Modified atmosphere packaging and 1-methylcyclopropene alleviate chilling injury of 'Youhou'sweet persimmon during cold storage. *Food Packaging and Shelf Life*, *24*, 100479. https://doi.org/10.1016/j.fpsl.2020.100479

Zhu, B., Liu, Y., Qin, Y., Chen, H., & Zhou, L. (2022). Release of clove essential oil loaded by mesoporous nano-silica in polylactic acid-based food packaging on postharvest preservation of white button mushroom. *International Journal of Food Science & Technology*, *57*(1), 457–465. https://doi.org/10.1111/ijfs.15440

Zhuang, H., Barth, M. M., & Cisneros-Zevallos, L. (2014). Modified atmosphere packaging for fresh fruits and vegetables. In *Innovations in food packaging* (pp. 445–473). Elsevier. https://doi.org/10.1016/B978-0-12-394601-0.00018-7

Vacuum Packaging Science and Postharvest Quality of Fruits and Vegetables

16

Hafiz Muhammad Shoaib Shah* Mahmood Ul Hasan and Sajid Ali

*Corresponding Author: shoaib.shah@ecu.edu.au

16.1 INTRODUCTION

Consuming 400 g of horticultural produce is suggested by WHO (2008) on an individual basis, which highlights these products's significance in reducing the risk of chronic diseases. Since 2006, there has been a remarkable reduction in the availability of fresh fruits and vegetables due to their abridged production and higher losses during the postharvest phase. Consequently, their consumption has been reduced by 13% (Baselice et al., 2017). Studies have evidenced about 45% loss in freshly harvested fruits and vegetables, mainly due to poor harvesting methods, inappropriate harvest handling techniques, packaging in poor quality materials, temperature fluctuations during postharvest storage and transportation, as well as unusual marketing delays, especially in developing countries (FAO, 2023; Irfan et al., 2023; Kitinoja & Kader, 2015; Molla et al., 2010). The perishability of fresh fruits and vegetables is the main constraint in their harvest, packaging, transportation, and storage. However, these losses also depend upon the type of commodity, with leafy vegetables reporting 70–80% losses, while roots and tuber crops experience approximately 45% losses (Kitinoja & Kader, 2015). Currently, the wastage of fruits and vegetables is about 40–50% globally, where 54% of losses occur at the farm gate, during postharvest storage and handling, with 24% wastage of fresh water and 30% of total agricultural land (Anand & Barua, 2022; FAO, 2023) causing malnutrition for 780 million people per annum and US$750 billion annually (Dos Santos et al., 2020; Porat et al., 2018).

The major factor limiting the availability of horticultural produce is their perishable nature, which is mainly contributed to by environmental conditions, pathogenic infections, and physiological and metabolic changes in their biochemical constituents (Li et al., 2017; Shah et al., 2023). Many alternative postharvest strategies, such as edible coatings, chemical dip treatments, thermal techniques, irradiations, and gaseous modifications, have been evidenced from the available literature to extend the postharvest window of fresh fruits and vegetables. The other possible way to minimize postharvest losses and to prolong the marketing window of horticultural produce is immediate processing to increase economic gain. However, consumers are more concerned with the availability of additive-free, natural, and fresh fruits and vegetables, which is the main aspiration for the development of non-chemical alternative postharvest technologies (Jiménez-Moreno et al., 2020). Consumer demand regarding the minimal use of chemicals for the conservation of horticultural produce enhanced reliance on physical quality preservation techniques, including heating, irradiation, freezing, and canning, which in turn results in the loss of health-promoting bioactive constituents. Moreover, modifications in color, texture, flavor, and aroma volatiles occur at exponential rates due to physical preservation techniques.

Increasing consumer concern about the use of chemicals and their food safety aspects as well as the deterioration of food bioactive compounds induced by thermal or irradiation treatments led to the foundation of pressure- and gaseous-modified storage of fruits and vegetables (Amsasekar et al., 2022). These technologies have been evidenced to produce minimally processed products during cold or even ambient conditions; mitigate microbial infestations causing cellular lysis or inactivation of microbial membranous protein as well as by increasing the period of lag phase; lessen the sensory and nutritional modifications in treated commodities due to instantaneous and uniform microclimate formation around the commodity; and lessening of chilling injury in tropical and subtropical fruits and vegetables (Rastogi et al., 2007). Pressure modifications may either be higher or lower than normal atmospheric pressure. This chapter focuses on the principles and functions of vacuum packaging (VP), its role in the preservation of

DOI: 10.1201/9781003370376-21

biochemical compounds, postharvest quality management, and the limitations of vacuum packaging in fresh fruits and vegetables.

16.2 PRINCIPLES AND FUNCTIONS OF VACUUM PACKAGING

VP is a form of modified atmosphere packaging (MAP) and is a traditional way to improve the shelf life of food products, creating an oxygen (O_2) deficient microclimate around the commodity by removing the air from the sealed package, which reduces the size of the package and leads to decreased pressure compared to atmospheric pressure. Low levels of O_2 inhibit ROS accumulation and lipid peroxidation and increase the lag phase of aerobic microbes inside the package. For VP, many packaging materials could be used depending upon the different fruits and vegetables, which would have certain qualities, including a barrier against light intensity. Oxygen levels within the packaging should be maintained to avoid enzymatic browning; modifications in flavor, aroma volatiles, and color variations; and reduction in AA levels. It should be economical and able to withstand vibration and heat shocks (Ravichandran & Upadhyay, 2019). The best results could be achieved by using a glass container, as dried fruits and vegetables may puncture plastic packaging after creating a vacuum around the commodity (Li et al., 2017). Modified vacuum packaging (MVP) is the most commonly adopted VP technique, in which a rigid container is used for the preservation of food products at ambient conditions with reduced air pressure (<40 kPa) and a deficit of O_2 leading to a reduction in aerobic bacterial and fungal infestations (Ravichandran & Upadhyay, 2019).

16.3 VACUUM PACKAGING FOR POSTHARVEST FRUIT AND VEGETABLE QUALITY PRESERVATION

16.3.1 Respiration and Ethylene Biosynthesis

VP shows static hypobaric storage conditions which increase the storage window of fresh, minimally processed, and fresh-cut horticultural produce owing to the suppression of respirational and metabolic catabolism by mitigating the levels of oxygen inside the packaging (Manju et al., 2007; Xu et al., 2022). Respiration and ethylene production during the postharvest phase cause a reduction in fruit quality by using sugars, polyphenols, and organic acids as the substrate to produce energy during the postharvest period, which ultimately lowers the quality of the fruit and reduces the marketing window of fresh commodities. VP alone or in combination with other postharvest strategies suppresses or delays the respiration rate in fresh produce (Table 16.1). For example, Rana et al. (2018) claimed that the VP of guava fruit exhibited a lower respiration rate during 21 days of cold storage. An exogenous AA application with subsequent storage in VP delayed respiration and ethylene biosynthesis in fresh-cut potatoes for up to 5 days (Xu et al., 2022). Similarly, VP delayed the climacteric ethylene peak in banana fruit for 48 hours without causing any off-flavor and browning of individual fingers. Nevertheless, the levels of AA and ethanol were significantly lessened by VP (Pesis et al., 2005). The efficacy of VP in the downregulation of ethylene production could be alleviated by a brief cold shock prior to packaging, possibly due to reduced oxidative stress and higher TPC (Lwin et al., 2020). However, there is sporadic information about the metabolic and enzymatic changes linked to ethylene production, which requires future research.

16.3.2 Enzymatic Browning Prevention

Enzymatic browning, caused by the transformation of phenolics to quinones by the oxidative activity of the PPO enzyme, is the most significant cause of the degradation of quality in fresh horticultural produce during storage (Ali et al., 2016; Ali et al., 2019; Ali et al., 2023; Luo et al., 2012; Zhang et al., 2023). Oxygen is the primary substrate for browning-related chemical reactions where it transforms into hydroxyl radicals ($OH^{\cdot}$) or superoxide anion ($O2^{-\cdot}$) during stress conditions. These free radicals catalyze the formation of quinones by the oxidation of phenolic compounds owing to activities of PPO and POD enzymes, which subsequently produce brown pigmentation in fruits and vegetables (Foyer & Noctor, 2005; Schieber & Chandel, 2014). VP proved to be effective in suppressing enzymatic browning in fresh fruits and minimally processed products due to a reduction in oxygen-dependent biochemical and phytochemical reactions by mitigating internal oxygen levels (Denoya et al., 2015; Min et al., 2019; Shah & Nath, 2008). The most accepted mechanism of VP-induced browning inhibition is suppressed respiratory rate, which is interlinked with a short supply of oxygen within the microclimate of the packaging. This may also lead to the formation of alcoholic fermented products, mainly formaldehyde or ethanol, in ordinary packaging but pressure processing lowers these constraints during storage (Araya et al., 2009). The alternative mechanistic inhibition of browning is associated with the expression of ethylene response factor (ERF) genes in climacteric fruits where the suppressed transcriptional expression of *NnPPOA*, *NnPAL1*, and *NnPOD2/3* genes delayed the activity of PAL, PPO, and POD enzymes in VP lotus slices, respectively, which in turn reduced the oxidative transformation of phenolics to quinones (Min et al., 2019). This suppression in gene expressions might be triggered by *NAC* transcription factor during cold storage, which is a major stress-mediated gene in plants (Min et al., 2020).

TABLE 16.1 Summary of Vacuum Packaging (VP) on the Quality Management of Horticultural Produce

HORTICULTURAL PRODUCE	*PACKAGING CONDITIONS*	*RESULTS*	*REFERENCES*
Asparagus (*Asparagus officinalis*)	LDPE (200 mm × 300 mm) with a thickness of 80 μm	VP suppressed ethylene production, mitigated color modifications, and reduced the activities of POD, CAD, and PAL enzymes while maintaining higher phenolics.	Lwin et al. (2020)
Banana (*Musa acuminata*)	LDPE (20 cm × 25 cm)	VP banana fruit exhibited higher fruit firmness while mitigating moisture loss and color modifications.	Othman et al. (2021)
Cauliflower (*Brassica oleracea*)	LDPE (25 μm) 23 x 18 cm, 6.673 ×10^{14} gm m^{-2} s^{-1} Pa^{-1}	VP maintained better sensory quality scores, AA, total sugars, and fruit firmness while reducing microbial content.	Nasrin et al. (2022)
Chinese water chestnut (*Eleocharis tuberosa*)	Polyamide/polyethylene (PA/PE) bags with a dimension of 20 × 15 cm (0.1 MPa)	VP conserved higher lightness and phenolics, and better sensory attributes. VP reduced the activities of PPO, POD, PLD, and LOX activities as well as MDA content.	Zhou et al. (2022)
Jujube (*Zizyphus jujube*)	LDPE bag (0.06 mm thick) oxygen 15–30 (ml × cm/cm^3), carbon dioxide 60–160 (ml × cm/cm^3)	VP maintained better sensory quality, fruit color, and firmness while reducing the browning index.	Moradinezhad & Dorostkar (2021)
Lotus (*Nelumbo nucifera*)	Polyethylene 200 × 280 mm	VP reduced the browning of lotus, and activities of PAL, PPO, and POD enzymes of lotus roots.	Min et al. (2020)
Peaches (*Prunus persica*)	Cryovac BB2800 bags O_2 transmission rate: 6–14 cm^3/m^2/24 h at 23°C, 101,325 Pa	VP inhibited browning, PPO activity, fermentation, and ethanol levels.	Denoya et al. (2015)
Pomegranate (*Punica granatum*)	LDPE bag (dimensions: 35 × 55 cm, with 50 μm thickness)	VP improved the marketability index, ripening index, and fruit color while reducing chilling injury, total anthocyanins, and electrolyte leakage and weight loss.	Moradinezhad et al. (2019)
Potato (*Solanum tuberosum*)	EVOH film (91.18 μm film thickness) with an O_2 permeability of 0.71 cm^3/m^2 24 h 0.1 MPa and polyethylene 8 μm.	VP maintained better visual quality, texture, and organic acid contents while lowering respiration, ethylene production, and fruit firmness and delaying the browning time and bacterial count.	Li et al. (2022); Xu et al. (2022)
Pumpkin (*Curcurbita maxima*)	Polyethylene (100-μm-thick) and chitosan (2%) coating	VP upregulated β-carotene and a^* values while lowering L^*, b^*, C^*, hue angle, and microbial quality.	Yüksel et al. (2022)
Sweet lemon (*Citrus limetta*)	LDPE bag (10 cm × 10 cm) with radish leaf extract coating	VP attenuated weight loss and fruit softening. VP maintained a higher AA level and better fruit color.	Zandi et al. (2021)
Sugar apple (*Annona squamosa*)	PVDC-nylon bags with a thickness of 25/50 μm	VP reduced PPO activity and maintained higher phenolics and chlorophyll content.	Wu (2000)
Taro (*Colocasia esculenta*)	Nylon/PE film bags (200 × 250 mm, 50 μm thickness)	VP reduced weight loss, fruit softening, sensory modifications, and microbial count.	Chang & Kim (2015)

Source: AA = ascorbic acid, CAD = cinnamyl alcohol dehydrogenase, LDPE = low-density polyethylene, LOX = lipoxygenase, MDA = malondialdehyde, PAL = phenylalanine ammonia-lyase, PLD = phospholipase D, POD = peroxidase, PPO = polyphenol oxidase

16.3.3 Moisture Loss and Fruit Firmness

Physiological moisture loss, often regarded as weight loss (WL), is the primary senescence-triggering mechanism in fresh horticultural produce and is associated with the significant rise in ROS accumulation and oxidative toxicity. Evaporation from fruit surface and higher respiration rate are primary causes of WL. However, VP has been reported to ameliorate WL, which is possibly due to reduced evaporation from the fruit surface, leading to the maintenance of firmness in potatoes (Xu et al., 2022). Asparagus exposed to short-term cold shock following VP mitigated WL (about 5.86%) compared to the control. Furthermore, fruit softening was also slowed by cold-shocked and VP asparagus for 16 days. The ameliorated vapor gradient due to VP acted as a barrier against moisture loss between the fruit surface and environment, thereby resulting in alleviated WL during storage. Moreover, the cold shock for short intervals causes the inactivation of oxidative enzymes due to alleviated ROS accumulation (Lwin et al., 2020). Gaseous modifications around the fruit stabilizes the microenvironment of fruit, and edible coatings have been extensively reported to improve this condition. The efficacy of VP to maintain optimum fruit quality with a prolonged marketing window may be increased by combining the edible coating on fruits and vegetables. The combination of radish leaf extract coating and VP has demonstrated better fruit firmness and reduced WL in sweet lemon. The combination treatments not only maintained a physical barricade between the fruit and the environment to reduce moisture loss but also alleviated metabolic activities due to reduced respiration resulting from lower oxygen levels (Zandi et al., 2021).

16.3.4 Role of Vacuum Packaging on Bioactive Compounds

Phenolics are aromatic compounds and secondary metabolites in plants that are associated with cellular energy metabolism and many metabolic pathways. Horticultural produce is a unique source of phenolics, which are not only the constituent of fruit quality but also play a substantial role in the antioxidant system (de la Rosa et al., 2019). VP effectively maintains higher total phenolic contents (TPC) during short- or long-term fruit and vegetable storage due to the inhibition of PPO activity, which produces brown pigments as a result of oxidation (Rocha et al., 2003; Zhou et al., 2022). Additionally, VP also upregulates TPC during storage due to the inhibition of phenylalanine ammonia-lyase (PAL) and peroxidase (POD) enzymes (Min et al., 2020). Xu et al. (2022) reported that VP conserved higher TPC by maintaining higher individual levels of cinnamate, ferulate, p-coumarate, chlorogenic, and gallic acids in fresh-cut lotus during low-temperature storage. However, sporadic information is available in the literature showing the role of VP on the phenolic metabolism of horticultural produce, which could be explored in future research.

Carbohydrates are primary compounds correlated with the postharvest quality of fruits and vegetables, which may be either due to reducing or non-reducing sugars (Shah & Nath, 2008). The main sugars in a horticultural commodity are fructose, glucose, and sucrose, with trace amounts of other sugars. VP conserves a higher amount of aforesaid sugars than unwrapped fruits (Xu et al., 2022). Overall reducing sugar levels were also maintained in potatoes stored for 7 days under VP. The reason for increasing reducing sugars stored in VP is overwhelmingly due to the interconversion of higher molecular weight sugars such as cellulose and hemicelluloses to low molecular weight sugars, which are a primary reason for higher levels of reducing sugars in potatoes (Rocha et al., 2003). Contrarily, Deepa et al. (2013) have observed a linear decline in the sugar contents of VP-stored chilies. They argued that this decline in sugar levels was caused by the consumption of low molecular weight sugar molecules during the respiratory activity of chilies during storage. However, the reduction in sugar levels was significantly lower than in chilies stored in jute bags. Overall, the total sugar contents exhibited a declining pattern in total sugars while higher reducing sugars were seen in VP-stored papaya fruit (Padmanaban et al., 2014). However, the sugar metabolism in fruits and vegetables during VP storage is still poorly elaborated in the literature, paving the way for further research in the future.

AA is one of the most reported bioactive constituents of fruits and vegetables and not only plays an important role in nutritional quality and medical benefits but also reduces the risk of oxidative injury in plants through the activation of the Halliwell-Asada pathway (Foyer & Noctor, 2011). The level of AA reduces with the accumulation of ROS and at the onset of senescence. The degradation of AA is exponential at ambient conditions, while cold storage maintains significantly higher levels of AA in fruits and vegetables. VP has been found to be effective in maintaining higher levels of AA in horticultural produce. Storage of vacuum-packed commodities in dark conditions further lowers the oxidation of AA, which indicates its sensitivity to temperature and light (Deepa et al., 2013). VP works to lower the levels of O_2 around the commodity, which may help retain the microbial spoilage and oxidative toxicity; however, the anaerobic conditions may cause slime mold infestations, modifications in flavor, and off-odor development in fresh horticultural produce (Nunez et al., 1986).

16.3.5 Microbial Deterioration

Microbial safety is the basic concern of postharvest technology in fresh or processed fruits and vegetables, as horticultural produce are good nutrient sources for microbes and also provide adequate moisture content for their colonial germination (Bico et al., 2009; Du et al., 2009). Microbial infestation may be either due to aerobic, mesophilic, yeast, or molds, which not only shortens the shelf life but also makes them unfit for human consumption. The reproductive and growth cycles are primarily correlated with the levels of oxygen around the fruit, and VP extends the duration of the lag phase during the reproductive cycle of microbes, leading to reduced microbial counts in fresh or minimally processed products during storage and marketing (Xing et al., 2012). However, the effectiveness of VP against microbial count may be increased by prethermal treatments. It was proposed by Chang and Kim (2015) that heat-treated (1 min at 60 °C) peeled taro exhibited alleviated anaerobic bacterial and coliform count during low-temperature storage for 12 days without causing any variations in organoleptic characteristics. Moderate VP has also been reported to mitigate 30–40% fruit decay by reducing microbial microflora in peeled onion for 30 days (Hong & Kim, 2004). However, the studies describing the influence of VP on disease incidence are poorly reported in the literature and require further investigation.

16.3.6 Sensory and Color Modification

Sensory attributes are one of the most imperative quality parameters for fresh horticultural produce and are mainly contributed by the soluble solids content (SSC), organic acids, aroma volatiles, and polyphenols. Ethylene production and respiration are the primary metabolic events which utilize the sugars and organic acids as an energy source, which in turn causes modifications in the sensory attributes of fresh produce (Rana et al., 2018; Xu et al., 2022). Padmanaban et al. (2014) have observed that VP maintained higher reducing sugars, which might be correlated with lower respiration rate due to delayed utilization of non-reducing sugars during glycolysis to maintain energy levels in cold-stored papaya. Color is the primary qualitative trait linked to the quality of fresh produce. Fruit darkening may be the result of many physical, physiological, and biochemical variations. VP has been reported to maintain the characteristic bright color for many fruits and vegetables during marketing, transportation, and storage. Lightness (L^*) and hue angle ($h°$) indicate the darkening of fruit skin (peel or pulp), and a higher L^* value is a symptom of browning in fresh fruits with characteristic bright colors. VP maintained lower L^* values for 21 days in peach fruit during cold storage (Denoya et al., 2015). However, sensory quality may deteriorate with higher ethanol production due to inappropriate packaging materials, lower oxygen levels, and pressure applied during storage, which may cause off-odor and a foul smell, making the commodity unacceptable for marketing (Ravichandran & Upadhyay, 2019).

16.4 EFFECT OF VACUUM PACKAGING ON MINIMALLY PROCESSED FRUITS AND VEGETABLES

The growing demand for convenience from consumers as well as the health advantages of fresh food has been cited as reasons for the appeal of minimally processed fruits and vegetables. Packaging conditions that limit the amount of oxygen within the package are known for extending the shelf life of produce by reducing the respiration rate and impeding the growth of aerobic bacteria. Vacuum packaging involves the elimination of oxygen from the container or package in order to remove microorganisms that can deteriorate the quality of produce. The VP prevents oxidative reactions and the growth of aerobic microorganisms, which often cause spoilage of fruits and vegetables during storage (Rocha et al., 2003).

Vacuum packaging is one of the key methods for quality preservation with proven potential to significantly prolong storability and enhance the overall quality of fruit and vegetables (Baselice et al., 2017). Combined application of VP (nylon or polyethylene films) and hot water treatment has proved to delay browning in fresh-cut taro along with maintenance of sensory quality (Chang & Kim, 2015). VP has been reported to maintain a better physical appearance of fresh-cut potato strips in comparison to modified atmospheric packaging (Beltrán et al., 2005). VP of minimally processed jackfruit bulbs resulted in preserved sensory quality for 3 weeks, 2 weeks, and 1 week in un-ripe, semi-ripe, and fully ripe bulbs, respectively (Taj et al., 2014). Likewise, Gana and Mini (2017) outlined that VP (polypropylene or low-density polyethylene) after treatment with potassium metabisulphite and citric acid is effective in enhancing shelf life, reducing fruit weight loss, and conserving the marketability of both soft-fleshed and firm-fleshed jackfruit bulbs. Buick and Damoglou (1987) reported that VP reduced microbial spoilage and extended the shelf life of fresh-cut carrots from 5 to 8 days at a temperature of 4 °C.

Later, Rocha et al. (2007) outlined that VP of fresh-cut carrots exhibited a positive effect on the color and total phenolics content when stored at a temperature of 4 °C. Similar results have been reported by Nasrin et al. (2022) in fresh-cut cauliflower where VP with LDPE resulted in a maximum sensory score compared with other packaging methods (polypropylene box, LDPE, sealed LDPE). Zudaire et al. (2020) also recommended a VP technique for the preservation of the phytochemical quality of minimally processed onions. Moreover, VP in combination with high hydrostatic pressure resulted in reduced fruit browning in fresh-cut peaches, along with reduced activity of polyphenol oxidase, ethylene production, and better textural properties (cohesiveness, chewiness, and firmness) of the fruit (Denoya et al., 2015).

16.5 CONCLUSIONS

The vacuum environment has demonstrated a considerable role in the quality management and preservation of fresh and processed horticultural commodities by reducing ethylene production, respiration rate, color modifications, and microbial infestations. Enzymatic discoloration is the leading constraint in fresh-cut storage, and VP is an alternative to alleviate the constraints associated with the use of anti-browning chemical and thermal treatments. However, the production of ethanol and other anaerobic products may limit the use of VP in the commercial horticultural industry. Moreover, commodity-specific VP requirements are other factors in the wider applicability of vacuum technology at the industrial level. However, on the basis of the literature available, the following prospects need to be addressed in future research in vacuum technology:

- Specific VP requirements, including the packaging materials, pressure during storage, and specific gaseous composition inside the sealed pack, should be optimized to reduce the risk of anaerobic respiration and the production of associated products.
- Although the antimicrobial role of VP has been reported in some studies, there is a need to explore the effectiveness of vacuum technology against the colonial growth and inhibition of foodborne pathogens such as *Salmonella* sp., *Erwinia* sp., *Pseudomonas* sp., hepatitis A virus, norovirus, *Bacillus* sp., *Escherichia coli*, *Yersinia enterocolitica*, and *Listeria monocytogenes*, causing different diseases in humans.
- The role of VP in the downregulation of chilling injury during cold storage should be evaluated both in fresh and minimally processed fruits and vegetables.
- A low-cost and eco-friendly packaging material needs to be developed for vacuum technology.

REFERENCES

Ali, S., Khan, A. S., Malik, A. U., Anjum, M. A., Nawaz, A., Shah, H. M. S. (2019). Modified atmosphere packaging delays enzymatic browning and maintains quality of harvested litchi fruit during low temperature storage. *Scientia Horticulturae, 254*, 14–20. https://doi.org/10.1016/j.scienta.2019.04.065

Ali, S., Khan, A. S., Malik, A. U., Shahid, M. (2016). Effect of controlled atmosphere storage on pericarp browning, bioactive compounds and antioxidant enzymes of litchi fruits. *Food Chemistry*, *206*, 18–29. https://doi.org/10.1016/j.foodchem.2016.03.021

Ali, S., Khan, A. S., Nawaz, A., Naz, S., Ejaz, S., Ullah, S. (2023). Glutathione application delays surface browning of fresh-cut lotus (*Nelumbo nucifera* Gaertn.) root slices during low temperature storage. *Postharvest Biology and Technology*, *200*, 112311. https://doi.org/10.1016/j.postharvbio.2023.112311

Amsasekar, A., Mor, R. S., Kishore, A., Singh, A., Sid, S. (2022). Impact of high pressure processing on microbiological, nutritional and sensory properties of food: a review. *Nutrition & Food Science*, *52*(6), 996–1017. https://doi.org/10.1108/NFS-08-2021-0249

Anand, S., Barua, M. (2022). Modeling the key factors leading to post-harvest loss and waste of fruits and vegetables in the agri-fresh produce supply chain. *Computers and Electronics in Agriculture*, *198*, 106936. https://doi.org/10.1016/j.compag.2022.106936

Araya, X. I. T., Smale, N., Zabaras, D., Winley, E., Forde, C., Stewart, C. M., Mawson, A. J. (2009). Sensory perception and quality attributes of high pressure processed carrots in comparison to raw, sous-vide and cooked carrots. *Innovative Food Science & Emerging Technologies*, *10*(4), 420–433. https://doi.org/10.1016/j.ifset.2009.04.002

Baselice, A., Colantuoni, F., Lass, D. A., Nardone, G., Stasi, A. (2017). Trends in EU consumers' attitude towards fresh-cut fruit and vegetables. *Food Quality and Preference*, *59*, 87–96. https://doi.org/10.1016/j.foodqual.2017.01.008

Beltrán, D., Selma, M. V., Tudela, J. A., Gil, M. I. (2005). Effect of different sanitizers on microbial and sensory quality of fresh-cut potato strips stored under modified atmosphere or vacuum packaging. *Postharvest Biology and Technology*, *37*(1), 37–46. https://doi.org/10.1016/j.postharvbio.2005.02.010

Bico, S., Raposo, M., Morais, R., Morais, A. (2009). Combined effects of chemical dip and/or carrageenan coating and/or controlled atmosphere on quality of fresh-cut banana. *Food Control*, *20*(5), 508–514. https://doi.org/10.1016/j.foodcont.2008.07.017

Buick, R. K., Damoglou, A. P. (1987). The effect of vacuum packaging on the microbial spoilage and shelf-life of, ready-to-use' sliced carrots. *Journal of the Science of Food and Agriculture*, *38*(2), 167–175. https://doi.org/10.1002/jsfa.2740380208

Chang, M.S., Kim, G.-H. (2015). Combined effect of hot water dipping and vacuum packaging for maintaining the postharvest quality of peeled taro. *Horticulture, Environment, and Biotechnology*, *56*, 662–668. https://doi.org/10.1007/s13580-015-0018-0

de la Rosa, L. A., Moreno-Escamilla, J. O., Rodrigo-García, J., Alvarez-Parrilla, E. (2019). Phenolic compounds. In E. M. Yahia & A. Carrillo-Lopez (Eds.), *Postharvest physiology and biochemistry of fruits and vegetables* (pp. 253–271). Elsevier. https://doi.org/10.1016/B978-0-12-813278-4.00012-9

Deepa, G., Chetti, M. B., Khetagoudar, M. C., Adavirao, G. M. (2013). Influence of vacuum packaging on seed quality and mineral contents in chilli (*Capsicum annuum* L.). *Journal of Food Science and Technology*, *50*, 153–158. https://doi.org/10.1007/s13197-011-0241-3

Denoya, G. I., Vaudagna, S. R., Polenta, G. (2015). Effect of high pressure processing and vacuum packaging on the preservation of fresh-cut peaches. *LWT-Food Science and Technology*, *62*(1), 801–806. https://doi.org/10.1016/j.lwt.2014.09.036

Dos Santos, S. F., Cardoso, R. d. C. V., Borges, Í. M. P., e Almeida, A. C., Andrade, E. S., Ferreira, I. O., do Carmo Ramos, L. (2020). Post-harvest losses of fruits and vegetables in supply centers in Salvador, Brazil: Analysis of determinants, volumes and reduction strategies. *Waste Management*, *101*, 161–170. https://doi.org/10.1016/j.wasman.2019.10.007

Du, J., Fu, Y., Wang, N. (2009). Effects of aqueous chlorine dioxide treatment on browning of fresh-cut lotus root. *LWT-Food Science and Technology*, *42*(2), 654–659. https://doi.org/10.1016/j.lwt.2008.08.007

FAO. (2023). *Food Loss and Waste Database*. Retrieved April 04, 2023 from www.fao.org/platform-food-loss-waste/flw-data/en/

Foyer, C. H., Noctor, G. (2005). Oxidant and antioxidant signalling in plants: a re-evaluation of the concept of oxidative stress in a physiological context. *Plant, Cell & Environment*, *28*(8), 1056–1071. https://doi.org/10.1111/j.1365-3040.2005.01327.x

Foyer, C. H., Noctor, G. (2011). Ascorbate and glutathione: the heart of the redox hub. *Plant Physiology*, *155*(1), 2–18. https://doi.org/10.1104/pp.110.167569

Gana, K., & Mini, C. (2017). Shelf life extension of minimally processed jackfruit (*Artocarpus heterophyllus* Lam.) portion. *Asian Journal of Dairy and Food Research*, 1–7.

Hong, S. I., Kim, D. (2004). The effect of packaging treatment on the storage quality of minimally processed bunched onions. *International Journal of Food Science & Technology*, *39*(10), 1033–1041. https://doi.org/10.1111/j.1365-2621.2004.00885.x

Irfan, M., Kumar, P., Ahmad, M. F., Siddiqui, M. W. (2023). Biotechnological interventions in reducing losses of tropical fruits and vegetables. *Current Opinion in Biotechnology*, *79*, 102850. https://doi.org/10.1016/j.copbio.2022.102850

Jiménez-Moreno, N., Esparza, I., Bimbela, F., Gandía, L. M., Ancín-Azpilicueta, C. (2020). Valorization of selected fruit and vegetable wastes as bioactive compounds: Opportunities and challenges. *Critical Reviews in Environmental Science and Technology*, *50*(20), 2061–2108. https://doi.org/10.1080/10643389.2019.1694819

Kitinoja, L., Kader, A. A. (2015). Measuring postharvest losses of fresh fruits and vegetables in developing countries. *PEF White Paper*, *15*, 26. https://doi.org/10.13140/RG.2.1.3921.6402

Li, L., Zhang, M., Adhikari, B., Gao, Z. (2017). Recent advances in pressure modification-based preservation technologies applied to fresh fruits and vegetables. *Food Reviews International*, *33*(5), 538–559. https://doi.org/10.1080/87559129.2016.1196492

Li, Z., Zhao, W., Ma, Y., Liang, H., Wang, D., Zhao, X. (2022). Shifts in the bacterial community related to quality properties of vacuum-packaged peeled potatoes during Storage. *Foods*, *11*(8), 1147. https://doi.org/10.3390/foods11081147

Luo, Z., Wu, X., Xie, Y., Chen, C. (2012). Alleviation of chilling injury and browning of postharvest bamboo shoot by salicylic acid treatment. *Food Chemistry*, *131*(2), 456–461. https://doi.org/10.1016/j.foodchem.2011.09.007

Lwin, W. W., Pongprasert, N., Boonyaritthongchai, P., Wongs-Aree, C., Srilaong, V. (2020). Synergistic effect of vacuum packaging and cold shock reduce lignification of asparagus. *Journal of Food Biochemistry*, *44*(11), e13479. https://doi.org/10.1111/jfbc.13479

Manju, S., Jose, L., Gopal, T. S., Ravishankar, C., Lalitha, K. (2007). Effects of sodium acetate dip treatment and vacuum-packaging on chemical, microbiological, textural and sensory changes of Pearlspot (*Etroplus suratensis*) during chill storage. *Food Chemistry*, *102*(1), 27–35. https://doi.org/10.1016/j.foodchem.2006.04.037

Min, T., Liu, E., Xie, J., Yi, Y., Wang, L., Ai, Y., Wang, H. (2019). Effects of vacuum packaging on enzymatic browning and ethylene response factor (ERF) gene expression of fresh-cut lotus root. *HortScience*, *54*(2), 331–336. https://doi.org/10.21273/HORTSCI13735-18

Min, T., Niu, L.F., Xie, J., Yi, Y., Wang, L.-m., Ai, Y.W., Wang, H.X. (2020). Effects of vacuum packaging on NAC gene expression in fresh-cut lotus root. *Journal of the American Society for Horticultural Science*, *145*(1), 36–44. https://doi.org/10.21273/JASHS04806-19

Molla, M., Islam, M., Nasrin, T., Bhuyan, M. (2010). Survey on postharvest practices and losses of litchi in selected areas of Bangladesh. *Bangladesh Journal of Agricultural Research*, *35*(3), 439–451. https://doi.org/10.3329/bjar.v35i3.6451

Moradinezhad, F., Dorostkar, M. (2021). Effect of vacuum and modified atmosphere packaging on the quality attributes and sensory evaluation of fresh jujube fruit. *International Journal of Fruit Science*, *21*(1), 82–94. https://doi.org/10.1080/15538362.2020.1858470

Moradinezhad, F., Khayyat, M., Ranjbari, F., Maraki, Z. (2019). Vacuum packaging optimises quality and reduces postharvest losses of pomegranate fruits. *Journal of Horticulture and Postharvest Research*, *2*, 15–26. https://doi.org/10.22077/jhpr.2018.1775.1030

Nasrin, T. A. A., Yasmin, L., Arfin, M. S., Rahman, M. A., Molla, M. M., Sabuz, A. A., Afroz, M. (2022). Preservation of postharvest quality of fresh cut cauliflower through simple and easy packaging techniques. *Applied Food Research*, *2*(2), 100125. https://doi.org/10.1016/j.afres.2022.100125

Nunez, M., Gaya, P., Medina, M., Rodríguez-Marín, M., Garcia-Aser, C. (1986). Changes in microbiological, chemical, rheological and sensory characteristics during ripening of vacuum packaged Manchego cheese. *Journal of Food Science*, *51*(6), 1451–1455. https://doi.org/10.1111/j.1365-2621.1986.tb13832.x

Othman, S. H., Abdullah, N. A., Nordin, N., Shah, N. N. A. K., Nor, M. Z. M., Yunos, K. F. M. (2021). Shelf life extension of Saba banana: Effect of preparation, vacuum packaging, and storage temperature. *Food Packaging and Shelf Life*, *28*, 100667. https://doi.org/10.1016/j.fpsl.2021.100667

Padmanaban, G., Singaravelu, K., Annavi, S. T. (2014). Increasing the shelf-life of papaya through vacuum packing. *Journal of Food Science and Technology*, *51*, 163–167. https://doi.org/10.1007/s13197-011-0468-z

Pesis, E., Arie, R. B., Feygenberg, O., Villamizar, F. (2005). Ripening of ethylene-pretreated bananas is retarded using modified atmosphere and vacuum packaging. *HortScience*, *40*(3), 726–731. https://doi.org/10.21273/HORTSCI.40.3.726

Porat, R., Lichter, A., Terry, L. A., Harker, R., Buzby, J. (2018). Postharvest losses of fruit and vegetables during retail and in consumers' homes: Quantifications, causes, and means of prevention. *Postharvest Biology and Technology*, *139*, 135–149. https://doi.org/10.1016/j.postharvbio.2017.11.019

Rana, S., Siddiqui, S., Gandhi, K. (2018). Effect of individual vacuum and modified atmosphere packaging on shelf life of guava. *International Journal of Chemical Studies*, *6*(2), 966–972.

Rastogi, N., Raghavarao, K., Balasubramaniam, V., Niranjan, K., Knorr, D. (2007). Opportunities and challenges in high pressure processing of foods. *Critical Reviews in Food Science and Nutrition*, *47*(1), 69–112. https://doi.org/10.1080/10408390600626420

Ravichandran, C., Upadhyay, A. (2019). Use of vacuum technology in processing of fruits and vegetables. In Khursheed Alam Khan, Megh R. Goyal, & A. A. Kalne (Eds.), *Processing of fruits and vegetables* (pp. 139–174). CRC Press.

Rocha, A. M., Coulon, E. C., Morais, A. M. (2003). Effects of vacuum packaging on the physical quality of minimally processed potatoes. *Food Service Technology*, *3*(2), 81–88. https://doi.org/10.1046/j.1471-5740.2003.00068.x

Rocha, A. M., Mota, C. C., Morais, A. M. (2007). Physico-chemical qualities of minimally processed carrot stored under vacuum. *Journal of Foodservice*, *18*(1), 23–30. https://doi.org/10.1111/j.1745-4506.2007.00043.x

Schieber, M., Chandel, N. S. (2014). ROS function in redox signaling and oxidative stress. *Current Biology*, *24*(10), R453-R462. https://doi.org/10.1016/j.cub.2014.03.034

Shah, H. M. S., Singh, Z., Afrifa-Yamoah, E., Hasan, M. U., Kaur, J., Woodward, A. (2023). Insight into the role of melatonin in mitigating chilling injury and maintaining the quality of cold-stored fruits and vegetables. *Food Reviews International*, 1–27. https://doi.org/10.1080/87559129.2023.2212042

Shah, N. S., Nath, N. (2008). Changes in qualities of minimally processed litchis: Effect of antibrowning agents, osmo-vacuum drying and moderate vacuum packaging. *LWT-Food Science and Technology*, *41*(4), 660–668. https://doi.org/10.1016/j.lwt.2007.04.012

Taj, F., Ranganna, K., Munishamanna, B. (2014). Vacuum packaging of minimally processed jackfruit bulbs for long distance transportation. *Mysore Journal of Agricultural Sciences*, *48*(3), 351–357.

WHO. (2008). *WHO European Action Plan for Food and Nutrition Policy 2007–2012*. https://apps.who.int/iris/handle/10665/349767

Wu, M. C. (2000). Effect of vacuum packaging on the color of frozen sugar apple (*Annona squamosa* L.). *Journal of Food Quality*, *23*(3), 305–315. https://doi.org/10.1111/j.1745-4557.2000.tb00216.x

Xing, Y., Li, X., Xu, Q., Yun, J., Lu, Y. (2012). Extending the shelf life of fresh-cut lotus root with antibrowning agents, cinnamon oil fumigation and moderate vacuum packaging. *Journal of Food Process Engineering*, *35*(4), 505–521. https://doi.org/10.1111/j.1745-4530.2010.00604.x

Xu, D., Chen, C., Zhou, F., Liu, C., Tian, M., Zeng, X., Jiang, A. (2022). Vacuum packaging and ascorbic acid synergistically maintain the quality and flavor of fresh-cut potatoes. *LWT*, *162*, 113356. https://doi.org/10.1016/j.lwt.2022.113356

Yüksel, Ç., Atalay, D., Erge, H. S. (2022). The effects of chitosan coating and vacuum packaging on quality of fresh-cut pumpkin slices during storage. *Journal of Food Processing and Preservation*, *46*(3), e16365. https://doi.org/10.1111/jfpp.16365

Zandi, M., Ganjloo, A., Bimakr, M., Moradi, N., Nikoomanesh, N. (2021). Effect of active coating containing radish leaf extract with or without vacuum packaging on the postharvest changes of sweet lemon during cold storage. *Journal of Food Processing and Preservation*, *45*(3), e15252. https://doi.org/10.1111/jfpp.15252

Zhang, C., Xin, Y., Wang, Z., Qin, L., Cao, l., Li, H., Ma, X., Yin, J., Zhao, Z., Liu, P. (2023). Melatonin-induced MYBs alleviates fresh-cut lotus root (*Nelumbo nucifera* Gaertn.) browning during storage by attenuating flavonoid biosynthesis and reactive oxygen species (ROS). *Journal of the Science of Food and Agriculture*. https://doi.org/10.1002/jsfa.12619

Zhou, X., Hu, W., Li, J., Iqbal, A., Murtaza, A., Xu, X., Pan, S. (2022). High-pressure carbon dioxide treatment and vacuum packaging alleviate the yellowing of peeled Chinese water chestnut (*Eleocharis tuberosa*). *Food Packaging and Shelf Life*, *34*, 100927. https://doi.org/10.1016/j.fpsl.2022.100927

Zudaire, L., Viñas, I., Abadias, M., Lafarga, T., Bobo, G., Simó, J., Aguiló-Aguayo, I. (2020). Effects of long-term controlled atmosphere storage, minimal processing, and packaging on quality attributes of calçots (*Allium cepa* L.). *Food Science and Technology International*, *26*(5), 403–412. https://doi.org/10.1177/108201321989100

Antimicrobial Packaging for Fruits and Vegetables

17

Asghar Ramezanian* and Hossein Shadmanfard

*Corresponding Author: ramezanian@shirazu.ac.ir

17.1 INTRODUCTION

Nowadays, increasing consumer desire for a less-processed, nutritional diet leads producers to unavoidable challenges to obtain healthier and higher-quality foods (Suppakul et al., 2003; Ju et al., 2019). Since packaging protects and maintains the sustainability of foods for long-term preservation, proper packaging is essential in order to keep foods fresh and prevent spoilage for longer periods (Han et al., 2014; Brockgreitens & Abbas, 2016; Ambaw et al., 2021). Food products, in fresh or processed form, must be packaged to protect them from external variables such as gas pollutants, microbial contamination or mechanical damage (Restuccia et al., 2010). Food products' quality may suffer physically, chemically and physiologically during distribution. Therefore, the transit and delivery of numerous varieties of fruits, vegetables and other fresh or processed organic substances after harvesting is significantly affected by packaging (Pathare & Opara, 2014; Defraeye et al., 2015). One of the most typical food deterioration factors, which makes food unsuitable for consumption and generates more food scrap, is microorganism intoxication (La Storia et al., 2012; Yousuf et al., 2018). As a result, innovations in new food packaging techniques are necessary (Restuccia et al., 2010; Han et al., 2018). The invention of active and antimicrobial packaging is an excellent example of a modern technological improvement in food packaging. The term "active packaging" refers to a new method of packaging where the product, its packaging and the surrounding area all collaborate to preserve the product's features while increasing the safety or sensory properties and prolonging the shelf life. This technology is particularly important for extending fresh foods' shelf life (Prasad & Kochhar, 2014; Kumar et al., 2018). Food shelf life is described as "the period of time, under specified environmental conditions, during which foods retain acceptable properties of color, aroma, flavor, texture, nutritional value, and safety" (Vasile & Baikan, 2021). The main reason to develop active packaging is to track and discover internal environment changes and react to changing the features of the packaging material in order to prolong the time that perishable products can be stored (Jideani & Vgot, 2016; Samanta et al., 2016). Active packaging (ACP) designs include the addition of certain compounds designed for transfer, dispersion and absorption through the food environment (Quintavalla & Vicini, 2002; Mangaraj et al., 2019). ACP has been classified into two main groups, i.e., active scavenging and active releasing. In addition, selected active-releasing emitters or complex substances, for instance antioxidant and antimicrobial compounds, along with gases such as CO_2, and odor to the internal headspaces of food packages could also be incorporated in ACP. It has also been reported that absorbers efficiently remove excessive or undesired gases/substances, including ethylene, moisture and ultraviolet radiations from the packaging environment, thereby improving the shelf-life potential of fresh produce (Viela et al., 2018; Yildirim et al., 2018). Antimicrobial packaging is a fascinating innovation in the development of ACP owing to an increasing consumer interest in foods with little or no synthetic additives (Fadiji et al., 2018; Opara & Fadiji, 2018). The main structure of antimicrobial packaging is formed by assisting biocide agents integrated into the packaging frame or polymers (Anwar & Warsiki, 2018). Therefore, the biocide feature of the antimicrobial packaging system leads to the delay and inhibition of microorganism growth by decreasing the number of living microbes and increasing the longevity of the product by reducing the quantity and development of microbial organisms (Malhotra et al., 2015). It also helps producers to inhibit spoilage and minimize pathogenic infections. Antimicrobial packaging attempts to combat infectious agents while enhancing the main objectives of ordinary food packaging, including extending shelf life, maintaining quality and providing safety (Malhotra et al., 2015; Sofi et al., 2018).

DOI: 10.1201/9781003370376-22

17.2 TYPES OF ANTIMICROBIAL PACKAGING SYSTEMS

There are numerous types of sachets, pads and polymers (either natural or artificial) that can carry volatile or non-volatile antimicrobials by forming ionic or covalent bondage. Coating with polymers containing natural antimicrobial agents, slow-releasing sachets and gas emitters/absorbers are some examples of coatings which can enable producers to satisfy consumers' demands for safe food products (Appendini & Hotchkiss, 2002; Marsh & Bugusu, 2007).

17.2.1 Packages Containing Microencapsulated Volatile Compounds

Loose or attached antimicrobial sachets, pouches and pads have been widely utilized in antimicrobial packaging for commercial applications (Karam et al., 2013). By definition, these pads are characterized as devices impregnated with volatile substances that exhibit antimicrobial properties. They are usually implanted within the food's surrounding area to let antimicrobial compounds gently leak out and prevent bacterial growth (Contreras et al., 2018). The amount of moisture inside the packages accelerates the slow release of active ingredients (Giannakourou & Tsironi, 2021). There are two types of sachets: sachets that shift and emit antimicrobial agents and sachets that produce antimicrobial substances and then release them *in situ* (Espitia et al., 2016). Some examples of typical commercial uses for antimicrobial packaging are oxygen and CO_2 scavengers and emitters, moisture absorbers and chlorine dioxide generators. Oxygen scavengers are commonly utilized in fresh fruit and vegetable packaging to stop oxidative damage, microbial development and food deterioration responses (Quintavalla & Vicini, 2002; Haghghi-Manesh & Azizi, 2017). Moisture absorbers are typically applied in foods like nuts, spices and dried fruits to prevent microbial growth by reducing water activity (Gaikwad et al., 2019). Due to silica gels' dry and free-flowing capacity, it is one of the most approved moisture absorbers. The excessive amount of CO_2 in a packaging system typically inhibits microbial growth. CO_2 generators are utilized in fresh produce packaging to increase CO_2 and decrease O_2 concentration, limiting the respiration rate and elongating the product's storability (Bhardwaj et al., 2019), and considering their preventing impact on multiple aerobic fungal and bacterial species, it gives them more antibacterial capability (Karam et al., 2013). For certain produce that exhales CO_2, an excessive content of CO_2 inside the packaging may dissolve into the product, which sometimes results in higher CO_2-related injury and off-odor formation. Due to poor CO_2 permeation, an increase in a package's pressure results in unwanted alteration in a product's texture and flavor, thus resulting in the probable collapse of the product packaging (Han et al., 2018; Suppakul et al., 2003). To avoid packaging rupture, particularly during storage preservation, absorbers for CO_2 can be utilized in order to evade the said types of adverse consequences (Lee, 2016). Another promising method for decontaminating fresh products is self-generating ClO_2 gas-releasing film patches (Ray et al., 2013). In order to decontaminate fresh products from microbial spoilage, chlorine dioxide (ClO_2), a potent oxidizing and sanitizing agent, can be applied in both gaseous and liquid forms (Aday et al., 2011; Lee et al., 2015). The ClO_2 has significant biocidal action against numerous microbial contaminators and is effective through an extensive pH range (3–8). The ClO_2 generators successfully eliminate disease-causing microbes and extend product shelf life (Trinetta et al., 2012). To be more specific, ClO_2 is more potent than ClO_2 or chlorinated water. Fresh fruits and vegetables can be disinfected with an aqueous or gaseous form of ClO_2 (Park et al., 2008). However, there are several limitations to aqueous ClO_2. The first and most important is the instability of ClO_2 and its explosive property, which makes the use of ClO_2 is inconvenient and expensive. Furthermore, aqueous ClO_2 cannot penetrate surface tension–resisting environments like stomata and fissures where microorganisms can attach. In addition, after washing, ClO_2 is no longer accessible to eliminate any remaining microorganisms. In order to solve these drawbacks, it would be ideal to design and develop a technology for food packaging systems that can self-generate ClO_2, which would decrease the cost of the manufacturing facility's production (Ray et al., 2013). Another positive aspect of gaseous ClO_2 is that it can access areas that aqueous ClO_2 cannot acquire. Furthermore, it does not generate carcinogenic chemicals like chloramines because it cannot combine with ammonia, a frequent byproduct of chlorine or chlorinated water treatment. Moreover, the efficiency of gaseous ClO_2 is not diminished by soil and/or other organic matter (Bellar et al., 1974; Woodworth & Jeng, 1990). Furthermore, through the integration of powdered citric acid and sodium chlorite with polymers such as polylactic acid (PLA), ClO_2-releasing packaging systems can be created in films or film patches. As anticipated, adding more reactants to PLA films and film patches can raise the volume of ClO_2 emitted from these substances considerably (Ray et al., 2013). Additionally, relative humidity (RH) is a crucial related factor in the formation of gaseous ClO_2 from films or film patches (Burg et al., 1999; Rasal et al., 2010). Moreover, unlike other petroleum-based polymers, PLA has significant ecological advantages since it is made from biomaterials and biodegradable ingredients. Nevertheless, the interior polymer matrix part may suffer from insufficient moisture availability due to PLA's relative hydrophobic nature (Yew et al., 2005). So, citric acid and sodium chlorite on the surface of films are limited to fresh produce limited to a certain amount of ClO_2. Therefore, polymers of a hydrophilic nature might be superior options for establishing a controlled-release-based system for RH-dependent processes. To develop a relatively complex packaging with a high capacity for moisture absorption, hydrophobic polymers, for instance PLA, can be modified with certain hydrophilic substances (Qin et al., 2011; Yew et al.,

2005). Thereafter, gaseous ClO_2 would be gently released in the controlled-release packaging to deliver permanent resistance to the persistence of select microorganisms (Chen et al., 2012; Zhu et al., 2012). Fruits and vegetables that can produce moisture inside packages are ideal for using this antimicrobial packaging system.

Although sachets, pouches and pads have numerous advantages, there are still some disadvantages. First, packaging takes longer, as sachets and pads are often manually placed into each container, which reduces productivity (Pereira et al., 2012; Contreras et al., 2018). A further challenge is that they cannot be used with liquid foods. The contents of the sachet may leak if liquids come into contact with the substance. Customer acceptance is yet another drawback. Since loose sachets are prone to disintegration, contamination and accidental ingestion, they should be handled cautiously (Contreras et al., 2018).

17.2.2 Polymers with Intrinsic Antimicrobial Properties

The only natural polymers with inherent antibacterial properties are chitosan and poly-lysine (Sayed & Jaedine, 2015; Sivakanthan et al., 2020). Due to the polycations that make up these polymers, microorganisms can be exterminated by their negative charge on their cell membranes (Santos et al., 2016). In other words, they prevent microbial growth by forcing intracellular components of microbial cells to leak out (Vasile & Baican, 2021). In composites or coatings, bioactive polymers, such as chitosan, have an inherent antibacterial effect (Perinelli et al., 2018; Wang et al., 2018). After cellulose, the second most frequent natural polymer is chitosan. Owing to its higher activity against antibacterial agents, zero toxicity, film-forming capabilities, biodegradable and biocompatible features and financial advantages, it is a promising material (Ahmad et al., 2014; Silberbauer & Schmid, 2017). Among other physicochemical factors, chitosan molecular weights and degree of deacetylation (DD) considerably affect its potential antibacterial efficacy (Verlee et al., 2017; Kumar et al., 2020). Chitosan can be used to develop sole films or films merged with certain other polymers. It can also be combined with other biopolymers to improve its functionality and use as a food packaging material (Zhu et al., 2019; Priyadarshi & Rhim, 2020). The aforesaid biopolymers include organic acids, proteins, certain plant-based extracts and some polysaccharides. In addition, it has demonstrated that the incorporation of nanomaterials into food packaging encompassing chitosan substances lessens the growth of bacterial pathogens and spoilage organisms, improves food quality characteristics and prolongs the shelf life of food (Kumar et al., 2020). For instance, silver is the most frequently used nanoparticle due to its noble reliability and safety. Silver nanoparticles (AgNPs) can be applied with non-biodegradable and edible polymers for ACP, as these exhibit antibacterial, antifungal, anti-yeast and antiviral activity (Table 17.1) (Sharma et al., 2013; Carbone et al., 2016). The methodology of green biochemistry is a form of chemical engineering and a subfield of chemistry which focuses on the formation of products and procedures that reduce or inhibit the production of harmful byproducts. In order to warrant an even distribution of chitosan and AgNPs in a polymer matrix, green chemistry techniques have been utilized. Another significant nanomaterial used as a food additive and generally accepted as safe is zinc oxide (ZnO) (Espitia et al., 2016; Aminzade et al., 2022). AnO can be added to promote antibacterial activity and enhance packaging properties in polymeric matrices (Espitia et al., 2012).

17.2.3 Antimicrobial Coatings and Films

In this case, the antibacterial ingredient is carried to the packaging through a coating matrix. Cast film coating is frequently used to apply volatile compounds and other heat-sensitive antimicrobials on packaging materials that cannot withstand polymer manufacturing temperatures (Appendini & Hotchkiss, 2002; Emamifar, 2011; Sung et al., 2013). Due to the simple coating process, it is the most common means of adding antifungal and antibacterial compounds to polymeric substrates. According to the definition, a coat is a "thin layer of material, generally thinner than 1 micron, applied onto a plastic or cellulose substrates" which can be used to increase surface properties, including wettability; improve water and oxygen barrier ability; or improve adhesion between two layers (Sung et al., 2013; Khaneghah & Hashemi, 2018). While coatings developed from edible biopolymers have increased packaging systems' biodegradability and antibacterial activity features, edible film coatings have gained popularity due to growing environmental concerns over packaging made from synthetic polymers (Bastarrachea et al., 2011; Malhotra et al., 2015). According to Bourtoom (2008), edible coatings and films are thin layers of material that can be eaten and act as a barrier to avoid the exchange of oxygen, solutes and moisture in food. The sensory qualities of the food product will not be affected by edible film coatings because they are an integral part of the food product (Guilbert et al., 1997). For food packaging purposes, edible films and coatings are required to be chosen based on the type of food to be packaged, the main contributors to quality loss and the purpose of application (Danijela et al., 2015). According to Donhowe and Fennema (1993), three ingredient types are utilized to make edible films: hydrocolloids (such as polysaccharides, proteins and alginate), lipids (such as fatty acids, waxes and acylglycerol) and composites. Materials that can form films can be used to make edible films. Film components must be mixed and dissolved in water, alcohol, a water–alcohol mixture or a combination of various solvents during production. It is possible to add plasticizers, antibacterial agents, coloring agents or flavors during this process. What is more, film solutions can be applied to food using various techniques, including dipping, spraying, brushing and panning, followed by drying (Bourtoom, 2008). Typically,

TABLE 17.1 Brief Examples of Nanoparticle Application in Antimicrobial Packaging

NANOPARTICLES	METHOD	FRUIT/VEGETABLE	RESULTS	REFERENCES
Calcium carbonate nanoparticles (Nano-$CaCO_3$) and low-density polyethylene (LDPE)	Packaging with refrigeration	Fresh-cut Chinese yams	Inhibited browning Preserved postharvest quality	Luo et al. (2015)
Preharvest nano-calcium spraying	Packaging	Apple cv. Red Delicious	maintained fruit quality and retarded degeneration over 4 months of storage	Ranjabr et al. (2018)
Silver nanoparticles	Cellulose paper packaging	Tomatoes	Preserved appearance, avoided spoilage and enhanced shelf-life stability	Sharma et al. (2023)
Selenium nanoparticles and chitosan nanoparticles	Coating	Oranges	Inhibited the growth of *Penicillium digitatum*	Salem at al. (2022)
Polyethylene film containing silver nanoparticles	Packaging with low-density polyethylene (LDPE) film	Fresh-cut carrots	Showed antimicrobial activity and maintained quality over 10 days	Bekaro et al. (2015)
Gum arabic enriched with zinc oxide nanoparticles (ZnO-NPs)	Commercial cartons	Mandarins	Preserved postharvest quality after 40 days of storage	Nxumalo et al. (2022)
Chitosan-loaded phenylalanine nanoparticles (Cs-Phe NPs)	Packaging	Persimmons	Delayed the negative effects of chilling stress	Nasr et al. (2021)
Selenium nanoparticles	plastic bags, cardboard, and brown paper	Strawberries (*Fragaria anannassa* Duch.)	strawberry fruit quality was maintained for up to 16 days at 6°C by preserving physiological characteristics	Aftab et al. (2023)
Chitosan, sodium tripolyphosphate and titanium dioxide nanoparticles	Packaging stored at 10°C	Cucumbers	Could prolong the shelf life and fruit quality for 21 days	Khojah et al. (2021)
cinnamaldehyde (CIN) and zinc oxide nanoparticles (ZnONPs)	Carboxy-Methylcellulose (CMC)-based film packaging	Cherry tomatoes	Exhibited antifungal performance against *Aspergillus niger and* decreased metabolic activities of cherry tomatoes	Guo et al. (2020)

this is done to make use of each component's capabilities to create a material with better qualities (Pascall & Lin, 2013). Combinations of multiple substances exhibit better permeability than single-component films and increase mechanical qualities (Elena et al., 2014). In order to preserve a high concentration of preservatives on food surfaces, edible coatings and films can act as a carrier for antibacterial and antioxidant substances. Their presence could prevent moisture loss during storage, decrease lipid and spoilage-induced browning, reduce the load of pathogenic and spoilage microorganisms on food surfaces and limit volatile flavor loss (Pérez-Pérez et al., 2016). Furthermore, these can increase mechanical integrity and carrier food constituents and can speed up product handling (Danijela et al., 2015). Edible coatings are becoming more popular due to environmental concerns, improved storage methods and developing markets for agricultural commodities that are not being used to their full potential (Pérez-Pérez et al., 2016). Since biopolymers do not have the same negative or detrimental environmental influences as packaging materials developed from non-renewable energy sources, edible coating materials and films are known as eco-friendly approaches (Danijela et al., 2015). In contrast with more conventional, non-environmentally-friendly packaging material types, use of edible films and coatings instead of synthetic-based polymer films can increase recycling rates and inhibit environmental pollution (Bourtoom, 2008).

17.2.4 Antimicrobial Agents and Polymers

Multiple antimicrobial substances, including coatings and films, can be directly added to the packaging materials (Karam et al., 2013). Selected compounds are incorporated into a polymer matrix to form antimicrobial film-forming materials, which can thereafter be released onto the food's surface to interact with pathogens, thereby resulting in their proliferation reduction. Polymeric packaging materials should be treated with antimicrobials to prevent food spoilage and contaminations from bacteria and other dangerous foodborne pathogens (Baldevraj et al., 2011; Giannakourou & Tsironi, 2021). Bacteria and fungicides, enzymatic agents, chelators, metal ions and organic acids are just a few examples of antimicrobial compounds that can be used to treat polymers (Appendini & Hotchkiss, 2002; Hanušová et al., 2009) and extend the shelf life of the packaged foods.

17.3 QUALITY MARKERS FOR SPOILAGE DETECTION

By minimizing oxidative processes, controlling decay and limiting water loss, modified atmosphere (MA) and modified humidity (MH) packaging can potentially increase the storage life of unprocessed or minimally processed products such as fresh fruits and vegetables. However, poor packaging design or temperature manipulation during handling can expose sealed products to extremely low levels of oxygen partial pressure (Cameron et al., 1993), where a lack of oxygen causes vegetables and fruits to experience harmful biosynthesis byproducts such as ethanol, acetaldehyde and ethyl acetate (Joles et al., 1994; Song et al., 1997). Furthermore, some spoilage agents like mold and yeast may pollute them. Even after harvesting, plants' vegetative, floral and fruiting parts breathe and release volatile metabolites. The number of volatiles produced increases in response to numerous biotic and abiotic stresses, including pathogen attacks (Roessner et al., 2001; Fiehn, 2002). Spoilage caused by microorganisms changes distinct aroma components and produces unpleasant volatile organic compounds (VOCs) in fresh fruits and vegetables (Shinde, 2014). For instance, the production of many volatiles, particularly simple ones such as acetone, promotes soft rot fungal disease caused by *Erwinia carotovora* (=*Pectobacterium carotovorum*) (Lui et al., 2005). Since there is a direct relationship between the headspace ethanol partial pressure and the quality decline brought on by low-O_2 injury, ethanol detection might be a valuable method to identify low-O_2 injury. To assess the quality of products immediately, researchers have created a variety of sensors, like ethanol biosensors, for detecting spoilage volatiles. In this case, in the presence of ethanol, the biosensors' color shifts from dull white to a bright bluish-green. This biosensor may also identify damaged products due to improper storage, distribution or marketing procedures by evaluating the degree of ethanol accumulation inside the MA packages. However, practical and financial limitations prevent continuous ethanol monitoring with a single sensor. Reichardt's betaine is a solvatochromic material that changes color depending on the solvent. It is another colorimetric indicator that is able to identify ethanol and acetone (Dickert et al., 2000).

Recently, technology has been developed, and more advanced VOC-responsive sensor devices that could be used as "electronic noses" have been introduced. These devices detect changes in responsive matrices' optical, electrochemical, electrical conductivity and chromic signals after exposure to VOCs. Polydiacetylene (PDA) could potentially be used as a volatile-detection tool, as it changes color from blue to red when exposed to air. It is fascinating to observe that the organic solvents in PDA-embedded electrospun fibers leads to a shift in color from blue to red, while the colorimetric responses might be connected to the organic compound concentrations (Yoon et al., 2007). Micropatterns based on PDAs have been used to produce multilayered structures using soft lithography approaches in conjunction with a sol-gel method. The smell system of mammals is mimicked by these sensors, providing suitable organic compound identification by employing cross-responsive arrays and reacting differently to different odors (Jiang et al., 2010). These methods might be helpful for spotting low-O_2 damage and microbiological agents in packaged foods and decreasing food safety concerns. Furthermore, nanotechnology has played an important role in the development of certain suitable technologies that can be used for microbial and foodborne toxins as well as spoilage detection. Nanosensors can be used directly as electronic tongues and noses, differentiating between healthy and contaminated produce by detecting specific chemicals released from foods which spoil. Also, there are other sensors, including microfluidic sensors, that can detect pathogens in real time. Silicon-based systems like NEMS (nanoelectromechanical systems) can be utilized in quality control devices to detect particular chemicals and biochemical signals like trans-fat content in foods. Nanocantilevers are another class of intelligent sensors that can detect microbes based on specific interactions such as antigen-antibody, enzyme-substrate or detection of mass change via electromechanical signals (Arora et al., 2011). These devices have already been used commercially in the market.

17.4 GAS-RELEASING FILMS

Considering the deficiencies of existing sanitizing methods, more efficient sanitizing techniques are required to guarantee the microbiological safety of fresh produce (Ray et al., 2013). Additionally, other technologies might be developed to capitalize upon the high humidity in packaged horticultural commodities and provide antimicrobial chemicals (Ayala-Zavala et al., 2008). Cyclodextrins (CDs) are one of the solutions to this problem. Antioxidant and antibacterial chemicals have been released from cyclodextrins, making them useful as antimicrobial transporters. Because of the host's amphipathic characteristics, adding up complexes, including hydrophobic antimicrobial guests, are possible. By intensifying the RH levels in the packaging headspace containing cyclodextrin compounds, host–guest interaction is weakened as a result of water and CDs collaboration. Then, the antimicrobial molecules release and save the product from microbial growth and the product is kept safe from microbial contamination and further degradation for the sake of the released antibacterial compounds. Cyclodextrins are glucose subunit compounds joined together in loops of different diameters of 6, 8 or 10 glucopyranosides linking with one another to create α-, β- and γ-cyclodextrin, respectively. Cyclodextrins form from starch through enzymatic conversion. They are utilized in healthcare and medication properties, chemical manufacturing and agricultural engineering (Del Valle, 2004). Typically, small or micro molecular compounds are often better complexed with α-CDs, which are made from 6 glucose subunits. At the same time, β-CDs are 7 glucose subunits, and this natural component could achieve better complexation with aromatic and heterocyclic compounds. Heavier compounds like macrocycles and steroids are well-suited to the γ-CDs because of their 8 glucose subunit structure and bigger ring diameter (Fujishima et al., 2001; Madene et al., 2006). However, the β-CDs generally have the most usage in the pharmaceutical and food industries due to their ease of accessibility and lower price than other cyclodextrins (Del Valle, 2004). In addition, the CDs' tendency to generate stable inclusion complex formations makes them appropriate for use in antimicrobial packaging systems (Bhardwaj et al., 2000; Lezcano et al., 2002).

17.5 ANTIMICROBIAL PACKAGING FOR FRUITS AND VEGETABLES

Fruits and vegetables are used fresh or cooked after harvest. Excess fruits and vegetables must be stored for the off season or delivered to consumers. Nevertheless, the shelf life of produce is too short, and it is vulnerable to injury, bruising and microbial contamination, considering of its high moisture level and perishable nature. Moreover, fruits and vegetables still undergo respiration after harvesting, which causes them to lose weight, wilt, shrivel and spoil and makes them vulnerable to contamination by microorganisms from harvest to distribution (Jung & Zhao, 2016; Kumar et al., 2020).

Proper packaging can save all energy and time spent on fruit and vegetable cultivation from farm to fork. The proper packaging for particular fruits and vegetables can increase their shelf life and reduce losses. Although conventional packaging can increase the shelf life of horticultural produce, vast losses of fruits and vegetables still occur throughout the year. Therefore, it seems necessary to modify this conventional packaging for the future. Due to the challenge of maintaining fruit and vegetable freshness and quality over long-term maintenance, minimally processed fruits and vegetables are the subject of extensive research (Mahmud & Khan, 2018), such as adding antimicrobial ingredients to fruit and vegetable packaging to reduce the impact of microbes, prolong shelf life and provide higher-quality products (Jung & Zhao, 2016; Giannakourou & Tsironi, 2021). While the respiration of fruits and vegetables continues after harvest, their moisture content, nutrients and other quality parameters are significantly reduced. Covering fruits and vegetables with waxes and other edible polymers can create an obstacle between the products' surface and the surrounding environment. These barriers can reduce moisture and gas transmission. In addition, coatings can give the fresh products a glossy appearance and make them more attractive to the consumer.

As previously mentioned, antimicrobial systems use artificial and/or organic chemical compounds such as chitosan, organic acids and their salts, essential oils (EO) and other plant substances in the packaging materials (Sharma et al., 2009). The inner surface of a packaging material is usually in touch with the surface of fresh produce, which can be a source of contamination. A potential solution is to use aseptic packaging. However, aseptic packaging is only practical for some food types. Recently, researchers have proven the antimicrobial properties of nanoparticles. Coating fruits and vegetables with nanoparticles not only helps to reduce moisture loss, fat oxidation and respiration rate, but also, compared to typical packaging, these films can serve as a carrier for colors, flavors, antioxidatives, nutrients and antimicrobial agents (Sadek et al., 2022). Studies have shown the oligodynamic effect of some metal nanoparticles, for instance copper (Cu), zinc (Zn), titanium (Ti), gold (Au) and silver (Ag), that could improve the shelf life of packaged fruits and vegetables. To be precise, heavy metals may have lethal effects on microbial organisms even at low concentrations. The term "oligodynamic effect" describes this trend. These types of metals can react with thiol or amine groups of enzymes of microorganisms. Then, the bacterial cell wall proteins are disrupted by copper ions, leading to bacterial precipitation and death (Nura, 2018).

Silver (Ag) and zinc oxide (ZnO) are two examples of metals and metal oxides that show great promise as antimicrobial ingredients to create less-expensive and safe methods of fruit and vegetable packing (Pérez-Gago & Palou, 2016; Aminzade et al., 2022).

Recently, the food and agricultural industries have shown a significant interest in extracting essential oils from aromatic plants to control the toxin production of microorganisms. Customers' demand for healthy and organic food products has challenged the food industries to look for innovative natural methods to satisfy consumers (Soliman & Badeaa, 2002; Tepe et al., 2005). Therefore, the food and agricultural industries have shown considerable interest in extracting essential oils from aromatic plants to control pathogen and toxin production of microorganisms (Ayala-Zavala et al., 2008).

Essential oils can kill different microorganisms that cause fruit and vegetable spoilage. However, the strong odor-flavor of EOs can negatively affect sensory acceptability and transmit from the oil to the packaging contents. Microencapsulation can suggest a solution by encapsulating antimicrobial compounds (Table 17.2). These slow-release sachets can constantly distribute odor and flavor into the packaging system's atmosphere and cover their pungent smell. CDs' ability to capture and preserve volatile molecules, including EOs, makes them an effective tool for long-term preservation. What is more, as volatile chemical compounds, EOs are able to flow over the packaging's headspace and any air gaps, unlike non-volatile substances that only flow above the surface area (Han, 2000). Therefore, microencapsulation can provide a way out of these impasses to avoid a decline in sensory acceptability, prolong antimicrobial features and extend the storage time of fruits and vegetables. For instance, the EO sachets were developed by Espitia et al. (2012) to be utilized in an antibacterial packaging system. Treated products with EOs exhibited a wider range of antifungal and antibacterial effects. Furthermore, Shemesh et al. (2016) developed a polyamides-based packaging film containing selected EOs against postharvest pathogens. The selected EOs are added to a variety of fresh vegetable and fruit packages, including lychees, grapes and cherry tomatoes. Substantial antifungal activity of the films was observed against the fungal molds, as well as incredible results in inhibiting decay and increasing the shelf life of horticultural products. Also, fruits and vegetables are subject to different treatments, such as pesticides, which may cause long-term health problems. Decontamination of foods and packaging equipment with the help of nanoemulsions has also been reported. For instance, a nanomicelle-based product containing natural glycerin cleans produce of any traces of harmful chemicals (Fadiji et al., 2023).

17.6 CONCLUSIONS

Antimicrobial packaging is a modern and promising technology which reduces microbial growth and extends the shelf life of harvested products. Various organic and inorganic antimicrobial agents are eligible for use in antimicrobial systems. Antimicrobial packaging can maintain the overall quality and extend the storage or shelf life by reducing microbial decay. This type of packaging can promote the new generation of smart packaging.

TABLE 17.2 Summarized Examples of Plant Essential Oil Utilization in Antimicrobial Packaging

ESSENTIAL OILS	*FRUIT/VEGETABLE*	*FINDINGS*	*REFERENCES*
Cinnamon essential oil incorporated in shellac	Oranges	incorporation of EOs into shellac could maintain citrus fruit quality and reduce the growth of *Penicillium digitatum*	Khorram and Ramezanian (2021)
Thyme (*Thymus vulgaris*) and vervain (*Verbena officinalis*)	Peaches	Vervain (1000 ppm) and thyme (500 ppm) significantly reduced the contamination of *Monilinia laxa*, *Monilinia fructigena*, and *Monilinia fructicola*	Elshafie et al. (2015)
Chitosan nanoparticles along with *Satureja hortensis* essential oil	Pomegranate arils	Encapsulation of savory EO in chitosan nanoparticles maintained phytochemicals and water content and reduced microbial activities	Amiri et al. (2021)
Basil, oregano, savory, thyme, lemon and fennel essential oils	Applese	EOs of thyme, savory and oregano, at all three concentrations, and basil, at 1.0% and 0.5%, inhibited the mycelial growth of *Penicillium expansum*.	*Buonsenso et al. (2023)*
Microencapsulation and pure oil of clove essential oil (CEO)	Navel oranges	CEO microencapsulation has shown a better ability to control *Penicillium digitatum*, both in vitro and on novel oranges	Wang et al. (2018)
Maltodextrin-based coating enriched with ajwain, neroli and rosemary essential oils	Tomatos	Coating enriched with ajwain oil reduced *Fusarium oxysporum* growth by 84.2% followed by rosemary and neroli with 66.7% and 24.6% inhibition, respectively	Khanjani et al. (2021)
Microemulsion of oregano, lemongrass and cinnamon essential oils	Iceberg lettuce	The 0.1% oregano oil microemulsion avoided the development of *Escherichia coli* and *Pseudomonas fluorescens*	Arellano et al. (2022)
Zanthoxylum bungeanum essential oil (10%) into polyvinyl alcohol/β-Cyclodextrin (PVA/β-CD, 6:1) nanofibers	Strawberries and sweet cherries	ZBEO/Films showed significant antifungal activities against *Aspergillus flavus* and *Botrytis cinerea* and both fruits kept their freshness	Zhang et al (2022)

REFERENCES

Aday, M.S., Caner, C. (2011). The applications of 'active packaging & chlorine dioxide for extended shelf life of fresh strawberries. *Packaging Technology & Science*, *24*, 123–136.

Aftab, A., Ali, M., Yousaf, Z., Binjawhar, D. N., Hyder, S., Aftab, Z. E., Maqbool, Z., Shahzadi, Z., Eldin, S. M., Iqbal, R., Ali, I. (2023). Shelf-life extension of *Fragaria* × *ananassa* Duch. using selenium nanoparticles synthesized from *Cassia fistula* Linn. leaves. *Food science & nutrition*, *11*(6), 3464–3484.

Ahmed, S., Ahmad, M., Ikram, S. (2014). Chitosan: A natural antimicrobial agent-a review. *Journal of Applied Chemistry*, *3*, 493–503.

Ambaw, A., Fadiji, T., Opara, U.L. (2021). Thermo-Mechanical Analysis in the Fresh Fruit Cold Chain: A Review on Recent Advances. *Foods*, *10*(6), 1357.

Aminzade, R., Ramezanian, A., Eshghi, S., Hosseini, S.M.H. (2022). Maintenance of pomegranate arils quality by zinc enrichment, a comparison between zinc sulfate & nano zinc oxide. *Postharvest Biology and Technology*, *184*, 111757.

Amiri, A., Ramezanian, A., Mortazavi, S.M.H., Hosseini, S.M.H., Yahia, E. (2021). Shelf-life extension of pomegranate arils using chitosan nanoparticles loaded with *Satureja hortensis* essential oil. *Journal of the Science of Food and Agriculture*, *101*(9), 3778–3786.

Anwar, R.W., Warsiki, E. (2018). The comparison of antimicrobial packaging properties with different applications incorporation method of active material. *IOP Conference* Series: *Earth* and *Environmental Science*, *141*, 012002.

Appendini, P., Hotchkiss, J.H. (2002). Review of antimicrobial food packaging. *Innovative Food Science* and *Emerging Technologies*, *3*, 113–126.

Arellano, S., Zhu, L., Kumar, G.D., Law, B., Friedman, M., Ravishankar, S. (2022). Essential oil microemulsions inactivate antibiotic-resistant bacteria on iceberg lettuce during 28-day storage at 4 °C. *Molecules*, *27*(19), 6699.

Arora, P., Sindhu, A., Dilbaghi, N., Chaudhury, A. (2011). Biosensors as innovative tools for the detection of food borne pathogens. *Biosensors & Bioelectronics*, *28*(1), 1–12.

Ayala-Zavala, J.F., Del-Toro-Sánchez, L., Alvarez-Parrilla, E., González-Aguilar, G.A. (2008). High relative humidity in-package of fresh-cut fruits and vegetables: Advantage or disadvantage considering microbiological problems and antimicrobial delivering systems. *Journal of Food Science*, *73*(4), 41–47.

Baldevraj, R.M., Jagadish, R.S. (2011). Incorporation of chemical antimicrobial agents into polymeric films for food packaging. In *Elsevier eBooks*, 368–420.

Bastarrachea, L., Dhawan, S., Sablani, S.S. (2011). Engineering properties of polymeric-based antimicrobial films for food packaging: A review. *Food Engineering Reviews*, *3*, 79–93

Bellar, T.A.L., Lichtenberg, J.J., Kroner, R.C. (1974). The occurrence of organohalides in chlorinated drinking waters. *American Water Works Association*, *66*, 703–706.

Bhardwaj, A., Alam, T., Talwar, N. (2019). Recent advances in active packaging of agri-food products: A review. *Journal of Postharvest Technology*, *7*, 33–62.

Bhardwaj, R., Dorr, R.T., Blanchard, J. (2000). Approaches to reducing toxicity of parenteral anticancer drug formulations using cyclodextrins. *PDA Journal of Pharmaceutical Science and Technology*, *54*, 233–239.

Bourtoom T. (2008). Edible films & coatings: characteristics & properties. *International Food Research Journal*, *15*(3), 237–248.

Brockgreitens, J., Abbas, A. (2016). Responsive food packaging: Recent progress & technological prospects. *Comprehensive Reviews in Food Science and Food Safety*, *15*, 3–15.

Buonsenso, F., Schiavon, G., Spadaro, D. (2023). Efficacy and mechanisms of action of essential oils' vapours against blue mould on apples caused by *Penicillium expansum*. *International Journal of Molecular Sciences*, *24*(3), 2900.

Burg, K.J., Holder, W.D., Culberson, C.R., Beiler, R.J., Greene, K.G., Loebsack, A.B., Roland, W. D., Mooney, D. J., Halberstadt, C. R. (1999). Parameters affecting cellular adhesion to polylactide films. *Journal of biomaterials science. Polymer Edition*, *10*(2), 147–161.

Cameron, A.C., Patterson, B.D., Talasila, P.C. Joles, D.W. (1993). Modeling the risk in modified-atmosphere packaging: a case for sense-&-respond packaging. *Proceeding of the 6th International Controlled Atmosphere Research Conference* (Cornell University, Ithaca, NY (15–17 June 1993)), *71*, 95–102.

Carbone, M., Donia, D.T., Sabbatella, G., Antiochia, R. (2016). Silver nanoparticles in polymeric matrices for fresh food packaging. *Journal of King Saud University*, *28*, 273–279.

Chen, X., Lee, D.S., Zhu, X., Yam, K.L. (2012). Release kinetics of tocopherol & quercetin from binary antioxidant controlled-release packaging films. *Journal of Agricultural and Food Chemistry*, *60*, 3492–3497.

Contreras, C.B., Charles, G., Toselli, R., Strumia, M.C. (2018). Antimicrobial active packaging. In *Biopackaging*, Masuelli, M.A., Ed., Taylor & Francis Group, LLC: Boca Raton, FL, pp. 36–58.

Danijela, Z.S., Vera, L.L., Senka, Z.P., Nevena, M.H. (2015). Edible films & coatings—sources, properties & application. *Food & Feed Research*, *42* (1), 11–22.

Defraeye, T., Cronje, P., Berry, T., Opara, UL, East, A., Hertog, M., Verboven, P., Nicolai, B. (2015). Towards integrated performance evaluation of future packaging for fresh produce in the cold chain. *Trends in Food Science & Technology*, *44*, 201–225.

Del Valle, E.M. (2004). Cyclodextrins & their uses: a review. *Process Biochemistry*, *39*, 1033–1046.

Dickert, F.L., Geiger, U., Lieberzeit, P., Reutner, U. (2000). Solvatochromic betaine dyes as optochemical sensor materials: detection of polar & non-polar vapours. *Sensors & Actuators, B: Chemical*, *70*, 263–269.

Donhowe, I.G., Fennema, O.R. (1993). The effects of plasticizers on crystallinity, permeability, & mechanical properties of methylcellulose films. *Journal of Food Processing & Preservation*, *17*, 247–257.

Elena, P., Roi, R., Shani, D., Batia, H., Victor, R. (2014). Gelatin-chitosan composite films & edible coatings to enhance the quality of food products: Layer-by-Layer vs. blended formulations. *Food* and *Bioprocess Technology*, *14*(1), 1–9.

Elshafie, H.S., Mancini, E., Camele, I.N., De Martino, L., De Feo, V. (2015). In vivo antifungal activity of two essential oils from Mediterranean plants against postharvest brown rot disease of peach fruit. *Industrial Crops & Products*, *66*, 11–15.

Emamifar, A. (2011). Applications of antimicrobial polymer nanocomposites in food packaging. In *Advances in Nanocomposite Technology*, Hashim, A., Ed., InTech: Rijeka, pp. 299–318.

Espitia, P.J.P, Avena-Bustillos, R.J., McHugh, T.H. (2016). Trends in antimicrobial food packaging systems: Emitting sachets & absorbent pads. *Food Research International*, *83*, 60–73.

Espitia, P.J.P., Soares, N.D.F.F., Botti, L.C.M., Melo, N.R.D., Pereira, O.L., Silva, W.A.D. (2012). Assessment of the efficiency of essential oils in the preservation of postharvest papaya in an antimicrobial packaging system. *Brazilian Journal of Food Technology*, *15*, 333–342.

Fadiji, T., Berry, T.M., Coetzee, C.J., Opara, U.L. (2018). Mechanical design & performance testing of corrugated paperboard packaging for the postharvest handling of horticultural produce. *Biosystems Engineering*, *171*, 220–244.

Fadiji, T., Rashvand, M., Daramola, M.O., Iwarere, S.A. (2023). A review on antimicrobial packaging for extending the shelf life of food. *Processes*, *11*(2), 590.

Fiehn, O. (2002). Metabolomics: the link between genotypes & phenotypes. *Plant Molecular Biology*, *48*, 155–171.

Fujishima, N., Kusaka, K., Umino, T., Urushinata, T., Terumi, K. (2001). Flour based foods containing highly branched cyclodextrins. *Japanese Patent*, *136*, 898.

Gaikwad, KK, Singh, S., Ajji, A. (2019). Moisture absorbers for food packaging applications. *Environmental Chemistry Letters*, *17*, 609–628.

Giannakourou, M.C., Tsironi, T.N. (2021). Application of processing & packaging hurdles for fresh-cut fruits & vegetables preservation. *Foods*, *10*, 830.

Guilbert S., Cuq, B. Gontard, N. (1997). Recent innovations in edible and/or biodegradable packaging materials. *Food Additives & Contaminants*, *14*, 741–751.

Guo, X., Chen, B., Xiaoling, W., Li, J., Hu, F.B. (2020). Utilization of cinnamaldehyde & zinc oxide nanoparticles in a carboxymethylcellulose-based composite coating to improve the postharvest quality of cherry tomatoes. *International Journal of Biological Macromolecules*, *160*, 175–182.

Haghighi-Manesh, S., Azizi, M.H. (2017). Active packaging systems with emphasis on its applications in dairy products. *Journal of Food Process Engineering*, *40*, 12542.

Han, J.H. (2000). Antimicrobial food packaging. *Food Technology*, *54*, 56–65.

Han, J.H., Patel, D., Kim, J.E., Min, S.C. (2014). Retardation of *Listeria monocytogenes* growth in mozzarella cheese using antimicrobial sachets containing rosemary oil & thyme oil. *Journal of Food Science*, *79*, 2272–2278.

Han, J.W., Ruiz-Garcia, L., Qian, J.P., Yang, X.T. (2018). Food packaging: A comprehensive review & future trends. *Comprehensive Reviews in Food Science and Food Safety*, *17*, 860–877.

Hanušová, K., Dobiáš, J., Klaudisová, K. (2009). Effect of packaging films releasing antimicrobial agents on stability of food products. *Czech Journal of Food Sciences*, *27*, 347–349.

Jiang, H., Wang, Y., Ye, Q., Zou, G., Su, W., Zhang, Q. (2010). Polydiacetylene-based colorimetric sensor microarray for volatile organic compounds. *Sensors and Actuators B: Chemical*, *143*(2), 789–794.

Jideani, V.A., Vogt, K. (2016). Antimicrobial packaging for extending the shelf life of bread: A review. *Critical Reviews in Food Science and Nutrition*, *56*, 1313–1324.

Joles, D.W., Cameron, A.C., Shirazi, A., Petracek, P.D., Beaudry R.M. (1994). Modified-atmosphere packaging of "Heritage" red raspberry fruit: respiratory response to reduced oxygen, enhanced carbon dioxide, & temperature. *Journal of the American Society for Horticultural Science*, *119*, 540–545.

Ju, J., Chen, X., Xie, Y., Yu, H., Guo, Y., Cheng, Y., Yao, W. (2019). Application of essential oil as a sustained release preparation in food packaging. *Trends in Food Science & Technology*, *92*, 22–32.

Jung, J., Zhao, Y. (2016). Antimicrobial packaging for fresh & minimally processed fruits & vegetables. In *Antimicrobial Food Packaging*, Barros-Velázquez, J., Ed., Academic Press: Cambridge, MA, pp. 243–256.

Karam, L., Jama, C., Dhulster, P., Chihib, N.E. (2013). Study of surface interactions between peptides, materials & bacteria for setting up antimicrobial surfaces & active food packaging. *Journal of Materials and Environmental Science*, *4*, 798–821.

Khaneghah, A.M., Hashemi, S.M.B., Limbo, S. (2018). Antimicrobial agents & packaging systems in antimicrobial active food packaging: An overview of approaches & interactions. *Food and Bioproducts Processing*, *111*, 1–19.

Khanjani, R., Dehghan, H., Sarrafi, Y. (2021). Antifungal edible tomato coatings containing ajwain, neroli, & rosemary essential oils. *Journal of Food Measurement & Characterization*, *15*(6), 5139–5148.

Khojah, E., Sami, R., Helal, M., Elhakem, A., Benajiba, N., Salamatullah, A. M., Salamatullah, A.M. (2021). Postharvest physicochemical properties & fungal populations of treated cucumber with sodium tripolyphosphate/titanium dioxide nanoparticles during storage. *Coatings*, *11*(6), 613.

Khorram, F., Ramezanian, A. (2021). Cinnamon essential oil incorporated in shellac, a novel bio-product to maintain quality of 'Thomson navel' orange fruit. *Journal of Food Science and Technology*, *58*(8), 2963–2972.

Kumar, S., Mukherjee, A., Dutta, J. (2020). Chitosan based nanocomposite films & coatings: Emerging antimicrobial food packaging alternatives. *Trends in Food Science & Technology*, *97*, 196–209

Kumar, K.V.P., Suneetha, J., Kumari, B.A. (2018). Active packaging systems in food packaging for enhanced shelf life. *Journal of Pharmacognosy and Phytochemistry*, *7*, 2044–2046.

La Storia, A., Ferrocino, I., Torrieri, E., Di Monaco, R., Mauriello, G., Villani, F., Ercolini, D. (2012). A combination of modified atmosphere & antimicrobial packaging to extend the shelf-life of beefsteaks stored at chill temperature. *International Journal of Food Microbiology*, *158*, 186–194.

Lee, D.S (2016). Carbon dioxide absorbers for food packaging applications. *rends in Food Science & Technology*, *57*, 146–155.

Lee, Y., Burgess, G., Rubino, M., Auras, R. (2015). Reaction & diffusion of chlorine dioxide gas under dark & light conditions at different temperatures. *Journal of Food Engineering*, *144*, 20–28.

Lezcano, M., Al-Soufi, W., Novo, M., Rodríguez-Núñez, E., Tato, J.V. (2002). Complexation of several benzimidazole-type fungicides with alpha- and beta-cyclodextrins. *Journal of Agricultural and Food Chemistry*, *50*(1), 108–112.

Lui, L.H., Ajit Vikram, Y., Abu-Nada, Kushalappa, A.C., Raghavan, G.S.V., Al-Mughrabi, K.I. (2005). Volatile metabolic profiling for discrimination of potato tubers inoculated with dry and soft rot pathogens. *American Journal of Potato Research*, *82*(1), 1–8.

Luo, Z., Wang, Y., Jiang, L., Xu, X. (2015). Effect of Nano-$CaCO_3$-LDPE packaging on quality and browning of fresh-cut Yam. *LWT—Food Science and Technology*, *60*, 1155–1161.

Madene, A., Jacquot, M., Scher, J., Desobry, S. (2006). Flavour encapsulation & controlled release: A review. *International Journal of Food Science & Technology*, *41*, 1–21.

Mahmud, J., Khan, R.A. (2018). Characterization of natural antimicrobials in food system. *Advances in Microbiology*, *8*, 894.

Malhotra, B., Keshwani, A., Kharkwal, H. (2015). Antimicrobial food packaging: Potential & pitfalls. *Frontiers in Microbiology*, *6*, 611.

Mangaraj, S., Yadav, A., Bal, L.M., Dash, S.K., Mahanti, N.K. (2019). Application of biodegradable polymers in food packaging industry: A comprehensive review. *Journal of Packaging Technology & Research*, *3*, 77–96.

Marsh, K., Bugusu, B. (2007). Food packaging Roles, materials, & environmental issues. *Journal of Food Science*, *72*, 39–55.

Nasr, F., Pateiro, M., Rabiei, V., Razavi, F., Formaneck, S.D., Gohari, G., Lorenzo, J.M. (2021). Chitosan-phenylalanine nanoparticles (Cs-Phe Nps) extend the postharvest life of persimmon

(*Diospyros kaki*) fruits under chilling stress. *Coatings*, *11*(7), 819.

Nura, A. (2018). Advances in food packaging technology- A review. *Journal of Postharvest Technology*, *6*(4), 55–64.

Nxumalo, K.A., Fawole, O.A., Oluwafemi, O.S. (2022). Evaluating the efficacy of gum arabic-zinc oxide nanoparticles composite coating on shelf-life extension of mandarins (cv. Kinnow). *Frontiers in Plant Science*, *13*.

Opara, U.L., Fadiji, T. (2018). Compression damage susceptibility of apple fruit packed inside ventilated corrugated paperboard package. *Scientia Horticulturae*, *227*, 154–161.

Park, E.J., Gray, P.M., Oh, S.W., Kronenberg, J., Kang, D.H. (2008). Efficacy of FIT produce wash and chlorine dioxide on pathogen control in fresh potatoes. *Journal of Food Science*, *73*(6), 278–282.

Pascall, M.A., Lin, S.J. (2013). The application of edible polymeric films and coatings in the food industry. *Journal* of *Food Processing & Technology*, *4*(2), 1–2.

Pathare, P.B., Opara, UL. (2014). Structural design of corrugated boxes for horticultural produce: A review. *Biosystems Engineering*, *125*, 128–140.

Pereira de Abreu, D.A., Cruz, J.M., Paseiro-Losada, P. (2012). Active & intelligent packaging for the food industry. *Food Reviews International*, *28*, 146–187.

Pérez-Gago, M.B., Palou, L. (2016). Antimicrobial packaging for fresh & fresh-cut fruits & vegetables. In *Fresh-Cut Fruits & Vegetables: Technology, Physiology, & Safety*, Pareek, S., Ed., CRC Press: Boca Raton, FL, pp. 403–452.

Pérez-Pérez, C., Regalado-González, C., Rodríguez-Rodríguez, C.A., Barbosa-Rodríguez, J.R., Villaseñor-Ortega, F. (2016). Incorporation of antimicrobial agents in food packaging films and coatings. *Research Signpost*, *37*(2), 193–216.

Perinelli, D.R., Fagioli, L., Campana, R., Lam, J.K., Baffone, W., Palmieri, G.F., Bonacucina, G. (2018). Chitosan-based nanosystems & their exploited antimicrobial activity. *European Journal of Pharmaceutical Sciences*, *117*, 8–20.

Prasad, P., Kochhar, A. (2014). Active packaging in food industry: A review. *Journal of Environmental Science Toxicology and Food Technology*, *8*, 1–7.

Priyadarshi, R., Rhim, J.W. (2020). Chitosan-based biodegradable functional films for food packaging applications. *Innovative Food Science and Emerging Technologies*, *62*, 102346.

Qin, L., Qiu, J., Liu, M., Ding, S., Shao, L., Lü, S., Zhang, G., Zhao, Y., Fu, X. (2011). Mechanical and thermal properties of poly(lactic acid) composites with rice straw fiber modified by poly(butyl acrylate). *Chemical Engineering Journal, 166, 772–778.*

Quintavalla, S., Vicini, L. (2002). Antimicrobial food packaging in meat industry. *Meat Science*, *62*, 373–380.

Rasal, R.M., Janorkar, A.V., Hirt, D.E. (2010). Poly (lactic acid) modifications. *Progress in Polymer Science*, *35*, 338–356.

Ray, S., Jin, T., Fan, X., Liu, L., Yam, K. L. (2013). Development of chlorine dioxide releasing film and its application in decontaminating fresh produce. *Journal of Food Science*, *78*(2), 276–284.

Restuccia, D., Spizzirri, U.G., Parisi, O.I., Cirillo, G., Curcio, M., Iemma, F., Picci, N. (2010). New E.U. regulation aspects & global market of active & intelligent packaging for food industry applications. *Food Control*, *21*, 1425–1435.

Roessner, U., Luedemann, A., Brust, D., Fiehn, O., Linke, T., Willmitzer, L., Fernie, A. (2001). Metabolic profiling allows comprehensive phenotyping of genetically or environmentally modified plant systems. *The Plant Cell*, *13*(1), 11–29.

Sadek, M.F., Shabana, Y.M., Sayed-Ahmed, K., Tabl, A.H.A. (2022). Antifungal activities of sulfur & copper nanoparticles against nucumber postharvest diseases caused by *botrytis cinerea & Sclerotinia sclerotiorum*. *Journal of Fungi*, *8*(4), 412.

Salem, M.L., Abd-Elraoof, W.A., Tayel, A.A., AlZuaibr, F.M., Abonama, O.M. (2022). Antifungal application of biosynthesized selenium nanoparticles with pomegranate peels & nanochitosan as edible coatings for citrus green mold protection. *Journal of Nanobiotechnology*, *20*(1). https://doi.org/10.1186/s12951-022-01393-x

Samanta, K.K., Basak, S., Chattopadhyay, S.K. (2016). Potentials of fibrous & nonfibrous materials in biodegradable packaging. In *Environmental Footprints of Packaging*, Springer: Singapore, pp. 75–113.

Santos, M.R., Fonseca, A.C., Mendonça, P.V., Branco, R., Serra, A.C., Morais, P.V., Coelho, J.F. (2016). Recent developments in antimicrobial polymers: A review. *Materials*, *9*, 599.

Sayed, S., Jardine, M.A. (2015). Antimicrobial biopolymers. In *Advanced Functional Materials*, John Wiley & Sons, Inc.: Hoboken, NJ, pp. 493–533.

Sharma, S., Kumar, S., Bulchini, B., Taneja, S., Banyal, S. (2013). Green synthesis of silver nanoparticles & their antimicrobial activity against gram-positive & gram-negative bacteria. *International Journal of Biotechnology and Bioengineering Research*, *4*, 711–714.

Sharma, S., Sharma, N., Kaushal, N. (2023). Utilization of novel bacteriocin synthesized silver nanoparticles (AgNPs) for their application in antimicrobial packaging for preservation of tomato fruit. *Frontiers in Sustainable Food Systems*, *7*.

Sharma, R.R., Singh, D., Singh, R. (2009). Biological control of postharvest diseases of fruits & vegetables by microbial antagonists: A review. *Biological Control*, *50*, 205–221.

Shemesh, R., Krepker, M., Nitzan, N., Vaxman, A., Segal, E. (2016). Active packaging containing encapsulated carvacrol for control of postharvest decay. *Postharvest Biology and Technology*, *118*, 175–182.

Shinde, R.L. (2014). Quality & aroma volatiles of fresh-cut melons: effects of genotype & M.A. packaging. MSc thesis, The Hebrew University of Jerusalem, Rehovot, Israel.

Silberbauer, A., Schmid, M. (2017). Packaging concepts for ready-to-eat food: Recent progress. *Journal of Packaging Technology & Research*, *1*, 113–126.

Sivakanthan, S., Rajendran, S., Gamage, A., Madhujith, T., Mani, S. (2020). Antioxidant & antimicrobial applications of biopolymers: A review. *Food Research International*, *136*, 109327.

Sofi, S.A., Singh, J., Rafiq, S., Ashraf, U., Dar, B.N., Nayik, G.A. (2018). A comprehensive review on antimicrobial packaging & its use in food packaging. *Current Nutrition & Food Science*, *14*, 305–312.

Soliman, K.M. Badeaa, R.I. (2002). Effect of oil extracted from some medicinal plants on different mycotoxigenic fungi. *Food and Chemical Toxicology*, *40*, 1669–1675.

Song, J., Gardner, B.D., Holland, J.F., Beaudry, R.M. (1997). Rapid analysis of volatile flavor compounds in apple fruit using SPME & G.C./time-of-flight mass spectrometry. *Journal of Agricultural and Food Chemistry*, *45*, 1801–1807.

Sung, S.Y., Sin, L.T., Tee, T.T., Bee, S.T., Rahmat, A.R., Rahman, W.A.W.A., Vikhraman, M. (2013). Antimicrobial agents for food packaging applications. *Trends in Food Science & Technology*, *33*, 110–123.

Suppakul, P., Miltz, J., Sonneveld, K., Bigger, S.W. (2003). Active packaging technologies with an emphasis on antimicrobial packaging & its applications. *Journal of Food Science*, *68*, 408–420.

Tepe, B., Daferera, D., Sokmen, A., Sokmen, M., Polissiou, M. (2005). Antimicrobial and antioxidant activities of the essential oil and various extracts of Salvia tomentosa Miller (Lamiaceae). *Food Chemistry*, *90*(3), 333–340.

Trinetta, V., Morgan, M., Linton, R. (2012). Chlorine dioxide for microbial decontamination of food. In *Microbial Decontamination in the Food Industry*, Woodhead Publishing: Cambridge, pp. 533–562.

Vasile, C., Baican, M. (2021). Progresses in food packaging, food quality, & safety—Controlled-release antioxidant and/or antimicrobial packaging. *Molecules*, *26*, 1263.

Verlee, A., Mincke, S., Stevens, C.V. (2017). Recent developments in antibacterial & antifungal chitosan & its derivatives. *Carbohydrate Polymers*, *164*, 268–283.

Vilela, C., Kurek, M., Hayouka, Z., Röcker, B., Yildirim, S., Antunes, M.D.C., Freire, C.S. (2018). A concise guide to active agents for active food packaging. *Trends in Food Science & Technology*, *80*, 212–222.

Wang, H., Qian, J., Ding, F. (2018). Emerging chitosan-based films for food packaging applications. *Journal of Agricultural* and *Food Chemistry*, *66*, 395–413.

Woodworth, A.J. Jeng, D.K. (1990). Chlorine dioxide gas sterilization under square-wave conditions. *Applied and Environmental Microbiology*, 56, 514–519.

Yew, G.H., Yusof, A.M., Ishak, Z.M., Ishiaku, U.S. (2005). Water absorption & enzymatic degradation of poly (lactic acid)/rice starch composites. *Polymer Degradation* and *Stability*, *90*, 488–500.

Yildirim, S., Röcker, B., Pettersen, M.K., Nilsen-Nygaard, J., Ayhan, Z., Rutkaite, R., Radusin, T., Suminska, P., Marcos, B., Coma, V. (2018) Active packaging applications for food. *Comprehensive Reviews in Food Science and Food Safety*, *17*, 165–199.

Yoon, J., Chae, S.K., Kim, J.M. (2007). Colourimetric sensors for volatile organic compounds (VOCs) based on conjugated polymer-embedded electrospun fibers. *Journal of American Chemical Society*, 129, 3038–3039.

Yousuf, B., Qadri, O.S., Srivastava, A.K. (2018). Antimicrobial packaging: Basic concepts & applications in fresh & fresh-cut fruits & vegetables. In *Innovative Packaging of Fruits & Vegetables: Strategies for Safety & Quality Maintenance*, Apple Academic Press Inc.: Oakville, ON, pp. 136–155.

Zhang, H., Zhang, C., Wang, X., Huang, Y., Xiao, M., Hu, Y., Zhang, J. (2022). Antifungal electrospinning nanofiber film incorporated with *Zanthoxylum bungeanum* essential oil for strawberry & sweet cherry preservation. *LWT - Food Science and Technology*, *169*, 113992.

Zhu, X., Schaich, K.M., Chen, X., Chung, D., Yam, K.L. (2012). Target release rate of antioxidants to extend induction period of lipid oxidation. *Food Research International*, *47*(1), 1–5.

Zhu, J., Wu, H., Sun, Q. (2019). Preparation of crosslinked active bilayer film based on chitosan & alginate for regulating ascorbate-glutathione cycle of postharvest cherry tomato (*Lycopersicon esculentum*). *International Journal of Biological Macromolecules*, *130*, 584–594.

PART VI

Irradiation and Water-Based Applications

Irradiation of Fresh Fruits and Vegetables

18

Qasid Ali, Adem Dogan and Mustafa Erkan*

*Corresponding Author: erkan@akdeniz.edu.tr

18.1 INTRODUCTION

Food irradiation was first used in 1905 by two British scientists, J. Appleby and A.J. Banks. Biomedical literature of the 1920s and 1930s reported that X-rays were applied on food. The research on food irradiation was increased in the 1940s and 1950s (Josephson, 1983). In 1963, the United States approved the use of irradiation to eliminate insects in wheat, while the European Economic Community allowed the usage of irradiation in 1987 for specific foods (Hadlington, 1987). Different organizations such as WHO, FAO, IAEA and Codex Alimentarius Commissions have been recommending and assisting with the application of radiations as a food safety procedure around the world. Currently, over 50 countries have approved food irradiations for over 60 foods (Ihsanullah & Rashid, 2017). Gamma rays target the nucleic acids and membrane lipids to affect the foods' biochemical functions. After irradiation, the reactive intermediate injures or damages the microorganism by modifying its nucleic acid and cell membrane structures, thereby hindering the metabolic activities of enzymes and cell division. The decrease in the number of cells is related to the entire irradiation dose acquired. Organisms vary in their sensitivity to irradiations because of the different sizes of their DNA molecules (Korkmaz & Polat, 2005). The smaller the microorganism the higher the irradiation dose required to damage it will be. Ordinary pathogenic bacteria can be damaged by irradiation doses lower than 10 kGy, for example, *Escherichia coli* (0.23–0.45 kGy), *E. coli* O157:H7 (0.24–0.47 kGy), Salmonella (0.37–0.80 kGy) and viruses (2.02–8.10 kGy) (Fellows, 2001; Monk et al., 1995). The nutritional acceptability of irradiated commodities is one of the most prominent debates regarding irradiation processed F&V. Irradiation, like other technologies, modifies the nutritional composition of F&V. The severity of those changes is related to the dose of irradiation, temperature, atmosphere composition, storage and packaging. Plums (1.2–1.5 kGy), kinnows (1.5 kGy), and papayas (0.5–1.0 kGy) irradiated with different doses maintained their sensory attributes (Hussain et al., 2013; Abdullah et al., 2018; Zhao et al., 1996). Like other technologies, irradiation technology also has some limitations, but five of the most important advantages of radiation of F&V are minimum postharvest losses, better human health, enhanced global trade, alternatice to fumigation and elevated energy efficiencies (Korkmaz & Polat, 2005). Irradiation technology has an important role in minimizing postharvest losses, especially in developing countries where these losses exceed more than 30%. Irradiated F&V can be transferred to markets from their origin of production without any significant losses through infection (Murray, 1990). Irradiation can be a good alternative to chemicals in reducing decay and postharvest losses, as they do not have negative impacts on human health and are ecofriendly. This chapter will give in-depth information about the effects of commonly used irradiation techniques such as UV-C, X-rays and gamma on fresh F&V.

18.2 UV-C IRRADIATION OF FRUIT AND VEGETABLE COMMODITIES

UV irradiation could be an inexpensive substitute to chemicals for the prevention of microorganism development. They are divided into UV-A, UV-B and UV-C with wavelength ranges of 315–400, 280–315 and 100–280 nm, respectively (Xu et al., 2019; González-Aguilar et al., 2010).

The 254 nm wavelength of UV-C irradiation showed significant effects on decreasing microbial effects, and DNA damage is the most common action mechanism of UV-C irradiation (Gayán et al., 2014). UV-C deactivates microorganisms by destroying genetic material in the nucleus of the cell or nucleic acid in a virus (Chang et al., 1985). The 250–270 nm range of UV-C is the most lethal because of its powerful assimilation by the nucleic acids of microorganisms. A 262

DOI: 10.1201/9781003370376-24

nm wavelength is known as the germicidal spectrum because it is the maximal germicidal wavelength (Gurzadyan et al., 1995). The dimerization of pyrimidine molecules induced by light destroys the DNA and RNA of microorganisms. In particular, cyclobutane dimers are produced by thymine (exist only in the DNA). The nucleic acids do not replicate due to the dimerization of thymine molecules, therefore a defect is produced which hinders the functionality of microorganisms (Dai et al., 2012). UV-C irradiation is a non-chemical surface sterilization method used to decrease microbial spoilage on the surface of F&V (Huang et al., 2017). Light intensity (irradiation source), irradiation period, space between the commodity and light source determine the dose of UV-C irradiation (Bintsis et al., 2000). A general view of a UV-C radiation instrument and UV-C radiation treatment in green beans is given in Fig. 18.1a and 18.1b, respectively. Different research has reported on the effects of UV-C illumination in fresh F&V.

Ustun et al. (2021) treated green beans with 0.1, 0.3 and 0.5 kJ m^{-2} UV-C irradiation doses. These researchers reported that all UV-C doses had the lowest respiration rate, while there were non-significant effects of UV-C doses on the amount of reducing sugar and chlorophyll. The 0.3 and 0.5 kJ m^{-2} doses displayed high total phenolic content. There was no surface damage from UV-C irradiation in cold storage; however, 0.3 and 0.5 kJ m^{-2} exhibited symptoms of surface damage at 15+3 days of shelf life. According to these researchers, 0.1 kJ m^{-2} UV-C irradiation had no symptoms of surface damage and preserved better storage quality of green beans in MAP for 25 days. Zucchini squash slices were treated for 1, 10 and 20 min with UV-C radiation from a germicidal lamp (Erkan et al., 2001). UV-C radiation applied for 10 and 20 min decreased decay at 5 and 10 °C. All the treatments resulted in a high respiration rate with no significant effects on ethylene production. After 12 days of storage at 10 °C, UV-C treatments for 10 and 20 min showed symptoms of irradiation damage. At 5 °C, no UV-C damage and non-significant effects of UV-C on chilling damage were recorded. These researchers mentioned that UV-C irradiation reduced the growth of microorganisms and delayed senescence and the deterioration of zucchini squash tissue slices. The influence of irradiation has also been performed on green beans (Figure 18.1).

Golden Bell sweet peppers were treated with 2.2, 4.4 and 6.6 kJm^{-2} UV-C irradiation doses and kept at 4 °C (Promyou & Supapvanich, 2013). UV-C treatment exhibited non-significant effects on the amount of total carotenoid in Golden Bell sweet peppers. The 6.6 kJm^{-2} UV-C irradiation dose decreased chilling damage, weight losses and firmness losses, and elevated antioxidant activity and catalase activity. Andrade Cuvi et al. (2011) applied 10 kJ m^{-2} UV-C radiation to red peppers during pre-storage at 0 °C. UV-C enhanced the activities of antioxidant enzymes such as superoxide dismutase, catalase, ascorbate peroxidase and guaiacol peroxidase and reduced chilling damage and weight losses in red peppers.

Broccoli heads were subjected to 4, 7, 10 and 14 kJ m^{-2} UV-C radiation doses. At 20 °C, all UV-C doses retarded yellowing and degradation of chlorophyll with 10 kJ m^{-2} had the maximum amount of chlorophyll. UV-C radiation reduced enzyme activities, i.e. chlorophyll peroxidase and chlorophyllase, and increased Mg-dechelatase activity. UV-C-treated broccoli had the lowest respiration rate and total phenolic and flavonoid contents, with the highest antioxidant activity. UV-C treatment could be a beneficial non-chemical treatment to retard the breakdown of chlorophyll, decrease damage and disruption of tissue and preserve the antioxidant potential of broccoli (Costa et al., 2006). Minimally processed broccoli florets were tested with UV-C radiation doses (4.4, 8.8 and 13.2 kJ m^{-2}) and kept under MAP (Dogan et al., 2018). These researchers reported that UV-C reduced weight losses and elevated total phenolic content in fresh-cut broccoli florets. The O_2 levels of UV-C-treated broccoli florets packed in MAP were lower, while CO_2 levels were higher than the control.

Allende and Artes (2003) used 0.4, 0.81, 2.44, 4.07 and 8.14 kJ m^{-2} UV-C doses on Red Oak Leaf lettuce kept in MAP packages at 5 °C. These researchers reported that UV-C decreased O_2 and increased CO_2 contents inside MAP. UV-C

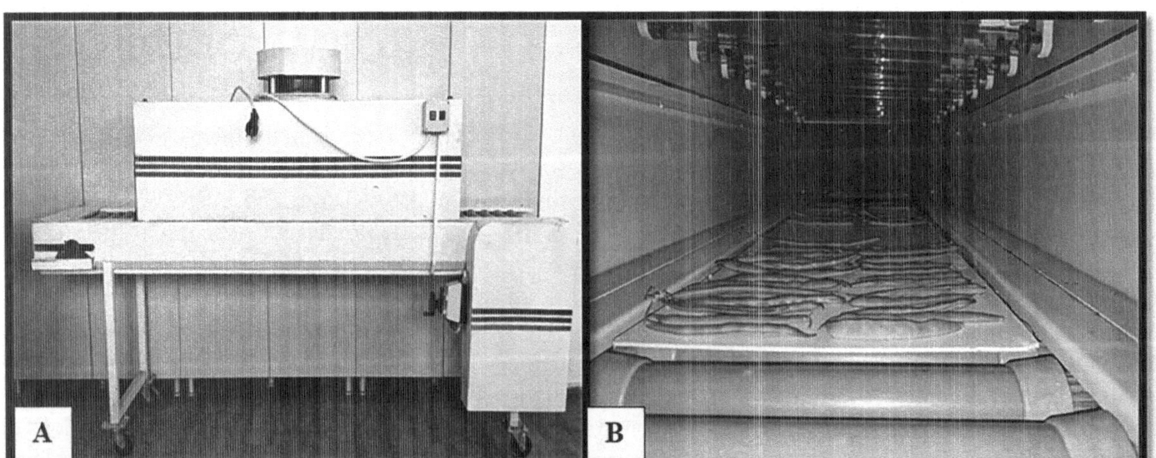

FIGURE 18.1 (A) General view of UV-C irradiation instrument; (B) UV-C irradiation treatment in green beans.

treatment combined with MAP effectively reduced psychrotrophic bacteria, coliform and yeast development, with no significant effects on lactic acid bacteria. The combined treatment of UV-C and MAP maintained better sensory quality of lettuce. Strawberries were irradiated with 0.43, 2.15 and 4.30 kJ m^{-2} UV-C doses for 1, 5 and 10 min (Erkan et al., 2008). UV-C treatment increased antioxidant activity and phenolic content and reduced decay. In particular, treatments for 5 and 10 min elevated antioxidant enzymes activiy, such as glutathione peroxidase, glutathione reductase, superoxide dismutate, ascorbate peroxidase, guaiacol peroxidase, monohydroascorbate reductase and dehydroascorbate reductase. These researchers recommended that UV-C treatment for 5 and 10 min maintained better postharvest quality of strawberries.

Collins and Bluecrop cultivars of blueberries were subjected to 0–4 kJ m^{-2} UV-C doses before storage at 5 °C (Perkins-Veazie et al., 2008). The 1–4 kJm^{-2} UV-C treatments reduced ripe rot (*Colletotrichum acutatum*, syn. *C. gloeosprioides*) by 10%. The 0 or 1 kJm^{-2} UV-C treatments had higher anthocyanin content, total phenolics and antioxidant activity in Collins fruit. The fruit treated with 2 or 4 kJ m^{-2} UV-C displayed maximum total anthocyanin content and ferric-reducing antioxidant power, with non-significant effects on phenolic content. UV-C treatment did show significant effects on weight loss and fruit firmness. These researchers suggested that UV-C radiation during postharvest can increase antioxidant activity and reduce ripe rot in blueberries.

Khademi et al. (2013) treated Karaj persimmons with UV-C radiation doses at 0, 1.5 and 3 kJm^{-2}. The findings obtained showed that UV-C irradiations of 1.5 and 3 kJm^{-2} decreased the decay incidence with no considerable effects on ethylene production, firmness and skin color. These researchers suggested that suitable firmness in Karaj persimmons can be maintained by combining UV-C treatments and other postharvest methods. González-Aguilar et al. (2007) mentioned that UV-C irradiation doses at 2.46 and 4.93 kJ m^{-2} were applied to mangoes (cv. Haden). UV-C irradiations showed maximum overall acceptance with lower decay incidence and enhanced shelf life. Furthermore, UV-C irradiation had a positive correlation with total phenolics, flavonoid content, lipoxygenase and phenylalanine ammonia-lyase activities. These researchers suggested that UV-C may enhance the shelf life of Haden mangoes.

Ustun and Dogan (2022) investigated the effects of UV-C (1.5, 3.0 and 4.5 kJ m^{-2}) on fresh fig fruit. Under cold and shelf storage, UV-C irradiations exhibited non-significant effects on firmness, total soluble solids (TSS) or titratable acidity (TA) content; however, UV-C treatments elevated total phenolics and antioxidant activity compared to untreated figs. The 3.0 and 4.5 kJ m^{-2} UV-C treatments were successful in controlling unmarketable fruit under both storage conditions.

The effects of a hot water dip at 55 °C for 30 s, 4.54 kJm^{-2} UV-C radiation, passive MAP and high O_2 active MAP on fresh-cut arils of Mollar de Elche pomegranates were studied (Maghoumi et al., 2013). All treatments showed less activity of superoxide dismutase with UV-C, and high O_2 individually or in combination maintained maximum superoxide dismutase activity. UV-C combined with high O_2 resulted in low catalase, peroxidase and polyphenol oxidase with higher anthocyanin and phenolic contents. These researchers suggested that the postharvest quality of fresh-cut arils was better preserved with UV-C + high O_2. Another study determined the effects of UV-C doses (12.47, 37.41 and 62.35 J mL^{-1}) on pomegranate juice showed maximum anthocyanin content, polymeric color of juice, total phenolics, total antioxidant activity, physical and chemical properties when compared to heat treatment. These researchers observed that UV-C irradiation has the ability to be a promising method for the disinfestation of pomegranate juice to enhance its shelf life (Pala & Toklucu, 2011).

18.3 EFFECTS OF X-RAY IRRADIATION ON FRUIT AND VEGETABLE COMMODITIES

X-ray, being a non-thermal technology, has shown excellent effects on the reduction of deterioration. Park and Ha (2019) reported that X-ray irradiation induces alteration in the cell membrane of pathogenic bacteria. X-rays are ionizing radiation that induces DNA damage because of reactive oxygen species produced through irradiation (Van Calenberg et al., 1999). Ionization radiation causes the loss of water molecules and produces hydroxyl (OH) radicals along with hydrogen peroxide (H_2O_2). OH and H_2O_2 are highly reactive substances associated with nucleic acids, hence vandalizing DNA, membrane-involved lipids, proteins and enzymes (Mahmoud, 2010). Electron rays can be negatively affected by the elimination of a 50 kilo electron volt dose of X-ray. The thickness of the processing product determines the selection of electron beams or X-rays (Van Calenberg et al., 1998). A dose of less than 1 kGy is sufficient to inactivate the insects. The US Food and Drug Administration reported that a 7.5 MeV dose of X-rays was enough to induce collision with electrons (Jung et al., 2015). The CDC (Centers for Disease Control and Prevention) and WHO (World Health Organization) suggested that radiation can protect humans against foodborne microorganisms. X-rays have been used to remove microorganisms in different products such as spinach leaves, lettuce, strawberries, shellfish, shrimp and dairy products (Mahmoud et al., 2010). X-rays have been successfully used to eliminate Mediterranean fruit flies (*Ceratitis capitata*) (Rojas-Argudo et al., 2012). Several researches have reported on the effects of X-rays in F&V.

Mahmoud (2012) applied different X-ray doses on inoculated whole cantaloupes for deactivation of *Escherichia coli* O157:H7, *Listeria monocytogenes*, *Salmonella enterica* and *Shigella flexneri*. Moreover, the effects of X-ray on the color, texture and numbers of microflora (mesophilic, psychrotrophic, yeast and mold) were also studied. Three strain mixtures from each organism (100 µL) were treated on 5 cm^2 of the cantaloupes' surface, dried at ambient temperature for 60 min and later at 22°C and 55% relative humidity X-rays were applied. For all the investigated pathogens a 5 log CFU decrease was obtained at a 2.0 kGy dose. Non-significant effects of X-ray on color and firmness of cantaloupes were observed. Moreover, X-ray minimized initial natural microflora on cantaloupes compared to the control during storage of 20 days.

Mahmoud et al. (2016) studied X-ray treatments (0, 0.1, 0.5, 1.0 and 1.5 kGy) on *Escherichia coli* O157:H7, *Listeria monocytogenes*, *Salmonella enterica* and *Shigella flexneri* pathogens in mangoes. Furthermore, these researchers studied the effects of X-rays on a number of mesophiles, psychrotrophic bacteria, yeast and mold for 30 days in storage at room temperature. The 0.5 kGy X-ray dose resulted in precisely 2.9, 1.8, 2.1 and 5.2 log CFU decreases. Moreover, the 1.5 kGy X-ray dose considerably decreased levels of *Escherichia coli* O157:H7, *Listeria monocytogenes*, *Salmonella enterica* and *Shigella flexneri* to undetectable levels, i.e. 2.0 log CFU cm^{-2}. The 1.5 kGy dose considerably lowered natural microflora on the peel of whole mangoes compared to the control at 22°C for 30 days of storage.

Alonso et al. (2007) studied X-ray irradiations in Clemenules mandarins as a quarantine treatment for fruit flies and compared it with standard cold temperature quarantine treatment. These researchers mentioned that no detrimental effects after 0.195 and 0.395 kGy X-ray doses were recorded on the fruits' quality. Findings showed that X-ray can be an efficient quarantine method in clementine mandarins, and it can be as beneficial as present cold treatments for Clemenules mandarins against fruit flies.

Jung et al. (2016) observed the effects of gamma rays, electron or X-rays on Fuji apples and Niitaka pears in terms of sensory characteristics, such as firmness, sugar levels and color. No considerable effects of the three types of radiations on sugar levels in Fuji apples and Niitaka pears were observed. Treatment doses between 200 and 1000 Gy had no significant effects on color. The radiation sources had no significant effects on firmness, while less hardness recorded in doses above 600 Gy. These researchers suggested that X-rays had homogenous effects to gamma rays and electron beam radiation in term of the sensorial characteristics of Fuji apples and Niitaka pears.

Mahmoud (2010) tested 0.1, 0.5, 0.75, 1.0 and 1.5 kGy X-ray doses to inactivate *Escherichia coli* O157:H7, *Listeria monocytogenes*, *Salmonella enterica* and *Shigella flexneri* pathogens in inoculated Roma tomato surfaces. The irradiation considerably lowered the amount of pathogens on whole Roma tomatoes compared to the control. The 0.75 kGy dose treatment had almost 4.2, 2.3, 3.7 and 3.6 log CFU decreases in *Escherichia coli* O157:H7, *Listeria monocytogenes*, *Salmonella enterica* and *Shigella flexneri*, respectively. X-ray treatments displayed no considerable effects on the color at the maximum dose of 2.0 kGy. The 1.0 and 1.5 kGy doses had more than a 5 log CFU decrease per tomato. Moreover, X-rays considerably lowered the natural microflora in Roma tomatoes.

Mahmoud et al. (2010) investigated different X-ray doses on the deactivation of *Escherichia coli* O157:H7, *Listeria monocytogenes*, *Salmonella enterica* and *Shigella flexneri* in spinach leaves. Furthermore, these researchers observed the quality of color and microflora numbers (mesophiles, psychrotrophic bacteria, yeast and mold) affected by X-rays. In all investigated pathogens, the 2 kGy X-ray dose decreased more than 5 log CFU/leaf. X-rays considerably decreased the initial natural microflora on spinach leaves, and natural microflora amounts were considerably lower than the control throughout the 30 days of refrigerated storage.

Moon et al. (2017) investigated different X-ray doses on the microbiological, physicochemical and organoleptic characteristics of fresh-cut vegetables, including carrots, green peppers, cherry tomatoes and paprika. Furthermore, these researchers evaluated the quality of food suitable for immune-depressed patients. Except cherry tomatoes, the total amount of aerobic bacteria in samples without irradiation was 1.63–3.34 log CFU/g. The minimum allowable dose regarding the sterilization of fresh-cut vegetables was 0.4 kGy. Firmness was reduced with the elevation of the absorbed dose, while non-significant differences were observed between the 0.4 kGy X-ray dose and samples without irradiation. The overall acceptance was maximum in irradiated fresh-cut samples than non-irradiated ones. These researchers suggested that an X-ray dose of 0.4 kGy can be used to make foods hygienically safe in fresh-cut vegetables, except carrots.

18.4 EFFECTS OF GAMMA IRRADIATION ON FRUIT AND VEGETABLE COMMODITIES

Gamma irradiations have been used as a promising technique to decrease microbial infection and increase the shelf life of fresh F&V (Prakash et al., 2000). Like X-rays, gamma irradiations are also ionizing radiations. Gamma rays directly or indirectly affect microorganisms (Bhatnagar et al., 2022). Microbial cell components, e.g. lipids, carbohydrates and DNA, are damaged directly through radiations in direct effects, while in indirect effects the free radicals and reactive oxygen species consisting of OH radicals, hydrogen atoms and hydrated electrons produced from water radiolysis react with the components of the cells (Fan & Wang, 2020). Furthermore, Wardman (2009) reported that cellular injury may occur due to the reactive nitrogen and additional species and it may be because of ionization on the essential molecules, i.e. DNA. Bisht et al. (2021) mentioned that the main mechanism in fresh products to inactivate microorganisms is an indirect effect, as water is the major substance in those products. Direct and indirect effects result in alterations of the physiologic, genetic, epigenetic and biologic makeup of microorganisms (Desouky et al., 2015). Pi et al. (2021) observed that a 5 MeV dose of X-ray from X-ray machines, a 10 MeV dose of electron beams from electron accelerators and gamma radiation of up to 5 MeV from cobalt-60 sources is thought to be radiologically safe. It is considered a safe technology for securing food because of its maximum penetration that can easily eliminate microorganisms. It is an ecofriendly technology which decreases the number of microorganisms (Zarbakhsh & Rastegar, 2019). Gamma irradiation was successfully used to decrease microbial load and maximize the shelf life of litchi, pomegranates, dates, bananas, strawberries, raspberries, potatoes, cucumbers, green onions, lettuce, and many other F&V (Bhatnagar et al., 2022).

Shahbaz et al. (2014) studied several gamma irradiation doses (0.4, 1 and 2 kGy) on the chemical and sensory properties of pomegranates. The 0.4 and 1 kGy doses did not significantly affect TSS, TA and pH values. Total anthocyanin and phenolic content were decreased by irradiation treatment. Antioxidant activity, total phenolics and anthocyanins were positively correlated. Irradiated fruit juice had higher preference scores. These researchers suggested that marketing and customer acceptance can be improved through gamma irradiation quarantine treatment.

Hajare et al. (2010) subjected two litchi varieties, Shahi and China, to 0.3 and 0.5 kGy gamma irradiation doses. Neither dose had considerable variations in pH. There were no significant differences in the TA of the Shahi cultivar until day 10, and increases in the TA were observed on day 20 in the 0.5 kGy dose and on day 28 in the 0.3 kGy dose. In China litchi the TA content was high on day 1 and reduced on day 10. The 0.3 and 0.5 kGy doses had similar TA after 20 days with an increase in TA on day 28, though this increase was lower as compared to day 1. The microbial load was reduced by the radiation treatment in a dose-dependent manner. These researchers suggested that gamma radiation treatment of 0.5 kGy combined with low temperature at 4°C had maximum shelf-life quality for 28 days in litchi fruit.

Mostafavi et al. (2012) applied different doses of gamma irradiation (0, 300, 600, 900 and 1200 Gy) on Red Delicious apples later stored at 1°C temperature. Prior to the irradiation treatment the fruit were inoculated with *Penicillium expansum*. The 900 and 1200 Gy doses significantly decreased the total phenolic content and antioxidant activity. A direct relationship between total phenolics and antioxidant activity was found. The moisture concentration was significantly affected by irradiation compared to storage duration and pathogen. After three months the lesion diameter in fruits with no radiation was higher compared to the radiated samples. An increase in storage time and dose decreased the firmness. These researchers recommended that low radiation doses of 300 and 600 Gy in combination with low storage temperature could be used to lessen losses in apple quality for 9 months of storage.

Majeed et al. (2014) subjected strawberries (cv. Corona) to 0.5, 1.0 and 1.5 kGy gamma irradiation doses to study the shelf life and chemical properties. The 1.0 and 1.5 kGy doses considerably enhanced the shelf life of strawberries for 5.75 and 7.75 days, respectively, compared to the untreated control group (3.25 days). The gamma radiation doses, especially the higher doses, significantly reduced the deterioration and weight losses compared to the control. Non-significant effects of radiation and storage time on TSS, TA and pH were recorded. These researchers suggested that 1.0 and 1.5 kGy doses may enhance shelf life, consumer acceptability, less weight loss and deterioration, with no negative effects on the chemical attributes of the strawberries.

Mahto and Das (2014) applied 0.04, 0.08, 0.12 and 1 kGy gamma radiation doses 5 and 30 days postharvest in potatoes (cv. Kufri Sindhuri) to investigate their effects on texture, microstructure, reducing and total sugars and losses in tubers at 22°C temperature and 85–90% relative humidity. The 0.04 kGy dose prevented browning in potatoes treated 5 days after harvest but not in potatoes treated 30 days after harvest. During storage, the appearance was maintained by the irradiated non-sprouted potatoes. The sensitivity to radiation-induced damage was higher in early irradiated potatoes, causing high tuber losses at 1 kGy; however, no increase in vulnerability to tuber rotting was observed in lower doses (0.04, 0.08 and 0.12 kGy). The blackening of bud tissues was observed in the 1 kGy dose group, resulting in high decay and unwanted texture. Low doses up to 0.12 kGy decreased deterioration in texture. Potatoes treated with 0.08–0.12 kGy had maximum textural scores. These researchers recommended that 0.08–0.12 kGy doses applied at 5 days after harvest in K. Sindhuri potatoes maintained the best textural properties. Furthermore, the shelf life of potatoes may be enhanced by the application of gamma radiation during non-refrigerated storage.

Khalili et al. (2017) investigated 2, 2.5 and 3 kGy gamma radiation doses on the shelf-life and quality parameters of cucumbers. These researchers infected some of the samples with white-*Sclerotinia sclerotiorum*, gray-*Botrytis cinerea* and olive-*Cladosporium cucumerinum* fungi. The irradiation treatment maintained the chlorophyll and ascorbic acid content by 1.4- to 3-folds with the shelf life being enhanced by 1 week. The chilling injuries were reduced, and the treatment delayed fungal growth on the samples for 1 week. These researchers mentioned that the 2 kGy irradiation dose maintained the best properties of cucumbers with low effect on taste, TSS and firmness.

Miladilari et al. (2020) subjected White-Qom, White-Neyshabour, Red-Ridge-Lump and Red-Ray-Corrugated onions to 0, 30, 60, 90, 120 and 150 Gy gamma radiation doses. The effects of irradiation on dry matter, protein, phynlyalanine ammonia lyase (PAL), peroxidase and PAL gene expression regarding irradiation dose and genotype were investigated. The peroxidase activity was enhanced by irradiation treatments, while protein concentration and PAL activity were reduced. The dry matter content was dependent on genotype. The increase in gamma irradiation decreased the activity of PAL, while the PAL gene expression rate was similar between the gamma-irradiated and control samples, showing the gene regulatory elements may not be directly linked with irradiation. These researchers mentioned that PAL activity can be reduced by gamma irradiation, limiting fungal and bacterial stresses.

Fan and Sokorai (2011) subjected cut iceberg lettuce to MAP and delayed gamma radiation doses of 0.5 and 1.0 kGy. The results obtained exhibited that tissue browning was observed in irradiated cut lettuce once kept in air while the irradiated cut lettuce kept in MAP maintained greater appearance scores due to lower O_2 levels than control cut lettuces. No significant effects of irradiation on the firmness of lettuce were observed. After storage, the control non-irradiated samples were below the marketable level with a score of ≈4, while both irradiated samples of 0.5 and 1 kGy had higher scores of 6.5 and 7.9, respectively, using a scale with a maximum of 9.

A delay of 24 h in irradiation of cut lettuce had no statistically significant differences in terms of cut or surface browning or visual quality compared to gamma-irradiated samples treated directly after preparation. These researchers suggested that using MAP is necessary to decrease deterioration in quality due to irradiation.

18.5 CONCLUSIONS

Postharvest losses in fresh F&V are very high, therefore different types of methods are used to overcome those losses. Irradiation treatment can be the safest non-chemical technique to control microbial decay. This technology damages the DNA of microorganisms, hence preventing them from multiplying. It serves as a surface sterilization technique to reduce the deterioration by enhancing antioxidant activity and defend against biotic and abiotic stresses. Irradiation technology is used as a quarantine treatment in different fresh F&V. As a conclusion, irradiation technology can be a promising method to enhance the shelf life of F&V by minimizing microbial load and activating defense-related enzymes.

REFERENCES

Abdullah, R., Rashid, S., Naz, S., Iqtedar, M., Kaleem, A. (2018). Postharvest preservation of citrus fruits (Kinnow) by gamma irradiation and its impact on physicochemical characteristics. *Progress in Nutrition*, 20, 133–145. https://doi.org/10.23751/pn.v20i1.5235

Allende, A., Artes, F. (2003). Combined ultraviolet-C and modified atmosphere packaging treatments for reducing microbial growth of fresh processed lettuce. *LWT-Food Science and Technology*, 36, 779–786. https://doi.org/10.1016/S0023-6438(03)00100-2

Alonso, M., Palou, L., Del Rio, M.A., Jacas, J.A. (2007). Effect of X-ray irradiation on fruit quality of clementine mandarin cv. 'Clemenules'. *Radiation Physics and Chemistry*, 76, 1631–1635. https://doi.org/10.1016/j.radphyschem.2006.11.015

Andrade Cuvi, M.J., Vicente, A. R., Concell´on, A., Chaves, A.R. (2011). Changes in red pepper antioxidants as affected by UV-C treatments and storage at chilling temperatures. *LWT-Food Science and Technology*, 44, 1666–1671. https://doi.org/10.1016/j.lwt.2011.01.027

Bhatnagar, P., Gururani, P., Bisht, B., Kumar, V., Kumar, N., Joshi, R., Vlaskin, M.S. (2022). Impact of irradiation on physic-chemical and nutritional properties of fruits and vegetables: A mini review. *Heliyon*, e10918. https://doi.org/10.1016/j.heliyon.2022.e10918

Bintsis, T., Litopoulou-Tzanetaki, E., Robinson, R.K. (2000). Existing and potential applications of ultraviolet light in the food industry—A critical review. *Journal of The Science of Food and Agriciculture*, 80, 637–645. https://doi.org/10.1002/(SICI)1097-0010(20000501)80:6<637::AID-JSFA603>3.0.CO;2-1

Bisht, B., Bhatnagar, P., Gururani, P., Kumar, V., Tomar, M.S., Sinhmar, R., Rathi, N., Kumar, S. (2021). Food irradiation: Effect of ionizing and non-ionizing radiations on preservation of fruits and vegetables- a review. *Trends in Food Science & Technology*, 114, 372–385. https://doi.org/10.1016/j.tifs.2021.06.002

Chang, J.C.H., Ossoff, S.F., Lobe, D.C., Dorfman, M.H., Dumais, C.M., Qualls, R.G., Johnson, J.D. (1985). UV inactivation of pathogenic and indicator microorganisms. *Applied and Environmental Microbiology*, 49(6), 1361–1365. https://doi.org/10.1128/aem.49.6.1361-1365.1985

Costa, L., Vicente, A.R., Civello, P.M., Chaves, A.R., Martínez, G.A. (2006). UV-C treatment delays postharvest senescence in broccoli florets. *Postharvest Biology and Technology*, 39, 204–210. https://doi.org/10.1016/j.postharvbio.2005.10.012

Dai, T., Vrahas, M.S., Murray, C.K., Hamblin, M.R. (2012). Ultraviolet C irradiation: an alternative antimicrobial approach to localized infections? *Expert Review of Anti-infective Theraphy*, 10(2), 185–195. https://doi.org/10.1586/eri.11.166

Desouky, O., Ding, N., Zhou, G. (2015). Targeted and non-targeted effects of ionizing radiation. *Journal of Radiation Research and Applied Sciences*, 8, 247–254. http://dx.doi.org/10.1016/j.jrras.2015.03.003

Dogan, A., Topcu, Y. Erkan, M. (2018). UV-C illumination maintains postharvest quality of minimally processed broccoli florets under modified atmosphere packaging. *Acta Horticulturae*, 1194, 537–544. https://doi.org/10.17660/ActaHortic.2018.1194.78

Erkan, M., Wang, C.Y., Krizek, D.T. (2001). UV-C irradiation reduces microbial populations and deterioration in *Cucurbita pepo* fruit tissue. *Environmental and Experimental Botany*, 45, 1–9. https://doi.org/10.1016/S0098-8472(00)00073-3

Erkan, M., Wang, S.Y., Wang, C.Y. (2008). Effect of UV treatment on antioxidant capacity, antioxidant enzyme activity and decay in strawberry fruit. *Postharvest Biology and Technology*, 48, 163–171. https://doi.org/10.1016/j.postharvbio.2007.09.028

Fan, X., Sokorai, K.J.B. (2011). Effects of gamma irradiation, modified atmosphere packaging, and delay of irradiation on quality of fresh-cut Iceberg lettuce. *HortScience*, 46(2), 273–277. https://doi.org/10.21273/HORTSCI.46.2.273

Fan, X., Wang, W. (2020). Quality of fresh and fresh-cut produce impacted by nonthermal physical technologies intended to enhance microbial safety. *Critical Reviews in Food Science and Nutrition*, 62(2), 362–382. https://doi.org/10.1080/10408398.2020.1816892

Fellows, P.J. (2001). *Food Processing Technology*. Cambridge: CRC Press, Woodhead Publishing.

Gayán, E., Condón, S., Álvarez, I. (2014). Biological aspects in food preservation by ultraviolet light: A review. *Food and Bioprocess Technology*, 7, 1–20. https://doi.org/10.1007/s11947-013-1168-7

González-Aguilar, G.A., Ayala-Zavala, J.F., Olivas, G.I., de la Rosa, L.A., Álvarez-Parrilla, E. (2010). Preserving quality of fresh-cut products using safe technologies. *Journal für Verbraucherschutz und Lebensmittelsicherheit*, 5, 65–72. https://doi.org/10.1007/s00003-009-0315-6

González-Aguilar, G.A., Zavaleta-Gatica, R., Tiznado-Hernandez, M.E. (2007). Improving postharvest quality of mango 'Haden' by UV-C treatment. *Postharvest Biology and Technology*, 45, 108–116. https://doi.org/10.1016/j.postharvbio.2007.01.012

Gurzadyan, G.G., Gorner, H., Schulte-Frohlinde, D. (1995). Ultraviolet (193, 216 and 254 nm) photoinactivation of *Escherichia coli* strains with different repair deficiencies. *Radiation Research*, 141(3), 244–251. https://doi.org/10.2307/3579001

Hadlington, S. (1987). Food irradiation legislation due despite public suspicion. *Nature*, 328, 751. https://doi.org/10.1038/328751a0

Hajare, S. N., Saxena, S., Kumar, S., Wadhawan, S., More, V., Mishra, B. B., Parte, M. N., Gautam, S., Sharma, A. (2010). Quality profile of litchi (*Litchi chinensis*) cultivars from India and effect of radiation processing. *Radiation Physics and Chemistry*, 79, 994–1004. https://doi.org/10.1016/j.radphyschem.2010.03.014

Huang, H., Ge, Z., Limwachiranon, J., Li, L., Li, W., Lup, Z. (2017). UV-C treatment affects browning and starch metabolism of minimally processed lily bulb. *Postharvest Biology and Technology*, 128, 105–111. https://doi.org/10.1016/j.postharvbio.2017.02.010

Hussain, P.R., Dar, M.A., Wani, A.M. (2013). Impact of radiation processing on quality during storage and post-refrigeration decay of plum (*Prunus domestica* L.) cv. Santaroza. *Radiation Physics and Chemistry*, 85, 234–242. https://doi.org/10.1016/j.radphyschem.2013.01.020

Ihsanullah, I., Rashid, A. (2017). Current activities in food irradiation as a sanitary and phytosanitary treatment in the Asia and the Pacific Region and a comparison with advanced countries. *Food Control*, 72, 345–359. https://doi.org/10.1016/j.foodcont.2016.03.011

Josephson, E.S. (1983). An historical review of food irradiation. *Journal of Food Safety*, 5(4), 161–189. https://doi.org/10.1111/j.1745-4565.1983.tb00469.x

Jung, K., Song, B.S., Kim, M.J., Moon, B.G., Go, S.M., Kim, J.K., Lee, Y.J., Park, J.H. (2015). Effect of X-ray, gamma ray, and electron beam irradiation on the hygienic and physicochemical qualities of red pepper powder. *LWT - Food Science and Technology*, *63*(2), 846–851. https://doi.org/10.1016/j.lwt.2015.04.030

Jung, K., Go, S.M., Moon, B.G., Song, B.S., Park, J.H. (2016). Comparative study on the sensory properties of Fuji apples and Niitaka pears irradiated by gamma rays, electron beams, or X-rays. *Food Science and Technology Research*, 22(1), 23–29. https://doi.org/10.3136/fstr.22.23

Khademi, O., Zamani, Z., Poor Ahmadi, E., Kalantari, S. (2013). Effect of UV-C radiation on postharvest physiology of persimmon fruit (*Diospyros kaki* Thunb.) cv. 'Karaj' during storage at cold temperature. *International Food Research Journal*, 20, 247–253.

Khalili, R., Ayoobian, N., Jafarpour, M., Shirani, B. (2017). The effect of gamma irradiation on the properties of cucumber. *Journal of Food Science and Technology*, 54(13), 4277–4283. https://doi.org/10.1007/s13197-017-2899-7

Korkmaz, M., Polat, M. (2005). Irradiation of fresh fruits and vegetables, In: *Improving the Safety of Fresh Fruits and Vegetables*, ed. W. Jongen, 387–428. Cambridge: Woodhead Publishing. https://doi.org/10.1533/9781845690243.3.387

Maghoumi, M., Gomez, P.A., Mostofi, Y., Zamani, Z., Artes-Hernandez, F., Artes, F. (2013). Combined effect of heat treatment, UV-C and superatmospheric oxygen packing on phenolics and browning related enzymes of fresh-cut pomegranate arils. *LWT - Food Science and Technology*, 54, 389–396. https://doi.org/10.1016/j.lwt.2013.06.006

Mahmoud, B.S., Bachman, G., Linton, R.H. (2010). Inactivation of *Escherichia coli* O157:H7, *Listeria monocytogenes*, *Salmonella enterica* and *Shigella flexneri* on spinach leaves by X-ray. *Food Microbiology*, 27(1), 24–28. https://doi.org/10.1016/j.fm.2009.07.004

Mahmoud, B.S.M. (2010). The effects of X-ray radiation on *Escherichia coli O157:H7, Listeria monocytogenes, Salmonella enterica and Shigella flexneri* inoculated on whole Roma tomatoes. *Food Microbiology*, 27, 1057–1063. https://doi.org/10.1016/j.fm.2010.07.009

Mahmoud, B.S.M. (2012). Effects of X-ray treatments on pathogenic bacteria, inherent microflora, color, and firmness on whole cantaloupe. *International Journal of Food Microbiology*, 156, 296–300. https://doi.org/10.1016/j.ijfoodmicro.2012.04.001

Mahmoud, B.S.M., Nannapaneni, R., Chang, S., Coker, R. (2016). Effect of X-ray treatments on *Escherichia coli O157:H7, Listeria monocytogenes*, *Shigella flexneri*, *Salmonella enterica* and inherent microbiota on whole mangoes. *Letters in Applied Microbiology*, 62(2), 138–144. https://doi.org/10.1111/lam.12518

Mahto, R., Das, M. (2014). Effect of gamma irradiation on the physico-mechanical and chemical properties of potato (*Solanum tuberosum* L.), cv. 'Kufri Sindhuri', in non-referigerated storage conditions. *Postharvest Biology and Technology*, 9, 37–45. https://doi.org/10.1016/j.postharvbio.2014.01.011

Majeed, A., Muhammad, Z., Majid, A., Shah, A.H., Hussain, M. (2014). Impact of low doses of gamma irradiation on shelf life and chemical quality of strawberry (*Fragaria* × *ananassa*) cv. 'Corona'. *The Journal of Animal & Plant Sciences*, 24(5), 1531–1536.

Miladilari, S., Ahmadi, M., Kashi, A., Mousavi, A., Mostofi, Y. (2020). Physiological responses of gamma-irradiated onion bulbs during storage. *Journal of Agricultural Sciences*, 26, 442–451. https://doi.org/10.15832/ankutbd.559604

Monk, J.D., Beuchat, L.R., Doyle, M.P. (1995). Irradiation inactivation of food borne microorganisms, *Journal of Food Protection*, 58, 197–208. https://doi.org/10.4315/0362-028X-58.2.197

Moon, B.G., Song, B.S., Park, J.H., Kim, J.K., Park, H.Y., Kim, D.H., Son, E.J., Im, D.S., Eun, J.B. (2017). Microbiological, physicochemical, and organoleptic evaluation of fresh-cut vegetables irradiated using X-rays. *Korean Journal of Food Preservation*, 24(1), 27–35. https://doi.org/10.11002/kjfp.2017.24.1.27

Mostafavi, H.A., Mirmajlessi, S.M., Mirjalili, S.M., Fathollahi, H., Askari, H. (2012). Gamma radiation effects on physic-chemical parameters of apple fruit during commercial post-harvest preservation. *Radiation Physics and Chemistry*, 81, 666–671. https://doi.org/10.1016/j.radphyschem.2012.02.015

Murray, D.R. (1990). *Biology of Food Irradiation*. Taunton: Research Studies Press.

Pala, C.U., Toklucu, A.K. (2011). Effect of UV-C light on anthocyanin content and other quality parameters of pomegranate juice. *Journal of Food Composition and Analysis*, 24, 790–795. https://doi.org/10.1016/j.jfca.2011.01.003

Park, J-S., Ha, J-W. (2019). X-ray irradiation inactivation of *Escherichia coli* O157:H7, *Salmonella enterica* Serovar Typhimurium, and *Listeria monocytogenes* on sliced cheese and its bactericidal mechanisms. *International Journal of Food Microbiology*, 289, 127–133. https://doi.org/10.1016/j.ijfoodmicro.2018.09.011

Perkins-Veazie, P., Collins, J.K., Howard, L. (2008). Blueberry fruit response to postharvest application of ultraviolet radiation. *Postharvest Biology and Technology*, 47, 280–285. https://doi.org/10.1016/j.postharvbio.2007.08.002

Pi, X., Yang, Y., Sun, Y., Wang, X., Wan, Y., Fu, G., Li, X., Cheng, J. (2021). Food irradiation: a promising technology to produce hypoallergenic food with high quality. *Critical Reviews in Food Science and Nutrition*, 62(24), 6698–6713. https://doi.org/10.1080/10408398.2021.1904822

Prakash, A., Inthajak, P., Huibregtse, H., Caporaso, F., Foley, D.M. (2000). Effects of low-dose gamma irradiation and conventional treatments on shelf life and quality characteristics of diced celery. *Journal of Food Science*, 65(6), 1070–1075. https://doi.org/10.1111/j.1365-2621.2000.tb09420.x

Promyou, S., Supapvanich, S. (2013). Chilling injury alleviation in "Golden Bell" sweet pepper caused by UV-C treatment. *Acta Horticulturae*, 1011, 357–362. https://doi.org/10.17660/ActaHortic.2013.1011.45

Rojas-Argudo, C., Palou, L., Bermejo, A., Cano, A., del Río, M.A., González-Mas, M.C. (2012). Effect of X-ray irradiation on nutritional and antifungal bioactive compounds of 'Clemenules' clementine mandarins. *Postharvest Biology and Technology*, 68, 47–53. https://doi.org/10.1016/j.postharvbio.2012.02.004

Shahbaz, H.M., Ahn, J.J., Akram, K., Kim, H.Y., Park, E.J. (2014). Chemical and sensory quality of fresh pomegranate fruits exposed to gamma radiation as quarantine treatment. *Food Chemistry*, 145, 312–318. https://doi.org/10.1016/j.foodchem.2013.08.052

Ustun, H., Ali, Q., Kurubas, M.S., Dogan, A., Balkhi, M., Peker, B., Erkan, M. (2021). Influence of postharvest UV-C illumination on biochemical properties of green beans. *Scientia Horticulturae*, 289, 110499. https://doi.org/10.1016/j.scienta.2021.110499

Üstün, H., Doğan, A. (2022). The effect of different UV-C illumination doses on postharvest quality of fresh fig. *GIDA*, 47(5), 744–753. https://doi.org/10.15237/gida.GD22063

Van Calenberg, S., Vanhaelewyn, G., Van Cleemput, O., Callens, F., Mondelaers, W., Huyghebaert, A. (1998). Comparison of the effect of X-ray and electron beam irradiation on some selected spices. *LWT- Food Science and Technology*, 31(3), 252–258. https://doi.org/10.1006/fstl.1997.0352

Van Calenberg, S., Van Cleemput, O., Mondelaers, W., Huyghebaert, A. (1999). Comparison of the effect of X-ray and electron beam irradiation on the microbiological quality of foodstuffs. *LWT-Food Science and Technology*, 32(6), 372–376. https://doi.org/10.1006/fstl.1999.0568

Wardman, P. (2009). The importance of radiation chemistry to radiation and free radical biology (The 2008 Silvanus Thompson Memorial Lecture). *The British Journal of Radiology*, 82, 89–104. https://doi.org/10.1259/bjr/60186130

Xu, Y., Charles, M.T., Luo, Z., Mimee, B., Tong, Z., Roussel, D., Rolland, D., Veronneau, P. (2019). Preharvest UV-C treatment affected postharvest senescence and phytochemicals alternation of strawberry fruit with the possible involvement of abscisic acid regulation. *Food Chemistry*, 299, 125138. https://doi.org/10.1016/j.foodchem.2019.125138

Zarbakhsh, S., Rastegar, S. (2019). Influence of postharvest gamma irradiation on the antioxidant system, microbial and shelf life quality of three cultivars of date fruits (*Pheonix dactylifera* L.). *Scientia Horticulturae*, 247, 275–286. https://doi.org/10.1016/j.scienta.2018.12.035

Zhao, M., Moy, J., Paull, R.E. (1996). Effect of gamma-irradiation on ripening papaya pectin. *Postharvest Biology and Technology*, 8, 209–222. https://doi.org/10.1016/0925-5214(96)00004-X

Electrolyzed Water-Based Technology for Fresh Fruits and Vegetables

19

Shafi Ahmed*, Sharmin Akther, and Md. Sakib Hossain
*Corresponding Author: shafi@just.edu.bd

19.1 INTRODUCTION

Fresh vegetables and fruits are rich in vitamins, polyphenols, organic acids, and dietary fiber, which benefit human health. Modern consumers prefer natural, healthy, and convenient foods that exhibit high quality and microbial safety (Ma et al., 2019). However, fresh produce can become contaminated by spoilage and pathogenic microorganisms in soil, water, and other surroundings during the growing, transporting, storing, and selling processes (Lu et al., 2022). Minimally processed fruit and vegetable commodities were found to be the most contaminated with microorganisms (Murray et al., 2017), mostly *Bacillus* spp., *Salmonella* spp., *Staphylococcus* spp., *Micrococcus* spp., *Pseudomonas* spp., *Escherichia coli*, and *Klebsiella* spp. (Finger et al., 2019). Thus, fresh fruit and vegetable disinfection is indispensable to eliminate pathogenic microorganisms for food safety and to combat fungal- and bacteria- based spoilage to prolong the products' shelf life (Qiang, et al., 2005; McKellar et al., 2004). Recently, the demand for antimicrobial agents and disinfectants has increased in food industries (Lopez-Gomez et al., 2009). An appropriate sanitizer must meet several criteria, including the ability to reduce microorganisms significantly, avoid cross-contamination, have minimal or no impact on quality, comply with available technical capabilities and processing practices, and be safe to use, affordable, and approved by regulatory agencies. A number of methods have been applied to enhance vegetable and fruit produce quality. These include controlled atmosphere storage, cold storage, edible coatings, ultraviolet light, and 1-methylcyclopropene treatment (Yoo et al., 2021), and some natural chemical excitons, such as nitric oxide, chitosan (Hesami et al., 2021), methyl jasmonate, and melatonin. Despite their advantages, these methods have a number of limitations, such as significant energy consumption and a convoluted application procedure (Zhang et al., 2021). Hence, a number of alternative disinfection techniques have been designed and their decontamination efficacy assessed for application in the food industry (Keskinen et al., 2009), including irradiation (Osaili & Al-Nabulsi, 2016), ozone (Kumar et al., 2016), electrolyzed water (Ding et al., 2015; Liu et al., 2021), high pressure (Li et al., 2016), ultrasound (Li et al., 2017), pulsed electric field (Zhao et al., 2013), and cold plasma (Liao et al., 2017; Xiang et al., 2018). In the last few years, electrolyzed water (EW) has gained substantial attention as an innovative decontamination and preservation method. The use of EW was reported for the very first time in Japan's soda industry during the 1980s (Rahman et al., 2016; Hricova et al., 2008). Recent technological innovations have led to further improvement of EW-based technology, which is now becoming gradually more popular and attracting substantially higher attention as a promising nonthermal-based technology, particularly in the food industry. In addition, EW has been widely accepted as safe and was previously accepted as one of the acceptable food additives in the USA, the Republic of Korea, and Japan (Xuan et al. 2017). Several studies have shown that EW can be utilized to decontaminate meats, cereals, as well as fruit and vegetable commodities (Qin et al., 2022; Akther et al., 2023). As a novel sanitizer, EW has demonstrated its efficacy for applications in fruit- and vegetable-based industries, as well as for preventing physiological diseases, enhancing postharvest storage life, and preserving quality (Graca et al., 2011; Yoon & Lee, 2018; Pinto et al., 2016).

The purpose of the current chapter is to summarize and discuss various EW types, their characteristics and production processes along with possible antimicrobial mechanisms and their impact on postharvest quality, and decontamination of fruit and vegetable commodities. Moreover, the key findings of several studies have been summarized and discussed, followed by vegetable and fruit disinfection along with the effect of EW on prolonging the shelf life of fresh produce.

DOI: 10.1201/9781003370376-25

19.2 ELECTROLYZED WATER

Electrolyzed water (EW) has commercially been used as an effective alternative to chlorine solution and has generally been produced electrolytically from a solution of NaCl at 50 g L^{-1} in electrolysis cells by applying a voltage across the anode and cathode. The two types of EW are concurrently produced at two electrodes. At the anode, acidic electrolyzed water (AEW) is produced with a pH of 2–3, and it possesses a strong oxidation-reduction potential (ORP) higher than 1100 mV, and the available chlorine concentration (ACC) in the range of 10–90 mg L^{-1} (Ahmed et al., 2022). In contrast, alkaline electrolyzed water (ALEW), which is produced at the cathode, has a pH of between 10 and 13, a lower ORP below zero, and a voltage between -800 and -900 mV in general (Hricova et al., 2008). Salt water's chloride is transformed into chlorine gas, which quickly creates hypochlorite in the solutions (Boal, 2009). Free chlorine species (such as HOCl, Cl_2, and ClO−) are often significant for the bactericidal action of EW (Dewi et al., 2017). However, it is greatly reliant on three properties of EW: pH, ORP, and ACC. The said fundamental characteristics of EW are stronger at 4 °C storage temperature and retain their effective germicidal potential for a full year (Robinson et al. 2012). Furthermore, EW pre-heating can further increase the probable antibacterial effectiveness and ACC value in contrast with post-production types of heating (Forghani et al. 2015). In EW, chloride-based compounds are determined on the basis of their pH values. The most vital form of the chloride compounds of the EW under the pH range of 5.0–6.5 is HOCl, but at a higher pH range it may convert into OCl^-, and at a lower pH it can change into Cl_2 (Rahman et al., 2010). Various reports have shown that EW is efficient against bacterial and viral microorganisms, molds, and some other forms of microorganisms (Dewi et al., 2017; Rahman et al., 2016). The leading reasons for the popularity of EW are its simple production, ease of application, lower cost production, and minimal health- and environment-associated detrimental effects.

19.3 VARIOUS TYPES OF ELECTROLYZED WATER (EW)

19.3.1 Acidic Electrolyzed Water (AEW)

Acidic EW (AEW) with higher ORP (1000–1200 mV), lower pH (2.5–3.5), and lower free chlorine (30–90 ppm) is produced at the anode, along with HOCl, chlorine gas, oxygen gas, and hydrochloride. In addition, an acidic solution with larger concentrations of free chlorine and ORP is produced by increasing voltage and amperage. AEW is also known as electrolyzed oxidizing water (EOW) owing to its strong oxidation-based capability. It has been recognized as a new disinfectant strategy (Jadeja et al., 2013), and it has been found that AEW has strong antimicrobial properties against certain types of microbes, for example *Salmonella*, *B. cereus*, *E. coli* O157:H7, and *L. monocytogenes* (Hao et al., 2017; Xiong et al., 2010; Ovissipour et al., 2016; Park et al., 2009). Furthermore, AEW revealed potent antifungal efficacy against *A. flavus* by rupturing cell structures and preventing Na+ and K+ from performing their regular duty (Xiong et al., 2014). In addition, it was discovered that on cabbages and green pepper fruits, the application of AEW and AlEW effectively reduced some pesticides, including dimethoate and methamidophos contents (Lin et al., 2006).

19.3.2 Alkaline Electrolyzed Water (AlEW)

The pH of AlEW produced from the cathode is characteristically 10–13, and its ORP value ranges from -800 to -900 mV (Rebezov et al., 2022). AlEW, which is also called electrolyzed reduced water (ERW), is used to eliminate substances which are hard to remove, such as protein and fat content (Hati et al., 2012). Up till now, limited researches have investigated AlEW's decontamination effectiveness (Ayebah et al., 2005; Rahman et al., 2011; Ovissipour et al., 2016). Han et al. (2017) investigated the efficacy of AlEW for eliminating six major pesticidal residual contents (chlorpyrifos, isoprocarb, bifenthrin, difenoconazole, azoxystrobin, and β-cypermethrin) in cowpeas. In contrast with previous treatments, the residual levels of chlorpyrifos, bifenthrin, isoprocarb, difenoconazole, β-cypermethrin, and azoxystrobin were diminished by 85%, 48%, 55%, 58%, 69%, and 75%, respectively, in cowpeas washed in AIEW solutions with pH 12.2 for about 45 min. Furthermore, as washing time treatment increased, progressive enhancement in the elimination of the said pesticide residual contents was observed.

19.3.3 Neutral Electrolyzed Water (NEW)

The anode is used to produce neutral EW (NEW), which has ORP in the range of 750–900 mV and a neutral pH of 7–8. In addition, it may also be generated within single cell–based chambers without separating membranes (Deza et al., 2007; Al-Haq et al., 2005). Sodium hypochlorite, hydrogen peroxide, hydroxyl radical, chlorine, oxygen free radicals, and certain types of disinfection by-products (DBPs) are the leading active ingredients in NEW and possess suitable antibacterial properties (Duan et al., 2016). It has been observed that NEW can effectively inactivate *Escherichia coli* (Jung et al., 2017b), *Salmonella* (Abadias et al., 2008), *Cronobacter sakazakii* (Santo et al., 2016), *Pseudomonas* (Ignat et al., 2016), *Listeria*

(Jung et al., 2017a), and *Yersinia enterocolitica* (Lehto et al., 2017) in fresh produce such as fruits and vegetables.

19.3.4 Slightly Acidic Electrolyzed Water (SAEW)

SAEW has a pH range of 5.0–6.5 and an ORP range of 800–900 mV. In chlorine-based sanitizers, hypochlorite ions (-OCl) are the most active chlorine compounds, although hypochlorous acid (HOCl) is the major effective bactericidal and sporicidal agent (Akther et al., 2023). Owing to its higher ORP, HOCl has 80 times greater bactericidal action in comparison with -OCl. The HOCl enters the cells and degrades cellular organelles by destroying cell membranes, disabling enzymes, and causing DNA damage, showing microbicidal activity (Dewi et al., 2017). In recent times, much research has been conducted regarding the use of SAEW for disinfection purposes because of its established antibacterial activity at even reduced available chlorine concentrations (ACC) of 10–30 mg L^{-1} (Wang et al., 2019; Calvo et al., 2019; Zhang et al., 2021).

19.4 GENERATION OF ELECTROLYZED WATER (EW)

A NaCl solution (KCl/$MgCl_2$ may also be used) must be run through an electrolytic cell, which has a diaphragm or septum separating the cathode and anode, which subsequently produces EW (Iram et al., 2021). SAEW, on the other hand, is generated by HCl or NaCl electrolysis in an electrolytic cell without the separating membranes (Forghani et al., 2015). Figures 19.1a and b depict the generator structure as well as the reactions on the electrodes for EW. When NaCl is dissolved in deionized water, it changes into Na^+ ions with a positive charge and Cl^- ions with a negative charge, and water molecules generate hydrogen (H^+) and hydroxide (OH^-) ions. On the anode's side, negatively charged ions ($Cl^-$and OH^-) release their electrons, leading to the creation of chlorine gas (Cl_2), oxygen gas (O_2), hypochlorous acid (HOCl), hypochlorite ion (ClO^-), and hydrochloric acid (HCl). Positively charged ions (H^+, Na^+) receive electrons from the cathode and create sodium hydroxide (NaOH) and hydrogen gas (H_2).

19.5 POSSIBLE ANTIMICROBIAL MECHANISM OF EW

ORP, pH, and the concentration as well as the form of active chlorine all have a substantial influence on the microbicidal action of EW (Rahman et al., 2016). Different forms of chlorine (HOCl, ClO^-, and Cl_2) work for microbial cell inactivation. The chlorine such as HOCl, $ClO^-$and Cl_2 species attack the cell walls, cell membranes, enzymes, cellular ribosomes, RNA, and associated components of the cell during the course of their germicidal activity (Liao et al., 2017). The anti-microbial efficacy of active chlorine species is shown in Figure 19.2. At pH levels of 5–6.5, hypochlorous acid (HOCl), the most potent form of all chlorine species, demonstrates an 80-fold increase in sanitizing power compared to hypochlorite (ClO^-) (Rahman et al., 2016). HOCl inactivates microbes by penetrating through their cell wall and membrane. The electrical neutrality and molecular size of HOCl ensure its penetration due to its comparable molecular size to water. Plasma membranes contain a hydrophobic lipid bilayer that prevents ionized ClO^- from penetrating microbial cells. It has been demonstrated by Deng et al. (2020) that ClO^- inactivates the key protein of the cell membrane and damages the outer membrane, resulting in

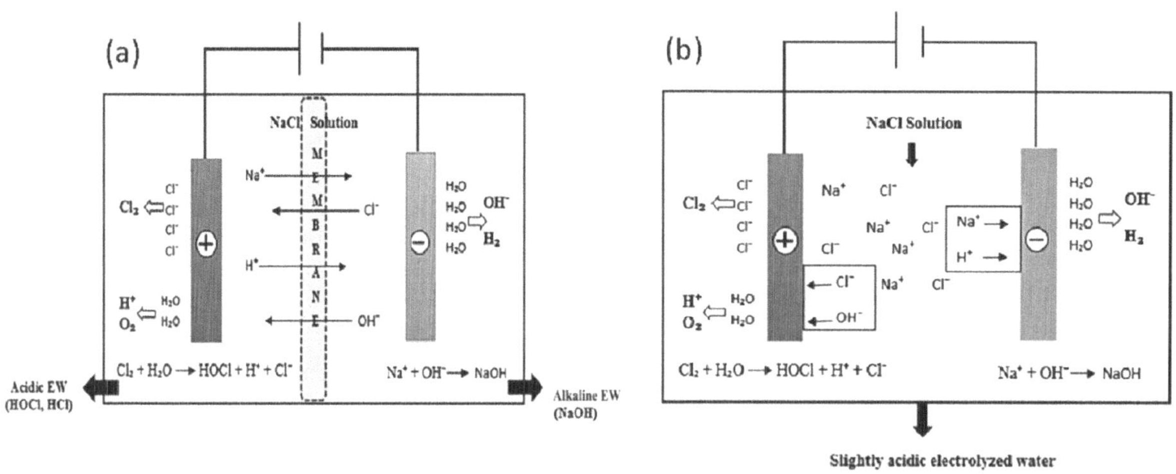

FIGURE 19.1 (a) Schematic of acidic and alkaline EW generation; (b) SAEW generation.

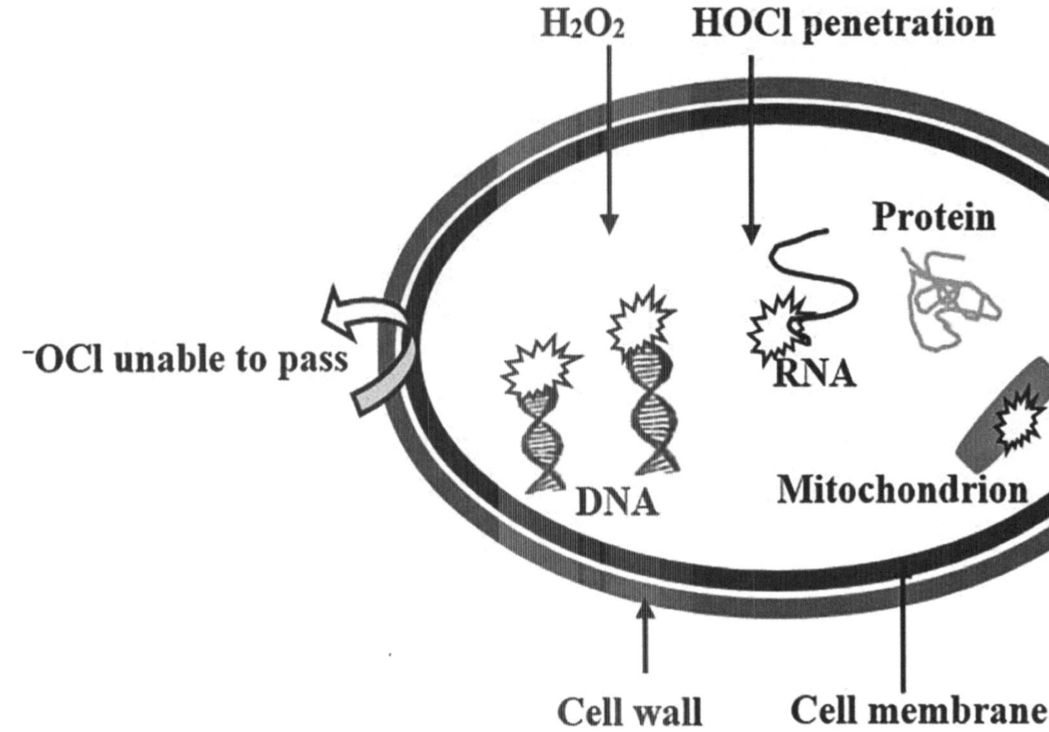

FIGURE 19.2 Antimicrobial mechanism of EW. Adapted from Lu et al. (2022).

cell wall and membrane disintegration. The germicidal activity of HOCl or ClO^- is supposed to be owing to damage to the membrane and DNA, the inactivation of enzymes essential for microbial growth, and the destabilization of the membrane transport system. Furthermore, the proteins and lipids in microbial cell membranes are broken down by additional oxidants such as H_2O_2 created during the electrolysis process, which causes the membranes to quickly inactivate (Zhao et al., 2021).

19.6 FRUIT AND VEGETABLE DISINFECTION

Fruits and vegetables contain a diversity of microorganisms that are capable of causing foodborne illnesses (Leff & Fierer, 2013; Graça et al., 2017). In raw materials where spoilage microbes or pathogens may accumulate, washing is a necessary step before processing (Gil et al., 2015). Pangloli and Hung (2013) reported that blueberries treated with AEW reduced *Escherichia coli* O157:H7 at a faster rate as treatment time increased (from 1 to 3 minutes). AEW can be utilized for treating leafy vegetables as a suitable alternative to chlorine (Yarahmadi et al., 2012). It was discovered that AEW treatment was effective against *S. aureus*, *Enterococcus faecalis*, *S. typhimurium*, *E. coli*, and *L. monocytogenes* when applied to spinach and lettuce (pH 6.3, ACC 100–120 ppm, 10 min exposure duration) (Guentzel et al., 2008). In spinach and lettuce, all bacterial species decreased by 2.43–3.81 and 4.0–5.0 log CFU/mL, respectively. After being treated with AEW (ACC 37.5 mg/L and pH 2.06) for 1 minute at 22 °C, the amount of *Escherichia coli* O157:H7, *Salmonella typhimurium*, and *Listeria monocytogenes* on tomatoes and green onions reduced by 4 log CFU/g, indicating the high microbicidal potential of AEW (Park et al., 2009). According to a different study, lettuce treated with AlEW considerably reduced the amount of *Listeria monocytogenes* throughout its life compared to lettuce that wasn't treated, with a maximum decrease of 1 log CFU/g (Ding et al., 2009). In apples, strawberries, and cherry tomatoes, SAEW has shown strong anti-microbial activity against microorganisms, reducing microbial populations by 2–3 log CFU of fungi and bacteria attached on the fruit's surface. This reduction level is evidently higher than that of conventional chlorine treatment, which achieves less than 2 log (Pinto et al., 2016). In comparison to the control, the SAEW treatment decreased the overall aerobic bacterial, yeast, and mold population in cauliflower by 2.21 log CFU/g and 1.17 log CFU/g (Akther et al., 2023). Other research demonstrated that SAEW might reduce the overall number of aerobic bacteria on celery surfaces by 4.33 log CFU (Zhang et al., 2016a, b). NEW treatment on tomatoes was shown to be efficient in reducing *L. monocytogenes*, *S. typhimurium*, *Salmonella enteritidis*, and *E. coli*, resulting in a higher than 5 log CFU/cm^2 reduction. Table 19.1 shows the disinfecting potential of several types of EW against fruits and vegetables.

TABLE 19.1 Applications of EW in Disinfecting Various Vegetables and Fruits

MICROORGANISM	*EW TYPE*	*FOOD COMMODITIES*	*PHYSIOCHEMICAL PROPERTIES OF EW (ACC-MG/L, ORP-MV)*	*WASHING CONDITIONS*	*REDUCTION (LOG CFU/G)*	*REFERENCE*
Cronobacter sakazakii	AEW	Apple, pear, and melon	pH-2.82 ACC-103 ORP-1121	5 min, 150 rpm	1.3–1.8	Santo et al. (2016)
Escherichia coli	AEW	Apple	pH-2.9 ACC-98 ORP-1128	30 min, 4°C	2.1	Graca et al. (2011)
Escherichia coli O157:H7	AEW	Strawberry	pH-2.7 ACC-55.5 ORP-1067.6	5 min, 4°C	1.58	Hung et al. (2010)
Escherichia coli O157:H7 *Listeria Monocytogenes* *Salmonella*	AEW	Lettuce	pH-2.89 ACC-13 ORP-1097	3 min, 15°C	3.43 2.55 3.72	Fishburn et al. (2012)
Total aerobic bacteria Yeasts and molds Coli-forming bacteria	AEW	Cilantro	pH-2.48 ACC-79.35 ORP-1134	5 min	2.58 2.04 1.31	Hao et al. (2011a, b)
E. coli O157:H7	AEW	Broccoli	pH-2.5 ACC-54.1 ORP-1108.6	5 min, 24°C	1.5	Hung et al., (2010)
Total aerobic bacteria *E. coli: 078*	AlEW	Cilantro	pH-11.6 ACC-NA ORP-824	5 min, 25°C	1.06 0.5	Hao et al. (2015a, b)
L. monocytogenes *E. coli* O157:H7	AlEW	Carrot	pH-11.3 ACC-NA ORP—(−810)	3min, 50°C	2.7 2.6	Rahman and others (2011)
L. monocytogenes *E. coli* O157:H7 Total bacteria count Yeast and mold	AlEW	Cabbage	pH-11 to 11.2 ACC-NA ORP- (−830 to 850)	5min, 50°C	2.6 2.6 3.0 2.7	Rahman and others (2010)
Escherichia coli	NEW	Mango	pH-7.95 ACC-101 ORP-797	5 min, 150 rpm	2.2	Santo et al. (2018)
Cronobacter sakazakii	NEW	Apple, pear, and melon	pH-8.18 ACC-94 ORP-762	5 min, 150 rpm	1–1.2	Santo et al. (2016)
A. acidoterrestris	NEW	Apple	pH-7.5 ACC-48 ORP-770	5 min, 20°C	<2/apple	Torlak (2014)
S. choleraesuis *L. innocua* *E. coli*	NEW	Apple	pH-8.3 ACC-49 ORP-753	30 min, 4°C	1.5	Graca et al. (2011)
Salmonella Enterica *Listeria Monocytogenes* *Escherichia coli O157:H7*	NEW	Lettuce	pH-6.83 ACC-43 ORP–	5 min	2.8–3.0 2.6–3.4 3.4–3.7	Jung et al. (2017a, b)
Salmonella Typhimurium *Escherichia coli O157:H7*	NEW	Tomato	pH-7.52 ACC-155 ORP-760	5 min, 65 rpm	8.0/tomato 4.4/tomato	Afari et al. (2015)

(Continued)

TABLE 19.1 (Continued)

MICROORGANISM	EW TYPE	FOOD COMMODITIES	PHYSIOCHEMICAL PROPERTIES OF EW (ACC-MG/L, ORP-MV)	WASHING CONDITIONS	REDUCTION (LOG CFU/G)	REFERENCE
Escherichia coli O157:H7 *Listeria Monocytogenes* Total aerobic bacteria	SAEW	Apple	pH-5.42 ACC-30 ORP-818–854	23°C, 3 min	2.28/fruit 2.30/fruit 1.89/fruit	Tango et al. (2017)
Total bacteria Yeasts and molds	SAEW	Strawberry	pH-6.49 ACC-34.33 ORP-853.7	25°C, 10 min	2.32 3.01	Ding et al. (2015)
Total bacteria Yeasts and molds	SAEW	Cherry tomato	pH-6.4 ACC-34.3 ORP-853.7	25°C, 10 min	1.4	Ding et al. (2015)
Total bacteria Yeasts and molds	SAEW	Cauliflower	pH-5.5–5.6 ACC-40 ± 1.5 ORP-(800–900)	25°C, 10 min	2.21 1.17	Akther S et al. 2023
Total microbial count	SAEW	Lettuce	pH-6.4 ACC-30 ORP-562	20°C, 10 min	1.9	Park et al. (2017)
Total microbial count	SAEW	Kale	pH-6.4 ACC-30 ORP-562	20°C, 10 min	1.7	Park et al. (2017)
Total viable count	SAEW	Lettuce	6.2–6.9 20 820–960	1 min, 20°C	1.9	Xuan et al. (2016)

19.7 PESTICIDE RESIDUE REMOVAL FROM FRUITS AND VEGETABLES

Agricultural pesticides are widely used globally to ensure crop quality and productivity by controlling weeds, eliminating pests, and eliminating insects and fungus (Cengiz et al., 2018). Their widespread use has led to them becoming significant food contaminants, posing a serious health risk to humans. Of all the ways that individuals might be exposed to pesticides, eating fresh fruit and vegetable products directly is the most likely one (Azam et al., 2020). Some pesticides have been linked with carcinogenesis, reproductive disorders, birth defects, endocrine disruption, and cardiovascular disease in laboratory research and epidemiological studies (Mostafalou & Abdollah, 2013). However, because plants are lipophilic, most pesticides do not reach their inner layers. The outer layers of the veggies are where the pesticides mainly adhere. Therefore, peeling or removing the skins of most vegetables can eliminate the majority of pesticides on them; nonetheless, certain fresh vegetables are often consumed without removing their skins (Pandiselvam et al., 2020). In order to eliminate pesticides from vegetables, washing is the most practical and effectice method for both home and commercial vegetable processing. Producers of fruits and vegetables depend on chlorine as one of the best sanitizers to guarantee the safety of their goods; the most widely used source of chlorine for a long time was sodium hypochlorite (NaOCl). Qi et al. (2018) reported that in comparison to conventional sodium hypochlorite–based sanitizers, electrolyzed oxidizing water, an improved chlorine-based solution, can efficiently remove pesticide residues from fresh fruit. Longer washing sessions typically result in the removal of pesticide residues more successfully. This is because longer washing durations cause more insecticides to break down (Han et al., 2017). In addition, EW possesses excellent emulsification properties. EW removes lipid-soluble pesticides easily since they contain hydrophobic groups that are poorly soluble in water (Bhilwadikar et al., 2019). Thermally unstable pesticides are removed more effectively as the temperature rises, since the washing effect increases. Pesticides that are thermally unstable break down faster at high temperatures.

19.8 DEGRADATION OF PESTICIDE RESIDUES

Various pesticide residues from fruits and vegetables can be reduced or removed by EW, as shown in Table 19.2. The findings of a research by Yuan Liu et al. (2021) suggested that

TABLE 19.2 Efficacy of Electrolyzed Water on Pesticide Removal of Fruits and Vegetables

PESTICIDES	EW TYPE	FOOD COMMODITIES	PHYSIOCHEMICAL PROPERTIES OF EW (ACC-MG/L, ORP-MV)	WASHING CONDITIONS	PERCENTAGE REDUCTION (%)	REFERENCE
Phorate, chlorpyrifos, lambda-cyhalothrin, cyfluthrin, procymidone, and chlorothalonil	AEW	Cabbage and broccoli florets	pH-2.9 ACC-100 ORP-1150	20 min	53.5–81.5 35.5–78.9	Liu, Y et al. (2021)
Diazinon, cyprodinil, and phosmet	AEW	Spinach and snap beans	pH-2.8 ORP-1151.5 ACC-120	1–15 min	37.1–85.7	Qi et al. (2018)
Chlorpyrifos	AlEW	Cowpeas	pH-10.1 ACC– OREP-(−541)	5 min	9.5	Han et al. (2017)
Lambdacyhalothrin	AlEW	Apples	pH-11.0±0.1 ACC– ORP-(−830)	90 min	67.9	Liu et al. (2015)
Carbendazim	NEW	Cherry tomatoes	pH-11.4 ACC-250 ORP-260	30 min	52.80	Pattanapo et al. (2020)
Azoxystrobin, bifenthrin, tebuconazole, buprofezin, etc.	NEW	Kumquats	pH-12.5 ACC- ORP-86	20 min	35.8–83.9	Yang et al. (2020)
Cyprodinil, tebuconazole, and iprodione	SAEW	Stone fruits	pH-6.7 ACC-165 ORP-775	15 min	<40	Calvo et al. (2019)
Imidacloprid	SAEW	Lettuce	pH-6.38 ACC-50 ORP-875	0.75 min	48.57	Cap et al. (2020)
Chlorpyrifos	SAEW	Leeks	pH-6.0 ACC-79.66 ORP-	15 min	72.55	Hu et al. (2016)

fresh-cut vegetables may successfully have fungicide, pyrethroid, and organophosphate residues removed without compromising their texture using EW treatment. AlEW treatment reduced carbendazim residue significantly (Pattanapo et al., 2020). Cherry tomatoes that were washed with 100, 250, 500, and 1000 mg/L AlEW for 30 minutes showed reductions in carbendazim residue of 60.36%, 80.49%, 67.32%, and 64.68%, respecticely. In a study by Yang et al. (2020), AlEW was found to be more effective than tap water at eliminating 10 types of pesticide residues from kumquats. The effect of AlEW on removing pesticides from apples was studied by Liu et al. (2015). The results showed that 67.9% of pesticide residues were effectively degraded by AlEW. According to Calvo et al. (2019), near-neutral EW may not be highly effective in degrading pesticide residues. SAEW reduced imidacloprid concentrations in lettuce by 48.57% after 0.75 minutes of treatment (Cap et al., 2020). Qi et al. (2018) reported that AEW removed pesticide residues in fresh produce more efficiently than AlEW. Hu et al. (2016) investigated the effect of SAEW on chlorpyrifos reduction in leeks. After soaking for 15 minutes, SAEW treatments effectively reduced pesticide residues by 72.55%. The use of EW reduced pesticide residues on fresh vegetable surfaces by 46–86%, which is better than detergents and tap water (Hao et al., 2011).

Although the aforementioned research suggests that EW has the capacity to eliminate pesticide residual constituents, the process by which pesticide residues degrade in fruits and vegetables remains unknown. The pH, ACC, and treatment duration of EW are the primary determinants of the efficiency of pesticide residue elimination. The structural variations across pesticides largely dictate how treatment duration affects the effectiveness of pesticide removal. Organophosphorus pesticides, such as acephate, generally consist of C=O and P=S double bonds. Because AlEW has a high pH and better emulsifying qualities than AEW, which possesses a lower pH and a higher ORP value, a nucleophilic reaction in an acidic or alkaline environment breaks double bonds (Bhilwadikar et al., 2019). Especially for pesticides with double bonds, EW has a great ability to eliminate them. Figure 19.3 depicts the mechanisms of pesticide residue degradation from fruits and vegetables. Therefore, without sacrificing quality, EW is a viable treatment technique for eliminating pesticide residue from fruits and vegetables. To assure future progress, more study is necessary to elucidate the precise process of pesticide breakdown by EW.

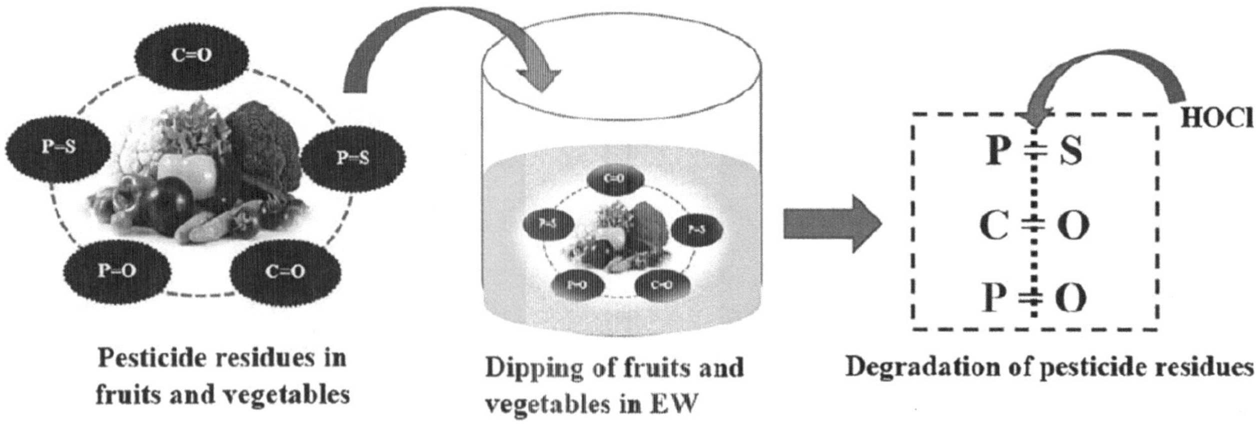

FIGURE 19.3 Degradation mechanism of pesticide residues.

19.9 ADVANTAGES AND DISADVANTAGES OF EW

EW has shown to be a successful alternative for hazardous pesticides in the degradation of pesticide residues and for the management of microbial contaminations and enhancement of fruit and vegetable quality-related attributes during postharvest. The advantages of EW are as follows: (1) Compared to chlorine, it is more efficient. (2) Because EW is made by electrolyzing distilled water with lower salt contents, it is simple to use. (3) EW is considered safer for the environment and human health, with few side effects on either (Chakka et al., 2021). (4) Using EW lowers the expenses and dangers involved in handling, storing, and applying chlorine-based solutions. (5) Because there are very few operating expenses beyond the initial cost of the electrolysis unit, using EW is highly cost-effective. (6) Since EW is a non-thermal method, fruits and vegetables retain their natural flavor, aroma, color, and texture (Rahman et al., 2016). (7) It is simple to modify the concentration of chlorine to meet the needs of a particular application. (8) After application, it may be changed back into ordinary water without producing any hazardous fumes. (9) On-site generation of EW reduces issues with storage and shipment. (10) It can shorten cleaning times by inactivating a wide range of non-selective bacteria (Hricova et al., 2008). However, consideration must be given to the disadvantages and potential drawbacks, which include: (1) EW may lose its antibacterial properties during storage at high temperatures or prolonged storage due to its self-decomposition. (2) If HOCl, Cl_2, and H^+ are not constantly provided by electrolysis, EW won't have antibacterial activity. (3) The initial costs of equipment purchasing and installation may be comparatively higher. (4) The organic matter presence (for example, proteins) will reduce the effectiveness of EW since it will react with free chlorine and decrease the availability of chlorine to disinfect. (5) EW with low pH and high ORP cause corrosion of equipment and can hurt hands (Lu et al., 2022).

19.10 CHALLENGES AND FUTURE PROSPECTS

The effectiveness of EW in obliterating pathogenic or spoilage microorganisms and pesticide residues on fruit and vegetable commodities without disturbing nutritional or sensory quality and extending shelf life has made it a promising choice. The United States Department of Agriculture (USDA) has approved EW as safe, and it may progressively gain worldwide acceptance in the near future. Many start-ups and industries around the globe have recently begun marketing and commercializing EW. For example, the American company Aqua Ox LLC makes two varieties of EW with varying quantities of hypochlorous acid that are used in food plants, in medicine, and to prevent and cure plant illnesses. Stores will soon carry EW for use as a fruit and vegetable cleaner and disinfectant in the kitchen. Much study has been conducted recently on the antibacterial properties of individual EW. Research is continuously being conducted to improve antibacterial effect of EW by integrating it with other technologies (UVC, O_3, ultrasound, and short heating time). Future research will focus on exploring certain other related approaches (modified atmosphere packaging and chitosan) and how EW can be combined to optimize quality, maximize efficiency, and maximize resources. Over the next decade, EW will also be broadly encouraged in certain other decontamination fields because of its numerous advantages (Lu et al., 2022). Given that it can lessen microbial and pesticide contamination, EW application for food safety is one of the most likelypotential uses of future sanitizers. Numerous research articles have suggested using AEW to disinfect surfaces (Moorman et al., 2017). As a result, EW may be effective at reducing exposure times in cases of novel coronaviruses (e.g., COVID-19). It has been discussed in scientific literature whether EW is capable of destroying the different virus structures, and it is recommended that research be done to determine how it affects COVID-19.

19.11 CONCLUSIONS

Electrolyzed water is a suitable technology with promising potential to increase the safety and shelf life of fresh produce. It can efficiently remove microorganisms and pesticide residues from most fruit and vegetable commodities. In general, EW has a modest impact on humans and the environment, leaving no hazardous residues in the food matrix. Furthermore, as an innovative nonthermal technology, it has high decontamination efficacy and simple operation over other traditional methods. The application potential of EW to inactivate spores of microbes, microbial toxins, and biofilms should be examined in detail. Finally, further investigation is required for the development of effective EW-integrated technologies for disinfecting, cleaning, preserving, and extending the shelf life of fruit- and vegetable-based produce while attaining the balance between quality assurance and safety status.

REFERENCES

Abadias, M., Usall, J., Oliveira, M., Alegre, I., Viñas, I. (2008). Efficacy of neutral electrolyzed water (NEW) for reducing microbial contamination on minimally-processed vegetables. *International Journal of Food Microbiology*, *123*(1–2), 151–158.

Afari, G.K., Hung, Y.C., King, C.H. (2015). Efficacy of neutral pH electrolyzed water in reducing *Escherichia coli* O157: H7 and *Salmonella Typhimurium* DT 104 on fresh produce items using an automated washer at simulated food service conditions. *Journal of Food Science*, *80*(8), M1815–M1822.

Ahmed, S., Akther, S., Alam, S.S., Ahiduzzaman, M., Islam, M.N., Azam, M.S. (2022). Individual and combined effects of electrolyzed water and ultrasound treatment on microbial decontamination and shelf life extension of fruits and vegetables: A review of potential mechanisms. *Journal of Food Processing and Preservation*, *46*(8), e16765.

Akther, S., Islam, M.R., Alam, M., Alam, M.J., Ahmed, S. (2023). Impact of slightly acidic electrolyzed water in combination with ultrasound and mild heat on safety and quality of fresh cut cauliflower. *Postharvest Biology and Technology*, *197*, 112189.

Al-Haq, M.I., Sugiyama, J., Isobe, S. (2005). Applications of electrolyzed water in agriculture and food industries. *Food Science and Technology Research*, *11*(2), 135–150.

Ayebah, B., Hung, Y.C, Frank, J.F. (2005). Enhancing the bactericidal effect of electrolyzed water on Listeria monocytogenes biofilms formed on stainless steel. *Journal of Food Protection*, *68*(7), 1375–1380.

Azam, S.R., Ma, H., Xu, B., Devi, S., Siddique, M.A.B., Stanley, S.L., Zhu, J. (2020). Efficacy of ultrasound treatment in the removal of pesticide residues from fresh vegetables: A review. *Trends in Food Science & Technology*, *97*, 417–432.

Bhilwadikar, T., Pounraj, S., Manivannan, S., Rastogi, N.K., Negi, P.S. (2019). Decontamination of microorganisms and pesticides from fresh fruits and vegetables: A comprehensive review from common household processes to modern techniques. *Comprehensive Reviews in Food science and Food Safety*, *18*(4), 1003–1038.

Boal, A.K. (2009). On-site generation of disinfectants. *National Environmental Services Center*, *9*(1), 1–4.

Calvo, H., Redondo, D., Remon, S., Venturini, M.E., Arias, E. (2019). Efficacy of electrolyzed water, chlorine dioxide and photocatalysis for disinfection and removal of pesticide residues from stone fruit. *Postharvest Biology and Technology*, *148*, 22–31.

Cap, M., Rojas, D., Fernandez, M., Fulco, M., Rodriguez, A., Soteras, T., Mozgovoj, M. (2020). Effectiveness of short exposure times to electrolyzed water in reducing salmonella spp. and imidacloprid in lettuce. *LWT-Food Science and Technology*, *128*, 109496.

Cengiz, M.F., Başlar, M., Basançelebi, O., Kılıçlı, M. (2018). Reduction of pesticide residues from tomatoes by low intensity electrical current and ultrasound applications. *Food Chemistry*, *267*, 60–66.

Chakka, A.K., Sriraksha, M.S., Ravishankar, C.N. (2021). Sustainability of emerging green non-thermal technologies in the food industry with food safety perspective: a review. *LWT—Food Science and Technology*, *151*, 112140.

Deng, L.Z., Mujumdar, A.S., Pan, Z., Vidyarthi, S.K., Xu, J., Zielinska, M., Xiao, H.W. (2020). Emerging chemical and physical disinfection technologies of fruits and vegetables: a comprehensive review. *Critical Reviews in Food Science and Nutrition*, *60*(15), 2481–2508.

Dewi, F.R., Stanley, R., Powell, S.M., Burke, C.M. (2017). Application of electrolysed oxidising water as a sanitiser to extend the shelf-life of seafood products: a review. *Journal of Food Science and Technology*, *54*(5), 1321–1332.

Deza, M.A., Araujo, M., Garrido, M.J. (2007). Efficacy of neutral electrolyzed water to inactivate Escherichia coli, Listeria monocytogenes, Pseudomonas aeruginosa, and Staphylococcus aureus on plastic and wooden kitchen cutting boards. *Journal of Food Protection*, *70*(1), 102–108.

Ding, T., Ge, Z., Shi, J., Xu, Y.T., Jones, C.L., Liu, D.H. (2015). Impact of slightly acidic electrolyzed water (SAEW) and ultrasound on microbial loads and quality of fresh fruits. *LWT-Food Science and Technology*, *60*(2), 1195–1199.

Ding, T., Jin, Y.G., Oh, D.H. (2009). Predictive model for growth of Listeria monocytogenes in untreated and treated lettuce with alkaline electrolyzed water. *World Journal of Microbiology and Biotechnology*, *26*(5), 863–869.

Duan, D., Liu, G., Yao, P., Wang, H. (2016). The effects of organic compounds on inactivation efficacy of Artemia salina by neutral electrolyzed water. *Ocean Engineering*, *125*, 31–37.

Finger, J.A.F.F., Baroni, W.S.G.V., Maffei, D.F., Bastos, D.H.M., Pinto, U.M., (2019). Overview of foodborne disease outbreaks in Brazil from 2000 to 2018. *Foods*, *8*(10), 434.

Fishburn, J.D., Tang, Y., Frank, J.F. (2012). Efficacy of various consumer-friendly produce washing technologies in reducing pathogens on fresh produce. *Food Protection Trends*, *32*(8), 456–466.

Forghani, F., Park, J.H., Oh, D.H. (2015). Effect of water hardness on the production and microbicidal efficacy of slightly acidic electrolyzed water. *Food Microbiology*, *48*, 28–34.

Gil, M.I., Gómez-López, V.M., Hung, Y.C., Allende, A. (2015). Potential of electrolyzed water as an alternative disinfectant agent in the fresh-cut industry. *Food and Bioprocess Technology*, *8*, 1336–1348.

Graca, A., Abadias, M., Salazar, M., Nunes, C. (2011). The use of electrolyzed water as a disinfectant for minimally processed apples. *Postharvest Biology and Technology, 61(2–3)*, 172–177.

Graça, A., Esteves, E., Nunes, C., Abadias, M., Quintas, C. (2017). Microbiological quality and safety of minimally processed fruits in the marketplace of southern Portugal. *Food Control*, *73*, 775–783.

Guentzel, J.L., Lam, K.L., Callan, M.A., Emmons, S.A., Dunham, V.L. (2008). Reduction of bacteria on spinach, lettuce, and surfaces in food service areas using neutral electrolyzed oxidizing water. *Food Microbiology*, *25*, 36–41.

Han, Y., Song, L., An, Q., Pan, C. (2017). Removal of six pesticide residues in cowpea with alkaline electrolyzed water. *Journal of the Science of Food and Agriculture*, *97*(8), 2333–2338.

Hao, J., Liu, H., Liu, R.U.I., Dalai, W., Zhao, R., Chen, T., Li, L. (2011). Efficacy of slightly acidic electrolyzed water (SAEW) for reducing microbial contamination on fresh-cut cilantro. *Journal of Food Safety*, *31*(1), 28–34.

Hao, J., Wu, T., Li, H., Liu, H. (2017). Differences of bactericidal efficacy on *Escherichia coli*, *Staphylococcus aureus*, and *Bacillus subtilis* of slightly and strongly acidic electrolyzed water. *Food and Bioprocess Technology*, *10*, 155–164.

Hati, S., Mandal, S., Minz, P.S., Vij, S., Khetra Y., Singh, B.P., Yadav, D. (2012). Electrolyzed oxidized water (EOW): non-thermal approach for decontamination of food borne microorganisms in food industry. *Food and Nutrition Sciences*, *3*(6), 760–768.

Hesami, A., Kavoosi, S., Khademi, R., Sarikhani, S. (2021). Effect of chitosan coating and storage temperature on shelf-life and fruit quality of *Ziziphus mauritiana*. *International Journal of Fruit Science*, *21*(1), 509–518.

Hricova, D., Stephan, R., Zweifel, C. (2008). Electrolyzed water and its application in the food industry. *Journal of Food Protection*, *71*(9), 1934–1947.

Hu, Z.H., Wu, T.J., Wan, Y.F., Hao, J., Liu, H. (2016). Study on the removal of dimethoate and chlorpyrifos in leek by slightly acidic electrolyzed water. *Science and Technology of Food Industry*, *37*(01), 49–52.

Ignat, A., Manzocco, L., Maifreni, M., Nicoli, M.C. (2016). Decontamination efficacy of neutral and acidic electrolyzed water in fresh-cut salad washing. *Journal of Food Processing and Preservation*, *40*(5), 874–881.

Iram, A., Wang, X.M., Demirci, A. (2021). Electrolyzed oxidizing water and its applications as sanitation and cleaning agent. *Food Engineering Reviews*, *13*, 41–27.

Jadeja, R., Hung, Y.C., Bosilevac, J.M. (2013). Resistance of various shiga toxin-producing Escherichia coli to electrolyzed oxidizing water. *Food Control*, *30*(2), 580–584.

Jung, Y., Jang, H., Guo, M., Gao, J., Matthews, K.R. (2017a). Sanitizer efficacy in preventing cross-contamination of heads of lettuce during retail crisping. *Food Microbiology*, *64*, 179–185.

Jung, Y., Gao, J., Jang, H., Guo, M., Matthews, K.R. (2017b). Sanitizer efficacy in preventing cross-contamination during retail preparation of whole and fresh-cut cantaloupe. *Food Control*, *75*, 228–235.

Keskinen, L.A., Burkea, A., Annous, B.A. (2009). Efficacy of chlorine, acidic electrolyzed water and aqueous chlorine dioxide solutions to decontaminate Escherichia coli O157:H7 from lettuce leaves. *International Journal of Food Microbiology*, *132*, 134–140.

Kumar, G.D., Williams, R.C., Sumner, S.S., Eifert, J. D. (2016). Effect of ozone and ultraviolet light on Listeria monocytogenes populations in fresh and spent chill brines. *Food Control*, *59*, 172–177.

Leff, J.W., Fierer, N. (2013). Bacterial communities associated with the surfaces of fresh fruits and vegetables. *PloS One*, *8*(3), e59310.

Lehto, M., Kuisma, R., Kymalainen, H.R. (2017). Neutral electrolyzed water (NEW), chlorine dioxide, organic acid based product, and ultraviolet-C for inactivation of microbes in fresh-cut vegetable washing waters. *Journal of Food Processing and Preservation*, *42*(1), 13354.

Li, J., Ding, T., Liao, X., Chen, S., Ye, X., Liu, D. (2017). Synergetic effects of ultrasound and slightly acidic electrolyzed water against Staphylococcus aureus evaluated by flow cytometry and electron microscopy. *Ultrasonics Sonochemistry*, *38*, 711–719.

Li, H., Xu, Z., Zhao, F., Wang, Y., Liao, X. (2016). Synergetic effects of high-pressure carbon dioxide and nisin on the inactivation of Escherichia coli and Staphylococcus aureus. *Innovative Food Science & Emerging Technologies*, *33*, 180–186.

Liao, X., Liu, D., Xiang, Q., Ahn, J., Chen, S., Ye, X., Ding, T. (2017). Inactivation mechanisms of non-thermal plasma on microbes: A review. *Food Control*, *75*, 83–91.

Lin, C.S., Tsai, P.J., Wu, C., Yeh, J.Y., Saalia, F.K. (2006). Evaluation of electrolysed water as an agent for reducing methamidophos and dimethoate concentrations in vegetables. *International Journal of Food Science & Technology*, *41*(9), 1099–1104.

Liu, H.J., Li, R.Z., Su, D.H. (2015). Degradation of Lambda-cyhalothrin in fruits and vegetable by alkaline electrolyzed water. *Food Science Technology*, *40*(2), 123–127.

Liu, Y., Wang, J., Zhu, X., Liu, Y., Cheng, M., Xing, W., Song, P. (2021). Effects of electrolyzed water treatment on pesticide removal and texture quality in fresh-cut cabbage, broccoli, and color pepper. *Food Chemistry*, *353*, 129408.

Lopez-Gomez, A., Fernández, P.S., Palop, A., Periago, P.M., Martinez-López, A., Marin-Iniesta, F., Barbosa-Cánovas, G. V. (2009). Food safety engineering: an emergent perspective. *Food Engineering Reviews*, *1*, 84–104.

Lu, L., Guo, H., Kang, N., He, X., Liu, G., Li, J., Yu, H. (2022). Application of electrolysed water in the quality and safety control of fruits and vegetables: a review. *International Journal of Food Science & Technology*, *57*(9), 5698–5711.

Ma, T., Lan, T., Ju, Y., Cheng, G., Que, Z., Geng, T., Fang, Y., Sun, X. (2019). Comparison of the nutritional properties and biological activities of kiwifruit (Actinidia) and their different forms of products: towards making kiwifruit more nutritious and functional. *Food & Function*, *10*(3), 1317–1329.

McKellar, R.C., Odumeru, J., Zhou, T., Harrison, A., Mercer, D.G., Young, J.C., Karr, S. (2004). Influence of a commercial warm chlorinated water treatment and packaging on the shelf-life of ready-to-use lettuce. *Food Research International*, *37*(4), 343–354.

Moorman, E., Montazeri, N., Jaykus, L.A. (2017). Efficacy of neutral electrolyzed water for inactivation of human norovirus. *Applied and Environmental Microbiology*, *83*(16), e00653–17.

Mostafalou, S., Abdollahi, M. (2013). Pesticides and human chronic diseases: evidences, mechanisms, and perspectives. *Toxicology and Applied Pharmacology*, *268*(2), 157–177.

Murray, K., Wu, F., Shi, J., Jun Xue, S., Warriner, K. (2017). Challenges in the microbiological food safety of fresh produce: Limitations of post-harvest washing and the need for alternative interventions. *Food Quality and Safety*, *1*(4), 289–301.

Osaili, T.M., Al-Nabulsi, A. (2016). Inactivation of stressed Escherichia coli O157:H7 in tahini (sesame seeds paste) by gamma irradiation. *Food Control*, *69*, 221–226.

Ovissipour, M., Al-Qadiri, H.M., Sablani, S.S., Govindan, B.N., Al-Alami, N., Rasco, B. (2016). Efficacy of acidic and alkaline electrolyzed water for inactivating Escherichia coli O104: H4, Listeria monocytogenes, Campylobacter jejuni, Aeromonas hydrophila, and Vibrio parahaemolyticus in cell suspensions. *Food Control*, *53*, 117–123.

Pandiselvam, R., Kaavya, R., Jayanath, Y., Veenuttranon, K., Lueprasitsakul, P., Divya, V., Ramesh, S. V. (2020). Ozone as a novel emerging technology for the dissipation of pesticide residues in foods–a review. *Trends in Food Science & Technology*, *97*, 38–54.

Pangloli, P., Hung, Y.C. (2013). Reducing microbiological safety risk on blueberries through innovative washing technologies. *Food Control*, *32*(2), 621–625.

Park, E.J., Alexander, E., Taylor, G.A., Costa, R., Kang, D.H. (2009). The decontaminative effects of acidic electrolyzed water for Escherichia coli O157: H7, Salmonella typhimurium, and Listeria monocytogenes on green onions and tomatoes with differing organic demands. *Food Microbiology*, *26*(4), 386–390.

Park, S.Y., Zhang, C.Y., Ha. S. (2017). Predictive modeling for the growth of salmonella enterica serovar typhimurium on lettuce washed with combined chlorine and ultrasound during storage. *Journal of Food Hygiene and Safety*, *34*(4), 374–379

Pattanapo, W., Whangchai, K., Uthaibutra, J., Pankasemsuk, T., Chanasut, U. (2020). Effect of electrolyzed reducing water with various concentration on carbendazim removal in cherry tomato. *International Journal of Geomate*, *18*, 54–59.

Pinto, L., Baruzzi, F., Ippolito, A. (2016). Recent advances to control spoilage microorganisms in washing water of fruits and vegetables: The use of electrolyzed water. In *III International Symposium on Postharvest Pathology: Using Science to Increase Food Availability 1144*, 379–384.

Qi, H., Huang, Q.G., Hung, Y.C. (2018). Effectiveness of electrolyzed oxidizing water treatment in removing pesticide residues and its effect on produce quality. *Food Chemistry*, *239*, 561–568.

Qiang, Z., Demirkol, O., Ercal, N., Adams, C. (2005). Impact of food disinfection on beneficial biothiol contents in vegetables. *Journal of Agricultural and Food Chemistry*, *53*, 9830–9840.

Qin, M., Fu, Y., Li, N., Zhao, Y., Yang, B., Wang, L., Ouyang, S. (2022). Effects of wheat tempering with slightly acidic electrolyzed water on the microbiota and flour characteristics. *Foods*, *11*(24), 3990.

Rahman, S.M.E., Jin, Y.G., Oh, D.W. (2010). Combined effects of alkaline electrolyzed water and citric acid with mild heat to control microorganisms on cabbage. *Journal of Food Science*, *75*, M111–M115.

Rahman, S.M.E., Jin, Y., Oh, D. (2011). Combination treatment of alkaline electrolyzed water and citric acid with mild heat to ensure microbial safety, shelf-life and sensory quality of shredded carrots. *Food Microbiology*, *28*(3), 484–491.

Rahman, S.M.E., Khan, I., Oh, D.H. (2016). Electrolyzed water as a novel sanitizer in the food industry: current trends and future perspectives. *Comprehensive Reviews in Food Science and Food Safety*, *15*(3), 471–490.

Rebezov, M., Saeed, K., Khaliq, A., Rahman, S.J.U., Sameed, N., Semenova, A., Lorenzo, J. M. (2022). Application of electrolyzed water in the food industry: a review. *Applied Sciences*, *12*(13), 6639.

Robinson, G., Thorn, R., Reynolds, D. (2012). The effect of long-term storage on the physiochemical and bactericidal properties of electrochemically activated solutions. *International Journal of Molecular Sciences*, *14*(1), 457–469.

Santo, D., Graça, A., Nunes, C., Quintas, C. (2016). Survival and growth of Cronobacter sakazakii on fresh-cut fruit and the effect of UV-C illumination and electrolyzed water in the reduction of its population. *International Journal of Food Microbiology*, *231*, 10–15.

Santo, D., Graça, A., Nunes, C., Quintas, C. (2018). Escherichia coli and Cronobacter sakazakii in 'Tommy Atkins' minimally processed mangos: Survival, growth and effect of UV-C and electrolyzed water. *Food Microbiology*, *70*, 49–54.

Tango, C.N., Khan, I., Kounkeu, P.F.N., Momna, R., Hussain, M.S., Oh, D.H. (2017). Slightly acidic electrolyzed water combined with chemical and physical treatments to decontaminate bacteria on fresh fruits. *Food Microbiology*, *67*, 97–105.

Torlak E. (2014). Inactivation of *Alicyclobacillus acidoterrestris* spores in aqueous suspension and on apples by neutral electrolyzed water. *International Journal of Food Microbiology*, *185*, 69–72.

Wang, H., Duan, D., Wu, Z., Xue, S., Xu, X., Zhou, G. (2019). Primary concerns regarding the application of electrolyzed water in the meat industry. *Food Control*, *95*, 50–56.

Xiang, Q., Kang, C., Niu, L., Zhao, D., Li, K., Bai, Y. (2018). Antibacterial activity and a membrane damage mechanism of plasma-activated water against Pseudomonas deceptionensis CM2. *LWT*, *96*, 395–401. doi:10.1016/j.lwt.2018.05.059

Xiong, K., Li, X.T., Guo, S., Li, L. T., Liu, H.J. (2014). The antifungal mechanism of electrolyzed oxidizing water against Aspergillus flavus. *Food Science and Biotechnology*, *23*, 661–669.

Xiong, K., Liu, H.J., Liu, R. (2010). Differences in fungicidal efficiency against Aspergillus flavus for neutralized and acidic electrolyzed oxidizing waters. *International Journal of Food Microbiology*, *137*(1), 67–75.

Xuan, X.T., Fan, Y.F., Ling, J.G., Hu, Y.Q., Liu, D.H., Chen, S.G., Ding, T. (2017). Preservation of squid by slightly acidic electrolyzed water ice. *Food Control*, *73*, 1483–1489.

Yang, J., Song, L., Pan, C.P., Han, Y.T. Kang, L. (2020). Removal of ten pesticide residues on/in kumquat by washing with alkaline electrolysed water. *International Journal of Environmental Analytical Chemistry*, *102*(15), 3638–3651.

Yarahmadi, M., Yunesian, M., Pourmand, M. R., Shahsavani, A., Mubedi, I., Nomanpour, B., Naddafi, K. (2012). Evaluating the efficiency of lettuce disinfection according to the official protocol in Iran. *Iranian Journal of Public Health*, *41*(3), 95.

Yoo, J., Win, N.M., Mang, H., Cho, Y.J., Jung, H.Y., Kang, I.K. (2021). Effects of 1-Methylcyclopropene treatment on fruit quality during cold storage in apple cultivars grown in Korea. *Horticulture*, *7*(10), 338.

Yoon, J.H., Lee, S.Y. (2018). Comparison of the effectiveness of decontaminating strategies for fresh fruits and vegetables and related limitations. *Critical Reviews in Food Science and Nutrition*, *58*(18), 3189–3208.

Zhang, C., Cao, W., Hung, Y.C., Li, B. (2016a). Application of electrolyzed oxidizing water in production of radish sprouts to reduce natural microbiota. *Food Control*, *67*, 177–182.

Zhang, W., Cao, J., Jiang, W. (2021). Application of electrolyzed water in postharvest fruits and vegetables storage: a review. *Trends in Food Science & Technology*, 114, 599–607.

Zhao, D., Zeng, X.A., Sun, D.W., Liu, D. (2013). Effects of pulsed electric field treatment on (+)-catechin–acetaldehyde condensation. *Innovative Food Science & Emerging Technologies*, *20*, 100–105.

Zhao, L., Li, S., Yang, H. (2021). Recent advances on research of electrolyzed water and its applications. *Current Opinion in Food Science*, *41*, 180–188.

Molecular Hydrogen and Hydrogen-Rich Water for Fresh Fruit and Vegetable Preservation

20

Nida Firdous*, Aliza Batool, Shabbir Ahmad, Muhammad Usman, Muhammad Sibt e Abbas, Umar Farooq, Farid Moradinezhad, and Fareena Jamil

*Corresponding Author: nida.firdous@mnsuam.edu.pk

20.1 INTRODUCTION

The most prevalent and lightest element in the universe, hydrogen (H_2), exists as a diatomic gas. It is tasteless, colorless, non-metallic, non-toxic, and highly flammable. The artificial creation of H_2 through chemical reactions involving acids and metals dates back to the 16th century. The literal translation of the word "hydrogen" from Greek is "water-former," reflecting its ability to burn and produce water. On Earth, hydrogen is primarily found in organic compounds or water molecules (Ohta, 2015). Hydrogen's unique ability to selectively reduce hydroxyl radicals makes it effective in cell protection. Studies have explored its protective effects in various areas, such as acute radiation syndrome, metabolic syndromes, inflammation, and ischemia/reperfusion (I/R) injury, with radioprotection being a notable focus. Ionizing radiation (IR) used in tumor therapy and medical diagnostics generates free radicals, and hydrogen has been identified as a scavenger of these radicals, providing radioprotective effects (Hu et al., 2020).

Molecular hydrogen, also known as H_2 or hydrogen gas, behaves as an inert gas in mammalian cells at body temperature (Li et al., 2021). However, recent research has highlighted the significant impact of chemical hydrogen on biological processes. Hydrogen therapy has been proposed for various disorders, including COVID-19 and neurodegenerative diseases. Deep-sea diving has utilized hydrogen concentrations of 49% in a mixture called Hydreliox, indicating its safety for human consumption. Hydrogen, being tasteless, odorless, and colorless, is considered suitable for food processing (Hancock et al., 2022). Numerous studies have explored the medicinal and preventative effects of hydrogen, covering various biological reactions to oxidative stress in nearly all organs. Previous review papers have documented its roles in different animal and cellular investigations, highlighting anti-allergic, anti-inflammatory, and anti-apoptotic properties in various model tissues (Ohta, 2014).

On the other side, the use of hydrogen (H_2) and hydrogen-rich water (HRW) has been reported on various fruits and vegetables, ranging from the production to postharvest stages (Yun et al., 2022; Dong et al., 2023a; Dong et al., 2023b; Liu et al., 2023). The postharvest application of H_2 and HRW improves quality (when applied during production) and preserves various attributes associated with the postharvest physiology of fruits and vegetables. The use of H_2 and HRW has been reported at various concentrations depending upon the types of fruits and vegetables (and cultivars), with varying efficacy to maintain quality by suppressing weight loss, reducing oxidative stress–associated factors, preserving higher antioxidative activities, and maintaining sensorial quality attributes under storage. This chapter reviews the potential role of H_2 and HRW in the regulation of postharvest quality of fresh fruits and vegetables during storage.

20.2 MOLECULAR HYDROGEN (H_2): A WORKABLE NEW PLANT THERAPY

Earth's atmosphere contains elevated concentrations of gases, including methane (CH_4), carbon dioxide (CO_2), and hydrogen (H_2). Molecular hydrogen, being a small, inert,

DOI: 10.1201/9781003370376-26

and insoluble molecule, has profound effects on both plants and animals. Numerous studies have confirmed its therapeutic properties for various neurodegenerative and infectious diseases, and it is also utilized as a sports supplement. H_2, whether applied to plants in the form of a gas or solution, can play diverse roles, such as influencing the quality and yield of horticultural plants. The application of HRW for irrigation or foliage spray has notable effects on the yield and quality of plants. H_2 has shown potential in enhancing the postharvest life of agricultural produce, especially in horticulture. Adding magnesium hydride (MgH_2), HRW, and H_2 nano-bubble water (HNW) to plants like roses, carnations, and lilies extends their shelf life. The safety and potential benefits of H_2 make it a promising option in the horticulture industry. HNW, with higher dissolved H_2 concentrations and longer residual time, exhibits advantages in preserving fruits and vegetables. H_2 gas has the potential to influence plant metabolites and their levels, particularly in pathways related to phenylpropanoid biosynthesis and metabolism, thus resulting in improved bioactive compounds and related attributes in fruits and vegetables (Russell et al., 2020; Hancock et al., 2022; Liu et al., 2023; Ding et al., 2023).

Understanding the cellular mechanisms of H_2 generation is crucial, considering its potential as an energy source supporting soil biogeochemical processes. Microbial growth in soil environments, often constrained by limited carbon sources, can benefit from H_2, which permeates microbial cells due to its abundance and low activation energy requirement. Plants can be exposed to H_2 through various means, including endogenous generation by enzymes like hydrogenases and external sources from the plant's microflora or specific treatments. H_2, categorized as a gasotransmitter along with molecules like ethylene, nitric oxide (NO), and hydrogen sulfide, could play a significant role in positively modulating the physiology and antioxidant capacity of the treated horticultural produce with improved shelf-life potential and quality.

H_2 exhibits both curative and preventive benefits associated with reactive oxygen/nitrogen species. Its effects on disease resistance in certain plants through the SA-dependent pathway highlight its potential applications. H_2 is now considered a strategy, and its concentration and presence may have significant impacts on plant cells, which may influence the yield and quality with improvement in shelf-life potential (Figure 20.1) (Russell et al., 2020). Ongoing debates surround the kinetics of H_2 reactions, and its efficacy against hydroxyl radicals (OH) and peroxynitrite. Despite these challenges, the exploration of H_2 roles in biological systems remains a fundamental area of study (Shao et al., 2023).

20.3 MANUFACTURING PROCESS OF H_2 (OTHER THAN ENDOGENOUS PRODUCTION IN PLANTS)

At present, more than 80% of the world's energy demands are met by fossil fuels, playing a crucial role in satisfying global energy needs. However, the sustainability of this heavy dependence is questionable in the long run due to the finite nature of fossil fuel resources (Das, 2001). These pollutants are emitted into the atmosphere during the combustion process of fossil

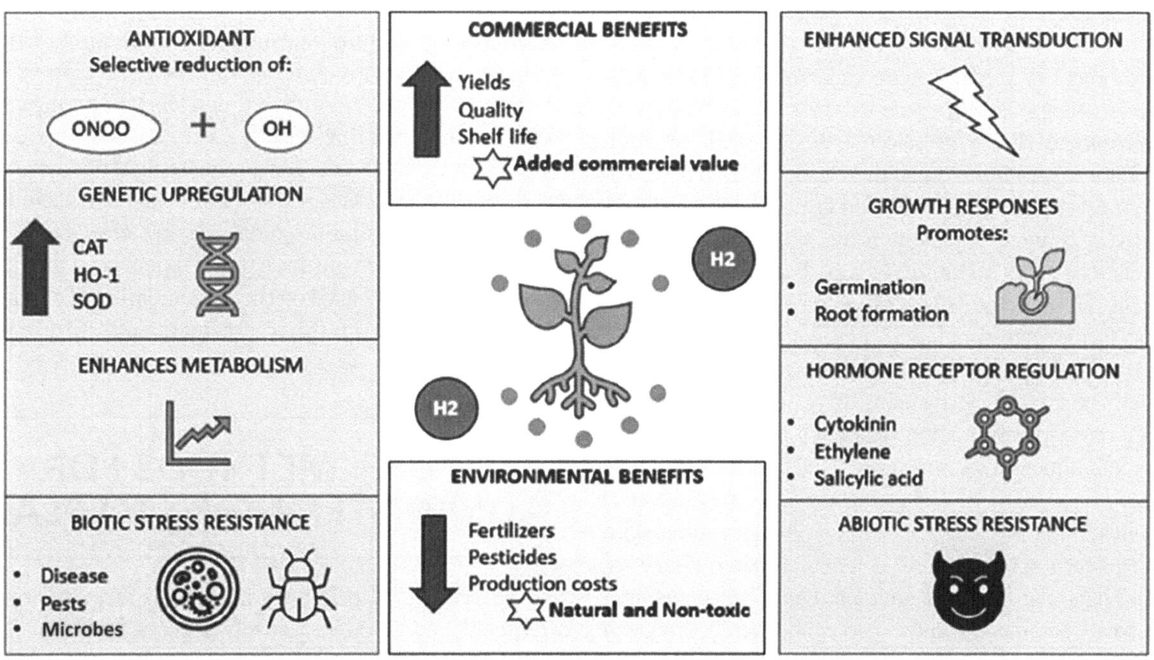

FIGURE 20.1 The potential mechanism by which H_2 gas acts as a protective agent in plants with possible beneficial effects. Russell et al. (2020).

fuels. In contrast, H_2 stands out for having the highest energy content per weight unit among known fuels. In terms of storage and usage, H_2 gas is recognized as a safer alternative to residential natural gas. It is increasingly acknowledged as a renewable energy source that can replace fossil fuels without contributing to the greenhouse effect, offering an environmentally friendly solution. H_2, unlike other fuels, contains no carbon, and its oxidation results in the formation of water. Whether used as a fuel for fuel cells or in internal combustion engines, it enables efficient energy conversion. While various applications for H_2 exist, the petroleum and fertilizer industries currently constitute the majority of users, accounting for approximately 37% and 50% of its global consumption, respectively. The annual growth rate of its exports has remained steady at around 6%, driven by stricter fuel-quality standards in refineries, leading to an increased demand for hydrogen in the refining process (Das & Veziroglu, 2008). The growing utilization of hydrogen underscores its significance as a potential clean energy source and its involvement in diverse industries. Presently, it is produced from fossil fuels using several methods. Fossil fuel–based production methods include natural gas thermal cracking, natural gas reforming by steam, coal gasification, and partial oxidation of hydrocarbons. Biomass-based production involves gasification, also known as pyrolysis, which yields a gas mixture containing hydrogen, methane, carbon dioxide, carbon monoxide, and nitrogen. Hydrogen can also be produced from water through various methods, including photolysis, electrolysis, thermolysis, thermochemical processes, and biological production (Federov et al., 1998).

20.4 PRODUCTION AND ROLE OF H_2 IN PLANTS

The identification of H_2 production traces its roots to 1931 in bacteria and 1942 in algae. In the same way, the biomedical sector extensively explored H_2 for its selective free radical reduction properties (Hu et al., 2014). Despite thorough studies on H_2 production in bacteria and algae, research on its occurrence in plants is comparatively limited. Typically, plants produce H_2 through symbiotic relationships with nitrogen-fixing bacteria in root nodules, releasing H_2 into the root surface vicinity. Additionally, abiotic stress factors can stimulate H_2 production (Zulfiqar et al., 2021). Seed germination, a pivotal stage in the plant life cycle, has been a subject of initial investigations. In winter rye seeds, the impact of H_2 on seed germination was first examined (Renwick et al., 1964). Subsequent studies revealed that H_2 positively affects *Oryza sativa* germination by enhancing amylase enzyme activity. The application of H_2 in agriculture is gaining attention, with increasing evidence supporting its potential benefits (Hancock et al., 2021). The use of H_2 gas is considered an effective method, as it can be easily obtained from the market and locally produced through the electrohydrolysis of water, simultaneously generating oxygen. HRW achieves this by modulating non-protein thiols and regulating various proteins (Li et al., 2018). The root, crucial for plant growth, is positively influenced by H_2. Moreover, water contributes to the development of adventitious roots in various plants, originating from plant parts other than primary roots, such as stems and leaves, as observed in cucumbers (Zhu et al., 2016). This application reduces the reliance on root-enhancing chemicals, contributing to environmental sustainability. HRW, a mixture of gas-rich water, is a commonly used form of H_2 in agriculture. An advanced approach, hydrogen nano-bubble water (HNW), has also been introduced to address the issue of H_2 quickly dissipating into the air in HRW solutions (Hancock et al., 2022). Plants face various stresses, both biotic and abiotic, and H_2 has emerged as a reactive signaling molecule that promotes plant health (Hancock & Russell, 2021). The molecular effects of H_2 have been observed in plants, influencing cell function and offering potential benefits for growth, productivity, and stress resistance. In addition, H_2 gas applications on harvested fruits and vegetables have been shown to maintain freshness, delay ripening, and prevent excessive softening owing to suppressed pectin solubilization (Hu et al., 2014).

20.5 CHARACTERISTICS OF H_2 GAS

H_2 functions as a mild yet potent antioxidant. Despite being the most prevalent element in the universe, constituting approximately 75% of the universe's mass, monoatomic hydrogen is not naturally occurring on Earth and is only present in water and various organic or inorganic compounds. H_2 is colorless, odorless, tasteless, and highly flammable, existing in the form of a diatomic gas. Under normal body temperature conditions and in the absence of catalysts, H_2 displays reduced reactivity, behaving as an inert gas. At room temperature, H_2 does not readily react with most substances, including O_2. Combustion of H_2 gas occurs only at temperatures exceeding 527 °C, and it becomes explosive within the concentration range of 4–75% (vol/vol) when reacting with O_2. Under atmospheric pressure, H_2 gas can dissolve in water up to 0.8 mM (1.6 ppm, wt/vol), and it easily permeates the walls of glass or plastic containers. However, aluminum containers are effective in securely containing hydrogen gas for extended periods (Ohta, 2011). So, care must be taken based on the aforementioned characteristics of H_2 during its uses postharvest treatments of fresh produce.

20.6 METHODS FOR ADMINISTERING H_2 TO PLANTS

The delivery of H_2, primarily in gaseous form, to organisms is often achieved through aerosolization, a common method in medical settings for animals that can inhale the gas. However, this approach is impractical for plants, requiring a controlled environment. Moreover, the flammability of gaseous

H_2 necessitates careful safety considerations. Therefore, an alternative method is essential for widespread applications in horticulture and other fields. The solubility of H_2 in water is low, and its tendency of quickly transitioning into the vapor phase may limit the maintenance of a high H_2 concentration in a saturated solution over time. Another method for generating H_2 is through HRW, which involves using commercially available magnesium-based tablets. Dissolving these tablets in water creates a hydrogen-enriched solution, applicable to plants through soil watering or foliage spraying. It is crucial to consider any residual by-products or additives introduced by these tablets during H_2 generation. Alternatively, H_2 can be introduced into a saline or salt solution. While effective in animal research, its applicability in plant treatments may be limited unless specific conditions, such as studies involving high salt concentrations, are considered. From a biological perspective, treating plants with H_2 is generally regarded as safe, with minimal or no detrimental effects on biological systems. However, practical considerations must address the highly explosive nature of H_2, particularly in the presence of atmospheric oxygen. Therefore, precautions and safety measures may be necessary when applying large-scale spray treatments.

20.7 ADVANTAGES OF H_2 APPLICATIONS IN PLANTS

H_2 is a promising antioxidant that has several benefits. Its efficient dispersion, which includes the capacity to pass through biological membranes and permeate into the cytosol, is one of its noteworthy features. Mitigating oxidative damages is essential since the respiratory chain of the mitochondria is a major source of reactive oxygen species (ROS) that cause it. Antioxidants have demonstrated poor therapeutic efficiency despite the clinical significance of oxidative damage in mitochondria, which may be related to the difficulty of targeted mitochondrial absorption (Smith & Murphy, 2011). H_2 demonstrates effective penetration into the nucleus and mitochondria, enhancing its ability to counteract oxidative damages. The measurement of H_2 content within distinct tissues provides evidence of its gaseous diffusion properties (Hayashida et al., 2008). These potentials of H_2 impart some advantages, and its use could be considered beneficial in increasing yield, quality at harvest, and shelf-life period after harvest of the produce.

20.8 H_2 TREATMENTS' EFFECTIVE CONCENTRATION RANGES

The effectiveness and safety of H_2 are directly correlated with its concentration range. Fresh HRW typically possesses an H_2 concentration ranging from approximately 220–860 M, often considered as 100% saturation, owing to the diverse H_2-producing characteristics of H_2 generators. It is noted that in certain plants, excessive H_2 concentration may lead to reduced benefits. The optimal HRW concentration is dependent on factors such as plant species, type, and various treatment durations. H_2 concentrations, such as 0.5% H_2 or 1% H_2, have been observed to markedly increase the vase life of cut roses, indicating the sensitivity of cut flowers to H_2 (Ren et al., 2017). Recent studies also suggested that an H_2-modified environment can extend the shelf life of fruits and vegetables. The potency and effective concentration of H_2 may vary significantly depending on the plant species.

20.9 H_2 USES IN FOOD TECHNOLOGIES

H_2, characterized by its colorless, tasteless, combustible, and non-toxic nature as a diatomic gas, holds a crucial role across diverse organisms. Its potential applications extend to the treatment of conditions such as COVID-19 (Ohta, 2015). Its utilization in the food industry, particularly in preserving fruits and vegetables, can contribute to enhancing postharvest shelf stability, transportation, and storage. Notably, H_2 is recognized as safe for human consumption and food processing, facing no regulatory challenges in the food sector (Ohno et al., 2012). Addressing the substantial challenge of food waste is crucial for realizing economic benefits and ensuring global food security (FAO, 2022), the issue of food waste continues to escalate, posing environmental, economic, and nutritional concerns for both farmers and consumers in terms of postharvest losses. Developing effective strategies to minimize food waste at various stages of the supply chain becomes paramount for bolstering food security without requiring additional resources. Therefore, there is a pressing need to explore alternative approaches and techniques for mitigating food wastes and losses (Nicastro & Carillo, 2021). In the agricultural domain, H_2 emerges as a critical factor influencing crop production (Cheng et al., 2021), potential yield, and quality of the produce at harvest and during storage.

20.10 USES OF H_2 IN HORTICULTURAL APPLICATIONS

The utilization of H_2 presents new opportunities for straightforward and affordable enhancements in horticultural production, leading to increased yield and superior product quality, all while minimizing CO_2 emissions. Numerous studies have illustrated the positive impacts of H_2 on horticultural crops, encompassing the extension of storage life and the enhancement of the postharvest quality of fruits and vegetables. The concept of "H_2 fertilization" was initially proposed when treated *Brassica napus* exhibited enhanced growth in response to H_2 application. Subsequent studies have explored the unique capabilities of H_2 to enhance postharvest preservation of fresh produce (along with some other effects). Its major use has been reported as (a) H_2 gas–treated soil, (b) H_2 gas fumigation, (c) postharvest treatments, and (d) use of magnesium hydroxide (MgH_2) in floriculture to enhance the vase life of flowers (among others). Nevertheless, the major focus of this chapter is its utilization in the postharvest preservation of fresh fruits and vegetables. H_2 applications in horticulture produce have been reported both during the preharvest and postharvest phases (Hu et al., 2018; An et al., 2021).

Several strategies have been employed to effectively harness H_2 gas, as outlined by Hu et al. (2018). In the context of kiwifruit, electro-hydrolysis of H_2O is utilized to obtain H_2, simultaneously producing O_2. It is crucial to separate the generated O_2 from H_2, resulting in an O_2/H_2 gas mixture. When using H_2 in gaseous form, particularly in confined spaces, adherence to safety precautions is imperative. Alternatively, H_2 can be utilized in its gaseous state within an enriched solution referred to as HRW (Dilisi, 2017). While the application of H_2 in food processing is somewhat restricted, one method involves bubbling H_2 gas into a saline solution to create hydrogen-rich saline (HRS). An advanced form of HRW, termed hydrogen nano-bubble water (HNW), involves generating small nano-bubbles of H_2 gas in water. The HNW holds promise across various industries, including agriculture and horticulture, potentially enhancing plant growth and productivity (Yang et al., 2012) as well as postharvest fresh produce preservation. Biological materials can also receive H_2 through donor molecules such as magnesium hydride (MgH_2), exemplified in a study on the vase life of carnations (Li et al., 2020). Donor molecules release H_2 gas at a slower and sustained rate, reducing the need for frequent applications compared to HRW (Quinn et al., 2015). Regardless of the delivery method, employing H_2 requires a comprehensive cost–benefit analysis, especially in cases where the primary objective is food preservation. In situations where the value of treated food does not outweigh the treatment costs, the feasibility of the treatment becomes questionable (Jain, 2009). Consequently, there is a growing demand for cost-effective H_2 gas production and distribution to enhance its applicability in the fresh produce–related food industry.

20.10.1 Potential Beneficial Effects of H_2 on Horticultrual Plants

Numerous studies have revealed that the production of H_2 increases in certain plants when exposed to abscisic acid, jasmonic acid, and ethylene. H_2 exhibits substantial impacts on antioxidant levels and gene expression, enhancing stress tolerance. This positive influence of H_2 extends to alleviating stresses, which triggers ROS accumulation, with the haem oxygenase system playing a role in mediating this effect. Another beneficial effect attributed to H_2 is the upregulation of antioxidant system–related genes. H_2 has been extensively studied in lower plant species like *C. reinhardtii*, which naturally produce significant quantities of H_2. Future investigations into H_2 enzymes and metabolism should consider this interplay and may necessitate the development of new fluorescent probes to monitor the spatial and temporal distribution of these signaling molecules. In the context of climate change–induced abiotic stress events, the use of non-toxic substances like H_2 is preferable to chemical fertilizers for the sustainable enhancement of horticultural crop yields. In the realm of plant science, it is imperative to explore effective implementations of H_2 applications in commercial settings, ensuring safety and sustainability given H_2's highly flammable nature. Spray treatments with HRW emerge as a practical and cost-effective method, offering convenience for delivery to soil or directly onto foliage and as dip treatments in the case of harvested fruits and vegetables. Future comprehensive research is necessary to fully comprehend the biochemical processes influenced by H_2 within plants cells. Various enzymes and regulatory proteins are sensitive to redox changes, necessitating exploration of the interactions between H_2 and the intracellular redox environment. Thus, forthcoming studies should concentrate on elucidating how H_2 affects gene expression and protein composition in various plant tissues (Russell et al., 2020).

20.10.2 Application of H_2 in the Postharvest Preservation of Fruits and Vegetables

The impact of hydrogen gas on kiwifruits was found to be mediated by suppressed ethylene biosynthesis and subsequent metabolism (Hu et al., 2018). A similar effect was also observed in bananas (Yun et al., 2022). Alterations in the enzymes associated with the metabolism of ethylene, particularly ACC synthesis and ACC oxidase, were observed in response to H_2 treatments. In a study involving kiwifruits treated with HRW (Zhao et al., 2021), delayed accumulation of soluble solid content, weight loss, and total microbial load were noted compared to control samples. The HRW treatment maintained firmness, color, and chlorophyll content during refrigeration, while also increasing flavonoids, total phenols, and ascorbic acid levels in kiwifruits. Notably, combining HRW with a small amount of electrolyzed water treatment enhanced these positive effects. In a study on strawberries, integrating 4% reducing hydrogen gas into packaging (RAP) was found to be more beneficial than conventional modified atmospheric packaging (MAP). H_2 in the packaging headspace effectively prevented fruit oxidation, preserving color, texture, and nutritional value (Table 20.1). The RAP also extended the best-before and expiration dates of strawberries compared to traditional MAP, offering an environmentally friendly method for prolonging shelf and storage life (Alwazeer, 2022).

The treatment of *Rosa sterilis* was significantly effective in maintaining its quality during storage owing to higher

TABLE 20.1 Possible Effects of H_2 and HRW-Based Applications on Fresh Produce during Storage

PRODUCE	*H_2 SOURCES*	*INFERENCES*	*REFERENCES*
Kiwifruit	H_2 fumigation	Improved fruit quality by ethylene metabolism suppression	Zhao et al. (2021)
Kiwifruit	HRW and SAEW	Reduced loss of antioxidants, such as flavonoids, and postponed breakdown of chlorophyll. Reduced oxidative stress indicators	Sun et al. (2023)
Rosa sterilis	HRW	Enhanced fruit quality facilitated by the modulation of reactive oxygen species (ROS) and energy metabolism	Yan et al. (2021)
Okra	HRW	Postponed softening of fruit with improved maintenance of cell wall stability	Dong et al. (2023b)
Litchi	HRW	Decreased pericarp browning and reduced indicators of oxidative stress	Yun et al. (2021)
Chinese water chesnut	HRW	Diminished tissue yellowing and decreased oxidative stress and impacts on the phenylpropanoid pathway	Li et al. (2022a)
Banana	HRW	Postponed ripening by ethylene metabolism suppression	Yun et al. (2022)
Strawberry	Within the package	Enhanced storage life with reduced fruit oxidation	Alwazeer and Özkan (2022)
Marsmoreus hypsizygus	HRW	Enhanced storage life by preserving antioxidants	Zhao et al. (2021)
Lily bulbs	HRW	Maintained overall eating quality of bulbs	Liu et al. (2023)

preservation of ascorbic acid levels in contrast with controls (Dong et al., 2022). HRW was also effective in preventing postharvest fruit softening in okra (*Abelmoschus esculentus* L.), influencing cell wall synthesis and gene expression related to cell disintegration (Dong et al., 2023b). The HRW treatment demonstrated antioxidant effects in Chinese water chestnut, maintaining color, reducing oxidative stress markers, and enhancing antioxidant capacity (Li et al., 2022a). Both HRW treatment and H_2 fumigation effectively reduced nitrite accumulation in tomatoes, with specific effects attributed to H_2 exposure (Zhang et al., 2019). H_2 has exhibited the potential to extend the storage life of fruits and preserve their nutritional value. Recent research indicated that HRW can slow the ripening process in bananas, delaying color changes and reducing the breakdown of cell walls and starch (Zhu et al., 2021). The HRW treatment was also found to regulate ethylene, a hormone involved in fruit ripening. Additionally, the HRW treatment influenced defense responses against *Botrytis cinerea*, a common disease affecting tomatoes (Lu et al., 2017).

H_2 exhibits highly beneficial effects across various organisms, from humans to plants, with applications ranging from treating neurodegenerative diseases to potential benefits in addressing harmful conditions (Hancock & Russell, 2021). In addition, its application in preserving harvested fruits and vegetables highlights its potential in improving food storage and transportation. Although the exact biochemical mechanisms of H_2 are not entirely elucidated, it is considered safe for use, leaving no harmful residues, as demonstrated in numerous studies. With advancing H_2 production and transportation technologies, the decreasing cost makes it increasingly advantageous to explore its potential in postharvest fresh produce conservation (Hancock et al., 2022; Yun et al., 2021).

The application of H_2 treatment has shown positive impacts on postharvest fruits and vegetables. Okras treated with HRW exhibited delayed fruit softening and maintained cell wall integrity (Dong et al., 2023b). Chinese water chestnuts treated with HRW resist tissue yellowing and experience reduced oxidative stress, particularly impacting the phenylpropanoid pathway (Li et al., 2022a). Gas treatment based on H_2 during strawberry packing reduced fruit deterioration and improved storage life (Alwazeer & Özkan, 2022). Lastly, HRW treatment acts as an antioxidant, enhancing storage in *Hypsizygus marmoreus* mushrooms (Zhao et al., 2021).

In the case of kiwifruit, HRW treatment resulted in the delayed ripening of fruits and the maintenance of firmness, with the most favorable outcomes observed at an 80% HRW concentration. This treatment led to reduced pectin solubilization, oxidative stress, and lipid peroxidation (Sun et al., 2023). It also enhanced the activity of the antioxidant enzyme superoxide dismutase (SOD) and preserved the integrity of the inner mitochondrial membrane. Another study utilized hydrogen gas fumigation on kiwifruits, increasing endogenous hydrogen levels and delaying fruit softening. This was achieved by inhibiting ethylene synthesis and reducing

1-aminocyclopropene-1-carboxylate (ACC) levels, accompanied by modifications in enzymes related to ethylene metabolism (Hu et al., 2018; Zhao et al., 2021).

For strawberries, incorporating 4% reducing hydrogen gas into the packaging (RAP) protected color, texture, and nutritional parameters. It preserved anthocyanin and phenolic content and extended the shelf life compared to conventional packaging methods (Alwazeer, 2022). The HRW treatment of *Rosa sterilis* reduced fruit weight loss, decay index, and oxidative stress, and increased glutathione, ascorbate levels, and antioxidant enzyme activity (Yan et al., 2021). Similarly, HRW treatment of litchi (Yun et al., 2021) and Chinese water chestnuts (Li et al., 2022a) maintained fruit quality by reducing pericarp browning and oxidative stress, and preserving antioxidant capacity. In tomatoes, HRW or H_2 gas fumigation decreased nitrite accumulation by affecting enzymes involved in nitrogen metabolism (Zhang et al., 2019). HRW treatment also delayed ripening in bananas by modulating ethylene metabolism (Yun et al., 2022). Overall, H_2 released from HRW exhibits the potential to maintain the quality and extend the shelf life of various postharvest plant materials, including fruits and vegetables under storage.

20.10.3 Mechanisms of H_2 Affecting Fresh Produce Quality and Shelf Life

Plant responses to various stressors often involve the generation of reactive oxygen species (ROS) and reactive nitrogen species (RNS). The overproduction of ROS can accelerate senescence in postharvest fruits and vegetables. Additionally, ROS and RNS play a crucial role as signal transducers in plant signaling networks related to stress and development (Ali et al., 2024b). Therefore, the metabolic control of ROS and RNS is essential for plants to respond to stress, grow, and develop. Endogenous H_2 may be produced under abiotic stress and plant senescence conditions. The exposure of a produce to H_2 helps to maintain freshness and quality (Hu et al., 2018).

20.11 HYDROGEN-RICH WATER

Hydrogen-rich water (HRW) has emerged as a novel functional beverage with potential positive impacts on human health. The production of HRW is critical and depends upon certain factors. Over a 160-minute duration at 25 °C, 300 mL of pure hydrogen gas per minute was introduced into 8 L of pure water. The freshly produced HRW underwent analysis using gas chromatography, revealing a stable H_2 level of 0.22 mM at 25 °C for a minimum of 12 hours. Various trials have demonstrated favorable effects of HRW consumption in people with non-alcoholic fatty liver disease and metabolic syndrome, contributing to improved performance and reduced inflammatory responses, as well as apoptosis prevention (Ostojic, 2021). HRW has demonstrated positive effects on anthocyanin formation in radish sprouts under UV-A irradiation (Su et al., 2014).

20.11.1 Potential Effects of HRW on Fruits and Vegetables

Different potential and beneficial effects of HRW have been reported in the literature, as mentioned next.

20.11.1.1 HRW Role in Improved Biosynthesis of Flavonoids

Plants employ flavonoids and their glycoconjugates, glycosides, as a defense mechanism against oxidative damage caused by ultraviolet (UV) light. The resilience of plants to UV-B stress was reduced through a combination of UV-B irradiation and treatment with HRW, leading to an enhancement in various flavonoid profiles. The HRW treatment upregulated the transcript levels of key genes involved in flavonoid synthesis. These genes play crucial roles in plant defense mechanisms against oxidative stress induced by UV radiation, highlighting the potential of HRW in modulating these processes (Su et al., 2014).

Anthocyanins, a prominent subclass of flavonoids, act as primary pigments and serve to attract pollinators and protect plants from UV radiation. The HRW treatment during UV-A irradiation led to the substantial accumulation of cyanidin, the primary anthocyanidin in radish sprout hypocotyls. Genes related to anthocyanin biosynthesis exhibited increased transcript levels under HRW regulation. Additionally, HRW-influenced anthocyanin production under UV-A irradiation was found to be significantly influenced by calcium signaling pathways that are InsP3-dependent. Moreover, HRW demonstrated the capability to enhance anthocyanidin levels under blue light exposure. Given the widespread applications of flavonoids in various industries, including food, pharmaceuticals, medicine, and cosmetics, HRW stands as a promising approach to elevate the quality of horticultural crops (Su et al., 2014).

20.11.1.2 HRW's Role in Maintaining the Quality of Fruits and Vegetables

HRW has a notable impact on the quality of fruits and vegetables, as these are perishable commodities owing to water loss, wilting, and softening (Dong et al., 2023a; Dong et al., 2023b). The HRW treatment, involving the infusion of H_2 into water, has emerged as a safe alternative due to its antioxidant capacity, effectively preserving the postharvest quality of fruits and vegetables and preventing diseases (Dong et al., 2022). The plant hormone abscisic acid, crucial in regulating fruit senescence and providing resilience against environmental challenges, has been a focus of experimentation. The HRW treatment has been found to inhibit abscisic acid production by reducing the expression of the corresponding gene, highlighting an interaction between HRW treatment and plant hormones and ultimately delaying the possible onset of senescence (Dong et al., 2023a). The application of an 80% HRW treatment to kiwifruit demonstrated inhibitory effects on endogenous ethylene synthesis, resulting in delayed ripening and extended shelf life (Hu et al., 2014). The HRW treatment positively influences the energy metabolism of plants, particularly in chloroplasts, mitochondria, and peroxisomes. Notably, hydrogen stimulates mitochondrial functions, leading to ATP production, which contributes to high energy levels that sustain plant vitality and freshness (Hancock et al., 2022). The energy level of a plant also affects fruit color, which is economically significant and attracts consumers. A low energy level makes plants more susceptible to physiological disorders (Li et al., 2019; Hancock et al., 2022). The short shelf life of litchi fruit (as it is prone to browning due to enzymatic and non-enzymatic reactions) has prompted the use of HRW treatment as a preservation technique. The hydrogen in HRW scavenges ROS responsible for browning and enhances the plant tissues' ability to resist stress-related effects (Yun et al., 2021). Although numerous studies supported HRW treatment as an effective and cost-efficient means of maintaining fruit quality (as it recognized as a non-toxic and safe method to boost the antioxidant capacity of fresh produce), further research is essential to validate its efficacy on a commercial scale.

20.11.1.3 Postharvest Senescence Alleviation of Fruits and Vegetables with HRW Applications

The issue of postharvest produce senescence, a source of food waste generation, emphasizes the urgent need for improved transportation and storage methods to ensure food security. Over 60% of food waste comprises cereals and vegetables, underscoring the necessity for effective solutions. Postharvest H_2 treatments have shown promise in addressing this challenge. Various studies illustrate the efficacy of H_2 treatments in preserving plant-based nutrients during postharvest. For instance, kiwifruit treated with 80% HRW concentration exhibited reduced oxidative stress, enhanced firmness, and decreased rot. H_2 fumigation in kiwifruit altered ethylene metabolism, resulting in delayed senescence and improved resistance against selected pathogens. In addition, the HRW-treated kiwifruits exbited a higher ability to scavenge radicals (such as DPPH, O_2^-, and OH^-) owing to the maintenance of stimulated activity of superoxide dismutase and the decreased the degree of lipid peroxidation. Furthermore, the mitochondrial inner membrane showed preserved integrity, which proved beneficial in increasing the storage period of treated kiwifruits in contrast with controls (Sun et al., 2023). Chinese chives treated with H_2 fumigation experienced extended storage time by mitigating oxidative stress, reducing weight loss, and preserving soluble protein content (Jiang et al., 2021). *Hypsizygus marmoreus* treated with 25% HRW concentration displayed reduced oxidative stress, decreased weight loss, and increased firmness (Zhao et al., 2021). Postharvest tomatoes treated with lower HRW concentrations exhibited delayed senescence along with maintained vitamin C levels (Zhang et al., 2019). These positive outcomes hold significant implications for reducing postharvest losses and enhancing food security. Continued research and cost-effectiveness assessments are crucial for broader adoption in industries such as food processing and postharvest horticultrue.

20.11.1.4 Postharvest Softening Suppression of Fruits and Vegetables with HRW Applications

The softening of fruits and vegetables, including okra, is primarily attributed to enzyme-mediated changes in the structure and composition of the cell wall, involving the solubilization of major cell wall polysaccharides such as pectins and cellulose (Ali et al., 2024a). Enzymes such as polygalacturonase (PG), cellulase (CX), and β-galactosidase (GAL) contribute to cell wall disassembly (Naveed et al., 2024). In the same way, maintaining the expression of genes related to cell wall synthesis, including galacturonosyltransferase (GAUT) and β-1,4-xylosyltransferase (IRX), can slow down the softening of fruits. The membrane lipids metabolism also plays a role in softening or pulp breakdown, as noted in longan fruits (Liu et al., 2022). Okra, scientifically termed *Abelmoschus esculentus* L., a member of the Malvaceae family, is a globally cultivated and highly nutritious vegetable known for its nutritional content and pleasant flavor. Despite their popularity, okras face several challenges in postharvest storage, marked by fruit softening and susceptibility to mechanical injuries, thus limiting their shelf life and resulting in declined consumer acceptance. Various technologies have been explored to extend the postharvest storage time of okras. However, there's a continuing need for simpler and safer techniques to preserve the freshness of okra fruit and delay its senescence (Dong et al., 2023a; Dong et al., 2023b). The treatment of HRW substantially delayed the softening by influencing the cell wall metabolism and interacting with certain phytohormones during the postharvest period of okra fruits (Dong et al., 2023a; Dong et al., 2023b). The HRW created by dissolving H_2 into pure water through a specific process offers a practical and safer method than hydrogen gas for application on postharvest fruits and vegetables. The HRW treatment has been reported to reduce rotting, maintain firmness, enhance free radical scavenging, and increase antioxidant activities in kiwifruits (Zhao et al., 2021). The treatment of pakchoi with HRW in combination with vacuum precooling markedly suppressed chlorophyll catabolism, thus resulting in suppressed color changes and improved capacity of antioxidative system during postharevst in comparison with controls (An et al., 2021). Interestingly, the preharvest irrigation of certain produce resulted in improved quality and shelf life owing to inhibited softening. For instance, H_2-based irrigation of strawberries resulted in improved cell wall components biosynthesis, which thus extended their shelf life and maintained their quality (Jin et al., 2023). In another study, H_2 nano-bubble water–based preharvest application was found effective in improving consumer accepance of strawberry fruits due to enhanced flavor profile (Li et al., 2022b).

20.11.1.5 Effect of HRW-Based Treatments on Physiological Attributes and Endogenous Phytohormones of Fruits and Vegetables

The application of HRW also markedly influences certain physiological attributes of fresh produce under storage. For instance, untreated okras exhibited a gradual loss of firmness during storage, whereas those subjected to HRW treatment maintained firmness throughout the storage period. Weight loss increased steadily with prolonged storage, and after 3 days HRW-treated fruits displayed significantly lower weight loss compared to the control group (Dong et al., 2023a; Dong et al., 2023b).

The research revealed a positive correlation between fruit senescence in HRW-treated postharvest okras and abscisic acid concentration on the endogenous level, along with an adverse correlation with melatonin, indole-3-acetic acid (IAA), and gibberellic acid (GA). By increasing melatonin, IAA, and GA contents decreased and inhibited ABA biosynthesis through the regulation of their metabolizing genes. HRW treatment effectively prolonged shelf life and delayed okra senescence owing to the positive regulation of anti-senescence-related phytohormones in contrast with controls (Dong et al., 2023a). HRW treatment has been associated with improved quality in fruits and vegetables along with overall freshness due to suppressed desiccation of the treated produce. For eample, HRW treatment prevented weight loss and retained markedly higher hardness in *Hypsizygus marmoreus* (Zhao et al., 2021).

Ascorbic acid (AsA) is considered a potent antioxidant possessing beneficial potential in declining the oxidative stress–associated metabolism in fresh produce. It is important that the senescence in fresh produce with higher AsA content is typically indicated by a decline in AsA levels under storage (Ali et al., 2020). The HRW treatment in a study significantly reduced endogenous AsA decline along with myo-inositol oxygenase (MIOX) mediated inositol conversion to glucuronic acid. This has suggested that MIOX might not be the key enzyme in the AsA production process. Notably, the expression patterns of numerous RrMYB transcription factors were found to be significantly associated with AsA levels, suggesting their potential role in transcriptionally influencing AsA biosynthesis and regeneration in postharvest *R. roxburghii* fruit (Figure 20.2). However, the fundamental regulatory system underlying this phenomenon requires further clarification. In general, HRW treatment emerged as a promising, safe, and environmentally favorable preservative for maintaining the quality of *R. roxburghii* fruit during storage (Ding et al., 2023).

The effect of HRW application and vacuum precooling was investigated on pakchoi senescence during storage (An et al., 2021). The study revealed that a 50% HRW treatment effectively preserved chlorophyll and its metabolic derivatives

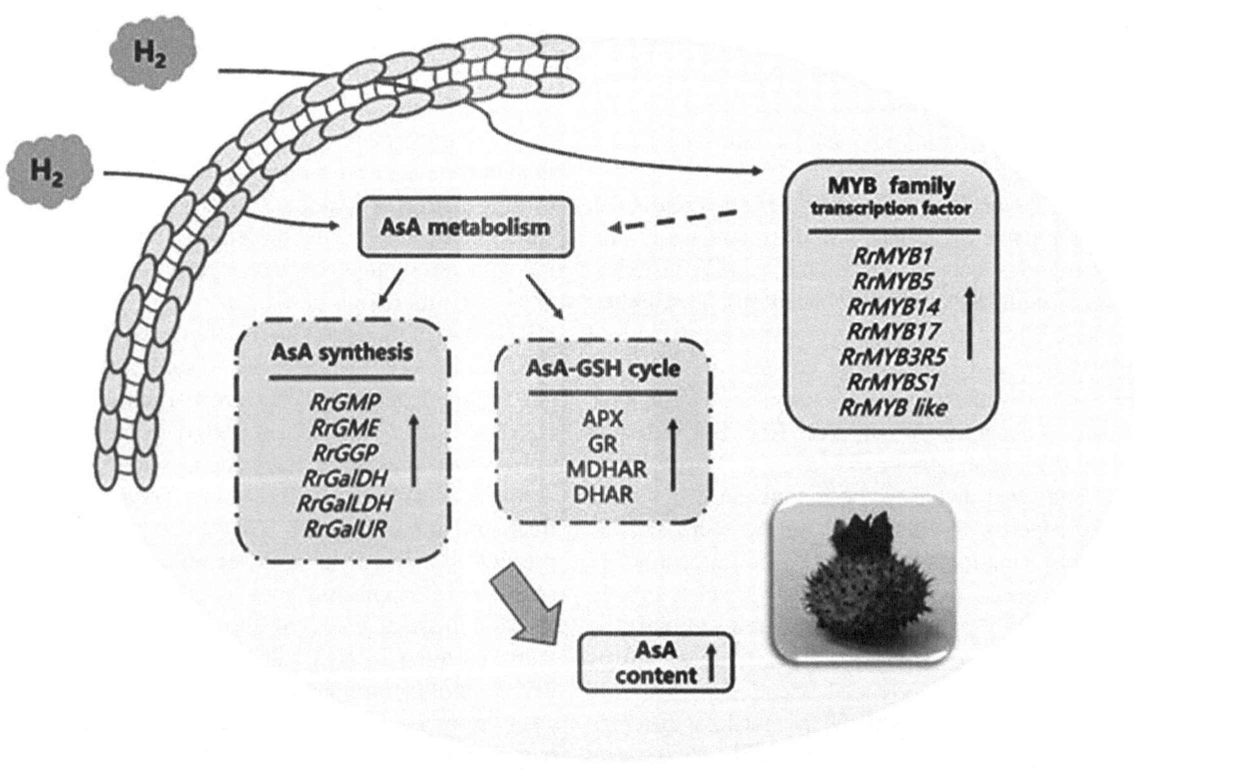

FIGURE 20.2 Expression levels.of regulation of AsA biosynthesis and regeneration in HRW-treated *R. roxburghii*. From (with permission) Ding et al. (2023).

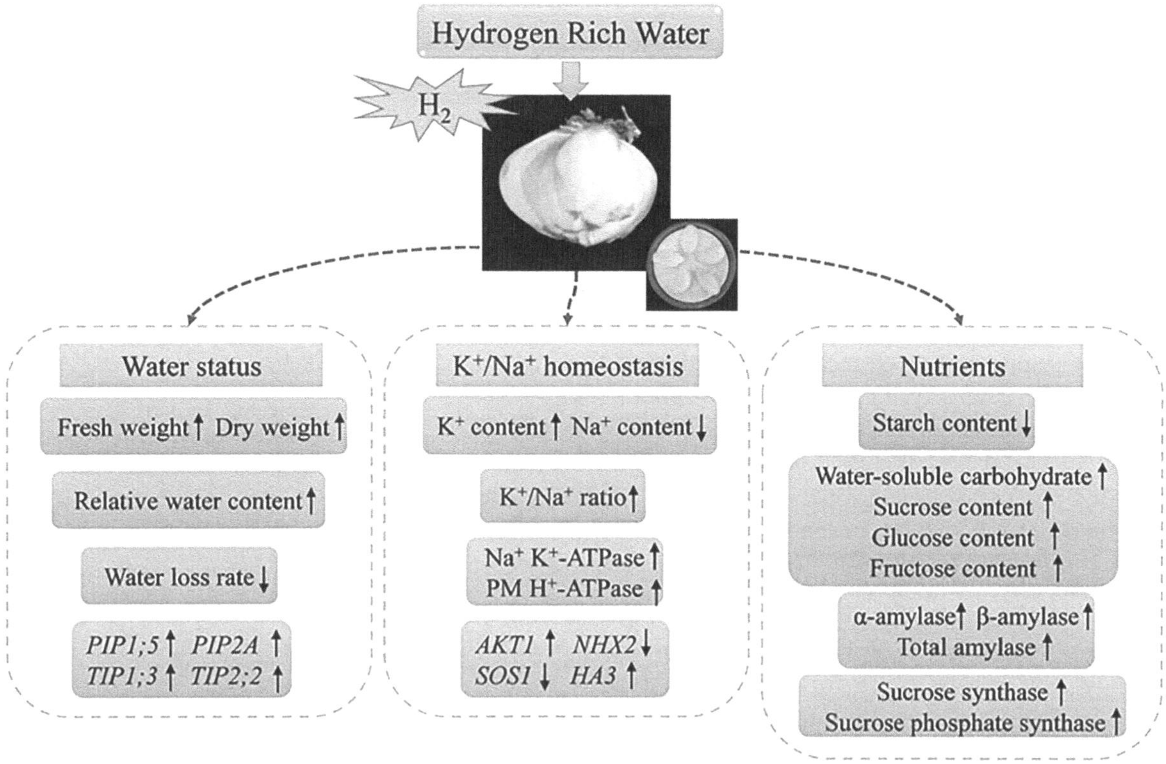

FIGURE 20.3 Effects of hydrogen-rich water treatment on lily bulb quality.during postharvest storage. Liu et al. (2023).

by reducing the activities of chlorophyll degradation–associated enzymes. Additionally, the combined treatment of HRW and vacuum precooling significantly reduced the rate of weight loss compared with other treatment groups. The synergistic effect preserved antioxidant substances, thus contributing to the delayed senescence of postharvest pakchoi (An et al., 2021). The said treatment actually suppressed the onset of oxidative stress by inhibiting free radicals generation, which reduced malondialdehyde generation and accumulation; thus all of the aforementioned factors were associated with HRW-induced suppression of color changes and subsequent senescence of pakchoi (An et al., 2021). Thus, the HRW emerged as a potential solution for enhancing the postharvest preservation of leafy vegetables, addressing the main challenges posed by rapid senescence and water loss during vacuum precooling of the said fresh produce. Nevertheless, further work is still required in this regard.

Fruits and vegetables undergo ripening and aging processes associated with the presence of ROS buildup. Maintaining a balance between ROS production and elimination is crucial for their quality and shelf-life enhancement. In fresh-cut kiwifruit, the application of HRW positively impacted ROS detoxification, specifically reducing superoxide anion and hydrogen peroxide contents, thereby alleviating oxidative stress during early storage (Sun et al., 2023). The reduced oxidative damages thus inhibited prompt senescence-related changes in the treated kiwifruits in comparison with the controls. These findings thus suggested that HRW treatment has the potential to influence the activity of cell wall degradation, thus preserving higher firmness and suppressing the prompt senescence of the treated fresh produce during postharvest storage (Sun et al., 2023). Recently, Liu et al. (2023) observed significantly maintained quality on lily bulbs treated with 50% HRW during 12 days of storage (Figure 20.3).

20.12 PRESERVATION POTENTIAL, SAFTEY REGULATIONS, AND FUTURE PERSPECTIVES OF H_2 APPLICATIONS IN POSTHARVEST FRESH PRODUCE

These studies collectively suggest the broad applicability of H_2-based (including HRW) treatments in preserving the quality and extending the shelf life of various fruits and vegetables. Implementing H_2-based treatments such as HRW holds enormous potential for reducing postharvest losses and enhancing food security but require ongoing research and cost-effectiveness considerations for widespread adoption of the said treatments (Hancock et al., 2021). As the global population expands, ensuring affordable and sustainable treatments becomes imperative to enhance the quality and shelf life of fresh produce, thereby averting food shortages. H_2 treatment exhibits promise in enhancing human health and agricultural outcomes, particularly in crop yield and health as well as postharvest storage. While the benefits of hydrogen treatments for flowers are evident, existing reports lack a thorough cost–benefit analysis, which is crucial for widespread adoption. The precise molecular mechanisms of hydrogen in plants remain a subject of debate, necessitating additional research to comprehend its interactions with various biomolecules. It is imperative to establish the safety and long-term consequences of H_2 treatments on both plant products and human health. In conclusion, H_2 displays substantial potential in enhancing plant tolerance, prolonging shelf life, and reducing nitrite accumulation in food items. Its application in the food industry, particularly in postharvest processing and packaging, merits further exploration. However, several questions persist regarding the specific food products that would benefit, safety considerations, its biochemical actions, its penetration into plant tissues, required treatment durations, cost-effectiveness assessments, integration with hydrogen transport systems, pathogen reduction, and compatibility with existing fresh produce industry practices. Despite these uncertainties, H_2 emerges as a promising preservative agent in postharvest management, urging its further consideration and exploration in future research.

20.13 CONCLUSIONS

In conclusion, H_2 has gained widespread recognition for its crucial role in the postharvest preservation of fresh produce quality and extension of shelf life. Its applications have been reported during both the pre- and postharvest stages on a number of fruits and vegetables. The available research suggests that H_2 and HRW-based treatments contribute to prolonging the shelf life of postharvest fresh produce by preserving firmness and delaying the onset of softening and subsequent senescence. In addition, its application activates antioxidant mechanisms, thus mitigating oxidative stress, which in turn delays senescence and improves the quality of fresh produce. In summary, harnessing H_2 and HRW in horticulture postharvest applications shows significant potential in delaying senescence and maintaining the overall quality of the treated fresh produce during postharvest storage.

REFERENCES

Ali, S., Anjum, M. A., Nawaz, A., Naz, S., Hussain, S., Ejaz, S., & Sardar, H. (2020). Effect of pre-storage ascorbic acid and Aloe vera gel coating application on enzymatic browning and quality of lotus root slices. *Journal of Food Biochemistry*, 44(3), e13136.

Ali, S., Ishtiaq, S., Nawaz, A., Naz, S., Ejaz, S., Haider, M.W., . . . & Javad, S. (2024a). Layer by layer application of chitosan and carboxymethyl cellulose coatings delays ripening of mango fruit by suppressing cell wall polysaccharides disassembly. *International Journal of Biological Macromolecules*, 256, 128429.

Ali, S., Nawaz, A., Naz, S., & Ejaz, S. (2024b). Application of ROS, RNS, and RSS for prolonging the shelf-life of horticultural crops via the control of postharvest bacterial infections. In *Oxygen, Nitrogen and Sulfur Species in Post-Harvest Physiology of Horticultural Crops* (pp. 341–367). Academic Press.

Alwazeer, D., & Özkan, N. (2022). Incorporation of hydrogen into the packaging atmosphere protects the nutritional, textural and sensorial freshness notes of strawberries and extends shelf life. *Journal of Food Science and Technology,* 59(10), 3951–3964.

An, R., Luo, S., Zhou, H., Zhang, Y., Zhang, L., Hu, H., & Li, P. (2021). Effects of hydrogen-rich water combined with vacuum precooling on the senescence and antioxidant capacity of pakchoi (*Brassica rapa* subsp. Chinensis). *Scientia Horticulturae,* 289, 110469.

Cheng, P., Wang, J., Zhao, Z., Kong, L., Lou, W., Zhang, T., . . . & Shen, W. (2021). Molecular hydrogen increases quantitative and qualitative traits of rice grain in field trials. *Plants,* 10(11), 2331.

Das, D., & Veziroglu, T.N. (2008). Advances in biological hydrogen production processes. *International Journal of Hydrogen Energy,* 33(21), 6046–6057.

Das, D., & Veziroğlu, T.N. (2001). Hydrogen production by biological processes: a survey of literature. *International Journal of Hydrogen Energy,* 26(1), 13–28.

DiLisi, G.A. (2017). The Hindenburg disaster: Combining physics and history in the laboratory. *The Physics Teacher,* 55(5), 268–273.

Ding, X., Yao, Q., Zhu, D., & Dong, B. (2023). Hydrogen-rich water maintains the quality of Rosa roxburghii by regulating AsA biosynthesis and regeneration. *Postharvest Biology and Technology,* 195, 112136.

Dong, B., Zhu, D., Yao, Q., Tang, H., & Ding, X. (2022). Hydrogen-rich water treatment maintains the quality of *Rosa sterilis* fruit by regulating antioxidant capacity and energy metabolism. *LWT,* 161, 113361.

Dong, W., Shi, L., Li, S., Xu, F., Yang, Z., Cao, S. (2023b). Hydrogen-rich water delays fruit softening and prolongs shelf life of postharvest okras. *Food Chemistry*, 399, 133997.

Dong, W., Cao, S., Zhou, Q., Jin, S., Zhou, C., Liu, Q., . . . & Shi, L. (2023a). Hydrogen-rich water treatment increased several phytohormones and prolonged the shelf life in postharvest okras. *Frontiers in Plant Science*, 14, 1108515.

FAO. 2022. *Sustainable Development Goal: Indicator 12.3.1 Global Food Losses*. Rome: Food and Agricultural Organization.

Fedorov, A.S., Tsygankov, A.A., Rao, K.K., & Hall, D.O. (1998). Hydrogen photoproduction by Rhodobacter sphaeroides immobilised on polyurethane foam. *Biotechnology Letters,* 20, 1007–1009.

Hancock, J.T., LeBaron, T.W., May, J., Thomas, A., & Russell, G. (2021). Molecular hydrogen: is this a viable new treatment for plants in the UK?. *Plants*, 10(11), 2270.

Hancock, J.T., & Russell, G. (2021). Downstream signalling from molecular hydrogen. *Plants*, 10(2), 367.

Hancock, J.T., Russell, G., & Stratakos, A.C. (2022). Molecular hydrogen: The postharvest use in fruits, vegetables and the floriculture industry. *Applied Sciences*, 12(20), 10448.

Hayashida, K., Sano, M., Ohsawa, I., Shinmura, K., Tamaki, K., Kimura, K., . . . & Fukuda, K. (2008). Inhalation of hydrogen gas reduces infarct size in the rat model of myocardial ischemia–reperfusion injury. *Biochemical and Biophysical Research Communications*, 373(1), 30–35.

Hu, H., Li, P., Wang, Y., & Gu, R. (2014). Hydrogen-rich water delays postharvest ripening and senescence of kiwifruit. *Food Chemistry*, 156, 100–109.

Hu, H., Zhao, S., Li, P., & Shen, W. (2018). Hydrogen gas prolongs the shelf life of kiwifruit by decreasing ethylene biosynthesis. *Postharvest Biology and Technology*, 135, 123–130.

Hu, Q., Zhou, Y., Wu, S., Wu, W., Deng, Y., & Shao, A. (2020). Molecular hydrogen: A potential radioprotective agent. *Biomedicine & Pharmacotherapy*, 130, 110589.

Jain, I. P. (2009). Hydrogen the fuel for 21st century. *International Journal of Hydrogen Energy*, 34(17), 7368–7378.

Jiang, K., Kuang, Y., Feng, L., Liu, Y., Wang, S., Du, H., & Shen, W. (2021). Molecular hydrogen maintains the storage quality of Chinese chive through improving antioxidant capacity. *Plants*, 10(6), 1095. https://doi.org/10.3390/plants10061095

Jin, Z., Liu, Z., Chen, G., Li, L., Zeng, Y., Cheng, X., . . . & Shen, W. (2023). Molecular hydrogen-based irrigation extends strawberry shelf life by improving the synthesis of cell wall components in fruit. *Postharvest Biology and Technology*, 206, 112551.

Li, C., Gong, T., Bian, B., & Liao, W. (2018). Roles of hydrogen gas in plants: A review. *Functional Plant Biology*, 45(8), 783–792.

Li, F., Hu, Y., Shan, Y., Liu, J., Ding, X., Duan, X., . . . & Jiang, Y. (2022a). Hydrogen-rich water maintains the color quality of fresh-cut Chinese water chestnut. *Postharvest Biology and Technology*, 183, 111743.

Li, L., Liu, Y., Wang, S., Zou, J., Ding, W., & Shen, W. (2020). Magnesium hydride-mediated sustainable hydrogen supply prolongs the vase life of cut carnation flowers via hydrogen sulfide. *Frontiers in Plant Science*, 11, 595376.

Li, L., Wang, J., Jiang, K., Kuang, Y., Zeng, Y., Cheng, X., . . . & Shen, W. (2022b). Preharvest application of hydrogen nanobubble water enhances strawberry flavor and consumer preferences. *Food Chemistry*, 377, 131953.

Li, L., Zeng, Y., Cheng, X., & Shen, W. (2021). The applications of molecular hydrogen in horticulture. *Horticulturae*, 7(11), 513.

Li, C., Zhang, J., Wei, M., Ge, Y., Hou, J., Cheng, Y., & Chen, J. (2019). Methyl jasmonate maintained antioxidative ability of ginger rhizomes by regulating antioxidant enzymes and energy metabolism. *Scientia Horticulturae*, 256, 108578.

Liu, Q., Xie, H., Chen, Y., Lin, M., Hung, Y.C., & Lin, H. (2022). Alleviation of pulp breakdown in harvested longan fruit by acidic electrolyzed water in relation to membrane lipid metabolism. *Scientia Horticulturae*, *304*, 111288.

Liu, X., Fang, H., Huang, P., Feng, L., Ye, F., Wei, L., . . . & Liao, W. (2023). Effects of Hydrogen-Rich Water on Postharvest Physiology in Scales of Lanzhou Lily during Storage. *Horticulturae*, 9(2), 156.

Lu, H., Wu, B., Wang, Y., Liu, N., Meng, F., Hu, Z., . . . & Zhao, H. (2017). Effects of hydrogen-rich water treatment on defense responses of postharvest tomato fruit to Botrytis cinerea. *Journal of Henan Agricultural Sciences*, 46(2), 64–68.

Naveed, F., Nawaz, A., Ali, S., & Ejaz, S. (2024). Xanthan gum coating delays ripening and softening of jujube fruit by reducing oxidative stress and suppressing cell wall polysaccharides disassembly. *Postharvest Biology and Technology*, 209, 112689.

Nicastro, R., & Carillo, P. (2021). Food loss and waste prevention strategies from farm to fork. *Sustainability*, 13(10), 5443.

Ohno, K., Ito, M., Ichihara, M., & Ito, M. (2012). Molecular hydrogen as an emerging therapeutic medical gas for neurodegenerative and other diseases. *Oxidative Medicine and Cellular Longevity*, 2012.

Ohta, S. (2011). Recent progress toward hydrogen medicine: potential of molecular hydrogen for preventive and therapeutic applications. *Current Pharmaceutical Design*, 17(22), 2241–2252.

Ohta, S. (2014). Molecular hydrogen as a preventive and therapeutic medical gas: initiation, development and potential of hydrogen medicine. *Pharmacology & Therapeutics,* 144(1), 1–11.

Ohta, S. (2015). Molecular hydrogen as a novel antioxidant: overview of the advantages of hydrogen for medical applications. *Methods in Enzymology*, 555, 289–317.

Ostojic, S.M. (2021). Hydrogen-rich water as a modulator of gut microbiota?. *Journal of Functional Foods*, 78, 104360.

Quinn, J.F., Whittaker, M.R., & Davis, T.P. (2015). Delivering nitric oxide with nanoparticles. *Journal of Controlled Release*, 205, 190–205.

Ren, P.J., Jin, X., Liao, W.B., Wang, M., Niu, L.J., Li, X.P., . . . & Zhu, Y.C. (2017). Effect of hydrogen-rich water on vase life and quality in cut lily and rose flowers. *Horticulture, Environment, and Biotechnology*, 58, 576–584.

Renwick, G.M., Giumarro, C., & Siegel, S.M. (1964). Hydrogen metabolism in higher plants. *Plant Physiology,* 39(3), 303.

Russell, G., Zulfiqar, F., & Hancock, J.T. (2020). Hydrogenases and the role of molecular hydrogen in plants. *Plants,* 9(9), 1136.

Shao, Y., Lin, F., Wang, Y., Cheng, P., Lou, W., Wang, Z., . . . & Shen, W. (2023). Molecular Hydrogen Confers Resistance to Rice Stripe Virus. *Microbiology Spectrum,* 11(2), e04417–e04422.

Smith, R.A., & Murphy, M.P. (2011). Mitochondria-targeted antioxidants as therapies. *Discovery Medicine*, *11*(57), 106–114.

Su, N., Wu, Q., Liu, Y., Cai, J., Shen, W., Xia, K., & Cui, J. (2014). Hydrogen-rich water reestablishes ROS homeostasis but exerts differential effects on anthocyanin synthesis in two varieties of radish sprouts under UV-A irradiation. *Journal of Agricultural and Food Chemistry*, 62(27), 6454–6462.

Sun, Y., Qiu, W., Fang, X., Zhao, X., Xu, X., & Li, W. (2023). Hydrogen-rich water treatment of fresh-cut kiwifruit with slightly acidic electrolytic water: influence on antioxidant metabolism and cell wall stability. *Foods*, 12(2), 426.

Yan, H., Liu, Y., Wu, Z., Yi, Y., & Huang, X. (2021). Phylogenetic relationships and characterization of the complete chloroplast genome of Rosa sterilis. *Mitochondrial DNA Part B*, 6(4), 1544–1546.

Yang, Y., Li, B., Liu, C., Chuai, Y., Lei, J., Gao, F., . . . & Cai, J. (2012). Hydrogen-rich saline protects immunocytes from radiation-induced apoptosis. *Medical Science Monitor: International Medical Journal of Experimental and Clinical Research*, 18(4), BR144.

Yun, Z., Gao, H., Chen, X., Chen, Z., Zhang, Z., Li, T., . . . & Jiang, Y. (2021). Effects of hydrogen water treatment on antioxidant system of litchi fruit during the pericarp browning. *Food Chemistry*, 336, 127618.

Yun, Z., Gao, H., Chen, X., Duan, X., & Jiang, Y. (2022). The role of hydrogen water in delaying ripening of banana fruit during postharvest storage. *Food Chemistry*, 373, 131590.

Zhang, Y., Zhao, G., Cheng, P., Yan, X., Li, Y., Cheng, D., . . . & Shen, W. (2019). Nitrite accumulation during storage of tomato fruit as prevented by hydrogen gas. *International Journal of Food Properties*, 22(1), 1425–1438.

Zhao, X., Meng, X., Li, W., Cheng, R., Wu, H., Liu, P., & Ma, M. (2021). Effect of hydrogen-rich water and slightly acidic electrolyzed water treatments on storage and preservation of fresh-cut kiwifruit. *Journal of Food Measurement and Characterization*, 15, 5203–5210.

Zhu, Y., Liao, W., Niu, L., Wang, M., & Ma, Z. (2016). Nitric oxide is involved in hydrogen gas-induced cell cycle activation during adventitious root formation in cucumber. *BMC Plant Biology*, 16, 1–13.

Zhu, L.S., Shan, W., Wu, C.J., Wei, W., Xu, H., Lu, W.J., . . . & Kuang, J.F. (2021). Ethylene-induced banana starch degradation mediated by an ethylene signaling component MaEIL2. *Postharvest Biology and Technology*, 181, 111648.

Zulfiqar, F., Russell, G., & Hancock, J. T. (2021). Molecular hydrogen in agriculture. *Planta*, 254(3), 56.

PART VII

Nanotechnology Applications

Potential of Nanoparticles for Postharvest of Fruits and Vegetables

21

Azri Shahir Rozman, Norhashila Hashim*, Bernard Maringgal, Intan Syafinaz Mohamed Amin Tawakkal, Khalina Abdan, and Akhmad Sabarudin

Corresponding Author: norhashila@upm.edu.my

21.1 INTRODUCTION

Fruits and vegetables contain high nutritional content, as well as health-promoting qualities over diseases such as diabetes, cancer, and heart disease, which leads to an increase in their demand (Islam & Kabir, 2019). However, postharvest losses of fruits and vegetables continue to be a major global issue and concern in the agricultural industries. Their metabolism speeds up after being harvested, which accelerates senescence and spoilage (Cid-López et al., 2021). The increase of water content, pH, and the presence of enzymes (α-galactosidase, polygalacturonase, and pectin methylesterase activities) during postharvest all contribute to the decrease of fruit and vegetable quality. The spoilage is also linked to physiological and microbiological processes, as well as the interactions between them (Perumal et al., 2022). Further, due to market trends in recent years towards minimally processed, fast produced, and "fresh" food products, the use of lower heat, high pressure, UV irradiation, or electric pulse treatments could increase the survival and proliferation of pathogenic microbes (Kowsalya et al., 2019).

Richard Feynman first proposed the idea of nanotechnology in 1959, and Norio Taniguchi later popularised the phrase in 1974. Nanotechnology primarily entails the creation, analysis, and manipulation of molecules on the nanoscale (< 100 nm). The design, production, processing, and application of polymer materials, including nanoparticles and/or nanoscale devices, constitute the application of nanotechnology to polymers (Sharma et al., 2017). The characteristics of nanoparticles are distinct from those of the same materials in bulk form. Due to their surface, quantum, and small-size effects, nanoparticles typically exhibit superior biocompatibility, great adsorption capacity, and ease of surface modification, making them desirable for applications in medical imaging, drug delivery, and disease detection (Tao, 2018). The emergence of nanotechnology has opened numerous possibilities for the creation of novel antibacterial substances. Due to their distinctive physical and chemical characteristics as well as their potent antibacterial effects, nanoparticles have attracted greater interest in recent years (Sardella et al., 2017).

Recently, because of the distinctive physicochemical features, regulated size, and form of metal nanoparticles, their usage in health engineering has greatly risen (Sivamaruthi et al., 2019). Also due to their special qualities, such as their higher surface-to-volume ratios, small size, and quantum effects, nanostructures and nanomaterials are well suited for a variety of uses in the food sector. For their inherent mesoporous particles, antibacterial activity, graphene, and carbon dots have particularly drawn attention in the foods business (Omerović et al., 2021). Inorganic nanomaterials include metal and metal oxide nanoparticles, such as silver, zinc oxide, copper/copper oxide, titanium dioxide, and magnesium oxide nanoparticles, as well as graphene and carbon dots. Many areas of the food industry, including processing, conditioning, packing, storage, and preservation, have been revolutionised by the recent advances in nanoscale research, which have boosted quality and usefulness. The studies have shown that food is protected from hazardous and spoilage-causing microorganisms by antimicrobial nanoparticles used in active packaging, allowing for longer-term freshness during storage and greater consumer convenience (Omerović et al., 2021). The potential for effective nanotechnology applications to alter current food science and the food business is widely acknowledged (Rodino et al., 2019).

DOI: 10.1201/9781003370376-28

Hence, in this chapter, the fundamental nanoparticle properties that have potential in fruit and vegetable postharvest application are elucidated. The concern over the adverse health effects of nanoparticles is highlighted. The challenges of applying nanoparticles as an advanced technology and its future trends are also discussed.

21.2 MECHANISM OF NANOPARTICLES FOR PRESERVING FRUIT AND VEGETABLE QUALITY

It is important to understand the essential properties of nanoparticles to produce materials that could be applied in any sector, especially agriculture. The mechanism of nanoparticles for preserving the quality of fruits and vegetables can be divided into its antimicrobial and antioxidant activities.

21.2.1 Antimicrobial Activity

Apparently, stored fruits and vegetables face a significant challenge of high decay rates caused by microorganisms, which lead to nutrient degradation (Xing et al., 2016). The spread of dangerous bacteria poses a serious threat to both human health and fruit quality, which has driven intensive study into nano-based products to overcome these threats (Xing et al., 2019). On the other hand, some manufacturers practice adding synthetic preservatives and antibiotics to food to prevent microorganism problems, which has led to increased community awareness of the effects of synthetic preservatives (Jadhav et al., 2022).

Numerous natural antimicrobial agents are available, such as plant-derived materials, peptides, and enzymes that have been used in films or coatings (Lopes & Brandelli, 2018). In addition, different types of nanoparticles have distinct antibacterial properties. Although the specific antibacterial activity of nanoparticles is unclear, it has been hypothesised that several different mechanisms may play a role in the antimicrobial mechanisms (Fernando et al., 2018). Some of the characteristics that affect the antimicrobial properties of nanoparticles are physical structure, ion release, surface-to-volume ratio, chemistry, particle size, particle shape, etc. (Wang et al., 2017).

Nanoparticles may infiltrate tissues more effectively due to their nanometric size, thus nanotechnology uses nanoparticles or nanomaterials as food additives for their beneficial effects (Cid-López et al., 2021). Relevant research has indicated that smaller microspheres with greater specific surface areas can engage with bacteria more closely and carry out the bacteriostatic function. Therefore, there has been a lot of attention given to regulating both the polydispersity and average size (Liang et al., 2019). Additionally, it has previously been documented that metallic nanoparticles made of silver, gold, copper, or zinc have antimicrobial characteristics. For instance, gold nanoparticles may adhere to the membrane of a bacterial species, allowing the contents of the bacterium to leak out, or they may pierce the outer membrane and the peptidoglycan layer, killing the bacteria (Zahran & Marei, 2019).

Due to their usage as cytotoxins, antioxidants, and antimicrobials, silver nanoparticles have attracted significantly more attention in recent years than any other type of nanomaterial (Akter et al., 2018). Also, Akter et al. (2018) reported that hypothesised particle size, particle shape, and concentration have influenced the cytotoxicity effects of silver nanoparticles, which means the ability to damage or kill cells. A study reported by Padilla-Camberos et al. (2021) concerning the synthesis of silver nanoparticles using pitaya fruit peel extract exhibited strong antimicrobial activity against gram-positive and gram-negative bacteria, fungi, and a methicillin-resistant strain of *Staphylococcus aureus*. According to the author, the antimicrobial activity shown was due to the contact between the negative charges on cell membranes of microorganism and functional groups on its surface and positive charges of nanoparticles.

Further, zinc oxide nanoparticles also demonstrated antimicrobial activity, which was reported by Amuthavalli et al. (2021) in their research on the antimicrobial properties of zinc oxide nanoparticles synthesised using plant *Lawsonia inermis* L. extracts. The authors reported that zinc oxide nanoparticles have inhibited several tested microbes, which included bacteria (*E. coli* and *B. subtilis*) and fungi (*A. alternate* and *A. flavus*), respectively. According to the authors, zinc oxide nanoparticles inhibited pathogen growth by disrupting cell walls, which may reduce the surface hydrophobicity of bacterial cells, and that their oxidative stress-resistance genes were down-regulated, resulting in cell degradation and death. Additionally, referring to Figure 21.1, zinc oxide nanoparticles were found secure and suitable for use in the human body. Numerous articles have discussed the potential antibacterial mechanisms of zinc oxide nanoparticles. These include (1) the generation of reactive oxygen species (ROS), such as the hydroxyl radical and peroxide, which cause oxidative stress, cell membrane disturbance, and DNA damage, ultimately leading to the death of the bacterial cell; and (2) dissolution of zinc oxide nanoparticles into Zn^{2+} ions, which interact with bacterial cells, particularly cytoplasm, cell membranes, and nucleic acids (Murali et al., 2021).

Another study on the antimicrobial activity of nanoparticles by Onitsuka et al. (2019) successfully developed gold nanoparticles using black tea extract of *Camellia sinensis* leaves. The authors stated that the nanoparticles were able to inhibit both gram-positive and -negative bacteria, *S. aureus* (NBRC 12732) and *K. pneumoniae* (NBRC 13277) respectively due to the small particles having high activity on the bacteria. In addition, a study by Chompunut et al. (2022) also showed the promising antimicrobial activity of copper

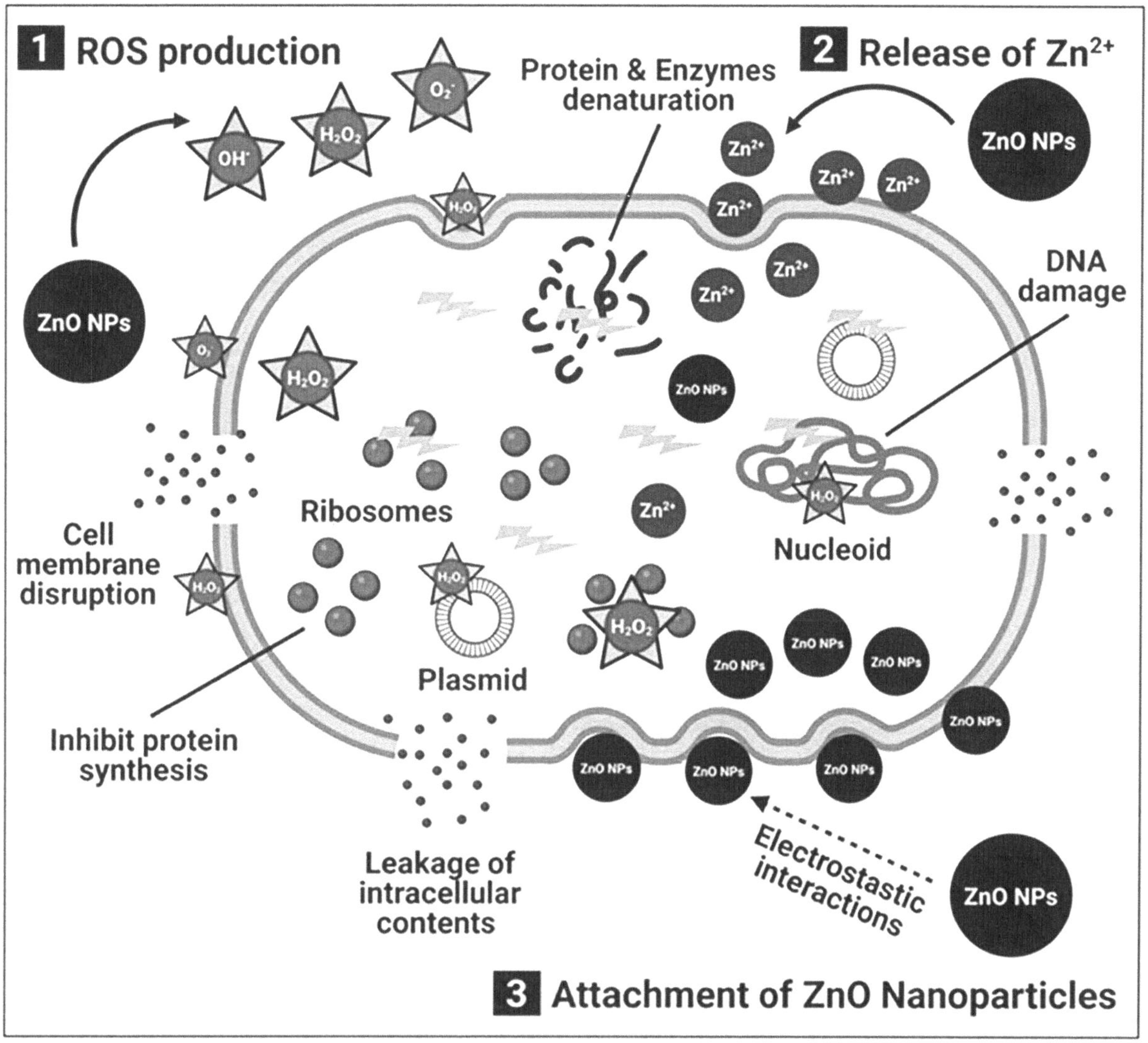

FIGURE 21.1 Possible antibacterial mechanisms of plant-mediated zinc oxide nanoparticles (Murali et al., 2021).

nanoparticles that were synthesised using *Cynodon dactylon* aqueous leaf extract. The nanoparticles were able to inhibit all tested bacteria and fungi due to the release of nanoparticles on the cell wall of microbes that hindered their growth and led to cell death. According to the author, the mechanism was also attributed to a high surface-to-volume ratio. Interestingly, a study on the antifungal activity of calcium oxide nanoparticles by Maringgal et al. (2020) that were synthesised using stingless bee honey also showed promising results in inhibiting *Colletotrichum brevisporum* at 15% concentration. According to the authors, the nanoparticles modified the permeability of the fungi cell membrane, which led to cell death. Table 21.1 comprises a summary of the recently reported antimicrobial activity of various nanoparticles.

21.2.2 Antioxidant Activity

Any agent or compound that could prevent the oxidation of a suitable substrate even at low concentrations is referred to as an antioxidant. The standard approach to determine the rate of oxygen consumption when fat is stored in an enclosed container with oxygen implies determining the antioxidant capacity of any component (Kumar et al., 2020a). The oxidative breakdown of organic materials, such as biological components like lipids and proteins, foods, and cosmetics, is induced by a radical-chain mechanism in which alkyl radicals are converted into peroxyl radicals (ROO) by atmospheric O_2 and utilised to establish new oxidative chains. As this occurs under benign conditions and for no apparent reason, this reaction is referred to as autoxidation. It is also referred to as peroxidation because the primary initially produced products are hydroperoxides (Ingold & Pratt, 2014). In addition to being highly reactive and capable of attacking even relatively stable compounds like DNA bases, hydroxyl (HO) and alkoxyl (RO) radicals can be produced when hydroperoxides are homolytically cleaved (Valgimigli et al., 2018). Further, oxidative stress is the accumulation of irreversible damage to lipids, proteins, and DNA that results in mutation and cell death. It is defined as the imbalance between the creation of ROS and the capacity of the

TABLE 21.1 A Summary of Nanoparticles' Antimicrobial Activity

NANOPARTICLES	*REMARKS*	*REFERENCE*
Silver nanoparticles	1. Nanoparticles were synthesised using pitaya fruit peel extract. 2. Tested microorganism: • Gram-negative (*E. coli* ATCC 11229, *S. enterica* ATCC 10708, *P. aeruginosa* ATCC 9027) • Gram-positive bacteria (*S. aureus* ATCC 6538, methicillin-resistant *S. aureus* (MRSA) ATCC 43300), and fungi (*C. albicans* ATCC 10231). 3. Antimicrobial activity of nanoparticles: • Interactions between negative charges on cell membranes of microorganism and functional groups on its surfaces and positive charges of nanoparticles. • Also, interruption of respiration and cycles of cell division causes a lysis due.	Padilla-Camberos et al. (2021)
Zinc oxide nanoparticles	1. Synthesised using plant *L. inermis* L. 2. Tested microorganism: • Bacteria (*E. coli* and *B. subtilis*) • Fungi (*A. alternate* and *A. flavus*) 3. Antimicrobial activity of nanoparticles: • Disrupting cell walls causing surface hydrophobicity of bacterial cell. • Down-regulated the oxidative stress–resistance genes.	Amuthavalli et al. (2021)
Gold nanoparticles	1. Synthesised using black tea extract of leaves of *C. sinensis*. 2. Tested microorganism: • *S. aureus* (NBRC 12732) • *K. pneumoniae* (NBRC 13277) 3. Antimicrobial action of nanoparticles: • Strongly reliant on the size, the smaller dimension of Ag NPs (<30 nm), imparting higher activity against both bacteria. • Assembly of functional organic molecule such as polyphenol, acquired from tea extract, onto the nanoparticles' surface.	Onitsuka et al. (2019)
Copper nanoparticles	1. Synthesised using *C. dactylon* aqueous leaf extract. 2. Tested microorganism: • *A. niger*, • Gram-negative (*Escherichia coli* and *K. pneumoniae*) • Gram-positive (*B. subtilis* and *S. aureus*) 3. Antimicrobial activity of nanoparticles: • Release of the nanoparticles on the cell wall of microbes eventually arrests its growth and causes the death of the microorganisms. • Smaller particle size and surface-to-volume ratio increase and help in release of more nanoparticles	Chompunut et al. (2022)
Calcium oxide nanoparticles	1. Synthesised using stingless bee honey 2. Tested microorganism: • *Colletotrichum brevisporum* 3. Antimicrobial activity of nanoparticles: • Nanoparticles modify the permeability of the fungi cell membrane, thus releasing the lipopolysaccharides and membrane proteins available. • Free radicals are generated, and the proton motive is forced to dissipate, which leads to cell death.	Maringgal et al. (2020)

TABLE 21.2 A Summary of Nanoparticles' Antioxidant Activity

NANOPARTICLES	*REMARKS*	*REFERENCE*
Silver nanoparticles	1. Nanoparticles were synthesised using *Citrus limon* zest. 2. Antioxidant activity of nanoparticles due to the chemical structure, redox properties, and particle shape and size.	Khane et al. (2022)
Zinc oxide nanoparticles	1. Nanoparticles were synthesised using mulberry fruit 2. Antioxidant activity of nanoparticles due to phytochemical compounds from the mulberry fruit such as polyphenols, flavonoids, proteins, and fatty acids.	Abbes et al. (2022)
Gold nanoparticles	1. Nanoparticles were synthesised using *Tecoma capensis* (L.) leaf extract. 2. The antioxidant activity of nanoparticles could be attributed to the high reducing capacity of these plant extracts which resulted in their high antioxidant efficiency.	Hosny et al. (2022)
Copper nanoparticles	1. Nanoparticles were synthesised using *Cissus vitiginea* leaf extract. 2. Antioxidant activity of nanoparticles due to extract from *C. vitiginea* leaves that contains abundant polyphenols and flavonoids.	Wu et al. (2020)

cell to trigger an effective antioxidative response (Valgimigli et al., 2018).

Plants, specifically edible fruits, vegetables, spices, and herbs that are high in vitamins, carotenoids, phenolic compounds, and microelements, are the primary sources of natural antioxidants. It should be mentioned, nonetheless, that the antioxidant activity varies for various natural resource types and morphological components (Flieger et al., 2021). It should be underlined that due to clear taste standards and requirement for EFSA (European Food Safety Authority) or US FDA (Food and Drug Administration) certification, the selection of bioactive chemicals for the food business is constrained. In the meantime, an intriguing development was observed in the approach to utilising processing industry byproducts (Lourenço et al., 2019).

Many types of nanomaterials have natural antioxidant effects that arise from the surface characteristics of the material rather than from their functionalisation with antioxidants. These materials mostly consist of inorganic metal nanoparticles (Valgimigli et al., 2018). Aside from natural extracts, several nanoparticles with antioxidant activity and the ability to remove reactive nitrogen (RNS) and reactive oxygen species (ROS) includes carbon nanotubes, metals, metal oxides, and various nanoparticle forms that are loaded with polymers (Eftekhari et al., 2018). Unfortunately, dangerous reducing and stabilising chemical compounds are commonly required to produce nanoparticles through chemical means. The harmful materials adsorbed on the nanoparticle surfaces restrict their uses in the biological domain (Flieger et al., 2021). Thus, natural biosynthesis techniques such as reduction of metallic cations by plant extracts, yeasts, fungi, and bacteria are being exploited to produce nanomaterials. Metal ions are reduced in the first stage, and then colloidal suspension aggregates to form oligomeric clusters in the second step, resulting in the creation of nanoparticles (Kumar et al., 2020b).

A study by Khane et al. (2022) regarding the synthesis of silver nanoparticles using aqueous *Citrus limon* zest extract showed significant antioxidant activity when compared to ascorbic acid and standard solution. The author stated that the significant differences in antioxidant activity may be due to the chemical structure, redox properties, and particle shape and size. Then, a study by Abbes et al. (2022) on the green synthesis of zinc oxide nanoparticles using mulberry fruit extract also showed significant antioxidant activity. Interestingly, the author stated that using the green method for producing zinc oxide nanoparticles could offer better antioxidant activity when compared to the conventional chemical method. The success in producing superior zinc oxide nanoparticles was due to phytochemical compounds from the mulberry fruit such as polyphenols, flavonoids, proteins, and fatty acids. Similar studies by Hosny et al. (2022) and Wu et al. (2020), in their study of gold nanoparticles and copper nanoparticles respectively, describe that the antioxidant activity of the nanoparticles was affected by the present phytochemical properties of the extracts used for synthesising. Table 21.2 briefly discusses recent reports of nanoparticles and their antioxidant activity.

21.3 APPLICATION OF NANOPARTICLES ON FRUITS AND VEGETABLES

Fruit and vegetable quality can be influenced by environmental conditions, such as storage temperature, humidity, the sharpness of the cutting tools, and chemical treatments, as well as internal factors like morphological, physiological, and biochemical defence mechanisms; fruit type; genotype; and stress-induced senescence programmes. Prolonging the shelf life of fruit is more difficult than enhancing the shelf life of vegetable crops due to the complex physiological makeups of fruits in terms of ethylene gas biosynthesis and ripeness stages (Yousuf et al., 2018). The applications of nanoparticles for fruit and vegetable quality preservation are usually incorporated through edible films and coatings as well as wrapping films.

21.3.1 Edible Films and Coatings

Any thin material that is used to wrap or coat a product to increase its shelf life and may be ingested together with the product is referred to as an edible film or coating. Edible films and coatings enhance the quality of foods like fruits and vegetables by preventing physical, chemical, and microbiological changes such as moisture loss, enzymatic browning reactions, and lipid oxidation (Kocira et al., 2021). The term edible coating refers to the thin layer of edible ingredients that covers the fruits' and vegetables; surface, whereas the term edible film refers to the layer that has already been made and applied to the raw materials. The raw material is dipped into a solution that contains the edible coating in liquid form. To apply edible films, the film is first formed into a solid sheet and then wrapped around the product (Vipan et al., 2018; Okcu et al., 2018). Compared to synthetic coatings, edible films and coatings offer many benefits, such as (1) they function as a gas and moisture barrier that alters the atmosphere inside the fruit, extending its shelf life and maintaining the freshness of the produce. (2) They serve as a barrier to microorganisms and promote good hygiene in this way. (3) The polymer matrix can be infused with a variety of active chemicals, including anti-browning agents, colours, fragrances, nutrients, and spices, to increase safety and even the nutritional and sensory characteristics of the fruit. (4) Due to their biodegradable nature, they assist in reducing synthetic packaging waste (Kocira et al., 2021).

In the case of edible coating, the food product is coated with the solution first, and then the films are developed by drying the applied solutions on the food products. In contrast, in the case of edible films, the film is first created by drying the solution, and then the food product is coated with the created film (Figure 21.2) (Bizymis & Tzia, 2021).

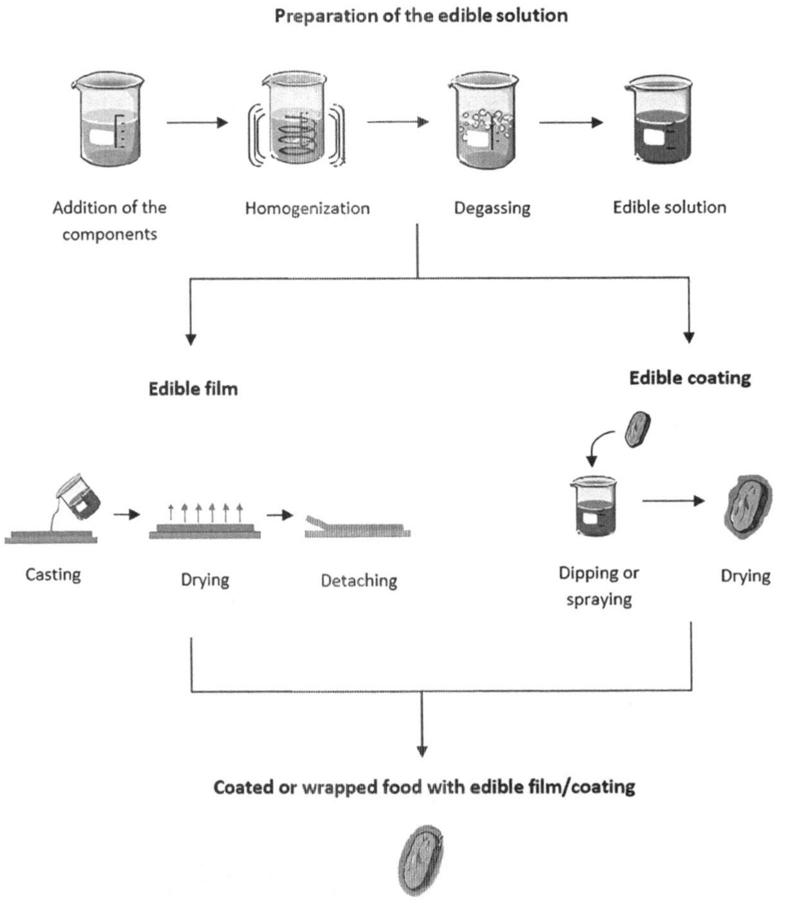

FIGURE 21.2 Preparation of edible films and edible coatings (Bizymis & Tzia, 2021).

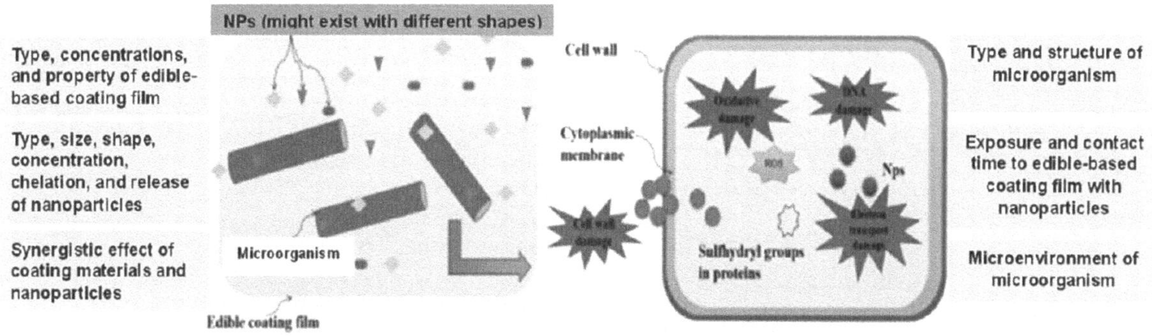

FIGURE 21.3 Factors that affect the antimicrobial activity of edible chitosan film with nanoparticles (Xing et al., 2019).

The ingredients that are frequently employed to make edible films and coatings can be divided into three fundamental groups: polysaccharides, proteins, and lipids. Edible films or coatings with just one basic ingredient frequently fall short in certain areas. However, mixing two or more elements, sometimes even ones from different classes, can enhance their qualities. It is difficult to conduct research on and apply these "composite edible films and coatings" to foods (Bizymis & Tzia, 2021). Recently, research has focused on the utilisation of nanomaterials to create edible films and coatings with better qualities that are not possible with traditional components (He et al., 2018; Seifari & Ahari, 2020). Additionally, the expected role of edible coatings or bio-nanocomposite films in protecting postharvest quality is widely established. Applying edible coatings is easy, safe, inexpensive, and environmentally benign, making it a viable method for preserving tropical fruits (Jafarzadeh et al., 2021).

Nanoscale antibacterial agents have drawn a lot of attention, as they could be added to the films to ensure food safety. Therefore, adding antibacterial agents to composite films is a direct technique especially given their high reactivity and effectiveness. Zinc oxide nanoparticles have been utilised widely as antibacterial treatments because they exhibit antibacterial action in a variety of microorganisms (Lian et al., 2021). As shown in Figure 21.3, the types, concentrations, and characteristics of the coating films as well as the types, sizes, shapes, concentrations, chelations, releases, and photocatalytic properties of the incorporated nanoparticle may play a major role in the effectiveness of these coatings in reducing microbial growth. Additionally, the types and structures of the studied microorganisms, the testing conditions, and the synergistic action of coating materials and nanoparticles in the barrier all affect the antibacterial effectiveness of edible coatings (Xing et al., 2019).

On the other hand, a recent study concerning edible film by Dong et al. (2022) showed that chitosan film integrated with silica nanoparticle contents significantly improved the mechanical, water vapour barrier property, and UV light barrier when compared to pure chitosan film. The author also stated that the incorporation of silica nanoparticles improved the antioxidant properties of the film, which were related to the phenolic compound present during the development of the nanoparticles. Then, a study by La et al. (2021) on chitosan/gum arabic edible coating incorporated with zinc oxide nanoparticles indicated positive results in preserving bananas for more than 17 days. The authors reported that the incorporation of zinc oxide nanoparticles significantly enhanced the performance of the chitosan/gum arabic edible coating when compared to the control sample in which the performance of the edible coating was evaluated by the changes in the physicochemical properties of the bananas. In Table 21.3, similar trends were reported by Jiang et al. (2021) and Hassan et al. (2021), whereby incorporating nanoparticles inside the film matrix of edible films and coatings could enhance the conservation of fruits by modifying the barrier properties and enhancing antimicrobial activity.

21.3.2 Wrapping Films

Without a doubt, the greatest invention of the 20th century is a polymer made from petroleum. In our modern daily life, it is normal and inevitable to use this non-biodegradable synthetic polymer that is derived from petroleum as a food packaging material. We indirectly contributed to global warming and resource depletion because of our over-reliance on food packaging for luxury and convenience. As an alternative to the non-biodegradable, synthetic polymer derived from petroleum, various forms of biopolymers have been used in the application of food packaging (Avella et al., 2005). Many synthetic and non-biodegradable polymers, including low- and high-density polyethylene, polyvinylchloride, and polyvinylpyrrolidone, have been utilised to create packaging materials to reduce fruit rotting losses during storage and transportation. After usage, improper disposal of these packing materials causes a persistent release of hazardous material into the environment, which has the potential to seriously harm ecosystems (Kalia et al., 2021). As a result, the growing problems with waste disposal brought on by the usage of synthetic packaging materials are driving research into the use of a potential substitute with environmentally acceptable biodegradable polymers (Kalia et al., 2021).

Researchers have therefore focused their efforts on the discovery of biopolymers or their combinations in the development of active and intelligent biofilms for food packaging due to the biodegradable, non-toxic, and renewable nature of

TABLE 21.3 A Summary of Nanoparticle Applications in Edible Films and Coatings

NANOPARTICLES	*FILMS/COATINGS*	*REMARKS*	*REFERENCE*
Silica nanoparticles	Chitosan edible films	1. Improve the mechanical, water vapour barrier property, and UV light barrier of chitosan film. 2. Improve the antioxidant properties of the film.	Dong et al. (2022)
Zinc oxide nanoparticles	Chitosan/gum arabic coatings	1. Nanoparticles significantly improve the functionality of coatings in colour change retention, weight loss retention, fruit firmness retention, organic acid retention, and reducing sugar retention. 2. The nanoparticles improve the mechanism of coatings by enhancing the barrier for oxygen and carbon dioxide penetration.	La et al. (2021)
Catechin/β-cyclodextrin inclusion complex nanoparticles	Zein films	1. Nanoparticles increase the barrier properties of films by occupying the pores of the film matrix. 2. Enhance the mechanical properties of films by improving intermolecular and intramolecular interactions. 3. A higher concentration of nanoparticles in films results in better antioxidant activity.	Jiang et al. (2021)
Chitosan nanoparticles	Chitosan/ thyme volatile oil films	1. Thyme volatile oil encapsulated with chitosan nanoparticles. 2. The shelf life of preserved basil leaves was improved significantly up to 140.61% compared to the control sample. 3. The nanoparticles improved the barrier property of the edible coating. 4. Thyme volatile oil enhanced the antimicrobial and antioxidative activity of edible coatings.	Hassan et al. (2021)

these biopolymers as a plastic substitute (Table 21.4). In order to avoid oxidation and spoilage, these biopolymers can also serve as a transporter for bioactive-like antioxidant or antimicrobial compounds (Thivya et al., 2021). Therefore, biodegradable polymer packaging must be developed for the food packaging business. Chitosan is one of the most prevalent polysaccharides in nature, among the various raw materials mentioned in the literature (Flórez et al., 2022). In addition, most of the natural-based wrapping film lacks any active antimicrobial properties for keeping food products. To enhance the antibacterial and antioxidant characteristics of the film, active ingredients such as plant extracts, essential oils, phenolic acids, chitosan, flavonoids, and metal nanoparticles are added (Mohamad et al., 2022).

The physical characteristics of biopolymer films have been greatly enhanced, and attributes like antioxidant and antibacterial activity have been increased, due to the incorporation of functional elements. As a result of the application of nanotechnology in the field of food packaging, active packaging materials with antibacterial capabilities have been created using a variety of functional nanomaterials, including nanoclays, metal or metal oxide nanoparticles, and organic nanoparticles (Ezati et al., 2022). Additionally, the strong surface reactivity of nanoparticles is what prevents some microorganisms from

TABLE 21.4 A Summary of Nanoparticle Applications in Wrapping Films

NANOPARTICLES	*FILMS*	*REMARKS*	*REFERENCE*
Zinc oxide nanoparticles	Agar film	1. 2% and 4% concentrations of nanoparticles enhanced the mechanical and physical characteristics of agar film when compared to control films. 2. Preservation of grapes for up to 21 days using a 4% concentration of zinc oxide nanoparticles.	Kumar et al. (2019)
Titanium dioxide and silver nanoparticles	Polylactic acid film	1. Nanoparticles improved the water barrier ability of packaging films. 2. The enzyme activity of tested fruits was inhibited by the reduction of water activity, resulting in preserving fruits' physicochemical properties. 3. The antimicrobial activity of the films also improved, which related to the small particle size of the nanoparticles, and it would generate many electron–hole pairs that induced redox reactions on those microorganisms.	Chi et al. (2019)
Copper oxide and zinc oxide nanoparticles	Chitosan film	1. Nanoparticles as a film filler showed significantly better antimicrobial activity and physical properties compared to pure chitosan film. 2. Fruits wrapped with a nanoparticle-filled film showed the lowest spoilage (6.3%), while fruits wrapped with pure chitosan film experienced 16.7% spoilage after 21 days of storage.	Kalia et al. (2021)
Gold nanoparticles and graphene oxide	Polyvinyl alcohol	1. Nanomaterials have improved the mechanical and physical properties of the film. 2. The presence of gold nanoparticles has also improved the antimicrobial activity of films better than the combination of both nanomaterials, graphene oxide alone, and control samples. 3. In the preservation of fruits, films with gold nanoparticles show the best result for the least changes in fruit properties.	Chowdhury et al. (2020)

growing while improving the mechanical and barrier properties of the resulting composite film by adding nanoparticles as a filler (Chowdhury et al., 2020).

Furthermore, when employed as filler, nanoparticles such as zinc oxide and titanium dioxide can act as antimicrobial agents and create antimicrobial food packaging materials. When antimicrobial food packaging is used, the antimicrobial agents are progressively released to the food surface to prevent bacteria development and increase the shelf life of the food (Shapi'i et al., 2020). Zinc oxide nanoparticles are currently regarded as the most promising class of nanofiller among all inorganic nanoscale particles due to their notable physicochemical characteristics. Zinc oxide nanoparticles are readily available, inexpensive to produce, and exhibit strong bioactivity and catalytic activity. Remarkably, the US FDA has classified zinc oxide nanoparticles as "Generally Recognised as Safe", and it shows greater biocompatibility than other nanoparticles (Guo et al., 2020). A study by Kumar et al. (2019) on developing agar film incorporated with zinc oxide nanoparticles (Figure 21.4) demonstrated that the films with 2% and 4% concentrations of zinc oxide nanoparticles improved the mechanical and physical characteristics of agar film when compared to the control film. The author also showed a promising result of the preservation of grapes for up to 21 days when applied with a 4% concentration of zinc oxide nanoparticles in the film. As reported by Chi et al. (2019), Kalia et al. (2021), and Chowdhury et al. (2020), nanoparticles also were able to enhance the barrier properties, physical properties, mechanical properties, and antimicrobial activities of the wrapping film.

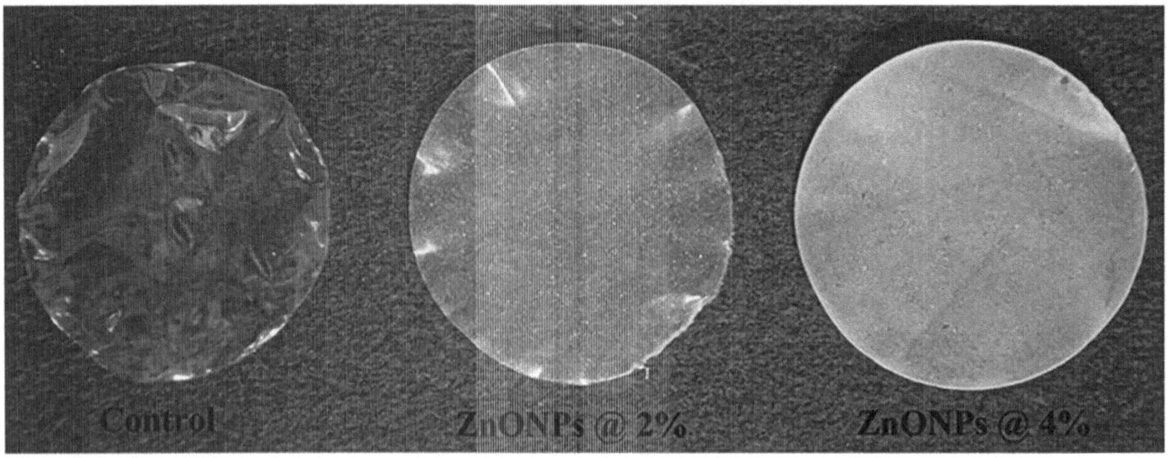

FIGURE 21.4 Photographs of the zinc oxide nanoparticles incorporated in agar nanocomposite films (Kumar et al., 2019).

21.4 CONCERNING THE ADVERSE HEALTH EFFECT OF NANOPARTICLES

Nanotechnology has had a significant impact on food packaging in the field of food science. Thus, due to the rapid development of nanotechnology and its effects in all areas of life in the 21st century, it is essential for consumers to have a thorough knowledge of the interactions between nanoparticles and cells, tissues, and organisms, particularly in relation to any potential risks to human health (Chaudhry & Castle, 2011). Despite many advantages, nanotechnology also poses certain risks and has unfavourable effects on both people and the environment. The persistent and non-degradable characteristics of nanomaterials are predominantly responsible for their toxicological effects, whereas the advantageous elements of the unique properties of nanoparticles are provided by their higher surface area and tiny size. But they also have negative side effects, such as high reactivity when interacting with biological elements (Chawla et al., 2021). Given their high reactivity, nanoparticles can easily bypass membrane barriers and blood arteries, which could consequently exhibit a number of hazardous effects (Nile et al., 2020).

However, due to their small size and strong ability to penetrate human tissue, nanoparticles are the subject of numerous hypotheses regarding their toxicity and consequent risk to human health. As an alternative, natural biopolymers such as cellulose, chitin, and chitosan have been employed to make nanoparticles that are produced for use as nanofillers. Generally, they are fit for human health and are edible. Chitosan nanoparticles are an example and have been employed in food coatings, packaging, and other applications because of their effective antibacterial qualities (Shapi'i et al., 2020). As discussed in the previous section, by improving material features such as mechanical and barrier properties, thermal resistance, processability, and durability, among others, nano-based food packaging systems offer a number of advantages and can address a number of the drawbacks of conventional packaging systems (Sharma et al., 2017). Additionally, the lack of adequate consumer knowledge, government regulations, guidelines, and policies, as well as the absence of reliable and efficient risk assessment detection methods, call for immediate action to develop a superior understanding of nano-toxicity, its characteristics, and the necessary regulatory procedures (Nile et al., 2020). To better understand the toxicological concerns related to these materials, it is crucial to assess various nanomaterials and their interactions with various living creatures in the ecosystem.

21.5 CHALLENGES AND FUTURE TRENDS

Nanotechnology has the potential to be extensively used in the food business in all aspects. This is based on minimal evidence obtained primarily through research. Given the limited ability to control the properties and interactions of materials at the nanoscale, and unclear environment-related impact and relatively scarce toxicity database, the practical applications of nanotechnology and the marketing of products based on nanomaterials remain dubious. Furthermore, this restricts the creation of new laws and regulations, which makes it harder to commercialise innovative items. The public is unaware of food nanotechnology, yet their attitudes can change depending on how the technology is applied and promoted. The issue appears to be the public's need for information regarding food nanotechnology. It is necessary to build up a suitable database and body of evidence to serve as logistical support for information for the public as well as for food makers.

The use of bioinspired approaches is growing in popularity in biological research and other fields since doing so will become vital for success in today's biotechnology industry. The development of nanoparticles from natural and safe sources using the "green method" is in line with the objectives set by most governments of the world, which are trying to accomplish sustainable developments goals (SDG), including good health and well-being, innovation, industry, and infrastructure, as well as ecological environments. Also, many products utilising innovative nanotechnology have been introduced all over the world, especially in the field of food contact technologies/materials, which is driven by food industries, a trillion-dollar sector. Due to the legislative branch of government power over safety codes, this will continue to be a battleground for manufacturers. The agricultural sector can be revolutionised with the use of nanotechnology. The nanotoxicological effects are based on entirely separate factors that are not based on the number of doses with respect to mass, albeit it exhibits different behaviours to their counterparts in bulk. Instead, the level of toxicity of nanoparticles, and hence the range of potential biochemical effects, is determined by factors like aggregation, shape, concentration, surface modification, and size.

21.6 CONCLUSIONS

In this chapter, a discussion of the fundamental properties of nanoparticles that could have potential in fruit and vegetable postharvest applications has been highlighted, including antimicrobial and antioxidant activity. Through the discussion of the antimicrobial and antioxidant activity of nanoparticles, it can be concluded that particle size, phytochemical properties of reducing agents, and types of precursors are significantly correlated. Additionally, the application of nanoparticles in fruit and vegetable postharvest has been majorly reported in edible films, edible coatings, and wrapping films. In terms of applications, various types of nanoparticles have been reported and have been used as fillers to enhance the properties of coatings or film matrixes while prolonging the shelf life of tested samples. However, health concerns regarding the usage of nanoparticles need to be critically evaluated to convince consumers of their safety. Proper regulations and procedures for producing, handling, application, and disposal of nanoparticles are also needed to protect the safety and health of every party that has direct interaction with the nanoparticles. The potential of nanoparticles in the application for fruit and vegetable postharvest is very promising and could be a breakthrough in various gaps that exist at present.

ACKNOWLEDGEMENTS

The authors wish to thank the Southeast Asian Regional Center for Graduate Study and Research in Agriculture (SEARCA) under the University Consortium (UC) Seed Fund for Collaborative Research Grants (Vot No. 6380085) for their financial support.

REFERENCES

Abbes, N., Bekri, I., Cheng, M., Sejri, N., Cheikrouhou, M., Xu, J. (2022). Green synthesis and characterization of zinc oxide nanoparticles using mulberry fruit and their antioxidant activity. *Materials Science*, *28*(2), 144–150.

Akter, M., Sikder, T., Rahman, M., Ullah, A.K.M.A., Fatima, K., Hossain, B., Banik, S., Hosokawa, T., Saito, T., Kurasaki, M. (2018). A systematic review on silver nanoparticles-induced cytotoxicity: Physicochemical properties and perspectives. *Journal of Advanced Research*, *9*, 1–16.

Amuthavalli, P., Hwang, J.S., Dahms, H.U., Wang, L. (2021). Zinc oxide nanoparticles using plant *Lawsonia inermis* and their anticancer applications showing moderate side effects. *Scientific Reports*, 1–13.

Avella, M., De Vlieger, J.J., Errico, M.E., Fischer, S., Vacca, P., Volpe, M.G. (2005). Biodegradable starch/clay nanocomposite films for food packaging applications. *Food Chemistry*, *93*(3), 467–474.

Bizymis, A., Tzia, C. (2021). Edible films and coatings: Properties for the selection of the components, evolution through composites and nanomaterials, and safety issues. *Critical Reviews in Food Science and Nutrition*, *0*(0), 1–16.

Chaudhry, Q., Castle, L. (2011). Food applications of nanotechnologies: An overview of opportunities and challenges for developing countries. *Trends in Food Science & Technology*, *22*(11), 595–603.

Chawla, R., Sivakumar, S., Kaur, H. (2021). Antimicrobial edible films in food packaging: Current scenario and recent nanotechnological advancements- a review. *Carbohydrate Polymer Technologies and Applications*, *2*, 100024.

Chi, H., Song, S., Luo, M., Zhang, C., Li, W., Li, L., Qin, Y. (2019). Effect of PLA nanocomposite films containing bergamot essential oil, TiO2 nanoparticles, and Ag nanoparticles on shelf life of mangoes. *Scientia Horticulturae*, *249*(January), 192–198.

Chompunut, L., Wanaporn, T., Anupong, W., Narayanan, M., Alshiekheid, M., Sabour, A., Karuppusamy, I., Lan Chi, N.T., Shanmuganathan, R. (2022). Synthesis of copper nanoparticles from the aqueous extract of *Cynodon dactylon* and evaluation of its antimicrobial and photocatalytic properties. *Food and Chemical Toxicology*, *166*(April), 113245.

Chowdhury, S., Teoh, Y.L., Ong, K.M., Rafflisman Zaidi, N. S., Mah, S.K. (2020). Poly(vinyl) alcohol crosslinked composite packaging film containing gold nanoparticles on shelf life extension of banana. *Food Packaging and Shelf Life*, *24*(January), 100463. https://doi.org/10.1016/j.fpsl.2020.100463

Cid-López, M.L., Soriano-Melgar, L.D.A.A., García-González, A., Cortéz-Mazatán, G., Mendoza-Mendoza, E., Rivera-Cabrera, F., Peralta-Rodríguez, R.D. (2021). The benefits of adding calcium oxide nanoparticles to biocompatible polymeric coatings during cucumber fruits postharvest storage. *Scientia Horticulturae*, *287*, 110285.

Dong, W., Su, J., Chen, Y., Xu, D., Cheng, L., Mao, L., Gao, Y. (2022). Characterization and antioxidant properties of chitosan film incorporated with modified silica nanoparticles as an active food packaging. *Food Chemistry*, *373*(PA), 131414.

Eftekhari, A., Maleki, S., Chodari, L., Sunar, S., Hasanzadeh, A. (2018). Biomedicine & Pharmacotherapy The promising future of nano-antioxidant therapy against environmental pollutants induced-toxicities. *Biomedicine & Pharmacotherapy*, *103*(February), 1018–1027.

Ezati, P., Riahi, Z., Rhim, J.W. (2022). CMC-based functional film incorporated with copper-doped TiO2 to prevent banana browning. *Food Hydrocolloids*, *122*(August 2021), 107104.

Fernando, S., Gunasekara, T., Holton, J. (2018). Antimicrobial Nanoparticles: applications and mechanisms of action. *Sri Lankan Journal of Infectious Diseases*, *8*(1), 2.

Flieger, J., Flieger, W., Baj, J., Maciejewski, R. (2021). Antioxidants: classification, natural sources, activity/capacity measurements, and usefulness for the synthesis of nanoparticles. *Materials*, *14*, 4135.

Flórez, M., Guerra-Rodríguez, E., Cazón, P., Vázquez, M. (2022). Chitosan for food packaging: Recent advances in active and intelligent films. *Food Hydrocolloids*, *124*.

Guo, X., Chen, B., Wu, X., Li, J., Sun, Q. (2020). Utilization of cinnamaldehyde and zinc oxide nanoparticles in a carboxymethylcellulose-based composite coating to improve the postharvest quality of cherry tomatoes. *International Journal of Biological Macromolecules*, *160*, 175–182.

Hassan, F.A.S., Ali, E.F., Mostafa, N.Y., Mazrou, R. (2021). Shelf-life extension of sweet basil leaves by edible coating with thyme volatile oil encapsulated chitosan nanoparticles. *International Journal of Biological Macromolecules*, *177*, 517–525.

He, X., Deng, H., Hwang, H. (2018). ScienceDirect: the current application of nanotechnology in food and agriculture. *Journal of Food and Drug Analysis*, *27*(1), 1–21.

Hosny, M., Fawzy, M., El-Badry, Y. A., Hussein, E.E., Eltaweil, A.S. (2022). Plant-assisted synthesis of gold nanoparticles for photocatalytic, anticancer, and antioxidant applications. *Journal of Saudi Chemical Society*, *26*(2), 101419.

Ingold, K.U., Pratt, D.A. (2014). Advances in radical-trapping antioxidant chemistry in the 21st century: A kinetics and mechanisms perspective. *Chemical Reviews*, *114*(18), 9022–9046.

Islam, J., Kabir, Y. (2019). Effects and mechanisms of antioxidant-rich functional beverages on disease prevention. In *Functional and Medicinal Beverages: Volume 11: The Science of Beverages*. Elsevier Inc.

Jadhav, R., Pawar, P., Choudhari, V., Topare, N., Raut-Jadhav, S., Bokil, S., Khan, A. (2022). An overview of antimicrobial nanoparticles for food preservation. *Materials Today: Proceedings*.

Jafarzadeh, S., Mohammadi Nafchi, A., Salehabadi, A., Oladzadabbasabadi, N., Jafari, S.M. (2021). Application of bio-nanocomposite films and edible coatings for extending the shelf life of fresh fruits and vegetables. *Advances in Colloid and Interface Science*, *291*, 102405.

Jiang, L., Jia, F., Han, Y., Meng, X., Xiao, Y., Bai, S. (2021). Development and characterization of zein edible films incorporated with catechin/β-cyclodextrin inclusion complex nanoparticles. *Carbohydrate Polymers*, *261*(February), 117877.

Kalia, A., Kaur, M., Shami, A., Jawandha, S.K., Alghuthaymi, M.A., Thakur, A., Abd-Elsalam, K. A. (2021). Nettle-leaf extract derived zno/cuo nanoparticle-biopolymer-based antioxidant and antimicrobial nanocomposite packaging films and their impact on extending the post-harvest shelf life of guava fruit. *Biomolecules*, *11*(2), 1–24.

Khane, Y., Benouis, K., Albukhaty, S., Sulaiman, G.M., Abomughaid, M.M., Al Ali, A., Aouf, D., Fenniche, F., Khane, S., Chaibi, W., Henni, A., Bouras, H.D., Dizge, N. (2022). Green synthesis of silver nanoparticles using aqueous citrus limon zest extract: characterization and evaluation of their antioxidant and antimicrobial properties. *Nanomaterials*, *12*(12).

Kocira, A., Kozłowicz, K., Panasiewicz, K., Staniak, M., Szpunar-Krok, E., Hortyńska, P. (2021). Polysaccharides as edible films and coatings: Characteristics and influence on fruit and vegetable quality—A review. *Agronomy*, *11*(5), 813.

Kowsalya, E., Mosa Christas, K., Balashanmugam, P., Tamil Selvi, A., aquline Chinna Rani, I. (2019). Biocompatible silver nanoparticles/poly(vinyl alcohol) electrospun nanofibers for potential antimicrobial food packaging applications. *Food Packaging and Shelf Life*, *21*(September 2018), 100379.

Kumar, H., Bhardwaj, K., Nepovimova, E., Kuča, K., Singh Dhanjal, D., Bhardwaj, S., Bhatia, S. K., Verma, R., Kumar, D. (2020b). Antioxidant functionalized nanoparticles: A combat against oxidative stress. *Nanomaterials*, *10*(7), 1334.

Kumar, H., Bhardwaj, K., Kuča, K., Kalia, A., Nepovimova, E., Verma, R., Kumar, D. (2020a). Flower-Based Green Synthesis of Metallic Nanoparticles: Applications beyond Fragrance. *Nanomaterials*, *10*, 766.

Kumar, S., Boro, J.C., Ray, D., Mukherjee, A., Dutta, J. (2019). Bionanocomposite films of agar incorporated with ZnO nanoparticles as an active packaging material for shelf life extension of green grape. *Heliyon*, *5*(6), e01867.

La, D.D., Nguyen-Tri, P., Le, K.H., Nguyen, P.T.M., Nguyen, M.D., Vo, A.T.K., Nguyen, M. T. H., Chang, S.W., Tran, L.D., Chung, W.J., Nguyen, D.D. (2021). Effects of antibacterial ZnO nanoparticles on the performance of a chitosan/gum arabic edible coating for post-harvest banana preservation. *Progress in Organic Coatings*, *151*(November 2020), 106057.

Lian, R., Cao, J., Jiang, X., Rogachev, A.V. (2021). Physicochemical, antibacterial properties and cytocompatibility of starch/chitosan films incorporated with zinc oxide nanoparticles. *Materials Today Communications*, *27*(March), 102265.

Liang, J., Wang, J., Li, S., Xu, L., Wang, R., Chen, R., Sun, Y. (2019). The size-controllable preparation of chitosan/silver nanoparticle composite microsphere and its antimicrobial performance. *Carbohydrate Polymers*, *220*, 22–29.

Lopes, N.A., Brandelli, A. (2018). Nanostructures for delivery of natural antimicrobials in food. *Critical Reviews in Food Science and Nutrition*, *58*(13), 2202–2212.

Lourenço, S.C., Mold, M., Alves, V.D. (2019). Antioxidants of natural plant origins: From sources to food industry applications. *Molecules*, *24*, 4132.

Maringgal, B., Hashim, N., Tawakkal, I.S.M.A., Hamzah, M.H., Mohamed, M.T.M. (2020). Biosynthesis of CaO nanoparticles using *Trigona* sp. Honey: Physicochemical characterization, antifungal activity, and cytotoxicity properties. *Journal of Materials Research and Technology*, *9*(5), 11756–11768.

Mohamad, N., Mazlan, M.M., Tawakkal, I.S.M.A., Talib, R.A., Kian, L.K., Jawaid, M. (2022). Characterization of active polybutylene succinate films filled essential oils for food packaging application. *Journal of Polymers and the Environment*, *30*(2), 585–596.

Murali, M., Kalegowda, N., Gowtham, H.G., Ansari, M.A., Alomary, M.N., Alghamdi, S., Shilpa, N., Adil, S.F., Hatshan, M.R., Amruthesh, K.N. (2021). Plant-mediated zinc oxide nanoparticles: Advances in the new millennium towards understanding their therapeutic role in biomedical applications. *Pharmaceutics*, *13*, 1662.

Nile, S.H., Baskar, V., Selvaraj, D., Nile, A., Xiao, J., Kai, G. (2020). Nanotechnologies in food science: applications, recent trends, and future perspectives. *Nano-Micro Letters*, *12*(1), 1–34.

Okcu, Z., Yavuz, Y., Kerse, S. (2018). Edible film and coating applications in fruits and vegetables. *Alınteri Journal of Agriculture Sciences*, *33*, 221–226.

Omerović, N., Djisalov, M., Živojević, K., Mladenović, M., Vunduk, J., Milenković, I., Knežević, N., Gadjanski, I., Vidić, J. (2021). Antimicrobial nanoparticles and biodegradable polymer composites for active food packaging applications. *Comprehensive Reviews in Food Science and Food Safety*, *20*(3), 2428–2454.

Onitsuka, S., Hamada, T., Okamura, H. (2019). Preparation of antimicrobial gold and silver nanoparticles from tea leaf extracts. *Colloids and Surfaces B: Biointerfaces*, *173*(September 2018), 242–248.

Padilla-Camberos, E., Sanchez-Hernandez, I.M., Torres-Gonzalez, O.R., Ramirez-Rodriguez, P., Diaz, E., Wille, H., Flores-Fernandez,

J.M. (2021). Biosynthesis of Silver Nanoparticles Using Stenocereus queretaroensis Fruit Peel Extract: Study of Antimicrobial Activity. *Materials 2021*, *14*, 4543.

Perumal, A.B., Huang, L., Nambiar, R.B., He, Y., Li, X., Sellamuthu, P.S. (2022). Application of essential oils in packaging films for the preservation of fruits and vegetables: A review. *Food Chemistry*, *375*(October 2021), 131810.

Rodino, S., Butu, M., Butu, A. (2019). Application of biogenic silver nanoparticles for berries preservation. *Digest Journal of Nanomaterials and Biostructures*, *14*(3), 601–606.

Sardella, D., Gatt, R., Valdramidis, V.P. (2017). Physiological effects and mode of action of ZnO nanoparticles against postharvest fungal contaminants. *Food Research International*, *101*(August), 274–279.

Seifari, F.K., Ahari, H. (2020). Active edible films and coatings with enhanced properties using nanoemulsion and nanocrystals. *Food & Health*, *3*(1), 15–22.

Shapi'i, R.A., Othman, S.H., Nordin, N., Kadir Basha, R., Nazli Naim, M. (2020). Antimicrobial properties of starch films incorporated with chitosan nanoparticles: *In vitro* and *in vivo* evaluation. *Carbohydrate Polymers*, *230*(August 2019), 115602.

Sharma, C., Dhiman, R., Rokana, N., Panwar, H. (2017). Nanotechnology: An untapped resource for food packaging. *Frontiers in Microbiology*, 8, 1735.

Sivamaruthi, B.S., Ramkumar, V.S., Archunan, G., Chaiyasut, C., Suganthy, N. (2019). Biogenic synthesis of silver palladium bimetallic nanoparticles from fruit extract of *Terminalia chebula—In vitro* evaluation of anticancer and antimicrobial activity. *Journal of Drug Delivery Science and Technology*, *51*(February), 139–151.

Tao, C. (2018). Antimicrobial activity and toxicity of gold nanoparticles: research progress, challenges and prospects. *Letters in Applied Microbiology*, *67*(6), 537–543.

Thivya, P., Bhosale, Y.K., Anandakumar, S., Hema, V., Sinija, V.R. (2021). International journal of biological macromolecules development of active packaging film from sodium alginate/carboxymethyl cellulose containing shallot waste extracts for anti-browning of fresh-cut produce. *International Journal of Biological Macromolecules*, *188*(July), 790–799.

Valgimigli, L., Baschieri, A., Amorati, R. (2018). Antioxidant activity of nanomaterials. *Journal of Materials Chemistry B*, *6*(14), 2036–2051.

Vipan, B., Mahajan, C., Tandon, R., Kapoor, S., Sidhu, M.K. (2018). Natural coatings for shelf-life enhancement and quality maintenance of fresh fruits and vegetables—a review. *Journal of Postharvest Technology*, *6*(1), 12–26.

Wang, L., Hu, C., Shao, L. (2017). The antimicrobial activity of nanoparticles: Present situation and prospects for the future. *International Journal of Nanomedicine*, *12*, 1227–1249.

Wu, S., Rajeshkumar, S., Madasamy, M., Mahendran, V. (2020). Green synthesis of copper nanoparticles using Cissus vitiginea and its antioxidant and antibacterial activity against urinary tract infection pathogens. *Artificial Cells, Nanomedicine and Biotechnology*, *48*(1), 1153–1158.

Xing, Y., Xu, Q., Yang, S.X., Chen, C., Tang, Y., Sun, S., Zhang, L., Che, Z., Li, X. (2016). Preservation mechanism of chitosan-based coating with cinnamon oil for fruits storage based on sensor data. *Sensors (Switzerland)*, *16*(7).

Xing, Y., Li, W., Wang, Q., Li, X., Xu, Q., Guo, X., Bi, X., Liu, X., Shui, Y., Lin, H., Yang, H. (2019). Antimicrobial nanoparticles incorporated in edible coatings and films for the preservation of fruits and vegetables. *Molecules*, *24*(9).

Yousuf, B., Qadri, O.S., Srivastava, A.K. (2018). Recent developments in shelf-life extension of fresh-cut fruits and vegetables by application of different edible coatings: A review. *LWT—Food Science and Technology*, *89*(June 2017), 198–209.

Zahran, M., Marei, A.H. (2019). Innovative natural polymer metal nanocomposites and their antimicrobial activity. *International Journal of Biological Macromolecules*, *136*, 586–596.

Nanocoatings for Fruits and Vegetables

22

Bushra Hussain, Sajid Ali*, Aamir Nawaz, Safina Naz, Waqar Shafqat, and Muhammad Moaaz Ali

**Corresponding Author:* sajidali@bzu.edu.pk

22.1 INTRODUCTION

Fruits and vegetables are an important component of the daily human diet. Their consumption defines dietary antioxidant capacity, which is a measure of the quality of the diet one is taking in order to remain healthy. The Food and Agricultural Organization (FAO) of the United Nations and the World Health Organization (WHO) have joined hands to make the public conscious about what they eat. Their food experts claim that at least 0.4 kg fresh consumption of fruits and vegetables every day is necessary for all human beings in order to reduce the odds of chronic diseases, obesity, heart problems and diabetes (WHO, 2005). It also reduces the production of carcinogens in the body, and daily consumption of greens can combat several kinds of cancers. Fruits and vegetables are rich sources of vitamins, add roughage to the diet, detoxify the body's cells and have medicinal values against several diseases (Nair et al., 2020). Nevertheless, the products are perishable and must be handled with care to preserve their natural nutrients after harvest.

The most feasible technique of storing fruits and vegetables is the use of palatable compounds applied as thin layers over the fruit or vegetable's surface. An applied layer that is 0.3 mm thick or less can either be eaten intact or can be washed away before consumption. Fruits and vegetables can be coated by dusting, plunging, sprinkling or brushing a suitable material over their surface (Yousuf et al., 2018). It is important that the coated material should not grant some off-flavor or toxic residues to the commodity under treatment. Commonly used coatings can be extracted from oils, starches, cellulose, seaweeds, chitosan, aloe vera, zein, casein and gums (Sharma et al., 2019). Edible coatings are gaining in popularity at an exponential rate. About 7 years ago the market value of edible coatings was almost 700 million USD, with the expected growth rate of 6.81%; it is now supposed to be almost 1.1 billion USD (Kondle et al., 2022).

Fruits and vegetables are living entities, and like all other living organisms they respire and transpire actively, even after harvest. Oxygen, which is the basic necessity of life, is considered the biggest foe of harvested commodities owing to the deadly effects it has, including loss of taste and nutritional value, oxidation of essential nutrients, triggering enzymatic browning and microbial growth (Rovera et al., 2020). Contrary to this, if oxygen concentration is too low the fresh produce can start accruing aldehydes, alcoholic compounds and CO_2. Such fermented fresh produce is not considered acceptable by consumers (Nor & Ding, 2020).

If we manage to regulate the amount of $O_2$and CO_2 around the fruits and vegetables in a suitable ratio, we can enhance their shelf life. Likewise, moisture loss is another issue concerning the quality of horticultural products. While monitoring the discussed issues, several studies were conducted to develop an aerobiotic intact layer of edible material that controls the amount of O_2 and H_2O exchange (Jafarzadeh et al., 2021). In short, the most vital quality of an edible coating is to maintain the physiochemical superiority of the coated commodity. According to several earlier studies on edible coatings and their effects on fruit and vegetable acceptability, coated products are aesthetically more appealing, shiny, clean and free of microbial or pathogenic invasion. Further, they are highly delicious and have a pleasing aroma (Chaple et al., 2017). This chapter covers different types of nanocoatings on the quality and postharvest life of harvested fresh fruits and vegetables.

22.2 LOSSES OF FRESH FRUITS AND VEGETABLES

Almost 45% of fresh vegetables and fruits are lost each year due to their perishable nature (Ömer Azabağaoğlu, 2018). Some of these perishable commodities hold over 80% water, and high water content leads to more than 30% of all the food losses around the globe (Zhou et al., 2021), causing an economical

DOI: 10.1201/9781003370376-29

loss of more than 940 million USD per year (Kirci et al., 2022). Another study states that of this large wastage of food, fruits and vegetables constitute around 50% (Sirohi et al., 2021). Moreover, changes in the eating habits of today's global population is putting high pressure on the supply of these commodities. Consumers are demanding large quantities of fresh products with high nutraceutical properties and minimum processing. Contrarily, fruits and vegetables are living entities which respire, transpire and age over time, being highly susceptible to deterioration owing to their high metabolic rates, amplified moisture contents, rapid changes in their physiology, heavy microbial attack and, of course, decline in taste with time due to biochemical changes (Sharma et al., 2019).

According to IFS&T, the time for which the edible items, stored under optimal storage conditions, can be consumed safely with acceptable taste and aroma and without compromising their mechanical, biochemical and antimicrobial qualities, is called the shelf life of that food. Being a living entity with 65–95% moisture content, fruits and vegetables decay after consuming their reserved nutrients (Rahiel et al., 2018). According to a survey by Schmidt-Traub et al. (2019), almost 0.8 billion people are facing food scarcity. On the other hand, 1.3 billion tons of food, which is more than 30% of the total food production, goes into waste bins. Food security is not only a need of the community but is also a serious question for biologists and biochemists to understand and solve it on scientific basis (Palit, 2020).

In order to meet the ever-increasing food demand, we need novel approaches which are friendly to the environment, easy to adopt, simple to perform and cost-effective. Besides, with the improvement in people's consumption models and eating habits, it's the need of the hour to supply fresh produce in their best possible condition to remote areas as well. It was reported that in America, around 9,000 people die every year because of consuming deteriorated fruits and vegetables (Wang et al., 2019). To address this stumbling block, we need to derive some strategies which will improve the shelf life and quality of fresh products so that they will be able to withstand postharvest changes, transportation stresses and handling problems. Before devising a method to improve the postharvest life of fruits and vegetables, we have to understand their biological nature and limits. This has been reported that different agricultural activities can emit almost 10% of all greenhouse gases, and if we consider these activities unavoidable then at least the products from these activities should be utilized. Also, some substitute technologies must be fashioned in order to save the environment and the earth (Gill, 2021; Abadi et al., 2021).

22.3 METHODS OF FRUIT AND VEGETABLE STORAGE DURING POSTHARVEST

According to FAO, perishable commodities have 50–70% water content (FAO, 2023) and United States Department of Agriculture (USDA) stated that a perishable commodity is only safe for consumption if it is either stored at a temperature lower than 40^{0}F or is frozen (USDA, 2023). It is not possible to access low-temperature storage for all harvested fruits and vegetables, hence it's not feasible in all cases to provide controlled atmospheric conditions. So there's a need to develop some alternative strategy in order to save the already produced food (Getahun et al., 2021). Additionally, during the glut season the produce is sold at cheap rates due to inadequate storage plans. This makes the farmers suffer serious economic damages, as their production is not in line with the marketing and handling facilities (Sharma et al., 2019). The postharvest life of fresh food commodities depends on both inherent (genetic and biochemical makeup) and extrinsic (handling and storage conditions) elements (Rajapaksha et al., 2021).

The biggest challenges faced by the food industry now are storage, preservation and supply of fresh food to consumers in its best possible form (Lopez-Polo et al., 2021). The main aim of the fresh food industry is to secure whatever fresh produce is at hand and then store it for enough time to meet the consumer. While securing the food, their main focus is on preserving nutrients and taste and even to increase the nutrients and flavor in some cases. These also have to control the production of any possible toxic compounds, e.g., aflatoxins, in the stored products. Thus, comprehensive technology is needed. Since ancient times, food has been stored by biological, mechanical and chemical methods like modified temperature and modified environment or by coating fruits with diverse materials (Yahia et al., 2019). The old style of packing material that is typically polyethylene, polyvinyl chloride, polyamide or polypropylene produces 350 M tons of non-biodegradable waste every year (Haghighi et al., 2020). However, the world population increases many-folds each year, and food production is not enough to cope with increasing needs; thus, for food storage, more environmentally friendly methods are needed.

To reduce the attack of microbes and to reduce pest contamination over fresh and delicate agricultural commodities, and to supplement the quality of the harvest, some common handling techniques are application of heat, radiation, low-pressure, MAP, CA storage, chemicals and hot water treatments. Sometimes fresh produce is also treated with fungicides to eliminate chances of contamination, but these have toxic residual effects on the health of consumers. Chilling injuries during low-temperature storage are common in fruits due to their delicate tissues. It thus reduces their consumer acceptance (Han et al., 2021). For MAP, non-biodegradable sheets are used, which leads to a serious problem for the environment as well as consumers. Also, this method requires high-cost vacuum technology, which is out of the reach of common farmers, particularly in developing and poor countries (Chiang et al., 2021). A more advanced technical approach is to utilize bio-based polyelectrolytes, forming reedy layers of polymers over the commodities. This approach is appreciated due to its recyclability, ease of application, flexible composition and restricted gaseous permissibility (Priolo et al., 2015).

One more innovative approach is the use of antagonistic microbes which kill food-deteriorating organisms over the surface of fresh produce. Genetic engineering can be employed for enhancing the life of fresh commodities, but the procedure is expensive and lacks the detailed knowledge for execution at the commercial scale (Jiang et al., 2021). Another more health-conscious approach is the use of edible coatings which are preferably organic in origin. These coatings can be proteinaceous or resinous. Some coatings also contain fats and starches. The main principle of their use is to check free gaseous and liquid exchange across the surface of the coated commodity. Some of the coated materials have certain medicinal or antimicrobial qualities. These are also safe for the environment and produce no hazardous waste. The coated commodity retains its taste, aroma, water and nutritional contents in general (Sharma et al., 2019; Ali et al., 2024a; Ali et al., 2024b).

22.4 APPLICATIONS OF DIFFERENT EDIBLE COATINGS IN POSTHARVEST FRUITS AND VEGETABLES

With the increase in the trend for eating more healthy food among the public, food habits are shifting toward plant-based diets. The same holds true for edible coatings. Plant-based materials are used to reduce senescence by controlling tissue browning and ethylene evolution. These also limit the escape of aromatic and flavoring volatile compounds from the subjected commodity, hence retaining the organoleptic qualities of the commodity under treatment (Hazrati et al., 2017; Anjum et al., 2020). Sometimes coating materials can add some off-flavor to the commodity, but the coating can be washed off before use or consumption (Ali et al., 2023). Yet most of the coatings are strong enough to bear transportation and handling shocks (Thakur et al., 2018).

With the intention of dealing with microbes, scientists have tried their best to encapsulate organic and inorganic nanoparticles into edible coatings via naoemulsions (Al-Tayyar et al., 2020). Nanoemulsions have proved to be helpful in retaining the excellence of the coated commodity by reducing the breakdown of vitamin C, antioxidants, color-granting pigments, lipids and other bioactive components (Govindappa et al., 2021). The agricultural industry produces huge amounts of by-products which are wasted and often go to landfills. It is highly probable that this material can be used in the assimilation of edible coatings due to its bioactive nature. The plant material is a rich source of lignocellulose, which can be processed into non-fiber and crystals (Teo et al., 2020). The said material is also rich in gums, pectin, chitin, carbohydrates and lignin, which can also be incorporated as edible coating. Another organic source of polymers is animal-based, including gelatin and collagen (Ranganathan et al., 2020). The advantage of using animal protein over plant-based starch coatings is the presence of both hydrophilic and hydrophobic peptides (Kouhi et al., 2020), which are responsible for lower penetrability regarding gaseous exchange and higher inbuilt mechanical stability (Nor & Ding, 2020).

Normally, organic coatings are carbohydrates, lipids or proteinaceous in nature. All of these have their own unique qualities. For example, lipids are hydrophobic, not allowing water from the commodity to escape into the environment. Proteinaceous coatings are elastic and are not easily prone to wear and tear during handling. Likewise, polysaccharides check the exchange of gases in and out of fresh produce. When combined, their utility improves and these coatings become more beneficial. Although the linkage between several components of the coating defines its value, the choice of coating material depends on indicators like organoleptic qualities, feasibility, availability and solubility. (Sharma et al., 2019). In the recent past these coatings have been used as vectors of other organic and inorganic nanomaterial to improve their performance (Zambrano-Zaragoza et al., 2018).

The benefits from the same coating can be doubled when incorporated as nanoparticles instead of large-sized particles because of their higher size compatibility with biomolecules and microbes (Petkoska et al., 2021). Due to their dimensional similarity, nanoparticles freely enter the bodies of microbes, produce ROS, cause cell oxidation and also oxidize heavy metal compounds (Vasile, 2018; Petkoska et al., 2021). Until now, Ag, ZnO, TiO_2, SiO_2, starches, chitosan and cellulose have been used to produce nanoparticles which lessen the extent of respiration in coated commodities (Ghosh & Singh, 2022). In another study, nanocoating containing essential oils (EO) extracted from amla was reported to reduce fungal attack on the fruits. The mentioned coating also reduced color and weight loss changes, retained fruit tissue firmness and alleviated tissue browning (Braich et al., 2022).

22.5 TYPES OF NANOCOATINGS FOR FRUITS AND VEGETABLES

Nanoparticle-based nanocoatings were developed for the first time in the 1980s by Gleiter in Germany, and now all over the world these are being utilized and further explored (Gleiter, 1989). In the 3D space of a nanometer, if a particle at its best is equal to 1D, that particle is claimed to be a nanoparticle (Tripathi & Chung, 2019). According to US FDA (Food and Drug Administration) standards, in 2006, nanoparticles were defined with dimensions <1 μm, but with some specified qualities (Dholariya et al., 2021). This tiny size grants certain magnetic, thermal, volumetric, mechanical, superficial and optical qualities that an ordinary particle cannot hold. With these miraculous properties, nanoparticles can form cross-links with the additional ingredients in order to make the medium impervious and prone to dissipation, as well as enhance heat resistance and physiological characteristics. Nanoparticles, even at a minute ratio, can do wonders and are

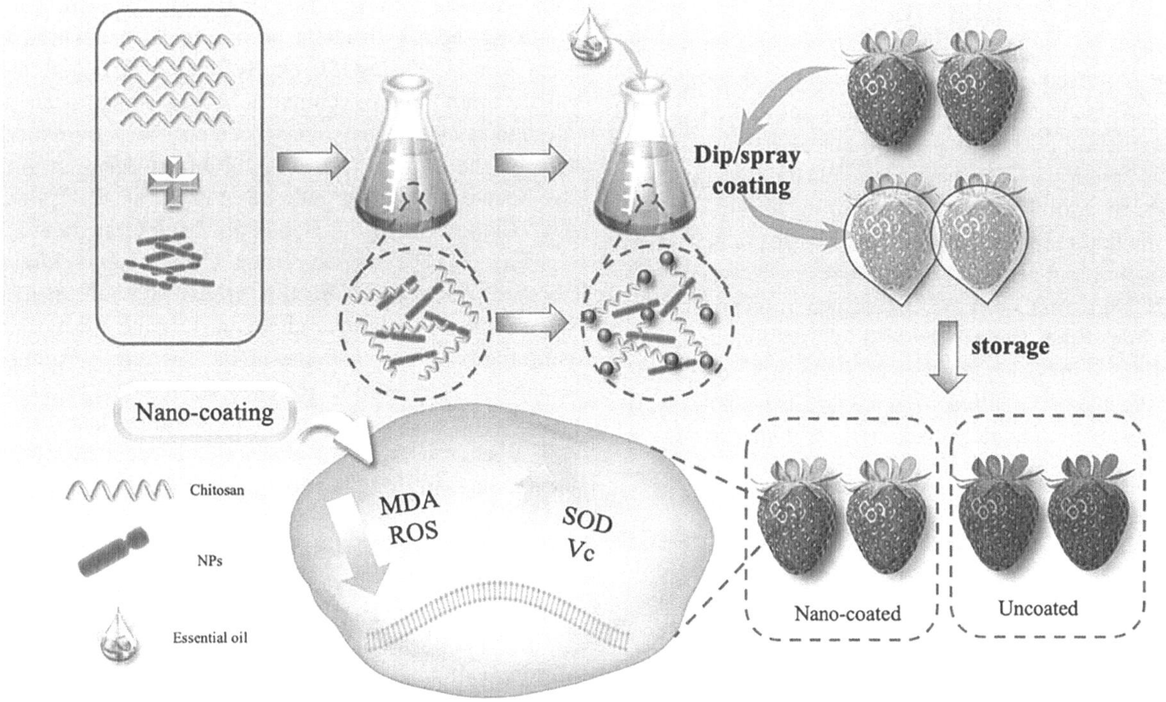

FIGURE 22.1 Preparation and application of nanocoatings on fruits and vegetables during postharvest storage. This figure is adapted from Shan et al. (2023).

thus claimed as a breakthrough technology (Gu et al., 2020). When nanoparticles are incorporated into edible coatings, they alter the physical, ocular, oxidative and resistive qualities of the coating (Arfat et al., 2017). These particles are a great choice for eliminating the possibility of pathogenic growth over fresh produce as well as for enhancing the antioxidative character of the coating against enzymatic deterioration. As when harvested products are stored for longer durations, due to biological reactions, the temperature of the commodities keeps on changing; these nanocoatings have certain ability to release the exact and required amounts of active agents in situ. Further, these combine well with water-soluble polymetrics. Moreover, nanoparticles are generally small enough to not alter the taste of the coated commodities (Luo et al., 2017). Nanoparticles provide a good portion of antioxidant and antimicrobial characteristics owing to their constituents, such as essential oils, phenols, etc., and due to the nano size of the particles, i.e. less than 1000 nm, their permeability against gases is excellent. Generally, the size of nanoparticles in edible coatings ranges from 100 to 500 nm. Nanoparticles are light in weight but also have excellent stretching abilities and are thus preferred over the use of conventional coatings alone. Despite their small size, these particles have active hydrophobic qualities and never compromise the visual quality of fresh produce. Along with having the benefit of dimension and size, the nanocoatings also have enhanced optical, electrical and thermal properties, which make them suitable to be applied in the fields of food and drug production, food storage, energy generation and mechanical material industries. Nanocoatings also have unique self-cleaning qualities along with their bactericidal efficiency; thus they have potential to overcome the inadequacies of older methods (Nair et al., 2020; Jagtiani, 2022). The preparation and subsequent application of selected nanocoatings have been presented in Figure 22.1.

In present times, the most commonly employed nanocoating involves nanoemulsions, solid nanoparticles and nanocomposites. For illustration purposes, we can have the example of thymol and carvacrol, obtained from the essential oils of oregano and thyme, respectively. These bioactive agents have the ability to kill microbes attacking the fresh produce. Correspondingly, their antioxidative nature can reduce enzymatic activities inside the commodity; thus better results are expected compared to when the main coating material is used alone. The pattern of nanoparticles and their structures varies greatly, and this variation is the cause of difference in interactions with different coating matrixes (Dang et al., 2017).

A nanosystem includes nanofibers, nanotubes, polymeric nanoparticles, solid nanoparticles, nanocrystals, nanoemulsions and nano-lipid carriers. When combined with a matrix, it becomes a nanocomposition. A nanocomposite can be defined as a matrix containing a single or double coating material, in which a nanomaterial is added to improve coating qualities. The mechanism of the nanosystem is the slow release of active agents from coating to the coated commodity for enhancing its shelf life, i.e. temporal distribution (Liu et al., 2017a).

22.5.1 Nanoemulsions

If lipids are dispersed in water in such a way that the extremely thin layer of aqueous emulsifier surrounds each and every lipid molecule, typically 50 nm to a maximum of 500 nm in size, the colloidal solution is called a 'nanoemulsion'. The emulsion can be made by adding water to a lipid or by adding a lipid to water, but the latter is considered better because the lipids have certain antimicrobial and antioxidant qualities (Salvia-Trujillo et al., 2017). Nanoemulsions are heat resistant, however, when kept for a long time; these can show alluviation, accretion, amalgamation, adhesion and Oswald's maturation (Öztürk, 2017). As the nanoemulsified particles are extremely small, Brownian motion is more dominant on them than gravity. This makes them relatively stable against gravitational separation techniques. Also, they can tolerate dilution without affecting the consistency of the emulsion. Due to the tiny size of lipid droplets, the coating becomes more bioactive, consumer acceptable and organoleptically more pleasing (McClements, 2015). The emulsifying agent can be any essential oil, fatty acid, carotenoid, antioxidant, plant-based sterol, quinone, etc. (Salvia-Trujillo et al., 2017).

22.5.2 Polymeric Nanoparticles

Nanoparticles ranging in size from 100 nm to 1000 nm are polymeric nanoparticles. Monomeric compounds can also be utilized for the same purpose, but these have noxious aftereffects. These colloidal structures are either in the form of nanospheres or nanocapsules. The former has a compact polymeric matrix, in which the bioactive agent is either dispersed or adsorbed; the latter has oil encapsulated inside a polymeric film and is more common (Yousuf et al., 2018). In both types, an active agent is released by diffusion, but in the case of some nanocapsules, sometimes the active agent is released by rupturing the membrane. Whichever type of polymer is used, the main thing is the bioactive agent, which defines when and how the active agent will be released. Poly-ε-caprolactone, chitosan, cellulose acetate phthalate, polylactic acid, alginate and poly-D,L-lactide-co-glycolide are generally used polymers (Kumar et al., 2015).

The discussed technology was invented for the drug industry, but then it was introduced in the food sector to enhance the organoleptic qualities and shelf life of fresh produce by loading the polymers with antioxidants and antimicrobials. The loaded compounds then release thereafter in situ to guarantee their uniform distribution over the fresh produce surface progressively. However, there is still a huge blank for research in this area. Polyphenols have been utilized in this technology. For example, curcumin, obtained from *Curcuma longa*, have long been used as an active agent in many edible nanoformulations (Liu et al., 2017b). Similarly, EO from *Jasminum officinale* was used by a coacervation method with gum arabic in combination with gelatin, EO from grape seed was collided with curcumin and resveratrol, starch from casava was unified with lycopene-PCL and PCL nanocapsules were also studied by fusing them with lutein by interfacial deposition and with β-carotene by using an emulsification-diffusion technique. Furthermore, hydroxycinnamic acids, epigallocatechin gallate and quercetin have also been successfully experimented with for the same purpose. The EO from *Mentha piperita* has also been used for said purpose (Assis et al., 2017; Brum et al., 2017; Liang et al., 2017; Hao et al., 2017; Granata et al., 2018).

The EOs extracted from *Cymbopogon citratus* and *Curcuma longa* are utilized in preparing nanocapsules due to the presence of alginate-chitosan. Despite the possibility of using many polymeric compounds, chitosan is often preferred because it is economically cheaper, not known to have any toxic residues, has good physical qualities, has compatibility with a generous range of material, has excellent film-forming ability, has antifungal and antipathogenic potential, allows limited gaseous exchange and is biodegradable. Strawberry fruits coated with chitosan nanocoating can be stored for 3 weeks (Pilon et al., 2015).

Love apple and other *Solanum* sp. provide cutin (Gómez-Patiño et al., 2020); yerba mate plant is a source of lignocellulose; queen's wreath and cacti plant excrete gelatin (Otálora et al., 2020); skeletons of insects, shells of crustaceans and some fungal and algal species are made of 2-acetamido-2-deoxy-β-D-glucose molecules and when nitrogen is deacetylated from them, chitosan is formed (Zambrano-Zaragoza et al., 2021). Under normal circumstances, chitosan is hydrophobic, but when nitrogen is attached as an α-amino it can be made soluble. The solubility of chitosan in water also depends on its molar mass. We can also get 2% weight by volume solution of chitosan by reducing the pH of the solvent below 6.3 (Kondle et al., 2022). The potential of chitosan-based films and coatings to act as natural preservatives is widely known. These materials contain antioxidative (Silva et al., 2018), anti-decaying (Farina et al., 2020b), antimicrobial (Balakrishnan et al., 2019), ultraviolet light blocking (Balasubramanian et al., 2019), anti-browning (Farina et al., 2020a), anti-spoilage (Fernández et al., 2010), antibacterial and antifungal properties (Kanikireddy et al., 2019). Thus, these can remarkably improve the shelf life (Severcan et al., 2020) and postharvest quality (Hasan et al., 2020) of fruits and vegetables. These also have water-holding ability (Santagata et al., 2018) and cytotoxic (Tibolla et al., 2019) qualities. These can conserve the ascorbate, hue and flavor of fruits, along with altering the extent of the deacetylation of treated fruits (Jiang et al., 2019). Until now the only limitation of using chitosan-based coatings is their high permeability to water, which is even worse if the external environment has high relative humidity. This factor can be controlled by adjusting the solvent ratio and the acidity of the solution and removing the acetyl group or supplementing it with essential oils, fats, carbohydrates, amino acids and surfactants. These can also be supplemented with hydroxycinnamic acid, genipapo extract, pentanedial, methanal, (E)-3-phenylprop-2-enal and Na-trimetaphosphate. There are several methods used for combining chitosan with other beneficial compounds, including heat, fumigation and MAP (Kondle et al., 2022).

Seaweeds contain carrageenan and alginate, and along with plants, certain fungal and bacterial species produce mycocel (Irbe et al., 2021) and carboxymethyl (Klunklin et al., 2020; Rachtanapun et al., 2021), respectively. Hydrophobic polymeric coatings can be formed by alginate, and when such a coating was tested over *Malus domestica* and *Prunus persica*, the results were extraordinarily positive (Nair et al., 2020). *Phaeophyceae algae*, *Pseudomonas aeruginosa* and Azotobacter provide rich amounts of alginate, which is a straight chain hybrid of (1–4) β-D-mannuronic (M) and α-L-guluronic (G) acid. Its common utilizable form is its salt made with alginic acid. The use of this material in the post-harvest edible coating sector is becoming popular (Kondle et al., 2022). Although the film-forming capacity of alginates is weaker (Medina-Jaramillo et al., 2020), it is easily found and has been declared to be safe for use in any food item by the FDA (Pirozzi et al., 2020). Argentum is also gaining fame in this sector owing to its antibacterial and microcidal qualities. Both chitosan and alginate coatings induce a little charge on nanoparticles which enhance the provision of healthy fats, carotenoids, provitamin A and vitamins D and E (Davoudi et al., 2021).

Pickering emulsion capsules, also known as colloidosomes, because these contain colloidal solutions, are also used in food storage. These capsules can be both water soluble and insoluble (Brossault et al., 2021). The method of preparation and the type of colloid used defines the physical forte, penetrability and combination ability with other materials. Almost the same qualities are offered by casein micelles, thus it is also considered a wise choice for preparing bioactive capsules to be used in edible nanocoatings. It also keeps the sensitive biochemical nutritional content of food safe (Sadiq et al., 2021).

22.5.3 Lipid-Based Solid Nanoparticles

An innovative approach to using lipid particles in nanocoatings is to use them in their solid state, in the 50–1000 nm size range, mixed with heat-resistant oil which is then dissolved in water with the help of surfactants or emulsifiers. The approach is not too old and was just started at the end of the 20th century. Solid particles containing bioactive agents are preferred in colloidal nanosolutions used as edible coatings because of their relatively minute size, solid physical state at room and body temperature, dimensions and certain other chemical properties, like the balanced and slow release of an oil- or water-loving bioagent. The homogenization of lipids in water is done at a temperature high enough to melt the solid lipids and dissolve them in the water as a nanoemulsion (Katouzian et al., 2017).

Nanoliposomes have also been tested with edible coatings to augment their mechanical and biochemical properties. These can also boost the antioxidative qualities of the coating. Nanoliposomes have antimicrobial properties as well (Lopez-Polo et al., 2021). These distribute the nutrients evenly across the coated fruit or vegetable while conveying genetic material, amino acids, chemical plant feed and pathocides (Kondle et al., 2022). Another lipid-based product is lipid nanocochleates, which transfer the nutrients without any loss of flavor and appearance in the commodity (Dholariya et al., 2021).

22.5.4 Nanocarriers

At the start of the 21st century, nanosystematic studies were extended and a new system, i.e. lipid-based nanocarriers, was introduced by Müller. These nanocarriers are not prone to crystallization, and along with being deliberately soluble (Göke & Bunjes, 2017), they have excellent defending qualities against oxidative and enzymatic damages (Sala et al., 2018). These nanocarriers are composed of 70% solid and 30% liquid fats (Sala et al., 2018), fashioned in an aqueous solvent with the help of emulsifiers, but they don't require an organic medium to make the solution.

Owing to being in a liquid state these are free of structure-related problems like surface area limit, melt at lower temperatures and are dispersed easily, thus ejecting the bioactive agents and nutraceuticals easily. Medium-sized linkages capturing lipid-friendly materials are further enclosed by a compact fatty layer containing lecithin. Like the other lipid-containing nanosystems, their major use was supposed to be in drug formation, and a little research has been performed on their use in the horticultural industry. Further, to use these nanocarriers in edible products, their grading is undeniable. Due to the critical selection criteria as a food additive, their use is limited. The concentration of the components in nanocarriers affects their properties. For example, if we enhance solid concentration while keeping the liquid the same, the nanocarriers will reduce in size. On the other hand, if we increase the amount of bioagent, it will change the entrapment efficiency of the system. The entrapped fat molecules define the compatibility, solubility, stability, consumer safety and biodegradability of the coating (Katouzian et al., 2017; Sala et al., 2018).

Another key feature of this nanosystem is choosing the solid-phase lipid with higher melting point so that it remains solid at room temperature, while the liquid-phase lipid has a melting point below room temperature to ensure that it remains liquid at body temperature (Katouzian et al., 2017). The main focus of researchers is to utilize such an organic material in nanocarriers that has excellent fighting ability against microbes and oxidation. It should be highly nutritious and have an aesthetically pleasing color.

Another leading example of a nanocarrier system is quercetin. The quercetin is compressed in glyceryl monostearate and linseed oil as a solid and a liquid, respectively, and used thereafter. Quercetin is known for its abilities against oxidation, carcinogenic agents, bacterial attack, inflammation and neurodegenerative diseases. The liquid-phase lipid is excellent for fighting heart problems, high blood pressure and also swelling and reddening, while supporting the nervous system and checking oxidation (Huang et al., 2017). Other researchers have claimed the use of maize oil, cocoa butter, sunflower oil, beeswax, essential oil from grape seeds, carnauba wax, soybean oil and oleic acid for the same purpose. The bioactive substance can be any water-soluble substance, e.g., antioxidants like lycopene, vitamins, β-carotene, carotenoids, curcumin and extracts from green tea or essential oils from berries (Katouzian et al., 2017).

22.5.5 Nanocomposites

Prominent progress in the field of nanosystems was made when the idea of inducing both organic and mineral substances in nanocompositions was introduced. These additives possess higher resistance against abrasive damage and gaseous exchange (Dang et al., 2017). Montmorillonite, argentum particles, oxides of silicon, ZnO and titanium dioxide have been experimented as nanoparticles in edible coatings (Shi et al., 2018). ZnO is well known for its antimicrobial qualities and is approved to be safe as a component of edible coatings by the FDA (Saba & Amini, 2017). Nano silicon dioxide (nano SiO_2-x, where x ranges between 0.4 and 0.8) has a 3D amorphous structure with stable bonds. The size of the molecule is small with a large surface area. Due to OH-group on the external surface it dissolves in larger compounds with no trouble (Xiong et al., 2015). CODEX-Alimentarius has approved it for being safe to utilize as an additive in edible items and has registered it with code E551. The major uses are as a thickener for pastes and a conditioner for grinded materials. In many edible and inedible products, it is also exploited as an aromatic and flavoring agent (Peters et al., 2012). Montmorillonite is a nanoclay comprising Si, with 2 dimensions of 100–150 nm and a 3rd dimension of only 1 nm (Klangmuang & Sothornvit, 2016). Besides the composition of the nanomaterial, the functionality of the nanoparticles also depends upon the way it merges into the coating material.

22.5.6 Nanotubes and Nanofibers

Nanotubes and nanofibers have recently been introduced in the postharvest industry due to their antioxidative and antimicrobial properties. In the recent past, nanotubes made of carbon were introduced and extensively applied for making food packaging materials because of their elasticity, ductility and easy mergence in starches, lipids, proteins and even in crystalline substances. Fractional hydrolysis of milk protein (α-lactalbumin) produces nanotubes with the ability to encapsulate edible substances in its cells, having a span of 8 nm. Contingent to the used polymer, these nanotubes slowly and meticulously release the active agent on the produce's surface (Ramos et al., 2017). A fibrous framework of particles ranging in size from 100 to 500 nm diameter, called a nanofiber, is also cast off to hinder the enzymatic movements and capture the active agent while altering the characteristics of the coating material in a beneficial way (Fahami & Fathi, 2018).

22.5.7 Nanobubbles

Nanobubbles are less than 1000 nm in size. Having such a minute size, the bubbles are not able to drift over water, so they burst immediately after touching the water's surface. These bubbles have a large volume, improved dissolution power and an interesting unique ability to generate a zeta charge on the surface in both acidic and basic media. On bursting, nanobubbles induce hydrolysis, generating OH- ions, which is the reason behind the antipathogenic behavior of these bubbles (Xiao et al., 2019). By adjusting the potential of the zeta charge, we can manage to defer the blast of the bubble for months. The use of microbubbles in edible coatings has also been extended to combine O_3, H_2, CO_2 and 1-MCP, and the bubbles are generated by twisted liquid flow, high-pressure nozzle, porous films made of ceramics, ultrasonic waves, gyratory cuts and a pumping divider or splitter, each having its own uses and defects. A pressure nozzle is good in producing extremely tiny bubbles but costs too much and requires high power for functioning. Industries cannot afford such expensive technology, but the cheaper ones, like the one using surfactants for rotary cutting, pose serious environmental damage. The splitter pump is by far the most adopted method to mix gases like O_3 with liquids without consuming chemicals and of course without generating harmful by-products (Patel et al., 2021). Due to the safety of their usage and ecofriendly nature, some of the nanotechnologies are considered green postharvest technologies (Aswani & Radhakrishnan, 2022). The preparation and potential applications of micro- and nanobubbles have been presented in Figure 22.2.

22.5.8 Montmorillonite Nanocomposites

Coatings containing nanosized montmorillonites (MMT) are categorized into 3 groups depending on their morphology. The first category is of elongated particles having less attraction for the polymers, and thus a true composite coating is impossible to be created. The second form is a multilayered structure having stretched clay particles sandwiched within the layers of the polymer, giving an excellent homogenized composite coating. The third form is alternate scales of polymer and clay nanoparticles within a thin sheet which is highly consistent in its formation. The clay particles principally increase the available area of the matrix and, if mixed with some surfactant or emulsifier, can additionally carry organic properties. Clay limits the exchange of gases and does not allow moisture to exhale from the fruit or vegetable under question. Commonly used MMT particles for the said purpose are Dellite HPS, Cloisite 93-A and Cloisite 30-B. These particles have excellent bactericidal effects against both gram-positive and -negative bacteria (Darie et al., 2014). While enhancing the physical strength against wear and tear and the external heat of the coating, the clay can reduce the aesthetic quality

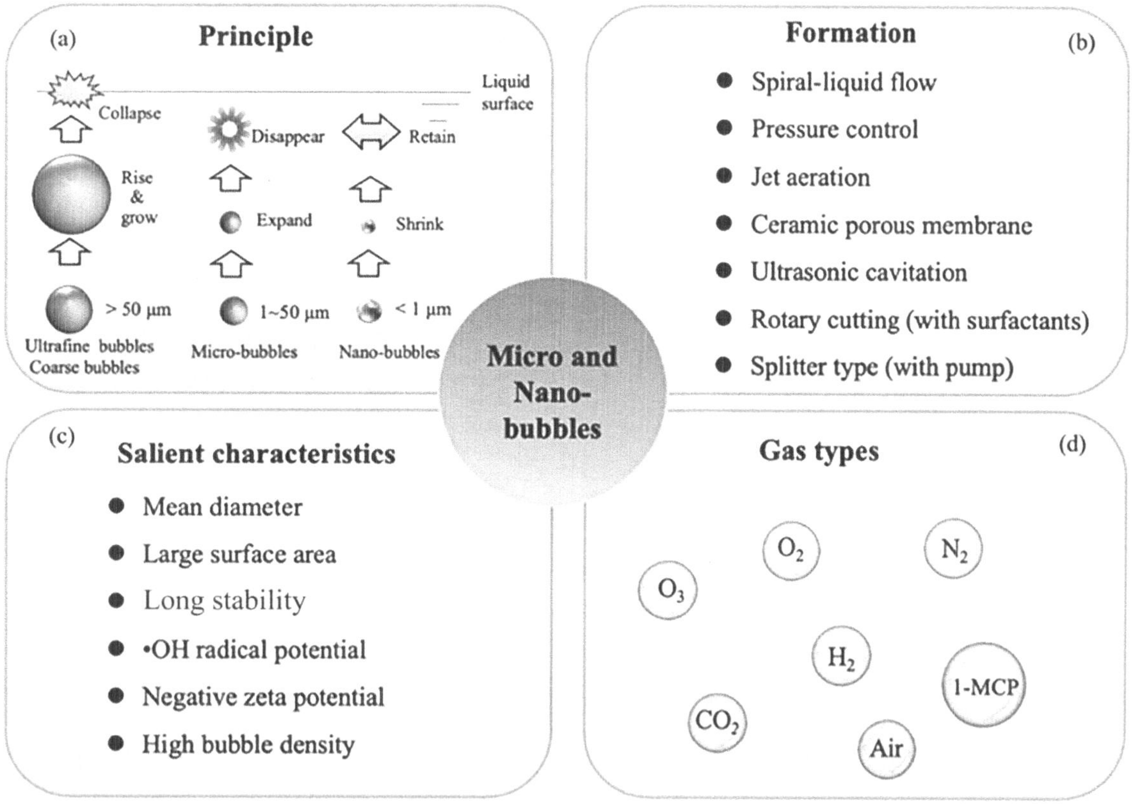

FIGURE 22.2 Advancement in micro/nanobubble-based technology for food applications: (a) schematic diagrams of macro, micro and nanobubbles, (b) formation of micro/nanobubbles, (c) salient properties of micro/nanobubbles and (d) applications of gas-based micro/nanobubble types. These figures are adapted from Shan et al. (2023).

of the coated commodity by reducing the shine and transparency of the coating (De Azeredo, 2009). The main advantages of using these coatings are their low molecular masses, mechanical strength, tolerance against high thermal exposure, blockage of gases, moisture and volatile aromatic compound escape and also their strength against microbes and pathogens (Aliofkhazraei, 2011).

22.6 PREPARATION OF NANOCOATINGS

Nanocomposite coatings can be prepared by mechanically stirring reverse charges, by altering temperature, by emulsifying oil in water, by steam condensation, by electrostatic interaction between polyions or by extreme stirring (Shapi'i et al., 2020). In a few studies, the stomatic movements of leaf cells have been imitated in nanocoatings so that a regular exchange of water vapors, oxygen and carbon dioxide can occur. For example, chitosan microspheres were merged in shellac with the help of tannic acid, and the results were outstanding (Zhou et al., 2021). Broadly speaking there are two procedures for producing nanoparticles, which are explained in the following sections.

22.6.1 Top-Down Method

With the use of physical and chemical changes, the complex compounds are converted into small submicron-sized particles. There are multiple methods of achieving this. The simplest is mechanical milling. The compound to be broken down is put in a mill, either alone (dry) or with a wet medium. Then with the help of a grinding stone the element is converted into nanosized particles. The efficiency of this assembly depends on mill type, ball adjustment, speed of the rolling ball, type of substrate (either dry or wet), time period of milling and heat. This is by far the cheapest method of preparing nanoparticles due to its feasibility for using both dry and wet substrate.

The second technique is nanolithography, which involves producing nanoparticles by nanofabrication. The size of the produced particles ranges between 1 and 100 nm. Lithography can be either masked or unmasked. The former embraces soft, nanoimprint and photolithography, involving the use of patterns for the preparation of fabricated nanoparticles. The latter technique embraces electron beams, focused ion beams and scanning probes, without using any mold.

The third one is sputtering, which counts on the use of dynamic plasma or gas striking a surface with heavy energy to eject a layer of nanoparticles over that surface. Nanoparticles formed by this method have the same composition as the targeted material. The several ways of carrying of this technique are the use of direct current, radio frequency, chemical reactions and magnetron. The next technology is laser excision. Extremely pure and distinguished nanoparticles are created when laser beams strike the aimed material. The composition of the target material, the focus of the laser beam and the pulse of the laser are the defining factors inthe success of this method. In the recent past, when a liquid target was tried and singly dispersed a colloidal end product was produced. The benefit of using it is that no chemical procedures are involved; further, there is no need for a mold or holding time. Also, it is safe ecologically.

The next technique is electrospinning, used in the preparation of nanofibers from polymers, metal and complex materials like ceramics. The extremely shrilled fibers, a few cm in length, were prepared by coaxial electrospinning. Last but not least is the formation of nanoparticles by the use of etching, which means chemical confiscation of the required material from the source. Etching can be done with a liquid medium, i.e. the wet method, or with gas and plasma, i.e. dry etching. The product is a metallic nanoparticle that can be utilized in any desired way.

22.6.2 Bottom-Up Method

Nanoparticles can be extracted from single or aggregated atoms by several methods, discussed next, collectively named bottom-up methods.

In chemical vapor deposition, nanoparticles are bonded as a slim film, at elevated temperature, over a previously designed surface by antecedent gases and vapors. Nanoparticles obtained by this technique are spotless, have eminence quality and have the lowest chance of blemishes. Yet this method is expensive and may have residual toxic effects due to some by-products generated during the process.

The second procedure is called the sol-gel process because first the source of the nanoparticles is dissolved in an aqueous medium (sol) and after breakdown of the water molecules and repeated condensation, a gel is formed. ZnO, titanium dioxide, stannic oxide and tungsten trioxide are generally prepared by this technique. These products are highly specific in terms of the magnitude and structure of nanosystems. The third scheme for preparing nanoparticles is carried out in a padlocked container at elevated temperature and under high pressure, using a hydro-solvent, and a solute as either a gas or solid. This is called the hydrothermal method and is close to the solvothermal method, in which a hydro-solvent is not used. Both of these techniques are useful in the preparation of nanosized structures, including sheets, rods and spherical or wire shapes.

The other technique is the reverse micelle technique, in which an emulsion, nanometric in size, is made by mixing lipids in an aqueous solvent or by mixing water as a solute in a lipid solvent. The resultant product has two poles. In the former case, the head is water-loving and remains outside, and the tail is intolerant to water and is directed toward the inside of the particle. In the latter case, the directions of the head and tail are reversed. The aqueous portion is called a water pool, and its size is adjustable and can be controlled by altering the proportions of water and other material in the emulsion. Surfactants are used to make the emulsion smooth and also to control the size and structure of the end product (Qiao et al., 2019; Zapata et al., 2019; Chaudhary et al., 2020; Nair et al., 2020; Pinzon et al., 2020; Bratovcic, 2021; de Oliveira Filho et al., 2021).

22.7 APPLICATIONS OF NANOCOATINGS ON FRUITS AND VEGETABLES

In an experiment, an aqueous solution of Na-caseinate was incorporated with curcumin and zein by using a nanoprecipitation technique. To advance the composition of this solution for film forming, a gelatin solution was added to water. This approach not only proved to be a wonder for controlling the attack of microorganisms and postharvest decay, but also enhanced the glossiness of grapes and maintained their biochemical qualities for 1 week (Lemes et al., 2017). One more study utilized polyvinyl alcohol along with silk fibroin in equal proportions to dip cut slices of apple. Ascorbic acid (AA) was added to improve antioxidative qualities. This bilayer coating was successful in retaining the weight, hue and moisture content of the slices. According to the images obtained from florescent microscope and SEM, the coating was a bilayer due to density differences and thus appreciably tensile to be used as an ultrathin flexible layer (Ruggeri et al., 2020). The effects of different nano-based coatings have been summarized in Table 22.1.

In the case of lipid nanoparticles, the common approaches to create them are the use of a surfactant in water and the homogenization of fats at a high temperature. Polyethylene glycol mixed with xanthan gum and incorporated with lipid nanoparticles (solid) when applied over guava fruits exhibited that the coating composition was evenly distributed and hydrophobic enough to maintain fruit quality (Zambrano-Zaragoza et al., 2013). Tannic acid paired with shellac also succeeds in maintaining mango fruit table quality. Shellac is appreciated for its resistance against moisture loss but is not tolerant to microbial attack, but that was compensated by using tannic acid. Overall results regarding coated mango fruits were outstanding (Ma et al., 2021).

Other organic waxes are also basic-alkanes in nature, which checks the escape of moisture from the fruit's surface, and owing to this quality these are being utilized as nanocoating components (Wang & Zhao, 2021). When jojoba oil and candelilla wax were emulsified by hot stirring with glycerol and Tween-80, they displayed excellent antioxidative characters, along with having uniform consistency (De León-Zapata et al., 2018). Hydroxypropyl methylcellulose when combined with beeswax controlled the water and gaseous movements from coated guava fruits by covering tiny openings like

TABLE 22.1 Application of Different Types of Nanotreatments on Fresh Fruits and Vegetables

PRODUCE	*TYPE*	*NANO-TYPE*	*NANOCOATING COMPOSITIONS*	*BENEFITS*	*REFERENCES*
Okra	Whole	Nanoemulsion	Alginate and basil oil	Retained firmness, color and taste. Reduced fungal attack	Gundewadi et al. (2018)
Strawberry	Whole	Nanoemulsion	Pullulan and cinnamon EO	Conserved fruit weight, reduced TA decline and exhibited antimicrobial effect against fungus	Chu et al. (2020)
Pineapple	Fresh cut	Nanoemulsion	Alginate and citral	More acceptable color, reduced respiration rate and less microbial attack	Prakash et al. (2020)
Apple	Juice	Nanoemulsion	Cinnamon EO and ascorbic acid	Inhibited PPO activity, enhanced oxidation rate and retained color	Xu et al. (2020)
Tomato	Whole	Nanoemulsion	Oregano EO	Enhanced shelf life and prevented microbial, yeast and fungal attacks	Pirozzi et al. (2020)
Spinach	Leaves	Nanobubbles	O_3 micro/nanobubbles	Improved antioxidative activities, reduced H_2O_2, ethylene, respiration rate, yellowing index and malondialdehyde content	Wang et al. (2020b)
Strawberry	Whole	Nanocapsules	Natamycin, nano-zein and CMC	Reduced attack of rot, mildew and gray mold	Lin et al. (2020)
Papaya	Fresh cut	Nanocomposite film	Chitosan and ZnO	Prevented microbial growth	Lavinia et al. (2020)
Strawberry	Whole	Nanocomposite film	CMC, guar gum and Ag-NPs	Presented antimicrobial activity on gram-positive and gram-negative bacteria and fungi. Revealed reduced weight loss	Kanikireddy et al. (2020)
Bell pepper	Fresh cut	Nanocomposite film	Polyvinyl pyrrolidone/ glycerosomes and Ag-NPs	Enhanced antibiotic activity and shelf life	Saravanakumar et al. (2020)
Guava	Whole	Nanocomposite film	Chitosan, Na-alginate and nano-ZnO	Inhibited rotting and microbial growth and improved quality and shelf life	Arroyo et al. (2020)
Strawberry	Whole	Nanocomposite film	Na-alginate and nano-ZnO	Reduced microbial attack and enhanced storage life	Emamifar & Bavaisi, 2020)
Sweet basil	Leaves	Nanoparticles	Thyme volatile oil and chitosan	Enhanced shelf life and reduced weight loss	Hassan et al. (2021)
Bell pepper	Whole	Nanoparticles	Chitosan and thyme EO	Reduced weight loss, maintained firmness and protected against *P. carotovorum*	Correa-Pacheco et al. (2021)
Cucumber	Fresh cut	Nanocapsules	Ionic gelation, alginate-pectin and lemon EO	Enhanced shelf life, reduced polyphenols, retained antioxidant capacity and reduced enzymatic activity	Zambrano-Zaragoza et al. (2021)
Strawberry, Apple	Whole, Fresh cut	Nanopolymer	Chitosan, grape seed EO and ascorbic acid	Retained postharvest quality and improved shelf life	Popescu et al. (2022)
Citrus	Whole	Nanocomposite film	Nano-Se, pomegranate peels and nano-chitosan	Protected from invasion and existing fungal infections and enhanced storage life and quality	Salem et al. (2022)
Banana, Apple	Whole, Cut slices	Nanofiber	Silk *fibroin*, VA, Honey and curcumin	Enhanced shelf life and retained weight, stiffness and general quality	Singh & Packirisamy, 2022)

(Continued)

TABLE 22.1 (Continued)

PRODUCE	TYPE	NANO-TYPE	NANOCOATING COMPOSITIONS	BENEFITS	REFERENCES
Strawberry, Mangoes, Apples, Peaches, Kiwifruit	Whole	Nanocapsules	Chitosan oligomers–hydroxyapatite–carbon nitride, *U. tomentosa* bark extract	Protected against *C. gloeosporioides*, *P. expansum, M. laxa* and *S. sclerotiorum*	Santiago-Aliste et al. (2023)
Pomegranate	Arils	Nanoparticles	Clove EO and chitosan	Enhanced shelf life	Hasheminejad & Khodaiyan (2020)
Pistachio	Whole	Nanopolymer	Chitosan and SA	Maintained weight, phenolics, enzymatic activities, color and organoleptic quality. Reduced oxidation of lipids and suppressed microbial growth	Molamohammadi et al. (2020)
Cherry tomato	Whole	Nanocomposite	Pectin and nano-Mg	Enhanced shelf life and quality	Kumar et al. (2020)
Lemon	Whole	Nanocomposite	Coconut oil and beeswax	Enhanced shelf life and quality	Nasrin et al. (2020)
Peach	Whole	Nanofiber	Zein, PEO and Hexanal	Maintained shelf life and quality	Ranjan et al. (2020)
Apple	Fresh cut	Nanopolymer	Chitosan, organic acids (acetic, lactic, and levulinic acids) and L-cysteine	Inhibited microbial growth	Jin et al. (2020)
Papaya	Fresh cut	Nanopolymer	Gellan gum, *A. vera* gel, α-carrageenan and Na-alginate	Strong oxygen barrier, slowed respiration and maintained texture	Farina et al. (2020)
Tomato	Whole	Nanotubes	Acrylic emulsion of EOs, B-cyclodextri and halloysite	Retained fruit freshness and reduced decay	Buendía et al. (2020)
Apple	Fresh cut	Nanofiber	Bacterial cellulose	Retained weight, firmness and TA. Showed reduced browning	Zhai et al. (2020)
Bell pepper	Whole	Biocomposite	Chitosan and α-Pinene	Reduced attack of *A. alternate* and retained physiochemical quality	Hernández-López et al. (2020)
Guava	Fresh cut	Nanocomposite	Chitosan and citric acid	Inhibited attack of *C. gloeosporioides* rot. Retained quality and sensorial attributes	Nascimento et al. (2020)
Fig	Whole	Nanocomposite	Nano-chitosan and nano-propolis	Decreased attack of *A. flavus*, delayed ripening and enhanced sensorial quality	Aparicio-García et al. (2021)
Mango	Whole	Nanocomposite	Chitosan and nano-TiO_2	Controlled attack of *C. gloeosporioides*, *C. oxysporum* and *P. steckii* on fruits	Xing et al. (2021)
Strawberry	Whole	Nanofiber, Nanocrystal	Cellulose and chitin/chitosan	Reduced attack of *B. cinerea*, preserved freshness and reduced decay, weight loss and color changes	Sun et al. (2021)
Blueberry	Whole	Nanocomposite	Chitosan and nano-TiO_2	Controlled attack of mesophilic aerobic bacteria, mold and yeasts and retained TSS, ripening index, pH and anthocyanins	Rokayya et al. (2021)

PRODUCE	*TYPE*	*NANO-TYPE*	*NANOCOATING COMPOSITIONS*	*BENEFITS*	*REFERENCES*
Strawberry	Whole	Nanocellulose	*Cellulose* nanofiber and cinnamon EO	Retained freshness and reduced fungal attack	Montero et al. (2021)
Pakchoi	Whole	Nanobubbles	H_2 micro/nanobubbles	Preserved chlorophlls, ROS scavenging ability and antioxidant activities along with reduced oxidative stress	An et al. (2021)
Pitaya	Whole	Nanopolymer	1-MCP, α-pinene and oleic acid	Reduced respiration and ethylene production. Retained texture	Espinal-Hernández et al. (2021)
Banana	Whole	Nanotubes	TiO_2, CuO and Cu_2O	Scavenged ethylene and retained firmness, freshness and weight along with delayed browning	Riahi et al. (2021)
Kiwi	Whole	Nanocellulose	Chitosan and curcumin	Maintained color, quality, weight, firmness and respiration. Showed reduced microbial count	Ghosh et al. (2021)
Cantaloupe	Fresh cut	Nanocellulose	Chitosan and *trans*-cinnamaldehyde	Maintained weight, firmness and color	Ozturk et al. (2021)
Rosa sterilis	Whole	Nanobubbles	H_2 micro/nanobubbles	Enhanced firmness, preserved biochemical and antioxidative activities with suppressed oxidative indicators	Dong et al. (2022)
Chinese water chestnut	Fresh cut	Nanobubbles	H_2 micro/nanobubbles	Enhanced antioxidant activity. Reduced hydrogen peroxide, improved phenylalanine ammonia lyase activity	Li et al. (2022)
Strawberry	Whole	Nanocomposite	Cinnamon essential oil, TiO_2 and chitosan	Retained firmness and fresh weight. Reduced mildew attack and enhanced shelf life	Yuan et al. (2023)
Rosa roxburghii	Whole	Nanobubbles	H_2 micro/nanobubbles	Enhanced firmness, biochemical quality and antioxidative activities and alleviated oxidative damages	Ding et al. (2023)
Parsley	Whole	Nanobubbles	O_3 micro/nanobubbles	Delayed leaf yellowing and chlorophyll degradation, reduced respiration and retained firmness, soluble solids concentration and antioxidative enzymatic activities	Shi et al. (2023)
Papaya	Cut slices	Nanoolymer	Nano-CuO, Nano-ZnO and guava extract	Extended shelf life and retained firmness and color	Sharma et al. (2023)
Cucumber	Whole	Nanofiber	Chitin	Reduced moisture loss and incidence of bacterial growth	Tanpichai et al. (2023)

lenticles and micropores on the fruit's peel (Formiga et al., 2019).

Metallic, inorganic nanoparticles have long been utilized in edible coatings, but their amalgamation with organic components is an innovative approach in horticulture. Xanthan gum was fabricated with ZnO (obtained by the sol-gel technique from Zn-acetate hydrate) by using a binary emulsion evaporative technique. Salicylic acid was added as a binder of suspension, and when this coating was applied over apple fruits, not only was the fruit quality maintained for a long time, but it was also safe for the consumer because hemolysis indicated the coating to be biofriendly. The same coating also eliminated the use of plastic packaging because of the antibacterial traits of ZnO (Joshy et al., 2020).

Likewise, when ZnO, as a nanoparticle, was mixed in the matrix of CMC along with cinnamaldehyde, the storage life and quality of cherry tomatoes were enhanced appreciably. The fruit's firmness, moisture content, nutritional quality and taste of coated fruits were exceptionally remarkable. There was no fungal attack on coated fruits, as the nanoparticles generated ROS to fight the invading microbes (Guo et al., 2020). When double layered alginate and AA matrix were injected with magnesium and aluminum hydroxide by using glycerol, the dip-coated strawberry fruits were able to be preserved for a long time. The treated fruits had higher TA, antioxidant content, less weight loss and firmer tissues. Also, no microbial attack was observed over the coated fruits (Madhusha et al., 2020).

Organic nanoparticles are apparently harmless and full of benefits, but the use of metallic particles has toxicity concerns. Thus, the present data and investigations are not enough, and detailed, long-term studies on their possible effects on human health are required. Also, the amount of metallic nanoparticles is critical. Another possible aspect for scientific studies is to find their resistance against viral attack (Ghosh & Singh, 2022). Further studies utilized egg cellulose and mixed it with polymer, also derived from egg. This organic nanocoating reduced the wastage of perishable products (Jung et al., 2020). In another investigation about nanocoatings, chitosan was used as cation and pectin was utilized as anion to form an organic polyelectrolyte nanocomplex. This experiment proved beneficial in reducing the use of polyethylene material for food packaging and to reduce handling losses of fruits and vegetables (Chiang et al., 2021).

The secret to getting full benefits from nanocoatings lies in finding the perfect ratio between the bioactive substance and the biopolymer which will encapsulate it. Some successful examples of such coatings are chitosan used for *Solanum lycopersicum* (Tampucci et al., 2021), quinoa protein for *Solanum lycopersicum* var. cerasiforme (Robledo et al., 2018), chitosan along with seed extract of *Citrus paradisi* (Won et al., 2018) and with zein protein on *P. gnaphalodes* (Mehdizadeh et al., 2021), A. vera on *Punica granatum* and other fruits (Pirzadeh & Aminlari, 2019) and alginate and chitosan encapsulating ZnO applied over *Psidium guava* (Arroyo et al., 2020). Other than the described coatings, other scientists have also used whey protein containing particles of TiO_2 along with EO of rosemary containing nanofibers of cellulose for coating several fruits and vegetables (Alizadeh-Sani et al., 2018), Co chelated biopolymers (Zhou et al., 2019), ZnO nanocomposites (Anugrah et al., 2020), chitosan-encapsulated salicylic acid (Lo'Ay & Taher, 2018), alginate with gellan (Tapia et al., 2007), whey protein isolate (Tapia et al., 2007), alginate glycerol citric acid biofilm (Rangel-Marrón et al., 2019), EOs from different resources (Guo et al., 2021) and several gum-based coatings (García-Cruz et al., 2021).

Nanospray is a state-of-the-art technique where nanocoatings are applied in the form of nanosprays containing submicron bioactive agents. Although very limited literature is available about this technique, still some describable research has been carried out. For example, it was studied that if a nano-mist is generated in the storage house, fruit weight loss can be reduced (Van Hung et al., 2011). In another study, the nano-mist contained encapsulated edible constituents (Díaz et al., 2019). Aerosol sprays with nanosized insecticidal particles were found to be successful against controlling fruit flies (Morozov & Kanev, 2015). In another study, fatty acids were loaded with nanoemulsified lactoferrin to keep the fruits fresh for longer durations (Vidinamo et al., 2021).

While storing fruits and vegetables in store houses, a very common and inevitable problem is wear and tear of the packaging material, especially cardboards, due to high relative humidity inside the storage cell. Nanosprays are helpful in maintaining and empowering the strength and shape of such materials. It has also been observed that nanosprays can control mold and yeast growth over the stored fruits and vegetables without compromising their aesthetic value (Aguirre-Güitrón et al., 2022). Nanosprays can also be useful in controlling the amount of nitric oxide evolving from the fruits (Zahedi et al., 2020). Nanosprays can increase the utilization of bioactive agents by improving their availability to the fruit and vegetable surface, hence relieving the commodity of possible oxidative loss (Cassi et al., 2022).

For making nanoparticles, another contemporary approach, sono-chemical nanoparticles, involves the use of ultrasound waves, via acoustic cavitation, to increase the rate of chemical reaction, leading to more intensified nanocomposites (Bodbodak & Moshfeghifar, 2016). Some of the applied sono-chemical nanocoatings include chitosan encapsulating ZnO (applied on *Abelmoschus esculentus*) (Al-Naamani et al., 2018), pectin (on *F. ananassa*) (Treviño-Garza et al., 2015) and CMC coating (on *Fragaria ananassa*) (Nadim et al., 2015).

Other than the mentioned nanoparticles, some other materials also offer antibacterial, antipathogenic and antidegradation properties when applied over *Spinacia oleracea*, *Lactuca sativa*, *Asparagus officinalis* and other greens (Kanikireddy et al., 2020). The $CaCl_2$ and ClO_2 have also been used for increasing the storage and utility span of fruits like *Fragaria ananassa* and *Prunus domestica*, respectively (Shahzad et al., 2020). Nanocoatings are an excellent solution for the control of postharvest decay in fruits and vegetables. Different kinds of whole fruits can be stored up to 60 days while cut slices of fruits treated with nanocoatings can be stored for around 6 weeks in good condition. However, the storage period is also reliant on the nature and biocomposition of fruit (Prabhu et al., 2022).

Nanosized zeolite particles with an outer layer of $KMnO_4$ when applied to peaches and nectarines lessened weight loss for some time and also removed ethylene from the fruits' surface to improve their storage life and quality (Emadpour et al., 2015). Results from another related study state that when nanosized alumina encumbered with $KMnO_4$ is applied over avocado fruit, it oxidizes more than 70% of the C_2H_4 within half an hour of application and maintains an ethylene-free environment for the next 5 days and improves the physical appearance of the fruit. The same coating applied over bananas held them in acceptable form for 1 additional week. This method is easy to carry, cheap and more acceptable for developing countries (Tirgar et al., 2018).

Titanium dioxide nanoparticles–based nanocoatings have also been reported to have the same properties when loaded over a chitosan coating. These also possess excellent mechanical strength and reduce the escape of moisture. If coated fruits are exposed to UV light, ROS, hydroxyl group and superoxide anion, radicals are generated which have bactericidal properties against *Staphylococcus aureus*, *Escherichia coli*, *Salmonella typhimurium* and *Pseudomonas aeruginosa*. It also kills *Aspergillus* and *Penicillium*. However, the possible side effects of their use on human health are still a mystery (Siripatrawan & Kaewklin, 2018). In another experiment, when packaging atmosphere was altered with the help of nanocoatings, it suppressed the exchange of O_2 completely but remained semipermeable to CO_2, which actually is useful in maintaining the quality of the product. With the production of ethylene, O_2 accumulation increased and delayed the opening of the head in mushrooms (Gholami et al., 2017).

PLA loaded with MgO and ZnO improved the organoleptic qualities of cucumbers and also reduced the incidence of yeast and fungal attack (Li et al., 2021). Likewise, Piel de Sapo melons, when coated with Ag nanoparticles loaded on cellulose film, were juicier even after storage for more than 1 week. Also, there was no microbial growth observed over cut slices of the fruits (Fernández et al., 2010). Loquat fruits coated with nanoparticles of silicon dioxide had high ROS scavenging ability along with higher amounts of AA, phenols, TSS and other nutrients after more than 10 days (Wang et al., 2020a). Likewise, titanium dioxide nanoparticles, Ag nanoparticles and MMT particles loaded over polyethylene sheet improved the storage life and quality of kiwifruits (Shan et al., 2023).

Plant-based nanoformulations are more appreciated than others because of their better action against microbes. The nano-edible coatings containing α-pinene, a type of terpene found in >400 plant species (Allenspach & Steuer, 2021), reduced the attack of *A. alternata* on stored sweet peppers (Hernández-López et al., 2020). Similarly, chitosan nanocoating, prepared in *B. crassifolia* extract, and EO from sea holly also retained the biochemical qualities of coated sweet peppers and cherries, respectively (Arabpoor et al., 2021). The EOs from other plants have also been recognized and used for their pleasing aroma, acceptable taste and antioxidative qualities (Vieira et al., 2022). Other than the stated EOs, other plants have also been used for extracting EO, later used for coating fruits and vegetables as nanocoatings, for example, Shirazi thyme, sweet worm-wort, cinnamon, lemongrass and great basil plant (Shan et al., 2023).

22.8 COMMERCIAL AVAILABILITY OF NANOCOATINGS

Nanotechnology is not limited to use in edible coatings; its use has also been reported in personal and skin care products, dyes and paints, greases and oils, medicines, surgeries, disease diagnosis, energy generation and other industries (Chaudhry, 2023). Half of the nanoproducts are availed in food storage, and the use is uninterruptedly increasing more than 11.5% each year (Anonymous, 2023b). Although a lot of research is expected to be carried out about several aspects of nanocoatings, these are still being used commercially on a large scale.

The most commonly used materials are nylon 6-nanoclay composite, cerium and iron oxides, oxidized polymers, Na-glycinate or Na_2CO_3, scavenging oxygen, AIT, ClO_2, enzyme-based TTI, lipase, nanoclay particles and composites, nanosilver, self-assembled nanoliquids and TiO_2. These materials can be available in the form of nylon resin barriers, sachets, labels, films, trays, tray pads, stickers, bags and sprays. These materials are sold in the market under different labels and brand names (Kondle et al., 2022).

On a commercial scale, to increase the lucidity and elasticity of the nanocoating, micro-sized pore plugs are used. A 550 nm thick coating has more than 5% elasticity, and a 400 nm thick coating can provide 86% lucidity (Vasile, 2018). Purdue University, USA, has introduced cellulose nanocrystals in wrapping the edible items. By the next year i.e. 2025 the food wrapping and packaging industry will increase in size by 6%, while the overall food coating industry is increasing in size by 45% each year; thus new techniques need to be adopted on a commercial scale to stay with the growing market trends (Anonymous, 2023a).

22.9 LIMITATIONS OF NANOCOATINGS

Regardless of the uncountable blessings we can get by exploiting nanocoatings, there are certain hurdles to their limitless uses. For starters, we can consider them as any other coating, which covers the fruit and limits gaseous movements in and out of coated produce, thus leading to quality degradation or flavor alteration (Mahela et al., 2020). Then, due to the novelty of nanosystems, their utilization is still ambiguous. There is still a huge blank space about their long-term consequences on consumer health. Until or unless their safety for consumption is not confirmed, their use will be in a challenging position. For instance, according to US food regulations, nanoparticles are still not included under the definition of Food Additives (1333/2008). On the other hand, if their use is supposed to be made legal, they must fall in the category of Novel Food (258/977). Hence, a lot of research is still needed in this sector (Gallocchio et al., 2015).

Some nanosystems can initiate allergies in consumers, especially those extracted from ground nuts, fish oil, milk products, soybeans and cereals, and such nanocoatings should be clearly labeled with their ingredients or should be banned. Another hurdle to using nanosystems is their high cost and the modern machinery and expertise required for their application and preparation. Being a novel technique, there is still so much to know about this system. There is no comprehensive data on effective making and application of these coatings. Their qualities, limitations, possible side effects and efficiency are still abstruse. The machinery required for industrial-scale production and application of nanocoatings is still not common and not even available in a lot of cases. Hence, for now, only highly precious, delicate and expensive products are being coated by this technique (Mahela et al., 2020).

According to the US EPA (Environmental Protection Agency), nanocoatings can have certain serious side effects on human health. For illustration we can consider the case of Ag nanoparticles, which are gaining popularity for their use in edible coatings. The issue with them is that they help gibberella fungus in mycotoxin production (Jian et al., 2022). Although other non-metallic nanoparticles, like selenium, have shown no residual effect on human health when applied over hazelnuts, it is still vague which elements will affect human health and which will not (Vera et al., 2016). Owing to their tiny size and excellent adsorbing power these particles can be inhaled by human beings and other living things. As, for now, we are not fully aware of the hazards they can cause to humans, we also need to develop a critical investigative method to detect the presence of these particles in food products (Shan et al., 2023).

The nanoparticles of nanocoatings can penetrate commodities due to their tiny size and can cause harm to the consumer's health, along with altering the taste and aroma of the commodity (Emamhadi et al., 2020). The extent of contamination can vary with variation in storage temperature, dimensions of the nanoparticle, physiological stress, the extent of bonding with other particles and the kind of main material applied in the coating (Emamhadi et al., 2020). If the nanoparticles slipping into the human body have toxic effects and their size range is more than 100 nm, it will assuredly lead to grave issues for the consumer. Even smaller ones, with sizes ranging from 50 nm to 70 nm, can enter the lungs, and particles with a 30 nm diameter can even enter the bloodstream and can also target the brain cells (Chau et al., 2007).

The Ag nanoparticles (Echegoyen & Nerín, 2013) and TiO_2-based (Yang et al., 2014) nanocoatings have been reported to contaminate the human body. Still, there is no reliable and comprehensive knowledge about the mode of entrance, the negative effects and safe limits regarding the use of nanoparticles. The US FDA has been working day and night to define the ultimate safe limits, the mode of usage and the possible bad outcomes of their usage. In Europe, however, if the concentration of nanoparticles found in food falls under their defined safe utility limit for that particle, it can be used in food, albeit with great caution (Paradise, 2019).

22.10 CONCLUSIONS

In conclusion, with the invention of nanocoatings, the food storage industry has seen a boost. Because of their capability to retain high antioxidant content in the coated fruits and vegetables and also because of their antimicrobial qualities, their use is highly appreciated for the storage of fresh produce. The choice of the material is important because it is contingent with the commodity to be treated and the commodity's own chemical composition. It is also subjected to how long we are planning to keep the said commodity in storage. Organic materials are ideal for this purpose because generally these are not considered to be toxic for human beings. Henceforth, if the coating is prepared properly, it will release the exact amount of active agent at exact time of requirement. As an outcome, the delicate and desired nutrients in the coated fruits and vegetables remain protected for a long time. Further research is required to extend the use of nanotechnology, especially if this technique will be able to modify the mechanical and heat-transporting abilities of the coated commodities according to the heat and moisture difference between the outside and inside of the commodity. This will bring a boom in the fresh produce postharvest industry in general. Using mineral or inert substances in nanocoatings is expected to bring a revolution in better regulation of O_2, CO_2 and vapor in and outside the fruits and vegetables. So, further scientific studies are needed in order to create a firm nanocoatings-based system that can be adopted on an industrial scale and is in favor of both the user and the industrialist. Another incontrovertible sector that needs to be studied in detail is the after-effects of nanocoatings. We have to be clear about what possible long-term outcomes we may face after the use of these nanoparticles in the human diet. We must know how these materials can affect our bodies after consumption, what their safe limits for consumption are and what precautions we must take before and after consuming such products.

REFERENCES

Abadi, B., Mahdavian, S., Fattahi, F. (2021). The waste management of fruit and vegetable in wholesale markets: Intention and behavior analysis using path analysis. *Journal of Cleaner Production*, *279*, 123802.

Aguirre-Güitrón, L., Calderón-Santoyo, M., Lagarón, J.M., Prieto, C., Ragazzo-Sánchez, J.A. (2022). Formulation of the biological control yeast Meyeroz yma caribbica by electrospraying process: effect on postharvest control of anthracnose in mango (*Mangifera indica* L.) and papaya (*Carica papaya* L.). *Journal of the Science of Food and Agriculture*, *102*(2), 696–706.

Ali, S., Khan, A.S., Nawaz, A., Naz, S., Ejaz, S., Shah, A.A., Haider, M.W. (2023). The combined application of Arabic gum coating and γ-aminobutyric acid mitigates chilling injury and maintains eating quality of 'Kinnow' mandarin fruits. *International Journal of Biological Macromolecules*, *236*, 123966.

Ali, S., Nawaz, A., Hussain, B., Ejaz, S., Sardar, H. (2024b). Carboxymethyl cellulose coating maintains quality of harvested aonla fruit by regulating oxidative stress and ascorbate-glutathione cycle. *Postharvest Biology and Technology*, *207*, 112621.

Ali, S., Ishtiaq, S., Nawaz, A., Naz, S., Ejaz, S., Haider, M.W., Shah, A.A., Ali, M.M., Javad, S. (2024a). Layer by layer application of chitosan and carboxymethyl cellulose coatings delays ripening of mango fruit by suppressing cell wall polysaccharides disassembly. *International Journal of Biological Macromolecules*, *256*, 128429.

Aliofkhazraei, M. (2011). Nanocoatings: Size Effect in Nanostructured Films; Springer: Berlin, p. 212.

Alizadeh-Sani, M., Khezerlou, A., Ehsani, A. (2018). Fabrication and characterization of the bionanocomposite film based on whey protein biopolymer loaded with TiO_2 nanoparticles, cellulose nanofibers and rosemary essential oil. *Industrial Crops and Products*, *124*, 300–315.

Allenspach, M., Steuer, C. (2021). α-Pinene: A never-ending story. *Phytochemistry*, *190*, 112857.

Al-Naamani, L., Dutta, J., Dobretsov, S. (2018). Nanocomposite zinc oxide-chitosan coatings on polyethylene films for extending storage life of okra (Abelmoschus esculentus). *Nanomaterials*, *8*(7), 479.

Al-Tayyar, N.A., Youssef, A.M., Al-Hindi, R.R. (2020). Edible coatings and antimicrobial nanoemulsions for enhancing shelf life and reducing foodborne pathogens of fruits and vegetables: A review. *Sustainable Materials and Technologies*, *26*, e00215.

An, R., Luo, S., Zhou, H., Zhang, Y., Zhang, L., Hu, H., Li, P. (2021). Effects of hydrogen-rich water combined with vacuum precooling on the senescence and antioxidant capacity of pakchoi (Brassica rapa subsp. Chinensis). *Scientia Horticulturae*, *289*, 110469.

Anjum, M.A., Akram, H., Zaidi, M., Ali, S. (2020). Effect of gum arabic and Aloe vera gel based edible coatings in combination with plant extracts on postharvest quality and storability of 'Gola' guava fruits. *Scientia Horticulturae*, *271*, 109506.

Anonymous (2023a). Purdue University. Manufacturing process provides low-cost, sustainable option for food packaging. https://phys.org/news/2018-06-low-cost-sustainable-option-food-packaging.html#jCp (Accessed: on 27 August, 2023).

Anonymous (2023b). European Commission, Regulation (EC) No 1935/2004 of the European Parliament and of the Council of 27 October 2004 on materials and articles intended to come into contact with food and repealing Directives 80/590/EEC and 89/109/EEC. *EEC*. www.interpack.com/en/Media_News/Tightly_Packed_Magazine/FOOD_INDUSTRY_PACKAGING/News/Food_Safety. (Accessed on: 26 August, 2023).

Anugrah, D.S.B., Alexander, H., Pramitasari, R., Hudiyanti, D., Sagita, C.P. (2020). A review of polysaccharide-zinc oxide nanocomposites as safe coating for fruits preservation. *Coatings*, *10*(10), 988.

Aparicio-García, P.F., Ventura-Aguilar, R.I., del Río-García, J.C., Hernández-López, M., Guillén-Sánchez, D., Salazar-Piña, D.A., Ramos-García, M.d.L., Bautista-Baños, S. (2021). Edible chitosan/propolis coatings and their effect on ripening, development of *Aspergillus flavus*, and sensory quality in fig fruit, during controlled storage. *Plants*, *10*(1), 112.

Arabpoor, B., Yousefi, S., Weisany, W., Ghasemlou, M. (2021). Multifunctional coating composed of Eryngium campestre L. essential oil encapsulated in nano-chitosan to prolong the shelf-life of fresh cherry fruits. *Food Hydrocolloids*, *111*, 106394.

Arfat, Y.A., Ahmed, J., Hiremath, N., Auras, R., Joseph, A. (2017). Thermo-mechanical, rheological, structural and antimicrobial properties of bionanocomposite films based on fish skin gelatin and silver-copper nanoparticles. *Food Hydrocolloids*, *62*, 191–202.

Arroyo, B.J., Bezerra, A.C., Oliveira, L.L., Arroyo, S.J., de Melo, E.A., Santos, A.M.P. (2020). Antimicrobial active edible coating of alginate and chitosan add ZnO nanoparticles applied in guavas (Psidium guajava L.). *Food Chemistry*, *309*, 125566.

Assis, R.Q., Lopes, S.M., Costa, T.M.H., Flôres, S.H., de Oliveira Rios, A. (2017). Active biodegradable cassava starch films incorporated lycopene nanocapsules. *Industrial Crops and Products*, *109*, 818–827.

Aswani, R., Radhakrishnan, E.K. (2022). Green approaches for nanotechnology. In Green Functionalized Nanomaterials for Environmental Applications (pp. 129–154). Elsevier: London.

Balakrishnan, P., Sreekala, M.S., Geethamma, V.G., Kalarikkal, N., Kokol, V., Volova, T., Thomas, S. (2019). Physicochemical, mechanical, barrier and antibacterial properties of starch nanocomposites crosslinked with pre-oxidised sucrose. *Industrial Crops and Products*, *130*, 398–408.

Balasubramanian, R., Kim, S.S., Lee, J., Lee, J. (2019). Effect of TiO_2 on highly elastic, stretchable UV protective nanocomposite films formed by using a combination of k-Carrageenan, xanthan gum and gellan gum. *International Journal of Biological Macromolecules*, *123*, 1020–1027.

Bodbodak, S., Moshfeghifar, M. (2016). Advances in controlled atmosphere storage of fruits and vegetables. In Eco-friendly Technology for Postharvest Produce Quality (pp. 39–76). Academic Press: London.

Braich, A.K., Kaur, G., Singh, A., Dar, B.N. (2022). Amla essential oil-based nano-coatings of Amla fruit: Analysis of morphological, physiochemical, enzymatic parameters, and shelf-life extension. *Journal of Food Processing and Preservation*, *46*(6), e16498.

Bratovcic, A. (2021). Physical–Chemical, Mechanical and Antimicrobial Properties of Bio-Nanocomposite Films and Edible Coatings. *International Journal for Research in Applied Sciences and Biotechnology*, *8*(5), 151–161.

Brossault, D.F., McCoy, T.M., Routh, A.F. (2021). Preparation of multicore colloidosomes: Nanoparticle-assembled capsules with adjustable size, internal structure, and functionalities for oil encapsulation. *ACS Applied Materials & Interfaces*, *13*(43), 51495–51503.

Brum, A.A.S., dos Santos, P.P., da Silva, M.M., Paese, K., Guterres, S.S., Costa, T.M.H., Pohlmann, A.R., Jablonski, A., Flôres, S.H., de Oliveira Rios, A. (2017). Lutein-loaded lipid-core nanocapsules: Physicochemical characterization and stability evaluation. *Colloids and surfaces A: Physicochemical and Engineering Aspects*, *522*, 477–484.

Buendía, L., Sánchez, M.J., Antolinos, V., Ros, M., Navarro, L., Soto, S., Martínez, G.B., López, A. (2020). Active cardboard box with a coating including essential oils entrapped within cyclodextrins and/or halloysite nanotubes. A case study for fresh tomato storage. *Food Control*, *107*, 106763.

Cassani, L., Marcovich, N.E., Gomez-Zavaglia, A. (2022). Valorization of fruit and vegetables agro-wastes for the sustainable production of carotenoid-based colorants with enhanced bioavailability. *Food Research International*, *152*, 110924.

Chaple, S., Vishwasrao, C., Ananthanarayan, L. (2017). Edible composite coating of methyl cellulose for post-harvest extension of shelf-life of finger hot indian pepper (*Pusa jwala*). *Journal of Food Processing and Preservation*, *41*(2), 12807

Chau, C.F., Wu, S.H., Yen, G.C. (2007). The development of regulations for food nanotechnology. *Trends in Food Science & Technology*, *18*(5), 269–280.

Chaudhary, P., Fatima, F., Kumar, A. (2020). Relevance of nanomaterials in food packaging and its advanced future prospects. *Journal of Inorganic and Organometallic Polymers and Materials*, *30*, 5180–5192.

Chaudhry, Q. (2023). Nanotechnology applications for food packaging. https://ilsi.eu/wp-content/uploads/sites/3/2016/06/S2.3-WSPM12-Chaudhry.pdf (Accessed on: 26 August, 2023).

Chiang, H.C., Eberle, B., Carlton, D., Kolibaba, T.J., Grunlan, J.C. (2021). Edible Polyelectrolyte Complex Nanocoating for Protection of Perishable Produce. *ACS Food Science & Technology*, *1*(4), 495–499.

Chu, Y., Gao, C., Liu, X., Zhang, N., Xu, T., Feng, X., Yang, Y., Shen, X., Tang, X. (2020). Improvement of storage quality of strawberries by pullulan coatings incorporated with cinnamon essential oil nanoemulsion. *LWT*, *122*, 109054.

Correa-Pacheco, Z.N., Corona-Rangel, M.L., Bautista-Baños, S., Ventura-Aguilar, R.I. (2021). Application of natural-based nanocoatings for extending the shelf life of green bell pepper fruit. *Journal of Food Science*, *86*(1), 95–102.

Dang, X., Cao, X., Ke, L., Ma, Y., An, J., Wang, F. (2017). Combination of cellulose nanofibers and chain-end-functionalized polyethylene and their applications in nanocomposites. *Journal of Applied Polymer Science*, *134*(42), 45387.

Darie, R.N., Pâslaru, E., Sdrobis, A., Pricope, G.M., Hitruc, G.E., Poiată, A., Hitruc, G.E., Poiată, A., Baklavaridis, A. & Vasile, C. (2014). Effect of nanoclay hydrophilicity on the poly (lactic acid)/clay nanocomposites properties. *Industrial & Engineering Chemistry Research*, *53*(19), 7877–7890.

Davoudi, Z., Peroutka-Bigus, N., Bellaire, B., Jergens, A., Wannemuehler, M., Wang, Q. (2021). Gut organoid as a new platform to study alginate and chitosan mediated PLGA nanoparticles for drug delivery. *Marine Drugs*, *19*(5), 282.

De Azeredo, H.M. (2009). Nanocomposites for food packaging applications. *Food research international*, *42*(9), 1240–1253.

De León-Zapata, M.A., Ventura-Sobrevilla, J.M., Salinas-Jasso, T.A., Flores-Gallegos, A.C., Rodríguez-Herrera, R., Pastrana-Castro, L., Rua-Rodríguez, M.L., Aguilar, C.N. (2018). Changes of the shelf life of candelilla wax/tarbush bioactive based-nanocoated apples at industrial level conditions. *Scientia Horticulturae*, *231*, 43–48.

de Oliveira Filho, J.G., Miranda, M., Ferreira, M.D., Plotto, A. (2021). Nanoemulsions as edible coatings: A potential strategy for fresh fruits and vegetables preservation. *Foods*, *10*(10), 2438.

Dholariya, P.K., Borkar, S., Borah, A. (2021). Prospect of nanotechnology in food and edible packaging: A review. *The Pharma Innovation Journal*, *10*(5), 197–203.

Díaz, D.I., Lugo, E., Pascual-Pineda, L.A., Jiménez-Fernández, M. (2019). Encapsulation of carotenoid-rich paprika oleoresin through traditional and nano spray drying. *Italian Journal of Food Science*, *31*(1).

Ding, X., Yao, Q., Zhu, D., Dong, B. (2023). Hydrogen-rich water maintains the quality of Rosa roxburghii by regulating AsA biosynthesis and regeneration. *Postharvest Biology and Technology*, *195*, 112136.

Dong, B., Zhu, D., Yao, Q., Tang, H., Ding, X. (2022). Hydrogen-rich water treatment maintains the quality of Rosa sterilis fruit by regulating antioxidant capacity and energy metabolism. *LWT*, *161*, 113361.

Echegoyen, Y., Nerín, C. (2013). Nanoparticle release from nanosilver antimicrobial food containers. *Food and Chemical Toxicology*, *62*, 16–22.

Emadpour, M., Ghareyazie, B., Kalaj, Y.R., Entesari, M., Bouzari, N. (2015). Effect of the potassium permanganate coated zeolite nanoparticles on the quality characteristic and shelf life of peach and nectarine. *Journal of Agricultural Technology*, *11*(5), 1263–1273.

Emamhadi, M.A., Sarafraz, M., Akbari, M., Fakhri, Y., Linh, N.T.T., Khaneghah, A.M. (2020). Nanomaterials for food packaging applications: A systematic review. *Food and Chemical Toxicology*, *146*, 111825.

Emamifar, A., Bavaisi, S. (2020). Nanocomposite coating based on sodium alginate and nano-ZnO for extending the storage life of fresh strawberries (*Fragaria × ananassa* Duch.). *Journal of Food Measurement and Characterization*, *14*(2), 1012–1024.

Espinal-Hernández, P., Colinas-León, M.T., Ybarra-Moncada, M.C., Méndez-Zúñiga, S.M., Corrales-García, J. (2021). Postharvest effects of 1-MCP and chitosan/oleic acid coating in pitaya (Stenocereus griseus H.). *Journal of the Professional Association for Cactus Development*, *23*, 43–57.

Fahami, A., Fathi, M. (2018). Fabrication and characterization of novel nanofibers from cress seed mucilage for food applications. *Journal of Applied Polymer Science*, *135*(6), 45811.

FAO (2023). Food and agricultural organization of the United Nations. www.fao.org/3/x5045e/x5045E04.htm#:~:text=Unlike%20the%20grains%20and%20similar,around%2060%25%20to%2070%25 (Accessed on: August 13, 2023)

Farina, V., Passafiume, R., Tinebra, I., Palazzolo, E., Sortino, G. (2020a). Use of aloe vera gel-based edible coating with natural anti-browning and anti-oxidant additives to improve post-harvest quality of fresh-cut 'fuji' apple. *Agronomy*, *10*(4), 515.

Farina, V., Passafiume, R., Tinebra, I., Scuderi, D., Saletta, F., Gugliuzza, G., Gallotta, A., Sortino, G. (2020). Postharvest application of aloe vera gel-based edible coating to improve the quality and storage stability of fresh-cut papaya. *Journal of Food Quality*, *2020*, 1–10.

Farina, V., Passafiume, R., Tinebra, I., Scuderi, D., Saletta, F., Gugliuzza, G., Gallotta, A., Sortino, G. (2020b). Postharvest application of aloe vera gel-based edible coating to improve the quality and storage stability of fresh-cut papaya. *Journal of Food Quality*, *2020*, 1–10.

Fernández, A., Picouet, P., Lloret, E. (2010). Cellulose-silver nanoparticle hybrid materials to control spoilage-related microflora in absorbent pads located in trays of fresh-cut melon. *International Journal of Food Microbiology*, *142*(1–2), 222–228.

Formiga, A.S., Junior, J.S.P., Pereira, E.M., Cordeiro, I.N., Mattiuz, B.H. (2019). Use of edible coatings based on hydroxypropyl methylcellulose and beeswax in the conservation of red guava 'Pedro Sato'. *Food Chemistry*, *290*, 144–151.

Gallocchio, F., Belluco, S., Ricci, A. (2015). Nanotechnology and food: brief overview of the current scenario. *Procedia Food Science*, *5*, 85–88.

García-Cruz, L., Guerra-Ramírez, D., Martínez-Damián, M.T., Zuleta-Prada, H., Valle-Guadarrama, S. (2021). Shelf Life of Pitaya (*Stenocereus pruinosus*) fruits affected by temperature and the use of biopolymers of Guar Gum. *Acta Agrícola y Pecuaria*, *7*(1).

Getahun, E., Delele, M.A., Gabbiye, N., Fanta, S.W., Demissie, P., Vanierschot, M. (2021). Importance of integrated CFD and product quality modeling of solar dryers for fruits and vegetables: a review. *Solar Energy*, *220*, 88–110.

Gholami, R., Ahmadi, E., Farris, S. (2017). Shelf life extension of white mushrooms (*Agaricus bisporus*) by low temperatures conditioning, modified atmosphere, and nanocomposite packaging material. *Food Packaging and Shelf Life*, *14*, 88–95.

Ghosh, M., Singh, A.K. (2022). Potential of engineered nanostructured biopolymer based coatings for perishable fruits with Coronavirus safety perspectives. *Progress in Organic Coatings*, *163*, 106632.

Ghosh, T., Nakano, K., Katiyar, V. (2021). Curcumin doped functionalized cellulose nanofibers based edible chitosan coating on kiwifruits. *International Journal of Biological Macromolecules*, *184*, 936–945.

Gill, V. (2021). Food waste: Amount thrown away totals 900 million tonnes. *BBC News*. www.bbc.com/news/science-environment-56271385.

Gleiter, H. (1989). Nanocrystalline materials. *Progress in Materials Science*, *33*, 223–315.

Göke, K., Bunjes, H. (2017). Drug solubility in lipid nanocarriers: Influence of lipid matrix and available interfacial area. *International Journal of Pharmaceutics*, *529*(1–2), 617–628.

Gómez-Patiño, M.B., Estrada-Reyes, R., Vargas-Diaz, M.E., Arrieta-Baez, D. (2020). Cutin from *Solanum myriacanthum* Dunal and *Solanum aculeatissimum* Jacq. as a potential raw material for biopolymers. *Polymers*, *12*(9), 1945.

Govindappa, M., Tejashree, S., Thanuja, V., Hemashekhar, B., Srinivas, C., Nasif, O., Pugazhendhi, A., Raghavendra, V.B. (2021). Pomegranate fruit fleshy pericarp mediated silver nanoparticles possessing antimicrobial, antibiofilm formation, antioxidant, biocompatibility and anticancer activity. *Journal of Drug Delivery Science and Technology*, *61*, 102289.

Granata, G., Consoli, G.M., Nigro, R.L., Geraci, C. (2018). Hydroxycinnamic acids loaded in lipid-core nanocapsules. *Food Chemistry*, *245*, 551–556.

Gu, Y., Xia, K., Wu, D., Mou, J., Zheng, S. (2020). Technical characteristics and wear-resistant mechanism of nano coatings: a review. *Coatings*, *10*(3), 233.

Gundewadi, G., Rudra, S.G., Sarkar, D.J., Singh, D. (2018). Nanoemulsion based alginate organic coating for shelf life extension of okra. *Food Packaging and Shelf Life*, *18*, 1–12.

Guo, Q., Du, G., Jia, H., Fan, Q., Wang, Z., Gao, Z., Yue, T., Yuan, Y. (2021). Essential oils encapsulated by biopolymers as antimicrobials in fruits and vegetables: A review. *Food Bioscience*, *44*, 101367.

Guo, X., Chen, B., Wu, X., Li, J., Sun, Q. (2020). Utilization of cinnamaldehyde and zinc oxide nanoparticles in a carboxymethylcellulose-based composite coating to improve the postharvest quality of cherry tomatoes. *International Journal of Biological Macromolecules*, *160*, 175–182.

Haghighi, H., Licciardello, F., Fava, P., Siesler, H.W., Pulvirenti, A. (2020). Recent advances on chitosan-based films for sustainable food packaging applications. *Food Packaging and Shelf Life*, *26*, 100551.

Han, J.W., Zuo, M., Zhu, W.Y., Zuo, J.H., Lü, E.L., Yang, X.T. (2021). A comprehensive review of cold chain logistics for fresh agricultural products: Current status, challenges, and future trends. *Trends in Food Science & Technology*, *109*, 536–551.

Hao, J., Guo, B., Yu, S., Zhang, W., Zhang, D., Wang, J., Wang, Y. (2017). Encapsulation of the flavonoid quercetin with chitosan-coated nano-liposomes. *LWT-Food Science and Technology*, *85*, 37–44.

Hasan, S.K., Ferrentino, G., Scampicchio, M. (2020). Nanoemulsion as advanced edible coatings to preserve the quality of fresh-cut fruits and vegetables: a review. *International Journal of Food Science & Technology*, *55*(1), 1–10.

Hasheminejad, N., Khodaiyan, F. (2020). The effect of clove essential oil loaded chitosan nanoparticles on the shelf life and quality of pomegranate arils. *Food Chemistry*, *309*, 125520.

Hassan, F., Ali, E., Mostafa, N., Mazrou, R. (2021). Shelf-life extension of sweet basil leaves by edible coating with thyme volatile oil encapsulated chitosan nanoparticles. *International Journal of Biological Macromolecules*, *177*, 517–525.

Hazrati, S., Kashkooli, A.B., Habibzadeh, F., Tahmasebi-Sarvestani, Z., Sadeghi, A.R. 2017. Evaluation of Aloe vera gel as an alternative edible coating for peach fruits during cold storage period. Gesunde Pflanzen, 69(3): 131–137.

Hernández-López, G., Ventura-Aguilar, R.I., Correa-Pacheco, Z.N., Bautista-Baños, S., Barrera-Necha, L.L. (2020). Nanostructured chitosan edible coating loaded with α-pinene for the preservation of the postharvest quality of Capsicum annuum L. and Alternaria alternata control. *International Journal of Biological Macromolecules*, *165*, 1881–1888.

Huang, J., Wang, Q., Li, T., Xia, N., Xia, Q. (2017). Nanostructured lipid carrier (NLC) as a strategy for encapsulation of quercetin and linseed oil: Preparation and in vitro characterization studies. *Journal of Food Engineering*, *215*, 1–12.

Irbe, I., Filipova, I., Skute, M., Zajakina, A., Spunde, K., Juhna, T. (2021). Characterization of novel biopolymer blend mycocel from plant cellulose and fungal fibers. *Polymers*, *13*(7), 1086.

Jafarzadeh, S., Nafchi, A.M., Salehabadi, A., Oladzad-abbasabadi, N., Jafari, S.M. (2021). Application of bio-nanocomposite films and edible coatings for extending the shelf life of fresh fruits and vegetables. *Advances in Colloid and Interface Science*, *291*, 102405.

Jagtiani, E. (2022). Advancements in nanotechnology for food science and industry. *Food Frontiers*, *3*(1), 56–82.

Jian, Y., Chen, X., Ahmed, T., Shang, Q., Zhang, S., Ma, Z., Yin, Y. (2022). Toxicity and action mechanisms of silver nanoparticles against the mycotoxin-producing fungus *Fusarium graminearum*. *Journal of Advanced Research*, *38*, 1–12.

Jiang, Y., Yu, L., Hu, Y., Zhu, Z., Zhuang, C., Zhao, Y., Zhong, Y. (2019). Electrostatic spraying of chitosan coating with different deacetylation degree for strawberry preservation. *International Journal of Biological Macromolecules*, *139*, 1232–1238.

Jiang, G., Zhang, D., Li, Z., Liang, H., Deng, R., Su, X., Jiang, Y., Duan, X. (2021). Alternative splicing of MaMYB16L regulates starch degradation in banana fruit during ripening. *Journal of Integrative Plant Biology*, *63*(7), 1341–1352.

Jin, T.Z., Chen, W., Gurtler, J.B., Fan, X. (2020). Effectiveness of edible coatings to inhibit browning and inactivate foodborne pathogens on fresh-cut apples. *Journal of Food Safety*, *40*(4), e12802.

Joshy, K.S., Jose, J., Li, T., Thomas, M., Shankregowda, A.M., Sreekumaran, S., Kalarikkal, N., Thomas, S. (2020). Application of novel zinc oxide reinforced xanthan gum hybrid system for edible coatings. *International Journal of Biological Macromolecules*, *151*, 806–813.

Jung, S., Cui, Y., Barnes, M., Satam, C., Zhang, S., Chowdhury, R.A., Adumbumkulath, A., Sahin, O., Miller, C., Sajadi, S.M., Ajayan, P.M. (2020). Multifunctional bio-nanocomposite coatings for perishable fruits. *Advanced Materials*, *32*(26), 1908291.

Kanikireddy, V., Kanny, K., Padma, Y., Velchuri, R., Ravi, G., Jagan Mohan Reddy, B., Vithal, M. (2019). Development of alginate-gum acacia-Ag0 nanocomposites via green process for inactivation of foodborne bacteria and impact on shelf life of black grapes (Vitis vinifera). *Journal of Applied Polymer Science*, *136*(15), 47331.

Kanikireddy, V., Varaprasad, K., Rani, M.S., Venkataswamy, P., Reddy, B.J.M., Vithal, M. (2020). Biosynthesis of CMC-Guar gum-Ag0 nanocomposites for inactivation of food pathogenic microbes and its effect on the shelf life of strawberries. *Carbohydrate Polymers*, *236*, 116053.

Katouzian, I., Esfanjani, A.F., Jafari, S.M., Akhavan, S. (2017). Formulation and application of a new generation of lipid nanocarriers for the food bioactive ingredients. *Trends in Food Science & Technology*, *68*, 14–25.

Kirci, M., Isaksson, O., Seifert, R. (2022). Managing perishability in the fruit and vegetable supply chains. *Sustainability*, *14*(9), 5378.

Klangmuang, P., Sothornvit, R. (2016). Combination of beeswax and nanoclay on barriers, sorption isotherm and mechanical properties of hydroxypropyl methylcellulose-based composite films. *LWT-Food Science and Technology*, *65*, 222–227.

Klunklin, W., Jantanasakulwong, K., Phimolsiripol, Y., Leksawasdi, N., Seesuriyachan, P., Chaiyaso, T., Insomphun, C., Phongthai, S., Jantrawut, P., Sommano, S.R., Rachtanapun, P. (2020). Synthesis, characterization, and application of carboxymethyl cellulose from asparagus stalk end. *Polymers*, *13*(1), 81.

Kondle, R., Sharma, K., Singh, G., Kotiyal, A. (2022). Using Nanotechnology for Enhancing the Shelf Life of Fruits. In Food Processing and Packaging Technologies-Recent Advances. IntechOpen: London, pp. 1–19.

Kouhi, M., Prabhakaran, M.P., Ramakrishna, S. (2020). Edible polymers: An insight into its application in food, biomedicine and cosmetics. *Trends in Food Science & Technology*, *103*, 248–263.

Kumar, N., Kaur, P., Devgan, K., Attkan, A.K. (2020). Shelf life prolongation of cherry tomato using magnesium hydroxide reinforced bio-nanocomposite and conventional plastic films. *Journal of Food Processing and Preservation*, *44*(4), e14379.

Kumar, V.D., Verma, P.R.P., Singh, S.K. (2015). Development and evaluation of biodegradable polymeric nanoparticles for the effective delivery of quercetin using a quality by design approach. *LWT-Food Science and Technology, 61(2)*, 330–338.

Lavinia, M., Hibaturrahman, S., Harinata, H., Wardana, A.A. (2020). Antimicrobial activity and application of nanocomposite coating from chitosan and ZnO nanoparticle to inhibit microbial growth on fresh-cut papaya. *Food Research*, *4*, 307–311.

Lemes, G.F., Marchiore, N.G., Moreira, T.F.M., Da Silva, T.B.V., Sayer, C., Shirai, M.A., Gonçalves, O.H., Gozzo, A.M., Leimann, F.V. (2017). Enzymatically crosslinked gelatin coating added of bioactive nanoparticles and antifungal agent: effect on the quality of Benitaka grapes. *LWT-Food Science and Technology*, *84*, 175–182.

Li, M., Yu, H., Xie, Y., Guo, Y., Cheng, Y., Qian, H., Yao, W. (2021). Effects of double layer membrane loading eugenol on postharvest quality of cucumber. *LWT-Food Science and Technology*, *145*, 111310.

Li, F., Hu, Y., Shan, Y., Liu, J., Ding, X., Duan, X., Zeng, J., Jiang, Y. (2022). Hydrogen-rich water maintains the color quality of fresh-cut Chinese water chestnut. *Postharvest Biology and Technology*, *183*, 111743.

Liang, J., Yan, H., Wang, X., Zhou, Y., Gao, X., Puligundla, P., Wan, X. (2017). Encapsulation of epigallocatechin gallate in zein/chitosan nanoparticles for controlled applications in food systems. *Food Chemistry*, *231*, 19–24.

Lin, M., Fang, S., Zhao, X., Liang, X., Wu, D. (2020). Natamycin-loaded zein nanoparticles stabilized by carboxymethyl chitosan: Evaluation of colloidal/chemical performance and application in postharvest treatments. *Food Hydrocolloids*, *106*, 105871.

Liu, R., Liu, D., Liu, Y., Song, Y., Wu, T., Zhang, M. (2017a). Using soy protein SiOx nanocomposite film coating to extend the shelf life of apple fruit. *International Journal of Food Science & Technology*, *52*(9), 2018–2030.

Liu, Y., Yu, S., Liu, L., Yue, X., Zhang, W., Yang, Q., Wang, L., Wang, Y., Zhang, D., Wang, J. (2017b). Inhibition of the double-edged effect of curcumin on Maillard reaction in a milk model system by a nanocapsule strategy. *LWT-Food Science and Technology*, *145, 84*, 643–649.

Lo'Ay, A.A., Taher, M.A. (2018). Effectiveness salicylic acid blending in chitosan/PVP biopolymer coating on antioxidant enzyme activities under low storage temperature stress of 'Banati' guava fruit. *Scientia Horticulturae*, *238*, 343–349.

Lopez-Polo, J., Monasterio, A., Cantero-López, P., Osorio, F.A. (2021). Combining edible coatings technology and nanoencapsulation for food application: A brief review with an emphasis on nanoliposomes. *Food Research International*, *145*, 110402.

Luo, X., Zhou, Y., Bai, L., Liu, F., Zhang, R., Zhang, Z., Zheng, B., Deng, Y., McClements, D.J. (2017). Production of highly concentrated oil-in-water emulsions using dual-channel microfluidization: Use of individual and mixed natural emulsifiers (saponin and lecithin). *Food Research International*, *96*, 103–112.

Ma, J., Zhou, Z., Li, K., Li, K., Liu, L., Zhang, W., Xu, J., Tu, X., Du, L., Zhang, H. (2021). Novel edible coating based on shellac and tannic acid for prolonging postharvest shelf life and improving overall quality of mango. *Food Chemistry*, *354*, 129510.

Madhusha, C., Munaweera, I., Karunaratne, V., Kottegoda, N. (2020). Facile mechanochemical approach to synthesizing edible food preservation coatings based on alginate/ascorbic acid-layered double hydroxide bio-nanohybrids. *Journal of Agricultural and Food Chemistry*, *68*(33), 8962–8975.

Mahela, U., Rana, D.K., Joshi, U., Tariyal, Y.S. (2020). Nano edible coatings and their applications in food preservation. *Journal of Postharvest Technology*, *8*(4), 52–63.

McClements, D.J. (2015). Food Emulsions: Principles, Practice and Techniques CRC Press. *New York: 2nd ed. Pp, 234*.

Medina-Jaramillo, C., Quintero-Pimiento, C., Gómez-Hoyos, C., Zuluaga-Gallego, R., López-Córdoba, A. (2020). Alginate-edible coatings for application on wild andean blueberries (*Vaccinium meridionale* swartz): Effect of the addition of nanofibrils isolated from cocoa by-products. *Polymers*, *12*(4), 824.

Mehdizadeh, A., Shahidi, S.A., Shariatifar, N., Shiran, M., Ghorbani-HasanSaraei, A. (2021). Evaluation of chitosan-zein coating containing free and nano-encapsulated *Pulicaria gnaphalodes* (Vent.) Boiss. extract on quality attributes of rainbow trout. *Journal of Aquatic Food Product Technology*, *30*(1), 62–75.

Molamohammadi, H., Pakkish, Z., Akhavan, H.R., Saffari, V.R. (2020). Effect of salicylic acid incorporated chitosan coating on shelf life extension of fresh in-hull pistachio fruit. *Food and Bioprocess Technology*, *13*, 121–131.

Montero, Y., Souza, A.G., Oliveira, É.R., dos Santos Rosa, D. (2021). Nanocellulose functionalized with cinnamon essential oil: A potential application in active biodegradable packaging for strawberry. *Sustainable Materials and Technologies*, *29*, e00289.

Morozov, V.N., Kanev, I.L. (2015). Knockdown of fruit flies by imidacloprid nanoaerosol. *Environmental Science & Technology*, *49*(20), 12483–12489.

Nadim, Z., Ahmadi, E., Sarikhani, H., Amiri Chayjan, R. (2015). Effect of methylcellulose-based edible coating on strawberry fruit's quality maintenance during storage. *Journal of Food processing and Preservation*, *39*(1), 80–90.

Nair, M.S., Tomar, M., Punia, S., Kukula-Koch, W., Kumar, M. (2020). Enhancing the functionality of chitosan-and alginate-based active edible coatings/films for the preservation of fruits and vegetables: a review. *International Journal of Biological Macromolecules*, *164*, 304–320.

Nascimento, J.I.G., Stamford, T.C.M., Melo, N.F.C.B., dos Santos Nunes, I., Lima, M.A.B., Pintado, M.M.E., Stamford-Arnaud, T.M., Stamford, N.P., Stamford, T.L.M. (2020). Chitosan–citric acid edible coating to control Colletotrichum gloeosporioides and maintain quality parameters of fresh-cut guava. *International Journal of Biological Macromolecules*, *163*, 1127–1135.

Nasrin, T.A.A., Rahman, M.A., Arfin, M.S., Islam, M.N., Ullah, M.A. (2020). Effect of novel coconut oil and beeswax edible coating on postharvest quality of lemon at ambient storage. *Journal of Agriculture and Food Research*, *2*, 100019.

Nor, S.M., Ding, P. (2020). Trends and advances in edible biopolymer coating for tropical fruit: A review. *Food Research International*, *134*, 109208.

Ömer Azabağaoğlu, M. (2018). Investigating fresh fruit and vegetables losses at contemporary food retailers. *Sosyal Bilimler Araştırma Dergisi*, *7*(4), 55–62.

Otálora, M.C., Camelo, R., Wilches-Torres, A., Cárdenas-Chaparro, A., Gómez Castaño, J.A. (2020). Encapsulation effect on the in vitro bioaccessibility of sacha inchi oil (*Plukenetia volubilis* L.) by soft capsules composed of gelatin and cactus mucilage biopolymers. *Polymers*, *12*(9), 1995.

Öztürk, B. (2017). Nanoemulsions for food fortification with lipophilic vitamins: Production challenges, stability, and bioavailability. *European Journal of Lipid Science and Technology*, *119*(7), 1500539.

Ozturk, S., Zhang, J., Singh, R.K., Kong, F. (2021). Effect of cellulose nanofiber-based coating with chitosan and trans-cinnamaldehyde on the microbiological safety and quality of cantaloupe rind and fresh-cut pulp. Part 2: Quality attributes. *LWT*, *147*, 111519.

Palit, S. (2020). Recent advances in the application of nanotechnology in food industry and the vast vision for the future. *Nanoengineering in the Beverage Industry*, 1–34.

Paradise, J. (2019). Regulating nanomedicine at the food and drug administration. *AMA Journal of Ethics*, *21*(4), 347–355.

Patel, A.K., Singhania, R.R., Chen, C.W., Tseng, Y.S., Kuo, C.H., Wu, C.H., Di Dong, C. (2021). Advances in micro-and nano bubbles technology for application in biochemical processes. *Environmental Technology & Innovation*, *23*, 101729.

Peters, B.L., Ramirez-Hernandez, A., Pike, D.Q., Müller, M., de Pablo, J.J. (2012). Nonequilibrium simulations of lamellae forming block copolymers under steady shear: A comparison of dissipative particle dynamics and Brownian dynamics. *Macromolecules*, *45*(19), 8109–8116.

Petkoska, A.T., Daniloski, D., D'Cunha, N.M., Naumovski, N., Broach, A.T. (2021). Edible packaging: Sustainable solutions and novel trends in food packaging. *Food Research International*, *140*, 109981.

Pilon, L., Spricigo, P.C., Miranda, M., de Moura, M.R., Assis, O.B.G., Mattoso, L.H.C., & Ferreira, M.D. (2015). Chitosan nanoparticle coatings reduce microbial growth on fresh-cut apples while not affecting quality attributes. *International Journal of Food Science & Technology*, *50*(2), 440–448.

Pinzon, M.I., Sanchez, L.T., Garcia, O.R., Gutierrez, R., Luna, J.C., Villa, C.C. (2020). Increasing shelf life of strawberries (Fragaria ssp) by using a banana starch-chitosan-Aloe vera gel composite edible coating. *International Journal of Food Science & Technology*, *55*(1), 92–98.

Pirozzi, A., Del Grosso, V., Ferrari, G., Donsì, F. (2020). Edible coatings containing oregano essential oil nanoemulsion for improving postharvest quality and shelf life of tomatoes. *Foods*, *9*(11), 1605.

Pirzadeh, M., Aminlari, M. (2019). Effects on the physiochemical properties of stored pomegranate seeds by the application of Aloe Vera Gel as coating agent. *Engineering, Technology & Applied Science Research*, *9*(4).

Popescu, P.A., Palade, L.M., Nicolae, I.C., Popa, E.E., Miteluț, A.C., Drăghici, M.C., Matei, F., Popa, M.E. (2022). Chitosan-based edible coatings containing essential oils to preserve the shelf life and postharvest quality parameters of organic strawberries and apples during cold storage. *Foods*, *11*(21), 3317.

Prabhu, K., Jayakumari, S., Devi, D.G., Rahe, R. (2022). A review of bio-nanocoatings used to extend fruits shelf life. *J Food Chem Nanotechnol*, *7*(4), 78–84.

Prakash, A., Baskaran, R., Vadivel, V. (2020). Citral nanoemulsion incorporated edible coating to extend the shelf life of fresh cut pineapples. *LWT*, *118*, 108851.

Priolo, M.A., Holder, K.M., Guin, T., Grunlan, J.C. (2015). Recent advances in gas barrier thin films via layer-by-layer assembly of polymers and platelets. *Macromolecular Rapid Communications*, *36*(10), 866–879.

Qiao, G., Xiao, Z., Ding, W., Rok, A. (2019). Effect of chitosan/nano-titanium dioxide/thymol and tween films on ready-to-eat cantaloupe fruit quality. *Coatings*, *9*(12), 828.

Rachtanapun, P., Jantrawut, P., Klunklin, W., Jantanasakulwong, K., Phimolsiripol, Y., Leksawasdi, N., Seesuriyachan, P., Chaiyaso, T., Insomphun, C., Phongthai, S., Ngo, T.M.P. (2021). Carboxymethyl bacterial cellulose from nata de coco: effects of NaOH. *Polymers*, *13*(3), 348.

Rahiel, H.A., Zenebe, A.K., Leake, G.W., Gebremedhin, B.W. (2018). Assessment of production potential and post-harvest losses of fruits and vegetables in northern region of Ethiopia. *Agriculture & Food Security*, *7*, 1–13.

Rajapaksha, L., Gunathilake, D.M.C.C., Pathirana, S.M., Fernando, T. (2021). Reducing post-harvest losses in fruits and vegetables for ensuring food security—Case of Sri Lanka. *MOJ Food Process Technols*, *9*(1), 7–16.

Ramos, O.L., Pereira, R.N., Martins, A., Rodrigues, R., Fucinos, C., Teixeira, J.A., Pastrana, L., Malcata, F.X., Vicente, A.A. (2017). Design of whey protein nanostructures for incorporation and release of nutraceutical compounds in food. *Critical Reviews in Food Science and Nutrition*, *57*(7), 1377–1393.

Ranganathan, S., Dutta, S., Moses, J.A., Anandharamakrishnan, C. (2020). Utilization of food waste streams for the production of biopolymers. *Heliyon*, *6*(9).

Rangel-Marrón, M., Mani-López, E., Palou, E., López-Malo, A. (2019). Effects of alginate-glycerol-citric acid concentrations on selected physical, mechanical, and barrier properties of papaya puree-based edible films and coatings, as evaluated by response surface methodology. *LWT-Food Science and Technology*, *145*, *101*, 83–91.

Ranjan, S., Chandrasekaran, R., Paliyath, G., Lim, L.T., Subramanian, J. (2020). Effect of hexanal loaded electrospun fiber in fruit packaging to enhance the post harvest quality of peach. *Food Packaging and Shelf Life*, *23*, 100447.

Riahi, Z., Priyadarshi, R., Rhim, J.W., Hong, S.I., Bagheri, R., Pircheraghi, G. (2021). Titania nanotubes decorated with Cu (I) and Cu (II) oxides: antibacterial and ethylene scavenging functions to extend the shelf life of bananas. *ACS Sustainable Chemistry & Engineering*, *9*(19), 6832–6840.

Robledo, N., Vera, P., López, L., Yazdani-Pedram, M., Tapia, C., Abugoch, L. (2018). Thymol nanoemulsions incorporated in quinoa protein/chitosan edible films; antifungal effect in cherry tomatoes. *Food Chemistry*, *246*, 211–219.

Rokayya, S., Jia, F., Li, Y., Nie, X., Xu, J., Han, R., Yu, H., Amanullah, S., Almatrafi, M.M., Helal, M. (2021). Application of nano-titanum dioxide coating on fresh Highbush blueberries shelf life stored under ambient temperature. *LWT*, *137*, 110422.

Rovera, C., Ghaani, M., Farris, S. (2020). Nano-inspired oxygen barrier coatings for food packaging applications: An overview. *Trends in Food Science & Technology*, *97*, 210–220.

Ruggeri, E., Kim, D., Cao, Y., Farè, S., De Nardo, L., Marelli, B. (2020). A multilayered edible coating to extend produce shelf life. *ACS Sustainable Chemistry & Engineering*, *8*(38), 14312–14321.

Saba, M.K., Amini, R. (2017). Nano-ZnO/carboxymethyl cellulose-based active coating impact on ready-to-use pomegranate during cold storage. *Food Chemistry*, *232*, 721–726.

Sadiq, U., Gill, H., Chandrapala, J. (2021). Casein micelles as an emerging delivery system for bioactive food components. *Foods*, *10*(8), 1965.

Sala, M., Diab, R., Elaissari, A., Fessi, H. (2018). Lipid nanocarriers as skin drug delivery systems: Properties, mechanisms of skin interactions and medical applications. *International Journal of Pharmaceutics*, *535*(1–2), 1–17.

Salem, M.F., Abd-Elraoof, W.A., Tayel, A.A., Alzuaibr, F.M., Abonama, O.M. (2022). Antifungal application of biosynthesized selenium nanoparticles with pomegranate peels and nanochitosan as edible coatings for citrus green mold protection. *Journal of Nanobiotechnology*, *20*(1), 182.

Salvia-Trujillo, L., Soliva-Fortuny, R., Rojas-Graü, M.A., McClements, D.J., Martín-Belloso, O. (2017). Edible nanoemulsions as carriers of active ingredients: a review. *Annual Review of Food Science and Technology*, *8*, 439–466.

Santagata, G., Mallardo, S., Fasulo, G., Lavermicocca, P., Valerio, F., Di Biase, M., Di Stasio, M., Malinconico, M., Volpe, M.G. (2018). Pectin-honey coating as novel dehydrating bioactive agent for cut fruit: enhancement of the functional properties of coated dried fruits. *Food Chemistry*, *258*, 104–110.

Santiago-Aliste, A., Sánchez-Hernández, E., Buzón-Durán, L., Marcos-Robles, J.L., Martín-Gil, J., Martín-Ramos, P. (2023). Uncaria tomentosa-loaded chitosan oligomers–hydroxyapatite–carbon nitride nanocarriers for postharvest fruit protection. *Agronomy*, *13*(9), 2189.

Saravanakumar, K., Hu, X., Chelliah, R., Oh, D.H., Kathiresan, K., Wang, M.H. (2020). Biogenic silver nanoparticles-polyvinylpyrrolidone based glycerosomes coating to expand the shelf life of fresh-cut bell pepper (*Capsicum annuum* L. var. grossum (L.) Sendt). *Postharvest Biology and Technology*, *160*, 111039.

Schmidt-Traub, G., Obersteiner, M., Mosnier, A. (2019). Fix the broken food system in three steps. *Nature*, *569*(7755), 181–183.

Severcan, S.S., Uzal, N., Kahraman, K. (2020). Clarification of apple juice using new generation nanocomposite membranes fabricated with TiO_2 and Al_2O_3 Nanoparticles. *Food and Bioprocess Technology*, *13*, 391–403.

Shahzad, S., Ahmad, S., Anwar, R., & Ahmad, R. (2020). Pre-storage application of calcium chloride and salicylic acid maintain the quality and extend the shelf life of strawberry. *Pakistan Journal of Agricultural Sciences*, *57*(2).

Shan, Y., Li, T., Qu, H., Duan, X., Farag, M. A., Xiao, J., Gao, H., Jiang, Y. (2023). Nano-preservation: An emerging postharvest technology for quality maintenance and shelf life extension of fresh fruit and vegetable. *Food Frontiers*, *4*(1), 100–130.

Shapi'i, R.A., Othman, S.H., Nordin, N., Basha, R.K., Naim, M.N. (2020). Antimicrobial properties of starch films incorporated with chitosan nanoparticles: In vitro and in vivo evaluation. *Carbohydrate Polymers*, *230*, 115602.

Sharma, H.P., Chaudhary, V., Kumar, M. (2019). Importance of edible coating on fruits and vegetables: A review. *Journal of Pharmacognosy and Phytochemistry*, *8*(3), 4104–4110.

Sharma, B., Nigam, S., Verma, A., Garg, M., Mittal, A., Sadhu, S.D. (2023). A biogenic approach to develop guava derived edible copper and zinc oxide nanocoating to extend shelf life and efficiency for food preservation. *Journal of Polymers and the Environment*, 1–14.

Shi, C., Wu, Y., Fang, D., Pei, F., Mariga, A.M., Yang, W., Hu, Q. (2018). Effect of nanocomposite packaging on postharvest senescence of *Flammulina velutipes*. *Food Chemistry*, *246*, 414–421.

Shi, J., Cai, H., Qin, Z., Li, X., Yuan, S., Yue, X., Sui, Y., Sun, A., Cui, J., Zuo, J. (2023). Ozone micro-nano bubble water preserves the quality of postharvest parsley. *Food Research International*, *170*, 113020.

Silva, W.B., Silva, G.M.C., Santana, D.B., Salvador, A.R., Medeiros, D.B., Belghith, I., da Silva, N.M., Cordeiro, M.H.M., Misobutsi, G.P. (2018). Chitosan delays ripening and ROS production in guava (*Psidium guajava* L.) fruit. *Food Chemistry*, *242*, 232–238.

Singh, A., Singh, R.K., Kumar, P., Singh, A. (2015). Mango biodiversity in eastern Uttar Pradesh, India: Indigenous knowledge and traditional products. *Indian Journal of Traditional Knowledge*, *14*, 258–264.

Singh, D.P., Packirisamy, G. (2022). Biopolymer based edible coating for enhancing the shelf life of horticulture products. *Food Chemistry: Molecular Sciences*, *4*, 100085.

Siripatrawan, U., Kaewklin, P. (2018). Fabrication and characterization of chitosan-titanium dioxide nanocomposite film as ethylene scavenging and antimicrobial active food packaging. *Food Hydrocolloids*, *84*, 125–134.

Sirohi, R., Gaur, V.K., Pandey, A.K., Sim, S.J., Kumar, S. (2021). Harnessing fruit waste for poly-3-hydroxybutyrate production: a review. *Bioresource Technology*, *326*, 124734.

Sun, X., Wu, Q., Picha, D.H., Ferguson, M.H., Ndukwe, I.E., Azadi, P. (2021). Comparative performance of bio-based coatings formulated with cellulose, chitin, and chitosan nanomaterials suitable for fruit preservation. *Carbohydrate Polymers*, *259*, 117764.

Tampucci, S., Castagna, A., Monti, D., Manera, C., Saccomanni, G., Chetoni, P., Zucchetti, E., Barbagallo, M., Fazio, L., Santin, M., Ranieri, A. (2021). Tyrosol-enriched tomatoes by diffusion across the fruit peel from a chitosan coating: a proposal of functional food. *Foods*, *10*(2), 335.

Tanpichai, S., Pumpuang, L., Srimarut, Y., Woraprayote, W., Malila, Y. (2023). Development of chitin nanofiber coatings for prolonging shelf life and inhibiting bacterial growth on fresh cucumbers. *Scientific Reports*, *13*(1), 13195.

Tapia, M.S., Rojas-Graü, M.A., Rodríguez, F.J., Ramírez, J., Carmona, A., Martin-Belloso, O. (2007). Alginate-and gellan-based edible films for probiotic coatings on fresh-cut fruits. *Journal of Food Science*, *72*(4), E190-E196.

Teo, H.L., Wahab, R.A. (2020). Towards an eco-friendly deconstruction of agro-industrial biomass and preparation of renewable cellulose nanomaterials: A review. *International Journal of Biological Macromolecules*, *161*, 1414–1430.

Thakur, R., Pristijono, P., Golding, J.B., Stathopoulos, C.E., Scarlett, C.J., Bowyer, M., Singh, S.P., Vuong, Q.V. 2018. Development and application of rice starch based edible coating to improve the postharvest storage potential and quality of plum fruit (*Prunus salicina*). *Scientia Horticulturae*, *237*, 59–66.

Tibolla, H., Pelissari, F.M., Martins, J.T., Lanzoni, E.M., Vicente, A.A., Menegalli, F.C., Cunha, R.L. (2019). Banana starch nanocomposite with cellulose nanofibers isolated from banana peel by enzymatic treatment: In vitro cytotoxicity assessment. *Carbohydrate Polymers*, *207*, 169–179.

Tirgar, A., Han, D., Steckl, A.J. (2018). Absorption of ethylene on membranes containing potassium permanganate loaded into alumina-nanoparticle-incorporated alumina/carbon nanofibers. *Journal of Agricultural and Food Chemistry*, *66*(22), 5635–5643.

Treviño-Garza, M.Z., García, S., del Socorro Flores-González, M., Arévalo-Niño, K. (2015). Edible active coatings based on pectin, pullulan, and chitosan increase quality and shelf life of strawberries (*Fragaria ananassa*). *Journal of Food Science*, *80*(8), M1823–M1830.

Tripathi, R.M., Chung, S.J. (2019). Biogenic nanomaterials: Synthesis, characterization, growth mechanism, and biomedical applications. *Journal of Microbiological Methods*, *157*, 65–80.

USDA, 2023. *United States Department of Agriculture, Food and Nutrition*. Food Waste Activities. www.usda.gov/foodwaste/activities (Accessed on: August 13, 2023)

Van Hung, D., Tong, S., Tanaka, F., Yasunaga, E., Hamanaka, D., Hiruma, N., Uchino, T. (2011). Controlling the weight loss of fresh produce during postharvest storage under a nano-size mist environment. *Journal of Food Engineering*, *106*(4), 325–330.

Vasile, C. (2018). Polymeric nanocomposites and nanocoatings for food packaging: A review. *Materials*, *11*(10), 1834.

Vera, P., Echegoyen, Y., Canellas, E., Nerín, C., Palomo, M., Madrid, Y., Cámara, C. (2016). Nano selenium as antioxidant agent in a multilayer food packaging material. *Analytical and Bioanalytical Chemistry*, *408*, 6659–6670.

Vidinamo, F., Fawzia, S., Karim, M.A. (2021). Effect of drying methods and storage with agro-ecological conditions on phytochemicals and antioxidant activity of fruits: a review. *Critical Reviews in Food Science and Nutrition*, *62*(2), 353–361.

Vieira, D.M., Pereira, C., Calhelha, R.C., Barros, L., Petrovic, J., Sokovic, M., Barreiro, M.F., Ferreira, I.C., Castro, M.C.R., Rodrigues, P.V., Machado, A.V. (2022). Evaluation of plant extracts as an efficient source of additives for active food packaging. *Food Frontiers*, *3*(3), 480–488.

Wang, L., Shao, S., Madebo, M.P., Hou, Y., Zheng, Y., Jin, P. (2020a). Effect of nano-SiO2 packing on postharvest quality and antioxidant capacity of loquat fruit under ambient temperature storage. *Food Chemistry*, *315*, 126295.

Wang, L., Wu, Z., Cao, C. (2019). Technologies and fabrication of intelligent packaging for perishable products. *Applied Sciences*, *9*(22), 4858.

Wang, T., Zhao, Y. (2021). Fabrication of thermally and mechanically stable superhydrophobic coatings for cellulose-based substrates with natural and edible ingredients for food applications. *Food Hydrocolloids*, *120*, 106877.

Wang, X., Zuo, J., Yan, Z., Shi, J., Wang, Q., Guan, W. (2020b). Effect of ozone micro-nano-bubble treatment on postharvest preservation of spinach. *Food Science*, *41*, 190–196.

WHO, 2005. *World Health Organization, Fruit and Vegetables for Health: Report of the Joint FAO*. WHO: Geneva, p. 07.

Won, J.S., Lee, S.J., Park, H.H., Song, K.B., Min, S.C. (2018). Edible coating using a chitosan-based colloid incorporating grapefruit seed extract for cherry tomato safety and preservation. *Journal of Food Science*, *83*(1), 138–146.

Xiao, Z., Aftab, T.B., Li, D. (2019). Applications of micro–nano bubble technology in environmental pollution control. *Micro & Nano Letters*, *14*(7), 782–787.

Xing, Y., Yi, R., Yang, H., Xu, Q., Huang, R., Tang, J., Li, X., Liu, X., Wu, L., Liao, X. (2021). Antifungal effect of chitosan/nano-TiO_2 composite coatings against *Colletotrichum gloeosporioides*, *Cladosporium oxysporum* and *Penicillium steckii*. *Molecules*, *26*(15), 4401.

Xiong, W., Yang, D., Zhong, R., Li, Y., Zhou, H., Qiu, X. (2015). Preparation of lignin-based silica composite submicron particles from alkali lignin and sodium silicate in aqueous solution using a direct precipitation method. *Industrial Crops and Products*, *74*, 285–292.

Xu, J., Zhou, L., Miao, J., Yu, W., Zou, L., Zhou, W., Liu, C., Liu, W. (2020). Effect of cinnamon essential oil nanoemulsion combined with ascorbic acid on enzymatic browning of cloudy apple juice. *Food and Bioprocess Technology*, *13*, 860–870.

Yahia, E.M., Gardea-Béjar, A., de Jesús Ornelas-Paz, J., Maya-Meraz, I.O., Rodríguez-Roque, M.J., Rios-Velasco, C., Ornelas-Paz, J., Salas-Marina, M.A. (2019). Preharvest factors affecting postharvest quality. In *Postharvest technology of perishable horticultural commodities* (pp. 99–128). Woodhead Publishing.

Yang, Y., Doudrick, K., Bi, X., Hristovski, K., Herckes, P., Westerhoff, P., Kaegi, R. (2014). Characterization of food-grade titanium dioxide: the presence of nanosized particles. *Environmental Science & Technology*, *48*(11), 6391–6400.

Yousuf, B., Qadri, O.S., Srivastava, A.K. (2018). Recent developments in shelf-life extension of fresh-cut fruits and vegetables by application of different edible coatings: A review. *LWT-Food Science and Technology*, *89*, 198–209.

Yuan, S., Xue, Z., Zhang, S., Wu, C., Feng, Y., Kou, X. (2023). The characterization of antimicrobial nanocomposites based on chitosan, cinnamon essential oil, and TiO_2 for fruits preservation. *Food Chemistry*, *413*, 135446.

Zahedi, S.M., Karimi, M., Teixeira da Silva, J.A. (2020). The use of nanotechnology to increase quality and yield of fruit crops. *Journal of the Science of Food and Agriculture*, *100*(1), 25–31.

Zambrano-Zaragoza, M.L., González-Reza, R., Mendoza-Muñoz, N., Miranda-Linares, V., Bernal-Couoh, T.F., Mendoza-Elvira, S., Quintanar-Guerrero, D. (2018). Nanosystems in edible coatings: A novel strategy for food preservation. *International Journal of Molecular Sciences, 19*(3), 705.

Zambrano-Zaragoza, M.L., Mercado-Silva, E., Ramirez-Zamorano, P., Cornejo-Villegas, M.A., Gutiérrez-Cortez, E., & Quintanar-Guerrero, D. (2013). Use of solid lipid nanoparticles (SLNs) in edible coatings to increase guava (*Psidium guajava* L.) shelf-life. *Food Research International*, *51*(2), 946–953.

Zambrano-Zaragoza, M.L., Quintanar-Guerrero, D., González-Reza, R.M., Cornejo-Villegas, M.A., Leyva-Gómez, G., & Urbán-Morlán, Z. (2021). Effects of UV-C and edible nano-coating as a combined strategy to preserve fresh-cut cucumber. *Polymers*, *13*(21), 3705.

Zapata, M.A.D.L., Castro, L.P., Rodríguez, M.L R., Pérez, O.B.A., Herrera, R.R. (2019). Candelilla wax nano emulsion coating with phytomolecules of tarbus improves the shelf life and quality of apples. *Scientia Horticulturae*, 231(32), 43–48.

Zhai, X., Lin, D., Li, W., Yang, X. (2020). Improved characterization of nanofibers from bacterial cellulose and its potential application in fresh-cut apples. *International Journal of Biological Macromolecules*, *149*, 178–186.

Zhou, S., Dai, F., Dang, C., Wang, M., Liu, D., Lu, F., & Qi, H. (2019). Scale-up biopolymer-chelated fabrication of cobalt nanoparticles encapsulated in N-enriched graphene shells for biofuel upgrade with formic acid. *Green Chemistry*, *21*(17), 4732–4747.

Zhou, Z., Ma, J., Li, K., Zhang, W., Li, K., Tu, X., Liu, L., Xu, J., Zhang, H. (2021). A plant leaf-mimetic membrane with controllable gas permeation for efficient preservation of perishable products. *ACS Nano*, *15*(5), 8742–8752.

PART VIII

Edible Film and Coating Applications

Edible Films for Postharvest Quality Management of Fresh Fruits and Vegetables

23

Farzana Rafeeq, Sami Ullah*, Kashif Razzaq, Uzman Khalil, Tanveer Hussain, and Seema Kanwal
*Corresponding Author: sami.ullah1@mnsuam.edu.pk

23.1 INTRODUCTION

Fruits and vegetables are a great source of vital nutrients which are essential to human nutrition and well-being. They provide plenty of nutrients, including vitamins, vital minerals, antioxidants, bioflavonoids, and dietary fiber. As living tissues and extremely perishable commodities, fruits and vegetables demand the best postharvest methods to extend their shelf life and assure storage stability (Mahajan et al., 2018). The sector is now compelled to create new, more effective procedures for maintaining the nutritional profile and prolonging the shelf life of fruits and vegetables due to the appeal for fresh commodities. Increased shelf life results in better fruit and vegetable quality, taste, and texture as well as reduced waste production and minimum related expenses to overcome the said waste management (Radev & Pashova, 2020). The use of biodegradable coatings, which might be an excellent substitute for industrial pesticides, constitutes a potential approach. They can successfully postpone fruit and vegetable degradation, increase their shelf life, and shield them from the environment's negative impacts. They are reliable, practical, and eco-friendly (Dhaka & Upadhyay, 2018).

Any thin material that is utilized to cover food or any other commodity in order to increase its storage life and which may be ingested together with the product is referred to as an edible film or coating. Coating prevents physical, chemical, and microbiological changes; enzymatic browning reactions; and lipid oxidation, and it can boost the nutrient content of produce like fruits and vegetables (Erkmen & Barzai, 2018). Edible films and coatings are a suitable storage technique, and monitoring quality standards like firmness and color, the production of ethylene, atmosphere composition, moisture content, and loss of weight, are essential for maintaining the shelf-life expectancy of fruits and vegetables (Swathi et al., 2017).

Generally, edible films are materials derived from carbohydrates, proteins, or lipids used for the packing of edible materials. From a technical point of view, an edible coating or film is made of polymers and additives having < 0.3 mm thickness dispersed in aqueous media. The use of plant-based coatings has become very common owing to numerous benefits, including protection against mechanical damage and extension of the freshness of the commodity and its antimicrobial characteristics (Guimarães et al., 2018; Salvia-Trujillo et al., 2017; Falguera et al., 2011). For a very long time, the most effective plant-based coating formulas comprise only one component. But recently, several investigations on two and many edible polymers with improved functional characteristics have been conducted. When designing materials with machinery, tangible, and barrier qualities that are better than those of an individual element substance, composite films or films composed of two or more monomers are formed (Salgado et al., 2015).

To increase or modify the material's fundamental functionality, a variety of compounds are added to film-forming preparations, including plasticizers (glycerol, polyoxymethylene emulsifiers, vegetable fatty acids), cross-linkers (glycol, propane glycol, and sorbitol), and enhancers. They could impair mechanical characteristics and increase water vapor permeability if incorporated in surplus (Galus et al., 2020).

23.2 FUNCTIONS OF EDIBLE FILMS

- They serve as a barrier to microbes and so support proper sanitation.
- They act as an oxygen and water barrier, altering the inside environment of the fruit to increase shelf life and preserve produce freshness.

DOI: 10.1201/9781003370376-31

- They contribute to reducing industrial waste associated with packaging because they are renewable.
- To increase safety and even nutritional and sensory qualities, multiple active substances can be added to the polymer base, including anti-browning agents, colors, fragrances, nutrients, and spices.

23.3 TYPES OF FILM-FORMING MATERIALS

Typically, edible film materials are categorized according to their form of cohesion and matrix. The main biopolymer components of edible films include lipids, proteins, and polysaccharides, as well as their mixtures (Figure 23.1). Among these materials are gums, pectin, cellulose derivates, starch, chitosan, carrageenan, and agar. Of the aforementioned biopolymers, polysaccharides have been widely used as an edible film on a wide range of fruits and vegetables. Mostly these are large molecules with numerous hydroxyl groups (Vilgis, 2012). A large number of derivatives of alginate alone and in combination with certain antimicrobial agents has been reported on many fruits and vegetables (Table 23.1 and Table 23.2).

23.4 POLYSACCHARIDE-BASED EDIBLE COATINGS

The special colloidal properties and reactivity of alginate with multivalent metal cations to generate strong gels or insoluble polymers make it a candidate for edible film and coating. Fruits' and vegetables' shelf lives have been extended using several polysaccharide-based coverings (Wang et al., 2022). Fruits and vegetables with edible coverings made of polysaccharides have demonstrated promising health benefits. Alginate, cellulose, starch, chitosan, as well as to a lesser degree gums (including guar, arabic, and xanthan gum), and carrageenan are the components that are used in forming these coatings (He et al., 2021).

23.4.1 Effect of Polysaccharide Films on the Physical Quality Attributes of Fruits and Vegetables

To evaluate the general nutritional profile of minimally processed papaya fruit, alginate (2%) or gellan-based (0.5%)

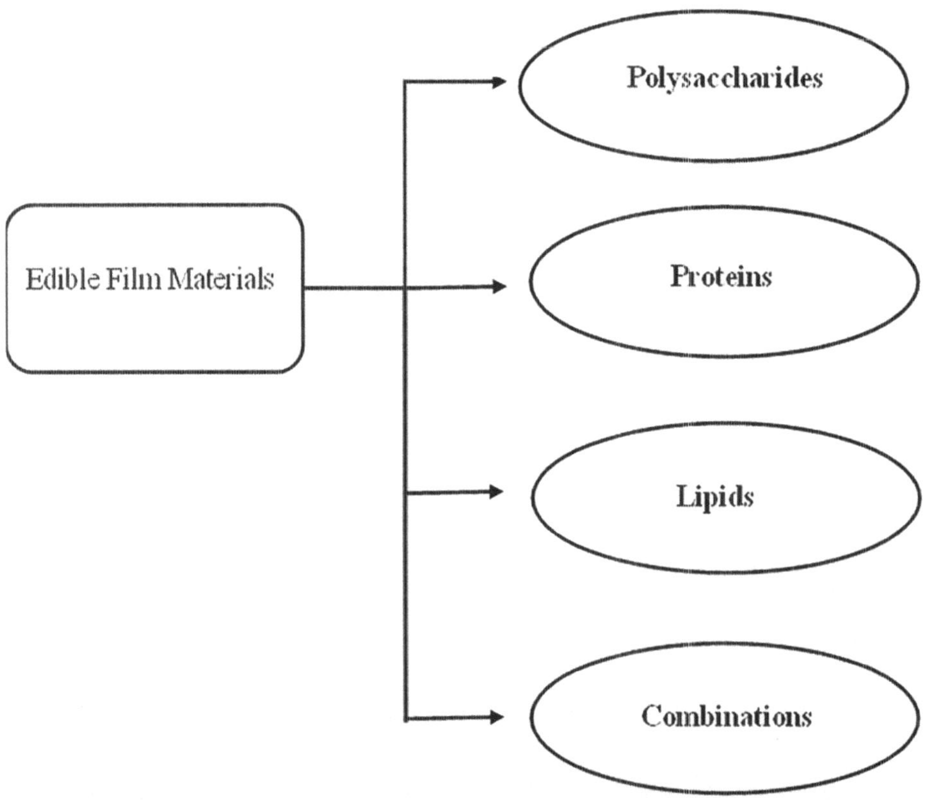

FIGURE 23.1 Types of edible film materials extensively used for making edible coatings and films.

TABLE 23.1 Effect of Alginate-Based Coatings with Incorporation of Antimicrobial Additives on Different Fruits and Vegetables

TREATMENT	*FRUIT*	*ANTIMICROBIAL AGENTS/ ANTIOXIDANTS*	*DOSE/METHOD*	*INFERENCE*	*REFERENCE*
Alginate/$CaCl_2$	Fresh-cut apple	Thyme oil	Alginate coating applied for 2 min, then 20 mL L^{-1} $CaCl_2$ with EOs (0.5, 3.5 and 6.5 μL mL^{-1}) for 2 min and kept at 4 °C for 16 days and evaluated at every 4th day.	Thyme oil significantly reduced growth of microbes, respiration rate, and weight loss and maintain browning index and firmness.	Sarengaowa et al. (2018)
Alginate/$CaCl_2$	Guava	Pomegranate peel extract	Chitosan (1%) and alginate (2%) combined application with 1% extract of pomegranate peel (PPE) on guava quality at 10 °C, changes were noted for 20 days.	Ascorbic acid, amount of phenolic compound and antioxidant capacity were analyzed with few losses. PPE coatings were effective at preserving quality.	Nair et al. (2018)
Alginate, sunflower oil/ $CaCl_2$	Fresh-cut papaya	Ascorbic acid	Glycerol+ ascorbic acid 2%+1% and glycerol + ascorbic acid 1%+1% with 0.025% of sunflower oil is incorporated in gellan- or alginate-based compositions.	Improved firmness and ascorbic acid and antioxidant contents. Maintained its overall nutritional profile during storage.	Tapia et al. (2008)
Alginate-apple puree/$CaCl_2$	Fresh-cut apple	Oregano, lemongrass, vanillin	Apple puree (26%), alginate solution (2%), and glycerol (1.5%) applied on apple pieces for 2 min and dipped for 2 min in the solution of calcium chloride and N-acetylcysteine stored at 23 °C and 0% RH for 21 days.	Reduced ethylene emission while maintaining color and firmness. After 2 weeks of storage, vanillin coatings (0.3%) were the most beneficial in terms of sensory quality.	Rojas-Graü et al. (2007)
Alginate	Strawberry	Cryptococcus laurentii	In addition to alginate (2%), glycerol (2%), palmitic acid (0.5%), glycerol monostearate (0.5%), and β-cyclodextrin (0.5%), the coating incorporated Cryptococcus laurentii.	Decreased loss of weight and weight decay, preservation of firmness, and increased quality of fruit and storage features over the duration of storage.	Fan et al. 2009
Alginate, sunflower oil/ $CaCl_2$	Fresh-cut pineapple	Lemongrass EO	Sodium alginate (1.29%), glycerol (1.16%), and sunflower oil (0.025%) were combined with lemongrass essential oil (0.1%, 0.3%, and 0.5%) subjected to freshly cut pineapple for 16 days of storage (10 °C, 65% RH).	Lemongrass (0.3%) preserves the fresh-cut pineapple's quality and shelf life. Lemongrass-coated samples with concentrations of 0.3% and 0.5% had lower levels of yeast, mold, and overall plate count than uncoated samples.	Azarakhsh, et al. (2014)
Alginate/calcium lactate	Fresh-cut watermelon	Trans-cinnamaldehyde	Solution contained sodium-alginate 0.5, 1, and 2 g/100 g supplemented with the natural antibacterial agent (trans-cinnamalde-hyde encapsulation powder at 2 g/100 g). Layer-by-layer coating was applied to the fruits, then stored at 4 °C for 15 days.	Freshly cut watermelon's shelf life is increased from 7 to 12 to 15 days while maintaining its sensory quality and increased firmness.	Sipahi et al. (2013)

(Continued)

TABLE 23.1 (Continued)

TREATMENT	*FRUIT*	*ANTIMICROBIAL AGENTS/ ANTIOXIDANTS*	*DOSE/METHOD*	*INFERENCE*	*REFERENCE*
Alginate, sunflower oil/ $CaCl_2$	Fresh-cut pears	N-acetylcysteine, glutathione	The impact of alginate (2%), pectin (2%), and gellan (0.5%) containing N-acetylcysteine (0.75%) and glutathione (0.75%) on Flor de Invierno pears was studied for 14 days at 4 °C.	Reduced ethylene rate and growth of microbes, prevented oxidation up to 2 weeks, and maintained firmness. Coatings with alginate, gellan, and pectin increased vitamin C and TPC and maintained antioxidant potential and sensory quality until day 14.	Oms-Oliu et al. (2008)
Sodium Alginate	Ber fruit	Ascorbic acid, Citric acid	To determine the impact of composite coating of 2% sodium alginate, 0.2% olive oil, and a mixture of 1% ascorbic acid and 1% citric acid on Ber fruit kept for 25 days at 2 °C and 65% RH.	The coatings increased the content of antioxidants and decreased the decay incidence, weight loss, and polygalacturonase activity of treated fruits.	Ramana et al. (2018)
Alginate-based coating	Fresh-cut apples		Solution of whey protein (65%), soy protein (90%) alginate, and glycerol (99%) was applied on fresh-cut apple pieces for 2 min and drained out for 5 min at ambient conditions.	Alginate-based coating was ideal in order to achieve maximum moisture content and retained firmness during storage.	Ghavidel et al. (2013)
Sodium alginate and methyl cellulose	Peach		MC 6g and 4g of sodium alginate were combined with 200 mL of water and ethyl alcohol to create the coating. By dipping the fruit in a 2% calcium chloride solution, an immediate cross-linking reaction was induced. At 15 °C, both coated and uncoated fruits were kept.	Treated fruits had decreased respiration rates and less moisture loss and quality alterations. Samples coated with sodium alginate and methyl cellulose kept their acceptability for 21 and 24 days, respectively.	Maftoonazad et al. (2008)
Alginate coating	Cherry		Sweet cherries treated with a sodium alginate edible coating in various concentrations (1%, 3%, and 5% w/v) were stored for 16 days.	Delayed ripening, color changes, softening, acidity loss, and respiratory rate. Maintained high TPC and antioxidants and increased storage period from 8 to 16 days.	Díaz-Mula et al. (2012)
Polysaccharide-based coatings	Mango	Ascorbic acid	Coatings with sodium alginate and high methoxyl pectin at 2% while cellulose chitosan at 0.5% in distilled water was tested for 14 days at 4 °C and 1 °C on freshly cut mango	Ascorbic acid enhanced the phenolics and carotenoids content as well as the sensory evaluations. It also delayed firmness loss.	Salinas-Roca et al. (2018)

TREATMENT	FRUIT	ANTIMICROBIAL AGENTS/ ANTIOXIDANTS	DOSE/METHOD	INFERENCE	REFERENCE
Alginate edible coating	Sliced carrots	Citric acid	High O_2 + high CO_2 modified atmosphere was investigated on sliced carrots.	Combined application of O_2 and CO_2 extended the shelf life by 2–3 days, while 0.1% citric acid and sodium alginate enhanced shelf life by 5–7 days.	Amanatidou, et al. (2000)
Alginate/zein edible coating	Tomato	Oleic acid	Zein (10%) in 75% ethanol and alginate (1%) in distilled water at 65 °C. Oleic acid (8%) was added as a plasticizer after cooling at 20 °C. 1 min of application, followed by 30 min of drying at 25 °C and 9 days of storage at 20 °C and 80% RH.	Higher fruit firmness, TSS, (TA), organic acids and sugars were achieved and rate of respiration, loss of weight and sensory quality of treated samples were reduced.	Zapata et al. (2008)
Alginate-based coating	Potato strips	Citric acid	Treatments; Blanching (85 °C, 3.5 min), sonicating (40 kHz, 5 min), and ultrasonic bath with 20 g L^{-1} of citric acid solution and 20 g L^{-1} alginate coating. Samples in vacuum packaging were stored for as long as 12 days at 31 °C.	After storage, the sonicated potato strips' visual quality improved in contrast to vacuum-packaged potato strips that had been blanched and coated with alginate.	Amaral et al. (2017)
Gellan and alginate coatings	Garlic bulbs	Mannuronic acid and guluronic acid	Coating was applied on garlic bulbs containing 2% sodium alginate, 1% or 2% gellan solution, 61% mannuronic acid, and 39% guluronic acid and then immersed in a 1% KC1 solution to induce gelation.	Coating was effective for mechanical and barrier properties.	Nussinovitch and Hershko (1996)
Alginate edible coating	Lettuce		One lot was treated with 1-MCP (500 µL/L) for 4 h at 20 °C. 2nd lot was immersed in alginate (0.5%) for 2 min. After that samples were immersed in a 1% $CaCl_2$ for 5 min to induce cross-linking reactions. Samples were refrigerated, at 2 °C and 6 °C.	1-methylcyclopropene reduced the samples' colour change, rate of respiration, and ethylene production.	Tay and Perera (2004)
Polysaccharide EC coating (alginate & chitosan)	Capsicum	Pomegranate peel extract	Capsicum was treated for 1 min either in conjunction with PPE or just using chitosan and alginate and dried for 45 min at (252 °C) stored for 25 days at 10 °C and 90–95% RH.	Whether used alone or in conjunction with PPE, chitosan decreased weight loss, hardness, color, ascorbic acid, and total chlorophyll. PPE maintained sensory quality, delayed microbiological growth, and increased shelf life by up to 25 days at 10 °C.	Nair et al. (2018)

TABLE 23.2 Effect of Plant-Based Gums Coating with Incorporation of Antimicrobial Additives on Different Fruits and Vegetables

TREATMENT	*VEGETABLES/ FRUITS*	*ANTIMICROBIAL AGENTS/ ANTIOXIDANTS*	*DOSE/METHOD*	*INFERENCE*	*REFERENCE*
Tragacanth gum	Mango	Glycerol	0.5%, 1%, and 1.5% of tragacanth gum (TCG) was applied to mango fruits, which were then kept at 20 ± 1°C and 80 ± 5% RH for 15 days.	Coated mangoes exhibited reduced ripening and softening with enhanced nutritional value.	Ali et al. (2022)
Tragacanth gum	Persimmon	Glycerol	0.5%, 1%, and 1.5% of tragacanth gum (TCG) was applied to persimmon fruits, which were then kept at 20 ± 2 °C and 80–85% RH for 20 days.	Treated fruits exhibited highest level of bioactive compounds, enzymatic activities, score of sensory quality and lowered the activities of softening enzymes.	Saleem et al. (2022)
Guar and arabic gum	Strawberry	Clove essential oil	Different ratios were applied 1% guar gum, 1% guar gum with 0.5% oil, gum arabic 1%, and gum arabic 1% with 0.5% oil. Fruits stored at 6 °C for 9 days.	Coating with guar gum 1% with essential was perform best to retained national value of fruit and enhanced shelf life.	Jodhani and Nataraj (2019)
Gum arabic	Cucumber	Glycerol monostearate	Gum arabic was applied to cucumbers in various concentrations (5, 10, 15, and 20%), and it was then kept at 10 and 25 degrees Celsius for up to 16 days.	Decreased weight loss, maintain firmness, postponed activity of softening enzymes and preserve sensory attributes at both temperatures.	Al-Juhaimi et al. (2012)
Guar gum	Sweet cherry	Glycerol	Coating based on guar gum, ginseng extract was applied on sweet cherry and stored at 20 °C and 70–75% RH for 8 days of storage period.	Compared to untreated fruit, coatings with GG-GSE prevented the loss of water, loss of firmness, titratable acid, ascorbic acid, and total phenols.	Dong and Wang (2018)
Guar gum and chitosan	Shiitake mushroom	Glycerol and acetic acid	Shiitake mushrooms (*Lentinus edodes*) were stored at 4 ± 1 °C for 16 days after the application edible coating made of chitosan (1%), guar gum (5, 15, and 25%), and other ingredients.	Shiitake mushrooms coated with CH 1% + GG 15% preserved firmness, ascorbic acid, malondialdehyde (MDA), decreasing sugar, electrolyte leakage, and total soluble solids.	Huang et al. (2019)
Basil seed gum	Fresh-cut apricot	Glycerol	The effects of coating on fresh apricot slices with edible films made of basil seed gum (BSG) and *Origanum vulgare* subsp. Viride essential oil (OEO) (1–6%) during cold storage at 4 °C for 8 days were studied.	Coated fruits incorporated with essential oil showed the higher number of phenolic compounds and enzymatic activity while reduced the incidence of microbial growth.	Hashemi et al. (2017)
Gum arabic	Apricot	Glycerol	The fruit were coated with 10% gum Arabic and stored at 20 ± 1 °C 8 days.	Reduced ripening and softening with higher eating quality	Ali et al. (2021)

TREATMENT	*VEGETABLES/ FRUITS*	*ANTIMICROBIAL AGENTS/ ANTIOXIDANTS*	*DOSE/METHOD*	*INFERENCE*	*REFERENCE*
Gum arabic	Tomato	Glycerol monostearate	Physiological mature tomatoes were preserved for 20 days at 20 °C and 80–90% relative humidity after being coated with an edible gum arabic coating in aqueous solutions of5,10,15, and 20%.	Tomatoes coated with 10% gum arabic showed prolong ripening, decreased ethylene generation and respiration while maintaining phenolic amount and activity of enzymes.	Ali et al. (2013)
Gum arabic, Ginger extract, garlic extract, and aloe vera gel.	Guava		The effects of the gum arabic, ginger extract, garlic extract, and the aloe vera gel coating (alone and in combination) were investigated on guava fruits for 15 days at 25 ±3 °C.	Combined application of gum arabic and garlic extract reduced weight loss, browning index. Retained phenolic and biochemical content enhanced shelf life.	Anjum et al. (2020); Zaidi et al. (2022)

coating preparations were tested on chunks of freshly cut papaya. During the time of the study, coatings increased the hardness of the fresh-cut product (Tapia et al., 2008). Using alginate- and gellan-based edible coatings and films for 23 days of preservation of freshly cut apples maintained fruit firmness and color during storage at a low temperature of 4 °C. Apples' shelf life is successfully extended by the application of coatings up to two weeks at the postharvest storage stage (Rojas-Graü et al., 2007). To assure microbiological safety and maintain the general fruit quality of strawberries, an antimicrobial edible alginate coating composed of 1.45% methyl cinnamate and 0.98% carvacrol was applied. A response surface approach was employed to increase the alginate coating's turbidity, viscosity, and color index, as well as its antibacterial efficacy against *Botrytis cinerea* and *E. coli*. Alginate edible covering decreased microbial development and preserved the freshness of the fruit (Peretto et al., 2014).

An alginate edible bio-film containing alginate, ß-cyclodextrin, glucose, palmitic acid as a component, and glycerol monostearate with *Cryptococcus laurentii* was investigated on strawberry fruit and stored for 20 days. Strawberry firmness and overall fruit quality and storage qualities were effectively preserved by a sodium alginate film entrapped with *C. laurentii* (Fan et al., 2009). Freshly cut pineapple had alginate-based (sodium and gellan) edible coatings applied to it, and it was kept there for 16 days at 5 °C. The findings also show that gellan-based coatings significantly contributed to keeping pineapples that have been freshly cut and frozen for 16 days at 5 °C. It may have to do with the advantageous outcome of an extra 2% of calcium chloride in the gellan-based coating's composition giving it the ability to retain the fruit's hardness (Azarakhsh, et al., 2014). A multilayered edible antimicrobial film composed of alginate was tested to see how well it would keep newly cut watermelon fresher longer without forgoing any of its excellent attributes. The coating was composed of sodium alginate. A coating consisting of beta-cyclodextrin, micro-encapsulated trans-cinnamaldehyde (a biological antimicrobial activity), and pectin with the mineral calcium lactate was created layer upon layer this way. The samples were covered layer by layer, stored at 4 °C for 15 days, and checked 3 days afterward. Application of the coating notably improved texture (firmness). Using 1 g/100 g of alginate to apply the multilayered edible coating will particularly enhance firmness, maintain the fresh-cut watermelon's quality and sensory appeal, and increase its shelf life (Sipahi et al., 2013).

Fresh-cut Flor de Invierno pears were covered in edible coatings made of alginate, pectin, and gellan and refrigerated for a period of 14 days at a low temperature of 4 °C (Oms-Oliu et al., 2008). These coatings contained N-acetylcysteine, glutathione, and antioxidants, and the amino acid have been included in the formulations to inhibit microbial development as well as the oxidation of freshly cut pear fruits and stored for a period of 2 weeks without allowing the texture of the fruit pieces to change. Alginate or pectin coatings best protected the pears' sensory properties for a duration of 14 days in storage (Oms-Oliu et al., 2008). Peaches were stored at 15 °C and 40% relative humidity, coated with coatings of sodium alginate along with methyl cellulose, and had their quality attributes evaluated at 3-day intervals. The edible coatings positively affected the physical attributes of the fruit. Peaches' firmness scores declined, indicating texture weakening as storage time increased for coated and uncoated fruit. However, fruit coating with methyl cellulose significantly improved firmness, displaying superior results to sodium alginate. At approximately 2 weeks of storage at ambient conditions the control fruits reached their full ripeness and softness (Maftoonazad et al., 2008). Citric acid (0.1%) is integrated into a sodium alginate edible coating applied on sliced carrots as a pretreatment, shelf life was increased by 5–7 days. The carrot had a glossy appearance thanks to the alginate coating, and its textural qualities persisted for at least 8 days. Even after 8 days, the edible coated-carrot root's hardness was slightly impacted by the addition of 0.1% or 0.5% citric acid (Amanatidou et al., 2000).

23.4.2 Effect of Polysaccharide Films on the Physiological Quality Attributes of Fruits and Vegetables

The impact of an edible film made of alginate and containing various concentrations of 15 different essential oils was examined on freshly cut Red Fuji apples for their *Salmonella typhimurium*, *Listeria monocytogenes*, *Escherichia coli*, *Staphylococcus aureus*, and *Listeria monocytogenes* antibacterial activity to increase the fresh-cut fruit's safety and quality. Findings revealed the superiority of thyme oil, cinnamon oil, and oregano oil to other essential oils in their ability to suppress the microorganisms. The outcome showed that the fresh-cut Red Fuji apples' microbial proliferation, respiration, weight loss, hardness, and browning could be significantly reduced by combining thyme oil with an alginate-based solution (Sarengaowa et al., 2018). The impact of chitosan (1%), alginate (2%), and extract of pomegranate peel (1%) upon guava fruit quality was investigated. Physiological changes were observed for 20 days at 10 °C, including respiration rate, ripening index, and pigment differences between treated and untreated samples. On freshly cut papaya pieces, formulas for alginate- or gellan-based (0.5% or 2.0%) coatings were used to see if they might increase water vapor resistance, modify gas exchange, and retain the general quality of the lightly processed fruit in comparison to the uncoated samples. The water barrier qualities of coatings based on either 1% glycerol + 1% ascorbic acid or 2% glycerol + 1% ascorbic acid was marginally stronger. The coated samples' capacity for vaporized water protection increased by 16% and 66%, respectively, when 0.025% sunflower oil was added to gellan- or alginate-based formulations. The qualities of the gas barrier of the developed solutions were not altered to permit appreciable distinctions in the respiratory rate of treated papayas and rate of ethylene production (Tapia et al., 2008).

The impact of alginate- and gellan-based films on the freshness of freshly cut apples were investigated by monitoring color changes and microbial development over a 23-day period of storage at 4 °C. The formation of ethylene in treated samples seems to be postponed for the duration of the short storage time. Furthermore, applying the edible films decreased the growth of microbes on fresh-cut apples (Rojas-Graü et al., 2007). Strawberry fruit was coated with a unique edible bio-film containing *Cryptococcus laurentii* together with cyclodextrin (0.5%), palmitic acid (0.5%), glycerol monostearate (0.5%), alginate (2%), and glycerol (2%) and kept for 20 days. According to Fan et al. (2009), a sodium alginate film entrapped with *C. laurentii* considerably reduced microbial decay, lowered weight loss, and enhanced strawberry fruit quality and storage capabilities. The coating contained sodium alginate (2%), glycerol (2%), and sunflower oil (0.025%). A gellan-based coating that included sunflower oil (0.025%), glycerol (1%), and gellan (1%) was applied on freshly cut pineapple pieces by using surface methodology. The alginate-based coating reduced the respiration rate, maintained color changes, and decreased the rate of weight loss. The loss of weight increased in both coated and uncoated samples as storage times progressed, but weight loss was lower in the gellan-based coated sample after 16 days of refrigeration at 5 °C when compared to uncoated samples (Azarakhsh et al., 2014).

Freshly cut pears were tested at 4 °C for 14 days along with edible coatings made of alginate (2%), pectin (2%), and gellan (0.5%) that also contained N-acetylcysteine (0.75%) and glutathione (0.75%). Freshly cut pear fruit treated with edible coatings composed of carbohydrates exhibited decreased ethylene production and enhanced resistance against water vaporization (Oms-Oliu et al., 2008). The effects of a synthesized edible coating consisting of 0.2% olive oil and 2% sodium alginate, along with 1% each of ascorbic acid and citric acid, on the nutritive value and storage duration of fruit kept at 65% relative humidity and 25 °C were evaluated. The coatings decreased the frequency of deterioration, weight loss, and polygalacturonase; the treated fruits displayed delayed pectate lyase and pectin methyl esterase activity as compared to the untreated fruit, suggesting a more gradual maturation and softening process (Ramana Rao et al., 2016). The effects of methyl cellulose and sodium alginate films were applied to peaches that were kept at 40% relative humidity and 15 °C with coated and uncoated control fruits, and the fruit quality attributes were evaluated at 3-day intervals. The edible coating lowered the rate of respiration, moisture loss, and modifications in qualitative characteristics. Treated samples with sodium alginate and methyl cellulose remained acceptable for 21 and 24 days, respectively, while uncoated samples lasted up to 15 days (Maftoonazad et al., 2008).

An edible coating made of sodium alginate was used on sweet cherry fruits that were obtained at the commercial maturity stage in different percentages (1%, 3%, or 5%). The coatings were successful at postponing the development of postharvest ripening-related factors, such as discoloration, softening, and acidity loss, as well as decreasing the rate of respiration (Díaz-Mula et al., 2012). On microbiological stability, the physicochemical characteristics, total phenol compounds and carotenoids, antioxidants, and sensual qualities of a freshly cut mango were examined. For storage of 14 days at 4 °C, edible coatings such as alginate, pectin, carboxymethyl cellulose, and chitosan were evaluated. The fresh mango with a chitosan coating showed the least amount of microbiological load (1log $CFUg^{-1}$) throughout the duration of preservation, keeping microbial levels below 6 log CFU g^{-1}. During preservation, alginate, pectin, and carboxymethyl cellulose coatings retained the fresh-cut mango's bright color (Salinas-Roca et al., 2018). Edible coatings comprised alginate (1%) in distilled water and zein (10%) in 75% ethanol at 65 °C. Oleic acid (8%) was cooled to 20 °C before being added as a plasticizer, fruits were submerged twice for 1 minute in solutions, and after drying at 25 °C for 30 minutes the samples were kept for 9 days at 20 °C and 80% relative humidity. According to Zapata et al. (2008), tomatoes that had been coated showed reduced respiration and ethylene rates, as well as fewer quality losses like weight reduction, color change, and softening than untreated fruits. MCP (1-methylcyclopropene) and an edible coating composed of alginate was applied on baby butterhead lettuce that had undergone minimal processing and kept for 12 days of storage at 2 °C and 6 °C. The lettuce treated with 1-MCP exhibited the lowest ethylene and respiration rates. The most delayed rusty spotting was found in the midribs of the lettuce leaves treated with 1-MCP (Tay & Perera, 2004).

23.4.3 Effect of Polysaccharide Films on the Biochemical Quality Attributes of Fruits and Vegetables

The impacts of 1% chitosan, 2% alginate, and 1% extract from pomegranate peel were investigated on the quality of guavas, and biochemical changes were recorded for 20 days or more at 10 °C. Chitosan-treated samples that had been enhanced with pomegranate peel turned out to be the most successful in preserving the general fruit quality (Nair et al., 2018). Alginate- or gellan-based coating incorporated with ascorbic acid on minimally processed papaya was applied to protect the fruit's nutritional value after minimal processing. Findings of the study revealed that fresh-cut papaya's natural ascorbic acid content was preserved using the coatings containing ascorbic acid as an antioxidant, helping to retain its nutritious quality over time (Tapia et al., 2008). How ber fruit's shelf life and nutrient contents are affected by a composite edible coating consisting of 2% sodium alginate, 0.2% olive oil, and a combination of 1% ascorbic acid and 1% citric acid were evaluated while stored at a temperature of 25 ± 2 °C and 65% relative humidity. The findings indicated that the ascorbic acid and overall total soluble solid of the ber fruit were preserved in the edible coatings (Ramana Rao et al., 2016).

Carrageenan, alginates, and isolates of soy protein and whey protein concentrations were used to create coatings. The emulsion coatings were applied to chunks of apple. The findings showed that edible films help to maintain the integrity of fresh-cut apples during storage and lengthen their shelf life (Ghavidel et al., 2013). The quality characteristics of peaches were assessed every 3 days while kept at 15 °C and 40% relatie humidity and coated with sodium alginate and methyl cellulose. The findings of the study showed that variation was discovered as a correlation between pH titration and storage period length, but coated fruits showed higher rates of acidity and pH, and there were no significant changes in the total soluble solids of peach fruit during storage (Maftoonazad et al., 2008). Alginate- or zein-based coatings were useful for retaining sugars, organic acids (particularly ascorbic acid), and assessments from the evaluation of sensations, which remained considerably higher in treated tomatoes as compared to untreated tomatoes toward the duration of the storage period (Zapata et al., 2008). During storage at 10 °C, the effects of extract of pomegranate peel on capsicum were examined in two edible polysaccharide-based coatings: chitosan and alginate. By characterizing the pomegranate peel extract, punicalagin, catechin, and rutin were found. According to Nair et al. (2018), PPE added to these two coatings can additionally preserve sensorial evaluations, inhibit microbiological development, and increase the shelf life at 10 °C by up to 25 days.

23.4.4 Effect of Polysaccharide Films on the Antioxidative Quality Attributes of Fruits and Vegetables

The quality of guava fruit (cv. Allahabad Safeda) was affected by the application of chitosan (1%), alginate (2%), and pomegranate peel extract (1%). According to study findings, following storage, the total amount of phenolic compounds and flavonoid content were measured, with negligible loss of antioxidant activity. Research revealed a more consistent association between several phytochemicals' antioxidant activities. Coatings coated with pomegranate peel extract were discovered to be helpful in preserving the nutritional value of guava fruits during a 20-day cold storage period (Kocira et al., 2021). Freshly sliced pineapple was coated with edible coatings made of alginate (sodium and gellan) and kept at low temperature of 5 °C in a refrigerator up to 16 days. Findings of this investigation showed that biochemical parameters did not significantly alter throughout the course of 16 days at a low temperature of 5 °C (Azarakhsh et al., 2014). Flor de Invierno pears that had been freshly cut were tested for 14 days at 4 °C with edible coatings made of alginate (2%), pectin (2%), and gellan (0.5%) that also contained N-acetylcysteine (0.75%) and glutathione (0.75%). Pear fruit pieces coated with alginate, gellan, and pectin had higher levels of ascorbic acid antioxidants (Oms-Oliu et al., 2008).

The effect of an edible coating consisting of 2% sodium alginate, 0.2% olive oil, and a combination of 1% ascorbic acid and 1% citric acid on the nutritional value and storage longevity of ber fruit after harvest while kept at ambient conditions was investigated. Results showed that edible coatings enhanced the level of antioxidants and phytochemicals in ber fruit. The study's findings may reduce the frequency of ber fruit decomposing, increase its shelf life, and enhance its beneficial nutritional properties (Ramana Rao et al., 2016). An edible sodium alginate-based coating was applied to sweet cherry fruits that were collected at a variety of physiologically mature stages (1%, 3%, and 5%). The edible coatings had a favorable impact on preserving greater total phenol concentrations and total antioxidant capacity, which had decreased due to the aging and over-ripening processes in untreated apple fruits. The storage period for control fruits was determined by quality parameters and antioxidant activity; cherries coated with alginate may be kept for 16 days at a low temperature of 2 °C and a further 2 days at 20 °C, whereas cherries kept for 8 days at 2 °C plus 2 days at 20 °C showed the best quality and improved antioxidant activity (Daz-Mula et al., 2012). Alginate, pectin, carboxymethyl cellulose, or chitosan and other edible coatings were applied to freshly chopped mango and kept at 4 °C for 14 days. The concentration of antioxidants in freshly cut mango increased when coated with alginate or chitosan, which have various monomers in their chains. Among those coated, the fresh-cut mangoes coated with alginate were the toughest after 14 days. Freshly cut mangoes coated in alginate (90.2%) had the highest level of consumer acceptability. The best coating to retain the quality of freshly cut mango's freshness during storage would be chitosan (Salinas-Roca et al., 2018).

23.5 EDIBLE FILMS/COATINGS BASED ON GUMS

Although naturally organic materials and their derived products are mostly used in food manufacturing, their application as palatable coatings to lengthen the shelf life of fresh fruits and vegetables has been recently investigated. When compared with artificially produced polymers, because these organic polymers' polysaccharides are harmless and beneficial to the ecosystem, they have several advantages, including being affordable and widely available in the environment. The partially synthetic modification of natural gums can result in derivatives that are easily competitive with the synthetic preservatives already on the market for food (Saha et al., 2017). Plant-derived gums are widely accessible, and due to the interaction of two or more polymers, gums can combine to form a coating that has combined advantages. Natural gums are now valuable ingredients for edible films and coatings because of these distinctive qualities. The effects of plant-based edible coatings on the freshness of fruits and vegetables has been documented in Table 23.1.

The gums that are naturally available are polysaccharides made up of sugars, besides glucose, and in very little amounts can improve a solution's consistency. They are easily accessible, inexpensive, biodegradable, environmentally friendly, odorless, and chemically inert. Because these gums dissolve in water, they are often referred to as hydrocolloids (Tahir et al., 2019). These characteristics allow gums to have an extensive variety of uses in numerous industries. They serve as stabilizers, thickeners, emulsifiers, integrating agents, and gelling compounds. These qualities result from their ability to bind water, their rheological characteristics, and their ability to encapsulate different compounds and produce films or gels, including flavors, fragrances, and nutraceutical components (Kocira et al., 2021). Naturally derived gums have recently attracted attention as a possible coating for edible fruits and vegetables and have shown that most polysaccharides act as good barriers when the overall humidity is low (Eghbalijoo et al., 2022).

23.5.1 Classification of Gums

Most gums contain sugars and are polysaccharides. Because of their texture, gums are utilized to produce attractive fruit and vegetable films. Only a few of these gums are well known. Plant-based gums of organic origin and their use as edible coatings includes gum arabic, guar gum, basil seed gum and xanthan gum, soybean gum, flaxseed gum, tamarind gum, and tara gum. Plant-based gums are classified on the basis of their source and chemical composition (Momin et al., 2021). There are three classes of gums:

1 Exudate gums, e.g., acacia gum.
2 Extractive gums, e.g., guar gum.
3 Fermented gums, e.g., xanthan gum (Kulkarni-Vishakha et al., 2012).

23.5.2 Effect of Gum-Based Film on Fruit and Vegetable Quality

Fruits and vegetables are a great source of antioxidants, which improve nutrition, lower the possibility of many diseases, and also prevent them. Additionally, these substances are linked to how plants react to biotic and abiotic stressors. Gum arabic proved successful in increasing the number of antioxidants overall in strawberry fruits, with the fruit's phenolic as well as anthocyanin contents rising as a result of low-temperature storage (Tahir et al., 2019). A coating made of tulsi flower extract, sodium caseinate, and gum arabic was applied on guava fruits stored for 7 days. At 28 ± 2 °C, the optimized coating exhibited at 7-day shelf life and greater general popularity. It also maintained an appropriate internal gas composition to delay ripening (Murmu & Mishra, 2017). A study was conducted to evaluate how ginseng extract and guar gum coating affected the quality of delicious cherries, which were kept at 20 °C for 8 days. Overall, coatings improved the nutritional value of sweet cherries up to 8 days (Dong & Wang, 2018). Mushrooms (*Lentinus edodes*) were covered with a guar gum and chitosan covering and kept at a low temperature for 16 days. The findings showed that ascorbic acid and soluble protein levels decreased at a rapid rate; on the other hand, total soluble content, lowering sugar, MDA, and leakage of electrolytes increased in mushrooms treated with CH 1% + GG 15%. The CH 1% + GG 15% coating was proposed as a potential commercial use for preserving the mushrooms' quality for a long storage duration (Huang et al., 2019).

A study was conducted to investigate how a coating based on cashew gum and carboxymethylcellulose affects freshly cut guava fruit. The coating solution was applied by a dipping technique and stored for 12 days at room temperature. Both coatings reduced mass loss, preserved stiffness, and delayed changes in skin color (Forato et al., 2015). A basil seed coating was applied on freshly cut apricot fruits, then kept at a low temperature of 4 °C for up to 18 days of storage. The film and coating could significantly improve odor and general acceptance and could also be used to preserve the freshness of cut apricot (Hashemi et al., 2017). The edible coating based on gum arabic was applied to physiologically mature tomatoes to prolong the ripening phenomena and retain the antioxidant activity by storing at 20 °C for up to 20 days. Findings indicated that tomato ripening can be postponed and antioxidant storage maintained up to 20 days of storage at 20 °C while preserving the nutritional profile through the application of 10% gum arabic, which is used as an edible coating (Ali et al., 2013).

Storage and marketing of guava fruit is associated with an enormous problem because of its limited shelf life and climacteric nature; that's why some effective treatments are required to maintain its overall quality. The goal of the study was to ascertain the effects of aloe vera gel, gum arabic, and ginger and garlic extracts, on guava fruit quality after harvest over a 15-day period of ambient storage. Combined, gum arabic and garlic extract dramatically decreased the degree of the disease, skin discoloration, and loss of weight while extending the storage life of guava fruits due of its greater titratable acidity than the control. This combination also lessened an excessive growth in sugar content until the completion of the storage period (Anjum et al., 2020). Persimmon fruits had postharvest losses and short shelf life due to their fast ripening process; therefore a plant-based hydrocolloid tragacanth gum coating was applied at various concentrations (0.5%, 1%, and 1.5%) then kept at 20 °C for 20 days of storage. The tragacanth gum coating lessened the rate of respiration, the production of ethylene, disease extent, and loss of weight. On the other hand, the treated fruits contained higher contents of bioactive compounds, including phytochemicals, flavonoids, tannins, ascorbic acid, carotenoids, and enzymatic activities; however, they experienced reduced activities of softening enzymes (Saleem et al., 2022).

The study was conducted applying a composite edible coating comprising guar gum, clove essential oil, and gum Arabic on strawberry fruits to enhance their shelf life and retain their nutritional profile. The fruits were stored at a low temperature (6 ± 2°C), and the results showed that the coated sample's shelf life was enhanced up to 9 days, which is quite noteworthy for delicate and perishable commodities. Fruits that had been treated lost less weight and experienced less decay incidence while maintaining their phenolic chemicals and the concentration of ascorbic acid. This proves that applying an edible coating made of polysaccharides after harvest, along with clove essential oil, increases the antioxidant potential of strawberries, prevents deterioration, and improves their overall nutritional profile (Jodhani & Nataraj, 2019). Gum arabic was applied as an edible coating on cucumber fruits at different concentrations (5%, 10%, 15%, and 20%) and kept at two distinct temperatures (10 °C and 25 °C) for the duration of the 16-day storage duration. Application of an edible gum coating reduced weight loss and prevented fruits from softening; retained their sensory attributes, including color, flavor, softness, and attractiveness; and maintained their overall acceptability for 16 days of storage at 10°C and 25°C (Al-Juhaimi et al., 2012).

23.6 LEGISLATION

Before the utilization of edible compounds as a coating for commercial applications, products that come into direct contact with food must obtain safety approval. Therefore, before the edible commodity can be consumed, the US Food and Drug Administration (FDA) must categorize it as Generally Regarded As Safe (GRAS) (Paidari et al., 2021). Furthermore, using authorized organic plant extracts along with essential oils may result in specific allergic reactions and have negligible negative effects based on the dosage (Dhall, 2013). In order to regularly confirm the harmful and allergic effects of extracts as well as essential oils employed in order to create edible coatings, the proper procedures must be put in place. Edible coating ingredients ought to be harmless, of food-safe quality, and processed with the highest level of hygiene (Nayak et al., 2019). Coatings that are edible are typically classified as food constituents, food components or products, and food-related packaging technologies. Given that they constitute a component of the meal that can be eaten, the edible coating materials for fresh foods must be produced using appropriate ingredients and in accordance with certain rules. The FDA has approved nutritional supplements or edible materials for coating that are GRAS for fresh fruits and vegetables (Kocira et al., 2021). When creating novel edible coating materials, caution must be given because different countries have different laws and regulations governing food additives and permitted edible coating biopolymers (Aayush et al., 2022).

23.7 FUTURE PROSPECTS

The quality of the product can be kept fresh with edible or biodegradable films and plant-based coatings during the product's market life. Fresh fruits and vegetables treated with alginate solutions containing additives have shown promising effects. New antibacterial, antioxidant, and anti-browning agents, however, could be added to existing ones to considerably improve food quality and safety. An excellent strategy is to use the film-formation method to improve its structural qualities in order to preserve the freshness of fruits and vegetables. There are currently very few commercial applications for alginate coatings, and the majority of research on their uses has been conducted at the research level. The industrial deployment of alginate-coated long-storage fruits and vegetables has to be the main priority of future research with real-world implications. Rresearchers should move toward the recycling process of coating materials that reduces the microbiological load of coatings during recycling to avoid the drawbacks of the methods, and that does not use excessive amounts of coating solution. Therefore, edible films and coatings based on sodium alginate could be utilized even more so than they already are.

23.8 CONCLUSIONS

Edible film is an ideal application to minimize the respiratory rate, water loss, aroma, and flavor score. It also protects against the growth of microbes and retains overall fruit and vegetable nutritional value characteristics. It is feasible to make the following deductions based on the evaluation of earlier studies on edible films and coatings made of alginate. These materials can effectively be employed to lengthen the shelf life of freshly cut fruits and vegetables. Researchers may use the knowledge compiled here to create effective coating applications. As edible films, save money and minimize food wastage, they are a smart choice for both producers and consumers of fresh fruits and vegetables.

REFERENCES

Aayush, K., McClements, D.J., Sharma, S., Sharma, R., Singh, G.P., Sharma, K., Oberoi, K. (2022). Innovations in the development and application of edible coatings for fresh and minimally processed Apple. *Food Control*, *141*, 109188.

Ali, A., Maqbool, M., Alderson, P.G., Zahid, N. (2013). Effect of gum arabic as an edible coating on antioxidant capacity of tomato (*Solanum Lycopersicon* L.) fruit during storage. *Postharvest Biology and Technology*, *76*, 119–124.

Ali, S., Akbar Anjum, M., Nawaz, A., Naz, S., Ejaz, S., Shahzad Saleem, M., Ul Hasan, M. (2021). Effect of gum arabic coating on antioxidative enzyme activities and quality of apricot (*Prunus armeniaca* L.) fruit during ambient storage. *Journal of Food Biochemistry*, *45*(4), e13656.

Al-Juhaimi, F., Ghafoor, K., Babiker, E.E. (2012). Effect of gum arabic edible coating on weight loss, firmness and sensory characteristics of cucumber (*Cucumis sativus* L.) fruit during storage. *Pakistan Journal of Botany*, *44*(4), 1439–1444.

Amanatidou, A., Slump, R.A., Gorris, L.G.M., Smid, E.J. (2000). High oxygen and high carbon dioxide modified atmospheres for shelf-life extension of minimally processed carrots. *Journal of Food Science*, *65*(1), 61–66.

Anjum, M.A., Akram, H., Zaidi, M., Ali, S. (2020). Effect of gum arabic and Aloe vera gel based edible coatings in combination with plant extracts on postharvest quality and storability of 'Gola' guava fruits. *Scientia Horticulturae*, *271*, 109506.

Azarakhsh, N., Osman, A., Ghazali, H.M., Tan, C.P., Adzahan, N.M. (2014). Lemongrass essential oil incorporated into alginate-based edible coating for shelf-life extension and quality retention of fresh-cut pineapple. *Postharvest Biology and Technology*, *88*, 1–7.

Dhaka, R., Upadhyay, A. (2018). Edible films and coatings: a brief overview. *The Pharma Innovation Journal*, *7*(7), 331–333.

Dhall, R.K. (2013). Advances in edible coatings for fresh fruits and vegetables: a review. *Critical Reviews in Food Science and Nutrition*, *53*(5), 435–450.

Díaz-Mula, H.M., Serrano, M., Valero, D. (2012). Alginate coatings preserve fruit quality and bioactive compounds during storage of sweet cherry fruit. *Food and Bioprocess Technology*, *5*, 2990–2997.

Dong, F., & Wang, X. (2018). Guar gum and ginseng extract coatings maintain the quality of sweet cherry. *LWT-Food Science and Technology*, *89*, 117–122.

Eghbaljoo, H., Sani, I.K., Sani, M.A., Rahati, S., Mansouri, E., Molaee-Aghaee, E., Jafari, S. M. (2022). Advances in plant gum polysaccharides; Sources, techno-functional properties, and applications in the food industry-A review. *International Journal of Biological Macromolecules*, *222*(B), 2327–2340.

Erkmen, O., Barazi, A. O. (2018). General characteristics of edible films. *Journal of Food Biotechnology Research*, *2*(1), 3.

Falguera, V., Quintero, J.P., Jiménez, A., Muñoz, J.A., Ibarz, A. (2011). Edible films and coatings: Structures, active functions and trends in their use. *Trends in Food Scieince & Technolology*, *22*, 292–303.

Fan, Y., Xu, Y., Wang, D., Zhang, L., Sun, J., Sun, L., Zhang, B. (2009). Effect of alginate coating combined with yeast antagonist on strawberry (*Fragaria× ananassa*) preservation quality. *Postharvest Biology and Technology*, *53*(1–2), 84–90.

Forato, L.A., de Britto, D., de Rizzo, J.S., Gastaldi, T.A., Assis, O.B. (2015). Effect of cashew gum-carboxymethylcellulose edible coatings in extending the shelf-life of fresh and cut guavas. *Food Packaging and Shelf Life*, *5*, 68–74.

Galus, S., Arik Kibar, E.A., Gniewosz, M., Kraśniewska, K. (2020). Novel materials in the preparation of edible films and coatings-A review. *Coatings*, *10*(7), 674.

Ghavidel, R.A., Davoodi, M.G., Asl, A.F.A., Tanoori, T., Sheykholeslami, Z. (2013). Effect of selected edible coatings to extend shelf-life of fresh-cut apples. *International Journal of Agriculture and Crop Sciences*, *6*(16), 11–71.

Guimarães, A., Abrunhosa, L., Pastrana, L.M., Cerqueira, M.A. (2018). Edible films and coatings as carriers of living microorganisms: A new strategy towards biopreservation and healthier foods. *Comprehensive Reviews in Food Science and Food Safety*, *17*, 594–614.

Hashemi, S.M.B., Khaneghah, A.M., Ghahfarrokhi, M.G., Eş, I. (2017). Basil-seed gum containing *Origanum vulgare* subsp. viride essential oil as edible coating for fresh cut apricots. *Postharvest Biology and Technology*, *125*, 26–34.

He, Y., Li, H., Fei, X., Peng, L. (2021). Carboxymethyl cellulose/cellulose nanocrystals immobilized silver nanoparticles as an effective coating to improve barrier and antibacterial properties of paper for food packaging applications. *Carbohydrate Polymers*, *252*, 117156.

Huang, Q., Qian, X., Jiang, T., & Zheng, X. (2019). Effect of chitosan and guar gum based composite edible coating on quality of mushroom (*Lentinus edodes*) during postharvest storage. *Scientia Horticulturae*, *253*, 382–389.

Jodhani, K.A., Nataraj, M. (2019). Edible coatings from plant-derived gums and clove essential oil improve postharvest strawberry (*Fragaria× ananassa*) shelf life and quality. *Environmental & Experimental Biology*, *17*(3).

Kocira, A., Kozłowicz, K., Panasiewicz, K., Staniak, M., Szpunar-Krok, E., Hortyńska, P. (2021). Polysaccharides as edible films and coatings: Characteristics and influence on fruit and vegetable quality—A review. *Agronomy*, *11*(5), 813.

Kulkarni-Vishakha, S., Butte-Kishor, D., Rathod Sudha, S. (2012). Natural polymers–A comprehensive review. *International Journal of Pharmaceutical and Bio-Medical Science, 3*(4), 1597–1613.

Maftoonazad, N., Ramaswamy, H.S., Marcotte, M. (2008). Shelf-life extension of peaches through sodium alginate and methyl cellulose edible coatings. *International Journal of Food Science & Technology*, *43*(6), 951–957.

Mahajan, B.V.C., Tandon, R., Kapoor, S., Sidhu, M.K. (2018). Natural coatings for shelf-life enhancement and quality maintenance of fresh fruits and vegetables-A review. *Journal of Postharvest Technology*, *6*(1), 12–26.

Momin, M.C., Jamir, A.R., Ankalagi, N., Henny, T., Devi, O.B. (2021). Edible coatings in fruits and vegetables: A brief review. *Pharma Innovation Journal*, *10*, 71–78.

Murmu, S.B., Mishra, H.N. (2017). Optimization of the arabic gum based edible coating formulations with sodium caseinate and tulsi extract for guava. *LWT-Food Science and Technology*, *80*, 271–279.

Nair, M.S., Saxena, A., Kaur, C. (2018). Characterization and antifungal activity of pomegranate peel extract and its use in polysaccharide-based edible coatings to extend the shelf-life of capsicum (*Capsicum annuum* L.). *Food and Bioprocess Technology*, *11*, 1317–1327.

Nayak, S.L., Sethi, S., Sharma, R.R., Prajapati, U. (2019). Active edible coatings for fresh fruits and vegetables. *Polymers for Agri-Food Applications*, 417–432.

Nussinovitch, A., Hershko, V. (1996). Gellan and alginate vegetable coatings. *Carbohydrate Polymers*, *30*(2–3), 185–192.

Oms-Oliu, G., Soliva-Fortuny, R., Martín-Belloso, O. (2008). Edible coatings with antibrowning agents to maintain sensory quality and antioxidant properties of fresh-cut pears. *Postharvest Biology and Technology*, *50*(1), 87–94.

Paidari, S., Zamindar, N., Tahergorabi, R., Kargar, M., Ezzati, S., Shirani, N., Musavi, S.H. (2021). Edible coating and films as promising packaging: a mini review. *Journal of Food Measurement and Characterization*, *15*(5), 4205–4214.

Peretto, G., Du, W.X., Avena-Bustillos, R.J., Berrios, J.D.J., Sambo, P., McHugh, T.H. (2014). Optimization of antimicrobial and physical properties of alginate coatings containing carvacrol and methyl cinnamate for strawberry application. *Journal of Agricultural and Food Chemistry*, *62*, 984–990.

Radev, R., Pashova, S. (2020). Application of edible films and coatings for fresh fruit and vegetables. *Qual. Access Success*, *21*, 108–112.

Ramana Rao, T.V., Baraiya, N.S., Vyas, P.B., Patel, D.M. (2016). Composite coating of alginate-olive oil enriched with antioxidants enhances postharvest quality and shelf life of Ber fruit (*Ziziphus mauritiana Lamk.* Var. Gola). *Journal of Food Science and Technology*, *53*, 748–756.

Rojas-Graü, M.A., Raybaudi-Massilia, R.M., Soliva-Fortuny, R.C., Avena-Bustillos, R.J., McHugh, T. H., Martín-Belloso, O. (2007). Apple puree-alginate edible coating as carrier of antimicrobial agents to prolong shelf-life of fresh-cut apples. *Postharvest Biology and Technology*, *45*(2), 254–264.

Saha, A., Tyagi, S., Gupta, R.K., Tyagi, Y.K. (2017). Natural gums of plant origin as edible coatings for food industry applications. *Critical Reviews in Biotechnology*, *37*(8), 959–973.

Saleem, M.S., Ejaz, S., Anjum, M.A., Ali, S., Hussain, S., Ercisli, S., Mlcek, J. (2022). Improvement of postharvest quality and bioactive compounds content of persimmon fruits after hydrocolloid-based edible coating application. *Horticulturae*, *8*(11), 10–45.

Salgado, P.R., Ortiz, C.M., Musso, Y.S., Di Giorgio, L., Mauri, A.N. (2015). Edible films and coatings containing bioactives. *Current Opinion in Food Science*, *5*, 86–92.

Salinas-Roca, B., Guerreiro, A., Welti-Chanes, J., Antunes, M.D., Martín-Belloso, O. (2018). Improving quality of fresh-cut mango using polysaccharide-based edible coatings. *International Journal of Food Science & Technology*, *53*(4), 938–945.

Salvia-Trujillo, L., Soliva-Fortuny, R., Rojas-Graü, M.A., McClements, D.J., Martín-Belloso, O. (2017). Edible nanoemulsions as carriers of active ingredients: A review. *Annual Review of Food Scieince and Technolology*, *8*, 439–466.

Sarengaowa, Hu, W., Jiang, A., Xiu, Z., Feng, K. (2018). Effect of thyme oil–alginate-based coating on quality and microbial safety of fresh-cut apples. *Journal of the Science of Food and Agriculture*, *98*(6), 2302–2311.

Sipahi, R.E., Castell-Perez, M.E., Moreira, R.G., Gomes, C., Castillo, A. (2013). Improved multilayered antimicrobial alginate-based edible coating extends the shelf life of fresh-cut watermelon (*Citrullus lanatus*). *LWT-Food Science and Technology*, *51*(1), 9–15.

Swathi, V., Gladvin, G., Babitha, B. (2017). Physico-chemical charectristics and applications of edible films for fruit preservation. *International Research Journal of Engineering and Technology*, *4*(02), 1954–1958.

Tahir, H E., Xiaobo, Z., Mahunu, G.K., Arslan, M., Abdalhai, M., Zhihua, L. (2019). Recent developments in gum edible coating applications for fruits and vegetables preservation: A review. *Carbohydrate Polymers*, *224*, 115141.

Tapia, M.S., Rojas-Graü, M.A., Carmona, A., Rodríguez, F.J., Soliva-Fortuny, R., Martin-Belloso, O. (2008). Use of alginate-and gellan-based coatings for improving barrier, texture and nutritional properties of fresh-cut papaya. *Food Hydrocolloids*, *22*(8), 1493–1503.

Tay, S.L., Perera, C.O. (2004). Effect of 1-methylcyclopropene treatment and edible coatings on the quality of minimally processed lettuce. *Journal of Food Science*, *69*(2), 131–135.

Vilgis, T.A. (2012). Hydrocolloids between soft matter and taste: Culinary polymer physics. *International Journal of Gastronomy and Food Science*, *1*(1), 46–53.

Wang, J., Euring, M., Ostendorf, K., Zhang, K. (2022). Biobased materials for food packaging. *Journal of Bioresources and Bioproducts*, *7*(1), 1–13.

Zaidi, M., Akbar, A., Ali, S., Akram, H., Ercisli, S., Ilhan, G., Anjum, M.A. (2022). Application of plant-based edible coatings and extracts influences the postharvest quality and shelf-life potential of "Surahi" guava fruits. *ACS Omega*, *8*(22), 19523–19531.

Zapata, P.J., Guillén, F., Martínez-Romero, D., Castillo, S., Valero, D., Serrano, M. (2008). Use of alginate or zein as edible coatings to delay postharvest ripening process and to maintain tomato (*Solanum Lycopersicon* Mill) quality. *Journal of the Science of Food and Agriculture*, *88*(7), 1287–1293.

Edible Coatings for Postharvest Fruit and Vegetable Storage

24

Sajid Ali*, Muhammad Hassan, Shaghef Ejaz, Muhammad Naveed Arshad and Irfan Ali Sabir
*Corresponding Author: sajidali@bzu.edu.pk

24.1 INTRODUCTION

Fresh fruits and vegetables (FV) are produced and utilized abundantly in the world due to their health-beneficial properties. The availability of the important nutrients of FV mainly depends upon their conservation after harvest (among others) because their storability has remained a great challenge and concern worldwide (Yousuf et al., 2018). Several factors affect the acceptance of FV nutrients and freshness, because these are generally processed and cooked before use. The conditions, safety and wholesomeness of FV often define their acceptance by consumers after harvest. Therefore, it is imperative to preserve the freshness of FV so that the reach consumers with proper safety (Slavin & Lloyd, 2012).

Different types of conservation techniques and approaches have been used and are still being practised in the world for the maintenance of FV quality after harvest (Werner et al., 2017; Olawuyi et al., 2021). The application of various types of edible packagings has been found effective for the postharvest conservation of FV during storage, because they are cheap, sustainable and easy to use (Olawuyi et al., 2021; Yaashikaa et al., 2023). Edible coating on FV is not a recent technique; it has long been used as a way to protect fresh produce and conserve its characteristic qualities (Ansorena et al., 2018). In the 20th century, different types and forms of coatings were tested to reduce weight loss and increase the shelf life of fresh commodities (Hassan et al., 2018).

A coating is a fine cover that is self-sustaining when applied on food items. These solutions are prepared in liquid phase either alone or in combination with other composites (Saberi & Golding, 2018). There are major dissimilarities between edible films and coatings as modes of treatment on fruits and vegetables. Coatings are applied by a dipping or misting technique, whereas films are applied in desiccated or moulded form (Falguera et al., 2011). Mostly, freshly consumed produce belongs to the horticulture industry, and FV are included. FV are known for their high perishability owing to their high water content, which is a major factor in their limited shelf life and is a leading issue if stored without a cold storage facility. Even though there has been much advancement in the preservation of fresh FV, a significant quantity of the said fresh produce is still wasted. So, it is direly needed to explore a suitable technique to overcome these FV losses.

Edible coatings used on fresh produce provide a partial covering that allows gaseous exchange and also acts as a barrier to foreign microbes, decreasing weight loss and extending the shelf life of a product. The FV are obtained from different plant parts and have different physiologies. So, edible coatings can be planned according to the nature of the produce. Application and type of coating material depends upon the aforementioned factors (Montero-Calderón et al., 2016). The application of coatings can be adjusted according to atmospheric conditions to increase the protection from microbes, thereby resulting in extended shelf life along with enhanced glossiness and preserved texture (Tapia-Blácido et al., 2018; Ali et al., 2023).

Edible packages, either in the form of stand-alone or combined/composite coatings, have been extracted and developed from different types of materials, including polysaccharides, lipids, proteins and mucilages, among others (Olawuyi et al., 2021; Yaashikaa et al., 2023). Among the different coating types, polysaccharide-based polymeric materials are of particular importance in the fresh FV industry for their appropriate preservation capabilities (Kumar et al., 2018; Olawuyi et al., 2021; Yaashikaa et al., 2023). Different types of polysaccharide-based coating materials include gum exudates, alginates, carrageenan, chitosan and mucilages, which

 DOI: 10.1201/9781003370376-32

have been much researched, developed, and applied in various types of edible packages for the preservation of fresh FV after harvest (Olawuyi et al., 2021; Yaashikaa et al., 2023). Polysaccharide-based edible packages have been applied to various FV for their preservation and to extend their storage/shelf-life potential (Ali et al., 2023; 2024b, 2024c; Naveed et al., 2024). Different factors affect the efficacy of edible coatings to preserve the characteristic qualities and prolong the storage and/or shelf life of fresh FV after harvest. In this chapter, various types of edible coatings, application methods, coating material characteristics and their safety status regulations have been summarized and discussed with potential future perspectives.

24.2 EDIBLE COATING FORMULATIONS

Edible coatings are usually prepared from edible compounds and distributed into different types, including hydrocolloids, phospholipids and their derivatives (Gutiérrez, 2018a). Polysaccharides and proteins (hydrocolloids) are mainly used as edible coating ingredients for edible coatings (Valencia & do Amaral Sobral, 2018). These are non-smelly, non-toxic and clear when applied as compared to lipid-based (non-transparent) coating materials. Edible coatings (such as polysaccharides among others) provide better tensile strength to fruits with minimum respiration rate and ethylene generation (Garrido et al., 2018; Ali et al., 2024b). But these are less protective against moisture due to their hydrophilic nature compared to lipids, which have more resistance to moisture. However, polysaccharide- or protein-based coating properties can be improved by adding lipids (Galus & Kadzińska, 2015). Recently, there has been progress in the use of edible constituents like polysaccharides, lipids and proteins alone or in combined form (Merino et al., 2019). The purpose of these studies was to develop a source that contains an effective recipe of coatings-based substances that are functionally more active and consistent. However, new combinations of constituents must be categorized and examined. For example, left-over FV flesh or its by-products in combination with hydrocolloids or other suitable coating types could be further investigated for more authentic results and study of their combined properties. Wang et al. (2011) found that a variety of materials was helpful in maintaining the freshness of carrots. He discovered that various polymeric materials (such as starch, carboxymethyl cellulose (CMC) and gelatine) can be used as edible coatings.

Proteins/lipids used in combination with polysaccharides enhanced the plasticizer ability of coatings by extending its elasticity (Álvarez et al., 2017). Plasticizers are used to minimize coating breakage or to increase its smoothness. Diverse kinds of plasticizers are commonly used in hydrocolloids (Medina-Jaramillo et al., 2016). Moreover, water also acts as a plasticizer in many combinations. Other constituents (binding and emulsifying agents) can also be used to increase the effectiveness of coatings (Collazo-Bigliardi et al. 2018). New studies have been done on different materials, including antioxidants, antimicrobials, nutraceuticals and other additives, being added to increase the characteristics of the coating to further improve its functioning as a smart packaging approach for fresh FV (Bracone et al., 2016; Gutiérrez, 2018b). The use of plastic containers can be reduced by using agricultural-based by-products or waste material containers, which may prove as biodegradable and safe materials for the environment (Salgado et al., 2015). There is a new trend to develop and investigate eco-friendly, cheap and humanly safe coatings/films for produce during postharvest storage (Ali et al., 2024c).

24.3 PROPERTIES OF EDIBLE COATINGS

The molecular structure of the coating matrix determines the various properties of an edible coating. Specific properties for edible coatings are described next. The advancement in coating development, especially based on polysaccharides, has significantly increased their efficiency and applicability with regards to the bulk of produce that can be coated (Olawuyi et al., 2021). Passive modified atmosphere created by an edible coating is the major characteristic of the coating material in extending the postharvest life of fresh FV. Sometimes high concentrations of CO_2 and low levels of O_2 have been associated with the increased incidence of certain disorders during storage. Therefore, such types of edible coating could be developed which should not deplete O_2 or accumulate excessive CO_2, as the presence of 1–3% of O_2 prevents the unwanted switch to anaerobic respiration from aerobic respiration (Dhall, 2013).

Similarly, in addition to gas permeability, a coating should also demonstrate reduced water vapor permeability. Moisture loss not only contributes to overall product loss but also deteriorates the appearance and cosmetic value of produce (Ali et al., 2022). The most important cause of postharvest losses is microbial infestation. Generally, harmful microbes reside on produce surfaces in the highest quantities. Therefore, a coating controlling the growth of these microbes is more efficient and desired than other pathogen-prone formulations (Abdollahzadeh et al., 2021; Yaashikaa et al., 2023). Mostly, a coating having the ability to carry antimicrobial agents is required. Similarly, the ability of a coating to carry other active agents, such as antioxidants, antibrowning agents and vitamins, is also preferred (Yaashikaa et al., 2023). The wettability of an edible coating is important in determining the functionality of active ingredients within the coating matrix. Generally, the water resistance of a coating maintains its integrity and intactness, which enables it to cover the produce properly. In addition to basic functions, a coating should improve the general appearance of the produce and also maintain its structural integrity during mechanical handling of the produce. The properties of non-stickiness and an efficient drying performance improve produce's appearance and handling. Generally, edible coatings are tasteless and odorless (Ali et al., 2024c). These do not interfere with the taste or aroma of fresh produce. During the commercial application of a coating on tons of a particular type of produce, ease of application is required; therefore, a coating formulation having low viscosity constitutes an efficient coating system and is preferred. Lastly, the overall coating procedure should be economically viable on a commercial scale (Ali et al., 2024c).

24.4 TYPES OF EDIBLE COATINGS

Edible coatings are categorized according to the matrix used to develop the coating. Coating matrices mostly consist of polysaccharides, proteins and lipids. The type of matrix and its concentration regulate the properties of a coating in terms of moisture permeability and gaseous exchange between the produce and its ambience. However, generally these biopolymers are combined, termed as a composite, by mixing or in different layers that provide the needed protection to a coated product. There are four types of coating matrices that are mostly used. These components are generally obtained from plants or a few other living organisms and are environmentally friendly and sustainable.

24.4.1 Edible Coatings Based on Polysaccharides

Various types of polysaccharide-based edible coatings have been reported in the literature for the preservation of freshness and eating quality of FV during storage. Most of the biopolymers have been extracted from aquatic or terrestrial animals and plants, particularly agricultural plants (Abdollahzadeh et al., 2021; Yaashikaa et al., 2023). Polysaccharides coating materials are considered hydrophilic in nature and provide a minimal barricade against moisture. Due to this nature, these coatings provide a gas barrier, thereby appropriately modifying the internal atmosphere and ultimately enhancing the storage life of the coated FV without developing severe anaerobic conditions. Nevertheless, the hydrophilic nature of polysaccharide-based coatings makes them a weak physical moisture barrier. However, their gas barrier properties, due to less permeability of O_2, helps polysaccharide-based coatings to preserve the quality of coated produce for a longer time without creating severe anaerobic conditions (Abdollahzadeh et al., 2021; Yaashikaa et al., 2023). Therefore, polysaccharide-based coatings have efficiently and effectively been used for coating various FV after harvest (Yousuf et al., 2018).

Different polymeric materials, such as cellulose, pectin, exudate gums, starches, dextrin, chitosan, carrageenan, alginate and gellan, have been grouped as polysaccharides (Yousuf et al., 2018). The coatings based on carrageenan and alginate have been used for FV preservation. These regulate various physico-chemical attributes, thereby ensuring better conservation owing to their higher tensile-based strength and reduced permeability to water vapours (Yousuf et al., 2018; Yaashikaa et al., 2023; Ali et al., 2021a, 2024b, 2024c). Gellan- and alginate-based coatings form insoluble gels or polymers and are also polysaccharides and possess good colloidal properties with certain cations such as calcium (Rojaz-Grau et al., 2008). Different types and antioxidative and antibrowning chemicals could be incorporated into these, which thereby reduce the surface browning and could be used to extend the shelf life of certain fresh-cut produce (Azarakhsh et al., 2012; Yousuf et al., 2018). Complex polysaccharides derived from fungi and bacteria, including xanthan, pullan, hyaluronic acid and curdlan, have also been gaining interest among researchers.

Chitosan has been recognized as one of the most efficient polysaccharides due to its properties, such as non-toxicity, safety and antimicrobial activity against potential storage microbiota. Several research articles have reported the application of chitosan-based coatings on fresh FV. Chitosan is also considered a biodegradable polysaccharide which has been found suitable for postharvest quality preservation and ensuring microbial safety of the coated produce (Dong et al., 2004). Saleem et al. (2021) reported improved postharvest quality of chitosan-coated strawberries during ambient storage. At the same time, it also has appropriate efficacy for preserving eating quality, suppressing chlorophyll catabolism, reducing oxidative stress and suppressing the cell wall disassembly either alone or combined with materials such as carboxymethyl cellulose (Ali et al., 2024b).

Pectins are a soluble compound of plant-based fiber obtained from certain cell wall materials of the plants. These have a poor barrier against moisture and thereby exhibit good potential for food materials with lower moisture content (Martinon et al., 2014). These are considered complex polysaccharides with high molecular weight which forms a highly efficient coating material. Generally, pectin used to develop an edible coating is extracted from food processing by-products. Pectin and its derivatives, including pectate and amidated pectin, form an edible coating that havs a good gas barrier and mechanical properties. Pectin-based edible coatings have been applied on apples, guavas, avocados, papayas, berries, tomatoes, carrots, apricots, chestnuts, melons, peaches and walnuts (Valdés et al., 2015). In addition, certain types of antibrowning compounds could be added into these coatings such as glutathione and N-acetylcysteine and was found to be effective for maintaining the quality of pears with suppressed microbial infestations (Oms-Oliu et al., 2008).

Cellulose is a naturally occurring biopolymer that is ubiquitous and abundant in nature. Its water solubility is increased by chemical derivation that yields carboxymethyl cellulose, methyl cellulose, hydroxypropyl methyl cellulose and/or hydroxypropyl cellulose. The coatings prepared from cellulose derivatives are gas permeable, tasteless, odorless, elastic, water soluble and form effective films around coated produce (Ali et al., 2023). Similarly, starch is a widely available polymeric carbohydrate. It is imperative that starch alone forms brittle films with inadequate mechanical properties such as tensile and flexural strength which need to be strengthened. Therefore, biodegradable plasticizers (glycerol) are blended into starch-based coatings to overcome the brittleness of starch. Rice starch reduced weight loss and respiration rate while maintaining total phenolics, flavonoids and antioxidants in coated plum fruits (Thakur et al., 2018). Similarly, modified and native starch-based coatings have been sourced from various materials and have been investigated and used as an edible coating. For instance, maize and cassava starches were investigated on pumpkin during their drying period (Yousuf et al., 2018).

The edible gums from certain plant exudates are also considered polysaccharides and have been found suitable to be used as coating materials. Various types of edible gums have been used for the postharvest quality preservation of fresh FV (Yousuf et al., 2018; Ali et al., 2023; Naveed et al., 2024). These could be applied alone or in combined form with other polysaccharide materials. In addition, the edible gums are considered nontoxic and ecofriendly in nature and applied on different FV during postharvest (Ali et al., 2021b; Shakir et al., 2022). Natural gums and mucilages from various agricultural plants have been derived and investigated as an efficient edible coating matrix. For example, guar gum (Dong & Wang, 2018), gum arabic (Alali et al., 2018), locust beam gum (Parafati et al., 2016), basil seed gum (Hashemi et al., 2017), gellan gum (Moreira et al., 2015), tragacanth gum (Ali et al., 2022) and almond gum (Rasool et al., 2023) have been tested on different FV.

Seaweed-derived alginates, extracted from brown seaweed, and carrageenans, extracted from red seaweed, have also been used to prepare edible coatings. For developing a thick and stable gel from sodium alginate, calcium ions are added to the alginate solution and form crosslinks between calcium ions and carboxylate ions of alginate guluronate units. The coating prepared from sodium alginate has shown to delay the ripening of stored FV such as plums (Valero et al., 2013), strawberries (Qamar et al., 2018) and sweet cherries (Diaz-Mula et al., 2012), as well as fresh-cut cantaloupe and strawberries (Senturk et al., 2019). Similarly, edible coatings based on carrageenan maintained the postharvest quality of papaya (Hamzah et al. 2013), prolonged the shelf life of Cavendish bananas (Dwivany et al., 2020) and reduced the decay incidence and physiological weight loss in strawberry fruits (Wani et al., 2021).

24.4.2 Edible Coatings Based on Proteins

Proteins are abundantly available as by-products from plants, such as wheat gluten, corn zein and proteins from soybeans, sunflowers and peanuts, whereas protein from animals includes whey protein, collagen, casein, gelatin and keratin. The most unique characteristics of proteins compared to polysaccharides and lipids are heat denaturation, electrostatic charges and amphiphilic character. Comparatively, protein-based coatings are better than coatings based on polysaccharides and lipids with regard to their physical, nutritional and functional activities (Iversen et al., 2022). Therefore, proteins demonstrate better coating/film-forming ability and adherence to hydrophilic surfaces. Although protein-based coatings exhibit good barrier properties to gases (O_2 and CO_2), they are poor in resisting moisture diffusion. For example, an *Idesia polycarpa* protein film-forming solution extended the shelf life of coated sweet cherries while also maintaining their colour and ascorbic acid content and reducing physiological weight loss (Yang et al., 2023). Similarly, Marelli et al. (2016) found that a silk fibroin coating slowed down fruit respiration rate, reduced fruit softening and inhibited the dehydration of stored strawberry fruits (at 22 °C), thus prolonging the fruit's freshness and marketability for up to 14 days. In an important finding reported by Liu et al. (2019), a fibroin coating increased the total soluble solids, decreased the respiration rate and mitigated chilling injury in banana fruits during storage. A nanocomposite coating based on soy protein isolate and nano-silicon oxide extended the shelf life of apple fruits by discharging active substances from the coating to the apple fruits (Liu et al., 2017).

Gelatin is commonly used in edible films, as it is highly economical, biocompatible and safe. It is an odor- and colourless water-soluble protein. Mannucci et al. (2017) demonstrated that a gelatin-based edible coating may extend the storage life of Fuji apples by slowing down the respiration rate and hence the ripening process. The addition of curcumin in the gelatin-based coating increased its antioxidant activity and antibacterial activity against *E. coli* infestation (Roy & Rhim, 2020).

Corn zein is the most important protein in corn that is insoluble in water while soluble in ethanol. Zein has an excellent film-forming ability and, therefore, may be applied as a biodegradable coating. An edible zein coating can be fabricated by drying its aqueous-ethanolic solution. Bai et al. (2003) developed a zein formulation to optimize the desirable modified atmosphere for apple fruit storage, which resulted in maintaining apple quality and extending its marketable life. Further, the application of zein has been found to significantly link with a lower grey mold severity, as noted in coated Cripps Pink apples (Magwebu et al., 2023). The protein albumin extracted from eggs can be utilized as an edible coating to maintain the quality of fresh produce (Poonia, 2018). Casein, derived from milk, is easy to process, highly stretchable and opaque in appearance. Casein, when applied in combination with turmeric on carrots retained colour, texture, carotenoid content and better overall acceptability (Villafañe, 2017).

Whey protein isolate is derived from cheese whey and has a high nutritional status with an adhesive coating–forming ability. Whey protein–based coatings are transparent and colourless, flexible, odorless and have low O_2 permeability. Soazo et al. (2015) coated strawberries with whey protein–based edible coating and suggested that whey protein coating-forming solutions can be used as an alternative to maintaining the quality attributes of rapidly frozen strawberries. Compared to uncoated fresh-cut broccoli, whey protein coatings retained a significantly higher amount of ascorbic acid, sulforaphane and total phenolics while showing lower weight loss (Elsayed et al., 2022). Wheat gluten, a water-insoluble protein of wheat and other cereals like rye and barley, is a mixture of polypeptide molecules. Chen et al. (2022) reported the preservative effect of edible coatings based on wheat gluten and lignocellulose nanofibres on bananas, grapes and persimmons. The authors reported that the coated bananas exhibited reduced weight loss, decay percentage and enzymatic browning. Soybeans have a higher protein content that varies from 38% to 44% than the protein in cereal grains (from 8% to 15%). Most of the soy proteins are water-insoluble but are soluble in diluted neutral salt solutions. Soy protein–based edible coatings have been found to improve the quality and storage life of fresh-cut apple slices during storage at 10 °C (Alves et al., 2017).

24.4.3 Edible Coatings Based on Lipids

Edible coatings based on lipids comprise acetic acid ester of monoglycerides, waxes, plant-based oils or minerals. Lipid-based coatings confer a shiny and glossy appearance to coated fresh produce. Lipid coatings, being hydrophobic, help to minimize the effect of O_2, water, light and humidity on produce quality during storage. Lipid-based coatings also decrease moisture loss by acting as a water barrier and mitigating chilling injury in certain produce during storage at low temperatures. There are various types of waxes used for coating different FV (Hassan et al., 2018; Milani & Nemati, 2022; Matloob et al., 2023). Nasirifar et al. (2018) studied the effect of a carnauba wax–based edible coating, with nontmorillonite nano-clay as an additive, on the storage life of blood oranges (at 7 °C). This wax coating improved antioxidant activity, titratable acidity, fruit firmness, total phenolics and the appearance (colour) of orange fruits. Similarly, paraffin wax reduced physiological weight loss and retained more marketable lime fruits, total soluble solids, titratable acidity, ascorbic acid and juice content than non-coated lime fruits (Bisen et al., 2012). Oregel-Zamudio et al. (2017) also highlighted the impact of a novel candelilla wax–based edible formulation containing *Bacillus subtilis HFC103* on extending strawberry storage life. Besides waxes, seed oils from castor, mustard and sesamum have been confirmed as efficient edible coatings (Bisen et al., 2012). Oil-based coatings have low permeability to water and act as a barrier between produce and the external environment, protecting the produce and reducing the postharvest losses of fresh FV, such as lime fruit (Bisen et al., 2012) and bell pepper (Panigrahi et al., 2018). The application of lipid-based emulsion coatings is a relatively recently developed concept. Nanoemulsion coatings have good water barrier properties and add a slight gloss as compared to other lipid-based coatings. The application of nanoemulsions has delayed epicarp maturation in avocados (Cenobio-Galindo et al., 2019); preserved moisture, colour, and firmness in okra (Gundewadi et al., 2018); and reduced postharvest decay in Fuji apples (Salvia-Trujillo et al., 2015).

24.4.4 Edible Coatings Based on Composites

Composite coatings have gained much interest from the research community in recent times. Composite coatings combine the strengths of different coating materials and fabricate a single more effective coating with significantly improved properties. These coatings are mostly heterogeneous in nature where a blend of polysaccharides, proteins and lipids may be formed. Composite coatings are applied as a single blended solution or successively layer-by-layer (multilayer coating) and have been reported in tomatoes (Shakir et al., 2022) and mangoes (Ali et al., 2024c) during storage.

A composite coating may consist of lipids (hydrophobic) and hydrocolloids (hydrophilic) combined to form a mixture or a layer-by-layer system (Tavassoli-Kafrani et al., 2016). A composite coating consisting of various polysaccharides, proteins and lipids combines the properties of each biopolymer to synergize their mechanical and barrier properties in the multicomponent coating system. Moreover, coatings based on emulsions are being formulated containing a lipid layer distributed uniformly within a biopolymer matrix. The number of biopolymers in a coating matrix determines whether it is a binary or ternary composite coating system. In these systems, numerous formulations of carbohydrate–carbohydrate, carbohydrate–lipid/protein and protein–protein have been investigated (Dhumal & Sarkar, 2018). Binary composite coating systems have been extensively studied as compared to ternary composite coating systems.

The fruits coated with composite edible coatings constituted from starch (extracted from banana pseudostem), chitosan and glycerol had significantly lower weight loss than uncoated fruits. Thus, the composite coating (starch-chitosan-glycerol) extended the shelf life of stored apples, mangoes and strawberries by potentially inhibiting microbial growth (Abera et al., 2023). Shakir et al. (2022) demonstrated that biocomposite coating developed with gum arabic and carboxymethyl cellulose delays senescence by increasing the accumulation of antioxidants and inhibiting cell wall degradation. Ali et al. (2024a) also showed that chitosan–chia seed mucilage-coated tomatoes had the lowest rates of respiration and ethylene production, thus maintaining the fruit quality of stored tomatoes.

Chitosan, along with carboxymethyl cellulose, in a layer-by-layer system lowered decay rate, weight loss and carotenoid accumulation and delayed fruit ripening and firmness loss in stored mangoes (Ali et al., 2024b). Similar findings have been reported in *Aloe vera* gel–gum arabic or *Aloe vera* gel–tragacanth gum (Saleem et al., 2023), chitosan–carrageenan (Nguyen et al., 2021), carboxymethyl chitosan–gelatin (Zhang et al., 2021), zein–chitosan nanowhiskers (Almeida et al., 2023) and cassava starch–gelatin–chitosan coatings (Silva et al., 2021).

24.4.5 Edible Coatings Based on Plant Extracts and Mucilages

Organic polymers are becoming famous and new coating materials are being developed. Java apples were treated under different levels of aloe vera gel as an edible coating to conserve their quality attributes (Mubarak & Engakanah, 2017). It was concluded that aloe vera gel at 100% concentration gave the most satisfying results in reducing weight loss and fruit softness. Extract of caldodes was used on prickly pear fruit, which not only reduced the weight loss but also delayed fruit softening issues. Metabolically coating conserved bioactive compounds, such as amino acids and sugars, thus preserving the quality of the fruit (Allegra et al., 2018). Robles-Flores et al. (2018) introduced a new material from pigeon pea seeds protein and used with a combination of a gum on strawberry fruits. Authors examined that coating markedly reduced the loss in weight, acidity and soluble solids. Additionally, no changes were observed in the organoleptic evaluation scores. Alginate was used in combination with Helianthus (sunflower) oil with the lactobacillus on sliced carrots. Using such a combination resulted in good water retention and conservation of colour and nutrition. The addition of acidophilus species in coatings reduced metabolic activity in sliced carrots (Shigematsu et al., 2018). Coating combination of three components of sodium-alginate, chitosan and different food fibres like apple, orange, oligo-fructose and inulin were investigated on blueberry fruit. A combination containing chitosan and fibre gave better quality, freshness and storage life compared with the additive of probiotics (Alvarez et al., 2018). Similarly, the impact of certain types of plant-based gels, such as *Aloe vera* gel (Qamar et al., 2018) and mucilage extract based on *Opuntia ficus indica* (Shinga & Fawole, 2023), have been investigated on strawberries and bananas, respectively.

24.4.6 Edible Coatings Made from Fruit and Vegetable Waste Materials

In the last few years, various purees of produce, flesh and extracts have been applied to develop edible films/coatings. The constituents being added act as a source of various active compounds and nutritive components and help in the improvement of the developed coatings (Gutiérrez, 2017b). Moreover, enormous waste material produced in the processing of FV is less valuable due to low demand, but it can be a meaningful addition to edible coatings. However, the main objective is to reduce the postharvest loss of produce by using novel raw materials produced in bulk for the development of consumable coatings. For this perspective, Torres-León et al. (2018) used mango skin and seeds to develop an ecofriendly coating for peach fruit. Mango fruit treated with the coating showed preserved texture, colour, absorbency, moisture-repellent capability and antioxidants. The coating containing mango by-product extract exhibited less CO_2 and ethylene emission along with reduced O_2 utilization compared to untreated peaches. This experimentation showed the capacity of the mango waste to be used as a biodegradable material for the extended storage life of peach fruit. The pulp of *Rubus fruticosus* was used as an additive material in *Maranta arundinacea* starch to make a smooth and stretchy coating. The prepared film preserved better aroma, colour, active metabolites and antioxidants that make it attractive to be used as a coating material on fresh produce (Nogueira et al., 2019). Rangel-Marrón et al. (2019) prepared a biodegradable, non-toxic packaging of alginate with papaya puree, glycerol and citrate. It is recommended that this ecofriendly packaging can be used for wrapping sushi or sliced horticultural produce with the preservation of a suitable colour. A double layer coating of an antioxidant nature developed from soybean production wastes has also been reported (Tkaczewska et al., 2023). In addition to wastes, edible coatings enriched with extracts of tropical fruits have also been reported with additional benefits compared to traditional coating types (Kupervaser et al., 2023).

24.5 EDIBLE COATING APPLICATION ADVANCES FOR FRESH PRODUCE

Biopolymers as a single constituent have been applied in the past but are now being applied in the form of composites, i.e. pectin and sodium alginate, to conserve the quality of minimally processed mango (Silva et al., 2018). Chitosan performed better in controlling the postharvest issues of fresh commodities either alone or in combination and gave promising results by preserving the quality parameters and protecting the coated produce against microbial attack (Gutiérrez, 2017a; Paul et al., 2018). Alginate-chitosan composite coating on fresh-cut melon significantly minimized the rate of gaseous exchange and ethylene gas release after a storage period of 13 days at 5 °C with maximum quality as compared to controls (Ortiz-Duarte et al., 2019). Oranges were preserved by treatment with chitosan in combination with pomegranate peel by-product and guar gum to control green mould. Results showed that the antifungal and bio-agent combination was helpful in controlling the postharvest deterioration of the fruits (Kharchoui et al., 2018). Layer-by-layer coating of chitosan and carboxymethyl cellulose on strawberry fruits reduced the loss of fruit firmness and preserved the organoleptic attributes. It was found that untreated fruits had maximum loss in primary metabolite (flavonoids, antioxidants, fatty acids) compared to treated strawberries (Yan et al., 2018).

The use of emulsification in edible coatings has more potential to hinder moisture absorbance/loss of FV quality. It is imperative to report that both nanoemulsions and coarse-emulsions have been found effective as carriers of target substances delivery, such as antimicrobials, flavouring agents or nutraceutical attributes, on the targeted produce (Galus & Kadzińska, 2015). Arnon-Rips et al. (2019) studied the effect of citral in nano- and coarse-emulsions with sunflower oil in combination with chitosan and carboxymethyl cellulose on melon. The prepared coating with nanoemulsion was consistent, so it acted as a potential obstruction to moisture and was mechanically resistant. Using nanoemulsions as a coating gave the melons a more protective shield against microbial activities and enhanced overall appearance with extended shelf life. However, sensory evaluations revealed that there was a presence of citrus-oil aroma which was not accepted by consumers, suggesting not to use this combination for citrus fruits. Starch, with the addition of k-carrageenan gel and sucrose esters, increased the storage life of bananas by 12 days in comparison with the untreated group (Thakur et al., 2019). For the last 10 years, there has been a trend of certain types of coatings from natural resources on food items, but there is still a need to study such materials at the commercial scale. For example, E-392 (European Commission, 2010a, 2010b) is a food preservative originated from rosemary (contains antioxidative properties) that resulted after vast researches. On the other side, advances in the form of application methods as double and/or multilayers have also been reported in edible coatings (Ali et al., 2024b). So, such types of advances for searching for the most appropriate and effective coatings for FV should be further researched to achieve the sustainable goal of global food security.

24.6 APPLICATION METHODS AND INNOVATIONS IN THE USE OF DIFFERENT APPROACHES COMBINED WITH EDIBLE COATINGS FOR FRUIT AND VEGETABLE STORAGE

There are various approaches to using edible coatings, but soaking and dipping are the most common. However, the use of vacuum pressure impregnation (VPI) is an alternative to the dipping method which was used in the case of an edible coating formulated with chitosan and chitosan+lauric-acid on pumpkin pieces. Results indicated that the VPI technique made coating effective and denser; however, parameters such as acidity, firmness and juice pH showed faster degradation compared to normal coating application methods (Soares et al., 2018). It is necessary to increase the shelf life of commodities due to current prevailing conditions like a growing population, waste generation (from lost FV), low production and market pressure. Current studies have been concerned with the development of novel recipes or blending to give a protective layer to FV. Concerning this perspective, El-Dein et al. (2018) developed a co-polymer coating of chitosan, polyvinyl alcohol and tannic acid compounds on dates to enhance their storage life. Film-developing constituents were treated with γ-rays prior to incorporation of the coating. Results indicated that the use of γ-rays and the addition of tannic acid extended the shelf life of dates from a week to a month without compromising the freshness and quality parameters. In another work, Muñoz-Labrador et al. (2018) investigated the impact of citrus-pectin which was prepared with the help of ultrasonication and incorporated on strawberry fruit; the treatment resulted in improved flavour and colour of strawberries along with suppressed weight loss compared to untreated fruits. The effect of coating constituents and hot air on the drying mechanism and properties of dehydrated red guava was studied. A coating containing pectin was applied to slices of guava prior to the dehydrating procedure. It was observed that the coating increased the phenolics and carotenoids content of the fruits and maintained good flavour and taste (Todisco et al., 2018). The effect of both alginate and light exposure and malic-acid with alginate were studied on pieces of mango fruits. The treated slices showed minimum browning and maximum reactive oxygen species production, showing that coating in combination can maintain nutritional attributes better for 14 days in storage (Salinas-Roca et al., 2018). A chitosan coating may reduce deterioration either due to the suppression of biotic or abiotic factors, as noted on kiwifruits slices which were treated with chitosan in combination with ultrasonication and sodium-hypochlorite separately that prolonged the storage life for 10 days in contrast with other treatments.

24.7 MAJOR CHALLENGES AND SAFETY STATUS OF EDIBLE COATINGS

It is well established that edible coatings generally enhance the safety, appearance, glassines and quality of the coated produce. Despite the potentials offered by the edible coatings by this innovative postharvest technology, several challenges which preclude their commercial application and feasibility are still limited in general (Yaashikaa et al., 2023). It is important that as the majority of the edible coatings are extracted from sustainable and natural sources, they are often used in coating materials and have been accompanied by objectionable colour and flavour, which thereby sometimes renders the coated produce unpalatable (Yaashikaa et al., 2023). The separation, isolation and subsequent characterization of certain constituents extracted from natural resources is quite expensive and exhibits fluctuations in the overall quality on the basis of batches extracted and impedes their application at the commercial level. In addition, some of the extracted compounds may exhibit potential toxicity and compromising safety of edible coatings as the coated produce is consumed by the consumers (Yaashikaa et al., 2023). The added surfactants, plasticizers and fillers may sometimes result in the development of reactions which may lead to the formation of potential toxicants which are considered unhealthy and toxic to consumers. The major hindrance which researchers face is the true identification of the ideal/most effective combinations of the biopolymers and potential additives which meets the packaging standards and safety of the intended commodities (Costa et al., 2018; Anis et al., 2021; Yaashikaa et al., 2023). The coatings should conform to particular guidelines about the use of certain food additives, as the coating may become a critical aspect of the edible fractions of the coated food product. The additives and chemicals used should be Generally Recognized as Safe and must completely be added at recommended levels to ensure the safety regulations of FV (Yaashikaa et al., 2023). Several coatings are based on casein, gluten, soy protein, whey protein and peanut protein in their formulations. The protein-derived components are declared allergens, as these substantially trigger allergic reactions in some people. Consequently, it is indispensable to specify the uses of such types of the allergic components in edible coating fabrication (Carpena et al., 2021; Yaashikaa et al., 2023).

24.8 CONCLUSIONS

In conclusion, edible coatings based on proteins, polysaccharides, lipids and/or composites have been significantly found to preserve fresh produce quality either alone or in combination. The coatings provide protection and preserve FV by modifying their internal atmosphere either by inhibiting oxygen uptake or due to the build-up of internal carbon dioxide, which thus ultimately delay quality changes and suppress the senescence of fresh produce. The major advancements regarding desirable formulations, extraction techniques, application methods and the addition of different types of antimicrobial agents in edible coatings have been found effective for fresh produce postharvest preservation. Future work should focus on new edible coatings, improved formulations, stability, functional ingredient addition and the development of multilayer-based combinations for fresh produce.

TABLE 24.1 The Effects of Various Types of Edible Coatings on the Preservation of Postharvest Quality Attributes of Fresh Fruits and Vegetables

PRODUCE	*COATING MATERIALS*	*FUNCTIONAL INGREDIENTS*	*INFERENCES*	*REFERENCES*
Strawberry	Chitosan, pigeon pea protein and gum	Shrimp protein, Roselle extract and cinnamon oil	Antifungal and antimicrobial. Maintained fruit weight by suppressing desiccation. Preserved the antioxidants, titratable acidity and fruit turgidity in general.	Hajji et al. (2018); Ventura-Aguilar et al. (2018); Robles-Flores et al. (2018)
Apple	CMC and gelatin	AsA, Aloe vera gel, $CaCl_2$, Cysteine, tea-extract	Acted as antimicrobial. Sustained colour and reduced browning and softening.	Kumar et al. (2018) Amiri et al. (2018)
Banana	Carrageenan and rice starch	–	Reduced fruit senescence and weight loss. Preserved freshness and overall appearance.	Thakur et al. (2019)
Melon	CMC and chitosan	Citral	Extended shelf life, preserved appearance and revealed antimicrobial activity	Ortiz-Duarte et al. (2019)
Bilberry cactus	Gelatin	Oily fluid	Declined TSS, suppressed weight loss and ripening along with stabilized turgidity in general	López-Palestina et al. (2018)
Apricot	Chitosan and soy protein	–	Decreased weight loss and pectin decomposition. Stabilized texture and turgidity status	Zhang et al. (2018)
Eggplant	Soy protein isolate	Cysteine	Inhibited weight loss and enzymatic browning	Ghidelli et al. (2018)
Tomato	Chitosan and cassava-starch	Oil, pomegranate peel extract	Reduced prompt senescence and weight loss, better quality	Araújo et al. (2018)
Carrots	Sunflower oil, sodium-alginate	*L. acidophilus*	Overall metabolism was decreased, colour remained intact and weight loss was minimized	Shigematsu et al. (2018)
Dates	PVA	Tannic acid	Enhanced the storage life of fresh fruit	El-Dein et al. (2018)

(Continued)

TABLE 24.1 (Continued)

PRODUCE	*COATING MATERIALS*	*FUNCTIONAL INGREDIENTS*	*INFERENCES*	*REFERENCES*
Kiwifruit	sodium-alginate	ε-Polylysine	Preserved the chlorophyll content, AsA, antioxidant activity and visual appearance	Li et al. (2017)
Cherry tomato	Chitosan	Grapefruit seed extract	Preserved fresh weight by suppressing desiccation and acted as antimicrobial	Won et al. (2018)
Guava	Gum arabic and casein-sodium	Lemongrass and cinnamon oil	Conserved the quality compounds and prolonged shelf life period	Murmu and Mishra (2018)
Capsicum	Sodium-alginate and chitosan	Peel extract of pomegranate	Reduced weight loss and microbial invasion. Conserved sensory attributes of coated groups	Sneha Nair et al. (2018)
Mandarin	Persian gum and gum arablic along with CMC	Carnauba and bee wax	Preserved the fruit's fresh weight and biochemical attributes	Khorram et al. (2017)
Avocado	CMC	Moringa leaf extract	Extended shelf life, preserved appearance and exhibited appropriate antimicrobial activity	Ali et al. (2021a)
Persimmon	Tragacanth gum	–	Decreased softening and preserved the bioactive compounds	Saleem et al. (2022)
Plum	Pectin	–	Higher antioxidants and ascorbic acid along with lower polyphenols	Panahirad et al. (2020)
Pear	Chitosan, chitooligosaccharide	–	Helped in wound healing and prolonged shelf life	Yu et al. (2022)
Tangerine	Alginate	Thyme oil	Conserved the organoleptic properties and extended the shelf life	Radi et al. (2023)
Aonla	CMC	–	Conserved juice pH, inhibited TSS increase and showed higher enzymes activity	Ali et al. (2024c)
Guava	Chitosan and alginate	ZnO	Prevented rotting and shelf life was increased	Arroyo et al. (2020)
Mango	Pirus banana starch	–	Lower weight loss and preserved flesh firmness	Abhirami et al. (2020)
Mandarin	Quince seed- mucilage	–	Shelf life was increased along with preserved metabolites	Kozlu and Elmacı (2020)
Pomegranate	Chitosan	Malic acid and oxalic acid	Reduced chilling injury and browning, maintained general eating quality	Ehteshami et al. (2020)
Litchi	Chitosan, pullulan	–	Higher phenolic content, improved shelf life and colour persistency	Kumar et al. (2020)
Mango	Aloe vera and chitosan	–	Discoloration was delayed and PPO activity decreased along with better quality attributes	Hajebi Seyed et al. (2021)
Banana	Aloe vera gel	Lemon peel extract	Decreased the microbial attack and maintained fresh weight, suppressed desiccation and lowered sugar accumulation	Jodhani and Nataraj (2021)
Starfruit	Maltodextrin and pectin	NaCl	Prevented entry of microbes and preserved higher TSS and fruits firmness	Suhaimi et al. (2021)

PRODUCE	COATING MATERIALS	FUNCTIONAL INGREDIENTS	INFERENCES	REFERENCES
Grape	Chitosan and gum ghatti	–	Reduced shattering of berries and suppressed browning, conserved quality	Eshghi et al. (2021)
Guava	Mango seed starch	Lemongrass oil	Conserved physiochemical properties along with reduced microbial attack	Yadav et al. (2022)
Broccoli	Chitosan and tea polyphenol	–	Reduced the curd yellowing and conserved nutrition-related attributes	Fang et al. (2022)
Pear	Pectin	Acetyl cysteine	Minimized browning and increased shelf life with preserved sensory aspects	Pleşoianu and Nour (2022)
Papaya	Carnauba-wax, *Cymbopogon martinii* oil	–	Delayed softening, loss of taste and colour changes, prolonged shelf life	Oliveira Filho et al. (2022)
Strawberry	Chitosan and Aloe vera gel	–	Reduced softening of berries, inhibited weight loss and slowed colour changes	Farida et al. (2023)
Mandarin	Chitosan and hydroxypropyl-methylcellulose	Bean gum	Preserved antioxidants, organic acids and taste	Jurić et al. (2023)
Strawberry	Agarwood bouya oil, yam bean starch	–	Increased storage life and preserved biochemical attributes	Wigati et al. (2023)
Cherry	*Thymus vulgaris*, *Satureja montana* extract	–	Preserved organoleptic properties and antioxidant activities, extended storage life	Afonso et al. (2023)
Grape	Lotus root starch	–	Exhibited suitable antifungal action, increased storage life and preserved concentrations of antioxidants	Zeng et al. (2024)
Mandarin	Chitosan and guar gum	Clove, cinnamon and coconut oils	Reduced postharvest decay, exhibited better firmness and inhibited rapid weight loss	Goswami et al. (2024)
Tomato	Chitosan and chia mucilage	Levan	Phenolic contents were increased and lower ethylene production was noted, which thereby reduced weight loss	Ali et al. (2024a)
Pear	$CaCl_2$ and Aloe vera gel	–	Reduced respiration rate and weight loss percentage, having improved and preserved enzyme activities of a antioxidative nature, which thus prolonged its shelf life	Ahmad et al. (2024)

AsA = Ascorbic acid, $CaCl_2$ = Calcium chloride, CMC = Carboxymethyl cellulose, NaCl = Sodium chloride, PPO = Polyphenol oxidase, TSS = Total soluble solids.

REFERENCES

Abdollahzadeh, E., Nematollahi, A., Hosseini, H. (2021). Composition of antimicrobial edible films and methods for assessing their antimicrobial activity: A review. *Trends in Food Science & Technology*, *110*, 291–303.

Abera, B., Duraisamy, R., Birhanu, T. (2023). Study on the preparation and use of edible coating of fish scale chitosan and glycerol blended banana pseudostem starch for the preservation of apples, mangoes, and strawberries. *Journal of Agriculture and Food Research*, 100916.

Abhirami, P., Modupalli, N., Natarajan, V. (2020). Novel postharvest intervention using rice bran wax edible coating for shelf-life enhancement of *Solanum lycopersicum* fruit. *Journal of Food Processing and Preservation*, *44*(12), e14989.

Afonso, S., Oliveira, I., Ribeiro, C., Vilela, A., Meyer, A. S., Gonçalves, B. (2023). Innovative edible coatings for postharvest storage of sweet cherries. *Scientia Horticulturae*, *310*, 111738.

Ahmad, F., Muhammad, A., Hashmi, M. S., Ahmad, A., Alam, S., Din, K. U., Siyab, A. (2024). Pre-storage calcium chloride and aloe vera gel coatings mitigate internal browning and senescence scald in 'Conference' pears. *Scientia Horticulturae*, *325*, 112684.

Alali, A. A., Awad, M. A., Al-Qurashi, A. D., Mohamed, S. A. (2018). Postharvest gum Arabic and salicylic acid dipping affect quality and biochemical changes of 'Grand Nain' bananas during shelf life. *Scientia Horticulturae*, *237*, 51–58.

Ali, Q., Kurubas, M. S., Mujtaba, M., Nazari, A. W., Dogan, A., Kaya, M., Oner, E. T., Yilmaz, B. A. Erkan, M. (2024a). Shelf life of cocktail tomato extended with chitosan, chia mucilage and levan. *Scientia Horticulturae*, *323*, 112500.

Ali, S., Anjum, M. A., Ejaz, S., Hussain, S., Ercisli, S., Saleem, M. S., Sardar, H. (2021a). Carboxymethyl cellulose coating delays chilling injury development and maintains eating quality of 'Kinnow' mandarin fruits during low temperature storage. *International Journal of Biological Macromolecules*, *168*, 77–85.

Ali, S., Ishtiaq, S., Nawaz, A., Naz, S., Ejaz, S., Haider, M. W., Shah, A. A., Ali, M. M. Javad, S. (2024b). Layer by layer application of chitosan and carboxymethyl cellulose coatings delays ripening of mango fruit by suppressing cell wall polysaccharides disassembly. *International Journal of Biological Macromolecules*, 128429.

Ali, S., Akbar Anjum, M., Nawaz, A., Naz, S., Ejaz, S., Shahzad Saleem, M., . . . & Ul Hasan, M. (2021b). Effect of gum arabic coating on antioxidative enzyme activities and quality of apricot (*Prunus armeniaca* L.) fruit during ambient storage. *Journal of Food Biochemistry*, *45*(4), e13656.

Ali, S., Nawaz, A., Hussain, B., Ejaz, S., Sardar, H. (2024c). Carboxymethyl cellulose coating maintains quality of harvested aonla fruit by regulating oxidative stress and ascorbate-glutathione cycle. *Postharvest Biology and Technology*, *207*, 112621.

Ali, S., Khan, A. S., Nawaz, A., Naz, S., Ejaz, S., Shah, A. A., Haider, M. W. (2023). The combined application of Arabic gum coating and γ-aminobutyric acid mitigates chilling injury and maintains eating quality of 'Kinnow' mandarin fruits. *International Journal of Biological Macromolecules*, *236*, 123966.

Ali, S., Zahid, N., Nawaz, A., Naz, S., Ejaz, S., Ullah, S., Siddiq, B. (2022). Tragacanth gum coating suppresses the disassembly of cell wall polysaccharides and delays softening of harvested mango (*Mangifera indica* L.) fruit. *International Journal of Biological Macromolecules*, *222*, 521–532.

Allegra, A., Gallotta, A., Carimi, F., Mercati, F., Inglese, P., Martinelli, F. (2018). Metabolic proiling and post-harvest behavior of "Dottato" Fig (*Ficus carica* L.) fruit covered with an edible coating from *O. ficus-indica*. *Frontiers in Plant Science*, *9*(1321). https://doi.org/10.3389/fpls.2018.01321.

Almeida, C. L., Figueiredo, L. R., Ribeiro, D. V., Santos, A. M., Souza, E. L., Oliveira, K. A., Medeiros, E. S. (2023). Antifungal edible coatings for fruits based on zein and chitosan nanowhiskers. *Journal of Food Science*. https://doi.org/10.1111/1750–3841.16831.

Álvarez, K., Famá, L., Gutiérrez, T.J. (2017). Chapter 12. Physicochemical, antimicrobial and mechanical properties of thermoplastic materials based on biopolymers with application in the food industry. In M. Masuelli & D. Renard (Eds.), Advances in physicochemical properties of biopolymers: Part 1 (pp. 358–400). Bentham Science Publishers. https://doi.org/10.2174/97816810845341170100l5.

Alvarez, M.V., Ponce, A.G., Moreira, M.R. (2018). Influence of polysaccharide-based edible coatings as carriers of prebiotic fibers on quality attributes of ready-to-eat fresh blueberries. *Journal of the Science of Food and Agriculture, 98*(7), 2587–2597. https://doi.org/10.1002/jsfa.8751.

Alves, M. M., Gonçalves, M. P., Rocha, C. M. (2017). Effect of ferulic acid on the performance of soy protein isolate-based edible coatings applied to fresh-cut apples. *LWT*, *80*, 409–415.

Amiri, S., Akhavan, H.R., Zare, N., Radi, M. (2018). Effect of gelatin-based edible coatings incorporated with aloe vera and green tea extracts on the shelf-life of fresh-cut apple. *Italian Journal of Food Science*, *30*(1), 61–74. https://doi.org/10.14674/1120-1770-ijfs699.

Anis, A., Pal, K., Al-Zahrani, S. M. (2021). Essential oil-containing polysaccharide-based edible films and coatings for food security applications. *Polymers*, *13*(4), 575. https://doi.org/10.3390/polym13040575

Ansorena, M. R., Pereda, M., Marcovich, N. E. (2018). Edible ilms. In T. J. Gutiérrez (Ed.), Polymers for food applications (pp. 5–24). Cham: Springer. https://doi.org/10.1007/978-3-319-94625-2_2.

Araújo, J.M. S., de Siqueira, A.C. P., Blank, A.F., Narain, N., de Aquino Santana, L.C. L. (2018). A cassava starch–chitosan edible coating enriched with *Lippia sidoides* Cham. Essential oil and pomegranate peel extract for preservation of italian tomatoes (*Lycopersicon esculentum* Mill.) stored at room temperature. *Food and Bioprocess Technology*, *11*(9), 1750–1760. https://doi.org/10.1007/s11947-018-2139-9.

Arnon-Rips, H., Porat, R., Poverenov, E. (2019). Enhancement of agricultural produce quality and storability using citral based edible coatings; the valuable effect of nano-emulsiication in a solid state delivery on fresh-cut melons model. *Food Chemistry*, *277*, 205–212. https://doi.org/10.1016/j.foodchem.2018.10.117.

Arroyo, B. J., Bezerra, A. C., Oliveira, L. L., Arroyo, S. J., de Melo, E. A., Santos, A. M. P. (2020). Antimicrobial active edible coating of alginate and chitosan add ZnO nanoparticles applied in guavas (*Psidium guajava* L.). *Food Chemistry*, *309*, 125566.

Azarakhsh, N., Osman, A., Ghazali, H. M., Tan, C. P., Mohd Adzahan, N. (2012). Optimization of alginate and gellan-based edible coating formulations for fresh-cut pineapples. *International Food Research Journal*, *19*(1).

Bai, J., Alleyne, V., Hagenmaier, R. D., Mattheis, J. P., Baldwin, E. A. (2003). Formulation of zein coatings for apples (*Malus domestica* Borkh). *Postharvest Biology and Technology*, *28*(2), 259–268.

Bisen, A., Pandey, S. K., Patel, N. (2012). Effect of skin coatings on prolonging shelf life of kagzi lime fruits (*Citrus aurantifolia* Swingle). *Journal of Food Science and Technology*, *49*(6), 753–759.

Bracone, M., Merino, D., González, J., Alvarez, V.A., Gutiérrez, T. J. (2016). Chapter 6. Nanopackaging from natural illers and biopolymers for the development of active and intelligent ilms. In S. Ikram & S. Ahmed (Eds.), Natural polymers: Derivatives, blends and composites (pp. 119–155). New York: Editorial Nova Science Publishers, Inc.

Carpena, M., Nuñez-Estevez, B., Soria-Lopez, A., Garcia-Oliveira, P., Prieto, M. A. (2021). Essential oils and their application on active packaging systems: A review. *Resources*, *10*(1), 7. https://doi.org/10.3390/resources10010007

Cenobio-Galindo, A. D. J., Ocampo-López, J., Reyes-Munguía, A., Carrillo-Inungaray, M. L., Cawood, M., Medina-Pérez, G., Fernández-Luqueño, F., Campos-Montiel, R. G. (2019). Influence of Bioactive Compounds incorporated in a nanoemulsion as coating on avocado fruits (*Persea americana*) during postharvest storage: Antioxidant activity, physicochemical changes and structural evaluation. *Antioxidants*, *8*(10), 500.

Chen, Y., Li, Y., Qin, S., Han, S., Qi, H. (2022). Antimicrobial, UV blocking, water-resistant and degradable coatings and packaging films based on wheat gluten and lignocellulose for food preservation. *Composite Part B, 238*, 109868.

Collazo-Bigliardi, S., Ortega-Toro, R., Chiralt, A. (2018). Properties of micro- and nano-reinforced biopolymers for food applications. In T.J. Gutiérrez (Ed.), Polymers for food applications (pp. 61–99). Cham: Springer. https://doi.org/10.1007/978-3-319-94625-2_4.

Costa, M. J., Maciel, L. C., Teixeira, J. A., Vicente, A. A., Cerqueira, M. A. (2018). Use of edible films and coatings in cheese preservation: Opportunities and challenges. *Food Research International*, *107*, 84–92.

Dhall, R.K. (2013). Advances in edible coatings for fresh fruits and vegetables: a review. *Critical Reviews in Food Science and Nutrition*, *53*(5), 435–450.

Dhumal, C. V., Sarkar, P. (2018). Composite edible films and coatings from food-grade biopolymers. *Journal of Food Science and Technology*, *55*, 4369–4383.

Díaz-Mula, H. M., Serrano, M., Valero, D. (2012). Alginate coatings preserve fruit quality and bioactive compounds during storage of sweet cherry fruit. *Food and Bioprocess Technology*, *5*, 2990–2997.

Dong, F., Wang, X. (2018). Guar gum and ginseng extract coatings maintain the quality of sweet cherry. *LWT-Food Science and Technology*, *89*, 117–122.

Dong, H., Cheng, L., Tan, J., Zheng, K., Jiang, Y. (2004). Effects of chitosan coating on quality and shelf life of peeled litchi fruit. *Journal of Food Engineering*, *64*(3), 355–358.

Dwivany, F. M., Aprilyandi, A. N., Suendo, V., Sukriandi, N. (2020). Carrageenan edible coating application prolongs Cavendish banana shelf life. *International Journal of Food Science*, *2020*, 8861610.

Ehteshami, S., Abdollahi, F., Ramezanian, A., Mirzaalian Dastjerdi, A. (2020). Maintenance of quality and bioactive compounds in pomegranate fruit (*Punica granatum* L.) by combined application of organic acids and chitosan edible coating. *Journal of Food Biochemistry*, *44*(9), e13393.

El-Dein, A.E., Khozemy, E.E., Farag, S.A., El-Hamed, N.A., Dosoukey, I.M. (2018). Effect of edible co-polymers coatings using γ-irradiation on Hyani date fruit behavior during marketing. *International Journal of Biological Macromolecules*, *117*, 851–857. https://doi.org/10.1016/j.ijbiomac.2018.05.170.

Elsayed, N., Hassan, A. A. M., Abdelaziz, S. M., Abdeldaym, E. A., Darwish, O. S. (2022). Effect of Whey Protein Edible Coating Incorporated with Mango Peel Extract on Postharvest Quality, Bioactive Compounds and Shelf Life of Broccoli. *Horticulturae*, *8*(9), 770.

Eshghi, S., Karimi, R., Shiri, A., Karami, M., Moradi, M. (2021). The novel edible coating based on chitosan and gum ghatti to improve the quality and safety of 'Rishbaba'table grape during cold storage. *Journal of Food Measurement and Characterization*, *15*(4), 3683–3693.

European Commission. (2010a). Commission directive 2010/67/EU of 22 October. Oficial Journal of the European Union, L277, 17–26.

European Commission. (2010b). Commission Directive 2010/69/EU of 22 October 2010. Oficial Journal of the European Union, L279, 22–31.

Falguera, V., Quintero, J.P., Jiménez, A., Muñoz, A., Ibarz, A. (2011). Edible ilms and coatings: Structures, active functions and trends in their use. *Trends Food Science and Technology*, *22*(6), 291–303. https://doi.org/10.1016/j.tifs.2011.02.004.

Fang, H., Zhou, Q., Yang, Q., Zhou, X., Cheng, S., Wei, B., Ji, S. (2022). Influence of combined edible coating with chitosan and tea polyphenol on the quality deterioration and health-promoting compounds in harvested broccoli. *Food and Bioprocess Technology*, *15*(2), 407–420.

Farida, F., Hamdani, J. S., Mubarok, S., Akutsu, M., Noviyanti, K., Nur Rahmat, B. P. (2023). Variability of Strawberry Fruit Quality and Shelf Life with Different Edible Coatings. *Horticulturae*, *9*(7), 741.

Galus, S., Kadzińska, J. (2015). Food applications of emulsion-based edible ilms and coatings. *Trends in Food Science and Technology*, *45*(2), 273–283. https://doi.org/10.1016/j.tifs.2015.07.011.

Garrido, T., Uranga, J., Guerrero, P., de la Caba, K. (2018) The potential of vegetal and animal proteins to develop more sustainable food packaging. In: Gutiérrez T.J. (Ed.). Polymers for food applications. Springer, Cham. pp 25–59. https://doi.org/10.1007/978-3-319-94625-2_3.

Ghidelli, C., Sanchís, E., Rojas-Argudo, C., Pérez-Gago, M.B., Mateos, M. (2018). Controlling enzymatic browning of fresh-cut eggplant by application of an edible coating and modified atmosphere packaging. *Acta Horticulturae*, *1209*, 239–245. https://doi.org/10.17660/ actahortic.2018.1209.34.

Goswami, M., Mondal, K., Prasannavenkadesan, V., Bodana, V., Katiyar, V. (2024). Effect of guar gum-chitosan composites edible coating functionalized with essential oils on the postharvest shelf life of Khasi mandarin at ambient condition. *International Journal of Biological Macromolecules*, *254*, 127489.

Gundewadi, G., Rudra, S. G., Sarkar, D. J., Singh, D. (2018). Nanoemulsion based alginate organic coating for shelf life extension of okra. *Food Packaging and Shelf Life*, *18*, 1–12.

Gutiérrez, T.J. (2017a). Chapter 8. Chitosan applications for the food industry. In S. Ahmed & S. Ikram (Eds.), Chitosan: Derivatives, composites and applications (pp. 183–232). New York: WILEY-Scrivener Publisher. https://doi.org/10.1002/9781119364849.ch8.

Gutiérrez, T.J. (2017b). Surface and nutraceutical properties of edible films made from starchy sources with and without added blackberry pulp. *Carbohydrate Polymers*, *165*, 169–179. https://doi.org/10.1016/j.carbpol.2017.02.016.

Gutiérrez, T.J. (2018a). Polymers for food applications: News. In T.J. Gutiérrez (Ed.), Polymers for food applications (pp. 1–4). Cham: Springer. https://doi.org/10.1007/978-3-319-94625-2_1.

Gutiérrez, T. J. (2018b). Active and intelligent films made from starchy sources/blackberry pulp. *Journal of Polymers and the Environment, 26*(6), 2374–2391. https://doi.org/10.1007/s10924-017-1134-y.

Hajebi Seyed, R., Rastegar, S., Faramarzi, S. (2021). Impact of edible coating derived from a combination of Aloe vera gel, chitosan and calcium chloride on maintain the quality of mango fruit at ambient temperature. *Journal of Food Measurement and Characterization*, *15*, 2932–2942.

Hajji, S., Younes, I., Affes, S., Boui, S., Nasri, M. (2018). Optimization of the formulation of chitosan edible coatings supplemented with carotenoproteins and their use for extending strawberries postharvest life. *Food Hydrocolloids*, *83*, 375–392. https://doi.org/10.1016/jfoodhyd.2018.05.013.

Hamzah, H. M., Osman, A., Tan, C. P., Ghazali, F. M. (2013). Carrageenan as an alternative coating for papaya (*Carica papaya* L. cv. Eksotika). *Postharvest Biology and Technology*, *75*, 142–146.

Hashemi, S. M. B., Khaneghah, A. M., Ghahfarrokhi, M. G., Eş, I. (2017). Basil-seed gum containing *Origanum vulgare* subsp. viride essential oil as edible coating for fresh cut apricots. *Postharvest Biology and Technology*, *125*, 26–34.

Hassan, B., Chatha, S. A. S., Hussain, A. I., Zia, K. M., & Akhtar, N. (2018). Recent advances on polysaccharides, lipids and protein based edible films and coatings: A review. *International Journal of Biological Macromolecules*, *109*, 1095–1107.

Iversen, L. J. L., Rovina, K., Vonnie, J. M., Matanjun, P., Erna, K. H., 'Aqilah, N. M. N., Funk, A. A. (2022). The emergence of edible and food-application coatings for food packaging: A review. *Molecules*, *27*(17), 5604.

Jodhani, K. A., Nataraj, M. (2021). Synergistic effect of Aloe gel (Aloe vera L.) and Lemon (*Citrus Limon* L.) peel extract edible coating on shelf life and quality of banana (Musa spp.). *Journal of Food Measurement and Characterization*, *15*, 2318–2328.

Jurić, S., Bureš, M. S., Vlahoviček-Kahlina, K., Stracenski, K. S., Fruk, G., Jalšenjak, N., Bandić, L. M. (2023). Chitosan-based layer-by-layer edible coatings application for the preservation of mandarin fruit bioactive compounds and organic acids. *Food Chemistry: X*, *17*, 100575.

Kharchoui, S., Parafati, L., Licciardello, F., Muratore, G., Hamdi, M., Cirvilleri, G., Restuccia C. (2018). Edible coatings incorporating pomegranate peel extract and biocontrol yeast to reduce *Penicillium digitatum* postharvest decay of oranges. *Food Microbiology*, *74*, 107–112. https://doi.org/10.1016/j.fm.2018.03.011.

Khorram, F., Ramezanian, A., Hosseini, S.M. H. (2017). Effect of different edible coatings on postharvest quality of 'Kinnow' mandarin. *Journal of Food Measurement and Characterization* *11*(4), 1827–1833. https://doi.org/10.1007/s11694-017-9564-8.

Kozlu, A., Elmacı, Y. (2020). Quince seed mucilage as edible coating for mandarin fruit; determination of the quality characteristics during storage. *Journal of Food Processing and Preservation*, *44*(11), e14854.

Kumar, N., Neeraj, Pratibha, Singla, M. (2020). Enhancement of storage life and quality maintenance of litchi (*Litchi chinensis* Sonn.) fruit using chitosan: Pullulan blend antimicrobial edible coating. *International Journal of Fruit Science*, *20*, S1662–S1680.

Kumar, P., Sethi, S., Sharma, R.R. (2018). Inhibition of browning in fresh-cut apple wedges through edible coatings and anti-browning agents. *Indian Journal of Horticulture*, *75*(3), 517–522. https://doi.org/10.5958/0974-0112.2018.00087.7.

Kupervaser, M. G., Traffano-Schiffo, M. V., Dellamea, M. L., Flores, S. K., & Sosa, C. A. (2023). Trends in starch-based edible films and coatings enriched with tropical fruits extracts: a review. *Food Hydrocolloids for Health*, 100138.

Li, S., Zhang, L., Liu, M., Wang, X., Zhao, G., Zong, W. (2017). Effect of poly-ε-lysine incorpo- rated into alginate-based edible coatings on microbial and physicochemical properties of fresh- cut kiwifruit. *Postharvest Biology and Technology*, *134*, 114–121. https://doi.org/10.1016/j postharvbio.2017.08.014.

Liu, J., Li, F., Li, T., Yun, Z., Duan, X., Jiang, Y. (2019). Fibroin treatment inhibits chilling injury of banana fruit via energy regulation. *Scientia Horticulturae*, *248*, 8–13.

Liu, R., Liu, D., Liu, Y., Song, Y., Wu, T., Zhang, M. (2017). Using soy protein SiOx nanocomposite film coating to extend the shelf life of apple fruit. *International Journal of Food Science and Technology*, *52*, 2018–2030.

López-Palestina, C. U., Aguirre-Mancilla, C. L., Raya-Pérez, J. C., Ramínez-Pimentel, J. G. Gutiérrez-Tlahque, J., Hernández-Fuentes, A. D. (2018). The effect of an edible coating with tomato oily extract on the physicochemical and antioxidant properties of garam bullo (*Myrtillocactus geometrizans*) fruits. *Agronomy*, *8*(11), 248. https://doi.org/10.3390 agronomy8110248.

Magwebu, S., Meitz-Hopkins, J. C., Pott, R. W., Lennox, C. L. (2023). Efficacy of the cyclolipopeptides fengycin and iturin A against postharvest pome fruit pathogens. *Frontiers in Horticulture*, *2*, 1175251.

Mannucci, A., Serra, A., Remorini, D., Castagna, A., Mele, M., Scartazza, A., Ranieri, A. (2017). Aroma profile of Fuji apples treated with gelatin edible coating during their storage. *LWT-Food Science and Technology*, *85*, 28–36.

Marelli, B., Brenckle, M. A., Kaplan, D. L., Omenetto, F. G. (2016). Silk fibroin as edible coating for perishable food preservation. *Scientific Reports*, *6*, 25263.

Martiñon, M. E., Moreira, R. G., Castell-Perez, M. E., Gomes, C. (2014). Development of a multilayered antimicrobial edible coating for shelf-life extension of fresh-cut cantaloupe (*Cucumis melo* L.) stored at 4 C. *LWT-Food Science and Technology*, *56*(2), 341–350.

Matloob, A., Ayub, H., Mohsin, M., Ambreen, S., Khan, F. A., Oranab, S., . . . & Ercisli, S. (2023). A review on edible coatings and films: Advances, composition, production methods, and safety concerns. *ACS Omega*, *8*(32), 28932–28944.

Medina-Jaramillo, C., Gutiérrez, T. J., Goyanes, S., Bernal, C., Famá, L. (2016). Biodegradability and plasticizing effect of yerba mate extract on cassava starch edible films. *Carbohydrate Polymers*, *151*, 150–159. https://doi.org/10.1016/j.carbpol.2016.05.025.

Merino, D., Gutiérrez, T. J., Alvarez, V.A. (2019). Potential agricultural mulch ilms based on native and phosphorylated corn starch with and without surface functionalization with chitosan. *Journal of Polymers and the Environment*, *27*(1), 97–105. https://doi.org/10.1007s10924-018-1325-1.

Milani, J. M., & Nemati, A. (2022). Lipid-based edible films and coatings: a review of recent advances and applications. *Journal of Packaging Technology and Research*, *6*(1), 11–22.

Montero-Calderón, M., Soliva-Fortuny, R., Martín-Belloso, O. (2016). Edible packaging for fruits and vegetables. In M. Â. P. R. Cerqueira R. N. C. Pereira, Ó. L. S. Ramos, J. A. C. Teixeira, & A. A. Vicente (Eds.), Edible food packaging: Materials and processing technologies (pp. 353–382). Boca Raton: CRC Press.

Moreira, M.R., Tomadoni, B., Martín-Belloso, O., Soliva-Fortuny, R. (2015) Preservation of fresh-cut apple quality attributes by pulsed light in combination with gellan gum-based prebiotic edible coatings. *LWT-Food Science and Technology*, *64*(2), 1130–1137.

Mubarak, A., Rao Engakanah, T. (2017). *Aloe vera* edible coating retains the bioactive compounds of wax apples (*Syzygium Samarangense*). *Malaysian Applied Biology*, *46*(4), 141–148.

Muñoz-Labrador, A., Moreno, R., Villamiel, M., Montilla, A. (2018). Preparation of citrus pec- tin gels by power ultrasound and its application as an edible coating in strawberries. *Journal of the Science of Food and Agriculture*, *98*(13), 4866–4875. https://doi.org/10.1002/jsfa.9018.

Murmu, S.B., Mishra, H.N. (2018). The effect of edible coating based on Arabic gum, sodium caseinate and essential oil of cinnamon and lemon grass on guava. *Food Chemistry*, *245*, 820–828. https://doi.org/10.1016/j.foodchem.2017.11.104.

Nasirifar, S. Z. Maghsoudlou, Y. Oliyaei, N. (2018) Effect of Active lipid-based coating incorporated with nanoclay and orange peel essential oil on physicochemical properties of Citrus sinensis. *Food Science and Nutrition*, 6(6), 1508–1518.

Naveed, F., Nawaz, A., Ali, S., Ejaz, S. (2024). Xanthan gum coating delays ripening and softening of jujube fruit by reducing oxidative stress and suppressing cell wall polysaccharides disassembly. *Postharvest Biology and Technology*, *209*, 112689.

Nguyen, T. H., Boonyaritthongchai, P., Buanong, M., Supapvanich, S., Wongs-Aree, C. (2021). Edible coating of chitosan ionically combined with κ-carrageenan maintains the bract and postharvest attributes of dragon fruit (*Hylocereus undatus*). *International Food Research Journal*, *28*(4), 682–694.

Nogueira, G.F., Soares, C.T., Cavasini, R., Fakhouri, F.M., de Oliveira, R.A. (2019). Bioactive films of arrowroot starch and blackberry pulp: Physical, mechanical and barrier properties and stability to pH and sterilization. *Food Chemistry*, *275*, 417–425. https://doi.org/10.1016/j foodchem.2018.09.054.

Olawuyi, I. F., Kim, S. R., Lee, W. Y. (2021). Application of plant mucilage polysaccharides and their techno-functional properties' modification for fresh produce preservation. *Carbohydrate Polymers*, *272*, 118371.

Oliveira Filho, J. G. D., Silva, G. D. C., Oldoni, F. C. A., Miranda, M., Florencio, C., Oliveira, R.M.D.D., Ferreira, M. D. (2022). Edible coating based on carnauba wax nanoemulsion and *Cymbopogon martinii* essential oil on papaya postharvest preservation. *Coatings*, *12*(11), 1700.

Oms-Oliu, G., Soliva-Fortuny, R., Martín-Belloso, O. (2008). Edible coatings with antibrowning agents to maintain sensory quality and antioxidant properties of fresh-cut pears. *Postharvest biology and Technology*, *50*(1), 87–94.

Oregel-Zamudio, E., Angoa-Pérez, M. V., Oyoque-Salcedo, G.; Aguilar-González, C. N., Mena-Violante, H.G. (2017) Effect of candelilla wax edible coatings combined with biocontrol bacteria on strawberry quality during the shelf-life. *Scientia Horticulturae*, *214*, 273–279.

Ortiz-Duarte, G., Pérez-Cabrera, L.E., Artés-Hernández, F., Martínez-Hernández, G.B. (2019) Ag-chitosan nanocomposites in edible coatings affect the quality of fresh-cut melon. *Postharvest Biology and Technology*, *147*, 174–184. https://doi.org/10.1016/j.postharvbio.2018.09.021.

Panahirad, S., Naghshiband-Hassani, R., Mahna, N. (2020). Pectin-based edible coating preserves antioxidative capacity of plum fruit during shelf life. *Food Science and Technology International*, *26*(7), 583–592.

Panigrahi, J., Patel, M., Patel, N., Gheewala, B., Gantait, S. (2018). Changes in antioxidant and biochemical activities in castor oil-coated *Capsicum annuum* L. during postharvest storage. *3 Biotech*, *8*(6), 280.

Parafati, L., Vitale, A., Restuccia, C., Cirvilleri, G. (2016). The effect of locust bean gum (LBG)-based edible coatings carrying biocontrol yeasts against *Penicillium digitatum* and *Penicillium italicum* causal agents of postharvest decay of mandarin fruit. *Food Microbiology*, *58*, 87–94.

Paul, S.K., Sarkar, S., Sethi, L.N., Ghosh, S.K. (2018). Development of chitosan based optimized edible coating for tomato (*Solanum lycopersicum*) and its characterization. *Journal of Food Science and Technology*, *55*(7), 2446–2456. https://doi.org/10.1007/s13197-018-3162-6.

Pleşoianu, A. M., Nour, V. (2022). Pectin-based edible coating combined with chemical dips containing antimicrobials and antibrowning agents to maintain quality of fresh-cut pears. *Horticulturae*, *8*(5), 449.

Poonia, A. (2018) Antimicrobial Edible Films and Coatings for Fruits and Vegetables. In Food science and nutrition: Breakthroughs in research and practice (pp 177–195). London: IGI Global.

Qamar, J., Ejaz, S., Anjum, M. A., Nawaz, A., Hussain, S., Ali, S., Saleem, S. (2018). Effect of Aloe vera gel, chitosan and sodium alginate based edible coatings on postharvest quality of refrigerated strawberry fruits of cv. Chandler. *Chandler. Journal of Horticulture Science and Technology*, *1*(8).

Radi, M., Shadikhah, S., Sayadi, M., Kaveh, S., Amiri, S., Bagheri, F. (2023). Effect of Thymus vulgaris essential oil-loaded nanostructured lipid carriers in alginate-based edible coating on the postharvest quality of tangerine fruit. *Food and Bioprocess Technology*, *16*(1), 185–198.

Rangel-Marrón, M., Mani-López, E., Palou, E., López-Malo, A. (2019). Effects of alginate-glycerol- citric acid concentrations on selected physical, mechanical, and barrier properties of papaya puree-based edible films and coatings, as evaluated by response surface methodology. *LWT–Food Science and Technology*, *101*, 83–91. https://doi.org/10.1016/j.lwt.2018.11.005.

Rasool, F., Zahoor, I., Ayoub, W. S., Ganaie, T. A., Dar, A. H., Farooq, S., Mir, T. A. (2023). Formulation and characterization of natural almond gum as an edible coating source for enhancing the shelf life of fresh cut pineapple slices. *Food Chemistry Advances*, *3*, 100366.

Robles-Flores, G., Abud-Arcgilla, M., Ventura-Canseco, L.M. C., Meza-Gordillo, R., Grajales-Lagunes, A., Ruiz-Cabrera, M.A., Gutiérrez-Miceli, F.A. (2018). Development and evaluation of a film and edible coating obtained from the *Cajanus cajan* seed applied to fresh strawberry fruit. *Food and Bioprocess Technology*, *11*(12), 2172–2181. https://doi.org/10.1007/s11947-018-2175-5.

Rojas-Graü, M. A., Tapia, M. S., Martín-Belloso, O. (2008). Using polysaccharide-based edible coatings to maintain quality of fresh-cut Fuji apples. *LWT-Food Science and Technology*, *41*(1), 139–147.

Roy, S., Rhim, J. W. (2020). Preparation of antimicrobial and antioxidant gelatin/curcumin composite films for active food packaging application. *Colloids and Surfaces B: Biointerfaces*, *188*, 110761.

Saberi, B., Golding, J.B. (2018). Postharvest application of biopolymer-based edible coatings to improve the quality of fresh horticultural produce. In T.J. Gutiérrez (Ed.), Polymers for food applications (pp. 211–250). Cham: Springer. https://doi.org/10.1007/978-3-319-94625-2_9.

Saleem, M. S., Anjum, M. A., Naz, S., Ali, S., Hussain, S., Azam, M., Ejaz, S. (2021). Incorporation of ascorbic acid in chitosan-based edible coating improves postharvest quality and storability of strawberry fruits. *International Journal of Biological Macromolecules*, *189*, 160–169.

Saleem, M. S., Ejaz, S., Mosa, W. F., Ali, S., Sardar, H., Ali, M. M., . . . & Anjum, M. A. (2023). Biocomposite coatings delay senescence in stored Diospyros kaki Fruits by regulating antioxidant defence mechanism and delaying cell wall degradation. *Horticulturae*, *9*(3), 351.

Saleem, M. S., Ejaz, S., Anjum, M. A., Ali, S., Hussain, S., Ercisli, S., Mlcek, J. (2022). Improvement of postharvest quality and bioactive compounds content of persimmon fruits after hydrocolloid-based edible coating application. *Horticulturae*, *8*(11), 1045.

Salgado, P.R., Ortiz, C.M., Musso, Y.S., Di Giorgio, L., Mauri, A.N. (2015). Edible films and coatings containing bioactives. *Current Opinion in Food Science*, *5*, 86–92. https://doi.org/10.1016/j.cofs.2015.09.004.

Salinas-Roca, B., Soliva-Fortuny, R., Welti-Chanes, J., Martín-Belloso, O. (2018). Effect of pulsed light, edible coating, and dipping on the phenolic proile and antioxidant potential of fresh-cut mango. *Journal of Food Processing and Preservation*, *42*(5), e13591. https://doi.org/10.1111/jfpp.13591.

Salvia-Trujillo, L., Rojas-Graü, M. A., Soliva-Fortuny, R., Martín-Belloso, O. (2015). Use of antimicrobial nanoemulsions as edible coatings: Impact on safety and quality attributes of fresh-cut Fuji apples. *Postharvest Biology and Technology*, *105*, 8–16.

Senturk Parreidt, T., Lindner, M., Rothkopf, I., Schmid, M., Müller, K. (2019). The development of a uniform alginate-based coating for cantaloupe and strawberries and the characterization of water barrier properties. *Foods*, *8*(6), 203.

Shakir, M. S., Ejaz, S., Hussain, S., Ali, S., Sardar, H., Azam, M., Canan, İ. (2022). Synergistic effect of gum Arabic and carboxymethyl cellulose as biocomposite coating delays senescence in stored tomatoes by regulating antioxidants and cell wall degradation. *International Journal of Biological Macromolecules*, *201*, 641–652.

Shigematsu, E., Dorta, C., Rodrigues, F.J., Cedran, M.F., Giannoni, J.A., Oshiiwa, M., Mauro, M.A. (2018). Edible coating with probiotic as a quality factor for minimally processed carrots. *Journal of Food Science and Technology*, *55*(9), 3712–3720. https://doi.org/10.1007/s13197-018-3301-0.

Shinga, M. H., Fawole, O. A. (2023). *Opuntia ficus indica* mucilage coatings regulate cell wall softening enzymes and delay the ripening of banana fruit stored at retail conditions. *International Journal of Biological Macromolecules*, 125550.

Silva, A.F., Finkler, L., Finkler, C.L.L. (2018). Effect of edible coatings based on alginate/pectin on quality preservation of minimally processed 'Espada' mangoes. *Journal of Food Science and Technology*, *55*(12), 5055–5063. https://doi.org/10.1007/s13197-018-3444-z.

Silva, O. A., Pellá, M. C. G., Friedrich, J. C., Pellá, M. G., Beneton, A. G., Faria, M. G., Dragunski, D. C. (2021). Effects of a native cassava starch, chitosan, and gelatin-based edible coating over guavas (*Psidium guajava* L.). *ACS Food Science & Technology*, *1*(7), 1247–1253.

Slavin, J.L., Lloyd, B. (2012). Health benefits of fruits and vegetables. *Advances in Nutrition*, *3*(4), 506–516.

Sneha Nair, M., Saxena, A., Kaur, C. (2018). Characterization and antifungal activity of pomegranate peel extract and its use in polysaccharide-based edible coatings to extend the shelf- life of capsicum (*Capsicum annuum* L.). *Food and Bioprocess Technology*, *11*(7), 1317–1327. https://doi.org/10.1007/s11947-018-2101-x.

Soares, A.S., Ramos, A.M., Vieira, É. N.R., Vanzela, E.S. L., Oliveira, P.M., Paula, D.A. (2018). Vacuum impregnation of chitosan-based edible coating in minimally processed pumpkin. *International Journal of Food Science and Technology*, *53*(9), 2229–2238. https://doi.org/10.1111/ijfs.13811.

Soazo, M., Pérez, L. M., Rubiolo, A. C., Verdini, R. A. (2015) Prefreezing application of whey protein-based edible coating to maintain quality attributes of strawberries. *International Journal of Food Science and Technology*, 50(3), 605–611.

Suhaimi, N.I.M., Ropi, A.A.M., Shaharuddin, S. (2021). Safety and quality preservation of starfruit (*Averrhoa carambola*) at ambient shelf life using synergistic pectin-maltodextrin-sodium chloride edible coating. *Heliyon*, *7*(2).

Tapia-Blácido, D.R., Maniglia, B.C., Tosi, M.M. (2018). Transport phenomena in edible ilms. In T.J. Gutiérrez (Ed.), Polymers for food applications (pp. 149–192). Cham: Springer. https://doi.org/10.1007/978-3-319-94625-2_7.

Tavassoli-Kafrani, E., Shekarchizadeh, H., Masoudpour-Behabadi, M. (2016) Development of edible films and coatings from alginates and carrageenans. *Carbohydrate Polymer*, *137*, 360–374

Thakur, R., Pristijono, P., Bowyer, M., Singh, S.P., Scarlett, C.J., Stathopoulos, C.E., Vuong, Q.V. (2019). A starch edible surface coating delays banana fruit ripening. *LWT–Food Science and Technology*, *100*, 341–347. https://doi.org/10.1016/j.lwt.2018.10.055.

Thakur, R., Pristijono, P., Golding, J. B., Stathopoulos, C. E., Scarlett, C. J., Bowyer, M. (2018). Development and application of rice starch based edible coating to improve the postharvest storage potential and quality of plum fruit (*Prunus salicina*). *Scientia Horticulturae*, *237*, 59–66.

Tkaczewska, J., Jamróz, E., Zając, M., Guzik, P., Gedif, H. D., Turek, K., & Kopeć, M. (2023). Antioxidant edible double-layered film based on waste from soybean production as a vegan active packaging for perishable food products. *Food Chemistry*, *400*, 134009.

Todisco, K.M., Janzantti, N.S., Santos, A.B., Galli, F.S., Mauro, M.A. (2018). Effects of tem- perature and pectin edible coatings with guava byproducts on the drying kinetics and quality of dried red guava. *Journal of Food Science and Technology*, *55*(12), 4735–4746. https://doi.org/10.1007/s13197-018-3369-6.

Torres-León, C., Vicente, A.A., Flores-López, M.L., Rojas, R., Serna-Cock, L., Alvarez-Pérez, O.B., Aguilar, C.N. (2018). Edible films and coatings based on mango (var. Ataulfo) by-products to improve gas transfer rate of peach. *LWT–Food Science and Technology*, *97*, 624–631. https://doi.org/10.1016/j.lwt.2018.07.057.

Valdés, A., Burgos, N., Jiménez, A., Garrigós, M. C. (2015) Natural Pectin Polysaccharides as Edible Coatings. *Coatings*, *5*(4), 865–886.

Valencia, G.A., do Amaral Sobral, P.J. (2018). Recent trends on nano-biocomposite polymers for food packaging. In T. J. Gutiérrez (Ed.), Polymers for food applications (pp. 101–130). Cham: Springer. https://doi.org/10.1007/978-3-319-94625-2_5.

Valero, D., Díaz-Mula, H. M., Zapata, P. J., Guillén, F., Martínez-Romero, D., Castillo, S., Serrano, M. (2013). Effects of alginate edible coating on preserving fruit quality in four plum cultivars during postharvest storage. *Postharvest Biology and Technology*, *77*, 1–6.

Ventura-Aguilar, R.I., Bautista-Baños, S., Flores-García, G., Zavaleta-Avejar, L. (2018). Impact of chitosan based edible coatings functionalized with natural compounds on *Colletotrichum fragariae* development and the quality of strawberries. *Food Chemistry*, *262*, 142–149. https:// doi.org/10.1016/j.foodchem.2018.04.063.

Villafañe, F. (2017). Edible coatings for carrots. *Food Reviews International*, *33*(1), 84–103.

Wang, X., Sun, X., Liu, H., Li, M., Ma, Z. (2011). Barrier and mechanical properties of carrot puree films. *Food and Bioproducts Processing*, *89*(2), 149–156. https://doi.org/10.1016/j.fbp.2010.03.012.

Wani, S. M., Gull, A., Ahad, T., Malik, A. R., Ganaie, T. A., Masoodi, F. A., Gani, A. (2021). Effect of gum Arabic, xanthan and carrageenan coatings containing antimicrobial agent on postharvest quality of strawberry: Assessing the physicochemical, enzyme activity and bioactive properties. *International Journal of Biological Macromolecules*, *183*, 2100–2108.

Werner, B. G., Koontz, J. L., Goddard, J. M. (2017). Hurdles to commercial translation of next generation active food packaging technologies. *Current Opinion in Food Science*, *16*, 40–48.

Wigati, L. P., Wardana, A. A., Tanaka, F., Tanaka, F. (2023). Strawberry preservation using combination of yam bean starch, agarwood *Aetoxylon bouya* essential oil, and calcium propionate edible coating during cold storage evaluated by TOPSIS-Shannon entropy. *Progress in Organic Coatings*, *175*, 107347.

Won, J.S., Lee, S. J., Park, H.H., Song, K.B., Min, S.C. (2018). Edible coating using a chitosan-based colloid incorporating grapefruit seed extract for cherry tomato safety and preservation. *Journal of Food Science*, *83*(1), 138–146. https://doi.org/10.1111/1750-3841.14002.

Yaashikaa, P. R., Kamalesh, R., Kumar, P. S., Saravanan, A., Vijayasri, K., Rangasamy, G. (2023). Recent advances in edible coatings and their application in food packaging. *Food Research International*, *173*, 113366.

Yadav, A., Kumar, N., Upadhyay, A., Singh, A., Anurag, R. K., Pandiselvam, R. (2022). Effect of mango kernel seed starch-based active edible coating functionalized with lemongrass essential oil on the shelf-life of guava fruit. *Quality Assurance and Safety of Crops & Foods*, *14*(3), 103–115.

Yan, J., Luo, Z., Ban, Z., Lu, H., Li, D., Yang, D., Aghdam, M.S., Li, L. (2018). The effect of the layer-by-layer (LBL) edible coating on strawberry quality and metabolites during storage. *Postharvest Biology and Technology*, *147*, 29–38. https://doi.org/10.1016/j.postharvbio.2018.09.002.

Yang, W., Zhang, Z., Chen, Y., Luo, K. (2023). Evaluation of the use of Idesia polycarpa Maxim protein coating to extend the shelf life of European sweet cherries. *Frontiers in Nutrition*, *10*, 1283086.

Yousuf, B., Qadri, O.S., Srivastava, A.K. (2018). Recent developments in shelf-life extension of fresh-cut fruits and vegetables by application of different edible coatings: A review. *LWT*, *89*, 198–209.

Yu, L., Zong, Y., Han, Y., Zhang, X., Zhu, Y., Oyom, W., Bi, Y. (2022). Both chitosan and chitooligosaccharide treatments accelerate wound healing of pear fruit by activating phenylpropanoid metabolism. *International Journal of Biological Macromolecules*, *205*, 483–490.

Zeng, Y. F., Chen, Y. Y., Deng, Y. Y., Zheng, C., Hong, C. Z., Li, Q. M., Zha, X. Q. (2024). Preparation and characterization of lotus root starch based bioactive edible film containing quercetin-encapsulated nanoparticle and its effect on grape preservation. *Carbohydrate Polymers*, *323*, 121389.

Zhang, Y.L., Cui, Q.L., Wang, Y., Shi, F., Fan, H., Zhang, Y.Q. (2021). Effect of edible carboxymethyl chitosan-gelatin based coating on the quality and nutritional properties of different sweet cherry cultivars during postharvest storage. *Coatings*, *11*, 396.

Zhang, L., Chen, F., Lai, S., Wang, H., Yang, H. (2018). Impact of soybean protein isolate-chitosan edible coating on the softening of apricot fruit during storage. *LWT-Food Science and Technology*, *96*, 604–611. https://doi.org/10.1016/j.lwt.2018.06.011.

Multilayer Edible Coatings Technology for Fresh Fruits and Vegetables

25

Somayeh Rastegar*, Soheila Aghaei Dargiri, and Emad Hamdy Khedr
*Corresponding Author: rastegarhort@gmail.com, s.rastegar@hormozgan.ac.ir.

25.1 INTRODUCTION

Fruits and vegetables maintain their respiration process, even after harvesting from their source, and consist of high moisture content, thus they are highly prone to microbial deterioration. In addition, fruit and vegetable commodities have been categorized as highly perishable and susceptible to dehydration, mechanical damage, pathological infestations, and enzymatic discoloration after harvest. Their restricted shelf life is caused by both intrinsic and external factors, like ethylene production, respiration rate, and transpiration, as well as storage conditions and environmental conditions (Banerjee et al., 2021). Postharvest losses, such as nutritional, functional, and sensory losses, as well as tasteless production following harvest, are serious challenges that result in a decrease in both the quantity and quality of fruits and vegetables. Because consumers always prefer to buy fresh fruits and vegetables, researchers are always looking for new advanced methods that can help conserve produce quality and extend its shelf life (Flores-López et al., 2016).

Edible coatings appear to be a novel postharvest technology that have been demonstrated to have a safe and positive approach to inhibiting deteriorative changes and enhancing commodities' shelf life. Owing to its mechanical, antimicrobial, thermal, and even antioxidant properties, edible coatings and films are a potential solution to postharvest losses (Yousuf et al., 2018). A thin layer of environmentally friendly and edible materials that protects foods from microbes, gases, and moisture is known as an edible coating (Dhall, 2013). The coatings have recently been widely applied to prevent fresh produce from dehydrating and increase their resistance to water vapor (Blancas-Benitez et al., 2022).

For normal CO_2/O_2 exchange, edible coatings applied to whole or cut fresh commodities have accurate and balanced gas permeability properties. Coatings, on the other hand, have the ability to limit water vapor permeability to prevent moisture escape because of the sensitivity of fresh goods to water loss, which results in texture impairment and size shrinkage (Panahirad et al., 2021). Additionally, the coatings should also have antimicrobial activity because of the susceptibility of fresh, particularly freshly cut, commodities to microbial damage (Sánchez-González et al., 2011). Coatings can be utilized to enhance fresh produce quality and extend its shelf life without causing anaerobiosis or decay. Various types of coating application methods, for instance brushing, spraying/dripping, and/or immersion (dipping in coating solutions), have been established so far. It is well established that these are accepted as ecofriendly, as they are generally based on biodegradable and/or biocompatible material types, and these coatings appropriately satisfy consumers' demand for ensuring their anticipated foods are categorized as safe and healthy (Raghav et al., 2016).

Adhesion to the surface, sensory quality, barrier property, and uniformity all play a significant role in the success of an edible coating. These coatings should adhere well when applied to a food's surface in order to suitably protect the coated commodity (Falguera et al., 2011). The coating formulation's viscosity, appearance, density, and surface tension must be adjusted based on the roughness and surface tension of the food product for achievement of good adhesion characteristics (Lin & Zhao, 2007). The use of hydrocolloid-based coatings on fruit and vegetable commodities with natural hydrophobic surface types poses additional challenges in achieving proper and efficient adhesion in general. Fresh-cut and/or minimally processed commodities, on the other hand, typically show surfaces which are substantially more hydrophilic in nature and also difficult to adhere with selected coatings. So, it is imperative that an edible coating possesses certain properties so it can be considered useful on a sustainable basis. It must also protect the coated commodities from environmental effects

DOI: 10.1201/9781003370376-33

and show adequate antimicrobial activity, gas barrier property, and enough water vapor permeability, along with balanced and suitable adhesion. It is highly unlikely that a single coating material could satisfy such a wide range of needs. So, composite edible coatings that can combine several benefits from different components has recently received much attention (Wang et al., 2022). The layer-by-layer (LBL) method is one of the main methods for developing composites/materials. This LBL approach involves successively layering several biopolymers, which thereby enhances certain chemical and physical characteristics and the functionality of the resultant edible coatings (Du et al., 2022; Ali et al., 2024). The multilayer coatings can alter or enhance a monolayer-based substance's effectiveness (Anukiruthika et al., 2020; Ali et al., 2024) and might improve its mechanical and gas barrier properties (Tanpichai et al., 2022). According to Dhall (2013), multilayers of coatings have shown the capacity to improve the quality and status of safety perspectives of certain produce. The multilayered edible coatings which have been developed for application on produce are normally applied to either whole and/or fresh-cut vegetable and fruit commodities. Chitosan-based multilayer coatings have been found ideally effective in enhancing safety protection and prolonging the shelf life of intact and ready-to-eat commodities in general (Zhang et al., 2019; Treviño-Garza et al., 2017). Thus, the primary purpose of the current chapter is to review and update the knowledge regarding multilayer coating types for edible horticultural produce.

25.2 BENEFITS OF EDIBLE COATINGS/FILMS FOR FRUITS AND VEGETABLES

25.2.1 Moisture Barrier

Weight loss due to moisture loss is essential in reducing the economic value of fruits and vegetables (Hailu & Derbew, 2015). It is imperative to describe that when poorly soluble or water-insoluble components are used in formulations of edible coatings, their water vapor permeability (WVP) is decreased, which in turn prevents weight loss. In order to regulate moisture flow among foods and associated environmental components, edible coatings are applied to various fruits, which thereby reduces mass loss of the treated commodities (Dhall, 2016).

25.2.2 Oxygen Scavengers

The rate of gas transfer is of the main factors in food coatings for fresh fruit and vegetable commodities (Dhall, 2016). The edible coating needs to keep a fine balance between the senescence processes and the prevention of over-maturity by reducing the gas exchange to allow for natural respiration and prevent harmful anaerobic conditions that can cause bad tastes and alcoholic fermentation. The gas transfer properties of numerous edible coatings are primarily influenced in response to two main factors, i.e., (a) permeability and (b) thickness. Permeability refers to the ability of gas molecules to pass through the applied coating/film, and it can be appropriately modified by arranging the coating layers in an appropriate spatial arrangement. By controlling the arrangement of a coating's components, its permeability could be adjusted to efficiently regulate the transfer of selected gases; thus it can alter the internal commodity atmospheres. Furthermore, the thickness of a coating, which is usually determined by the concentrations of specific components, also affects gas transfer properties. By changing the concentration and composition of coating components, its thickness can also be suitably modified, thus impacting gas permeability and transfer rates through the applied/deposited coating (Mishra et al., 2010).

25.2.3 Antimicrobial Properties

The primary causes of fresh produce deterioration are microbial infestations. The absence of appropriate packaging for fruit and vegetable produce is one of the leading causes of their contamination. Food quality is harmed by spore-forming microorganisms (Balali et al., 2020). Exposure to microorganisms that are human pathogens compromises the safety of food commodities because it can result in serious health problems. The most susceptible micro wounds and cracks to microbial attacks are covered by edible coatings, which are applied on the surfaces of food products. Such mechanical protection reduces microbial attacks to a certain extent (Campos et al., 2011). Additionally, edible coatings frequently contain antimicrobial components that give them the ability to eliminate harmful microorganisms. Based on where they come from, antimicrobial compounds for edible coating materials can be divided into four major groups. They include inorganic antimicrobial agents, animal sources, microbial sources, and plant extracts. Organic acids, such as lactic and citric acids, are among the leading antimicrobial substances (Chawla et al., 2021).

25.2.4 Inhibition of Texture Modification

One of the main important consumer quality attributes is firmness and crispness of a food's texture. It is imperative that microbial damage, moisture loss, and in the case of fresh commodities, over-ripening all have an effect on the texture of a commodity during storage. Edible coatings can help food products maintain a satisfying texture by (a) utilizing texture-improving components (such as Ca^{+2} ions), (b) lowering the water vapor permeability (WVP) by using components that are either water-insoluble or poorly soluble, and (c) incorporating antimicrobial components to reduce microorganism-caused textural damage (Olivas et al., 2008; Dhall, 2013).

25.2.5 Improve the Food Product's Appearance

Appearance is an important aspect of food products, which includes factors such as glossiness, color, and the absence of visible flaws. However, during storage, certain fresh horticultural commodities are susceptible to oxidation-based reactions, thus resulting in discoloration owing to enzymatic browning. These processes can lead to undesirable color changes, moisture loss, and a reduction in the size and gloss of food products. It is imperative to implement proper storage conditions and employ appropriate preservation techniques to minimize these effects and preserve the desired appearance of the food products (Pott et al., 2020). The fresh fruit and vegetable commodities may lose their marketable potential as a result of mechanical damages, microbial infections, and certain other factors. It is well reported that edible coatings can protect fresh produce against adverse effects during storage. Furthermore, suitable coating components can further enhance a food product's glossiness and visual appeal (Falguera et al., 2011). By producing semi-permeable barriers, edible coating treatments enhance texture, aid in product appearance improvement, and extend the commodity's shelf life (Garg et al., 2021). Enzymatic browning has correlated with the oxidation of phenolics in fruit and vegetable commodities, which are subjected to minimal processing operations. This browning happens when a polyphenol oxidase (PPO) enzyme transforms polyphenolic contents into dark pigments (known as quinones) accompanied with O_2 (Taranto et al., 2017). PPO activity can be controlled by edible coating applications (as these impede O_2 uptakes) specifically incorporated with effective anti-browning substances, which serve as strong oxygen barriers. Carboxylic acids (citric and oxalic acid), ascorbic acid, cinnamic acid, selected thiol-containing agents (glutathione and cysteine), and phenolic acids have been reported as the most commonly tested anti-browning chemicals. PPO enzymes brought these reduced o-quinone compounds back to corresponding phenolic contents (Lee et al., 2003; Kumar et al., 2018).

25.3 TYPES OF EDIBLE COATINGS

There are various types of edible coating types which considerably extend the shelf life of fresh commodities or products by reducing lipid oxidation, lowering water activity, improving product appearance, and retaining native volatile flavors. Edible coatings may also comprise suitable components like antimicrobials, antioxidants, and a number of components and molecules that are embedded in a polymer matrix and express their barrier and surface protection qualities.

25.3.1 Single Edible Coating

Edible coatings can be broadly categorized into three key types: polysaccharide-based, protein-based, and lipid-based.

25.3.2 Polysaccharide-Based Edible Coatings

Natural polymers called polysaccharides are often used to make edible coatings or films. Polysaccharides, also known as polyglucans, are biopolymers composed of monosaccharide units with a hydroxyl group. These can interact with water or other molecules through hydrogen bonds within and between the molecules. Polysaccharides that are used in these edible coatings include chitosan, cellulose, pectin, alginates, pullulan and starch (Valdés et al., 2015). Polysaccharide coatings have good gas barrier properties, especially when the relative humidity (RH) is low, but owing to their hydrophilic nature, these typically have poor water barrier properties. Water vapor permeability (WVP) values tend to be lower in coatings made from pure polysaccharides than in coatings made from proteins. However, all protein-based films exhibit elevated WVP values at elevated RH. As a result, variations in moisture affect these hydrophilic materials' selectivity. Nevertheless, monomer compounds and substituent groups of polysaccharides also have an influence on their behavior, permitting for a wider range of functional and molecular characteristics (Felicia et al., 2022).

25.3.3 Protein-Based Edible Coatings

Protein-based coatings are derived from both animals and plants. Gluten (derived from wheat), zein (derived from maize), soy protein, and other plant-based proteins are used as protective coatings, whereas animal-based proteins are derived from egg albumin, whey protein, gelatin, and milk protein casein. It has been reported that proteins have favorable mechanical and gas barrier properties. Casein, gluten, and soy protein are all examples of protein-based coatings that help protect food products from deterioration by acting as effective oxygen blockers (Hassan et al., 2018). Owing to their hydrophilic properties, despite having a very high potential to create a barrier against mechanical strengths and organoleptic and aroma preservation, and higher permeability of oxygen, protein coatings lack waterproof properties. However, its hydrophilic characteristics can be enhanced by incorporating hydrophobic substances, for instance, lipid compounds (Armghan Khalid et al., 2022).

25.3.4 Lipid-Based Edible Coatings

Because lipid contents are naturally hydrophobic, they are excellent materials for edible coatings, as they can prevent moisture movement into fresh foods, which can lead to significant quality deterioration. Due to their relatively low polarity, lipids can be used to coat foods because they primarily prevent moisture transport.

Because lipid coatings are hydrophilic, they can help reduce the rate of water evaporation from foods and limit the effects of water, light, oxygen, and other external factors on the quality of a product during preservation. Additionally, these protect from and prevent chilling-related damages, which usually occur during cold storage (Dhall, 2013). Some examples of lipid-based edible coatings include acetylated monoglycerides, vegetable oils, paraffin, and waxes. These coatings give fresh fruit a glossy, shiny appearance. In the food industry, waxes are commonly used lipid materials. Natural waxes include carnauba wax, beeswax, and candelilla wax; synthetic waxes include paraffin wax and oxidized polyethylene wax. Natural waxes contain complex organic molecules such as long-chain-based fatty acids, ketones, long-chain alcohols, long alkyl chains, aldehyde compounds, and fatty acid–based ester contents. They may also have aromatic components (Valencia-Chamorro et al., 2009). A diverse collection of molecules, lipids include waxes, phospholipids, fats, fatty acids, and others. However, the formed coatings are fragile due to their hydrophobic nature. To improve the coating properties of lipids, other coating agents like proteins or polysaccharides must be combined with them. Polysaccharides or proteins, for instance, provide selective gas (O_2 or CO_2) permeability, durability, integrity, and structural cohesion. Lipids, on the other hand, increase the mixture's resistance to water vapor. However, reports claim that coated food products may lose their gloss and appearance as a result of lipid-containing coatings or films (Bravin et al., 2004; Perez-Gago et al., 2002)

25.4 MULTILAYER EDIBLE COATINGS

Adhesion issues are caused by an exceptionally hydrophobic surface of waxed fruit or vegetable commodities or a hydrophilic-based surface of fresh-cut produce. So, it can be challenging to apply edible coatings (Olivas & Barbosa-Cánovas, 2009). Considering the role of edible coatings in creating barriers on the surfaces of the fruit, producing a transparent and natural coating, controlling respiration, inhibiting enzyme-based degradation of cell walls, and increasing the firmness of fruit tissue, these properties could be further enhanced by using multilayer-based coatings which can be made from natural components like carbohydrates, proteins, or a combination of these. In this technique, compounds with opposite charges are used, and the LBL coating that is created extends the efficacy of the monolayer-based constituent and might increase its mechanical strength. In addition, it is possible to increase the adhesion of the edible coating and therefore the shelf life of commodities could be further extended (Zhang et al., 2022).

Lipids can also be a part of edible coatings. It has been demonstrated that multilayered coatings made up of hydrocolloids and lipids perform better than composite coatings made up of the same blended components. This is because the coatings' gas permeability and water vapor properties can be more precisely controlled through the alternations of lipids and hydrocolloids films (Yousuf et al., 2022).

The said method uses the alternate layers of various polymers on the surface of commodities as a simple and low-cost preservation technique. The double or multilayer-based coatings have complete structural and stable properties, and these can increase major advantages of several single layers of applied coating. These can also commendably preserve the color, general freshness, and firmness of fruits, as well as extend their shelf life after harvesting (Arnon-Rips & Poverenov, 2018). Multilayer-based coatings have a complete structure and stable effectiveness and can merge the various merits of the single layers of used coatings. These can also efficiently preserve fruit and vegetable produce color, textural stability, and freshness, along with prolonged storage life (Zhang et al., 2019).

Extensive research and applications regarding LBL self-assembly-based approaches as an ecofriendly way has been reported to make functional multilayer-based coatings. During the postharvest coating of fresh produce, the LBL deposition and utilization have effectively been used and successfully tested to preserve the general quality of fresh fruit and vegetable commodities. The LBL method has stimulated a lot of considerations owing to its efficacy to control the thickness of a coating at the nanoscale level (Arnon-Rips & Poverenov, 2018). In order to create antimicrobial-oriented composites, it is possible to incorporate selected components of an antimicrobial nature into matrixes of intended polymers. The use of the LBL strategy, which is based (mostly) on chitosan/pectin and/or chitosan/alginate alternate multilayers, has resulted in enhanced protection along with extension of the shelf life of numerous fruits and vegetables (Brasil et al., 2012). In the same way chitosan/alginate (Nair et al., 2020) and chitosan/carboxymethyl cellulose (CMC) (Ali et al., 2024) as alternate multilayer and double layer, respectively, have also been reported. The LBL-based application of edible coatings made of mucilages, pullulan, and chitosan preserved the eating quality of fresh-cut pineapple and extended its shelf life in comparison with controls (Treviño-Garza et al., 2017).

25.4.1 Layer-by-Layer (LBL) Edible Coatings

A method for preparing multilayer films based on the LBL fabrication technique was proposed previously by Decher and Hong (1991). The LBL technique uses alternate deposition of polyelectrolyte-based constituents with opposite charges to better control the coating's properties and functionality. In general, this method involves serially depositing a number of film-forming materials onto a commodity surface to form multilayer-based films or coatings, which then subsequently help to increase the edible coating's effectiveness. The LBL technique involves the consecutive depositions of two or more suitable materials onto the commodity's surface. The said materials can include inorganic nanomaterials, polyelectrolytes, biomolecules, organic small molecules, and several more having appropriate properties. The deposition technique relies on several intermolecular forces, comprising hydrogen bonds, covalent bonds, electrostatic forces, host–guest interfaces, and certain others, thus eventually facilitating the assembly of targeted layers (Zhang et al., 2022).

A common method to form these kinds of films or coatings is to utilize the opposite attractions between two targeted layers. It involves dipping substrates with a negative charged surface in a solution which has positive charged cations to attract the said constituents to the substrates' surface layers. These substrates with a positive charge are then dipped in other solutions that have substances with negative charges, which make new layers after the extra solution is removed. For the production of films or coatings with multiple layers, the process can be repeated multiple times. It may be essential to do a washing or drying step (Figure 25.1) between each dipping step prior to putting the objects in the next dipping solution, to discard the extra solution from the surface (McClements et al., 2009). The electrostatic LBL depositions process typically applies materials which are water-dispersible and possess an electric charge, for instance colloidal particles or biopolymers. The aforementioned characteristics of coating materials which are formed, for instance their thickness, surface charges, and compositions, can be altered by changing the number, types, and order of charged substances that are considered compulsory to form the desired coating layers in general (Wang et al., 2022). In addition, electrostatic attractions, selected intermolecular interactions, including covalent bonding, hydrogen bonding, and hydrophobic interfaces, can be efficiently utilized for the construction of a LBL film or coating. The said interfaces are involved in the development of environmentally friendly and biodegradable coatings which have been reported to depend on the properties of the substances utilized. Proteins often create connections with neighboring molecules by combinations of covalent, hydrophobic, electrostatic, and hydrogen bonds (Wang et al., 2022; Anukiruthika et al., 2020).

During the preparation of multilayer-oriented packaging/coatings, the LBL method, as reported previously, exhibits several peculiar advantages among alternative fabrication techniques (Wang et al., 2022; Anukiruthika et al., 2020). These include:

- Applications of edible coatings to minimally processed fresh produce can present numerous technical difficulties, including poor adhesion to fruits which have hydrophilic surfaces, degradation of antimicrobial compounds, and poor diffusion rates. Nevertheless, the said challenges can be managed by utilizing the LBL electrodepositions technique. The method based on the LBL approach involves creating multilayers of coating structure by sequentially depositing more than two materials onto the surface of the targeted produce. By employing LBL, the adhesion issues can be addressed, and controlled incorporations of antimicrobial compounds could further help to prevent their probable degradation. Furthermore, optimizing the layer composition and thickness enhances the diffusion rates of active ingredients within the coating. Using LBL electrodeposition, these technical challenges can be effectively resolved, leading to improved coating adhesion, sustained antimicrobial effectiveness, and enhanced diffusion rates of bioactive components. By selecting various numbers, types, and sequences of coating-forming materials, it is possible to create coatings with various functional attributes by modifying their structural and physicochemical properties (such as thickness, composition, and charge).
- The enhanced gas exchange properties of the LBL coating, which were better than monolayer-based coating types and resulted in less ethanol

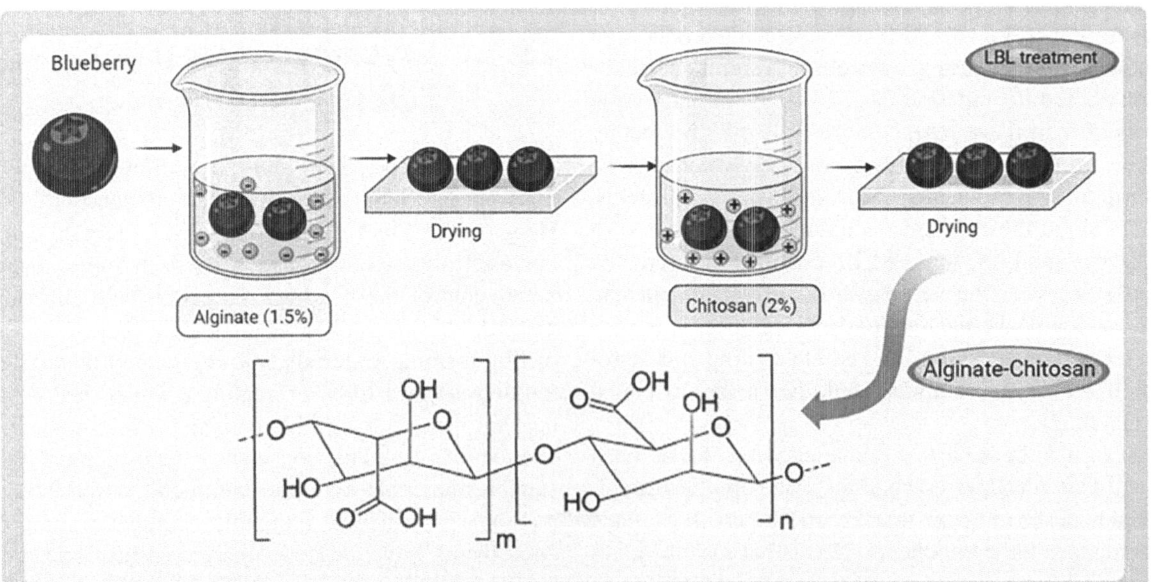

FIGURE 25.1 Schematic illustration of an edible coating applied via the LBL electrostatic deposition method.

production, were an unexpected advantage. The accumulation of carbon dioxide showed the same tendency. Ethanol is a well-known off-flavor volatile marker, and carbon dioxide may lead to anaerobic conditions. The accumulation of these compounds is undesirable because it is linked to a decline in quality.

- Its preparation is simple, inexpensive, and does not require any special equipment. From a practical standpoint, there are no special tools or equipment required for using LBL edible coatings. Nowadays, certain coatings, mostly waxes, are applied industrially to citrus, apples, and other fruits and vegetables.
- Multilayer films or coatings can be applied to foods of all shapes and can be prepared for flat surfaces as well.

25.4.2 Multilayered LBL Edible Coating Based on Polysaccharides

Alternate deposition of polyelectrolytes with opposing charges onto the fruit surface is the LBL method, which may enable effective control of coating properties and functionality. Recently, polysaccharide-based coatings, such as pectin, chitosan, alginate, and gellan coatings, are increasingly being used to prevent fresh produce from dehydrating and resisting water vapor. Polysaccharide-based coatings are becoming more and more popular due to their low cost, easy application, high biodegradability, and availability. Consequently, organic solvents are not required prior to or during application (Kumar, 2019). Polysaccharides can adhere to the surface of fresh produce and create a uniform coating with the assistance of Ca^{2+} bridges and intra-strands cross-linking. These interactions occur naturally between the polysaccharides and pectin content present on the fruit's surface. Additionally, the well-defined chemical structure of polysaccharides permits the modification of their coating properties. Polysaccharides can be grouped into several categories based on the magnitude of their electric charge, regardless of their origin or composition: amphoteric, anionic, cationic, and non-ionic.

25.4.2.1 Anionic Polysaccharides

Anionic polysaccharides show a negative charge when the pH is higher than their pKa value, while they become neutral when the pH is much lower than their pKa value. Commonly used natural anionic polysaccharides in multilayer edible coatings include alginate, carrageenan, pectin, gellan gum, xanthan, and gum arabic. Modified anionic polysaccharides such as carboxymethyl chitin (CMCH) and sodium carboxymethyl cellulose (CMC) are also frequently employed. The electrical charge of polysaccharides can be influenced by the presence of ionic species in the surrounding environment. The negative groups on the biopolymer chain can interact with multivalent or monovalent ions like Ca2+ or Na+, thereby altering the complete charge properties. The anionic polysaccharide contents' gelation requires interaction between opposite group charges on the chain polymers. For instance, alginate and pectin can be gelled using divalent metal ions like Ca^{2+} (Matalanis et al., 2011).

25.4.2.2 Cationic Polysaccharides

When well above their pKa value, cationic polysaccharide contents tend to be neutral, while those below their pKa value tend to be positive. Chitosan is the only naturally occurring cationic polysaccharide (Matalanis et al., 2011). It generally consists of (1,4)-linked 2-amino-2-deoxy-β-D-glucan, which is the result of chitin's partial deacetylation (Gruber, 1999). Insect exoskeletons and a variety of bacterial parasites naturally contain chitosan, but in insufficient quantities for commercial use. Chitosan has been mixed with other agents and coatings, like xanthan gum, alginate, starch, and whey protein, for multilayer edible coatings, despite its broad antimicrobial spectrum (Pech-Canul et al., 2020).

The preparation of biodegradable edible coating types based on polysaccharides using the LBL electrostatic deposition method in recent years has significantly improved fresh produce quality. Alginate and chitosan-based edible coatings applied layer by layer recently significantly improved the microbial and physiological qualities of fresh-cut melon without incorporation of any active agent (Poverenov et al., 2014). In addition, Hira et al. (2022) suggested that a shelf life extended by a chitosan/alginate-based LBL coating may primarily be attributable to the management of the equilibrium between anaerobic and aerobic metabolisms. The primary common and suitable combinations are alternate depositions of polyelectrolyte constituents of a biocompatible nature with opposite charges, for instance chitosan (CH) and CMC (Yan et al., 2019; Ali et al., 2024). Additionally, Liu et al. (2017) demonstrated that LBL electrostatic deposition of a CMC and CS polyelectrolyte multilayer (PEM) coating inhibited weight loss, maintained firmness, and protected fresh-cut apple fruit from surface browning occurrence. The LBL-applied coatings made from the combination of two polysaccharides, for instance CMC (as the inner layer) and CH (as the outer layer), exhibited encouraging results in enhancing numerous quantitative attributes of Rishon and Michal mandarin fruit (Arnon et al., 2015). Figure 25.2 shows the schematic of the formation of an edible coating based on CMC and CH on fruits' surface. According to observations made by Yan et al. (2019), the LBL treatment considerably reduced the primary and secondary metabolites content of carbohydrates, amino acids, and fatty acids. The LBL electrostatic application using CH and CMC was used to coat strawberries in a study. The findings demonstrated that the LBL edible coating had an impact on soluble solids content or acidity content while significantly reducing strawberry firmness and aroma volatiles loss. Brasil et al. (2012) created the first edible multilayer coatings made of CH and pectins

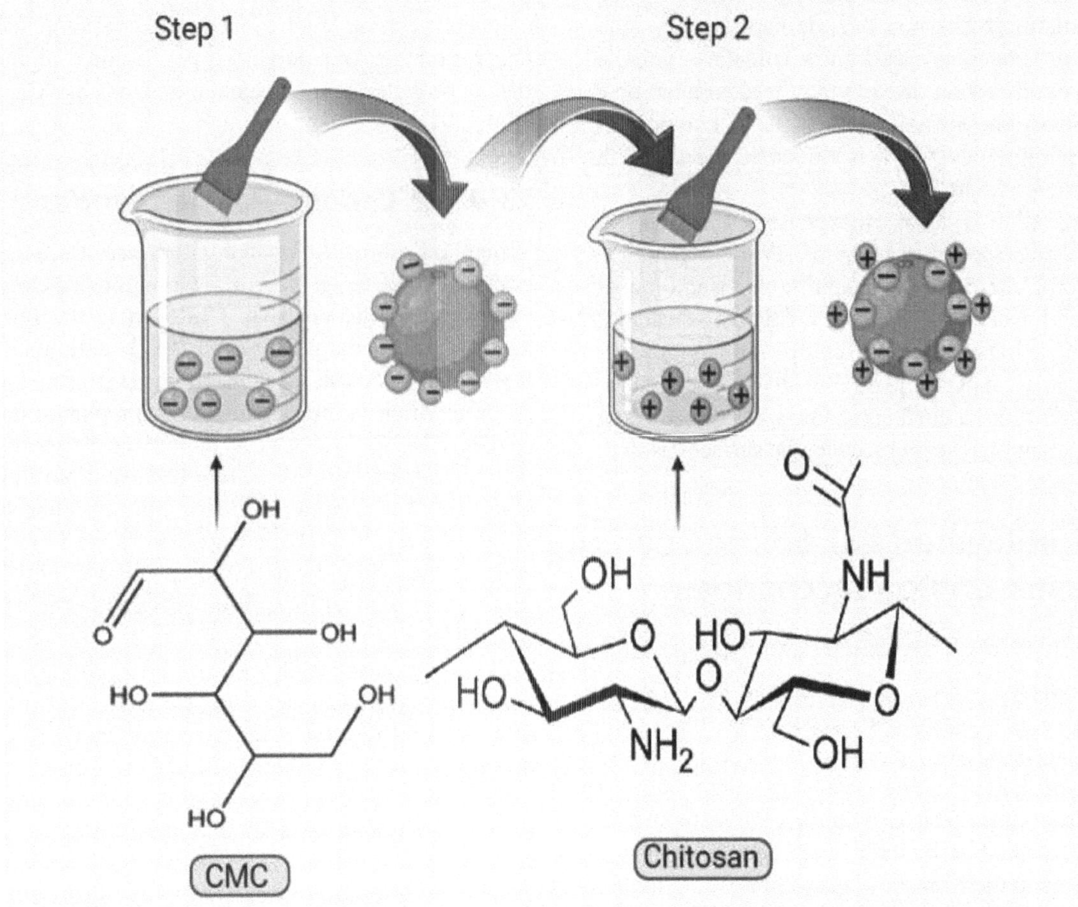

FIGURE 25.2 Schematic of the formation of an LBL edible coating made of CMC and chitosan on fruit.

with $CaCl_2$ as a cross-linking agent to prolong the shelf life of freshly cut papayas for 15 days. Treviño-Garza et al. (2017) also carried out another study which enhanced the storage life of fresh-cut pineapples for 18 days using LBL coatings based on the combination of chitosan, pullulan (nonionic polymer), and mucilage.

25.4.3 The Physicochemical Principles behind Electrostatic LBL Edible Coatings

The most frequently utilized chemical forces for the preparation of LBL structures are electrostatic interactions. In the original LBL method, the process involved depositing polyelectrolytes, specifically polyanions and polycations, in alternating sequences onto a charged surface. Promptly after each deposition, a rinsing step was performed (Decher, 2012). The overall electroneutrality of an LBL structure is achieved through the overcompensation of charges, which relies on competitive ion pairing as well as intrinsic and extrinsic charge compensations. Combining polyelectrolytes with opposing charges is a process known as intrinsic charge compensation. Eventually, the bulk of the multilayer contains no counter ions, and the surface contains all exchangeable charges. Ion pairing with charged counter ions is the foundation of extrinsic compensation. This extrinsic charge compensation transforms into intrinsic charge compensation during LBL assembly, resulting in the release of counter ions and solvating water molecules, as well as an entropic driving force for LBL deposition (Mateos-Maroto et al., 2022). The LBL process will proceed in a linear growth mode when polyelectrolytes are strongly attracted to one another, which is primarily driven by intrinsic charge compensation. In this mode, the number of deposited bilayers and the number of polyelectrolytes increase linearly. The concentration of the polyelectrolyte makes it simple to adjust the thickness of materials, which is characteristically a few nanometers per bilayer (Campbell & Vikulina, 2020). Both extrinsic and intrinsic charge compensation drive the process when two polyelectrolytes have a weak attraction to one another. Therefore, methods of LBL deposition could employ exponential increases, where the numbers of bilayers and adsorbed polyelectrolyte-oriented components will increase further exponentially. As a result, the subsequent materials will have limited numbers of bilayers or multilayers and possess an efficient layer of thickness on a microscale (Joseph et al., 2014).

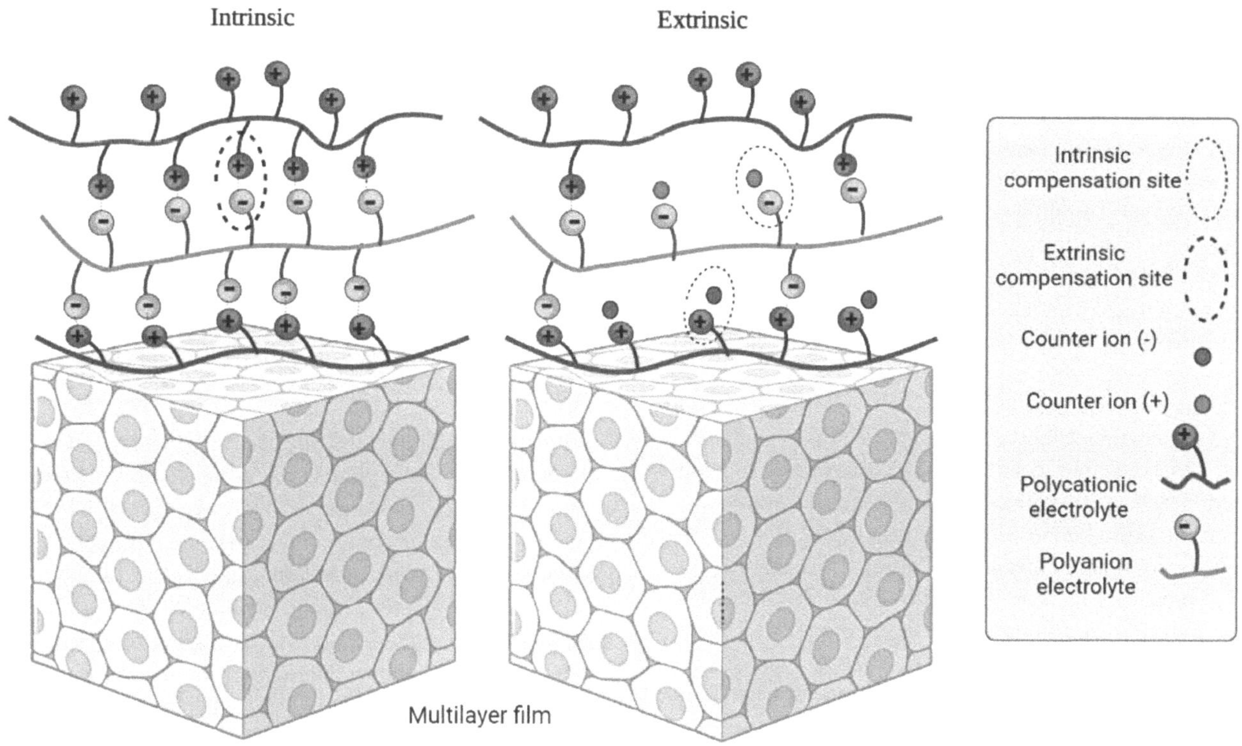

FIGURE 25.3 In multilayer polyelectrolytes, there are two charge compensation mechanisms.

Two leading aspects which have an impact on the complex equilibrium of interactions which leads to the formation of electrostatic-based LBL as a self-assembled coating include (a) factors of electrostatic and (b) entropy, as well as the quality of solvents (Alkekhia et al., 2020; Alotaibi et al., 2018). During the formation of the self-assembled film, the phenomenon of charge inversion plays an imperative role in driving the depositions of charged substances onto target surfaces with an opposite charge. The said phenomenon involves a problem of surface charge, resulting in attraction and adsorption of charged species onto an originally based, oppositely charged surface. This process thereby contributes to the formation and stability of a self-assembled film in general (Guzman et al., 2020). The multilayer structures therefore become unstable owing to charge inversion–induced excess charges. As a result, the charge compensation mechanism must be utilized to keep the structures of the overall system of electrical neutrality throughout electrostatic-based LBL self-assembly deposition (Ghoussoub et al., 2018). Figure 25.3 illustrates the compensation mechanism for the polyelectrolyte multilayer film. The intrinsic compensation mechanism involves a direct compensation between the two layers generated during the LBL assembly technique (Guzmán et al., 2017). In this mechanism, counter ions are released into the solution by the multilayer film, and the increase in entropy resulting from the counter ion release acts as an additional driving force for the formation of the LBL self-assembled film (Dubas & Schlenoff, 1999). The compensation mechanism has been influenced by various factors, for instance the solvent quality of polyelectrolyte and interactive forces. One imperative factor is the ionic strength, which plays a substantial role in the compensation mechanism. The ionic solution strength also substantially affects overall charge distribution and interactions among polyelectrolyte layers, subsequently influencing the anticipated release of counter ions and the development of a self-assembled structure. The intrinsic compensation produced owing to an increase in entropy occurs when a majority of the counter ions are released into the solutions when the ionic strength is lower. Counter ion release is irrelevant when ionic strength is high. It is currently a compensatory extrinsic mechanism (Guzmán et al., 2016; Mateos-Maroto et al., 2021). Therefore, the electrostatic LBL self-assembled film can be altered by adjusting the compensation mechanism and the solvent quality. In addition, electrostatic interactions play a significant role as a driving force behind assembly; however, neither intrinsic nor extrinsic compensation is differentiated. Hydrogen bonding and hydrophobic interactions may drive the LBL assembly process in addition to electrostatic interactions.

25.5 METHODS OF MULTILAYER COATING APPLICATION TO FOOD PRODUCTS

Various techniques are used to cover fruits and vegetables, which depend on various factors such as: (a) the properties of the produce that will be coated, particularly their surface roughness and hydrophobicity; (b) the coating's physical properties, like its surface tension, density, and viscosity; (c) the intended effects of coatings; (d) the available method of drying; (e) the intended application in industry; and (f) production cost. The most common way to prepare for LBL electrostatic self-assembly is by dipping. The template is rinsed in the middle after being alternately submerged in solutions comprising oppositely charged substances to remove substances with weak adsorptive charges (Vander Straeten et al., 2020). The dipping method consists of three imperative stages:

1 Immersion and holding. After immersing the substrate in the coating solution, the time of holding is allowed for the substrate to interact with the coating solution for a sufficient amount of time to complete wetting.
2 Drainage and deposition. The thin layers of the coating solution are entrained by pulling the substrate in an upward direction, thereby resulting in film deposition. The additional liquid is then removed from the substrate's surface.
3 Drying and/or evaporation. By drying with heated air or evaporating at room temperature, the excessive diluent evaporates from the food's surface, leaving behind a thin film of the coating solution (Brinker, 2013).

Spin coating and spray-based methodologies have been widely investigated in various fields. These methods offer numerous advantages, such as uniform and efficient deposition. However, their practical application in food-related industries and active packaging types are comparatively restricted owing to certain potential challenges and concerns. Factors such as control of the deposited coating's thickness, potential contamination risks, and difficulties in attaining precise and controlled coating applications may further restrict their utilization in active packaging systems for fresh fruit and vegetable produce. So, some alternative coating techniques with higher efficacy are often explored and preferred in the context of active packaging systems. Recently, fluidic and electromagnetic-based assemblies have also received a lot of attention in order to explore their benefits (Guzmán et al., 2022).

25.6 ENHANCEMENT OF THE FUNCTIONALITY OF MULTILAYER EDIBLE COATINGS/FILMS

Researchers have demonstrated clear evidence of the substantial role played by selected additives in prolonging the shelf life and maintaining the nutritional values of various fruits and vegetables. When combined with edible coatings, these functional molecules work synergistically, effectively preserving the moisture content, antioxidant potential, antioxidant enzyme activity, browning enzyme activity, and antimicrobial properties of fresh fruits and vegetables. As a result of these properties, the coated fruits and vegetables retain their appearance, experience reduced senescence, and enjoy an extended shelf life (Oyom et al., 2022).

25.7 ENHANCING THE FUNCTIONALITY OF MULTILAYER EDIBLE COATINGS/FILMS USING PLANT SOURCES

Strong antioxidant and antifungal properties are found in compounds derived from plants, according to some studies (Ong et al., 2021). The presence of hydroxyl groups and the high content of phenolic compounds, terpenes, aliphatic alcohols, iso-flavonoids, ketones, aldehydes, and acids in plant substances are the primary factors that contribute to their antimicrobial activity (Tiwari et al., 2009). By scavenging free radicals, phenolics prevent oxidative chain reactions, making them antioxidants. The individual phenolic constituents collaborate to improve the overall antimicrobial and antioxidant properties of coating formulations. The active compounds inhibit the spoilage organism by destroying protein translocation, releasing cellular components, and breaking down the cytoplasmic membrane (Amensour et al., 2010). In addition, plant extracts can easily diffuse and disperse when they come into contact with microorganisms. Using selected plant-based antimicrobials in edible coatings, on the otherhand, necessitates an in-depth knowledge of the minimal concentrations required to attain the desired results and outcomes. According to a comparative recent study, the concentration of phenolic extracts and the duration of exposure to test certain microorganisms determines their probable antimicrobial properties (Nguyen et al., 2020). The binding of polyphenol content to polysaccharides is influenced in response to various factors, including temperature, structure, pH, ionic strength, and native charge. These factors play an imperative role in determining the interaction between phenol and polysaccharide components. Major structural characteristics which affect binding between phenols and polysaccharides include the degree of steric hindrances caused by side chains existing in polysaccharides, the molecular flexibility of polyphenols, and the presence of certain hydroxyl and hydrogen bonding groups which can interact with corresponding binding sites on selected polysaccharides. All these factors collectively influence the strengths and specificity of binding interactions among certain polyphenol- and polysaccharide-based constituents (Dobson et al., 2019). Essential oils (EOs) are volatile-based secondary metabolites which have strong antimicrobial, antioxidative, and aromatic properties and are typically extracted from aromatic plants. As important antimicrobial agents, EOs are able to interact with bacterial membranes and destroy selected fungal pathogens. It was observed that the effectiveness of a CH coating was increased when it was combined with certain EOs (Li et al., 2021).

25.8 MULTILAYER EDIBLE COATING/FILM FUNCTIONALITY ENHANCEMENT WITH SELECTED INORGANIC SUBSTANCES

Various innovations and advancements have been reported for the preservation of fresh fruit and vegetable produce after harvest. Recently, nanoparticles (NPs) have been investigated as critical components of LBL coating assembly with improved overall performance. Nano-silver, titanium dioxide, nano-cupric oxide, zinc dioxide (ZnO), chlorine dioxide, and cinnamic acids are some major examples of produced inorganic nanoparticles which are frequently utilized. ZnO nanoparticles are the subject of extensive investigations due to their antimicrobial potential with limited constraints in food systems (La et al., 2021). The ROS formation on the particle's surface could explain the antimicrobial activity of ZnO, as observed earlier (Ma et al., 2019). In past works, edible coatings along with some inorganic antimicrobial agents were utilized to prolong the shelf life of agricultural commodities and their byproducts. Studies have established the fact that these could extend the shelf life of fresh produce, as noted in the case of fresh-cut papaya (Lavinia et al., 2019) and shiitake mushrooms (Jiang et al., 2013) with formulations enriched with certain nanoparticles. In addition to incubation or dipping, spin-coatings, spin-spray assembly, and spray-assisted alignments along with cross-linking after infiltration have all been exploited for the LBL assembling process incorporated with appropriate nanoparticles (Gittleson et al., 2015). However, the tendency of such compounds to migrate into the food product from the coating remains a leading obstacle which is impeding their probable commercialization (Chhetri et al., 2019).

25.9 CONCLUSIONS

In conclusion, edible multilayer coatings have been utilized to extend the shelf life of vegetable- and fruit-based commodities. One of the most inventive ways to enhance the efficacy of barrier properties and extend the shelf life of an edible coating is through LBL coating. The LBL method is characteristically based on polyelectrolyte self-assembling electrostatic interactions. As was mentioned in this review, the most prevalent LBL coating is polysaccharides. Nevertheless, lipids and proteins can also be utilized in this manner. There are numerous benefits to using LBL edible coatings in food products. They can prolong storability, improve physiological quality, improve appearance, and prevent the growth of harmful microorganisms. The LBL coating strategy has the potential to facilitate the efficient delivery of active agents to food products. However, the complicated coatings deposition method, which relies on alternate uses of various coating types and necessitates intermediate washing to eliminate excess solution, limits its use in industrial applications. Additional research is also needed to determine the optimal storage temperature for fruit coated with LBL and the processes that play a part in LBL-mediated shelf life extension.

REFERENCES

Ali, S., Ishtiaq, S., Nawaz, A., Naz, S., Ejaz, S., Haider, M. W., . . . & Javad, S. (2024). Layer by layer application of chitosan and carboxymethyl cellulose coatings delays ripening of mango fruit by suppressing cell wall polysaccharides disassembly. *International Journal of Biological Macromolecules*, *256*, 128429. https://doi.org/10.1016/j.ijbiomac.2023.128429

Alkekhia, D., Hammond, P.T., Shukla, A., (2020). Layer-by-layer biomaterials for drug delivery. *Annual Review of Biomedical Engineering*, *22*, 1–24.

Alotaibi, H.F., Al Thaher, Y., Perni, S., Prokopovich, P., (2018). Role of processing parameters on surface and wetting properties controlling the behaviour of layer-by-layer coated nanoparticles. *Current Opinion in Colloid & Interface Science*, *36*, 130–142.

Amensour, M., Bouhdid, S., Fernández-López, J., Idaomar, M., Senhaji, N.S., Abrini, J., (2010). Antibacterial activity of extracts of Myrtus communis against food-borne pathogenic and spoilage bacteria. *International Journal of Food Properties*, *13*, 1215–1224.

Anukiruthika, T., Sethupathy, P., Wilson, A., Kashampur, K., Moses, J.A., Anandharamakrishnan, C., (2020). Multilayer packaging: Advances in preparation techniques and emerging food applications. *Comprehensive Reviews in Food Science and Food Safety*, *19*, 1156–1186.

Armghan Khalid, M., Niaz, B., Saeed, F., Afzaal, M., Islam, F., Hussain, M., Mahwish, Muhammad Salman Khalid, H., Siddeeg, A., Al-Farga, A., (2022). Edible coatings for enhancing safety and quality attributes of fresh produce: A comprehensive review. *International Journal of Food Properties*, *25*, 1817–1847.

Arnon, H., Granit, R., Porat, R., Poverenov, E., (2015). Development of polysaccharides-based edible coatings for citrus fruits: A layer-by-layer approach. *Food Chemistry*, *166*, 465–472.

Arnon-Rips, H., Poverenov, E., (2018). Improving food products' quality and storability by using Layer by Layer edible coatings. *Trends in Food Science & Technology*, *75*, 81–92.

Balali, G.I., Yar, D.D., Afua Dela, V.G., Adjei-Kusi, P., (2020). Microbial contamination, an increasing threat to the consumption of fresh fruits and vegetables in today's world. *International Journal of Microbiology*, *2020*.

Banerjee, S., Haldar, S., Kumar, B.N.S., Mitra, J., (2021). Storage of fruits and vegetables: an overview. *Packaging and Storage of Fruits and Vegetables*, 157–181.

Blancas-Benitez, F.J., Montaño-Leyva, B., Aguirre-Güitrón, L., Moreno-Hernández, C.L., Fonseca-Cantabrana, A., del Carmen Romero-Islas, L., González-Estrada, R.R., (2022). Impact of edible coatings on quality of fruits: A review. *Food Control*, 109063.

Brasil, I.M., Gomes, C., Puerta-Gomez, A., Castell-Perez, M.E., Moreira, R.G., (2012). Polysaccharide-based multilayered antimicrobial edible coating enhances quality of fresh-cut papaya. *LWT-Food Science and Technology*, *47*, 39–45.

Bravin, B., Peressini, D., Sensidoni, A., (2004). Influence of emulsifier type and content on functional properties of polysaccharide lipid-based edible films. *Journal of Agricultural and Food Chemistry*, *52*, 6448–6455.

Brinker, C.J., 2013. Dip coating. Chemical Solution Deposition of Functional Oxide Thin Films 233–261.

Campbell, J., Vikulina, A.S., (2020). Layer-by-layer assemblies of biopolymers: Build-up, mechanical stability and molecular dynamics. *Polymers*, *12*, 1949.

Campos, C.A., Gerschenson, L.N., Flores, S.K., (2011). Development of edible films and coatings with antimicrobial activity. *Food and Bioprocess Technology*, *4*, 849–875.

Chawla, R., Sivakumar, S., Kaur, H., (2021). Antimicrobial edible films in food packaging: Current scenario and recent nanotechnological advancements-a review. *Carbohydrate Polymer Technologies and Applications*, *2*, 100024.

Chhetri, V.S., Janes, M.E., King, J.M., Doerrler, W., Adhikari, A., (2019). Effect of residual chlorine and organic acids on survival and attachment of Escherichia coli O157: H7 and Listeria monocytogenes on spinach leaves during storage. *LWT-Food Sciene and Technolog*, *105*, 298–305.

Decher, G., (2012). Layer-by-layer assembly (putting molecules to work). Multilayer thin films: *Sequential Assembly of Nanocomposite Materials*, 1–21.

Decher, G., Hong, J., (1991). Buildup of ultrathin multilayer films by a self-assembly process, 1 consecutive adsorption of anionic and cationic bipolar amphiphiles on charged surfaces, in: *Makromolekulare Chemie. Macromolecular Symposia*. Wiley Online Library, pp. 321–327.

Dhall, R.K., (2013). Advances in edible coatings for fresh fruits and vegetables: a review. *Critical Reviews in Food Science and Nutrition*, *53*, 435–450.

Dhall, R.K., (2016). Application of edible coatings on fruits and vegetables. *Biobased and Environmental Benign Coatings*, 87–119.

Dobson, C.C., Mottawea, W., Rodrigue, A., Pereira, B.L.B., Hammami, R., Power, K.A., Bordenave, N., (2019). Impact of molecular interactions with phenolic compounds on food polysaccharides functionality. *Advances in Food and Nutrition Research*, *90*, 135–181.

Du, Y., Yang, F., Yu, H., Yao, W., Xie, Y., (2022). Controllable fabrication of edible coatings to improve the match between barrier and fruits respiration through layer-by-layer assembly. *Food and Bioprocess Technology*, *15*, 1778–1793.

Dubas, S.T., Schlenoff, J.B., (1999). Factors controlling the growth of polyelectrolyte multilayers. *Macromolecules*, *32*, 8153–8160.

Falguera, V., Quintero, J.P., Jiménez, A., Muñoz, J.A., Ibarz, A., (2011). Edible films and coatings: Structures, active functions and trends in their use. *Trends in Food Science & Technology*, *22*, 292–303.

Felicia, W.X.L., Rovina, K., Nur'Aqilah, M.N., Vonnie, J.M., Erna, K.H., Misson, M., Halid, N.F.A., (2022). Recent advancements of polysaccharides to enhance quality and delay ripening of fresh produce: A review. *Polymers*, *14*, 1341.

Flores-López, M.L., Cerqueira, M.A., de Rodríguez, D.J., Vicente, A.A., (2016). Perspectives on utilization of edible coatings and nano-laminate coatings for extension of postharvest storage of fruits and vegetables. *Food Engineering Reviews*, *8*, 292–305.

Garg, M., Kaur, P., Sadhu, S.D., (2021). Edible Coating for improvement of horticulture crops. in: *Packaging and Storage of Fruits and Vegetables*. Apple Academic Press, pp. 89–107.

Ghoussoub, Y.E., Zerball, M., Fares, H.M., Ankner, J.F., von Klitzing, R., Schlenoff, J.B., (2018). Ion distribution in dry polyelectrolyte multilayers: A neutron reflectometry study. *Soft Matterials*, *14*, 1699–1708.

Gittleson, F.S., Hwang, D., Ryu, W.-H., Hashmi, S.M., Hwang, J., Goh, T., Taylor, A.D., (2015). Ultrathin nanotube/nanowire electrodes by spin–spray layer-by-layer assembly: A concept for transparent energy storage. *ACS Nano*, *9*, 10005–10017.

Gruber, J. V, (1999). Polysaccharide-based polymers in cosmetics. *Cosmetic Science and Technology Series*, 325–390.

Guzmán, E., Mateos-Maroto, A., Ruano, M., Ortega, F., Rubio, R.G., (2017). Layer-by-Layer polyelectrolyte assemblies for encapsulation and release of active compounds. *Advances in Colloid and Interface Science*, *249*, 290–307.

Guzmán, E., Ortega, F., Rubio, R.G., (2016). Comment on "Formation of polyelectrolyte multilayers: ionic strengths and growth regimes" by K. Tang and AM Besseling, Soft Matter, 12, 1032. Soft Matter 12, 8460–8463.

Guzmán, E., Ortega, F., Rubio, R.G., (2022). Layer-by-layer materials for the fabrication of devices with electrochemical applications. *Energies*, *15*, 3399.

Guzman, E., Rubio, R.G., Ortega, F., (2020). A closer physico-chemical look to the Layer-by-Layer electrostatic self-assembly of polyelectrolyte multilayers. *Advances in Colloid and Interface Science*, *282*, 102197.

Hailu, G., Derbew, B., (2015). Extent, causes and reduction strategies of postharvest losses of fresh fruits and vegetables–A review. *Journal of Biology, Agriculture and Healthcare*, *5*, 49–64.

Hassan, B., Chatha, S.A.S., Hussain, A.I., Zia, K.M., Akhtar, N., (2018). Recent advances on polysaccharides, lipids and protein based edible films and coatings: A review. *International Journal of Biological Macromolecules*, *109*, 1095–1107.

Hira, N., Mitalo, O.W., Okada, R., Sangawa, M., Masuda, K., Fujita, N., Ushijima, K., Akagi, T., Kubo, Y., (2022). The effect of layer-by-layer edible coating on the shelf life and transcriptome of 'Kosui'Japanese pear fruit. *Postharvest Biology and Technology*, *185*, 111787.

Jiang, T., Feng, L., Wang, Y., (2013). Effect of alginate/nano-Ag coating on microbial and physicochemical characteristics of shiitake mushroom (*Lentinus edodes*) during cold storage. *Food Chemistry*, *141*, 954–960.

Joseph, N., Ahmadiannamini, P., Hoogenboom, R., Vankelecom, I.F.J., (2014). Layer-by-layer preparation of polyelectrolyte multilayer membranes for separation. *Polymer Chemistry*, *5*, 1817–1831.

Kumar, N., (2019). Polysaccharide-based component and their relevance in edible film/coating: A review. *Nutrition & Food Science*, *49*, 793–823.

Kumar, P., Sethi, S., Sharma, R.R., Singh, S., Varghese, E., (2018). Improving the shelf life of fresh-cut 'Royal Delicious' apple with edible coatings and anti-browning agents. *Journal of Food Science and Technology*, *55*, 3767–3778.

La, D.D., Nguyen-Tri, P., Le, K.H., Nguyen, P.T.M., Nguyen, M.D.-B., Vo, A.T.K., Nguyen, M.T.H., Chang, S.W., Tran, L.D., Chung, W.J., (2021). Effects of antibacterial ZnO nanoparticles on the performance of a chitosan/gum arabic edible coating for postharvest banana preservation. *Progress in Organic Coatings*, *151*, 106057.

Lavinia, M., Hibaturrahman, S.N., Harinata, H., Wardana, A.A., (2019). Antimicrobial activity and application of nanocomposite coating from chitosan and ZnO nanoparticle to inhibit microbial growth on fresh-cut papaya. *Food Research*, *4*, 307–311.

Lee, J.Y., Park, H.J., Lee, C.Y., Choi, W.Y., (2003). Extending shelf-life of minimally processed apples with edible coatings and antibrowning agents. *LWT-Food Science and Technology*, *36*, 323–329.

Li, S., Sun, J., Yan, J., Zhang, S., Shi, C., McClements, D.J., Liu, X., Liu, F., (2021). Development of antibacterial nanoemulsions incorporating thyme oil: Layer-by-layer self-assembly of whey protein isolate and chitosan hydrochloride. *Food Chemistry*, *339*, 128016.

Lin, D., Zhao, Y., (2007). Innovations in the development and application of edible coatings for fresh and minimally processed fruits and vegetables. *Comprehensive Reviews in Food Science and Food Safety*, *6*, 60–75.

Liu, X., Tang, C., Han, W., Xuan, H., Ren, J., Zhang, J., Ge, L., (2017). Characterization and preservation effect of polyelectrolyte multilayer coating fabricated by carboxymethyl cellulose and chitosan. *Colloids and Surfaces A: Physicochemical and Engineering Aspects*, *529*, 1016–1023.

Ma, J., An, W., Xu, Q., Fan, Q., Wang, Y., (2019). Antibacterial casein-based ZnO nanocomposite coatings with improved water resistance crafted via double in situ route. *Progress in Organic Coatings*, *134*, 40–47.

Matalanis, A., Jones, O.G., McClements, D.J., (2011). Structured biopolymer-based delivery systems for encapsulation, protection, and release of lipophilic compounds. *Food Hydrocolloids*, *25*, 1865–1880.

Mateos-Maroto, A., Fernández-Peña, L., Abelenda-Núñez, I., Ortega, F., Rubio, R.G., Guzmán, E., (2022). Polyelectrolyte multilayered capsules as biomedical tools. *Polymers*, *14*, 479.

Mateos-Maroto, A., Abelenda-Núñez, I., Ortega, F., Rubio, R.G., Guzmán, E., (2021). Polyelectrolyte multilayers on soft colloidal nanosurfaces: A new life for the layer-by-layer method. *Polymers*, *13*, 1221.

McClements, D.J., Decker, E.A., Park, Y., Weiss, J., (2009). Structural design principles for delivery of bioactive components in nutraceuticals and functional foods. *Critical Reviews in Food Science and Nutrition*, *49*, 577–606.

Mishra, B., Khatkar, B.S., Garg, M.K., Wilson, L.A., (2010). Permeability of edible coatings. *Journal of Food Science and Technology*, *47*, 109–113.

Nair, M.S., Tomar, M., Punia, S., Kukula-Koch, W., Kumar, M., (2020). Enhancing the functionality of chitosan-and alginate-based active edible coatings/films for the preservation of fruits and vegetables: A review. *International Journal of Biological Macromolecules*, *164*, 304–320.

Nguyen, T.T., Dao, U.T.T., Bui, Q.P.T., Bach, G.L., Thuc, C.N.H., Thuc, H.H., (2020). Enhanced antimicrobial activities and physiochemical properties of edible film based on chitosan incorporated with *Sonneratia caseolaris* (L.) Engl. leaf extract. *Progress in Organic Coatings*, *140*, 105487.

Olivas, G.I.I., Barbosa-Cánovas, G., (2009). Edible films and coatings for fruits and vegetables. *Edible Films and Coatings for Food Applications*, 211–244.

Olivas, G.I., Dávila-Aviña, J.E., Salas-Salazar, N.A., Molina, F.J., (2008). Use of edible coatings to preserve the quality of fruits and vegetables during storage. *Stewart Postharvest Review*, *3*, 1–10.

Ong, G., Kasi, R., Subramaniam, R., (2021). A review on plant extracts as natural additives in coating applications. *Progress in Organic Coatings*, *151*, 106091.

Oyom, W., Zhang, Z., Bi, Y., Tahergorabi, R., (2022). Application of starch-based coatings incorporated with antimicrobial agents for preservation of fruits and vegetables: A review. *Progress in Organic Coatings*, *166*, 106800.

Panahirad, S., Dadpour, M., Peighambardoust, S.H., Soltanzadeh, M., Gullón, B., Alirezalu, K., Lorenzo, J.M., (2021). Applications of carboxymethyl cellulose-and pectin-based active edible coatings in preservation of fruits and vegetables: A review. *Trends in Food Science & Technology*, *110*, 663–673.

Pech-Canul, A. de la C., Ortega, D., García-Triana, A., González-Silva, N., Solis-Oviedo, R.L., (2020). A brief review of edible coating materials for the microencapsulation of probiotics. *Coatings*, *10*, 197.

Perez-Gago, M.B., Rojas, C., DelRio, M.A., (2002). Effect of lipid type and amount of edible hydroxypropyl methylcellulose-lipid composite coatings used to protect postharvest quality of mandarins cv. fortune. *Journal of Food Science*, *67*, 2903–2910.

Pott, D.M., Vallarino, J.G., Osorio, S., (2020). Metabolite changes during postharvest storage: Effects on fruit quality traits. *Metabolites*, *10*, 187.

Poverenov, E., Danino, S., Horev, B., Granit, R., Vinokur, Y., Rodov, V., (2014). Layer-by-layer electrostatic deposition of edible coating on fresh cut melon model: Anticipated and unexpected effects of alginate–chitosan combination. *Food and Bioprocess Technology*, *7*, 1424–1432.

Raghav, P.K., Agarwal, N., Saini, M., (2016). Edible coating of fruits and vegetables: a review. *International Journal of Scientific Research and Modern Education*, *1*, 188–204.

Sánchez-González, L., Vargas, M., González-Martínez, C., Chiralt, A., Chafer, M., (2011). Use of essential oils in bioactive edible coatings: a review. *Food Engineering Reviews*, *3*, 1–16.

Tanpichai, S., Srimarut, Y., Woraprayote, W., Malila, Y., (2022). Chitosan coating for the preparation of multilayer coated paper for food-contact packaging: Wettability, mechanical properties, and overall migration. *International Journal of Biological Macromolecules*, *213*, 534–545.

Taranto, F., Pasqualone, A., Mangini, G., Tripodi, P., Miazzi, M.M., Pavan, S., Montemurro, C., (2017). Polyphenol oxidases in crops: biochemical, physiological and genetic aspects. *International Journal of Molecular Sciences*, *18*, 377.

Tiwari, B.K., Valdramidis, V.P., O'Donnell, C.P., Muthukumarappan, K., Bourke, P., Cullen, P.J., (2009). Application of natural antimicrobials for food preservation. *Journal of Agricultural and Food Chemistry*, *57*, 5987–6000.

Treviño-Garza, M.Z., García, S., Heredia, N., Alanís-Guzmán, M.G., Arévalo-Niño, K., (2017). Layer-by-layer edible coatings based on mucilages, pullulan and chitosan and its effect on quality and preservation of fresh-cut pineapple (*Ananas comosus*). *Postharvest Biology and Technology*, *128*, 63–75.

Valdés, A., Burgos, N., Jiménez, A., Garrigós, M.C., (2015). Natural pectin polysaccharides as edible coatings. *Coatings*, *5*, 865–886.

Valencia-Chamorro, S.A., Pérez-Gago, M.B., del Río, M.Á., Palou, L., (2009). Effect of antifungal hydroxypropyl methylcellulose (HPMC)–lipid edible composite coatings on postharvest decay development and quality attributes of cold-stored 'Valencia'oranges. *Postharvest Biology and Technology*, *54*, 72–79.

Vander Straeten, A., Lefèvre, D., Demoustier-Champagne, S., Dupont-Gillain, C., (2020). Protein-based polyelectrolyte multilayers. *Advances in Colloid and Interface Science*, *280*, 102161.

Wang, Q., Chen, W., Zhu, W., McClements, D.J., Liu, X., Liu, F., (2022). A review of multilayer and composite films and coatings for active biodegradable packaging. *NPJ Science of Food*, *6*, 18.

Yan, J., Luo, Z., Ban, Z., Lu, H., Li, D., Yang, D., Aghdam, M.S., Li, L., (2019). The effect of the layer-by-layer (LBL) edible coating on strawberry quality and metabolites during storage. *Postharvest Biology and Technology*, *147*, 29–38.

Yousuf, B., Qadri, O.S., Srivastava, A.K., (2018). Recent developments in shelf-life extension of fresh-cut fruits and vegetables by application of different edible coatings: A review. *LWT-Food Science and Technology*, *89*, 198–209.

Yousuf, B., Sun, Y., Wu, S., (2022). Lipid and lipid-containing composite edible coatings and films. *Food Reviews International*, *38*, 574–597.

Zhang, C., Li, C., Aliakbarlu, J., Cui, H., Lin, L., (2022). Typical application of electrostatic layer-by-layer self-assembly technology in food safety assurance. *Trends in Food Science & Technology*, *129*, 88–97.

Zhang, L., Huang, C., Zhao, H., (2019). Application of pullulan and chitosan multilayer coatings in fresh papayas. *Coatings*, *9*, 745.

PART IX

Natural Products, Antagonistic Microorganisms and Preharvest Applications

Role of Active Packaging in the Preservation of Fresh Fruits and Vegetables

26

Shaghef Ejaz*, Sajid Ali, Rehaba Abid, Neha Khizar and Laraib Amjad

*Corresponding Author: shaghef.ejaz@bzu.edu.pk

26.1 INTRODUCTION

Food packaging is essential for preserving the quality of agricultural produce during the transportation, storage and sale/marketing chains. Conventionally, packaging is generally intended to safeguard food items from mechanical injury, damaging radiation from light and gases that encourage unwanted reactions and also from contamination by pathogenic/decaying bacteria or noxious chemicals (Wang et al., 2017). It can also serve additional functions, such as informing consumers about the nature, origin, manufacturing and nutritional value of the items contained in it (Yam & Dong, 2012). Thus, the information on packaging frequently acknowledges how sustainable they are for both humans and the environment. As a result, food packaging supplies must be very adaptable, suited to being created in a variety of sizes and forms and possess various functional properties (Cheng et al., 2020; Yam & Dong, 2012).

Consumers' demand for information on food safety and additives has gained a significant amount of recognition in recent decades. As a result, there has been increased interest and curiosity in the designing and manufacturing of packaging products based on environmentally friendly components, such as edible proteins, polysaccharides and lipids separated from food waste sources (Cazón et al., 2017). The preventative measures adopted for packaging are mainly passive, working as a barrier between the surface of the food, the ambient atmosphere and external climate shifts. But there are certain exceptions, for example fresh products such as fruits and vegetables, which require extremely low gas permeability or porous packaging to facilitate the exchange of gases through the wrapping (Hussein et al., 2015; Lee & Paik, 1997).

Active packaging may consist of O_2 barriers, antimicrobial and antioxidants agents or light blockers that can aid in extending the shelf life of food commodities. The US market has spent more than $54 billion for active, regulated and intelligent packaging for foods and drinks. Active and intelligent packaging may have a promising future. In this regard, legislation such as the Regulation of Europe in 2004/EC and the new Regulation in 2009/EC provide a new foundation for general regulations as well as particular safety and marketing considerations relating to active and intelligent packaging (Restuccia et al., 2010). The main purpose for the invention of novel active packaging technology is due to the increasing demand for reliable, durable and transparent packaging which prolongs the shelf life of fresh food (Yildirim et al., 2018). The major purpose of active packaging is to prolong the shelf life of fresh produce while preserving its characteristic qualities. So, this chapter covers various aspects of active packaging and its utilization in the fresh produce industry.

26.2 ACTIVE PACKAGING

Advances in the production, distribution, storage and retailing of foods in response to the growing customer requirement for higher quality and extended shelf life introduced active and intelligent packaging concepts (Brody et al., 2001). These packaging types have numerous advantages compared to traditional packaging materials.

The inclusion of active compounds like antibacterial and antioxidizing agents through active packaging instead of directly adding them to the food may decrease the amount of such necessary compounds. Introducing active compounds through active packaging could be more efficient than adding them to most food items. Active packaging methods tested in a model system may not necessarily act similar to a real-world food application. The complex structure of the food may have

DOI: 10.1201/9781003370376-35

an impact on the activity of the packaging. Active substance release rates and absorption rates may be modified (Almasi et al., 2021). Furthermore, active components or the carrier's substance may interact or bind to dietary components, limiting the desired effect. Therefore, it is crucial to carefully evaluate active packaging research relevant to particular foods, allowing nutrition and packaging experts to fully appreciate the advantage of such types of systems and perhaps hasten their acceptance in commercial applications.

Applications of active packaging have a diverse and emerging scope of use. To increase the durability of most O_2-sensitive items, O_2 scavengers and antioxidant-producing substances can be used. Desiccant substances have been widely employed in the treatment of dry and mold-sensitive foods. Antibacterial-releasing compounds may be employed with bakery items, milk products and other food items, and C_2H_4-scavengers are utilized for the horticulture produce market (Mohan & Ravishankar, 2019).

26.3 CLASSIFICATION OF ACTIVE PACKAGING

Active packaging techniques are classified into two groups:

- Non-migratory active packaging or active scavenging systems (absorbers)
- Migratory active packaging or active releasing systems (emitters)

Both absorbing and releasing systems are used to increase the durability and/or the nutritional value of food. Oxygen (O_2), extra liquid, ethylene, carbon dioxide (CO_2), contaminants, and other non-significant food components are all removed through scavenging mechanisms (Figure 26.1), while the releasing systems add substances such as CO_2, water, antioxidants and preservatives to packed food (Kruijf et al., 2002). The principal approaches of active packaging are concerned with compounds that absorb O_2, ethylene, humidity, CO_2, flavors/odors, antibacterial agents and antioxidants (Restuccia et al., 2010). Active packaging has diverse and expanding applications. O_2 scavengers and antioxidant-producing mechanisms are utilized to increase the shelf life of various O_2-sensitive products (Mohan & Ravishankar, 2019; Vermeiren et al., 1999). Active packaging methods and techniques serve as an inert barrier against environmental and unfavorable conditions, as well as tasks related to the storage of food, property removal, microbiological control and maintaining the quality of the food (Alves et al., 2023).

Active packaging systems have two ingredients: i) the absorbing system absorbs components from the packaging of food or the headspace, i.e., O_2, ethylene, off-flavors or ethanol; and ii) the releasing system adds substances such as CO_2, antioxidants, antimicrobials, ethylene or aroma (Kruijf et al., 2002).

26.3.1 Non-migratory Active Packaging (Absorbers)

The preservation parameters for fresh produce, such as the levels of oxygen and carbon dioxide, are critical to extend a product's shelf life (Herrera et al., 2020). These parameters can be regulated to a tolerable extent by active packaging, which largely depend on the choice of materials, their geometric arrangement, improved gas removal capability, the quantity of packed product and the presence of active elements (Garavito et al., 2022). Utilizing absorbents, such as CO_2, O_2 and ethylene absorbers, as the active packaging method for fresh-cut fruits and vegetables in modified atmospheric packaging (MAP) may increase the product's shelf life and preserve its freshness (Siddiqui, 2020).

26.3.1.2 Oxygen Scavenger

Oxygen, which is being continuously consumed by freshly packaged fruits and vegetables, is the primary factor in fruit senescence (Singh et al., 2019); phenolic browning (Cichello, 2015); changing color, flavor and aroma; and the growth of aerobic bacteria and molds (Mir et al., 2018). One of the primary active packaging techniques to regulate oxygen level is the use of oxygen scavengers (OS), which work as an active barrier or eliminate any remaining oxygen from the food package. OS slows down food senescence by reducing respiration rate, inhibiting microbial growth, decreasing oxidation reactions (Yildirim et al., 2018) and inhibiting lipid oxidation or enzymatic browning (Leneveu et al., 2021). Generally, during produce packing, oxygen levels are reduced to maximize the effects of oxygen scavenging. Nowadays, the majority of commercially used OS are flexible sachets with gas permeability containing reduced iron particles (iron that is not totally oxidized) that are placed inside of food packaging (Bhardwaj et al., 2019).

26.3.1.3 Types of Oxygen Scavengers

26.3.1.3.1 Iron-based Oxygen Scavengers

When an OS is in contact with the moisture from the produce, it is turned into iron oxide by a process called rusting, which involves an iron and oxygen reaction and results in a stable ferric oxide trihydrate complex (Yildirim et al., 2018), thus consuming oxygen in the headspace (Gupta, 2023). In theory, 300 cc of oxygen are absorbed by 1 g of iron-based scavenger, resulting in the formation of non-toxic iron oxide (Mir et al., 2018). OS are non-toxic and fully safe to use with no hazardous gas emission. OS come in the form of monolayer or multilayer packaging materials, self-adhesive labels, sachets or self-adhesive labels. In high-permeability films like plasticized PVC or polyethylene, OS chemicals may be spread or blended in order to make them completely available to oxygen (Nwakaudu et al., 2015).

Certain commercially accessible oxygen-scavenging solutions are Ageless Omacr (OS film, Mitsubishi Gas Chemical

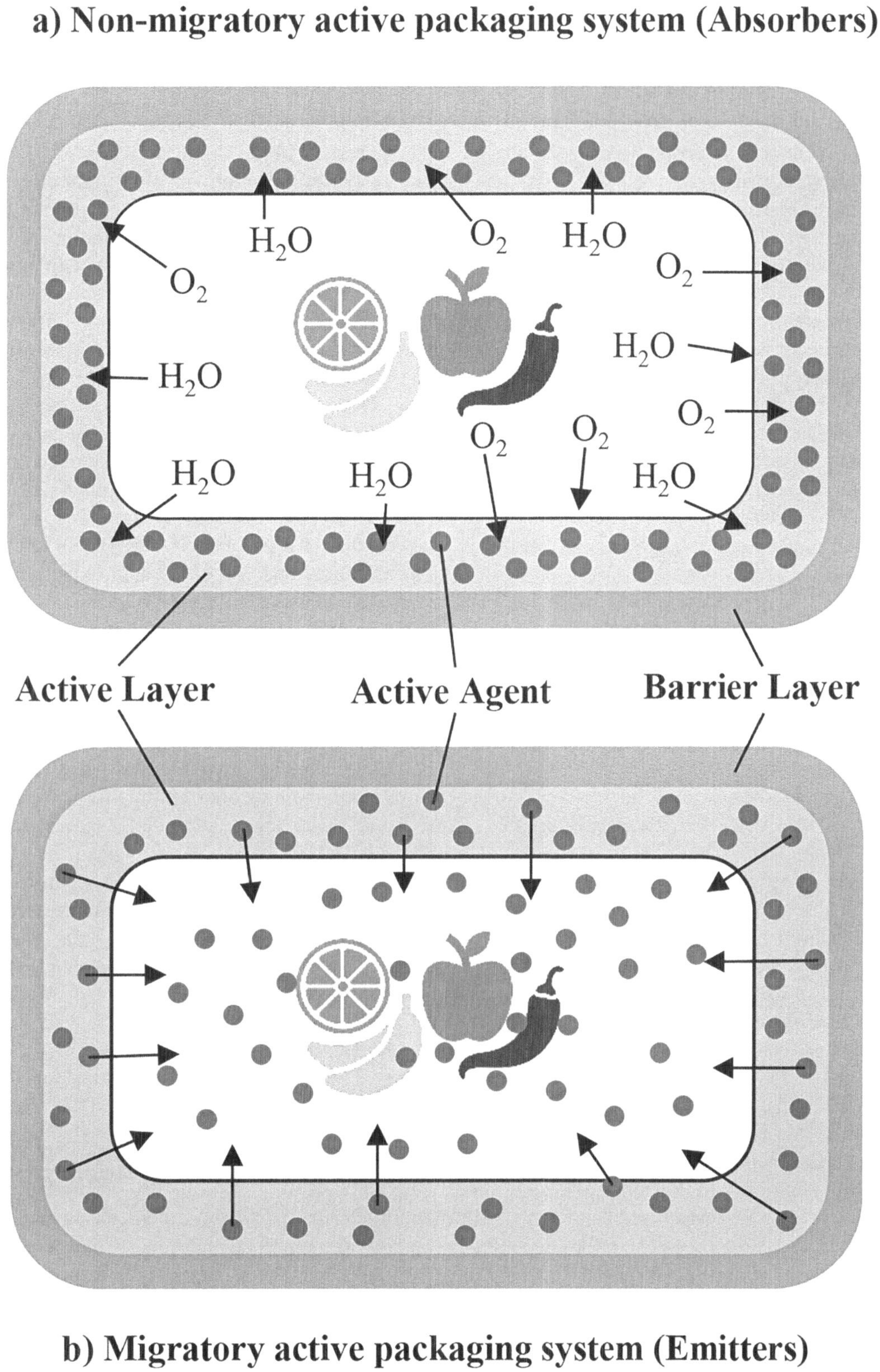

FIGURE 26.1 Schematic representations of (a) non-migratory active packaging and (b) migratory active packaging systems.

Inc., USA), OxyRx® (OS container, Mullinix Packages Inc., USA), Shelfplus® O_2 (Albis Plastic GmbH, USA) Aegis® Oxce (Honeywell International Inc., USA), Amosorbtm ColorMatrix™ (PolyOne™, UK), Cryovac® (Sealed Air Corporation, USA) (Yildirim et al., 2018).

Lulo fruits packed with a polylactic acid film and iron filings as an oxygen absorber showed that active materials extended the postharvest life of the fruits by 25 days (Escobar et al., 2023). Yellow bell peppers were stored in LDPE packaging with oxygen absorbers (ferrous iron oxide, finely powdered) for active modification and without oxygen absorbers for the passive modification of the headspace. The packed commodity had a shelf life of 28 days, much longer than the control (Devgan et al., 2019). Similarly, the barberries in the OS packages retained good sensory qualities regarding biochemical characteristics, texture and color because the oxygen level decreased, thereby extending its shelf life (Niazmand & Yeganehzad, 2020). Jamun fruits stored with iron powder sachets (1 g/kg of fruits) had the lowest respiration rate, the maximum retention of titratable acidity, and the highest levels of phenols and antioxidant activity over 6 days of storage time (Pattar et al., 2021). Bartlett pears packed in LDPE film bags (50.8 μm) including 15% iron powder sachets had the highest ascorbic acid, titratable acidity and total soluble solids (TSS) and the least fruit softening, spoilage, physiological weight loss, rate of respiration and activities of pectin methylesterase and polygalacturonase enzymes. Further, a rise in fruit Ca content and premier level of general acceptance after cold storage and ambient shelf-life intervals was also noticed (Kumar & Thakur, 2020).

26.3.1.3.2 Non-metallic Oxygen Scavengers

Non-metallic OS like ascorbate salts, glutathione, ascorbic acid, catechol, oleic and linoleic acids and enzymes (for example, glucose oxidase and ethanol oxidase) have also been utilized and are generally non-hazardous. However, non-metallic OS are less effective at capturing oxygen compared to metallic scavengers (Nayik & Muzaffar, 2014).

Certain following parameters must be met by oxygen scavengers (Bhardwaj, Alam & Talwar, 2019), such as OS:

- should not be harmful to the human body
- should timely absorb oxygen
- should not produce harmful gases or odors
- should have consistent quality and performance
- should be small in size
- should be able to absorb the most oxygen
- should be fairly priced

However, several factors that must be considered for oxygen absorbers' selection for food packaging systems include certain environmental variables like temperature and pH (Damaj, Joly & Guillon, 2015), humidity (Yildirim et al., 2018) and UV radiation (Zerdin et al., 2003). Moreover, familiarity with preliminary headspace oxygen content, the packaging's barrier qualities and headspace volume are required. In addition, knowledge about the oxygen concentration, packaging volume and permeability, and the product's minimum shelf life are crucial in deciding the needed OS rate and capacity as well as the cost of the OS packaging (Yildirim et al., 2018).

26.3.1.4 Carbon Dioxide Scavengers

Fresh fruits and vegetables chiefly generate carbon dioxide during respiration. An excessive amount of CO_2 inside the packaging can have negative consequences on both the packaging and the food (for example: flavor alterations, color fading and changes in the texture) (Alves et al., 2023). Several fruits and vegetables, such as potatoes, lettuce, onions, cucumbers, cauliflower, artichokes, apricots, peaches, apples and carrots, are particularly vulnerable to quality deterioration when suffering from prolonged exposure to CO_2 (Drago et al., 2020).

CO_2 may be absorbed or removed by physical adsorption, chemical reaction (involving alkaline solution), membrane separation or cryogenic condensation (Lee, 2016). But the latter two technologies are inappropriate for usage in food packaging, as they are associated with high pressure and refrigeration equipment which are frequently used in industrial and power plants. So, the two main techniques for CO_2 scavenging in food packaging are non-harmful chemicals and physical absorbers (Han et al., 2018).

CO_2 scavenging substances can be designed as a sheet or coating or may be contained in a sachet placed within the food packaging. Absorbers, either physical or chemical, can be coupled in a formulation with a catalyst or synergistic ingredient (Lee, 2016). The chemical absorbers include calcium oxide, silica gel, KOH, NaOH and $Ca(OH)_2$ (Bhardwaj et al., 2019). The chemical reaction of calcium oxide and water produces $Ca(OH)_2$, which then interacts with CO_2 to produce $CaCO_3$, while zeolites and activated carbon fall under the category of physical absorbers (Alves et al., 2023). The $Ca(OH)_2$ system in films presents a potential way to use its capacity for CO_2 adsorption, and the results revealed outstanding CO_2 adsorption capability (Lee et al., 2020). It was discovered that strawberries were adequately preserved when they were exposed to MAP while being stored at 4 °C for 4 weeks using two different CO_2 scavengers and one O_2 scavenger (Aday, Caner & Rahvali, 2011).

When choosing the type of CO_2 absorber to be utilized, the following things should be taken into account (Lee, 2016):

- The food product's qualities (e.g., the amount of CO_2 it produces)
- The intended level of CO_2
- The packaging factors
- The CO_2 absorption capability and speed of CO_2 scavenging

26.3.1.5 Antimicrobial Agents

Antimicrobial packaging methods involve the incorporation of antimicrobial substances into a packaging system. Antimicrobial agents serve the purpose of controlling or limiting the growth of undesirable microbes on the surface of produce. These agents are frequently transferred from the packaging material to the surface of the food. Antimicrobials are used as a coating on a polymer or in the matrix of a polymeric material. As the microbes encounter the antimicrobial agent (either embedded in the packaging material or emitted or gradually diffused from the packaging to the food), the microbes are eliminated (Giannakourou & Tsironi, 2021; Appendini & Hotchkiss, 2002).

Antimicrobial substances employed in active packaging are believed to prolong the lag phase and limit microbe growth rates, hence increasing the postharvest life and ensuring produce safety (Ribeiro-Santos et al., 2017). Food packaging performance is improved by using antimicrobial packaging (Atef et al., 2015). Numerous types of antimicrobial agents can be incorporated into packaging materials, such as phenolic compounds, isothiocyanates, inorganics, organic acids, mineral acids and inorganics.

These antimicrobial agents are classified as either natural or synthetic. Inorganic antimicrobial agents (metals and metal oxides), organic agents (such as organic acids, peptides, enzymes, etc.) and naturally occurring compounds derived from plants (polyphenols, essential oil constituents and complex plant extracts) make up the core group of antimicrobial agents. Their use is frequently determined by the packaging material. A few studies, for example, suggested to add potassium salt of sorbic acid and nisin peptide compounds to a chitosan-based coating solution for forming an active packaging film with increased flexibility (Remedio et al., 2019).

Nanocomposite films comprise a matrix (a bio-based polymer) that may be mixed with fillers (such as antimicrobial compounds). The fillers enhance the material's functional qualities. Biopolymers are natural (including different proteins, polysaccharides, lignins and lipids) and synthetic (including partially bio-based polymers like polyethylene, polyethylene terephthalate and polyamide). Metal or metal oxides are frequently used as nanofillers (Jamroz et al., 2019). Nanoparticles, generally <100 nm-sized particles, interact with metal ions, surfactants and polymers. The three types of nanoparticles are i) organic, ii) inorganic and iii) nanoparticles based on carbon. Nanoparticles are added to food packaging materials that boost their functional characteristics and, therefore, are part of diverse packaging applications. Nano-ZnO, silver (Ag)-nanoparticles, TiO_2-nanoparticles, carbon nanotubes, cellulose-nanowhiskers and starch nanocrystals are a few of the nanoparticles that have been incorporated in some active packaging systems owing to their distinct physical and chemical properties (Sharma et al., 2023). It is possible to use metal oxide nanoparticles as nanofillers. ZnO, TiO_2, silica (SiO_2), aluminum oxide (Al_2O_3), iron oxide (Fe_2O_3) and copper oxide (CuO) are the most popular materials utilized in this method. TiO_2 NPs incorporated with various organic and inorganic matrixes (like poly vinyl alcohol, polyethylene glycol and polyurethane) resulted in bactericidal property of TiO_2 coatings against *E. Coli* O157:H7 (Yemmireddy et al., 2015).

The most often utilized metal NPs with an antibacterial function are those made of copper, silver, selenium and gold. The polymer matrix can be supplemented with metal NPs. Not all metal NPs, however, are included in food coatings because they can occasionally be employed alone or in combination with other antimicrobials. An LDPE film modified with 1% (weight basis) silver NPs showed antifungal and antibacterial activities in packaged barberries (Motlagh et al., 2012). Similarly, Youssef et al. (2014) reported a chitosan film modified with silver and gold NPs showed promising antibacterial activities against *Staphylococcus aureus* and *Pseudomonas aeruginosa* and antifungal activities against *Aspergillus niger* and *Candida albicans*.

Numerous organic acids, including lactic acid, butyric acid, valeric acid, sodium formate, monopropionin, monobutyrin, monovalerin and monolaurin, have been shown to have antibacterial action (Kovanda et al., 2019). However, only a small number of these acids are utilized in food packaging. In one study, it was noticed that *Candida albicans*, *Escherichia coli*, *Staphylococcus aureus* and *Bacillus mycoides* all showed inhibitory activity when sodium nitrite and sodium benzoate were combined, while *B. mucoides*, *P. aeruginosa* and *E. coli* all showed antimicrobial effects when potassium sorbate and sodium nitrite were combined (Stanojevic et al., 2009).

There are tens of thousands of antimicrobial peptides (AMPs). Bacteriocins were the first AMPs characterized that were isolated from bacteria. The well-recognized bacteriocins include bifidin, enterocin, lactocin, lactococcin A, leucocin, nisin (a Class 1 bacteriosin) and pediocin. Other AMPs include cathelicidins, cecropins, defensins, pleurocidin, LL-37, plectasin, protegrins and magainins (Kumar et al., 2018; Rydlo et al., 2006; Santos et al., 2018). The majority of AMPs that have been discovered and researched are cationic and amphiphilic α-helical peptides. AMPs have been integrated directly into a packaging polymer or coated on the surface of the polymeric.

26.3.1.6 Moisture Absorbers

Active components in a moisture absorber reduce moisture and condensation within food packaging to prevent the growth of microorganisms (Belay et al. 2018). These systems incorporate a moisture absorber based on the weight of the product to boost the packaging system's capacity to remove moisture, improving the product's preservation for a longer period of time. Further, to facilitate the process of adsorption from the active element, the kinetics and adsorption equilibrium must be determined based on the moisture content and the storage temperature (Agudelo-Rodríguez et al., 2020). The moisture absorbers principle is based on the fact that hygroscopic substrates or materials may draw and hold water molecules from their surroundings (Yildirim et al., 2018).

26.3.1.7 Ethylene Scavengers

Over the course of the different phases of the supply chain (i.e., production, packaging, storage, marketing and consumption), 45–55% of the globally produced fruits and vegetables are wasted due to the ripening phytohormone, ethylene (Navarro-Martínez et al., 2021), initiating physiological processes, such as fruit ripening, vegetable yellowing, respiration, textural changes (e.g., softening and senescence) and shortening the shelf life of several fruits and vegetables (Singh et al., 2021); it also hastens chlorophyll breakdown, particularly in leafy vegetables (Yildirim et al., 2018). Therefore, an ethylene scavenging technology that removes ethylene from a package's headspace helps to preserve fresh produce and avoid food spoiling (Sultana et al., 2022). For example, individually packaged carambola fruits in LDPE bags with or without a sachet of commercial oxygen and ethylene absorbers were more firm, retained fruit color with limited losses of weight and ascorbic acid contents over the storage period of 35 days at 10 °C (Nimitkeatkai et al., 2022).

Unwanted ethylene within the package's headspace removal can be achieved using absorption, adsorption or scavenging (Siddiqui, 2016). However, the three basic methods used for ethylene control are (Gaikwad et al., 2020):

1. Ethylene reduction by modified atmospheric packaging by means of the exchange of different gases in the headspace
2. Usage of perforated packaging materials which minimally allow passage of gases through tiny holes in the packaging material
3. Ethylene removal via ethylene scavenger

Scavenging is a chemical reaction (takes place between two substances as a result of water absorption), while absorption is a physical reaction (between substances that physically absorb and confine ethylene molecules).

Ethylene scavenging practices include chemical compounds, i.e., potassium permanganate, sodium permanganate and titanium dioxide, and their irreversible reaction involves the oxidation of ethylene into water and carbon dioxide. On the other hand, the absorption/adsorption method typically includes zeolites, clays, charcoal, activated carbon and metal–organic frameworks to adsorb ethylene (Bhat et al., 2023). However, potassium permanganate, titanium dioxide and ozone fall under the category of ethylene catalytic oxidants, while 1-Methylcyclopropene is an ethylene inhibitor (Wei et al., 2021). Other ethylene remover materials (adsorbents) are Japanese Oya stone, palladium (Gaikwad et al., 2020), silica gel, zeolites, carbon nanospheres, activated carbon, halloysite nanotubes, porous metal oxide and montmorillonite (Wei et al., 2021). Absorbents like activated carbon or zeolite are replaced after these materials have achieved absorbent capability/saturation (Fang & Wakisaka, 2021).

Certain commercial $KMnO_4$-based scavengers include MM-1000 MULTIMIX® MEDIA (Circul-Aire Inc., Montreal, Canada), Bi-On® SORB (Bioconservacion S.A., Barcelona, Spain), Sofnofil™ (Molecular Products Limited, Essex, UK) and Chemisorbant (Purafil, Inc., Doraville, GA, USA) (Gaikwad, Singh & Negi, 2020). Usually, climacteric fruit packages contain sachets of $KMnO_4$ powder. The toxicity of $KMnO_4$ and its purple color prevent it from being allowed to come into direct contact with food (Sharma & Ghoshal, 2018). The effectiveness of $KMnO_4$ in reducing ethylene levels in mango (*Mangifera indica* L.) has been demonstrated by Fatima et al. (2023). These researchers packed mango fruits with $KMnO_4$ (wrapped in muslin cloth) and kept these fruits at 25 °C with 75% relative humidity (RH). They found that mangoes packed with $KMnO_4$ reached the ripening stage later than the control (Fatima et al., 2023). Tomato fruits packed in polypropylene trays with $KMnO_4$ showed the lowest ethylene production rate (Muhammad et al., 2023). Flowers can be protected against ethylene damaging effects such as flower wilting or abscission, epinasty of buds and flowers and leaf shedding by using ethylene scavenging products based on $KmnO_4$, thereby keeping their aesthetic appeal and increasing their shelf life (Álvarez-hernández et al., 2019). Ripe bananas were packed in LDPE film–based MAP bags with an ethylene absorber 'Green Keeper' and stored at 12 °C and 85–90% RH. This resulted in an extended shelf life of bananas up to 28 days (Kudachikar et al., 2011).

26.3.1.8 Titanium Dioxide (TiO_2)

Titanium dioxide is one of the most widely used photocatalysts for the breakdown of ethylene (Wei et al., 2021). TiO_2 is light-sensitive; TiO_2 becomes activated whenever exposed to visible light, photocatalyst or UV-light, and breaks down ethylene to O_2 and H_2O, producing reactive oxygen species (ROS) and hydroxyl radicals (OH•) that efficiently stop microbial development, slow down fruit's ripening, as well as increase the fresh product's shelf life (Phothisarattana et al., 2021). Alonso-Salinas et al. (2023) utilized the mechanism of ethylene elimination through UV light radiation on TiO_2. They found that a combination of $KMnO_4$ + UV-C + TiO_2 lessened ethylene levels, improved organoleptic quality and lengthened the shelf life of broccoli (Alonso-Salinas et al., 2023). The shelf life and physiological and physiochemical characteristics of fresh strawberries were maintained for 12 days when stored at 4 °C in an improved polyethylene mixed with nano-Ag, TiO_2 and kaolin clay compared to the control sample (Yang et al., 2010). Papayas wrapped with PE foam nets incorporated with gelatin-TiO_2 (by dip coating) exhibited delayed ripening and maintained their green color and firmness over a period of 4 days under UV-A light (Fonseca et al., 2021).

26.3.1.9 1-Methylcyclopropene

1-Methylcyclopropene (1-MCP) is a small ring hydrocarbon that has been extensively utilized in fruit and vegetable preservation due to its non-toxicity, great efficacy and ease of handling (Lv et al., 2020). 1-MCP competes with ethylene to bind with ethylene receptors and inhibits the expression of a number of transcription factors to prevent the over-ripeness of harvested fruits (Liu et al., 2019). Giné-Bordonaba et al. (2020) explained the physiology of pear scald with 1-MCP after 4 months of cold storage and discovered that 1-MCP treatments reduced the formation of scald by repressing the gene for the enzyme polyphenol oxidase and limiting the generation of ethylene. It is suggested to release 1-MCP into the package headspace at a predictable rate using heat-pressed polymer films to counteract the effects of ethylene. A film was created using three layers: an inner layer of hydrophilic hydrogel, a middle layer of 1-MCP and palladium on carbon and an outside layer of hydrophobic ethyl cellulose. At 20 °C, the film's effects on *Agaricus bisporus* mushrooms showed that the film might prevent the softening, browning and weight loss of mushrooms while maintaining cell membrane integrity and stopping the activity of enzymes involved in ethylene formation (Wu et al., 2021).

26.3.1.10 Zeolite

The potential of several clays to act as ethylene adsorbents was studied. Zeolite is a porous material that is frequently utilized because it has huge cavities for seizing ethylene soon after adsorption (Ichiura et al., 2003). Zeolite-based products work very well because of their porous, three-dimensional structure and cation exchange, adsorption and molecular sieving capabilities (Yildirim et al., 2018). Such ethylene scavengers are appropriate for commercial uses because of their characteristics of being abundant, economical, nontoxic and environmentally acceptable, mainly in the agro-food industry (Mariah et al., 2022). Esturk et al. (2014) made zeolite-based (active ingredient) low-density polyethylene bags that increased the shelf life of broccoli by 20 days compared to 5 days for the control (unpacked). A novel and environmentally friendly active material packaging was developed by the inclusion of silver-impregnated zeolite with chitosan-coated kraft paper. Cherry tomatoes were packed to test the effectiveness of active packaging materials in terms of ethylene scavenging for 30 days at 25 ± 2 °C. It was revealed that these novel active packaging techniques had the ability to slow the rate at which the fruit ripened by regulating the ethylene content in the headspace, as the fruits exhibited a reduction in weight loss while retaining their texture (Soares et al., 2023).

26.3.1.11 Activated Carbon

Ethylene can also be successfully removed using scavengers made of activated carbon with different metal catalysts (Ebrahimi et al., 2022). Activated carbon comes with some benefits, such as its hydrophobic nature, high surface area and comparatively low production costs (Gaikwad et al., 2020). Such absorbents have been utilized as sachets inside of packaging of fresh products and as part of paper produce storage bags (Ebrahimi et al., 2022). Biocomposite foam with activated carbon significantly reduced the ethylene levels in the air, thus delaying Hom Thong banana ripening and reducing changes in color, total soluble solids and flesh firmness during ambient storage (Nooun et al., 2023). According to research, acidified activated carbon powder along with a $PdCl_2$–$CuSO_4$ ethylene scavenger reduced ethylene concentration in the storage room at 20 °C, thereby prolonging the shelf life of broccoli and delaying yellowing and quality loss (Cao et al., 2015).

26.3.1.12 Palladium

Palladium-based scavengers have been found to have significant ethylene adsorption capability; however, they are typically evaluated as sachets in packages or in a storage area rather than in the framework of packaging films. Palladium's industrial use is restricted by its high price (Yildirim et al., 2018). Bananas ripened slowly as a result of palladium being added to synthetic activated carbon (made from sugarcane bagasse) to boost the effectiveness of ethylene and carbon dioxide removal (Tepamatr, 2023). It has been reported that tomatoes packed in MAP [infused palladium catalyst plus granular-carbon activated (GAC-Pd)] showed decreased ethylene concentration, which postponed ripening compared to the control group (MAP+ GAC without palladium) (Bailén et al., 2006). Sapota fruits were first packed in $PdCl_2$-encapsulated poly vinyl alcohol nanomats and then placed in corrugated fiber board boxes. This technique revealed greater quality traits, as well as reduced weight loss and respiration rate over the course of 8 days of storage (Gundewadi et al., 2022).

26.3.1.13 Nanocomposites

Nanocomposites can be used for a variety of purposes. Their uses include enhancing the mechanical, antibacterial and barrier qualities of food products and the development of innovative food packaging materials that improve food flavor, appearance and texture (Ebrahimi et al., 2022). A chitosan/nano-sized TiO_2 nanocomposite-developed film coated on tomatoes promoted photodegradation of ethylene, thereby delaying ripening as well as maintaining the fruits' quality (Kaewklin et al., 2018). The $KMnO_4$-impregnated nanoparticles (nanosilica and nano-clay) based on polyolefin elastomer (POE)–created nanocomposite films scavenged more ethylene, had less water vapor permeability and extended the bananas' shelf life at room temperature (Ebrahimi et al., 2021).

26.3.1.14 Biopolymers

Biopolymers have been employed to regulate ethylene in packaging film systems, as well as in the formulation of edible coatings for fresh fruits and vegetables. Biopolymer films are a potentially useful, inventive and environmentally friendly material for controlling ethylene and offer another possibility for fruit and vegetable storage and transportation with an extended shelf life. However, for ethylene control, biopolymer films have not received much research (de Moraes et al., 2020). Paper products, which are entirely made of cellulose fibers, can also be used to package food. Paper products' efficacy in the preservation of fruits can be further enhanced by the addition of ethylene scavengers, antibacterial agents, or other useful active compounds. Reasonably priced, sufficient airflow, preventing products from drying out and great mechanical strength are some advantages of paper packaging in contrast to polyolefin-based plastics (Wei et al., 2021). A multifunctional (bactericidal, water-locking and ethylene eliminating) packing biopolymer-based film (i.e., CMCGF) was formed (a modified metal-organic framework, MOF; i.e., Cer@MHKUST-1) and embedded within a conjugated matrix of chitosan quaternary ammonium salt (CQAS)/gelatin]. Researchers looked into the packaging's properties and preservative effects. In comparison to the control film (i.e., CGF), the results showed that the physicochemical features of CMCGF (such as mechanical, gas/light obstruction and wettability) were upgraded compared to CGF (7.8 cm^3/g STP), and CMCGF has better ethylene adsorption ability (65–69 cm^3/g STP). The death of the bacteria was caused by copper ions released from CMCGF and against *E. coli* and *S. aureus*, and CMCGF's antibacterial effectiveness was 97–100% and 98–100%, respectively.

Winter jujube and tomato postharvest storage experiments supported the CMCGF packaging's highly effective preservation effect (Nian et al., 2023). An ethylene scavenging paper based on pine needle (PN) waste was made by incorporating nanomaterials halloysite nanotube (HNT) and microfibrillated cellulose (MFC). The ability of the developed nanocomposite paper to scavenge ethylene was assessed using banana storage. The ripening and senescence process was postponed by the 30% PN/MFC/HNT packaging solution and the color of the banana, the TSS and the pH remained steady throughout the time of storage (Kumar et al., 2022).

26.3.2 Migratory Active Packaging (Emitters)

Fresh horticultural produce is extremely perishable, and the commercial and nutritional quality of these types of produces decrease during storage due to microbial contamination or metabolic processes like respiration (da Rosa et al., 2020; Ramesh et al., 2018). To overcome these problems, active and intelligent packaging techniques are being developed to preserve highly perishable produce while they are being stored and transported, as well as to provide information about their quality condition (Pavinatto et al., 2020; Taqi et al., 2014).

Active packaging materials, which may consist of antioxidants, antimicrobials and light or oxygen barriers can aid in extending the shelf life of fruits and vegetables. These compounds can migrate to the produce or not. Packaging that allows controlled non-volatile agent movement or volatile compound emissions into the atmosphere around food is known as migratory active packaging (Nwakaudu et al., 2015).

By releasing active ingredients through a smart regulated diffusion system from a packaging material to the produce surface or in the headspace of the package, this novel active food packaging idea extends the shelf life of packed produce (Bastarrachea et al., 2015; Goddard & Hotchkiss, 2007). Active agents used in active packaging may include antioxidants, CO_2 emitters, O_2 scavengers, ethylene scavengers and antimicrobial agents (Table 26.1).

TABLE 26.1 Active Agents Used in Active Packaging Systems

ACTIVE AGENTS	*ACTIVE COMPOUNDS*
Antioxidants	Polyphenols, essential oils, α-tocopherols, plant extracts, lignin
CO_2 emitters	Ferrous carbonate, Na_2CO_3, citric acid
O_2 scavengers	Iron, pyrogallol, palladium, gallic acid, glucose oxidase, laccase
Ethylene scavengers	$KMnO_4$, metal oxides, activated carbon, titanium dioxide
Antimicrobial agents	Metals, essential oils, chitosan, nisin, lactoferrin, lysozyme

Source: Vilela et al. (2018)

26.3.2.1 *Antioxidants*

A number of antioxidants of natural or synthetic origin have been investigated to be incorporated in innovative active packaging systems and coating products. These active compounds in the packing films move to the produce and absorb oxidative radicals from it, thus improving the product's nutritional quality and shelf life (Rangaraj et al., 2021). Antioxidants often prevent oxidation through the neutralization of singlet oxygen, the reduction of H_2O_2 and scavenging free radicals (Yildirim et al., 2018; Gomez-Estaca et al., 2014). Antioxidants mitigate oxidative stress directly by interacting with ROS. Antioxidants may indirectly up-regulate the activity or expression of intracellular antioxidant enzymes or suppress the activity/expression of free radical–producing enzymes (Lu et al., 2010).

Antioxidants are classified into two broad categories based on their mode of action within the active packaging system: release-type and scavenger-type.

Release-type antioxidants are designed to slowly release active compounds from the film during the entirety of a product's shelf life. For the development of antioxidant active packaging for a well-regulated release, antioxidants are embedded into a polymer base using an extrusion technique (Tian et al., 2014). Scavenger-type or non-migratory antioxidant coatings, on the other hand, are designed to hunt for oxidants (free radicals, ROS and transition metals), thus extending the shelf life of fruits or vegetables showing higher oxidation sensitivity. To ensure an extended shelf life, uniform antioxidant levels are maintained during produce storage. Scavenger-type antioxidants have a preventative mechanism that reduces or inhibits the oxidation process of the fruit or vegetable (Gomez-Estaca et al., 2014; Islam et al., 2017; Licciardelloa et al., 2015).

Antioxidant products are available in the form of a sachet, pad or label. These products are incorporated into the packaging monolayer or multilayer polymers. Currently and commonly used primary synthetic antioxidants for active packaging include butylhydroxyanisole, butylhydroxytoluene, tert-butylhydroquinone and propyl gallate (Xu et al., 2021a; Lu et al., 2010). Although these antioxidants are effective at reducing lipid oxidation, there is evidence that their consumption can have negative effects on humans. As a result, consumers are increasingly choosing antioxidant alternatives from natural sources, such as those found in compounds from various plant extracts or essential oils (Shahidi & Ambigaipalan, 2015; Yashin et al., 2017; Cruz et al., 2018).

26.3.2.2 *Natural Antioxidants*

Various botanicals from plants, including garlic, saffron, potatoes, cabbage and tomatoes, have been incorporated as natural antioxidant additives in active packaging systems (Yildirim et al., 2018; Pereira et al., 2015). These include numerous antioxidant substances, including flavonoids, phenolics (phenol and polyphenols), steroids, thiol compounds and carotenoids. Lipid oxidation and protein denaturation can be effectively delayed by antioxidants. They also disrupt oxidation chain reactions, preventing further oxidation progress and limiting or inactivating enzyme activities to catalyze the oxidation reactions.

Polyphenols (phenolic acids, flavonoids and lignans), vitamins (C and E) and carotenoids (carotenes and xanthophylls) are the three ubiquitous sub-classes of natural antioxidants. Carotenoids are renowned for their capacity to scavenge ROS, particularly $^1O_2$ (Edge & Truscott, 2021; Ramel et al., 2012). Enzymes that catalyze the breakdown of ROS and radicals are also considered antioxidants. While catalase converts H_2O_2 to water and oxygen, peroxidase causes the reduction of hydrogen peroxide, organic hydroperoxides and peroxynitrite and superoxide dismutase (SOD) dismutases $O_2^{-\bullet}$ (Waisundara, 2021). The regeneration of glutathione (GSH) from the oxidized form (GSSG) via GSSG reductase is another activity performed by enzymes that act as antioxidants (Lushchak, 2012). Aside from its capacity to give an electron or hydrogen atom, vitamin C can also regenerate other antioxidants, such as α-tocopherol (Hamza, 2017; Morelli et al., 2020). Additionally, some antioxidants work by suppressing the Fenton and Haber-Weiss reactions by chelating metal ions (such as Fe and Cu) (Akbari et al., 2016). This prevents the generation of ROS.

Fruit peels are among the most widely used agricultural by-products. Due to their antioxidative properties, the peels of fruits such as mangoes, oranges, lemons, tomatoes and pomegranates have frequently been utilized in the development of active packaging (Adilah et al., 2018; Hanani et al., 2018; Jridi et al., 2019; Marino-Cortegoso et al., 2022). Antioxidant packaging systems utilizing plant-based volatiles, extracts and essential oils have been created to maintain the quality and extend the shelf life of different horticultural products.

Plant-based antioxidants, which are added to edible active films, are complex molecules extracted largely as an essential oil or an extract (Tuberoso et al., 2013). Natural antioxidants derived from plant extracts release active ingredients from the packaging to defend the produce against free radicals, prooxidants and secondary breakdown products. Furthermore, these chemicals have antibacterial and antifungal activities that help protect fruit or vegetable quality. As a result, there has recently been a lot of interest in extracting new antioxidant molecules from diverse plant sources.

Polyphenolic compounds have been extracted from various sources (plant leaves, roots, fruits or vegetables, etc.) like acerola fruit (Teixeira & Aranha, 2022), green tea (Colon & Nerin, 2012), barley husk (Sanches-Silva et al., 2014) and rosemary (Andrade et al., 2023). Anthocyanins, a class of water-soluble color pigments, are present in large amounts in a variety of plant products, such as flowers, cereals, vegetables and fruits (Yildirim et al., 2018). Active packaging materials that use anthocyanins as a reducing agent and an oxygen suppressor have been found to benefit from their potential antioxidant activities. The type of anthocyanin utilized, the makeup of the biopolymer matrix, and the technique employed to generate the film all have an impact on how effective these antioxidants are. These naturally occurring phenolic compounds also possess outstanding antioxidant and antibacterial capacity (Yong & Liu, 2020; Alizadeh-Sani et al., 2021).

One of the most well-documented and noteworthy examples of plant components with dual effects that have been utilized in industrial practice are essential oils. Essential oils are plant secondary metabolites that inhibit pathogens like *Escherichia coli*, *Vibrio parahaemolyticus*, *Clostridium perfringens*, *Staphylococcus aureus*, *Listeria monocytogenes*, *Shigella spp.*, *Salmonella spp.*, *Pseudomonas spp.* and other microorganisms which otherwise cause food deterioration and associated financial losses (Tajkarimi et al., 2010; Soylu et al., 2010).

26.3.2.3 Carbon Dioxide (CO_2) Emitters

Carbon dioxide, at a higher concentration, preserves the quality and extends the shelf life of fresh produce by lowering its respiration rate. Therefore, owing to its advantageous qualities, CO_2 regulation has extensive application in the food packaging industry. Optimizing CO_2 levels in a packaging system is effective in prolonging the postharvest life of fresh agricultural produces (Jiang et al., 2022). Modified atmosphere packaging (MAP) is one accomplishment that is garnering more and more attention (Olveira-Bouzas et al., 2021). The purpose of MAP is to alter the composition of the gases surrounding preserved goods in order to impact how fruits and vegetables respire. By incorporating CO_2 emitters inside MAP packages, it is possible to raise fill levels, decrease package sizes, increase transport effectiveness and minimize environmental impact overall. A modified atmosphere of low O_2 and high CO_2 is recommended for storing fresh fruits and vegetables and has been demonstrated to reduce respiration rate and slow down senescence.

Numerous research reports have been compiled on the efficiency of a CO_2 release system. In a study by Huynh et al. (2023), the effect of a CO_2 emitter on extending the shelf life of raspberries was investigated. Modified atmosphere and modified humidity technology incorporated into packaging bags fitted with cardboard trays were tested to store raspberries along with vapor-releasing sanitization techniques. The initial trial investigated two modified atmosphere (14–15% O_2 and 7–8% CO_2) packaging variants (mineral-clay impregnated and microperforated), as well as H_2O_2 and SO_2 as sanitizers. After 20 days, compared to unpacked fruits, raspberry fruits packed in microperforated bags showed less than 2% moldy fruit by weight and slower anthocyanin accumulation. In another study that used a similar methodology, two types of microperforated bags and H_2O_2 with different levels of perforation were studied. The testing showed that the pre-retail shelf life of raspberries increased in response to a modified environment of 14–15% O_2 and 7–8% CO_2. Several studies have shown that low concentrations of O_2 and high concentrations of CO_2 and N_2 extend the shelf life of cilantro, iceberg lettuce, pears, potatoes, fresh-cut papaya and lily bulbs (Jiang et al., 2022; McManamon et al., 2019; Shen et al., 2019; Waghmare & Annapure, 2013; Waghmare & Annapure, 2015; Wang et al., 2021). Similarly, fresh-cut honeydew melon, eggplants, artichokes, iceberg lettuce and garlic scape had longer shelf lives when packed with active MAP (high content CO_2

and low content O_2) developed from a polypropylene plastic bag (Chen et al., 2019; Ghidelli et al., 2014 and 2015; Paillart et al., 2017; Zhang et al., 2013).

26.3.2.4 Natural Antimicrobials

Although several chemical antimicrobial agents have regulatory approval for their use in the food industry, these compounds have shown to be a risk to the health of consumers (Kuorwel et al., 2011; Mahmud & Khan, 2018). The list of such hazardous antimicrobials includes sodium benzoates, potassium sorbates, sodium salt of propionic acid, triclosan, nisin, sorbic acid, Na/K pyrosulfites, chlorides, tartaric acid and nitrites. Therefore, natural antimicrobial compounds are garnering much attention owing to their capacity to inhibit the growth and activity of microbes and boost the shelf life of food products (Gyawali & Ibrahim, 2014). Natural antimicrobial additives are derived from organic matter from a diverse range of sources, including plants, animals, algae and fungi, which results in reducing humans' exposure to certain health concerns (Mahmud & Khan, 2018).

The most significant natural antimicrobial agents, according to several studies, are organic acids (e.g., sorbic acid, propionic acid and citric acid), essential oils extracted from plants (e.g., thyme, cinnamon, clove, rosemary, sage, oregano, basil and vanillin) and naturally occurring polymers (such as chitosan). These compounds also include enzymes from animal sources (such as lysozyme and lactoferrin) and bacteriocins from microbial sources (nisin, natamycin, lactocin and pediocin) (Mahmud & Khan, 2018; Gutierrez et al., 2010). In the food industry, natural compounds such as spices and essential oils have been used to increase foods' shelf life. Food-borne pathogens like *Salmonella*, *Listeria monocytogenes*, *Escherichia coli*, *Bacillus cereus* and *Staphylococcus aureus* are eliminated, and the overall quality of food products is improved, using a variety of plant- and spice-based antimicrobials. A total of 30,000 components and more than 1,340 plants with known antibacterial properties have been extracted from plant-oil compounds with phenolic groups that are used in the food industry (Mahmud & Khan, 2018). Polyphenols, alkaloids, glucosinolates, isothiocyanates, polyamines, saponins, terpenoids and thiosulfinates are natural antimicrobial agents used mostly in active packaging (Cowan, 1999).

26.3.2.5 Essential Oils

Biosourced antimicrobials, notably essential oils (EOs) and their active components, have attracted significant attention in recent years, emerging as powerful and convenient natural alternatives to synthetic antimicrobials (Yammine et al., 2022). EOs are mixes of volatile compounds generated by the secondary metabolism of aromatic and other plant types (Bassole & Juliani, 2012). EOs have a wide range of biological and antimicrobial activities (Falleh et al., 2020).

EOs are a different class of plant compounds with potent antioxidant and antibacterial capabilities (Alizadeh-Sani et al., 2020). They are made up of a complex blend of phenolics, terpenes (α-pinene, limonene), phenylpropanoids (linalool, carvacrol) and terpenoids (cinnamaldehyde, eugenol) (Bakkali et al., 2008; Hyldgaard et al., 2012). Because of their Generally Recognized as Safe (GRAS) status, EOs have been utilized in active food packaging matrices as functional components. Distillation can be used to extract oils from botanical sources like citrus, cinnamon, thyme, rosemary, carvacrol, tea and other EOs (Djenane, 2015; Maisanaba et al., 2017). Cinnamon, citron, clove, coriander, garlic, lemongrass, oregano, rosemary, parsley and sage are the most popular EOs, and they are utilized in smart or bioactive packaging materials to stop microbial growth on food surfaces (Tajkarimi et al., 2010). The physicochemical, mechanical, antibacterial and antioxidant properties of packing materials can all be improved by adding these EOs (Alizadeh-Sani et al., 2018).

Numerous EOs, such as peppermint, tea tree, thyme, lemon, clove, lavender, mint and bergamot, as well as active EO components like thymol, carvacrol and eugenol, have been shown to have significant antimicrobial effects against a variety of microorganisms, including *Staphylococcus aureus*, *Listeria monocytogenes*, *Candida spp.*, *Escherich coli*, *Salmonella spp.*, *Proteus spp.*, *Klebsiella spp.* and *Pseudomonas Aeruginosa* (Cakmak et al., 2020; Valkova et al., 2021; Ed-Dra et al., 2021; Orlo et al., 2021; El-Darier et al., 2018; Rajkowska et al., 2016).

Lipid oxidation and microbial growth have been shown to be considerably reduced by the addition of rosemary oil and TiO_2 nanoparticles to a biopolymer packaging material (Alizadeh-Sani et al., 2020). Similar to this, it has been demonstrated that adding EOs contained in cyclodextrins to packaging materials will improve cherry tomato quality and shelf life (Buendía-Moreno et al., 2019). According to Da Rocha Neto et al. (2019), antimicrobial packaging materials using star anise or palmarosa EOs have been demonstrated to extend the shelf life of and inhibit *Penicillium expansum* development in apples.

In comparison to EOs from thyme, lemon and rosemary, tea tree EO has been shown to have better antibacterial and antioxidant properties (Xu et al., 2021b). In a study conducted by Tian et al. (2023) to assess the effect of tea tree EO on antimicrobial activity, sustained-release (tea tree EO) solid preservative (SP) was applied to fresh pineapple slices in MAP. The result showed that SP increased the sensory quality of fresh-cut pineapple while reducing nutritional loss and microbiological deterioration, extending its shelf life to 4 days. Furthermore, antioxidant ability was increased by boosting antioxidant enzyme activity, antioxidant content and 2,2-diphenyl-1-picrylhydrazine scavenging capacity, whereas MDA buildup was decreased.

Recently, cinnamon EO has gained popularity due to its potential antioxidant, anti-inflammatory and antitumor properties (Bertacchi et al., 2021). In one study, pectin modified with 0.7% v/v of cinnamon EO, glycerol and Tween 80 resulted in shelf-life extension of Golden Delicious apples (*Mauls domestica*) (Naqash et al., 2022). Cinnamon EO inhibits the growth of gray mold and *Rhizopus* rot that otherwise results in the deterioration of strawberry fruit quality (Yousef et al., 2019). Similarly, potato starch, when combined with cumin EO, obtained from *Cuminum cyminum*, was effective in maintaining the internal quality of pear fruit (Oyom et al., 2022). The starch in combination with 0.2% and 0.4% cumin EO efficiently reduced change in color, fruit firmness, water loss and *Alternaria alternata*–induced decay in pear fruits. The shelf life of pears increased by 28 days. The effect of thymol EO with chitosan was studied on the postharvest quality of peaches. The fruits coated with chitosan and thymol had significantly less weight loss, TSS and fungal decay and a maximum storage life of 28–30 days (Rahimi et al., 2019). In one study, it was revealed that a cassava starch–based edible covering integrated with clove or cinnamon EOs improved papaya preservation for longer periods of time. The addition of clove or cinnamon EOs efficiently decreased counts of mold and yeast, hence lowering microbial activity (do-Nascimento et al., 2023). Similarly, lemongrass EO increased the antifungal efficiency of a chitosan-based film, which led to an extension in the storage life of coated strawberry fruits (Perdones et al., 2012). The shelf life of strawberries was extended by adding lavender EO in a chitosan film that inhibited the infestation of *Botrytis cinerea* against spoiling fresh strawberries (Rattanapitigorn et al., 2006; Sangsuwan et al., 2016).

26.3.2.6 Encapsulated Essential Oils

Some studies have reported that directly applying EOs on fresh fruits and vegetables adversely affects the sensory properties of produce, mainly owing to off-flavor or high reactivity. However, regulated release of EOs and biologically active compounds have been achieved through a microencapsulation technique. Two approaches are adopted to prepared microencapsulated EOs: top-down and bottom-up (Guo et al., 2021). Mechanical procedures like shearing, emulsification, extrusion, homogenization and grinding are used for size-reduction and structural reshaping of large materials in top-down methods. Self-assembly and self-organization reactions construct the large particles from small molecules in bottom-up approaches. These methods utilize techniques like electro-spinning, electro-spraying, spray-drying, nano-precipitation and coacervation.

EO encapsulation reduces undesirable sensory degradation while extending the postharvest life of freshly harvested fruits and vegetables. Cherry tomatoes were packed in active cardboard trays with their inner surface coated with β-cyclodextrins enriched with encapsulated EOs of carvacrol, oregano and cinnamon (Buendía-Moreno et al., 2019) and were stored in the coated cardboard at 8 °C. This active cardboard tray packaging maintained the firmness and color of cherry tomatoes up to 24 days. Dextrin-derived nanosponges incorporated with encapsulated coriander EO not only had greater crystallinity but also had higher antimicrobial activity, mainly owing to controlled oil release (Silva, Caldera, Trotta, Nerin & Domingues, 2019).

Rusková et al. (2023) encapsulated lemongrass and oregano EOs into a poly(lactic acid) and poly(3-hydroxybutyrate) packaging to assess the biochemical and quality traits of stored strawberry fruits at 4 ±0.5 °C and 85 ±5% RH. They found that EO-incorporated active packaging significantly maintained fruit quality and extended the shelf life of strawberries. Similarly, Navarro-Martínez et al. (2021) found that active paper sheets releasing vapors of encapsulated EOs of citral, carvacrol, oregano and cinnamon significantly reduced ethylene production in broccoli florets stored at 2 °C (optimum storage temperature) or 22 °C (abusive commercialization temperatures).

26.4 CONCLUSIONS

Active packaging systems are being increasingly used to package fresh fruits and vegetables. These systems have been proven to be highly beneficial for the fresh food industry due to their effectiveness in prolonging the shelf life of fresh fruits and vegetables. The use of active packaging mechanisms and systems may vary with the kind of fruit or vegetable that is being packaged. The most promising active packaging systems for fresh fruits and vegetables are based on oxygen scavengers, carbon dioxide absorbers and emitters and ethylene absorbers. Additionally, antioxidants and antimicrobial agents are also added to the main packaging systems, mostly to increase the efficiency of the main active packaging system. Research on active packaging systems is being conducted, with numerous commercial products available or under development. Recently, concepts of nanotechnology, biodegradability, recycling and edibility have been introduced to active packaging systems to minimize production costs and protect the environment. However, more research is needed in this area to extend the postharvest life of fresh fruits and vegetables while improving food safety and security. This will enable the fresh fruit and vegetable industry to focus on food safety, affordability, traceability and environmental concerns linked to evolving active food packaging systems.

REFERENCES

Aday, M. S., Caner, C., Rahvalı, F. (2011). Effect of oxygen and carbon dioxide absorbers on strawberry quality. *Postharvest Biology and Technology*, *62*(2), 179–187.

Adilah, A. N., Jamilah, B., Noranizan, M. A., Hanani, Z. A. N. (2018). Utilization of mango peel extracts on the biodegradable films for active packaging. *Food Packaging and Shelf Life*, *16*, 1–7.

Agudelo-Rodríguez, G., Moncayo-Martínez, D., Castellanos, D. A. (2020). Evaluation of a predictive model to configure an active packaging with moisture adsorption for fresh tomato. *Food Packaging and Shelf Life*, *23*, 100458.

Akbari, A., Jelodar, G., Nazifi, S., Sajedianfard, J. (2016). An overview of the characteristics and function of vitamin c in various tissues : Relying on its antioxidant function. *Zahedan Journal of Research in Medical Sciences*, *18*(11), e4037.

Alizadeh-Sani, M, Khezerlou, A., Ehsani, A. (2018). Fabrication and characterization of the bionanocomposite film based on whey protein biopolymer loaded with TiO_2 nanoparticles, cellulose nanofibers and rosemary essential oil. *Industrial Crops and Products*, *124*, 300–315.

Alizadeh-Sani, M. Kia, E. M., Ghasempour, Z., Ehsani, A. (2021). Preparation of active nanocomposite film consisting of sodium caseinate, ZnO nanoparticles and rosemary essential oil for food packaging applications. *Journal of Polymers and the Environment*, *29*, 588–598.

Alizadeh-Sani, M, Mohammadian, E, Rhim, J. W., Jafari, S. M. (2020). pH-sensitive (halochromic) smart packaging films based on natural food colorants for the monitoring of food quality. *Trends in Food Science & Technology*, *105*, 93–144.

Almasi, H., Jahanbakhsh Oskouie, M., Saleh, A. (2021). A review on techniques utilized for design of controlled release food active packaging. *Critical Reviews in Food Science and Nutrition*, *61*(15), 2601–2621.

Alonso-Salinas, R., López-Miranda, S., González-Baidez, A., Pérez-López, A. J., Noguera-Artiaga, L., Núñez-Delicado, E., Carbonell-Barrachina, A., Acosta-Motos, J. R. (2023). Effect of potassium permanganate, ultraviolet radiation and titanium oxide as ethylene scavengers on preservation of postharvest quality and sensory attributes of broccoli stored with tomatoes. *Foods*, *12*(12), 2418.

Álvarez-Hernández, M. H., Martínez-Hernández, G. B., Avalos-Belmontes, F., Castillo-Campohermoso, M. A., Contreras-Esquivel, J. C., Artés-Hernández, F. (2019). Potassium permanganate-based ethylene scavengers for fresh horticultural produce as an active packaging. *Food Engineering Reviews*, *11*, 159–183.

Alves, J., Gaspar, P. D., Lima, T. M., Silva, P. D. (2023). What is the role of active packaging in the future of food sustainability? A systematic review. *Journal of the Science of Food and Agriculture*, *103*, 1004–1020.

Andrade, M. A., Barbosa, C. H., Cerqueira, M. A., Azevedo, A. G., Barros, C., Machado, A. V., Coelho, A., Furtado, R., Correia, C. B., Saraiva, M., Vilarinho, F., Silva, A. S., Ramos, F. (2023). PLA films loaded with green tea and rosemary polyphenolic extracts as an active packaging for almond and beef. *Food Packaging and Shelf Life*, *36*, 101041.

Appendini, P., Hotchkiss, J. H. (2002). Review of antimicrobial food packaging. *Innovative Food Science and Emerging Technologies*, *3*, 113–126.

Atef, M., Rezaei, M., Behrooz, R. (2015). Characterization of physical, mechanical, and antibacterial properties of agar-cellulose bionanocomposite films incorporated with savory essential oil. *Food Hydrocolloids*, *45*, 150–157.

Bailén, G., Guillén, F., Castillo, S., Serrano, M., Valero, D., Martínez-Romero, D. (2006). Use of activated carbon inside modified atmosphere packages to maintain tomato fruit quality during cold storage. *Journal of Agricultural and Food Chemistry*, *54*, 2229–2235.

Bakkali, F., Averbeck, S., Averbeck, D., Idaomar, M. (2008). Biological effects of essential oils—A review. *Food and Chemical Toxicology*, *46*(1), 446–475.

Bassolé, I. H. N., Juliani, H. R. (2012). Essential oils in combination and their antimicrobial properties. *Molecules*, *17*(4), 3989–4006.

Bastarrachea, L. J., Wong, D. E., Roman, M. J., Lin, Z., Goddard, J. M. (2015). Active packaging coatings. *Coatings*, *5*, 771–791.

Belay, Z. A., Caleb, O. J., Mahajan, P. V., Opara, U. L. (2018). Design of active modified atmosphere and humidity packaging (MAHP) for 'wonderful'pomegranate arils. *Food and Bioprocess Technology*, *11*, 1478–1494.

Bertacchi, S., Pagliari, S., Cant, C., Bruni, I., Labra, M., Branduardi, P. (2021). Enzymatic hydrolysate of cinnamon waste material as feedstock for the microbial production of carotenoids. *International Journal of Environmental Research and Public Health*, *18*, 11–46.

Bhardwaj, A., Alam, T., Talwar, N. (2019). Recent advances in active packaging of agri-food products: a review. *Journal of Postharvest Technology*, *7*(1), 33–62.

Bhat, S. A., Rizwan, D., Mir, S. A., Wani, S. M., Masoodi, F. A. (2023). Advances in apple packaging: a review. *Journal of Food Science and Technology*, *60*(7), 1847–1859.

Brody, A. L., Strupinsky, E. P., Kline, L. R. (2001). *Active packaging for food applications*. CRC Press.

Buendía-Moreno, L., Soto-Jover, S., Ros-Chumillas, M., Antolinos, V., Navarro-Segura, L., Jose Sanchez-Martinez, M., et al. (2019). Innovative cardboard active packaging with a coating including encapsulated essential oils to extend cherry tomato shelf life. *Lwt-Food Science and Technology*, 116, 108584.

Çakmak, H., Özselek, Y., Turan, O. Y., Fıratlıgil, E., Karbancioğlu-Güler, F. (2020). Whey protein isolate edible films incorporated with essential oils: Antimicrobial activity and barrier properties. *Polymer Degradation and Stability*, *179*, 109–285.

Cao, J., Li, X., Wu, K., Jiang, W., Qu, G. (2015). Preparation of a novel $PdCl_2$–$CuSO_4$–based ethylene scavenger supported by acidified activated carbon powder and its effects on quality and ethylene metabolism of broccoli during shelf-life. *Postharvest Biology and Technology*, *99*, 50–57.

Cazón, P., Velazquez, G., Ramírez, J. A., Vázquez, M. (2017). Polysaccharide-based films and coatings for food packaging: A review. *Food Hydrocolloids*, *68*, 136–148.

Chen, J., Hu, Y., Yan, R., Hu, H., Chen, Y. and Zhang, N. (2019). Modeling the dynamic changes in O_2 and CO_2 concentrations in MAP-packaged fresh-cut garlic scapes. *Food Packaging and Shelf Life*, *22*, 00432.

Cheng, D., Ngo, H. H., Guo, W., Chang, S. W., Nguyen, D. D., Liu, Y., Wei, D. (2020). A critical review on antibiotics and hormones in swine wastewater: Water pollution problems and control approaches. *Journal of Hazardous Materials*, *387*, 121682.

Cichello, S. A. (2015). Oxygen absorbers in food preservation: a review. *Journal of Food Science and Technology*, *52*(4), 1889–1895.

Colon, M., Nerin, C. (2012). Role of catechins in the antioxidant capacity of an active film containing green tea, green coffee, and grapefruit extracts. *Journal of Agricultural and Food Chemistry*, *60*(39), 9842–9849.

Cowan, M.M. (1999). Plant products as antimicrobial agents. *Clinical Microbiology Reviews*, *12*(4), 564–582.

Cruz, R. G. da, Beney, L., Gervais, P., Lira, S. P. de, Vieira, T. M. F. de S., Dupont, S. (2018). Comparison of the antioxidant property of acerola extracts with synthetic antioxidants using an in vivo method with yeasts. *Food Chemistry*, *277*, 698–705.

Da Rocha Neto, A. C., Beaudry, R., Maraschin, M., Di Piero, R. M., Almenar, E. (2019). Vitamin c and cardiovascular packaging for apple shelf-life extension. *Food Chemistry*, *279*, 379–388.

Da Rosa, C. G., Sganzerla, W. G., Maciel, M. V. D. O. B., de Melo, A. P. Z., da Rosa Almeida, A., Nunes, M. R., Bertoldi, F. C., Barreto, P. L. M. (2020). Development of poly (ethylene oxide) bioactive nanocomposite films functionalized with zein nanoparticles. *Colloids and Surfaces A: Physicochemical and Engineering Aspects*, *586*, 124268.

Damaj, Z., Joly, C., Guillon, E. (2015). Toward new polymeric oxygen scavenging systems: formation of poly (vinyl alcohol) oxygen scavenger film. *Packaging Technology and Science*, *28*, 293–302.

de Moraes, M. A., da Silva, C. F., Vieira, R. S. (2020). Biopolymer Membranes and Films: Health, Food, Environment, and Energy Applications. Elsevier.

Devgan, K., Kaur, P., Kumar, N., Kaur, A. (2019). Active modified atmosphere packaging of yellow bell pepper for retention of physico-chemical quality attributes. *Journal of Food Science and Technology*, *56*(2), 878–888.

Djenane, D. (2015). Chemical profile, antibacterial and antioxidant activity of algerian citrus essential oils and their application in *Sardina pilchardus*. *Foods*, *4*(2), 208–228.

do-Nascimento, A., Toneto, L. C., Lepaus, B. M., Valiati, B. S., Faria-Silva, L., de São José, J. F. B. (2023). Effect of edible coatings of cassava starch incorporated with clove and cinnamon essential oils on the shelf life of papaya. *Membranes*, *13*(9), 772.

Drago, E., Campardelli, R., Pettinato, M., Perego, P. (2020). Innovations in smart packaging concepts for food: an extensive review. *Foods*, *9*, 1628 1–42.

Ebrahimi, A., Khajavi, M. Z., Mortazavian, A. M., Asilian-Mahabadi, H., Rafiee, S., Farhoodi, M., Ahmadi, S. (2021). Preparation of novel nano–based films impregnated by potassium permanganate as ethylene scavengers: An optimization study. *Polymer Testing*, *93*, 106934.

Ebrahimi, A., Khajavi, M. Z., Ahmadi, S., Mortazavian, A. M., Abdolshahi, A., Rafiee, S., Farhoodi, M. (2022). Novel strategies to control ethylene in fruit and vegetables for extending their shelf life: A review. *International Journal of Environmental Science and Technology*, *19*, 4599–4610.

Ed-Dra, A., Nalbone, L., Filali, F. R., Trabelsi, N., El Majdoub, Y. O., Bouchrif, B., Giarratana, F., Giuffrida, A. (2021). Comprehensive evaluation on the use of thymus vulgaris essential oil as natural additive against different serotypes of salmonella enterica. *Sustainability (Switzerland)*, *13*(8), 4594.

Edge, R., Truscott, T. G. (2021). The reactive oxygen species singlet oxygen, hydroxy radicals, and the superoxide radical anion—examples of their roles in biology and medicine. *Oxygen*, *1*(2), 77–95.

El-Darier, S. M., El-Ahwany, A. M. D., Elkenany, E. T., Abdeldaim, A. A. (2018). An in vitro study on antimicrobial and anticancer potentiality of thyme and clove oils. *Rendiconti Lincei*, *29*(1), 131–139.

Escobar, H. J., Garavito, J., Castellanos, D. A. (2023). Development of an active packaging with an oxygen scavenger and moisture adsorbent for fresh lulo (*Solanum quitoense*). *Journal of Food Engineering*, *349*, 111484.

Esturk, O., Ayhan, Z., Gokkurt, T. (2014). Production and application of active packaging film with ethylene adsorber to increase the shelf life of broccoli (*Brassica oleracea* L. var. Italica). *Packaging Technology and Science*, *27*(3), 179–191.

Falleh, H., Ben Jemaa, M., Saada, M., Ksouri, R. (2020). Essential oils: A promising eco-friendly food preservative. *Food Chemistry*, *330*, 127268.

Fang, Y., Wakisaka, M. (2021). A review on the modified atmosphere preservation of fruits and vegetables with cutting-edge technologies. *Agriculture*, *11*, 992.

Fatima, F., Basit, A., Younas, M., Shah, S.T., Sajid, M., Aziz, I., Mohamed, H.I. (2023). Trends in potassium permanganate (ethylene absorbent) management strategies: towards mitigating postharvest losses and quality of mango (*Mangifera indica* L.) fruit. *Food and Bioprocess Technology*, *16*, 2172–2183.

Fonseca, J., Pabón, N. Y. L., Nandi, L. G., Valencia, G. A., Moreira, R. D. F. P. M., Monteiro, A. R. (2021). Gelatin-TiO_2-coated expanded polyethylene foam nets as ethylene scavengers for fruit postharvest application. *Postharvest Biology and Technology*, *180*, 111602.

Gaikwad, K. K., Singh, S., Negi, Y. S. (2020). Ethylene scavengers for active packaging of fresh food produce. *Environmental Chemistry Letters*, *18*, 269–284.

Garavito, J., Mendoza, S. M., Castellanos, D. A. (2022). Configuration of biodegradable equilibrium modified atmosphere packages, including a moisture absorber for fresh cape gooseberry (*Physalis peruviana* L.) fruits. *Journal of Food Engineering*, *314*, 110761.

Ghidelli, C., Mateos, M., Rojas-Argudo, C., Pérez-Gago, M. B. (2014). Extending the shelf life of fresh-cut eggplant with a soy protein-cysteine based edible coating and modified atmosphere packaging. *Postharvest Biology and Technology*, *95*, 81–87.

Ghidelli, C., Mateos, M., Rojas-Argudo, C., Pérez-Gago, M. B. (2015). Novel approaches to control browning of fresh-cut artichoke: Effect of a soy protein-based coating and modified atmosphere packaging. *Postharvest Biology and Technology*, *99*, 105–113.

Giannakourou, M. C., Tsironi, T. N. (2021). Application of processing and packaging hurdles for fresh-cut fruits and vegetables preservation. *Foods*, *10*, 1–23.

Giné-Bordonaba, J., Busatto, N., Larrigaudière, C., Lindo-García, V., Echeverria, G., Vrhovsek, U., Farneti, B., Biasioli, F., De Quattro, C., Rossato, M., Delledonne, M., Costa, F. (2020). Investigation of the transcriptomic and metabolic changes associated with superficial scald physiology impaired by lovastatin and 1-methylcyclopropene in pear fruit (cv. "blanquilla"). *Horticulture Research*, *7*, 49.

Goddard, J. M. and Hotchkiss, J. H. (2007). Polymer surface modification for the attachment of bioactive compounds. *Progress in Polymer Science*, *32*(7), 698–725.

Gomez-Estaca, J., Lopez-de-Dicastillo, C., Hernandez-Munoz, P., Catala, R., Gavara, R. (2014). Advances in antioxidant active food packaging. A review. *Trends in Food Science and Technology*, *35*, 42–51.

Gundewadi, G., Rudra, S.G., Prasanna, R., Banerjee, T., Singh, S.K., Dhakate, S.R., Gupta, A., Anand, A. (2022). Palladium encapsulated nanofibres for scavenging ethylene from sapota fruits. *Frontiers in Nutrition*, *9*, 1–14.

Guo, Q., Du, G., Jia, H., Fan, Q., Wang, Z., Gao, Z., Yue, T., Yuan, Y. (2021). Essential oils encapsulated by biopolymers as antimicrobials in fruits and vegetables: A review. *Food Bioscience*, *44*, 101367.

Gupta, P. (2023). Role of oxygen absorbers in food as packaging material, their characterization and applications. *Journal of Food Science and Technology*, https://doi.org/10.1007/s13197-023-05681-8.

Gutiérrez, L., Batlle, R., Sánchez, C., Nerín, C. (2010). New approach to study the mechanism of antimicrobial protection of an active packaging. *Foodborne Pathogens and Disease*, *7*(9), 1063–1069.

Gyawali, R., Ibrahim, S. A. (2014). Natural products as antimicrobial agents. *Food Control*, *46*, 412–429.

Hamza, A. H. (2017). Vitamin C: An antioxidant agent. IntechOpen.

Han, J. W., Ruiz-Garcia, L., Qian, J. P., Yang, X. T. (2018). Food packaging: a comprehensive review and future trends. *Comprehensive Reviews in Food Science and Food Safety*, *17*(4), 860–877.

Hanani, Z. A., Yee, F. C., Nor-Khaizura, M. A. R. (2018). Effect of pomegranate (*Punica granatum L.*) peel powder on the antioxidant and antimicrobial properties of fish gelatin films as active packaging. *Food Hydrocolloid*, *89*, 253–259.

Herrera, A.O., Castellanos, D.A., Mendoza, R., Patiño, L.S. (2020). Design of perforated packages to preserve fresh produce considering temperature, gas concentrations and moisture loss. *Acta Horticulturae*, *1275*, 185–192.

Hussein, Z., Caleb, O. J., Opara, U. L. (2015). Perforation-mediated modified atmosphere packaging of fresh and minimally processed produce—A review. *Food Packaging Shelf Life*, *6*, 7–20.

Huynh, N. K., Wilson, M. D., Stanley, R. A. (2023). Extending the shelf life of raspberries in commercial settings by modified atmosphere/modified humidity packaging. *Food Packaging and Shelf Life*, *37*, 101069.

Hyldgaard, M., Mygind, T., Meyer, R. L. (2012). Essential oils in food preservation: Mode of action, synergies, and interactions with food matrix components. *Frontiers in Microbiology*, *3*, 1–24.

Ichiura, H., Kitaoka, T., Tanaka, H. (2003). Removal of indoor pollutants under UV irradiation by a composite TiO_2–zeolite sheet prepared using a papermaking technique. *Chemosphere*, *50*, 79–83.

Islam, T., Yu, X., Badwal, T. S., Xu, B. (2017). Comparative studies on phenolic profiles, antioxidant capacities and carotenoid contents of red goji berry (*Lycium barbarum*) and black goji berry (*Lycium ruthenicum*). *Chemistry Central Journal*, *11*(1), 59.

Jamróz, E., Kulawik, P., Kopel, P. (2019). The effect of nanofillers on the functional properties of biopolymer-based films: A review. *Polymers*, *11*(4), 675.

Jiang, F., Zhou, L., Zhou, W., Zhong, Z., Yu, K., Xu, J., Zou, L., Liu, W. (2022). Effect of modified atmosphere packaging combined with plant essential oils on preservation of fresh-cut lily bulbs. *LWT-Food Science and Technology*, *162*, 113513.

Jridi, M., Boughriba, S., Abdelhedi, O., Nciri, H., Nasri, R., Kchaou, H. (2019). Investigation of physicochemical and antioxidant properties of gelatin edible film mixed with blood orange (*Citrus sinensis*) peel extract. *Food Packaging and Shelf Life*, *21*, 100342.

Kaewklin, P., Siripatrawan, U., Suwanagul, A., Lee, Y. S. (2018). Active packaging from chitosan-titanium dioxide nanocomposite film for prolonging storage life of tomato fruit. *International Journal of Biological Macromolecules*, *112*, 523–529.

Kovanda, L., Zhang, W., Wei, X., Luo, J., Wu, X., Atwill, E. R., Vaessen, S., Li, X., Liu, Y. (2019). In vitro antimicrobial activities of organic acids and their derivatives on several species of gram-negative and gram-positive bacteria. *Molecules*, *24*(20), 3770.

Kruijf, N. D., Beest, M. V., Rijk, R., Sipiläinen-Malm, T., Losada, P. P., Meulenaer, B. D. (2002). Active and intelligent packaging: applications and regulatory aspects. *Food Additives Contaminants*, *19*, 144–162.

Kudachikar, V. B., Kulkarni, S. G., Prakash, M. N. K. (2011). Effect of modified atmosphere packaging on quality and shelf life of 'robusta' banana (*Musa* sp.) stored at low temperature. *Journal of Food Science and Technology*, *48*(3), 319–324.

Kumar, A., Deshmukh, R. K., Gaikwad, K. K. (2022). Quality preservation in banana fruits packed in pine needle and halloysite nanotube-based ethylene gas scavenging paper during storage. *Biomass Conversion and Biorefinery*. https://doi.org/10.1007/s13399-022-02708-6.

Kumar, P., Kizhakkedathu, J. N., Straus, S. K. (2018). Antimicrobial peptides: Diversity, mechanism of action and strategies to improve the activity and biocompatibility in vivo. *Biomolecules*, *8*(1), 4.

Kumar, S., Thakur, K.S. (2020). Active packaging technology to retain storage quality of pear cv "Bartlett" during shelf-life periods under ambient holding after periodic cold storage. *Packaging Technology and Science*, *33*(7), 239–254.

Kuorwel, K. K., Cran, M. J., Sonneveld, K., Miltz, J., Bigger, S. W. (2011). Essential oils and their principal constituents as antimicrobial agents for synthetic packaging films. *Journal of Food Science*, *76*(9). R164-R177.

Lee, D.S. (2016). Carbon dioxide absorbers for food packaging applications. *Trends in Food Science & Technology*, *57*, 146–155.

Lee, D. S., Paik, H. D. (1997). Use of a pinhole to develop an active packaging system for kimchi, a Korean fermented vegetable. *Packaging Technology Science: International Journal*, *10*(1), 33–43.

Lee, H. G., Cho, C. H., Kim, H. K., Yoo, S. (2020). Improved physical and mechanical properties of food packaging films containing calcium hydroxide as a CO_2 adsorbent by stearic acid addition. *Food Packaging and Shelf Life*, *26*, 100558.

Leneveu-Jenvrin, C., Apicella, A., Bradley, K., Meile, J.C., Chillet, M., Scarfato, P., Incarnato, L., Remize, F. (2021). Effects of maturity level, steam treatment, or active packaging to maintain the quality of minimally processed mango (*Mangifera indica* cv. José). *Journal of Food Processing and Preservation*, *45*, e15600.

Licciardello, F., Wittenauer, J., Saengerlaub, S., Reinelt, M., Stramm, C. (2015). Rapid assessment of the effectiveness of antioxidant active packaging—Study with grape pomace and olive leaf extracts. *Food Packaging and Shelf Life*, *6*, 1–6.

Liu, J., Yang, J., Zhang, H., Cong, L., Zhai, R., Yang, C., Wang, Z. Ma, F., Xu, L. (2019). Melatonin inhibits ethylene synthesis via nitric oxide regulation to delay postharvest senescence in pears. *Journal of Agricultural and Food Chemistry*, *67*, 2279–2288.

Lü, J., Lin, P. H., Yao, Q., Chen, C. (2010). Chemical and molecular mechanisms of antioxidants : Experimental approaches and model systems. *Molecular Medicine*, *14*(4), 840–860.

Lushchak, V. I. (2012). Glutathione homeostasis and functions: Potential targets for medical interventions. *Journal of Amino Acids*. *2012*, 736837.

Lv, J., Zhang, M., Bai, L., Han, X., Ge, Y., Wang, W., Li, J. (2020). Effects of 1-methylcyclopropene (1-MCP) on the expression of genes involved in the chlorophyll degradation pathway of apple fruit during storage. *Food Chemistry*, *308*, 125707.

Mahmud, J., Khan, R. A. (2018). Characterization of natural antimicrobials in food system. *Advances in Microbiology*, *8*(11), 894–916.

Maisanaba, S., Pichardo, S., Prieto, A. I., Jos, A., Cameán, A. M. (2017). New advances in active packaging incorporated with essential oils or their main components for food preservation. *Food Reviews International*, *33*(5), 447–515.

Mari˜no-Cortegoso, S., Stanzione, M., Andrade, M. A., Restuccia, C., Quir´os, A. R.-B. de, Buonocore, G. G., Barbosa, C. H., Vilarinho, F., Silva, A. S., Ramos, F., Khwaldia, K., Send´on, R., Barbosa-Pereira, L. (2022). Development of active films utilizing antioxidant compounds obtained from tomato and lemon by-products for use in food packaging. *Food Control*, *140*, 109–128.

Mariah, M. A. A., Vonnie, J. M., Erna, K. H., Nur'Aqilah, N. M., Huda, N., Abdul Wahab, R., Rovina, K. (2022). The emergence and impact of ethylene scavenger's techniques in delaying the ripening of fruits and vegetables. *Membranes*, *12*, 117.

McManamon, O., Kaupper, T., Scollard, J., Schmalenberger, A. (2019). Nisin application delays growth of Listeria monocytogenes on fresh-cut iceberg lettuce in modified atmosphere packaging, while the bacterial community structure changes within one week of storage. *Postharvest Biology and Technology*, *147*, 185–195.

Mir, A. A., Sood, M., Bandral, J. D. (2018). Effect of active packaging on quality and shelf life of peach fruits. *The Pharma Innovation Journal*, *7*(4), 1076–1082.

Mohan, C. and Ravishankar, C. (2019). Active and intelligent packaging systems-application in seafood. *World Journal of Aquaculture Research & Development*, *1*, 10–16.

Morelli, M. B., Gambardella, J., Castellanos, V., Trimarco, V. and Santulli, G. (2020). Vitamin C and cardiovascular disease: An update. *Antioxidants*, *9*(12), 1227.

Motlagh, N. V., Mosavian, M. T. H., Mortazavi, S. A. (2012). Effect of polyethylene packaging modified with silver particles on the microbial, sensory and appearance of dried barberry. *Packaging and Technology and Science*, *29*, 399–412.

Muhammad, A., Dayisoylu, K. S., Khan, H., Khan, M. R., Khan, I., Hussain, F., Basit, A., Ali, M., Khan, S., Idrees, M. (2023). An integrated approach of hypobaric pressures and potassium permanganate to maintain quality and biochemical changes in tomato fruits. *Horticulturae*, *9*(1), 9.

Naqash, F., Masoodi, F. A., Ayob, O., Parvez, S. (2022). Effect of active pectin edible coatings on the safety and quality of fresh-cut apple. *International Food Science Technology*, *57*, 57–66.

Navarro-Martínez, A., López-Gómez, A., Martínez-Hernández, G. B. (2021). Potential of essential oils from active packaging to highly reduce ethylene biosynthesis in broccoli and apples. *ACS Food Science & Technology*, *1*(6), 1050–1058.

Nayik, G. A., Muzaffar, K. (2014). Developments in packaging of fresh fruits-shelf life perspective: a review. *American Journal of Food Science and Nutrition Research*, *1*(5), 34–39.

Nian, L., Xie, Y., Sun, X., Wang, M., Cao, C. (2023). Chitosan quaternary ammonium salt/gelatin-based biopolymer film with multifunctional preservation for perishable products. *International Journal of Biological Macromolecules*, *228*, 286–298.

Niazmand, R., Yeganehzad, S. (2020). Capability of oxygen-scavenger sachets and modified atmosphere packaging to extend fresh barberry shelf life. *Chemical and Biological Technologies in Agriculture*, *7*, 28.

Nimitkeatkai, H., Techavuthiporn, C., Boonyaritthongchai, P., Supapvanich, S. (2022). Commercial active packaging maintaining physicochemical qualities of carambola fruit during cold storage. *Food Packaging and Shelf Life*, *32*, 100834.

Nooun, P., Chueangchayaphan, N., Ummarat, N., Chueangchayaphan, W. (2023). Fabrication and properties of natural rubber/rice starch/activated carbon biocomposite-based packing foam sheets and their application to shelf life extension of 'hom thong' banana. *Industrial Crops and Products*, *195*, 116409.

Nwakaudu A. A., Nwakaudu M. S., Owuamanam C. I., Iheaturu N. C. (2015). The use of natural antioxidant active polymer packaging films for food preservation. *Applied Signals Reports*, *2*(4), 38–50.

Olveira-Bouzas, V., Pita-Calvo, C., Vazquez-Od´eriz, M., Lourdes, Romero-Rodríguez, M. (2021). Evaluation of a modified atm packaging system in pallets to extend the shelf-life of the stored tomato at cooling temperature. *Food Chemistry*, *364*, 130309.

Orlo, E., Russo, C., Nugnes, R., Lavorgna, M., Isidori, M. (2021). Natural methoxyphenol compounds: Antimicrobial activity against foodborne pathogens and food spoilage bacteria, and role in antioxidant processes. *Foods*, *10*(8), 1807.

Oyom, W., Xu, H., Liu, Z., Long, H., Li, Y., Zhang, Z., Bi, Y., Tahergorabi, R., Prusky, D. (2022). Effects of modified sweet potato starch edible coating incorporated with cumin essential oil on storage quality of 'early crisp.' *LWT-Food Science and Technology*, *153*, 112475.

Paillart, M. J. M., van der Vossen, J. M. B. M., Levin, E., Lommen, E., Otma, E. C., Snels, J. C. M. A., Woltering, E. J. (2017). Bacterial population dynamics and sensorial quality loss in modified atmosphere packed fresh-cut iceberg lettuce. *Postharvest Biology and Technology*, *124*, 91–99.

Pattar, A., Kukanoor, L., Naik, K. R., Naik, N., Jholgikar, P., Koulagi, S., Naika, B. N. M. (2021). Influence of post-harvest treatments on quality and shelf life of jamun (*Syzygium cuminii* Skeels.) fruits under ambient storage. *The Pharma Innovation Journal*, *10*(1), 523–528.

Pavinatto, A., de Almeida Mattos, A. V., Malpass, A. C. G., Okura, M. H., Balogh, D. T., Sanfelice, R. C. (2020). Coating with chitosan-based edible films for mechanical/ biological protection of strawberries. *International Journal of Biological Macromolecules*, *151*, 1004–1011.

Perdones, A., Sánchez-González, L., Chiralt, A., Vargas, M. (2012). Effect of chitosan–lemon essential oil coatings on storage-keeping quality of strawberry. *Postharvest Biology and Technology*, *70*, 32–41.

Pereira, V. A., de Arruda, I. N. Q., Stefani, R. (2015). Active chitosan/PVA films with anthocyanins from *Brassica oleraceae* (Red Cabbage) as time-temperature indicators for application in intelligent food packaging. *Food Hydrocolloids*, *43*, 180–188.

Phothisarattana, D., Wongphan, P., Promhuad, K., Promsorn, J., Harnkarnsujarit, N. (2021). Biodegradable poly (butylene adipate-co-terephthalate) and thermoplastic starch-blended TiO_2 nanocomposite blown films as functional active packaging of fresh fruit. *Polymers*, *13*(23), 4192.

Rahimi, R., ValizadehKaji, B., Khadivi, A., Shahrjerdi, I. (2019). Effect of chitosan and thymol essential oil on quality maintenance and shelf life extension of peach fruits cv. "Zaferani." *Journal of Horticulture and Postharvest Research*, *2*(2), 143–156.

Rajkowska, K., Nowak, A., Kunicka-Styczyńska, A., Siadura, A. (2016). Biological effects of various chemically characterized essential oils: Investigation of the mode of action against: Candida albicans and HeLa cells. *RSC Advances*, *6*(99), 97199–97207.

Ramel, F., Birtic, S., Ginies, C., Soubigou-Taconnat, L., Triantaphylidès, C., Havaux, M. (2012). Carotenoid oxidation products are stress signals that mediate gene responses to singlet oxygen in plants. *Proceedings of the National Academy of Sciences of the United States of America*, *109*(14), 5535–5540.

Ramesh, M., Narendra, G., Sasikanth, S. (2018). A review on biodegradable packaging materials in extending the shelf life and quality of fresh fruits and vegetables. *Octa Journal of Biosciences*, *6*(2), 42–45.

Rangaraj, V. M., Rambabu, K., Banat, F., Mittal, V. (2021). Natural antioxidants-based edible active food packaging: An overview of current advancements. *Food Bioscience*, *43*, 101251.

Rattanapitigorn, P., Arakawa, M., Tsuro, M. (2006). Vanillin enhances the antifungal effect of plant essential oils against *Botrytis cinerea*. *International Journal of Aromatherapy*, *16*(3–4), 193–198.

Remedio, L. N., Silva dos Santos, J. W., Vieira Maciel, V. B., Yoshida, C. M. P., Aparecida de Carvalho, R. (2019). Characterization of active chitosan films as a vehicle of potassium sorbate or nisin antimicrobial agents. *Food Hydrocolloids*, *87*, 830–838.

Restuccia, D., Spizzirri, U. G., Parisi, O. I., Cirillo, G., Curcio, M., Iemma, F., Picci, N. (2010). New EU regulation aspects and global market of active and intelligent packaging for food industry applications. *Food Control*, *21*(11), 1425–1435.

Ribeiro-Santos, R., Andrade, M., de Melo, N. R., Sanches-Silva, A. (2017). Use of essential oils in active food packaging: Recent advances and future trends. *Trends in Food Science & Technology*, *61*, 132–140.

Rusková, M., Opálková Šišková, A., Mosnáčková, K., Gago, C., Guerreiro, A., Bučková, M., Puškárová, A., Pangallo, D. and Antunes, M. D. (2023). Biodegradable active packaging enriched with essential oils for enhancing the shelf life of strawberries. *Antioxidants*, *12*(3), 755.

Rydlo, T., Miltz, J., Mo, A. (2006). Eukaryotic antimicrobial peptides : Promises and premises in food safety. *Journal of Food Science*, *71*(9), R125-R135.

Sanches-Silva, A., Costa, D., Albuquerque, T. G., Buonocore, G. G., Ramos, F., Castilho, M. C., Machado, A. V., Costa, H. S. (2014). Trends in the use of natural antioxidants in active food packaging: A review. *Food Additives and Contaminants—Part A, 31*(3), 374–395.

Sangsuwan, J., Pongsapakworawat, T., Bangmo, P., Sutthasupa, S. (2016). Effect of chitosan beads incorporated with lavender or red thyme essential oils in inhibiting *Botrytis cinerea* and their application in strawberry packaging system. *LWT-Food Science and Technology, 74*, 14–20.

Santos, J. C. P., Sousa, R. C. S., Otoni, C. G., Moraes, A. R. F., Souza, V. G. L., Medeiros, E. A. A., Espitia, P. J. P., Pires, A. C. S., Coimbra, J. S. R., Soares, N. F. F. (2018). Nisin and other antimicrobial peptides: Production, mechanisms of action, and application in active food packaging. *Innovative Food Science and Emerging Technologies, 48*, 179–194.

Shahidi, F., Ambigaipalan, P. (2015). Phenolics and polyphenolics in foods, beverages and spices: Antioxidant activity and health effects—A review. *Journal of Functional Foods, 18*, 820–897.

Sharma, A., Ranjit, R., Pratibha, Kumar, N., Kumar, M., Giri, B. S. (2023). Nanoparticles based -nanosensors: Principles and their applications in active packaging for food quality and safety detection. *Biochemical Engineering Journal, 193*, 108861.

Sharma, R., Ghoshal, G. (2018). Emerging trends in food packaging. *Nutrition & Food Science, 48*(5), 764–779.

Shen, X., Zhang, M., Devahastin, S., Guo, Z. (2019). Effects of pressurized argon and nitrogen treatments in combination with modified atmosphere on quality characteristics of fresh-cut potatoes. *Postharvest Biology and Technology, 149*, 159–165.

Siddiqui, M. W. (2016). *Eco-friendly technology for postharvest produce quality*. Academic Press.

Siddiqui, M. W. (2020). *Fresh-Cut Fruits and Vegetables: Technologies and Mechanisms for Safety Control*. Academic Press.

Silva, F., Caldera, F., Trotta, F., Nerin, C., Domingues, F. C. (2019). Encapsulation of coriander essential oil in cyclodextrin nanosponges: A new strategy to promote its use in controlled-release active packaging. *Innovative Food Science & Emerging Technologies, 56*, 102177.

Singh, A. K., Ramakanth, D., Kumar, A., Lee, Y. S., Gaikwad, K. K. (2021). Active packaging technologies for clean label food products: a review. *Journal of Food Measurement and Characterization, 15*, 4314–4324.

Singh, S., Gaikwad, K. K., Lee, Y. S. (2019). Development and application of a pyrogallic acid-based oxygen scavenging packaging system for shelf life extension of peeled garlic. *Scientia Horticulturae, 256*, 108548.

Soares, T. R. P., Reis, A. F., dos Santos, J. W. S., Chagas, E. G. L., Venturini, A. C., Santiago, R. G., Bastos-Neto, M., Vieira, R. S., Carvalho, R. A., da Silva, C. F., Yoshida, C. M. P. (2023). NaY-Ag zeolite chitosan coating kraft paper applied as ethylene scavenger packaging. *Food and Bioprocess Technology, 16*, 1101–1115.

Soylu, E. M., Kurt, S., Soylu, S. (2010). In vitro and in vivo antifungal activities of the essential oils of various plants against tomato grey mould disease agent *Botrytis cinerea*. *International Journal of Food Microbiology, 143*(3), 183–189.

Stanojevic, D., Comic, L., Stefanovic, O., Solujic-Sukdolak, S. (2009). Antimicrobial effects of sodium benzoate, sodium nitrite and potassium sorbate and their synergistic action in vitro. *Bulgarian Journal of Agricultural Science, 15*(4), 307–311.

Sultana, A., Kathuria, A., Gaikwad, K. K. (2022). Metal–organic frameworks for active food packaging. a review. *Environmental Chemistry Letters, 20*, 1479–1495.

Tajkarimi, M. M., Ibrahim, S. A., Cliver, D. O. (2010). Antimicrobial herb and spice compounds in food. *Food Control, 21*(9), 1199–1218.

Taqi, A., Mutihac, L., Stamatin, I. (2014). Physical and barrier properties of apple pectin/cassava starch composite films incorporating *Laurus* nobilis oil and oleic acid. *Journal of Food Processing and Preservation, 38*(4), 1982–1993.

Teixeira, B. F., Aranha, J. B., Vieira, T. M. F. de S. (2022). Replacing synthetic antioxidants in food emulsions with microparticles from green acerola (*Malpighia emarginata*). *Future Foods, 5*, 100130.

Tepamatr, P. (2023). Efficacy of a palladium-modified activated carbon in improving ethylene removal to delay the ripening of gros michel banana. *Journal of Agriculture and Food Research, 12*, 100561.

Tian, Y., Zhou, L., Liu, J., Yu, K., Yu, W., Jiang, H., Zhong, J., Zou, L., Liu, W. (2023). Effect of sustained-tea tree essential oil solid preservative on fresh-cut pineapple storage quality in modified atmospheres packaging. *Food Chemistry, 417*, 135898.

Tian, J., Zeng, X., Zhang, S., Wang, Y., Zhang, P., Lü, A., Peng, X. (2014). Regional variation in components and antioxidant and antifungal activities of *Perilla frutescens* essential oils in China. *Industrial Crops and Products, 59*, 69–79.

Tuberoso, C. I. G., Boban, M., Bifulco, E., Budimir, D., Pirisi, F. M. (2013). Antioxidant capacity and vasodilatory properties of Mediterranean food: The case of Cannonau wine, myrtle berries liqueur and strawberry-tree honey. *Food Chemistry, 140*(4), 686–691.

Valková, V., ˇDúranová, H., Galoviˇcová, L., Vukovic, N. L., Vukic, M., Kaˇcániová, M. (2021). In vitro antimicrobial activity of lavender, mint, and rosemary essential oils and the effect of their vapours on growth of *Penicillium spp.* in a bread model system. *Molecules, 26*, 38–59.

Vermeiren, L., Devlieghere, F., van Beest, M., de Kruijf, N., Debevere, J. (1999). Developments in the active packaging of foods. *Trends in Food Science Technology, 10*(3), 77–86.

Vilela, C., Kurek, M., Hayouka, Z., Röcker, B., Yildirim, S., Antunes, M. D. C., Nilsen-Nygaard, J., Pettersen, M. K., Freire, C. S. R. (2018). A concise guide to active agents for active food packaging. *Trends in Food Science and Technology, 80*, 212–222.

Waghmare, R. B., Annapure, U. S. (2013). Combined effect of chemical treatment and/or modified atmosphere packaging (MAP) on quality of fresh-cut papaya. *Postharvest Biology and Technology, 85*, 147–153.

Waghmare, R. B., Annapure, U. S. (2015). Integrated effect of sodium hypochlorite and modified atmosphere packaging on quality and shelf life of fresh-cut cilantro. *Food Packaging and Shelf Life, 3*, 62–69.

Waisundara, V. (2021). *Antioxidants—Benefits, Sources, Mechanisms of Action*. IntechOpen.

Wang, D., Ma, Q., Li, D., Li, W., Li, L., Aalim, H., Luo, Z. (2021). Moderation of respiratory cascades and energy metabolism of fresh-cut pear fruit in response to high CO_2 controlled atmosphere. *Postharvest Biology and Technology, 172*, 111379.

Wang, Y., Liu, B., Wen, X., Li, M., Wang, K., Ni, Y. (2017). Quality analysis and microencapsulation of chili seed oil by spray drying with starch sodium octenylsuccinate and maltodextrin. *Powder Technology, 312*, 294–298.

Wei, H., Seidi, F., Zhang, T., Jin, Y., Xiao, H. (2021). Ethylene scavengers for the preservation of fruits and vegetables: a review. *Food Chemistry, 337*, 127750.

Wu, W., Ni, X., Shao, P., Gao, H. (2021). Novel packaging film for humidity-controlled manipulating of ethylene for shelf-life extension of *Agaricus bisporus*. *LWT-Food Science and Technology, 145*, 111331.

Xu, X., Liu, A., Hu, S., Ares, I., Martínez-Larrañaga, M. R., Wang, X., Martínez, M., Anadón, A., Martínez, M. A. (2021a). Synthetic phenolic antioxidants: Metabolism, hazards and mechanism of action. *Food Chemistry*, *353*, 129488.

Xu, Y., Wei, J., Wei, Y., Han, P., Dai, K., Zou, X., Jiang, S., Xu, F., Wang, H., Sun, J., Shao, X. (2021b). Tea tree oil controls brown rot in peaches by damaging the cell membrane of *Monilinia fructicola*. *Postharvest Biology and Technology*, *175*, 111474.

Yam, K. L., Dong, S. L. (2012). Emerging Food Packaging Technologies. Elsevier.

Yammine, J., Chihib, N. E., Gharsallaoui, A., Dumas, E., Ismail, A., Karam, L. (2022). Essential oils and their active components applied as: Free, encapsulated and in hurdle technology to fight microbial contaminations. A review. *Heliyon*, *8*(12), e12472.

Yang, F. M., Li, H. M., Li, F., Xin, Z. H., Zhao, L. Y., Zheng, Y. H., Hu, Q. H. (2010). Effect of nano-packing on preservation quality of fresh strawberry (*Fragaria ananassa* Duch. cv fengxiang) during storage at 4 °C. *Journal of Food Science*, *75*(3), C236–C240.

Yashin, A., Yashin, Y., Xia, X., Nemzer, B. (2017). Antioxidant activity of spices and their impact on human health: A review. *Antioxidants*, *6*(3), 70.

Yemmireddy, V. K., Hung, Y. C. (2015). Effect of binder on the physical stability and bactericidal property oftitanium dioxide (TiO_2) nanocoatings on food contact surfaces. *Food Control*, *57*, 82–88.

Yildirim, S., Röcker, B., Pettersen, M. K., Nilsen-Nygaard, J., Ayhan, Z., Rutkaite, R., Radusin, T., Suminska, P., Marcos, B., Coma, V. (2018). Active packaging applications for food. *Comprehensive Reviews in Food Science and Food Safety*, *17*(1), 165–199.

Yong, H., Liu, J. (2020). Recent advances in the preparation, physical and functional properties, and applications of anthocyanins-based active and intelligent packaging films. *Food Packaging and Shelf Life*, *26*, 100550.

Yousef, N., Niloufar, M., Elena, P. (2019). Antipathogenic effects of emulsion and nanoemulsion of cinnamon essential oil against Rhizopus rot and grey mold on strawberry fruits. *Foods and Raw Materials*, *7*(1), 210–216.

Youssef, A. M., Abdel-Aziz, M. S., El-Sayed, S. M. (2014). Chitosan nanocomposite films based on Ag-NP and Au-NP biosynthesis by *Bacillus Subtilis* as packaging materials. *International Journal of Biological Macromolecules*, *69*, 185–191.

Zerdin, K., Rooney, M.L., Vermuë, J. (2003). The vitamin C content of orange juice packed in an oxygen scavenger material. *Food Chemistry*, *82*, 387–395.

Zhang, B. Y., Samapundo, S., Pothakos, V., Sürengil, G., Devlieghere, F. (2013). Effect of high oxygen and high carbon dioxide atmosphere packaging on the microbial spoilage and shelf-life of fresh-cut honeydew melon. *International Journal of Food Microbiology*, *116*(3), 378–390.

27 Natural Plant Extracts and Postharvest Quality of Fruits and Vegetables

Ambreen Naz, Kashif Razzaq*, Sami Ullah, Misbah Sharif, Ali Hamza, and Syed Bilal Hussain
*Corresponding Author: kashif.razzaq@mnsuam.edu.pk

27.1 INTRODUCTION

Plants have been a resilient source to cater basic human needs such as food, shelter, and clothing for many centuries. They were also used for arrows and poison darts for hunting, murder, and ritualistic and hallucinogenic purposes, as well as for medicines and inebriants. Various medicinal plants have been extensively used in crude or pure form to treat various health issues. Plant-derived drugs serve as the initial point to prepare the formulation for modern bioactive ingredients. Humans have used phytonutrients for different purposes since prehistoric times, and their knowledge was widespread among ancient civilizations. Since early times, the world has relied on medicinal plant extracts as traditional medicines or complementary alternative medicine. Presently, these extracts are also under discussion for the preservation of various edible commodities, including fruits and vegetables. Various plant metabolites that are further derivatives of primary compounds (such as lipids, amino acids, or polysaccharides) are directly or indirectly used for the development of such chemical classes (alkaloids, tannins, terpenoids, and phenolics) that can act as preservatives for fresh commodities.

Fresh horticultural commodities maintain biological processes even after harvest. However, postharvest changes can sometimes result in undesirable outcomes, ultimately contributing to an overall deterioration in quality. These changes may manifest as mass loss, a shriveled appearance, reduced shelf life, and diminished organoleptic properties. Such limitations can restrict the transportation of fruits and vegetables to long-distance markets. Ripening is a highly synchronized process that triggers a series of physical and chemical alterations, including color loss, changes in color, astringency loss, seed maturation, tissue wilting, and other related issues. These factors collectively determine the ultimate quality of the horticultural products. Plant extracts are considered a promising source of various biochemical compounds that can be obtained through various extraction methods (Abdullahi et al., 2022).

Natural extracts obtained from plants are considered promising biomolecules which can be extracted from different parts of the plants. In recent developments, the food packaging industry has been exploring plants-based materials of a sustainable nature to meet consumers' health needs and their personal standards regarding consumption of organic commodities. Among these solution types, active techniques of different packagings stand out owing to their sustainable potential to improve fresh produce's quality characteristics and sensorial profiles, and to ensure food safety status globally. In the present chapter, a thorough review of the literature has been carried out to assess the impacts of plant-derived botanical extracts, both in the form of independent active ingredients and as a part of the formulations in extending freshly harvested vegetable- and fruit-based produce quality. Horticultural produce is frequently considered vulnerable to different diseases and pests, particularly after harvest, with certain infections often caused due to the attack of certain types of viruses, fungi, and bacteria. Plant extract–based additives have gained enormous popularity in the food industry, primarily owing to their nature of having higher concentrations of polyphenolic and carotenoid-based compounds. Such extracts possess antioxidative and antimicrobial properties, efficiently impeding damages caused by the free radicals produced after harvest, thus preventing off-flavor development, enhancing color stability, and prolonging the shelf life of products. These natural compounds have replaced synthetic counterparts in edible coating development for fresh horticultural products. Some of the plant extracts currently used in edible coatings to extend the shelf life of postharvest produce include *Allium sativum* (garlic), *Azadirachta indica* (neem), *Eucalyptus globulus* (eucalyptus),

DOI: 10.1201/9781003370376-36

Curcuma longa (turmeric), *Nicotiana tabacum* (tobacco), and *Zingiber officinale* (ginger). Next, we provide brief details about these botanical extracts and their applications.

27.2 ESSENTIAL OILS, HERBAL EXTRACTS, AND VOLATILE COMPOUNDS

Botanical extracts, such as those from neem, thyme, eucalyptus, lemongrass oil, rue, and tea tree, contain essential oils that are extracted to combat postharvest pathogens. The application of these extracted essential oils also offers fungistatic and aromatic terpenoids. In addition to essential oils, plants serve as excellent sources of volatile compounds—organic molecules with lower molecular weights and appreciable vapor pressure at room temperature. These volatile complexes significantly contribute to the defense systems of living organisms while inhibiting pathogen growth. Volatile compounds can be isolated through steam or hydrodistillation processes. Not only that, but some well-known sources of volatiles include black pepper, nutmeg, clove, thyme, and oregano. These volatile compounds have been recognized as safe for industrial use and are GRAS (Generally Recognized as Safe).

Spices and condiments are a major source of antioxidants and have long been under use, starting their journey in India and traveling to other parts of the world. The use of spices is always encouraged owing to their allied health benefits. Not only used in Indian cuisine, spices are admired on a global scale, as they contribute many essential oils, e.g., vanillin. The aromatic/essential oils are extracted from various herbs, spices, and other plant-based contents that are further used as supplements in packing wraps or coating materials to extend the shelf life of fresh horticultural products, but they also enhance the qualities of such products. These essential oils reduce odors as well. The US FDA (Food and Drug Administration) has also declared their status as GRAS (Serrano et al., 2005). Additionally, *P. citrophthora* infections were prevented by the extracted aromatic oils from *Origanum vulgare* and *Verbena officinalis*. The existence of substances with distinctive chemical structures, particularly the occurrence of functional groups containing hydrophilic functional contents, i.e., hydroxyl groups from phenolic compounds and/or some of the lipophilic essential oils, are considered to have antimicrobial activity (Dorman & Deans, 2000).

The impact of natural antibacterial agents was checked on freshly cut melons after applying essential oils and malic acid of cinnamon, lemongrass, and palmarosa as natural antimicrobial components. An extension of shelf life was noticed in malic acid–coated samples compared to the fresh-cut control samples while considering physicochemical and microbial standpoints. Owing to the raised antibacterial qualities of various malic acids and essential oils, the integration of these active components into the coating occasionally increased the antimicrobial stability by >21 days. The essential oils from thyme (*Thymus vulgaris*) worked well against *Penicillium italicum*; nevertheless, it proved ineffective when checked against fruit rot caused by *Rhizopus stolonifer*, *Botrytis cinerea*, and *Phytophthora citrophthora* (Camele et al., 2010; Raybaudi-Massilia et al., 2008). Fresh-cut peaches treated with an edible pectin coating plus cinnamon leaf extract presented a longer shelf life, as the coating had raised the content of antioxidants, thus less microbial growth was observed (Ayala-Zavala et al., 2008). Another compound known as menthol, extracted from peppermint, has been documented as a useful compound that prevents various fruit diseases during storage. The volatiles extracted from *Zingiber officinale*, *Ocimum canum*, and *Metha arvensis* were successful in controlling blue mold rot of oranges, indicating a synergistic interaction among the elements (Tripathi, 2001).

Microorganisms can enter fruits from various sources, including the plant's surface, the air, the soil, or a combination of these factors. Fruits are often exposed to a multitude of microorganisms. These microorganisms invade via lenticels, stomatal apertures, surface damage, or a growth crack, which ultimately leads to spoilage of harvested crops. Numerous plant extracts and plant-based products exhibit broad-spectrum antibacterial capabilities. These serve as easily recognizable bio-preservatives without any negative impact on human health. Therefore, the use of herbal extracts to enhance the quality of fruits holds promise due to their non-toxic and safe characteristics. The utilization of herbal extracts has promised new avenues for controlling spoilage. Their application is straightforward, and they retain their effectiveness even at room temperature. Extracts of ginger and garlic, when applied at a 10% concentration, have proven beneficial against most microbial isolates, with the exception of *Aspergillus* and *Rhizopus*.

Tulsi (*Ocimum sanctum*) leaf extracts have polyamine biosynthesis inhibitors that hinder the ornithine decarboxylase pathway, ultimately reducing fruit rots (Patil et al., 1992; Shivpuri et al., 1997). Ginger and garlic extracts with 10% dilution were applied via spray over tomatoes to hinder the occurrence of rot. Research has shown that the tomatoes sprayed with garlic extract (10%) followed by ginger extract (10%) exhibited a pronounced decline in weight percentage. In one of the studies, fresh-cut tomatoes presented a longer shelf life after treatment with natural herbal extracts with antimicrobial properties (Ayala-Zavala et al., 2008). Applying ginger and garlic extract is an inexpensive technique that is highly effective in preserving fresh tomato produce. Various herbal plant compounds have shown antimicrobial capacity. Aqueous extracts and freshly squeezed ginger and turmeric have shown antifungal properties against the fungi *Penicillium digitatum* and *Aspergillus niger* (Kapoor, 1997). Natural plant products have also been effective against mango storage rot, with mango fruit dipped in plant extracts showing a decrease in the occurrence of disease (Hasabnis & Souza, 1988).

Azadirachtin, the active ingredient of neem oil, has the capacity to strengthen the pectin molecules by altering the galacturonic acid (methyl group remnants). By this process, the pectin molecules are broken down during storage (Kleeberg, 1996). During a study on mango, the neem oil presented a substantial reduction in weight loss compared to the control sample (Singh et al., 2000). The NLE (neem leaf extract) behaved as a reasonable component in maintaining most of the biological and chemical attributes of pulpy and juicy fruits as compared to neem and castor oils (Singh et al., 2003).

In one of the studies, freshly cut tomatoes were subjected to treatment with natural volatile substances, e.g., ethanol, garlic oil, methyl jasmonate (MeJA), and tea tree oil, for a period of 2 weeks at 5 °C. During this tenure, the microbial proliferations were controlled effectively by MeJA and ethanol together more than by either substance alone. Compared with other antibacterials, the mixture also preserved better color index and firmness. Furthermore, enhanced levels of ascorbic acid, lycopene, and phenolic components were retained thanks to the MeJA. The quality of fresh chunks of Fuji apples and the impact of oregano oil, lemongrass, and vanillin when incorporated with alginate-coated apple were examined.

The vanillin coating with 0.3% (w/w) was reported as being comparatively better on the basis of sensory perspective during the 3-week storage study at 4 °C. The proliferation of psychrophilic yeasts, aerobes, and molds was strongly hindered by every other antimicrobial covering that had been investigated. The MeJA, a significant amount of jasmonic acid found in plants, is essential for promoting resistance to fungi.

MeJA is a naturally occurring substance that is crucial for plant development, growth, fruit ripening, and responses to environment-based challenges (Creeman & Mullet, 1997). According to reports, MeJA therapy can successfully prevent postharvest illnesses in a variety of loquats (Cao et al., 2008). Furthermore, it is documented that MeJA treatment after harvesting preserved the quantity of bioactive compounds and antioxidant capacity in berries such as strawberries, raspberries, and blackberries (Kaituo et al., 2009).

MeJA can reduce the softness and microbial population after 6 and 12 days of storage when added to sliced pineapple. Peach fruits having a MeJA volatile treatment experienced slower rates of decay and better quality for 8 days after treatment than untreated fruit (Jin et al., 2006). MeJA prevented chilling injury when given 25 M as a dip or 100 M as gas to grapefruits, avocados, and peppers before chilling (Fung et al., 2004). Additionally, it has been demonstrated that MeJA keeps papayas' postharvest quality intact and lessens deterioration (Gonzalez-Aguilar et al., 2003).

Plant extracts are simple and effective to be utilized in various industries as antimicrobials, antioxidants, flavoring compounds, enzymes, nutrient enhancers, and part of packaging materials. Based on plant parts, the extracts can be derived from various components such as roots, leaves, and stems.

27.2.1 Leaves

The leaves are a crucial part of plants that are responsible for the synthesis of carbohydrates and the generation of oxygen. According to various scientific communities, leaves are a rich reservoir of health-promoting bioactives. Leaves contain anthocyanin and carotenoids, responsible for red to purple and orange to yellow colors, in addition to chlorophyll, which imparts a green color to plants. These coloring compounds are responsible for the high antioxidative activities of leaves. Additionally, some leaf extracts impart antimicrobial activity owing to the presence of flavonoids. Leaf extracts show more free radical scavenging activity compared to seed, stem, and bark polyphenols (Table 27.1). Leaf extracts are popular for their utility as a food additive and in preservative activities.

TABLE 27.1 Various Important Plants and Their Industrial Applications

PLANT	*SOURCE*	*INDUSTRIAL APPLICATION*	*FUNCTION*
Green tea(*Camellia sinensis*)	Leaves	Packaging films	Mechanical strength, water barrier, delayed spoilage by microbes, and rancidity
Olive	Fruit, leaves	Preservative impact	Reduced lipid rancidity in cooked patties, antimicrobial activity during meat storage
Mint	Leaves	Preservative impact	Antimicrobial activity and antioxidant potential, shelf life extension of meat, minimized rancidity in meat products
Black mulberry	Fruit	Preservative impact	Reduced lipid oxidation and microbial invasion during storage
Jabuticaba (*Myrciaria jaboticaba*)	Fruit	Bakery products	Improved nutritional value and acted as a color enhancer
Blueberry(*Vaccinium vitis-idaea*)	Fruit	Lard and tallow	Acted as an antioxidant for packaged tallow and lard
Pomegranate(*Punica Granatum*)	Fruit	Raw pork,ground goat meat and nuggets	Decreased numbers of disease-causing pathogenic bacteria and color degradation in raw meat, reduced TBARS value
Kiwifruit(*Actinidia chinensis*)	Fruit	Chicken meat	Retarded oxidation of lipids in chicken meat
Pitaya(*Hylocereus costaricensis*)	Fruit	Pork patties	Improved oxidative stability

PLANT	*SOURCE*	*INDUSTRIAL APPLICATION*	*FUNCTION*
Baobab(*Adansonia digitate*)	Seed s	Preservative impact	Improved lipid and microbial stability in beef patties
Cumin(*Nigella sativa*)	Seeds	Beef patties	Retarded oxidative rancidity
Guarana(*Paullinia cupana*)	Seeds	Meat patties	Enhanced shelf life of baked items and meat, protected against lipid and protein
Litchi(*Litchi chinensis*)	Fruit, leaves, seeds	Meatballs	Reduced lipid rancidity and degradation of protein
Broccoli(*Brassica oleracea*)	Bark	Goat meat nuggets	Reduced TBARS, enhanced phenolic contents
Bitter orange(*Citrus aurantium*)	Bark	Rice pudding	Preservative effect against food spoilage agents
Grapefruit(*Citrus paradisi*)	Bark	Sausage	Minimize lipid peroxidation and enhance microbial stability
Cork bark(*Quercus suber*)	Bark	Packaging films	Improve flexibility and resistance to light of films, Enhanced antimicrobial and antioxidant properties
Cinnamon(*Cinnamomum verum*)	Bark	Raw meat	Reduce the number of disease-causing pathogenic micro-organisms in raw meat
Beetroot(*Beta vulgaris*)	Bark	Packaging films	Protect the polymer against light degradation, and enhance plasticizing potential
Carrot(*Daucus carota*)	Roots	Packaging films	Protect the polymer against light-degradation, and enhance plasticizing potential
Licorice(*Glycyrrhiza glabra*)	Roots	Packaging films for gram balls	Protection against microbial degradation and lipid oxidation

27.2.2 Fruits and Vegetables

Fruits and vegetables have a plethora of functional ingredients, such as phenolic acids, lignans, stilbenes, proanthocyanidins, and tannins. The most prevalent phenolic acids are flavonoids, cinnamic, and benzoic acids, and they have a number of applications in food-related industries. The extracts that are produced from blueberries, grapes, pomegranates, raspberries, chestnuts, and murtas are prepared by using different extraction technologies and are used in products such as cheese, meat, fish, and poultry.

Peels from fruits and vegetables are considered one of the major processing waste materials that have the potential to incorporate higher values of phytonutrients in the food system and are considered superior to pulp in the case of natural antioxidants. In the case of seeds, grape seed extract has wide application in the food industry owing to its rich phenolic contents, such as epicatechin, catechin, proanthocyanidins, flavonols, and gallate. The nature of flower extract is also antioxidative and antimicrobial with promising health benefits. The flower extracts also present antimicrobial, color enrichment, and flavoring properties. These extracts are incorporated in various food products such as beverages, meat, packing films, edible coatings, and sausages. The bark of some plants has a higher phenolic content compared to wood, which is important in the defense mechanisms of the body. The Chinese chive root is rich in polyphenols as well as antimicrobial constituents.

27.3 EXTRACTION TECHNIQUES FOR PLANT EXTRACTS

The separation of plant extracts from plant material involves extracting medicinally active bio-components using inert or inactive components and selective solvents (Table 27.2). For extraction, traditional and modern methods are employed. Solid-phase micro-extraction, liquid-liquid, and solid-phase methods are conventionally used. However, ultrasound-assisted, pressurized liquid, subcritical water, supercritical fluid, microwave-assisted, and instant-controlled

TABLE 27.2 Common Solvents Used for Active Component Extraction

EXTRACTION MEDIUM	*COMPOUNDS*
Acetone	Flavonols
Ether	Coumarins, terpenoids, alkaloids
Dichloro-methanol	Terpenoids
Chloroform	Flavonoids, terpenoids
Methanol	Flavones, tannins, saponins, terpenoids
Ethanol	Flavonols, terpenoids, tannins, alkaloids
Water	Terpenoids, saponins, tannins

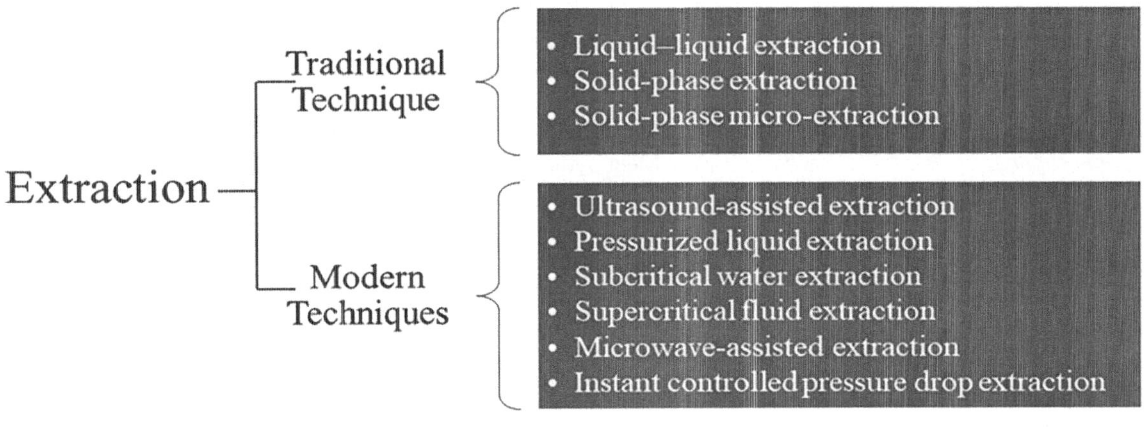

FIGURE 27.1 Extraction techniques for plant extracts.

pressure drop methods are traditionally used. The solvent diffuses into the plant tissue sample and dissolves the compounds of the same polarity. The quality of the phyto-extract depends on the target plant material, the selected solvent, and the method of extraction.

The variations in extraction protocols are often due to the length of the extraction procedure, the particle size of the plant material, the pH of the solvent, the temperature, and the sample-to-solvent ratio. The governing principle is the grinding of plant material, whether wet or dry, to a fine consistency that enhances the surface area of the sample (Figure 27.1). The extraction of finely crushed particles with a size 10 µm for a period of 5 min has higher extraction efficiency compared to roughly crushed particles subjected to extraction for a period of 24 hrs. Earlier studies have shown that a ratio of solvent-to-sample v/w @ 10:1 to dry weight is an ideal ratio (Green, 2004). Fresh plant samples, either wet or dry, can be ground into fine particles using a blender and vigorously shaken for 5–10 min, or mild shaking can be done for 24 hrs followed by filtration for efficient extraction.

27.4 BOTANICALS AS NATURAL PRESERVATIVES FOR FRESH HORTICULTURAL PRODUCE

Fruits and vegetables are nutrient-dense and have a pleasant taste due to organic acids, vitamins, and polyphenolic compounds, making them preferred for human consumption. During storage, these products undergo various changes due to environmental fluctuations. These changes are further represented in their nutritional value and taste, and result in the loss of quality (Zhang et al., 2019). Senescence is a postharvest stage when the anabolic process stops and the anabolic phase starts, leading to the deterioration of fresh produce (Figure 27.2). Various undesirable changes occur in the final postharvest produce, and its progression is influenced by both exogenous and endogenous factors. A significant loss of fresh vegetables and fruits occur in terms of mass due to water loss and shrinkage, resulting in poor retail sales and financial losses (Mbili et al., 2017). Because fruits are perishable in nature, postharvest technologists strive to find environmentally-friendly treatments to increase the quality of the final product (Zhang et al., 2019).

Plant extracts have gained significant global consideration owing to several health benefits and other potential industrial applications. Numerous studies have shown that treating fruits and vegetables with plant extracts increases their durability and postharvest shelf life. Postharvest treatments of fruits and vegetables reduce non-enzymatic antioxidant activity and enhance hormone biosynthesis regulation along with delayed degradation of cell walls (Bose et al., 2021). Plant extracts are commercially used as bio-preservatives, which are biological compounds or constituents employed to extend the shelf life of several products. They are used to enhance the safety of foods by utilizing antimicrobial compounds or controlled microbiota (Ananou et al., 2007).

Plant extracts have found various direct applications in the form of additives in the food industry owing to their contents of carotenoid and polyphenol compounds. Research activities have recognized various plant extracts that can be used in the form of coatings, sprays, and emulsion to decrease postharvest-based losses without resulting in some potential health hazards (Dawlatana, 2019). Plants possess the ability to produce aromatic metabolites, such as oxygen alternatives and phenols. Important subclasses of these compounds include phenols, flavones, flavonoids, tannins, quinones, flavonols, and coumarins. These compounds possess antimicrobial activities and serve as bioactive phytochemicals in the defense of plants against pathogenic infestations. Simple phenolics and phenols are bioactive phytonutrients composed of single-substituted rings of phenolic compounds. Quinones are notably reactive, colored compounds with ketonic-substituted phenolic rings.

Different flavonoids, flavones, and flavonols are types of phenolic compounds known for their potential antimicrobial

FIGURE 27.2 Behavior variation of coated and un-coated fruits and vegetables.

properties. Such compounds are biosynthesized by plants in response to microbial contaminations and exhibit antimicrobial activity against a wide range of bacterial and fungal microbes. Tannins are polymeric phenolic constituents known for their astringent properties, which also act as antimicrobials. These are soluble in both water and alcohol and form precipitates when mixed with proteins. The coumarin-based compounds are another group of antimicrobial phenols characterized by fused rings of α-pyrone and benzene. These are found in essential oils, raw sap, and volatile constituents extracted from various plant parts, including stem parts (bark and cambium), roots, flowers, leaves, seeds, and fruits. These compounds are widely utilized to prepare antimicrobial formulations for horticultural commodities, protecting them against diseases and pathogens. There are six broad categories of natural chemicals, including (i) saponins, (ii) flavonoids and isoflavonoids, (iii) steroids, (iv) tannins, (v) phenolic acids (including chlorogenic acid, caffeic acid, protocatechuic acid, ferulic acid, phenolic acids, and other phenols), and (vi) pyrones and coumarins.

The plant extracts have the ability to preserve the quality of harvested fruits and vegetables via spraying, brushing, or coating; however, the most reliable technology is edible coating. Fresh horticultural products, including fruits and vegetables, maintain biological activity even after harvesting, which results in water loss and solute concentration. Moreover, the gaseous exchange also deteriorates the product quality. This presents a significant loss in postharvest quality owing to moisture loss, shriveled texture, poor quality attributes, and limited shelf life. Immediate coating of edible material reduces postharvest quality losses. This process reduces microbial invasion and delays senescence and water loss. Some coatings include "Edible coatings based on plant extracts" (ECPE) labels on fruits and vegetables, providing relief to vegan customers. The ECPE is derived from essential oils and volatiles and can be applied by different methods such as spraying, brushing, dipping, and panning. The ECPE performs various functions, such as improving aesthetic look, hindering minor injuries, and improving shine. ECPEs are categorized into two different types based on their composition: lipids (acylglycerol, fatty acids, and waxes) and hydrocolloids (polysaccharides and proteins). There are a few functional properties that an ECPE must perform, i.e., it should not alter the taste and color, and it should not degrade the final quality owing to physiological reactions. Moreover, it should be resistant to moisture and must create partial separation between the internal and external atmosphere, sufficiently giving coverage after the final application, but it should also be a bit porous enough to permit fruit/vegetable respiration.

The food manufacturing industry is constantly up against the difficulty of food storage issues and extending the shelf life of fresh produce after harvest. Foodborne infections are one of the main culprits, causing food to decay and consumer health to deteriorate (Al-Tayyar et al., 2020; Yousuf et al., 2018). To overcome the difficulties of food preservation with minimal or no impact on consumer health, there is a growing knowledge regarding the utilization of more natural and safer plant-based preservers for fruit and vegetable commodity shelf life prolongation as an effective substitute for certain dangerous chemicals (Hosseini et al., 2020). Research is moving toward the age of plant-based extracts as a replacement for dangerous synthetic pesticides, as the demand for pesticide-free or organic foods has greatly increased (Raherison et al., 2019). Due to their efficient biodegradability, biocompatibility, and compliance with chemical and biological alterations, natural polymer–based compounds have numerous benefits over synthetic polymers.

Foliar spray of aloe-based gels can be used to control and ensure the suppression of banana fruit metabolism, thereby resulting in resistance of the treated group against anthracnose incidence. Aloe gel treatment serves as a synthetic pesticide alternative in inducing resistance to anthracnose and preserving the banana fruit's postharvest quality (Khaliq et al., 2019). The shelf life of blueberry fruits was extended by up to 5 days upon spraying with aloe vera–based extract compared with chitosan, indicating that a combination of chitosan treatment and aloe vera mucilage can considerably extend a product's shelf life. Postharvest quality and storage life of grapes are both positively impacted by aloe gel. The application of aloe vera extracts has proven to be a healthy substitute for synthetic chemical–based treatments in preserving the postharvest quality attributes of litchi fruit. Surface browning in aloe vera gel–treated litchi fruit was reduced, along with extended storage life and overall quality preservation (Ali et al., 2019; Shah & Hashmi, 2020). The quality of postharvest papaya fruits was preserved in storage for up to 15 days thanks to the aloe vera extract coating that was applied (Ali et al., 2019; Mendy et al., 2019). Additionally, the antifungal property of aloe vera offers an appropriate potential for increasing the shelf life of blueberry fruits without the using of synthetic fungicides in edible coatings or films. The aloe vera mucilage may provide a healthy alternative to potentially dangerous pesticides and fungicides when dealing with fungal and certain pest-based infestations on fruits and vegetables, thereby protecting humans' health and environmental sustainability.

At different concentration levels of plant extract, the pathogen's mycelial growth and sporulation could be inhibited, and guava fruits treated with lemongrass extract showed a significant reduction in *C. gloeosporioides*–based disease incidence (Mazón-Suástegui et al., 2019; Jagana et al., 2018). Orange *Penicillium* infection might be efficiently suppressed by using vaporized lemongrass-based essential oil at different concentrations. The antifungal properties of lemongrass extracts were also effective against *A. niger* and *C. musae* on orange fruits (Pinto et al., 2021; Nyamath et al., 2018).

Neem extract extended the storage life of banana fruits and enhanced their quality preservation. The shelf life of bananas was significantly extended by using neem extract at 40% concentration (Siddiqua et al., 2018). This approach is not limited to bananas; it can also benefit some other tropical and subtropical fruits by impeding postharvest losses caused by different types of fungal diseases. Neem extract was found effective in inhibiting the growth of *Neofusicoccum parvum*, *Lasiodiplodia theobromae*, *Aspergillus flavus*, *B. cinerea*, and *A. niger*. The application of neem extract on mango fruit after harvest showed a significant impact in improving its storage life and preserving quality in comparison with controls.

Different studies have demonstrated that the quality and shelf life of harvested strawberries could be effectively conserved by applying various sprays of plant protein–based extracts. The treatment helped to maintain the ascorbic acid content of strawberries, which is considered a major factor in determining antioxidant activity and free radical scavenging potential (Mbili et al., 2017). For various avocado fruit varieties, such as Hass and Gem, the application of moringa seed/leaf extracts with 1% carboxymethyl cellulose (CMC) significantly preserved postharvest quality and improved shelf life. Compared with uncoated fruit, the treatment combination reduced ethylene production and respiration rate while increasing fruit firmness under storage. Additionally, moringa ethanol extracts exhibited considerably higher antibacterial activity and inhibited foodborne microorganisms. An edible spray containing moringa leaf/seed extracts at 1% CMC could be utilized as an organic postharvest treatment for Hass and Gem avocados. It was observed that at a concentration of 10 g, crude extracts of dukung anak or turmeric powder can be employed as a fungicide to inhibit postharvest anthracnose in dragon fruits. Furthermore, the treatment of dragon fruits with ginger and turmeric extracts conserved their natural color and flavor (Bordoh et al., 2020). In the case of guava fruits, a combination of gum arabic (10%) and garlic extract effectively prevented weight loss, skin browning, and disease incidence, and extended the shelf life of the fruit compared to the controls. Overall, the combination of gum arabic and garlic extract preserved higher physiological, biochemical, and sensory qualities of the guavas (Anjum et al., 2020). In addition, sole treatments of ginger and garlic extracts, as well as aloe vera and gum arabic, were also found to be effective in preserving the quality of harvested Surahi guavas during ambient storage (Zaidi et al., 2023).

27.4.1 Use of Plant Extracts for Fruit and Vegetable Disease Control during Storage

After harvest, horticultural produce undergo a wide range of biological and chemical processes and changes that result in the growth of certain biological diseases, quality degradation, and ultimately higher losses. The deterioration of certain horticultural commodities is mostly attributed to the attack of postharvest diseases from molds, *Penicillium spp.*, and *B. cinerea*. The utilization of synthetic fungicide chemicals remains the most effective approach in managing postharvest infections due to their quick control. Nevertheless, the utilization of these chemicals has progressively reduced on a global scale owing to their detrimental effects on the health of consumers and customers' increased preference for organic substitutes. This problem has incited researchers to explore the possibility of natural flora as alternatives for crop protection and preservation. A number of phytochemicals and antioxidative compounds have been reported to conserve quality and control disease during storage. Plant extracts from *Ruta chalepensis* and *Eucalyptus globulus* have shown higher efficiency in managing postharvest diseases of horticulture produce and prolonging shelf life, thereby serving as potential alternatives to synthetic chemicals. It has been reported that neem is known to possess substantial insecticidal and antifungal properties. The utilization of botanical extracts from medicinal plants as an alternative to synthetic chemicals in the context of preserving and maintaining the quality of horticultural commodities after harvest holds potential advantages (Nxumalo et al., 2021).

Despite the existence of extensive ethnobotanical literature, there is a lack of scientific evidence about the effectiveness and phytochemistry of native medicinal plants and their excerpts in safeguarding and preserving horticulture commodities after harvest. Fungi are identified as the primary factor responsible for the occurrence of postharvest losses in fresh fruits and vegetables, whether it is during transportation or storage. During the commercialization phase, these events lead to substantial economic losses and render the products unfit for human consumption (D'Aquino et al., 2010).

About one-third (33%) of the world's yearly human food output goes to waste (FAO, 2020). Physical damage; poor transportation and storage methods; improper packing; postharvest infections like *Colletotrichum gloeosporioid*, *Botrytis cineria*, *Penicillium digitatum*, *Alternaria alternata*, and *Rhizoctonia solani*; and natural senescence processes all contribute to postharvest losses (FAO, 2017: FAO, 2020). Capital-intensive machinery and fungicides are employed in the postharvest phase of the horticultural industry to reduce postharvest crop failure or decomposition (Ncama et al., 2018: Riva et al., 2020). Global food security and sustainability generally rely heavily on the protection and conservation of food crops. Food deterioration due to diseases has been reduced because of advances in food preservation techniques (Pimentel, 2009; Ogunnupebi, 2020; Sharif et al., 2017). Researchers have been looking toward safer and commercially feasible plant-based alternatives which could be considered safe for consumers and the environment (FAO, 2013). The food, pharmaceutical, and agrochemical sectors have made significant advancements in the development and utilization of natural preservatives and pesticides (Santos-Sánchez et al., 2017).

Horticultural produce protection and preservation heavily rely on the application of various pesticides and fungicides (Thorat et al., 2017; Masarirambi et al., 2020). Several types of phytochemicals, including cyanogenic glycosides, alkaloids, phenylpropanoids, polyketides, carbohydrates, lipids, nucleic acids, anthocyanins, amino acids, flavonoids, terpenoids, tannins, saponins, and phenols, have been reported for this purpose. The antibacterial, antifungal, and antioxidant properties of plant extracts have been related to a variety of phytochemicals in the extracts. *Geophagus brasiliensis* contains fukugetin; *Lepidium meyenii* and *fukugiside* have carnosic acid, rosmanol, and carnosol; and *Cymbopogon citratus* has citral and aspilactonol B, 8-methyl-6-prenylquercetin. Therefore, it is crucial to assess medicinal plants' biological activities (antimicrobial and antioxidant properties), as well as their safety features (like cytotoxic and genotoxic effects), by conducting a thorough investigation of their phytochemical constituents (active ingredients, nutritional components, and mineral contents). Assessment is also essential for determining whether these can replace synthetic pesticides and fungicides in the postharvest preservation (Arnold et al., 2002; Williams et al., 2013) of fruits and vegetables.

There were more than 360,000 species of flowering plants and over 400,000 known plant species as of 2015. In addition, reports suggest that over 2,000 new species are identified yearly. Primary and secondary metabolites from these plants have been shown to have antibacterial and antifungal effects (Lee et al., 2007). Therapeutic plant extracts such as those from turmeric (*Curcuma longa*) leaf, neem (*Azadirachta indica*) leaf, and lemongrass (*Cymbopogon citratus*) plant have all been demonstrated to be particularly efficient against fungi. Neem leaf extract's efficacy has been connected to a wide variety of chemicals (Saxena et al., 2013), including phytol, nonanoic acid, dibutyl phthalate, tritriacontane, and 1,2-benzenedicarboxylic acid. Some plants may create secondary metabolites, including essential oils and biocidal chemicals. Fungicides such as tea tree oil, jojoba oil, neem oil, and aloe vera gel are widely available in stores. Pesticides like nicotine and pyrethrum are available for purchase. Medicinal plant extracts have recently been proposed as a feasible alternative to synthetic fungicides for the postharvest control of rot fungi on fruit and vegetable produce. Table 27.3 provides a comprehensive overview of the literature on medicinal plant extracts, which can be used to maintain and preserve crops after harvest.

TABLE 27.3 Herbal Extracts and Their Potential Usage

SR	PLANT NAME	PLANT PART USED	FOCUS OF THE STUDY	TREATMENT APPLICATION	REFERENCE
1	*Azadiracta indica*	Aqueous leaf extracts	Reduce cabbage damage caused by *Pieris brassicae*.	Cabbage leaves were sprayed with aqueous solutions of several concentrations (10%, 5%, 2.5%, and 1%).	Sharma and Gupta (2009)
2	*Zataria multiflora*	Essential oil from leaves	Keep button mushrooms (*Agaricus bisporus*) from becoming brown.	The treatments included a placebo (water), tragacanth gum (0.6% of the coating), tragacanth gum plus 0.01% ethanol oxide (ZEO) (100 parts per million), tragacanth gum plus 0.06% ethanol oxide (ZEO) (500 parts per million), tragacanth gum plus 0.06% ethanol oxide (ZEO) (1000 parts per million), and sodium metabisulphite (1000 parts per million). Button mushrooms were dipped for 5 minutes in each solution and then analyzed for browning after 16 days under refrigeration at 4 degrees Celsius.	Nasiri et al. (2017)

(Continued)

TABLE 27.3 (Continued)

SR	PLANT NAME	PLANT PART USED	FOCUS OF THE STUDY	TREATMENT APPLICATION	REFERENCE
3	*Azadiracta indica*	Kernel aqueous extract	Eradicate the red slug caterpillar.	Neem kernel aqueous extracts (NKAE) of various concentrations (2%, 4%, 6%, and 8%) were sprayed on tea leaves. As concentrations increased, so did the anti-feedant action. The maximum leaf area devoured was 1158.6 + 254.79 cm^2 by 2% 5th instar larvae, while the minimum was 92.2 cm^2 by 8% 1st instar larvae.	(Das et al., 2010)
4	*Bobgunnia madagascariensis*	Dried pods	Brassica napus ladybird beetle suppression using aqueous extracts of dried pods.	Various aqueous extracts were tested at the concentrations of 5, 10, 15, 20, and 25% w/v in a controlled environment. At 24, 48, and 72 hours after exposure, H. variegata death was recorded.	Mazhawidza et al. (2018)
5	Aloe vera	Leaves	Protect nectarines from *Penicillium digitatum*, *Botrytis cinerea*, and *Rhizopus stolonifera* via aloe vera gel solution.	Aloe vera gel solution was used to coat the fruit, which was then dried at room temperature for 10 minutes. To inoculate the nectarine cultivars with *P. digitatum*, *B. cinerea*, or *R. stolonifer* we created a 2 mm long, 2 mm wide, and 2 mm deep incision and left it open for 6 days at room temperature.	Navarro et al. (2011)
6	*Moringa olifera*	Leaf extracts	Coatings made of gum arabic (GA) and moringa (M) leaf extract are effective against *Colletotrichum gloeosporioides* on Maluma avocados.	The fruit samples were exposed to GA 10%, CMC 1% + M, GA 15% + M, and GA 15%, GA 10% + M. The fruits were kept at 5.5 degrees Celsius and 95% relative humidity (RH) for 21 days before being transferred to 21.1 degrees Celsius and 60% RH for 7 days to imitate an environment.	Kubheka et al. (2020)
7	*Lippia javanica*	The essential oil of leaves	*F. graminearum* must be contained in sweet corn.	Experimental concentrations of 0, 87, 65, 43, 22, 11, 054, and 0.027 mg of essential oil per milliliter were used in the bioassays.	Philemon et al. (2015)
8	*Solanum incanum*	Aqueous crude fruit sap extract	Preventing the spread of *Myzus persicae*, or green peach aphids, on kale.	Kale was regularly sprayed with an extract of *S. incanum* at concentrations of 10, 25, 50, and 75%.	Umar et al. (2015)
9	*Thymus vulgaris* L.	Leaves	Examining the impact of thyme oil and edible coatings on the development and spread of anthracnose in avocados that have been artificially infected.	To compare the potency of various plant extract concentrations, the 'poisoned food technique' was adopted. Chitosan (CH), aloe (AL), thyme oil (TO), chitosan + thyme oil (3:1), and aloe + thyme oil (3:1) were all used to treat the contaminated fruit before it was stored at room temperature for 5 days.	Bill et al. (2014)
10	*Melia azedarach*	Aqueous leaf extracts	Management of the cabbage pest *Pieris brassicae*.	Cabbage leaves were sprayed with aqueous solutions of several concentrations (10%, 5%, 2.5%, and 1%).	Sharma and Gupta (2009)

27.4.2 Preventive Behavior of Biological Extracts

The limited frequency of infectious diseases in medicinal plants serves as a key sign of the existence of robust defensive systems (Ugboko et al., 2020). Limited research has been conducted on the microbial interactions occurring on the surface of fruits and within wounds, mostly due to the intricate nature of researching the interplay between the pathogen, antagonist, host, and other microorganisms. Understanding the mechanism of action is essential for effectively inhibiting pathogens in their hosts. It is crucial to comprehend the underlying mechanism via which biocontrol of fruit diseases takes place in order to facilitate the future improvement and broadening of the application of medicinal plant extracts for disease control (Talibi et al., 2014). Within the fructoplane, antagonistic interactions encompass a wide range of processes, such as nutrition and induced resistance, space competition, antibiosis, and parasitism. These applied extracts may function independently or in combination with one another. In order to establish effective postharvest control measures on the basis of plant extracts, it is necessary to comprehensively understand pathogens, antagonistic bacterial strains, produce, and the surrounding environments (Liu et al., 2018).

Meskin et al. (2002) have identified different mechanisms by which phytochemicals from plant extracts can exhibit antimicrobial effects. These mechanisms encompass the inhibition of enzymatic activities, the disruption of bacterial membranes, the suppression of virulence factors, the formation of biofilms, the inhibition of protein synthesis, and quorum quenching. The mode of action of tannins involves their ability to bind proteins, resulting in the inhibition of cell protein biosynthesis (Saxena et al., 2013). On the other hand, phenolic compounds have been observed to modify the pathogens' membrane functionality. Nevertheless, additional research is required to fully understand the mode of action of these plant-based products before they can be suggested for the management of postharvest diseases in horticultural crops (Lanciotti et al., 2004).

The availability of a wide range of bioactive chemicals in extracts from medicinal plants implies that their antimicrobial activities are not attributed to a factor (Talibi et al., 2014). According to one study, essential oils (EOs) obtained from medicinal plants have been found to contribute to plant defense mechanisms against phytopathogenic microbes. Additionally, the combined action of the many components found in these EOs effectively hinders the formation of fungal strains that are resistant to their effects. Green mold can be inhibited by increasing the host's immune response, and it has been shown that extracts from the winter cherry and the shittah tree are beneficial for this purpose (Lanciotti et al., 2004).

According to Mekbib et al. (2007), the defensive mechanism resulted in the production of the cell walls. This cell wall can function as a protective barrier, both physically and biologically, to prevent the entry of pathogens. Additionally, it may lead to an elevation in the overall concentration of soluble phenolic compounds in orange peels. Numerous phytochemical substances present in medicinal plants participate in the process of membrane rupture, the suppression of crucial metabolic functions, and the imposition of stress on pH homeostasis by depositing anions within the cell and triggering the activation of fruit defenses (Youssef et al., 2014).

In *in vitro* situations, these phytochemicals demonstrate the ability to impede the growth of certain phytopathogenic fungi. Fungal species exhibit a preference for acidic to neutral conditions rather than alkaline environments in fruits such as citrus. In instances of this kind, pathogens that necessitate an acidic milieu, such as *P. digitatum*, allocate a greater amount of energy toward the synthesis of fungal acids rather than the elongation of hyphae, resulting in the suppression of their proliferation. Extracts from medicinal plants are effective against *P. italicum*, *P. digitatum*, and *G. citri-aurantii* because they disrupt oxidative phosphorylation in mitochondria, limit enzyme and protein production, and modify membrane and transport function in the cells (Nunes, 2011; El-Mougy et al., 2008).

The pH of medicinal extracts plays a vital role in controlling postharvest citrus illnesses, as it has a direct impact on the germination of conidia (Talibi et al., 2014; Youssef et al., 2014) and regulates the aggressiveness of pathogens during their colonization of the host tissue. Yusoff et al. (2021) utilized a scanning electron microscope (SEM) to investigate the antibiotic mode of action exhibited by antifungal compounds derived from the extract of bitter leaf (*V. amygdalina*), specifically in terms of its inhibitory effects on fungus. The structural characteristics of the hyphae of *B. cinerea* were modified upon exposure to the phytochemical substances present in an extract derived from a bitter leaf. The mycelia exhibited contortions and creases, characterized by a serrated edge. Following the treatment, certain mycelia exhibited agglutination, characterized by dried hyphae end and conidia that had undergone shrinkage. The abundant and easily distributed asexual spores of *B. cinerea* may serve as a potential barrier against the transmission of gray mold illness to adjacent fruits. The alteration in the morphology of fungi occurred as a consequence of the presence of antifungal secondary metabolites within the plant extract, which effectively impeded the growth of the fungus. The combination of antifungal drugs has shown synergistic properties in inhibiting the growth and proliferation of *Botrytis cinerea*. In the context of an oxidative sequence, Yang et al. (2015) have identified several mechanisms through which antioxidants can exert their effects. These mechanisms include (i) reduction of localized oxygen concentrations; (ii) inhibition of chain initiation by scavenging starting radicals; (iii) sequestration of metal ions that have the potential to decompose lipid peroxides into peroxyl and alkoxyl radicals; and (iv) decomposition of peroxides by converting them into non-radical products. Phenols or aromatic amines are commonly seen as antioxidants that possess the ability to interrupt chain reactions. In the context of the postharvest processing of horticultural crops and the food business, an antioxidant is characterized as a chemical that effectively inhibits or retards the oxidation process of easily oxidizable compounds, such as lipids, when present in small amounts (Miguel, 2010).

The hydroxyl groups in alkaloids found in medicinal plants have been shown to interact with reactive species. Additionally, their nitrogen moiety enables them to chelate ferrous ions (Gali & Bedjou, 2019; Gomaa et al., 2019). Terpenoids and saponins isolated from various medicinal plants have demonstrated the ability to reduce the production of reactive oxygen species, resulting in the inhibition of lipid oxidation and the reduction of oxidative stress (Dini et al., 2009; Nzowa et al., 2010).

27.5 CONCLUSIONS

Natural plant extracts consist of biomolecules that can be extracted from various parts of plants. Researchers have explored that plant-based extracts can meet consumer demand by decreasing the detrimental effects of synthetic compounds on human health. Fresh vegetables and fruits are living commodities. However, postharvest changes can sometimes lead to undesirable outcomes, ultimately contributing to an overall decline in quality. These changes manifest as mass loss, a withered appearance, reduced shelf life, and diminished organoleptic properties. Medicinal plants contain polyphenols, including flavonoids and phenolic compounds, which exhibit various biological activities, such as scavenging reactive species, inhibiting lipoxygenase, and reducing metmyoglobin formation. While synthetic fungicides and other chemicals have traditionally been effective in controlling postharvest diseases, concerns about their overuse and potentially negative effects have led to increased demand for organic alternatives. All these factors underscore the importance of research into plant extracts and their application in controlling postharvest diseases, medicinal uses, and as viable substitutes for chemically synthesized fungicides and insecticides in crop protection and preservation. Plant extracts represent natural resources and valuable tools for maintaining the optimal postharvest condition of fruits and vegetables.

REFERENCES

Abdullahi, A., Tijjani, A., Abubakar, A. I., Khairulmazmi, A., Ismail, M. R. (2022). Plant biomolecule antimicrobials: An alternative control measures for food security and safety. In *Herbal Biomolecules in Healthcare Applications* (pp. 381–406). Academic Press.

Ali, S., Khan, A.S., Nawaz, A., Anjum, M.A., Naz, S., Ejaz, S., Hussain, S. (2019). Aloe Vera gel coating delays postharvest browning and maintains quality of harvested litchi fruit. *Postharvest Biology and Technology*, *157*, 110960.

Al-Tayyar, N. A., Youssef, A. M., Al-Hindi, R. R. (2020). Edible coatings and antimicrobial nanoemulsions for enhancing shelf life and reducing foodborne pathogens of fruits and vegetables: A review. *Sustainable Materials and Technologies*, *26*, e00215.

Ananou, S., Maqueda, M., Martínez-Bueno, M., Valdivia, E. (2007). Biopreservation, an ecological approach to improve the safety and shelf-life of foods. In Mendez-Vilas, A. (ed), *Communicating Current Research and Educational Topics and Trends in Applied Microbiology* (pp. 475–486). Formatex.

Anjum, M. A., Akram, H., Zaidi, M., Ali, S. (2020). Effect of gum arabic and Aloe vera gel based edible coatings in combination with plant extracts on postharvest quality and storability of 'Gola' guava fruits. *Scientia Horticulturae*, *271*, 109506.

Arnold, T.H., Prentice, C.A., Hawker, L.C., Snyman, E.E., Tomalin, M., Crouch, N.R., Pottas-Bircher, C. (2002). *Medicinal and Magical Plants of Southern Africa: An Annotated Checklist* (p. 203). National Botanical Institute.

Ayala-Zavala, J.F., Oms-Oliu, G., Odriozola-Serrano, I., Gonzalez-Aguilar, G.A., Alvarez-Parrilla, E., Martin-Belloso, O. (2008). Biopreservation of fresh-cut tomatoes using natural antimicrobials. *European Food Research and Technology*, *226*, 1047–1055.

Banon, S., Díaz, P., Rodríguez, M., Garrido, M. D., Price, A. (2007). Ascorbate, green tea and grape seed extracts increase the shelf life of low sulphite beef patties. *Meat Science*, *77*(4), 626–633.

Bill, M., Sivakumar, D., Korsten, L., Thompson, A.K. (2014). The efficacy of combined application of edible coatings and thyme oil in inducing resistance components in avocado (*Persea americana* Mill.) against anthracnose during post-harvest storage. *Crop Protection*, *64*, 159–167.

Bordoh, P. K., Ali, A., Dickinson, M., Siddiqui, Y. (2020). Antimicrobial effect of rhizome and medicinal herb extract in controlling postharvest anthracnose of dragon fruit and their possible phytotoxicity. *Scientia Horticulturae*, *265*, 109249.

Bose, S. K., Howlader, P., Wang, W., Yin, H. (2021). Oligosaccharide is a promising natural preservative for improving postharvest preservation of fruit: A review. *Food Chemistry*, *341*, 128178.

Camele, I., De Feo, V., Altieri, L., Mancini, E., De Martino, L., Luiqi Rana, G. (2010). An attempt of postharvest orange fruit rot control using essential oils from mediterranean plants. *Journal of Medicinal Food*, *13*, 1515–1523.

Cao, S.F., Zheng, Y.H., Yang, Z.F., Tang, S.S., Jin, P., Wang, K.T., Wang, X.M. (2008). Effect of methyl jasmonate on the inhibition of *Colletotrichum acutatum* infection in loquat fruit and the possible mechanisms. *Postharvest Biology and Technology*, *49*, 301–307.

Creeman, R.A., Mullet, J.E. (1997). Biosynthesis and action of jasmonate in plants. *Annual Review of Plant Physiology*, *48*, 355–381.

D'Aquino, S., Palma, A., Schirra, M., Continella, A., Tribulato, E., La Malfa, S. (2010). Influence of film wrapping and fludioxonil application on quality of pomegranate fruit. *Postharvest Biology and Technology*, *55*, 121–128.

Das, K., Tiwari, R.K.S., Shrivastava, D.K. (2010). Techniques for evaluation of medicinal plant products as antimicrobial agent: Current methods and future trends. *Journal of Medicinal Plants Research*, *4*, 104–111.

Dawlatana, M. (2019). Science and Technology for Sustainable Development. *Bangladesh Journal of Scientific and Industrial Research*, *54*, 1–134.

Dini, I. Tenore, G.C., Dini, A. (2009). Saponins in *Ipomoea batatas* tubers: Isolation, characterization, quantification and antioxidant properties. *Food Chemistry*, *113*, 411–419.

Dorman, H.J., Deans, S.G. (2000). Antimicrobial agents from plants: antibacterial activity of plant volatile oils. *Journal of Applied Microbiology*, *88*, 308–316.

El-Mougy, N.S., El-Gamal, N.G., Abd-El-Kareem, F. (2008). Use of organic acids and salts to control postharvest diseases of lemon fruits in Egypt. *Archives of Phytopathology and Plant Protection*, *41*(7), 467–476.

Food and Agriculture Organization of the United Nations (FAO). (2017). *SAVE FOOD: Global Initiative on Food Loss and Waste Reduction*. FAO.

Food and Agriculture Organization of the United Nations (FAO). (2020). *The State of Agricultural Commodity Markets 2020. Agricultural Markets and Sustainable Development: Global Value Chains; Smallholder Farmers and Digital Innovations*. FAO.

Food and Agriculture Organization (FAO). (2013). *World Health Organisation [WHO] of the United Nations. Pesticide Residues in Food*. WHO.

Fung, R., Wang, C., Smith, D., Gross, K., Tian, M. (2004). Mesa and Meja increase steady-state transcript levels of alternative oxidase and resistance against chilling injury in sweet peppers (Capsicum annuum L.). *Plant Science*, *166*, 711–719.

Gali, L., Bedjou, F. (2019). Antioxidant and anticholinesterase effects of the ethanol extract, ethanol extract fractions and total alkaloids from the cultivated *Ruta chalepensis*. *South African Journal of Botany*, *120*, 163–169.

Gomaa, A.A., Makboul, R.M., El-Mokhtar, M.A., Abdel-Rahman, E.A., Ahmed, I.A., Nicola, M.A. (2019). Terpenoid-rich Elettaria cardamomum extract prevents Alzheimer-like alterations induced in diabetic rats via inhibition of the GSK3β activity, oxidative stress and pro-inflammatory cytokines. *Cytokine*, *113*, 405–416.

Gonzalez-Aguilar, G.A., Wang, C.Y., Buta, J.G. (2003). Methyl jasmonate and modified atmosphere packaging (MAP) reduce decay and maintain postharvest quality of papaya "Sunrise". *Postharvest Biology and Technology*, *28*, 361–379.

Green, R.J. (2004). Antioxidant activity of peanut plant tissues. Master Thesis, North Carolina State University

Hasabnis, S.N., Souza, T.F. (1988). Use of natural plant products in the control of the storage rot of alphanso mango fruits. *Journal of Maharastra Agricultural University*, *12*, 105–106.

Hosseini, H., Jafari, S. M. (2020). Introducing nano/microencapsulated bioactive ingredients for extending the shelf-life of food products. *Advances in Colloid and Interface Science*, *282*, 102210.

Jagana, D., Hegde, Y. R., Lella, R. (2018). Bioefficacy of essential oils and plant oils for the management of banana anthracnose-a major post-harvest disease. *International Journal Current Microbiology and Applied Sciences*, *7*(4), 2359–2365.

Jin, P., Zheng, Y.H., Gao, H.Y., Cheng, C.M., Chen, W.X., Chen, H.J. (2006). Effects of methyl jasmonate treatment on fruit decay and quality in peaches during storage at ambient temperature. *Acta Horticulture*, *712*, 711–716.

Kaituo, W., Peng, J., Shifeng, C., Haitao, S., Zhenfeng, Y., Yonghua, Z. (2009). Methyl jasmonate reduces decay and enhances antioxidant capacity in Chinese bayberries. *Journal of Agricultural and Food Chemistry*, *57*, 5809–5815.

Kapoor, A. (1997). Antifungal activities of fresh juice and aqueous extracts of turmeric and ginger. *Journal of Phytological Research*, *10*, 59–62.

Khaliq, G., Ramzan, M., Baloch, A.H. (2019). Effect of Aloe Vera Gel Coating Enriched with Fagonia Indica Plant Extract on Physicochemical and Antioxidant Activity of Sapodilla Fruit during Postharvest Storage. *Food Chemistry*, *286*, 346–353.

Kleeberg, H. (1996). The Neem Azal conception: Future possibilities of the use of neem in biological and integrated pest management. In Singh, R.P., Chavi, M. S., Raheja, R.K. (eds.), *Neem and Environment* (pp. 875–882). IBH publication Company Pvt. Ltd.

Kubheka, S.F., Tesfay, S.Z., Mditshwa, A., Magwaza, L.S. (2020). Evaluating the efficacy of edible coatings incorporated with moringa leaf extract on postharvest of 'Maluma' avocado fruit quality and its biofungicidal effect. *HortScience*, *55*, 410–415.

Lanciotti, R., Gianotti, A., Patrignani, F., Belletti, N., Guerzoni, M., Gardini, F. (2004). Use of natural aroma compounds to improve the shelf-life and safety of minimally processed fruits. *Trends in Food Science and Technology*, *15*, 201–208.

Lee, S.H., Chang, K.S., Su, M.S., Huang, Y.S., Jang, H.D. (2007). Effects of some Chinese medicinal plant extracts on five different fungi. *Food Control*, *18*, 1547–1554.

Liu, J., Sui, Y., Chen, H., Liu, Y., Liu, Y. (2018). Proteomic analysis of kiwifruit in response to the postharvest pathogen, Botrytis cinerea. *Frontiers in Plant Science*, *9*, 302202.

Masarirambi, M.T., Nxumalo, K.A., Kunene, E.N., Dlamini, D.V., Mpofu, M., Manwa, L., Earnshaw, D.M., Bwembya, G.C. (2020). Traditional/Indigenous vegetables of the kingdom of eswatini: biodiversity and their importance: a review. *Journal o f Experimental Agriculture International*, *42*, 204–215.

Mazón-Suástegui, J.M., Salas-Leiva, J., Teles, A., Tovar-Ramírez, D. (2019). Immune and antioxidant enzyme response of longfin yellowtail (*Seriola rivoliana*) juveniles to ultra-diluted substances derived from phosphorus, silica and pathogenic Vibrio. *Homeopathy*, *108*(01), 043–053.

Mbili, N. C., Opara, U. L., Lennox, C. L., Vries, F. A. (2017). Citrus and lemongrass essential oils inhibit Botrytis cinerea on 'Golden Delicious', 'Pink Lady' and 'Granny Smith' apples. *Journal of Plant Diseases and Protection*, *124*, 499–511.

Mekbib, S.B., Regnier, T.J.C. (2007). Korsten, L. Control of *Penicillium digitatum* on citrus fruit using two plant extracts and study of their mode of action. Phytoparasitica, 35, 264–276.

Mendy, T. K., Misran, A., Mahmud, T. M. M., Ismail, S. I. (2019). Application of Aloe vera coating delays ripening and extend the shelf life of papaya fruit. *Scientia Horticulturae*, *246*, 769–776.

Meskin, M.S., Bidlack, W.R., Davies, A.J., Omaye, S.T. (2002). *Phytochemicals in Nutrition and Health* (p. 224). CRC Press.

Miguel, M.G. (2010). Antioxidant activity of medicinal and aromatic plants. A review. *Flavour and Fragrance Journal*, *25*, 291–312.

Nasiri, M., Barzegar, M., Sahari, M.A., Niakousari, M. (2017). Tragacanth gum containing *Zataria multiflora* Boiss. essential oil as a natural preservative for the storage of button mushrooms (*Agaricus bisporus*). *Food Hydrocolloids*, *72*, 202–209.

Ncama, K., Magwaza, L.S., Mditshwa, A., Tesfay, S.Z. (2018). Plant-based edible coatings for managing postharvest quality of fresh horticultural produce: A review. *Food Packaging and Shelf Life*, *16*, 157–167.

Nunes, C.A. (2011). Biological control of postharvest diseases of fruit. *European Journal of Plant Pathology*, *133*, 181–196.

Nxumalo, K. A., Aremu, A. O., Fawole, O. A. (2021). Potentials of medicinal plant extracts as an alternative to synthetic chemicals in postharvest protection and preservation of horticultural crops: A review. *Sustainability*, *13*(11), 5897.

Nyamath, S., Karthikeyan, B. (2018). In vitro antifungal activity of lemongrass (*Cymbopogon citratus*) leaf extracts. *Journal of Pharmacognosy and Phytochemistry*, *7*(3), 1148–1151.

Nzowa, L.K., Barboni, L., Teponno, R.B., Ricciutelli, M., Lupidi, G., Quassinti, L., Bramucci, M., Tapondjou, L.A. (2010). Rheediinosides A and B, two antiproliferative and antioxidant *Triterpene saponins* from *Entada rheedii*. *Phytochemistry*, *71*, 254–261.

Ogunnupebi, T.A., Oluyori, A.P., Dada, A.O., Oladeji, O.S., Inyinbor, A.A., Egharevba, G.O. (2020). Promising natural products in crop protection and food preservation: basis, advances, and future prospects. *International Journal of Agronomy*, *2020*, 1–28.

Patil, R.K., Patel, K.D., Sharma, A., Pathak, V.N. (1992). Inhibition effects of *Ocimum sanctum* extract on fruit rot fungi. *Indian Journal of Mycology and Plant Pathology*, *22*, 199–200.

Philemon, Y.K., Matasyoh, J.C., Wagara, I.N. (2015). Chemical composition and antifungal activity of the essential oil from *Lippia javanica* (Verbenaceae). *International Journal of Biotechnology and Food Science*, *4*, 1–6.

Pimentel, D. (2009). Pesticides and pest control. In Peshin, R., Dhawan, A.K. (eds.), *Integrated Pest Management: Innovation-Development Process* (pp. 83–87). Dordrecht: Springer.

Pinto, L., Cefola, M., Bonifacio, M. A., Cometa, S., Bocchino, C., Pace, B., Baruzzi, F. (2021). Effect of red thyme oil (Thymus vulgaris L.) vapours on fungal decay, quality parameters and shelf-life of oranges during cold storage. *Food Chemistry*, *336*, 127590.

Raherison, C., Baldi, I., Pouquet, M., Berteaud, E., Moesch, C., Bouvier, G., Canal-Raffin, M. (2019). Pesticides exposure by air in vineyard rural area and respiratory health in children: a pilot study. *Environmental research*, *169*, 189–195.

Raybaudi-Massilia, R.M., Mosqueda-Melgar, J., Martín-Belloso, O. (2008). Edible alginate-based coating as carrier of antimicrobials to improve shelf-life and safety of fresh-cut melon. *International Journal of Food Microbiology*, *121*, 313.

Riva, S.C., Opara, U.O., Fawole, O.A. (2020). Recent developments on postharvest application of edible coatings on stone fruit: A review. *Scientia Horticulturae*, *262*, 109074.

Santos-Sanchez, N.F., Salas-Coronado, R., Valadez-Blanco, R., Hernández-Carlos, B., Guadarrama-Mendoza, P.C. (2017). Natural antioxidant extracts as food preservatives. Acta *Scientiarum Polonorum Technologia Alimentaria*, *16*, 361–370.

Saxena, M., Saxena, J., Nema, R., Singh, D., Gupta, A. (2013). Phytochemistry of medicinal plants. *Journal of Pharmacognacy and Phytochemistry*, *1*, 168–182.

Serrano, M., Martinez-Romero, D., Castillo, S., Guillén, F., Valero, D. (2005). The use of natural antifungal compounds improves the beneficial effect of MAP in sweet cherry storage. *Innovative Food Science and Emerging Technologies*, *6*, 115–123.

Shah, S., Hashmi, M. S. (2020). Chitosan–aloe vera gel coating delays postharvest decay of mango fruit. *Horticulture, Environment, and Biotechnology*, *61*, 279–289.

Sharif, Z., Mustapha, F., Jai, J., Yusof, N.M., Zaki, N. (2017). Review on methods for preservation and natural preservatives for extending the food longevity. *Chemical Engineering Research Bulletin*, *19*, 145.

Sharma, A., Gupta, R. (2009). Biological activity of some plant extracts against *Pieris brassicae* (Linn.). *Journal of Biopestic*, *2*, 26–31.

Shivpuri, A., Sharma, O.P., Jhamaria, S.L. (1997). Fungitoxic properties of plant extracts against pathogenic fungi. *Journal of Mycology and Plant Pathology*, *70*, 13–17.

Siddiqua, M., Khan, S. A. K. U., Tabassum, P., Sultana, S. (2018). Effects of neem leaf extract and hot water treatments on shelf life and quality of banana: Effect of plant extract and hot water on banana. *Journal of the Bangladesh Agricultural University*, *16*(3), 351–356.

Singh D., Thakur, R.K., Singh, D. (2003). Effect of pre-harvest sprays of fungicides and calcium nitrate on postharvest of kinnow in low temperature storage. *Plant Disease Research*, *18*, 9–11.

Singh, J. N., Pinaki, A., Singh, B. B. (2000). Effect of GA_3 and plant extracts on storage behavior of mango (*Mangifera indica* L.) cv. Langra. *Haryana Journal of Horticultural Sciences*, *29*(3/4), 199–200.

Siripatrawan, U., Harte, B. R. (2010). Physical properties and antioxidant activity of an active film from chitosan incorporated with green tea extract. *Food Hydrocolloids*, *24*(8), 770–775.

Talibi, I., Boubaker, H., Boudyach, E., Ben Aoumar, A.A. (2014). Alternative methods for the control of postharvest citrus diseases. *Journal of Applied Microbiology*, *117*, 1–17.

Thorat, P., Kshirsagar, R., Sawate, A., Patil, B. (2017). Effect of lemongrass powder on proximate and phytochemical content of herbal cookies. *Journal of Pharmacognacy and Phytochemistry*, *6*, 155–159.

Tripathi, P. (2001). Evaluation of some plant products against fungi causing postharvest diseases of some fruits. Ph.D. thesis, Department of Botany, Banaras Hindu University.

Ugboko, H.U., Obinna, C.N., Solomon, U.O., Toluwase, H.F., Conrad, A.O. (2020). Antimicrobial importance of medicinal plants in Nigeria. *Scientific World Journal*, 7059323.

Umar, A., Piero, N.M., Mgutu, A.J., Ann, N.W., Maina, G.S., Maina, M.B., Njagi, J.M., Mworia, J.K., Ngure, J.K., Mwonjoria, J.K. (2015). Bio efficacy of aqueous crude fruit sap extract of *Solanum incanum* against green peach aphids *Myzus persicae* Sulzer (Homoptera: Aphididae). *Entomology, Ornithology & Herpatology*, *5*, 169, 2161-0983.

Williams, V., Victor, J., Crouch, N. (2013). Red Listed medicinal plants of South Africa: Status, trends, and assessment challenges. *South African Journal of Botany*, *86*, 23–35.

Yang, X.-N., Khan, I., Kang, S.C. (2015). Chemical composition, mechanism of antibacterial action and antioxidant activity of leaf essential oil of *Forsythia koreana* deciduous shrub. *Asian Pacific Journal of Tropical Medicine*, *8*, 694–700.

Youssef, K., Sanzani, S.M., Ligorio, A., Ippolito, A., Terry, L.A. (2014). Sodium carbonate and bicarbonate treatments induce resistance to postharvest green mould on citrus fruit. *Postharvest Biology & Technology*, *87*, 61–69.

Yousuf, B., Qadri, O. S., Srivastava, A. K. (2018). Recent developments in shelf-life extension of fresh-cut fruits and vegetables by application of different edible coatings: A review. *LWT*, *89*, 198–209.

Yusoff, S., Haron, F., Asib, N., Mohamed, M., Ismail, S. (2021). Development of Vernonia amygdalina Leaf Extract Emulsion Formulations in Controlling Gray Mold Disease on Tomato (*Lycopersicon esculentum* Mill.). *Agronomy*, *11*, 373.

Zaidi, M., Akbar, A., Ali, S., Akram, H., Ercisli, S., Ilhan, G., Sarkar, E., Marc, R.A., Sonmez, D.A., Ullah, R., Bari, A., Anjum, M.A. (2023). Application of Plant-Based Edible Coatings and Extracts Influences the Postharvest Quality and Shelf Life Potential of "Surahi" Guava Fruits. *ACS Omega*, *8*, 19523–19531.

Zhang, W., Jiang, W. (2019). UV treatment improved the quality of postharvest fruits and vegetables by inducing resistance. *Trends in Food Science & Technology*, *92*, 71–80.

Antagonistic Microorganism Technology for Horticultural Produce

28

Soheila Aghaei Dargiri*, Somayeh Rastegar, and Emad Hamdy Khedr
*Corresponding Author: s.aghaei6418@gmail.com; S.aghaei.phD@hormozgan.ac.ir.

28.1 INTRODUCTION

Fruits and vegetables contribute a substantial portion of the nutrients that go into human nutrition and have a big impact on people's safety and quality of life (Lapuente, Estruch, Shahbaz, & Casas, 2019). Plant diseases, which include bacteria, fungi, viruses, and nematodes, successfully harm and destroy agricultural crops all over the world and drastically diminish their variety and production. These fatalities pose a severe threat to the world's agricultural output each year (Dean, 2012; O'Brien, 2017; Singh, 2014). Health issues for people and farm animals can arise from pathogenic infections in the field or during post-harvest storage, particularly if the pathogen creates pollutants that end up in or on consumer items (Menzler-Hokkanen, 2006).

Several strategies and approaches are used to control plant diseases, including the application of genetically modified organisms (GMOs), the utilization of certain chemicals and physical processes (such as UV irradiations, heat treatments, an altered or governed atmosphere, cold storage, and intent to induce opposition through the application of elicitors), the use of biocontrol agents, and excellent agronomic and horticultural practices (Droby, 2006; Gupta, 2014; O'Brien, 2017; Singh, 2014; Singh & Chawla, 2012; Stevens et al., 1997).

Biological defense strategies against plant diseases also include decreasing the number of pathogens or their influence (disease-causing behavior) that is attained through the introduction of biochemical pathways or activities of common antagonists, which happens by modification of the microenvironment to prefer the action of antagonists (Baker, 1987; Stirling & Stirling, 1997). Microbial biocontrol agents (BCAs) are typically bacterial or fungal species that have been isolated from the phyllosphere, endosphere, or rhizosphere in order to control plant-pathogenic organisms. Microbial antagonists or biocontrol agents stop the pathogen from infecting the host plant or from becoming established there. It has been presumed that these are the primary pathogen-targeting control systems. The antagonists may display a variety of direct or indirect related illness regulatory systems. These processes involve starting the war (inducing a plant's innate immunity against plant pathogens), pathogenicity (where the antagonist originates a portion or all of its nutrient content from the fungal host), antibiosis (where the antagonist produces an inhibiting metabolite or antibiotic), and tissue regeneration (BCAs encourage crop growth via microorganisms' hormones like indoleacetic acid and gibberellic acid, as well as where the disease's impacts are being mitigated). Biological infection control also involves the antagonist secreting intracellularly hydrolytic enzymes, competing with other microbes for nutrients and space, and detoxifying virulence genes (Chandrashekara et al., 2012; Deketelaere et al., 2017; Heydari & Pessarakli, 2010; Punja, 1997; Singh, 2014; Wilson et al., 1991; Zhang et al., 2014).

In order to reduce the amount of decomposition in harvested goods and to control plant diseases in the field, which result in significant financial failures, trends are moving away from agrochemical applications to treat plant diseases. At the moment, the focus of most research is on finding effective, environmentally responsible, and safe alternatives to synthetic fungicides (Chandrashekara et al., 2012; Droby, 2006; Stirling & Stirling, 1997). In this chapter, features, methods, applications, and antagonistic mechanisms in horticultural products have been discussed.

28.2 BIOCONTROL OF MICROORGANISMS AND RESILIENT BIOTECHNOLOGIES

Numerous biotic and abiotic stresses brought on by climate change are affecting the products of agricultural production. Pathogens, nutrient deficiency, and weather extremes are a few that should

DOI: 10.1201/9781003370376-37

be highlighted. Some of these are promoting increased chemical consumption. (Naamala & Smith, 2020). The promotion of environmentally friendly agriculture has received a lot of attention recently because it allows agricultural crop productivity to be increased while still maintaining yields and only slightly affecting the environment (Umesha et al., 2018). Agriculture manufacturing is anticipated to rise by at least 70% by 2050. Altogether, people are becoming aware of how crucial sustainable farming is to meeting future agricultural needs (Altieri, 2004). Enhancing the beneficial plant-associated microbiome is one strategy for creating crop production methods that are sustainable. The features of the phyto microbiome are significantly influenced by this association, or halobiont (Hartmann et al., 2014).

28.2.1 What Is Microbial Antagonism?

The struggle for resources such as food and a home are referred to as microbial antagonism. So, if one organism outperforms another, the organism that didn't succeed in the environment is inhibited.

28.2.2 Definition of Antagonistic

An organism that inhibits or significantly interferes with a plant pathogen's normal growth and activity is said to be acting antagonistically in the field of phytopathology; examples include the major components of bacteria or fungi. Antagonizing organisms can suppress or hinder the typical development, spread, and activity of neighboring phytopathogens, which is a phenomenon known as antagonism. These organisms are referred to as biological control agents because of their capacity to control the pathogens and insect pests that harm horticultural crops (Heydari & Pessarakli, 2010).

28.2.3 Antagonist Types

Both antagonists and the pathogens they fight are taxonomically diverse. The main purpose of fungi and bacteria is biological control of bacterial and fungal plant pathogens. Both fungi and bacteria can be controlled by the other, and vice versa.

28.2.3.1 The Qualities of the Perfect Microbial Antagonist

The characteristics preferred in a microbial antagonist in the disease-controlling procedure have been presented across different reviews (Sharma et al., 2009; Droby et al., 2009). Wilson and Wisniewski (1989) suggested the following criteria to choose the perfect antagonist:

1 it should be steadfast.
2 It ought to work well at small doses.
3 Nutrient requirements must not make it difficult.
4 it should be capable of withstanding challenging environmental situations.
5 It must be efficient against a variety of goods and pathogens in a variety of settings.
6 It must be capable of being produced on negative growth press.
7 Long-lasting preparations must be adaptable to this.
8 It must be simple to use and safe for people's health.
9 Chemicals utilized in the surroundings after harvest should not be a problem.
10 It should be environmentally friendly.
11 it should be appropriate for advert preparation techniques.
12 It shouldn't have an adverse effect on fruits and vegetables.

28.2.3.2 Selection and Isolation of Opportunistic Microbes

In the biocontrol procedures of postharvest degradation in fruits and vegetables, the separation of the antagonistic microorganism is a crucial step. Since there are few antagonistic microbes sold commercially in so many parts of the world and multiple attempts to develop specific products have had only patchy achievement, separation before biocontrol action is crucial (Droby et al., 2009). Microorganisms are typically removed from the leaf and surrounding environment of organically grown fruits and vegetables. Antagonizing microorganisms have occasionally been isolated from samples collected beneath fruit and vegetable crops.

28.2.3.3 Preharvest and Postharvest Use of Antagonistic Microorganisms

Choosing a prospective microbial antagonist is the first process; the next is finding an application strategy that can effectively repress or control the disorder pathogens (Sharma et al., 2009). As of right now, antagonists are used either pre- or postharvest. There is compelling evidence that some fruit and vegetable commodities are infected by pathogens in the ground and that these infections play a substantial role in decomposition during the transport or storage of the commodities. As a result, many researchers have claimed that preharvest uses of microbial antagonistic cultures are frequently beneficial in preventing postharvest deterioration in fruits and vegetables (Ippolito & Nigro, 2000; Ippolito et al, 2004; Irtwange, 2006; Janisiewicz & Korsten, 2002). Typically, sprays during preharvest are utilized to pre-colonize the fruit exterior with antagonistic bacteria so that any damage incurred at harvest could be defeated by the antagonists before being colonized by the disease (Ippolito & Nigro, 2000).

28.2.3.4 Preliminary of Microbial Antagonists

Fruit and vegetable postharvest illnesses must be efficiently controlled with a microbial antagonist. A microbial agent should be given to the wound site before the pathogen does, according to various studies (Barkai-Golan, 2001; Droby,

2005; El Ghaouth et al., 2004; Singh & Sharma, 2007; Smilanick, 1994).

28.2.3.5 Characteristics of the Perfect Antagonist

To be the ideal bioagent, a beneficial microbial antagonist must possess definite admirable characteristics (Barkai-Golan, 2001; Wilson & Wisniewski, 1989): The antagonist must fulfill the following requirements: (a) be efficient at small doses; (b) not be picky about what it eats; (c) possess the capacity to endure adverse environmental conditions; (d) be effective against numerous pathogens and various harvested goods; (e) possess a pesticide resistance; (f) not generate metabolites that are potentially harmful to humans; and (g) possess an adaptive benefit over particular pathogens in addition to not being pathogenic to the host (Wilson & Wisniewski, 1989).

28.2.3.6 Antagonistic Arrangement

Only a small number of microbial antagonists that were discovered to prevent postharvest illnesses of fruit and vegetable produce in experimental settings were made widely accessible. Numerous factors could be at play, but the two main obstacles that stood in the way of this were (a) comparative inefficacy of the antagonist treatments in contrast with chemical control techniques and (b) the lack of financial incentives (Wilson & Wisniewski, 1989). Nevertheless, after an efficient antagonist is found, research into its preparation, storage, and use procedures begins.

28.3 TECHNIQUES FOR APPLYING MICROBIAL ANTAGONISTS

It is necessary to look for a technique that appears to apply an efficient antagonist to suppress or control the pathogen once one has been identified or chosen. Microbial antagonists are typically applied in one of two ways, namely, preharvest application or postharvest usage. Technology assistance as well as strain characteristics affect how well prospective antagonistic microorganisms can control the pathogens in harvested fruits and vegetables. Biocontrol agents can generally be used at either the pre- or postharvest stage.

Utilization of pre-harvest. Nevertheless, "hidden" infection can emerge under storage of goods and act as a considerable degrading factor, causing major food losses. During the growing season, pathogens regularly infect produce in the field and live asymptomatic carriers in plant cells (Coates & Johnson, 1997). According to studies concerning the use of antagonistic microorganisms as beneficial microbes, their usage minimizes the impact of stress and improves crop yield when stored (Gao et al., 2016; Maksimov et al., 2011).

Utilization of post-harvest. Researchers have attempted to introduce a suitable and efficient method to control disease, including the use of microbial antagonists for fruits and vegetables. According to this technique, biological products based on antagonistic microorganisms can be sprayed on or dissolved in the crop after harvest (Figure 28.1) (Barkai-Golan, 2001; Irtwange, 2006).

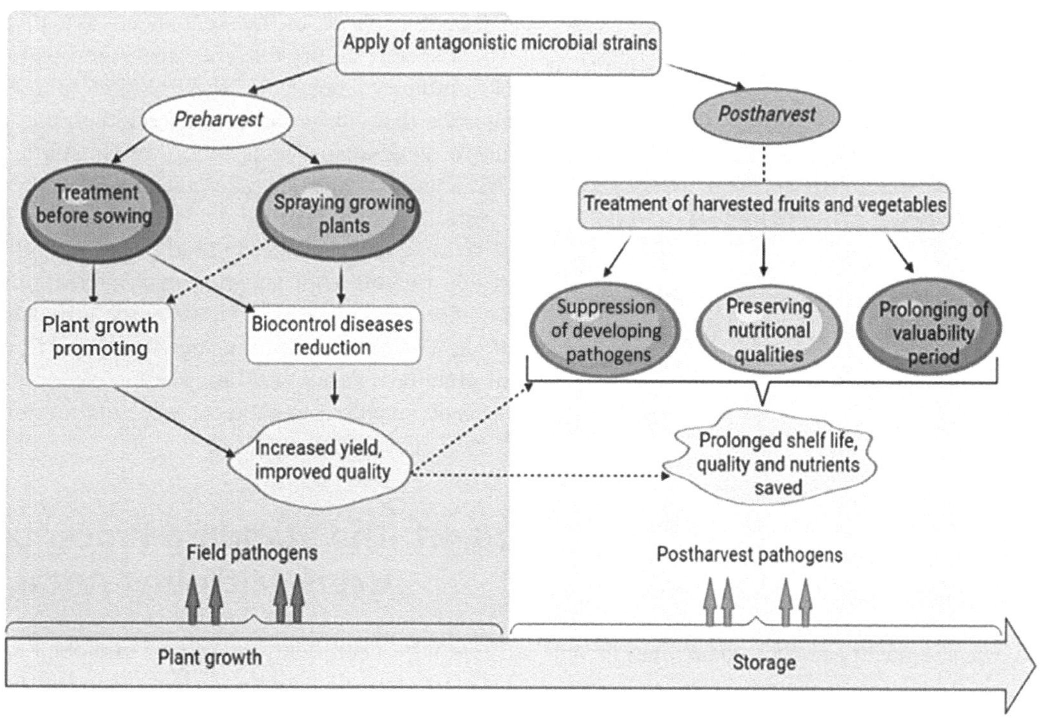

FIGURE 28.1 The diagram for methods of antagonistic microorganisms' application in the control of fruit/vegetable diseases during storage.

28.3.1 Preharvest Application

Fruits and vegetables could become infected with pathogens in the field in a number of situations, and these infections of a latent nature become a major cause of decay when the produce is transported or stored. In order to prevent fruit and vegetable postharvest decomposition, microbial antagonistic cultures made prior to harvest are frequently efficient (Ippolito & Nigro, 2000; Ippolito et al., 2004; Irtwange, 2006; Janisiewicz & Korsten, 2002). Pre-colonizing the fruit surface as soon as possible before it is actually harvested is the goal of the preharvest user so that any wounds sustained during harvest can be treated by the antagonist before contracting a disease (Ippolito & Nigro, 2000). This approach has proven to be very successful in some circumstances, despite the fact that it was not economically feasible because of the poor ground conditions that allowed microbial antagonists to persist.

28.3.2 Postharvest Application

According to the already available research, utilizing microbial antagonists postharvest is a more advantageous, practical, and efficient strategy to stop postharvest illnesses of fruits and vegetables. In this technique, microbial cultures are harvested and then sprayed or dipped in a solution of an antagonist agent (Barkai-Golan, 2001; Irtwange, 2006).

28.3.2.1 Improving the Biological Effectiveness of Microbial Antagonists

Postharvest illnesses of fruits and vegetables are rarely entirely eradicated by microbial antagonists when applied alone. The effective steps have proven to be beneficial for boosting their bio-efficacy and performance.

28.3.2.2 Physical and Chemical Manipulations in the Storage Environment

In order to preserve the performance and satisfy consumer requirements, fruits and vegetables are typically kept for varying lengths of time at predefined temperatures, comparative humidity levels, and gas mixtures. Only those microbial antagonists are selected that are able to develop quickly under the necessary storage duration after being examined to ensure they adhere to the basic criteria. However, because it is probable to change the chemical and physical surroundings to the advantage of the microbial antagonist in storage, changing the storage area may be a practical method for increasing the efficiency of microbial antagonists (Janisiewicz & Korsten, 2002). Nevertheless, these adjustments must be made in a way that does not compromise the quality of the product and is suitable for the founding of the microbial antagonist (Droby et al., 2016). The horticultural produce is regularly examined and/or treated in water prior to, during, and after storage, which presents a significant opportunity to change the environment.

28.4 USING MULTIPLE CULTURES

Over the past 20 years, a lot of progress has been done in the problematic subject of biological management of postharvest diseases employing microbial antagonists. Finding a particular microbial strain that works well on fruits and vegetables while also being effective against the main postharvest pathogens has proven to be difficult. Therefore, in order to offer the necessary spectrum of activities for the suitable postharvest control of diseases of fruits and vegetables, it is necessary to utilize appropriate strains (Barkai-Golan, 2001; El Ghaouth et al., 2004; Singh & Sharma, 2007).

There are certain benefits to applying blends of microbial antagonists:

1. A broader range of microbial activity that controls two or more postharvest diseases.
2. Improving performance in a variety of environments, including cultivars, stages of sophistication, and sites.
3. Increasing the effectiveness and dependability of biocontrol because the mixture's constituent parts work through a range of techniques, including host genetic initiation, predation, and antagonism.
4. Combining various biocontrol traits without genetically modifying them to introduce foreign genes.

The improvement in the bio-efficacy of microbial antagonists may be owing to the emergence of a more stable microbial population that may exclude particular bacteria, such as pathogens. These include enhanced oxidative metabolism, which speeds up growth, the other microbial agent's release of inhibitory chemicals, the creation of nutrients by one microbe that can be utilized by another, and enhanced oxidative metabolism (Janisiewicz, 1998). Additionally, certain characteristics must be taken into account when selecting the constituents of antagonistic mixtures, such as: (a) the absence of rivalry between microbial antagonists and (b) the choice of constituents with advantageous interactions (mutualism), which permit more efficient resource utilization. Evaluation of the biocontrol agents using a variety of antagonists and elimination of any that are ineffective or incompatible is the most practical method for selecting mixture components (Fukui et al., 1999).

28.4.1 The Standard Procedures for Using Microbial Antagonists

There have been many reports of microbial antagonists that have antagonistic activity against pathogens both before and after harvest. These microorganisms create metabolites, which are pathogen-specific antifungals that impede the metabolism and growth of the pathogen. These organisms preclude further deterioration in fruit quality during storage by inhibiting the viruses' capacity to spread while beginning to grow on fruit. Antifungal microbial agents use both direct and indirect inhibitory processes to biologically suppress fungal pathogen growth. A small number of taxonomic groupings, including yeast, bacteria, and filamentous fungi, make up the antagonistic microbial agents that are utilized to suppress fungal infections (Dukare et al., 2019). There are two primary ways to use microbial antagonists to safeguard fruit and vegetable produce from postharvest diseases. One has two options for dealing with postharvest pathogens: (a) using microorganisms that are already on the product and can be incentivized and handled, or (b) artificially introducing them.

28.4.2 Microbial Antagonism Agents Found in Nature

Most microbial antagonists can be naturally found on the surfaces of fruits and vegetables, where they seem to be widespread. They have come into contact with a wide range of antagonists, several of which have been found to be effective biocontrol agents for the management of postharvest infections (Janisiewicz et al., 2013; Vero et al., 2013).

Organic antibodies are those that are found on the outside of fruits and vegetables and are used to prevent postharvest illnesses after they have already been separated (Janisiewicz, 1987; Sobiczewski et al., 1996).

28.4.3 Synthetic Microbial Antagonists

The first reports of the Trichoderma species of fungus as a microbial antagonist for the management of strawberry (*Fragaria* x *ananassa* Duch.) botrytis rot were published by Tronsmo and Dennis (1977). A number of other antagonists have been found and are currently being used to treat postharvest issues in a selection of fruits and vegetables. When it comes to controlling postharvest illnesses of fruits and vegetables, synthetic emergence of microbial antagonists is far more effective than other biocontrol strategies.

28.5 ANTAGONISM AGENTS' PROCESSES OF ACTIVITY

Microbial antagonists' precise mechanisms of action and mode of action against pathogens are still unknown. It is essential to comprehend how microbial antagonists work because doing so will make it possible to create new strategies and tools to improve the effects of antagonists currently in use. Additionally, it will make it possible to choose more desirable and effective antagonists or antagonist strains (Wilson & Wisniewski, 1989; Wisniewski & Wilson, 1992). However, it is still considered that the fundamental mechanism by which microbial agents suppress pathogens responsible for postharvest breakdown is pathogen and antagonist competition for nutrients and space (Droby et al., 1992; Filonow, 1998; Ippolito & Nigro, 2000; Wilson et al., 1993).

Although the inhibit manner of postharvest pathogens by hostile bacteria is not entirely understood, several mechanisms of action have been put forth, including (a) competition for resources with pathogenic bacteria and the availability of colonization niches (Beneduzi, Ambrosini, & Passaglia, 2012; W. Janisiewicz, 1998; Jiang, Chen, Li, & Liu, 1997); (b) making of antibiosis metabolites, such as hydrogen cyanide, biosurfactants, siderophores, and antibiotics (Baindara & Korpole, 2016; Cawoy et al., 2015; El Ghaouth et al., 2004; Fan, Ru, Zhang, Wang, & Li, 2017; Jijakli, Grevesse, & Lepoivre, 2001); (c) creation of compounds and hydrolytic enzymes that can kill pathogenic fungus cells, including chitinases, glucanases, proteases, and lipases (IV Maksimov et al., 2011; Van der Ent, Van Wees, & Pieterse, 2009); and (d) activity of the elicitors and systemically induced resistance (ISR and SAR) (De Vleesschauwer & Höfte, 2009; Van der Ent et al., 2009). Salicylic acid, abscisic acid, jasmonic acid, and ethylene are examples of host plant hormones that primarily control how these compounds are metabolized. Additionally, bacteria predictors such as flagellins, lipopolysaccharides, siderophores, antibiotics, biosurfactants, and volatile chemicals lead them to manifest (Cawoy et al., 2015; De Vleesschauwer & Höfte, 2009; Igor Maksimov & Khairullin, 2016).

Direct exposure to molecular components of host-pathogen and microbial antagonism dialogues, however, is now made easier by recent omics breakthroughs and developments. This contributes to the development of more precise and insightful understanding about the mechanisms of microbial biocontrol at the extension of pre- or postharvest storage (Morales-Cedeño et al., 2021).

In order to prevent disease or slow the spread of pathogens during postharvest storage, microbial antagonists employ a variety of mechanisms. In addition to producing siderophores, mucolytic enzymes, and volatiles as well as synthesizing antibiotics, this involves competition for nutrients and available space (Dukare et al., 2019; Liu et al., 2013; Safdarpour & Khodakaramian, 2018). ACC deaminase activity and biofilm development are also essential for regulating phytopathogen growth during pre- or postharvest storage, in addition to this method. In order to control several postharvest disease infections of fruits and vegetables, it is mentioned that a number of antagonists, including yeasts, fungi, and bacteria, are utilized (Figures 28.2 and 28.3).

28.5.1 Postharvest Applications

The microbial antagonists are implemented using either a spray application or dipping in a solution in this method. In several biocontrol studies, this strategy performed better than

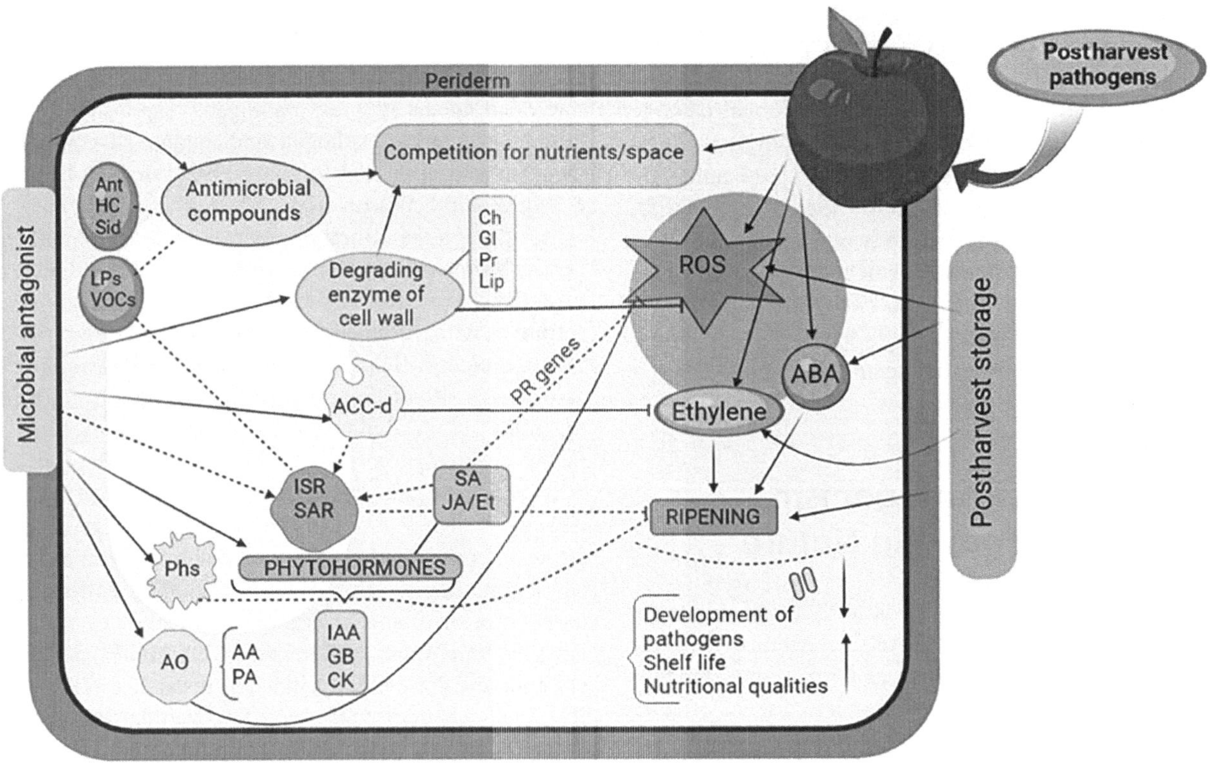

FIGURE 28.2 The illustration shows how pathogenic infections are fought off by microbial antagonists in stored fruits and vegetables. Ascorbic acid, abscisic acid, ACCd-1-aminocyclopropane-1-carboxylate deaminase, antibacterials, antioxidants, chitinases, cytokinin, ethylene, gibberellins, gluconates, and hydrogen cyanide are examples of these compounds. ISR-induced systemic resistance; IAA-indole-3-acetic acid; jasmonic acid; lipases; peroxidase; proteases; phytoalexins; reactive oxygen species (ROS); SAR, or systemic acquired resistance, stands for salicylic acid; siderophores; LPs-lipopolysaccharide; and VOCs-volatile compounds.

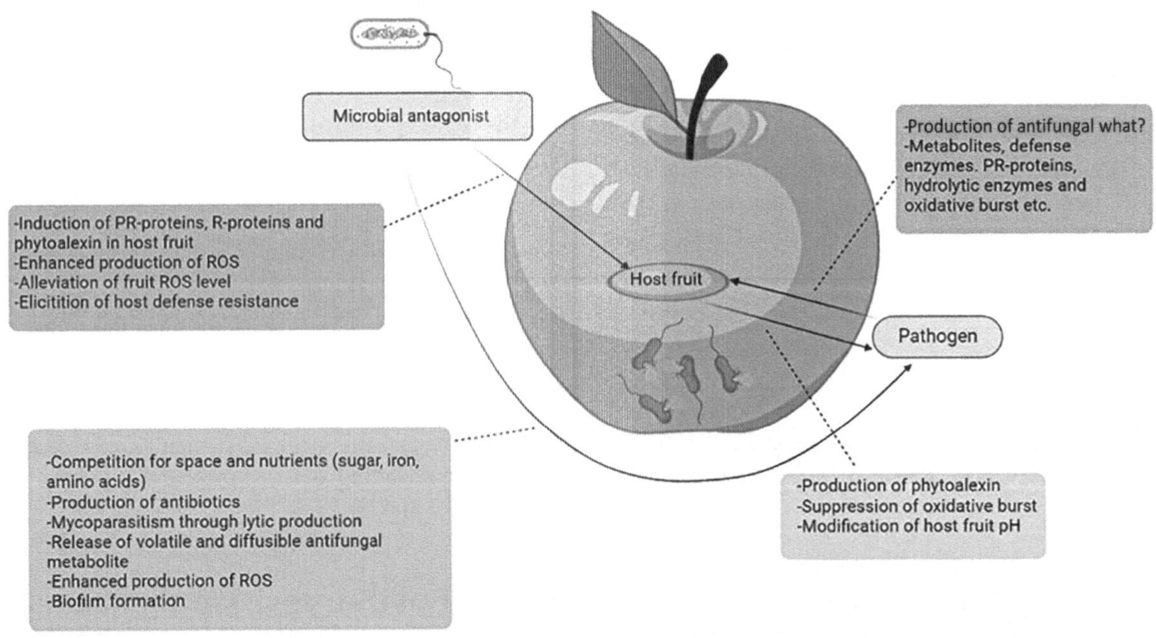

FIGURE 28.3 Diagrammatic representation of the potential biocontrol mechanisms involved in the tritrophic system, including interactions between microbial antagonists, the pathogen, and the host fruit.

preharvest application at suppressing storage pathogens. The likelihood of fungal contamination is significantly reduced more frequently when potential antagonists are applied postharvest. The management of fungal infections during postharvest handling entails the postharvest, exogenous introduction of microbial bioagents, according to research findings on postharvest biocontrol.

28.5.2 Commercial Applications

The achievement of the highest and most dependable level of disease control is necessary for the commercial success of biocontrol calculations of hostile bacteria. Therefore, semi-commercial, in pilot, and large-scale commercial research in varied wrapping situations is required to assess the therapeutic efficacy of biocontrol agents (Droby et al., 2009). Once the aforementioned processes have been successfully completed, the following step is getting regulatory licensing and regulatory agency approval. The ability of the formulation to successfully control disease and on unbiased evaluations of the product's safety are typically the two criteria used by regulatory agencies to approve biocontrol formulations. For the prevention of postharvest disease, numerous antagonistic microbes have been discovered over the past few decades, but only a few have been produced and made available for purchase. Here are some illustrations of commercially available formulations for antagonistic microbe-based biocontrol products. To combat several postharvest fungal infections, these products may be applied to horticultural produce (Dukare et al., 2019).

28.6 MECHANISMS USED BY ANTAGONISTIC MICROORGANISMS TO CONTROL DISEASE

It is still unknown how pathogens in fruits and vegetables are controlled by antagonistic microorganisms. Over time, it has been reported that the antagonistic microbes fight disease incidence brought on by fungi in six main ways (Sharma et al., 2009). The biocontrol techniques work by competing for resources and space, producing antibiotics, attaching, harboring microbial antagonists, directly parasitizing organisms, and inducing resistance. In the absence of a hostile environment, pathogens that cause disease expand more slowly and can withstand brutal environments. The biocontrol ability of microbial antagonists is said to enhance with antagonistic concentration levels, which in turn reduces pathogen accumulation (Zhang et al., 2014). Numerous investigations have revealed that a variety of microbial antagonists have the ability to combat postharvest fungal infections (Gbadeyan, Orole, & Gerard, 2016; Nunes, 2012; M. Wisniewski, Droby, Norelli, Liu, & Schena, 2016). A greater understanding of the interactions between microbial antagonists and hosts and pathogens, as well as their methods of biocontrol actions, has been made possible by the application of effective technologies based on genomic perspectives. There are numerous potential mechanisms that might be operating in a tritrophic interaction system to reduce pathogen infection, as shown in Figure 28.1.

28.7 IMPLEMENTATION OF MICROORGANISMS AS BIOCONTROL POSTHARVEST

Despite the fact that postharvest losses from microbial infections have been significantly reduced by modern food preservation procedures, which make up around 20–25% of the manufacturing process, considerable product losses are still a problem (Blakeney, 2019; Sellitto et al., 2021).

In experimental, semi-commercial, and commercial research in this area, several microbial antagonistic agents (yeasts, fungi, and bacteria) which can be utilized on fruits and vegetables during and after harvest have been found. There are currently a variety of these antagonists in the industry, though their implementation is primarily focused on worsening microorganisms (primarily fungal pathogens) that harm fruit quality when in the field ripening (preharvest) stage. Since widespread adoption of a single product hasn't yet been attained, the post-collection situation is actually more challenging despite thousands of reports identifying possible economically viable adversaries. A few product lines were released but eventually dropped from the market, while others were profitable in specific countries (Droby et al., 2016). This is due to a number of factors, such as inconsistent results, a lack of industrial acceptance, a higher cost in contrast with synthetic pesticides, difficulties with certification, and formulation challenges (M. Wisniewski et al., 2016). This complementary approach combines a number of methods that each significantly reduces decay incidence to achieve prevention goals (using controlled or modified atmosphere packaging, or dipping; exposing to wet/dry heat treatments; applying microbial antagonists; sanitizing and/or carefully harvesting and handling). The different tactics come together or add up to significant levels (97–99%) of disease control (M. Wisniewski et al., 2016). The application of microbial antagonist agents in this strategy is an efficient tool for decreasing spoiling microorganisms and extending the shelf life of produce as part of an overall sustainable approach (Spadaro & Gullino, 2019). This may also lead to future advances in the pre- through postharvest phases of an integrated approach. In reality, if the many actors handling them (both post- and prior to harvest) have a better understanding of how much they may gain from it, this strategy can be applied throughout the full lifespan of fruit and vegetable produce. In actuality, no single procedure can completely rid a product of hazardous bacteria and the decays they bring about (M. Wisniewski et al., 2016).

28.8 MICROBIOLOGICAL CONTROL OF POSTHARVEST OPERATIONS

It is generally recognized that changes in the microbiome's diversity and composition will affect the host phenotype. The insertion of microbial strains or microbial consortia that carry out specific activities for the host plant and aid in the production of a favored phenotypic expression is the basis of microbial interference (Kumar, 2022). The addition of microbes or microbial consortia to an innate microbial community for their advantageous and effective usage in pre- or postharvest pathogen administration is a contemporary development. The artificial microbial population provides the natural microbiome with a difference in the incidence and nutrient status that prevents pathogen incursion (Eng & Borenstein, 2019).

28.9 MAJOR CHALLENGES TO THE COMMERCIAL EXPLOITATION OF MICROBIAL ANTAGONISTIC PRODUCTS

1. There are few companies developing biocontrol products.
2. Market networks that are established and insufficient funding.
3. The microorganisms must manage a variety of pathogens on various fruits and vegetables because the postharvest market is small.
4. The expense and duration of registering a biological control agent.
5. The reliability of the registration package, which must include fundamental toxicological tests on the formulated products for ingestion, contact with the skin and the eyes, as well as a history of proven environmental and human health safety.
6. The difficulties some nations have registering biocontrol agents. Consider Europe.
7. The current safety and health issues surrounding the addition of antagonists to our diet.
8. The conundrum of post-rehydration cell physiology and metabolism.
9. The biocontrol agents are not packaged with high water holding technology that can prevent contamination.
10. The inability to determine the acceptability of the products using uniform quality assurance standards.
11. Incapability of the biocontrol to achieve 95–98% control efficacy when used by itself.
12. Farms' negative attitudes toward using microbes to physiologically control postharvest illnesses in fruits and vegetables (Abano & SAM, 2001).

28.10 A PARADIGM SHIFT IN MICROBIAL ANTAGONISTIC BIOCONTROL SCIENTIFIC KNOWLEDGE

The review of the new paradigm shifts to successfully manage deterioration in the fruit and vegetable industry by Droby et al. (2009) is instructive for the scientific community studying biocontrol. One of the many recommendations made is the necessity of incorporating lower-risk chemical-based pesticides, natural antimicrobial compounds, and physical methods like hot water, infrared, microwave, and ultrasonic treatment into the biocontrol procedure. Other recommendations involve the requirement to perceive organic methods as a procedure instead of the control of one organism: (a) the implementation of more effective DNA-based techniques to track the destiny and action of biocontrol agents after proposal. (b) The use of proteomics and genetic analysis to snoop on the effects of environmental stress and the biological properties of biocontrol agents; (c) as also, during mass manufacture, formulation, and storage, as well as in reaction to exposure to and contact with host plant tissue following application, genes from several biocontrol agents may be combined or their expression may be raised. Additionally, (d) the use of biocontrol agents derived from genetically altered organisms. Since only a small percentage of the earth's microflora has been identified and documented, studies in this area aim to discover new antagonists in substitution of those that are now utilized in operations.

28.11 DEVELOPMENTS IN THE RESEARCH OF POSTHARVEST BIOCONTROL PROCESSES

Recent developments in DNA- and proteomics-based technology coupled with bioinformatics have enabled a comprehensive understanding of potential molecular interactions between the pathogen, the host, and the microbial antagonist (An et al., 2014). Additionally, advances and developments in a range of "omics" technologies, such as deep metagenomics, sequencing, functional genomics, comparative genomics, proteomics, and transcriptomics, may be used more successfully for the in-depth elucidation of the suppression mechanism of disease-controlling biocontrol agents. These innovative techniques can also be used to investigate alterations in the expression level of "biocontrol genes" during bulk production, its preparation and storage, the physiological state of microbial biocontrol agents, and the effects of different environment-associated stresses on its intracellular machineries (Hershkovitz et al., 2013). Furthermore, the host fruits' differentially up- and down-regulated genes in response to the injection of microbes of a biocontrol nature can all be assessed, as can changes in the transcriptome and/or proteome threshold. Each of these studies has the potential to advance our understanding of the microbial control of fungal infections after harvest.

28.12 ENVIRONMENTAL CONCERNS TO BE TAKEN INTO ACCOUNT

The very same species may act as a pathogen for both plants and people as well as a biological defense against disease. Even identical strains can benefit plants while also posing a hazard to the environment (Sobiczewski & Iakimova, 2022). As a consequence, sometimes it can be difficult to distinguish between good and bad bacteria. It's also important to note that because horizontal gene transfer (HGT) is a possibility, it's possible for bacteria to acquire undesirable traits from one another, such as the ability to cause disease (Arnold, Huang, & Hanage, 2022; Soucy et al., 2015). Examples of HGT include the advancement of antibiotic resistance in organisms from the genera *Pseudomonas*, *Erwinia*, and *Xanthomonas*, or the acquisition of pathogenicity (Sundin & Wang, 2018) by nonpathogenic Agrobacterium bacteria (Platt et al., 2014). A comparative study of genomic sequences revealed that some bacteria share targeted genomic structures, implying a chance for them to adapt their lifestyle to fit in with a particular ecological specialty and prospective host (Bulgari et al., 2019). According to this theory, using bacteria to control plant pathogens can lead to the rapid deterioration of disease and even the creation of new pathogens. It is important to consider the prospect that antagonists could discharge metabolites that are harmful for plant pathogens and may pollute the environment to some large extent, putting the lives of nearby species in danger.

28.13 CONCLUSIONS AND FUTURE PROSPECTS

The primary problems with utilizing synthetic pesticides to prevent postharvest disease are the growth of pathogen-resistant strains and the accumulation of hazardous residues in fruit and vegetable commodities. Eco-friendly techniques that employ little to no synthetic fungicides are therefore given preference. Significant progress has been made in postharvest biological and integrative disease management techniques during the last few decades. However, since optimal conditions for biocontrol agents' production are rare, it is irrational to believe that the use of bio-fungicides alone will always result in comprehensive control of postharvest illnesses. As a result, microbial antagonists' biocontrol efficacy is inferior to that of synthetic fungicides. After harvest, a few bio-fungicides have been approved and introduced to the market to control fungal infections on a range of commodities. In order to permanently reduce the usage of fungicides, it is necessary to include microbe-mediated biological disease inhibition as a crucial component of an integrated disease control approach. In the near future, it is projected that the usage of bio-based fungicides will increase along with social acceptance of them as a component of a comprehensive postharvest disease strategy. To create a microbial biocontrol product that is economically viable, commercially successful, and effective, however, a number of obstacles must be overcome. Further research is required on the microbial antagonistic organisms that possess a wide range of antifungal potential on various yields, their development, the fundamentals of postharvest biocontrol processes, and environmental effects. Another significant challenge is the creation of mass multiplication techniques that are affordable and the conceptualization of microbial antagonists on a large scale.

REFERENCES

Abano, E., SAM, A. L. (2001). Application of antagonistic microorganisms for the control of postharvest decays in fruits and vegetables. *International Journal of Advanced Biological and Biochemical Research*, *2*, 1–8.

Altieri, M. A. (2004). Linking ecologists and traditional farmers in the search for sustainable agriculture. *Frontiers in Ecology and the Environment*, *2*(1), 35–42.

An, B., Chen, Y., Li, B., Qin, G., Tian, S. (2014). Ca^{2+}–CaM regulating viability of Candida guilliermondii under oxidative stress by acting on detergent resistant membrane proteins. *Journal of Proteomics*, *109*, 38–49.

Arnold, B. J., Huang, I.T., Hanage, W. P. (2022). Horizontal gene transfer and adaptive evolution in bacteria. *Nature Reviews Microbiology*, *20*(4), 206–218.

Baindara, P., Korpole, S. (2016). Lipopeptides: Status and strategies to control fungal infection. *Recent Trends in Antifungal Agents and Antifungal Therapy*, 97–121.

Baker, K. F. (1987). Evolving concepts of biological control of plant pathogens. *Annual Review of Phytopathology*, *25*(1), 67–85.

Barkai-Golan, R. (2001). *Postharvest diseases of fruits and vegetables: development and control*: Elsevier.

Beneduzi, A., Ambrosini, A., Passaglia, L. M. (2012). Plant growth-promoting rhizobacteria (PGPR): their potential as antagonists and biocontrol agents. *Genetics and Molecular Biology*, *35*, 1044–1051.

Blakeney, M. (2019). *Food loss and food waste: Causes and solutions*: Edward Elgar Publishing.

Bulgari, D., Montagna, M., Gobbi, E., Faoro, F. (2019). Green technology: bacteria-based approach could lead to unsuspected microbe–plant–animal interactions. *Microorganisms*, *7*(2), 44.

Cawoy, H., Debois, D., Franzil, L., De Pauw, E., Thonart, P., Ongena, M. (2015). Lipopeptides as main ingredients for inhibition of fungal phytopathogens by B acillus subtilis/amyloliquefaciens. *Microbial Biotechnology*, *8*(2), 281–295.

Chandrashekara, K., Manivannan, S., Chandrashekara, C., Chakravarthi, M. (2012). Biological control of plant diseases. *Chapter• January*.

Coates, L., Johnson, G. (1997). Postharvest diseases of fruit and vegetables. *Plant pathogens and Plant Diseases*, 533–548.

De Vleesschauwer, D., Höfte, M. (2009). Rhizobacteria-induced systemic resistance. *Advances in Botanical Research, 51*, 223–281.

Dean, R. (2012). Jan a. *L. Van Kan, Zacharias A. Pretorius, Kim E. Hammond Kosack, Antonio Di Pietro, Pietro D. Spanu, Jason J. Rudd, et al*, 414–430.

Deketelaere, S., Tyvaert, L., França, S. C., Höfte, M. (2017). Desirable traits of a good biocontrol agent against Verticillium wilt. *Frontiers in Microbiology, 8*, 1186.

Droby, S. (2005). *Improving quality and safety of fresh fruits and vegetables after harvest by the use of biocontrol agents and natural materials*. Paper presented at the I International Symposium on Natural Preservatives in Food Systems 709.

Droby, S. (2006). Biological control of postharvest diseases of fruits and vegetables: difficulties and challenges. *Phytopathologia Polonica, 39*, 105–117.

Droby, S., Chalutz, E., Wilson, C., Wisniewski, M. (1992). Biological control of postharvest diseases: a promising alternative to the use of synthetic fungicides. *Phytoparasitica, 20*, S149–S153.

Droby, S., Romanazzi, G., Tonutti, P. (2016). Alternative approaches to synthetic fungicides to manage postharvest decay of fruit and vegetables: Needs and purposes of a Special Issue. *Postharvest Biology and Technology, 100*(122), 1–2.

Droby, S., Wisniewski, M., Macarisin, D., Wilson, C. (2009). Twenty years of postharvest biocontrol research: is it time for a new paradigm? *Postharvest Biology and Technology, 52*(2), 137–145.

Dukare, A. S., Paul, S., Nambi, V. E., Gupta, R. K., Singh, R., Sharma, K., Vishwakarma, R. K. (2019). Exploitation of microbial antagonists for the control of postharvest diseases of fruits: a review. *Critical Reviews in Food Science and Nutrition, 59*(9), 1498–1513.

El Ghaouth, A., Wilson, C., Wisniewski, M. (2004). Biologically-based alternatives to synthetic fungicides for the control of postharvest diseases of fruit and vegetables. *Diseases of Fruits and Vegetables: Volume II: Diagnosis and Management*, 511–535.

Eng, A., Borenstein, E. (2019). Microbial community design: methods, applications, and opportunities. *Current Opinion in Biotechnology, 58*, 117–128.

Fan, H., Ru, J., Zhang, Y., Wang, Q., Li, Y. (2017). Fengycin produced by Bacillus subtilis 9407 plays a major role in the biocontrol of apple ring rot disease. *Microbiological Research, 199*, 89–97.

Filonow, A. (1998). Role of competition for sugars by yeasts in the biocontrol of gray mold of apple. *Biocontrol Science and Technology, 8*(2), 243–256.

Fukui, R., Fukui, H., Alvarez, A. (1999). Comparisons of single versus multiple bacterial species on biological control of anthurium blight. *Phytopathology, 89*(5), 366–373.

Gao, H., Xu, X., Dai, Y., He, H. (2016). Isolation, identification and characterization of Bacillus subtilis CF-3, a bacterium from fermented bean curd for controlling postharvest diseases of peach fruit. *Food Science and Technology Research, 22*(3), 377–385.

Gbadeyan, F. A., Orole, O. O., Gerard, G. (2016). Study of naturally sourced bacteria with antifungal activities. *International Journal of Microbiology, and Mycology, 4*(1), 9–16.

Gupta, S. (2014). *Approaches and trends in plant disease management*: Scientific Publishers (India).

Hartmann, A., Rothballer, M., Hense, B. A., Schröder, P. (2014). Bacterial quorum sensing compounds are important modulators of microbe-plant interactions. In (Vol. 5, p. 131): Frontiers Media SA.

Hershkovitz, V., Sela, N., Taha-Salaime, L., Liu, J., Rafael, G., Kessler, C., Droby, S. (2013). De-novo assembly and characterization of the transcriptome of Metschnikowia fructicola reveals differences in gene expression following interaction with Penicillium digitatumand grapefruit peel. *BMC Genomics, 14*(1), 1–13.

Heydari, A., Pessarakli, M. (2010). A review on biological control of fungal plant pathogens using microbial antagonists. *Journal of Biological Sciences, 10*(4), 273–290.

Ippolito, A., Nigro, F. (2000). Impact of preharvest application of biological control agents on postharvest diseases of fresh fruits and vegetables. *Crop Protection, 19*(8–10), 715–723.

Ippolito, A., Nigro, F., Schena, L. (2004). Control of postharvest diseases of fresh fruits and vegetables by preharvest application of antagonistic microorganisms. *Crop Management and postharvest Handling of Horticultural Products. Volume IV: Diseases and Disorders of Fruits and Vegetables*, 1–30.

Irtwange, S. (2006). Application of biological control agents in pre-and postharvest operations. *Agricultural Engineering International: CIGR Journal, 8*, 1–12.

Janisiewicz, W. (1987). Postharvest biological control of blue mold on apples. *Phytopathology, 77*(3), 481–485.

Janisiewicz, W. (1998). Biocontrol of postharvest diseases of temperate fruits. Challenges and opportunities. *Plant Pathology*, 171–198.

Janisiewicz, W. J., Korsten, L. (2002). Biological control of postharvest diseases of fruits. *Annual Review of Phytopathology, 40*(1), 411–441.

Janisiewicz, W. J., Jurick II, W. M., Vico, I., Peter, K. A., Buyer, J. S. (2013). Culturable bacteria from plum fruit surfaces and their potential for controlling brown rot after harvest. *Postharvest Biology and Technology, 76*, 145–151.

Jiang, Y., Chen, F., Li, Y., Liu, S. (1997). A Preliminary Study on the Biological Control of the Postharvest Disease of Litchi Fruit. *Journal of Fruit Science, 14*, 185–186.

Jijakli, M., Grevesse, C., Lepoivre, P. (2001). Modes of action of biocontrol agents of postharvest diseases: challenges and difficulties. *Bulletin OILB/SROP, 24*(3).

Kumar, A. (2022). *Microbial Biocontrol: Sustainable Agriculture and Phytopathogen Management: Volume 1* (Vol. 1): Springer Nature.

Lapuente, M., Estruch, R., Shahbaz, M., Casas, R. (2019). Relation of fruits and vegetables with major cardiometabolic risk factors, markers of oxidation, and inflammation. *Nutrients, 11*(10), 2381.

Liu, J., Sui, Y., Wisniewski, M., Droby, S., Liu, Y. (2013). Utilization of antagonistic yeasts to manage postharvest fungal diseases of fruit. *International Journal of Food Microbiology, 167*(2), 153–160.

Maksimov, I., Abizgil'Dina, R., Pusenkova, L. (2011). Plant growth promoting rhizobacteria as alternative to chemical crop protectors from pathogens. *Applied Biochemistry and Microbiology, 47*, 333–345.

Maksimov, I., Khairullin, R. (2016). The role of Bacillus bacterium in formation of plant defense: Mechanism and reaction. *The Handbook of Microbial Bioresources*, 56–80.

Menzler-Hokkanen, I. (2006). Socioeconomic significance of biological control. *An Ecological and Societal Approach to Biological Control*, 13–25.

Morales-Cedeño, L. R., del Carmen Orozco-Mosqueda, M., Loeza-Lara, P. D., Parra-Cota, F. I., de Los Santos-Villalobos, S., Santoyo, G. (2021). Plant growth-promoting bacterial endophytes as biocontrol agents of pre-and post-harvest diseases: Fundamentals, methods of application and future perspectives. *Microbiological Research, 242*, 126612.

Naamala, J., Smith, D. L. (2020). Relevance of plant growth promoting microorganisms and their derived compounds, in the face of climate change. *Agronomy, 10*(8), 1179.

Nunes, C. A. (2012). Biological control of postharvest diseases of fruit. *European Journal of Plant Pathology*, *133*, 181–196.

O'Brien, P. A. (2017). Biological control of plant diseases. *Australasian Plant Pathology*, *46*, 293–304.

Platt, T. G., Morton, E. R., Barton, I. S., Bever, J. D., Fuqua, C. (2014). Ecological dynamics and complex interactions of *Agrobacterium megaplasmids*. *Frontiers in Plant Science*, *5*, 114700.

Punja, Z. K. (1997). Comparative efficacy of bacteria, fungi, and yeasts as biological control agents for diseases of vegetable crops. *Canadian Journal of Plant Pathology*, *19*(3), 315–323.

Safdarpour, F., Khodakaramian, G. (2018). Assessment of antagonistic and plant growth promoting activities of tomato endophytic bacteria in challenging with Verticillium dahliae under in-vitro and in-vivo conditions. *Biological Journal of Microorganism*, *7*(28), 77–90.

Sellitto, V. M., Zara, S., Fracchetti, F., Capozzi, V., Nardi, T. (2021). Microbial biocontrol as an alternative to synthetic fungicides: Boundaries between pre-and postharvest applications on vegetables and fruits. *Fermentation*, *7*(2), 60.

Sharma, R., Singh, D., Singh, R. (2009). Biological control of postharvest diseases of fruits and vegetables by microbial antagonists: A review. *Biological Control*, *50*(3), 205–221.

Singh, D., Sharma, R. (2007). *Postharvest Diseases of Fruit and Vegetables and Their Management* (Ed. D. Prasad). New Delhi: Sustainable Pest Management, Daya Publishing House.

Singh, H. (2014). *Management of plant pathogens with microorganisms*. Paper presented at the Proc Indian Natl Sci Acad.

Singh, V. K., Chawla, S. (2012). Cultural Practices: An Ecofriendly Innovative Approach in Plant Disease Management. *International Book Publishers and Distributers, Dehradun–248*, *1*, 01–20.

Smilanick, J. (1994). Strategies for the isolation and testing of biocontrol agents. *Biological Control of Postharvest Diseases Theory and Practice*.

Sobiczewski, P., Bryk, H., Berczynski, S. (1996). Evaluation of epiphytic bacteria isolated from apple leaves in the control of postharvest apple diseases. *Journal of Fruit and Ornamental Plant Research*, *4*(1).

Sobiczewski, P., Iakimova, E. T. (2022). Plant and human pathogenic bacteria exchanging their primary host environments. *Journal of Horticultural Research*, *30*(1), 11–30.

Soucy, S. M., Huang, J., Gogarten, J. P. (2015). Horizontal gene transfer: building the web of life. *Nature Reviews Genetics*, *16*(8), 472–482.

Spadaro, D., Gullino, M. L. (2019). Sustainable management of plant diseases. *Innovations in Sustainable Agriculture*, 337–359.

Stevens, C., Khan, V., Lu, J., Wilson, C., Pusey, P., Igwegbe, E., Chalutz, E. (1997). Integration of ultraviolet (UV-C) light with yeast treatment for control of postharvest storage rots of fruits and vegetables. *Biological Control*, *10*(2), 98–103.

Stirling, M., Stirling, G. (1997). Disease management: biological control. *Plant Pathogens and Plant Diseases*, 427–439.

Sundin, G. W., Wang, N. (2018). Antibiotic resistance in plant-pathogenic bacteria. *Annual Review of Phytopathology*, *56*, 161–180.

Tronsmo, A., Dennis, C. (1977). The use of Trichoderma species to control strawberry fruit rots. *Netherlands Journal of Plant Pathology*, *83*, 449–455.

Umesha, S., Singh, P. K., Singh, R. P. (2018). Microbial biotechnology and sustainable agriculture. In *Biotechnology for Sustainable Agriculture* (pp. 185–205). London: Elsevier.

Van der Ent, S., Van Wees, S. C., Pieterse, C. M. (2009). Jasmonate signaling in plant interactions with resistance-inducing beneficial microbes. *Phytochemistry*, *70*(13–14), 1581–1588.

Vero, S., Garmendia, G., González, M. B., Bentancur, O., Wisniewski, M. (2013). Evaluation of yeasts obtained from Antarctic soil samples as biocontrol agents for the management of postharvest diseases of apple (Malus× domestica). *FEMS Yeast Research*, *13*(2), 189–199.

Wilson, C. L., Wisniewski, M. E. (1989). Biological control of postharvest diseases of fruits and vegetables: an emerging technology. *Annual Review of Phytopathology*, *27*(1), 425–441.

Wilson, C. L., Wisniewski, M. E., Biles, C. L., McLaughlin, R., Chalutz, E., Droby, S. (1991). Biological control of post-harvest diseases of fruits and vegetables: alternatives to synthetic fungicides. *Crop Protection*, *10*(3), 172–177.

Wilson, C. L., Wisniewski, M. E., Droby, S., Chalutz, E. (1993). A selection strategy for microbial antagonists to control postharvest diseases of fruits and vegetables. *Scientia Horticulturae*, *53*(3), 183–189.

Wisniewski, M., Droby, S., Norelli, J., Liu, J., Schena, L. (2016). Alternative management technologies for postharvest disease control: The journey from simplicity to complexity. *Postharvest Biology and Technology*, *122*, 3–10.

Wisniewski, M. E., Wilson, C. L. (1992). Biological control of postharvest diseases of fruits and vegetables: recent advances. *HortScience*, *27*(2), 94–98.

Zhang, Q., Zhang, J., Yang, L., Zhang, L., Jiang, D., Chen, W., Li, G. (2014). Diversity and biocontrol potential of endophytic fungi in Brassica napus. *Biological Control*, *72*, 98–108.

The Effects of Preharvest Treatments on the Postharvest Storage Quality of Different Horticultural Products

29

María E. García-Pastor*, Fernando Garrido-Auñón, Jenifer Puente-Moreno, Huertas M. Díaz-Mula, María Serrano and Daniel Valero
*Corresponding Author: m.garciap@umh.es

29.1 INTRODUCTION

The growth of the population and consumption levels have generated a dizzying increase in the demand for all types of resources in the last two centuries, which is causing great global impacts on the planet. The largest impact is that caused by the increase in "greenhouse gases" (GHG), the most important emissions being carbon dioxide (CO_2), nitrous oxide (N_2O) and methane (CH4) and, therefore, altering the normal composition of the atmosphere. The emission of these gases has been the main cause of a totally unusual warming of 1.07 °C during the last 150 years (IPPC, 2021). This process, known as "climate change", is negatively affecting the planet's atmosphere. Climate change is an unavoidable phenomenon that will increasingly impact all human activities. Agriculture will be affected by both an excess and deficit of precipitation, variations in temperature and other associated effects. These events can constitute serious threats to achieve food security worldwide.

Among cultivated plants, it is possibly the woody species, and particularly fruit trees, that are the most dependent on climatic conditions, especially the temperature of the environment where they are grown. Two types of factors condition fruit cultivation: those related to crop management (water and nutrients, mainly), and environmental factors (light and temperature). Fruit trees carry out their productive cycle in two years, with winter, spring and summer temperatures being those that can condition the productivity and quality of fruits. Therefore, during the winter season cold temperatures are necessary since all fruit trees need it for the cessation of growth. Furthermore, during spring, adequate temperatures must be recorded for the physiological processes of photosynthesis and flowering, thus guaranteeing that the tree can express its maximum productivity and the cultural techniques applied to it are effective (Zalasiewicz & Williams, 2016).

The variation in temperatures has a negative impact on fields and orchards around the world and forces farmers to take drastic and costly measures to deal with droughts, high temperatures or frosts. The countries of the Mediterranean basin show a greater risk of global warming and, in the case of Spain, about 75% of the land shows symptoms of desertification. Warmer climates and rising CO_2 levels negatively impact food supply, safety and quality, apart from food losses and waste that, on average, account for 25–30% of the food produced worldwide (FAO, 2019).

Crop production requires optimal conditions related to appropriate soil characteristics, water quality, light, and moderate heat. The current conditions of global warming will alter the normal cycle of fruit development in terms of advancement of flowering and harvest periods, leading to reductions in crop yields. Moreover, there is an enhancement in the occurrence of weeds and pets mainly due to changes in rainfall and, consequently, reductions in crop efficacy, also affecting the nutritional quality, while food security and safety are severely compromised (Wheeler & Von Braun, 2013).

DOI: 10.1201/9781003370376-38

In general, there is a certain consensus that the climate conditions will be somewhat more hostile in the near future so that risk-control strategies must be implemented, such as stress reduction or, simply, relocating the species and thus avoiding the most extreme risks. The challenge is to adapt the production systems that cause loss of competitiveness to the fruit sector with a minimal increase in costs (García-Barreda et al., 2021). The production of fruit depends on the daily interaction of trees with the climate, soil and technological interventions, which are allowing plants to develop their maximum growth potential and crop yield and, in turn, mitigating the causes of different stresses.

Apart from the climate-related abiotic stresses, such as global warming temperatures; increases in salinity, drought, cold, freezing, heavy metals and flooding; and the decrease in water availability, crop productivity will be also affected by biotic stresses, including microbes, pests and diseases. To cope with these stresses, plants have two types of defence mechanisms: the first are physical barriers (cuticles, trichomes and waxes), and the second are chemical substances being synthesized before any occurrence of infection by pathogens, including alkaloids, saponins, antifungal proteins and inhibitory enzymes (Figure 29.1).

On the other hand, plants can produce endogenously, or after an exogenous application, chemical compounds known as "elicitors", whose objective is to act as activators of defensive reactions, that is, to induce the production of phytoalexins or stimulate any defence mechanism in plants to protect themselves (Savvides et al., 2016). These inducers are capable of promoting different modes of plant defence, the main one being systemic acquired resistance (SAR).

In a broad sense, the term "elicitor" refers to chemical substances of different origins triggering several responses that alter a plant's physiological and morphological characteristics. The application of exogenous elicitors mimics the role of plant signalling compounds and thus generates different reactive oxygen species (ROS) that induce the synthesis of plant defence hormones, as well as stimulates the enzymatic or non-enzymatic antioxidant mechanisms to counter the lethal effects of the ROS molecules (Das & Roychoudhury, 2014).

The interest in the use of elicitors has grown exponentially in recent years due to the benefits they trigger in the plant when they are used in crops. Elicitors are substances of natural origin that, when they are applied to plants preventively, help reduce or avoid damage caused by diseases, pests or adverse abiotic factors (Hilker & Schmülling, 2019). Although there are different classifications, the most accepted are according to their nature (Figure 29.2): 1) biotic elicitors, molecules of pathogenic origin synthetized by pathogenesis-related (PR) protein genes that can induce defence responses in plants, such as the accumulation of phytoalexins or the hypersensitive response; and 2) abiotic elicitors, compounds generated by physical factors such as temperature, UV light and ultrasound; chemical moieties (heavy metals, ozone and pH); and plant hormones, namely, jasmonic acid (JA) and its derivative methyl jasmonate (MeJA), salicylic acid (SA) and its derivatives methyl salicylate (MeSa) and acetylsalicylic acid (ASA), oxalic acid (OA) and polyamines (PAs), among others.

Therefore, there is a need to investigate the responses of crops to different types of abiotic stresses, as well as to evaluate the different responses in the plant, especially those related to the role of the plant hormones. This chapter is focused on the use of plant growth regulators (PGR), applied as preharvest treatments, in different horticultural products and their role in growth, development and ripening, crop yield and horticultural product quality both at harvest and under postharvest storage.

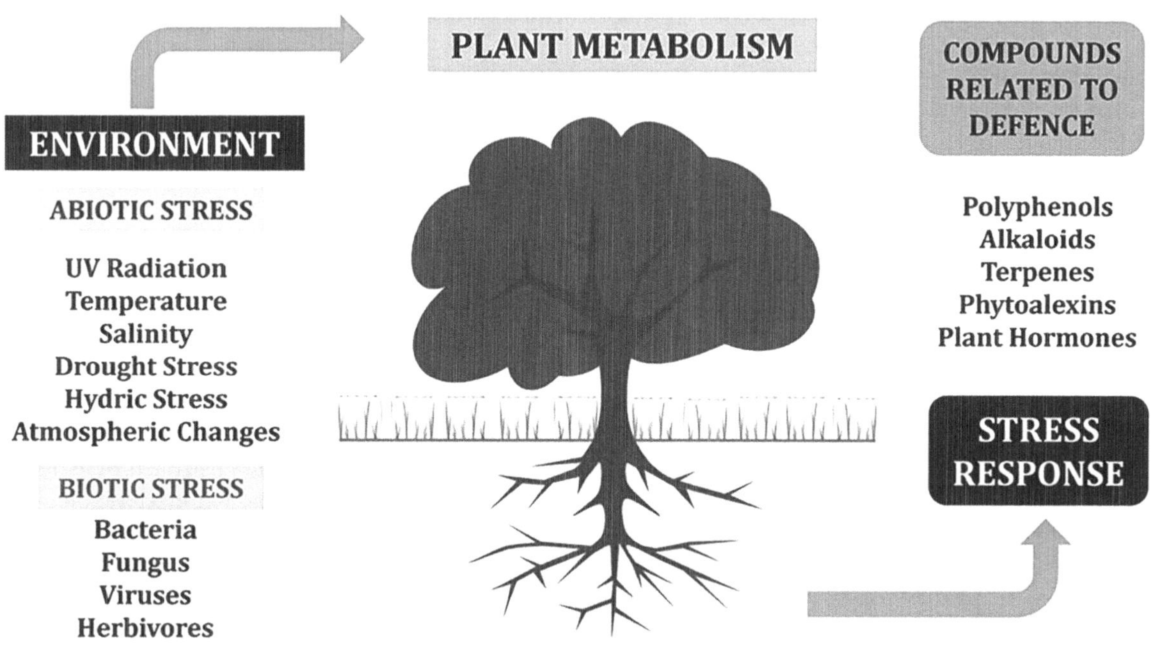

FIGURE 29.1 Biotic and abiotic stresses of plants and their responses.

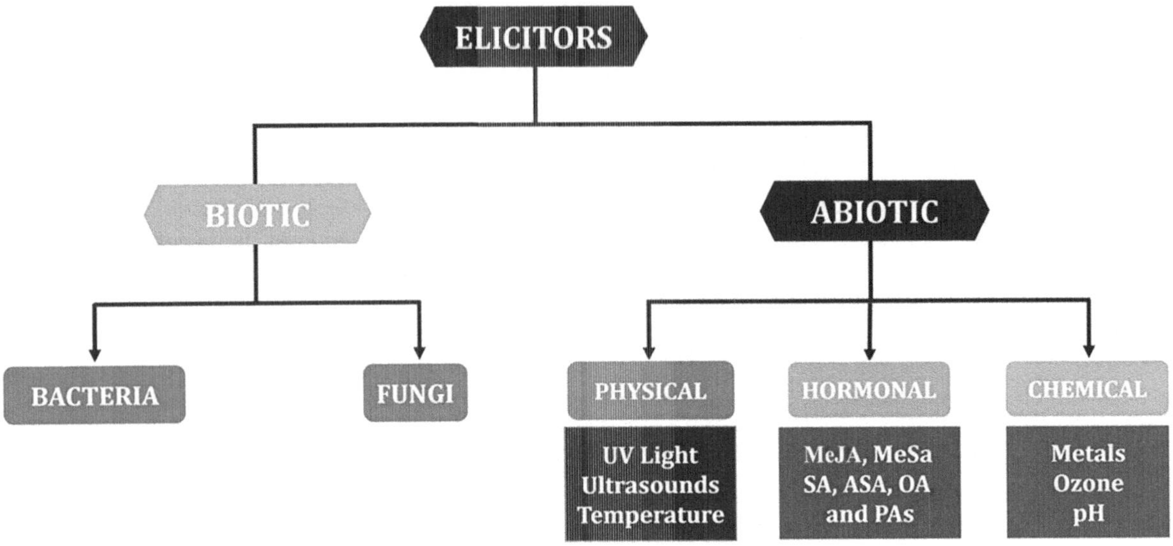

FIGURE 29.2 Classification of elicitors. Abbreviations: MeJA (methyl jasmonate), MeSa (methyl salicylate), SA (salicylic acid), ASA (acetylsalicylic acid), OA (oxalic acid) and PAs (polyamines).

29.2 GROWTH AND DEVELOPMENT OF HORTICULTURAL PRODUCTS

During plant development, a series of physiological events take place, such as germination, vegetative development and the formation of flowers and fruits. These events are regulated by a genetic order of the species or cultivars, which indicate to the plant the specific time and site of synthesis of phytohormones so that these in turn regulate the physiological event in question and the appearance or not of the specific event.

The genetics of the plant species or cultivars will always be influenced by the environment and crop management and, therefore, this defines the type of genetic order and the type of compound that must be formed at that moment. Each hormone produced by the plant fulfils several functions, some very specific, and in other cases two or more hormones must act to regulate a single event. The hormonal presence is critical for the event to occur. On the contrary, if it is absent, in small quantities or in a different location, the event does not occur or manifests itself very poorly.

Like any living organism, the life of a horticultural product goes through different stages that conventionally have been structured in three periods: growth, maturation and senescence, although a clear distinction among them has not been established. While the growth stage includes cell division and elongation, maturation usually begins before the completion of the growth phase and extends until senescence. During this last period, the anabolic processes give way to the catabolic ones, producing ageing and final death of the horticultural product. It is precisely during fruit maturation when a series of biochemical and structural events take place in the vegetable that cause changes in its chemical composition that make the horticultural product attractive for consumption. But unfortunately, once this moment has been reached, senescence and degradation of the cells is generally rapid, negatively affecting the organoleptic characteristics.

The life cycle of a horticultural product is characterized by a sequence of continuous metabolic changes. Thus, after pollination, the growth and development of fruit can be categorized into three fundamental physiological stages: growth, maturation and senescence, without knowing when one ends and the next begins. The maturation process is considered the most important and complex stage of fruit development. Furthermore, this stage can be divided into two phases: 1) the physiological maturation and 2) the organoleptic maturation.

29.3 QUALITY TRAITS OF FRUITS

The most palpable changes during the process of fruit ripening are colour, flavour, total soluble solids (TSS), firmness and total acidity (TA). Others, such as the number of days since full flowering, the respiration rate and ethylene production, are most suitable for studying the physiological characteristics (Valero & Serrano, 2010). These changes are the result of the profound and restructuring chemical metabolism that is triggered within the developmental process. In climacteric fruits, this process is fundamentally controlled by ethylene production and its action, while in non-climacteric ones the ripening process is regulated by other plant hormones.

There are different concepts about the maturation process in response to the different needs consumers have about a product. Therefore, it is necessary to establish some definitions to interpret maturation with respect not only to the anatomical and physiological part of the fruit but also to its proposed use. The definitions are as follows:

29.3.1 Physiological Maturity

Physiological maturity refers to the fruit development stage in which maximum growth and development have occurred. The fruit is fully developed and has all the biochemical elements that will allow it to start the production of aromas, flavours and colour changes.

29.3.2 Harvest Maturity

Harvest maturity is known as the moment in which the fruit still on the plant presents a series of visually noticeable qualities through which it can be marketed, such as size, shape, and colour.

29.3.3 Commercial Maturity

Commercial maturity is the stage at which the fruit is commonly marketed because it has the required attributes to satisfy the consumer.

29.3.4 Consumer Maturity

Consumer maturity is the state in which the fruit has acquired its own characteristics of appearance, consistency, firmness, flavour and aroma.

Once these differences are established, it is necessary to determine the internal characteristics of the fruit for which maturity indices are used in order to decide when a given product can be harvested in such a way that the different markets can be supplied and ensure that the quality of the product is accepted by the consumer.

Maturity indices are also external quality characteristics (visual appearance), which are related to factors such as the colour and the absence of defects or deterioration. It is important to keep in mind that maturity indices must be objective to ensure the acceptability of their quality to the consumer. These indices should be included in the standards for fresh fruits and vegetables since they allow the quality criteria for the products to be defined. The commonly used maturity indices are:

29.3.4.1 External Colour

Depending on the type of synthesis, class, final concentration of the pigments and the conversion of chloroplasts to chromoplasts, the following colour changes may occur, according to the fruits studied: from green to orange (tangelo, pineapples, oranges and mandarin), to yellow (lemons, passion fruit, and yellow pitahaya) and to red or purple (tomato, table grape, sweet cherry, blackberry and strawberry) (Valero & Serrano, 2010).

29.3.4.2 Content of Pulp or Juice

As the ripening process progresses in the fruit, the pulp or juice content of most fruits increases. This increase is due to the degradation of starch and pectin from the first stages and to the accumulation of water, which makes the fruit softer and juicier when ripe. Depending on the characteristics of each fruit, it is expressed as pulp or juice content (Valero & Serrano, 2010).

29.3.4.3 Firmness

Like the increase in juiciness, fruit ripening is usually associated with a softening of the tissues due to degradation of the cell wall components and loss of turgor. In particular, degradation of pectic substances make the cells less cohesive and lead to their separation. Therefore, the hardness of the fruit is reduced (Valero & Serrano, 2010).

29.3.4.4 Content of Sugars

As well as other characteristics of the fruit, the sugar content depends on the cultivar, the nutritional status of the plants and the state of development of the fruit. Many fruit species accumulate starch (in chloroplast or chromoplast) during their development, which creates simpler sugars before or during ripening when it is hydrolysed. During development, bananas accumulate large amounts of starch (ranging from 15% to 35% w/w of their fresh weights based on cultivar or genotype), while during ripening the softening of pulp and the sweetening of the fruit are ascribed to changes in cell wall hydrolase activity and to starch-to-sugar metabolism to reach their optimum postharvest quality (Cordenunsi-Lysenko et al., 2019). Other fruit species accumulate sugars, such as fructose, sucrose and glucose, of which fructose is considered sweeter and glucose less sweet. The amounts of these sugars vary in each species and cultivar, being identifiable by analytical methods such as high-performance liquid chromatography (HPLC). In sweet cherries, total soluble sugar contents (as glucose and fructose) start to significantly accumulate at stage III, gradually increasing during the ripening of the fruit (Kovács et al., 2008; Teribia et al., 2016). Glucose was the most abundant among the soluble sugars in sweet cherry fruit, and concentrations of both glucose and fructose progressively increased during ripening on the tree, reaching the highest level at the stage of commercial maturity (Teribia et al., 2016).

29.3.4.5 Content of Acidity

Most fruits are acidic, although there are differences in acid concentrations. Similarly, the content of acid generally decreases during ripening, although with some exceptions. This decrease is owing to the use of organic acids during the respiration process or their conversion into sugars. To determine the acid content in fruits, total acidity (TA) can be used as a measurement method, which relates the acidity perceived by consumers. This method measures the proportion of acids present in fruits, through their total neutralization with a base solution. The result of this analysis is expressed as the percentage of the acid present in a greater quantity over the other acids, of which citric, malic and tartaric acid are fully identified as the most abundant, depending on the fruit. For instance, the major acids in apples are malic, citric and tartaric acids, and their concentrations generally depend on the nature of the cultivar and the extent of the fruit's ripeness (Cunha et al., 2002). During ripening, the concentration of acids in apple fruits is reduced owing to endogenous enzyme activity (Fagundes et al., 2013) and exposure to oxygen (Fonseca et al., 2002). Conversely, the initial growth phases between stages I and II in sweet cherries is linked with a reduction in pH value and an increase in TA (Teribia et al., 2016). Afterward, malic acid concentration remains elevated up to the commercial maturity stage (Teribia et al., 2016), while pH values markedly decrease at stage III due to the biosynthesis of certain other organic acids, for instance ascorbic or citric acids (Mahmood et al., 2012) and remain lower and constant until stage VIII.

29.4 PREHARVEST FACTORS AFFECTING THE QUALITY OF HORTICULTURAL PRODUCTS

Over the last few years, there has been a growing awareness of consumers' need to have at their disposal edible fruits that have reached a level of satisfactory maturity and show their true organoleptic characteristics. The concept of quality has evolved over time; at first the perception of quality was different according to the particular interest of each of the agents who intervened in the production process (producer, consumer or retailer). Nevertheless, there is increasingly more alignment among the different sectors involved since all of them tend to bring their criteria towards those imposed by the consumers, in which the state of maturity of the fruit plays an important role.

29.4.1 Climate Change: Global Temperature

Earth's climate has evolved over many millions of years and has shown that atmospheric CO_2 concentrations and global surface temperature are strongly correlated. Global warming is one of the most important manifestations of climate change. The increases in the frequency and intensity of temperature waves will affect the most sensitive populations, such as older adults and people suffering from severe illnesses. Temperature changes will also increase chemical processes and will trigger contaminants that are detrimental to human health (Moreti et al., 2010). Global warming and the associated changes in the climate cycle, as well as the rise in sea level, can cause negative impacts on ecosystems, natural resources, human health and economies. Climate change is expected to alter ecosystems and will lead to the loss of species diversity, as many species will not be able to adapt to the rapid changes in environmental conditions (Ganopolski, 2012).

29.4.2 Environmental Conditions Inducing Abiotic Stress

Abiotic stress refers to certain environment-related factors that alter different metabolic and physiological processes of plants. It is reported as the leading cause of the loss of more than 50% of the key agricultural crops worldwide, as well as the decrease in plant growth, development and yield (Sah et al., 2016). This condition is further aggravated by climate changes, which deeply influence the frequency of fires, rapid temperature increases and disturbances in rainfall patterns, among others (Cramer et al., 2011). Plants have managed to evolve to develop in environments where there are various types of stresses and associated factors, owing to the development of particular mechanisms that allow the plants to sense the changes in their environment and respond to them, reducing damages and at the same time preserving different useful resources that allow plants to continue with their sustainable development (Atkinson & Urwin, 2012).

The increase in temperatures would constitute a relevant factor for food production due to its possible effects on plant behaviour. Likewise, this increase in temperature directly affects photosynthesis, causing alterations in fruit sugar, organic acid and flavonoid contents; firmness; and antioxidant activity (Moretti et al., 2010). It is known that climate change can possibly cause certain postharvest quality alterations in fruits and vegetables. Although numerous researchers have focused on climate change–related aspects in the past, the information on the use of plant elicitors as a preharvest treatment is still scarce. This information is approached in the next section of this chapter.

29.5 PLANT ELICITORS AS TOOLS TO IMPROVE THE QUALITY OF HORTICULTURAL PRODUCTS

Plants are autotrophic organisms that transform light energy into biochemical energy, such as glucose, through the process of photosynthesis. In addition to their energy value, their nutrient content makes them the base of the planet's trophic pyramid. Plants are constantly changing and adapting to the environment. Among the main adaptations are the formations of various organic molecules, collectively named as plant elicitors. However, under abiotic stress situations, the endogenous levels could not reach the necessary concentrations to cope with the symptoms or consequences of the stress. The role of these elicitors will be addressed in the following sections.

29.5.1 Methyl Jasmonate (MeJA)

Jasmonic acid (JA) is characterized by acting as a "negative" growth regulator in vegetables, because it generally represses organ growth and accelerates both the processes of senescence and abscission and those of flowering and fruit ripening. JA and its derivatives, for instance methyl jasmonate (MeJA) (Figure 29.3A), have a role in root growth, the maturation of fruits and senescence, as well as in plant defence mechanisms.

The JA family, including its derivatives, comprises compounds of lipid origin with a molecular structure like those of prostaglandins in animals. They act as signal molecules for plant responses to various stress situations (wounds, attack by pathogens and pests, exposure to drought and ozone) and participate in various growth- and development-related aspects (Creelman & Mullet, 1997). During the 1980s, JA and some of its derivatives were initially discovered as growth inhibitors and senescence promoters in several plant species, both monocotyledonous and dicotyledonous.

The biosynthesis pathway, also called the octadecanoid pathway, has been studied extensively, in which several enzymes are involved in various steps (Berger, 2002; Agrawal et al., 2004). JA is formed from the unsaturated fatty acids, specifically linoleic and linolenic acids, that are released from the phospholipids of cell membranes by the action of lipases. From linolenic acid and through a series of steps which include cyclization, lipoxidation, and β-oxidation, (+)-7-isojasmonic acid is formed, which is isomerized under natural conditions into (-)-jasmonic acid. Among these steps, it is worth distinguishing the action of the allene oxide synthetase (AOS) enzyme, which catalyses the first limiting step of this pathway, involving linolenic acid in the JA synthesis (Schaller et al., 2005). JA can be esterified with methyl groups to form MeJA, which is highly volatile. Although the most abundant forms are (-)-jasmonic acid and (-)-methyl-jasmonic acid, the isomers (+)- 7-isojasmonate and methi-(+)-7-isojasmonate also are present in certain species (Creelman & Mullet, 1997). JAs can be synthesized throughout the plant, but they have a greater activity in growing tissues, such as apices of stems, roots, young leaves and immature fruits.

MeJA is a volatile derivative of JA, which has been used successfully at a concentration of 0.2 mM to reduce rot developed by *Monilinia fructicola* in cherries, being more effective than applying it during postharvest (Yao & Tian, 2005). The effective dose of MeJA to induce systemic acquired resistance (SAR) varies with the fruit, the minimum concentration applied with effectiveness being 0.045 mM for the control of anthracnose diseases in tomatoes (Tzortzakis, 2007), while the highest was 10 mM to control *Botrytis cinerea* in tomatoes (Zhu & Tian, 2012).

The preharvest application of MeJA has an important impact on the ripening of fruits based on its applied concentration, in both climacteric and non-climacteric fruit. In plums, MeJA at 0.5 mM postponed postharvest ripening processes by lowering the respiration rate, reducing ethylene emission and delaying the softening process (Zapata et al., 2014). Interestingly, MeJA increased the crop yield in table grape and pomegranate fruits (Table 29.1) and enhanced the weight and size of the fruits (García-Pastor et al., 2019; 2020a). In addition, MeJA at 0.5, 1.0 and 2 mM led to plums with improved quality parameters at harvest, for instance firmness, colour and total acidity. MeJA also increased the concentrations of bioactive compounds at harvest, such as total phenolics and hydrophilic total antioxidative activity, although the effect was dose-dependent. Moreover, MeJA induced the expression of antioxidant enzymes, including catalase (CAT), peroxidase (POD), ascorbate peroxidase (APX) and superoxide dismutase (SOD), as measured in pomegranates (García-Pastor et al., 2020a; 2020b), table grapes (García-Pastor et al., 2019), blood oranges (Habibi et al., 2019), and plums (Zapata et al., 2014).

Chilling injury (CI) is one of the major limitations for the cold storage of harvested fruits and vegetables, leading to quality deterioration and substantial economic losses (Zhang et al., 2021). Recently, studies showed that MeJA is pivotal for the regulation of the possible potential of harvested fruit and vegetable produce tolerance against CI incidence. Min et al. (2023) noted that MeJA application substantially suppressed CI in postharvest fresh produce, proposing that the function mechanisms of MeJA on fruit and vegetable CI tolerance could be: 1) improving membrane integrity and sustained energy supply; 2) enhancing antioxidative activity; 3) increasing the arginine pathway; 4) improving the metabolism of sugars; 5) regulating phenolics metabolism; 6) stimulating the pathway of C-repeat/dehydration response–related element binding factor CBFs, which play an imperative role in the responses of plants to low temperature–related stresses; 7) regulating the expression and accumulation of heat shock proteins (HSPs), which are a group of proteins that play a critical role in responding to different stresses, such as cold, heat and oxidative stress; and (8) crosstalk with phytohormones. MeJA treatments play a substantial role in conserving membranes integrity and decreasing membrane damages caused by CI stress in various postharvest crops, effectively inhibiting electrolyte leakage and reducing MDA accumulation in harvested fruit and vegetable commodities, as listed in Table 29.1. Recently, Bagheri and Esna-Ashari (2022) reported that MeJA at 8, 16 and 24 μL L^{-1} decreased the electrolyte leakage and malondialdehyde (MDA) content in persimmon.

Thus, MeJA favored the activation of antioxidative systems, both at the enzymatic level and increasing the phenolic and anthocyanin compounds through the activation of the phenylalanine ammonium lyase (PAL) enzyme (García-Pastor et al., 2019). Positive effects enhancing crop yields, quality traits and bioactive compounds have been observed in other fruits (Table 29.1) after the preharvest application of different concentrations of MeJA in fruits, such as lemons (Serna-Escolano et al., 2019; 2021a), table grapes (García-Pastor et al., 2019) and pomegranate fruits (García-Pastor et al., 2020a; 2020b). During the growth of radish sprouts (Baenas et al., 2019) or artichokes (Martínez-Esplá et al., 2017a), MeJA treatment induced a higher accumulation of phytochemicals with antioxidant activity.

TABLE 29.1 A Review of Several Preharvest Treatments with Effects on the Postharvest Storage of Different Horticultural Products

TREATMENT	*HORTICULTURAL PRODUCT*	*CONCENTRATION*	*EFFECT*	*REFERENCE*
Methyl Jasmonate (MeJa)	Plum	0.5, 1.0 and 2.0 mM	Increased fruit size and weight, firmness, color, total acidity, total phenolics and hydrophilic total antioxidative activity at harvest, although the optimal concentration was cultivars dependent.	Martínez-Esplá et al. (2014a)
	Plum	0.5, 1.0 and 2.0 mM	MeJA at 0.5 mM suppressed the ripening process due to decreased ethylene production, respiration rate and softening. For this dose, total phenolics and total antioxidative activity and antioxidant enzyme activities were found higher in treated plums than controls.	Zapata et al. (2014)
	Peach	10 μM	Enhanced activity and expressions of SPS and concentrations of sucrose and sorbitol; decreased glucose, activity and transcript level of AI, SD H, PFK, G6PDH.	Yu et al. (2016)
	Tomato	50 μM	Enhanced arginase activity and expression of SlARG2.	Zhang et al. (2016)
	Artichoke	0.5 mM	Improved yield and artichoke quality, as well as phytochemical content with antioxidant activity at harvest and under storage.	Martínez-Esplá et al. (2017a)
	Tomato	50 μM	Decreased electrolyte leakage and MDA content and enhanced SOD, POD, APX and CAT activities.	Min et al. (2018)
	Radish sprouts	125, 250 and 350 μM	Induction in bioactive compounds in almost all combination treatments (seeds priming and spray), contributing to human health.	Baenas et al. (2019)
	Table grape	1, 0.1 and 0.01 mM	The concentration of 10 mM suppressed the ripening of the berry, while 1, 0.1 and 0.01 mM accelerated it. In this sense, total yield was enhanced by 0.1 and 0.01 mM MeJA doses, which thus improved berry quality and bioactive compounds.	García-Pastor et al. (2019)
	Lemon fruit	0.1, 0.5 and 1.0 mM	Increased antioxidant compounds content, such as phenolics, with the most effective level being 0.1 mM. Additionally, antioxidant enzyme activities were also increased, without influencing yield or fruit quality.	Serna-Escolano et al. (2019)
	Blood orange	50 μM	Decreasing electrolyte leakage and MDA content and enhancing SOD, APX and CAT activities.	Habibi et al. (2019)
	Hot pepper	50 μM	Enhanced fructose, sucrose and glucose content and increased transcript level of CaLOX3 and CaAOC and JA contents.	Seo et al. (2020)
	Pomegranate fruit	1, 5 and 10 mM	Increase on pomegranate crop yield. The concentrations of 1 and 5 mM enhanced on-tree fruit ripening, whereas this was postponed in 10 mM treatment. The firmness, weight and organic acid content losses during storage at 10°C were suppressed, leading to maintenance of quality. In addition, there was an improvement on arils colour owing to increased anthocyanins concentration.	García-Pastor et al. (2020a)
	Pomegranate fruit	5 mM	Reduction in both external and internal CI symptoms and ion leakage, mainly due to higher concentrations of unsaturated fatty acids at harvest, which was preserved under storage, along with maintained stability of membrane and bioactive compounds with antioxidative activity.	García-Pastor et al. (2020b)

	Lemon	0.1 mM	Improvement of total antioxidative activity, total phenolics content, main individual phenolics (eriocitrin and hesperidin) and activity of antioxidant enzymes.	Serna-Escolano et al. (2021a)
	Tomato	50 μM	Decreasing $O2^{-}$ production, H_2O_2 content, DHA, GSSG; Increasing contents of AsA and GSH, activities of APX, DHAR, MDHAR, GR, GST.	Li et al. (2021)
	Tomato	50 μM	Decreasing electrolyte leakage and MDA content.	Zhou et al. (2021)
	Pomegranate fruit	100 μM	Decreasing electrolyte leakage and malondialdehyde (MDA) content.	Chen et al. (2021)
	Persimmon	8, 16, 24 μL L^{-1}	Decreasing electrolyte leakage and MDA content.	Bagheri and Esna-Ashari (2022)
	Sweet cherry	0.5 mM	Reduction of fruit cracking, improvement of abiotic stresses tolerance and induction of an overall fruit ripening delay on the tree.	Ruiz-Aracil et al. (2023)
Salicylic acid (SA)	Tomato	1.0 mM	Improving antioxidant capacity and phenolics and delaying senescence process.	Zhang et al. (2011)
	Sweet cherry	0.5, 1.0 and 2.0 mM	Increase on fruit weight, improvement of quality attributes at harvest, and led to cherries with higher concentrations in bioactive compounds, as well as higher antioxidant properties, being 0.5 mM the most effective dose.	Giménez et al. (2014)
	Table grape	1.0, 1.5 and 2.0 mM	Enhancing total phenolics, delaying ripening process, reducing incidence of browning and decay percentage of grapes.	Champa et al. (2014)
	Sweet cherry	0.5 mM	Enhancement of color, total soluble solids, firmness, total phenolic contents, total anthocyanin contents, total hydrophilic antioxidative activity and antioxidant enzymes activities.	Giménez et al. (2017)
	Plum	0.5 mM	Increase on quality and functional attributes, such as weight, firmness, total acidity, total phenolic contents, including anthocyanins, total carotenoids, antioxidative activity and antioxidant enzymes activities at harvest and during storage, delaying postharvest ripening and senescence processes and extending the shelf-life.	Martínez-Esplá et al. (2017b)
	Plum	1.0 mM	Increase on total phenolics and total carotenoids at harvest, along with increase in antioxidative activity, without influencing on-tree ripening process of the fruits.	Serrano et al. (2018)
	Plum	0.5 mM	Increased firmness, weight, individual sugars, organic acids, phenolics, total carotenoids and anthocyanins at harvest. During storage, softening, color changes and acidity losses were delayed, which could be attributed to the effect of treatment in decreasing ethylene production.	Martínez-Esplá et al. (2018)
	Daikon sprouts	0.5, 1.0 and 2.0 mM	Enhanced antioxidative capacity and phenolics. Preserved chlorophyll pigments and total carotenoids of sprouts.	Supapvanich et al. (2019)
	Radish sprouts	50, 100 and 200 μM	Induction in bioactive compounds contents in almost all combined treatments (priming seeds and spray), contributing to human health.	Baenas et al. (2019)

(Continued)

TABLE 29.1 (Continued)

TREATMENT	*HORTICULTURAL PRODUCT*	*CONCENTRATION*	*EFFECT*	*REFERENCE*
	Pomegranate fruit	1, 5 and 10 mM	The best results were attained with 10 mM dose of three salicylates studied, SA being the most effective treatment improving quality traits, mainly in terms of red colour, and maintaining concentrations of anthocyanins, phenolics, and ascorbic acid during prolonged storage at 10°C.	García-Pastor et al. (2020c)
	Table grape	1, 0.1 and 0.01 mM	Acceleration of ripening and increment on yield, as well as improvement in colour of berries owing to improved total and individual anthocyanins, the highest effects were noted in 0.1 mM MeSa.	García-Pastor et al. (2020d)
	Table grape	0.01 mM	Increase in total acidity, bioactive compound content and activity of antioxidant enzymes, inducing a resistance to *Botrytis cinerea* spoilage, though 0.1 mM MeSa was most effective.	García-Pastor et al. (2020e)
	Jujube fruit	1 and 2 mM	Improved antioxidant capacity and total phenolics of Indian jujube.	Shanbehpour et al. (2020)
	Table grape	1, 2, 3 and 4 mmol L^{-1}	Improved antioxidant capacity, total phenolics and bioactive compounds of grape.	Gomes et al. (2021)
	Green pepper fruit	0.5, 1 and 5 mM	Increase in yield and quality attributes at harvest, specifically for the 0.5 mM dose, as well as higher phenolics and total antioxidative activity were observed in treated pepper fruit.	Dobón-Suárez et al. (2021)
Acetyl salicylic acid (ASA)	Tomato	1.0 mM	Improving antioxidant capacity and phenolics and delaying senescence process.	Zhang et al. (2011)
	Sweet cherry	0.5, 1.0 and 2.0 mM	Increase on fruit weight, improvement of quality attributes at harvest, and led to cherries with higher concentration in bioactive compounds, as well as higher antioxidant properties, being 1 mM the most effective dose.	Giménez et al. (2014)
	Sweet cherry	1.0 mM	Enhancement of firmness, colour, total phenolics, total soluble solids, total anthocyanin contents, hydrophilic total antioxidant activity and antioxidant enzymes activities.	Giménez et al. (2017)
	Plum	1.0 mM	Increase on quality and functional parameters, like weight, firmness, total acidity, total phenolics, such as total carotenoids, anthocyanins, antioxidant activity and antioxidant enzymes activities at harvest and during storage, delaying postharvest ripening and senescence processes and prolonging the shelf-life.	Martínez-Esplá et al. (2017b)
	Plum	1.0 mM	Increase on total carotenoids and total phenolics concentration at harvest, as well as an increase in the antioxidative activity, without influencing the on-tree ripening process of the fruits.	Serrano et al. (2018)
	Plum	1.0 mM	Increase on weight, firmness, individual organic acids, sugars, phenolics, anthocyanins and total carotenoids at harvest. During storage, softening, color changes and acidity losses were delayed, which could be attributed to the effect of treatment in decreasing ethylene production.	Martínez-Esplá et al. (2018)

	Pomegranate fruit	1, 5 and 10 mM	The best results were accomplished with 10 mM level of the three salicylates studied, being SA the most effective treatment improving quality traits, mainly in terms of red color, and maintaining higher phenolics, anthocyanins and ascorbic acid concentrations during prolonged storage at 10 °C.	García-Pastor et al. (2020c)
	Table grape	1, 0.1 and 0.01 mM	Acceleration of ripening and increment on yield, as well as improvement of berry color owing to enhanced concentrations of total and individual anthocyanin contents, the highest effects were noted in 0.1 mM MeSa.	García-Pastor et al. (2020d)
	Table grape	1 and 0.1 mM	Increase on total acidity, bioactive compounds content and activity of antioxidant enzymes, inducing a resistance to *Botrytis cinerea* spoilage, although the best results were noted in 0.1 mM MeSa treatment.	García-Pastor et al. (2020e)
	Loquat fruit	2 mM spermidine (Spd)	Spermidine did not show any significantly beneficial effect on loquat fruit quality parameters.	Hadjipieri et al. (2021)
Methyl salicylate (MeSa)	Sweet cherry	1.0 mM	Total phenolic and anthocyanin contents were significantly increased in MeSa-treated than in controls at harvest and during storage, leading to fruit with higher hydrophilic total antioxidant activity and antioxidant enzymes activities.	Valverde et al. (2015)
	Sweet cherry	0.5, 1.0 and 2.0 mM	MeSa at 1 mM improved fruit size and quality attributes at harvest and after postharvest storage, showing a delay on the postharvest ripening process, exhibited by reduced degree in color changes, and lower loss of acidity and firmness.	Giménez et al. (2015)
	Plum	0.5 mM	Increase on quality and functional parameters, including weight, firmness, total acidity, total phenolics, such as anthocyanins, total carotenoids, antioxidative activity and antioxidant enzymes activities at harvest and during storage, delaying postharvest ripening and senescence processes and extending the shelf life.	Martínez-Esplá et al. (2017b)
	Plum	0.5 mM	Increase in firmness, weight, individual organic acids, phenolics, sugars, total carotenoids and anthocyanins at harvest. During storage, softening, colour changes and acidity losses were delayed, which could be ascribed to the effect of treatment in decreasing ethylene production.	Martínez-Esplá et al. (2018)
	Pomegranate fruit	1, 5 and 10 mM	The 10 mM was most effective dose among the three salicylates studied, SA being the most effective treatment improving quality traits, mainly in terms of red colour, and maintaining the concentrations of anthocyanins, phenolics, and ascorbic acid during prolonged storage at 10°C.	García-Pastor et al. (2020c)
	Table grape	1, 0.1 and 0.01 mM	Acceleration of ripening and increment on yield, as well as improvement of berry colour owing to enhanced total and individual anthocyanins concentration and 0.1 mM MeSa was the most effective.	García-Pastor et al. (2020d)

(Continued)

TABLE 29.1 (Continued)

TREATMENT	*HORTICULTURAL PRODUCT*	*CONCENTRATION*	*EFFECT*	*REFERENCE*
	Table grape	0.1 mM	Increase on total acidity, bioactive compound content and activity of antioxidant enzymes, inducing a resistance to *Botrytis cinerea* spoilage, and 0.1 mM MeSa was the most effective.	García-Pastor et al. (2020e)
	Apricot	0.05, 0.1 and 0.2 mmol L^{-1}	Enhancing antioxidant capacity, alleviating CI incidence, maintaining soluble solid and organic acid, and decreasing decay percentage of apricot.	Fan et al. (2021)
Oxalic acid (OA)	Sweet cherry	0.5, 1.0 and 2.0 mM	Increase in fruit size and other parameters related to fruit quality, such as firmness and colour, but did not affect the normal ripening process on-tree, the higher effect being found with 2.0 mM. In addition, a stimulation on the bioactive phenolic content and the antioxidant capacity was observed.	Martínez-Esplá et al. (2014b)
	Kiwifruit	5 mM	Lower incidence of blue mould rot and lesion diameter and higher defence-related enzymes.	Zhu et al. (2016)
	Peach	5 mM	Higher fruit firmness levels, total phenolic content and total antioxidants.	Razavi and Hajilou (2016)
	Sweet cherry	2 mM	Higher fruit weight, firmness and content of soluble solids.	Giménez et al. (2017)
	Pineapple	5 mM	Lower internal browning and higher ascorbic acid.	Youryon et al. (2017)
	Coriander	1 mM	Higher concentration of phenolic compounds.	El-Zaeddi et al. (2017)
	Strawberry	1 mM	Higher fruit number, sensory attributes and ascorbic acid levels.	Anwar et al. (2018)
	Plum	1.0 mM	No effect on on-tree ripening process, although led to fruit with high bioactive compounds and antioxidant potential at commercial harvest.	Serrano et al. (2018)
	Plum	2.0 mM	Enhancement of crop yield and weight of fruit, although the on-tree process of ripening and the ripening process during cold storage were delayed. In addition, the antioxidant compounds were improved.	Martínez-Esplá et al. (2019)
	Kiwifruit	5 mM	Lower off-flavour, acetaldehyde and ethanol content and higher ascorbic acid.	Ali et al. (2019)
	Pomegranate fruit	1, 5 and 10 mM	Increase in yield, the on-tree ripening process being promoted in a concentration-dependent manner. The concentration of 10 mM showed the best results in terms of firmness, external colour, respiration rate and sensory attributes.	García-Pastor et al. (2020f)
	Apricot	1 mM	Lower water loss, ethylene production and respiration rate.	Üzümcü et al. (2020)
	Apricot	2 mM	Higher levels of firmness and ascorbic acid, fruit weight, juice yield and lower water loss and antioxidant capacity.	Ahmed et al. (2021)
	Lemon fruit	0.1, 0.5, and 1 mM	Reduction of weight loss and maintenance of firmness, soluble solids content and total acidity. The activity of antioxidative enzymes and the total phenolic content were enhanced.	Serna-Escolano et al. (2021b)

	Table grape	5 mM	Increase on abscisic acid (ABA) and ABA glucose ester (ABA-GE) content, being regulated by ABA biosynthesis pathway changes, specially by upregulation of the 9-cis-epoxycarotenoid dioxygenase (*Vv*NCED1) gene. The treatment showed a delay on senescence process, which was also associated with enzymatic antioxidative system stimulation.	García-Pastor et al. (2021)
	Blueberry	2 mM	Higher firmness, total anthocyanins and antioxidant activity.	Retamal-Salgado et al. (2023)
Polyamines (PAs)	Fresh pistachio	1 and 2 mM putrescine (Put) and spermidine (Spd)	Increasing the fresh pistachio storability due to their effect on delaying the softening and decreasing loss of weight, and fungal infection.	Mirdehghan et al. (2013)
	Table grape	1 and 2 mM putrescine (Put) and spermidine (Spd)	Higher firmness levels, lower microbial infection and higher phenolics and anthocyanins content, as well as antioxidant activity.	Mirdehghan and Rahimi (2016)
	Apricot	0.1 mM spermidine (Spd)	Increasing shelf-life up to 30 days with good quality in MAP conditions.	Erbaş et al. (2018)
	Jujube fruit	1 and 2 mM putrescine (Put)	Improved antioxidant capacity and total phenolics of Indian jujube.	Shanbehpour et al. (2020)
	Loquat fruit	2 mM spermidine (Spd)	Spermidine did not show any significantly beneficial effect on loquat fruit quality parameters.	Hadjipieri et al. (2021)
	Pear fruit	0.1 mM and 0.25 mM putrescine (Put), 0.05 and 0.25 mM spermine (Spm), and 0.05 and 0.25 mM spermidine (Spd)	Spd (0.05 mM) and Put (0.25 mM) resulted in a significant increase in June fruit set percentage. The highest maturity index was noted in Put (0.25 mM), followed by Spd (0.05 mM). Spd at 0.25 mM showed enhanced total sugars, antioxidants, anthocyanins, and phenols content.	Sayyad-Amin et al. (2022)

29.5.2 Salicylates (SA, ASA and MeSa)

Salicylic acid (SA), or 2-hydroxybenzoic acid, is an endogenous plant growth regulator of a phenolic nature having an aromatic ring with a hydroxyl group or other functional derivatives (Figure 29.3B).

Acetyl salicylic acid (ASA) is an analogue of SA which is transformed spontaneously when it is hydrolysed. Another derivative is called methyl salicylate (MeSa), which is volatile and can diffuse across membranes, so it acts as a signal molecule (Figure 29.3C).

SA plays a role in the transmission of signals, since its most relevant physiological activity has been demonstrated as a signal that intervenes in the initiation of systemic acquired resistance (SAR), an immunological type of response in the event of an infection by pathogens, which is related to plant response mechanisms to both abiotic and biotic stress. SA is synthesized and transported by the phloem to the entire plant, protecting it from future infections. These chemically very different products are now recognized as having hormonal action in the plant and whose function at the cell level has been possible to recognize only with progress in the knowledge of molecular biology and the availability of mutants. Although they are different in chemical structure, it is estimated that there would be several interrelationships between them, leading to a specific response.

Studies carried out in rice proposed that SA is biosynthesized from cinnamic acid, presenting two probable biosynthetic routes; one of them involves the decarboxylation of the cinnamic acid side chain, forming benzoic acid, which is hydroxylated two more times to form SA, which is then regulated by the enzyme benzoic acid 2-hydrolase. The other route proposes that cinnamic acid undergoes two hydroxylations, giving rise to *o*-coumaric, which is then decarboxylated and forms SA (Silverman et al., 1995). More recent studies have presented the shikimate pathway, proposing two routes from the chorismate. One is mediated by the enzyme isochorismate synthetase in chloroplasts, and the other by the intervention of the PAL enzyme in the cytoplasm, which converts benzoic acid to acid *o*- coumaric acid, the precursor of SA (Dempsey et al., 2011).

These SA derivatives have shown important physiological roles from the stimulation of seed germination, flowering, fruit growth, development and ripening. Preharvest ASA, SA and MeSa treatments have been used in various fruit commodities, such as sweet cherries (Giménez et al., 2014; 2015; 2017; Valverde et al., 2015), pomegranates (García-Pastor et al., 2020c), plums (Martínez-Esplá et al., 2017b; 2018; Zapata et al., 2014; Serrano et al., 2018) and table grapes (García-Pastor et al., 2020d; 2020e). For all fruit species, the treatment with either SA, ASA or MeSa led to a positive effect in crop productivity by increasing the kg tree^{-1}, the fruit weight and size, which was ascribed to an enhancement of net photosynthesis rate (Table 29.1). These treatments were also effective in improving the quality attributes at harvest, with fruits having higher firmness, TSS, TA and colour. In addition, the reduction of respiratory rate as influenced in response to SA and its aforementioned derivatives depends on the suppression of ethylene biosynthesis and closer stomata (Champa et al., 2014). The contents of bioactive components, including anthocyanins, phenolic contents and antioxidative activity, have been improved with the application of these treatments. Furthermore, during postharvest storage, both quality traits and phytochemicals remained at higher concentrations in salicylate-treated horticultural products. The application of SA or MeSa has been found to improve antioxidative capacities in apricots (Fan et al., 2021), grapes (Gomes et al., 2021) and daikon sprouts (Supapvanich et al., 2019) under storage.

In chilling-sensitive fruits (plums and pomegranate fruit), salicylate preharvest treatments were also effective in reducing CI symptoms, which was attributed to stimulating the activity of antioxidative enzymes such as CAT, POD, SOD and APX. These enzymes together with the phenolics would alleviate the chilling stress by scavenging the reactive oxygen species (ROS) and preserving the integrity of the membrane (Martínez-Esplá et al., 2018; García-Pastor et al., 2020b).

29.5.3 Oxalic Acid (OA)

Oxalic acid (OA) is an organic acid whose synthesis has not been well documented in plants (Figure 29.3D). Nevertheless, it has been suggested that oxalate is biosynthesized from three major precursors: ascorbate, glyoxylate/glycolate and oxaloacetate. During photorespiration, oxalate is released from glycolate/glyoxylate through the oxidation of the glycolate oxidase enzyme (Cai et al., 2018). It has also been suggested that oxidative oxaloacetate degradation would be another source of oxalate, a process in which the oxaloacetate acetylhydrolase enzyme intervenes.

The OA preharvest treatment delays senescence and preserves the quality of fruit and vegetable commodities, as in the case of cherries (Martínez-Esplá et al., 2014b), peaches (Razavi & Hajilou., 2016) or apricots (Üzümcü et al., 2020). Regarding the mode of action, it has been established that it intervenes in ROS accumulation (Ding et al., 2007), decreases the activity of alcohol dehydrogenase (ADH), enhances proline accumulation, conserves higher ATP levels (Li & Zheng, 2015), regulates unsaturated and saturated fatty acid content proportions (Jin et al., 2014) and mediates sugar metabolisms (Wang u al., 2016). In contrast, OA application has been associated with increased activity of enzymes associated with resistance against biotic (certain pathogens) and abiotic stress, such as POD, APX, SOD and CAT. In this perspective, improved accumulation of phenolic contents has been observed (Serrano et al., 2018; García-Pastor et al., 2020f, 2021; Serna-Escolano et al., 2021b), as well as an enhancement of ascorbic acid levels in strawberries (Anwar et al., 2018). The OA treatments could favour a response in the plant and the fruit that improves cell homeostasis by reducing oxidative stress. Studies carried out with *Arabidopsis thaliana* show that treatments with OA up to a concentration of 5 mM did not affect plant development and induced gene expression related to defence, which protects the plant from possible infections. Similarly, OA applied at 5 mM in kiwifruit reduced the blue mould rot incidence, since increased the defence-related enzymes (Zhu et al., 2016). However, concentrations higher than 6 mM promoted programmed apoptosis of cells and facilitated the development of possible pathogenic fungi (Lehner et al., 2008).

The OA concentrations may also have an effect on the quality characteristics of postharvest fruit and vegetable produce. Table 29.1 exhibits that the concentration of OA from 1 mM to 10 mM as a spray application during preharvest stages has been found to exhibit better outcomes in contrast with other concentrations irrespective of crop, cultivar or number of spray applications. Nevertheless, the spray applications of 2 mM OA as a preharvest treatment showed substantial outcomes in increasing fruit size, antioxidative potential, delay of ripening and preserving quality attributes under postharvest storage (Ahmed et al., 2021; Martínez-Esplá et al., 2014b; 2019; Giménez et al., 2017; Retamal-Salgado et al., 2023). In addition, the impact of concentrations may be unpredictable with regard to crops, as previously observed by García-Pastor et al. (2021), who described that a spray treatment at 5 mM concentration on table grapes could be considered the optimum level for improved yield, berry maturation and bioactive content accumulation among 1 mM, 5 mM and 10 mM concentrations investigated over two growing periods.

29.5.4 Polyamines (PAs)

Polyamines (PAs) are cationic compounds with several amino groups. Unlike classic hormones, these are generally found in higher concentrations (Figure 29.3E). They are involved in cell division stimulation, the development of some fruits and vegetables and in the delay of senescence.

PAs are low molecular weight nitrogenous aliphatic compounds that have various effects on cell growth, division and differentiation, and these are present in bacteria, animals and plants. In plants, these are normally found at somewhat higher levels than classic hormones. However, these are also considered regulators of plant growth or sometimes as mediators of the action of other hormones. Due to their polycationic character, PAs easily bind to anion-based molecules, like phosphate groups of nucleic acids, with some proteins, and with phospholipids or wall components, such as pectin-type polysaccharides. Many plants can produce the three main types of PAs that exist, namely, Put, Spd and Spm, although this may vary depending on the vegetable species. Each of these PAs generally is derived from the amino acid arginine and is biosynthesized through two metabolic routes: arginine dexcarboxylase (ADC) and ornithine dexcarboxylase (ODC) (Figure 29.4).

PAs are synthesized from three basic amino acids through reactions of decarboxylation. Arginine, lysine and ornithine provide the carbon base, while methionine contributes an aminopropyl group. The smallest PA, putrescine (containing two amino groups), is formed from arginine through an intermediate (agmatine) and by the action of arginine decarboxylase or ornithine by the action of ornithine decarboxylase. Spermidine (with three amino groups) and spermine (with four amino groups) are formed from putrescine with the addition of aminopropyl groups given by S-adenosyl methionine, which is also a precursor of the ethylene hormone. The said reactions are catalysed by Spd synthase and Spm synthase, respectively (Kakkar & Sawhney, 2002).

Previous studies suggest that the decrease in PA levels is related to ageing and senescence in vegetables, while the accumulation of various free PAs, such as spermidine and spermine, can be associated with tissue growth and organogenesis. A few studies have focused on the effects of PAs applied in

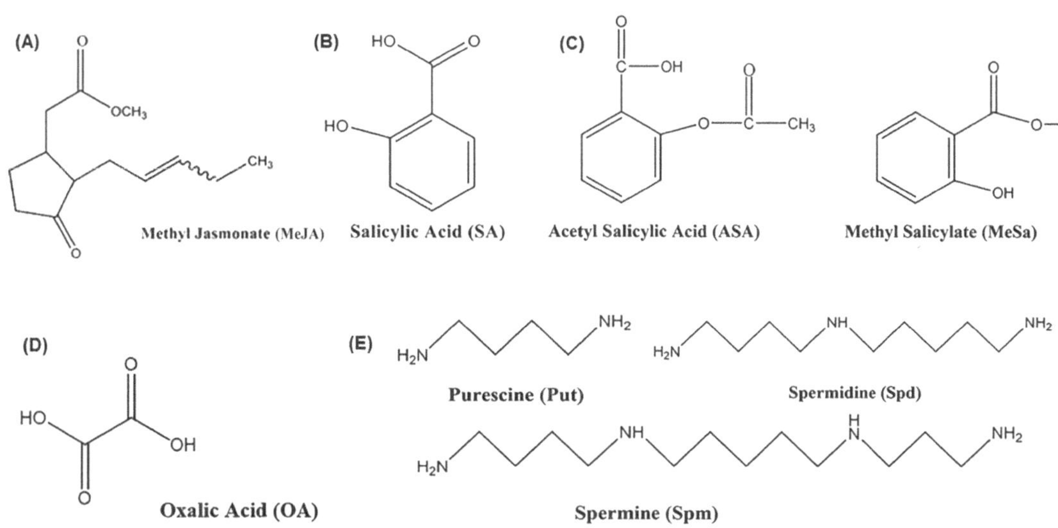

FIGURE 29.3 Chemical structure of (A) methyl jasmonate (MeJA), (B) salicylic acid (SA), (C) acetyl salicylic acid (ASA) and methyl salicylate (MeSa), (D) oxalic acid (OA) and (E) main polyamines: putrescine (Put), spermidine (Spd) and spermine (Spm).

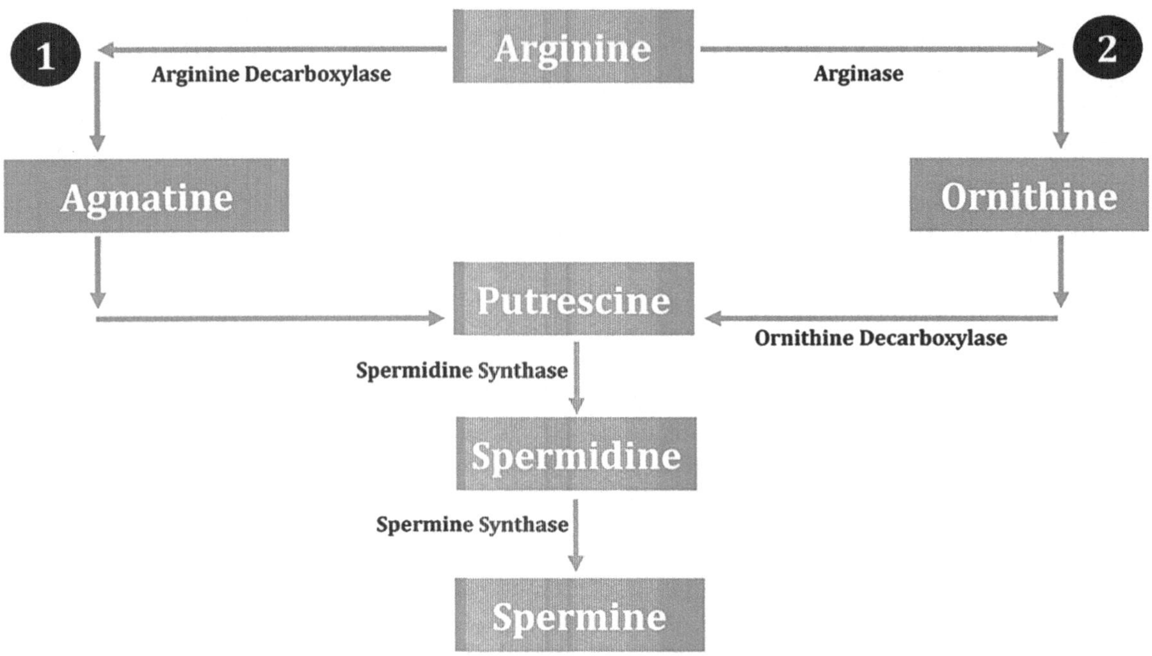

FIGURE 29.4 Simplified biosynthesis pathway of polyamines.

preharvest on fruit and vegetable quality, as can be observed in Table 29.1. Putrescine and spermidine at 1 and 2 mM concentrations demonstrated a significant increase on shelf life due to a delay of quality losses in fresh pistachio and apricot (Mirdehghan et al., 2013; Erbaş et al., 2018). Furthermore, an improvement in the antioxidant system has also been observed as a response to preharvest treatments with putrescine and spermidine in table grapes and jujube fruit (Mirdehghan & Rahimi, 2016; Shanbehpour et al., 2020). In the same way, Sayyad-Amin et al. (2022) explored the effects of the three PAs as preharvest treatments in pear fruit and observed that 0.25 mM spermidine improved total sugar content and antioxidants in comparison to control fruits.

29.6 CONCLUSIONS AND FUTURE PROSPECTS

Although climate change is negatively affecting the crop yield and quality of horticultural products, several preharvest treatments, such as MeJA, SA, ASA, OA and PAs could effectively alleviate this abiotic stress in plants, improving productivity and maintaining product quality during postharvest conditions. Interestingly, the applications of such preharvest treatments, considered as natural compounds and to be environmentally friendly and sustainable, had an additional advantage in improving the content of bioactive compounds and antioxidative activities. The chapter provides evidence for several beneficial influences of exogenous elicitor applications applied preharvest on fruit and vegetable quality and functional attributes. However, the anticipated commercial applications are nowadays limited since no specific regulation exists about the use of these treatments in Europe. Finally, comprehensive metabolic and transcriptomic approaches are required to know about the deeper insights into the nutritional and functional traits of elicitor-treated horticultural products. Furthermore, new elicitors (melatonin, gamma-aminobutyric acid or brassinosteroids) will be become the new research lines.

ACKNOWLEDGEMENTS

The authors greatly appreciate the project "Innovative and eco-friendly pre- and postharvest strategies with natural compounds to improve quality of fruits" funded by Conselleria d'Innovació, Universitats, Ciència i Societat Digital (Generalitat Valenciana) through the Prometeo Program (PROMETEO/2021/089).

REFERENCES

Agrawal, G., Tamogami, S., Han, O., Iwahashi, H., Rakwal, R. (2004). Rice octadecanoid pathway. *Biochemical and Biophysical Research Communications*, *317*, 1–15. DOI: 10.1016/j.bbrc.2004.03.020.

Ahmed, M., Ullah, S., Razzaq, K., Rajwana, I.A., Akhtar, G., Nazb, A., Aminc, M., Khalidd, M.S., Khalidd, S. (2021). Pre-harvest oxalic acid application improves fruit size at harvest, physicochemical and sensory attributes of 'Red Flesh' apricot during fruit ripening. *Journal of Horticultural Science & Technology*. DOI: 10.46653/jhst2142048.

Ali, M., LIU, M.M., Wang, Z.E., Li, S.E., Jiang, T.J., Zheng, X.L. (2019). Pre-harvest spraying of oxalic acid improves post-harvest quality associated with increase in ascorbic acid and regulation of ethanol fermentation in kiwifruit cv. Bruno during storage. *Journal of Integrative Agriculture*, *18*, 2514–2520. DOI: 10.1016/S2095–3119(19)62791–7.

Anwar, R., Gull, S., Nafees, M., Amin, M., Hussain, Z., Khan, A., Malik, A. (2018). Preharvest foliar application of oxalic acid improves strawberry plant growth and fruit quality. *Journal of Horticultural Science & Technology*, *1*, 35–41. DOI: 10.46653/jhst180101035.

Atkinson, N.J., Urwin, P.E. (2012). The interaction of plant biotic and abiotic stresses: From genes to the field. *Journal of Experimental Botany, 63 (10)*, 3523–3543. DOI: 10.1093/jxb/ers100.

Baenas, N., Fusari, C., Moreno, D.A., Valero, D., García-Viguera, C. (2019). Biostimulation of bioactive compounds in radish sprouts (*Raphanus sativus* 'Rambo') by priming seeds and spray treatments with elicitors. *Acta Horticulturae*, DOI: 10.17660/ActaHortic.2019.1256.94.

Bagheri, M., Esna-Ashari, M. (2022). Effects of postharvest methyl jasmonate treatment on persimmon quality during cold storage. *Scientia Horticulturae*, *294*, 110756. DOI: 10.1016/j.scienta.2021.110756.

Berger, S. (2002). Jasmonate-related mutants of *Arabidopsis* as tools for studying stress signaling. *Planta*, *214*, 497–504. DOI: 10.1007/s00425–001–0688-y.

Cai, X., Ge, C., Xu, C., Wang, X., Wang, S., Wang, Q. (2018). Expression Analysis of Oxalate Metabolic Pathway Genes Reveals Oxalate Regulation Patterns in Spinach. *Molecules*, *23*, 1286. DOI: 10.3390/molecules23061286.

Champa, W.A.H., Gill, M.I.S., Mahajan, B.V.C., Arora, N.K. (2014). Preharvest salicylic acid treatments to improve quality and postharvest life of table grapes (*Vitis vinifera* L.) cv. Flame Seedless. *Journal of Food Science and Technology*, *6*, 3607–3616. DOI: 10.1007/s13197-014-1422-7.

Chen, L., Pan, Y.F., Li, H.D., Jia, X.Y., Guo, Y.L., Luo, J.S., Li, X.H. (2021). Methyl jasmonate alleviates chilling injury and keeps intact pericarp structure of pomegranate during low temperature storage. *Food Science and Technology International, 27 (1)*, 22–31. DOI: 10.1177/1082013220921597.

Cordenunsi-Lysenko, B.R., Nascimento, J.R.O., Castro-Alves, V.C., Purgatto, E., Fabi, J.P., Peroni-Okyta, F.H.G. (2019). The Starch Is (Not) Just Another Brick in the Wall: The Primary Metabolism of Sugars During Banana Ripening. *Frontiers in Plant Science*, *10*, 391. DOI: 10.3389/fpls.2019.00391.

Cramer, G.R., Urano, K., Delrot, S., Pezzotti M., Shinozaki, K. (2011). Effects of abiotic stress on plants: A systems biology perspective. *BMC Plant Biology*, *11*, 163. DOI: 10.1186/1471-2229-11-163.

Creelman, R.A., Mullet, J.E. (1997). Biosynthesis and action of jasmonates in plants. *Annual Review of Plant Physiology and Plant Molecular Biology*, *48*, 355–381. DOI: 10.1146/annurev.arplant.48.1.355.

Cunha, S.C., Fernandes, J.O., Ferreira, I.M.P.L.V.O. (2002). HPLC/UV determination of organic acids in fruit juices and nectars. *European Food Research and Technology*, *214(1)*, 67–71. DOI: 10.1007/s002170100412.

Das, K., Roychoudhury, A. (2014). Reactive oxygen species (ROS) and response of antioxidants as ROS-scavengers during environmental stress in plants. *Frontiers in Environmental Science*, *2*, 53. DOI: 10.3389/fenvs.2014.00053.

Dempsey, D.A., Vlot, A.C., Wildermuth, M.C., Klessig, D.F. (2011). Salicylic acid biosynthesis and metabolism. Arabidopsis Book, 9, e0156. DOI: 10.1199/tab.0156.

Ding, Z.S., Tian, S.P., Zheng, X.L., Zhou, Z.W., Xu, Y. (2007). Responses of reactive oxygen metabolism and quality in mango fruit to exogenous oxalic acid or salicylic acid under chilling temperature stress. *Physiologia Plantarum*, *30*, 112–121. DOI: 10.1111/j.1399–3054.2007.00893.x.

Dobón-Suárez, A., Giménez, M.J., García-Pastor, M.E., Zapata, P.J. (2021). Salicylic Acid Foliar Application Increases Crop Yield and Quality Parameters of Green Pepper Fruit during Postharvest Storage. *Agronomy*, *11*, 2263. DOI: 10.3390/agronomy11112263.

El-Zaeddi, H., Calín-Sánchez, Á., Nowicka, P., Martínez-Tomé, J., Noguera-Artiaga, L., Burló, F., Wojdyło, A., Carbonell-Barrachina, Á.A. (2017). Preharvest treatments with malic, oxalic, and acetylsalicylic acids affect the phenolic composition and antioxidant capacity of coriander, dill and parsley. *Food Chemistry*, *226*, 179–186. DOI: 10.1016/j.foodchem.2017.01.067.

Erbaş, D., Koyuncu, M.A., Onursal, C.E. (2018). Effects of pre- and postharvest spermidine treatments on storage life and quality of 'Aprikoz' apricot. *Acta Horticulturae*, *1194*, 785–791. DOI: 10.17660/ActaHortic.2018.1194.111.

Fagundes, C., Carciofi, B.A.M., Monteiro, A.R. (2013). Estimate of respiration rate and physicochemical changes of fresh-cut apples stored under different temperatures. *Food Science and Technology, 33 (1)*, 60–67. DOI: 10.1590/S0101–20612013005000023.

Fan, X., Du, Z., Cui, X., Ji, W., Ma, J., Li, X., Wang, X., Zhao, H., Liu, B., Guo, F., Gong, H. (2021). Preharvest methyl salicylate treatment enhances the chilling tolerance and improve the postharvest quality of apricot during low temperature storage. *Postharvest Biology and Technology*, *177*, 111535. DOI: 10.1016/j.postharvbio.2021.111535.

FAO (Food and Agriculture Organization of the United Nations). (2019). The State of Food and Agriculture 2019. Moving forward on food loss and waste reduction. Rome, Lucense: CC BY-NC-SA 3.0 IGO.

Fonseca, S.C.; Oliveira, F.A.R.; Brecht, J.K. (2002). Modelling respiration rate of fresh fruits and vegetables for modified atmosphere packages: a review. *Journal of Food Engineering, 52 (2)*, 99–119. DOI: 10.1016/S0260–8774(01)00106–6.

Ganopolski, A. (2012). Climate Change Models. Ecological models 6, pp. 603–612.

García-Barreda, S., Sangüesa-Barreda, G., Madrigal-González, J., Seijo, F., González de Andrése, E., Camarero, J. (2021). Reproductive phenology determines the linkages between radial growth, fruit production and climate in four Mediterranean tree species. *Agricultural and Forest Meteorology*, *307*. DOI: 10.1016/j.agrformet.2021.108493.

García-Pastor, M.E., Giménez, M.J., Valverde, J.M., Guillén, F., Castillo, S., Martínez-Romero, D., Serrano, M., Valero, D., Zapata, P.J. (2020f). Preharvest Application of Oxalic Acid Improved Pomegranate Fruit Yield, Quality, and Bioactive Compounds at Harvest in a Concentration-Dependent Manner. *Agronomy, 10 (10)*, 1522. DOI: 10.3390/agronomy10101522.

García-Pastor, M.E., Giménez, M.J., Serna-Escolano, V., Guillén, F., Valero, D., Serrano, M., García-Martínez, S., Terry, L.A., Alamar, M.C., Zapata, P.J. (2021). Oxalic acid preharvest treatment improves colour and quality of seedless table grape 'Magenta' upregulating on-vine ABA metabolism and relative *Vv*NCED1 gene expression and the antioxidant system in berries. *Frontiers in Plant Science*, *12*, 740240. DOI: 10.3389/fpls.2021.740240.

García-Pastor, M.E., Serrano, M., Guillén, F., Castillo, S., Martínez-Romero, D., Valero, D., Zapata, P.J. (2019). Methyl jasmonate effects on table grape ripening, vine yield, berry quality and bioactive compounds depend on applied concentration. *Scientia Horticulturae*, *247*, 380–389. DOI: 10.1016/j.scienta.2018.12.043.

García-Pastor, M.E., Serrano, M., Guillén, F., Giménez, M.J., Martínez-Romero, D., Valero, D., Zapata, P.J. (2020a). Preharvest application of methyl jasmonate increases crop yield, fruit quality and bioactive compounds in pomegranate 'Mollar de Elche' at harvest and during postharvest storage. *Journal of the Science of Food and Agriculture, 100 (1)*, 145–153. DOI: 10.1002/jsfa.10007.

García-Pastor, M.E., Giménez, M.J., Zapata, P.J., Guillén, F., Valverde, J.M., Serrano, M., Valero, D. (2020e). Preharvest application of methyl salicylate, acetyl salicylic acid and salicylic acid alleviated disease caused by *Botrytis cinerea* through stimulation of antioxidant system in table grapes. *International Journal of Food Microbiology*, *334*, 108807. DOI: 10.1016/j.ijfoodmicro.2020.108807.

García-Pastor, M.E., Zapata, P.J., Castillo, S., Martínez-Romero, D., Guillén, F., Valero, D., Serrano, M. (2020c). The Effects of Salicylic Acid and Its Derivatives on Increasing Pomegranate Fruit Quality and Bioactive Compounds at Harvest and During Storage. *Frontiers in Plant Science*, *11*, 668. DOI: 10.3389/fpls.2020.00668.

García-Pastor, M.E., Zapata, P.J., Castillo, S., Martínez-Romero, D., Valero, D.; Serrano, M., Guillén, F. (2020d). Preharvest Salicylate Treatments Enhance Antioxidant Compounds, Color and Crop Yield in Low Pigmented-Table Grape Cultivars and Preserve Quality Traits during Storage. *Antioxidants, 9 (9)*, 832. DOI: 10.3390/antiox9090832.

García-Pastor, M.E., Serrano, M., Guillén, F., Zapata, P.J., Valero, D. (2020b). Preharvest or a combination of preharvest and postharvest treatments with methyl jasmonate reduced chilling injury, by maintaining higher unsaturated fatty acids, and increased aril colour and phenolics content in pomegranate. *Postharvest Biology and Technology*, *167*, 111226. DOI: 10.1016/j.postharvbio.2020.111226.

Giménez, M.J., Valverde, J.M., Valero, D., Díaz-Mula, H.M., Zapata, P.J., Serrano, M., Moral, J., Castillo, S. (2015). Methyl salicylate treatments of sweet cherry trees improve fruit quality at harvest and during storage. *Scientia Horticulturae*, *197*, 665–673. DOI: 10.1016/j.scienta.2015.10.033

Giménez, M.J., Valverde, J.M., Valero, D., Guillén, F., Martínez-Romero, D., Serrano, M., Castillo, S. (2014). Quality and antioxidant properties on sweet cherries as affected by preharvest salicylic and acetylsalicylic acids treatments. *Food Chemistry*, *160*, 226–232. DOI: 10.1016/j.foodchem.2014.03.107.

Giménez, M.J., Serrano, M., Valverde, J.M., Martínez-Romero, D., Castillo, S., Valero, D., Guillén, F. (2017). Preharvest salicylic acid and acetylsalicylic acid treatments preserve quality and enhance antioxidant systems during postharvest storage of sweet cherry cultivars. *Journal of the Science of Food and Agriculture, 97 (4)*, 1220–1228. DOI: 10.1002/jsfa.7853.

Gomes, E.P., Vanz Borges, C., Monteiro, G.C., Filiol Belin, M.A., Minatel, I.O., Pimentel Junior, A., Tecchio, M.A., Lima, G.P.P. (2021). Preharvest salicylic acid treatments improve phenolic compounds and biogenic amines in 'Niagara Rosada' table grape. *Postharvest Biology and Technology*, *176*, 111505. DOI: 10.1016/j.postharvbio.2021.111505.

Habibi, F., Ramezanian, A., Rahemi, M., Eshghi, S., Guillen, F., Serrano, M., Valero, D. (2019). Postharvest treatments with γ-aminobutyric acid, methyl jasmonate, or methyl salicylate enhance chilling tolerance of blood orange fruit at prolonged cold storage. *Journal of the Science of Food and Agriculture, 99 (14)*, 6408–17. DOI: 10.1002/jsfa.9920.

Hadjipieri, M., Georgiadou, E.C., Drogoudi, P., Fotopoulos, V., Manganaris, G.A. (2021). The efficacy of acetylsalicylic acid, spermidine and calcium preharvest foliar spray applications on yield efficiency, incidence of physiological disorders and shelf-life performance of loquat fruit. *Scientia Horticulturae*, *289*, 110439. DOI: 10.1016/j.scienta.2021.110439.

Hilker, M., Schmülling, T. (2019). Stress priming, memory, and signalling in plants. *Plant, Cell, and Environment*, *42*, 753–776. DOI: 10.1111/pce.13526.

IPCC (Intergovernmental Panel on Climate Change). (2021). Summary for Policymakers. *In:* Climate Change 2021: The Physical Science Basis. Contribution of Working Group I to the Sixth Assessment Report of the Intergovernmental Panel on Climate Change [Masson-Delmotte, V., Zhai, P., Pirani, A., Connors, S.L., Péan, C., Berger, S., Caud, N., Chen, Y., Goldfarb, L., Gomis, M.I., Huang, M., Leitzell, K., Lonnoy, E., Matthews, J.B.R., Maycock, T.K., Waterfield, T., Yelekçi, O., Yu, R.; Zhou, B. (eds.)]. Cambridge University Press, Cambridge, United Kingdom and New York, NY, USA, pp. 3–32. DOI: 10.1017/9781009157896.001.

Jin, P., Zhu, H., Wang, L., Shan, T., Zheng, Y. (2014). Oxalic acid alleviates chilling injury in peach fruit by regulating energy metabolism and fatty acid contents. *Food Chemistry*, *161*, 87–93. DOI: 10.1016/j.foodchem.2014.03.103.

Kakkar, R.K., Sawhey, V.P. (2002). Polyamine research in plants–a changing perspective. *Physiologia Plantarum*, *116*, 281–292. DOI: 10.1034/j.1399–3054.2002.1160302.x.

Kovács, E., Kristóf, Z., Perlaki, R., Szőllősi, D. (2008). Cell wall metabolism during ripening and storage of nonclimacteric sour cherry (*Prunus cerasus* L., cv. Kántorjánosi). *Acta Alimentaria*, *37*, 415–426. DOI: 10.1556/aalim.2008.0011.

Lehner, A., Meimoun, P., Errakhi, R., Madiona, K., Barakate, M., Bouteau, F. (2008). Toxic and signalling effects of oxalic acid: oxalic acid-Natural born killer or natural born protector? *Plant Signaling & Behavior*, *3*, 746–748. DOI: 10.4161/psb.3.9.6634.

Li, P., Zheng, X. (2015). Prestorage Application of Oxalic Acid to Alleviate Chilling Injury in Mango Fruit. Proceedings of the Florida State Horticultural Society, 128, 190–195. Downloaded from https://cabidigitallibrary.org by 85.217.156.17, on 12/24/23.

Li, Z.L., Min, D.D., Fu, X.D., Zhao, X.M., Wang, J.H., Zhang, X.H., Li, F.J., Li, X.A. (2021). The roles of SlMYC2 in regulating ascorbate-glutathione cycle mediated by methyl jasmonate in postharvest tomato fruits under cold stress. *Scientia Horticulturae*, *288*, 110406. DOI: 10.1016/j.scienta.2021.110406.

Mahmood, T., Anwar, F., Abbas, M., Boyce, M.C., Saari, N. (2012). Compositional variation in sugars and organic acids at different maturity stages in selected small fruits from Pakistan. *International Journal of Molecular Sciences*, *13*, 1380–1392. DOI: 10.3390/ijms13021380.

Martínez-Esplá, A., Serrano, M., Valero, D., Martínez-Romero, D., Castillo, S., Zapata, P.J. (2017b). Enhancement of antioxidant systems and storability of two plum cultivars by preharvest treatments with salicylates. *International Journal of Molecular Sciences, 18 (9)*, 1911. DOI: 10.3390/ijms18091911.

Martínez-Esplá, A., Valero, D., Martínez-Romero, D., Castillo, S., Giménez, M.J., García-Pastor, M.E., Serrano, M., Zapata, P.J. (2017a). Preharvest application of methyl jasmonate as an elicitor improves the yield and phenolic content of artichoke. *Journal of Agricultural and Food Chemistry, 65 (42)*, 9247–9254. DOI: 10.1021/acs.jafc.7b03447.

Martínez-Esplá, A., Zapata, P.J., Castillo, S., Guillén, F., Martínez-Romero, D., Valero, D., Serrano, M. (2014a). Preharvest application of methyl jasmonate (MeJA) in two plum cultivars.

1. Improvement of fruit growth and quality attributes at harvest. *Postharvest Biology and Technology, 98*, 98–105. DOI: 10.1016/j.postharvbio.2014.07.011.

Martínez-Esplá, A., Zapata, P.J., Valero, D., García-Viguera, C., Castillo, S., Serrano, M. (2014b). Preharvest application of oxalic acid increased fruit size, bioactive compounds, and antioxidant capacity in sweet cherry cultivars (*Prunus avium* L.). *Journal of Agricultural and Food Chemistry, 62 (15)*, 3432–3437. DOI: 10.1021/jf500224g.

Martínez-Esplá, A., Zapata, P.J., Valero, D., Martínez-Romero, D., Díaz-Mula, H.M., Serrano, M. (2018). Preharvest treatments with salicylates enhance nutrient and antioxidant compounds in plum at harvest and after storage. *Journal of the Science of Food and Agriculture, 98 (7)*, 2742–2750. DOI: 10.1002/jsfa.8770.

Martínez-Esplá, A., Serrano, M., Martínez-Romero, D., Valero, D., Zapata, P.J. (2019). Oxalic acid preharvest treatment increases antioxidant systems and improves plum quality at harvest and during postharvest storage. *Journal of the Science of Food and Agriculture, 99 (1)*, 235–243. DOI: 10.1002/jsfa.9165.

Min, D.D., Li, F.J., Zhang, X.H., Cui, X.X., Shu, P., Dong, L.L., Ren, C.T. (2018). SlMYC2 involved in methyl jasmonate-induced tomato fruit chilling tolerance. *Journal of Agricultural and Food Chemistry, 66 (12)*, 3110–7. DOI: 10.1021/acs.jafc.8b00299.

Min, D., Li, F., Ali, M., Zhang, X., Liu, Y. (2023). Application of methyl jasmonate to control chilling tolerance of postharvest fruit and vegetables: a meta-analysis and eliciting metabolism review. *Critical Reviews in Food Science and Nutrition*, DOI: 10.1080/10408398.2023.2258201.

Mirdehghan, S.H., Khanamani, Z., Shamshiri, M.H., Hokmabadi, H. (2013). Preharvest foliar application of putrescine and spermine on postharvest quality of fresh pistachio. *Acta Horticulturae, 1012*, 299–304. DOI: 10.17660/ActaHortic.2013.1012.37.

Mirdehghan, S.H., Rahimi, S. (2016). Pre-harvest application of polyamines enhances antioxidants and table grape (*Vitis vinifera* L.) quality during postharvest period. *Food Chemistry, 196*, 1040–1047. DOI: 10.1016/j.foodchem.2015.10.038.

Moretti, C.L., Matos, L.M., Calvo, A.G., Sargent, S.A. (2010). Climate changes and potential impacts on postharvest quality of fruit and vegetable crops: A review. *Food Research International, 43*, 1824–1832. DOI: 10.1016/j.foodres.2009.10.013.

Razavi, F., Hajilou, J. (2016). Enhancement of postharvest nutritional quality and antioxidant capacity of peach fruits by preharvest oxalic acid treatment. *Scientia Horticulturae, 200*, 95–101. DOI: 10.1016/j.scienta.2016.01.011.

Retamal-Salgado, J., Adaos, G., Cedeño-García, G., Ospino-Olivella, S.C., Vergara-Retamales, R., Lopéz, M.D., Olivares, R., Hirzel, J., Olivares-Soto, H., Betancur, M. (2023). Preharvest applications of oxalic acid and salicylic acid increase fruit firmness and polyphenolic content in blueberry (*Vaccinium corymbosum* L.). *Horticulturae, 9*, 639. DOI: 10.3390/horticulturae9060639.

Ruiz-Aracil, M.C., Valverde, J.M., Lorente-Mento, J.M., Carrión-Antolí, A., Castillo, S., Martínez-Romero, D., Guillén, F. (2023). Sweet Cherry (*Prunus avium* L.) Cracking during Development on the Tree and at Harvest: The Impact of Methyl Jasmonate on Four Different Growing Seasons. *Agriculture, 13*, 1244. DOI: 10.3390/agriculture13061244.

Sah, S.K., Reddy, K.R., Li, J. (2016). Abscisic Acid and Abiotic Stress Tolerance in Crop Plants. *Frontiers in Plant Science 7*, 571. DOI: 10.3389/fpls.2016.00571.

Savvides, A., Ali, S., Tester, M., Fotopoulos, V. (2016). Chemical priming of plants against multiple abiotic stresses: Mission possible? *Trends in Plant Science 21*, 329–340. DOI: 10.1016/j.tplants.2015.11.003.

Sayyad-Amin, P., Davarynejad, G., Abedi, B. (2022). Effects of Preharvest Application of Chemical Compounds on Pear (*Pyrus communis* cv. 'Shekari') Fruit Traits. *Erwerbs-Obstbau, 64*, 657–662. DOI: 10.1007/s10341-022-00756-w.

Schaller, F., Schaller, A., Stintzi, A. (2005). Biosynthesis and metabolism of jasmonates. *Journal of Plant Growth Regulation, 23*, 179–199. DOI: 10.1007/BF02637260.

Seo, J., Yi, G., Lee, J.G., Choi, J.H., Lee, E.J. (2020). Seed browning in pepper (*Capsicum annuum* L.) fruit during cold storage is inhibited by methyl jasmonate or induced by methyl salicylate. *Postharvest Biology and Technology, 166*, 111210. DOI: 10.1016/j.postharvbio.2020.111210.

Serna-Escolano, V., Martínez-Romero, D., Giménez, M.J., Serrano, M., García-Martínez, S., Valero, D., Valverde, J.M., Zapata, P.J. (2021a). Enhancing antioxidant systems by preharvest treatments with methyl jasmonate and salicylic acid leads to maintain lemon quality during cold storage. *Food Chemistry, 338*, 128044. DOI: 10.1016/j.foodchem.2020.128044.

Serna-Escolano, V., Valverde, J.M., García-Pastor, M.E., Valero, D., Castillo, S., Guillén, F., Martínez-Romero, D., Zapata, P.J., Serrano, M. (2019). Pre-harvest methyl jasmonate treatments increase antioxidant systems in lemon fruit without affecting yield or other fruit quality parameters. *Journal of the Science of Food and Agriculture, 99 (11)*, 5035–5043. DOI: 10.1002/jsfa.9746.

Serna-Escolano, V., Giménez, M.J., Castillo, S., Valverde, J.M., Martínez-Romero, D., Guillén, F., Serrano, M., Valero, D., Zapata, P.J. (2021b). Preharvest treatment with oxalic acid improves postharvest storage of lemon fruit by stimulation of the antioxidant system and phenolic content. *Antioxidants, 10*, 963. DOI: 10.3390/antiox10060963.

Serrano, M., Martínez-Esplá, A., Zapata, P., Castillo, S., Martínez-Romero, D., Guillén, F., Valverde, J.M., Valero, D. (2018). Effects of methyl jasmonate treatment on fruit quality properties. *In*: Barman, K., Sharma, S., and Siddiqui, M.W. (Eds.), Emerging Postharvest Treatment of Fruits and Vegetables. Apple Academic Press, Oakville, Canada, pp. 85–106 (Chapter 4).

Shanbehpour, F., Rastegar, S., Ghasemi, M. (2020). Effect of preharvest application of calcium chloride, putrescine, and salicylic acid on antioxidant system and biochemical changes of two Indian jujube genotypes. *Journal of Food Biochemistry, 44 (11)*. DOI: 10.1111/jfbc.13474.

Silverman, P., Seskar, M., Kanter, D., Schweizer, P., Metraux, J.P., Raskin, I. (1995). Salicylic acid in rice. Plant Physiology, 108, 633–639. DOI: 10.1104/pp.108.2.633.

Supapvanich, S., Anan, W., Chimsonthorn, V. (2019). Efficiency of combinative salicylic acid and chitosan preharvest-treatment on antioxidant and phytochemicals of ready to eat daikon sprouts during storage. *Food Chemistry, 284*, 8–15. DOI: 10.1016/j.foodchem.2019.01.100.

Teribia, N., Tijero, V., Munné-Bosch, S. (2016). Linking hormonal profiles with variations in sugar and anthocyanin contents during the natural development and ripening of sweet cherries. *New Biotechnology, 33 (6)*, 824–833. DOI: 10.1016/j.nbt.2016.07.015.

Tzortzakis, N.G. (2007). Maintaining postharvest quality of fresh produce with volatile compounds. *Innovative Food Science & Emerging Technologies, 8 (1)*, 111–116. DOI: 10.1016/j.ifset.2006.08.001.

Üzümcü, S.S., Koyuncu, M.A., Onursal, C.E., GüneylÍ, A., Erbas, D. (2020). Effect of pre-harvest oxalic acid treatment on shelf-life of apricot cv. 'Roxana. *Nevsehir Journal of Science and Technology 9*, 73–80. DOI: 10.17100/nevbiltek.561704.

Valero, D.; Serrano, M. (2010). Postharvest Biology and Technology for Preserving Fruit Quality (1st ed.). CRC Press, 287. DOI: 10.1201/9781439802670.

Valverde, J.M., Giménez, M.J., Guillén, F., Valero, D.; Martínez-Romero, D., Serrano, M. (2015). Methyl salicylate treatments of sweet cherry trees increase antioxidant systems in fruit at harvest and during storage. *Postharvest Biology and Technology*, *109*, 106–113. DOI: 10.1016/j.postharvbio.2015.06.011.

Wang, Z., Cao, J., Jiang, W. (2016). Changes in sugar metabolism caused by exogenous oxalic acid related to chilling tolerance of apricot fruit. *Postharvest Biology and Technology*, *114*, 10–16. DOI: 10.1016/j.postharvbio.2015.11.015.

Wheeler, T., von Braun, J. (2013). Climate change impacts on global food security. *Science*, 341, 508–513. DOI: 10.1126/science.1239402.

Yao, H., Tian, S. (2005). Effects of pre- and post-harvest application of salicylic acid or methyl jasmonate on inducing disease resistance of cherry fruit in storage. *Postharvest Biology and Technology*, *35(3)*, 253–262. DOI: 10.1016/j.postharvbio.2004.09.001.

Youryon, P., Chimphakdee, N., Supapvanich, S. (2017). Effects of preharvest salicylic acid or oxalic acid on quality of 'Queen' pineapple fruit harvested at different month stored at room temperature. *Acta Horticulturae*, *1256*, 489–494. DOI: 10.17660/ActaHortic.2019.1256.69.

Yu, L.N., Liu, H.X., Shao, X.F., Yu, F., Wei, Y.Z., Ni, Z.M., Xu, F., Wang, H.F. (2016). Effects of hot air and methyl jasmonate treatment on the metabolism of soluble sugars in peach fruit during cold storage. *Postharvest Biology and Technology*, *113*, 8–16. DOI: 10.1016/j.postharvbio.2015.10.013.

Zalasiewicz, J., Williams, M. (2016). A geological history of climate change, pp. 3–17. *In:* Letcher, T.M. (ed.). Climate change: observed impacts on planet Earth. Elsevier, Amsterdam, Netherlands.

Zapata, P.J., Martínez-Esplá, A., Guillén, F., Díaz-Mula, H.M., Martínez-Romero, D., Serrano, M., Valero, D. (2014). Preharvest application of methyl jasmonate (MeJA) in two plum cultivars. 2. Improvement of fruit quality and antioxidant systems during postharvest storage. *Postharvest Biology and Technology*, *98*, 115–122. DOI: 10.1016/j.postharvbio.2014.07.012.

Zhang, W.L., Jiang, H.T., Cao, J.K., Jiang, W.B. (2021). Advances in biochemical mechanisms and control technologies to treat chilling injury in postharvest fruits and vegetables. *Trends in Food Science & Technology*, *113*, 355–65. DOI: 10.1016/j.tifs.2021.05.009.

Zhang, X., Shen, L., Li, F., Meng, D., Sheng, J. (2011). Methyl salicylate-induced arginine catabolism is associated with up-regulation of polyamine and nitric oxide levels and improves chilling tolerance in cherry tomato fruit. *Journal of Agricultural and Food Chemistry*, *59*, 9351–9357. DOI: 10.1021/jf201812r.

Zhang, X., Li, F., Ji, N., Shao, S., Wang, D., Li, L., Cheng, F. (2016). Involvement of arginase in methyl jasmonate–induced tomato fruit chilling tolerance. *Journal of the American Society for Horticultural Science, 141 (2)*, 139–45. DOI: 10.21273/JASHS.141.2.139.

Zhou, J. X., Min, D.D., Li, Z.L., Fu, X.D., Zhao, X.M., Wang, J.H., Zhang, X.H., Li, F.J., Li, X.A. (2021). Effects of chilling acclimation and methyl jasmonate on sugar metabolism in tomato fruits during cold storage. *Scientia Horticulturae*, *289*, 110495. DOI: 10.1016/j.scienta.2021.110495.

Zhu, Y., Yu, J., Brecht, J.K., Jiang, T.J., Zheng, X.L. (2016). Preharvest application of oxalic acid increases quality and resistance to *Penicillium expansum* in kiwifruit during postharvest storage. *Food Chemistry*, *190*, 537–543. DOI: 10.1016/j.foodchem.2015.06.001.

Zhu, Z., Tian, S. (2012). Resistant responses of tomato fruit treated with exogenous methyl jasmonate to *Botrytis cinerea* infection. *Scientia Horticulturae*, *142*, 38–43. DOI: 10.1016/j.scienta.2012.05.002.

Index

Note: Page numbers in *italics* indicate a figure and page numbers in **bold** indicate a table on the corresponding page.

D

E

F

G

H

I

J

L

O

P